D1217902

GRAPHS of ELEMENTARY and SPECIAL FUNCTIONS

HANDBOOK

N.O. Virchenko
I.I. Lyashko

BEGELL HOUSE, INC. PUBLISHERS
New York

Typeset by Kalashnikoff & Co., Ltd. (Moscow, Russia)

Library of Congress Cataloging-in-Publication Data

Virchenko, N. O. (Nina Opanasivna)
[Grafiki elementarnykh i spetsial'nykh funktsii. English]
Graphs of elementary and special functions : handbook / N.O. Virchenko, I.I. Lyashko.
p. cm.
Includes bibliographical references and index.
ISBN 1-56700-156-4
1. Functions—Graphic methods. 2. Functions, Special—Graphic methods.
I. Liashko, Ivan Ivanovich. II. Title.
QA231 .V57 2001
515—dc21 2001025293

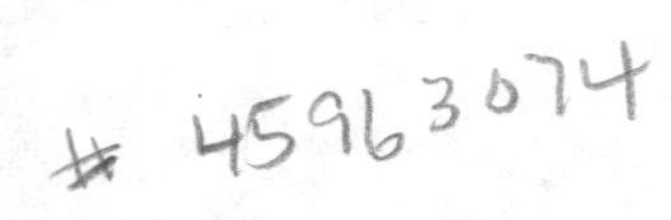

Contents

Part I. Plots of graphs with elementary methods

Chapter 1. Basic notions of numbers, variables and functions

Chapter 2. Analysis of functions for constructing graphs

Preface

The notion of a functional dependence is one of the most important notions in modern mathematics. In studying processes in nature and solving engineering problems, one encounters a situation in which one quantity varies depending on another quantity.

Among the many ways to define the function (analytical, tabular, descriptive), defining it by its graph is widely used because of its visual presentation. In some cases, this method of defining the function is the only one possible. However, it is surprising that so far there are not many reference books that describe different aspects of studying and plotting graphs of various functions, in particular, with the use of methods of calculus.

This reference book gives essentials on functions and methods of plotting their graphs. Special attention is paid to methods used for plotting graphs of functions that are defined implicitly and in a parametric form, methods for making plots in polar coordinates, and plotting graphs of remarkable curves, special functions, etc.

The book consists of two parts. The first part gives general information about the notion of a function and methods of plotting graphs of functions without the use of derivatives. The second part deals with methods of studying functions and plotting their graphs with the use of calculus. The book contains many examples of functions that are more complicated than commonly considered, graphs of important curves, and properties, and graphs of common special functions (gamma function, integral exponential functions, Fresnel integrals, Bessel functions, orthogonal polynomials, elliptic functions, Mathieu functions, etc.).

The book contains 760 figures and uses material found in existing references, textbooks, articles, and other sources on the subject.

Part I

Plots of graphs with elementary methods

Chapter 1

Basic notions of numbers, variables, and functions

1.1. Numbers. Variables. Functions

In this book we will occasionally use certain symbols to denote logical expressions. Symbol (generality quantifier) will replace the expressions "for arbitrary," "for any," etc., symbol $\exists$ (existential quantifier) — the expressions "there exists," "one can find," and so on.

The notation $A \Rightarrow B$ (implication symbol) means that statement A implies statement B.

The notation $A \Leftrightarrow B$ (equivalence symbol) means that $A \Leftarrow B$ and $B \Rightarrow A$.

The notation $A \wedge B$ (conjunction quantifier) means that both statements A and B hold.

If one of the statements A or B holds, then we use notations $A \vee B$ (disjunction quantifier).

1.1.1. Real numbers

All infinite decimal fractions (positive, negative, and zero) make up the set of *real numbers* R (continuum of numbers). This set can be subdivided into two parts (subsets): the subset of rational numbers (integers, fractions, both positive and negative, and the number 0) and the set of irrational numbers. Every rational number can be written as a fraction $\pm p/q$, where p and q are natural numbers, i.e., 1, 2, ... , n, ... A rational number is a periodic decimal fraction.

An irrational number is a nonperiodic infinite decimal fraction. For example, $1/3 = 0.333$ ..., $5 = 5/1$ are rational numbers; $\pi = 3.14159...$, $\sqrt{2} = 1.4142...$ are irrational.

Basic properties of the set of real numbers

The set of real numbers is *ordered*, i.e any two numbers from this set a, b are either equal or one is greater than the other. The order relation is denoted by $a < b$ (a is less than b), $a \leq b$ (a is not greater than b), $a > b$ (a is greater than b), $a \geq b$ (a is not less than b).

The set of real numbers is *dense*, i.e. for any two arbitrarily close real numbers a and b there are infinitely many real numbers (both rational and irrational) lying between these numbers.

The set of real numbers is *continuous*, i.e. any Dedekind cut defines a unique real number a which separates the numbers belonging to the class A from the numbers belonging to the class

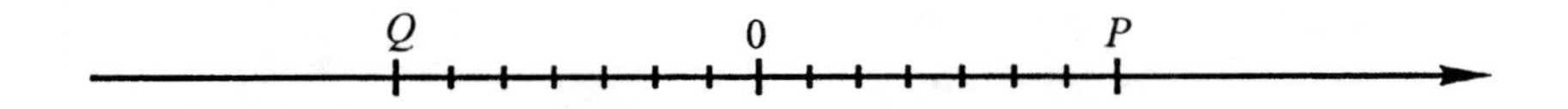

Figure 1.1. Number axis. O is the origin; points P and Q lie on the axis.

B, that is, a is the greatest number in the class A (and then the class B does not contain the smallest number), or it is the smallest number in the class B (then class A does not have the greatest number).

A *Dedekind cut* D of the set of real numbers is a splitting of all real numbers into two classes, the lower class A and the upper class B, in such a way that each real number belongs to only one class and each number from class A is less than an arbitrary number from class B.

It is convenient to draw real numbers as points on the real axis. The *real axis* is a line with an initial point O, direction, which is taken to be positive, and a scale (a line segment the length of which is taken as one). To every real number there corresponds a unique point on the real axis. The number 0 corresponds to the point O. If a number x is positive, then the corresponding point P lies to the right of the point O with the distance OP equal to x. If a number x is negative, then it corresponds to a point Q which lies to the left of O with the distance OQ equal to $-x$ (Figure 1.1). The converse statement is also true: every point in the real axis corresponds to a unique real number. This means that there is a one-to-one correspondence between real numbers and points on the real axis. This allows us in many instances to use interchangeably the notions "a number x" and "a point x." The following statement will be of practical use: every irrational number can be arbitrarily well approximated by rational numbers.

Real numbers can be subdivided into *real algebraic numbers* (real roots of algebraic equations with integer coefficients) and *real transcendental numbers* (the rest of the real numbers).

Any finite or infinite collection of real numbers is called a *number set*. Sets will be denoted by capital letters $X, Y, Z, \ldots$, and we use small letters $x, y, z, \ldots$ to denote numbers that belong to a set. For example $x \in X$ (x belongs to X). If an element x does not belong to a set X, we write this as $x \notin X$. The notation $X = \{x\text{:condition } K \text{ holds}\}$ is used to denote the set of all elements which verify condition K.

Let us consider two sets A and B. If each element of the set A is an element of the set B, then we say that A is a *subset* of the set B and denote this by $A \subset B$ (or $B \supset A$). If $A \subset B \wedge B \subset A$, then the sets A and B are called equal. This fact is denoted by $A = B$. The set which does not contain any element is called the *empty set* and is denoted by $\varnothing$. Any subset contains the empty set as a subset.

The simplest operations one can perform with sets are taking unions (sums), intersections (products), and differences.

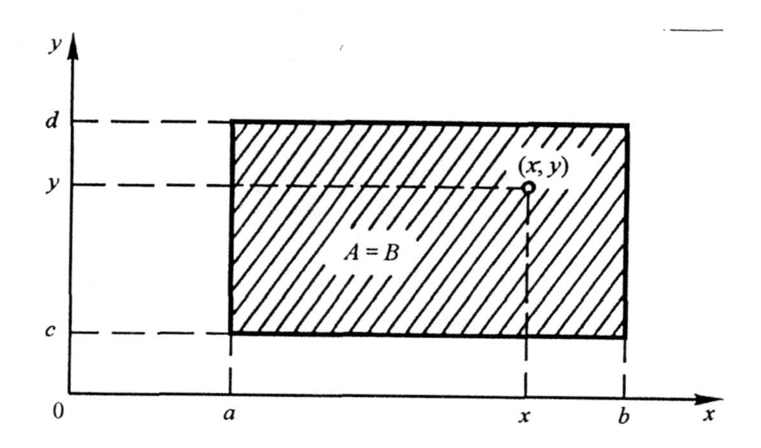

Figure 1.2. Direct product of sets A and B.

By the *union* (or sum) of sets A and B, we call the set $C = A \cup B = \{x : x \in A \vee x \in B\}$.

The *intersection* (or product) of sets A and B is the set $D = A \cap B = \{x : x \in A \wedge x \in B\}$.

The *difference* of sets A and B is the set $F = A \setminus B = \{x : x \in A \wedge x \notin B\}$.

The *symmetric difference* of two sets A and B is the set $L = A \Delta B = (A \cup B) \setminus (A \cap B)$.

A pair of elements a and b is called the *ordered pair* and is denoted by (a,b) if the order in which they appear is given. By this definition, $(a,b) = (c,d) \Leftrightarrow (a = c) \wedge (b = d)$.

Let $A \subset M$, $B \subset M$. The totality of all possible ordered pairs (a,b), where $a \in A$, $b \in B$, is called the *direct product* of the sets A and B, and is denoted by $A \times B$. For example, let $A = \{x : a \le x \le b\}$, $B = \{y : c \le y \le d\}$. Then the direct product $A \times B$ is in one-to-one correspondence with the set of points (x, y) that make the rectangle (see Figure 1.2).

Among number sets, the following sets will be of importance.

A bounded closed interval (line segment) $X = [a;b]$. The set X contains all the numbers x which satisfy the inequality $a \le x \le b$. On the real axis this set forms the line segment ab, with the endpoints included (Figure 1.3 a).

A bounded open interval (line segment) $X = (a;b)$. The set X contains all numbers x which satisfy the inequality $a < x < b$. On the real axis, this set is the line segment ab without the endpoints (Figure 1.3 b).

Bounded half-open (or half-closed) intervals $[a;b)$ and $(a;b]$. These are sets X which contain numbers x for which $a \le x < b$ and, respectively, $a < x \le b$. The corresponding interval on the real axis contains the left endpoint and does not contain the right endpoint, and, respectively, does not contain the left endpoint and contains the right endpoint.

One-sided infinite intervals $(b; \infty)$, $[b; \infty)$, $-\infty; a)$, $(-\infty; a]$. These are sets X of numbers x for which $x > b, x \ge b, x < a, x \le a$ respectively. On the real line these sets correspond to respective infinite half-axes.

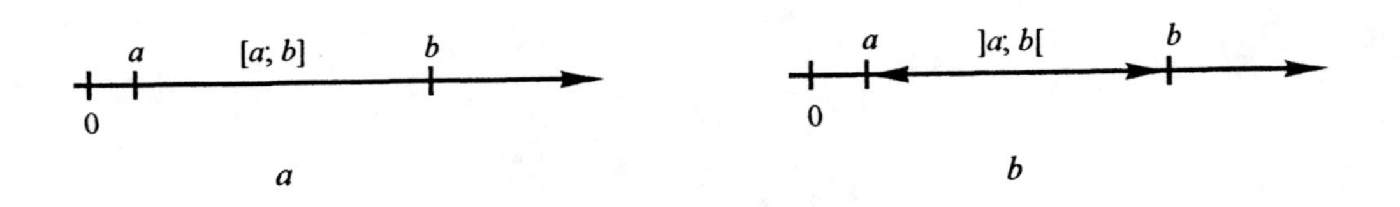

Figure 1.3. (*a*) segment $[a;b]$, (*b*) interval $(a;b)$.

The set of all real numbers (the real axis) is also considered as an interval and is denoted by $(-\infty; +\infty)$.

In the sequel, we will be using the following notations. The set of all natural numbers will be denoted by **N**, the set of all nonnegative integers — $\mathbf{Z}_0$, the set of all integers — **Z**, the set of all rational numbers — **Q**, the set of all real numbers — **R**, the set of all complex numbers — **C**.

Let a be an arbitrary number. A *δ-neighborhood* of the point $x = a$ is the open interval $(a-\delta;a+\delta)$ containing the point a. In other words, a δ-neighborhood of a point a is the set of all x for which $|x - a| < \delta$, where δ is a positive number (Figure 1.4). A *neighborhood* of a point $x = a$ is an arbitrary set containing a δ-neighborhood of this point for some $\delta > 0$.

The *absolute value* of a real number a, denoted by $|a|$, is the nonnegative number defined by:

$$|a| = \begin{cases} a, & \text{if } a > 0, \\ -a, & \text{if } a < 0, \\ 0, & \text{if } a = 0. \end{cases}$$

The main properties of the absolute value are the following:

$$|a| \geq 0,$$
$$|a| = 0 \quad \text{implies} \quad a = 0,$$
$$\big||a| - |b|\big| \leq |a + b| \leq |a| + |b|,$$
$$\big||a| - |b|\big| \leq |a - b| \leq |a| + |b|,$$
$$|ab| = |a|\,|b|, \qquad \left|\frac{a}{b}\right| = \frac{|a|}{|b|} \quad (b \neq 0),$$
$$\text{if} \quad |a| \leq A \quad \text{and} \quad |b| \leq B, \quad \text{then} \quad |a + b| \leq A + B, \quad |ab| \leq AB.$$

Consider a nonempty set X. An arbitrary number M such that $x \leq M$ for all $x \in X$ is called the *upper bound* of the set X. The *lower bound* of a set X is a number m such that $x \geq m$ for all $x \in X$.

Sets that have an upper bound are called *bounded from above*. If a set has a lower bound, then it is called *bounded from below*. A set which is bounded from below and above is called *bounded*.

The smallest among the numbers M is called the *least upper bound*, and the largest among the numbers m is called the *greatest lower bound*. More precisely, a number

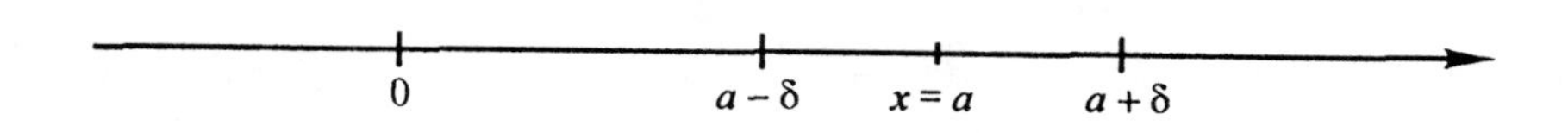

Figure 1.4. δ-neighborhood of point *a*.

$M^* = \sup X = \sup\{x\}$ is called the least upper bound of a set $X = \{x\}$ if for all $x \in X, x \le M^*$, and all $\varepsilon > 0$ there exists $x_0 \in X$ such that $M^* - \varepsilon < x_0 \le M^*$. The number $m^* = \inf X = \inf\{x\}$ is called the greatest lower bound of a set $X = \{x\}$ if for all $x \in X$, we have $x \ge m^*$, and for every $\varepsilon > 0$ there exists $x \in X$ such that $m^* \le x < m^{*+}\varepsilon$.

If a set X is bounded from above (below), then it has the least upper (greatest lower) bound. This means that if a set is bounded, then it has the least upper bound and the greatest lower bound.

1.1.2. Constants and variables

While observing and studying processes in nature, one sees that some quantities vary, i.e. their numeric values change, whereas the others remain constant. For example, in a uniform motion, the time and distance change, but the velocity remains constant. If a gas is heated in a closed vessel, then the gas pressure and temperature change, but the mass and the volume of the gas are unchanged.

A *variable* is a quantity that can assume different values in a problem under consideration. A quantity that does not change in a problem under consideration is called *constant*. Quantities that remain constant in all problems are called *absolute constants*. An example of an absolute constant is $\pi = 3.141...$.

Variables can change in various ways. Some variables can take only integer positive values, others — infinite, negative values, etc. One also considers *discrete* variables, — those which take values from a finite or infinite set of isolated values, and *continuous* variables that, assuming two values $x = a$ and $x = b$, take also all the values x with $a < x < b$.

The set of numbers from which a variable x takes its values is called the *value domain* of the variable x. A variable is called *increasing* if each of its next values is greater than the previous one. A variable is called *decreasing* if each of its next values is less than the previous one. Increasing and decreasing variables are called *monotone*. A variable is *bounded* if its value domain is a bounded set.

1.1.3. Concept of a function

Often we have to consider not variables by themselves but a relationship between them, a dependence of one variable on another. Actually there are no variables in nature that would change independently from other physical variables. For example, in studying a motion, one can consider the distance as a variable that changes with time. The distance then will be a function of time. The distance that a projectile flies and the accuracy with which it hits the target depend on the mass, the initial velocity of the projectile, the pitch angle of the barrel, the wind velocity and its direction, and other effects.

In mathematics the concept of functional dependence or a function is introduced irrespective of a particular nature of the dependence of variables.

Consider two sets X and Y, elements of which could have an arbitrary nature, and suppose that there is a rule that, to each element x from the set X, assigns an element y from the set Y. We write this as $y = f(x)$. Then f is called a *function* from X into Y (or a mapping from X into Y)*

Thus if there is a function f from a set X into a set Y, then we say that the function f is defined in the set X and takes values $y = f(x)$ in the set Y. The set X is called the *domain* of the function f; the set $f(X)$ is called the *range* of the function. It is clear that $f(X) \subset Y$. The variable x is called *independent* or *argument*. The equality $y = f(x)$ means that, by applying the rule f to a value of the variable x, we get the value of y that corresponds to this value of x (Figure 1.5).

It is important to clearly see the difference between the variable x, which is an element of the set X, the value $f(x)$, which is an element of the set Y, and the mapping f itself, which is a correspondence between elements $x \in X$ and $f(x) \in Y$.

Let $f: X \to Y$ and $D \subset X$. The set of those elements of Y which are images under f of at least one element of D is called the *image* of D and denoted by $f(D)$. It is clear that $f(D) = \{f(x) \in Y : x \in D\}$.

If $f(X) = Y$, then we say that f *maps* X *onto* Y. If $f(X) \subset Y$, then f *maps* X *into* Y.

A mapping $f: X \to Y$ is called *injective* (or *injection*, or *one-to-one* mapping of the set X *into* the set Y) if $(x \neq x') \Rightarrow (f(x) \neq f(x'))$, or else if for every y the equation $f(x) = y$ has at most one solution. A mapping $f: X \to Y$ is called *bijective* (or *bijection* or *one-to-one* mapping of the set X *onto* the set Y) if it is injective and onto the set Y, that is for all $y \in Y$ the equation $f(x) = y$ has a solution and it is unique.

An injective mapping f from a set X into a set Y is a bijective mapping from X onto $f(X)$.

1.1.4. Methods of defining a function

To define a function means to give a law by which, for a given value of the independent variable, one finds its corresponding values. Consider a few methods to define a function.

Table method

The table method consists of writing out, in a certain order, values of the independent variable $x_1, x_2, \ldots, x_n$ and the values of the function $y_1, y_2, \ldots, y_n$ that correspond to them.

---

* Definition of a function was introduced by Leibniz in 1692. The notation $f(x)$ is due to Euler.

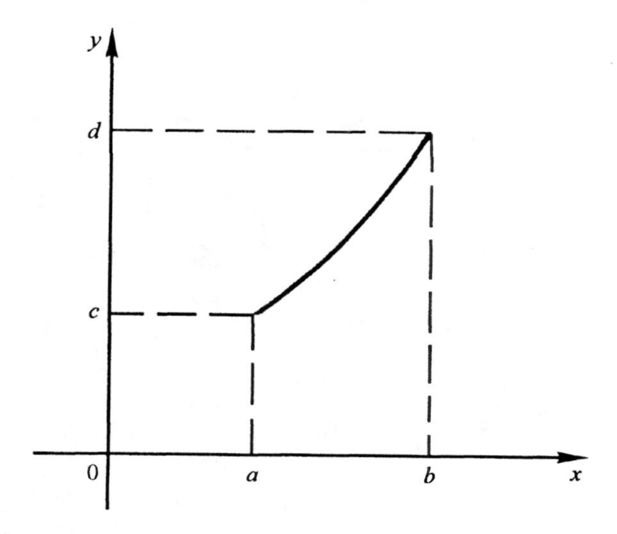

Figure 1.5. Domain of the function $y=f(x)$ (segment $[a; b]$) and range of the function (segment $[c; d]$).

Examples are tables of logarithms, trigonometric functions, etc. The table method is widely used in engineering, the natural sciences, and so on. Numerical results of observations are written in the form of a table. For example, readings of the temperature taken over a day can be written as the table:

t	1	2	3	4	5	6	7	8	9	10
T	–1	–2	2.5	–2	–0.5	1	3	3.4	3.6	3.9

This table gives the temperature T as a function of t, $T=f(t)$.

The table method has the advantage that it allows one to determine particular values of the function without further calculations. The drawback of this method is that it does not define the function completely but only for certain values of the argument, and it does not show the complete behavior of the function as the argument changes.

Graphical method

If a method for determining the position of a point on the plane is given, then we say that the plane is equipped with a coordinate system.

Various coordinate systems are used for constructing graphs. The most widely used is the *Cartesian coordinate system*, that is, two mutually perpendicular number axes Ox and Oy (Figure 1.6) intersecting at the point O.

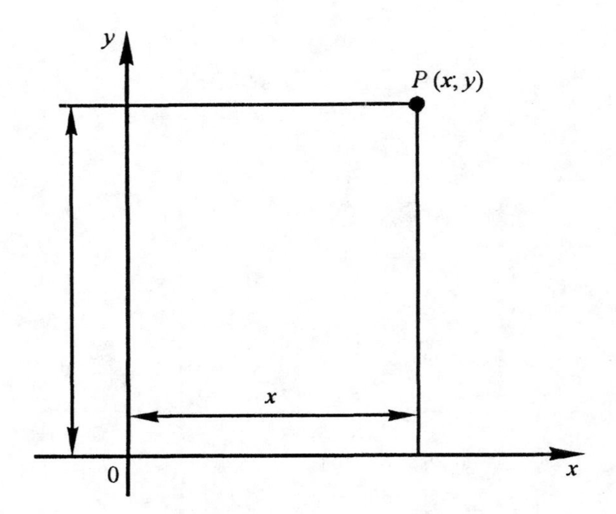

Figure 1.6. Rectangular coordinate system.

The *graph* of a function $f: X \to Y$ is the set of all ordered pairs $(x; f(x))$, where $x \in X$, $f(x) \in Y$. It is clear that the graph is a subset of the direct product $X \times Y$ (see Figure 1.2). In other words, the graph of a function $y = f(x)$ is the set of the points in the coordinate plane Oxy, coordinates of which satisfy the equality $y = f(x)$.

If the graph of a function is given, then the function is said to be defined graphically. Any curve in the plane can be regarded as the graph of a function. In particular, a straight line or a set of isolated points could be graphs of functions.

Graphical definition of a function does not allow one to find precise values of the function at particular points. Its advantage, however, is the visualization of the function.

Analytical method

The analytical method of defining a function in terms of a formula or an expression permits one to find the precise or approximate value of the function for any numeric value of its argument x.

Analytical expression of the function can be given by various formulas. For example, the function

$$f(x) = \begin{cases} \cos x, & \text{if} \quad -\pi \le x \le 0, \\ 1, & \text{if} \quad 0 < x < 1, \\ 1/x, & \text{if} \quad 1 \le x \le 2 \end{cases}$$

is defined in the interval $[-\pi; 2]$ in terms of three formulas.

If the dependence between x and y is given by a formula, defining y in terms of x, i.e., $y = f(x)$, then it is said that the function is defined *explicitly*. The functions $y = 6x - 2$, $y = x^2 \ln x$ are examples of functions defined explicitly.

Let on a set $P = A \times B$ there be given a function $F : P \to Q$, where the set Q contains zero. Suppose there are sets $X \subset A$, $Y \subset B$ such that, for each fixed $x \in X$, the equation $F(x, y) = 0$ has a unique solution. In this case there is a mapping $f: X \to Y$ defined on the set X by assigning, to every $x \in X$, the corresponding $y \in Y$ which is a solution of the equation $F(x,y) = 0$. In this case the function $y = f(x)$ is said to be defined by equation $F(x, y) = 0$ *implicitly*. In other words, the function $y = f(x)$ is defined implicitly if the values of x and y satisfy an equation $F(x, y) = 0$, i.e. this formula is not solved with respect to y.

For example, let $F(x, y) = x^2 - y^2 - x + y = 0$, $A = \{x \in \mathbf{R}: -\infty < x < +\infty\}$, $B = \{y \in \mathbf{R}: -\infty < y < +\infty\}$. Because for every $x \in X$, where $X \subset A$, $X = \{x \in \mathbf{R}: 1/2 \le x < +\infty\}$, the equation $F(x, y) = 0$ has a unique solution $y = x$, $x \in Y$, $Y \subset B$, $Y = \{y \in \mathbf{R}: 1/2 \le y < +\infty\}$, we see that on X, the equation implicitly defines the function $y = x$ with values belonging to the set Y. It should be remarked that not every implicitly defined function can be written in an explicit form $y = f(x)$. Any explicitly defined function can be written implicitly: $y - f(x) = 0$.

Consider one more method of analytical definition of a function, namely the method where both the argument x and the function y are functions of a third variable, a parameter t,

$$x = \varphi(t), \quad y = \psi(t), \tag{1.1}$$

where t runs over a set T. Here, for every value of t, t_0, from its value domain T, one assigns to x and y numerical values $x_0 = \varphi(t_0)$ and $y_0 = \psi(t_0)$. This correspondence defines y as a function of x (or x as a function of y). Such a method of defining a function is called *parametric*.

If x and y are considered as coordinates of a point in the coordinate plane Oxy, then each value of t corresponds to a point in the plane. As the value of t varies from t_1 to t_2, the point in the plane describes certain curve. For example, the equations $x = a\cos t$, $y = a\sin t$, where $t \in [0, 2\pi]$, describe a circle of radius a centered at the origin $O(0,0)$.

If t is eliminated from equations (1.1), then we obtain a function $y = f(x)$ given explicitly. Parametric form of functions is widely used in theoretical studies in mechanics, geometry and other sciences.

The analytical form of defining a function is the one used the most. This is the main form of functions one works with in calculus. Compactness, the possibility of calculating values of the function for any value of the argument from the domain, and the possibility of applying the machinery of mathematical analysis — these are the main advantages of the analytical definition of functions. Its shortcomings are poor visualization and the necessity to perform sometimes cumbersome calculations.

Descriptive method

This method consists of just describing the function. For example, the function $E(x)$ is the integer part of the number x.

In general, by $E(x) = [x]$, we denote the greatest integer, which is less than the number x, that is if $x = n + r$, where n is an integer (possibly negative) and $0 \le r < 1$, then $[x] = n$. The function $E(x) = [x]$ is constant in the interval $[n; n + 1)$. For example $E(2.3) = 2$, $E(-\pi) = -4$.

The Dirichlet function is defined as follows:

$$\chi(x) = \begin{cases} 1, & \text{if } x \text{ is rational,} \\ 0, & \text{if } x \text{ is irrational.} \end{cases}$$

The Dirichlet function is defined in the set of all real numbers.

The function that gives the fractional part of the number is denoted by $y = \{x\}$, more precisely,

$$y = \{x\} = x - [x],$$

where $[x]$ is the integer part of x. This function is defined for all real x.

If x is an arbitrary number, then writing it as $x = n + r$ $(n = [x])$, where n is such an integer that $0 \le r < 1$, we get that $\{x\} = n + r - n = r$. For example, $\{0\} = 0$, $\{1\} = \{2\} = \ldots = \{n\} = \ldots = 0$, $\{1.37\} = 0.37$, $\{-1,3\} = -1.3 + 2 = 0.7$.

The Riemann function is given by:

$$f(x) = \begin{cases} \dfrac{1}{n}, & \text{if } x = \dfrac{m}{n}, \text{ where } x \text{ and } y \text{ are mutually prime;} \\ 0, & \text{if } x \text{ is irrational.} \end{cases}$$

This function is defined by describing the relationship between x and y.

Many so-called "arithmetic functions" also fall into this class. Examples are integer-valued functions of integer argument: the function $n! = 1 \cdot 2 \cdot 3 \ldots \cdot n$; the function $\tau(n)$ which gives the number of divisors of the number n, for example, $\tau(16) = 5$, $\tau(12) = 6$, etc.; the function $\varphi(n)$ that gives the number of positive integers mutually prime with n, for example, $\varphi(12) = 4$, $\varphi(16) = 8$.

1.2. Classification of functions

1.2.1. Composition of functions

Let $f: X \to Y$ and $g: Y \to Z$. Because $f(X) \subset Y$, the function g maps each element $f(x) \in Y$ into an element $g(f(x)) \in Z$. Thus, to every element $x \in X$ there is a corresponding element $(g \circ f)(x) = g(f(x))$ in Z, that is the functions f and g define a new function (mapping) $g \circ f$, which is called the *composition* of functions.

The domain of the composition of functions, $z = g(f(x))$, is a part of the domain of the function $y = f(x)$ such that y belongs to the domain of the function $z = g(y)$.

Example 1.2.1. Let $y = \sin z$, where $z = x^3$. The functions $y = \sin z$ and $z = x^3$ are defined on the entire real axis. Then the composition of these functions is $y = \sin x^3$ and is defined on $(-\infty, +\infty)$.

Example 1.2.2. Let $y = \sqrt{4 - x^2}$, where $y = \sqrt{z}$, $z = 4 - x^2$. The function $y = \sqrt{z}$ is defined for $z \ge 0$, i.e. $4 - x^2 \ge 0$ or $|x| \le 2$. Although the function $z = 4 - x^2$ is defined for all x, the domain of the function $y = \sqrt{4 - x^2}$ is the interval $[-2; 2]$.

Example 1.2.3. Let $y = \sqrt{1 - x^2}$, where $x = 1 + \sin^2 \pi t$. It is clear that the domain of the function $y(x)$ is the interval $[-1; 1]$. The values of the function $x = 1 + \sin^2 \pi t$ belong to this interval if $\sin \pi t = 0$, that is for $t = 0, \pm 1, \pm 2, \ldots$. So the composition $y = y(x(t))$ is defined on the set of integers Z.

Working with the composition of functions one must to take care of the domains of the functions. For example, one cannot take composition of the functions $y = \arccos z$ and $z = 3 + x^4$. Indeed, the expression $y = \arccos(3 + x^4)$ does not make sense because the function $y = \arccos z$ is defined on the segment $[-1; 1]$ but values of the function $3 + x^4$ do not lie in $[-1; 1]$ for any x.

It is possible to consider compositions of an arbitrary number of functions. For example, let $f(x) = \dfrac{1}{1 - x}$. Then

$$f(f(f(x))) = \frac{1}{1 - f(f(x))} = \frac{1}{1 + \frac{1-x}{x}} = x, \quad x \neq 0 \quad x \neq 1.$$

1.2.2. Inverse functions

Let f be a mapping from a set X onto a set Y. If for any element y from the set Y there exists a unique element $x = g(y)$ from the set X such that $f(x) = y$, then the mapping f is called *invertible*. The mapping g is called the *inverse* of f and is denoted by f^{-1}.

The domain of the inverse function f^{-1} is the range of the function f; the range of the function f^{-1} is the domain of f. For any $x \in X$ we have the following identity:

$$f^{-1}(f(x)) \equiv x, \quad x \in X.$$

Note that if the function $x = f^{-1}(y), y \in Y$, is the inverse of the function $y = f(x), x \in X$, then the function $y = f(x)$, $x \in X$, is the inverse of the function $x = f^{-1}(y)$, $y \in Y$, and $f(f^{-1}(y)) \equiv y$, $y \in Y$.

A necessary and sufficient condition for a function to have an inverse is that the function takes different values at different points of the domain. If $f: X \to Y$ is such that $f(x_1) \neq f(x_2)$ for $x_1 \neq x_2$ and $x_1, x_2 \in X$, then $f^{-1}: Y \to X$ and $f^{-1}(f): X \to X$, $f(f^{-1}): Y \to Y$, moreover $f^{-1}(f(x)) \equiv x, x \in X$, and $f(f^{-1}(y)) \equiv y, y \in Y$. The pair of functions f and f^{-1} are inverses of each other.

To prove that a function does not have an inverse, it is enough to find two values of the argument $x_1 \neq x_2$ such that $f(x_1) = f(x_2)$. For example, the function $f(x) = x^2$, $x \in (-\infty; +\infty)$, does not have an inverse (Figure 1.7), because $-3 \in (-\infty; +\infty)$ and $3 \in (-\infty; +\infty)$, and $9 = f(-3) = f(3)$. But the function $y = f(x) = x^2$, $x \in [0; +\infty)$, does have the inverse $x = \sqrt{y}$, because for any two points $x_1 \neq x_2$ from $[0, +\infty)$, the points $y_1 = x_1^2$ and $y_2 = x_2^2$ are distinct. Indeed, $y_1 - y_2 = x_1^2 - x_2^2 = (x_1 - x_2)(x_1 + x_2) \neq 0$.

It is easy to determine whether a function $y = f(x)$ has an inverse or not by using its graph. If any straight line parallel to the axis Ox intersects the graph of the function in a single point, then the inverse function $x = f^{-1}(y)$ exists. If one of such lines intersects the graph in two or more points, then there is no inverse function.

If a function $y = f(x)$ is strictly monotone on a set X, then the inverse function $x = f^{-1}(y)$ exists and is strictly monotone on the set Y.

When studying functions inverse to each other, f and f^{-1}, it is usual to denote independent variables by the same letter x, whereas the letter y is used to denote values of the function; in other words, for a function $y = f(x)$, $x \in X$, the inverse function is written as $y = f^{-1}(x)$, $x \in Y$. In these notations, we have the identities: $f^{-1}(f(x)) \equiv x$, $x \in X$, and $f(f^{-1}(x)) \equiv x$, $x \in Y$. For

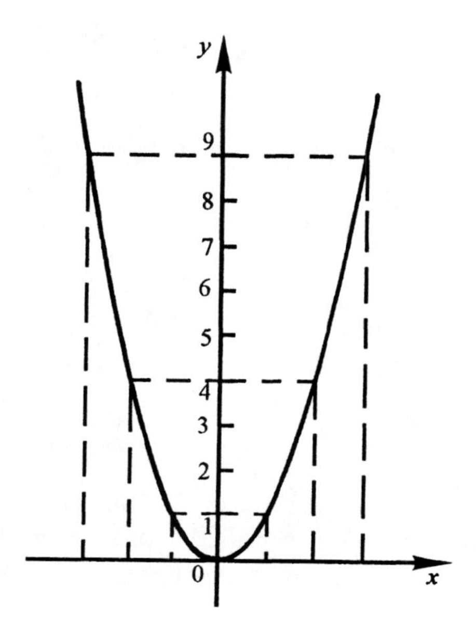

Figure 1.7. Function $y = x^2$.

example, the function $y = x + 2, x \in \mathbf{R}$, and $y = x - 2, x \in \mathbf{R}$, are inverses of each other. The same holds true for the functions $y = 3^x, x \in \mathbf{R}$, and $y = \log_3 x, x \in (0; +\infty)$.

1.2.3. Elementary functions

Principal elementary functions are the following: the power function $y = x^\alpha$ for $\alpha \in \mathbf{R}$, the exponential function $y = a^x$ for $a > 0$ and $a \neq 1$, the logarithmic function $y = \log_a x$, for $a > 0$, $a \neq 1$, the trigonometric functions $y = \sin x$, $y = \cos x$, $y = \tan x$, $y = \cot x$, and the inverse trigonometric functions $y = \arcsin x, y = \arccos x, y = \arctan x, y = \operatorname{arccot} x$.

Principal elementary functions and functions that can be constructed by using a finite number of operations of addition, subtraction, multiplication, division, and compositions are called *elementary*. For example, the functions

$$y = \sqrt{1 + 3\sin^3 x},$$

$$y = \frac{\log_4 x + 3\sqrt[3]{x} + 3\tan x}{10^{2x} + 2x + 4}$$

are elementary.

There are several types of elementary functions.

The *integer rational function*, or *polynomial* is the function

$$y = a_0x^n + a_1x^{n-1} + \dots + a_n,$$

where n is a nonnegative integer (degree of the polynomial), $a_0,$, $a_1, \dots, a_n$ are constants (coefficients).

The *fractional rational function* is the function that can be written as a ratio of two integer rational functions, i.e.

$$y = \frac{a_0x^n + a_1x^{n-1} + \dots + a_{n-1}x + a_n}{b_0x^m + b_1x^{m-1} + \dots + b_{m-1}x + b_m}.$$

For example, $y = \frac{4x^3 - 2x^2 + 5}{2x + 1}$, $y = \frac{x}{x^4 - 1}$ are fractional rational functions.

Integer rational and fractional rational functions form a class of *rational functions.*

An *irrational function* is obtained by using compositions of rational functions and power functions with rational non-integer exponents. For example, such are the functions $y = \sqrt{x}$, $y = \frac{3x^2 + \sqrt{x^2 - 1}}{\sqrt[3]{1 + 6x} - 4}$, $y = \sqrt[3]{x + 1}$.

Let us prove that the function $y = \sqrt{x}$ is irrational. Suppose it is rational, i.e.

$$\sqrt{x} = \frac{P_n(x)}{Q_m(x)}, \quad x \geq 0, \tag{1.2}$$

where $P_n(x)$ is a polynomial of degree $n \geq 0$, $Q_m(x)$ — polynomial of degree $m \geq 0$. We can assume that the polynomials $P_n(x)$ and $Q_m(x)$ do not have common multiplies in the form x^k, $k > 0$. Consider identity (1.2) on a segment $[a; b]$, $b > a > 0$. Then

$$Q_m^2(x)x = P_n^2(x). \tag{1.3}$$

This implies that the polynomial $P_n^2(x)$, and hence the polynomial $P_n(x)$ itself, is divisible by x. Consequently, the degree of $P_n(x)$ is at least one, and $P_n(x) = xS_{n-1}(x)$ for some polynomial of degree $(n - 1)$. By substituting xS_{n-1} into (1.3) and dividing both sides by x, we find that $Q_m^2(x) = xS_{n-1}^2(x)$. Applying the same reasoning to $Q_m(x)$, we see that the degree of $Q_m(x)$ is also at least one and that it is divisible by x. Thus, the polynomials $P_n(x)$ and $Q_m(x)$ have the common divisor x. This means that equality (1.3) does not hold on any interval.

Rational and irrational functions make a subclass in the class of explicit algebraic functions.

The *algebraic function* is an arbitrary function $y = f(x)$ which satisfies an equation of the form

$$A_0(x)y^n + A_1(x)y^{n-1} + \dots + A_{n-1}(x)y + A_n(x) = 0,$$

where n is a positive integer, $A_0(x), A_1(x), \ldots, A_{n-1}(x), A_n(x)$ are integer rational functions of x and $A_0(x) \not\equiv 0$. For example, $y = \sqrt[3]{x^2+1}$ is an algebraic function, because it satisfies the equation $y^3 - x^2 - 1 = 0$; the function defined by the equation $y^5 + y - x^3 = 0$ is algebraic, but it is not an explicit algebraic function, because this equation cannot be solved in radicals. Hence it is an implicit algebraic function.

Elementary functions which are not algebraic are called *transcendental*. For example, the functions $y = \cos x$, $y = 2^x$, $y = \log_3 x$ are transcendental.

Let us prove that the function $y = 4^x$ is transcendental. Suppose, by contradiction, that the function $y = 4^x$ is algebraic. This means that it satisfies some algebraic equation

$$P(x, y) \equiv A_0(x)y^n + A_1(x)y^{n-1} + \ldots + A_{n-1}y + A_n(x) = 0.$$

Let $A_0(x) = \alpha x^m + \ldots$, where $\alpha \neq 0$, $m \geq 0$. Clearly, we can assume that $\alpha > 0$. Rewrite now this identity, $P(x, 4^x) \equiv 0$, in the form

$$A_0(x)\cdot 4^{nx} + A_1(x)\cdot 4^{(n-1)x} + \ldots + A_{n-1}\cdot(x)4^x + A_n(x) = 0$$

or

$$A_0(x) = -A_1(x)\cdot 4^{-x} - \ldots - A_{n-1}(x)\cdot 4^{-(n-1)x} - A_n(x)\cdot 4^{-nx}.$$

If x approaches $+\infty$, the polynomial $\alpha x^m + \ldots$ approaches $+\infty$ if $m > 0$, or remains constant if $m = 0$. The right-hand side of this identity is the sum of a finite number of terms, each of which has the form $\beta x^r 4^{-ix}$, $i > 0$ and hence approaches zero. This means that the right-hand side also approaches zero. This contradiction proves that the function 4^x is transcendental.

1.2.4. Single-valued and multivalued functions

If in the given definition of function, we let the law assign to every element x not one but several values of y from a set Y, then such a function will be called *multivalued* as opposed to the *single-valued* function defined above. For example, the function $y = \pm\sqrt{x}$ is a two-valued function. Sometimes, instead of a two-valued function, one considers two single-valued functions, the branches of the two-valued function, e.g. $y = \sqrt{x}$ and $y = -\sqrt{x}$. The function $\arcsin x$ is multivalued.

1.2.5. Bounded and unbounded functions

A function f defined in a set X is called *bounded* on a set $X_1 \subset X$ if $f(X_1)$, i.e. the set of its values on X_1, is bounded. In other words, the function f is called bounded on a set X_1 if there are constants m and M such that $m \leq f(x) \leq M$ for all x in X_1. Otherwise, it is called *unbounded*.

The number $m_0 = \inf_{x \in X_1} \{f(x)\}$ is called the *greatest lower bound* of the function $f(x)$ on the set X_1; the number $M_0 = \sup_{x \in X_1} \{f(x)\}$ is the *least upper bound* of $f(x)$ on X_1. The difference $M_0 - m_0$ is called the *variation* of the function f on the set X_1. For example, the function $f(x) = \sin x$ is bounded on the whole real axis, because, for any x, $|\sin x| \le 1$. The function $f(x) = a^x$ is bounded from below, because for any x we have $0 < a^x$.

The function $y = \tan x$ is bounded on the set $[-\pi/4; \pi/4]$, because for x from this interval, the function satisfies $|\tan x| \le 1$. It is also bounded on any interval $[a; b]$ if $-\pi/2 < a < b < \pi/2$. However, the function is not bounded on either of the intervals $[0; \pi/2)$, $(-\pi/2; \pi/2)$.

A function $f(x)$ is *bounded at a point* x_0, if it is bounded on a neighborhood of the point x_0. For a function $y = f(x)$ to be bounded on an interval $[a; b]$, it is necessary and sufficient that it be bounded at every point of this interval.

Bounded functions possess the following properties:

1. If functions $f(x)$ and $\varphi(x)$ are defined and bounded on the same set X, then the functions $f(x) \pm \varphi(x)$, $f(x)\cdot\varphi(x)$, $|f(x)|$ and, in particular, the function $C\cdot f(x)$, where C is a constant, are defined and bounded on this set.
2. If functions $f(x)$ and $\varphi(x)$ are defined on a set X, the function $f(x)$ is bounded on this set and $\varphi(x)$ satisfies $|\varphi(x)| > M > 0$, then the function $f(x)/\varphi(x)$ is bounded on the set X.
3. If a function $f(x)$ is bounded, then the functions $a^{f(x)}$, $\sqrt[n]{f(x)}$, $\cos f(x)$, $\sin f(x)$, $\arcsin f(x)$, $\arccos f(x)$, $\arctan f(x)$, $\operatorname{arccot} f(x)$ are bounded on the set where they are defined.

1.2.6. Monotone functions

A function $f(x)$ is called *increasing* on a set X if for arbitrary x_1, x_2 from X, $x_1 < x_2$, the function satisfies $f(x_1) < f(x_2)$, i.e. greater values of the argument correspond to greater values of the function.

A function $f(x)$ is called *decreasing* on a set X if for arbitrary x_1, x_2 from X, $x_1 < x_2$, we have $f(x_1) > f(x_2)$, i.e. greater values of the argument correspond to smaller values of the function.

Functions that are increasing or decreasing are called *strictly monotone*.

If for arbitrary x_1 and x_2 from X such that $x_1 < x_2$, the function satisfies $f(x_1) \le f(x_2)$ ($f(x_1) \ge f(x_2)$), then the function $f(x)$ is called *nondecreasing* (*nonincreasing*) on X. Nondecreasing and nonincreasing functions are called *monotone*.

A function $f(x)$ is called *piecewise monotone* on a set X if the set X can be subdivided into a finite number of subsets such that the function is monotone on each of these subsets.

The sum of two increasing (decreasing) functions on X is increasing (decreasing) on X.

The product of two positive increasing (decreasing) functions on X is an increasing (decreasing) function on X. The function $Cf(x)$, where C is a positive constant and $f(x)$ is increasing (decreasing), is again an increasing (decreasing) function on X.

If a function $y = f(x)$ is increasing on X, then the function $y = -f(x)$ will be decreasing on X, and vice versa.

If a function $y = f(x)$ is increasing on X and $f(x) \neq 0$, then the function $1/f(x)$ is decreasing on X, and vice versa.

If a function $y = f(x)$ is strictly monotone on X, then the inverse single-valued function $x = f^{-1}(y)$ is strictly monotone on Y.

If a function $x = f(t)$ is increasing on $[\alpha; \beta]$, and a function $y = F(x)$ is increasing on $[f(\alpha); f(\beta)]$, then the function $y = F(f(t))$ is increasing on $[\alpha; \beta]$.

If a function $x = f(t)$ is decreasing on $[\alpha; \beta]$, and a function $y = F(x)$ is decreasing on $[f(\beta); f(\alpha)]$, then the function $y = F(f(t))$ is increasing on $[\alpha; \beta]$.

If a function $x = f(t)$ is increasing on $[\alpha; \beta]$, and a function $y = F(x)$ is decreasing on $[f(\alpha); f(\beta)]$, then the function $y = F(f(t))$ is decreasing on $[\alpha; \beta]$.

Let $\varphi(x)$, $\psi(x)$, and $f(x)$ be monotone increasing functions on X. Then if $\varphi(x) \leq f(x) \leq \psi(x)$, we have that $\varphi(\varphi(x)) \leq f(f(x)) \leq \psi(\psi(x))$.

Note that a nonmonotone function $y = f(x)$, $-\infty < x < +\infty$, can have a single-valued inverse function, if the equation $y = f(x)$ has a unique solution for every fixed $y \in (-\infty; +\infty)$.

All principal elementary functions are monotone either on the entire domain or on some subsets of the domain. In particular, if a function $f(x)$ is increasing (decreasing) on X, then the function $a^{f(x)}$ is increasing (decreasing) on X for $a > 1$, and is decreasing (increasing) on X for $0 < a < 1$; if $f(x) > 0$, the function $\log_a f(x)$ is increasing (decreasing) on X for $a > 1$, and is decreasing (increasing) on X for $0 < a < 1$; if $f(x) \geq 0$, then the function $\sqrt{f(x)}$ is increasing (decreasing) on X.

1.2.7. Even and odd functions

A number set X is called *symmetric* if for an arbitrary $x \in X$, the number $(-x)$ belongs to X. For example, the set of integers, interval $[-a; a]$, interval $(-a; a)$, interval $(-\infty; +\infty)$ are symmetric with respect to the origin.

A function f defined on a symmetric set X is called *even* on X if for all $x \in X$, the equality $f(-x) = f(x)$ holds true.

A function f defined on a symmetric set X is called *odd* on X if for all $x \in X$, we have $f(-x) = -f(x)$.

The functions $y = x^2$, $y = \cos x$, $y = f(|x|)$ are even, the functions $y = x^3$, $y = \tan x$, $y = |x|/(2x)$ are odd.

Main properties of even (odd) functions defined on the same symmetric set X are the following.

- The sum of a finite number of even (odd) functions is an even (odd) function.
- The product of odd functions is an even function if the number of terms in the product is even, and it is an odd function if the number of terms is odd. In particular, the product of two even or odd functions is even.
- The product of an even and an odd function is odd.

- If $y = f(x)$, $x = \varphi(t)$ are odd functions, then the composition $y = f(\varphi(t))$ is odd.
- If the function $y = f(x)$ is even and $x = \varphi(t)$ is odd, then the function $y = f(\varphi(t))$ is even.
- If a function $x = \varphi(t)$ is even, then the function $f(\varphi(t))$ is even.
- If $y = f(x)$ is even and $f(x) \neq 0$, then the function $1/f(x)$ is even.
- For both even and odd functions, we have $|f(x)| = |f(-x)|$.

It follows from the definition that it does not make sense to say that a function is even or odd if it is defined on a set which is not symmetric. Thus the symmetry of the domain is a necessary condition for a function to be even or odd. However, this condition is not sufficient. For example, domains of the functions $y = x + 3^x$, $y = 3^x$ are symmetric (it is the whole real axis) but the functions are neither even nor odd.

Functions that are neither even nor odd are called *general type functions*.

It should be noted that an arbitrary function defined on a symmetric set can be expressed as a sum of an even and an odd function,

$$f(x) = \frac{f(x) + f(-x)}{2} + \frac{f(x) - f(-x)}{2},$$

where the first term is an even function, the second term is odd. Such a representation is unique. For example, we have $e^x = \frac{e^x + e^{-x}}{2} + \frac{e^x - e^{-x}}{2}$ for the general type function e^x defined on the interval $(-\infty; +\infty)$.

Note that if $f(x)$ is an even function, and $\varphi(x)$ is odd and not identically zero, and both functions are defined on the whole real axis, then each of the functions $|f(x)|$, $|\varphi(x)|$, $f(-x) + \varphi(|x|)$ is even, and all of the functions $\varphi(-x)$, $xf(x) + x^2\varphi(x)$, $\varphi(x|x|)$ are odd.

The notion of even functions can be extended to a definition of functions that are symmetric with respect to a vertical line. A function $y = f(x)$ is called *symmetric with respect to a vertical line* that passes through a point $(a; 0)$ if the domain of the function is symmetric with respect to the point $(a; 0)$ and, for every x from the domain, we have $f(2a - x) = f(x)$.

For example, the function $y = \sin x$ is symmetric with respect to the vertical line that passes through the point $(\pi/2; 0)$.

1.2.8. Periodic functions

Let f be defined on a set X. If there exists $\omega \neq 0$ such that for every $x \in X$, the numbers $x + \omega$ and $x - \omega$ also belong to the set X and $f(x + \omega) = f(x)$ ($f(x - \omega) = f(x)$), then the function is called *periodic* with period ω. For example, functions $\sin x$ and $\cos x$ are periodic and have periods $2\pi n$, where $n = \pm 1, \pm 2, \pm 3, \ldots$, the functions $\tan x$ and $\cot x$ — have periods $n\pi$, $n = \pm 1, \pm 2, \pm 3, \ldots$.

If a periodic function has period ω, then it also has an infinite set of periods: $2\omega, 3\omega, \ldots, -\omega, -2\omega, -3\omega, \ldots$.

If ω_1 and ω_2 are periods of a function $f(x)$, $x \in X$, and $\omega_1 \neq \omega_2$, then $\omega_1 \pm \omega_2$ will also be a period of this function;

If ω_1 is a period of a function $f(x)$, then $n\omega_1$ is also a period of f for any $n \in Z \{0\}$.

If ω_1 and ω_2 are periods of a function $f(x)$, $n\omega_1 + k\omega_2 \neq 0$ for arbitrary integers n and k, then $n\omega_1 + k\omega_2$ are also periods of the function.

The smallest positive period (if it exists) is called *principal.* It can happen that no smallest positive period exists. For example, the function $f(x) = 3$ has each real number as a period, and there is no least positive real number. In general, any function $f(x) = \text{const}$ is periodic because it has the same value for any x, i.e. $f(x + \alpha) = f(x)$, $\alpha \neq 0$. Thus, any number α will be a period of the function, which means that there is no principal period. If a function $f(x)$ is continuous, periodic, and not a constant function, then it always has a principal period.

A function is called *antiperiodic* if $f(x + \omega) = -f(x)$, $\omega > 0$. In this case, $f(x)$ will be periodic with the period 2ω. Indeed,

$$f(x + 2\omega) = f(x + \omega + \omega) = -f(x + \omega) = -(-f(x)) = f(x).$$

It follows from the definition of a periodic function that the sum and product of two functions that are periodic with the same period ω are also periodic with the period ω. If ω is a principal period of each of the functions, then it need not be a principal period of the sum or product of these functions. For example, the principal period of the functions $f_1(x) = 3\sin x + 2$ and $f_2(x) = 2 - 3\sin x$ is 2π, but the sum is $f_1(x) + f_2(x) = 4$ does not have minimal positive periods. The principal period of the functions $\varphi_1(x) = \sin x + 1$ and $\varphi_2(x) = 1 - \sin x$ is 2π, the product $\varphi_1(x)\cdot\varphi_2(x) = \cos^2 x = \frac{1 + \cos 2x}{2}$ has the smallest positive period equal to π.

The following statements can be used to determine the period of a function:

- if $f(x)$ is periodic with period ω, then the function $y(x) = f(ax)$ is periodic with period τ, where $\tau = \frac{\omega}{|a|}$. It is assumed that $a \neq 0$ (a is an arbitrary real number), and the numbers x and ax belong to the domain of $f(x)$;
- if $f(x)$ is periodic with period ω, then the function $f(ax + b)$ is periodic with period $\frac{\omega}{|a|}$, $a \neq 0$;
- if for all x and some ω, $f(x + \omega) = \frac{1}{f(x)}$, then the function $f(x)$ is periodic with period 2ω;
- if the function $u = \varphi(x)$ is periodic, then the composition $y = f(\varphi(x))$ is also periodic and has the same period as in the case where the function $y = f(x)$ is strictly monotone; the period of the composition can be less than the period of the function $u = \varphi(x)$, if $y = f(u)$ is not strictly monotone. For example, the periods of the functions $y = 3^{\sin x}$, $y = \log_4 \cos x$ is $\omega = 2\pi$.

The period of an algebraic sum of periodic functions is the least common multiple of all periods of all the terms in the sum, excluding those terms the sum of which equals zero. In particular, periods of the functions:

$$f(x) = \sum_{i=1}^{n} a_i \sin b_i x + \sum_{k=1}^{m} c_k \cos d_k x, \tag{1.4}$$

$$f(x) = \sum_{i=1}^{n} a_i \tan b_i x + \sum_{k=1}^{m} c_k \cot d_k x, \tag{1.5}$$

where a_i, c_k are real numbers different from zero, b_i, $d_k \in \mathbf{N}$, can be found from the corresponding formulas:

$$\omega = \frac{2\pi}{D(b_1, b_2, \ldots, b_n, d_1, d_2, \ldots, d_m)}, \quad \omega = \frac{\pi}{D(b_1, b_2, \ldots, b_n, d_1, d_2, \ldots, d_m)},$$

where $D(\ldots)$ is the greatest common divisor.

If b_i, d_k are rationals ($b_i = \frac{p_i}{q_i}$, $d_k = \frac{r_k}{s_k}$), then periods of functions (1.4) and (1.5) can be respectively found from the formulas:

$$\omega = \frac{2\pi K(q_1, q_2, \ldots, q_m, s_1, s_2, \ldots s_m)}{D(p_1, p_2, \ldots, p_n, r_1, r_2, \ldots r_m)},$$

$$\omega = \frac{\pi K(q_1, q_2, \ldots, q_m, s_1, s_2, \ldots s_m)}{D(p_1, p_2, \ldots, p_n, r_1, r_2, \ldots r_m)},$$

where $D(\ldots)$ is the greatest common divisor, $K(\ldots)$ is the least common multiple.

The function which is an algebraic sum, product, or quotient of two or more functions that have different periods need not be periodic. For example, the functions $f(x) = \sin \alpha x + \cos \beta x$, $f(x) = \tan \alpha x + \cot \beta x$ are periodic if the number α/β is rational, and are not periodic if α/β is irrational.

Periodic functions possess the following properties.

1. If a point x belongs to the domain of a periodic function with period ω, then all points $x + n\omega$, $n \in \mathbf{Z}$, also belong to the domain of the function.
2. A periodic function cannot have a finite number of points of discontinuity in its domain.
3. A periodic function takes each of its values at infinitely many points x, among which there are positive and negative numbers, and numbers with arbitrarily large absolute value. In particular, a periodic function can not be strictly monotone on its domain.
4. If $f(x)$ is a periodic function defined on the whole real axis, then the equation $f(x + \omega) = f(x)$ with respect to ω, and x being a parameter, has at least one positive solution $\omega = \tilde{\omega}$, which makes the equation hold for every value of x. In particular, if for a function $f(x)$ there are numbers $x = a$ and $x = b$ such that the equations $f(a + \omega) = f(a)$ and $f(b + \omega) = f(b)$ do not have a common positive solution $\omega = \tilde{\omega}$, then the function $f(x)$ is not periodic.
5. If a function $f(x)$ with period ω is bounded on an interval $[a; a + |\omega|]$ that belongs to the domain of the function, then the function is bounded on the whole domain. In particular, if a periodic function $f(x)$ is continuous on the whole real axis, then there exists a number

$M > 0$ such that $f(x) \leq M$ for all $x \in \mathbf{R}$. If a function is continuous on the whole real axis but not bounded, then it is not periodic.

6. If functions $f_1(x)$ and $f_2(x)$ are defined on the whole real axis, periodic with periods $\omega_1 > 0$ and $\omega_2 > 0$ correspondingly, with ω_1/ω_2 — a rational number, then the functions $f_1(x) \pm f_2(x)$ and $f_1(x) \cdot f_2(x)$ are periodic.

In practice, it is easy to determine whether the function is periodic by using the definition and properties of periodic functions.

Remark 1.2.1. A function, which is defined on the whole real axis but at one point, will not be periodic. The sum of the functions $f_1(x) = x + \sin x$ and $f_2(x) = -x + \sin x$, each of which is not periodic, will be periodic. Product of the functions

$$f_1(x) = \begin{cases} \sin x, & x \neq \pi/2, \\ 1/2, & x = \pi/2, \end{cases} \quad f_2(x) = \begin{cases} \sin x, & x \neq \pi/2, \\ 2, & x = \pi/2. \end{cases}$$

will be a periodic function.

1.3. Limit of a function. Continuity of a function

1.3.1. Limit of a number sequence

Let a function $f(x)$ be defined on the set $N = \{1,2,3,...,n,...\}$ of all natural numbers. Such a function is called the *infinite number sequence.* Denoting $f(n)$ by x_n, the number sequence is written as $\{x_n\}$, or, in a more explicit form, as $x_1, x_2, \ldots, x_n, \ldots$.

A sequence is considered to be given if there is a law by which any member of the sequence can be calculated.

A number a is called the *limit of a number sequence* $\{x_n\}$ if for every $\varepsilon > 0$ there exists $N(\varepsilon)$ such that $|x_n - a| < \varepsilon$ for all x_n with $n > N(\varepsilon)$. It is denoted by $\lim\limits_{n \to \infty} x_n = a$ or $x_n \to a, n \to \infty$.

The geometric interpretation of the limit of a sequence is the following: if a is the limit of a sequence $\{x_n\}$, then all members of the sequence, starting with $n = N + 1$, belong to the ε-neighborhood of the point a (Figure 1.8).

A number sequence which has a limit is called *converging*; if it does not have a limit, then it is called *divergent.*

A sequence $\{x_n\}$ is convergent if for every $\varepsilon > 0$ there exists $n_0 \in \mathbf{N}$ such that if $m,n > n_0$, then $|x_m - x_n| < \varepsilon$. One can consider $m > n$. Then $m = n + p$, where p is a positive integer, and the last inequality becomes $|x_{n+p} - x_n| < \varepsilon$ for all $n > n_0$ and $p \in \mathbf{N}$.

Main theorems about the limit of a sequence

1. A sequence has only one limit.
2. A sequence that has a finite limit is bounded. A sequence that has an infinite limit is unbounded.
3. A monotone bounded sequence has a finite limit; if the sequence is increasing, then $\lim\limits_{n \to \infty} x_n \geq x_n$; if the sequence is decreasing, then $\lim\limits_{n \to \infty} x_n \leq x_n$.
4. A monotone unbounded sequence has an infinite limit: if this sequence is increasing, then $\lim\limits_{n \to \infty} x_n = +\infty$; if the sequence is decreasing, then $\lim\limits_{n \to \infty} x_n = -\infty$.

1.3.2. Limit of a function

Let a function $f(x)$ be defined on some neighborhood of a point a. A number b is called the *limit* of the function $y = f(x)$ for $x \to a$ if for every $\varepsilon > 0$ there exists $\delta(\varepsilon) > 0$ such that, for all x different from a and satisfying $|x - a| < \delta$, we have $|f(x) - b| < \varepsilon$. This is denoted by $b = \lim\limits_{x \to a} f(x)$.

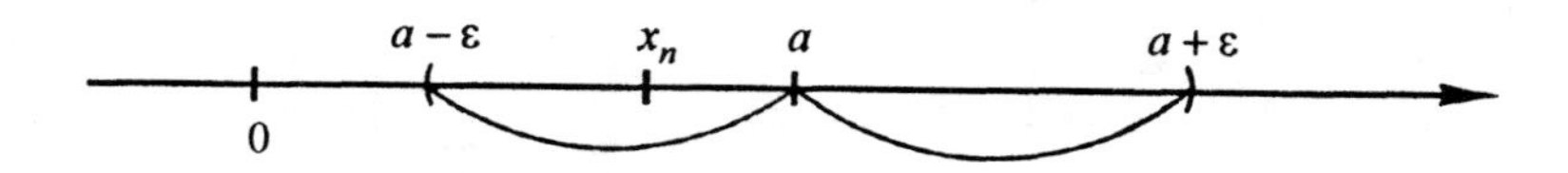

Figure 1.8. Geometric meaning of the limit of the sequence $\{x_n\}$.

Because $|x - a| < \delta \Rightarrow |f(x) - b| < \varepsilon$, geometrically this means that, for all points x lying from a at a distance not exceeding δ, the points on the graph $y = f(x)$ lie inside the strip of width 2ε and which is bounded by straight lines $y = b - \varepsilon$, $y = b + \varepsilon$ (Figure 1.9).

Criteria for existence of a limit

1. The number b is the limit of a function $f(x)$ at $x = a$ if for arbitrary sequence of x (x_1, x_2, ..., x_n, ...), which belongs to domain of the function and has a as its limit, the sequence of the corresponding values $f(x_1), f(x_2), ..., f(x_n), ...$ converges to b.
2. Another limit existence criterion is the *Cauchy criterion*. For a function $f(x)$ to have a finite limit at $x = a$, it is necessary and sufficient that for every $\varepsilon > 0$ there exist $\delta = \delta(\varepsilon) > 0$ such that, for any x_1, x_2 from the domain of the function and $|x_1 - a| < \delta$, $|x_2 - a| < \delta$, we have $|f(x_1) - f(x_2)| < \varepsilon$.

One-sided limits

A number b is called the *right-hand limit* of a function $f(x)$ at a point a if for every $\varepsilon > 0$ there exists $\delta > 0$ such that $|f(x_1) - b| < \varepsilon$ for all x, $a < x \leq a + \delta$. The right-hand limit of a function $f(x)$ at a point a is denoted by $\lim\limits_{x \to a+0} f(x)$ or $f(a + 0)$.

A number b is called the *left-hand limit* of a function $f(x)$ at a point a if for every $\varepsilon > 0$ there exists $\delta > 0$ such that $|f(x) - b| < \varepsilon$ for all x, $a - \delta \leq x < a$. The left-hand limit is denoted by $\lim\limits_{x \to a-0} f(x)$ or $f(a - 0)$.

For a number b to be a limit of a function $f(x)$ for $x \to a$, it is necessary and sufficient that both the left-hand and the right-hand limits $f(a - 0)$, $f(a + 0)$ exist and equal b.

A function $f(x)$ has $+\infty$ as its limit for $x \to a$, if for an arbitrarily large positive M there exists $\delta > 0$ such that $f(x) > M$ for all x satisfying $|x - a| < \delta$.

A function $f(x)$ has the limit $-\infty$ for $x \to a$, if for an arbitrarily large positive M there exists δ, $\delta > 0$, such that $f(x) < -M$ for all x, $|x - a| < \delta$.

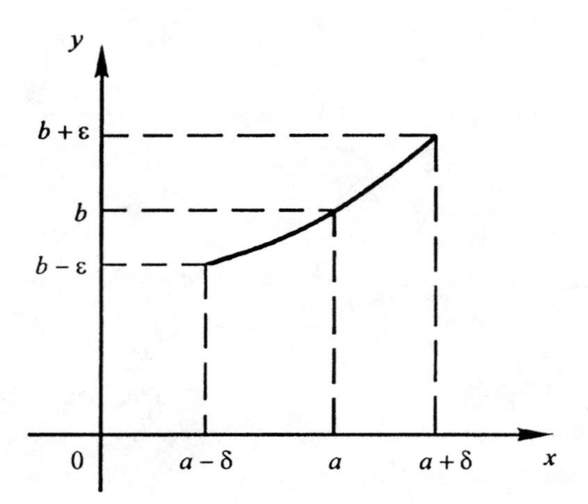

Figure 1.9. Geometric meaning of the limit of function $\lim_{x \to a} f(x) = b$.

A function $f(x)$ has the limit ∞, if for an arbitrarily large positive M there exists $\delta > 0$ such that $|f(x)| > M$ for all x, $|x - a| < \delta$.

A number b is a limit of a function $f(x)$ for $x \to +\infty$ if for every $\varepsilon > 0$ there is an arbitrarily large positive number N such that $|f(x) - b| < \varepsilon$ for all x whenever $x > N$.

A number b is a limit of a function $f(x)$ for $x \to -\infty$ if for every $\varepsilon > 0$ there exists an arbitrarily large positive number N such that $|f(x) - b| < \varepsilon$ for all x, $x < -N$.

A number b is a limit of a function $f(x)$ for $x \to \infty$ if for every $\varepsilon > 0$ there exists an arbitrarily large positive number N such that $|f(x) - b| < \varepsilon$ for all x, $|x| > N$.

Theorems about limits

1. The limit of a constant function equals the value of the constant, $\lim a = a$.
2. The limit of an algebraic sum of a finite number of functions that have limits equals the algebraic sum of the limits of these functions,

$$\lim_{x \to a} \left(f_1(x) \pm f_2(x) \pm \ldots \pm f_n(x)\right) = \lim_{x \to a} f_1(x) \pm \lim_{x \to a} f_2(x) \pm \ldots \pm \lim_{x \to a} f_n(x).$$

3. The limit of the product of a finite number of functions equals the product of the limits if they exist,

$$\lim_{x \to a} \left(f_1(x) f_2(x) \ldots f_n(x)\right) = \lim_{x \to a} f_1(x) \lim_{x \to a} f_2(x) \ldots \lim_{x \to a} f_n(x).$$

4. The limit of the quotient of two functions equals the quotient of the limits of these functions if the limits of the numerator and denominator exist and the latter is different from zero,

$$\lim_{x \to a} \frac{f(x)}{\varphi(x)} = \frac{\lim_{x \to a} f(x)}{\lim_{x \to a} \varphi(x)}, \qquad \lim_{x \to a} \varphi(x) \neq 0.$$

5. If a function $f(x)$ satisfies the inequality $\varphi(x) \le f(x) \le \psi(x)$ and $\lim_{x \to a} \varphi(x) = b$, $\lim_{x \to a} \psi(x) = b$, then $\lim_{x \to a} f(x) = b$.

6. A monotone function of a continuous argument has a limit (finite or infinite) for any value of x (finite or infinite). A monotone bounded function has a finite limit for any value of x.

We list some important limits:

1) $\lim_{x \to 0} \frac{\sin x}{x} = 1$, where x is the length of the arc or angle expressed in radians;

2) the number e (irrational): $\lim_{x \to \infty} \left(1 + \frac{1}{x}\right)^x = e = 2.71828...$;

3) the number C (Euler's constant): $\lim_{n \to \infty} \left(1 + \frac{1}{2} + \frac{1}{3} + ... + \frac{1}{n} - \ln n\right) = C = 0.5772...$;

4) for two functions $f(x)$ and $g(x) > 0$, which are defined on a domain D, the symbol $f(x) = O(g(x))$ (the Landau symbol) means that there exists a constant $K > 0$ such that $|f(x)| \le Kg(x)$ for all $x \in D$.

Infinitely small functions

A function $f(x)$ is called *infinitely small* as $x \to x_0$, if $\lim_{x \to x_0} f(x) = 0$. If $\lim_{x \to a} f(x) = b$, then it is always possible to write $f(x) = b + \alpha(x)$, where $\alpha(x)$ is an infinitely small function as $x \to x_0$.

Let $\alpha(x)$ and $\beta(x)$ be infinitely small functions as $x \to x_0$. If $\lim_{x \to x_0} \frac{\alpha(x)}{\beta(x)} = L$, L = const, $L \neq 0$, then the infinitely small functions $\alpha(x)$ and $\beta(x)$ are said to have the same *order of smallness*, and it is denoted by $\alpha = O(\beta)$, $\beta = O(\alpha)$.

Let $\alpha(x)$ and $\beta(x)$ be infinitely small functions as $x \to x_0$, and k be a fixed positive number.

If $\alpha(x)$ and $\beta^k(x)$ are infinitely small functions of the same order of smallness, then $\alpha(x)$ is called an infinitely small function of order k with respect to the infinitely small function $\beta(x)$.

If $\lim_{x \to x_0} \frac{\alpha(x)}{\beta(x)} = 0$, then $\alpha(x)$ is said to have a higher order of smallness than $\beta(x)$; this is denoted by $\alpha = o(\beta)$. We say then that the order of smallness of the function β is lower than the order of the function $\alpha(x)$.

If $\lim\limits_{x\to x_0} \dfrac{\alpha(x)}{\beta(x)} = 1$, i.e. $\alpha(x) - \beta(x) = o(\alpha)$, then the infinitely small functions $\alpha(x)$ and $\beta(x)$ are called *equivalent* and this is denoted by $\alpha \sim \beta$.

If the limit $\lim\limits_{x\to x_0} \alpha(x)/\beta(x)$ does not exist, then the functions $\alpha(x)$ and $\beta(x)$ are called *incomparable*.

When finding a limit of two infinitely small functions, each of them can be replaced by an equivalent infinitely small function.

1.3.3. Continuity of a function

A function $y = f(x)$ is called *continuous* at a point $x = a$ if a belongs to the domain of the function, $\lim\limits_{x\to a} f(x)$ exists and equals the value of the function at the point a, that is the function $f(x)$ is defined at $x = a$, and for all $\varepsilon > 0$ there exists $\delta = \delta(\varepsilon, a) > 0$ such that $|f(x) - f(a)| < \varepsilon$ for all x, $|x - a| < \delta$.

The notion of continuity can be reformulated by using increments.

The *increment* of the argument x at a point x_0 is the difference $\Delta x = x - x_0$. The increment of a function $f(x)$ at a point x_0 is the difference $\Delta f(x_0) = f(x) - f(x_0)$ (see Figure 1.10).

The increments of the argument and the function could be both positive and negative.

A function $f(x)$ defined in a neighborhood of a point a is called *continuous* at the point a if infinitely small increments of the argument correspond to infinitely small increments of the function.

Continuity criterion

For a function $f(x)$ which is defined in a neighborhood containing a point a, to be continuous at the point a, it is necessary and sufficient that, for any sequence of points x_n, $n = 1,2,...$, which lies in this neighborhood and converges to the point a, the sequence $f(x_n)$, $n = 1, 2, ...$, converge to $f(a)$.

A function which is continuous at every point of a set X is said to be *continuous* on this set.

If $f(a - 0) = f(a)$ ($f(a + 0) = f(a)$), then the function $f(x)$ is called *right (left) continuous* at the point a. For a function to be continuous at a point $x = a$, it is necessary and sufficient that $f(a - 0) = f(a + 0) = f(a)$.

That value of a belonging to the interior or the boundary of the domain of the function, at which the function is not defined or the value $f(a)$ does not equal the value $\lim\limits_{x\to a} f(x)$, or the limit $\lim\limits_{x\to a} f(x)$ does not exist, is called the *point of discontinuity*.

There are points of discontinuity of the first and second kind. A *point of discontinuity of the first kind* is the point where the one-sided limits $f(a - 0) = \lim\limits_{x\to a-0} f(x)$ and $f(a + 0) = \lim\limits_{x\to a+0} f(x)$ exist. If a is a point of discontinuity of the first kind of a function $f(x)$ and the left-hand limit of

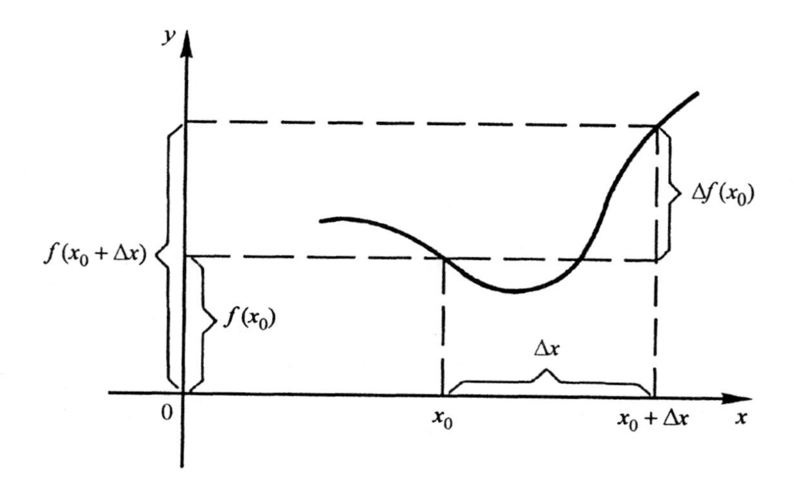

Figure 1.10. Argument and function increments.

the function is not equal to the right-hand limit, then the absolute value of the difference between these limits is called the *jump* of the function $f(x)$ at the point a.

A point a is called the *point of discontinuity of the second kind* if at least one of the one-sided limits does not exist.

A *removable point* is a point of discontinuity of the function $f(x)$ if $\lim_{x \to a} f(x)$ exists, i.e. $f(a-0) = f(a+0)$, but the function is not defined at $x = a$ or $f(a) \neq \lim_{x \to a} f(x)$.

If at least one of the limits $f(a-0)$ or $f(a+0)$ equals ∞, then the point $x = a$ is called the *point of infinite discontinuity*.

Theorem on continuity of a monotone function. If the set of values that a monotone increasing (decreasing) function $f(x)$ takes as x runs over a set X is contained in some set Y and fills this set completely, then the function $f(x)$ is continuous on X.

By using theorems about properties of limits of functions, one deduces a *theorem on algebraic operations with continuous functions*: if functions $f(x)$ and $g(x)$ are continuous at a point a, then the functions $f(x) \pm g(x), f(x)g(x), \dfrac{f(x)}{g(x)}$ (if $g(a) \neq 0$) are continuous}. The *theorem on continuity of elementary functions* follows from this and the previous theorem: *any elementary function is continuous in its domain*.

Theorem on composition of continuous functions. Let a function $\varphi(y)$ be defined on a set Y, a function $f(x)$ — on a set X such that its values are contained in Y as x ranges over X. If $f(x)$ is continuous at a point $x = a \in X$ and $\varphi(y)$ is continuous at the corresponding point $y_0 = f(a) \in Y$, then the composition $\varphi(f(x))$ is continuous at the point $x = a$.

Main properties of functions continuous on a closed interval

First Weierstrass theorem. A function continuous on a closed interval $[a; b]$ is bounded on this interval.

Second Weierstrass theorem. If a function $f(x)$ is continuous on a closed interval $[a; b]$, then it attains the greatest and the least value on this interval.

First Bolzano—Cauchy theorem. If a function $f(x)$ is continuous on a closed interval $[a; b]$ and the values the function takes at the ends of this interval have opposite signs, then there exists at least one point $c \in (a; b)$ at which the value of the function is zero, $f(c) = 0$.

Second Bolzano—Cauchy theorem. If a function $f(x)$ is continuous on a closed interval $[a; b]$ and $f(a) \neq f(b)$, then for any number A between the numbers $f(a)$ and $f(b)$ there exists at least one point $c \in (a; b)$ at which the function takes the value A, $f(c) = A$.

Chapter 2

Analysis of functions for constructing graphs

2.1. Coordinate systems

2.1.1. Cartesian coordinate system

The plane with a coordinate system is called the *coordinate plane*. The *Cartesian* or *rectangular coordinate system* is defined by giving a scale and two mutually orthogonal number axes. The point of intersection of the axes is called the *origin*; the axes are called *coordinate axes*. One of these axes is called the *abscissa axis* (x-axis), the other is the *ordinate axis* (y-axis). They are denoted by Ox and Oy respectively. The positive directions are chosen so that looking from the origin in the positive direction along the x-axis, the positive direction in the y-axis will be to the left.

To every point P in the plane with a coordinate system, one can associate two numbers obtained in the following way. Draw two straight lines that pass through P and such that one is orthogonal to the x-axis, and the other — to the y-axis. Let P_x be the point at which one of the straight lines intersects the axis Ox, P_y — at which the other straight line intersects Oy (if the point P lies in the x-axis, then $P_x = P$; if it lies in the y-axis, then $P_y = P$). To the point P_x, which lies in the number axis, there is a corresponding real number x; similarly, there is a real number y corresponding to the point P_y. The numbers x and y are called *coordinates* of the point P; x is the abscissa of the point P and y is its ordinate (Figure 2.1). Each pair of real numbers can be regarded as coordinates of a point in the plane. To draw this point by using its coordinates x_0, y_0, one needs to mark two points on the axes Ox and Oy at the distance equal to x_0 and y_0 respectively. Then two straight lines perpendicular to corresponding axes are constructed. The point of intersection of these straight lines will be the needed point.

The coordinates are written as follows: first we write the x-coordinate, then a semicolon, and then the y-coordinate. The coordinates are written in parentheses. So a point P is written as $P(x_0; y_0)$. If in the plane a coordinate system is chosen, then there is a one-to-one correspondence between points in the plane and ordered pairs of real numbers.

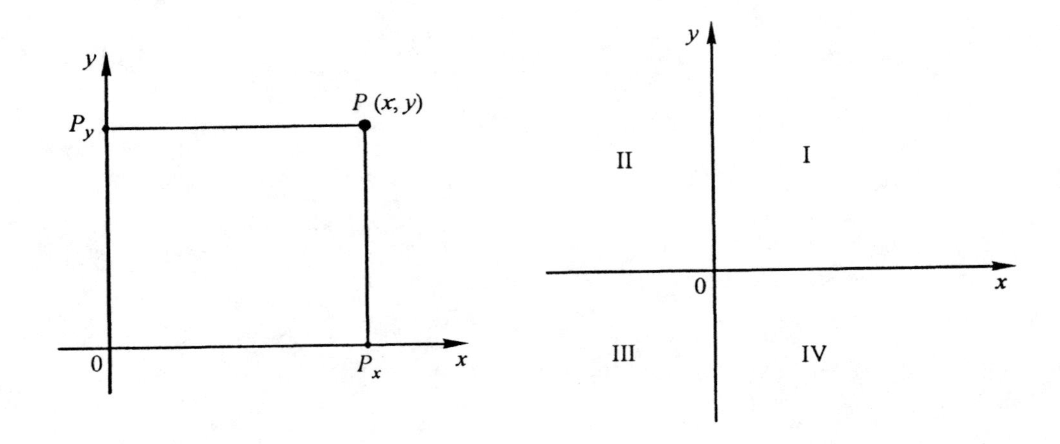

Figure 2.1. Coordinates of the point P in the plane Oxy.

Figure 2.2. Coordinate plane Oxy.

Coordinate axes divide the plane into four parts, which are called the quadrants. Each quadrant is bounded by a straight angle with the vertex at the origin (Figure 2.2). The signs of the coordinates of the points in each quadrant do not change. These are $(+;+)$ in the first quadrant, $(-;+)$ in the second, $(-;-)$ in the third, and $(+;-)$ in the forth quadrant.

2.1.2. Polar coordinate system

Polar coordinates of a point are two numbers that define the position of this point with respect to some fixed point O, the *pole*, and a certain fixed ray Ox, the *polar axis*. The first coordinate ρ of a point $M(\rho;\varphi)$ is the *polar radius*. It measures the distance from the point to the pole, $OM = \rho$. The second coordinate φ is the *polar angle*. It is the angle, positive or negative, at which the axis Ox needs to be rotated in order to coincide with the ray OM (Figure 2.3). The polar angle is positive if the axis Ox was rotated counterclockwise, and negative otherwise.

If the length of the polar radius is taken to be nonnegative, $0 \le \rho < +\infty$, and the polar angle is taken to be in the interval $0 \le \varphi < 2\pi$, then this establishes a one-to-one correspondence between the points of the plane and the polar coordinates, with an exception of the pole, to which it is impossible to assign a definite polar angle. It is convenient to use this correspondence in problems related to the construction of points in the plane, determining the distance between points, subdividing an interval in a certain proportion, etc.

While studying functions and constructing graphs, it often happens that the polar radius assumes negative values, and the polar angle — negative values or values greater than 2π. For

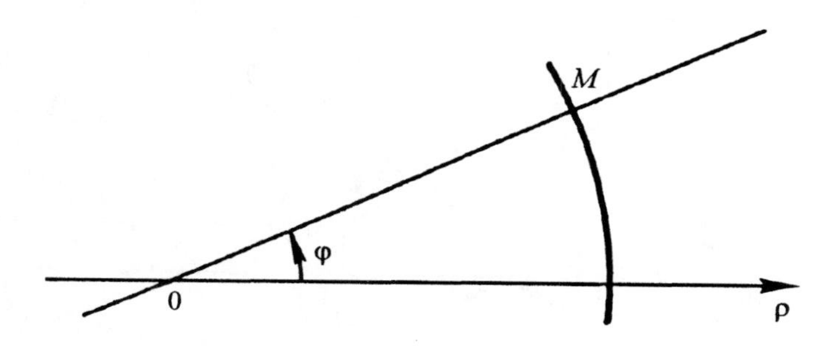

Figure 2.3. Polar coordinates of point M.

example, the function $\rho = \sin 2\varphi$ yields $\rho = -\sqrt{3}/2$ for $\varphi = 2/3\pi$; for the function $\rho = 2\varphi$, the polar angle can take any value.

It is sometimes useful to consider the values of the polar radius and polar angle ranging from $-\infty$ to $+\infty$. Negative values of φ are obtained by rotating the axis Ox clockwise; for negative values of ρ the points in the plane are constructed by continuing the ray OQ and plotting the point on the opposite side of the ray with respect to the pole. In such a case, the values of φ that lie between 0 and 2π or between $-\pi$ and π are called *principal*. In this case, an arbitrary pair of the coordinates $(\rho; \varphi)$ defines a unique point in the plane, and a point in the plane determines a set of ordered pairs of numbers. Indeed, as the value of φ runs from $-\infty$ to $+\infty$, the ray (Figure 2.4) can take the same position infinitely many times: the position of the ray Q_1Q is the same for $\varphi = \alpha$, $\varphi = \alpha + 2\pi$, $\varphi = \alpha + 4\pi$, . . . , $\varphi = \alpha + 2k\pi$, where $k \in \mathbf{Z}$. The point lying in the ray and having the polar radius $\rho = a$ is determined by one pair of the coordinates $(a; \alpha)$, $(a; \alpha + 2\pi)$, ..., $(a; \alpha + 2k\pi)$. This fact should be taken into consideration while constructing graphs of functions given in the polar coordinate system.

The transition from the Cartesian to polar coordinates and vice versa is carried out according to the formulas:

$$x = \rho \cos \varphi, \quad y = \rho \sin \varphi,$$

and

$$\rho = \sqrt{x^2 + y^2}, \quad \varphi = \arctan\frac{y}{x} \quad (x \neq 0),$$

where x, y are the Cartesian coordinates of the point $M(x; y)$, and ρ, φ are its polar coordinates. Here the origin O of the Cartesian system coincides with the pole, and the axis Ox coincides with the polar axis.

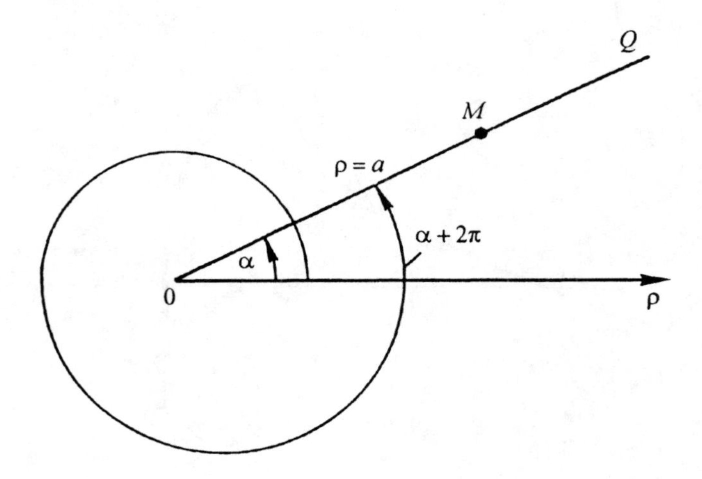

Figure 2.4. Drawing point in polar coordinates.

2.1.3. Transformations of the Cartesian coordinate system

The position of a point in the plane is determined by two coordinates in some coordinate system. The coordinates of the point change if another coordinate system is chosen. The transformation is known if coordinates of a point in one coordinate system are expressed in terms of coordinates in another system.

Translation of the origin

Under such a transformation, the position of the origin changes but the direction of the axes and the scale remain constant. Suppose we have two coordinate systems with the origins O and O_1 and having the same direction of the coordinate axes (Figure 2.5). Let a and b denote the values of coordinates of the origin O_1 in the first coordinate system, and let x, y and x', y' denote the coordinate of an arbitrary point M in the first and second coordinate systems respectively. The coordinates x and y of the point M are expressed in terms of the coordinates x', y' by $x = a + x', y = b + y'$, where $BM = y, B'M = y'$, and the coordinates x', y' of M can be calculated as $x' = x - a, y' = y - b$.

Rotation of coordinate axes

This is a transformation of the Cartesian coordinate system such that the origin and the scale do not change, but the axes are rotated by the same angle.

Let M be a point in the plane with the coordinates x and y in the first coordinate system (Figure 2.6), and let the angle between the axis Ox' of the second coordinate system and the axis Ox of the first system be α (it is the same as the angle between Oy and Oy'). The rotation angle

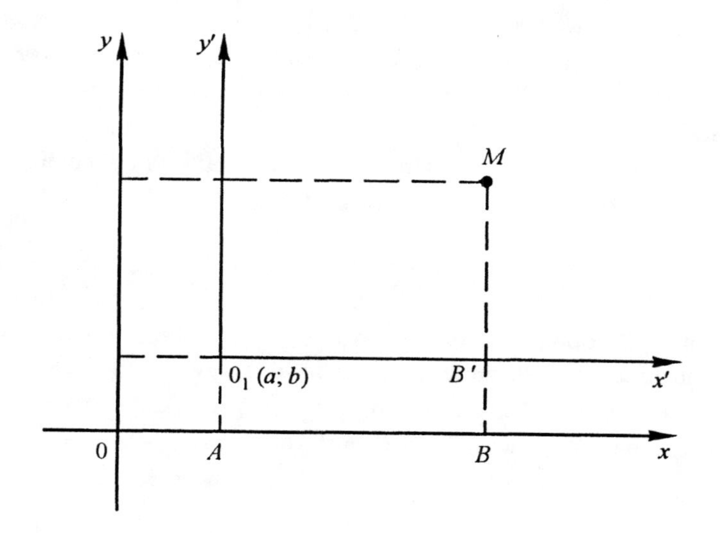

Figure 2.5. Translation of the origin.

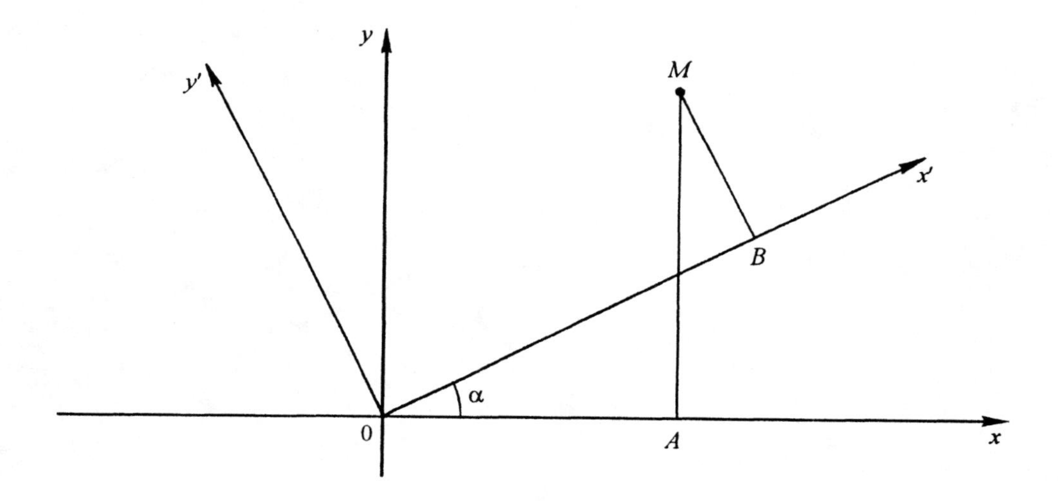

Figure 2.6. Rotation of coordinates.

is measured counterclockwise. Denote by x', y' the coordinates of the point M in the second coordinate system. The coordinates in the first system are expressed in terms of the coordinates in the second system rotated by α as: $x = x' \cos\alpha - y' \sin\alpha$, $y = x' \sin\alpha + y' \cos\alpha$. Here $OA = x$, $OB = x'$, $AM = y$, $MB = y'$.

The coordinates in the second system, expressed in terms of the coordinates in the first system, are $x' = x \cos\alpha + y \sin\alpha$, $y' = -x \sin\alpha + y \cos\alpha$.

General transformation (translation of the origin and rotation of axes)

By combining the formulas for translation of the origin and rotation of the axes, we get the following formulas for the general case, which includes, as particular cases, the translation ($\alpha = 0$) and rotation ($a = b = 0$): $x = x' \cos\alpha - y' \sin\alpha + a$, $y = x' \sin\alpha + y' \cos\alpha + b$. Note that these formulas are linear with respect to x' and y', i.e. $x = Ax' + By' + C$, $y = A_1 x' + B_1 y' + C_1$. The expressions for the coordinates in the second system are: $x' = (x - a) \cos\alpha + (y - b) \sin\alpha$, $y' = -(x - a) \sin\alpha + (y - b) \cos\alpha$. These formulas are also linear in x and y.

2.2. Analysis of functions in the Cartesian coordinate system

2.2.1. Domain of a function

The *domain of a function*, $D(f)$, is the set of all values of the argument at which the function exists and takes real values.

When a function is defined by a table, its domain is the set of all the values of x listed in the table. For x not in the table, the function may not be defined. It should be noted that the function may also be defined for all x lying between its largest and smallest value given in the table. But to calculate the value of the function for x not from the table, the calculation rule must be given (most often the linear interpolation is used).

In the graphical method of defining a function, the domain is clear from the graph; in the descriptive method, the domain is explicitly described.

If a function is given analytically, the domain of the function is understood to be all the values of x such that the formula that defines the function makes sense. Often such a domain is called natural. Additional conditions may decrease the size of the natural domain.

The domain of the function may consist of isolated points. For example, the function $f(x) = \sqrt{-(x-1)^2}$ has the domain consisting of the only point $x = 1$.

If a function is defined analytically, then the expression that gives the function may contain:

- a fraction with a denominator that becomes zero for some x;
- roots of an even degree of expressions that may become negative.

To find the domain of a function in these two cases, it is necessary to exclude all the values of x that make the denominator of fractions zero or the expressions under the roots of even degree negative. While finding the domain of a function, it is advisable not to make transformations in the formulas that define the function.

In general cases, the domains of the functions

$y = (f(x))^r$, where $r = 2k, 2k+1, \dfrac{1}{2k+1}, \dfrac{2n}{2k+1} > 1$ $k, n \in \mathbf{N}$,

$y = a^{f(x)}$, where $a|a-1| > 0$,

$y = \sin f(x), y = \cos f(x), y = \arctan f(x), y = \operatorname{arccot} f(x), y = \operatorname{sign} f(x)$

coincide with the corresponding domain of the function $f(x)$. The domain of the function $y = (f(x))^r$, where $r = \dfrac{1}{2k}, \dfrac{2n+1}{2k} > 1$, $k, n \in \mathbf{N}$, is found by using the condition $f(x) \geq 0$. For the function $y = \log_a f(x)$, where $a|a-1| > 0$, the domain is found from the condition $f(x) > 0$. The domain of the function $y = \log_{\varphi(x)} f(x)$ is determined from $f(x) > 0$, $\varphi(x) > 0$, $\varphi(x) \neq 1$. To find the domain of the function $y = \tan f(x)$, we use the condition $f(x) \neq (2k+1)\dfrac{\pi}{2}$; for the function $y = \cot f(x)$ the condition is $f(x) \neq k\pi$, where $k \in \mathbf{Z}$. For the functions $y = \arcsin f(x)$, $y = \arccos f(x)$, the domain is found from the condition $-1 \leq f(x) \leq 1$.

2.2.2. Range of a function. Graph of a bounded function

The *range of a function* $E(f)$ is the set of values of the function taken as x ranges over its domain (see Figure 5). The range of a function may be a set of isolated points, a single point, one or several intervals, etc. It is clear that for a bounded function, the range cannot be unbounded, and for an unbounded function, the range can not be bounded. For example, the range of the function $y = x^4$ is the set $[0; +\infty)$, the range of the function $y = \cos x$ is the interval $[-1; 1]$.

If a function is given graphically or by a table, then its range is determined from the graph or the table. To find the range of a function given analytically, $y = f(x)$, it is necessary to find all the values of y such that the equation $f(x) = y$ has a real solution. If the general solution of the equation $f(x) = y$ can be written as $x = F(y)$, then the range of the function f consists of those y for which the expression $F(y)$ makes sense.

If a function is defined on several intervals in different ways, then the range is found by considering each interval separately. Sometimes for finding the range of a function, it is convenient to calculate $\lim\limits_{x \to \pm\infty} y = \lim\limits_{x \to \pm\infty} f(x)$. This method is often used when the function is linear-fractional. In many instances to determine the range of the function is not an easy task: one needs to analyze the function, draw its graph.

If the range of the function is known, then it is easy to find the largest and smallest values that the function takes.

The graph of a bounded function, $|f(x)| \leq M$, is contained in the strip formed by two straight lines parallel to the x-axis and lying on both sides of the x-axis at the distance M. A function bounded below, $f(x) \geq m$, has its graph lying above the line parallel to the x-axis and drawn at the distance m from Ox. The graph of a function bounded above, $f(x) < M$, lies below the line parallel to Ox at the distance M. Thus the graph of the function $y = \sin x$ lies between the lines $y = -1$, $y = 1$, because $|\sin x| \leq 1$. The function $y = a^x$ is bounded below ($a^x > 0$), and its graph lies above the x-axis. The function $y = -\frac{1}{2}x^2$ is bounded above ($-\frac{1}{2}x^2 \leq 0$), and its graph lies below the x-axis.

Functions $y = f(x)$ and $y = \varphi(x)$ are called *identical* or *equal* on a set X, if they are defined on X and for every $x_0 \in X, f(x_0) = \varphi(x_0)$. For example, the functions $f(x) = \sqrt{x^2}$ and $\varphi(x) = |x|$ are identical on the set $\mathbf{R}$ of all real numbers.

Of course, the graphs of two identical functions coincide. And so to analyze a function or constructe its graph, it is possible to replace the function with an identical function. For example, the function $f(x) = \frac{x^2}{x}$ can be replaced with the function $f(x) = x$, $x \in \mathbf{R} \setminus \{0\}$; the function $f(x) = 3^{\log_3 x}$ — with the function $f(x) = x$, $x \in (0; +\infty)$; the function $f(x) = \sin \arcsin x$ — with $f(x) = x$, $x \in [-1; 1]$.

2.2.3. Even and odd functions. Properties of graphs

To check that a function is even or odd, one uses the corresponding definition (see Section 2.2). For example, the function $f(x) = \sqrt{4 - x^2}$, which is defined on the symmetric with respect to the origin set $|x| \leq 2$, is even. Indeed, $f(-x) = \sqrt{4-(-x)^2} = \sqrt{4-x^2} = f(x)$ for all $x \in [-2; 2]$.

The function $f(x) = \log_a(x + \sqrt{1+x^2})$, $a \neq 1$, $a > 0$, defined on the symmetric with respect to the origin set $(-\infty; +\infty)$, is odd. Indeed,

$$f(-x) = \log_a(-x + \sqrt{1+x^2}) = \log_a \frac{1}{x + \sqrt{1+x^2}} = -\log_a(x + \sqrt{1+x^2}) = -f(x),$$

that is $f(-x) = -f(x)$ for all $x \in (-\infty; +\infty)$.

The function $f(x) = \log_a x$ is defined on the set $(0; +\infty)$, and this set is not symmetric. So, by the definition, this function cannot be either even or odd.

Proposition 2.2.1. The graph of an even function is symmetric with respect to the axis Oy.

Indeed, let a point $(\tilde{x}; \tilde{y})$ lie on the graph of an even function $y = f(x)$. Then $\tilde{y} = f(\tilde{x})$. Because the function is even, $\tilde{y} = f(-\tilde{x})$, which means that the point $(-\tilde{x}; \tilde{y})$, which is symmetric with respect to Oy, also belongs to the graph of the function $y = f(x)$.

Proposition 2.2.2. The graph of an odd function is symmetric with respect to the origin.

Indeed, let a point $(\tilde{x}; \tilde{y})$ belong to the graph of an odd function. Then $\tilde{y} = f(\tilde{x})$. Because the function is odd, $-\tilde{y} = f(-\tilde{x})$, and hence the point $(-\tilde{x}; -\tilde{y})$, which is symmetric with respect to the origin, also lies on the graph of the function $y = f(x)$.

It follows that the graph of an even function is a curve such that the axis Oy is a symmetry axis. The graph of an odd function is a curve such that the origin is the center of symmetry.

These properties allow one to simplify the procedure of constructing graphs of even and odd functions, namely, it is enough to construct the graph for positive values of the argument from the domain, and then extend it by using the symmetry with respect to Oy, if the function is even, or the origin, if the function is odd (Figures 2.7a, 2.7b).

2.2.4. Types of symmetry

1. Symmetry of the graph of a function with respect to a vertical line

The graph of a function $y = f(x)$ is *symmetric with respect a vertical line* $x = x_0$, if

$$f(x_0 - x) \equiv f(x_0 + x)$$

for all x in the domain of the function $f(x)$. In particular, the graph of an even function is symmetric with respect to the line $x = 0$.

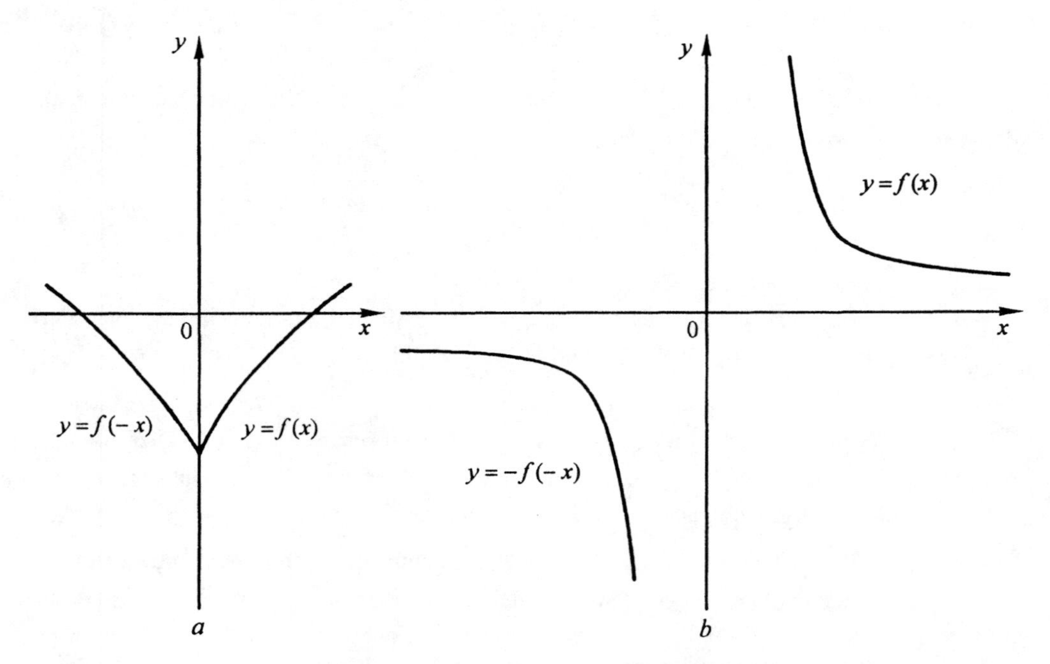

Figure 2.7. Graph of an even (a) and odd (b) function.

2. Symmetry of the graph of a function with respect to a point

A point $M(x_0, y_0)$ is the *center of symmetry* of the function $y = f(x)$, if for all x from the domain of the function, we have the following: $y_0 \equiv \frac{1}{2}(f(x_0 - x) + f(x_0 + x))$. In particular, the graph of an odd function is symmetric with respect to the origin $O(0; 0)$.

Analyzing a function, it is useful to keep in mind that:

- if the graph of a function $y = f(x)$ is symmetric with respect to two vertical lines $x = a$ and $x = b$ $(b > a)$, then the function is periodic, and its period is $\omega = 2(b - a)$;
- if the graph of a function $y = f(x)$ is symmetric with respect to two points $A(a; y_0)$ and $B(b; y_1)$ $(b > a)$, then the function $f(x)$ can be written as the sum of a linear function and a periodic function with the period $\omega = 2(b - a)$; in particular, if $y_0 = y_1$, then the function $f(x)$ is periodic;
- if the graph of a function $y = f(x)$ $(-\infty < x < +\infty)$ is symmetric with respect to a point $A(a; y_0)$ and a vertical line $x = b$, $b \neq a$, then the function $f(x)$ is periodic, and its period equals $4(b - a)$.

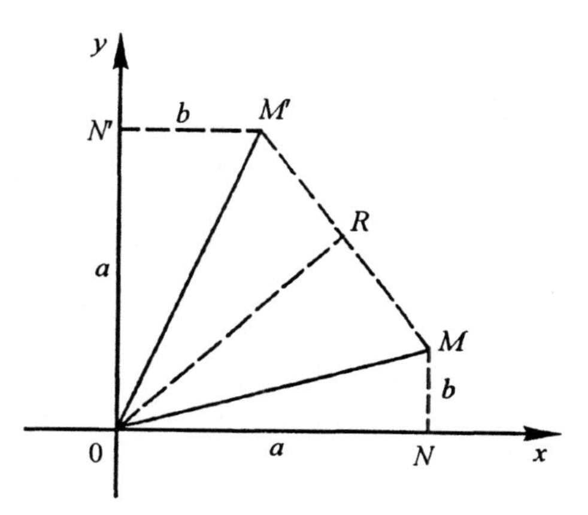

Figure 2.8. Construction of graph of the inverse function.

2.2.5. Graph of the inverse function

If values of y, the independent variable of the inverse function $x = f^{-1}(y)$, are marked on the axis Oy and the values of the initial function — on the axis Ox, then the graph of the inverse function will coincide with the graph of the initial function $y = f(x)$. Denote by x the independent variable of the inverse function, and use letter y to denote the values of the function, that is, instead of the function $x = f^{-1}(y)$, consider the function $y = \varphi(x)$. Note that the inverse function $y = \varphi(x)$, x runs over the set Y, and the function $x = f^{-1}(y)$, y runs over Y, are the same, since they have the same domain.

To draw the graph of the inverse function $y = \varphi(x)$, the graph of the function $y = f(x)$ must be symmetrically reflected with respect to the bisectrix of the first and third quadrants. Indeed, let $x = a$ and $y = f(a) = b$. The point $M(a; b)$ belongs to the graph of the function $y = f(x)$, and the point $M'(B; a)$ belongs to the graph of the inverse function, because $a = \varphi(b)$. We have (Figure 2.8) that $\Delta ON'M' = \Delta ONM$. This implies that $\angle N'OM' = \angle NOM$ and $OM' = OM$. Then the bisectrix of the first quadrant, OR, will be the bisectrix of the angle MOM' and, because $\Delta MOM'$ is isosceles, the bisectrix OR is a symmetry axis of $\Delta MOM'$. So, to every point M lying in the graph of the function $y = f(x)$ there is a corresponding point M', lying on the graph of the inverse function, symmetric with respect to the bisectrix of the first and third quadrants.

Remark 2.2.1. There is another way to construct the graph of the inverse function. Construct the graph of the function $y = f(-x)$ (by reflecting the graph of the function $y = f(x)$ about the axis Oy), and then rotate the graph clockwise about the origin by 90°.

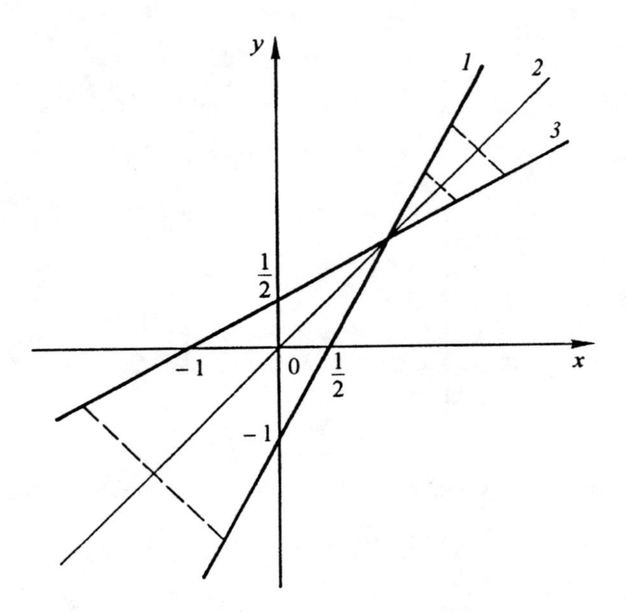

Figure 2.9. Graphs of functions $y = 2x - 1$ (*1*), inverse function $y = \frac{x+1}{2}$ (*3*), symmetry axis $y = x$ (*2*).

Example 2.2.1. Let us find the function which is the inverse to the function $y = 2x - 1$, and draw its graph.

We find that $x = \frac{y+1}{2}$. Interchange the variables x and y, $y = \frac{x+1}{2}$. The graph of the function and its inverse is shown in Figure 2.9.

2.2.6. Periodicity of a function

Knowing that a function is periodic makes it easer to study and construct its graph, because a periodic function can be studied only within one period. To construct the graph of a periodic function with a period ω, it is sufficient to construct its graph on an interval $(x; x + \omega)$, and then periodically continue this graph. The *periodic continuation* with a period ω of a function $y = f(x)$ defined on some interval $[a; b]$ or $(a; b)$ is the construction of a periodic function $F(x)$ with the period ω such that it coincides with $f(x)$ on $[a; b]$ or $(a; b)$.

Let us look at some examples of periodic continuations.

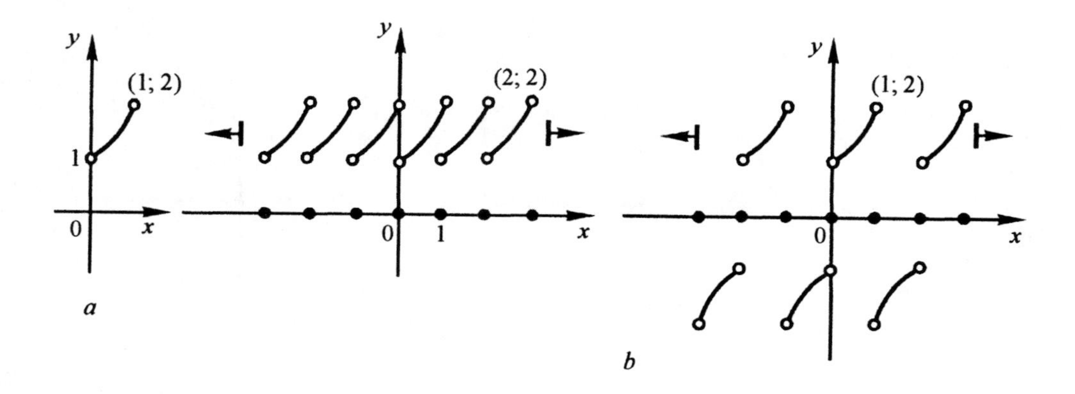

Figure 2.10. (*a*) Graph of function $f(x) = x^2 + 1$ on $(0; 1)$; (*b*) its periodic continuation on R.

Example 2.2.2. Let a function $y = f(x)$, $x \in (0,\omega)$, be given, $\omega > 0$. Then this function can periodically be defined over the whole real line, so that the resulting function will be periodic with the period ω.

Indeed, if for every $x \in (n\omega; \omega + n\omega)$, $n \in \mathbf{Z}$, we set $\varphi(x) = f(x - n\omega)$, and for every $x \in \{n\omega\}$, set $\varphi(x) = C$, where C is a fixed number, then the function $y = \varphi(x)$, $x \in \mathbf{R}$, will be periodic with the principal period ω, i.e. it is a periodic continuation on $\mathbf{R}$ of the given function.

The graph of the function

$$\varphi(x) = \begin{cases} (x-n)^2 + 1, & x \in (n; n+1) \quad n \in \mathbf{Z}, \\ 0, & x = n, \quad n \in \mathbf{Z}, \end{cases}$$

is a periodic continuation on R of the function $f(x) = x^2 + 1$, $x \in (0; 1)$, with the principal period $\omega = 1$. The graph is shown in Figure 2.10.

Example 2.2.3. Let a function $y = f(x)$, $x \in (0; \omega)$, be given, $\omega > 0$. This function can be defined over the whole real axis such that the resulting function will be odd with the principal period 2ω, i.e. will be the *odd periodic continuation* over $\mathbf{R}$.

Indeed, to every $x \in (0; \omega)$ we can put into the correspondence $\varphi(x) = f(x)$, and to every $x \in (-\omega; 0)$ — $\varphi(x) = -f(x)$, and set $\varphi(0) = 0$. Then the function $y = \varphi(x)$, $x \in (-\omega; \omega)$, will be odd. If for every $x \in (-\omega + 2\omega n; \omega + 2\omega n)$, $n \in \mathbf{Z}$, we set $\psi(x) = \varphi(x - 2\omega n)$, and to every $x \in \{2n\omega + \omega\}$, set $\psi(x) = 0$, then the function $y = \psi(x)$ will be odd with the principal period 2ω.

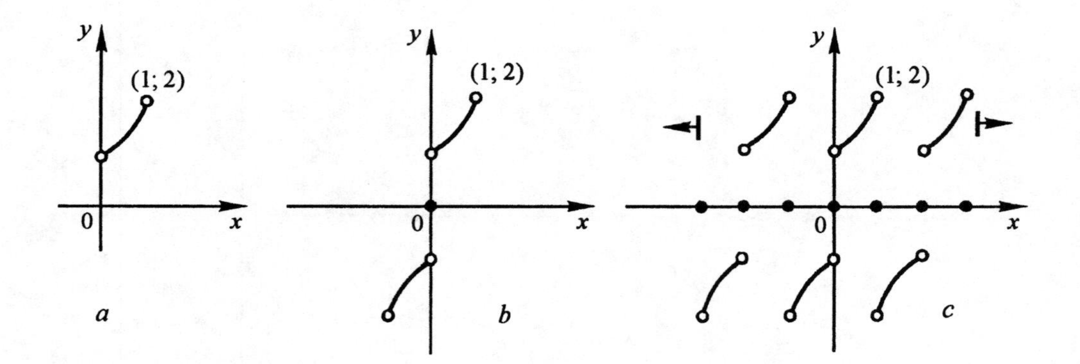

Figure 2.11. (*a*) Graph of $f(x) = x^2 + 1$ on (0; 1); (*b*) its odd continuation on $(-1; 1)$; (*c*) its odd continuation on R.

The graph of the function

$$\varphi(x) = \begin{cases} x^2 + 1, & x \in (0; 1), \\ 0, & x \in \{0\}, \\ -x^2 - 1, & x \in (-1; 0), \end{cases} .$$

is shown in Figure 2.11 *b*. The graph of its periodic continuation, i.e. of the function

$$\psi(x) = \begin{cases} (x - 2n)^2 + 1, & x \in (2n; 1 + 2n), \quad n \in \mathbf{Z}, \\ 0, & x \in \{n\}, \quad n \in \mathbf{Z}, \\ -(x - 2n)^2 - 1, & x \in (-1 + 2n; 2n), \quad n \in \mathbf{Z}, \end{cases}$$

is shown in Figure 2.11 *c*.

Example 2.2.4. Consider a function $y = f(x), x \in (0; \omega), \omega > 0$. This function can be continued over the real line so that it becomes an *even periodic continuation* over **R**.

Indeed, to every $x \in (0; \omega)$, we set $\varphi(x) = f(x)$, and to every $x \in (-\omega; 0)$ set $\varphi(x) = f(-x)$, and $\varphi(0) = C$ for some fixed C. The function $\varphi(x)$, $x \in (-\omega; \omega)$, will be even. Now we set $\varphi(x - 2\omega n) = \psi(x)$ if $x \in (-\omega + 2n\omega; \omega + 2n\omega)$, $n \in \mathbf{Z}$, and $\psi(x) = A$ if $x \in \{\omega + 2n\omega\}$, $n \in \mathbf{Z}$. The function $y = \psi(x)$, $x \in \mathbf{R}$, is an even periodic function with the principal period 2ω.

The graph of the function

$$\varphi(x) = \begin{cases} x^2 + 1, & x \in (0; 1), \\ -1, & x \in \{0\}, \\ x^2 + 1, & x \in (-1; 0), \end{cases}$$

which is an even continuation of the function $f(x) = x^2 + 1$, $x \in (0; 1)$, over $(-1; 1)$, is shown in Figure 2.12*a*, 2.12*b*. Its periodic continuation is the function

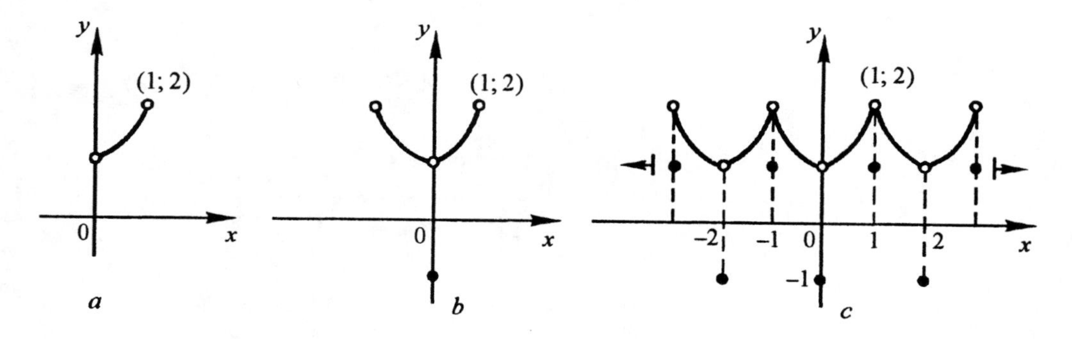

Figure 2.12. (*a*) Graph of $f(x) = x^2 + 1$ on (0; 1); (*b*) its even continuation on (− 1; 1); (*c*) its odd continuation on *R*.

$$\psi(x) = \begin{cases} (x - 2n)^2 + 1, & x \in (2n;\ 1 + 2n), \quad n \in \mathbf{Z}, \\ -1, & x \in \{2n\}, \quad n \in \mathbf{Z}, \\ (x - 2n)^2 + 1, & x \in (-1 + 2n;\ 2n), \quad n \in \mathbf{Z}, \\ 1, & x \in \{1 + 2n\}, \quad n \in \mathbf{Z}, \end{cases}$$

its graph is shown in Figure 2.12*c*.

Remark 2.2.2. Let $f(x)$, $x \in \mathbf{R}$, be a periodic function with the principal period ω. Then the function $\varphi(x) = f(\frac{\omega}{a}x)$, $x \in \mathbf{R}$, will be periodic with the principal period a.

2.2.7. Zeroes and signs of functions

A *zero* of a function is the real value of x at which the value of the function equals zero.

The geometric interpretation of a zero is the x coordinate of the point at which the graph of the function intersects or touches the axis *Ox*. At these points, the function may change sign (Figure 2.13).

Note that the function may also change sign at points of discontinuity. If the graph of a function only touches the axis *Ox*, then it does not change sign at this point.

To find zeros of a function $f(x)$, it is necessary to solve the equation $f(x) = 0$. Real roots of this equation are zeros of the function $y = f(x)$. Assume that the equation $f(x) = 0$ has roots x_1, x_2, ..., x_n. These numbers subdivide the domain of the function $f(x)$ into intervals such that the function is defined and does not have zeros on each of the intervals.

The intervals in which a function does not have either zeros or points of discontinuity, i.e. the intervals where the sign of the function does not change, are called *intervals of constant sign*.

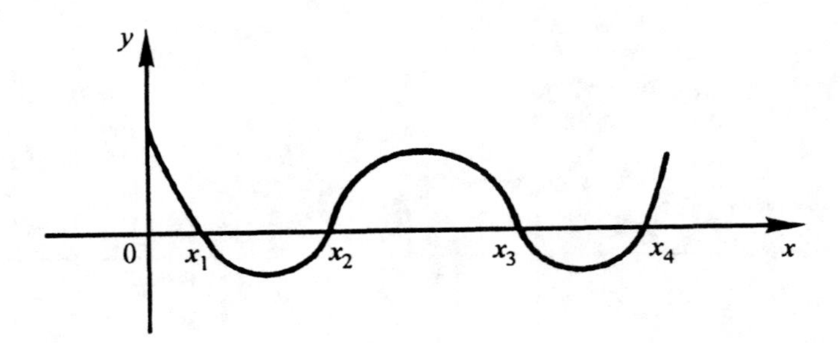

Figure 2.13. Zeros of function $y = f(x)$.

To find out what the sign of a function is on an interval where the function has a constant sign, it is sufficient to determine the sign at any point of this interval. If the domain of a function consists of several constant sign intervals, then the sign of the function is determined on each interval separately.

2.2.8. Monotonicity of functions

Monotonicity of the function is one of the most important properties. For the analysis of a function, it is necessary to know how to determine the monotonicity intervals of the function. One can make use of the following proposition. *If a function is strictly monotone on an interval* $[a; b]$, *then the function takes each of its values only at one* x. And conversely, *if a function defined on* $[a; b]$ *takes each of its values only once, then the function is strictly monotone.* When performing the elementary analysis of monotonicity, one mainly uses the definition and main properties of monotone functions (see Section 1.2.6).

Suppose that the equation $f(x) = y$ yields several single-valued solutions of x as a function of y, $x = \varphi_1(y)$, $x = \varphi_2(y)$, ..., $x = \varphi_n(y)$. If $x \in (a; b)$, then for every value of $f(x) = y$ there is a unique corresponding value of x from $(a; b)$, namely, $x = \varphi_i(y)$, $(i = 1,2,...,n)$. To find all monotonicity intervals, it is necessary to find all the intervals which are domains of the functions $\varphi_1(y)$, ..., $\varphi_n(y)$.

If a function is monotone on an interval $[a; b]$, then:

1) if the function is positive and increasing, then its graph departs from the axis Ox as x increases;
2) if the function is positive and decreasing, then its graph approaches the axis Ox as x increases;
3) if the function is negative and decreasing, then its graph departs (downwards) from the axis Ox as x increases;
4) if the function is negative and increasing, then its graph approaches the axis Ox as x increases.

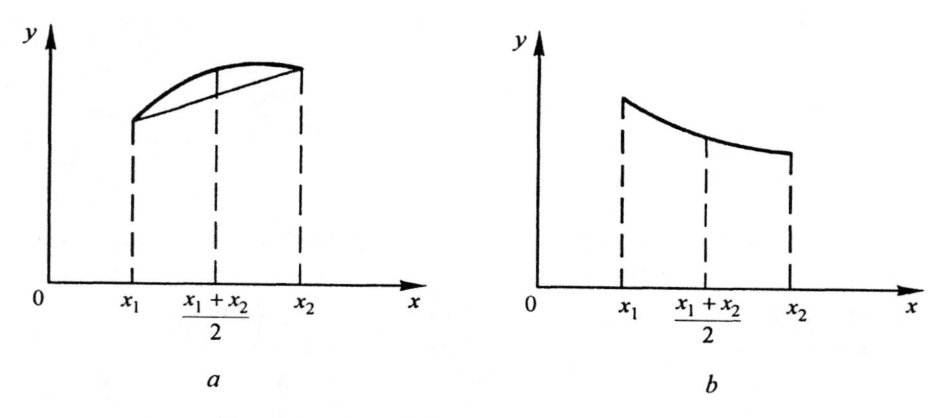

Figure 2.14. Functions convex (*a*) and concave (*b*).

2.2.9. Convexity of functions

Consider a function $f(x)$ defined on an interval $[a; b]$. The function is *convex* on this interval, if

$$\frac{f(x_1)+f(x_2)}{2} \leq f\left(\frac{x_1+x_2}{2}\right) \quad (x_1 \neq x_2)$$

for all x_1, x_2 from $[a; b]$ (Figure 2.14 *a*). The function is *concave* on an interval $[a; b]$, if

$$\frac{f(x_1)+f(x_2)}{2} \geq f\left(\frac{x_1+x_2}{2}\right) \quad (x_1 \neq x_2)$$

for all x_1, x_2 from $[a; b]$ (Figure 2.14 *b*).

This definition of convexity (concavity) is a particular case (with $\alpha = 1/2$) of a more general definition. A function $f(x)$, continuous on X, is called *convex* (*concave*) if for any x_1, x_2 from this interval and any $\alpha \in (0; 1)$, we have the inequality $\alpha f(x_1) + (1-\alpha)f(x_2) \leq$ $\leq f(\alpha x_1 + (1-\alpha)x_2)$ $(\alpha f(x_1) + (1-\alpha)f(x_2) \geq f(\alpha x_1 + (1-\alpha)x_2))$.

We recall some properties of convex functions.

The product of a convex (concave) function by a positive constant is a convex (concave) function. The product of a convex (concave) function by a negative constant is a concave (convex) function.

The sum of two convex functions is a convex function.

If a function $f(x)$ is concave and increasing, and $x = \varphi(t)$ is concave, then the composition $f(\varphi(x))$ is a concave function.

If a function $f(x)$ is concave and decreasing, and $x = \varphi(t)$ is convex, then the composition $f(\varphi(x))$ is concave.

If a function $f(x)$ is convex and increasing, and $x = \varphi(t)$ is convex, then the composition $f(\varphi(x))$ is convex.

If a function $f(x)$ is convex and decreasing, and $x = \varphi(t)$ is concave, then the composition $f(\varphi(x))$ is convex.

If functions $y = f(x)$ and $y = g(x)$ are inverses to each other (in corresponding intervals), then:

a) if the function $f(x)$ is concave and increasing, then the function $g(x)$ is convex and increasing;

b) if the function $f(x)$ is convex and decreasing, then the function $g(x)$ is convex and decreasing;

c) if the function $f(x)$ is convex and increasing, then the function $g(x)$ is concave and increasing;

d) if the function $f(x)$ is concave and decreasing, then the function $g(x)$ is concave and decreasing.

A concave (convex) function $f(x)$ defined on $[a; b]$, which is not identically constant, cannot take its maximum (minimum) in the interior of this interval.

If functions $f(x)$ and $\varphi(x)$ are positive and convex, then the function $y = \varphi(x)/f(x)$ will be convex only if one of the functions is increasing and the other is decreasing.

If a function $y = f(x)/\varphi(x)$ is convex, and the function $f(x)$ is negative, convex, and increasing, then the function $\varphi(x)$ is positive, concave, and increasing.

If a function $y = f(x)/\varphi(x)$ is concave, and $f(x)$, $\varphi(x)$ are convex and negative, then one of these functions is increasing, and the other is decreasing.

If a function $f(x)$ is convex and positive, then the function $y = 1/f(x)$ is concave.

If a function $f(x)$ is convex and positive, then the function $y = \sqrt[n]{f(x)}$ is convex (n is a natural number).

If functions $f(x)$ and $\varphi(x)$ are positive, increasing, and concave, then the function $y = f(x)\varphi(x)$ is positive, increasing, and concave.

If functions $f(x)$ and $\varphi(x)$ are positive, decreasing, and concave, then the function $y = f(x)\varphi(x)$ is positive and concave.

If functions $f(x)$ and $\varphi(x)$ are negative, increasing, and convex, then the function $y = f(x)\varphi(x)$ is positive and concave.

If functions $f(x)$ and $\varphi(x)$ are negative, decreasing, and convex, then the function $y = f(x)\varphi(x)$ is positive and concave.

If a function $f(x)$ is positive, decreasing, and convex, and a function $\varphi(x)$ is positive, increasing and convex, then the function $y = f(x)\varphi(x)$ is positive and convex.

If a function $f(x)$ is negative, increasing, and concave, and a function $\varphi(x)$ is negative, decreasing, and concave, then the function $y = f(x)\varphi(x)$ is positive and convex.

If a function $f(x)$ is positive, increasing, and convex, and a function $\varphi(x)$ is negative, increasing, and concave, then the function $y = f(x)\varphi(x)$ is negative and concave.

If a function $f(x)$ is positive, decreasing, and convex, and a function $\varphi(x)$ is negative, decreasing, and concave, then the function $y = f(x)\varphi(x)$ is negative and concave.

If a function $f(x)$ is positive, increasing, and concave, and a function $\varphi(x)$ is negative, increasing, and convex, then the function $y = f(x)\varphi(x)$ is negative and convex.

If a function $f(x)$ is positive, decreasing, and concave, and a function $\varphi(x)$ is negative, increasing, and convex, then the function $y = f(x)\varphi(x)$ is negative and convex.

The following *Jensen's inequality* is useful in applications: if a concave function $f(x)$ is defined on an interval $(a; b)$, and $x_1, x_2, ..., x_n$ are points of this interval, $\alpha_1, \alpha_2, ..., \alpha_n$ are nonnegative numbers such that $\alpha_1 + \alpha_2 + ... + \alpha_n = 1$, then $f(\alpha_1 x_1 + \alpha_2 x_2 + ... + \alpha_n x_n) \leq$ $\leq \alpha_1 f(x_1) + \alpha_2 f(x_2) + ... + \alpha_n f(x_n)$.

Notice that if there are at least two numbers among $\alpha_1, \alpha_2, ..., \alpha_n$ different from zero, then the equality holds if and only if $x_1 = x_2 = ... = x_n$.

For a convex function the Jensen's inequality is:

$$f(\alpha_1 x_1 + \alpha_2 x_2 + ... + \alpha_n x_n) \geq \alpha_1 f(x_1) + \alpha_2 f(x_2) + ... + \alpha_n f(x_n).$$

Jensen's inequalities permit to obtain other inequalities.

Example 2.2.6. Consider the function $f(x) = x^p, x \geq 0, p > 1$. Because the function is concave on $[0; +\infty)$, the Jensen's inequality gives $(\alpha_1 x_1 + \alpha_2 x_2 + ... + \alpha_n x_n)^p \leq \alpha_1 x_1^p + \alpha_2 x_2^p + ... + \alpha_n x_n^p$ if $\alpha_1 + \alpha_2 + ... + \alpha_n = 1, \alpha_1 > 0, \dots, \alpha_n > 0$.

Example 2.2.7. Because the function $f(x) = \ln x$ is convex on the set of positive numbers, by Jensen's inequality we have: $\alpha_1 \ln x_1 + \alpha_2 \ln x_2 + ... + \alpha_n \ln x_n \leq \ln(\alpha_1 x_1 + \alpha_2 x_2 + ... + \alpha_n x_n)$, where $x_i > 0, \alpha_i > 0, 1 \leq i \leq n$, and $\alpha_1 + \alpha_2 + ... + \alpha_n = 1$.

In particular, if $\alpha_1 = \alpha_2 = ... = \alpha_n = 1/n$, then we get a well known inequality between the geometric and arithmetic means, $\sqrt[n]{x_1 x_2 ... x_n} \leq \dfrac{x_1 + x_2 + ... + x_n}{n}$.

If $n = 2$, $\alpha_1 = 1/p$, $\alpha_2 = 1/q$, $x_1 = a$, $x_2 = b$, $1/p + 1/q = 1$, we obtain the Young's inequality, $a^{1/p} b^{1/q} \leq a/p + b/q$.

Example 2.2.8. If a function $f(x)$ is concave on $(a; b)$, and $x_1, x_2, ..., x_n$ belong to the interval $(a; b), p_1, p_2, \dots, p_n$ are positive numbers, then

$$f\left(\frac{p_1 x_1 + p_2 x_2 + ... + p_n x_n}{p_1 + p_2 + ... + p_n}\right) \leq \frac{p_1 f(x_1) + ... + p_n f(x_n)}{p_1 + p_2 + ... p_n}.$$

Example 2.2.9. Let $x_1, x_2, ..., x_n$ be positive, and $k > 1$, then

$$\left(\frac{x_1 + x_2 ... + x_n}{n}\right)^k \le \frac{1}{n}(x_1^k + x_2^k + ... + x_n^k).$$

These properties are useful for studying convexity of functions.

The study of convexity can easily be carried out by using methods of calculus (these methods will be discussed below). There are different methods of elementary mathematics that allow to study convexity. For example, one can study the sign of $D = f(x_1) + f(x_2) - 2f\left(\frac{x_1 + x_2}{2}\right)$ in the domain of the function. If this sign changes, then the domain is subdivided into subintervals where the sign does not change. If the value of D is positive on an interval, then the function is concave on this interval. If D is negative, then the function is convex.

The point at which the convexity of the function changes is called the *inflection point*. The tangent to the graph at the inflection point intersects the graph.

2.2.10. Characteristic points of graph

Characteristic points of graphs are:

1) points where the graph intersects the axes;
2) boundary values of the function, i.e. the values the function takes at boundary points;
3) points at which the function takes extremal values, i.e. y_{max}, y_{min}, and the corresponding values of x, inflection points, etc.

Points at which the graph intersects the axis Oy are points $x = 0$, $y = f(0)$; the axis Ox is intersected at points where $y = 0$. The x coordinate is found from the equation $f(x) = 0$.

Example 2.2.10. The function $y = \log_2(x + 8) - 1$ intersects the axis Oy at $x = 0$, $y(0) =$ $= \log_2(0 + 8) - 1 = \log_2 8 - 1 = 2$, i.e. the point of intersection is unique and has the coordinates $(0; 2)$. For the point at which the graph intersects the axis Ox, we have $y = \log - 2(x + 8) - 1 = 0$, $\log_2(x + 8) = 1$, $x + 8 = 2$, $x = -6$. Hence the point is unique and has the coordinates $(-6; 0)$.

Boundary values of a function

If the domain of a function is a closed interval, then the boundary values of the function are obtained by substituting the values of the argument at the boundary points.

If the domain is an open interval, then the boundary values are found as limits (if they exist) with x approaching the value at the boundary, $\lim\limits_{x \to a} f(x)$, where a is the x-coordinate of the endpoint. This limit may exist and be finite, and it is also possible that the limit approaches plus or minus infinity.

If the function has an infinite limit at the point a, then the graph of the function approaches the vertical line $x = a$, and as x approaches a, the graph infinitely departs from the axis Ox upward or downward depending on the sign of the function.

If the function has a finite limit y_0 as x approaches a, then the graph approaches the point $(a; y_0)$. This is indicated by drawing an arrow on the graph to the point $(a; y_0)$.

If the limit (finite or infinite) does not exist, then it is impossible to draw the graph of the function in the vicinity of the point a.

Extremum. The greatest and the least value of function

A point x_0 is called a point of *relative maximum* (*relative minimum*) of a function $f(x)$, $x \in X$, if there exists an interval $(x_0 - \delta; x_0 + \delta)$, $\delta > 0$, contained in X, such that $f(x) \leq f(x_0)$ $(f(x) \geq f(x_0))$ for $x \in (x_0 - \delta; x_0 + \delta)$.

Points of relative maximum and relative minimum are called the points of *relative extremum* of the function.

If in the above definition the inequalities are strict, $f(x) < f(x_0)$ $(f(x) > f(x_0))$, $x \neq x_0$, then the point x_0 is the point of *strict relative maximum* (*minimum*).

Sufficient condition for relative extremum

If a function $y = f(x)$, $x \in X$, is increasing (decreasing) on some interval $(x_0 - \delta; x_0] \subset X$, and is decreasing (increasing) on $[x_0; x_0 + \delta) \subset X$, then the point x_0 is the point of relative maximum (minimum) of the function $f(x)$.

If there exists a point x_0 in the set A, $A \subset eqX$, such that, for all $x \in A$, $f(x) \geq f(x_0)$ $(f(x) \leq f(x_0))$, then the function $y = f(x)$ is said to have its minimum (maximum) on A at the point $x = x_0$. Notice that such points may not exist at all, or there may be a finite or infinite number of such points.

If a function $y = f(x)$ is increasing (decreasing) on a closed interval $[a; b]$, then it reaches its minimum (maximum) at the point $x = a$, and the maximum (minimum) — at $x = b$.

If a function $y = f(x)$, $x \in X$, is unbounded from above (below) on $A \subseteq X$, then it cannot reach maximum (minimum) on A.

A bounded function also may not reach a maximal (minimal) value (see Figure 2.15).

To find a maximal (minimal) value of a continuous function $y = f(x)$ on an interval $[a; b]$ that has a finite number of relative maximums (minimums), it is necessary to calculate the values of the function at every point of relative maximum (minimum) and at the endpoints, and then to choose the largest (smallest) value.

To facilitate finding maximums (minimums) of a function, it is sometimes convenient to use a transformation of the expression that defines the function.

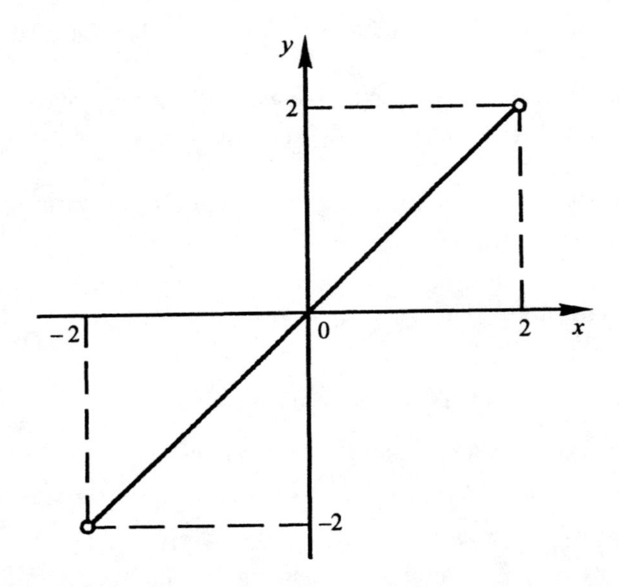

Figure 2.15. Bounded function that does not take its maximum and minimum.

The maximum of a function $y=f(x)$ on an interval $[a; b]$ is denoted by $\max\limits_{x \in [a;\, b]} f(x)$, the minimum — by $\min\limits_{x \in [a;\, b]} f(x)$.

Let a function $y=f(x)$ be defined on a set X, and let it take a maximum at a point $x=x_0$, $x_0 \in X$.

1. Let C be a constant. Then a) if $C>0$, then at the point $x=x_0$, the function $Cf(x)$ reaches its maximum on the set X; b) if $C<0$, then at the point $x=x_0$, the function $Cf(x)$ reaches its minimum on X.
   In particular, the function $-f(x)$ reaches on X its minimum at the point $x=x_0$.
2. The function $f(x)+C$ reaches its maximum on X at $x=x_0$.
3. If $f(x)>0$ (or $f(x)<0$) on X, then the function $1/f(x)$ achieves its minimum (maximum) on the set X.
4. Let a function $\varphi(x)$ be defined on the set X and take the maximal value on X at the point $x=x_0$. The function $af(x)+b\varphi(x)$, $a, b>0$, reaches its maximum on X at $x=x_0$.
5. If $f(x)\geq 0$ on X and $n \in \mathbf{N}$, $a>1$, then each of the functions $f^n(x)$, $\sqrt[n]{f(x)}$, $a^{f(x)}$, $\log_a(1+f(x))$ reaches its maximum on X at the point $x=x_0$.

Similar properties hold for the minimal value of a function $y=f(x)$ on X, as well as for its relative extremums.

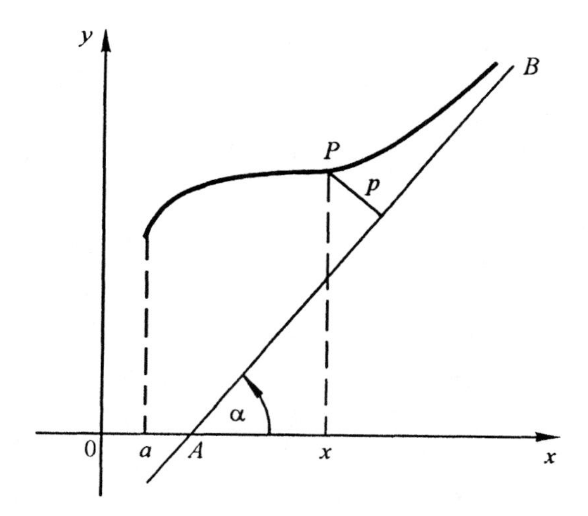

Figure 2.16. Asymptote of curve $y = f(x)$.

Asymptotes of graph of a function

When studying properties of a function, it is very important to find out its behavior as the (absolute) value of the argument x goes to infinity, as well as find the points at which the absolute value of the function $f(x)$ increases to infinity. Geometric interpretation of these ideas leads to the notion of an asymptote of the graph.

A straight line $y = kx + b$ is called the *asymptote* of the curve $y = f(x)$ if the distance from the point $P(x; f(x))$ on this curve to the straight line goes to zero as $x \to +\infty$ $(x \to -\infty)$ (Figure 2.16).

There are vertical, horizontal, and oblique asymptotes.

A curve $y = f(x)$ has a *horizontal asymptote* $y = b$ if the function $y = f(x)$ has a finite limit as $x \to +\infty$ or $x \to = -\infty$ and this limit equals b, i.e. $\lim_{x \to +\infty} f(x) = b$ or $\lim_{x \to -\infty} f(x) = b$.

If a function $f(x)$ can be written as $f(x) = kx + b + \varepsilon(x)$, where $\varepsilon(x) \to 0$ as $x \to +\infty$ or $x \to -\infty$, then $y = kx + b$ is an asymptote. This means that if $f(x)$ is a ratio of integer in x polynomials and the degree of the numerator is equal to or greater than the degree of the denominator by one, then the asymptote can be found by calculating the integer part of the ratio. So, if $f(x) = P_{n+1}(x)/Q_n(x) = kx + b + \varepsilon(x)$, where $\varepsilon(x) \to 0$ as $x \to \infty$, then $y = kx + b$ is an asymptote of the graph. In the case where $f(x) = P_n(x)/Q_n(x)$, the integer part of the ratio is a constant, $f(x) = b + \varepsilon(x)$, and hence the asymptote $y = b$ is parallel to Ox. In the case where the degree of the numerator is less than the degree of the denominator, the asymptote is the axis Ox, because $\lim_{x \to \infty} f(x) = 0$ in this case.

So, if

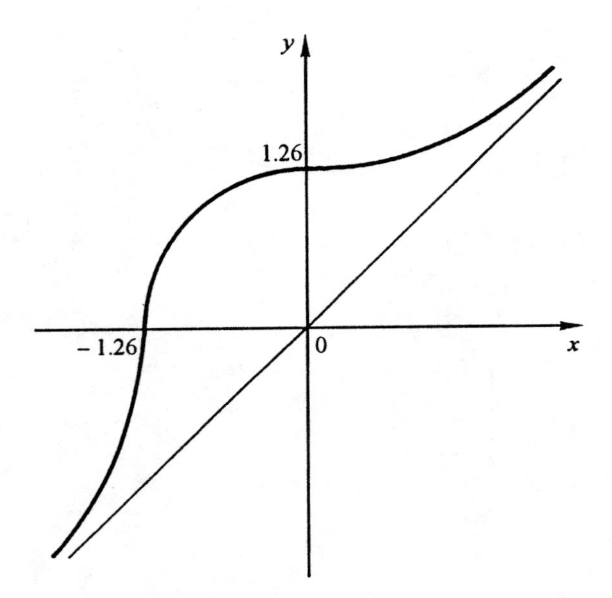

Figure 2.17. Two-sided asymptote of the function $y = \sqrt[3]{x^3 + 2}$.

$$f(x) = \frac{a_0x^n + a_1x^{n-1} + ... + a_n}{b_0x^k + b_1x^{k-1} + ... + b_k},$$

where $n, k \in N$, the graph has

a) the oblique asymptote $y = (a_0/b_0)x + (a_1 - b_1a_0/b_0)$, if $n = k + 1$;

b) the horizontal asymptote $y = a_0/b_0$, if $n = k$;

c) the horizontal asymptote $y = 0$, if $n < k$;

d) the vertical asymptote $x = p$, if $a_0p^n + a_1p^{n-1} + ... + a_n \neq 0$ and $b_0p^k + b_1p^{k-1} + ... + b_0 = 0$.

A curve $y = f(x)$ has the vertical asymptote $x = a$ if $f(x) \to +\infty$ ($f(x) \to -\infty$) for $x \to a$, $x \to a - 0$, or $x \to a + 0$. To find vertical asymptotes, it is necessary to find values of x in a neighborhood the absolute value of the function tends to infinity. If the values of such x are a_1, $a_2, \dots, a_n$, then the asymptotes will be $x = a_1, x = a_2, \dots, x = a_n$.

A curve $y = f(x)$ has the oblique asymptote $y = kx + b$ if and only if there exist finite limits $k = \lim\limits_{x \to \infty} \frac{f(x)}{x}$, $b = \lim\limits_{x \to \infty} (f(x) - kx)$ (cases $x \to +\infty$ and $x \to -\infty$ should be treated separately).

An oblique asymptote is called *right* (*left*) if the graph of the function approaches this asymptote as $x \to +\infty$ ($x \to -\infty$); it is called *two-sided* if it is simultaneously left and right (Figures 2.17, 2.18 and 2.19).

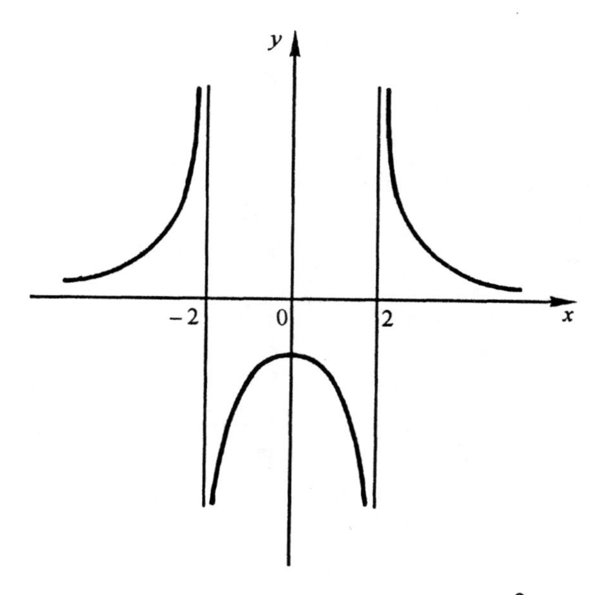

Figure 2.18. Graph of the function $y = \frac{3}{x^2 - 4}$.

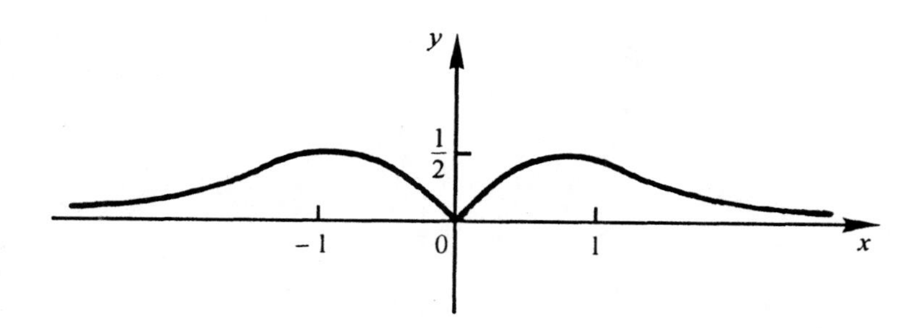

Figure 2.19. Graph of the function $y = \frac{|x|}{x^2 + 1}$.

We remark that an asymptote of a curve $y = f(x)$ may intersect the graph of this curve in a finite or infinite number of points. To determine the intervals on which the curve is above or below the asymptote, one studies the difference $\delta = y_{\text{curve}} - y_{\text{asymptote}}$. The intervals on which $\delta > 0$, the curve lies above the asymptote; if $\delta < 0$ — below the curve.

Procedure for studying the curve and constructing its graph

To study the curve and to construct its graph, it is useful to follow these steps:

1) determine the domain of the function;
2) determine the range of the function;
3) determine whether the function is even, odd, and periodic;
4) determine characteristic points of the graph, monotonicity intervals, and intervals of convexity and concavity;
5) determine asymptotes of the function;
6) draw the graph of the function.

To make the drawing more precise, it is useful to calculate several values of the function in those intervals of the domain which are large and where the function does not have characteristic points.

Chapter 3

Graphs of principal elementary functions

3.1. Power function

Power function is the function $y = x^{\alpha}$, where $\alpha \in \mathbf{R}$. The nature of this function essentially depends on arithmetic properties of the number α. If α is a rational number, then the function $y = x^{\alpha}$ is algebraic; if α is irrational, then the function $y = x^{\alpha}$ is transcendental.

Power function with natural exponent

Let $\alpha = m$, where $m \in N$. The domain of the function is $(-\infty; +\infty)$. If $x \to +\infty$, then the function unboundedly increases. Indeed, for an arbitrary number M, $x^m > M$, if $x > \sqrt[m]{M}$.

If the number m is even, the function $y = x^m$ is even and its graph is symmetric with respect to the axis Oy; if m is odd, the function $y = x^m$ is odd, and its graph is symmetric with respect to the origin. Indeed, let $m = 2n$. Then $f(-x) = (-x)^{2n} = x^{2n} = f(x)$. If $m = 2n + 1$, then $f(-x) = (-x)^{2n+1} = -x^{2n+1} = -f(x)$.

The function monotonically increases on $[0; +\infty)$, where it takes the values from 0 to $+\infty$. Indeed, if $0 \le x_1 < x_2$, then

$$y(x_2) - y(x_1) = x_2^m - x_1^m = (x_2 - x_1)(x_2^{m-1} + x_2^{m-2}x_1 + \ldots + x_1^{m-1}) > 0.$$

If $m = 2n$, then the function $y = x^m$ monotonically decreases on $(-\infty; 0]$ from $+\infty$ to 0; if $m = 2n + 1$, the function monotonically increases on $(-\infty; 0]$ from $-\infty$ to 0.

If $m > 1$, the function $y = x^m$ is concave on $(0; +\infty)$. Indeed, if $x_1 \ne x_2$, then $f\left(\frac{x_1 + x_2}{2}\right) \le \frac{f(x_1) + f(x_2)}{2}$, because $\left(\frac{x_1 + x_2}{2}\right)^m \le \frac{x_1^m + x_2^m}{2}$. The latter inequality is proved by induction.

The inequality holds for $m = 2$, since

$$\left(\frac{x_1+x_2}{2}\right)^2 = \frac{x_1^2+x_2^2+2x_1x_2}{4} \le \frac{x_1^2+x_2^2+x_1^2+x_2^2}{4} = \frac{x_1^2+x_2^2}{2}.$$

Suppose that the inequality holds for $m = k$,

$$\left(\frac{x_1+x_2}{2}\right)^k \le \frac{x_1^k+x_2^k}{2}.$$

By multiplying both sides of this inequality by $\frac{x_1+x_2}{2}$, we obtain:

$$\left(\frac{x_1+x_2}{2}\right)^{k+1} \le \frac{x_1^{k+1}+x_2^{k+1}+x_1^kx_2+x_2^kx_1}{4} \le$$

$$\le \frac{x_1^{k+1}+x_2^{k+1}+x_1^{k+1}+x_2^{k+1}}{4} = \frac{x_1^{k+1}+x_2^{k+1}}{2}.$$

This means that the inequality holds for $m = k + 1$ if it holds for $m = k$. By induction, this means that the inequality is true for all $m \in \mathbf{N}$.

For $m = 2n$, the function $y = x^m$ is concave on $(-\infty; 0)$; for $m = 2n + 1$, the function is convex on this interval. The graph of the function $y = x^m$ passes through the points (0; 0) and (1; 1).

If $m = 2$, the graph of the function is called the *quadratic parabola*, if $m = 3$, the graph is called the *qubic parabola*. For $m = 4, 5, \ldots$, the graph is called parabola of the corresponding degree.

As can be seen from Figure 3.1, the graphs of the curves $y = x^m$, for positive even m, have the parabola form. All of them have three points in common: O, P, Q, where O is the vertex of the parabolas, the points P and Q have $x = 1$ and $x = -1$ correspondingly. All parabolas have two symmetric with respect to Oy branches that tend to $+\infty$. As m increases, the branches of the parabolas tend towards Ox for $x < 1$, and get closer to Oy for $x > 1$.

For positive odd values of m, the graphs of the curves $y = x^m$ consist of two branches, which are symmetric with respect to the origin. The point O is an inflection point. All such parabolas have three points in common: Q, P, and R (Figure 3.1).

The function $y = ax$, where a is a real number, defines the proportional relation between x and y. Its graph is a straight line that passes through the origin. The graph lies in the first and third quadrants if $a > 0$ (Figure 3.2 a), and it lies in the second and forth quadrant if $a < 0$ (Figure 3.2 b).

Graphs of some functions are shown in Figure 3.4.

Power function with integer negative exponent

Let $\alpha = -r$, $r \in \mathbf{N}$. The domain of the function $y = 1/x^r$ is $(-\infty; 0) \cup (0; +\infty)$. If r is even, the function is even and its graph is symmetric with respect to the axis Oy. If r is odd, the function is odd and its graph is symmetric with respect to the origin (Figure 3.3).

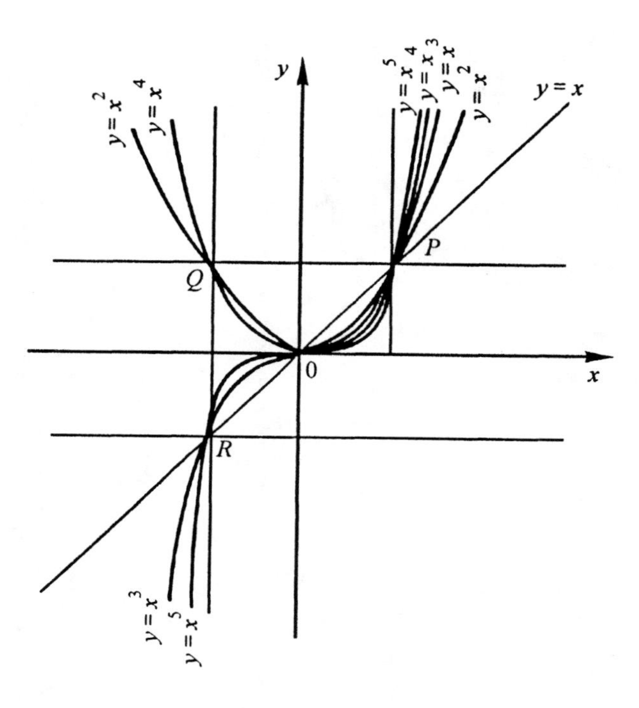

Figure 3.1. Curves $y = x^m$, $m = 1,2,3,4,5$.

Indeed, $f(-x) = 1/(-x)^{2n} = 1/x^{2n} = f(x)$, and $f(-x) = 1/(-x)^{2n+1} = -1/x^{2n+1} = -f(x)$. The function $y = 1/x^r$ decreases from $+\infty$ to 0 on the interval $(0; +\infty)$. This follows from the fact that the function $y = x^r$ increases from 0 to $+\infty$ on $(0; +\infty)$ (if $x_1 < x_2$, then $\frac{1}{x_1^r} > \frac{1}{x_2^r}$). By using the property of the function of being even or odd (for corresponding r), we conclude that the function $y = 1/x^r$ monotonically increases from 0 to $+\infty$ on the interval $(-\infty; 0)$ for even r, and monotonically decreases from 0 to $-\infty$ on this interval for odd r.

For $x \in (0; +\infty)$, the function $y = 1/x^r$ is concave. This follows from the following inequalities:

$$\left(\frac{x_1 + x_2}{2}\right)^r \le \frac{x_1^r + x_2^r}{2}, \quad \frac{1}{\frac{x_1 + x_2}{2}} \le \frac{1}{2}\left(\frac{1}{x_1} + \frac{1}{x_2}\right).$$

On the interval $(-\infty; 0)$, the function $y = 1/x^r$ is concave if r is even, and convex if r is odd. The graph of the function does not intersect the coordinate axes, but passes through the point $(1; 1)$. The horizontal asymptote is given by the equation $y = 0$. Indeed,

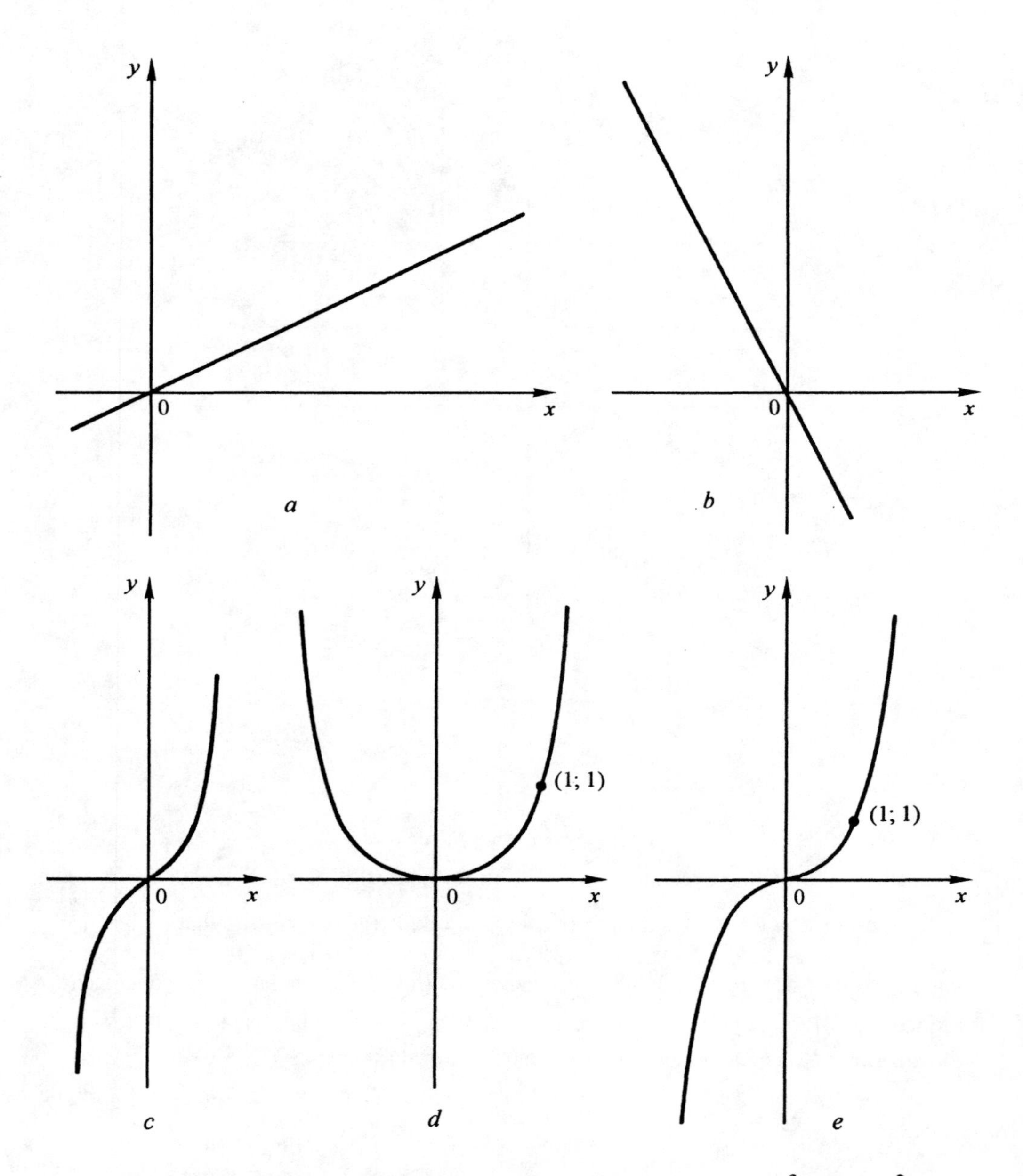

Figure 3.2. Graphs of functions: (*a*) $y = 0.5x$, (*b*) $y = -2x$, (*c*) $y = x^3$, (*d*) $y = x^{2n}$, (*e*) $y = x^{2n+1}$, $n \in \mathbf{N}$.

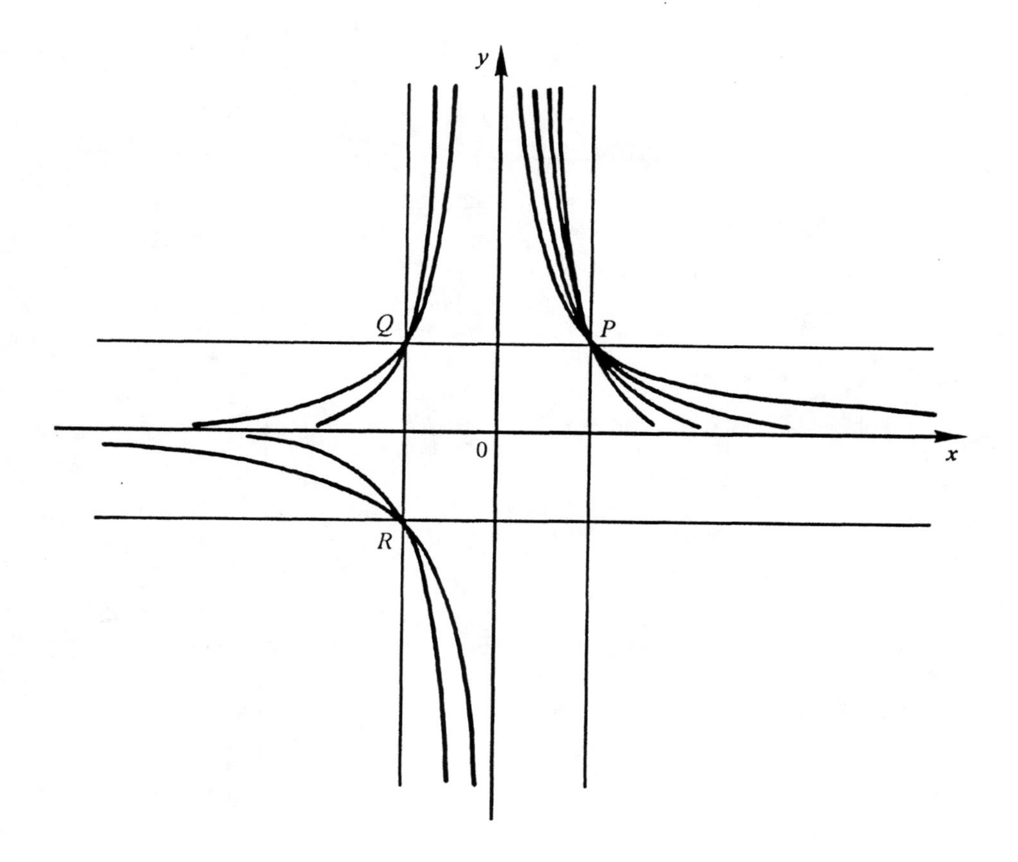

Figure 3.3. Curves $y = \frac{1}{x^m}$, $m = 1,2,3,4$.

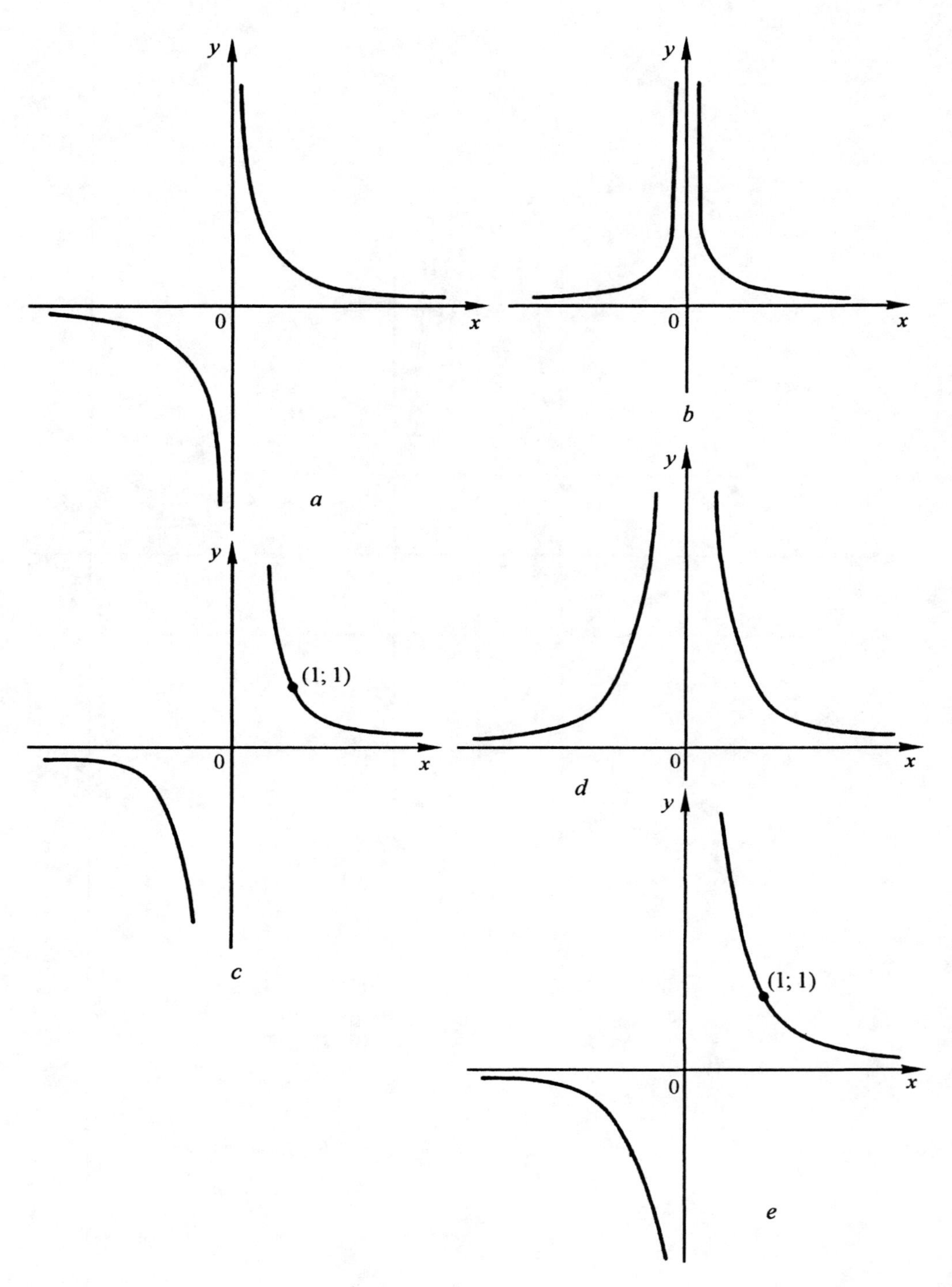

Figure 3.4. Curves $(a)\ y = \frac{1}{x}$, $(b)\ y = \frac{1}{x^2}$, $(c)\ y = \frac{1}{x^3}$, $(d)\ y = \frac{1}{x^{2n}}$, $(e)\ y = \frac{1}{x^{2n+1}}$.

$$k = \lim_{x \to \pm\infty} \frac{f(x)}{x} = \lim_{x \to \pm\infty} \frac{1}{x^{r+1}} = 0,$$

$$b = \lim_{x \to \pm\infty} (f(x) - kx) = \lim_{x \to \pm\infty} \frac{1}{x^r} = 0.$$

The vertical asymptote is $x = 0$, because the absolute value of the function $y = 1/x^r$ increases to infinity in a neighborhood of the point $x = 0$. If the exponent of the power function is even and negative, the graph consists of two hyperbolic type branches symmetric with respect to the axis Oy (Figure 3.3). The graph lies above the axis Ox. For different even r, the curves form two bundles, one of which passes through the point P, and another one — through the point Q (Figure 3.3). For odd negative integer r, the graphs are hyperbolic type curves, but lying in the first and third quadrants. They form branches that pass through the points P and R (Figure 3.3).

Notice that the function $y = k/x$, where $k \in \mathbf{R}$, defines the inverse proportion between the variables x and y. Its graph is a hyperbola. If $k > 0$, two branches of the hyperbola lie in the first and third quadrants, respectively. If $k < 0$, the two branches lie in the second and fourth quadrants (Figures 3.5 and 3.6).

Power function with rational exponent

Let us consider the function $y = x^{\alpha}$, where $\alpha = r$ is a rational number.

1. Let $r = 1/m$, where $m \in \mathbf{N}$. Suppose that m is odd. As can be seen from the graph of the function $y = x^{2n+1}$ (see Figure 3.2*d*), to every $y \in (-\infty; +\infty)$ there corresponds a unique value of $x = \sqrt[2n+1]{y}$. This means that for odd m the function $y = x^m$ has an inverse, $x = \sqrt[m]{y}$, which is defined for all y. By denoting the argument by x and the independent variable by y, we obtain the inverse function $y = \sqrt[m]{x}$. Its graph is obtained by symmetrically reflecting the graph of the function $y = x^m$ with respect to the bisectrix of the first and third quadrants (Figure 3.7).

Let now m be even. Consider the graph of the function $y = \sqrt[m]{x}$ (Figure 3.8). To every positive y, there exist two values of x, $x = \sqrt[m]{y}$ and $x = -\sqrt[m]{y}$. So, if m is even, the inverse function is two-valued. However, if the function $y = x^m$ is considered on the interval $[0; +\infty)$, on which it is increasing, then to every value of y, there corresponds the unique $x = \sqrt[m]{y}$ $(x > 0)$. Hence, on the interval $[0; +\infty)$, the function $y = x^m$ has the inverse function $x = \sqrt[m]{y}$, which is defined on $[0; +\infty)$.

In the sequel for even m, $\sqrt[m]{x}$ is taken to be a nonnegative number. The graph of the function $y = \sqrt[m]{x}$ is constructed by symmetrically reflecting the graph of the function $y = x^m$ with respect to the bisectrix of the first and third quadrants.

Remark 3.1.1. The shaded regions in Figure 3.9 show the parts of the plane where graphs of the function $y = x^{\alpha}$ lie for different α.

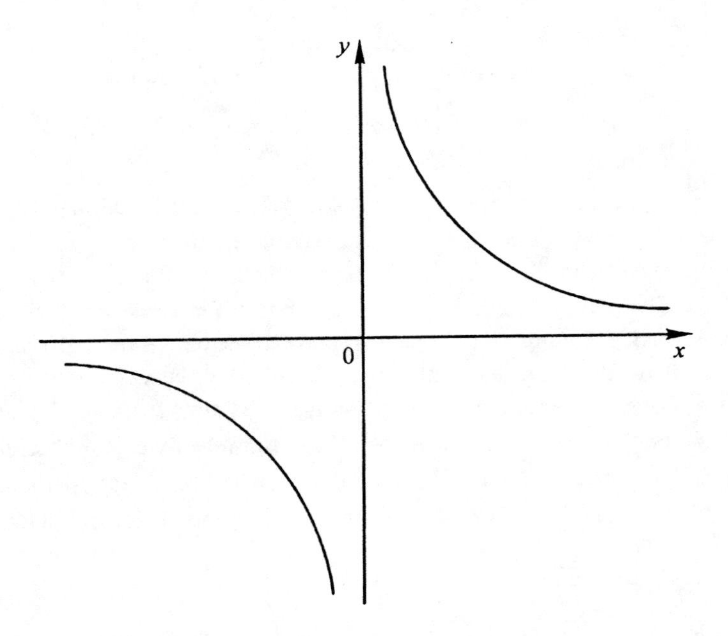

Figure 3.5. Function $y = \frac{k}{x}$, $k > 0$.

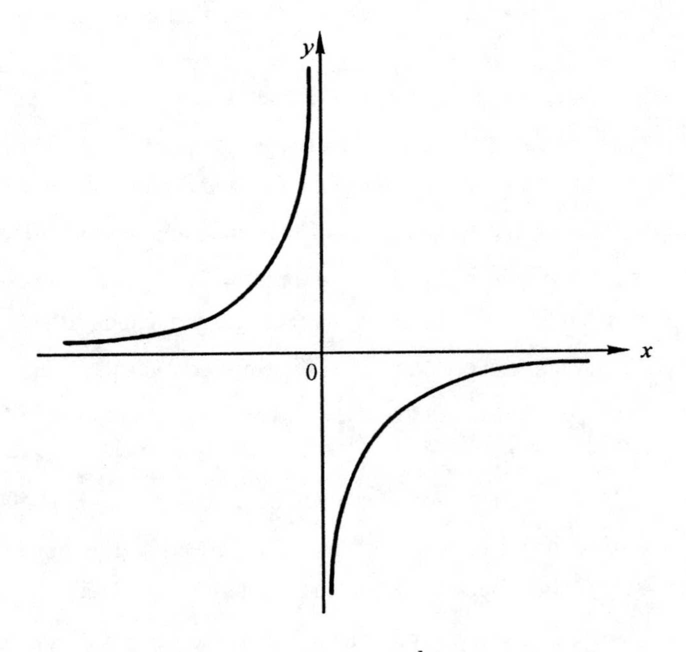

Figure 3.6. Function $y = \frac{k}{x}$, $k < 0$.

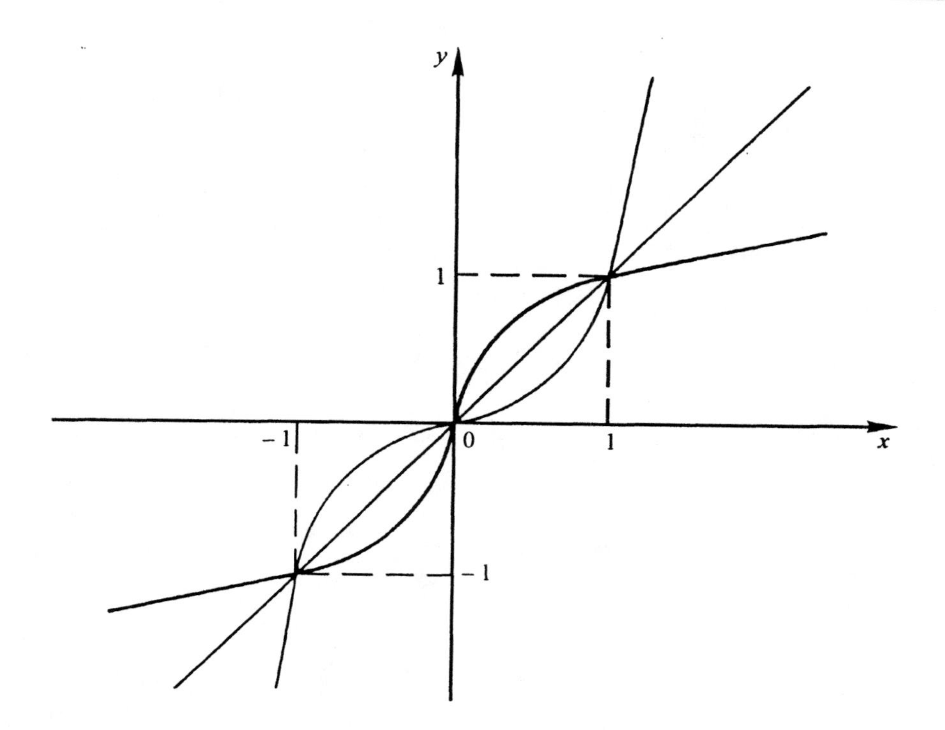

Figure 3.7. Power function $y = x^{1/m}$, m is odd.

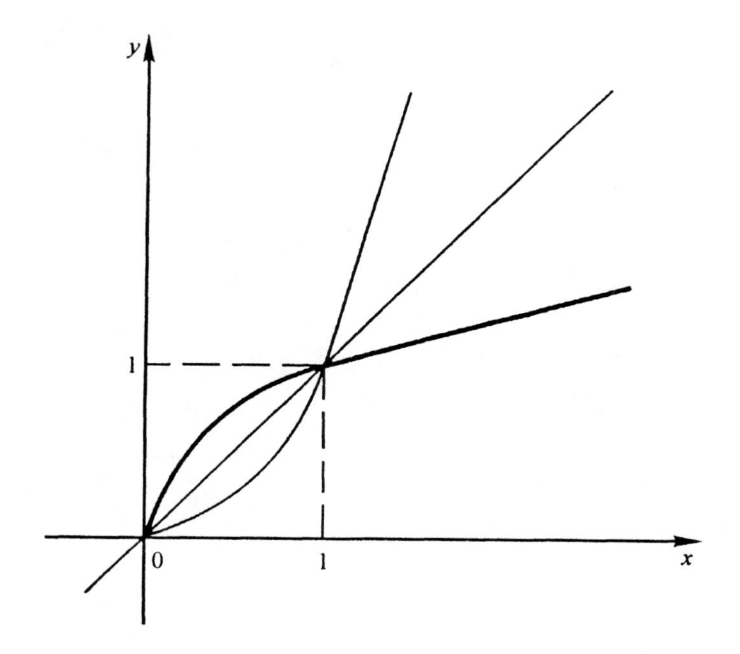

Figure 3.8. Power function $y = x^{1/m}$, m is even, positive.

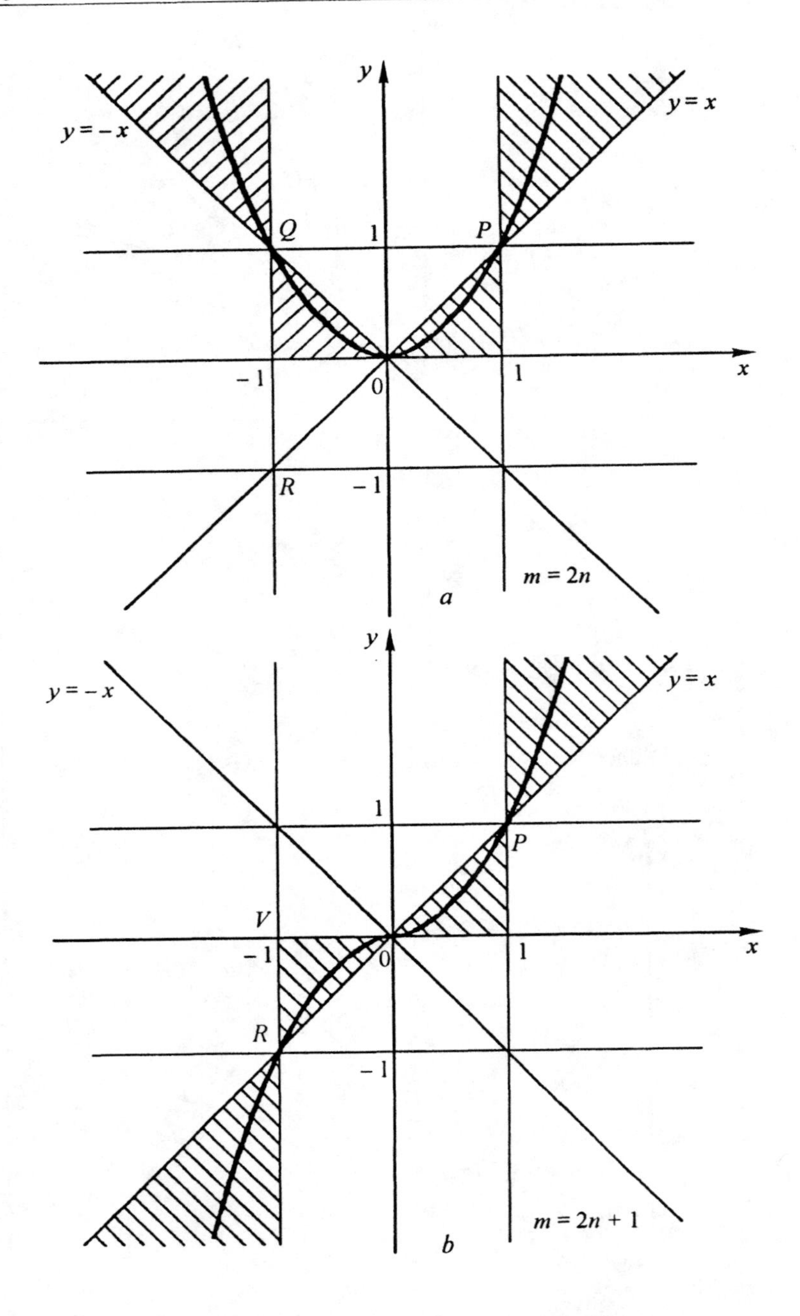

Figure 3.9. Graphs of functions: (*a*) $y = x^{2n}$, (*b*) $y = x^{2n-1}$.

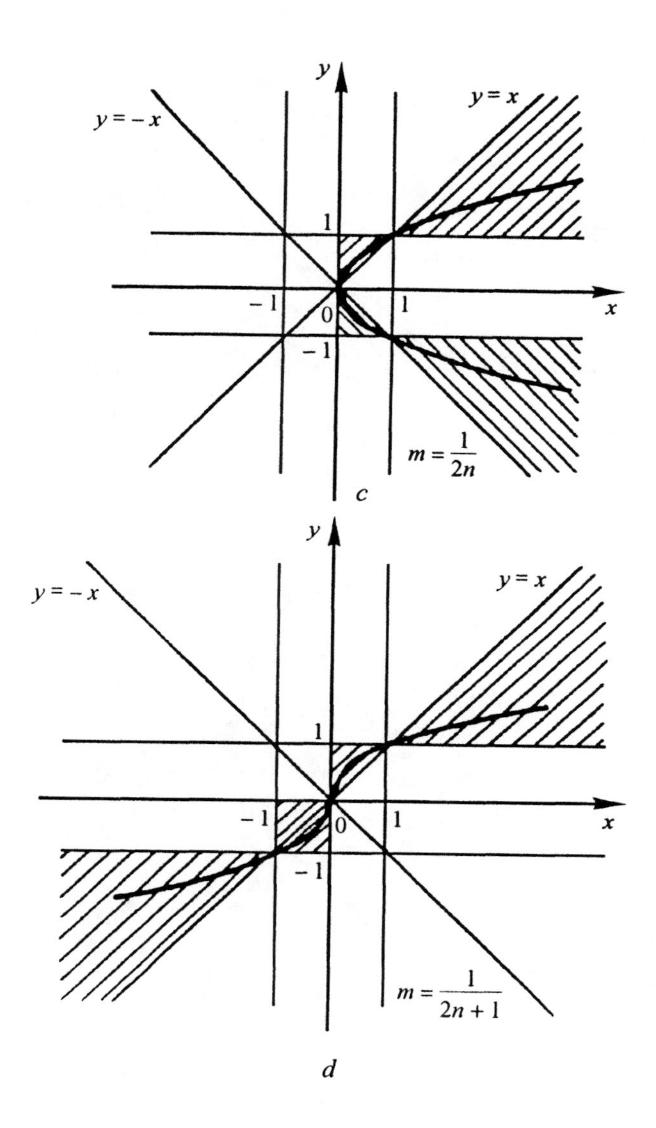

Figure 3.9. Graphs of functions: (*c*) $y = x^{1/(2n)}$, (*d*) $y = x^{1/(2n-1)}$.

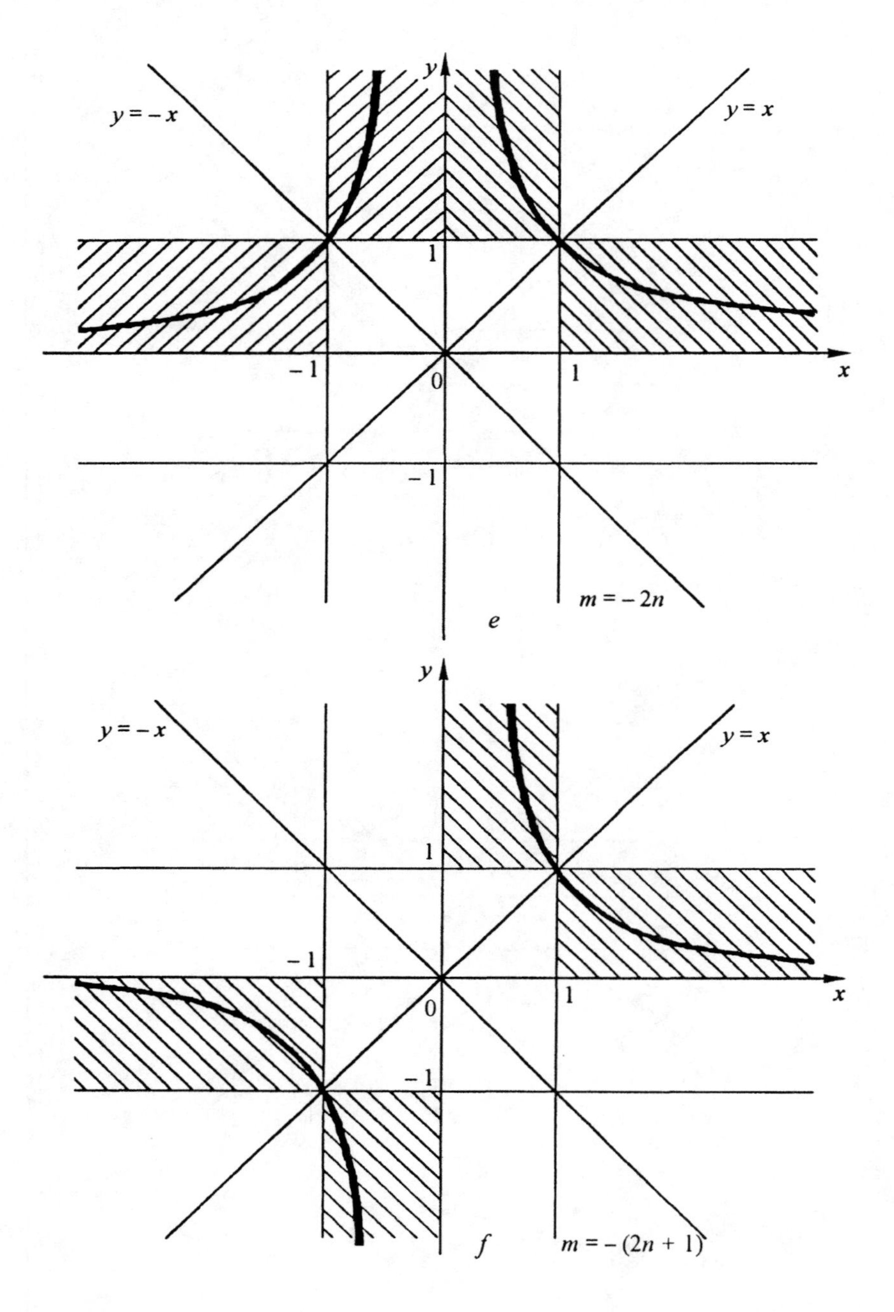

Figure 3.9. Graphs of functions: (*e*) $y = x^{-2n}$, (*f*) $y = x^{-2n+1}$.

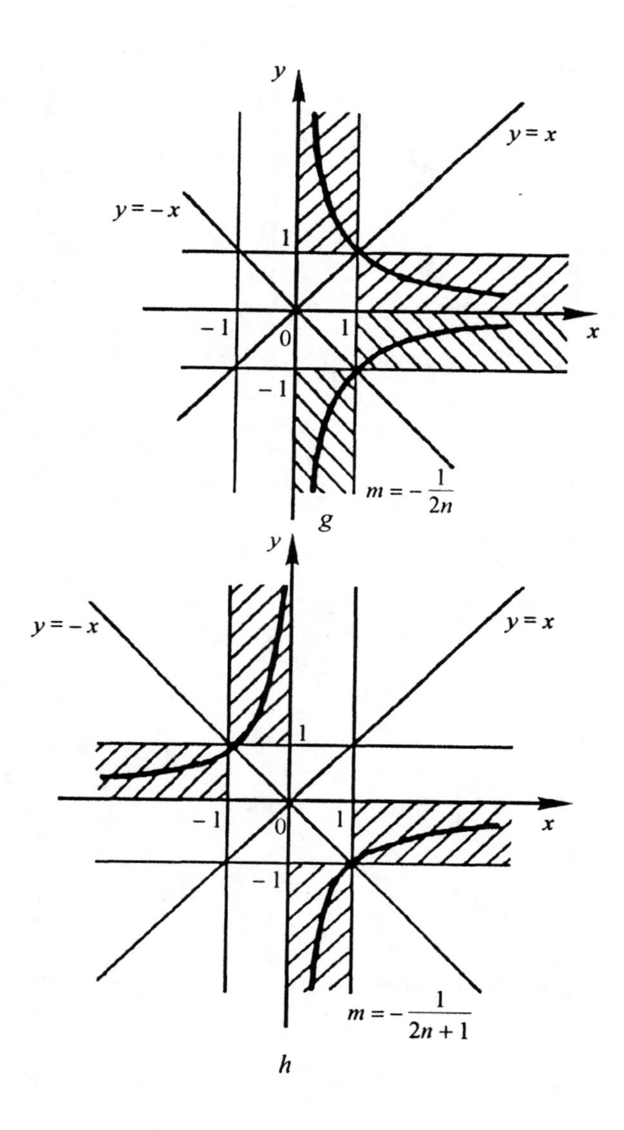

Figure 3.9. Graphs of functions: (g) $y = x^{-1/(2n)}$, (h) $y = x^{-1/(2n-1)}$.

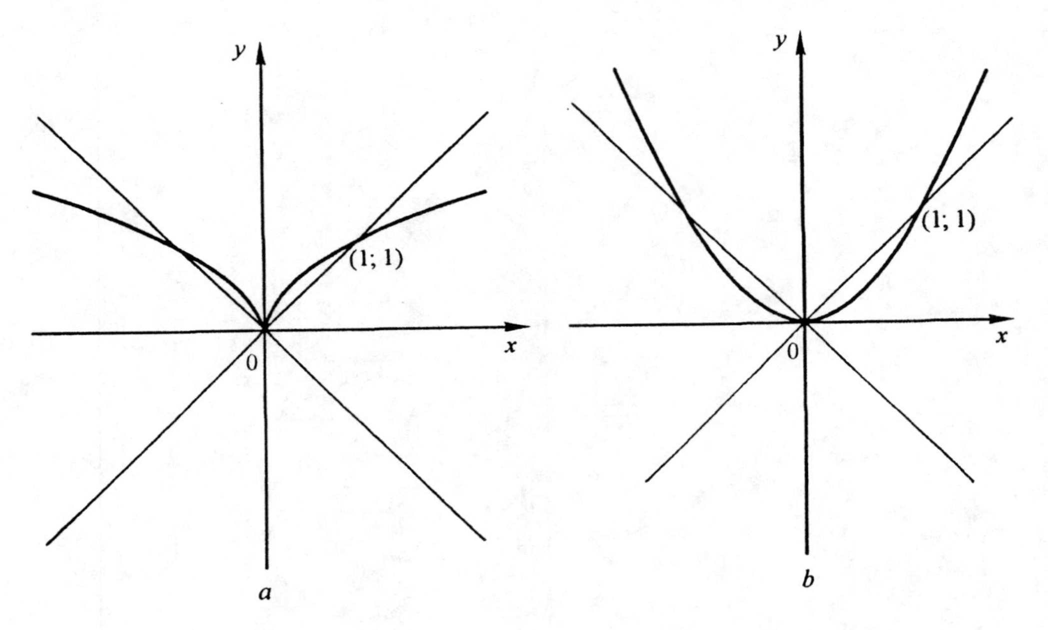

Figure 3.10. Function $y = x^{p/q}$, p — even, q — odd. (a) $0 < \frac{p}{q} < 1$, (b) $\frac{p}{q} > 1$.

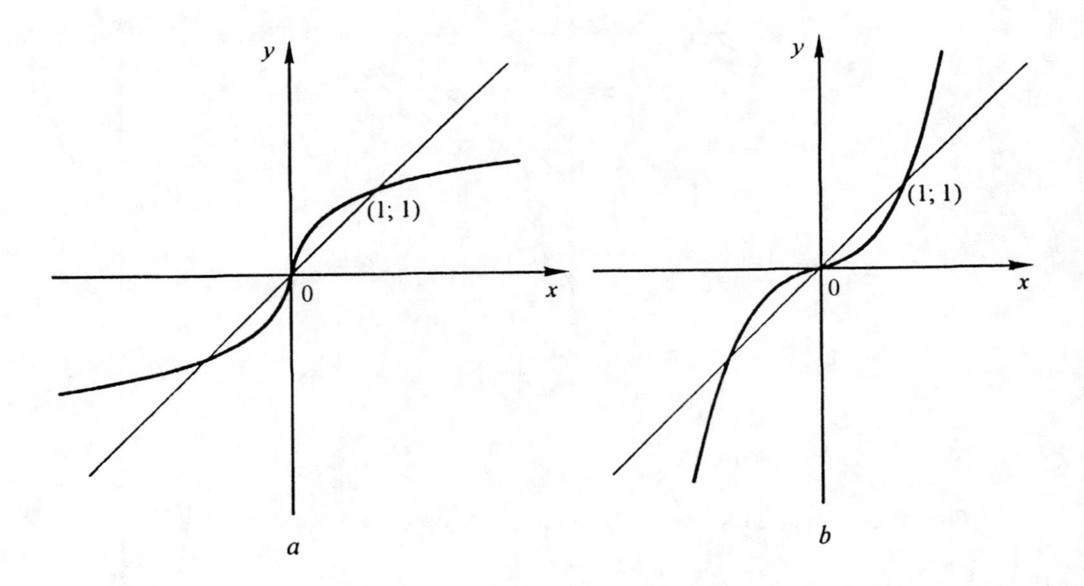

Figure 3.11. Function $y = x^{p/q}$, p,q — odd. (a) $0 < \frac{p}{q} < 1$, (b) $\frac{p}{q} > 1$.

2. Let $r = \frac{p}{q}$, where p and q are mutually prime numbers. The graph $y = \sqrt[q]{x^p}$ depends on p and q. Consider several cases.

Case 1. Let p be even, q — odd. The function $y = \sqrt[q]{x^p}$ is defined on $(-\infty; +\infty)$. Since $\sqrt[q]{(-x)^p} = \sqrt[q]{x^p}$, the function is even, i.e. its graph is symmetric with respect to the axis Oy.

Because the function is monotone increasing on $[0; +\infty)$ and $y = \sqrt[q]{x^p} \geq 0$ for all x, the graph lies above the axis Ox and passes through the points $(0; 0)$ and $(1; 1)$.

The function $y = \sqrt[q]{x^p}$ is convex on $[0; +\infty)$ if $0 < \frac{p}{q} < 1$, and is concave if $\frac{p}{q} > 1$. If $0 < \frac{p}{q} < 1$, the graph lies above the bisectrix $y = x$ for $x \in (0; 1)$, and below the bisectrix for $x \in (1; +\infty)$. In the case where $\frac{p}{q} > 1$, the graph lies below the bisectrix in the interval $(0; 1)$, and above it in the interval $(1; +\infty)$ (Figure 3.10).

Case 2. Let p and q be odd. The function $y = \sqrt[q]{x^p}$ is defined on $(-\infty; +\infty)$ and is odd there, since $\sqrt[q]{(-x)^p} = -\sqrt[q]{x^p}$. Its graph is symmetric with respect to the origin.

The function monotonically increases on $(-\infty; +\infty)$ from $-\infty$ to $+\infty$. The graph passes through the points $(0; 0)$ and $(1; 1)$. The function $y = \sqrt[q]{x^p}$ is convex on $(0; +\infty)$ if $0 < \frac{p}{q} < 1$, and is concave on this interval if $\frac{p}{q} > 1$ (Figures 3.11).

Case 3. Let p be odd and q — even. The function $y = \sqrt[q]{x^p} = (x^p)^{1/q}$ is defined on $[0; +\infty)$. The graph passes through the points $(0; 0)$ and $(1; 1)$ (Figure 3.12).

3. Let $r = -p/q$, where p and q are mutually prime natural numbers. The power function with negative rational exponent can be written as $y = x^{-p/q} = 1/x^{p/q} = 1/\sqrt[q]{x^p}$.

Case 1. Let p be even, and q — odd. The domain of the function is $(-\infty; 0) \cup (0; +\infty)$. The point $x = 0$ is a point of discontinuity. The function is even; its graph is symmetric with respect to the axis Oy. On the interval $(0; +\infty)$, the function $y = x^{-p/q}$ is decreasing, because the function $y = x^{p/q}$ is increasing on this interval. For $x \in (0; +\infty)$, $y = x^{-p/q} > 0$ and its graph lies above the axis Ox. The graph passes through the point $(1; 1)$. The line $y = 0$ is a horizontal asymptote (Figure 3.13).

Case 2. Let p and q be odd. The domain of the function is $(-\infty; 0) \cup (0; +\infty)$. The function is odd; its graph is symmetric with respect to the origin. The function is decreasing. The graph has the horizontal asymptote $y = 0$ and vertical asymptote $x = 0$ (Figure 3.14).

Case 3. Let p be odd, q — even. The domain of the function is $(0; +\infty)$. The horizontal asymptote is $y = 0$, the vertical asymptote — $x = 0$ (Figure 3.15).

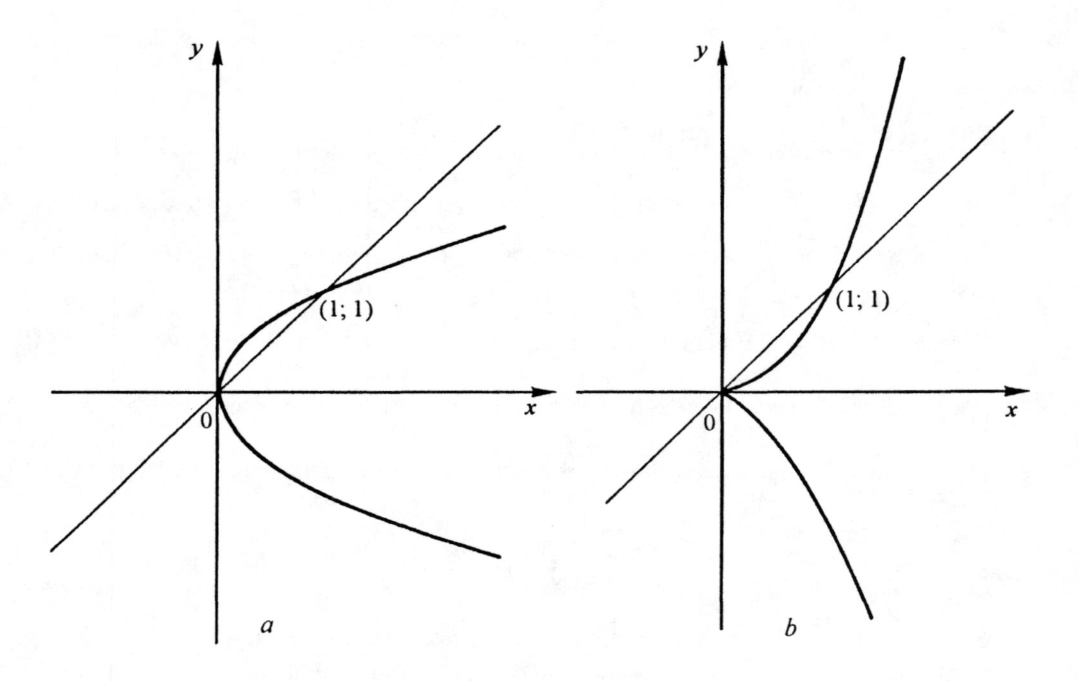

Figure 3.12. Function $y = x^{p/q}$, p — odd, q — even. (a) $0 < \frac{p}{q} < 1$, (b) $\frac{p}{q} > 1$.

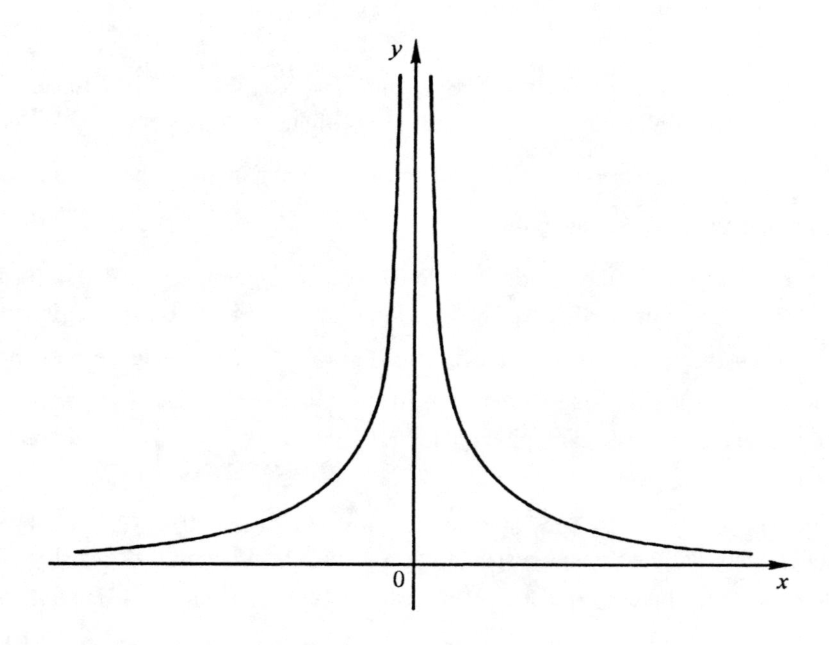

Figure 3.13. Function $y = x^{-p/q}$, p — even, q — odd, $\frac{p}{q} > 0$.

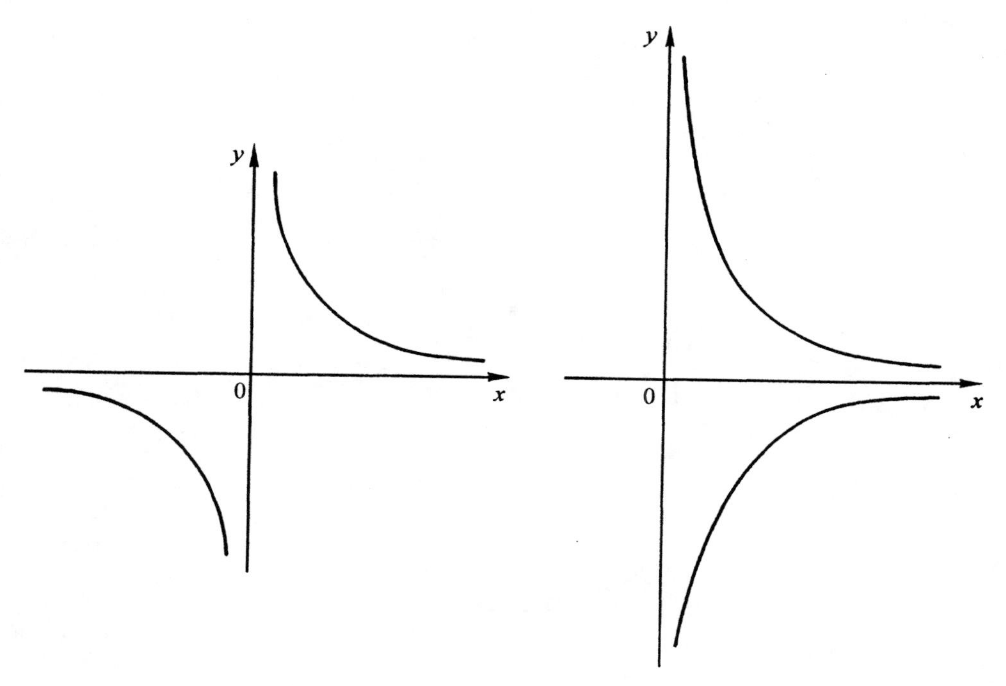

Figure 3.14. Function $y = x^{-p/q}$, p, q — odd, $\frac{p}{q} > 0$.

Figure 3.15. Function $y = x^{-p/q}$, p — odd, q — even, $\frac{p}{q} > 0$.

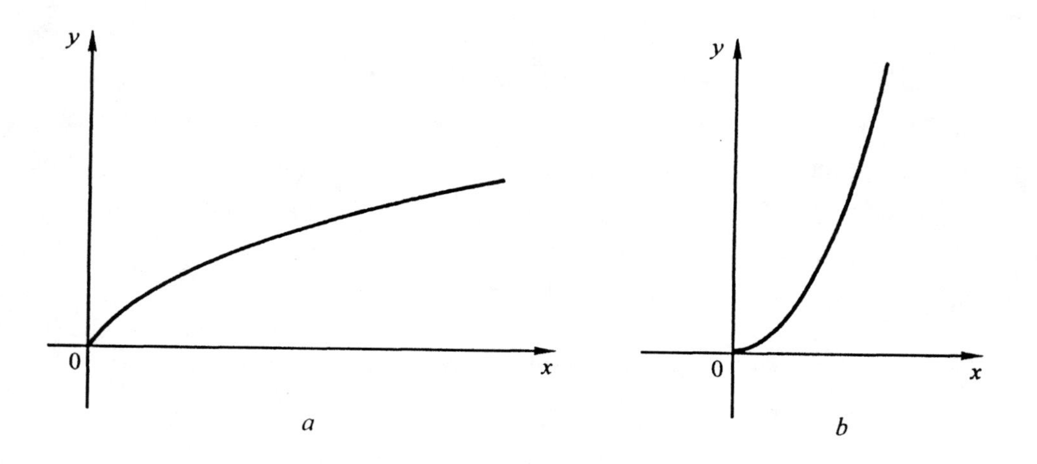

Figure 3.16. Function $y = x^{\alpha}$, α — positive irrational. (a) $\alpha < 1$, (b) $\alpha > 1$.

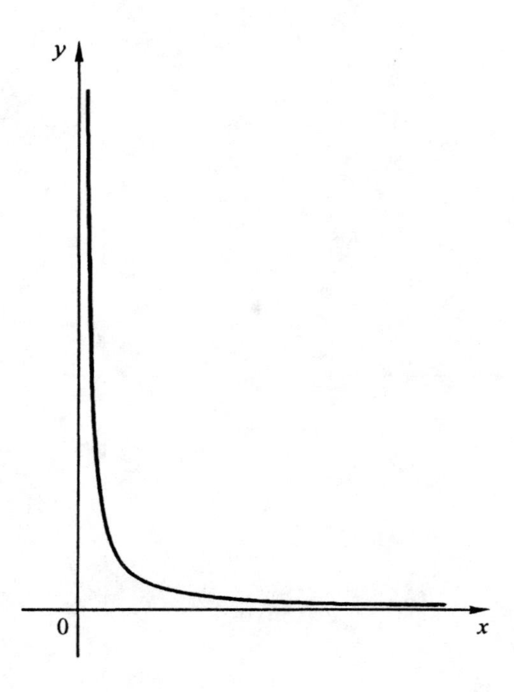

Figure 3.17. Function $y = x^{-\alpha}$, α — irrational.

Power function with irrational exponent

The power with irrational exponent α of a positive number x is the limit of a sequence of rational powers of the number x when the rational exponents approach α, i.e. $x^{\alpha} = \lim\limits_{r_n \to \alpha} x^{r_n}$.

If x is negative, then x^r makes sense only for some rationals r.

For α irrational, we will consider $y = x^{\alpha}$ as a function on $[0; +\infty)$. The function is increasing on $[0; +\infty)$ (as the limit of increasing functions $y = x^{r_n}$, $r_n \to \alpha$) and varies from 0 to $+\infty$.

For all $x \in [0; +\infty)$, the function $x^{\alpha} \geq 0$, that is the graph lies above the x-axis. The function is concave for $\alpha > 1$, and convex for $0 < \alpha < 1$. The graph passes through the point $(1; 1)$ and the origin (Figure 3.16).

The graph of the function $y = x^{-\alpha}$ for positive irrational α, is shown in Figure 3.17.

3.2. Exponential function

Exponential function is the function $y = a^x = e^{bx}$, where $a > 0$ and $a \neq 1$, $b = \ln a$. This function is one of the transcendental functions. The domain is $(-\infty; +\infty)$, the range is $(0; +\infty)$. Its graph lies above the x-axis.

If $x = 0$, then $y = 1$, that is the graph intersects the y-axis at the point $(0; 1)$.

If $a > 1$, then the function $y = a^x$ is monotone increasing, and it is decreasing if $a < 1$.

Let us calculate the value of the function at points x_1, x_2 $(x_1 < x_2)$:

$$f(x_1) = a^{x_1}, \quad f(x_2) = a^{x_2}.$$

Consider the difference $a^{x_1} - a^{x_2} = a^{x_1}(a^{x_2 - x_1} - 1)$, and determine its sign. Because a^{x_1} is always greater than zero, the sign of the expression coincides with the sign of the second factor.

If $x_2 - x_1 > 0$, then $a^{x_2 - x_1} > 1$ for $a > 1$, and $a^{x_2 - x_1} < 1$ for $a < 1$. Hence $f(x_2) - f(x_1) > 0$ $(a > 1)$, and $f(x_2) - f(x_1) < 0$ $(a < 1)$, that is, if $x_2 > x_1$, then $f(x_2) > f(x_1)$ $(a > 1)$, and $f(x_2) < f(x_1)$ $(a < 1)$.

If $x \to +\infty$, then $a^x \to +\infty$ for $a > 1$, and $a^x \to 0$ for $a < 1$. The x-axis is an asymtote of the graph as $x \to -\infty$ if $b > 0$, and as $x \to +\infty$ if $b < 0$. The greater the value of $|b|$ the faster the graph approaches the asymptote.

To study the convexity of the graph, consider the sign of the expression $\dfrac{f(x_1) + f(x_2)}{2} - f\left(\dfrac{x_1 + x_2}{2}\right)$:

$$\frac{a^{x_1} + a^{x_2}}{2} - a^{(x_1 + x_2)/2} = \frac{(a^{x_1/2} - a^{x_2/2})^2}{2} \geq 0.$$

This means that the graph is concave.

Note that the graphs of the functions a^x and a^{-x} are symmetric with respect to the y-axis. Examples of exponential functions are shown in Figures 3.18 and 3.19.

3.3. Logarithmic function

Logarithmic function is the function $y = \log_a x$, where $a > 0$, $a \neq 1$. The domain of the function is $(0; +\infty)$.

For $a > 1$, the logarithmic function takes positive values for $x > 1$, and negative values for $0 < x < 1$. At the point $x = 1$, the function equals zero.

For $a < 1$, the logarithmic function is positive for $0 < x < 1$, negative for $x > 1$, and zero at $x = 1$.

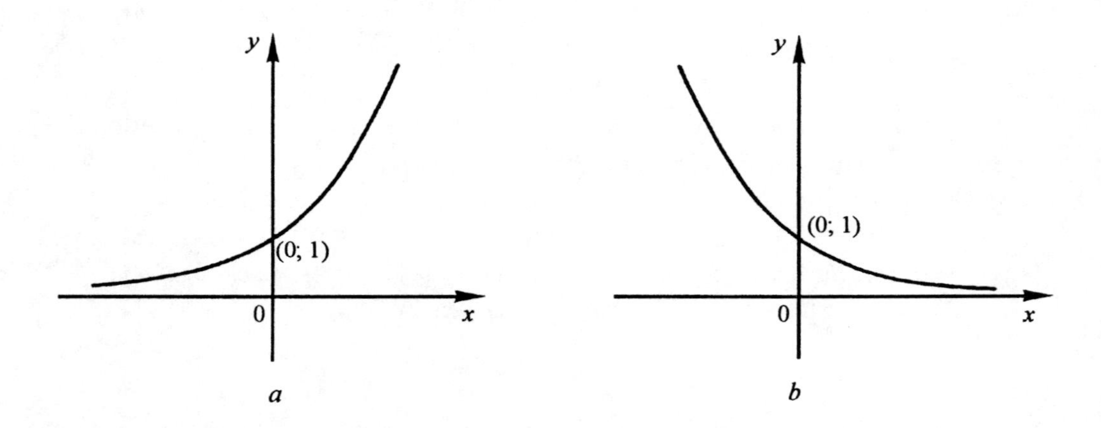

Figure 3.18. Exponential function $y = a^x$, (*a*) $a > 1$, (*b*) $a < 1$.

For $a > 1$, the function monotonically increases from $-\infty$ to $+\infty$. For $a < 1$, the function is monotone decreasing with values ranging from $+\infty$ to $-\infty$, and the greater the value of $|\ln a|$ the more rapidly the function decreases.

If $a > 1$, then $\lim\limits_{x \to +\infty} \log_a x = +\infty$, $\lim\limits_{x \to 0+0} \log_a x = -\infty$.

If $a < 1$, then $\lim\limits_{x \to +\infty} \log_a x = -\infty$, $\lim\limits_{x \to 0+0} \log_a x = +\infty$.

The function is convex for $a > 1$, and concave for $a < 1$.

The graph passes through the point (1; 0) and asymptotically approaches the y-axis at negative y if $a > 1$, and at positive y if $a < 1$. Because the logarithmic function $y = \log_a x$ is the inverse of the exponential function $y = a^x$, it is symmetric to the graph of the exponential function with respect to the bisectrix of the first and third quadrants.

Graphs of some logarithmic functions are shown in Figures 3.20, 3.21.

3.4. Trigonometric functions

Trigonometric functions are the functions $y = \sin x$, $y = \cos x$, $y = \tan x$, $y = \cot x$.

Function $y = \sin x$. (Figure 3.22). The domain of the function is $(-\infty; +\infty)$, the range is $[-1; 1]$. The function is odd, periodic with the period 2π. The graph intersects the x-axis at points $n\pi$, $n \in \mathbf{Z}$. These points are also inflection points.

The function increases on $(0; \pi/2)$ from 0 to 1, decreases on $(\pi/2; 3/2\pi)$ from 1 to -1, and increases on $(3/2\pi; 2\pi)$ from -1 to 0. The points $x = \pi/2 + 2\pi m$, $m \in \mathbf{Z}$, are points of relative maximum; the points $x = -\pi/2 + 2\pi m$, $m \in \mathbf{Z}$, are points of relative minimum.

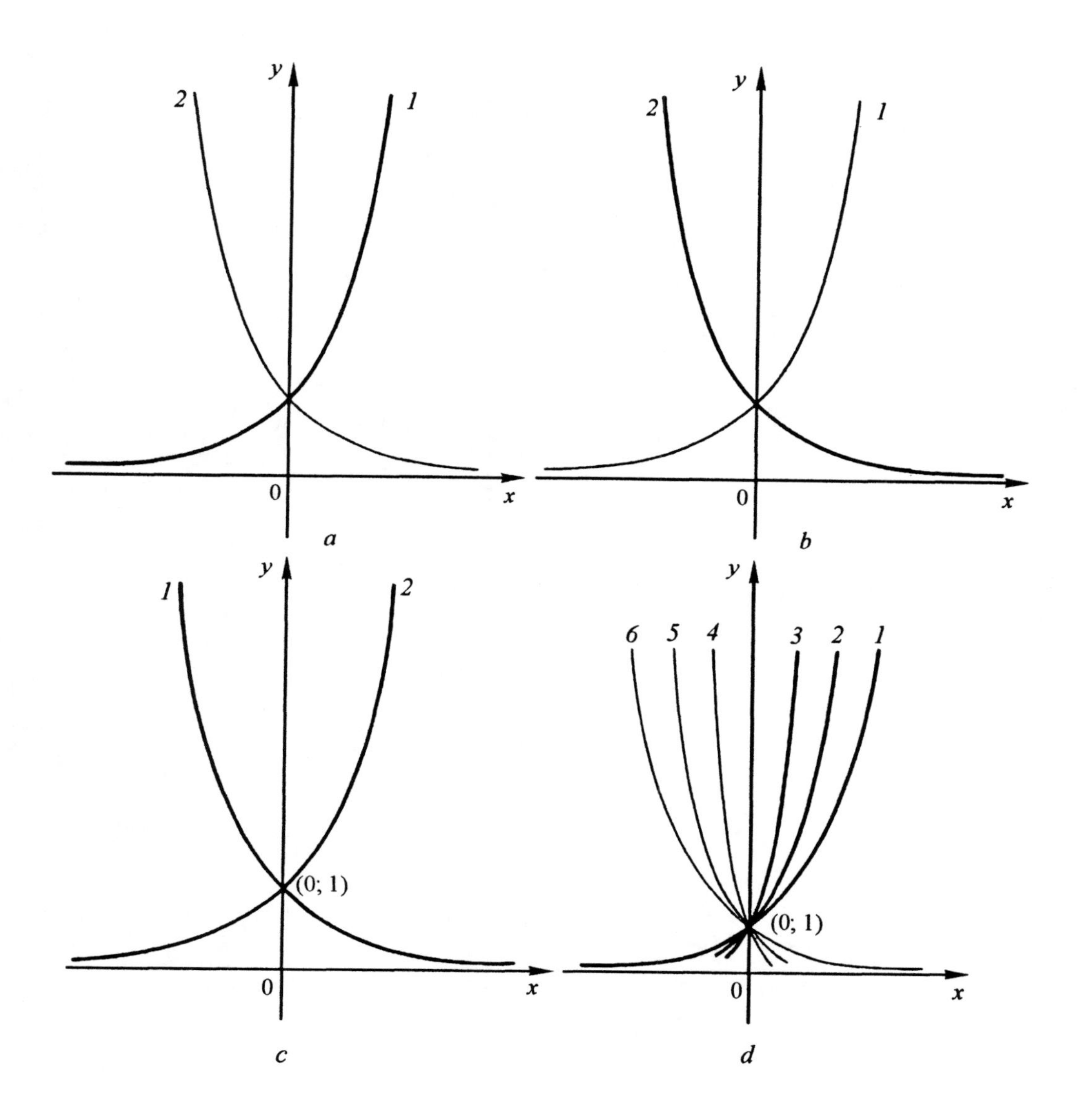

Figure 3.19. Exponential functions: (*a*) $y = a^x$, $a > 1$ (*1*), $a < 1$ (*2*); (b) $y = a^{-x}$, $a < 1$ (*1*), $a > 1$ (*2*); (*c*) $y = e^{-x}$ (*1*), $y = e^x$ (*2*); (*d*) $y = a^x$, $a = 2$ (*1*), $a = e$ (*2*), $a = 10$ (*3*), $a = \frac{1}{10}$ (*4*), $a = \frac{1}{e}$ (*5*), $a = \frac{1}{2}$ (*6*).

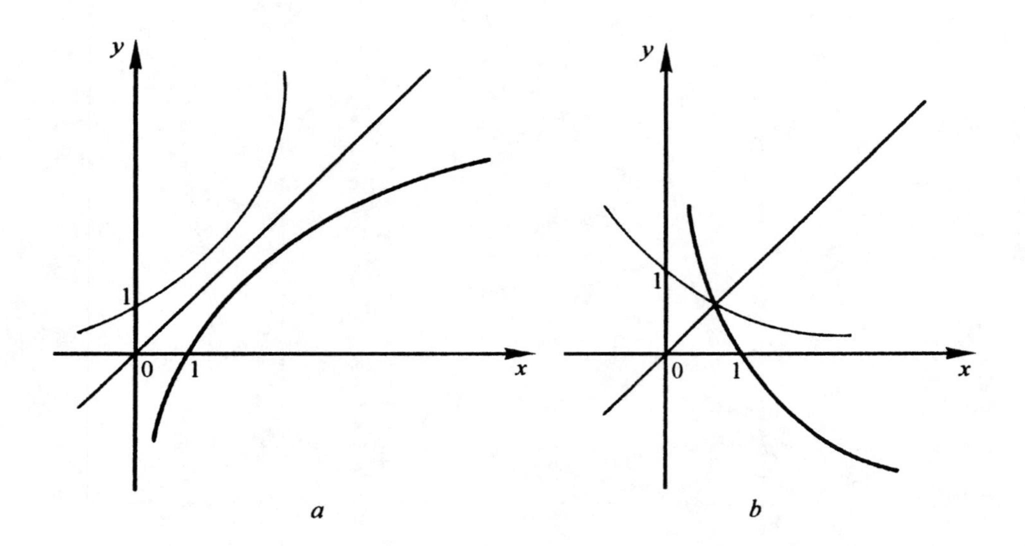

Figure 3.20. Logarithmic function $y = \log_a x$: (*a*) $a > 1$, (*b*) $a < 1$.

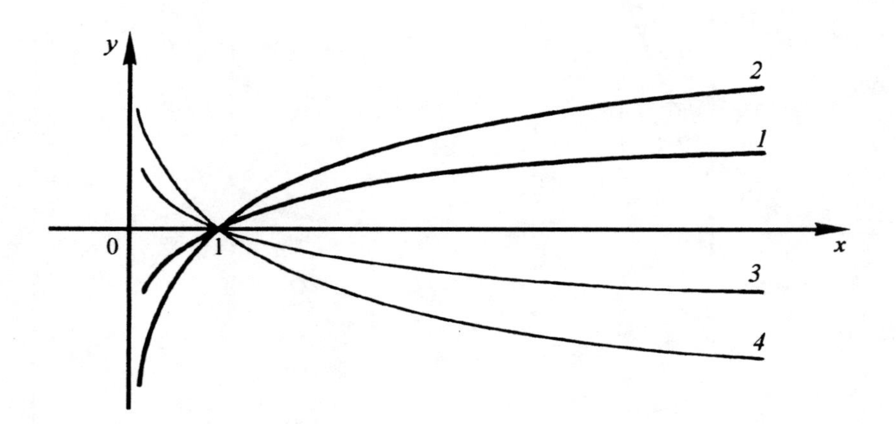

Figure 3.21. Logarithmic functions $y = \log_a x$: $a = e$ (*1*); $a = 2$ (*2*); $a = \frac{1}{e}$ (*3*); $a = \frac{1}{2}$ (*4*).

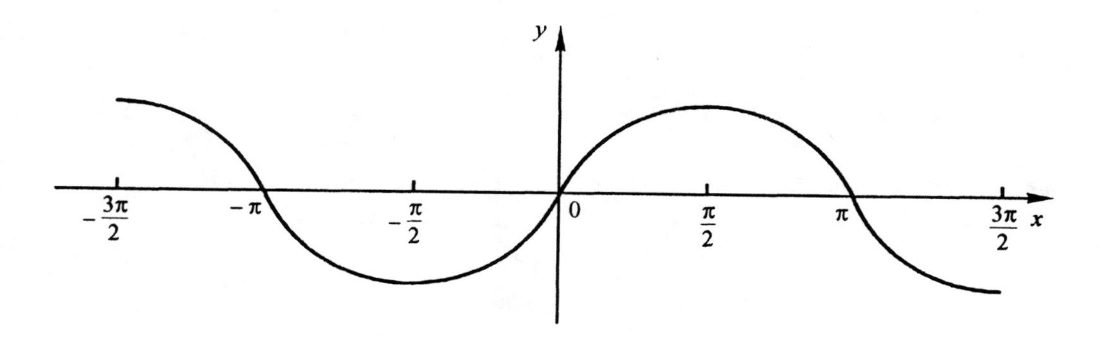

Figure 3.22. Function $y = \sin x$.

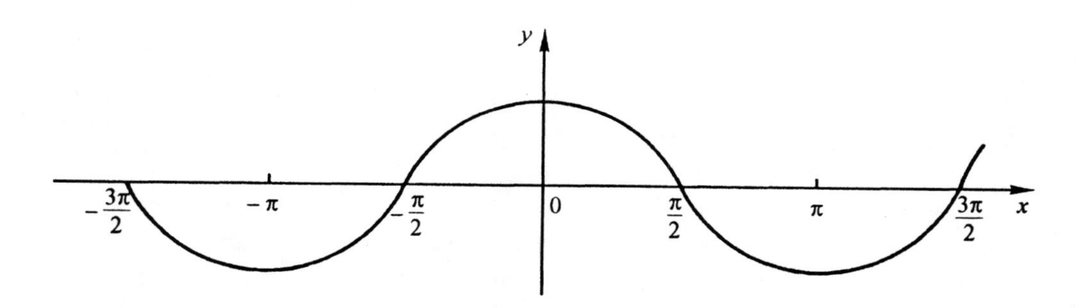

Figure 3.23. Function $y = \cos x$.

The function is convex on $[0; \pi]$, and concave on $(\pi; 2\pi)$.

The graph of $y = \sin x$ is called the *sine curve*.

Function $y = \cos x$. (Figure 3.23). The domain of the function is $(-\infty; +\infty)$; its range is $[-1; 1]$. The function is even, periodic with the period 2π. The points where the curve intersects the x-axis are $x = (2n+1)\pi/2$, $n \in \mathbf{Z}$. These points are the inflection points.

The function decreases on $(0; \pi)$ from 1 to -1, and increases on $(\pi; 2\pi)$ from -1 to 1. The function is convex on $[0; \pi/2]$, and concave on $(\pi/2; 3/2\pi)$. The points $x = 2\pi k$, $k \in \mathbf{Z}$, are points of relative maximum, the points $x = (2k+1)\pi$, $k \in \mathbf{Z}$, are points of relative minimum.

The graph of the function $y = \cos x$ is called the *cosine curve*.

Function $y = \tan x$. (Figure 3.24). The domain of the function is the whole real axis excluding the points $x = \pi/2 + k\pi$, $k \in \mathbf{Z}$. The range of the function is $(-\infty; +\infty)$. The function is odd periodic with the period π. The points of intersection of the graph and the x-axis are $x = k\pi$, $k \in \mathbf{Z}$. These points are inflection points.

The function increases on $(-\pi/2; \pi/2)$ from $-\infty$ to $+\infty$. The function is concave on $(0; \pi/2)$, and convex on $(\pi/2; 0)$. The vertical asymptotes are $x = (k + 1/2)\pi$, $k \in \mathbf{Z}$.

The graph of $y = \tan x$ is called the *tangent curve*.

Function $y = \cot x$. (Figure 3.25). The domain of the function is the set of all real numbers excluding the points $x = k\pi$, $k \in \mathbf{Z}$. The range of the function is $(-\infty; +\infty)$. The function is odd, periodic with the period π. The curve intersects the x-axis at the points $x = (k + 1/2)\pi$, $k \in \mathbf{Z}$. These points are inflection points.

The function decreases on $(0; \pi)$ from $+\infty$ to $-\infty$. The function is concave on $(0; \pi/2)$, and convex on $(\pi/2; \pi)$. The vertical asymptotes are $x = k\pi$, $k \in \mathbf{Z}$.

The graph of $y = \cot x$ is called the *cotangent curve*.

3.5. Inverse trigonometric functions

The *principal inverse trigonometric functions* are the functions $y = \arcsin x$, $y = \arccos x$, $y = \arctan x$, $y = \operatorname{arccot} x$. Note that the functions $y = \sin x$ and $y = \operatorname{Arcsin} x$, $y = \cos x$ and $y = \operatorname{Arccos} x$, $y = \tan x$ and $y = \operatorname{Arctan} x$, $y = \cot x$ and $y = \operatorname{Arccot} x$ are inverse to each other.

To construct graphs of the inverse trigonometric functions, one usually uses properties of the graphs of inverse functions (see Section 3.2). The construction is done by symmetrically reflecting the corresponding graph with respect to the bisectrix of the first and third quadrants.

Function $y = \operatorname{Arcsin} x$. (Figure 3.26). The function $y = \operatorname{Arcsin} x$ is multivalued and defined for $|x| \le 1$. The function that corresponds to the branch $y \in [-\pi/2; \pi/2]$ is called the *principal value* of $\operatorname{Arcsin} x$ and is denoted by $\arcsin x$. Other values of the function $y = \operatorname{Arcsin} x$ are calculated by $\operatorname{Arcsin} x = \arcsin x + 2k\pi$, $k \in \mathbf{Z}$. So, the domain of the function $y = \arcsin x$ is $[-1; 1]$; its range is $[-\pi/2; \pi/2]$.

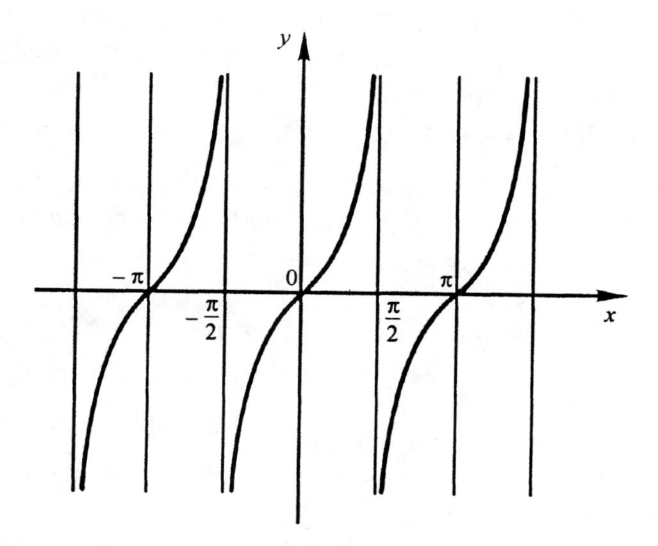

Figure 3.24. Function $y = \tan x$.

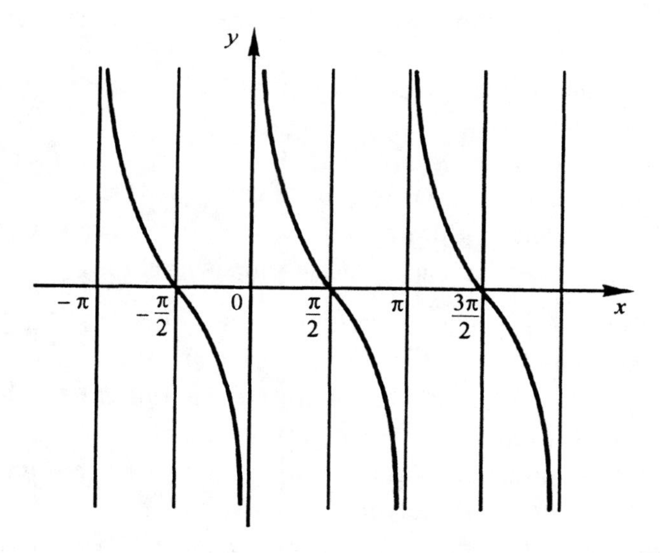

Figure 3.25. Function $y = \cot x$.

The function $y = \arcsin x$ is odd on $[-1; 1]$. It monotonically increases on this interval from $-\pi/2$ to $\pi/2$. The graph of this function (Figure 3.27) passes through the point $(0; 0)$. This point is also an inflection point. On the interval $[-1; 0]$, the function is convex, it is concave on $(0; 1]$. Graphs of the functions $y = \text{Arcsin}\, x$ and $y = \arcsin x$ are shown in Figures 3.26 and 3.27.

Function $y = \text{Arccos}\, x$. (Figure 3.28). The function $y = \text{Arccos}\, x$ is defined for $|x| \le 1$. The function corresponding to the branch $y \in [0; \pi]$ is called the *principal value* of Arccos x, and is denoted by $\arccos x$. Other values of the function $y = \text{Arccos}\, x$ are calculated by $\text{Arccos}\, x = 2\pi k - \arccos x$, $k \in \mathbf{Z}$.

The domain of the function $y = \arccos x$ is $[-1; 1]$. Its range is $[0; \pi]$. The function monotonically decreases on $[-1; 1]$ from π to 0. The function $y = \arccos x$ satisfies the identity $\arccos(-x) = \pi - \arccos x$.

The graph of the function $y = \arccos x$ (Figure 3.29) passes through the point $(0; 0)$ which, at the same time, is an inflection point and the center of symmetry of the curve.

The function is concave on $[-1; 0]$, and convex on $[0; 1]$.

Graphs of the functions $y = \text{Arccos}\, x$ and $y = \arccos x$ are shown in Figures 3.28 and 3.29.

Function $y = \text{Arctan}\, x$. (Figure 3.30). The function $y = \text{Arctan}\, x$ is multivalued and is defined on $(-\infty; +\infty)$. The function corresponding to the values $y \in (-\pi/2; \pi/2)$ is called the *principal value* of $y = \text{Arctan}\, x$ and is denoted by $y = \arctan x$. Other values of $y = \text{Arctan}\, x$ are calculated by the formula $\text{Arctan}\, x = \arctan x + k\pi$, $k \in \mathbf{Z}$.

The domain of the function $y = \arctan x$ is $(-\infty; +\infty)$. Its range is $(-\pi/2; \pi/2)$. The function is odd.

The function $y = \arctan x$ monotonically increases on $(-\infty; +\infty)$ from $-\pi/2$ to $\pi/2$. (The values $-\pi/2$ and $\pi/2$ are not attained.)

The graph of the function $y = \arctan x$ (Figure 3.31) passes through the point $(0; 0)$, which is an inflection point and a center of symmetry of the curve.

The function $y = \arctan x$ is concave on $(-\infty; 0]$ and is convex on $(0; +\infty)$. Horizontal asymptotes of the graph of the function $y = \text{Arctan}\, x$ are straight lines $y = \pm(2k+1)\pi$, $k \in \mathbf{Z}$.

Function $y = \text{Arccot}\, x$. (Figure 3.32). The function $y = \text{Arccot}\, x$, defined on $(-\infty; \infty)$, is multivalued. The function corresponding to the values $y \in (0; \pi)$ of the function $y = \text{Arccot}\, x$ is called the *principal value* of the function $y = \text{Arccot}\, x$ and is denoted by $y = \text{arccot}\, x$. All other values of the function are calculated by $\text{Arccot}\, x = \text{arccot}\, x + k\pi$, $k \in \mathbf{Z}$.

The domain of the function $y = \text{arccot}\, x$ is $(-\infty; +\infty)$. Its range is $(0; \pi)$. The function satisfies the identity $\text{arccot}(-x) = \pi - \text{arccot}\, x$.

The function monotonically decreases on $(-\infty; +\infty)$ from π to 0 (the values π and 0 are not attained).

The graph of the function $y = \text{arccot}\, x$ (Figure 3.33) passes through the point $(0; 0)$. This point is an inflection point and a center of symmetry of the curve.

The function $y = \text{arccot}\, x$ is convex on $(-\infty; 0]$ and is concave on $(0; +\infty)$. Horizontal asymptotes of the graph $y = \text{Arccot}\, x$ are given by $y = \pm k\pi$, $k \in \mathbf{Z}$.

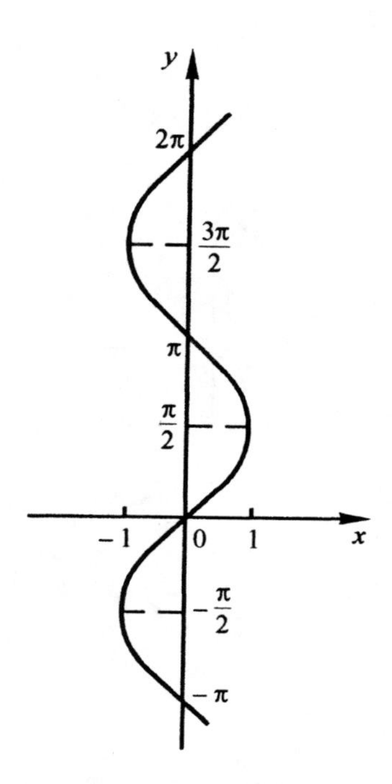

Figure 3.26. Function $y = \operatorname{Arcsin} x$.

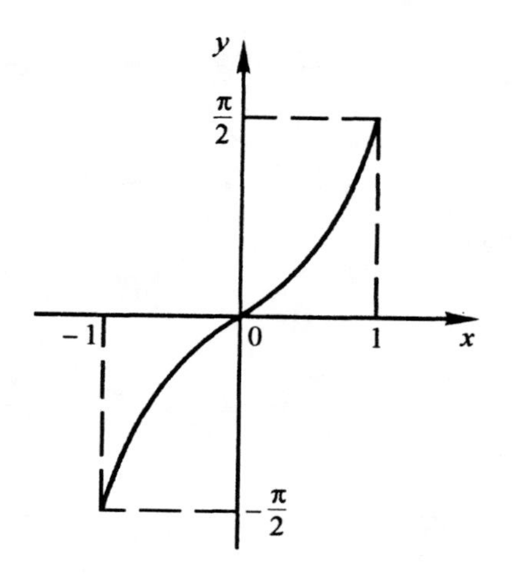

Figure 3.27. Function $y = \arcsin x$.

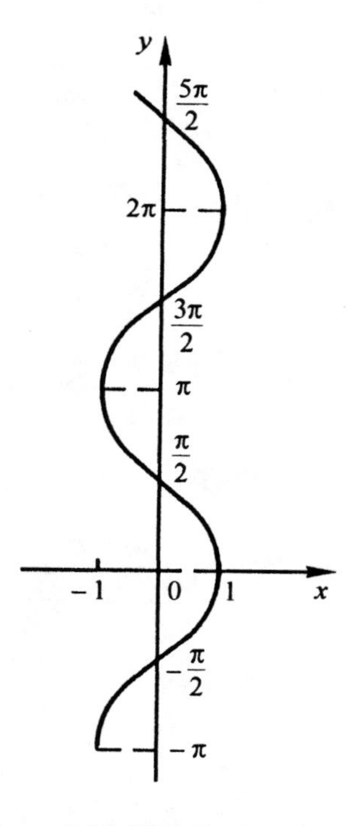

Figure 3.28. Function $y = \operatorname{Arccos} x$.

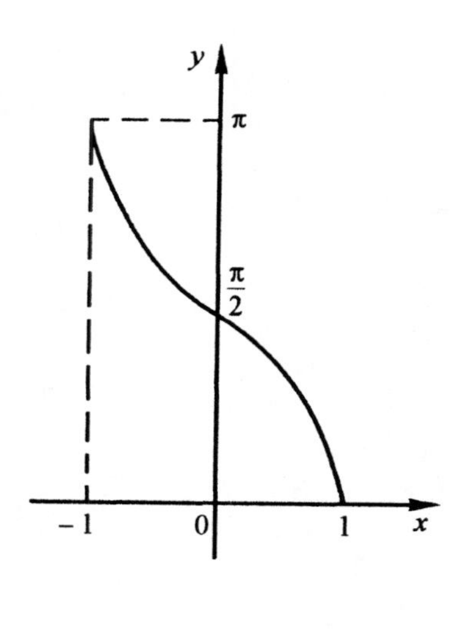

Figure 3.29. Function $y = \arccos x$.

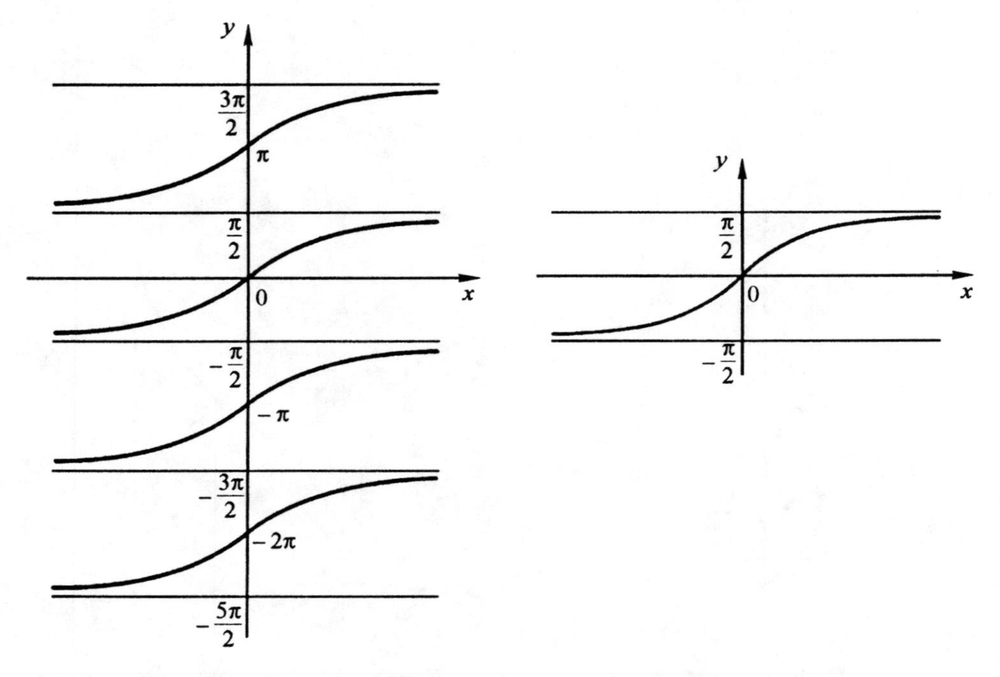

Figure 3.30. Function $y = \text{Arctan}\, x$.

Figure 3.31. Function $y = \arctan x$.

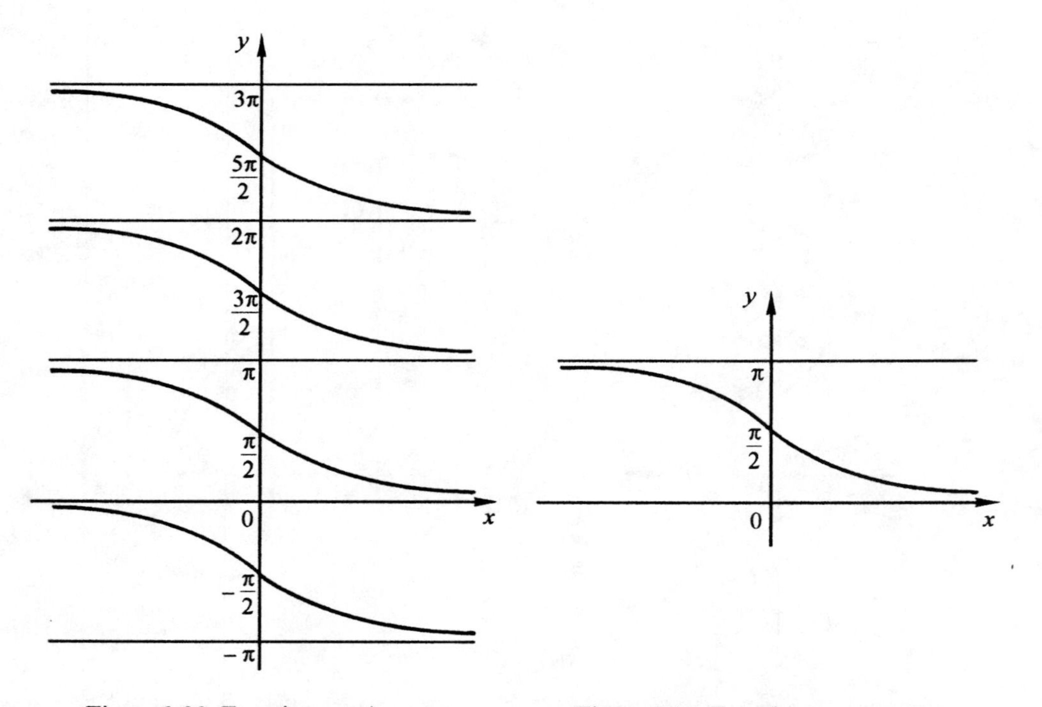

Figure 3.32. Function $y = \text{Arccot}\, x$.

Figure 3.33. Function $y = \text{arccot}\, x$.

Chapter 4

METHODS FOR PLOTTING GRAPHS BY USING ARITHMETIC TRANSFORMATIONS. TRANSFORMATIONS OF GRAPHS IN THE CARTESIAN COORDINATE SYSTEM

4.1. Arithmetic transformations of graphs

4.1.1. Addition and subtraction of graphs

A general method for plotting graphs of the sum or difference of two functions consists in plotting the graph of each function and then adding or subtracting the y-coordinates of the points of the curves that have the same x-coordinates. (It is convenient to do this for characteristic points of the graphs.) The graph is then drawn using the constructed points, and is verified at several check points.

In certain cases, the graph of the sum or difference of two functions is constructed as follows.

1. To plot the graph of the sum of two functions, first the graph of a more simple function is plotted. Then, the graph of the sum is constructed by adding the y-coordinates of the second function to the y-coordinates of the points of the plotted curve.
2. To plot the graph of the difference of two functions, first the graph of the minuend is plotted. Then from the y-coordinate of points on the curve, the y-coordinate of the subtrahend is subtracted. Sometimes it convenient to draw an auxiliary graph of the subtrahend taken with the minus sign and then add the two graphs.

When constructing the graphs of functions $y = f(x) \pm \varphi(x)$, one should use properties that the functions have: evenness, oddness, monotonicity, periodicity, and other properties of the functions $f(x)$, $\varphi(x)$. The graph is plotted in the intersection of the domains of the functions.

If $f(x) > 0$ and $\varphi(x) > 0$, then the graph of $y = f(x) + \varphi(x)$ will lie above the graphs of the functions $y = f(x)$ and $y = \varphi(x)$, and if $f(x) < 0$ and $\varphi(x) < 0$, then the graph of $y = f(x) + \varphi(x)$ will lie below the graphs of $y = f(x)$ and $y = \varphi(x)$.

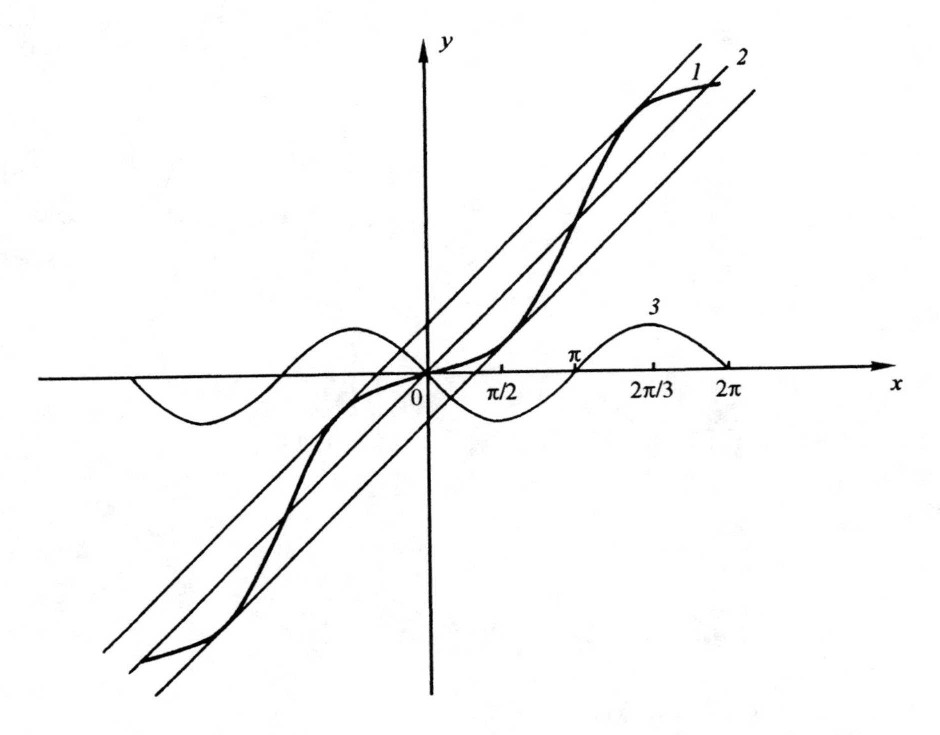

Figure 4.1. The graphs of functions: $y = x - \sin x$ (*1*), $y = x$ (*2*), $y = -\sin x$ (*3*).

Remark 4.1.1. This method can be used for constructing graphs of a finite algebraic sum of simple functions. If the functions are complicated, then it is better to use the procedure given in Section 2.2, and still better is to use methods of calculus described in Part II.

Example 4.1.1. Plot the graph of the function $y = x - \sin x$.

We have two functions $y_1 = x$ and $y_2 = -\sin x$. Plot the function $y_1 = x$, and then from this graph (not from the x-axis), draw the y-coordinates of the second function. Because $|\sin x| \le 1$, it is useful to draw two additional lines $y = x + 1$ and $y = x - 1$.

At the points where $\sin x = 0$, i.e. at $x = k\pi$, $k \in \mathbf{Z}$, we have that $y = x$, and this means that the points of the graph lie on the line $y = x$.

At the points where $\sin x = 1$, i.e. at $x = \pi/2 + 2k\pi$, we have that $y = x - 1$, and this means that the corresponding points of the plotted graph lie on the line $y = x - 1$.

At the points where $\sin x = -1$, i.e. at $x = -\pi/2 + 2k\pi$, the value of the function is $y = x + 1$. Such points lie on the line $y = x + 1$.

So the vertices of the sine curve lie on the lines $y = x - 1$ and $y = x + 1$. The graph is shown in Figure 4.1.

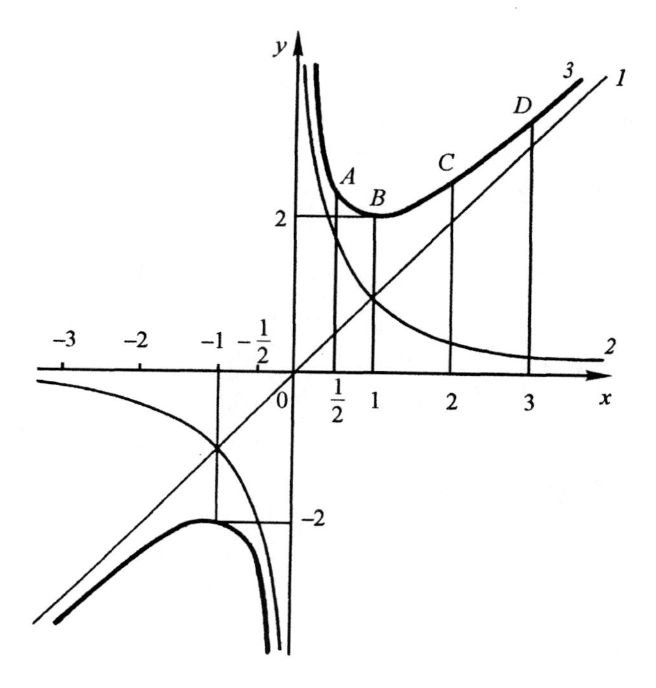

Figure 4.2. The graphs of functions: $y = x + 1/x$ (*1*), $y = 1/x$ (2), $y = x$ (*3*).

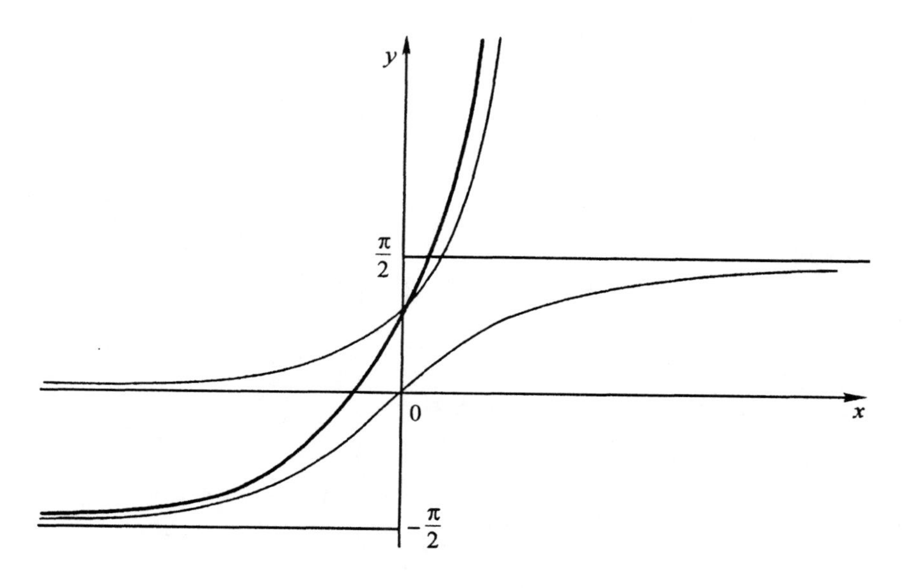

Figure 4.3. The graph of the function $y = \arctan x + 2^x$.

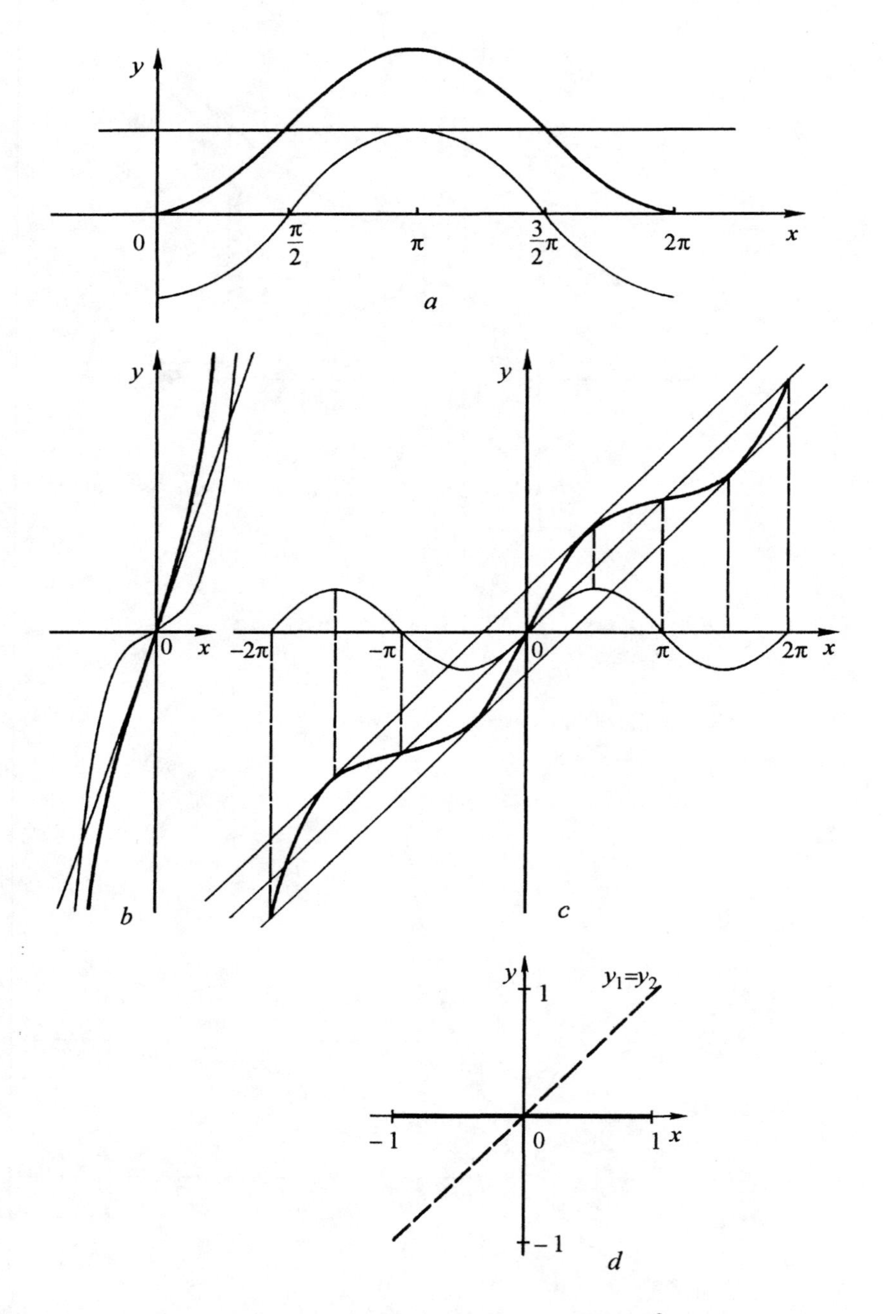

Figure 4.4. The graph of functions: (*a*) $y = 1 - \cos x$; (*b*) $y = x^3 + 3x$; (*c*) $y = x + \sin x$; (*d*) $y = \sin(\arcsin x) - x$.

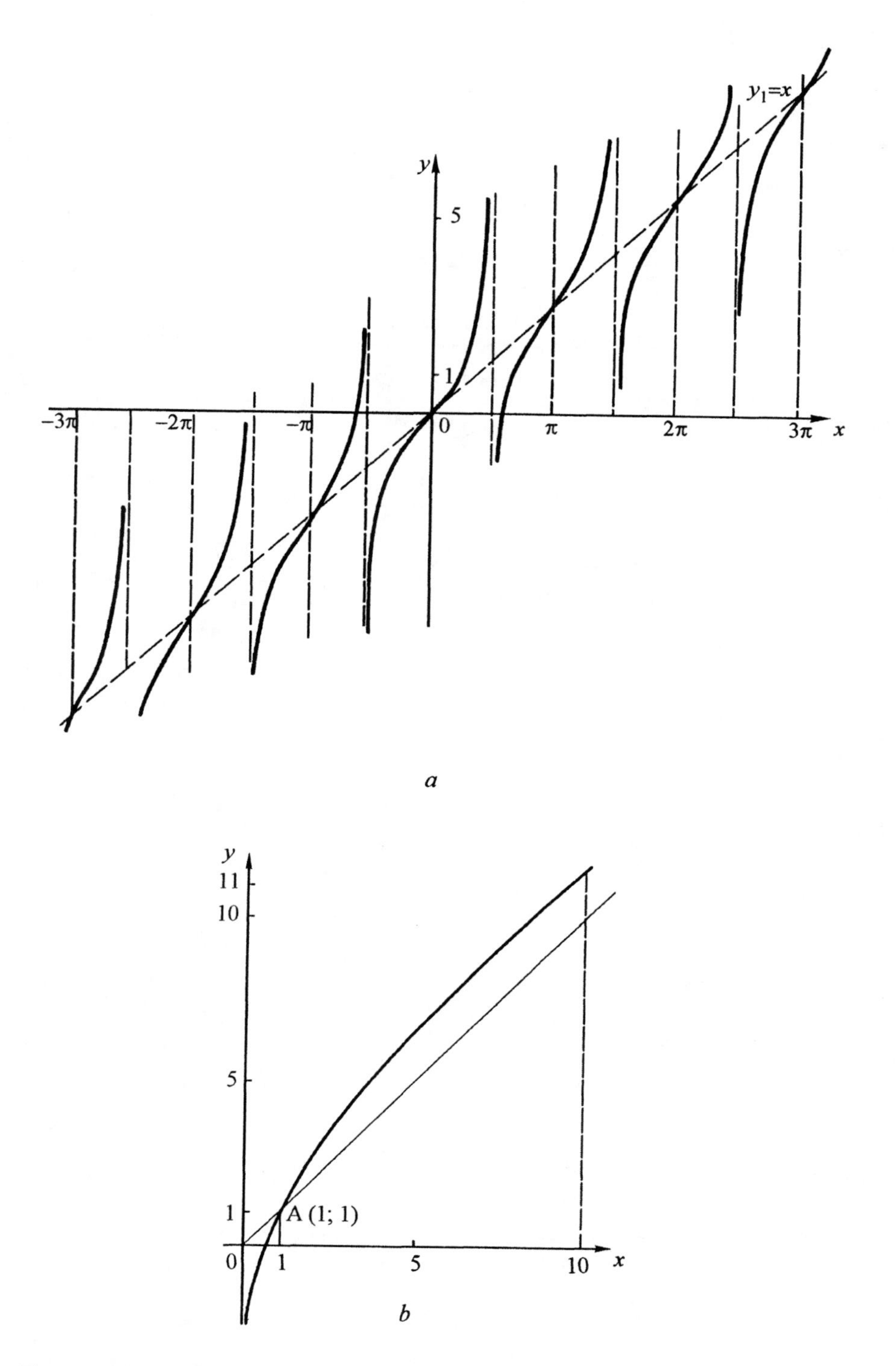

Fig. 4.5. The graph of functions: (*a*) $y = x + \tan x$; (*b*) $y = x + \ln x$.

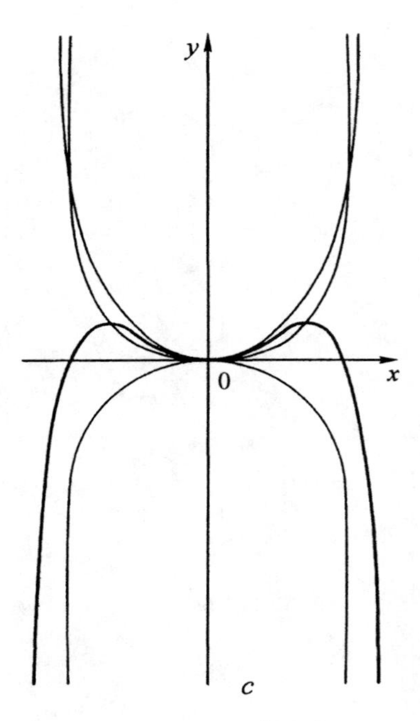

Figure 4.5. The graph of functions: (*c*) $y = x^2 - x^4$.

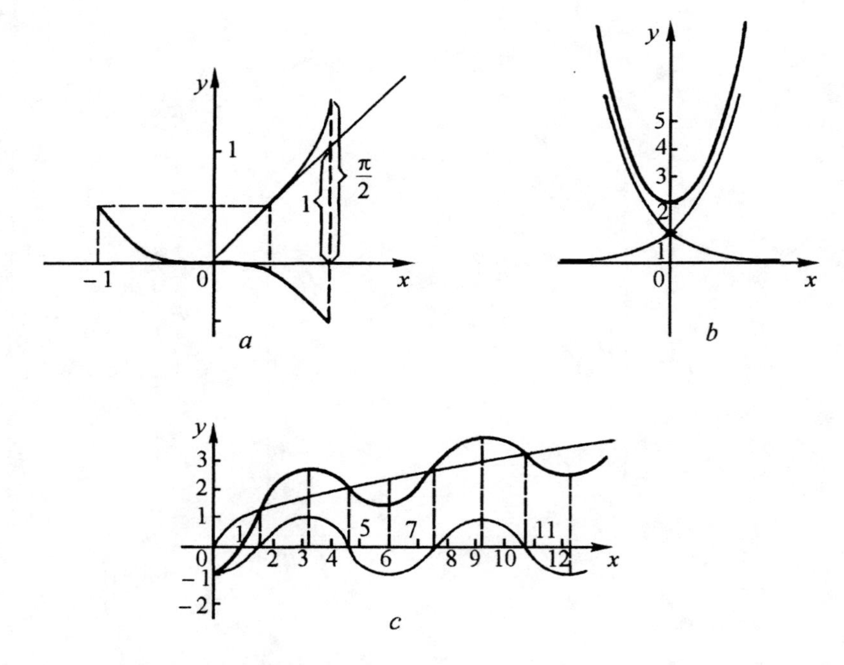

Figure 4.6. The graphs of functions: (*a*) $y = x - \arcsin x$; (*b*) $y = a^x + a^{-x}$; (*c*) $y = \sqrt{x} - \cos x$.

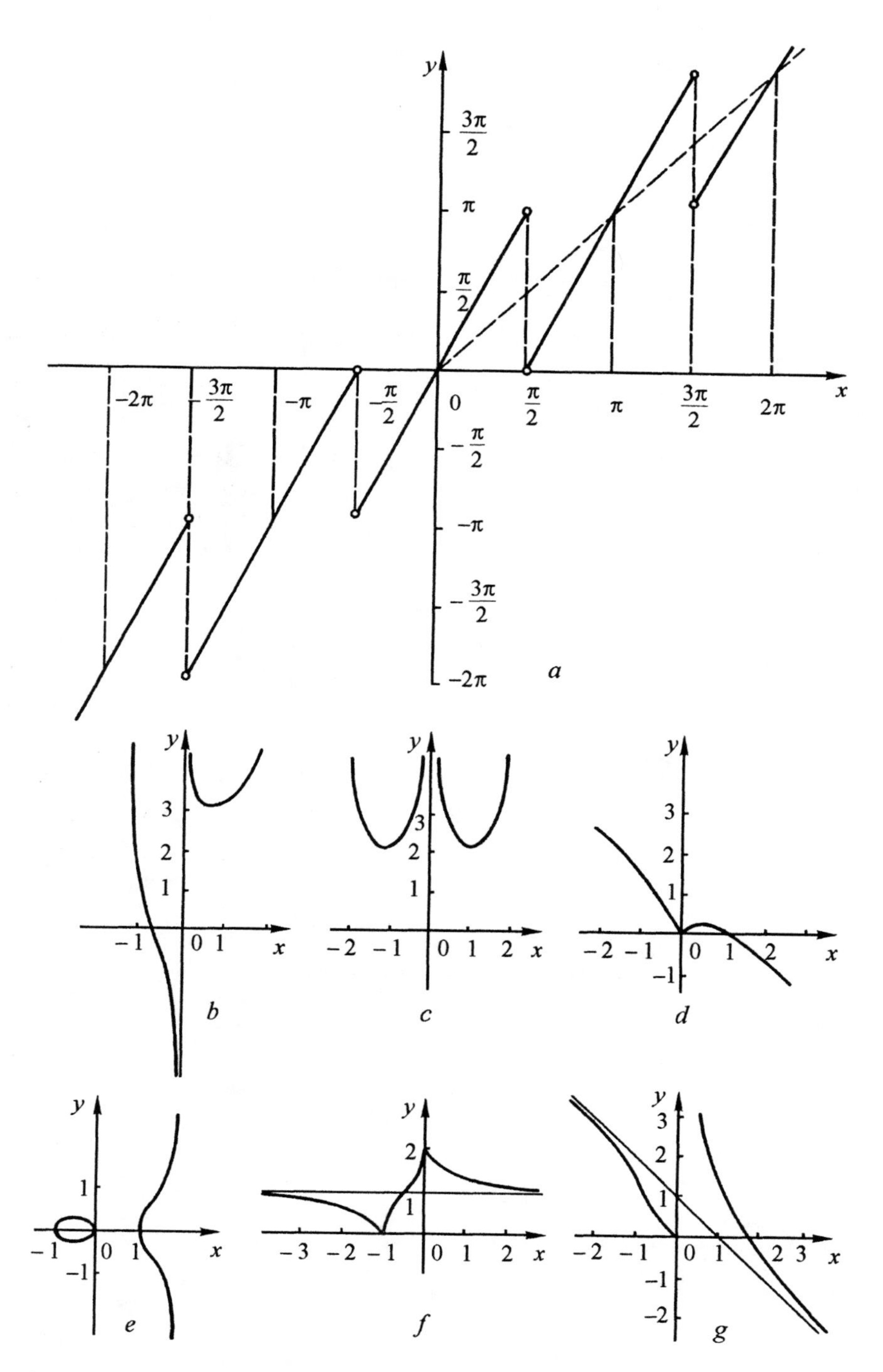

Figure 4.7. The graph of functions: (*a*) $y = x + \arctan(\tan x)$; (*b*) $y = 1/x + 4x^2$; (*c*) $y = x^2 + 1/x^2$; (*d*) $y = \sqrt[3]{x^2} - x$; (*e*) $y^2 = x^{2-x}$; (*f*) $y = \sqrt[3]{(x+1)^2} - \sqrt[3]{x^2} + 1$; (*g*) $y = e^{1/x} - x$.

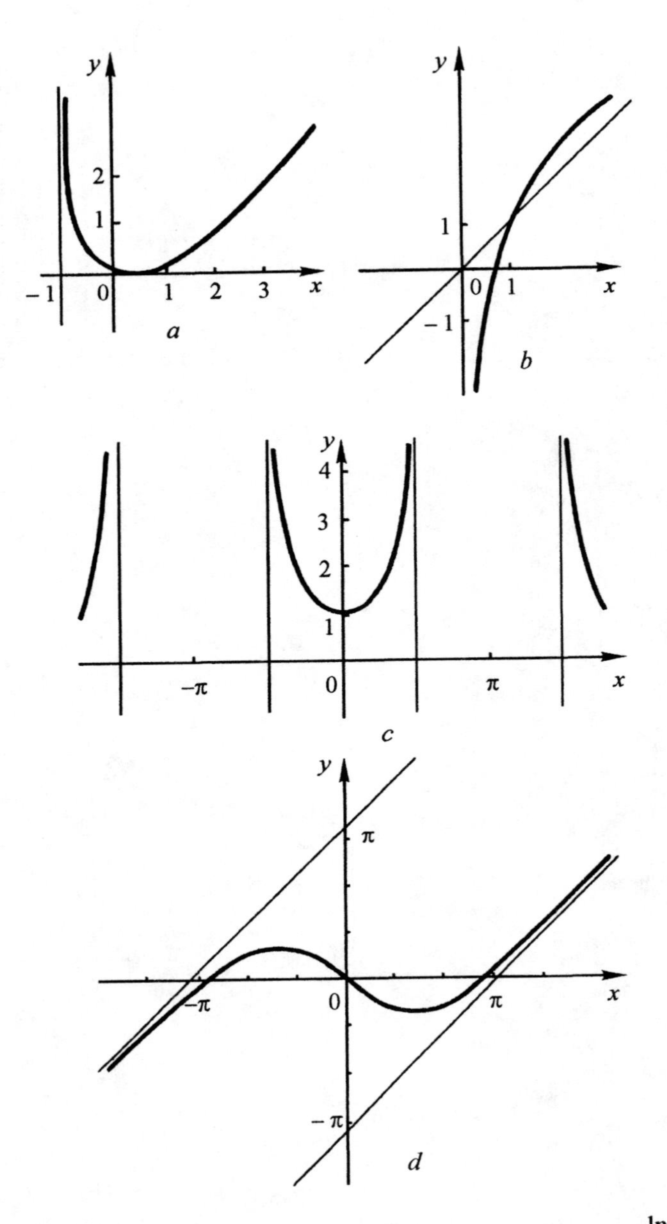

Figure 4.8. The graph of functions: (*a*) $y = x - \ln(x+1)$; (*b*) $y = x + \frac{\ln x}{x}$; (*c*) $y = \cos x - \ln \cos x$; (*d*) $y = x - \arctan x$.

Example 4.1.2. Plot the graph of the function $y = x + 1/x$.

First we plot the graphs of the functions $y_1 = 1/x$ and $y_2 = x$. Then we add the y-coordinates of points that have the same x-coordinate. Let us take $x = 1/2, 1, 2, 3$. By adding the y-coordinates, we obtain points A, B, C, and D. Drawing a smooth curve that passes through these points, we get a branch of the graph of the function (for $x > 0$). By noting that the function is odd, and hence that its graph is symmetric with respect to the origin, we construct the other branch of the graph for $x < 0$ by the symmetry (Figure 4.2).

Graphs of functions of type $y = f(x) \pm \varphi(x)$ are shown in Figures 4.3 – 4.8.

4.1.2. Multiplication and division of graphs

If the graphs of functions $y_1 = f_1(x)$ and $y_2 = f_2(x)$ are given, it is then possible to plot the graphs of the functions $y = y_1(x) \cdot y_2(x)$ and $y = \frac{y_1(x)}{y_2(x)}$, $y_2(x) \neq 0$, by using pointwise multiplication and division. Note that if adding or subtracting graphs, one can use a compass to add or subtract coordinates; however, to plot the product or ratio of functions it is necessary first to find the numerical value of the line segment representing the y-coordinates and then multiply (divide) these values. The division can be reduced to multiplication by writing $y = y_1(x) \frac{1}{y_2(x)}$.

A better way to study the product or ratio of two functions is to use the scheme described in Section 2.2. Sometimes the construction can be facilitated if the expressions of the functions are first simplified.

This method of plotting graphs of the product or ratio of two functions is not efficient and is rarely used. A better way to plot such graphs is to use methods of calculus (see Part 2).

Example 4.1.3. Plot the graph of the function $y = x\sin x$.

Let us plot the graphs of the functions $y_1 = x$, $y_2 = \sin x$ (Figure 4.9).

Multiplication of graphs of these functions is simplified in this case, because the function $y_2 = \sin x$ periodically takes the values 0, 1, −1.

If $x = k\pi$ for $k \in \mathbb{Z}$, $\sin x = 0$ and $x\sin x = 0$, which means that the corresponding point lies in the x-axis.

If $x = \pi/2 + 2k\pi$ for $k \in \mathbb{Z}$, $\sin(\pi/2 + 2k\pi) = 1$ and $y = \pi/2 + 2k\pi$, so that the corresponding point of the graph of the function $y = x\sin x$ lies on the line $y = x$.

If $x = -\pi/2 + 2k\pi$ for $k \in \mathbb{Z}$, $\sin(-\pi/2 + 2k\pi) = -1$ and the point lies on the line $y = -x$.

Because the function $y = x\sin x$ is even, we need to construct the graph only for $x \geq 0$, and plot the remaining part of the graph symmetrically with respect to the y-axis.

Graphs of products and quotients of functions are shown in Figures 4.10–4.22.

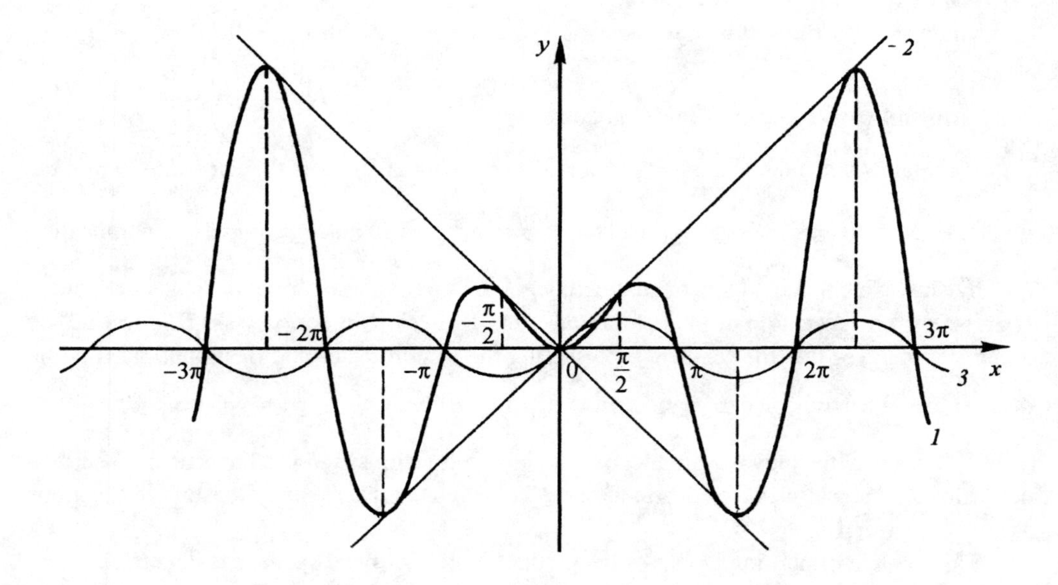

Figure 4.9. The graph of functions: *1*. $y = x \sin x$; *2*. $y = x$; *3*. $y = \sin x$.

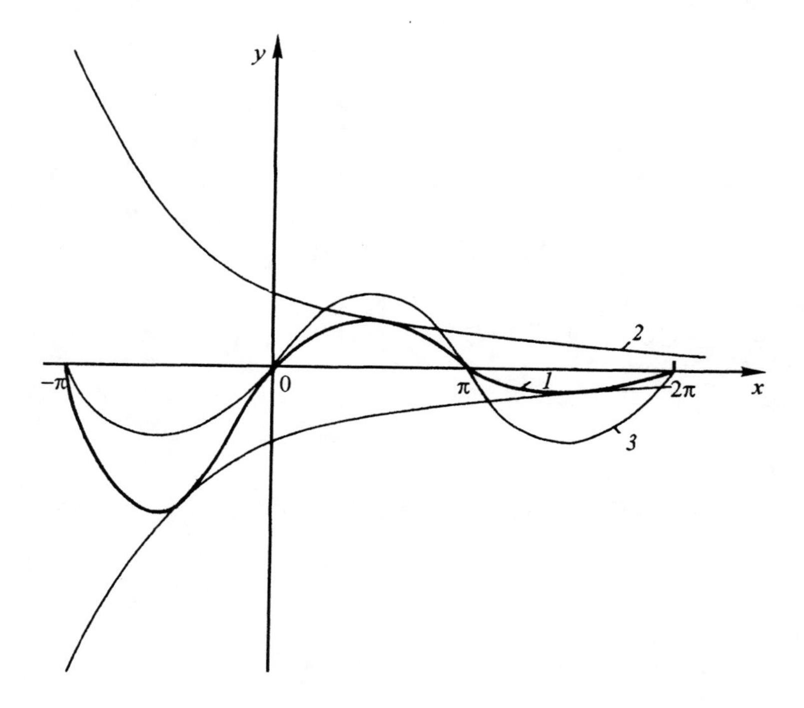

Figure 4.10. The graph of functions: *1.* $y = e^{-x} \sin x$; *2.* $y = e^{-x}$; *3.* $y = \sin x$.

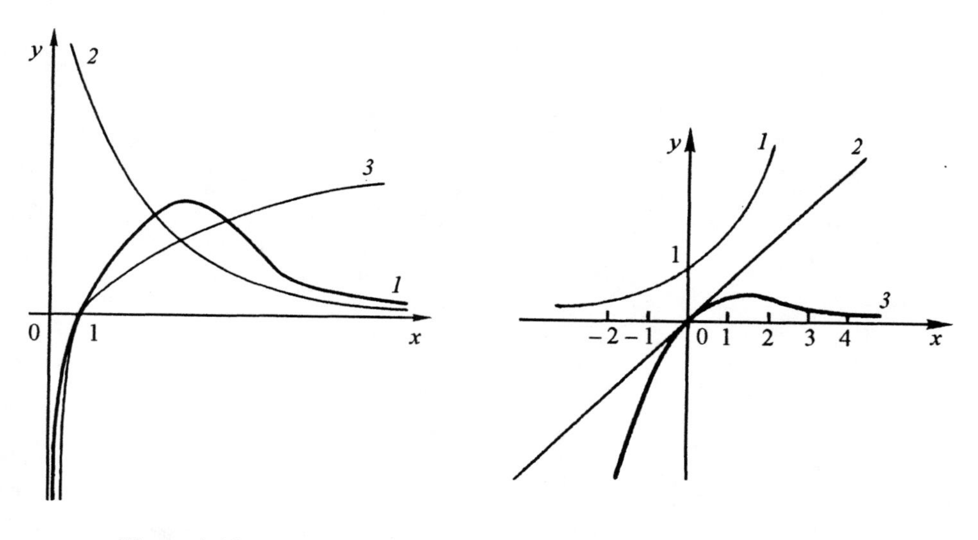

Figure 4.11

Figure 4.12

Figure 4.11. The graph of functions: *1.* $y = \dfrac{\ln x}{x}$; *2.* $y = 1/x$; *3.* $y = \ln x$.

Figure 4.12. Graph of functions: *1.* $y = 2^x$; *2.* $y = x$; *3.* $y = x/2^x$.

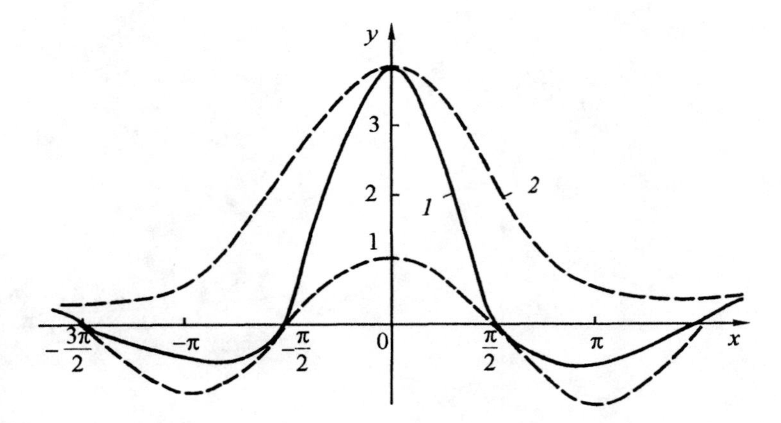

Figure 4.13. The graph of functions: *1*. $y = \dfrac{4\cos x}{1+x^2}$; *2*. $y = \dfrac{4}{1+x^2}$.

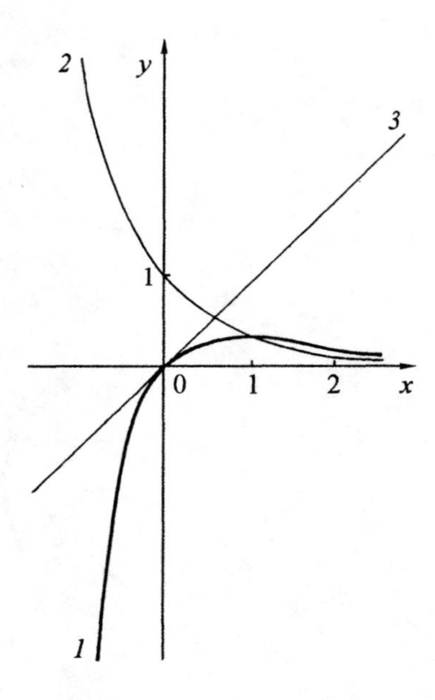

Figure 4.14. Graph of functions: *1*. $y = xe^{-x}$; *2*. $y = e^{-x}$; *3*. $y = x$.

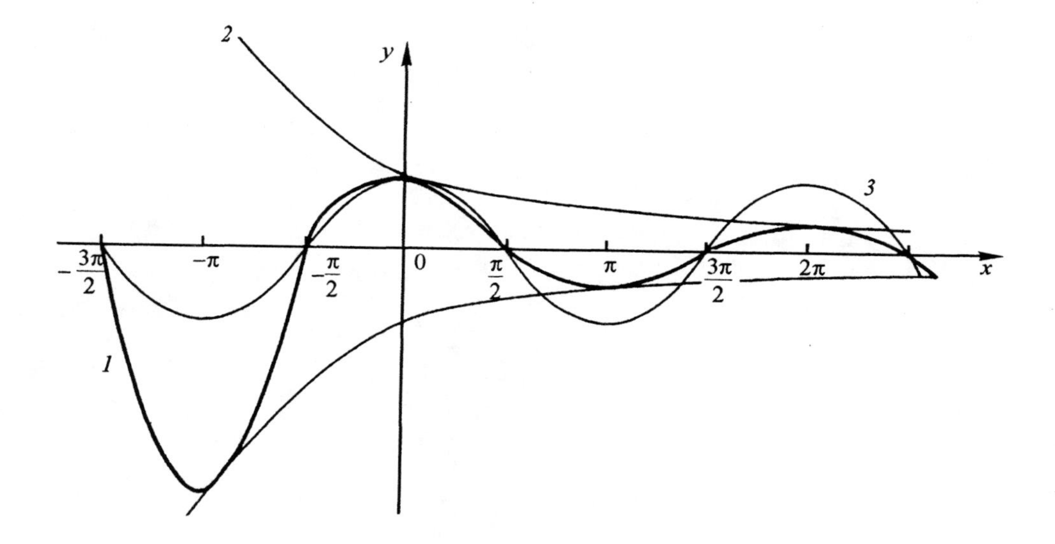

Figure 4.15. The graph of functions: *1.* $y = e^{-x}\cos x$; *2.* $y = e^{-x}$; *3.* $y = \cos x$.

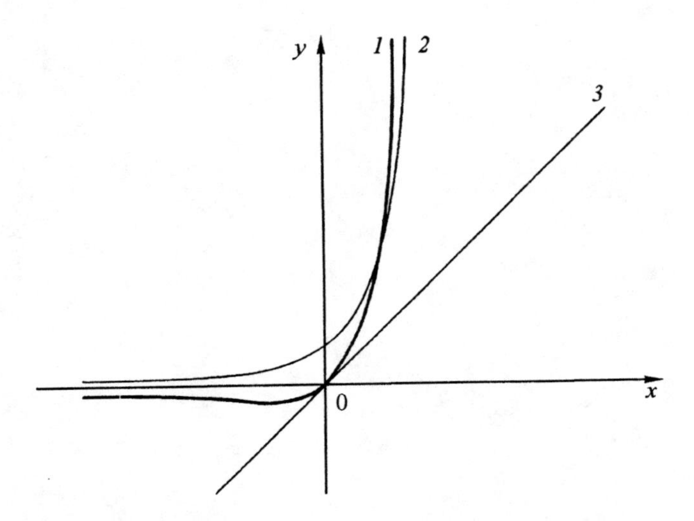

Figure 4.16. The graph of functions: *1.* $y = xe^x$; *2.* $y = e^x$; *3.* $y = x$.

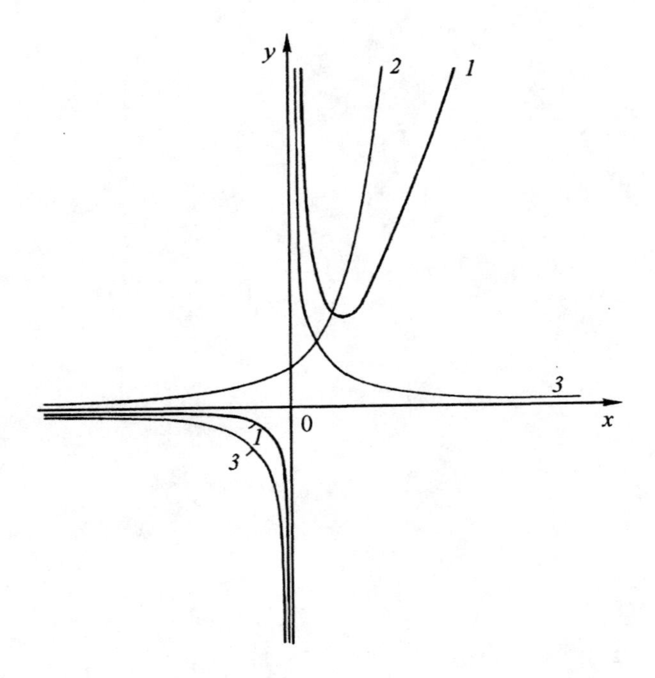

Figure 4.17. The graph of functions: *1.* $y = \frac{e^x}{x}$; *2.* $y = e^x$; *3.* $y = 1/x$.

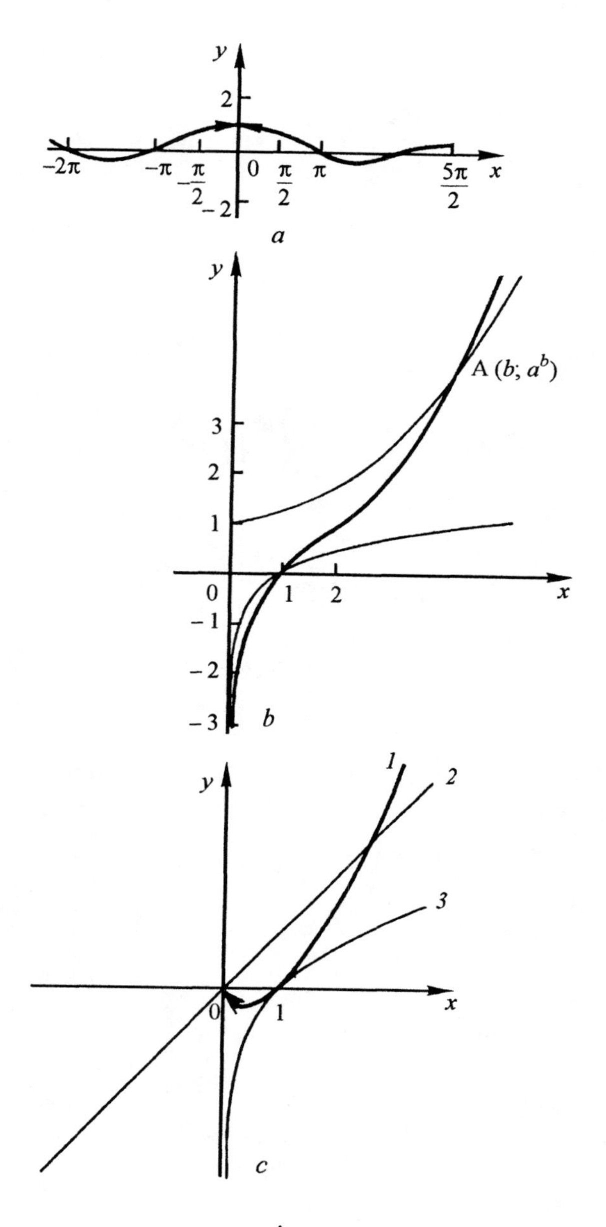

Figure 4.18. The graph of functions: (*a*) $y = \dfrac{\sin x}{x}$; (*b*) $y = a^x \log_b x$, $a > 1$, $b > 1$; (*c*) *1*. $y = x \sin x$, *2*. $y = x$, *3*. $y = \ln x$.

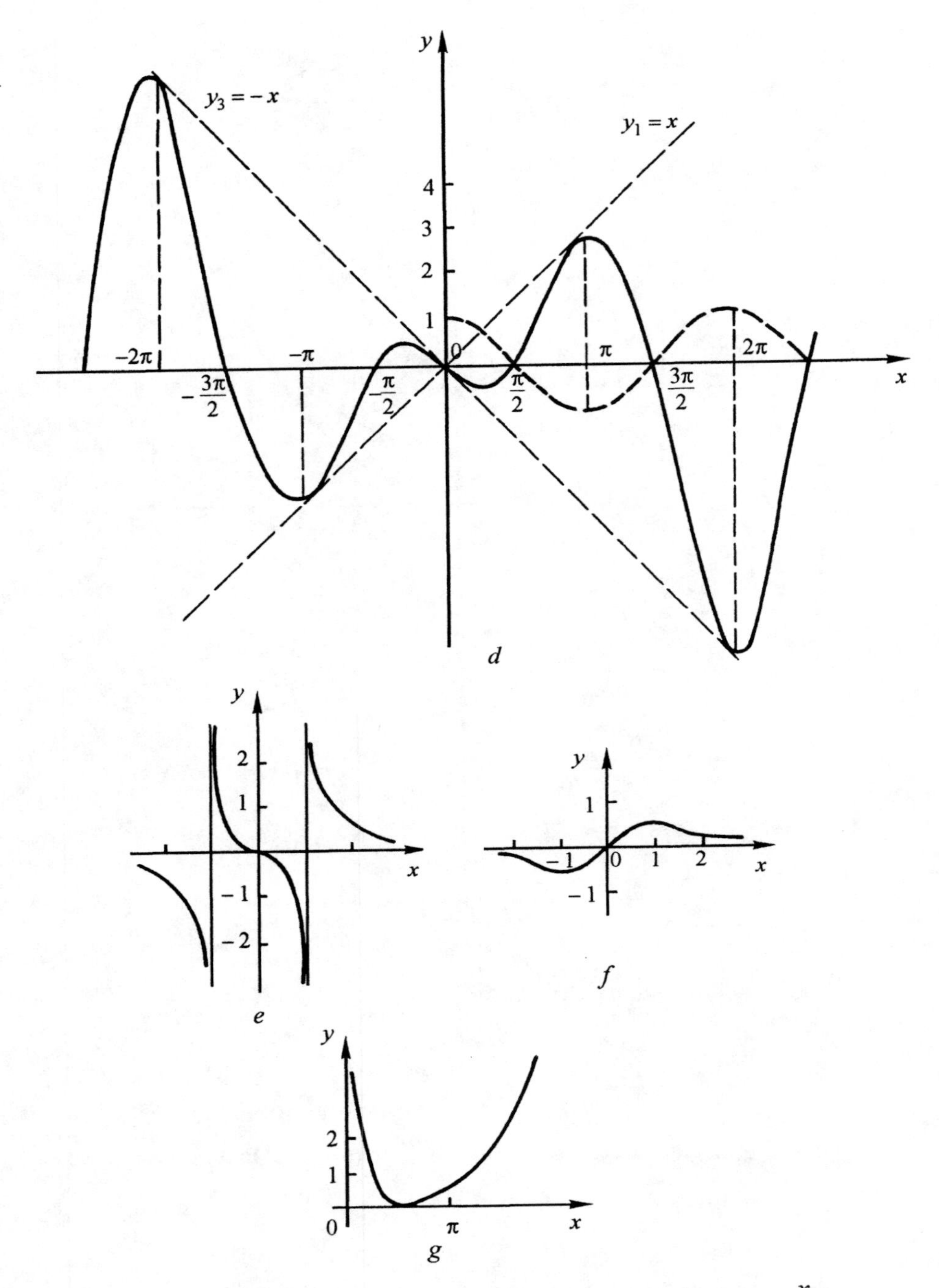

Figure 4.18. The graph of functions: (*d*) $y = -x \cos x$; (*e*) $y = \dfrac{x}{x^2 - 1}$; (*f*) $y = \dfrac{x}{1 + x^2}$; (*g*) $y = \dfrac{1 - \sin x}{1 - \cos x}$.

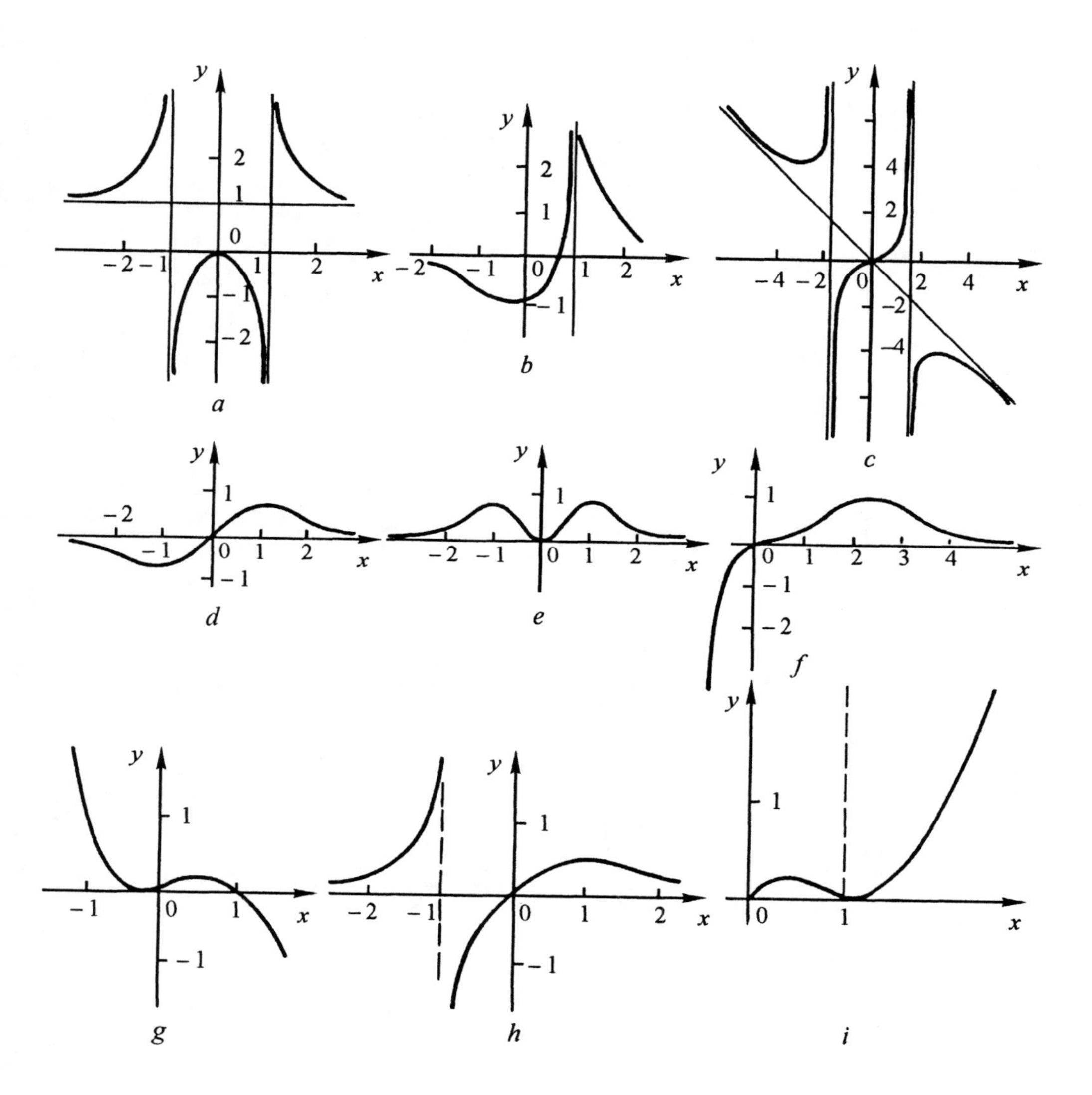

Figure 4.19. The graph of functions: (*a*) $y = \dfrac{x^2}{x^2 - 1}$; (*b*) $y = \dfrac{2x - 1}{(x-1)^2}$;

(*c*) $y = \dfrac{x^3}{3 - x^2}$; (*d*) $y = xe^{-x^2/2}$; (*e*) $y = x^2 e^{x^2}$; (*f*) $y = x^3 e^{-x}$; (*g*) $y = x^2(1 - x)$;

(*h*) $y = \dfrac{x}{1 + x^3}$; (*i*) $y = (1 - x)^2\sqrt{x}$.

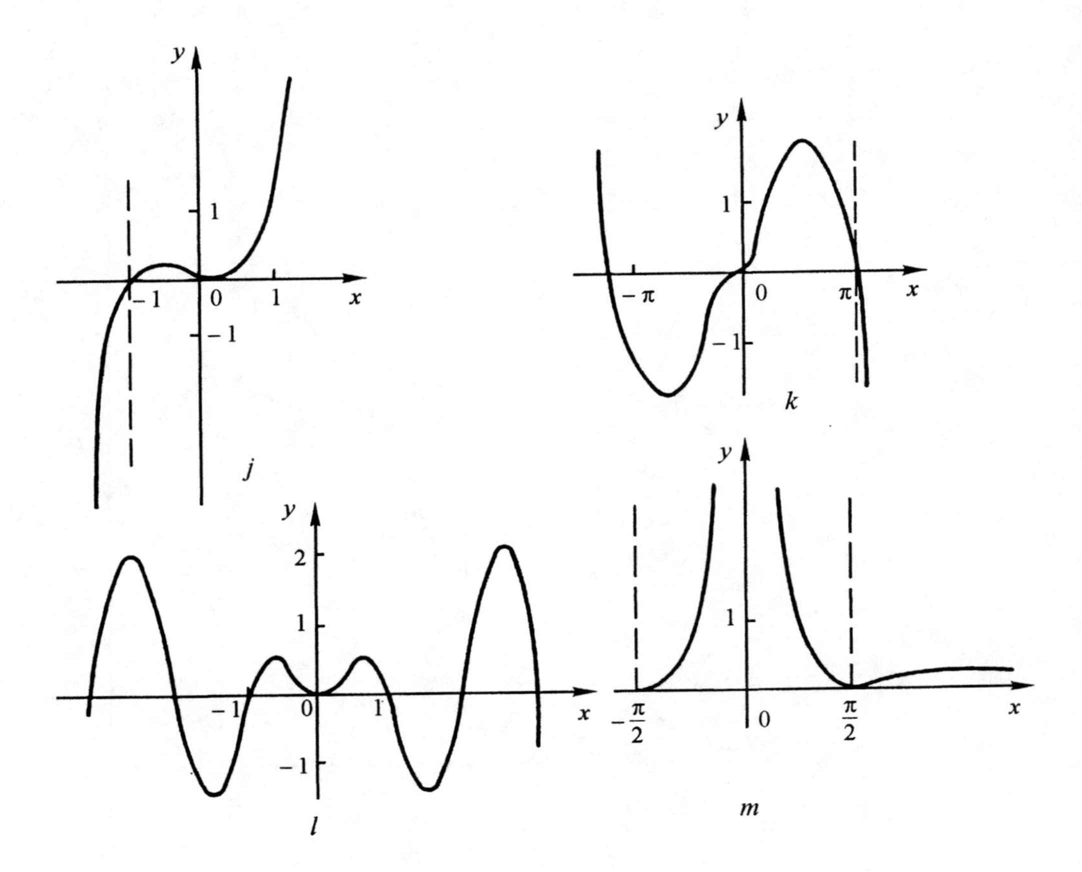

Figure 4.19. The graph of functions: (*j*) $y = x^2\sqrt[3]{1+x}$; (*k*) $y = x^2 \sin x$;

(*l*) $y = x \sin 3x$; (*m*) $y = \dfrac{\cos^2 x}{x^2}$.

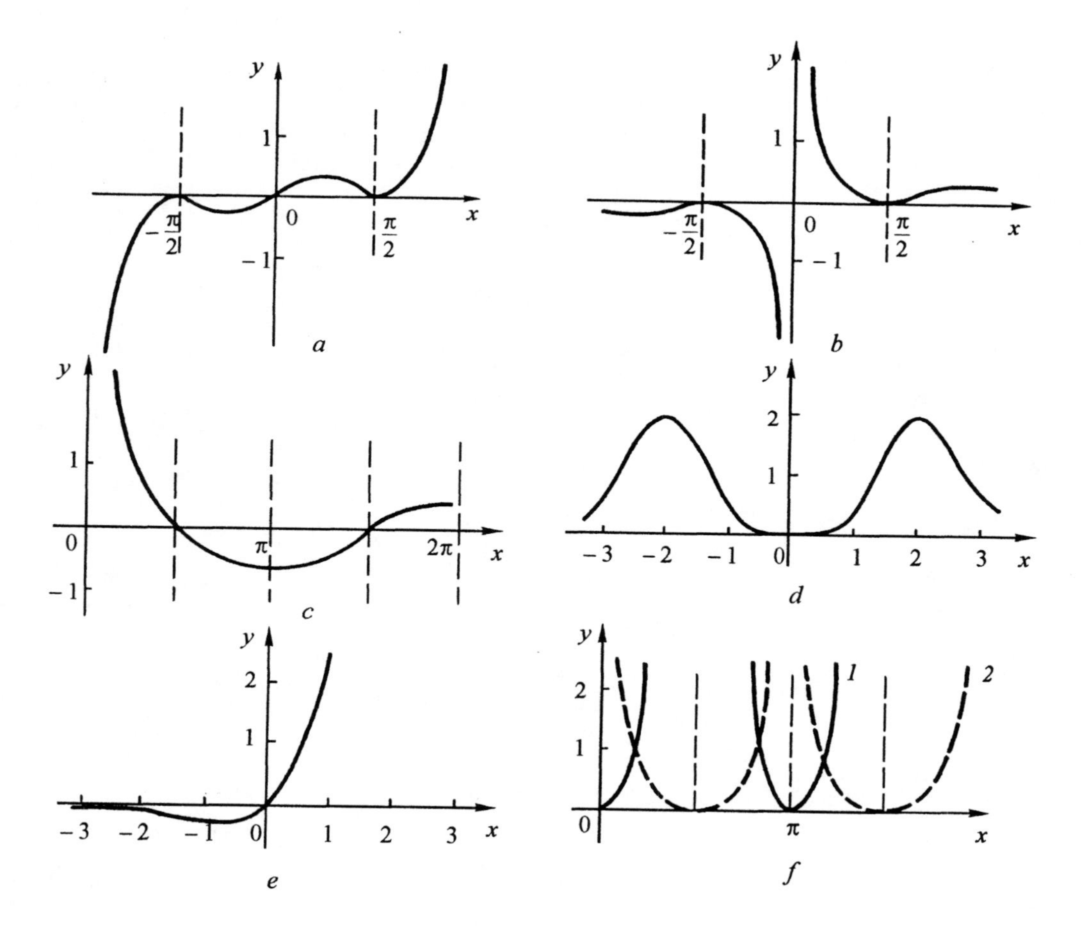

Figure 4.20. The graph of functions: (*a*) $y = x\cos^2 x$; (*b*) $y = \dfrac{\cos^2 x}{x}$; (*c*) $y = \cos\dfrac{x}{\sqrt{x}}$;

(*d*) $y = x^4 e^{-x^2/2}$; (*e*) $y = e^x\sqrt[3]{x}$; (*f*) (*1*) $y = e^x\sin^2 x$, (*2*) $\dfrac{1}{e^x\sin^2 x}$.

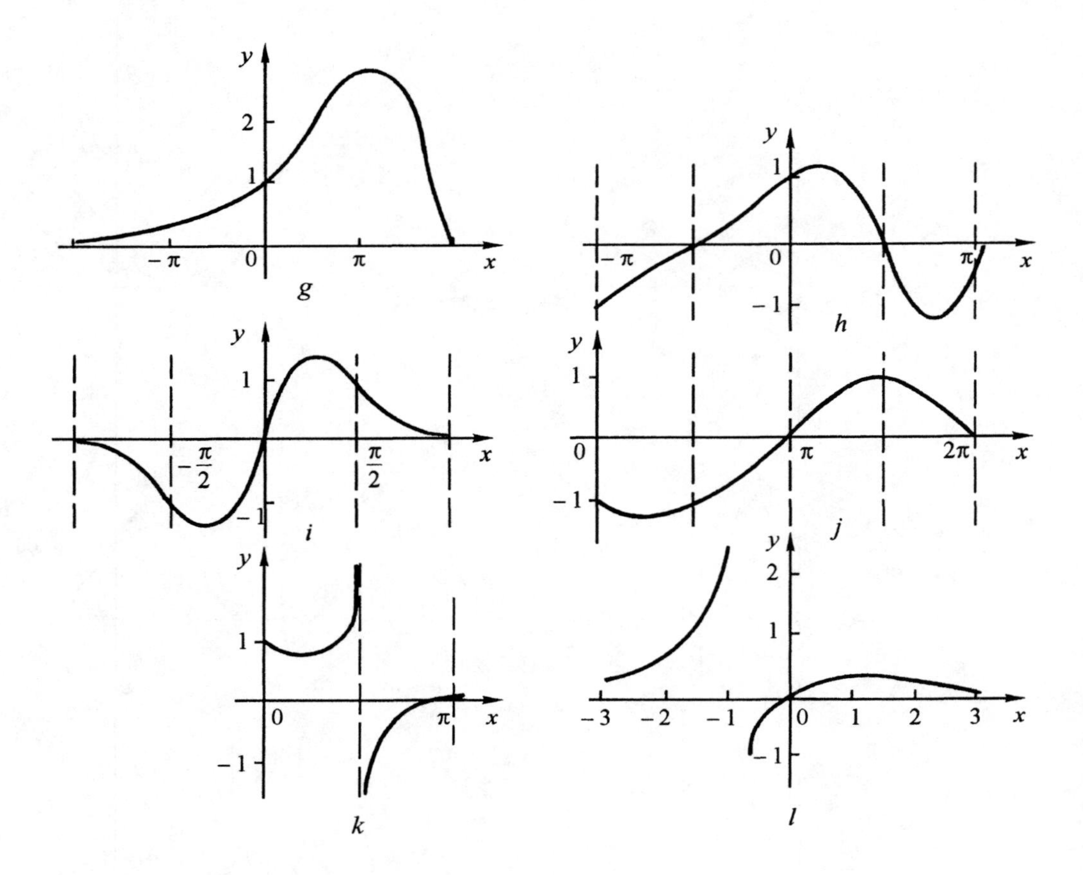

Figure 4.20. The graph of functions: (*g*) $y = \dfrac{e^x \sin^2 x}{x}$; (*h*) $y = \dfrac{e^x \sin x}{x}$;

(*i*) $y = e^{\cos x} \sin x$; (*j*) $y = \dfrac{\sin x}{e^{-x} - 1}$; (*k*) $y = \dfrac{e^{-x} \tan x}{x}$; (*l*) $y = \dfrac{\sinh x}{\sinh + 1}$.

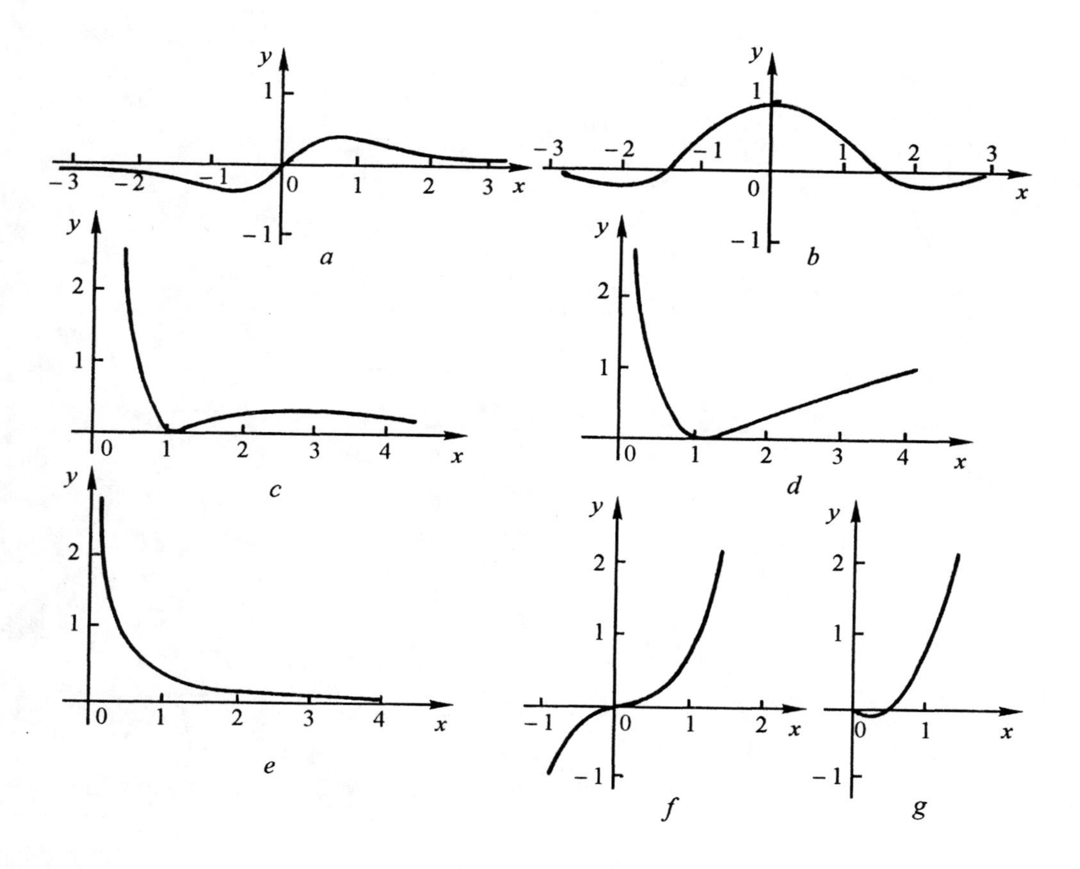

Figure 4.21. The graph of functions: (*a*) $y = \dfrac{\sinh x}{\cosh^3 x}$; (*b*) $y = \dfrac{\cos x}{\cosh x}$; (*c*) $y = \dfrac{\ln^2 x}{x^3}$; (*d*) $y = \dfrac{\ln^2 x}{\sqrt{x}}$; (*e*) $y = \dfrac{\ln x}{x^2} - 1$; (*f*) $y = x^2 \ln(1 + x)$; (*g*) $y = x(1 + \ln x)$.

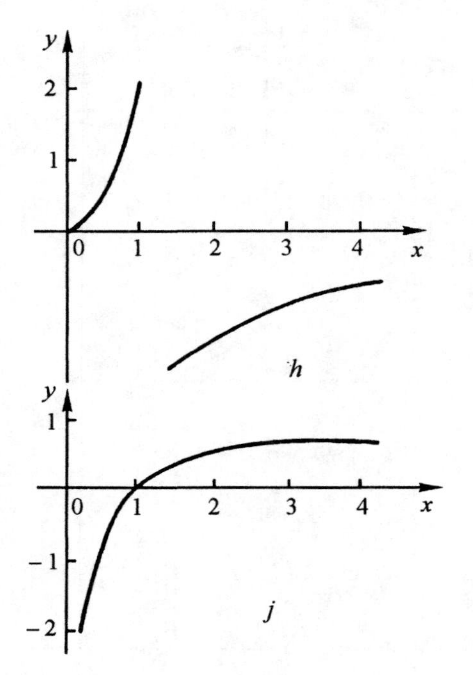

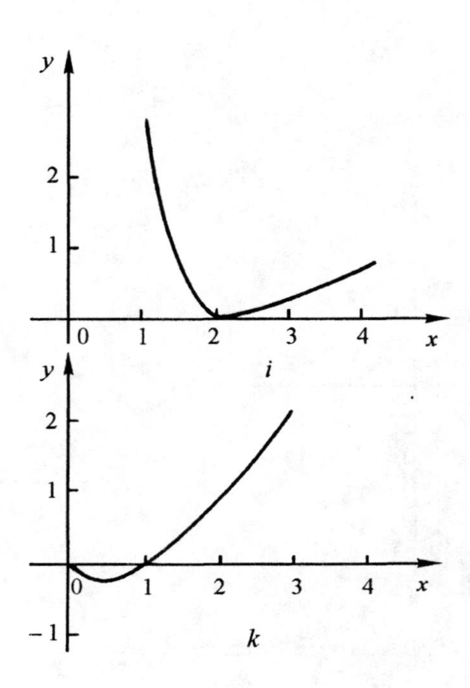

Figure 4.21. The graph of functions: (*h*) $y = \dfrac{\ln(1+x^2)}{1-x}$; (*i*) $y = \dfrac{\ln^4(x-1)}{x}$; (*j*) $y = \dfrac{\ln x}{\sqrt{x}+1}$; (*k*) $y = x \ln \sqrt{x}$.

4.2. Plotting graphs by using simplest transformations

4.2.1. Transformations that do not change the scale

Consider symmetry transformations induced by the property of the function's being even or odd.

Plots of graphs of the functions $y=-f(x)$, $y=f(-x)$, $y=-f(-x)$ by using the graph of the function $y=f(x)$.

1. Let $P(x_0; y_0)$ be a point in the graph of a function $y=f(x)$, i.e. $y_0=f(x_0)$. Take the point $P_1(x_0; -y_0)$, which is symmetric to the point $P(x_0; y_0)$ with respect to the x-axis. The coordinates of the point $P_1(x_0; -y_0)$ satisfy the equation $y_0=-f(x_0)$; hence the point $P_1(x_0; -y_0)$ belongs to the graph of the function $y=-f(x)$. Since the point $P(x_0; y_0)$, which belongs to the graph of the function $y=f(x)$, is arbitrary and the functions $y=f(x)$, $y=-f(x)$ have the same domain, the graph of the function $y=-f(x)$ is obtained from the graph of the function $y=f(x)$ by symmetrically reflecting it about the x-axis (Figure 4.23). Examples of graphs of such functions are given in Figure 4.23.

2. Let $P(x_0; y_0)$ be an arbitrary point in the graph of a function $y=f(x)$, i.e. $y_0=f(x_0)$. Take the point $P_2(-x_0; y_0)$, which is symmetric to the point $P(x_0; y_0)$ with respect to the y-axis. Coordinates of the point $P_2(-x_0; y_0)$ satisfy the equality $y_0=f(-(-x_0))$, and so the point $P_2(-x_0; y_0)$ belongs to the graph of the function $y=f(-x)$. Since the point $P(x_0; y_0)$ of the graph of the function $y=f(x)$ is arbitrary and the functions $y=f(x)$ and $y=f(-(-x))$ have the same domains, the graph of the function $y=f(-x)$ is obtained by symmetrically reflecting the graph of the function $y=f(x)$ about the y-axis (Figure 1.17a). Examples of this construction are given in Figures 4.24, 4.25.

Similarly, it is easy to show that the graph of the function $y=-f(-x)$ is symmetric to the graph of the function $y=f(x)$ with respect to the origin (Figure 1.17b).

Translations parallel to the x-axis.

The graph of the function $y=f(x+a)$ is obtained by translating the graph of the function $y=f(x)$ at the distance equal to $|a|$ parallel to the axis Ox and in the direction opposite to the sign of a. In practice, it is convenient to plot the graph of the function $y=f(x)$, with the y-axis drawn as a dotted line. Then the y-axis is translated by $|a|$ in the same direction as the sign of a. This is the new y-axis.

Examples of such constructions are shown in Figures 4.26 – 4.28.

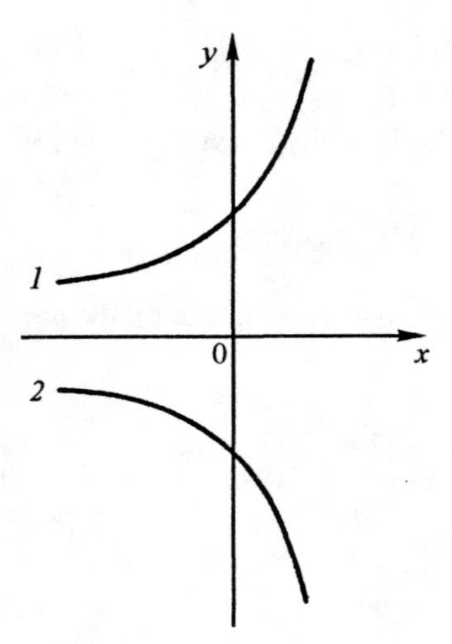

Figure 4.22. The graphs of functions *1.* $y = f(x)$; *2.* $y = -f(x)$.

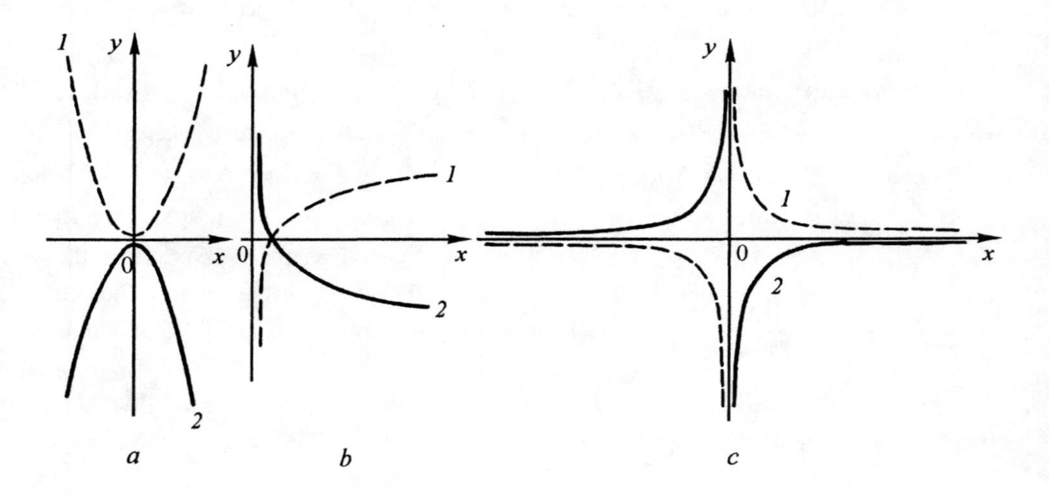

Figure 4.23. The graphs of functions (*a*) *1.* $y = x^2$, *2.* $y = -x^2$; (*b*) *1.* $y = \log_2 x$, *2.* $t = -\log_2 x$; (*c*) *1.* $y = \frac{1}{x}$, *2.* $y = -\frac{1}{x}$.

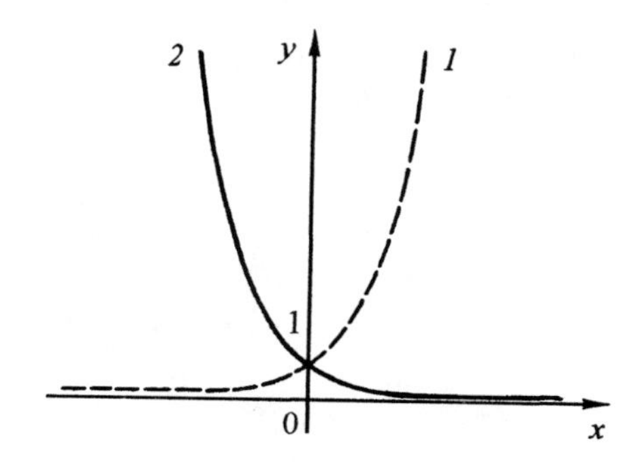

Figure 4.24. The graphs of functions *1.* $y = 2^x$, *2.* $y = 2^{-x}$.

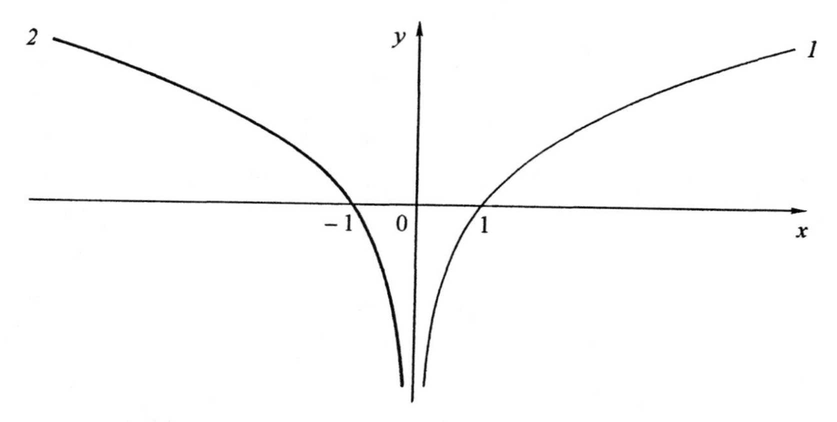

Figure 4.25. The graphs of functions *1.* $y = \log_2 x$, *2.* $y = \log_2 (-x)$.

Translations parallel to y-axis

The graph of the function $y = f(x) + b$ is obtained from the graph of the function $y = f(x)$ by translating it by $|b|$ along the y-axis in the same direction as the sign of b. Practically, this can be done as follows: draw the graph of the function $y = f(x)$, with the x-axis being shown as a dotted line. Then translate the x-axis parallel to itself by $|b|$ in the direction opposite to the sign of b.

Examples of graphs of such functions are shown in Figures 4.29 – 4.31.

4.2.2. Transformations that change the scale

Expansion or contraction along the x-axis

The graph of the function $y = f(kx)$, $k \neq 0$, is obtained from the graph of the function $y = f(x)$ by scaling it by k along the x-axis: if $k > 1$, then the graph is contracted by k; if $0 < k < 1$, then the graph is expanded by $1/k$; if $k = 1$, then the graph remains unchanged; if

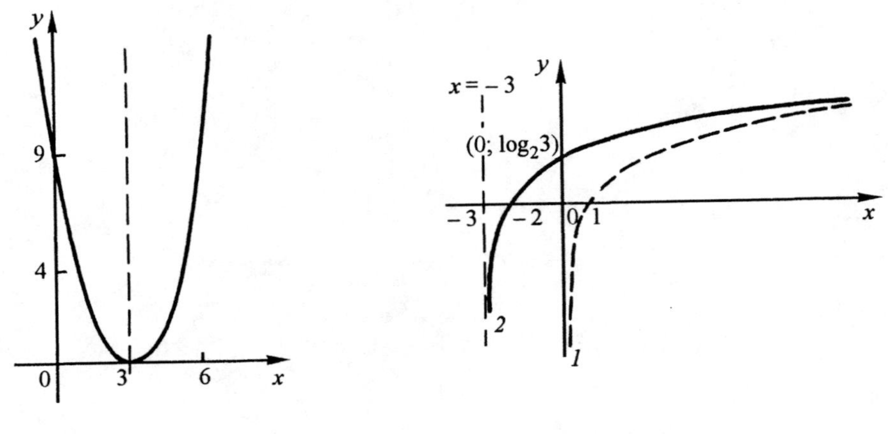

Figure 4.26 Figure 4.27

Figure 4.26. The graph of function $y = (x - 3)^2$.

Figure 4.27. The graphs of functions *1*. $y = \log_2 x$, *2*. $y = \log_2 (x + 3)$.

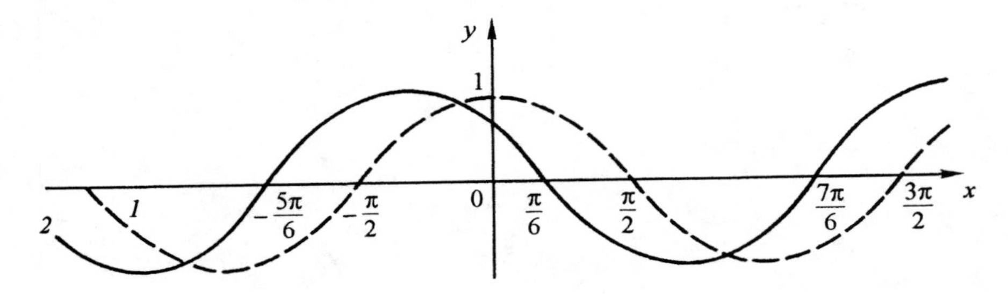

Figure 4.28. The graphs of functions *1*. $y = \cos x$; *2*. $y = \cos(x + \frac{\pi}{2})$.

$k < 0$, then we first plot the graph $y = f(|k|x)$, and then reflect it symmetrically with respect to the y-axis.

In practice, expansion or contraction of the graph along the x-axis is carried out as follows: plot the graph of the function $y = f(x)$, and then, if $k > 1$, contract the scale of the x-axis by k, and if $0 < k < 1$, expand the scale of the x-axis by $1/k$, leaving the scale of the y-axis unchanged.

Examples of these constructions are shown in Figures 4.32 – 4.35.

Expansion or contraction along the y-axis

The graph of the function $y = mf(x)$, $m \neq 0$, is obtained from the graph of the function $y = f(x)$ by scaling it by m along the y-axis: if $m > 1$, then the graph is expanded by m; if $0 < m < 1$, then the graph is contracted by $1/m$; if $m = 1$, then the graph remains unchanged; if

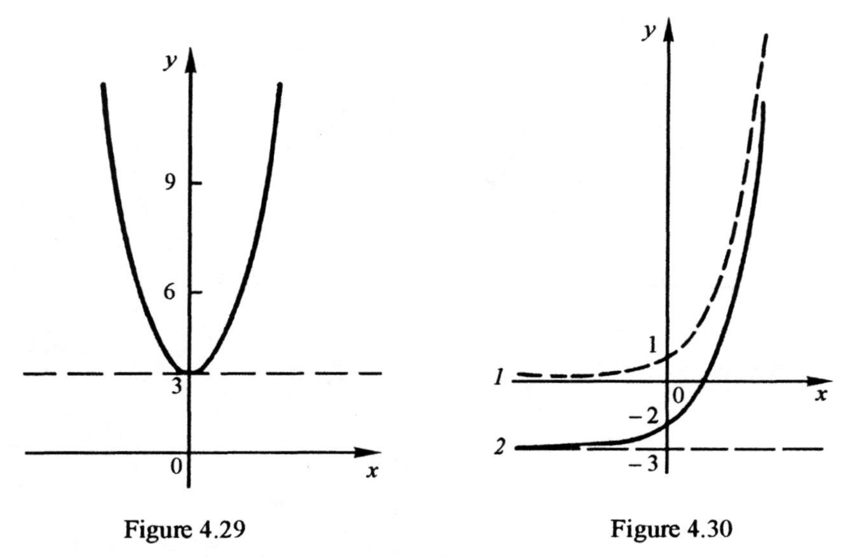

Figure 4.29

Figure 4.30

Figure 4.29. The graph of function $y = x^2 + 3$.

Figure 4.30. The graphs of functions $y = 2^x$ (*1*); $y = 2^x - 3$ (*2*).

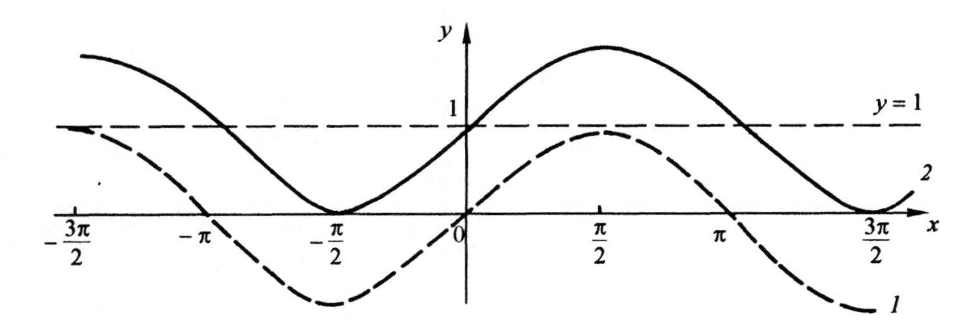

Figure 4.31. The graphs of functions $y = \sin x$ (*1*); $y = \sin x + 1$ (*2*).

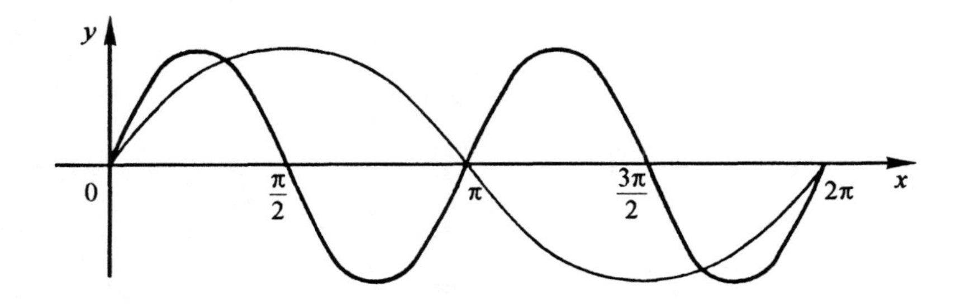

Figure 4.32. The graph of function $y = \sin 2x$.

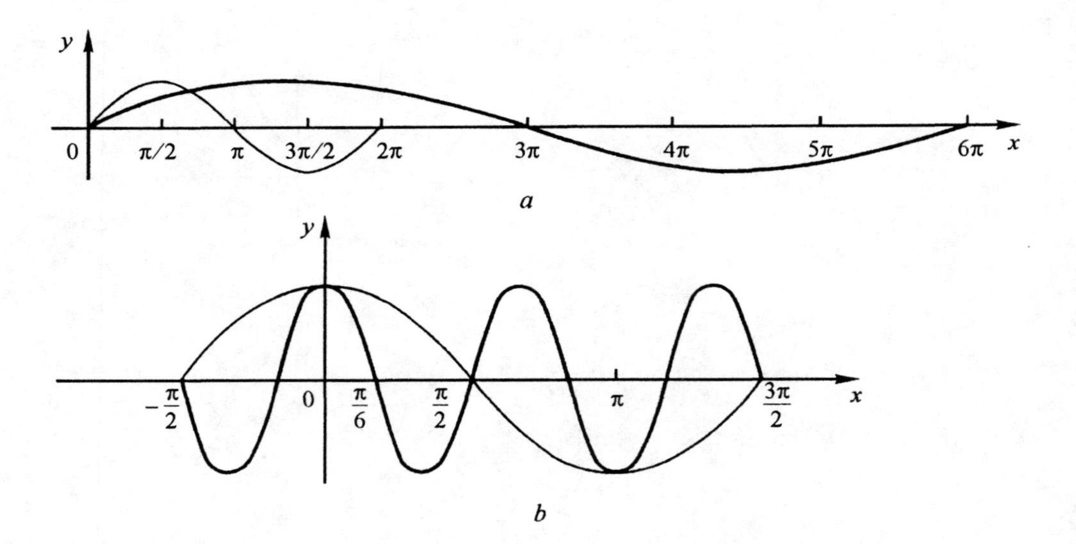

Figure 4.33. The graphs of functions: (*a*) $y = \sin\frac{x}{3}$; (*b*) $y = \cos 3x$.

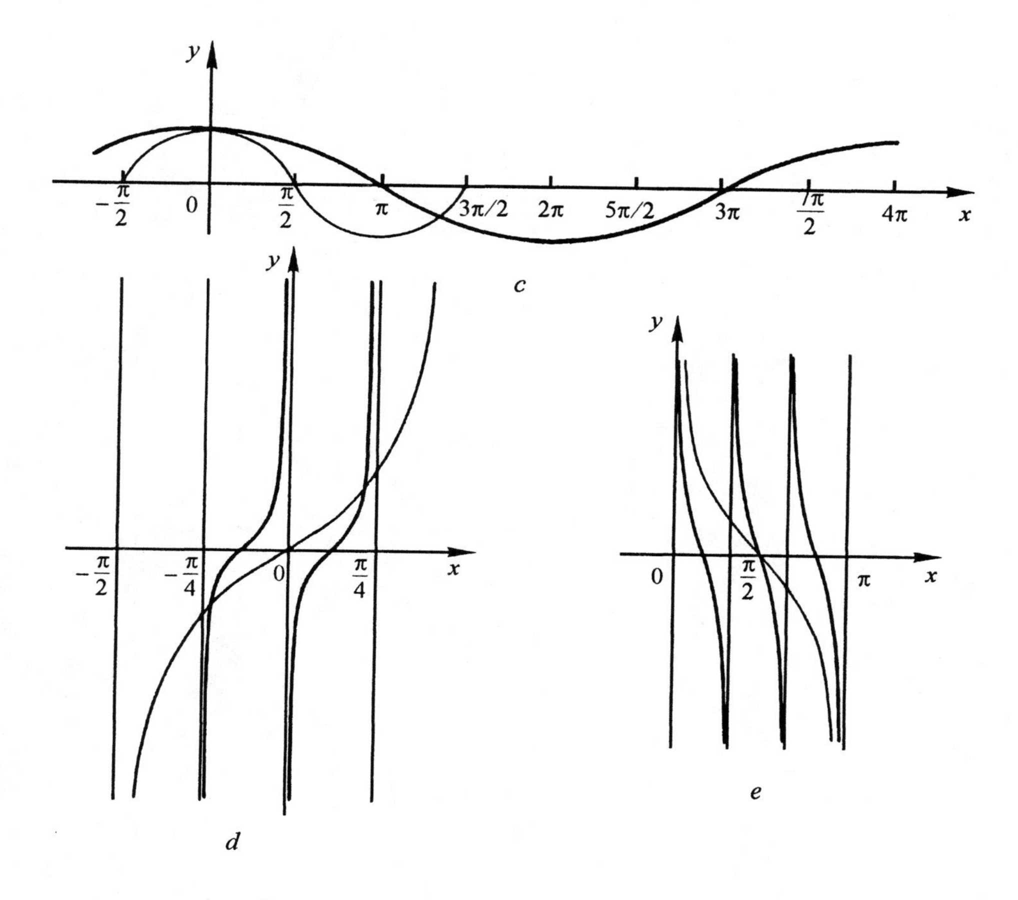

Figure 4.33. The graphs of functions: (*c*) $y = \cos\frac{x}{2}$; (*d*) $y = \tan 2x$; (*e*) $y = \cot 3x$.

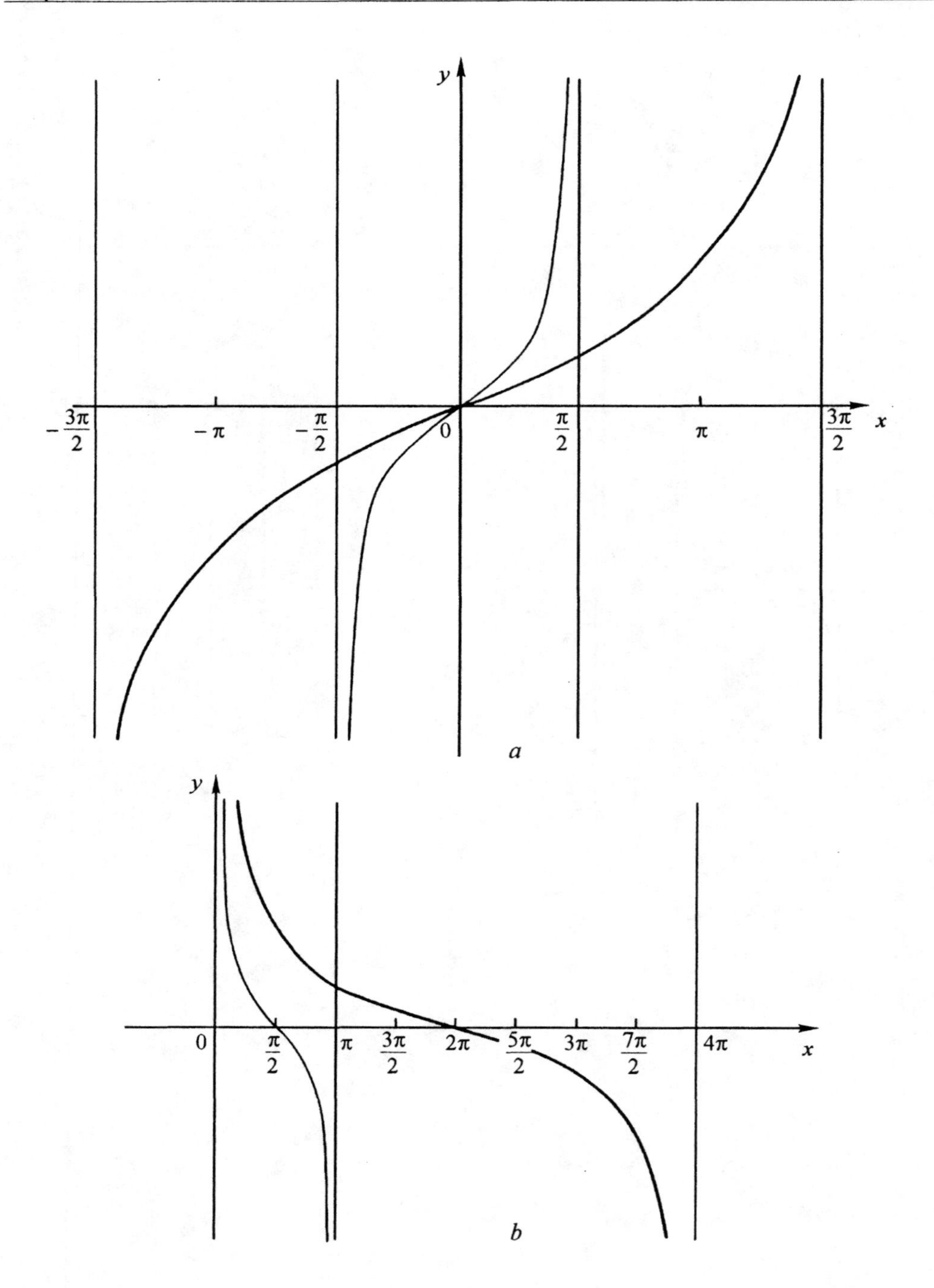

Figure 4.34. The graphs of functions: (*a*) $y = \tan \frac{x}{3}$; (*b*) $y = \cot \frac{x}{4}$.

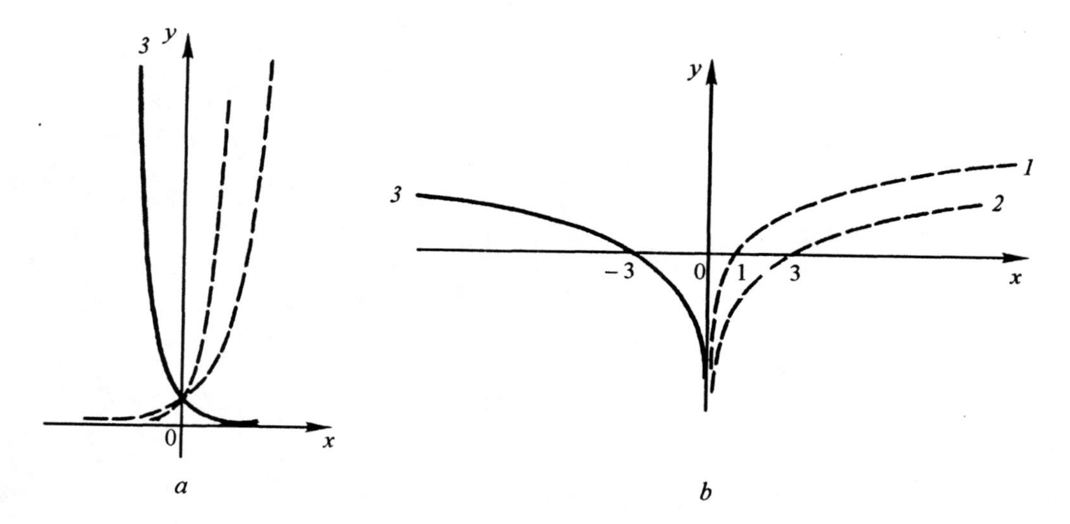

Figure 4.35. The graphs of functions: (*a*) *1.* $y = 2^x$, *2.* $y = 2^{2x}$, *3.* $y = 2^{-2x}$;
(*b*) *1.* $y = \log_2 x$, *2.* $y = \log_2 (\frac{x}{3})$, *3.* $y = \log_2(-\frac{1}{3}x)$.

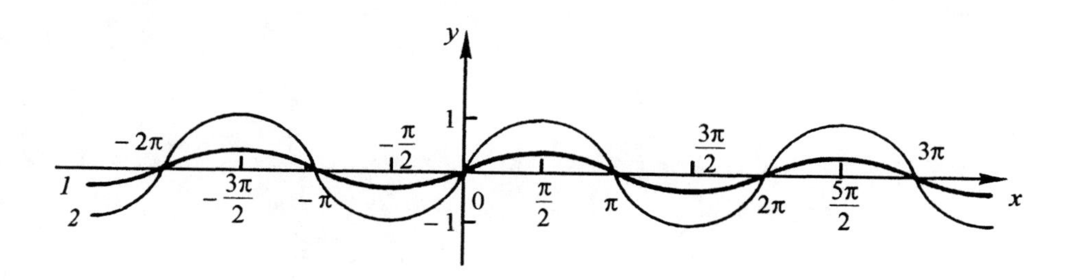

Figure 4.36. The graphs of functions: *1.* $y = \frac{1}{3}\sin x$, *2.* $y = \sin x$.

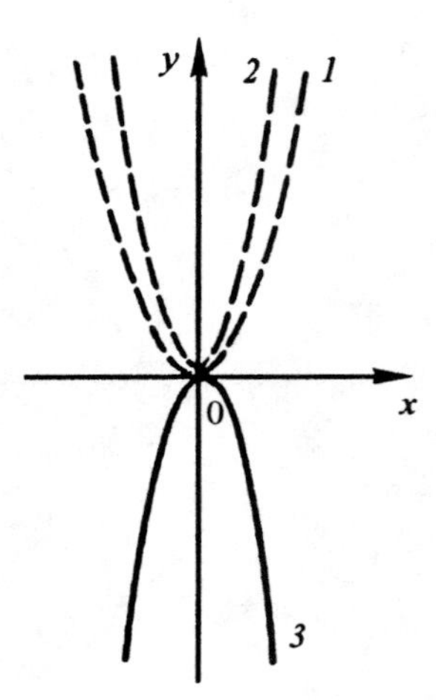

Figure 4.37. The graphs of functions: *1*. $y = x^2$, *2*. $y = 2x^2$, *3*. $y = -2x^2$.

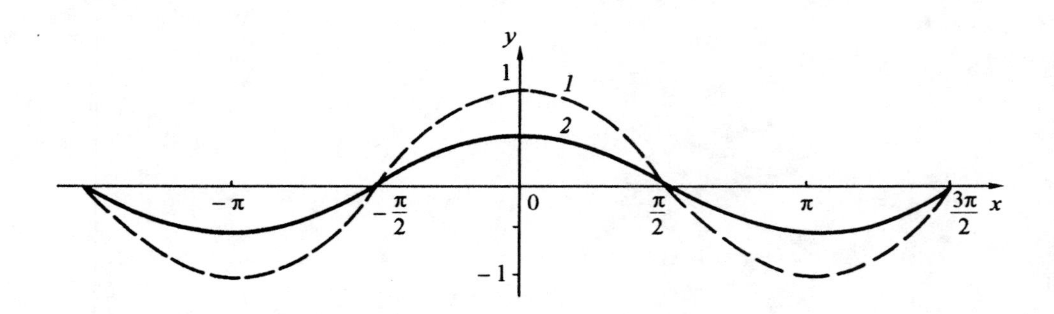

Figure 4.38. The graphs of functions: $y = \cos x$ (*1*), $y = \frac{1}{2}\cos x$ (*2*).

$m < 0$, then first the graph of the function $y = |m|f(x)$ is plotted, and then this graph is reflected about the x-axis.

In practice, the scaling of the graph can be carried out as follows: plot the graph of the function $y = f(x)$; if $m > 1$, expand the scale of the y-axis by m, and if $0 < m < 1$, contract the scale of the y-axis by $1/m$. The scale of the x-axis is unchanged.

Examples of this construction are shown in Figures 4.36 – 4.38.

Example 4.2.1. Plot the graph of the function $y = mf(kx + a) + b$.

The graph of this function is constructed by performing the above transformations in a certain order. For example, plot the graph of the function $y = f(x + a)$ and then plot the graph of the function $y = f(kx + a)$ (note that this transformation involves only the multiplication of the

x-coordinate by k). Then construct the graph of the function $y = mf(kx + a)$. Finally, construct the graph of the function $y = mf(kx + a) + b$.

The same transformations could be performed in another order. First plot the graph of the function $y = f(kx)$, then translate it to the right (left) by a/k, lift or lower the graph by b/m, and finally, expand or contract it along the y-axis by m. Depending on the signs of k and m, reflections about Ox and Oy may have to be performed.

Notice that these transformations could be carried out in any order, but this order determines the values by which the graph is translated along the axes.

These transformations are shown in Tables 1 and 2.

Plots constructed by using transformations shown in Tables 1 and 2 are given in Figures 4.39 – 4.47.

4.2.3. Plotting graphs of functions the analytic expression of which contains the absolute value sign

Plotting the graph of the function $y = f(|x|)$

The function $y = f(|x|)$ can be written as

$$y = f(|x|) = \begin{cases} f(x), & \text{if } x \geq 0, \\ f(-x), & \text{if } x < 0. \end{cases}$$

Whence it follows that the domain of the function is the set containing points x from the domain of the function $f(x)$ such that $x \geq 0$, and the points which are symmetric to them with respect to the y-axis. Since $f(|-x|) = f(x)$, the function $y = f(|x|)$ is even, and its graph is symmetric with respect to the y-axis.

Hence, to construct the graph of the function $y = f(|x|)$, it is necessary to plot the graph of the function $y = f(x)$ for $x \geq 0$, and then symmetrically reflect it with respect to the y-axis. These two steps yield the graph of the function $y = f(|x|)$ (Figure 4.48). It should be remarked that if the domain of the function is a set each element of which is negative, then $f(|x|)$ does not define a function. For example, for $f(x) = \log_{10}(-2x)$, $x < 0$, $f(|x|) = \log_{10}(-2|x|)$ does not make sense.

Example 4.2.2. Plot the graph of the function $y = \sin|x|$.

First, plot the graph of the function $y = \sin x$ for $x \geq 0$, and then symmetrically reflect this graph with respect to the axis Oy (Figure 4.49).

Example 4.2.3. Draw the graph of the function $y = 2^{|x|}$.

Plot the graph of the function $y = 2^x$ for $x \geq 0$, and then symmetrically reflect it with respect to the y axis. The resulting graph is the graph of the needed function (Figure 4.50).

Graphs of functions $y = f(|x|)$ are shown in Figure 4.51.

Table 1

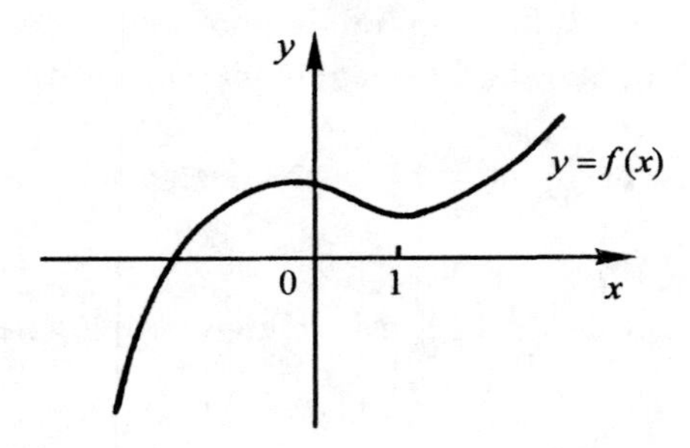
y
y = f(x)
0
1
x

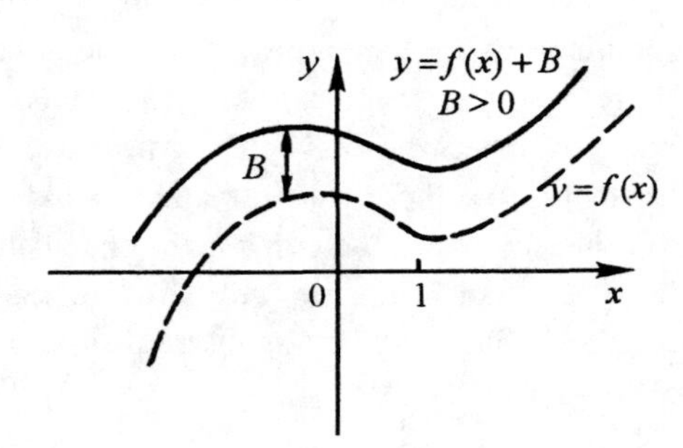
y
y = f(x) + B
B > 0
B
y = f(x)
0
1
x

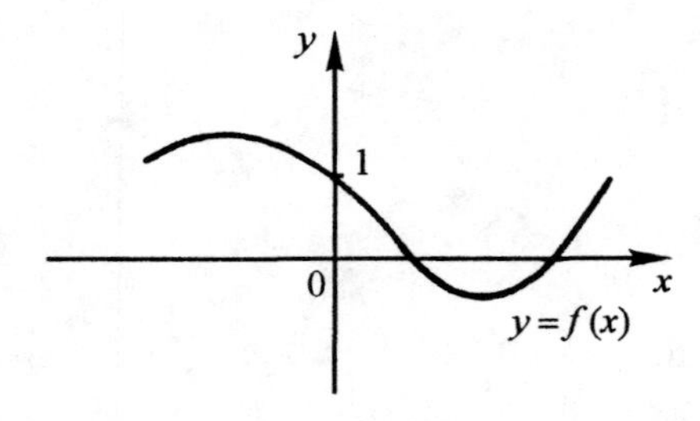
y
1
0
x
y = f(x)

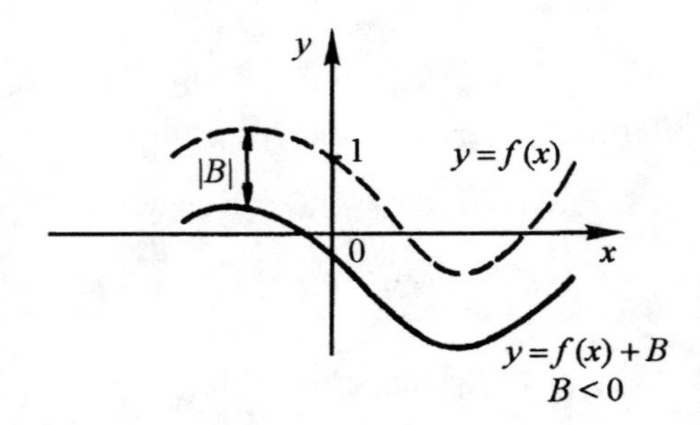
y
|B|
1
y = f(x)
0
x
y = f(x) + B
B < 0

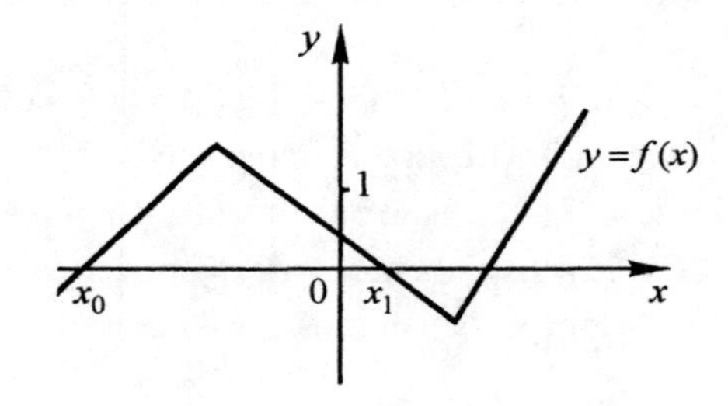
y
y = f(x)
1
x0
0
x1
x

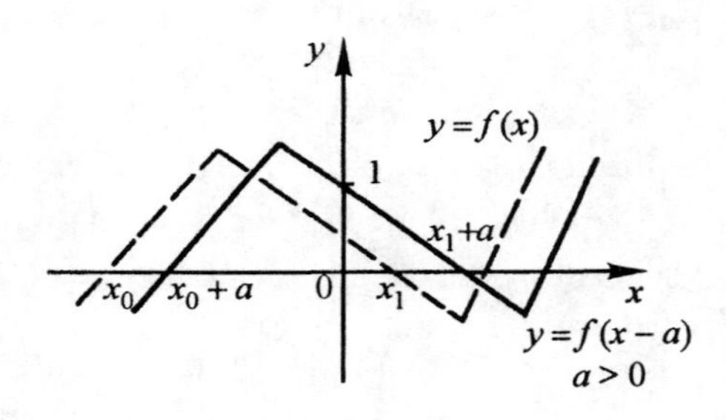
y
y = f(x)
1
x1+a
x0
x0 + a
0
x1
x
y = f(x − a)
a > 0

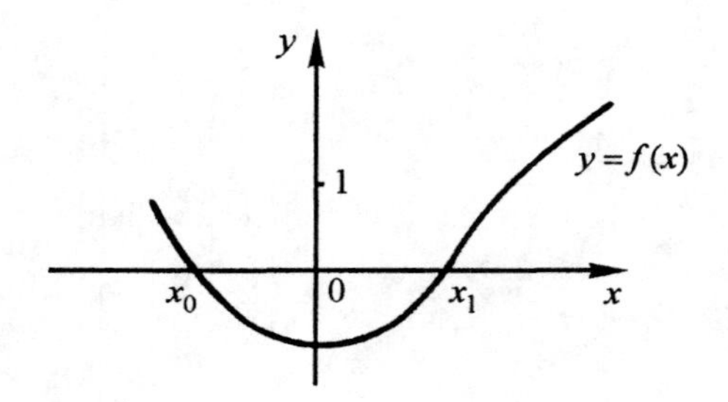
y
y = f(x)
1
x0
0
x1
x

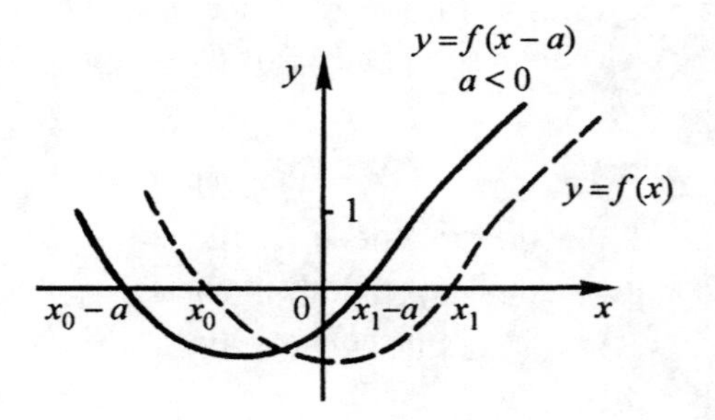
y = f(x − a)
a < 0
y
1
y = f(x)
x0 − a
x0
0
x1−a
x1
x

Table 2

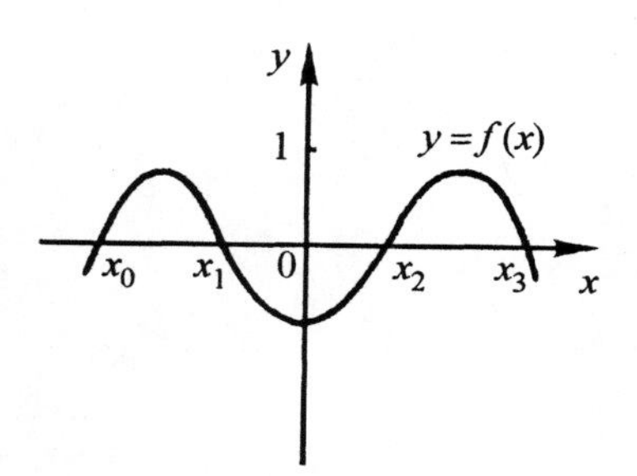
y
1
y = f(x)
x0
x1
0
x2
x3
x

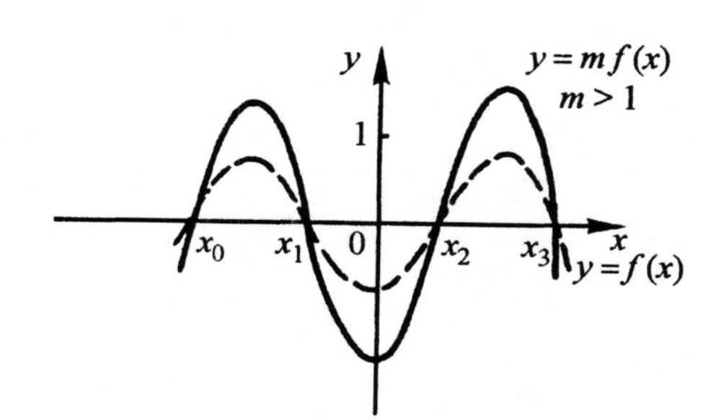
y
y = m f(x)
m > 1
1
x0
x1
0
x2
x3
x
y = f(x)

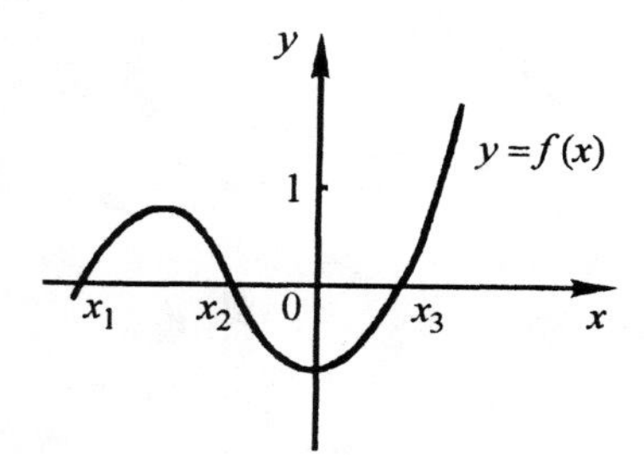
y
y = f(x)
1
x1
x2
0
x3
x

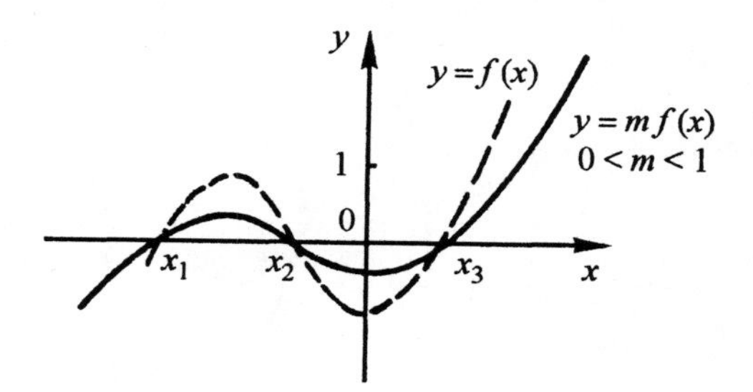
y
y = f(x)
y = m f(x)
0 < m < 1
1
0
x1
x2
x3
x

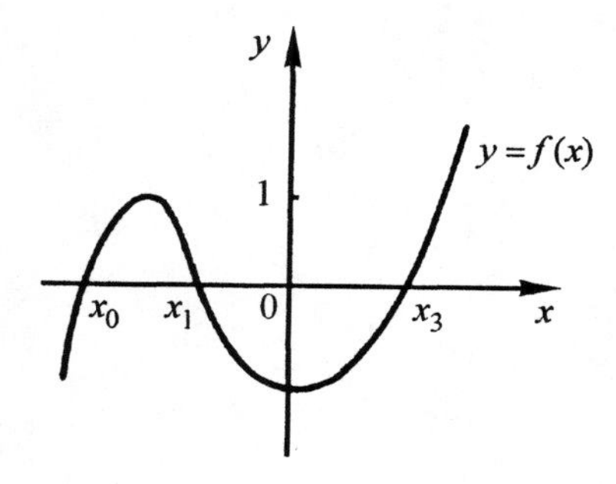
y
y = f(x)
1
x0
x1
0
x3
x

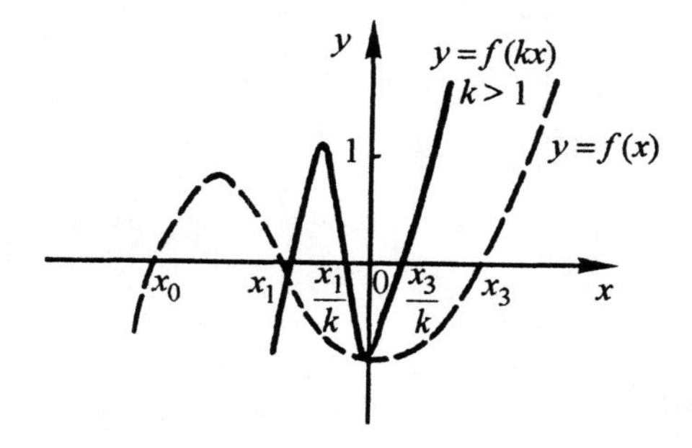
y
y = f(kx)
k > 1
y = f(x)
1
x0
x1
x1/k
0
x3/k
x3
x

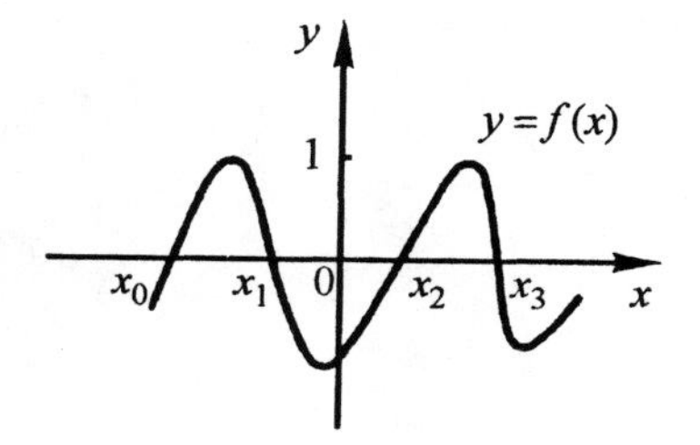
y
y = f(x)
1
x0
x1
0
x2
x3
x

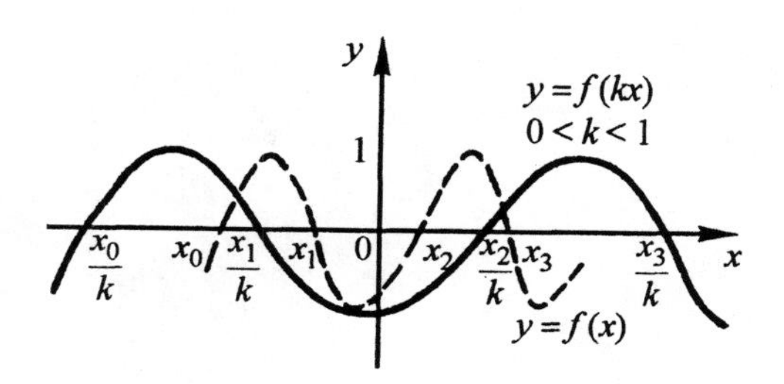
y
y = f(kx)
0 < k < 1
1
x0/k
x0
x1/k
x1
0
x2
x2/k
x3
x3/k
x
y = f(x)

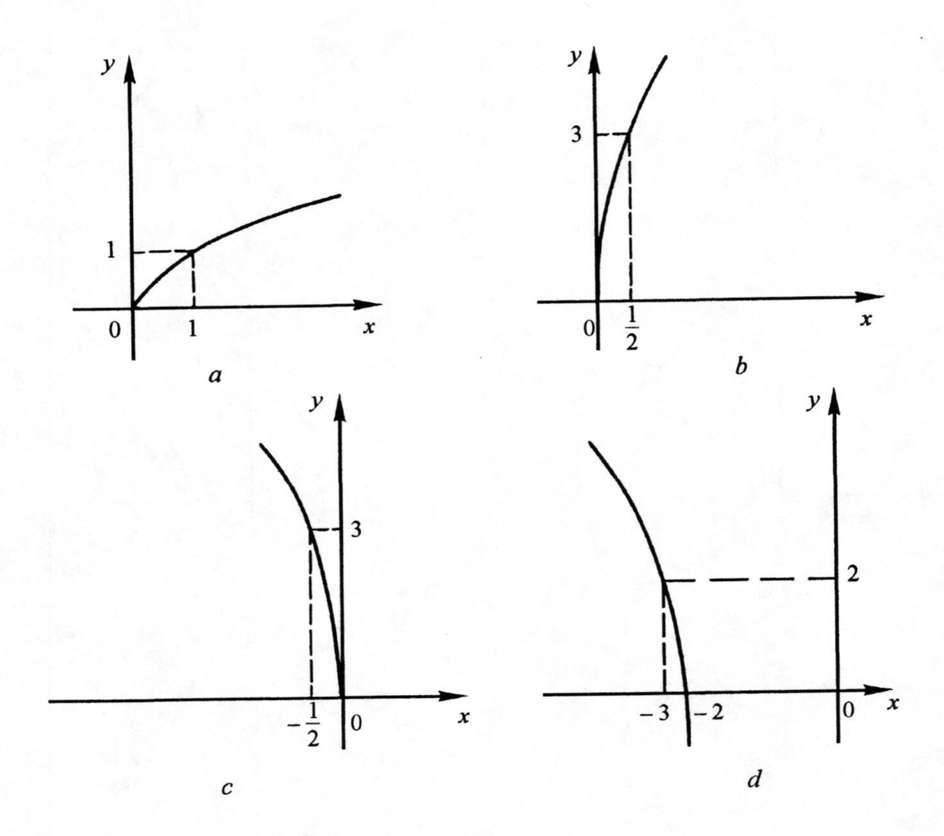

Figure 4.39. The graphs of functions: (*a*) $y = \sqrt{x}$; (*b*) $y = 3\sqrt{2x}$; (*c*) $y = 3\sqrt{-2x}$; (*d*) $y = 4\sqrt{-2(x+2)} - 0.7$.

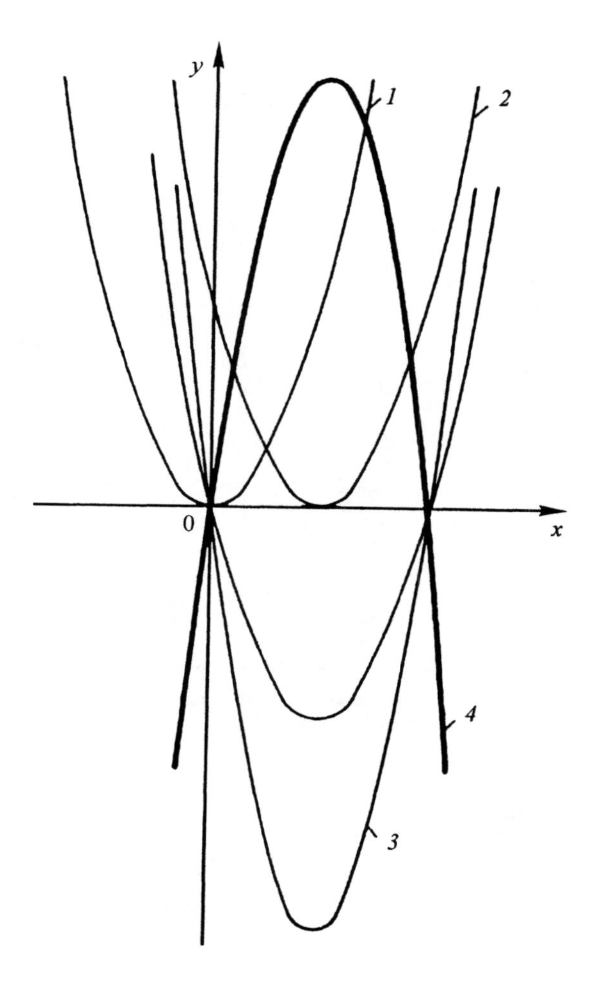

Figure 4.40. The graphs of functions: *1.* $y = x^2$, *2.* $y = (x-2)^2$, *3.* $y = 2(x-2)^2 - 4$, *4.* $y = -2((x-2)^2 - 4)$.

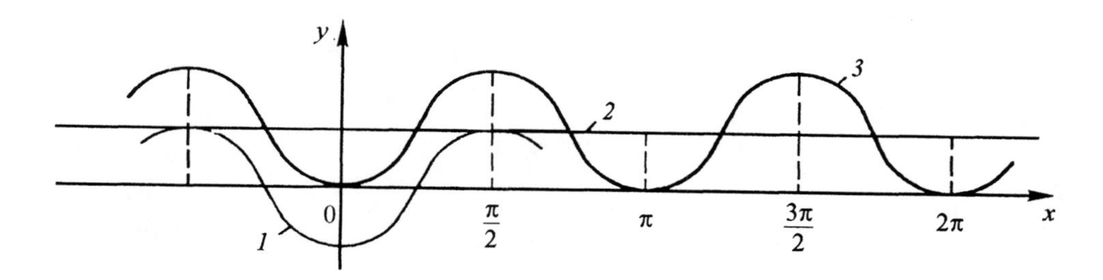

Figure 4.41. The graphs of functions: *1.* $y = -\frac{1}{2}\cos x$, *2.* $y = \frac{1}{2}$, *3.* $y = -\frac{1}{2}\cos 2x + \frac{1}{2}$.

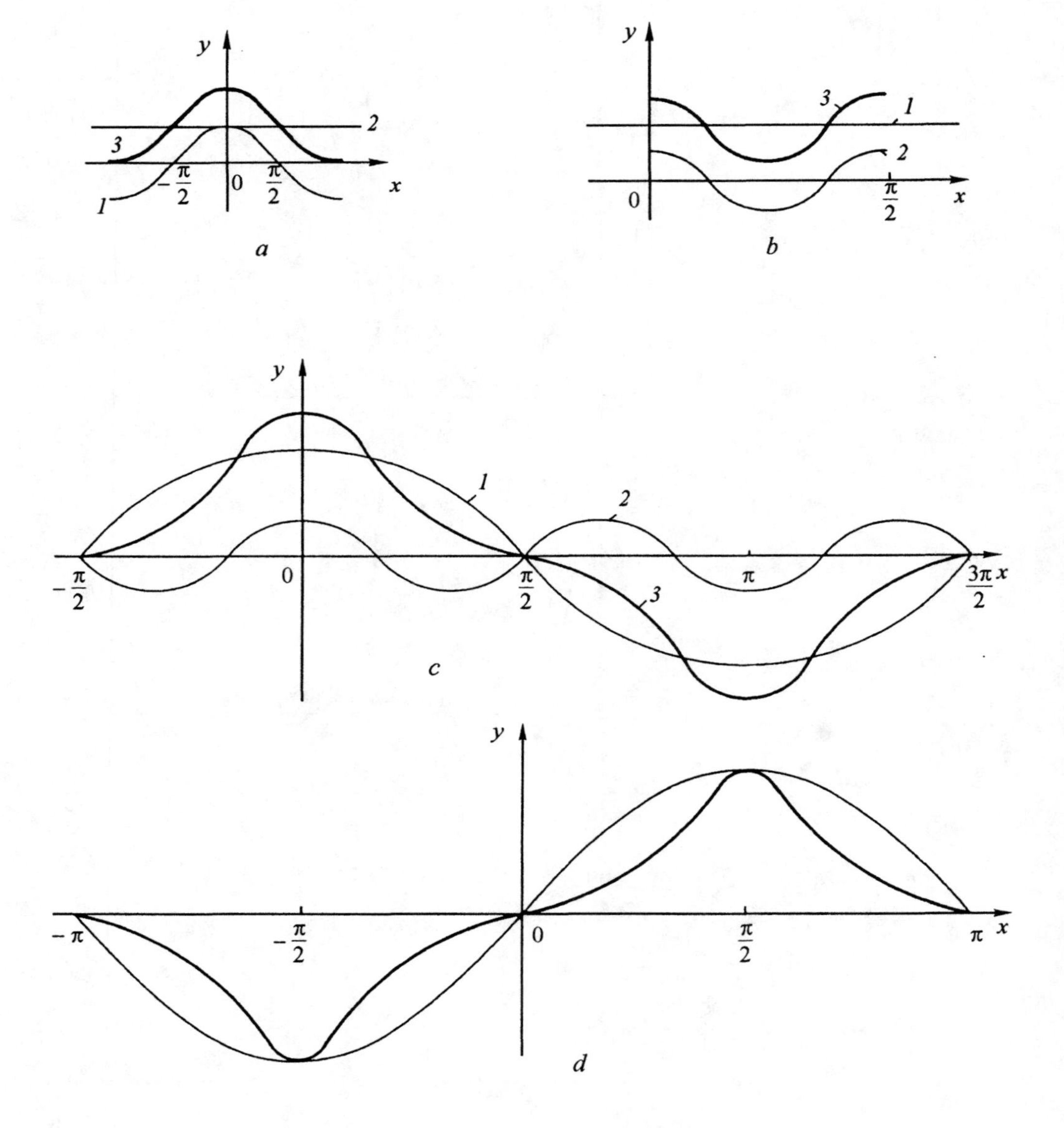

Figure 4.42. The graphs of functions: (*a*) *1.* $y = \frac{1}{2}\cos 2x$, *2.* $y = \frac{1}{2}$, *3.* $y = \cos^2 x$; (*b*) *1.* $y = \frac{3}{4}$, *2.* $y = \frac{1}{4}\cos 4x$, *3.* $y = \frac{1}{4}\cos 4x + \frac{3}{4}$; (*c*) *1.* $y = \frac{3}{4}\cos x$, *2.* $y = \frac{1}{4}\cos 3x$, *3.* $y = \cos^3 x$; (*d*) $y = \sin^3 x$.

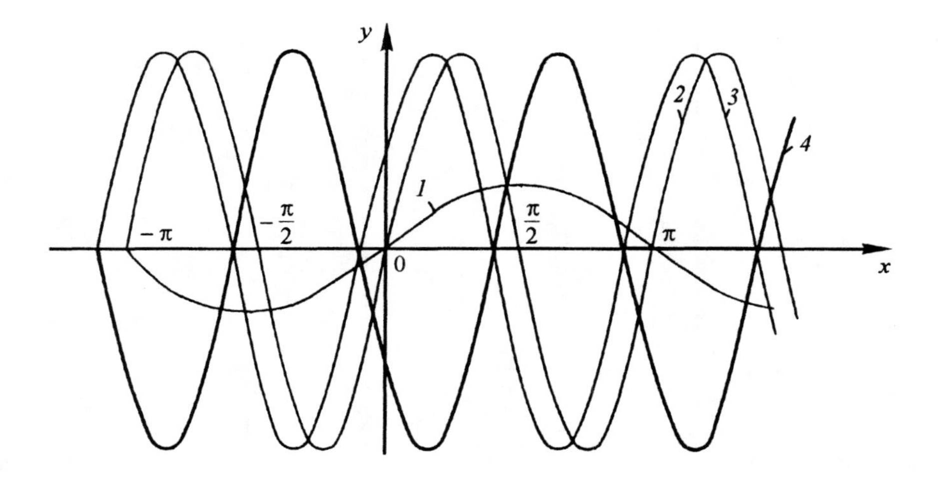

Figure 4.43. The graphs of functions: *1.* $y = \sin x$, *2.* $y = 3\sin 2x$, *3.* $y = 3\sin(2x + 8)$, *4.* $y = -3\sin(2x + 8)$.

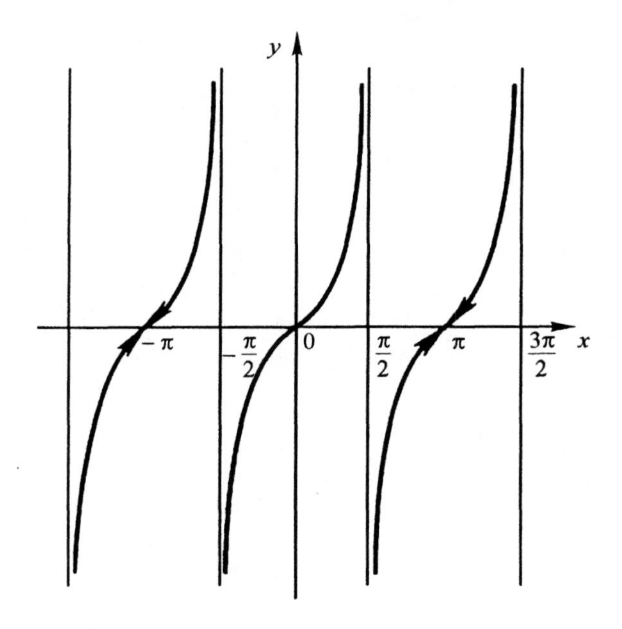

Figure 4.44. The graphs of function $y = \dfrac{2\tan(x/2)}{1 - \tan^2(x/2)}$.

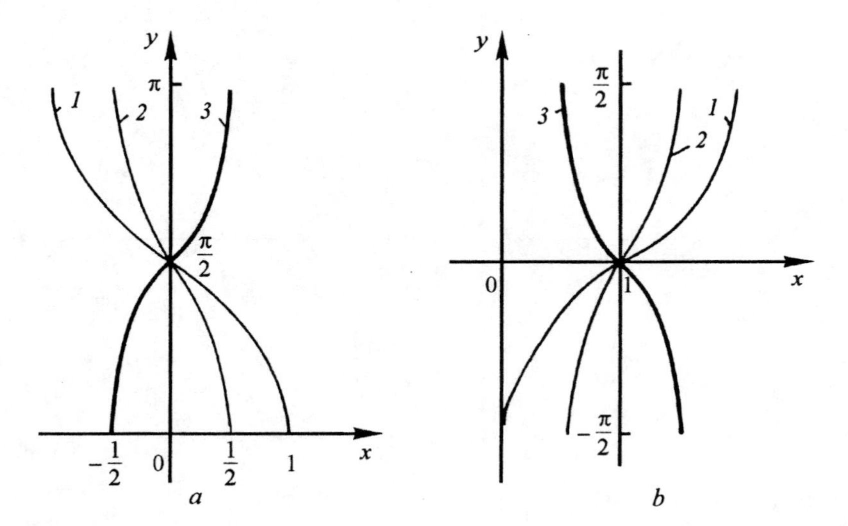

Figure 4.45. The graphs of functions: (*a*) *1.* $y = \arccos x$, *2.* $y = \arccos 2x$, *3.* $y = \arccos(-2x)$; (*b*) *1.* $y = \arcsin(x-1)$, *2.* $y = \arcsin 2(x-1)$, *3.* $y = \arcsin 2(x-1)$.

Plot of the graph of the function

Let us write the function $y = |f(x)|$ in a more explicit form:

$$f(|x|) = \begin{cases} f(x), & \text{if } x \text{ is such that } f(x) \geq 0, \\ -f(x), & \text{if } x \text{ is such that } f(x) < 0. \end{cases}$$

This formula shows that to plot the graph of the function $y = |f(x)|$, one needs to construct the graph of the function $y = f(x)$, leave the part of the graph lying above the x-axis unchanged, and reflect symmetrically with respect to the axis Ox the parts of the graph that lie below it (Figure 4.52).

Graphs of functions $y = |f(x)|$ are shown in Figures 4.53 – 4.55.

Graph of the function $y = |f(|x|)|$

The graph of the function $y = |f(|x|)|$ is constructed from the graph of the function $y = f(|x|)$ by leaving the parts of the graph lying above the x-axis unchanged and symmetrically reflecting about Ox the parts of the graph lying below the axis.

Example 4.2.4 Plot the graph of the function $y = |\log_2|x||$.

We plot the graph of the function $y = \log_2|x|$. Then draw the absolute value of this function, and thus get the plot of the function $y = |\log_2|x||$ (Figure 4.56).

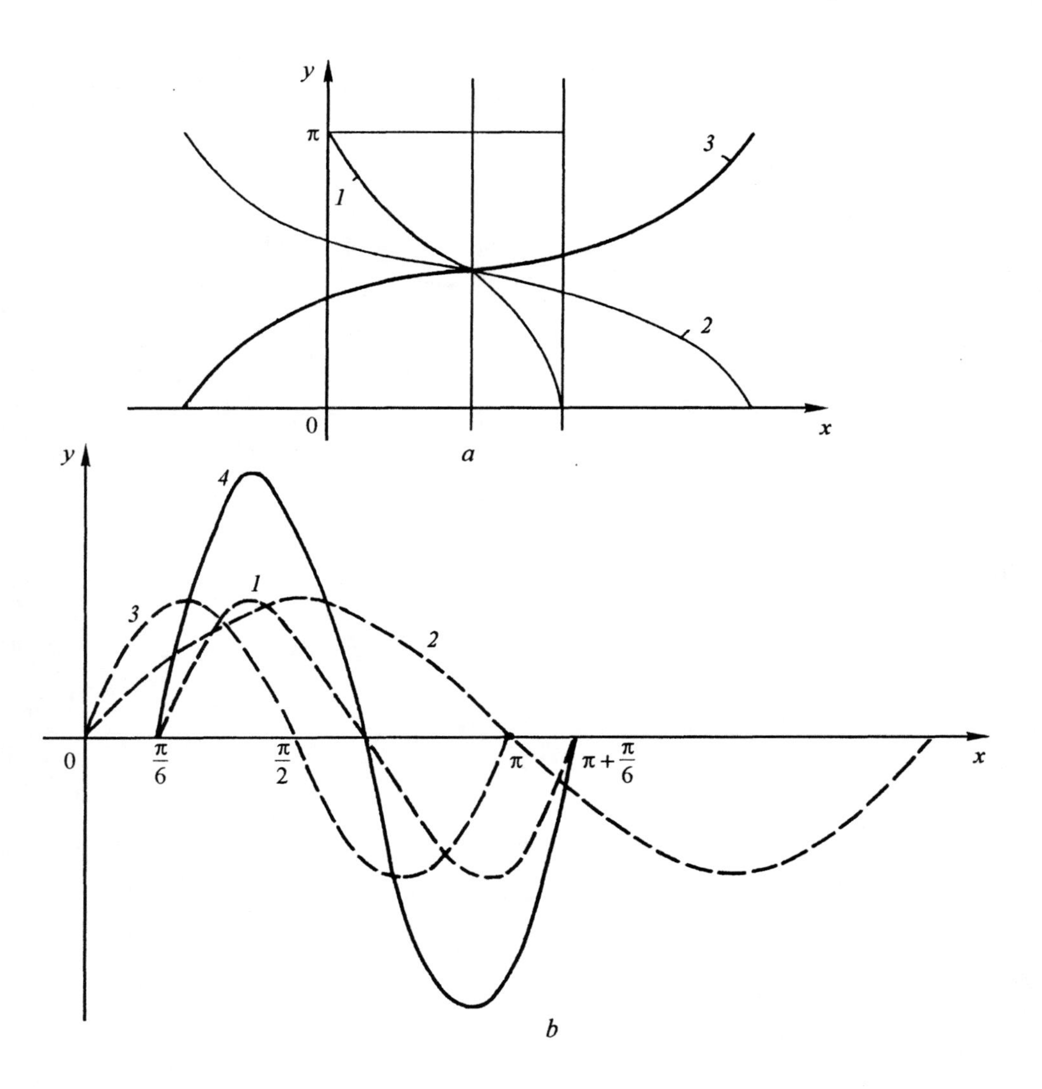

Figure 4.46. The graphs of functions: (*a*) *1*. $y = \arccos x$, *2*. $y = \arccos \frac{1}{3}(x - \frac{3}{2})$, *3*. $y = \arccos(-\frac{1}{3}x + \frac{1}{2})$; (*b*) *1*. $y = \sin x$, *2*. $y = \sin 2x$, *3*. $y = \sin 2(x - \frac{\pi}{6})$, *4*. $y = 2\sin 2(x - \frac{\pi}{6})$.

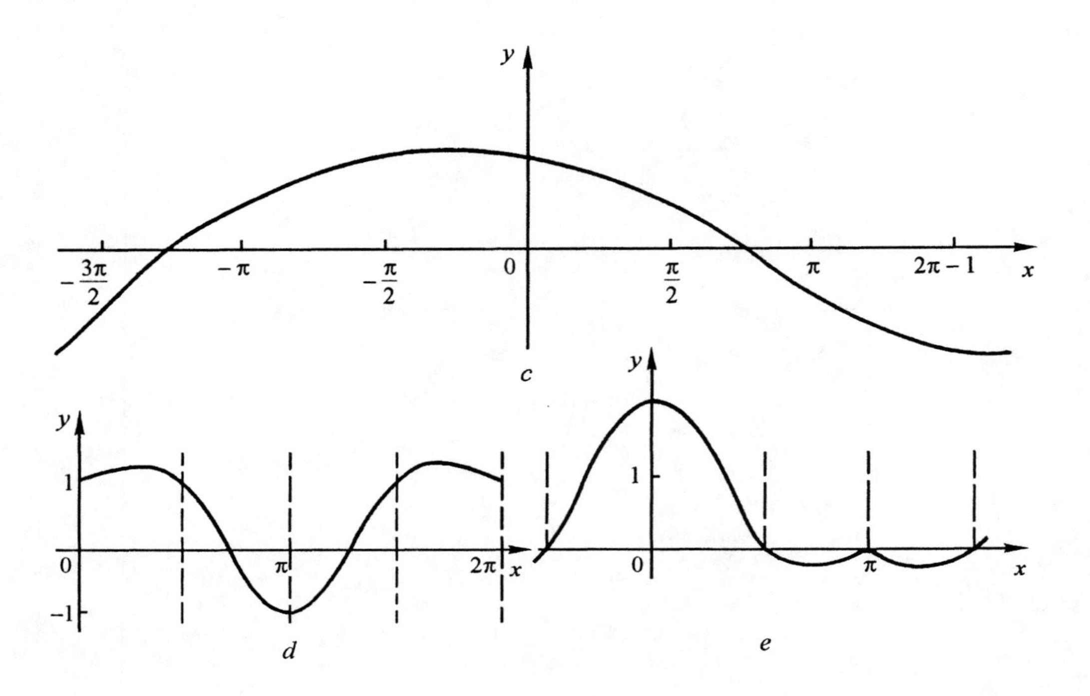

Figure 4.46. The graphs of functions: (*c*) $y = \cos\frac{x+1}{2}$; (*d*) $y = \sin^2 x + \cos x$; (*e*) $y = (1 + \cos x)\cos x$.

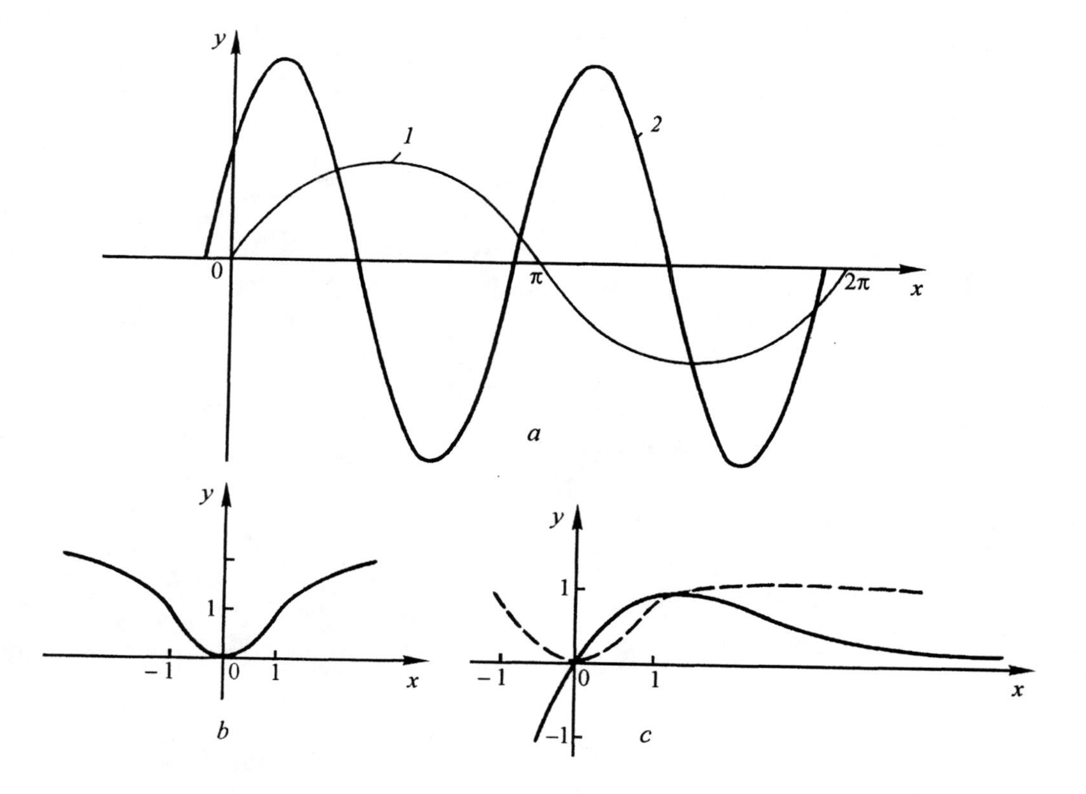

Figure 4.47. The graphs of functions: (*a*) *1*. $y = \sin x$, *2*. $y = 2\sin\left(2x + \frac{\pi}{6}\right)$;

(*b*) *1*. $y = \ln(x^2 + 1)$; (*c*) $y = \frac{x}{\sqrt{1 + x^4}}$.

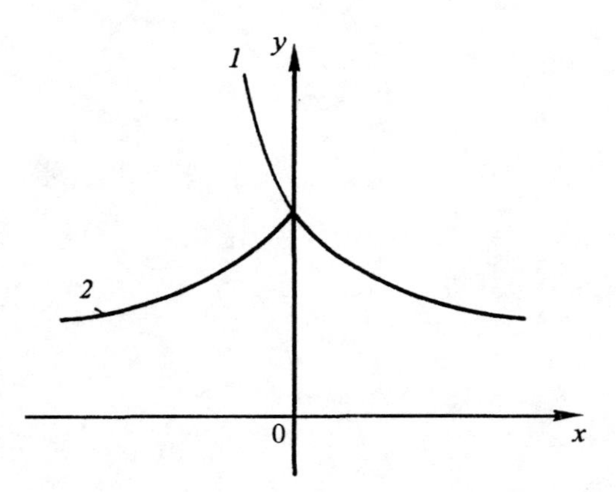

Figure 4.48. The graphs of functions: *1*. $y = f(x)$, *2*. $y = f(|x|)$.

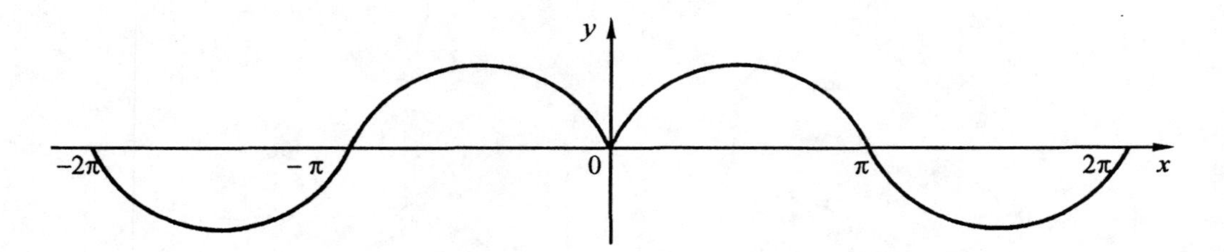

Figure 4.49. The graph of function $y = \sin|x|$.

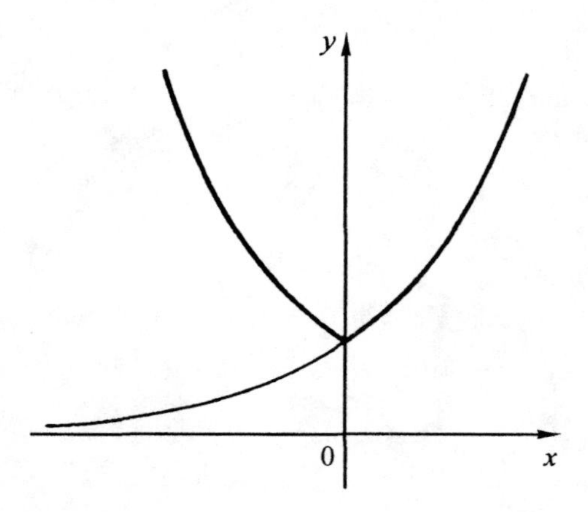

Figure 4.50. The graph of function $y = 2^{|x|}$.

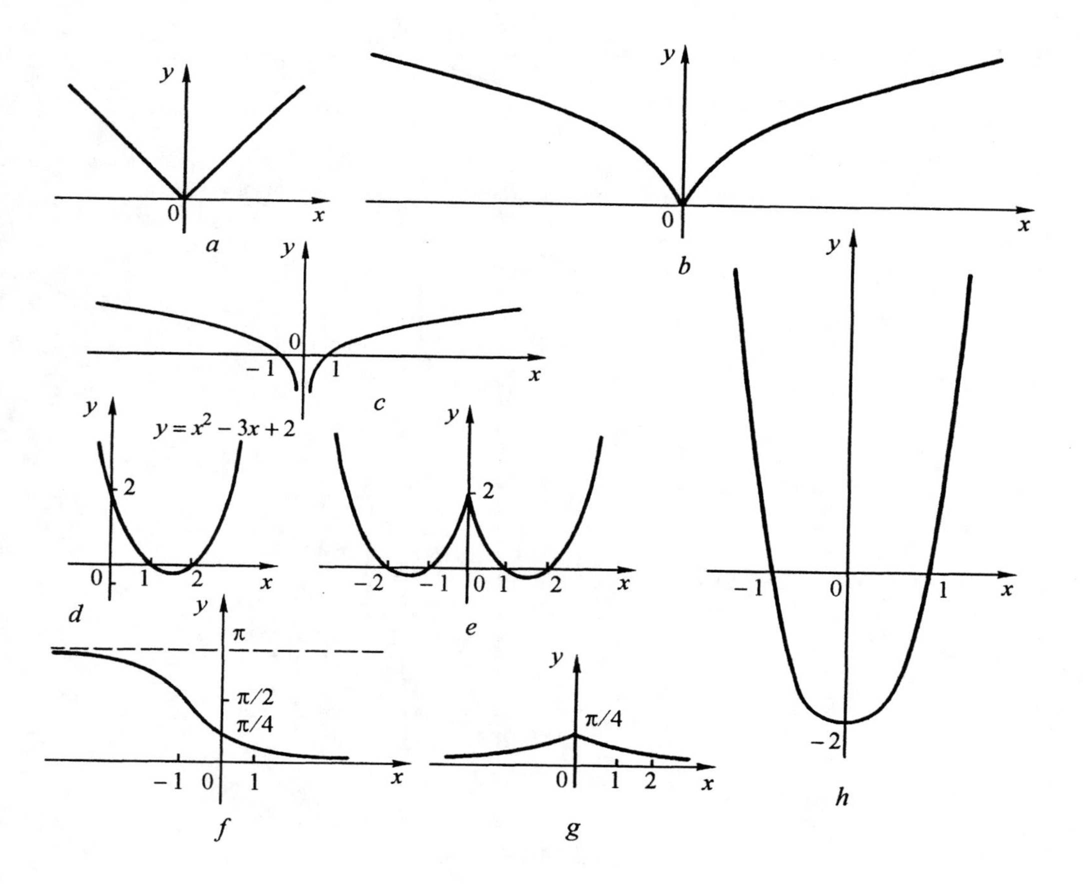

Figure 4.51. The graphs of functions: (*a*) $y = |x|$; (*b*) $y = \sqrt{|x|}$; (*c*) $y = \log_2 |x|$; (*d*) $y = x^2 - 3x + 2$; (*e*) $y = x^2 - 3|x| + 2$; (*f*) $y = \operatorname{arccot}(x + 1)$; (g) $y = \operatorname{arccot}(|x| + 1)$; (h) $y = x^2 + |x| - 2$.

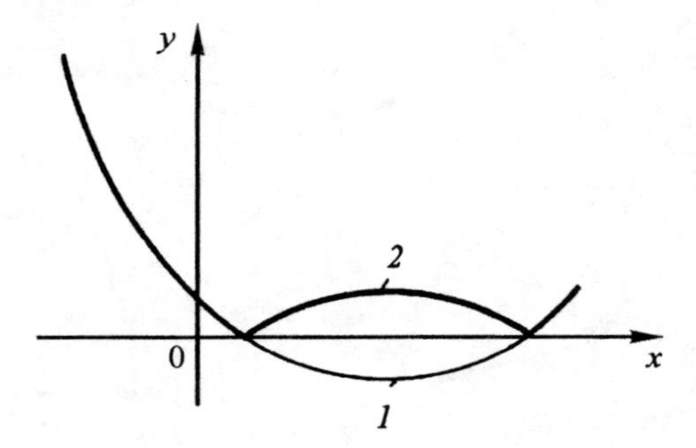

Figure 4.52. The graph of functions: *1.* $y = f(x)$, *2.* $y = |f(x)|$.

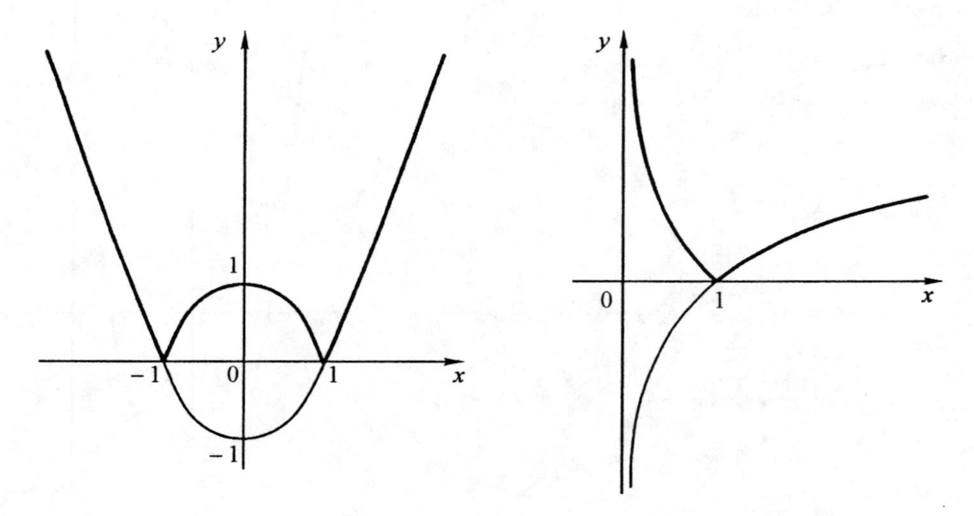

Figure 4.53 Figure 4.54

Figure 4.53. The graph of the function $y = |x^2 - 1|$.

Figure 4.54. The graph of the function $y = |\log_a x|$, $a > 1$.

Example 4.2.5. Plot the graph of the function $||x| - 1|$.

The construction of the graph is carried out as follows: draw the graph of the function $y = |x|$, move the graph downwards by 1, and symmetrically reflect the part located under the x-axis about the axis. The graph is shown in Figure 4.57.

Graphs of functions $y = |f(|x|)|$ are shown in Figures 4.58 and 4.59.

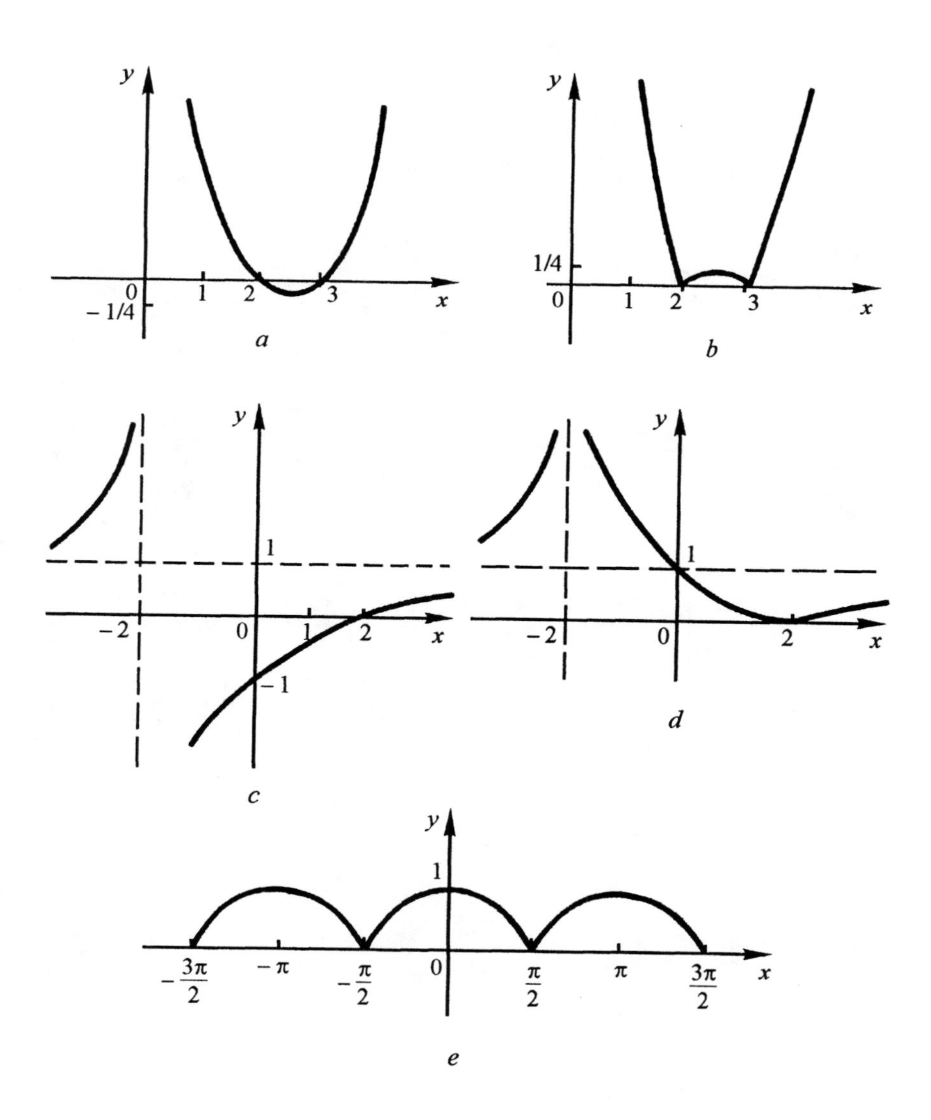

Figure 4.55. The graphs of functions: (*a*) $y = x^2 - 5x + 6$; (*b*) $y = |x^2 - 5x + 6|$; (*c*) $y = \frac{x-2}{x+2}$; (*d*) $y = \left|\frac{x-2}{x+2}\right|$; (*e*) $y = |\cos x|$.

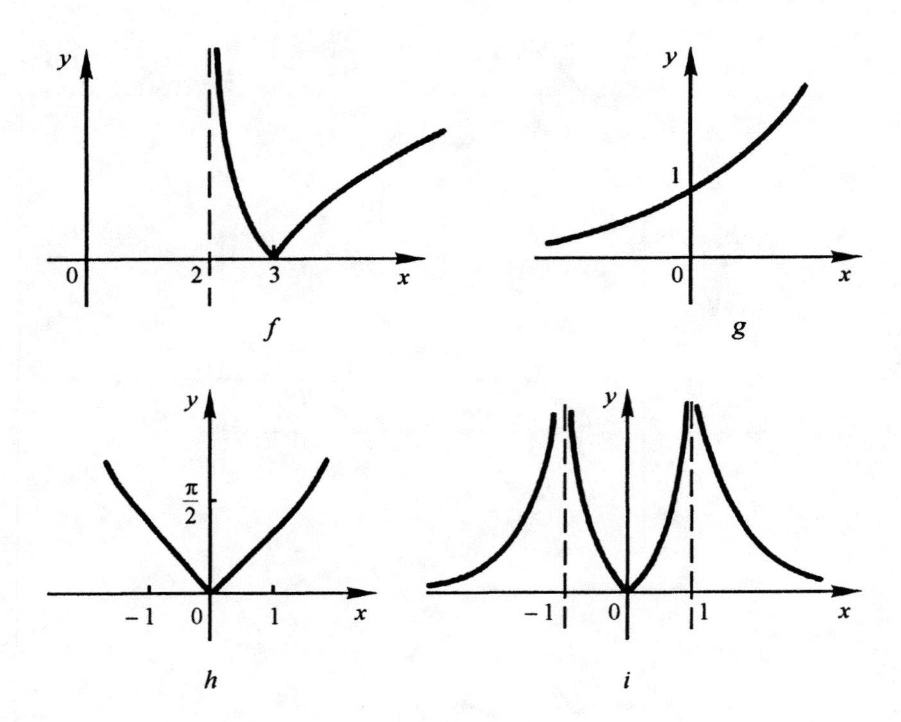

Figure 4.55. The graphs of functions: (*f*) $y = |\log_3 (x-2)|$; (*g*) $y = |3^x|$; (*h*) $y = |\arcsin x|$; (*i*) $y = \left|\frac{x}{x^2-1}\right|$.

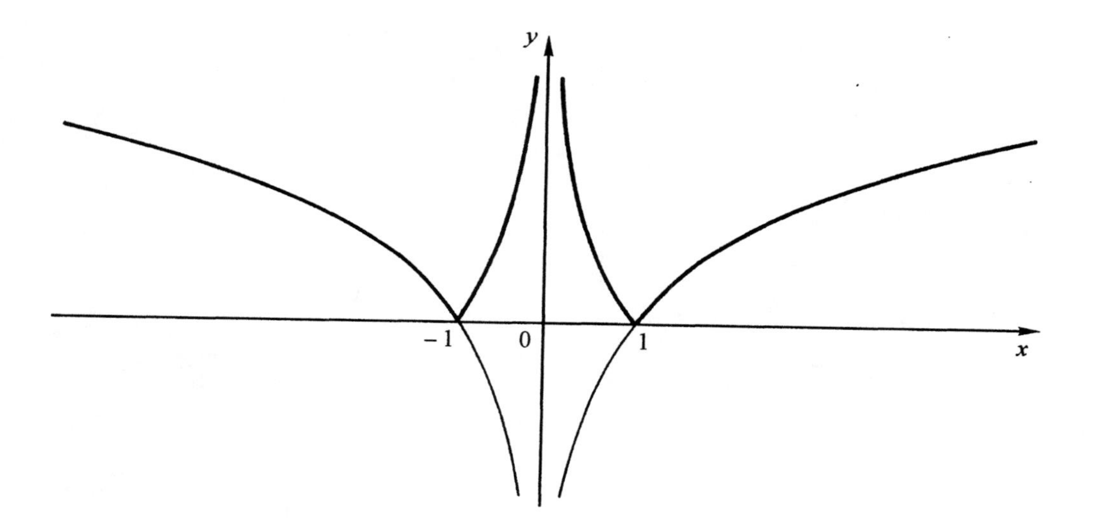

Figure 4.56. The graph of function $y = |\log_2 |x||$.

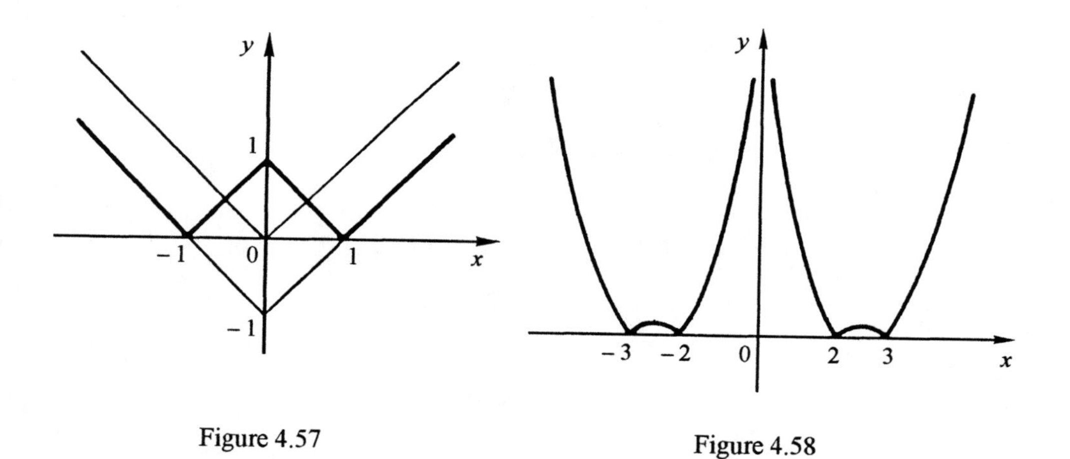

Figure 4.57

Figure 4.58

Figure 4.57. The graph of the function $y = ||x| - 1|$.

Figure 4.58. The graph of the function $y = |x^2 - 5|x| + 6|$.

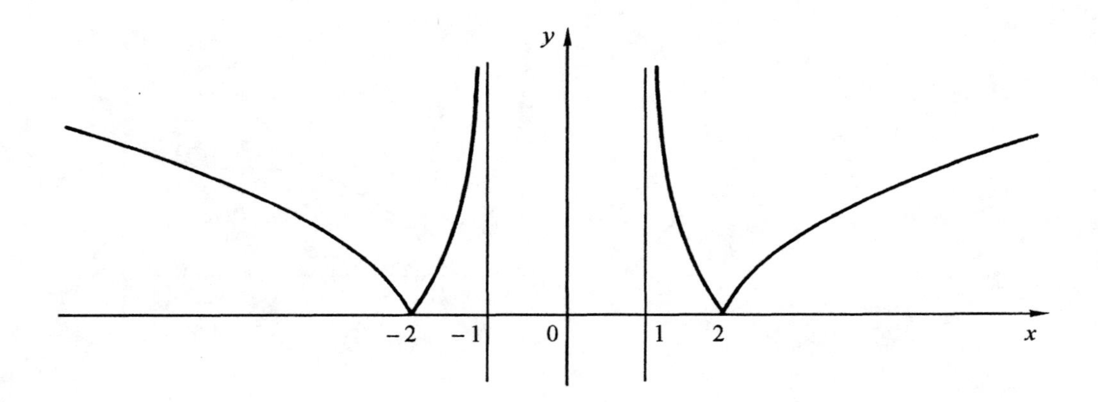

Figure 4.59. The graph of function $y = |\log_2 (|x| - 1)|$.

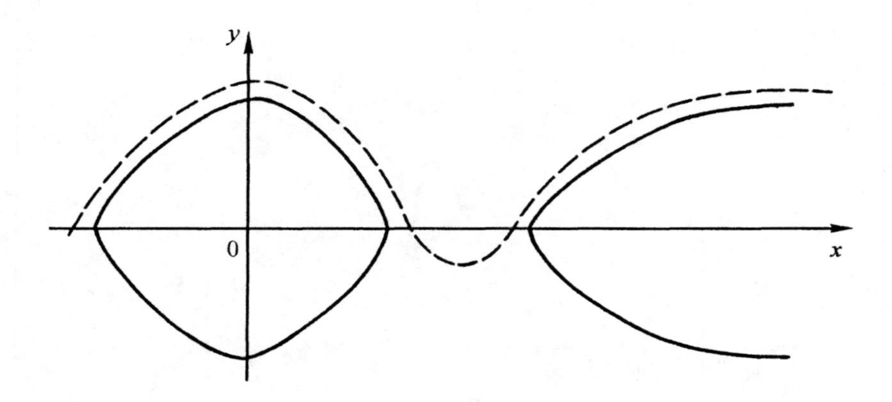

Figure 4.60. The graph of function $|y| = f(x)$.

Graph of the function $|y| = f(x)$

Since the left-hand side contains the absolute value sign, only those values of x from the domain of the function $y = f(x)$ should be considered for which $f(x) \geq 0$. For these x, the equality $|y| = f(x)$ can be written as $y = \pm f(x)$, i.e. we have two functions. If for all x from the domain of the function $y = f(x)$, we have that $f(x) < 0$, then no curve is defined by the equality $|y| = f(x)$. So the graph of the function $|y| = f(x)$ is plotted by taking the part of the graph of the function $y = f(x)$ which lies above the x-axis together with its symmetric reflection with respect to Ox (Figure 4.60).

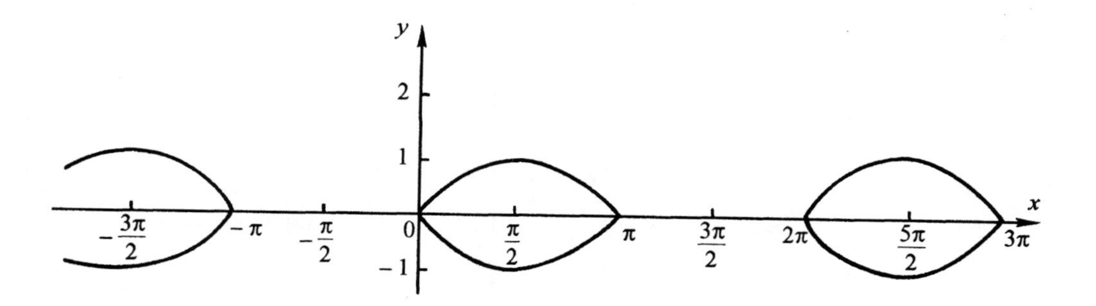

Figure 4.61. The graph of function $|y| = \sin x$.

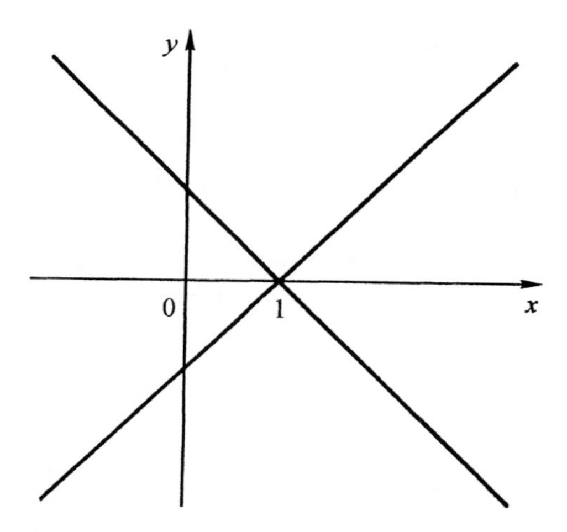

Figure 4.62. The graph of function $|y| = |x - 1|$.

Example 4.2.6. Plot the graph of the function $|y| = \sin x$. The domain is found from the condition $0 \le \sin x \le 1$. This yields $2\pi n \le x \le \pi(2n + 1)$, $n \in \mathbf{Z}$.

Draw the portion of the graph of the function $y = \sin x$ for these x, that is for x such that $\sin x \ge 0$. The lower part is constructed symmetrically (Figure 4.61).

Graph of the function $|y| = |f(x)|$

The graph of this function is plotted by drawing the graph of the function $y = |f(x)|$ and then drawing its symmetric reflection with respect to the x-axis.

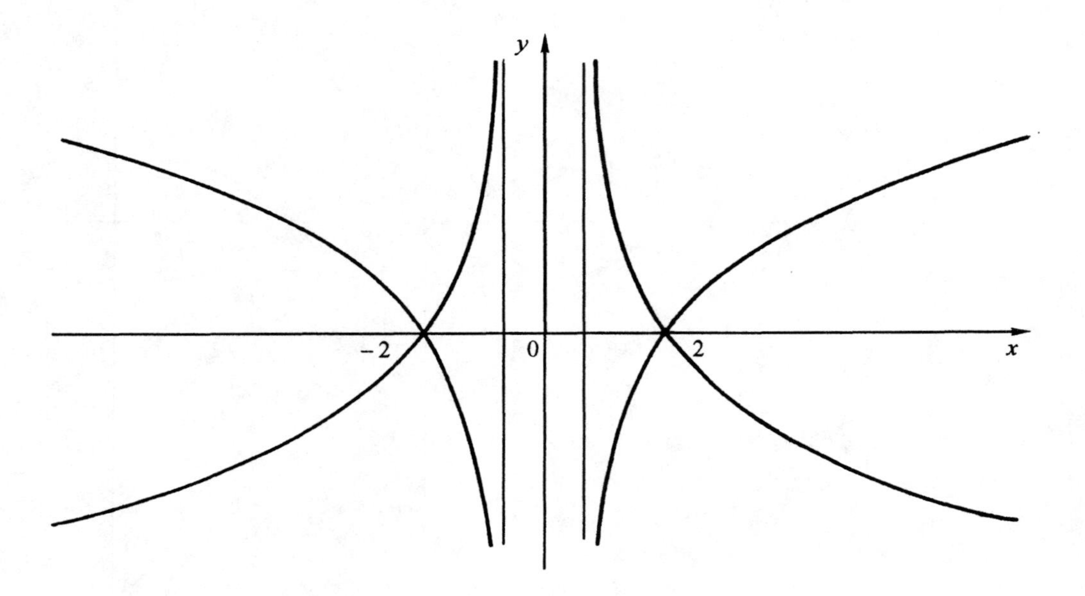

Figure 4.63. The graph of the function $|y| = |\log_2 (|x| - 1)|$.

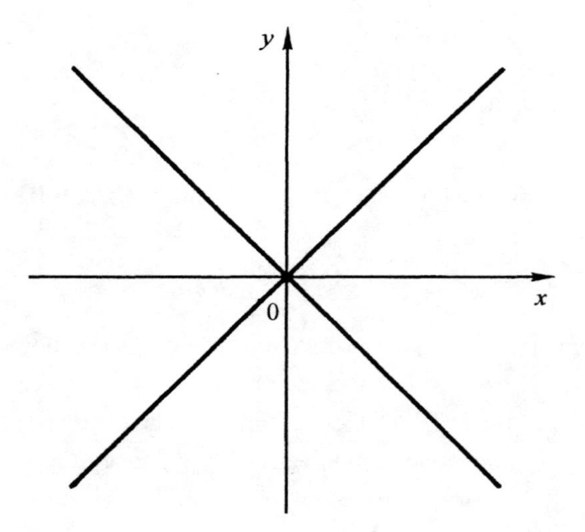

Figure 4.64. The graph of the function $|y| = |x|$.

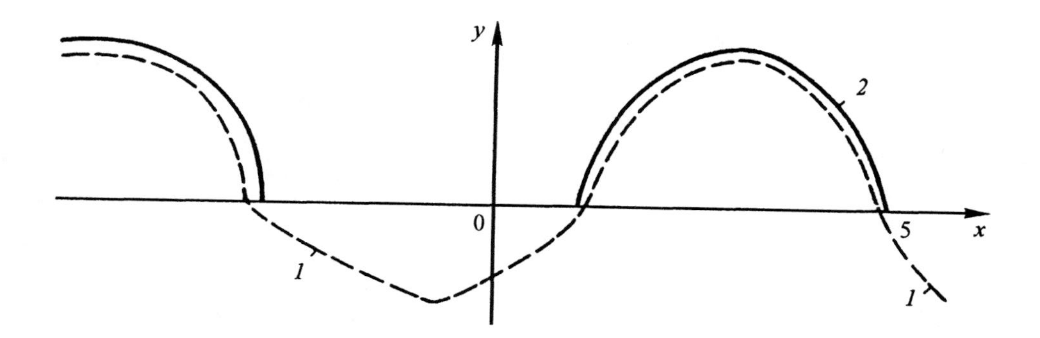

Figure 4.65. The graph of functions: *1.* $y = f(x)$, *2.* $y = \frac{1}{2}(|f(x)| + f(x))$.

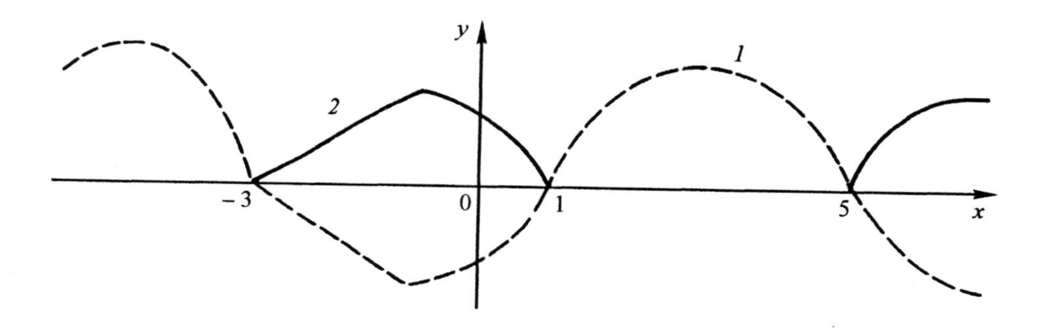

Figure 4.66. The graph of functions: *1.* $y = f(x)$, *2.* $y = \frac{1}{2}(|f(x)| - f(x))$.

Example 4.2.7. Plot the graph of the function $|y| = |x - 1|$.

First, plot the graph of the function $y = |x - 1|$, and then plot another portion of the graph which is symmetric with respect to the x-axis (Figure 4.62).

Graph of the function $|y| = |f(|x|)|$

To plot the graph of the function $|y| = |f(|x|)|$, we first plot the graph of the function $y = |f(|x|)|$, and then draw the other portion of the graph symmetrically with respect to the x-axis. Examples of such graphs are shown in Figures 4.63 and 4.64.

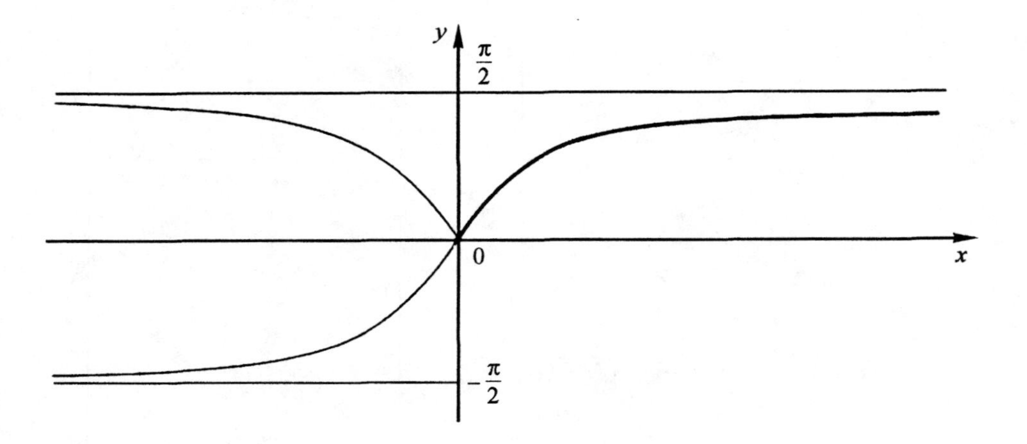

Figure 4.67. The graph of the function $y = \frac{1}{2}(|\arctan x| + \arctan x)$.

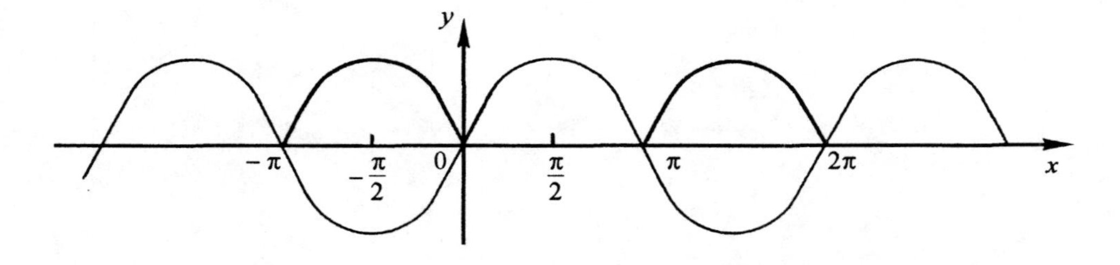

Figure 4.68. The graph of the function $y = \frac{1}{2}(|\sin x| + \sin x)$.

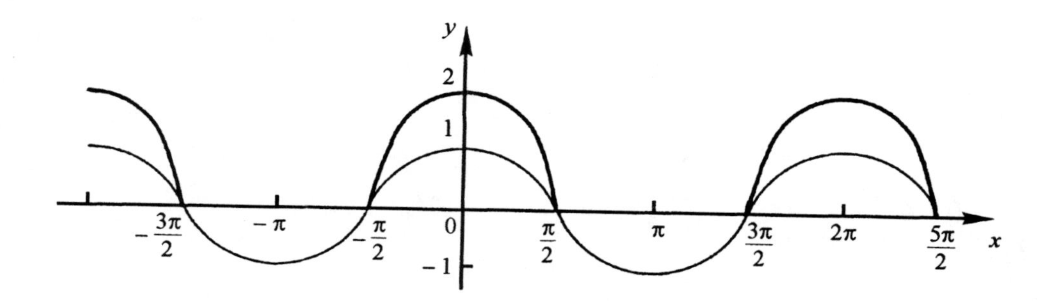

Figure 4.69. The graph of the function $y = |\cos x| + \cos x$.

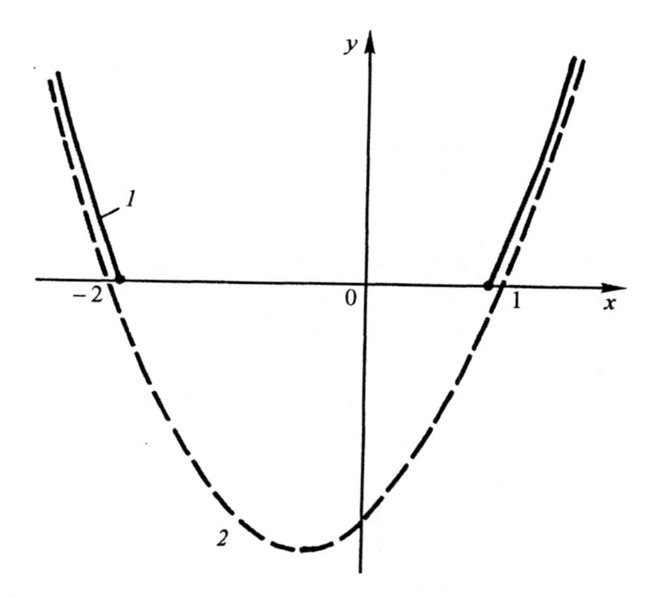

Figure 4.70. The graph of the function $y = \frac{1}{2}(|x^2 + x - 2| + x^2 + x - 2)$.

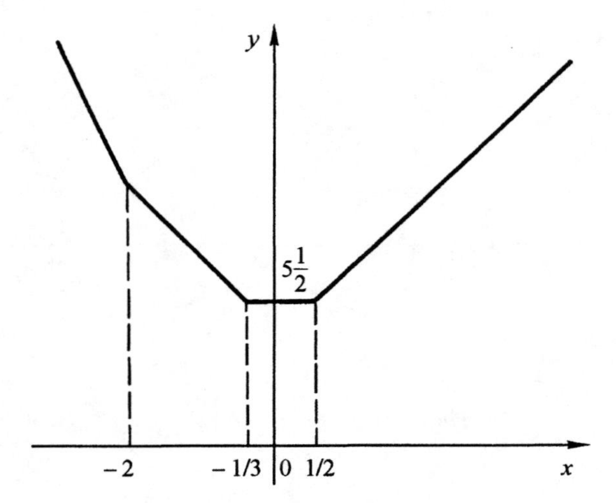

Figure 4.71. The graph of the function $y = |x+2| + |3x+1| + |x - \frac{1}{2}| - 3x + 2$.

Graph of the function $y = \frac{1}{2}(f(|x|) \pm f(x))$

1. Let $y = \frac{1}{2}(f(|x|) + f(x))$. The domain of this function is the same as the domain of the function $f(x)$. Writing the function as

$$y = \frac{1}{2}(|f(x)| + f(x)) = \begin{cases} f(x), & \text{if } f(x) \geq 0, \\ 0, & \text{if } f(x) < 0, \end{cases}$$

we conclude that on the sets where the graph of the function $f(x)$ is above the x-axis, the needed graph coincides with the graph of the function $f(x)$. On the intervals where the graph of $f(x)$ is below the x-axis, the graph of the function coincides with the x-axis (Figure 4.65).

2. Let $y = \frac{1}{2}(f(|x|) - f(x))$. Then

$$y = \frac{1}{2}(|f(x)| - f(x)) = \begin{cases} 0, & \text{if } f(x) \geq 0, \\ |f(x)|, & \text{if } f(x) < 0, \end{cases}$$

This implies that the graph of the function coincides with the x-axis on the intervals where $f(x) \geq 0$, and coincides with the graph of the function $y = |f(x)|$ on the intervals where $f(x) < 0$ (Figure 4.66).

Examples of graphs of the function $y = \frac{1}{2}(|f(x)| \pm f(x))$ are shown in Figures 4.67 – 4.70.

Remark 4.2.1. To plot the graph of a function which is a sum, product, ratio, or a more complex function which includes the absolute value sign, the domain of the function is subdivided into intervals on which the sign of the expressions under the absolute value sign does not change.

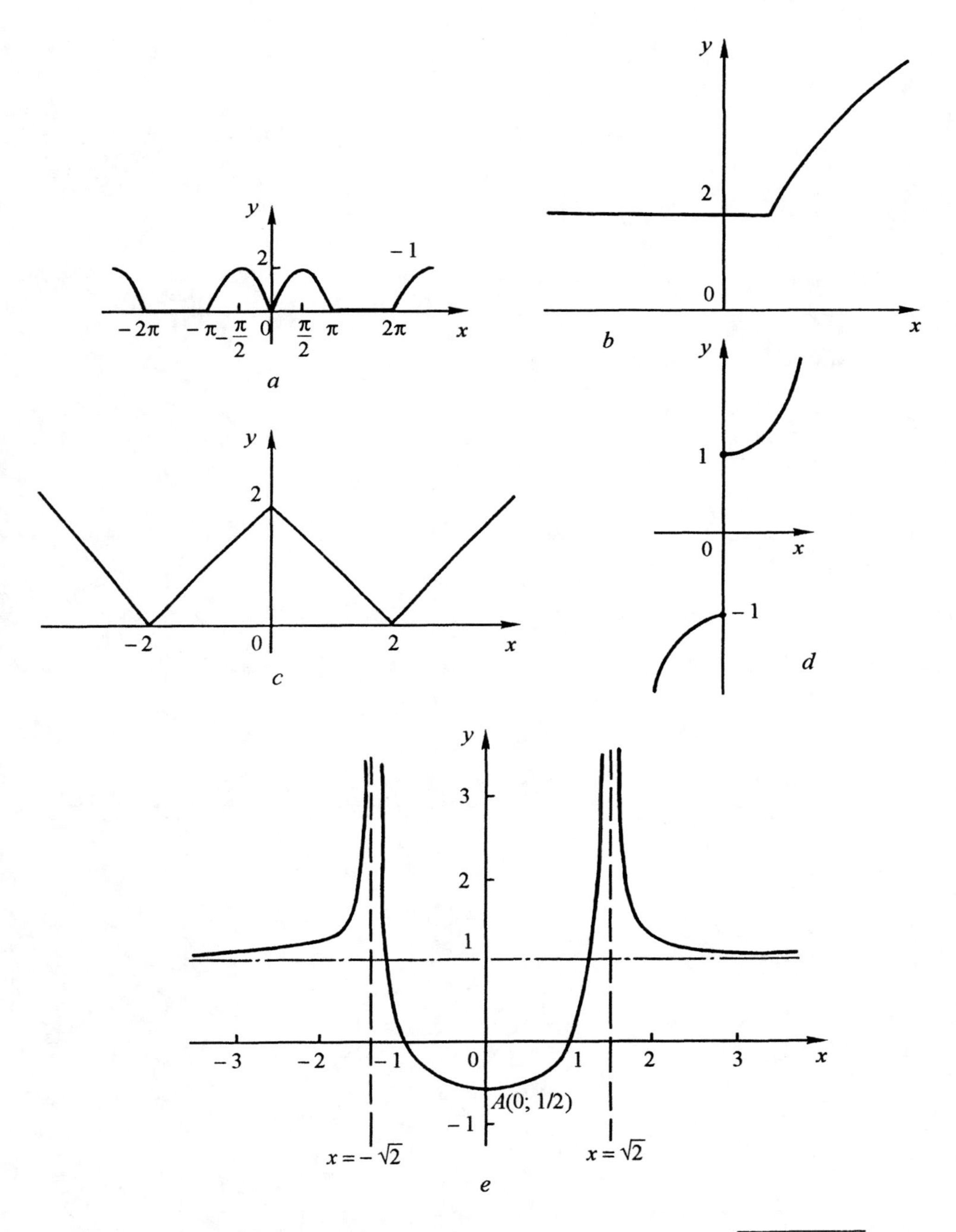

Figure 4.72. The graph of functions: (*a*) $y = |\sin x| + \sin|x|$; (*b*) $y = \sqrt{2(x + |x - 2|)}$; (*c*) $y = |2 - |x||$; (*d*) $y = \dfrac{x(1 + x^2)}{|x|}$, $x \neq 0$; (*e*) $y = \dfrac{x^2 - 1}{|x^2 - 2|}$.

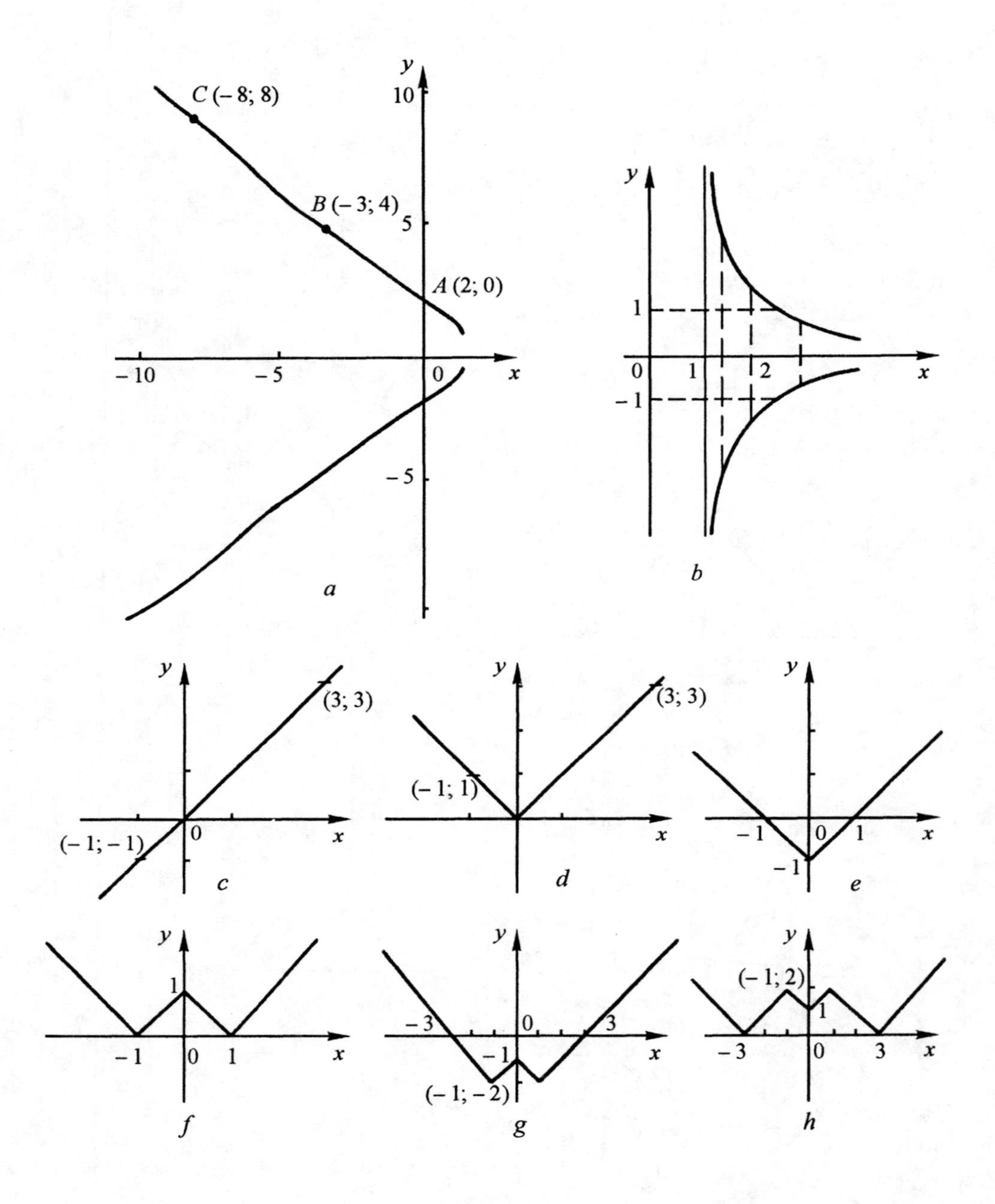

Figure 4.73. The graph of functions: *(a)* $|y| = 2^{\sqrt{1-x}}$; *(b)* $|y| = \dfrac{1}{x-1}$; *(c)* $y = x$; *(d)* $y = |x|$; *(e)* $y = |x| - 1$; *(f)* $y = |x - 1|$; *(g)* $y = |\,|x| - 1|$; *(h)* $y = |\,|x - 1| - 2|$.

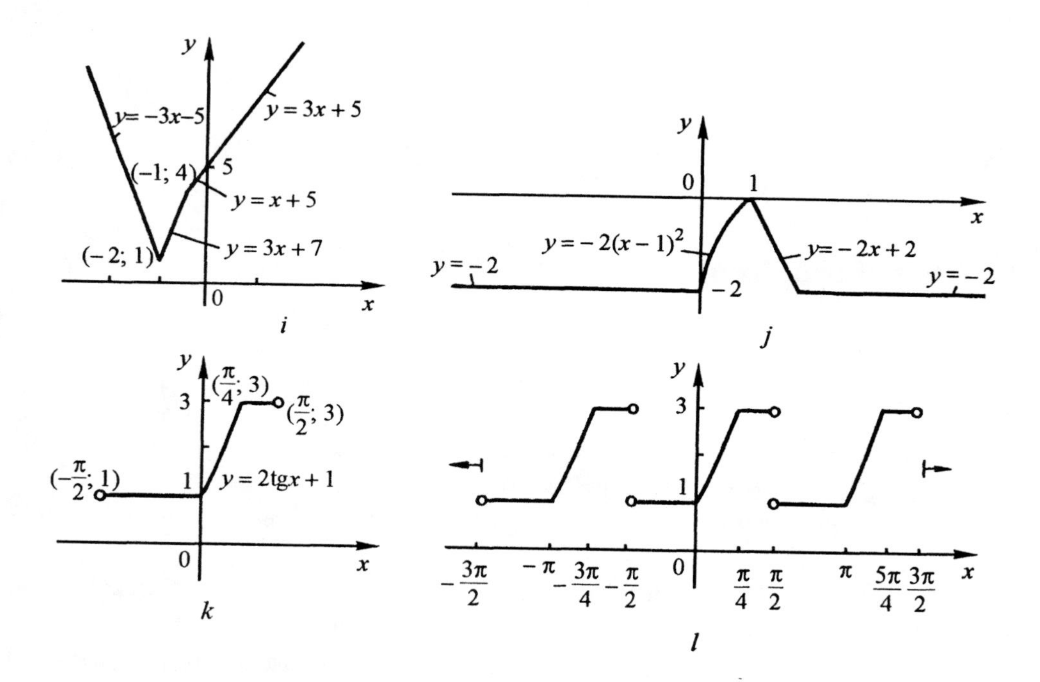

Figure 4.73. The graph of functions: (i) $y = |x| - |x + 1| + 3|x + 2|$;
(j) $y = |x^2 - 2x| - |x^2 - 3x + 2| - x$; ($k$, l) $y = |\tan x| - |\tan x - 1| + 2$.

The graph of the function is then plotted on those intervals, where the function is given by different analytic expressions. For example, if the function is of the form

$$y = |f_1(x)| \pm |f_2(x)| \pm \ldots \pm |f_n(x)| + r(x),$$

then we find all x which satisfy at least one of the equations: $f_1(x) = 0, f_2(x) = 0, \ \ldots \ , f_n(x) = 0$. These values of x subdivide the domain of the function under consideration into intervals such that the signs of all the functions $y_1 = f_1(x), y_2 = f_2(x), \ \ldots \ , y_n = f_n(x)$ do not change on any of the intervals in the subdivision. On each of these intervals, the absolute value signs can be removed from the expression of the function. The graph of the resulting function is then plotted on this interval.

Graphs of such functions are shown in Figures 4.71 – 4.73.

Chapter 5

Plotting of graphs of elementary functions

5.1. Plots of compositions of functions

Consider a composition $y = f(\varphi(x))$, where the function y depends on x not directly but via the function $\varphi(x)$. Denoting $\varphi(x)$ by u, we get $y = f(u)$, $u = \varphi(x)$.

Sometimes the auxiliary function $u = \varphi(x)$ is called the *inner function*, and the function $y = f(x)$, the *outer function*.

Graphs of compositions can be constructed on the basis of a general study of the function (see Section 2.2). In such a case, properties of the functions $y = f(u)$ and $u = \varphi(x)$ must be used. In particular, the expression $f(\varphi(x))$ makes sense for x such that $\varphi(x)$ exists and assumes the values of u for which the expression $f(u)$ is defined. Properties of being even, odd, periodic, and monotone should be used. It is also necessary to find the value of the composition at the endpoints of the monotonicity intervals, or to determine the behavior of the function in neighborhoods of such points, if they (or one of them) does not belong to the domain of the function.

Let us describe a geometric method for constructing the graph of the composition $y = f(\varphi(x))$. The construction is done in one figure.

Denote by A, A_1, A_2, A_3, A_4 the points $(x; 0)$, $(x; \varphi(x)$, $(\varphi(x); \varphi(x))$, $(\varphi(x); f(\varphi(x))$, $(x; f(\varphi(x)))$, correspondingly.

The construction is carried out in the following order.

1. Plot the graphs of the functions $y = f(x)$ and $y = \varphi(x)$.
2. Draw the bisectrix of the first and third quadrants ($y = x$).
3. Draw a straight line passing through the point A parallel to the axis Oy until it intersects the graph of the function $y = \varphi(x)$ (the point of intersection is A_1).
4. Draw a straight line passing through the point A_1 parallel to the axis Ox until it intersects the bisectrix (the point of intersection is A_2).
5. Draw a straight line through the point A_2 parallel to the axis Oy until it intersects the graph of the function $y = f(x)$ (the point of intersection is A_3).

6. Draw a straight line passing through the point A_3 parallel to the axis Ox until it intersects the line obtained by continuing the line segment AA_1. The point of intersection $A_4(x; f(\varphi(x)))$ will be a point of the graph of the function $y = f(\varphi(x))$ (Figure 5.1).

Other points of the graph are constructed similarly.

If the construction is carried out with a ruler and compass, then this method could be simplified. Draw a perpendicular line through the point $P(x; 0)$ until it intersects the graph of $y = \varphi(x)$. We get the point Q with $PQ = \varphi(x)$. Then take on the x-axis a point T such that $OT = PQ$. Drawing a perpendicular through the point T until it intersects the graph of the function $y = f(x)$ at a point S, we get that $TS = f(x)$ (Figure 5.2).

Remark 5.1.1. The graph of the function $y = f(\varphi(x))$ can also be constructed as follows. First construct the graph of the function $u = \varphi(x)$. Then, using the y-values of this function and main properties of the function $y = f(u)$, construct the graph of the composition.

Example 5.1.1. Plot the graph of the curve $y = e^{-x^2}$ (*Gaussian curve*). Rewrite the function as $y = e^{-u}$, $u = x^2$. Let us analyze the curve $y = e^{-x^2}$ as described in Section 2.2. The domain of the function is $(-\infty; +\infty)$. The range is $(0; 1]$. The function is even, because the inner function $u = x^2$ is such. Hence, it suffices to construct the graph for $x \geq 0$. The function $y = e^{-u}$ is decreasing on $(0; +\infty)$, and the function $u = x^2$ is increasing on $(0; +\infty)$. This means that the composition $y = e^{-x^2}$ is decreasing on $(0; +\infty)$ (Figure 5.3).

The graph passes through the point $(0; 1)$. The limit values of the function, as $x \to \pm\infty$, equal zero. The horizontal asymptote is $y = 0$.

Let us now do the plotting of the graph of the function $y = e^{-x^2}$ in another way (Figure 5.4).

As an example, we construct an arbitrary point M_2 of the graph of the function $y = e^{-x^2}$. Take the point $P_2(2; 0)$. Draw a straight line perpendicular to Ox until it intersects the graph of the function $y = x^2$ at the point $Q_2(2; 4)$. Draw the line segment P_2Q_2 on the axis Ox with one endpoint coinciding with the origin. The second endpoint of the segment gives the point T_2. Draw a straight line perpendicular to Ox and passing through the point T_2, and continue it until its intersection with the curve e^{-x^2}. We get the point S_2. Through the point S_2 draw a straight line parallel to Ox until it intersects the straight line P_2Q_2. We get the point M_2. Several other points of the graph of the function $y = e^{-x^2}$ are constructed similarly (Figure 5.4).

Well-classified graphs of compositions of functions and ways of their constructions can be found in [6]. We recall some facts from this book.

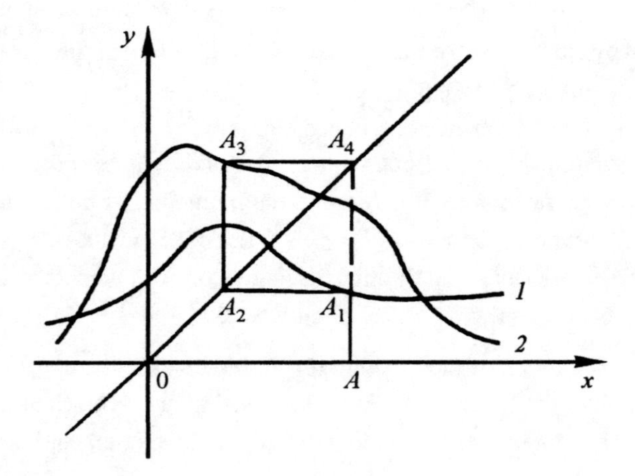

Figure 5.1. Geometric construction of composition $y = f(\varphi(x))$.

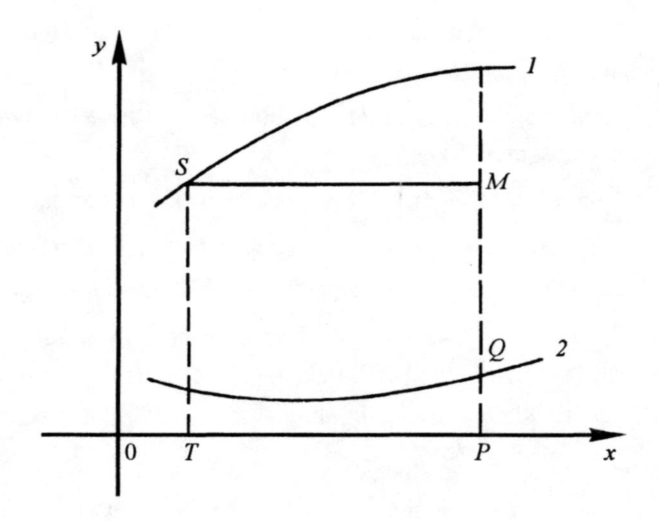

Figure 5.2. Construction of the graph of composition of functions by using a rule and compass: *1*. $y = f(x)$; *2*. $y = \varphi(x)$.

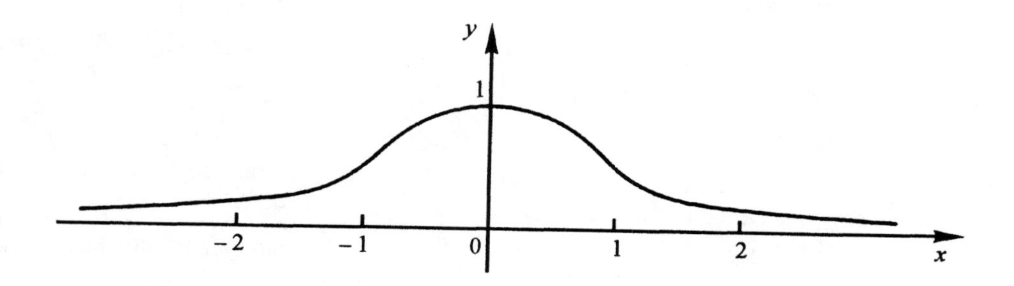

Figure 5.3. The graph of the function $y = e^{-x^2}$ (Gaussian curve).

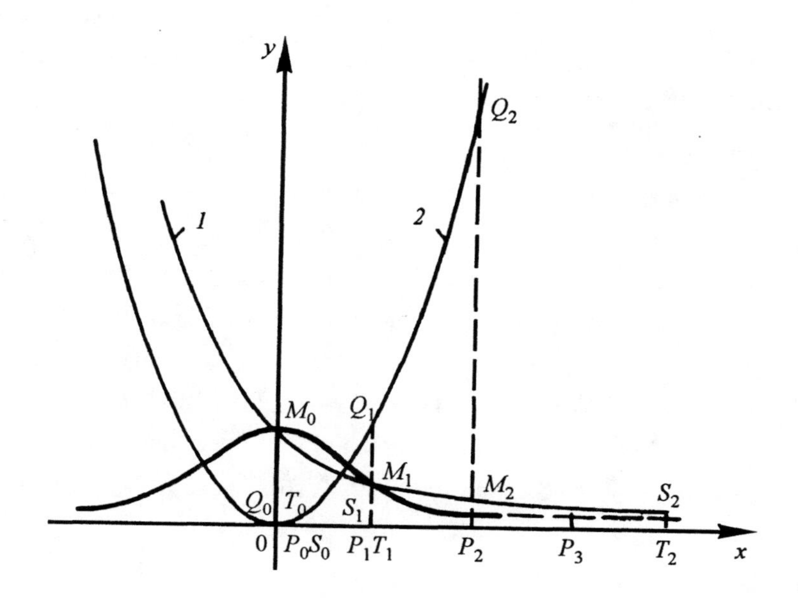

Figure 5.4. Another way of constructing the graph of the function $y = e^{-x^2}$.
1. $y = e^{-x}$. 2. $y = x^2$.

5.1.1. Plotting the graphs of functions $y = (f(x))^r$, where $r \in \mathbf{Q}$.

1. Let $r = 2k$, $k \in \mathbf{N}$. The domain of the function $y = (f(x))^{2k}$ coincides with the domain of the function $y = f(x)$.

If the function $f(x)$ is even or odd, then the function $(f(x))^{2k}$ is even.

If the function $f(x)$ is periodic, then so is the function $(f(x))^{2k}$. Its principal period could be less than the principal period of the function $f(x)$.

If $f(x) > 0$, then the intervals on which the functions $f(x)$ and $(f(x))^{2k}$ are increasing (decreasing) are the same. If $f(x) < 0$, then the intervals on which the function $f(x)$ is increasing (decreasing) are the intervals on which the function $(f(x))^{2k}$ is decreasing (increasing).

Because the graphs of both functions intersect at those x at which $f(x) = 0$ or $f(x) = 1$, we find these points on the graph of the function $(f(x))^{2k}$ by drawing the line $y = 1$ and finding its intersection with the graph of the function $f(x)$.

Because $(f(x))^{2k} \geq 0$, the graph of this function lies entirely above the x-axis. For values of x such that $f(x) < 0$ or $f(x) > 1$, the graph of the function $(f(x))^{2k}$ will be above the graph of the function $f(x)$; for values of x such that $0 < f(x) < 1$, the graph of $y = (f(x))^{2k}$ lies below the graph of the function $y = f(x)$ (Figure 5.5).

2. Let $r = 2k + 1$, where $k \in \mathbf{N}$.

The domain of the function $y = (f(x))^{2k+1}$ coincides with the domain of the function $y = f(x)$. If $f(x)$ is even (odd), then $(f(x))^{2k+1}$ is even (odd). If the function $y = f(x)$ is periodic, then the function $y = (f(x))^{2k+1}$ is also periodic with the same principal period. The intervals on which the functions $y = f(x)$ and $y = (f(x))^{2k+1}$ are increasing (decreasing) are the same.

The graphs of the functions $y = (f(x))^{2k+1}$ and $y = f(x)$ have common points with x such that $f(x) = 0$, $f(x) = \pm 1$. For x such that $-\infty < f(x) < -1$ or $0 < f(x) < 1$, the graph of the function $(f(x)^{2k+1})$ lies below the graph of the function $y = f(x)$. For x such that $-1 < f(x) < 0$ or $1 < f(x) < +\infty$, the graph of the function $y = (f(x))^{2k+1}$ lies above the graph of the function $y = f(x)$.

3. Let $r = 1/2k$, where $k \in \mathbf{N}$.

The domain of the function $y = \sqrt[2k]{f(x)}$ is found from the condition $f(x) \geq 0$.

If the function $y = f(x)$ is even, then the function $\sqrt[2k]{f(x)}$ is even. If $f(x)$ is periodic, then the function $\sqrt[2k]{f(x)}$ is also periodic. Their periods are equal. The intervals on which the functions $y = \sqrt[2k]{f(x)}$ and $y = f(x)$ are increasing (decreasing) are the same. For x such that $f(x) = 0$ or $f(x) = 1$, the graphs of the functions coincide. For x such that $0 < f(x) < 1$, the graph of $y = \sqrt[2k]{f(x)}$ lies above the graph of the function $y = f(x)$. For x such that $f(x) > 1$, the graph of the function $y = \sqrt[2k]{f(x)}$ lies below the graph of the function $y = f(x)$ (Figure 5.6).

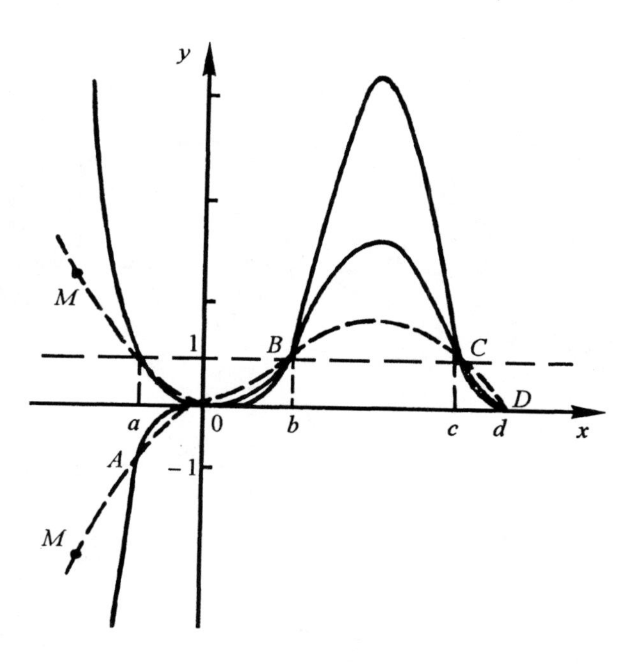

Figure 5.5. The graph of the function $y = (f(x))^{2k}$.

4. Let $r = \frac{1}{2k+1}$, $k \in \mathbf{N}$.

The domains of the functions $y = \sqrt[2k+1]{f(x)}$ and $y = f(x)$ are the same. If the function $y = f(x)$ is even (odd), then so is the function $y = \sqrt[2k+1]{f(x)}$. If the function $y = f(x)$ is periodic, then $y = \sqrt[2k+1]{f(x)}$ is periodic. The intervals on which the functions are increasing (decreasing) coincide.

The graphs of the functions coincide for x such that $f(x) = 0$ or $f(x) = \pm 1$. For x such that $f(x) < -1$ or $0 < f(x) < 1$, the graph of the function $y = \sqrt[2k+1]{f(x)}$ lies above the graph of the function $y = f(x)$. For x such that $-1 < f(x) < 0$ or $f(x) > 1$, the graph of $y = \sqrt[2k+1]{f(x)}$ lies above the graph of the function $y = f(x)$.

Note that: (a) for $r = \frac{2n}{2k+1>1}$, $r = \frac{2n+1}{2k+1}$, where $n,k \in \mathbf{N}$, the method used for plotting the graph of the function is the same as for plotting the function $y = (f(x))^{2k}$; (b) for $0 < r = \frac{2k-1}{2n} < 1$, $0 < r = \frac{2k}{2n+1} < 1$, where $k,n \in \mathbf{N}$, the construction is carried out similarly to the construction of the function $y = \sqrt[2k]{f(x)}$.

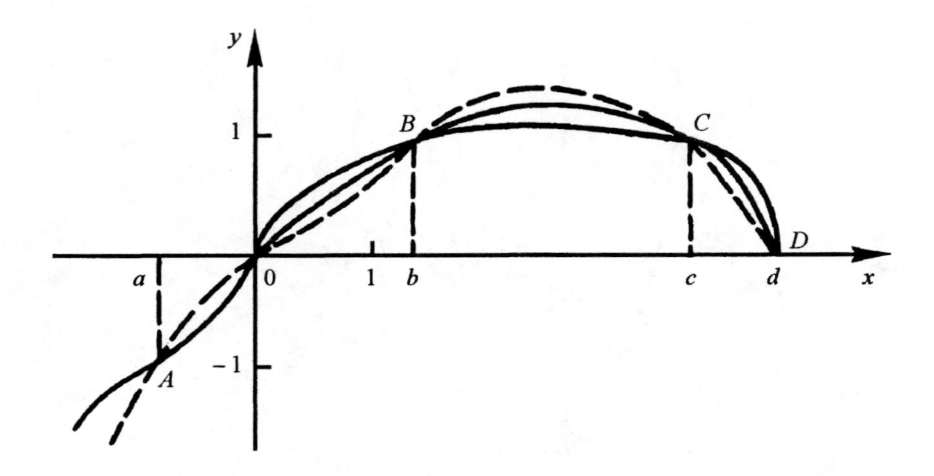

Figure 5.6. The graph of the function $y = \sqrt[2k]{f(x)}$.

5.1.2. Plotting the graph of the function $y = 1/f(x)$ from the graph of $y = f(x)$.

The analysis of the function $y = 1/f(x)$ follows the procedure described in Section 2.2 and uses properties of the function.

The function $y = 1/f(x)$ is defined for x such that $f(x) \neq 0$. Zeros of the function $y = f(x)$ are points of discontinuity of the function $y = 1/f(x)$. Denote these points by $x_1, x_2, \ldots, x_n$. The straight lines $x = x_1$, $x = x_2$, ... , $x = x_n$ are vertical asymptotes of the graph of the function $y = 1/f(x)$.

The range of the function $y = 1/f(x)$ is determined from the range of the function $y = f(x)$. If the function $y = f(x)$ is even (odd), then the function $y = 1/f(x)$ is even (odd). If the function $y = f(x)$ is periodic with the period ω, then the function $y = 1/f(x)$ is also periodic and its period is ω. On intervals where the function $y = f(x)$ is increasing (decreasing), the function $y = 1/f(x)$ is decreasing (increasing). If the function $y = f(x)$ is convex on some interval, then the function $y = 1/f(x)$ is concave on this interval.

Oblique asymptotes of the graph of the function $y = 1/f(x)$ are found from the formulas in Section 2.2.

Note that if $0 < f(x) < 1$, then $1 < 1/f(x) < +\infty$; if $-\infty < f(x) < -1$, then $-1 < 1/f(x) < 0$ and the graph of the function $1/f(x)$ lies above the graph of the function $f(x)$; if $1 < f(x) < +\infty$, then $) < 1/f(x) < 1$. If $-1 < f(x) < 0$, then $-\infty < 1/f(x) < -1$. The graphs of the functions $f(x)$ and $1/f(x)$ have common points with x such that $f(x) = \pm 1$ (Figure 5.7).

To plot the graph of the function $y = 1/(f(x))^r$, first the graph of the function $y = (f(x))^r$ is plotted, and then the described scheme is applied to construct the needed graph.

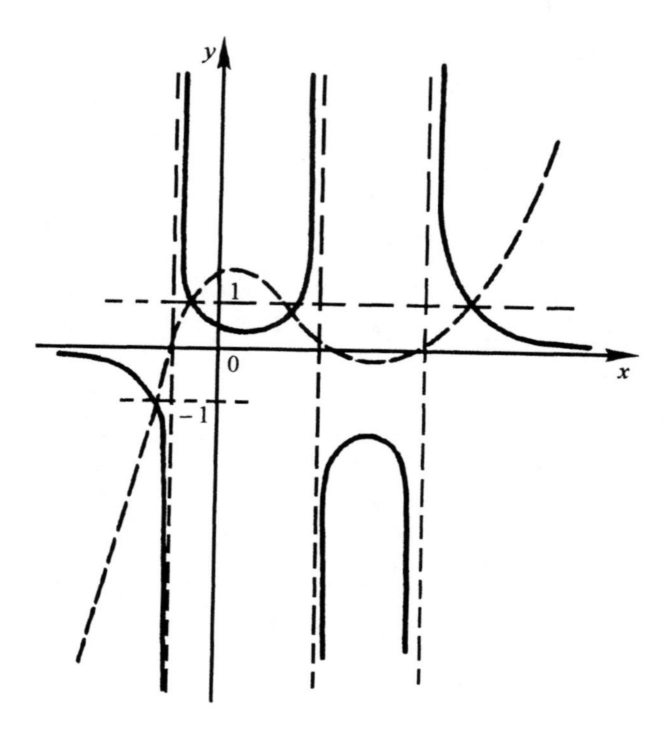

Figure 5.7. The graph of the function $y = \dfrac{1}{f(x)}$.

Remark 5.1.2. To construct the graph of the function $y = 1/f(x)$ from the graph of $y = f(x)$, one can use the transformation of inversion with respect to the axis Ox applied to the graph of the function $f(x)$.

The graphs of functions of type $y = (f(x))^r$ are given in Figures 5.8 – 5.14.

5.1.3. Plotting the graphs of the function $y = a^{f(x)}$, where $a|a-1| > 0$

The domain of the function $y = a^{f(x)}$ coincides with the domain of the function $y = f(x)$. The graph lies in the upper half of the plane, since $a^{f(x)} > 0$.

If the function $f(x)$ is even, then the function $a^{f(x)}$ is even.

If the function $f(x)$ is periodic, then the function $a^{f(x)}$ is also periodic and has the same principal period as the function $f(x)$.

If the function $y = f(x)$ has a horizontal asymptote $y = b$, then $y = a^b$ is a horizontal asymtote of the function $y = a^{f(x)}$.

If $\lim_{x \to d} f(x) = +\infty$ and $a > 1$, then $\lim_{x \to d} a^{f(x)} = +\infty$ and the straight line $x = d$ is a vertical asymptote. The straight line $x = d$ is an asymptote if $\lim_{x \to d} f(x) = -\infty$ and $0 < a < 1$.

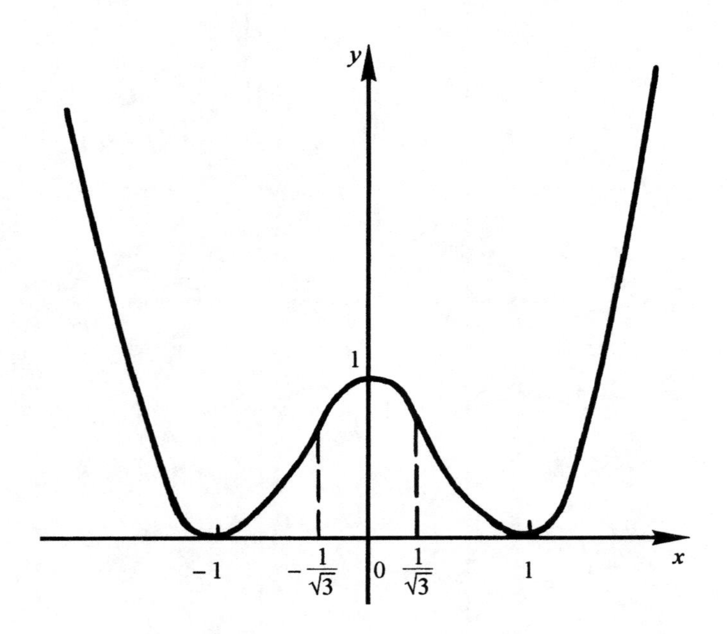

Figure 5.8. The graph of the function $y = (x^2 - 1)^2$.

If $a > 1$, then the intervals on which the functions $a^{f(x)}$ and $f(x)$ are increasing (decreasing) are the same; if $0 < a < 1$, then the function $a^{f(x)}$ is increasing on the interval on which $f(x)$ is decreasing, and vice versa.

If $f(x) = 0$, then $y = a^{f(x)} = 1$.

If $a > 1$, then the points at which the function $y = f(x)$ reaches maximum (minimum) are points of maximum (minimum) of the function $y = a^{f(x)}$. If $0 < a < 1$, then points of maximum (minimum) of the function $f(x)$ are points of minimum (maximum) of the function $a^{f(x)}$.

The graphs of functions $y = a^{f(x)}$ are shown in Figures 5.15 – 5.21.

5.1.4. Plotting the graphs of the function $y = \log_a f(x)$, where $a|a - 1| > 0$

The domain of the function $y = \log_a f(x)$ is found from the condition $f(x) > 0$.

Zeros of the function are found by setting $f(x) = 1$. For those x such that $f(x) > 1$, the graph of the function lies in the upper (lower) half of the plane for $a > 1$ $(0 < a < 1)$.

The function $y = \log_a f(x)$ is even, if the function $y = f(x)$ is such.

If the function $y = f(x)$ is periodic, then the function $y = \log_a f(x)$ is also periodic with the same principal period.

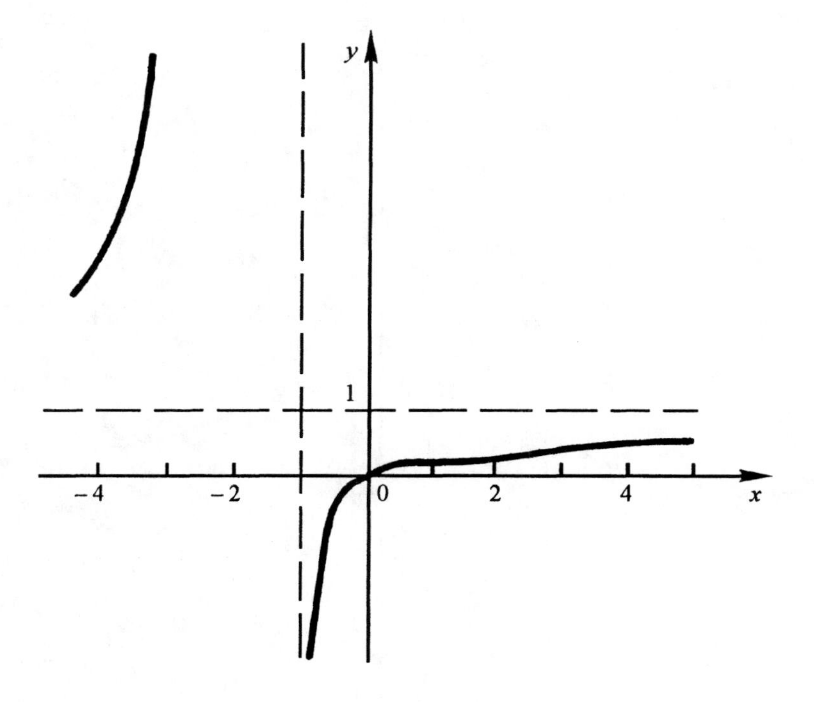

Figure 5.9. The graph of the function $y = \left(\frac{x}{x+1}\right)^5$.

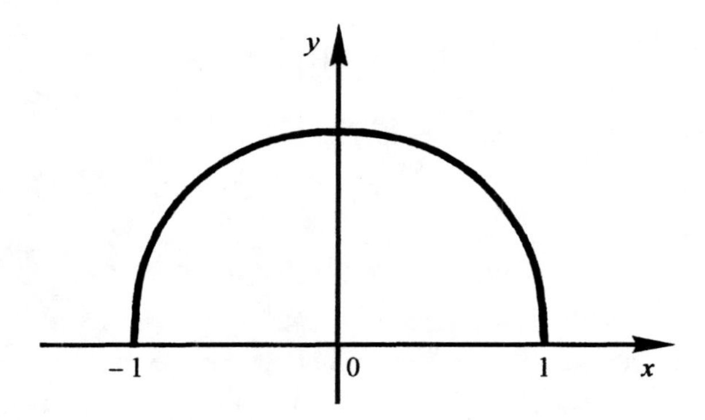

Figure 5.10. The graph of the function $y = \sqrt[3]{1 - x^2}$.

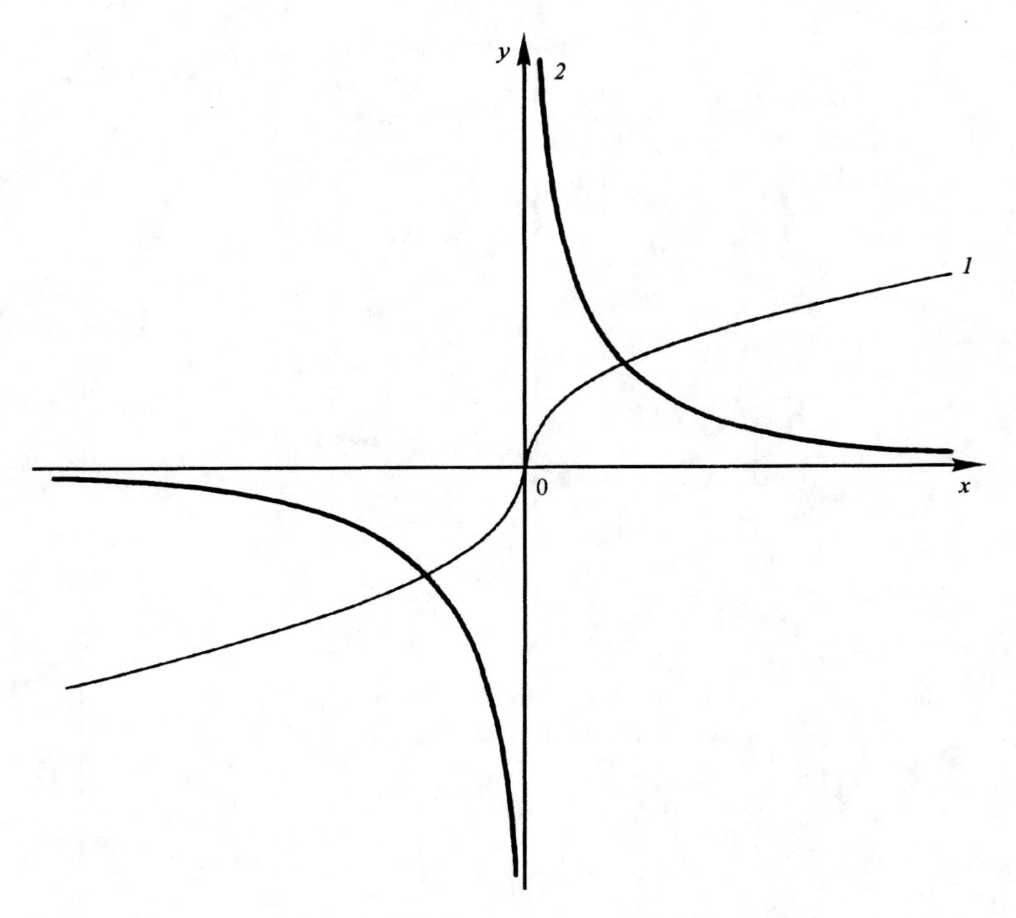

Figure 5.11. The graphs of functions *1.* $y = \sqrt[3]{x}$, *2.* $y = 1/\sqrt[3]{x}$.

If $a > 1$ and $f(x) > 0$, the intervals on which the functions $y = \log_a f(x)$ and $y = f(x)$ are increasing (decreasing) are the same; if $0 < a < 1$, they are opposite.

If the function $y = f(x)$ reaches maximum (minimum) at a point $M(x_0; n)$, where $n > 0$, then, at this point, the function $y = \log_a f(x)$ will reach maximum (minimum) for $a > 1$ and minimum (maximum) for $0 < a < 1$.

If $a > 1$ and $\lim f(x) = +\infty$, then $\lim \log_a f(x) = +\infty$;

if $\lim f(x) = 0$, then $\lim \log_a f(x) = -\infty$.

In the case where $0 < a < 1$, $\lim f(x) = +\infty$ implies that $\lim \log_a f(x) = -\infty$,

and $\lim f(x) = 0$ implies that $\lim \log_a f(x) = +\infty$.

The graphs of functions $y = \log_a f(x)$ are shown in Figures 5.22 – 5.27.

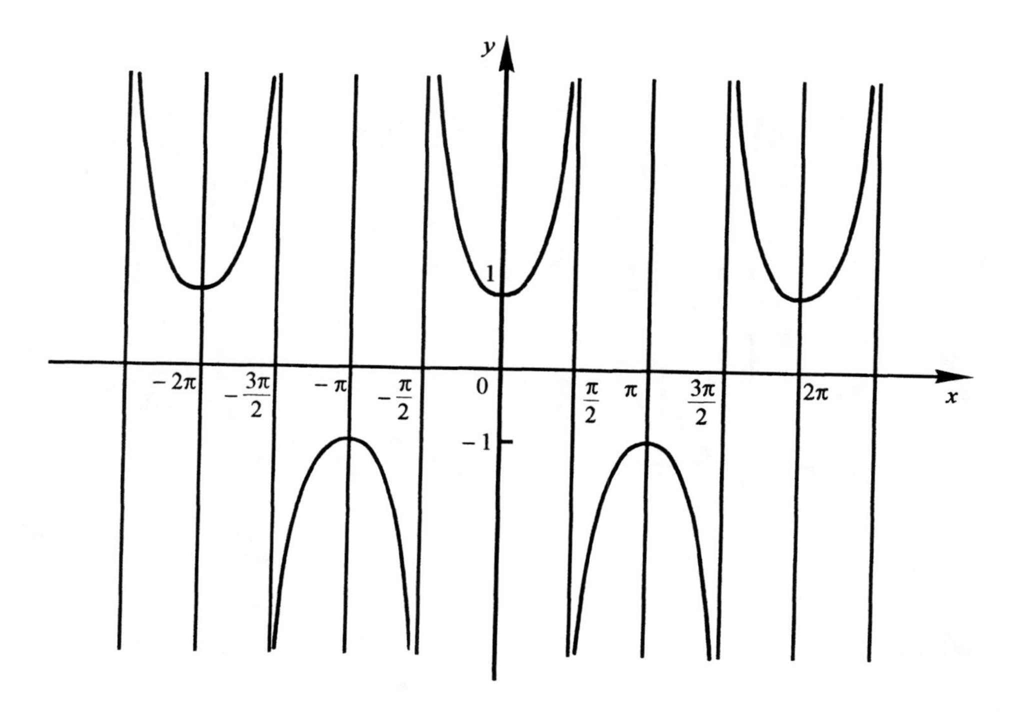

Figure 5.12. The graph of function $y = \frac{1}{\cos x}$.

5.1.5. Plotting the graphs of the functions $y = \sin f(x)$, $y = \cos f(x)$

The domains of the functions $y = \sin f(x)$, $y = \cos f(x)$ coincide with the domains of the function $y = f(x)$.

Because $|\sin x| \le 1$, $|\cos x| \le 1$, the graphs of the functions lie inside the strip formed by the straight lines $y = -1$, $y = 1$. The functions do not have vertical or oblique asymptotes. If the line $y = d$ is a horizontal asymptote of the function $y = f(x)$, then the line $y = \sin d$ (correspondingly, $y = \cos d$) is an asymptote of the function under consideration.

If the function $y = f(x)$ is periodic, then the functions $y = \sin f(x)$, $y = \cos f(x)$ are also periodic.

The intervals on which the function $y = \sin f(x)$ is monotonic increasing (decreasing) are found from the inequalities:

$$-\frac{\pi}{2} + 2k\pi < f(x) < \frac{\pi}{2} + 2k\pi \quad \left(\frac{\pi}{2} + 2k\pi < f(x) < \frac{3}{2}\pi + 2k\pi\right), \quad k \in \mathbf{Z},$$

for the function $y = \cos f(x)$, the respective conditions are:

$$\pi + 2k\pi < f(x) < 2\pi + 2k\pi, \quad (2k\pi < f(x) < \pi + 2k\pi), \quad k \in \mathbf{Z}.$$

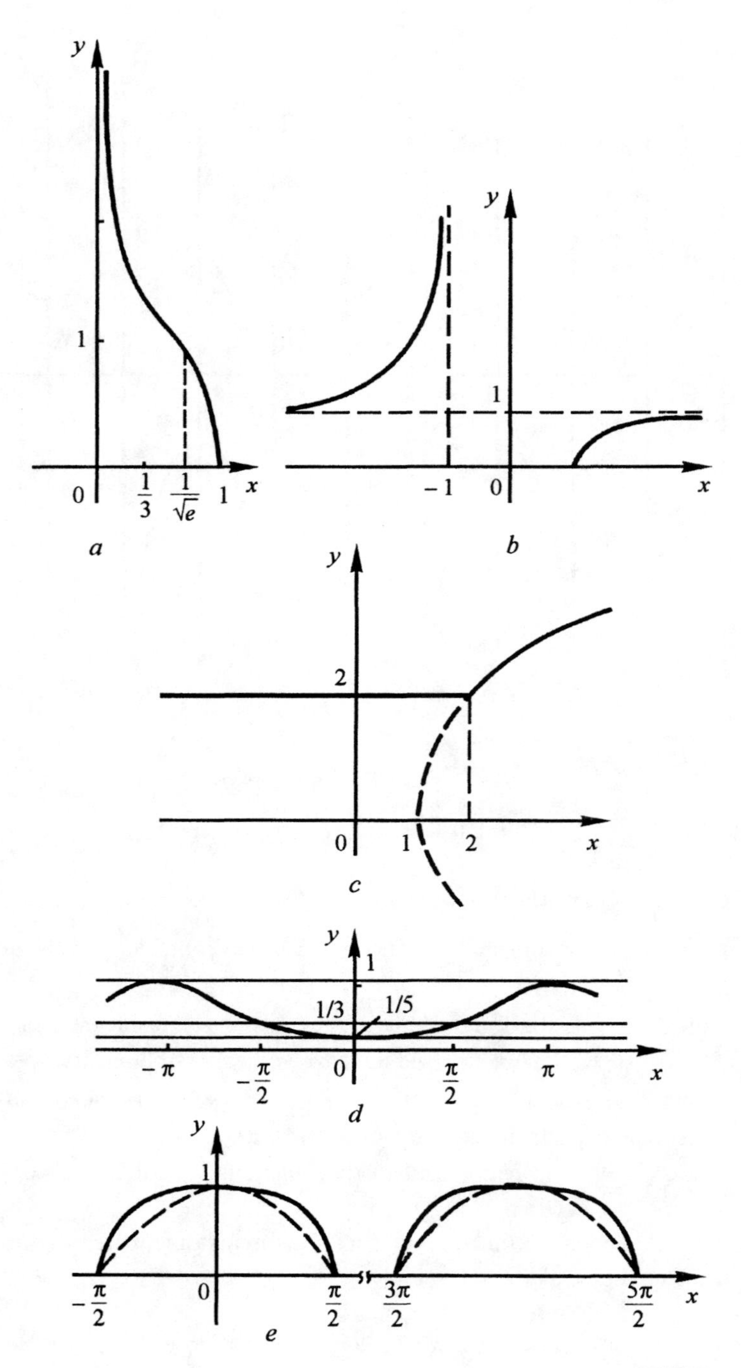

Figure 5.13. The graphs of functions: (*a*) $y = \sqrt{\log_{1/3} x}$; (*b*) $y = \sqrt{\frac{x-1}{x+1}}$; (*c*) $y = \sqrt{2(x + |x-2|)}$; (*d*) $y = \frac{1}{3 + 2\cos x}$; (*e*) $y = \sqrt{\cos x}$.

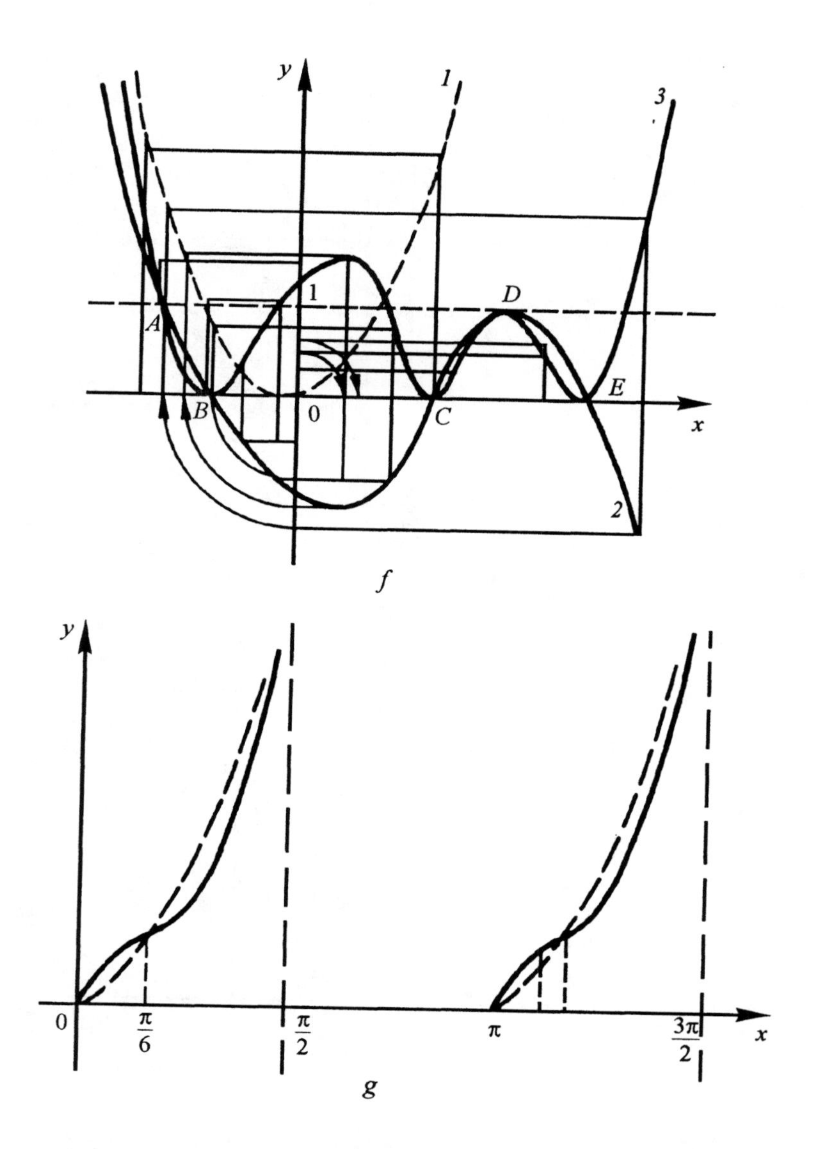

Figure 5.13. The graphs of functions: (*f*) *1.* $y = x^2$, *2.* $y = f(x)$, *3.* $y = (f(x))^2$; (*g*) $y = \sqrt{\tan x}$.

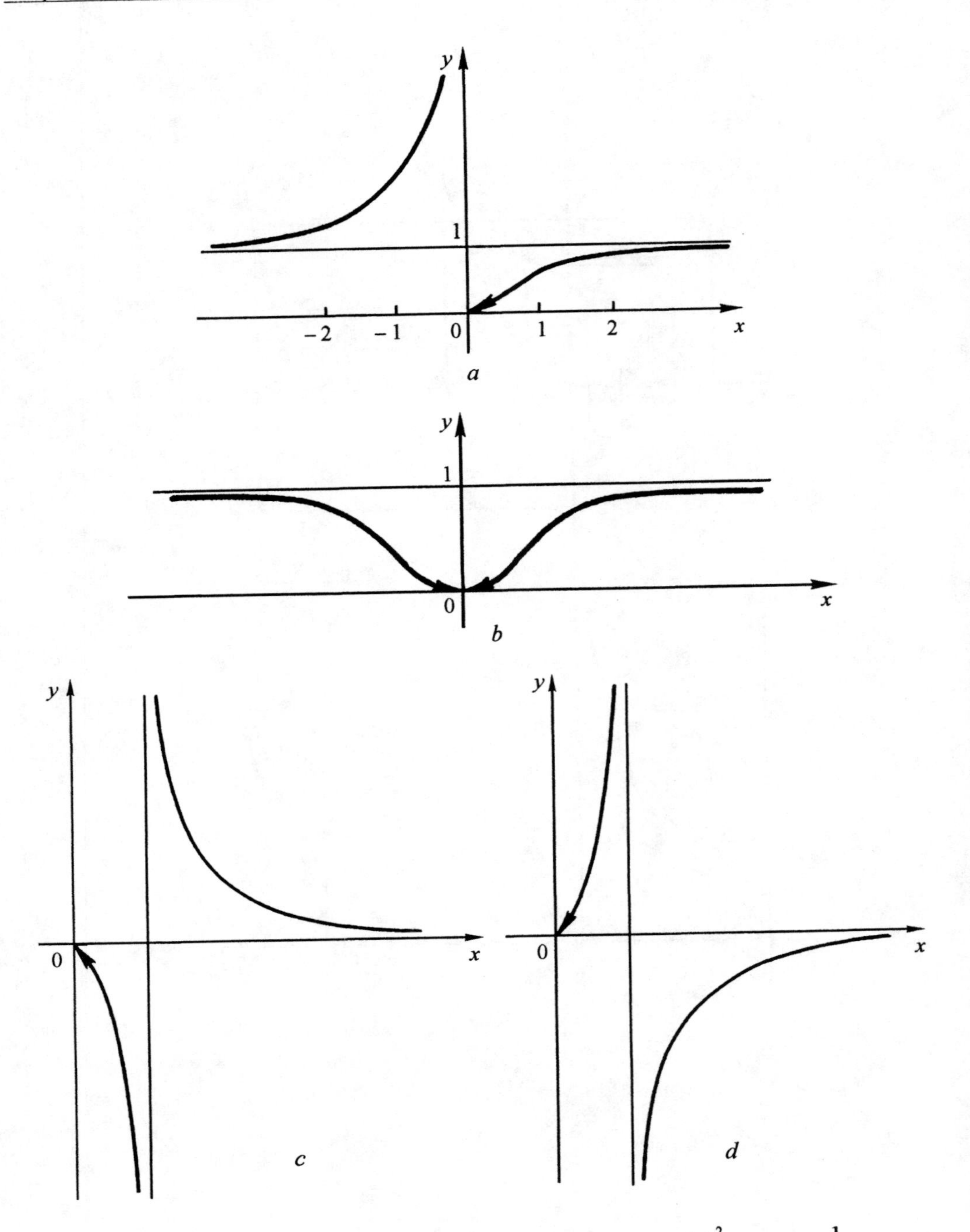

Figure 5.14. The graphs of functions: (*a*) $y = e^{-1/x}$; (*b*) $y = e^{-1/x^2}$; (*c*) $y = \dfrac{1}{\log_a x}$, $a > 1$; (*d*) $y = \dfrac{1}{\log_a x}$, $a < 1$.

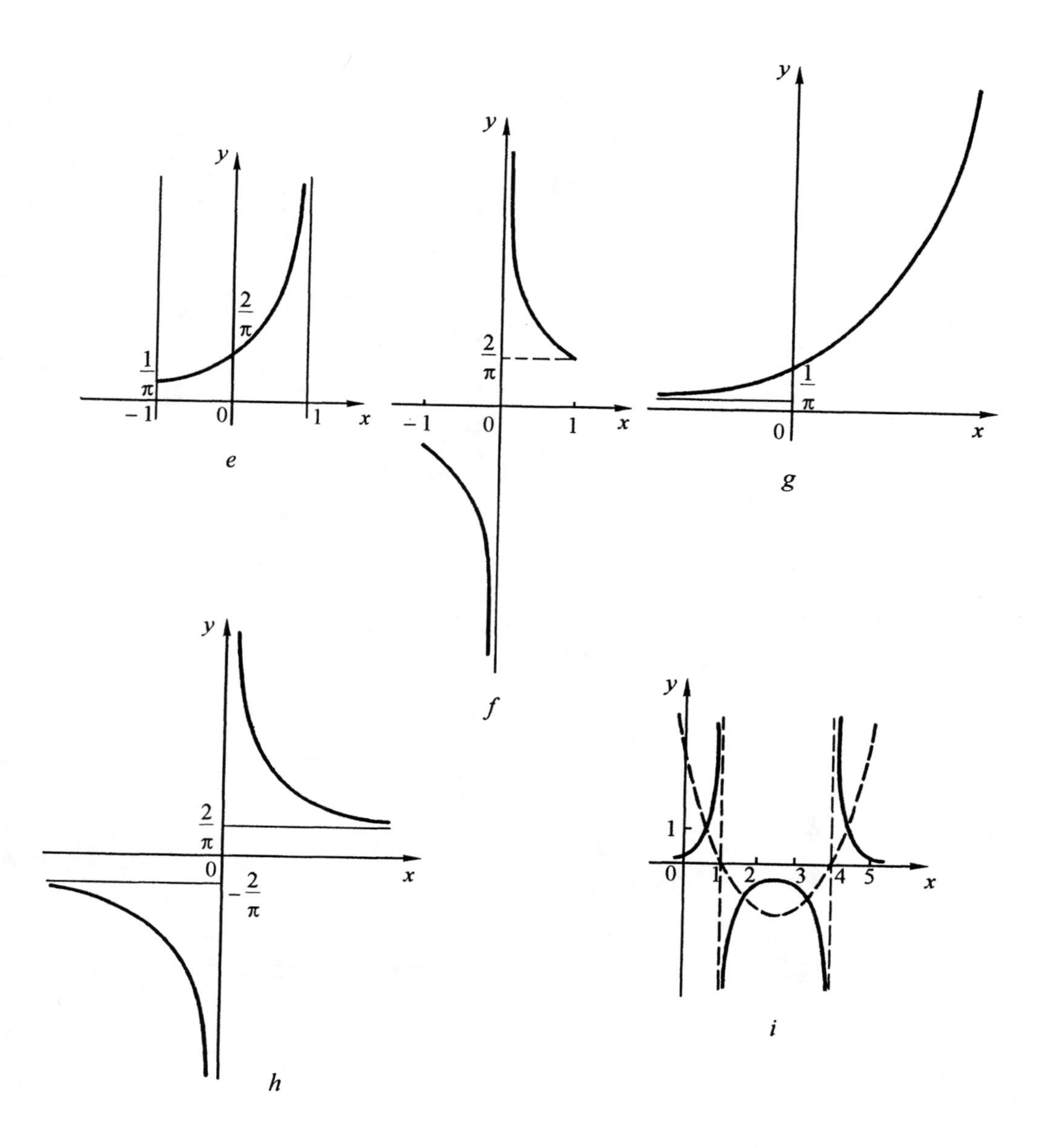

Figure 5.14. The graphs of functions: (*e*) $y = \dfrac{1}{\arccos x}$; (*f*) $y = \dfrac{1}{\arcsin x}$; (*g*) $y = \dfrac{1}{\operatorname{arccot} x}$; (*h*) $y = \dfrac{1}{\arctan x}$; (*i*) $y = \dfrac{1}{x^2 - 5x + 4}$.

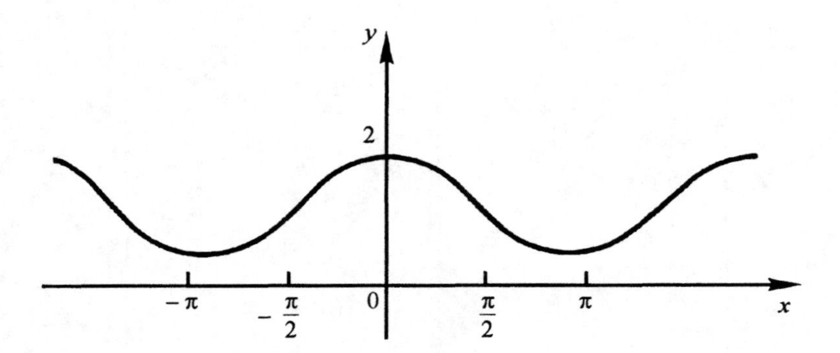

Figure 5.15. The graph of the function $y = 2^{\cos x}$.

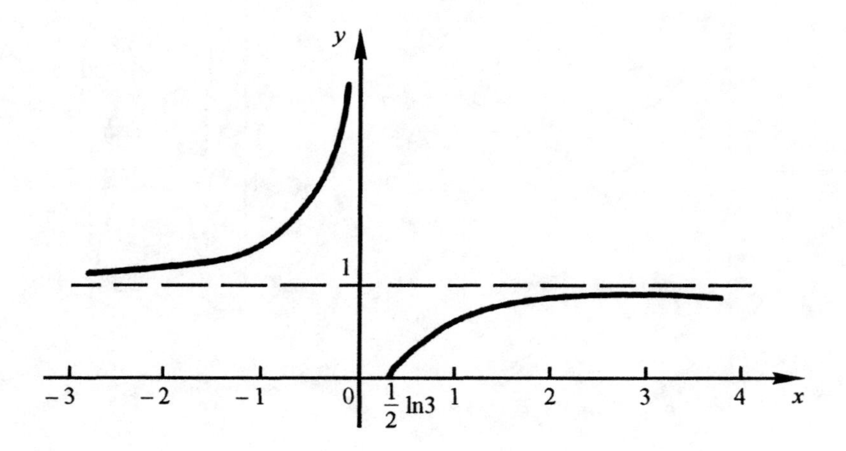

Figure 5.16. The graph of the function $y = (1/3)^{1/x}$.

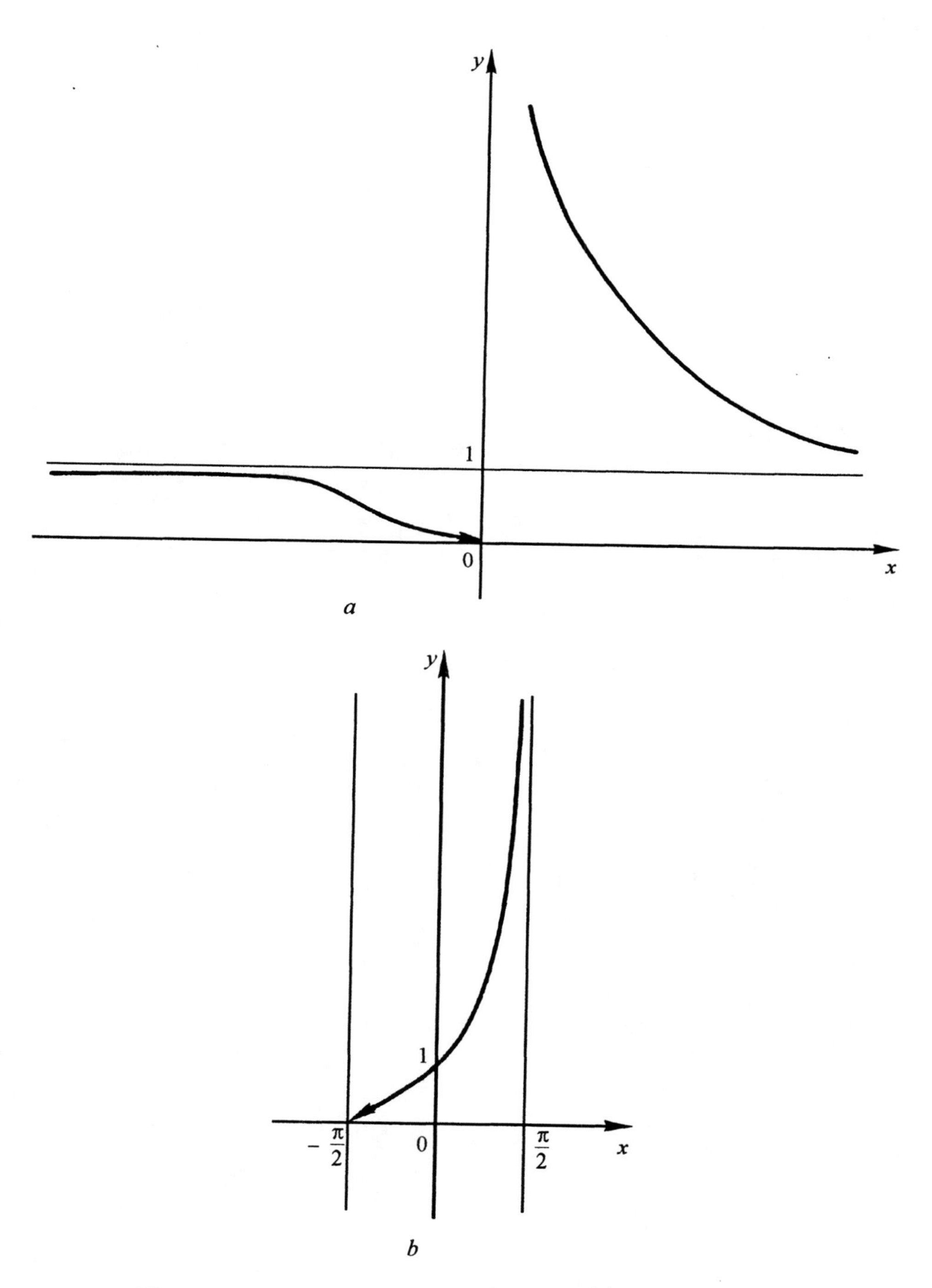

Figure 5.17. The graphs of functions: (*a*) $y = e^{1/x}$; (*b*) $y = 2^{\tan x}$.

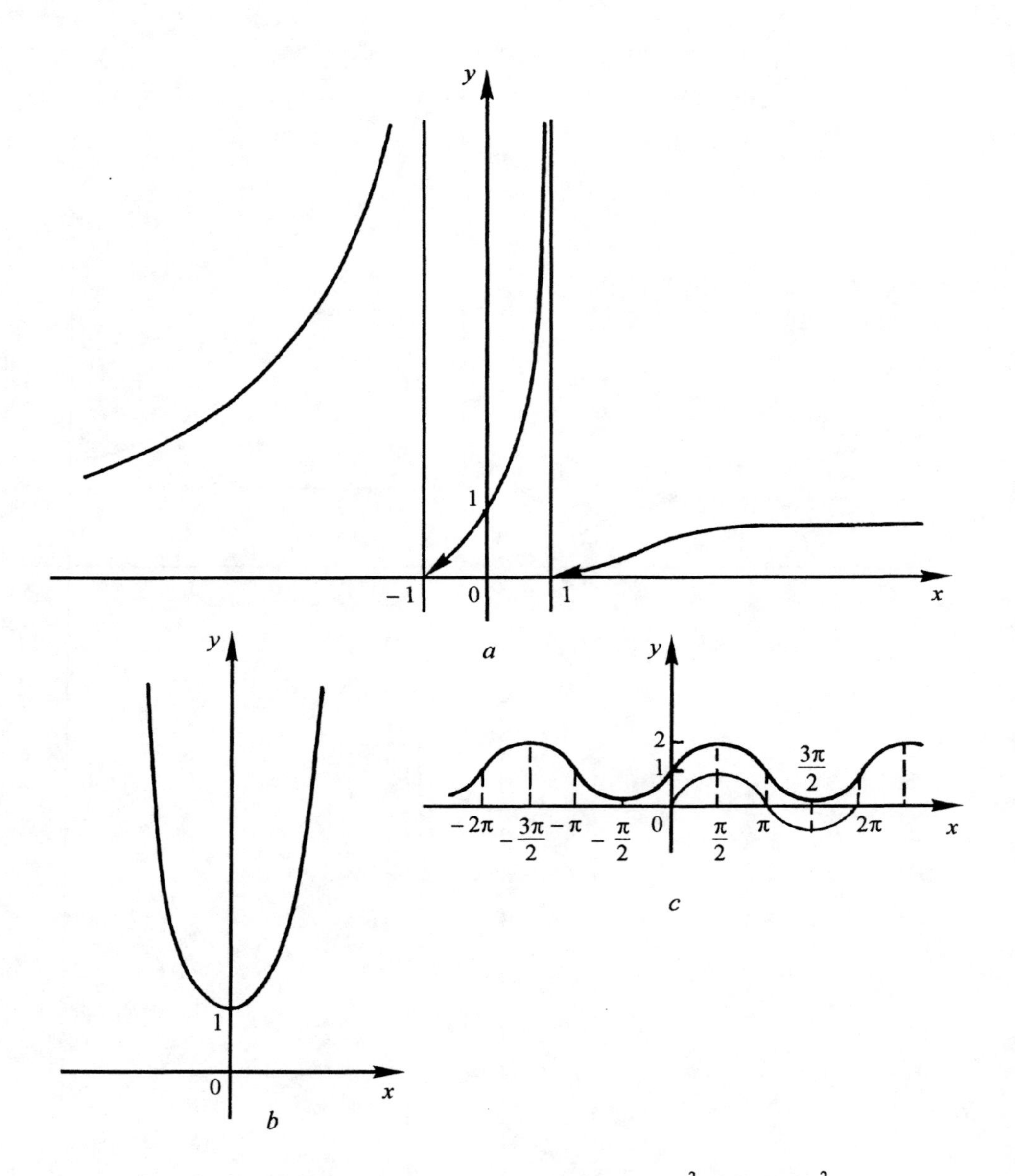

Figure 5.18. The graphs of functions: (*a*) $y = e^{2x/(1-x^2)}$; (*b*) $y = e^{x^2}$;
(*c*) $y = 2^{\sin x}$.

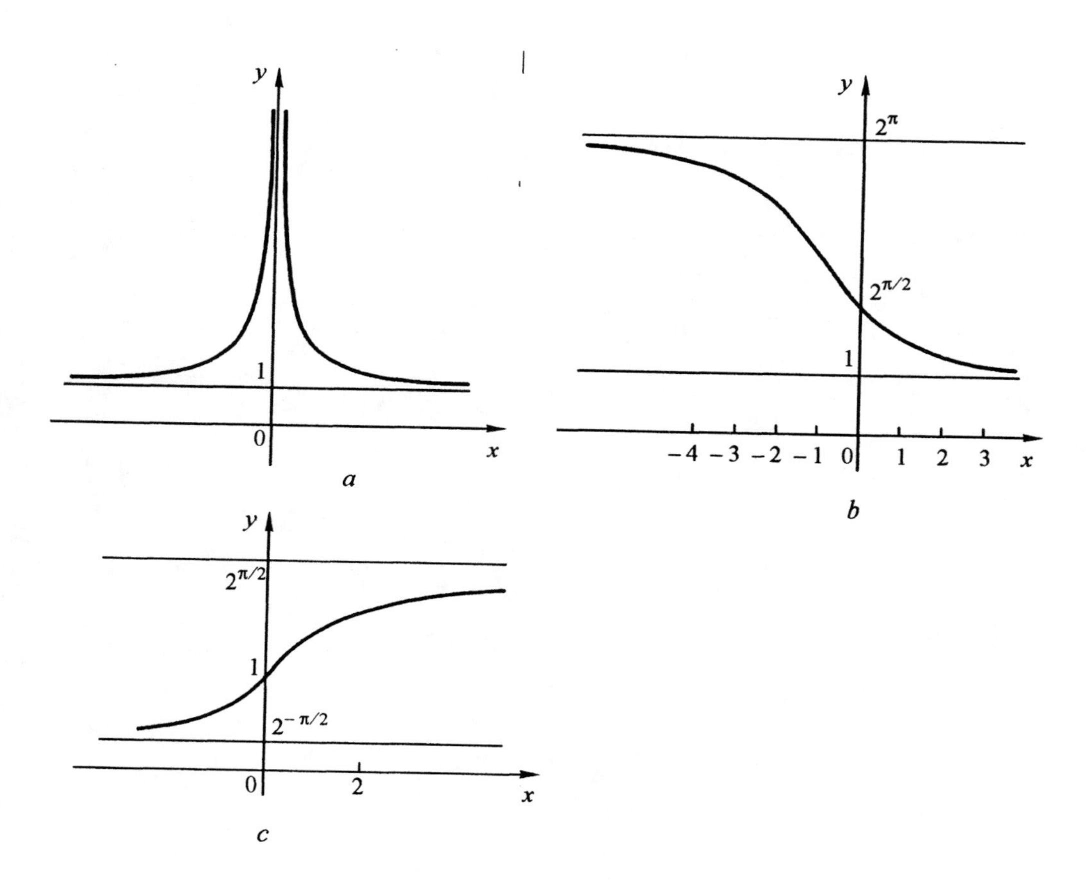

Figure 5.19. The graphs of functions: *(a)* $y = e^{1/x^2}$; *(b)* $y = 2^{\operatorname{arccot} x}$; *(c)* $y = 2^{\arctan x}$.

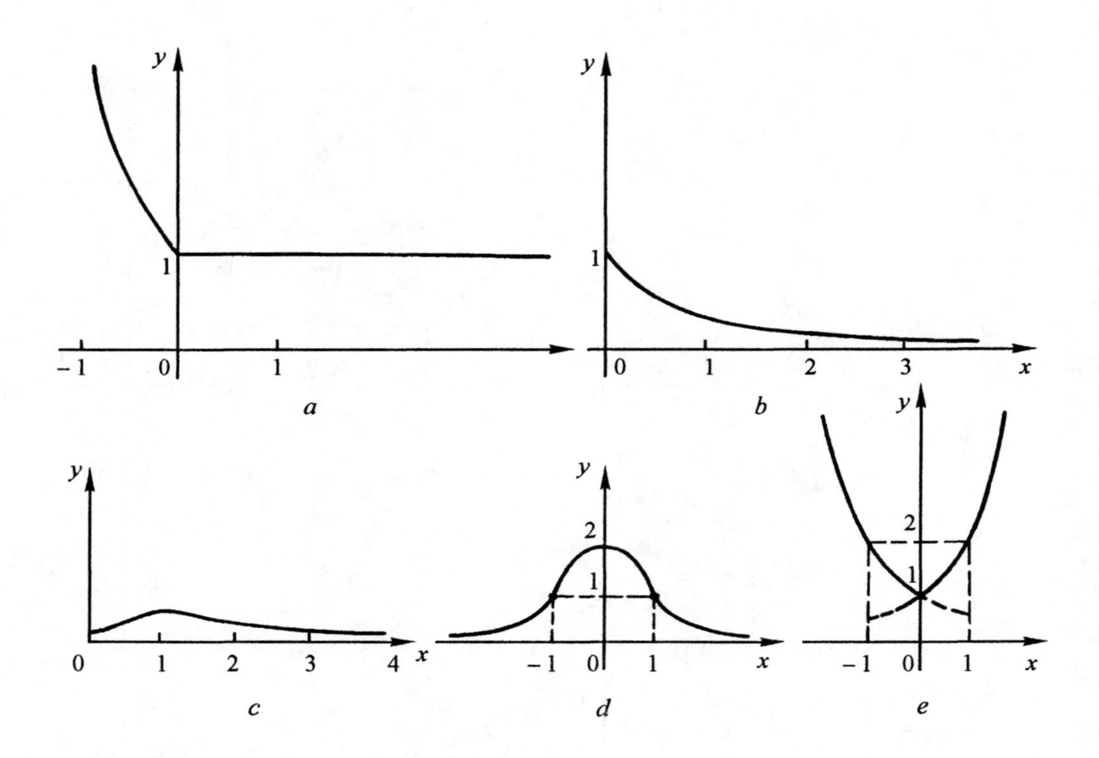

Figure 5.20. The graphs of functions: (*a*) $y = 2^{|x| - x}$; (*b*) $y = (1/3)^{\sqrt{x^3}}$; (*c*) $y = \frac{e^{-(x-1)^2}}{\sqrt{2\pi}}$; (*d*) $y = 2^{1 - x^2}$; (*e*) $y = 2^{x^2/|x|}$.

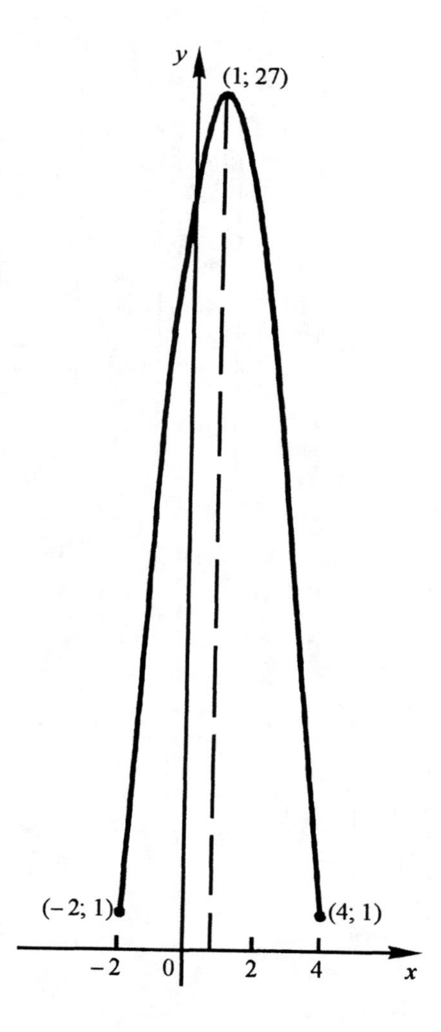

Figure 5.21. The graph of the function $y = 3^{\sqrt{-x^2+2x+8}}$.

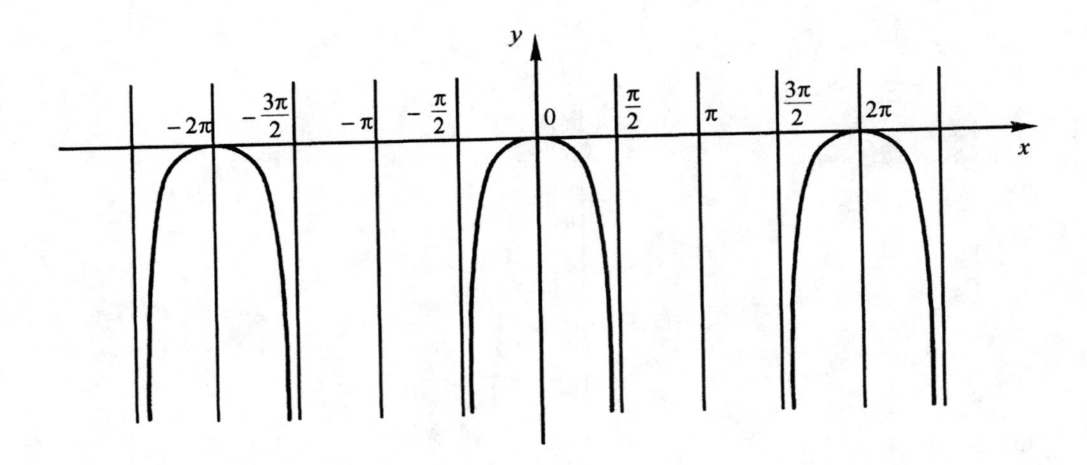

Figure 5.22. The graph of the function $y = \log_{10} \cos x$.

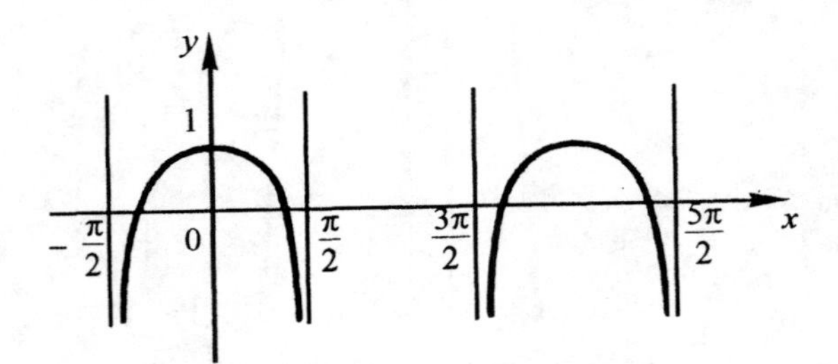

Figure 5.23. The graphs of function $y = \ln \cos x + \cos x$.

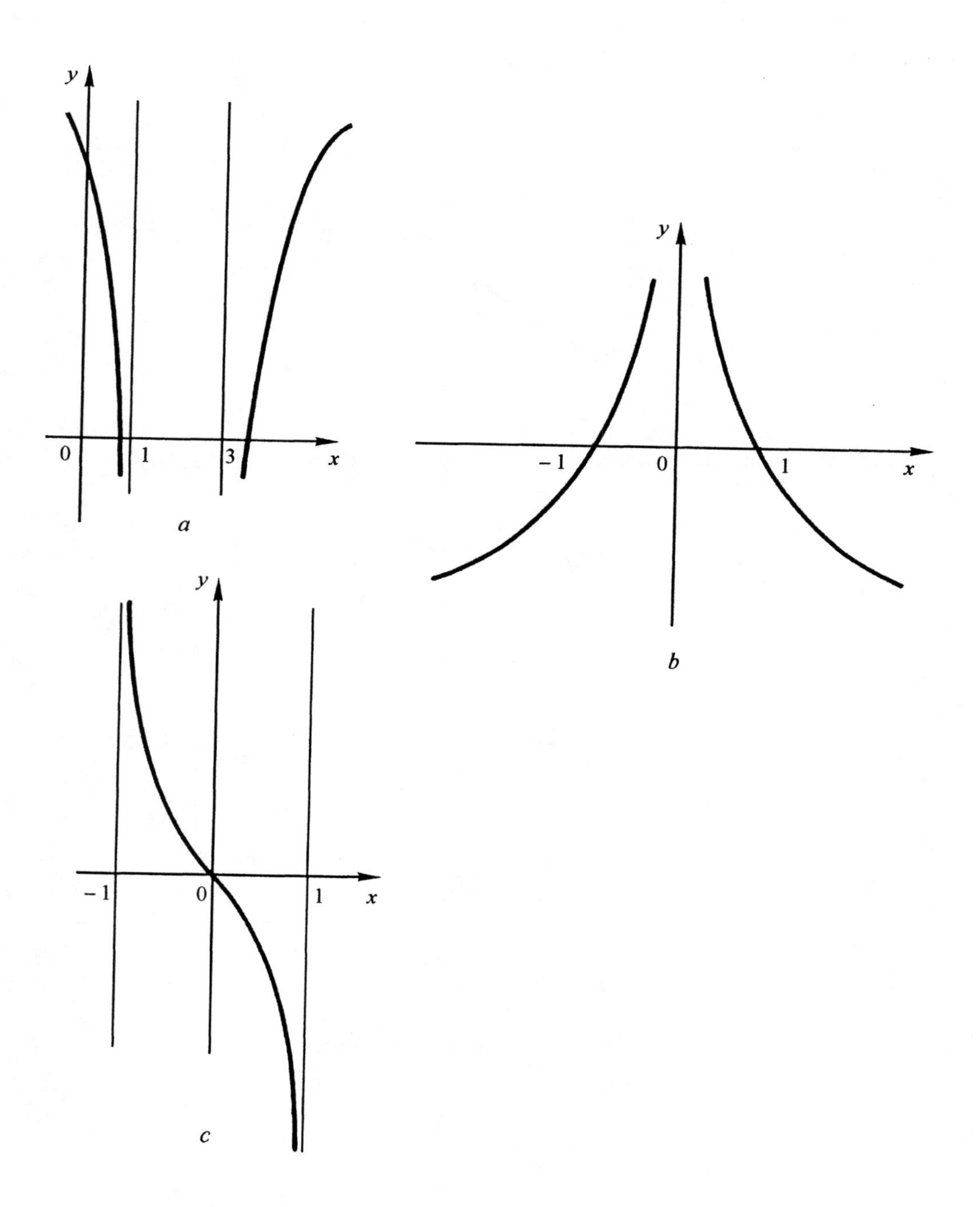

Figure 5.24. The graphs of functions: (*a*) $y = \ln((x-1)(x-2)^2(x-3)^3)$; (*b*) $y = \ln 1/x^2$; (*c*) $y = \ln \frac{1-x}{1+x}$.

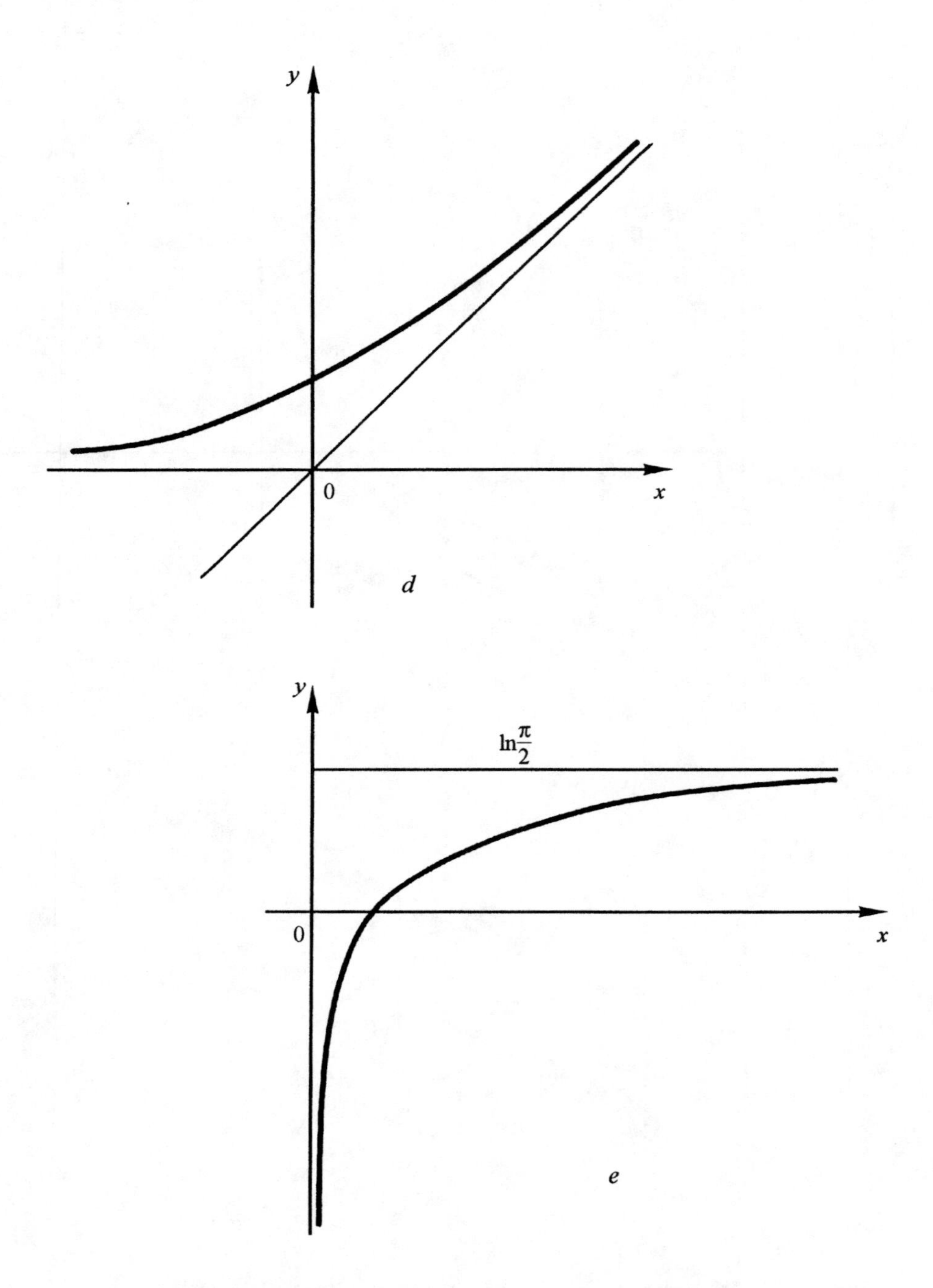

Figure 5.24. The graphs of functions: (*d*) $y = \ln(1 + e^x)$; (*e*) $y = \ln(\arctan x)$.

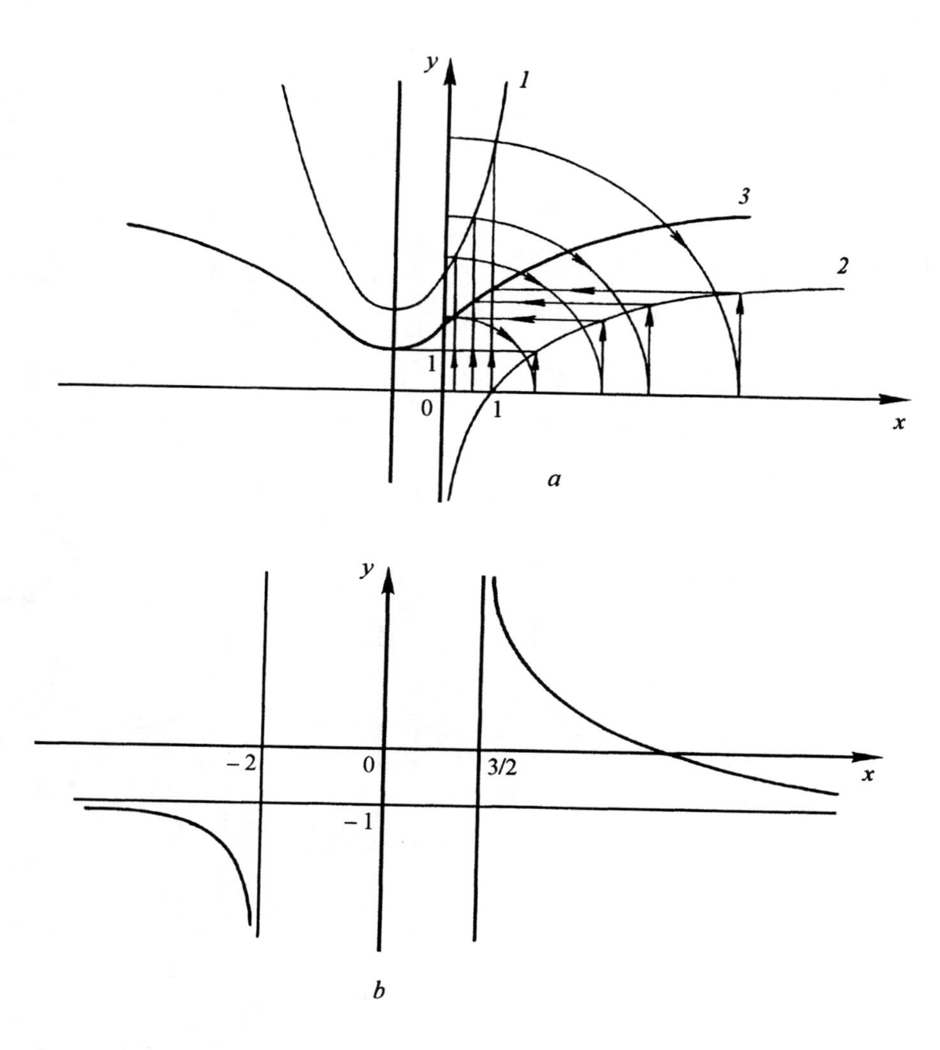

Figure 5.25. The graphs of functions: (*a*) *1.* $y = x^2 + 2x + 3$, *2.* $y = \log_2 x$, *3.* $y = \log_2 (x^2 + 2x + 3)$; (*b*) $y = \log_2 \dfrac{2 + x}{2x - 3}$.

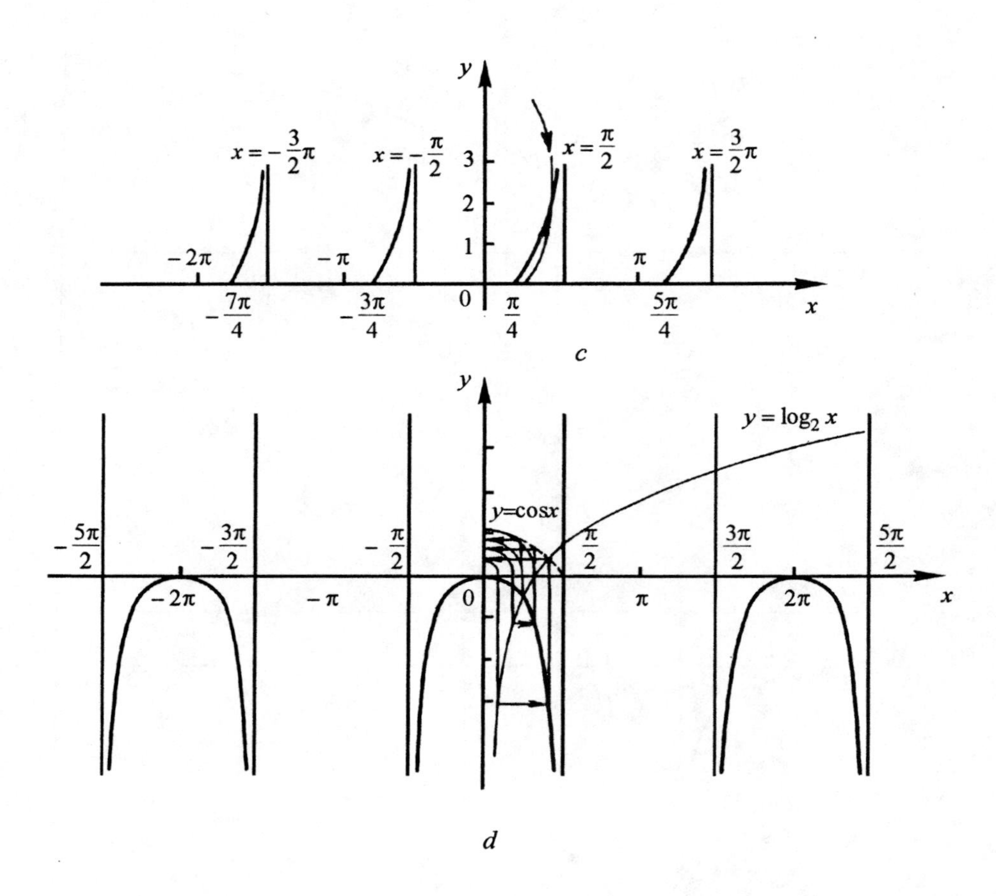

Figure 5.25. The graphs of functions: (*c*) $y=\sqrt{\log_{10}\tan x}$; (*d*) $y=\log_2\cos x$.

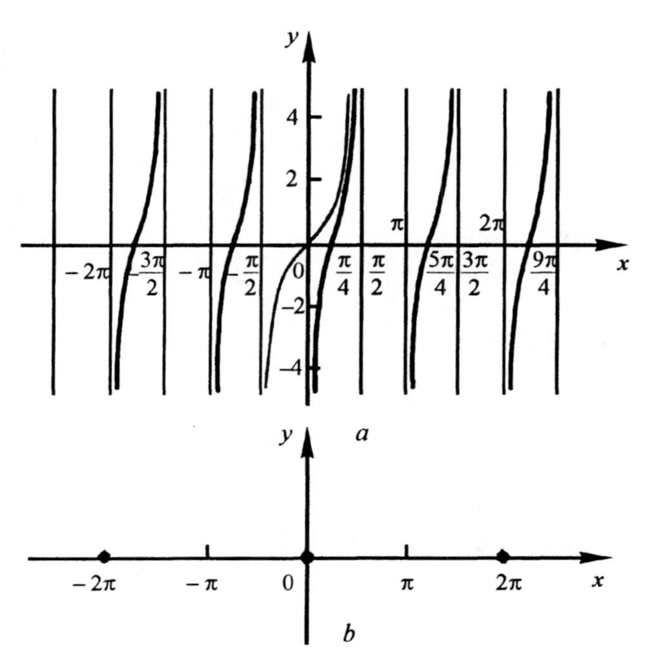

Figure 5.26. The graphs of functions: (*a*) $y = \log \tan x$; (*b*) $y = \sqrt{\log \cos x}$.

The graphs of the functions $y = \sin f(x), y = \cos f(x)$ are shown in Figures 5.28 – 5.32.

5.1.6. Plotting the graphs of functions $y = \tan f(x), y = \cot f(x)$

The domain of the function $y = \tan f(x)$ can be found from the condition $f(x) \neq (2k + 1)\frac{\pi}{2}$, the domain of the function $y = \cot f(x)$ — from the condition $f(x) \neq k\pi$, where $k \in \mathbf{Z}$.

The function is even (odd) if the function $y = f(x)$ is even (odd).

On intervals where the function $y = f(x)$ is increasing (decreasing), the function $y = \tan f(x)$ is increasing (decreasing), and the function $y = \cot f(x)$ is decreasing (increasing).

Vertical asymptotes for the function $y = \tan f(x)$ are found from the equality $f(x) = (2k + 1)\frac{\pi}{2}$; for the function $y = \cot x$, — from the equality $f(x) = k\pi$, where $k \in \mathbf{Z}$.

Graphs of functions $y = \tan f(x)$ and $y = \cot f(x)$ are shown in Figures 5.33 – 5.35.

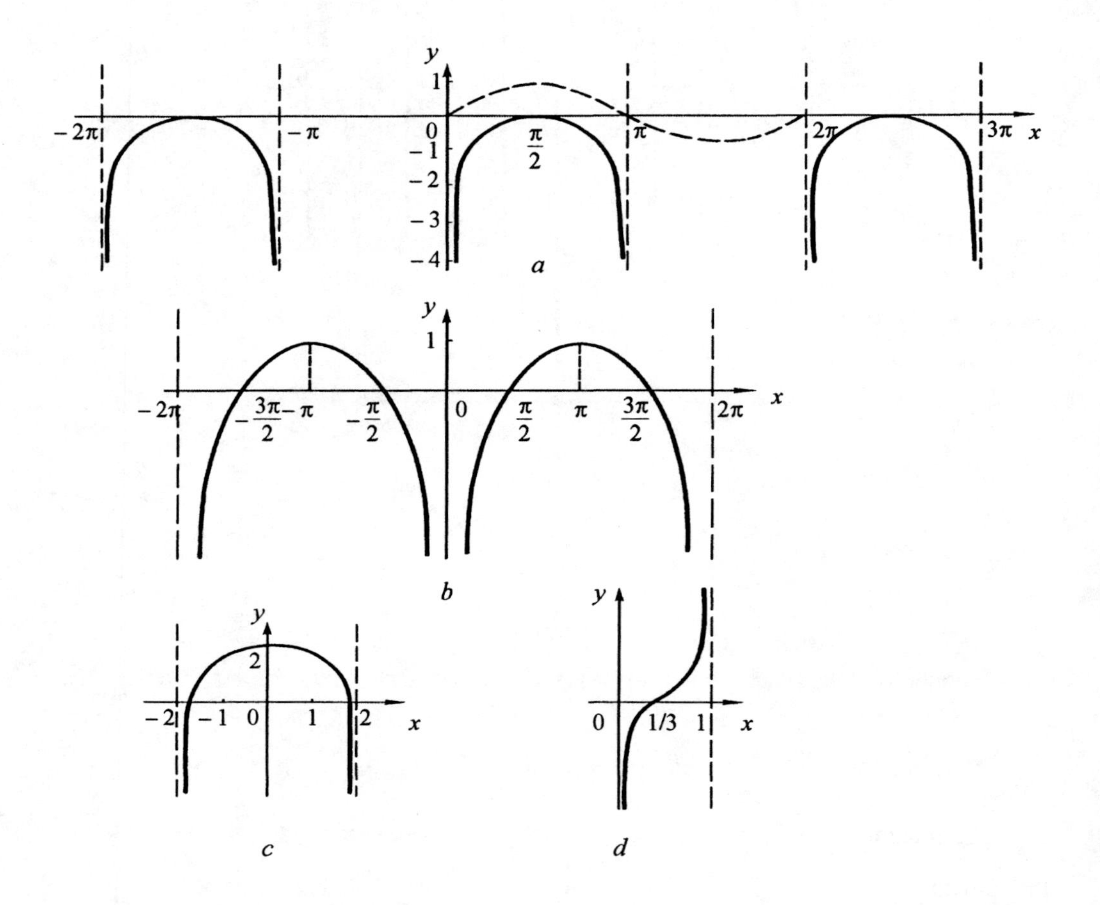

Figure 5.27. The graphs of functions: (*a*) $y = 2\log_a \sin x$; (*b*) $y = \log_2 (1 - \cos x)$; (*c*) $y = \log_2 (4 - x^2)$; (*d*) $y = \log_3 (3^x + 1)$.

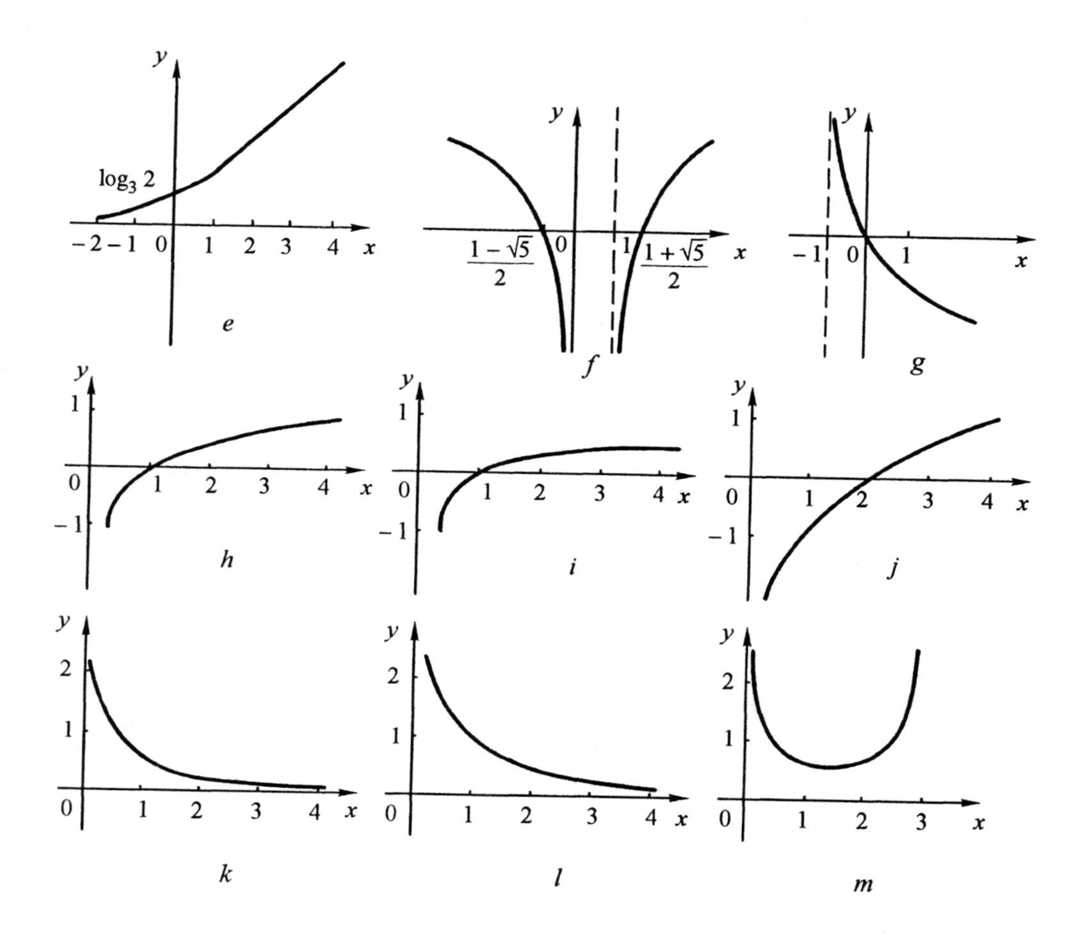

Figure 5.27. The graphs of functions: (*e*) $y = \log_{1/3}(\log_{1/3} x)$;
(*f*) $y = \log_{10}(x^2 - x)$; (*g*) $y = \log_{1/2}(x + 1)$; (*h*) $y = \ln \sqrt{x}$; (*i*) $y = \ln \sqrt[3]{x}$;
(*j*) $y = \ln \frac{x}{2}$; (*k*) $y = \ln\left(1 + \frac{1}{x}\right)$; (*l*) $y = \ln \frac{1 + \sqrt{x^2 + 1}}{x}$; (*m*) $y = \ln \frac{1 + \sin x}{\sin x}$.

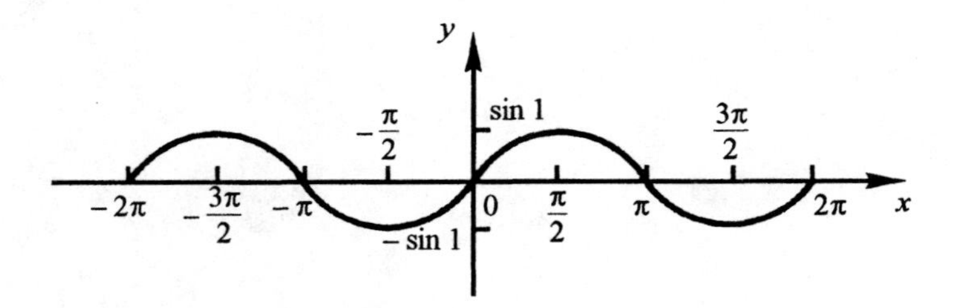

Figure 5.28. The graph of the function $y = \sin(\sin x)$.

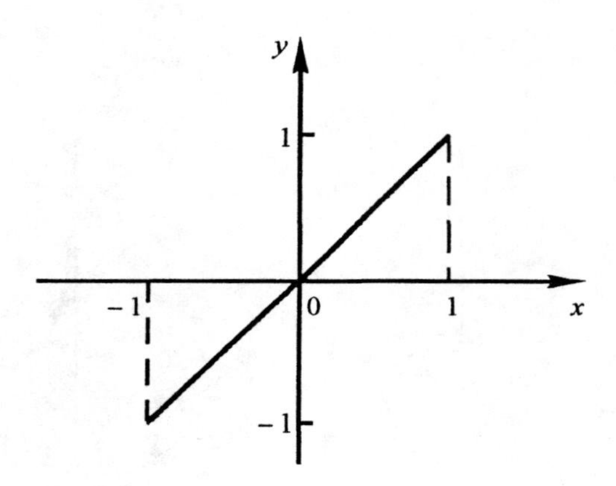

Figure 5.29. The graph of the function $y = \sin(\arcsin x)$.

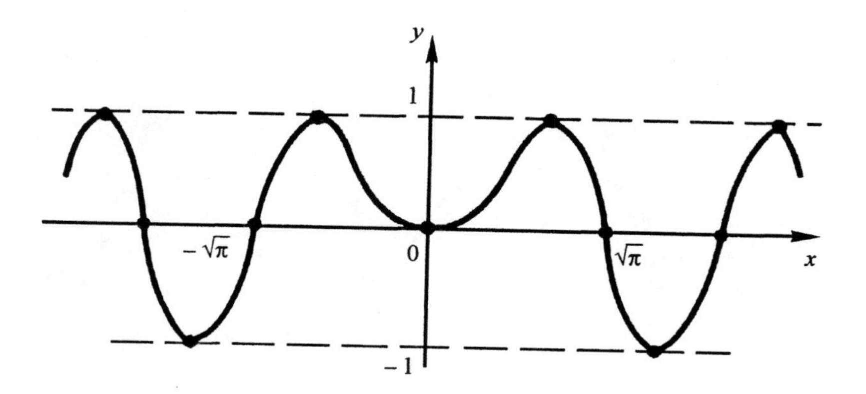

Figure 5.30. The graph of the function $y = \sin x^2$.

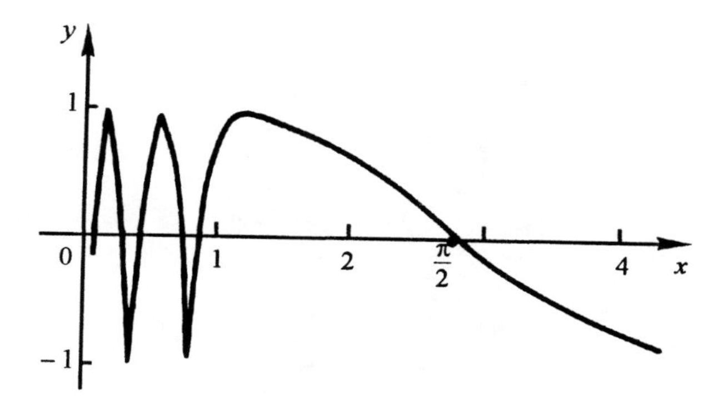

Figure 5.31. The graph of the function $y = \cos(\log_2 x)$.

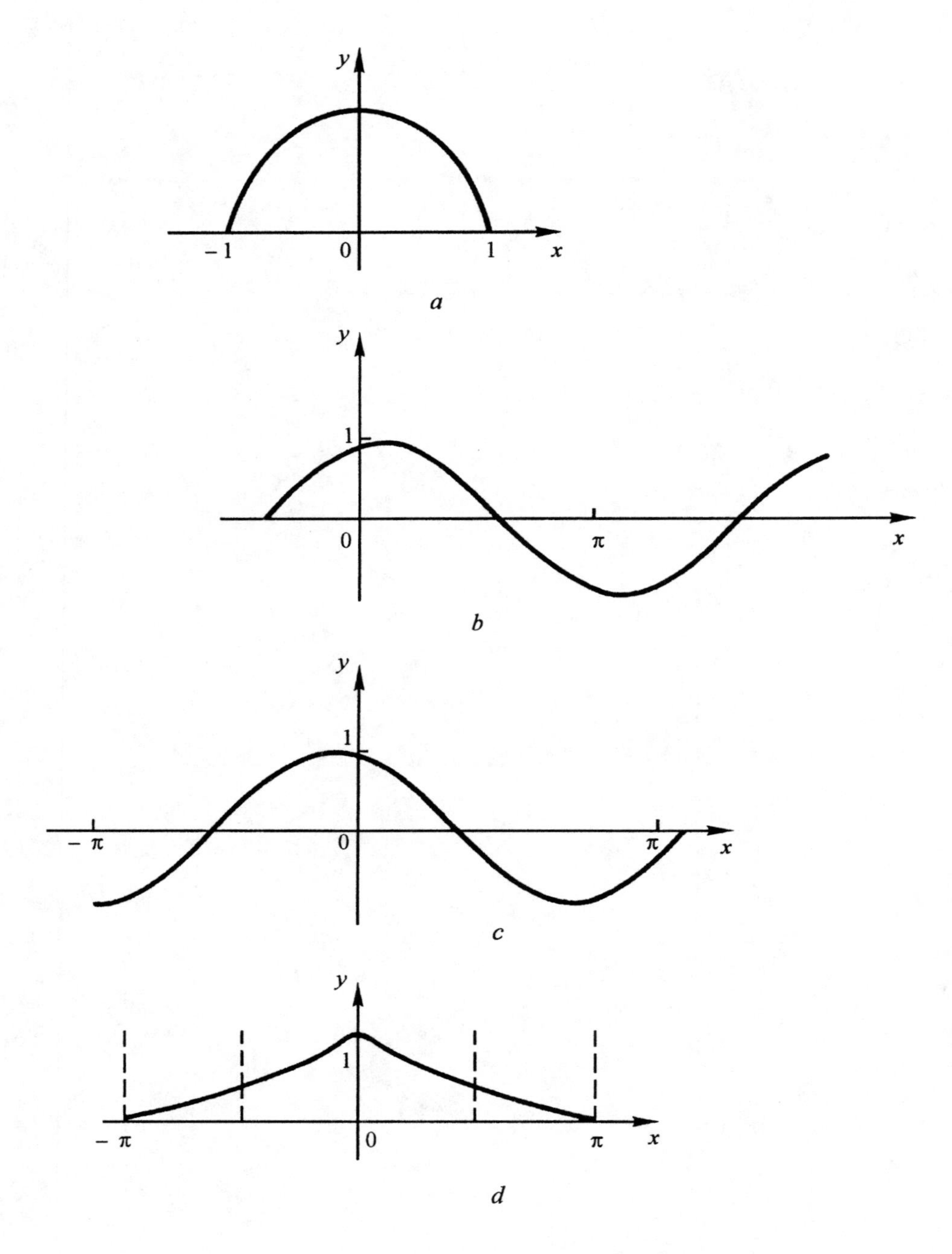

Figure 5.32. The graphs of functions: (*a*) $y = \sin(\arccos x) = \sqrt{1-x^2}$, $y = \cos(\arcsin x) = \sqrt{1-x^2}$; (*b*) $y = \sin(1+x)$; (*c*) $y = \sin(1-x)$; (*d*) $y = \cos\sqrt[3]{x}$.

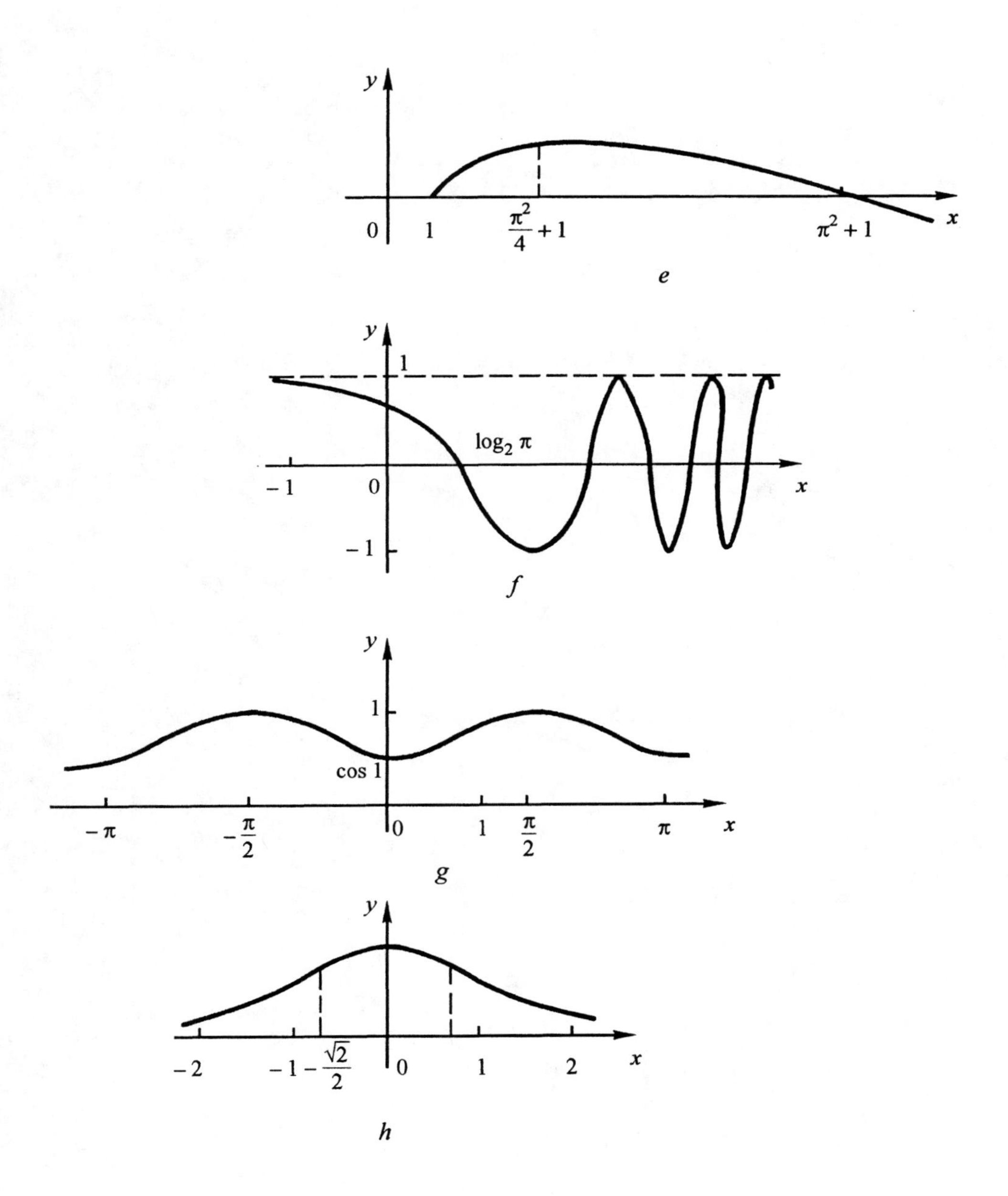

Figure 5.32. The graphs of functions: *(e)* $y = \sin\sqrt{x-1}$; *(f)* $y = \cos 2^x$; *(g)* $y = \cos(\cos x)$; *(h)* $y = \sin(\arctan x)$.

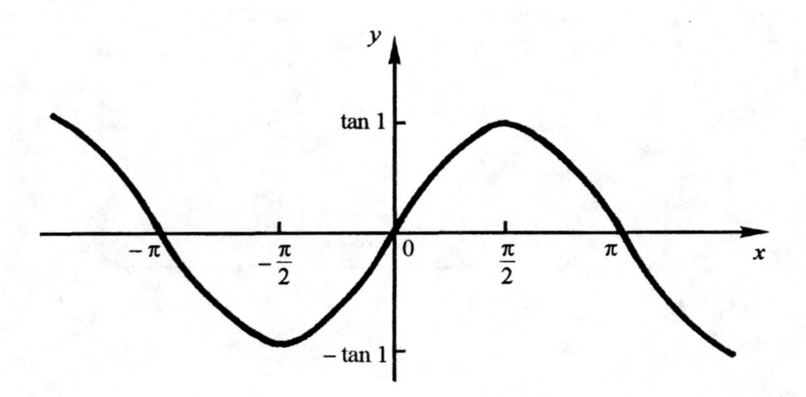

Figure 5.33. The graph of the function $y = \tan(\sin x)$.

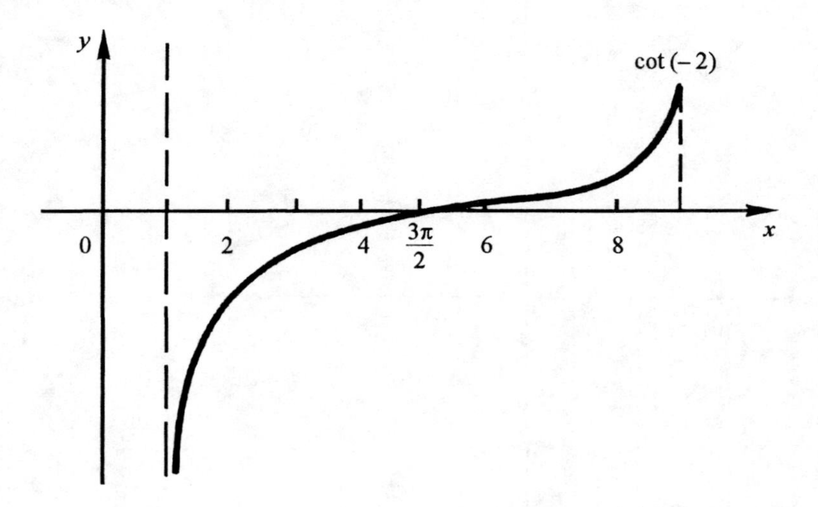

Figure 5.34. The graph of the function $y = \cot \log_{1/3} x \ (1 \le x \le 9)$.

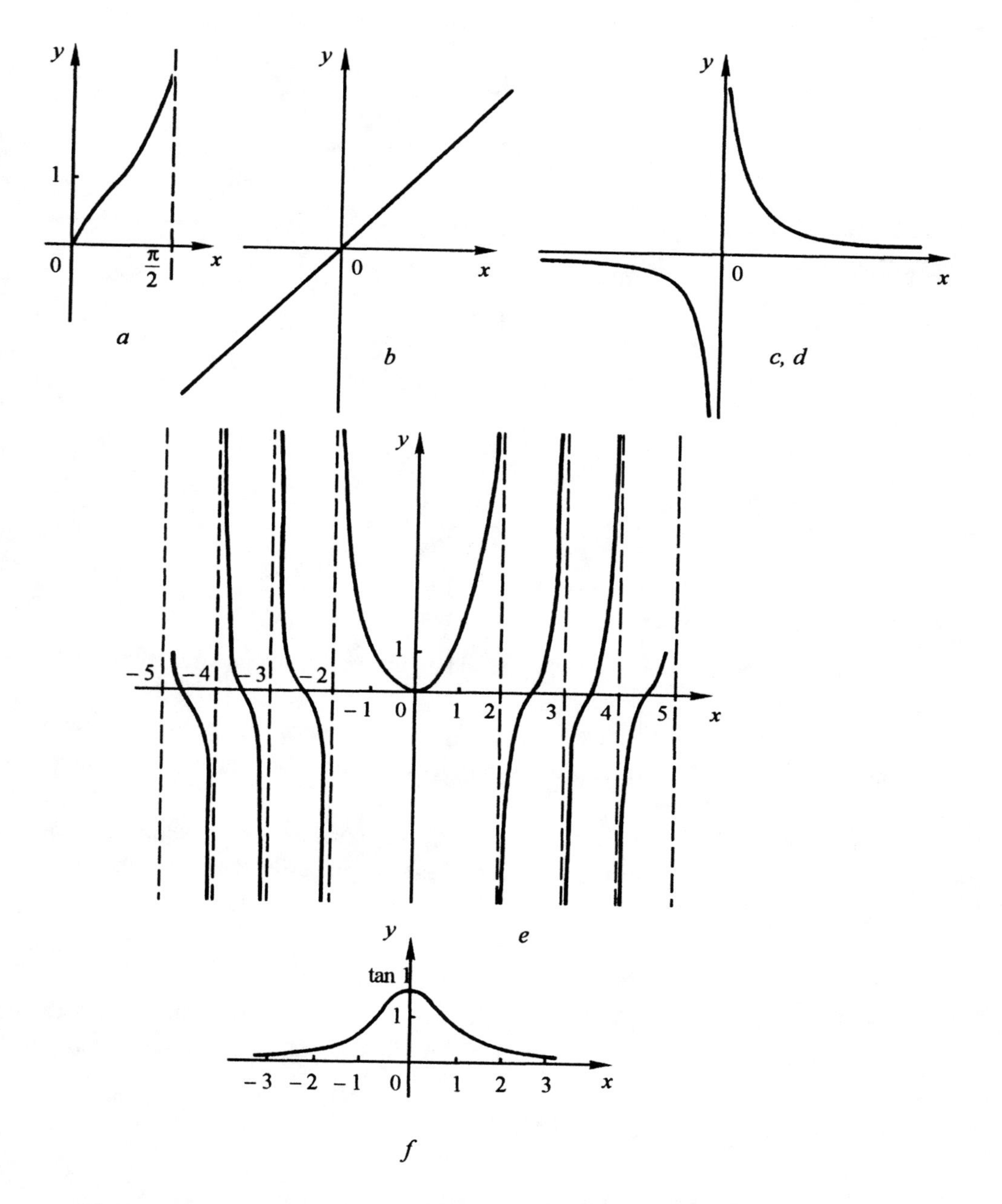

Figure 5.35. The graphs of functions: (*a*) $y = \tan\sqrt{x}$; (*b*) $y = \tan(\arctan x)$; (*c*) $y = \tan(\operatorname{arccot} x) = \frac{1}{x}$; (*d*) $y = \cot(\arctan x)$; (*e*) $y = \tan\frac{x^2}{2}$; (*f*) $y = \tan\frac{1}{1+x^2}$.

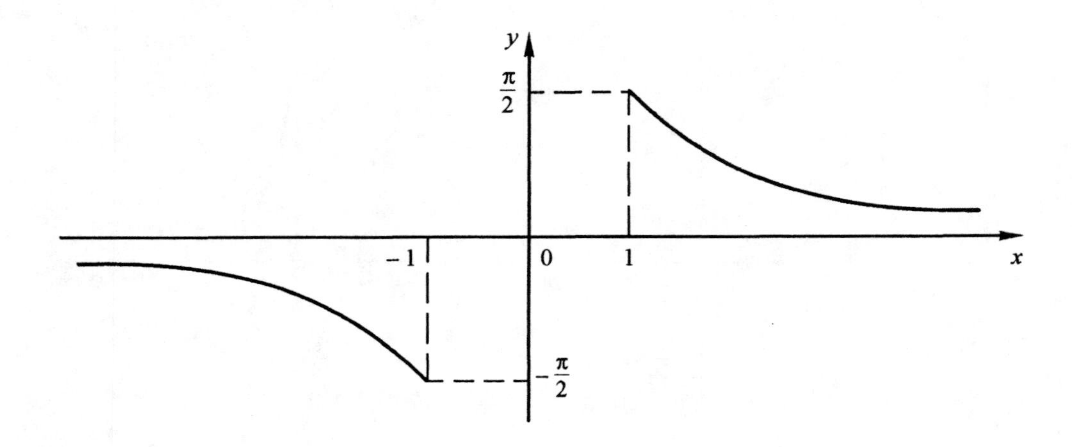

Figure 5.36. The graphs of the function $y = \arcsin \frac{1}{x}$.

5.1.7. Plotting the graphs of the functions $y = \arcsin f(x)$, $\arccos f(x)$

The domain of the functions $y = \arcsin f(x)$ and $y = \arccos f(x)$ is found from the conditions $-1 \le f(x) \le 1$.

The functions are even (odd) if the function $y = f(x)$ is even (odd). The functions are periodic if the function $y = f(x)$ is periodic; the principal periods of the functions under consideration and the function $y = f(x)$ are the same.

The intervals on which the function $y = \arcsin f(x)$ is increasing (decreasing) is the interval where the function $y = f(x)$ is increasing (decreasing); the opposite is true for the function $y = \arccos f(x)$.

Example 5.1.2. Plot the graph of the function $y = \arcsin 1/x$.

The domain of the function is $|1/x| \le 1$, or $(-\infty; -1] \cup [1; +\infty)$. The range of the function is $[-\pi/2; \pi/2]$, with $y(-1) = -\pi/2$, $y(1) = \pi/2$. The function is odd, monotone decreasing in its domain.

Note that $\lim_{x \to \pm\infty} y = 0$. The horizontal asymptote is $y = 0$ (Figure 5.36).

Example 5.1.3. Plot the graph of the function $y = \arccos (\cos x)$

The function is even and periodic with the period 2π. The range of the function is $[0; 2\pi]$.

If $\cos y = \cos x$, then $y = x + 2k\pi$ or $y = -x + 2k\pi$, $k \in \mathbf{Z}$.

On the interval $[0; 2\pi]$, the function is written as:

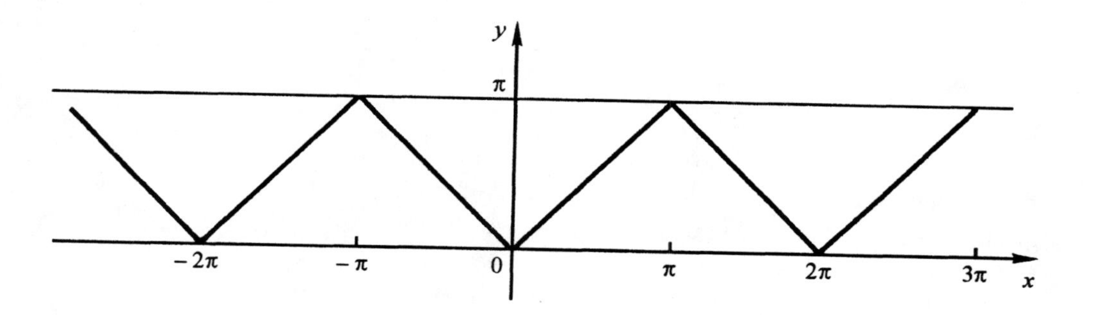

Figure 5.37. The graphs of the function $y = \arccos(\cos x)$.

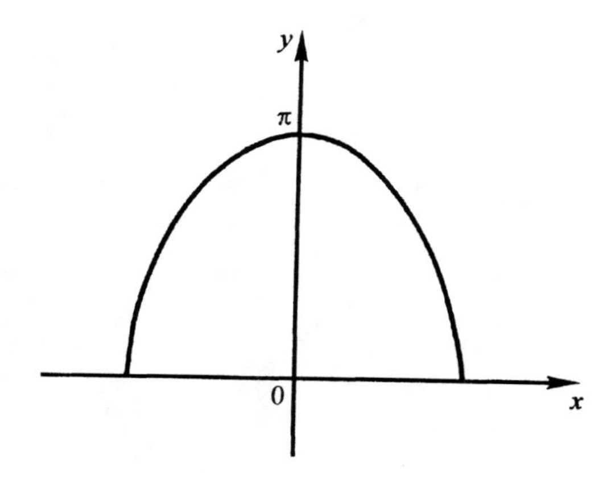

Figure 5.38. The graphs of the function $y = \arccos x^2$.

$$y = \begin{cases} x, & \text{if } 0 \le x \le \pi \\ -x + 2\pi, & \text{if } \pi \le x \le 2\pi \end{cases}$$

The graph is plotted with the use of periodicity (Figure 5.37).

Example 5.1.4. Plot the graph of the function $y = \arccos x^2$.

The domain of the function is $[-1; 1]$. The function is even. On the interval $[0; 1]$, the function decreases from $\pi/2$ to 0; on the interval $[-1; 0]$, the function increases from 0 to $\pi/2$. The graph is shown in Figure 5.38.

Graphs of the functions $y = \arcsin f(x)$ and $y = \arccos f(x)$ are also shown in Figures 5.39 – 5.41.

5.1.8. Plotting the graphs of the functions $y = \arctan f(x)$, $y = \operatorname{arccot} f(x)$

The domain of the functions $y = \arctan f(x)$, $y = \operatorname{arccot} f(x)$ is the same as the domain of the function $f(x)$.

The functions are even (odd) if the function $y = f(x)$ is even (odd). The functions are periodic if $y = f(x)$ is periodic.

The intervals on which the function $y = \arctan f(x)$ is increasing (decreasing) and the function $y = \operatorname{arccot} f(x)$ is decreasing (increasing) coincide with the intervals where the function $y = f(x)$ is increasing (decreasing).

Example 5.1.5. Plot the graph of the function $\arctan x^2$.

The domain of the function is $(-\infty; +\infty)$. The function is even. It increases from 0 to $\pi/2$ as x ranges from 0 to $+\infty$. The line $y = \pi/2$ is a horizontal asymptote. The graph of the function is shown in Figure 5.42.

Example 5.1.6. Plot the graph of the function $y = \arctan \frac{1}{\sin x}$.

The domain of the function is $(k\pi; (k+1)\pi)$, $k \in \mathbf{Z}$. The function is odd and periodic with $\omega = 2\pi$. On the interval $(0; \pi/2]$, the function decreases from $\pi/2$ to $\pi/4$, and increases on the interval $[\pi/2; \pi)$ from $\pi/4$ to $\pi/2$ (Figure 5.43).

Example 5.1.7. Plot the graph of the function $y = \operatorname{arccot} \frac{1}{x}$.

The domain of the function is $(-\infty; 0) \cup (0; +\infty)$. The function is odd. If $|x| \to \infty$, then $\operatorname{arccot} \frac{1}{x} \to 0$; if $x \to -0$, then $\operatorname{arccot} \frac{1}{x} \to \pi - 0$, and if $x \to +0$, then $\operatorname{arccot} \frac{1}{x} \to +0$ (Figure 5.44).

Graphs of functions $y = \arctan f(x)$ and $\operatorname{arccot} f(x)$ are shown in Figures 5.45 – 5.46.

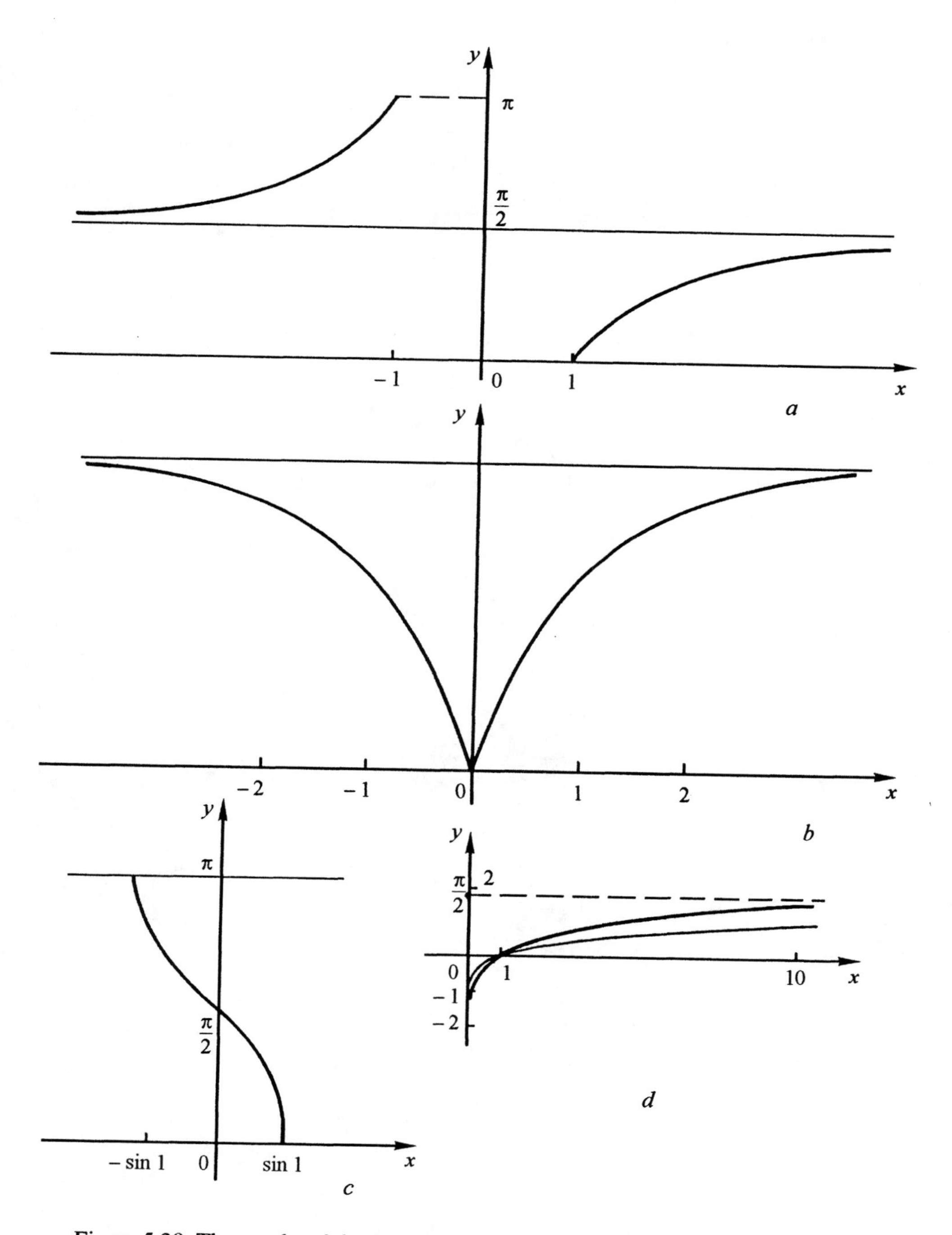

Figure 5.39. The graphs of the functions: (*a*) $y = \arccos 1/x$; (*b*) $y = \arccos \dfrac{1-x^2}{1+x^2}$; (*c*) $y = \arccos(\arcsin x)$; (*d*) $y = \arcsin \log_{10} x$.

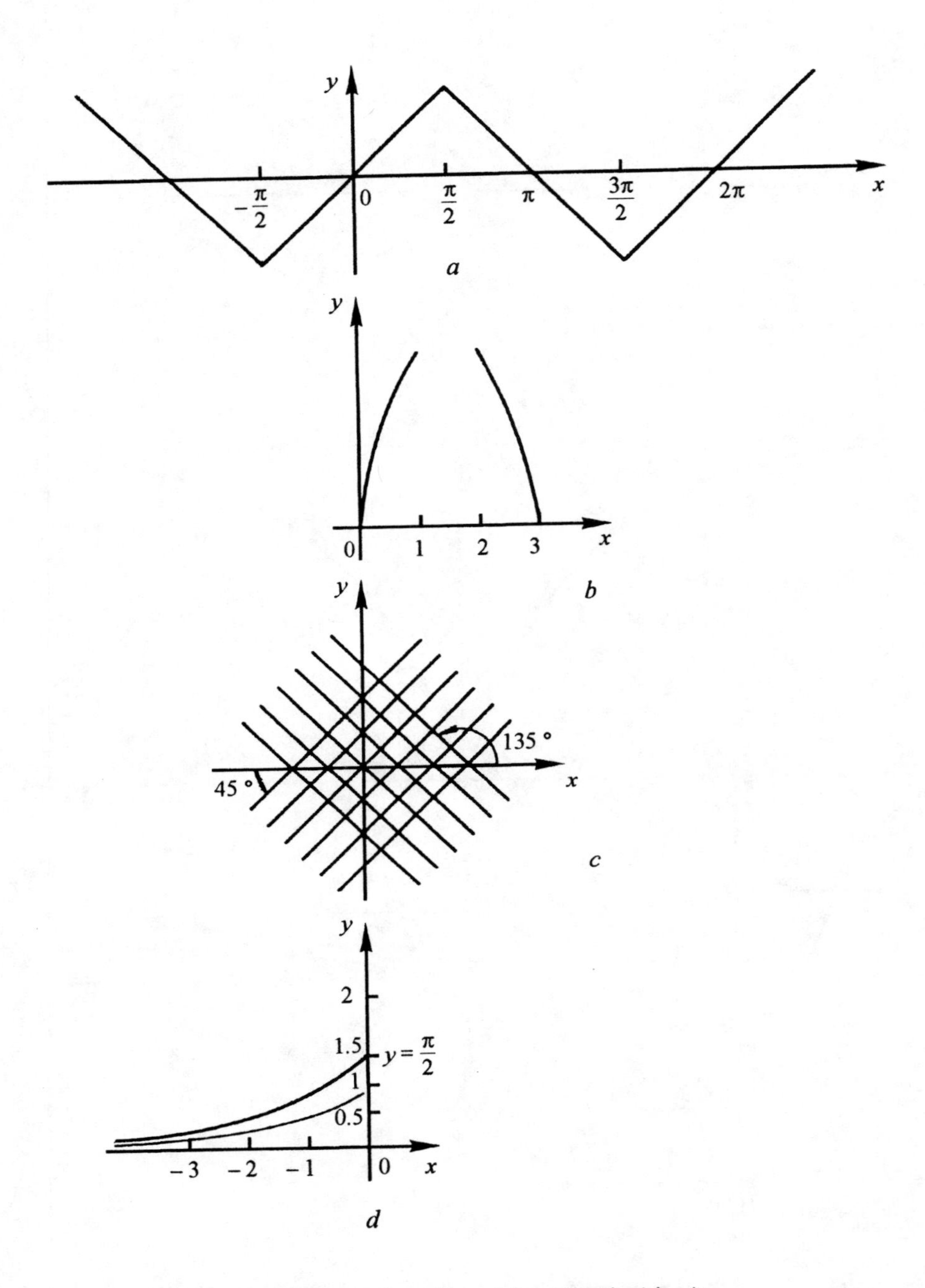

Figure 5.40. The graphs of the functions: (*a*) $y = \arcsin(\sin x)$; (*b*) $\arccos(x^2 - 3x + 1)$; (*c*) $y = \arccos(\cos x)$; (*d*) $y = \arcsin a^x$, $a > 1$.

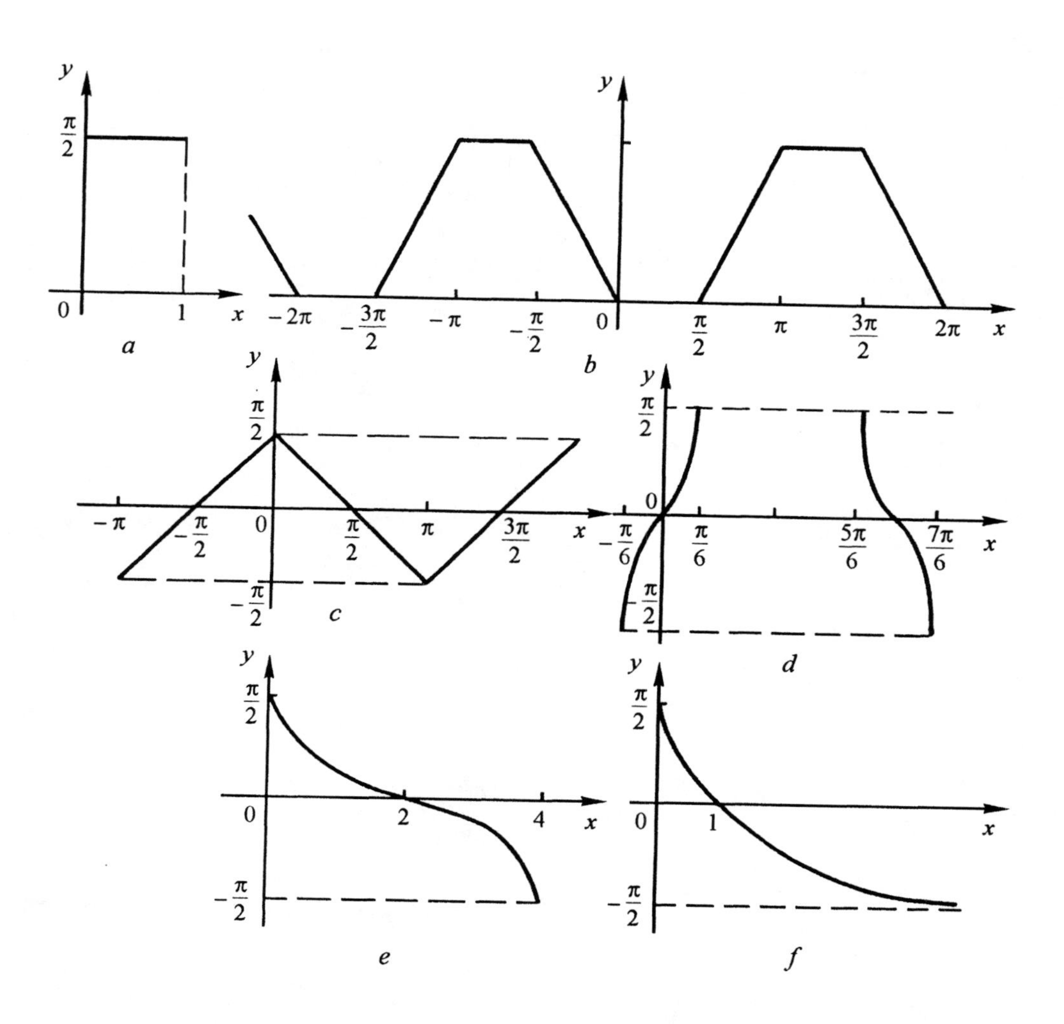

Figure 5.41. The graphs of the functions: (*a*) $y = \arcsin\sqrt{1-x} + \arcsin\sqrt{x}$; (*b*) $y = \arccos(\cos x) - \arcsin(\sin x)$; (*c*) $y = \arcsin(\cos x)$; (*d*) $y = \arcsin(2\sin x)$; (*e*) $y = \arcsin(1 - x/2)$; (*f*) $y = \arcsin\frac{1-x}{1+x}$.

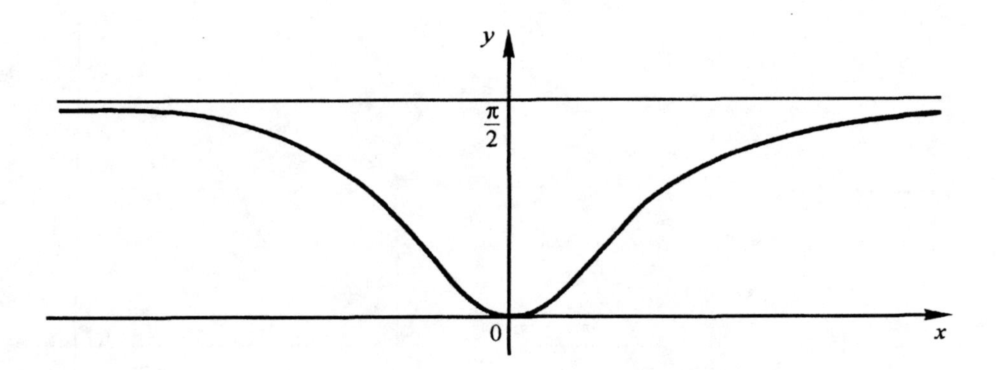

Figure 5.42. The graphs of the function $y = \arctan x^2$.

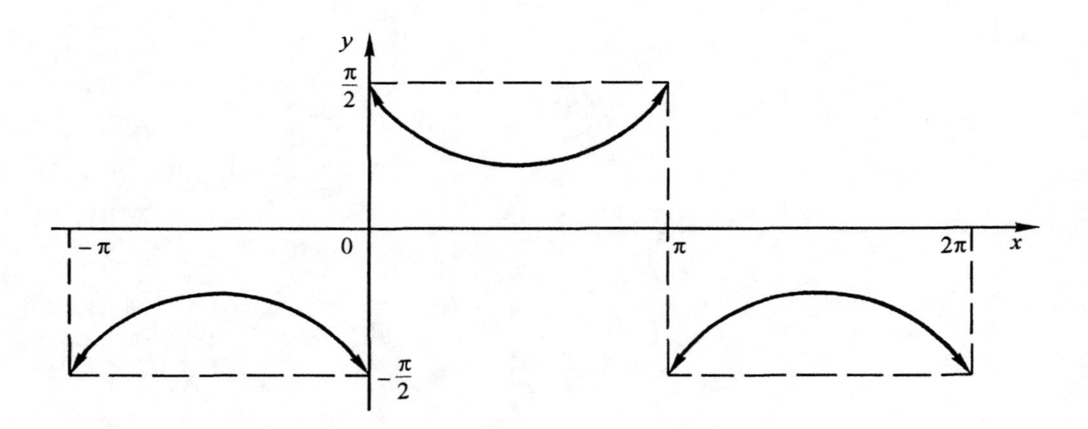

Figure 5.43. The graphs of the function $y = \arctan \frac{1}{\sin x}$.

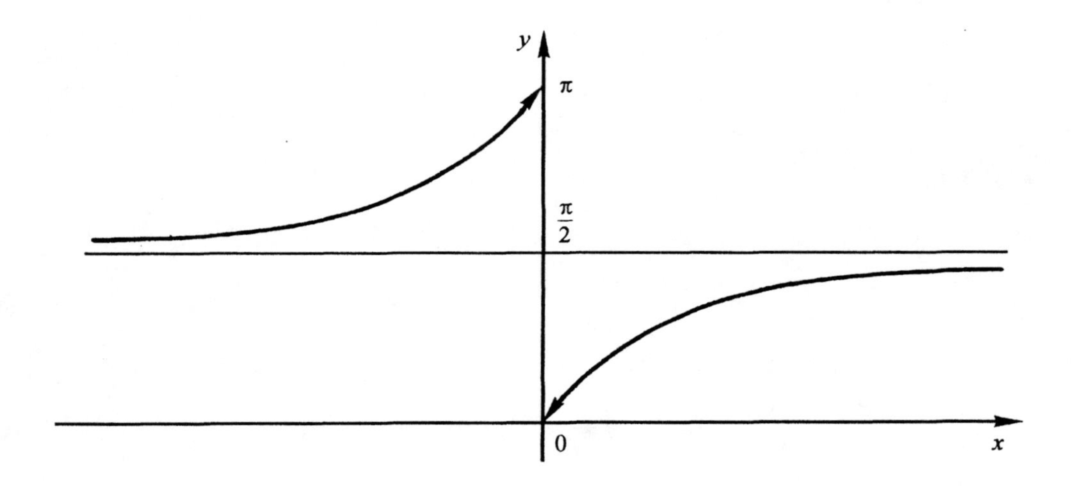

Figure 5.44. The graph of the function $y = \operatorname{arccot} \frac{1}{x}$.

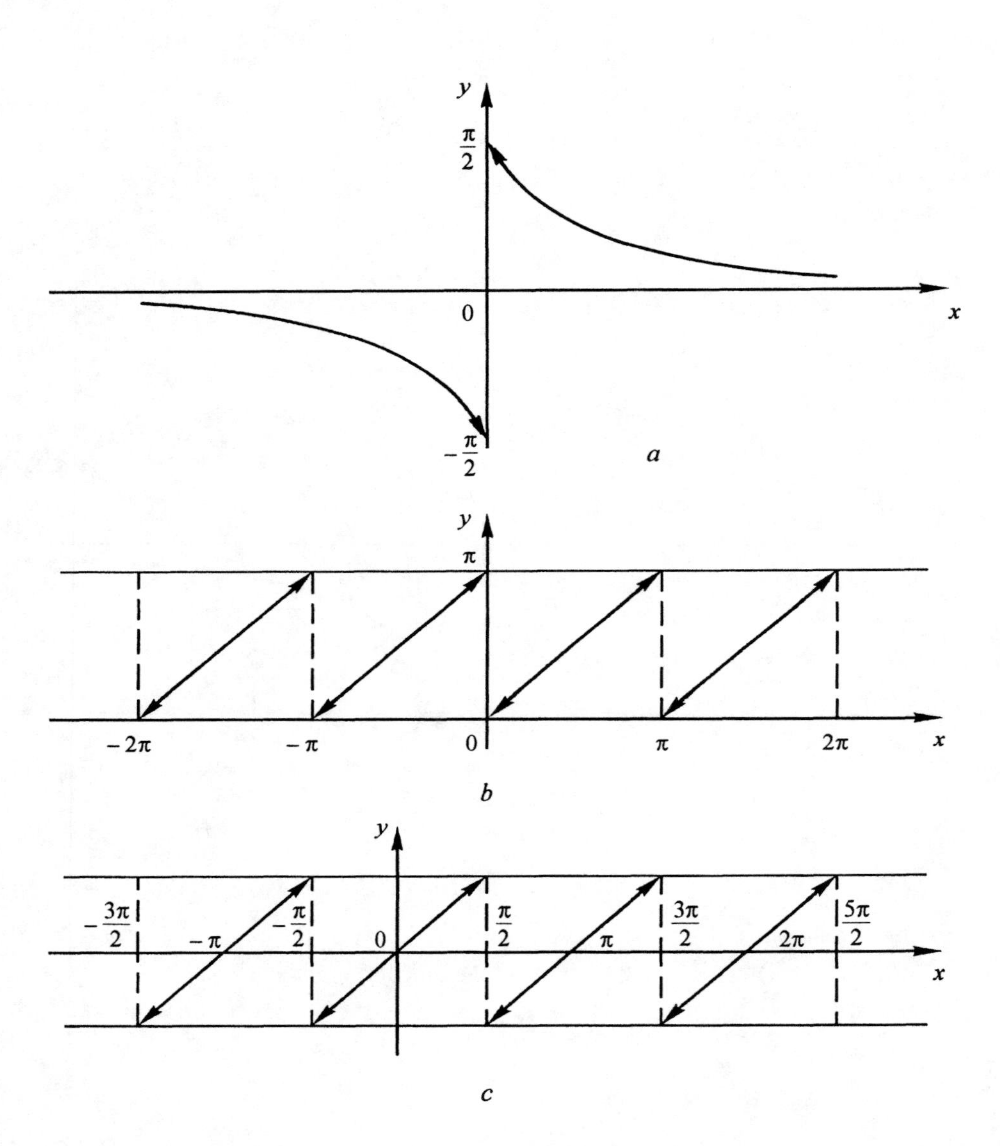

Figure 5.45. The graphs of the functions: (*a*) $y = \arctan 1/x$; (*b*) $y = \operatorname{arccot}(\cot x)$; (*c*) $y = \arctan(\tan x)$.

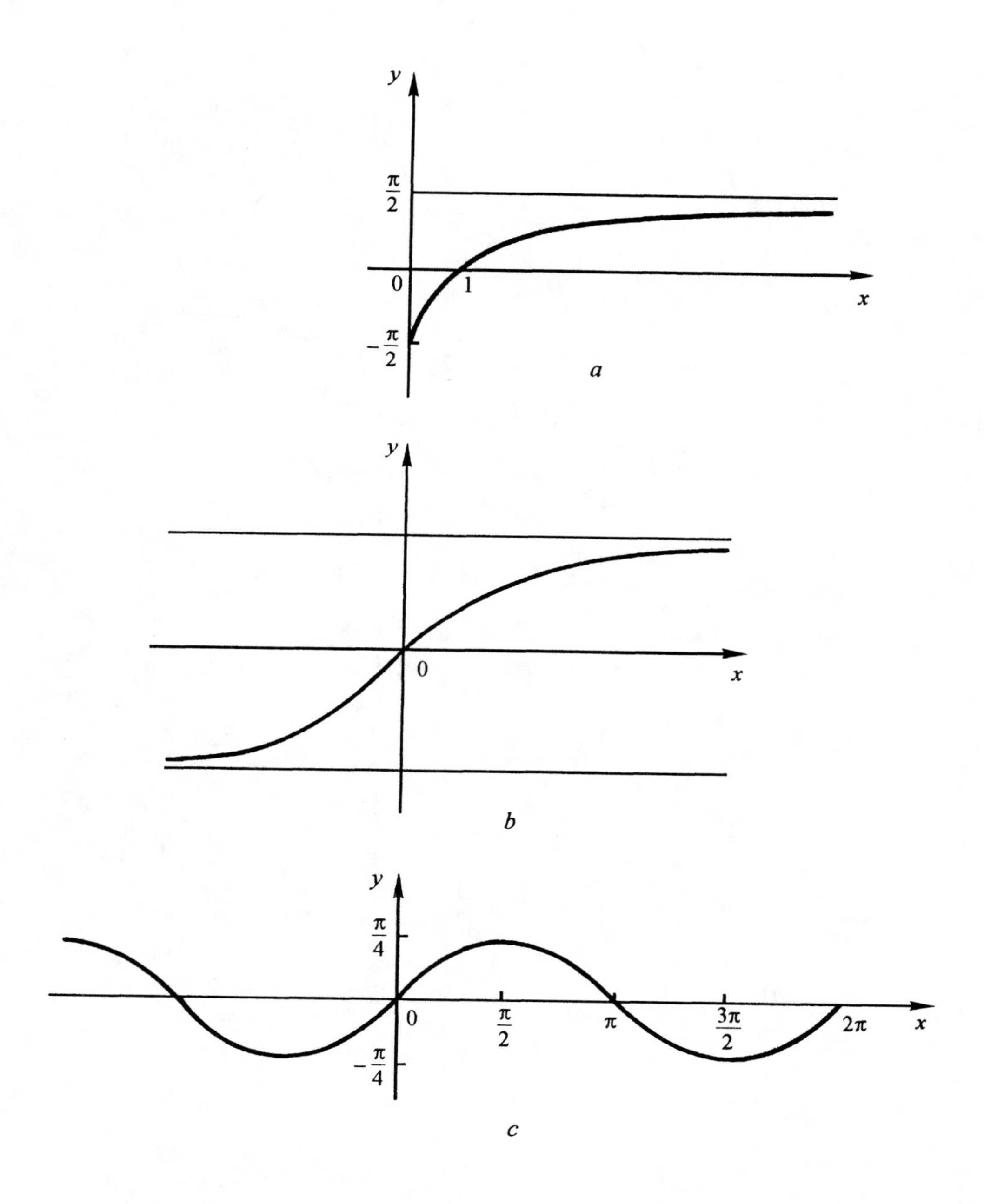

Figure 5.46. The graphs of the functions: (*a*) $y = \arctan(\log_{10}/x)$; (*b*) $y = \arctan(\cot x)$; (*c*) $y = \arctan(\sin x)$.

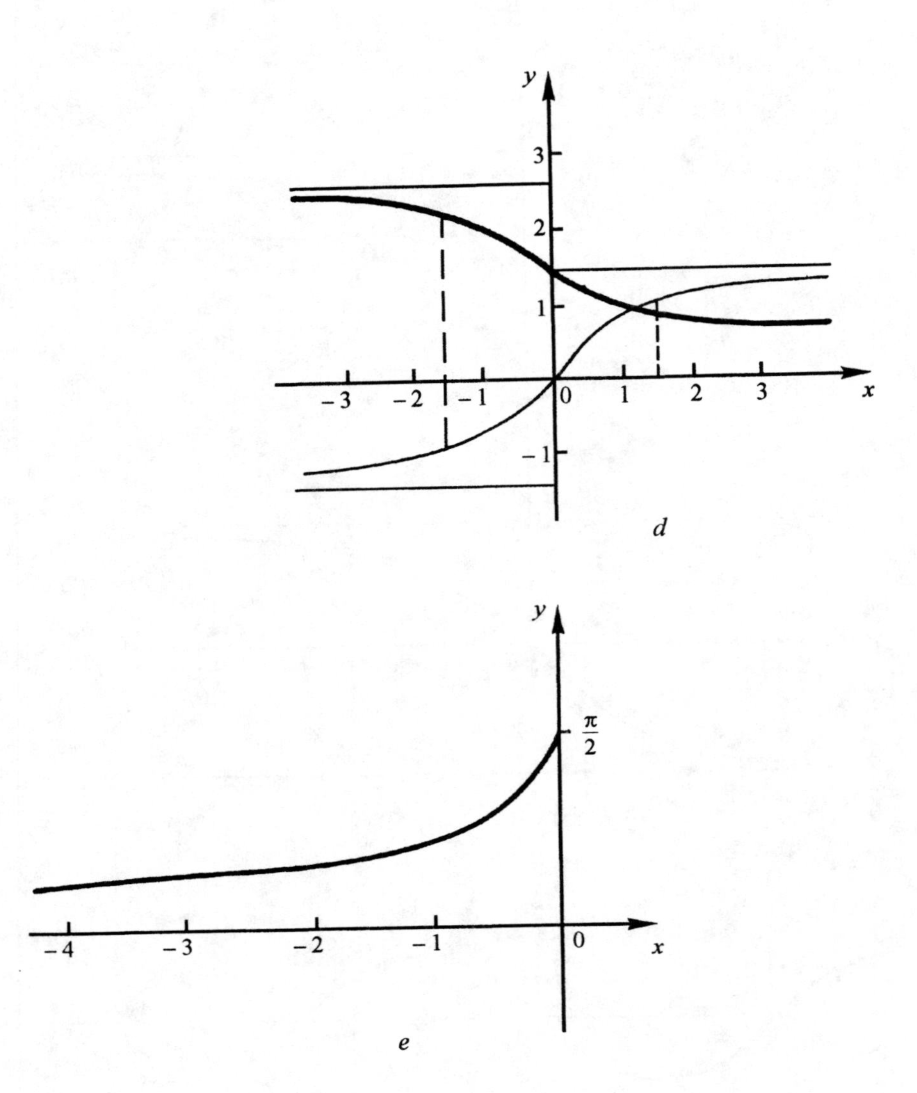

Figure 5.46. The graphs of the functions: (*d*) $y = \text{arccot}\,(\arctan x)$; (*e*) $y = \text{arccot}\,\sqrt{-x}$.

5.2. Graphs of algebraic functions

In this section we consider the graphs of integer rational functions (polynomials), fractional rational functions, and irrational functions. Of course, construction of their graphs can be carried out by following the general methods for graph plotting of compositions of functions (Section 5.1) with the use of transformations of graphs (Chapter 4). We consider the most important and widely used groups of such functions and study specifics of constructing graphs of these functions.

5.2.1. Graphs of integer functions

An integer rational function, or a polynomial, is a function of the form

$$y = a_0x^n + a_1x^{n-1} + \dots + a_{n-1}x + a_n,$$

where n is a nonnegative integer, and $a_0, a_1, \dots, a_n$ are real constants.

Linear function

The most simple integer rational function is a linear function ($n = 1$). We write it as $y = ax + b$. The graph of this function is a straight line.

The function is monotone increasing if $a > 0$, decreasing if $a < 0$, and is a constant if $a = 0$. The points of intersection with the coordinate axes are $(-b/a;\ 0)$ and $(0;\ b)$. The graph of the function depends on the signs of a and b, see Figure 5.47.

Quadratic function (quadratic trinomial)

The general form of a quadratic trinomial is $y = ax^2 + bx + c$, where a, b, c are some real numbers. The graph of the quadratic trinomial is a parabola with the vertical symmetry axis $x = -b/2a$ (Figure 5.48).

Note that the form of the graph of the function $y = ax^2 + bx + c$ depends on the value of the coefficient a: the greater the absolute value of a, the steeper the branches of the parabola (Figure 5.49).

There are two methods to plot the graph of the quadratic trinomial: 1) to separate a complete square, and to do the construction following the corresponding transformations of graphs (Chapter 4); and 2) to study the quadratic trinomial and use this to plot the graph.

Cubic function (polynomial of degree three)

The general form of a polynomial of degree three is $y = ax^3 + bx^2 + cx + d$, where $a \neq 0$, b, c, d are some real numbers.

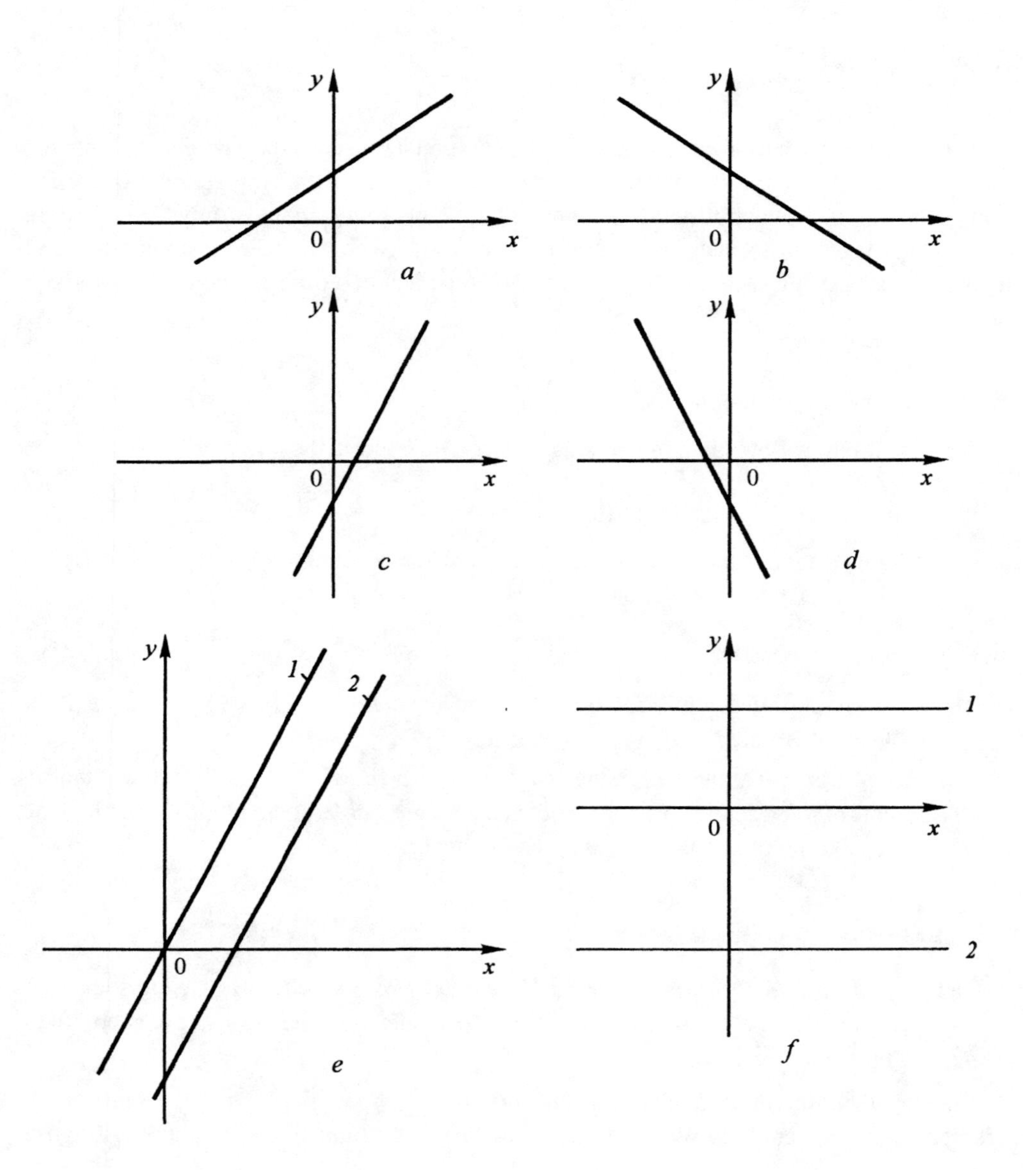

Figure 5.47. The linear function $y = ax + b$: (*a*) $a > 0, b > 0$; (*b*) $a < 0, b > 0$; (*c*) $a > 0, b < 0$; (*d*) $a < 0, b < 0$; (*e*) $b = 0$, *1*. $b \neq 0$, *2*. $y = 2x - 3$; (*f*) $a = 0$, *1*. $y = 2$, *2*. $y = -3$.

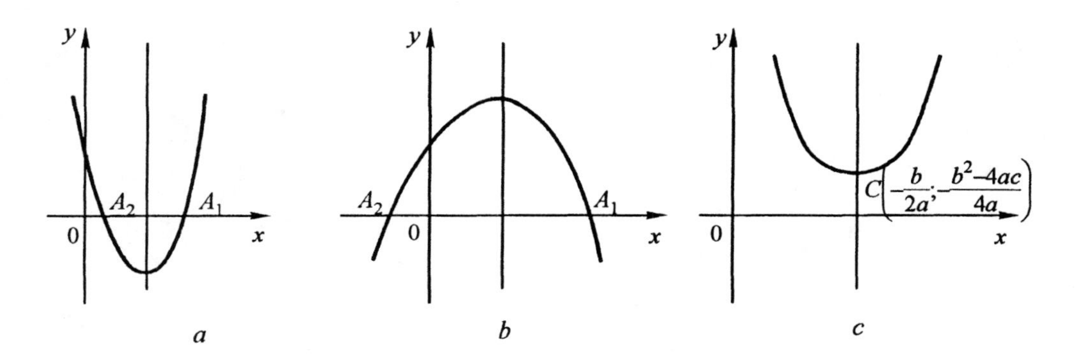

Figure 5.48. The quadratic function $y = ax^2 + bx + c$: (*a*) $a > 0$, $b < 0$; (*b*) $a > 0$, $b > 0$; (*c*) $y = ax^2 + bx + c$.

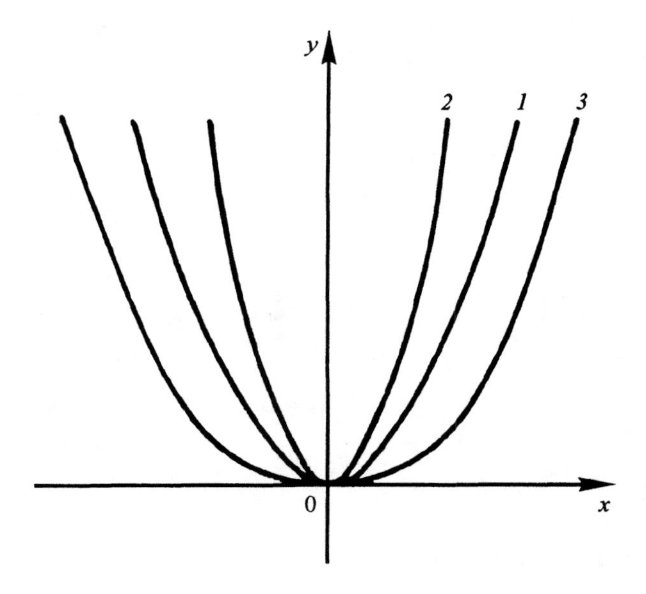

Figure 5.49. The quadratic function $y = ax^2$: *1*. $a = 1$, *2*. $a = 2$, *3*. $a = \frac{1}{2}$.

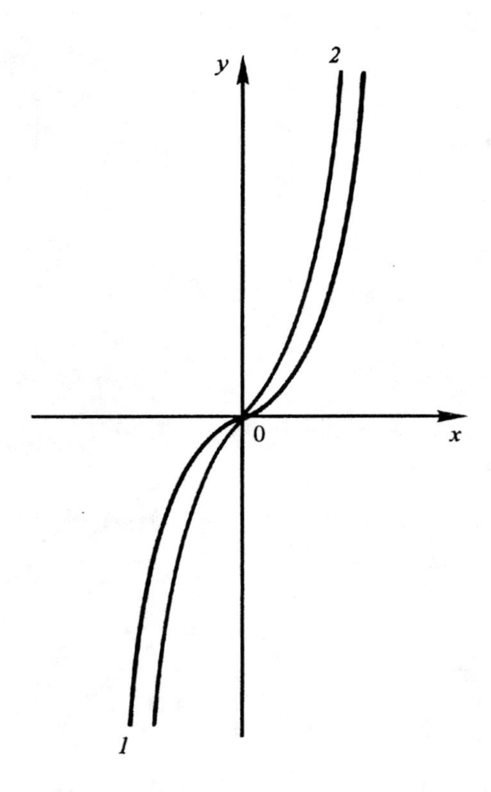

Figure 5.50. Qubic functions: *1.* $y = x^3$, *2.* $x^3 + x$.

1. If $a = 1$, $b = c = d = 0$, then the polynomial has the form $y = x^3$. This function has already been studied in Section 3.1.

2. If $a = 1$, $c = k \neq 0$, $b = d = 0$, then the function $y = ax^3 + bx^2 + cx = d$ has the form $y = ax^3 + kx$.

The domain of this function is $(-\infty; +\infty)$. The function is odd.

If $k > 0$, the graph passes through the point (0; 0). Since the functions x^3 and kx are increasing, the function $y = x^3 + kx$ is increasing and does not have extremums. The function is concave for $x \in (0; +\infty)$.

Note that adding a linear function to any function does not influence its concavity. The point (0; 0) is an inflection point. It is clear that the graph of the function $y = x^3 + kx$ is steeper than the graph of the function $y = x^3$ (Figure 5.50).

If $k < 0$, then the graph of the function $y = x^3 + kx$ passes through the point (0; 0), $(\sqrt{-k}; 0)$, $(-\sqrt{-k}; 0)$.

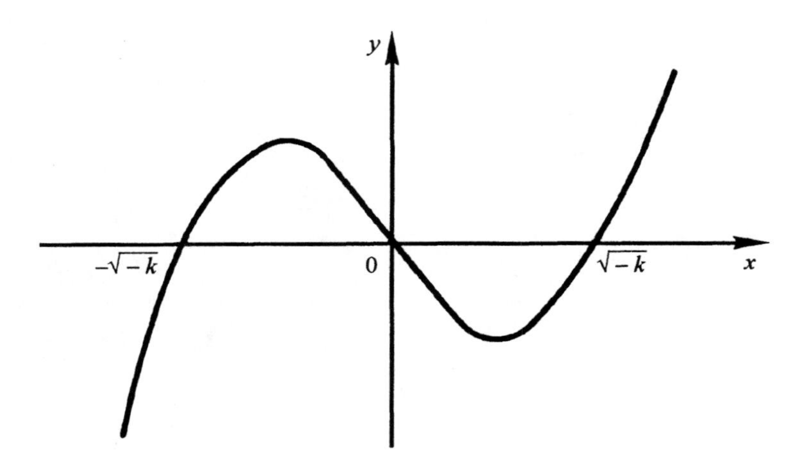

Figure 5.51. Qubic function $y = y^3 + kx$, $k < 0$.

For $x \in (-\infty; -\sqrt{(-k/3)}) \cup (\sqrt{-k/3}; +\infty)$, the function is increasing, and for $x \in (-\sqrt{-k/3}; \sqrt{-k/3})$, it is decreasing. The function attains a maximum at $x = -\sqrt{-k/3}$ ($y_{max} = 2k\sqrt{-k/3}$) and a minimum at $x = \sqrt{-k/3}$ ($y_{min} = -2k\sqrt{-k/3}$), see Figure 5.51.

3. The graph of the function $y = x^3 + kx + b$ is obtained from the graph of the function $y = x^3 + bx$ by parallel translation of the graph along the y-axis by $|b|$ in the same direction as the sign of b.

4. To construct the graph of the function $y = x^3 + bx^2 + cx + d$, we make a parallel translation of the origin, $x = x' - b/3$, $y = y'$. After that the function becomes $y = x'^3 + k_1x' + b_1$. Hence the graph of the function $y = x^3 + bx^2 + cx + d$ is the same as the graph of the function $y = x^3 + kx + b$, axcept only for the translation of its center of symmetry along the x-axis by $-b/3$.

5. Plotting the graph of the function $y = ax^3 + bx^2 + cx + d$ is reduced to plotting the graph of the function $y = x^3 + kx + b$.

Indeed, write the function $y = ax^3 + bx^2 + cx + d$ as $y = a(x^3 + \frac{b}{a}x^2 + \frac{c}{a}x + \frac{d}{a})$. Then plot the graph of the function $y = x^3 + \frac{b}{a}x^2 + \frac{c}{a}x + \frac{d}{a}$, and then stretch it along the y-axis by a.

The graph of the function $y = ax^3 + bx^2 + cx + d$ is called the *cubic parabola.*

Note that the behavior of the function $y = ax^3 + bx^2 + cx + d$ depends on the signs of a and $\Delta = 3ac - b^2$. In the case where $\Delta \geq 0$ (Figure 5.52), the function is monotonic increasing if $a > 0$, and monotonic decreasing if $a < 0$. In the case where $\Delta < 0$, the function has one maximum and one minimum,

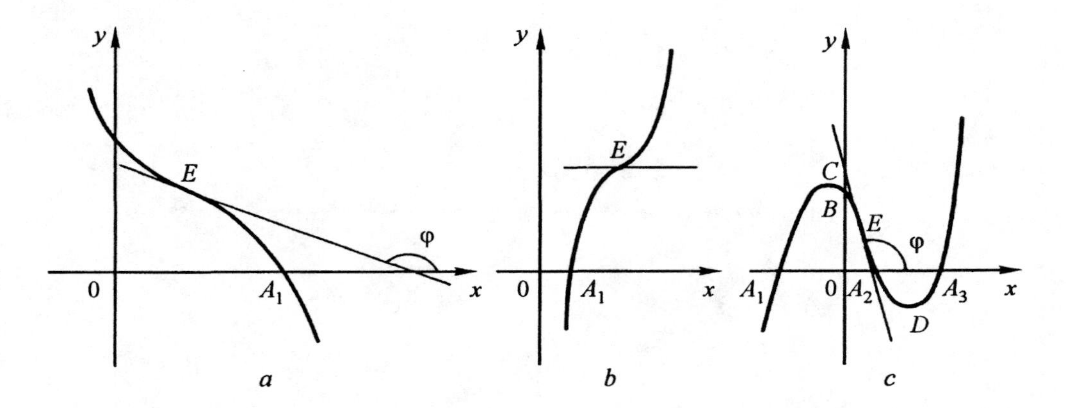

Figure 5.52. Quadratic function $y = ax^3 + bx^2 + cx + d$: (*a*) $\Delta > 0$, $a < 0$; (*b*) $\Delta = 0$, $a > 0$; (*c*) $\Delta < 0$, $a > 0$.

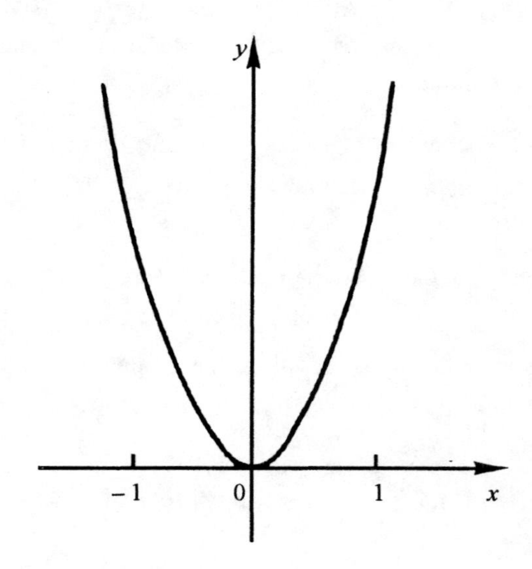

Figure 5.53. Biquadratic function $y = ax^4 + x^2$.

$$C\left(-b+\frac{\sqrt{-\Delta}}{3a};\ \frac{2b^3-9abc-(6ac-2b^3)\sqrt{-\Delta}}{27a^2}+d\right),$$

$$D\left(-b-\frac{\sqrt{-\Delta}}{3a};\ \frac{2b^3-9abc-(6ac-2b^3)\sqrt{-\Delta}}{27a^2}+d\right).$$

The x-coordinates of points of intersection of the graph with the axis Ox are calculated by solving the equation $ax^3+bx^2+cx+d=0$. The number of roots could be 1, 2, or 3. The inflection point, which at the same time is the symmetry point of the curve, is $E\left(-\frac{b}{3a};\ \frac{2b^3-9abc}{27a^2}+d\right)$.

Biquadratic function

A function of the form $y=ax^4+bx^2+c,\ a\neq 0$, is called *biquadratic*.

A study and construction of the graph of the function is carried out following the general scheme (see Section 2.2). Sometimes one can plot the graph of this function by using certain transformations of the graphs of the functions $y=x^4, y=x^2$. Indeed, since

$$y=ax^4+bx^2+c=\frac{b^2}{c}\left(\frac{a^2}{b^2}x^4+\frac{a}{b}x^2\right)+c=$$

$$=\frac{b^2}{a}\left(\frac{x^4}{b^2/a^2}+\frac{x^2}{b/a}\right)+c=\frac{b^2}{a}\left[\left(\frac{x}{\sqrt{|b/a|}}\right)^4\pm\left(\frac{x}{\sqrt{|b/a|}}\right)^2\right]+c$$

it is clear that the graph of the function $y=ax^4+bx^2+c$ can be obtained from the graph of the function $y=x^4+x^2$ or $y=x^4-x^2$ by using scaling along both coordinate axes and parallel translation along the y-axis by $|c|$.

The graph of the function $y=x^4+x^2$ is the sum of the graphs of the functions $y=x^4$ and $y=x^2$ (Figure 5.53). The graph of the function $y=x^4-x^2$ is shown in Figure 5.54.

Polynomial of degree *n*

The polynomial of degree n has the form $y=a_0x^n+a_1x^{n-1}+\ldots+a_n$. Its graph is a parabolic type curve of degree n.

1. If n is odd, then y continuously changes from $-\infty$ to $+\infty$ for $a_0>0$, and from $+\infty$ to $-\infty$ for $a_0<0$. The curve can intersect (or touch) the x-axis from 1 to n times. The function either does not have extremums at all or has an even number of them (from 2 to $n-1$), with minima and maxima being alternated. The number of inflection points is odd and could be from 1 to $n-2$.

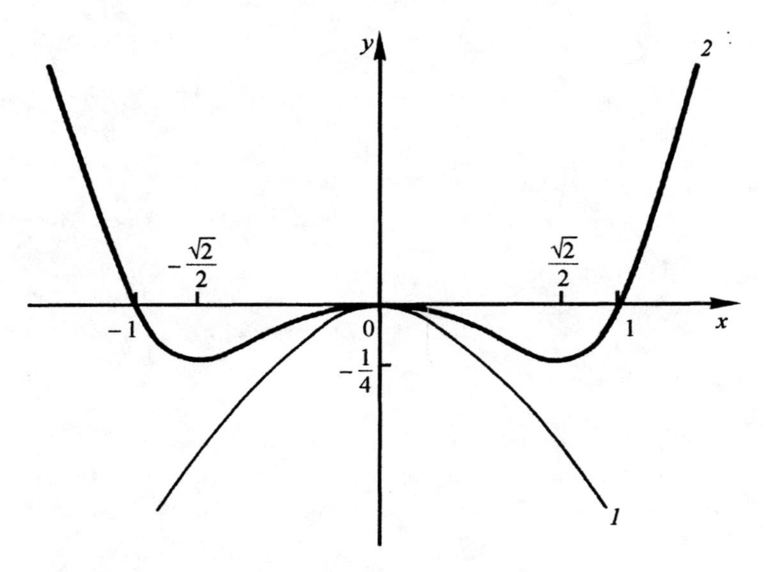

Figure 5.54. Functions *1*. $y = -x^2$; *2*. $y = x^4 - x^2$.

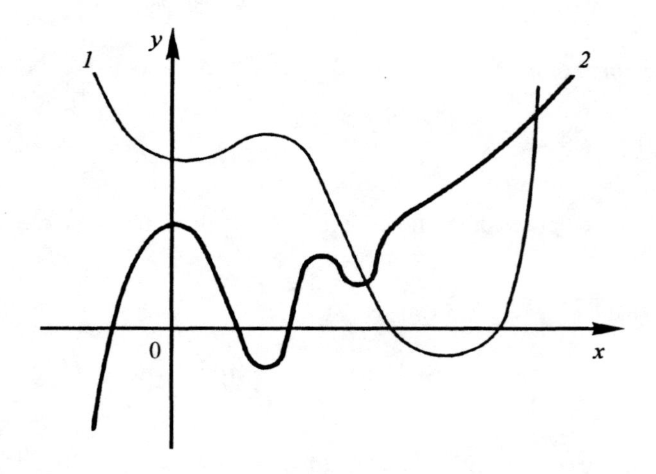

Figure 5.55. Polynomial of degree n: *1*. n is even, *2*. n is odd.

2. In the case where n is even, y continuously ranges from $+\infty$ to $+\infty$ if $a_0 > 0$, and from $-\infty$ to $-\infty$ if $a_0 < 0$. The curve either does not intersect the x-axis or intersects it (or touches it) from 1 to n times.

The function has an odd number of extremums with maxima and minima alternating. The number of inflection points is even (from 0 to $n-2$). The curves do not have asymptotes (Figure 5.55).

It should be noted that it is better to perform an analysis and construction of the graph of a polynomial of degree n by using methods of calculus.

Graphs of some polynomials are shown in Figures 5.56 – 5.63.

Functions of the form $y = (ax^2 + bx + c)^n$, $n \in Z_0$

If the quadratic trinomial $ax^2 + bx + c$ has equal real roots, then the function $y = (ax^2 + bx + c)^n$ can be written as $y = a^n(x - \alpha)^{2n}$, where α is a real root of the trinomial. The graph is constructed by using transformations of the graph of the power function $y = x^{2n}$ (see Section 4.2).

If the quadratic trinomial $ax^2 + bx + c$ has two distinct real roots, then the function $y = (ax^2 + bx + c)^n$ is written as $y = a^n(x - \alpha)^n(x - \beta)^n$.

Plotting of the graph of this function is reduced to drawing the graphs of the functions $y = a^n(x - \alpha)^n$, $y = (x - \beta)^n$ and then multiplying the graphs.

If the quadratic trinomial $ax^2 + bx + c$ has complex roots, then the graph of the function $y = (ax^2 + bx + c)^n$ is constructed as the graph of the composition of functions (see Section 5.1).

Remark 5.2.1. A study and construction of the graph of the function $y = f(ax^2 + bx + c)$ is conducted by regarding the function as the composition of functions (see Section 5.1).

Graphs of this type of function are shown in Figures 5.64 and 5.65.

Note. Let us briefly describe a method to construct the graph of an integer rational function

$y = a_0x^n + a_1x^{n-1} + \ldots + a_{n-1}x + a_n$, which is due to Segner (1761).

This method is based on constructing a chain of binomials of the same type for a given value of x_i:

$$y_1 = a_0x_i = a_1,$$

$$y_2 = a_0x_i^2 + a_1x_i + a_2 = y_1x_i + a_2,$$

$$y_3 = a_0x_i^3 + a_1x_i^2 + a_2x_i + a_3 = y_2x_i + a_3,$$

$$\ldots \quad \ldots$$

$$y_n = y = a_0x_i^n + a_1x_i^{n-1} + a_2x_i^{n-2} + \ldots + a_{n-1}x_i + a_n = y_{n-1}x_i + a_n,$$

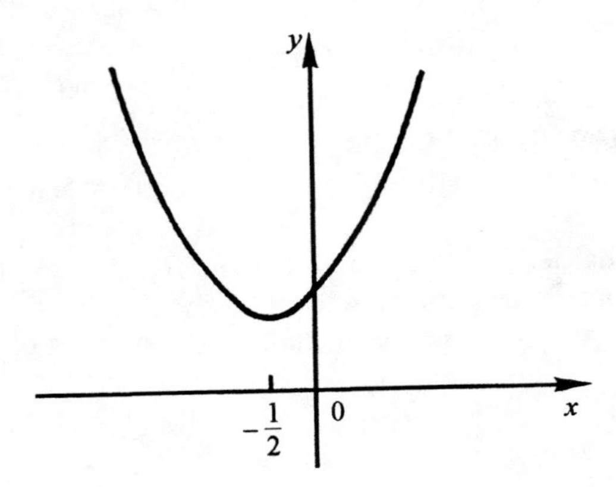

Figure 5.56. Quadratic function $y = x^2 + x + 1$.

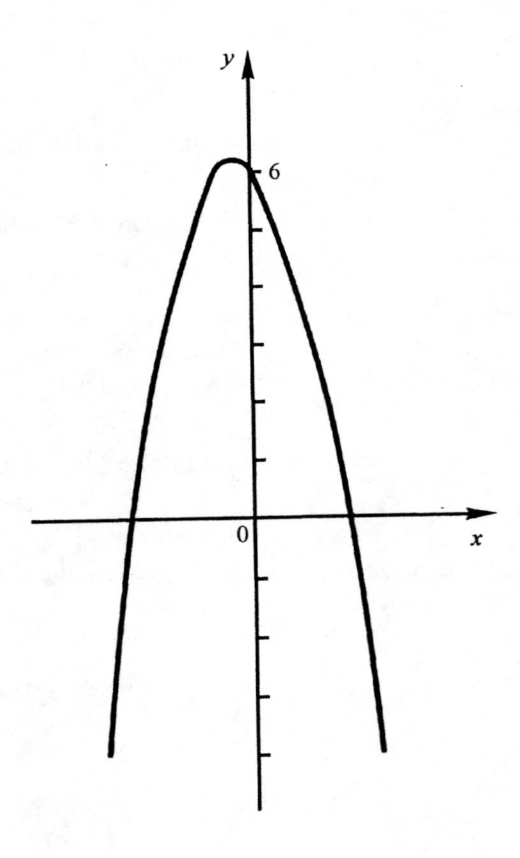

Figure 5.57. Quadratic function $y = -2x^2 - x + 6$.

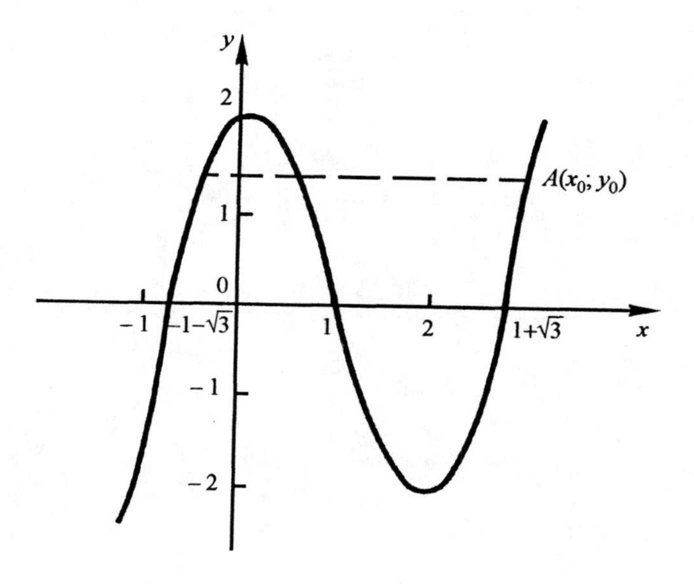

Figure 5.58. Cubic function $y = x^3 - 3x^2 + 2$.

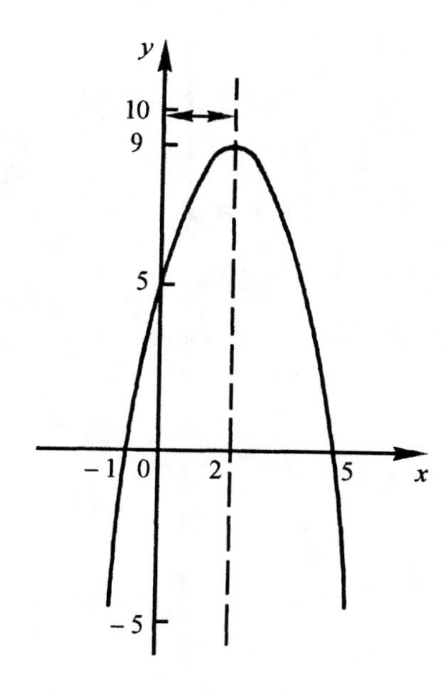

Figure 5.59. Function $y = (1 + x)(5 - x)$.

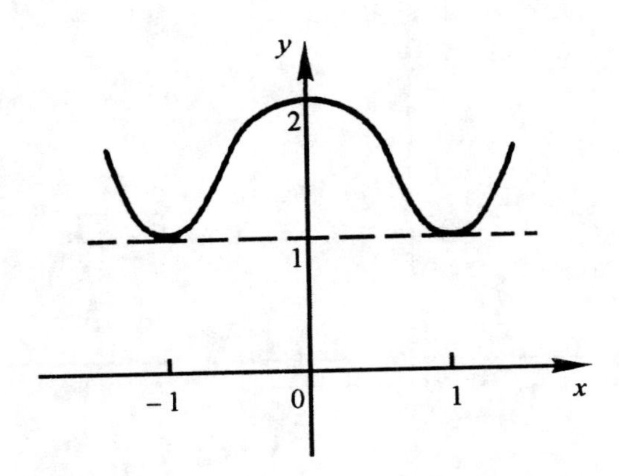

Figure 5.60. Biquadratic function $y = x^4 - 2x^2 + 2$.

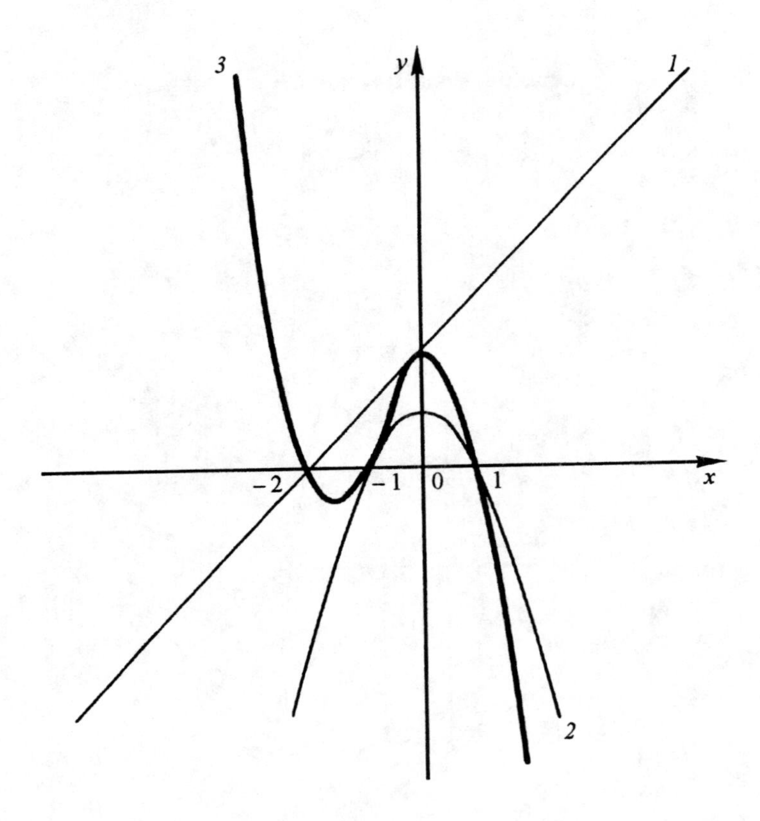

Figure 5.61. The graphs of polynomials: *1*. $y = 2 + x$, *2*. $1 - x^2$, *3*. $y = (1 - x^2)(2 + x)$.

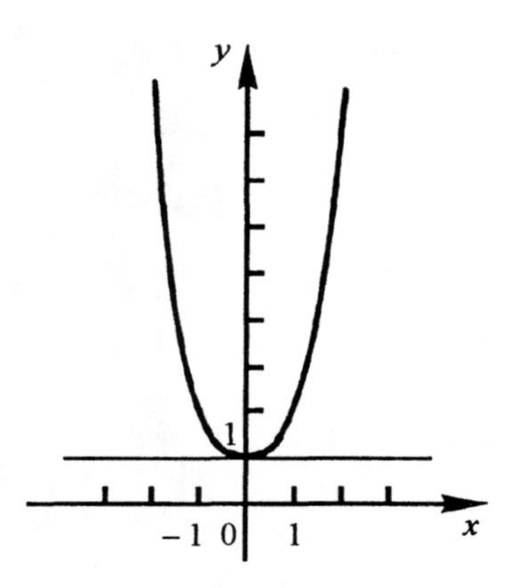

Figure 5.62. The graph of the function $y = 1 + |-x^3|$.

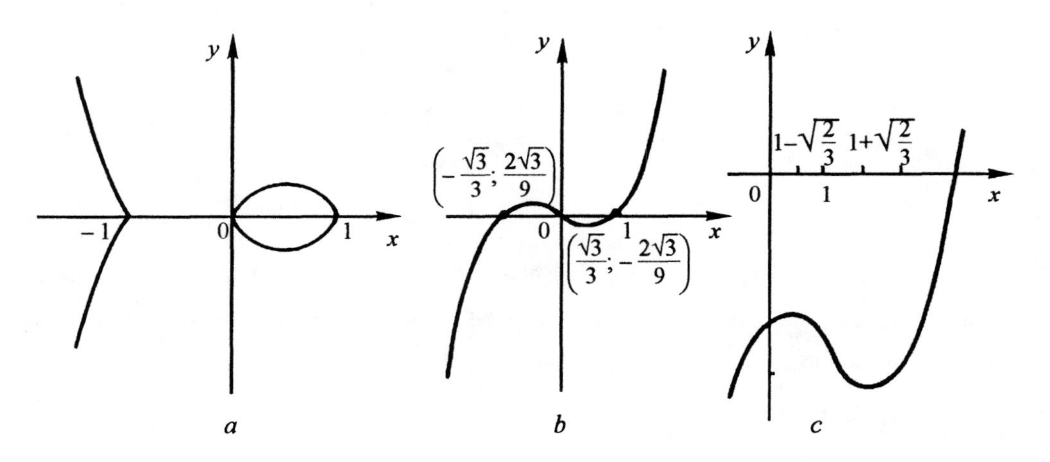

Figure 5.63. The graphs of functions: (*a*) $|y| = x - x^3$; (*b*) $y = x^3 - x$;
(*c*) $y = x^3 - 3x^2 + x - 3$.

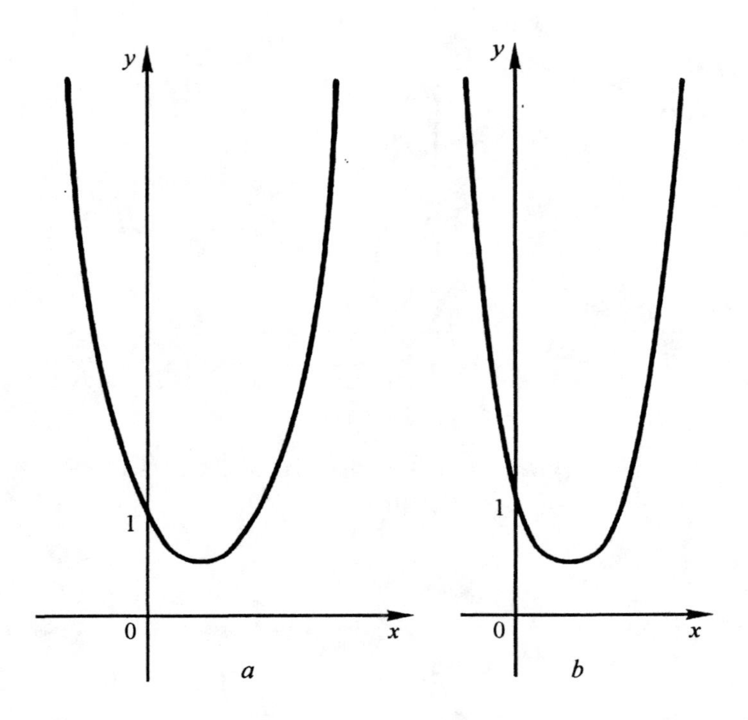

Figure 5.64. The graphs of functions: (*a*) $y=(x^2-x+1)^2$; (*b*) $y=(x^2-x+1)^3$.

Let us consider the case where $x_i<1$ (Figure 5.66) in more detail. To construct the first binomial $y_1=a_0x_i+a_1$, mark off, on one side of the right angle yOx, the line segment $On=\alpha 1$, $O_{k_0}=\alpha x_i$, and a line segment that is perpendicular to it, $nq_0=\alpha a_0$. The straight line Oq_1 and the perpendicular P give a point p_1 such that $\frac{kOp_1}{\alpha a_0}=\frac{\alpha x}{\alpha}$ or $\overline{kOp_1}=\alpha a_0 x_i$. If now the line segment $OO_1=\alpha a_1$ is drawn from the point O downward in the direction of the axis Oy and then a perpendicular straight line O_1n_1 is constructed, then the intersection with the line P gives a point k_1 such that $k_1p_1=\alpha(a_0x+a_1)=\overline{y}_1$. So that the first binomial y_1 has been constructed for $x=x_i$.

Let us construct the second binomial y_2. To do that, we draw from the point P a perpendicular p_{1q_1}. Then $\overline{n_1q_1}=k_1p_1=\overline{y}_1$. The intersection of the inclined line O_1q_1 and the straight line P gives a point p_2, where, as before, $k_1p_2=\alpha y_1x_i$. If we mark off the line segment $\overline{O_1O_2}=\alpha a_2$ from the point O_1 downwards and draw a perpendicular a_2n_2 we get the line segment $k_2p_2=\alpha(y_1x_i+a_2)=\alpha(a_0x_i^2+a_1x_i+a_2)$.

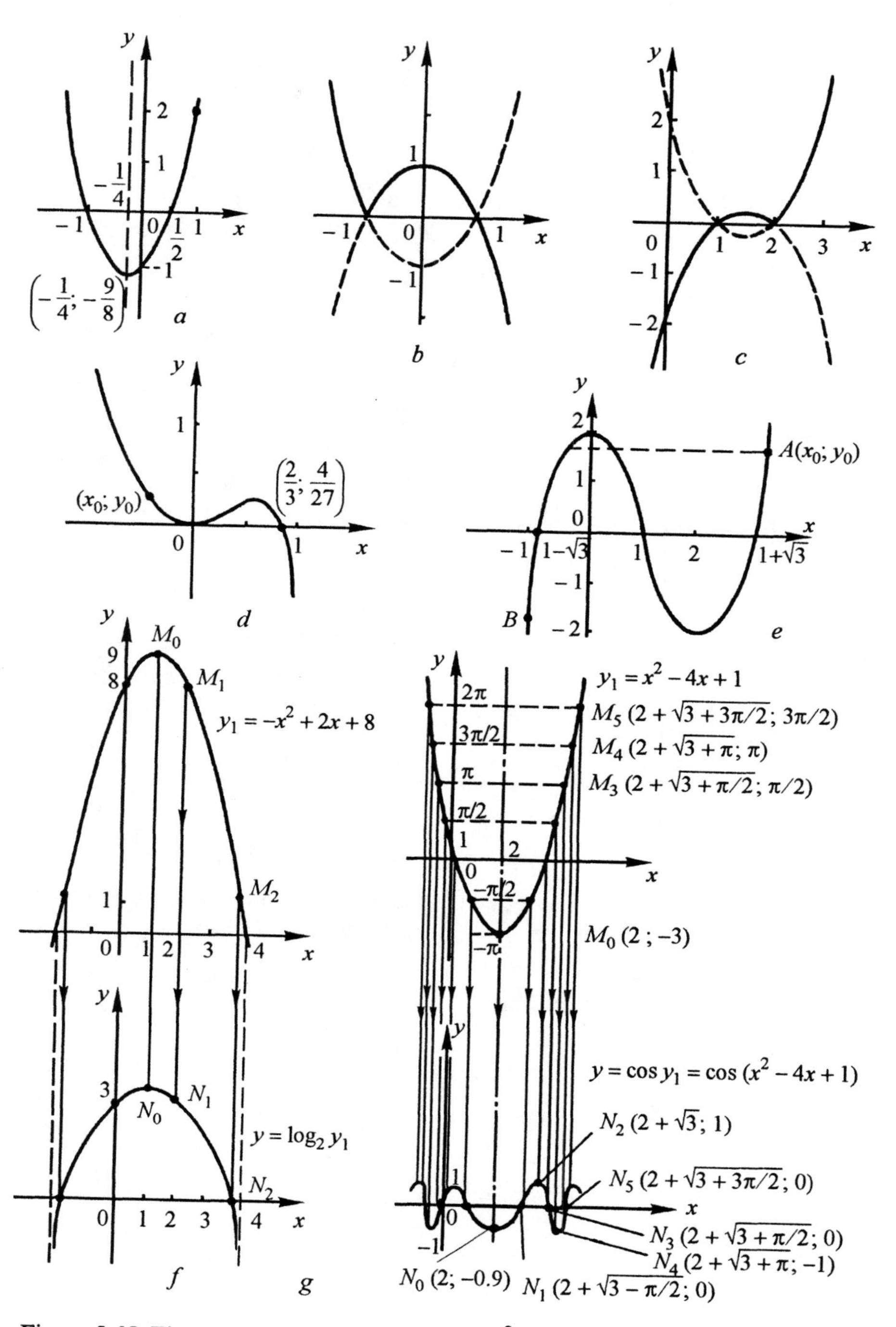

Figure 5.65. The graphs of functions: *(a)* $y = 2x^2 + x - 1$; *(b)* $y = (1 - x)|x + 1|$; *(c)* $y = (x - 1)|x - 2|$; *(d)* $y = x^2 - x^3$; *(e)* $y = x^3 - x^2 - x + 1$; *(f)* $y = \log_2(-x^2 + 2x + 8)$; *(g)* $y = \cos(x^2 - 4x + 1)$.

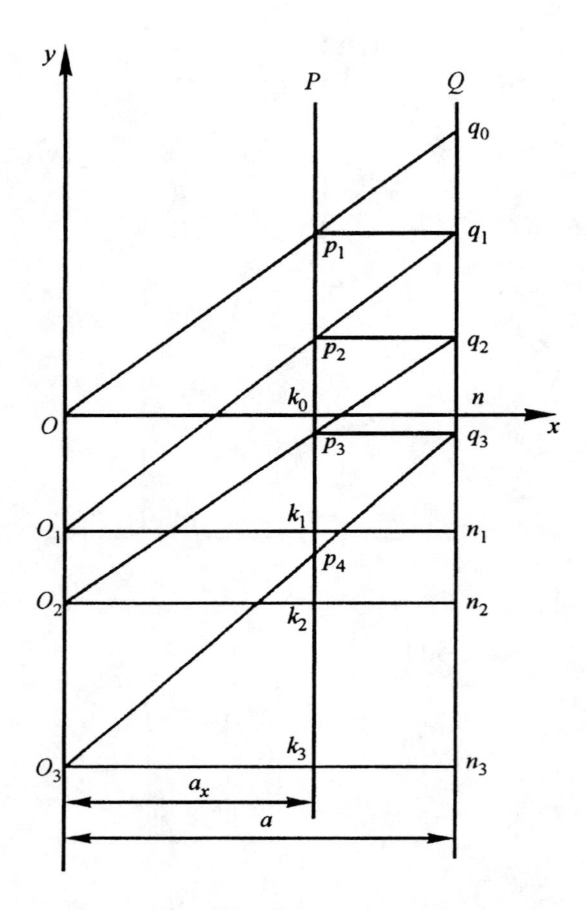

Figure 5.66. Segner's method of constructing the graph of an integer rational function.

In the same way one gets $y_3 = \alpha(y_2 x_i + a_3)$, etc. In the above construction we assumed all the coefficients a_i to be positive. If some of a_i, say a_1, is negative, then the line segment OO_1 is turned upwards from the point O.

If $x_1 > 1$, then the construction remains the same; however, the straight line P will be located from the axis y further than the line Q. For a negative x_i, the construction is similar.

5.2.2. Graphs of fractional rational functions

Fractional linear function

The *fractional linear function* is a function of the form

$$y = \frac{ax+b}{cx+d} \quad (ad - bc \neq 0),$$

where the numerator and denominator are linear functions. By dividing the numerator by the denominator, we get

$$y = \frac{a}{c} + \frac{bc-ad}{c^2\left(x+\frac{d}{c}\right)}.$$

It is now clear that the graph of the fractional linear function can be obtained from the graph of the function $y = 1/x$ by using the translation along the x-axis by $|d/c|$ in the direction opposite to the sign of d/c, stretching along the y-axis by $\frac{bc-ad}{c^2}$, and translation along the y-axis by $|a/c|$ in the same direction as the sign of a/c.

Remark 5.2.2. The construction of the graph of any fractional linear function can be carried out by using a direct study of the function as described in Section 2.2.

The graph of a fractional linear function is an isosceles hyperbola with asymptotes parallel to the coordinate axis, $x = -d/c$ and $y = a/c$.

To plot the graph of a fractional linear function, it is sufficient to determine the "cross" of the asymptotes and its position with respect to one branch of the hyperbola, since the other branch is symmetric to the first one with respect to the point of intersection of the asymptotes.

Example 5.2.1. Plot the graph of the function $y = \frac{2x-1}{x+1}$.

Write the function as $y = \frac{2x-1}{x+1} = 2 - \frac{3}{x+1}$, which shows the transformations needed to be applied to the graph of the function $1/x$ (Figure 5.67):

$$\frac{1}{x} \to 3\frac{1}{x} \to -3\frac{1}{x} \to -3\frac{1}{x+1} \to -3\frac{1}{x+1} + 2 = \frac{2x-1}{x+1}.$$

Graphs of fractional linear functions are shown in Figures 5.68 and 5.69.

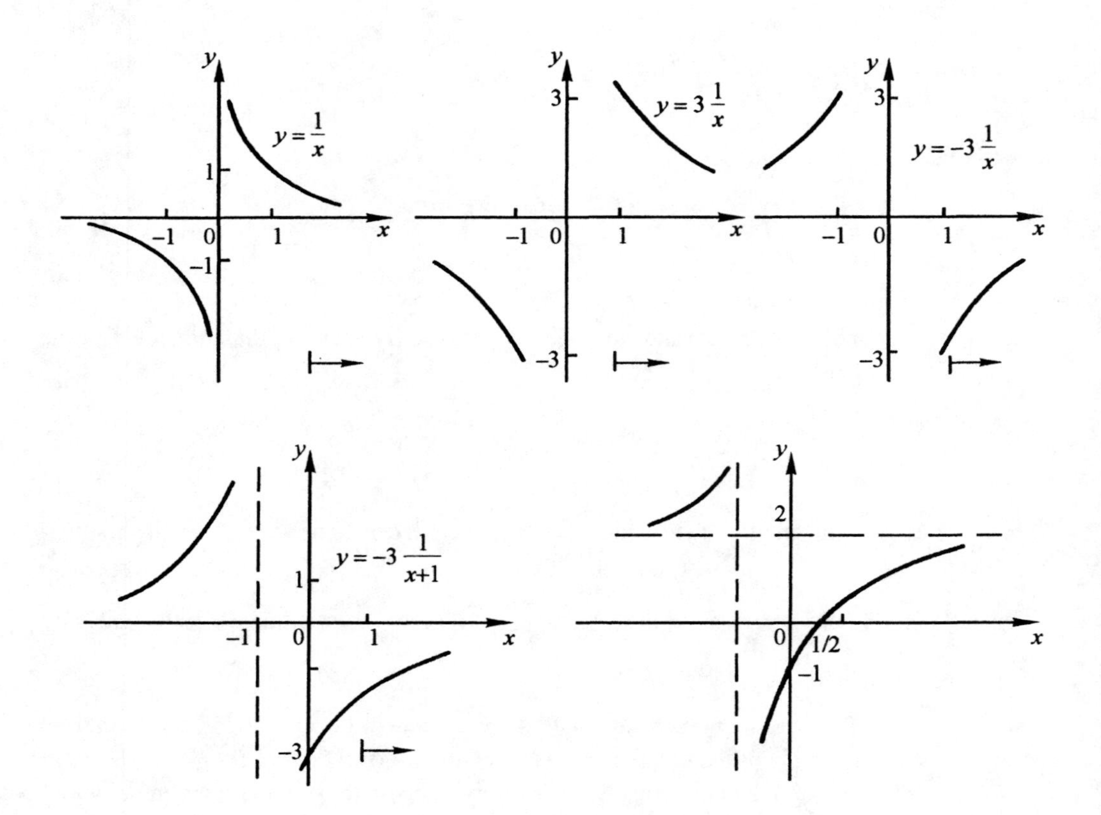

Figure 5.67. The graph of the function $y=\frac{2x-1}{x+1}$.

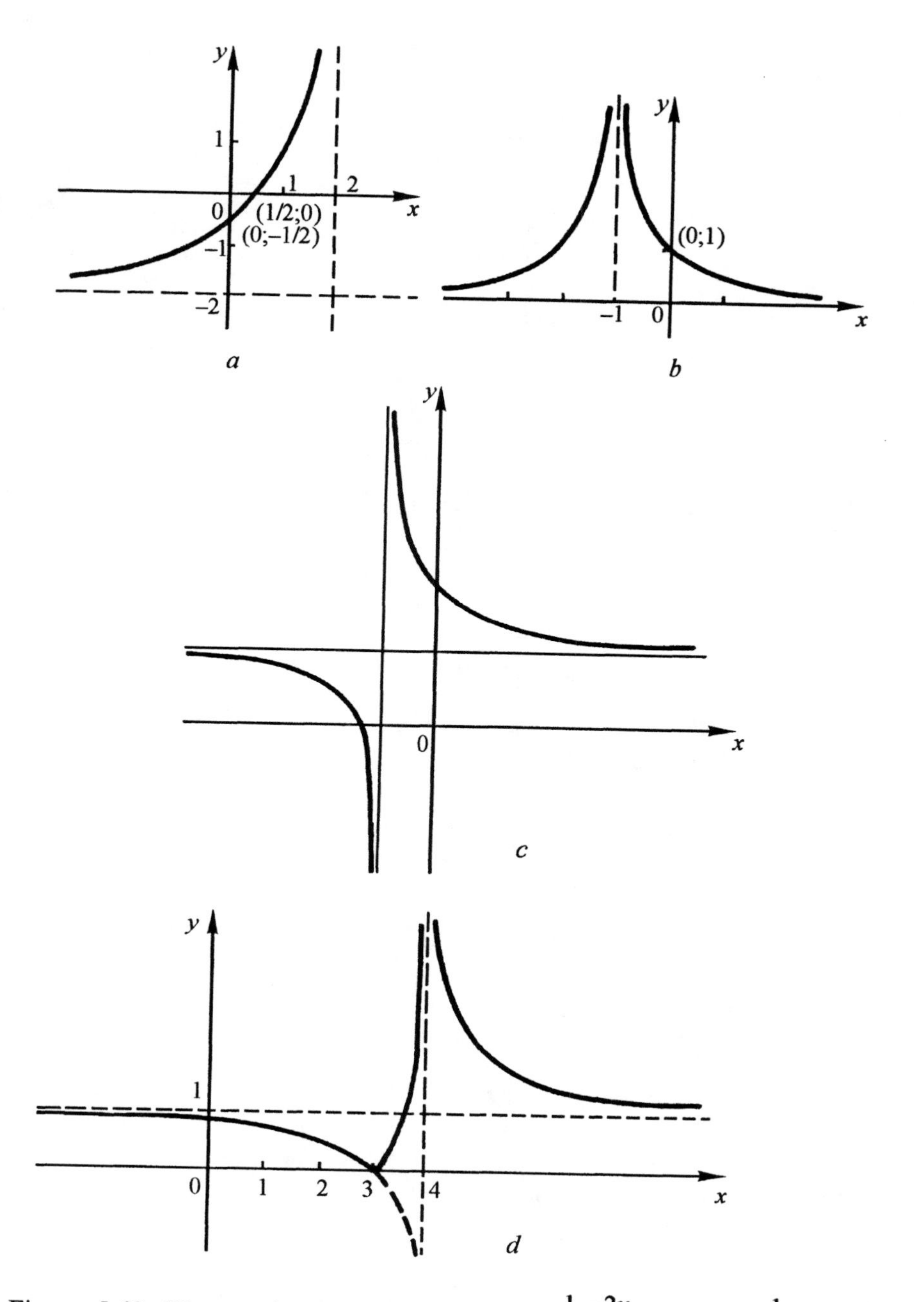

Figure 5.68. The graphs of functions: (*a*) $y=\dfrac{1-2x}{x-2}$; (*b*) $y=\dfrac{1}{|x+1|}$; (*c*) $y=\dfrac{3x+4}{2(x+1)}$; (*d*) $y=\left|\dfrac{x-3}{x-4}\right|$.

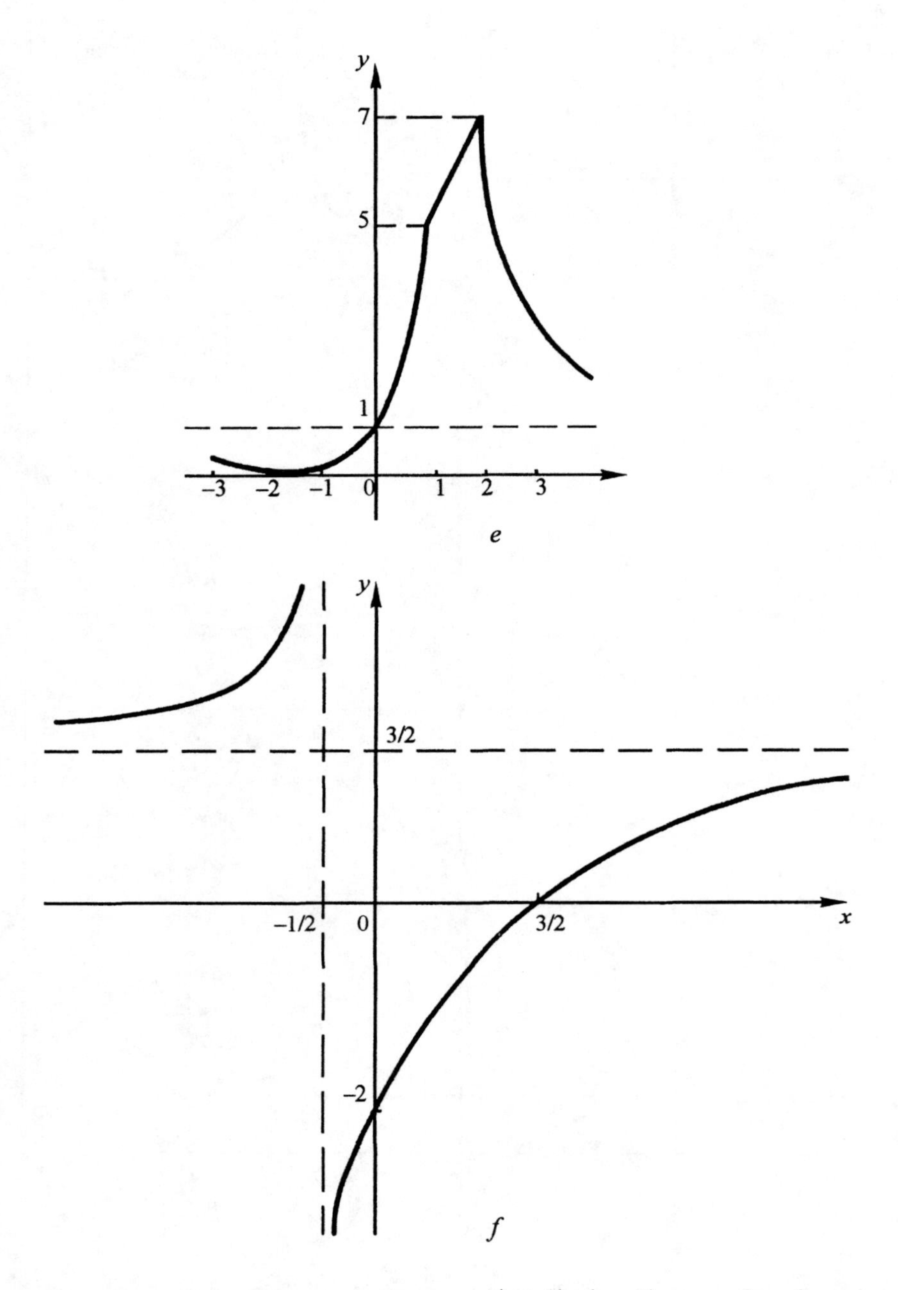

Figure 5.68. The graphs of functions: (*e*) $y = \dfrac{|x+1| + |x+2|}{|x-1| + |x-2|}$; (*f*) $y = \dfrac{3x-2}{2x+1}$.

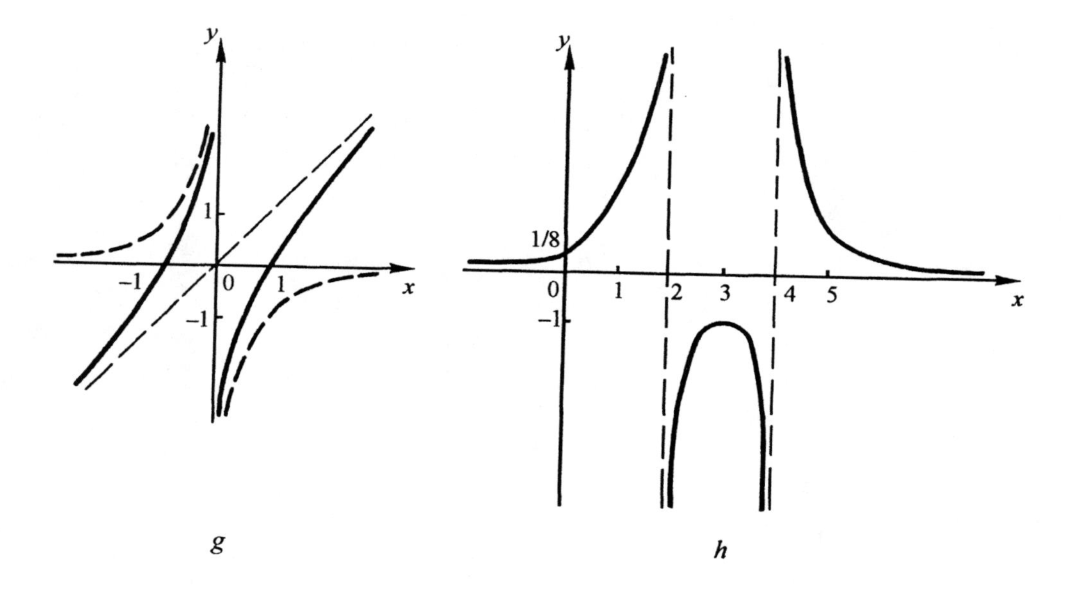

Figure 5.68. The graphs of functions: (g) $y = x - \frac{1}{x}$; (h) $y = \frac{1}{x^2 - 6x + 8}$.

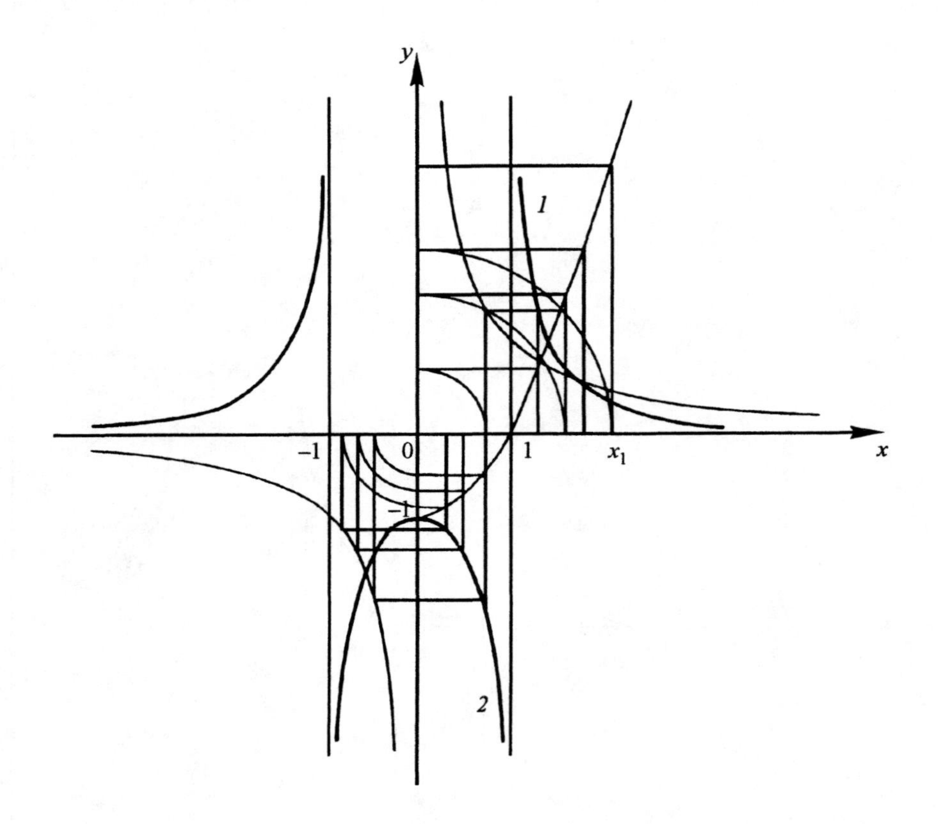

Figure 5.69. The graphs of functions: *1*. $y = \frac{1}{x}$, *2*. $y = \frac{1}{x^2 - 1}$.

Fractional rational function

Consider a fractional rational function

$$y = \frac{a_0x^n + a_1x^{n-1} + \ldots + a_{n-1}x + a_n}{b_0x^m + b_1x^{m-1} + \ldots + b_{m-1}x + b_m} = \frac{P(x)}{Q(x)},$$

where the numerator and denominator are polynomials of degree n and m, correspondingly. Let the fraction $P(x)/Q(x)$ be proper, i.e. $n < m$. It is well known that any proper irreducible rational fraction can be uniquely represented as the sum of a finite number of elementary fractions which are determined from the decomposition

$$Q(x) = a_0(x - k - 1)^{m_1} \cdot \ldots \cdot (x - k_s)^{m_s}(x^2 + p_{1x} + q_1)^{\mu_1} \cdot \ldots \cdot (x^2 + p_tx + q_t)^{\mu_t}$$

of the denominator $Q(x)$ into the product of real factors, where $k_1, \ \dots \ , k_s$ are real roots of the polynomial $Q(x)$ with multiplicities $m_1, \ \dots \ , m_s$, and the trinomials correspond to pairs of complex conjugate roots of $Q(x)$ with multiplicities $\mu_1, \dots, \mu_r$. The decomposition of the fraction is

$$\frac{P(x)}{Q(x)} = \frac{A_1}{(x-k_1)^{m_1}} + \frac{A_2}{(x-k_1)^{m_1-1}} + \dots + \frac{A_{m_1}}{x-k_1} + \dots +$$

$$+ \frac{L_1}{(x-k_s)^{m_s}} + \frac{L_2}{(x-k_s)^{m_s-1}} + \dots + \frac{L_{m_s}}{x-k_s} + \dots +$$

$$+ \frac{B_1x+C_1}{(x^2+p_1x+q_1)^{\mu_1}} + \dots + \frac{B_{\mu_1}x+C_{\mu_1}}{x^2+p_1x+q_1} + \dots +$$

$$+ \frac{M_1x+N_1}{(x^2+p_tx+q_t)^{\mu_1}} + \dots + \frac{M_{\mu_1}x+N_{\mu_1}}{x^2+p_tx+q_t}.$$

The fractions of the form

$$\frac{A}{x-k}, \quad \frac{A}{(x-k)^m}, \quad \frac{Bx+C}{x^2+px+q}, \quad \frac{Bx+C}{(x^2+px+q)^{\mu}}$$

are called *elementary rational fractions* of the first, second, third, and fourth type, correspondingly. Here A, B, C, k are real numbers, m and μ are integers, $m, \mu > 1$, the trinomial x^2+px+q has complex roots.

It is clear that the graph of a fractional linear function can obtained as the sum of the graphs of elementary fractions.

The graph of the function $y = \dfrac{1}{(x-k)^m}$, $m \in \mathbf{N}$, is obtained from the graph of the function $1/x^m$ by translating it along the x-axis to the right by $|k|$.

The graph of the function $y = \dfrac{1}{x^2+px+q}$ is easily constructed by separating a complete square in the denominator, $y = \dfrac{1}{(x+p/2)^2+q-p^2/4}$, and applying the corresponding transformations to the graph of the function $1/x^2$.

Plotting the graph of the function $y = \dfrac{Bx+C}{x^2+px+q}$ is reduced to constructing the graphs of two functions: $y = Bx + C$ and $y = \dfrac{1}{x^2+px+q}$.

Remark 5.2.3. Plotting the graphs of the functions $y=\frac{P(x)}{Q(x)}$, $y=\left(\frac{ax+b}{cx+d}\right)^n$, where $ad-bc\neq 0$, $y=\frac{1}{(ax^2+bx+c)^n}$, $y=\left[\frac{P(x)}{Q(x)}\right]^n$, where n is an integer, can be performed according to the general scheme (see Section 2.2). In some particular cases, one can successfully use transformations of graphs (see Section 2.2). The best way to construct graphs of this type of functions is to use methods of calculus (see Part II).

Examples of graphs of fractional rational functions are shown in Figures 5.70 – 5.75.

5.2.3. Graphs of irrational functions

Consider some simple cases.

Functions of the form $y=\pm\sqrt{ax+b}$

The graph of this function can easily be obtained from the graph of the function $y=\sqrt{x}$ by using corresponding transformations (see Chapter 4). The graph of the function $y=\pm\sqrt{ax+b}$ is a parabola. Its axis coincides with the x-axis, the vertex is $A(-b/a;0)$, and the parameter is $p=a/2$. The graph is shown in Figure 5.76.

Function of the form $y=\pm\sqrt{ax^2+bx+c}$

The analysis and construction of the curve is done according to the scheme described in Section 2.2.

The form of the graph of the function $y=\pm\sqrt{ax^2+bx+c}$ depends on the signs of a and $\delta=4ac-b^2$. If $a<0$, $\delta<0$, then the graph is an ellipse; if $a>0$, $\delta<0$, then it is a hyperbola. Its axes are $y=0$, $x=-b/2a$, the vertices — $A(\frac{b+\sqrt{-\delta}}{2a};0)$, $C(-\frac{b-\sqrt{-\delta}}{2a};0)$, $B(-b/2a;\sqrt{\delta/4a})$, $D(-b/2a;-\sqrt{\delta/4a})$ (Figure 5.77).

If $a<0$ and $\delta>0$, there are no real x satisfying the equation.

Remark 5.2.4. It is easier to construct the graphs of the functions

$$y=(ax^2+bx+c)^{p/q},\quad y=\left(\frac{ax+b}{cx+d}\right)^{p/q},\quad y=\left[\frac{P(x)}{Q(x)}\right]^{p/q},$$

where p and q are mutually prime, if methods of calculus are used (see Part II).

Graphs of other irrational functions are shown in Figures 5.78 – 5.81.

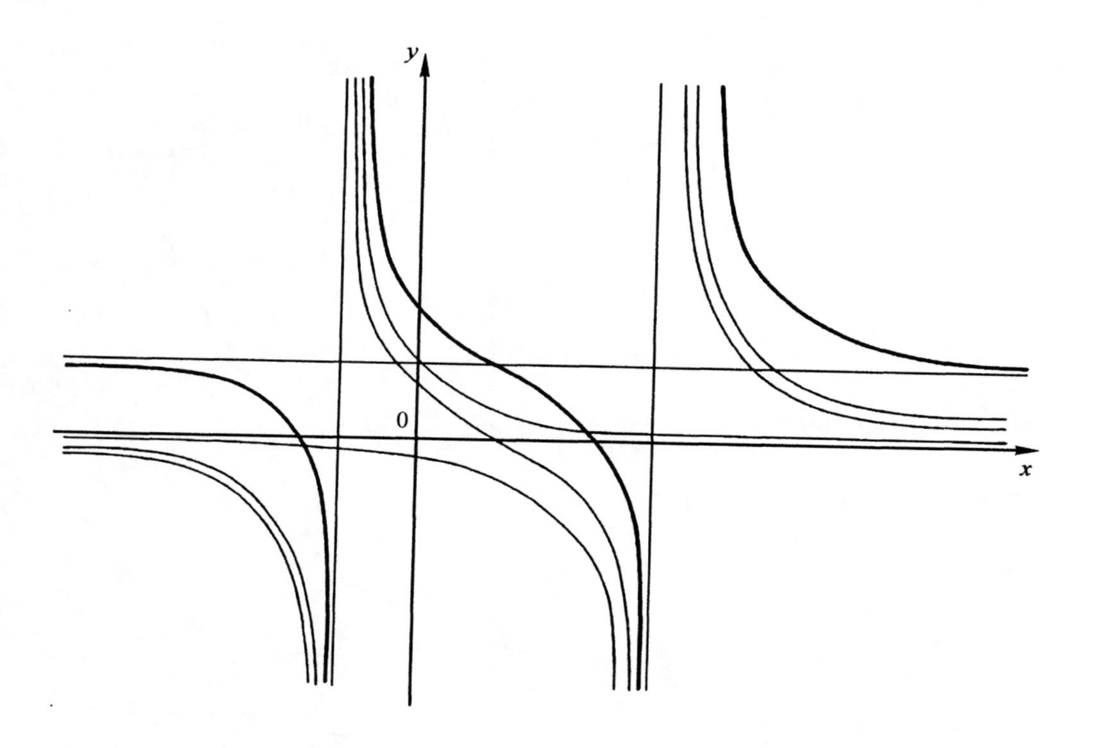

Figure 5.70. Fractional rational function $y = \dfrac{x^2 - 4}{x^2 - 2x - 3}$.

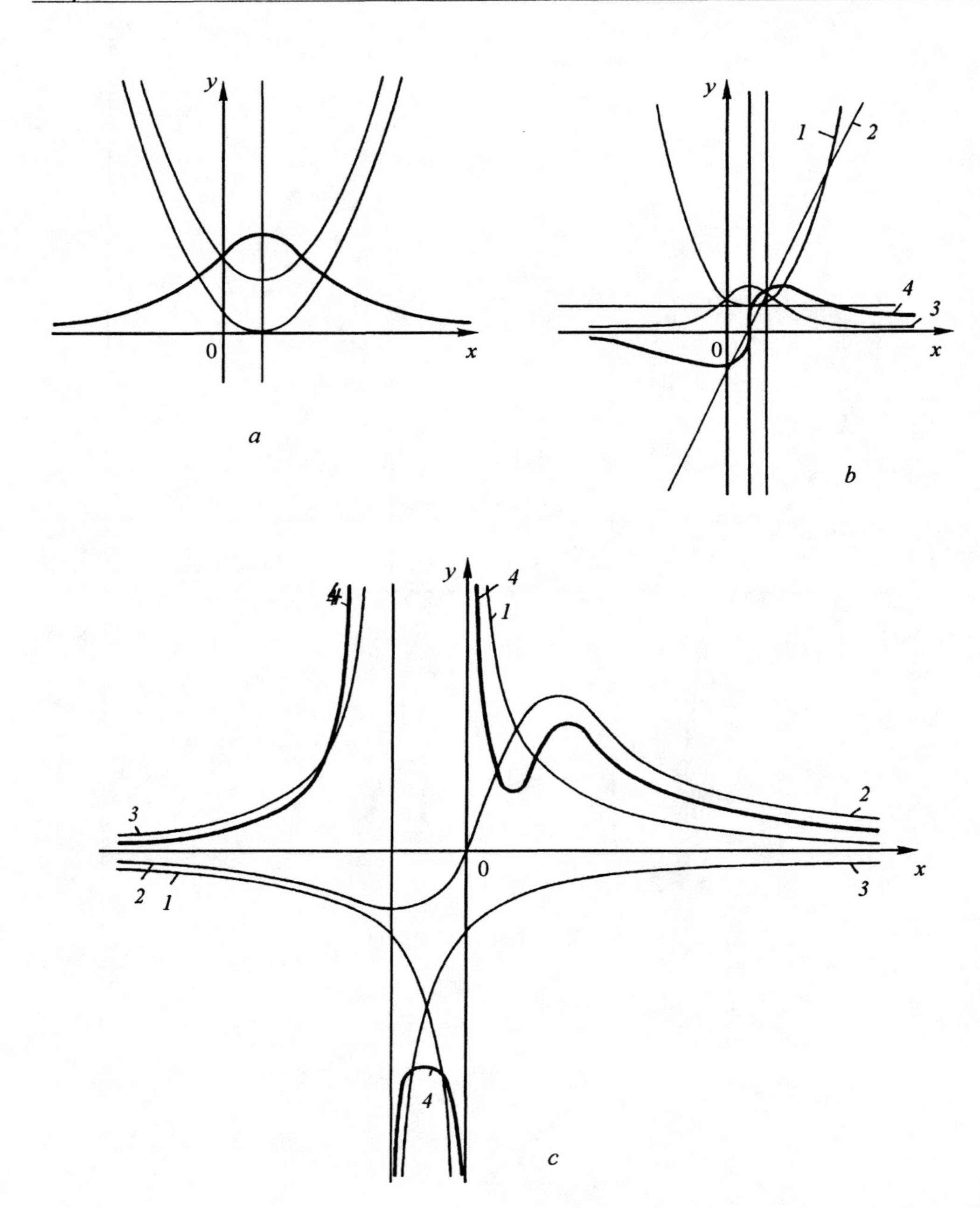

Figure 5.71. The graphs of functions: *(a)* $y = \dfrac{1}{x^2 - x + 1}$; *(b)* *1.* $y = x^2 - x + 1$, *2.* $y = 2x - 1$, *3.* $y = \dfrac{1}{x^2 - x + 1}$, *4.* $y = \dfrac{2x - 1}{x^2 - x + 1}$; *(c)* *1.* $y = \dfrac{1}{x}$, *2.* $y = \dfrac{2x}{x^2 - x + 1}$, *3.* $y = -\dfrac{2}{x + 1}$, *4.* $y = \dfrac{x^3 + 4x^2 - 2x + 1}{x(x + 1)(x^2 - x + 1)}$.

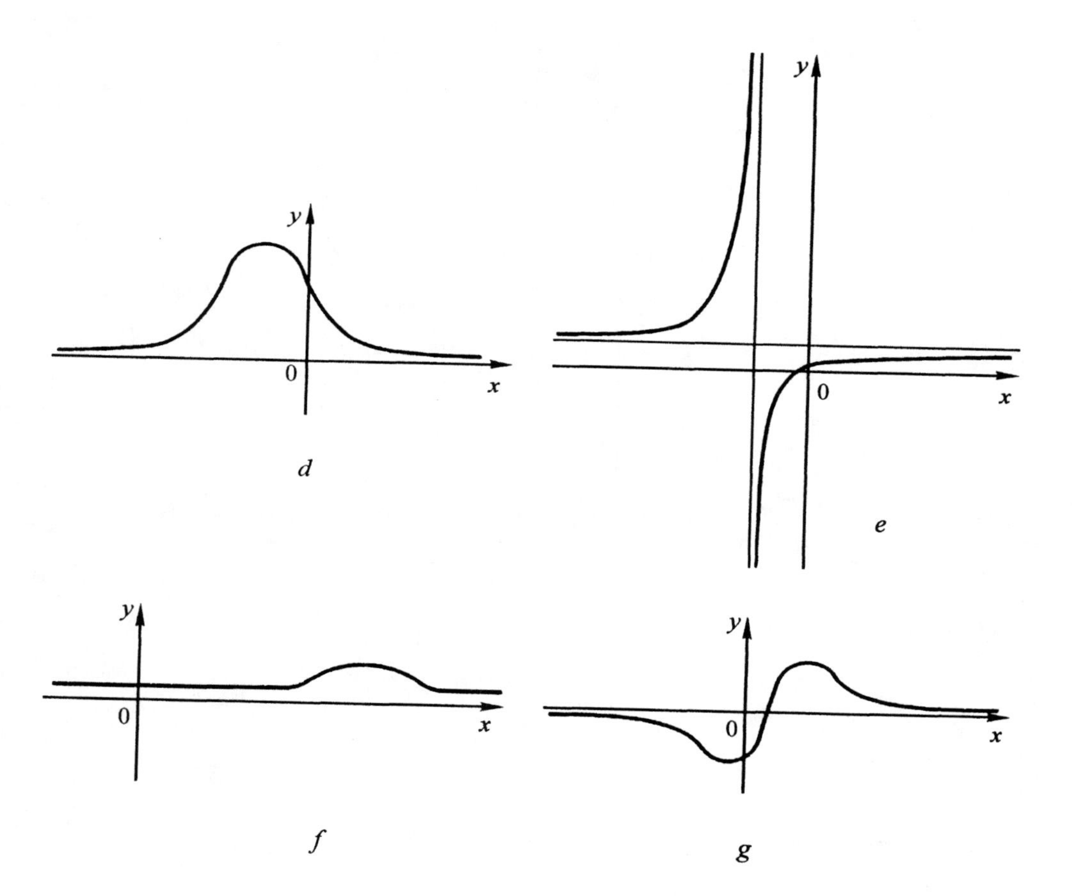

Figure 5.71. The graphs of functions: (*d*) $y = \dfrac{1}{(1+x+x^2)^2}$; (*e*) $y = \left(\dfrac{2x+1}{3x+4}\right)^3$;

(*f*) $y = \dfrac{1}{(x^2-4x+7)^2}$; (*g*) $y = \left(\dfrac{2x-1}{x^2-x+1}\right)^3$.

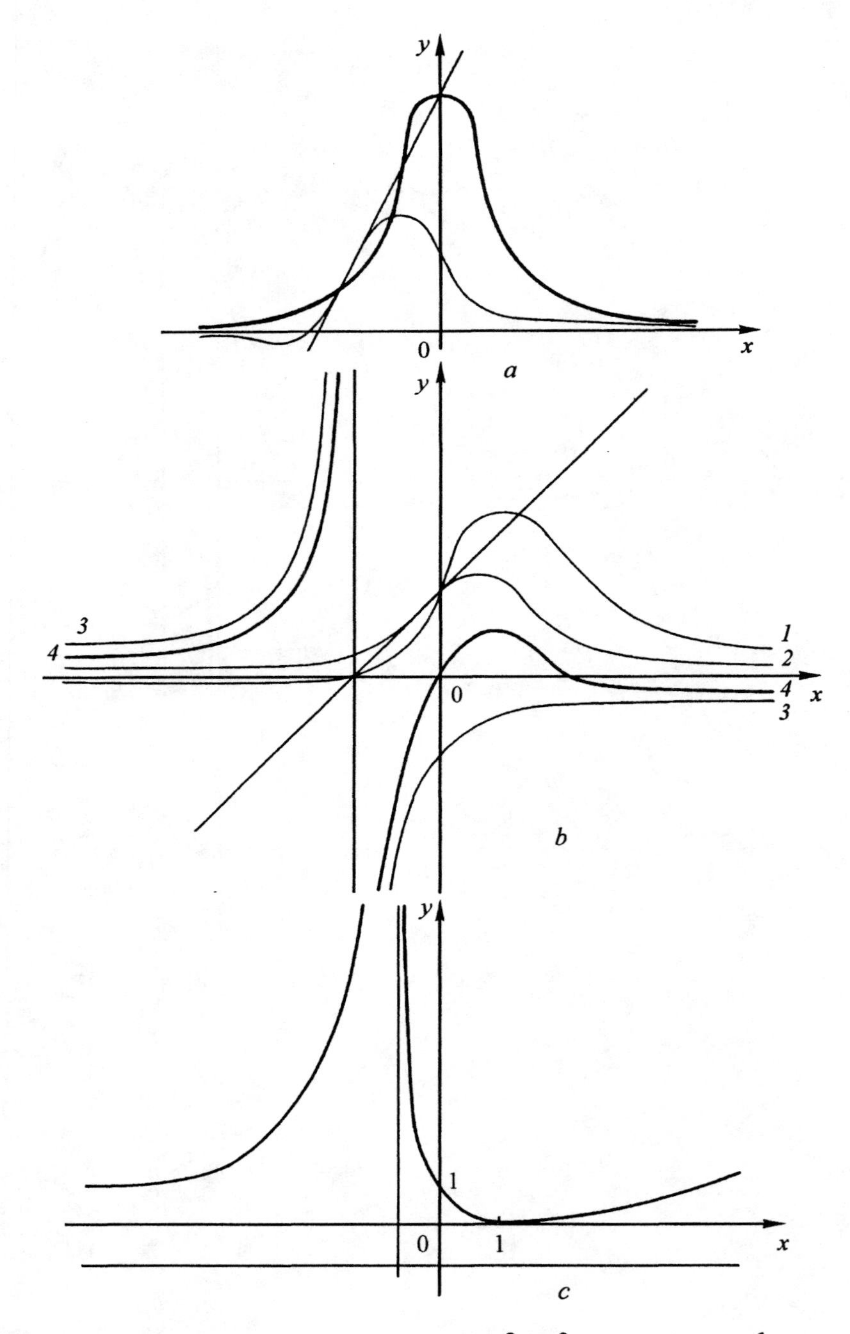

Figure 5.72. The graphs of functions: (*a*) $y = \dfrac{2x+3}{1+x+x^2}$; (*b*) *1*. $y = \dfrac{1}{x^2-x+1}$, *2*. $y = \dfrac{x+1}{x^2-x+1}$, *3*. $y = -\dfrac{1}{x+1}$, *4*. $y = \dfrac{3x}{1+x^3}$; (*c*) $y = \left(\dfrac{1-x}{1+x}\right)^2$.

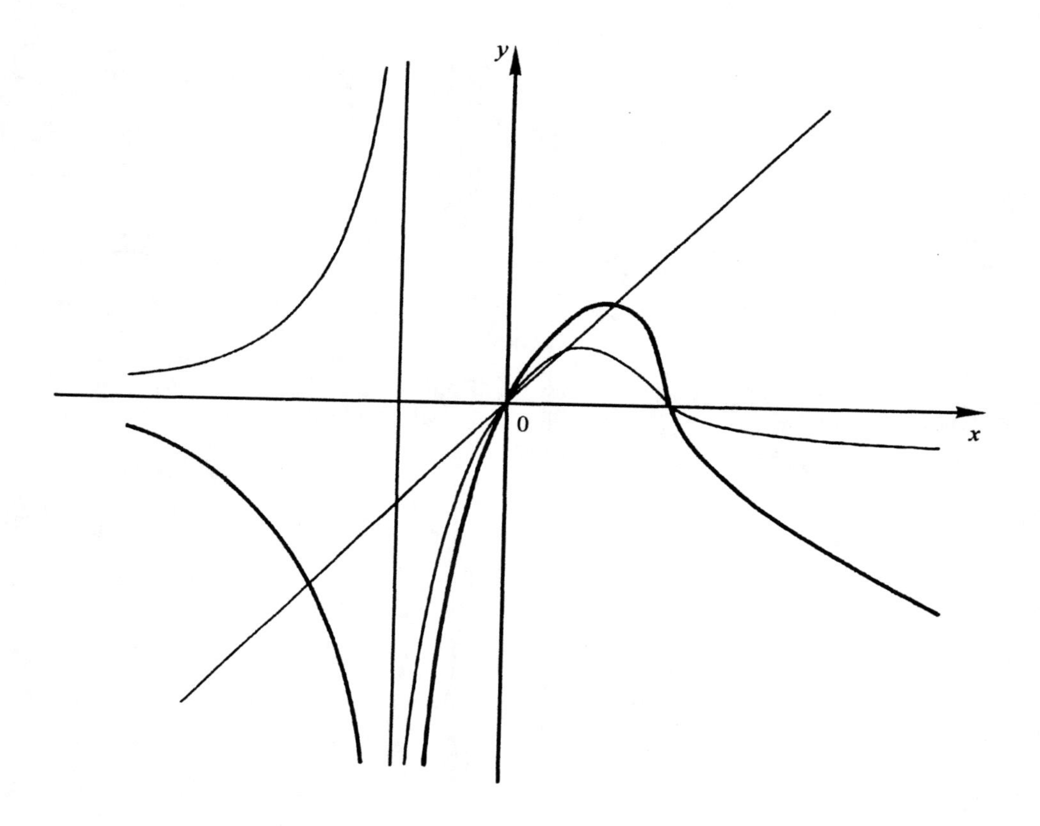

Figure 5.73. Fractional linear function $y = \frac{3x^2}{1+x^3}$.

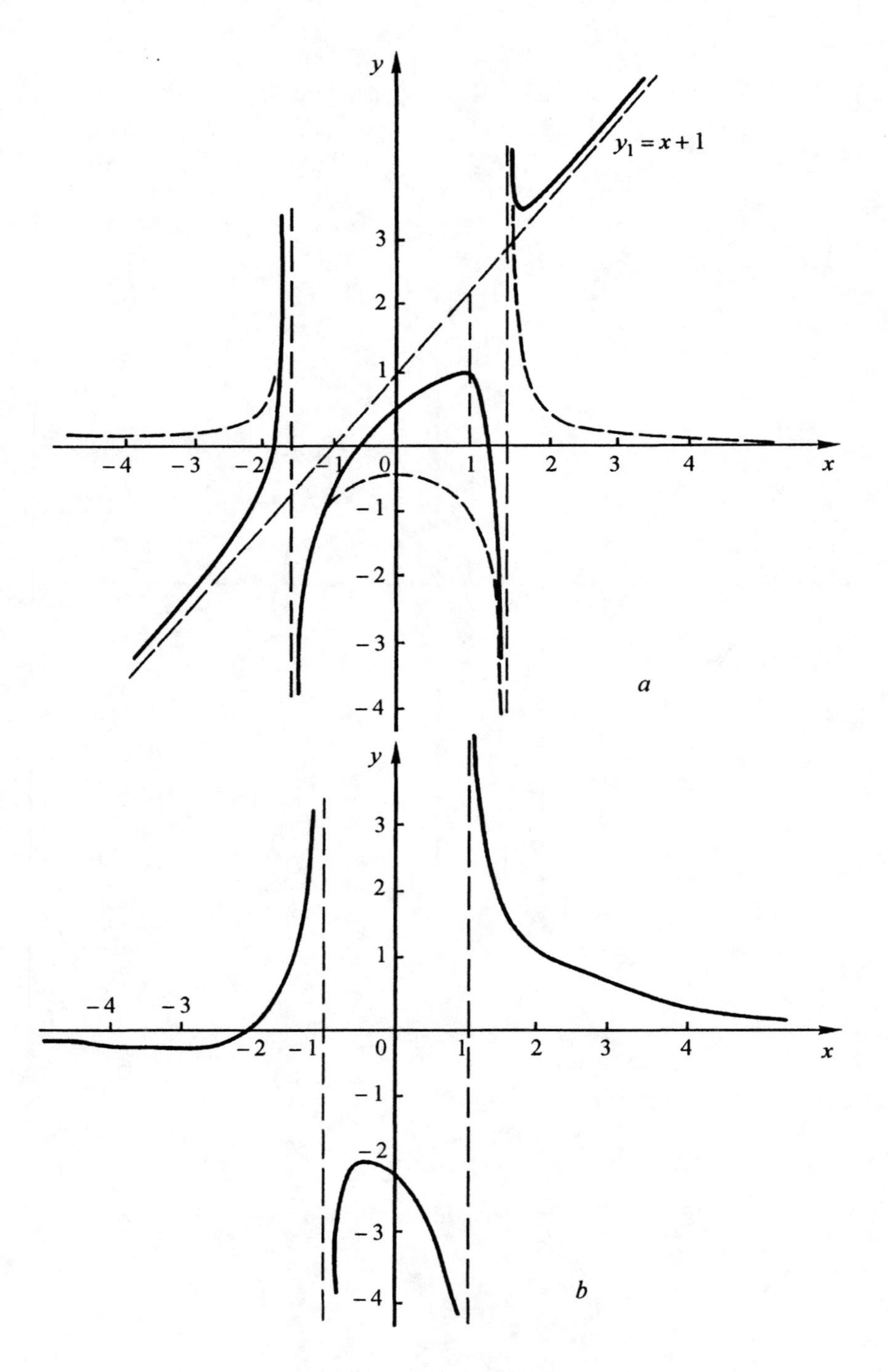

Figure 5.74. The graphs of functions: (*a*) $y = \dfrac{x^3 + x^2 - 2x - 1}{x^2 - 2}$; (*b*) $y = \dfrac{x + 2}{x^2 - 1}$.

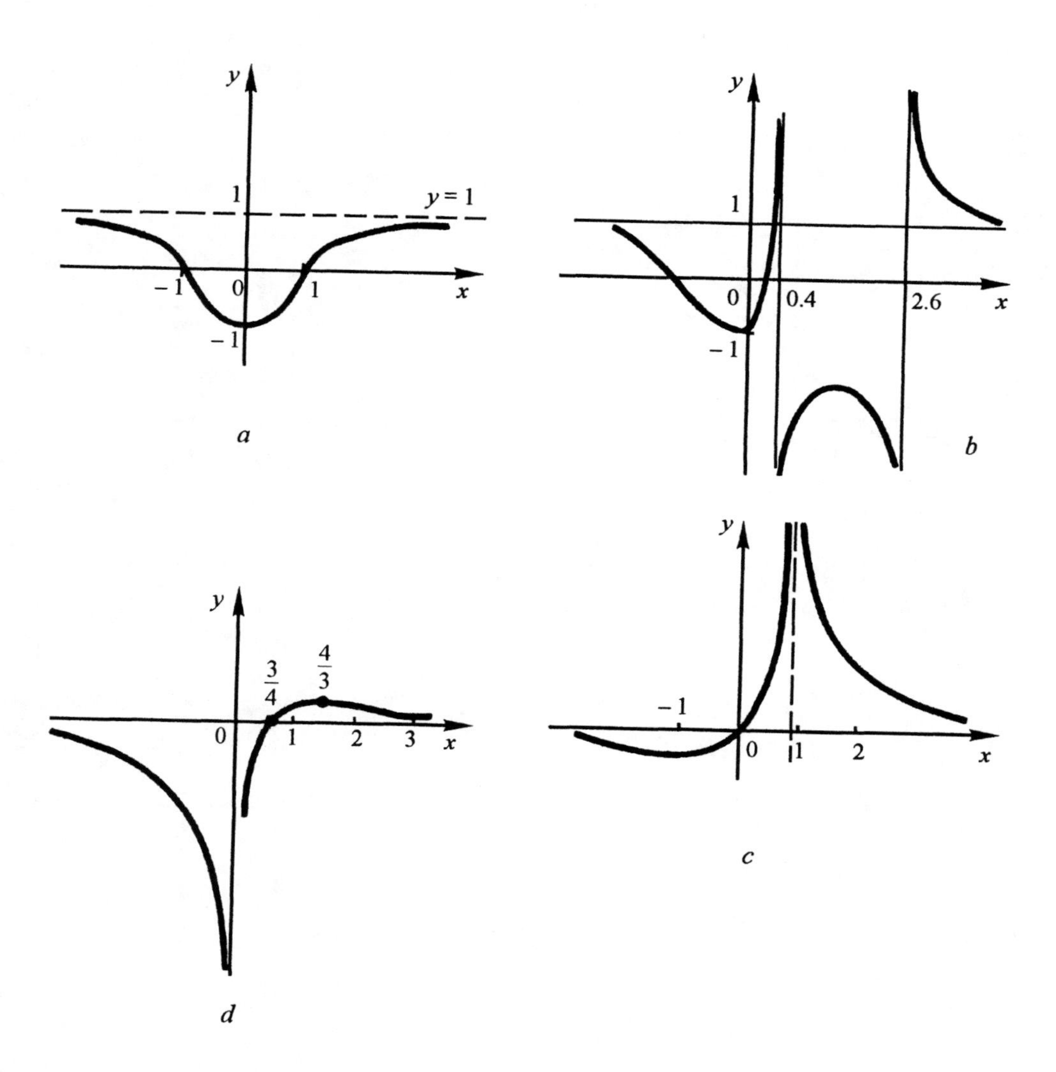

Figure 5.75. The graphs of functions: $(a)\ y = \dfrac{x^2-1}{x^2+1}$; $(b)\ y = \dfrac{x^2+3x-1}{x^2-3x+1}$; $(c)\ y = \dfrac{x}{(x-1)^2}$; $(d)\ y = \dfrac{3x-2}{5x^2}$.

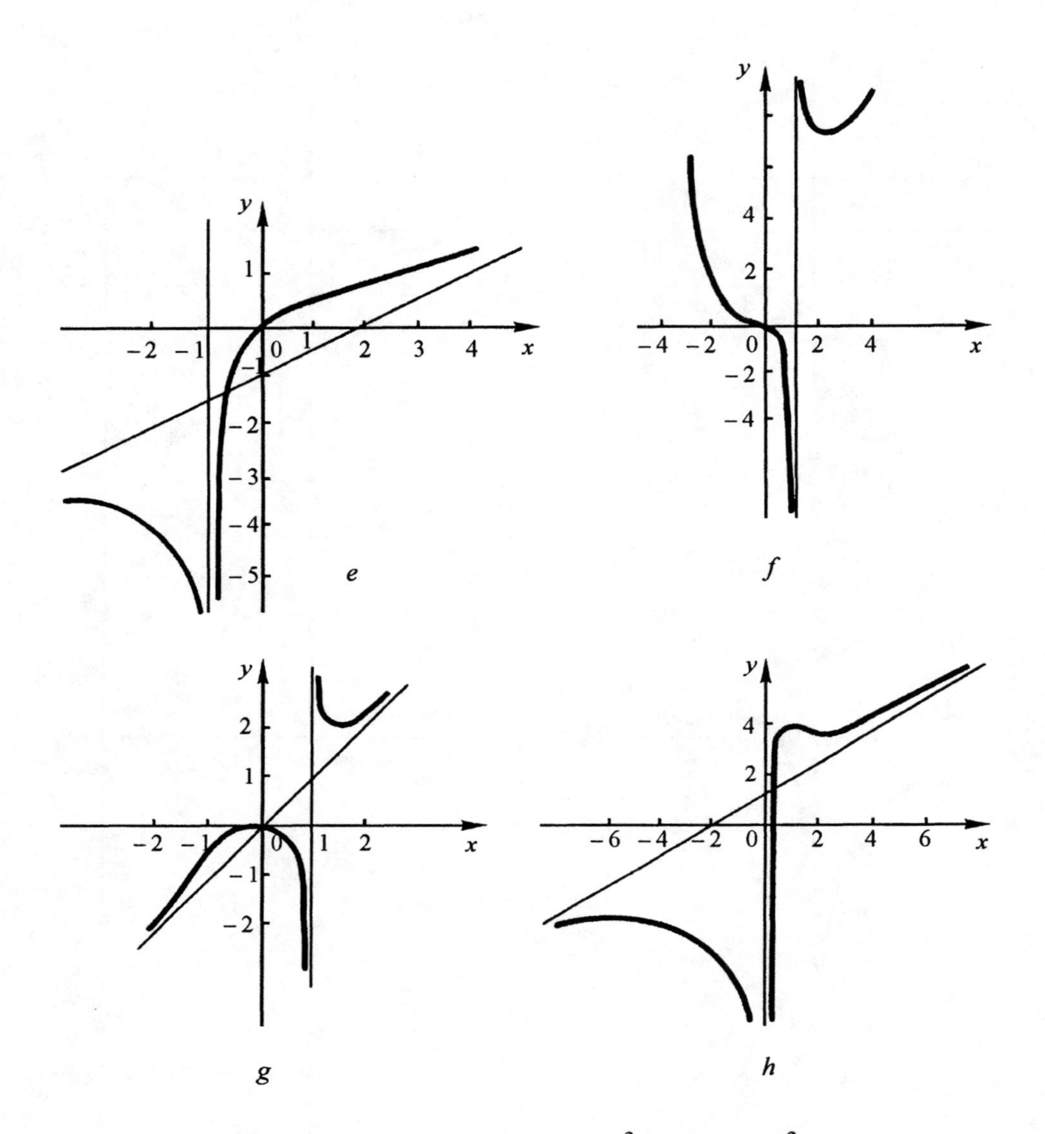

Figure 5.75. The graphs of functions: (*e*) $y = \dfrac{x^3}{2(x+1)^2}$; (*f*) $y = \dfrac{x^3}{x-1}$; (*g*) $y = \dfrac{x^4}{x^3-1}$; (*h*) $y = \dfrac{x^3+2x^2+7x-3}{2x^2}$.

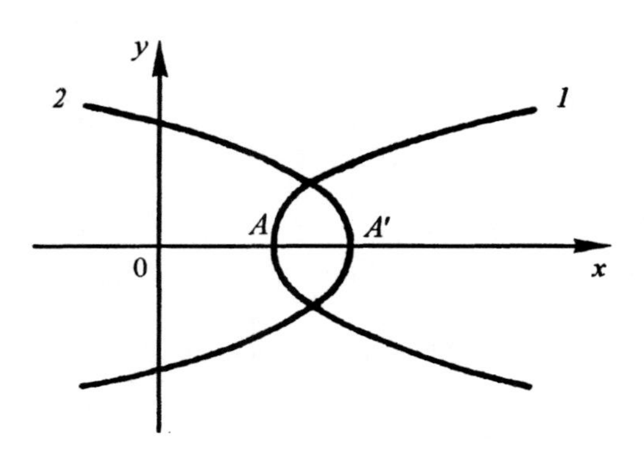

Figure 5.76. Irrational function $y = \pm\sqrt{ax+b}$. *1*. $A > 0$, *2*. $a < 0$.

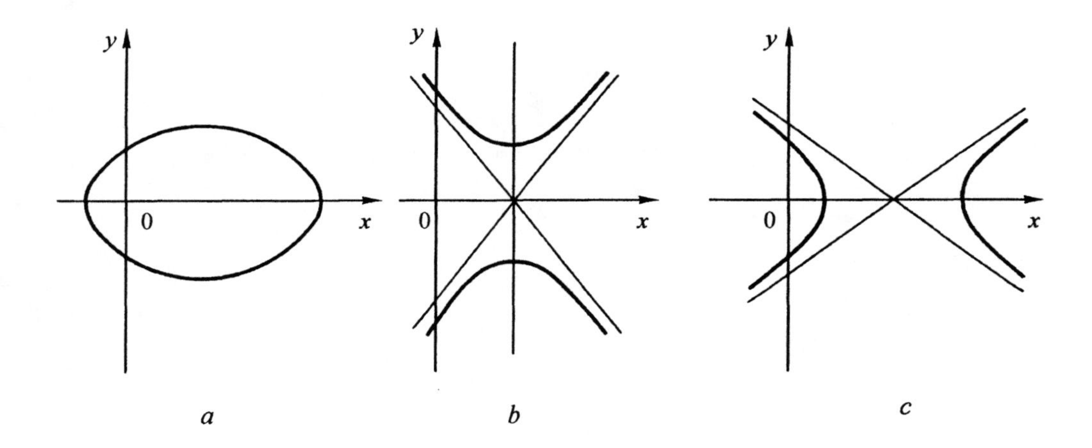

Figure 5.77. Function $y = \pm\sqrt{ax^2+bx+c}$: (*a*) $a < 0$, $\delta < 0$; (*b*) $a > 0$, $\delta > 0$; (*c*) $a > 0, \delta < 0$.

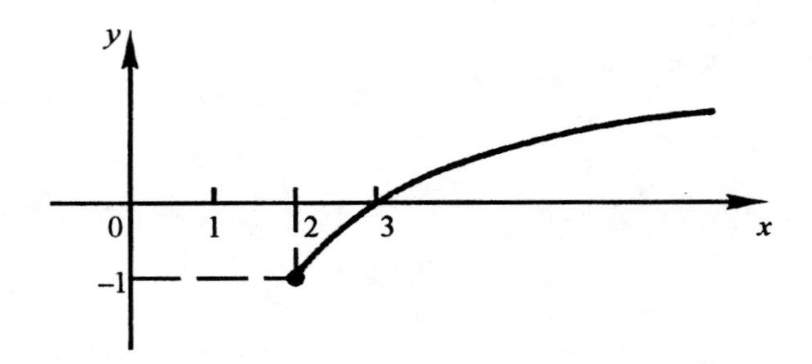

Figure 5.78. The graph of the function $y = \sqrt{x-2} - 1$.

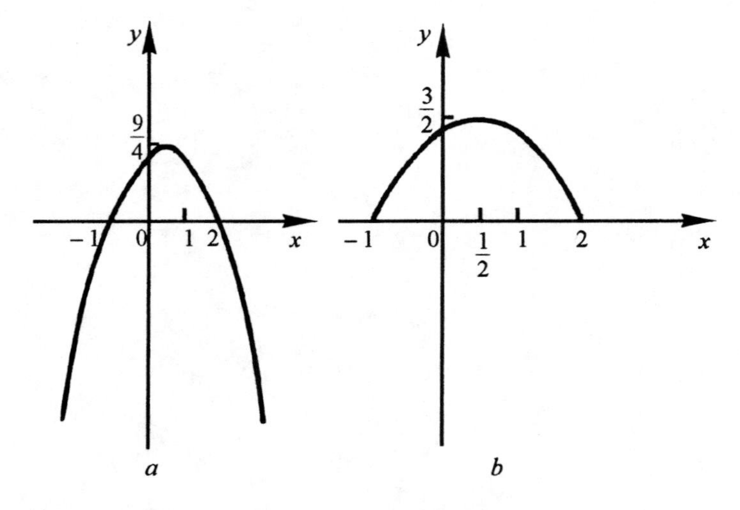

Figure 5.79. The graphs of functions: (*a*) $y = -x^2 + x + 2$; (*b*) $y = \sqrt{-x^2 + x + 2}$.

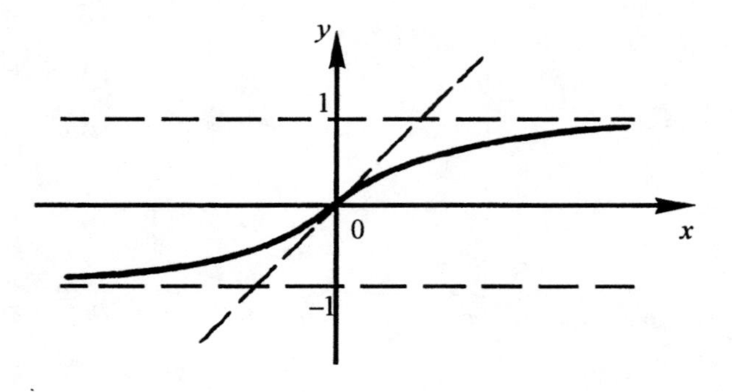

Figure 5.80. The graph of the function $y = \sqrt{x^2 + x + 1} - \sqrt{x^2 - x + 1}$.

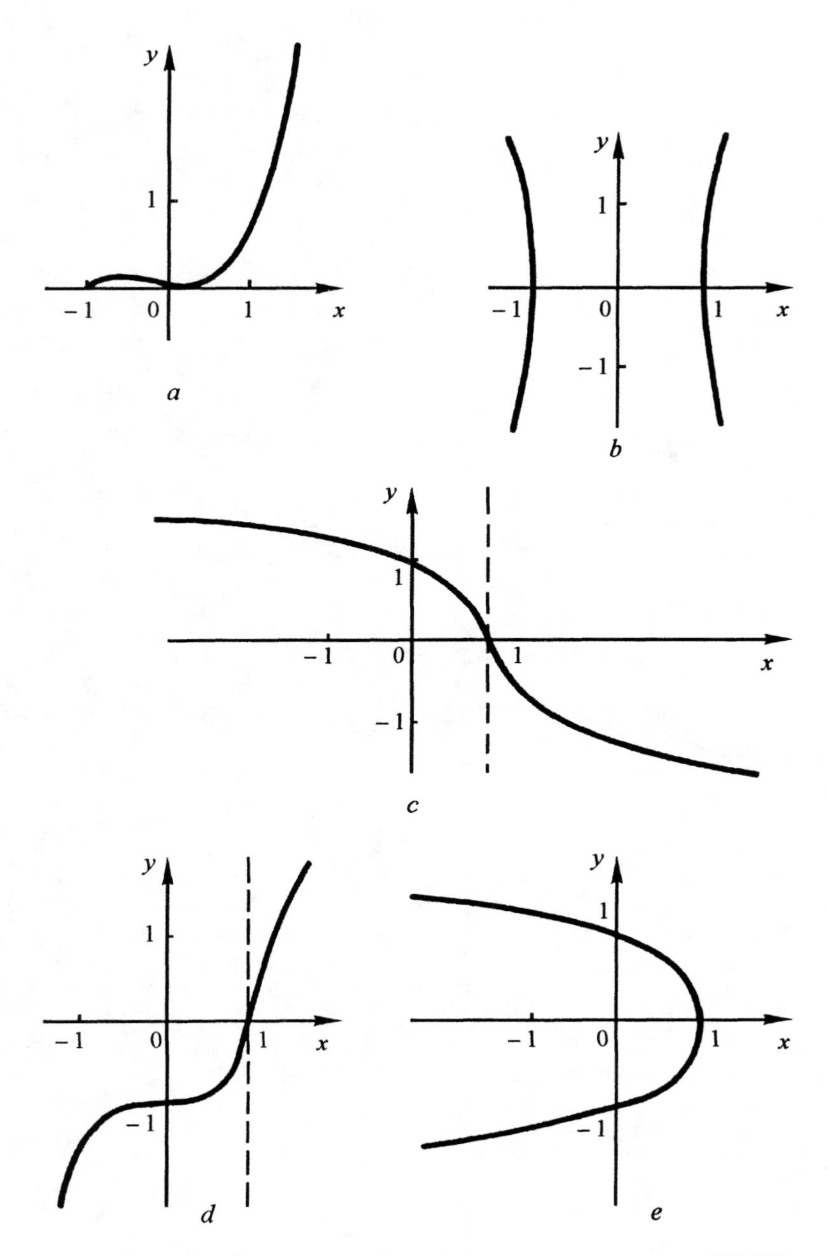

Figure 5.81. The graphs of functions: (*a*) $y = x^2\sqrt{1+x}$; (*b*) $y = \pm x^2\sqrt{x^2-1}$; (*c*) $y = \sqrt[3]{1-x}$; (*d*) $y = -\sqrt[3]{1-x}$; (*e*) $y = \pm\sqrt[4]{1-x}$.

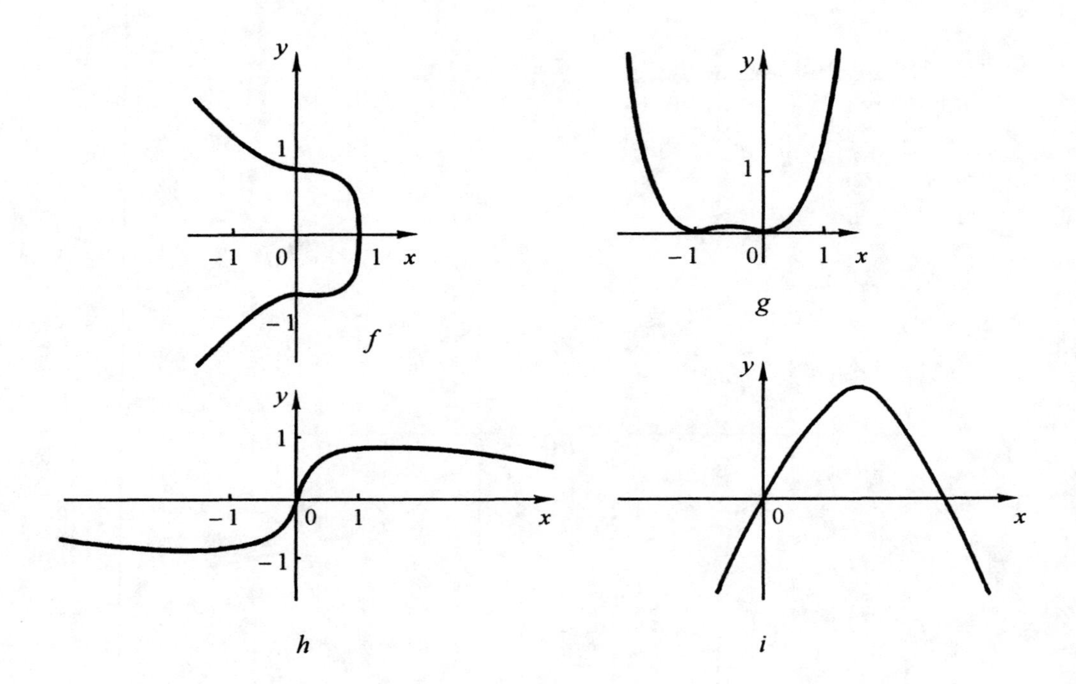

Figure 5.81. The graphs of functions: *(f)* $y = \pm\sqrt{1 - x^3}$; *(g)* $y = x^2\sqrt[3]{(1 + x)^2}$; *(h)* $y = \dfrac{x}{\sqrt[3]{1 + x^4}}$; *(i)* $y = \sqrt[3]{8x - 2x^2}$.

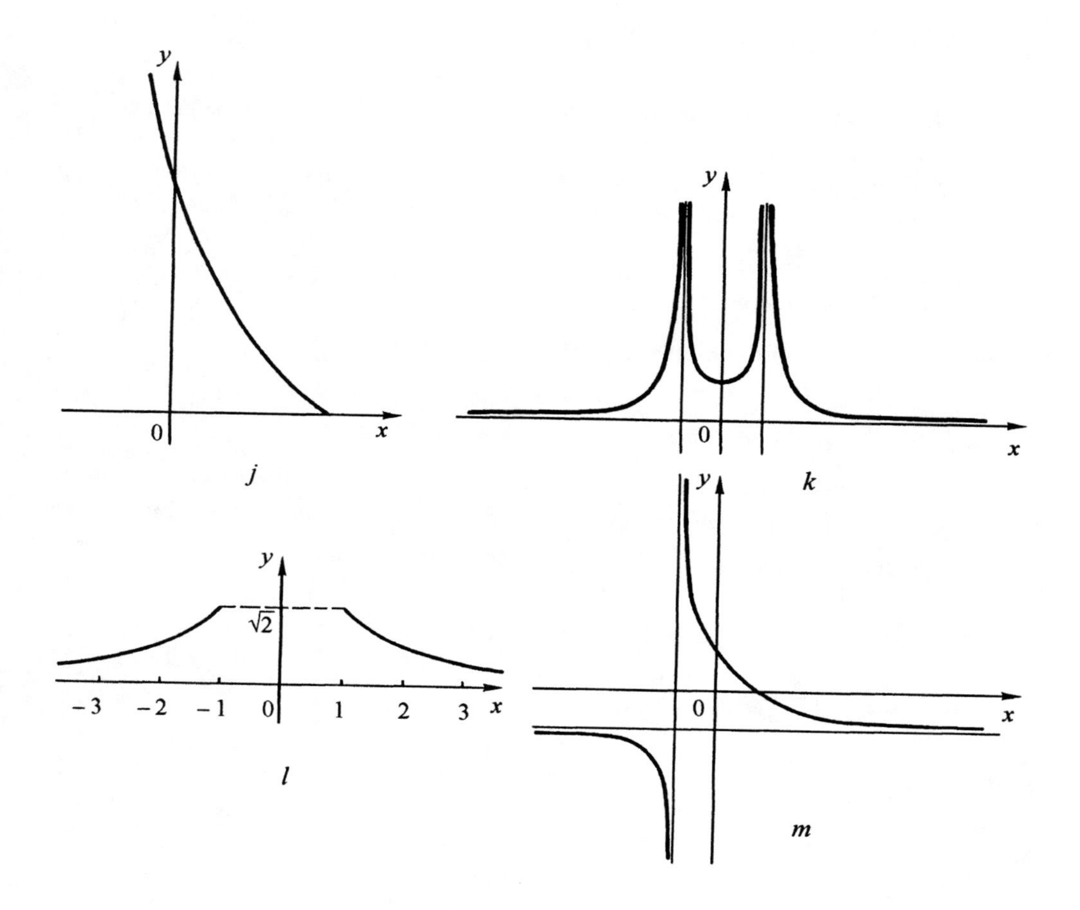

Figure 5.81. The graphs of functions: (*j*) $y = \sqrt{(2-x)^3}$; (*k*) $y = \frac{1}{\sqrt[3]{(1-x^2)^2}}$;

(*l*) $y = \sqrt{x^2+1} - \sqrt{x^2-1}$; (*m*) $y = \sqrt[3]{\frac{1-x}{1+x}}$.

5.3. Graphs of transcendental functions

Graphs of the most simple trancendental functions were considered in Sections 3.2 – 3.5. More complex transcendental functions will be considered in this section. Plotting of such graphs can be carried out using the scheme given in Section 2.2. A better way is to use methods of calculus (see part II). Graphs of some transcendental functions are shown in Figures 5.82 – 5.87.

5.3.1. Hyperbolic functions

Hyperbolic sine function, sinhx, is given by the formula

$$y = \sinh x = \frac{e^x - e^{-x}}{2.}.$$

The domain of the function is $(-\infty; +\infty)$. The range is $(-\infty; +\infty)$. The function is odd. The point (0; 0) is the center of symmetry; at the same time it is an inflection point. The function does not have asymptotes. The graph can be plotted similarly to the graph of the function $y = e^x$ and $y = e^{-x}$.

Hyperbolic cosine function is defined by the formula

$$y = \cosh x = \frac{e^x + e^{-x}}{2}.$$

Its domain is $(-\infty; +\infty)$. The function is even. It reaches the minimum at the point (0; 1). The graph does not have asymptotes. Graphs of the functions $y = \sinh x$ and $\cosh x$ are shown in Figure 5.88.

Hyperbolic tangent $\tanh x$ is defined by the formula

$$y = \tanh x = \frac{e^x - e^{-x}}{e^x + e^{-x}}.$$

The domain of the function is $(-\infty; +\infty)$, its range is $(-1; 1)$. The function is odd. The point (0; 0) is the center of symmetry and, at the same time, is an inflection point. Horizontal asymptotes are straight lines $y = \pm 1$ (Figure 5.89).

Hyperbolic cotangent $\coth x$ is given by

$$\coth x = \frac{1}{\tanh x} = \frac{e^x + e^{-x}}{e^x - e^{-x}}.$$

The domain of the function is the set $(-\infty; 0) \cup (0; +\infty)$. The range of the function is $(-\infty; -1) \cup (1; +\infty)$. The function is odd. The graph has horizontal asymptotes $y = \pm 1$, and the vertical asymptote is $x = 0$ (Figure 5.90).

Hyperbolic secant function, sech x, is defined by the formula

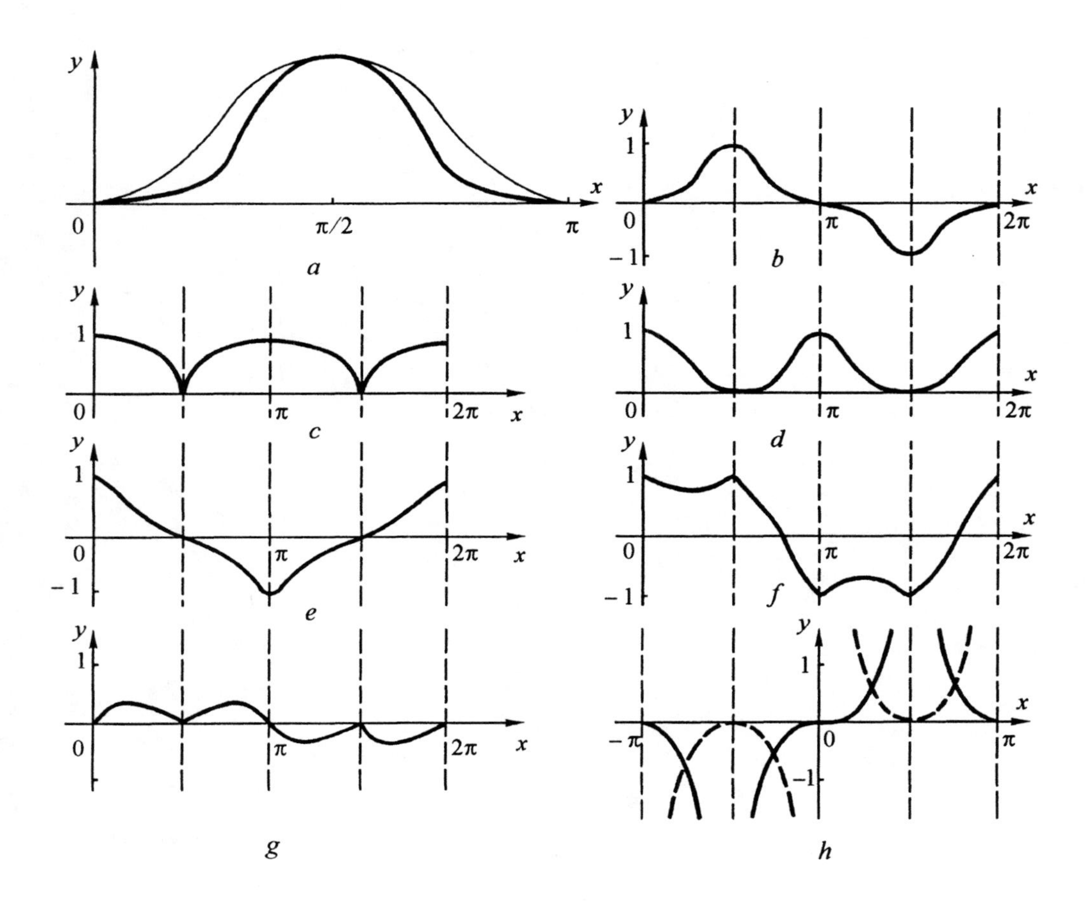

Figure 5.82. The graphs of trancendental functions: (*a*) $y = \sin^4 x$; (*b*) $y = \sin^5 x$; (*c*) $y = \sqrt[4]{1 - \sin^4 x}$; (*d*) $y = \cos^4 x$; (*e*) $y = \cos^5 x$; (*f*) $y = \sin^3 x + \cos^3 x$; (*g*) $y = \sin x \cos^2 x$; (*h*) *1.* $y = \dfrac{\sin^{3x}}{\cos^2 x}$, *2.* $y = \dfrac{\cos^2 x}{\sin^3 x}$.

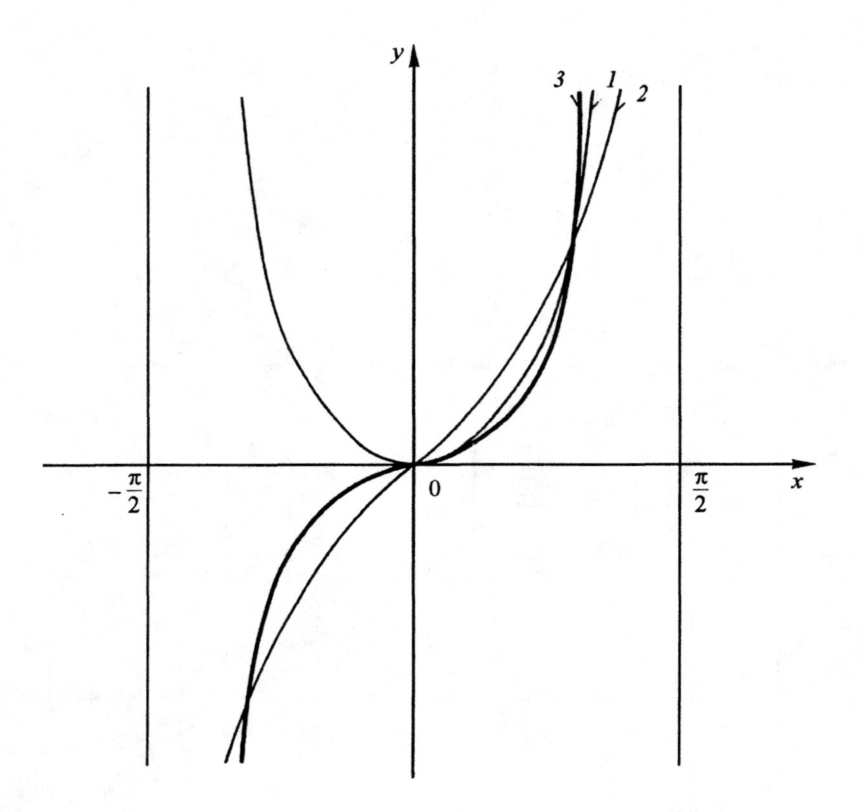

Figure 5.83. The graphs of the functions: *1*. $y = \tan^2 x$, *2*. $y = \tan x$, *3*. $y = \tan^3 x$.

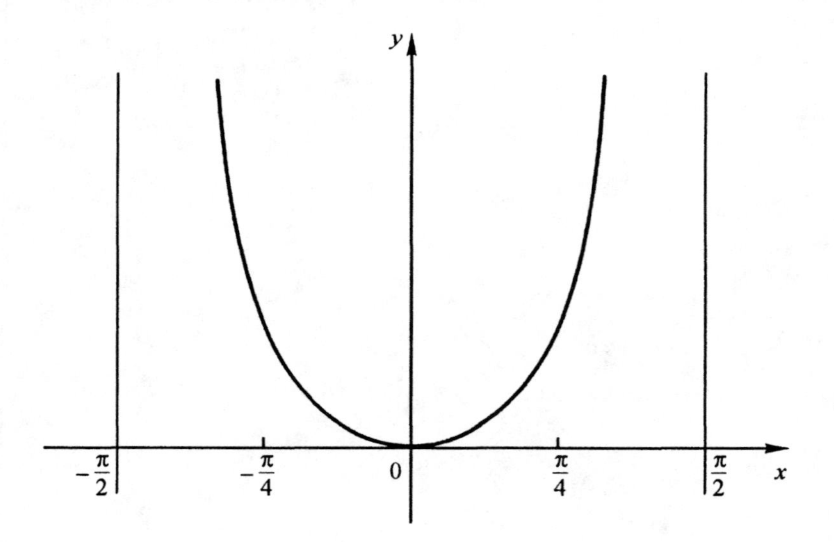

Figure 5.84. The graphs of the function $y = \tan^4 x$.

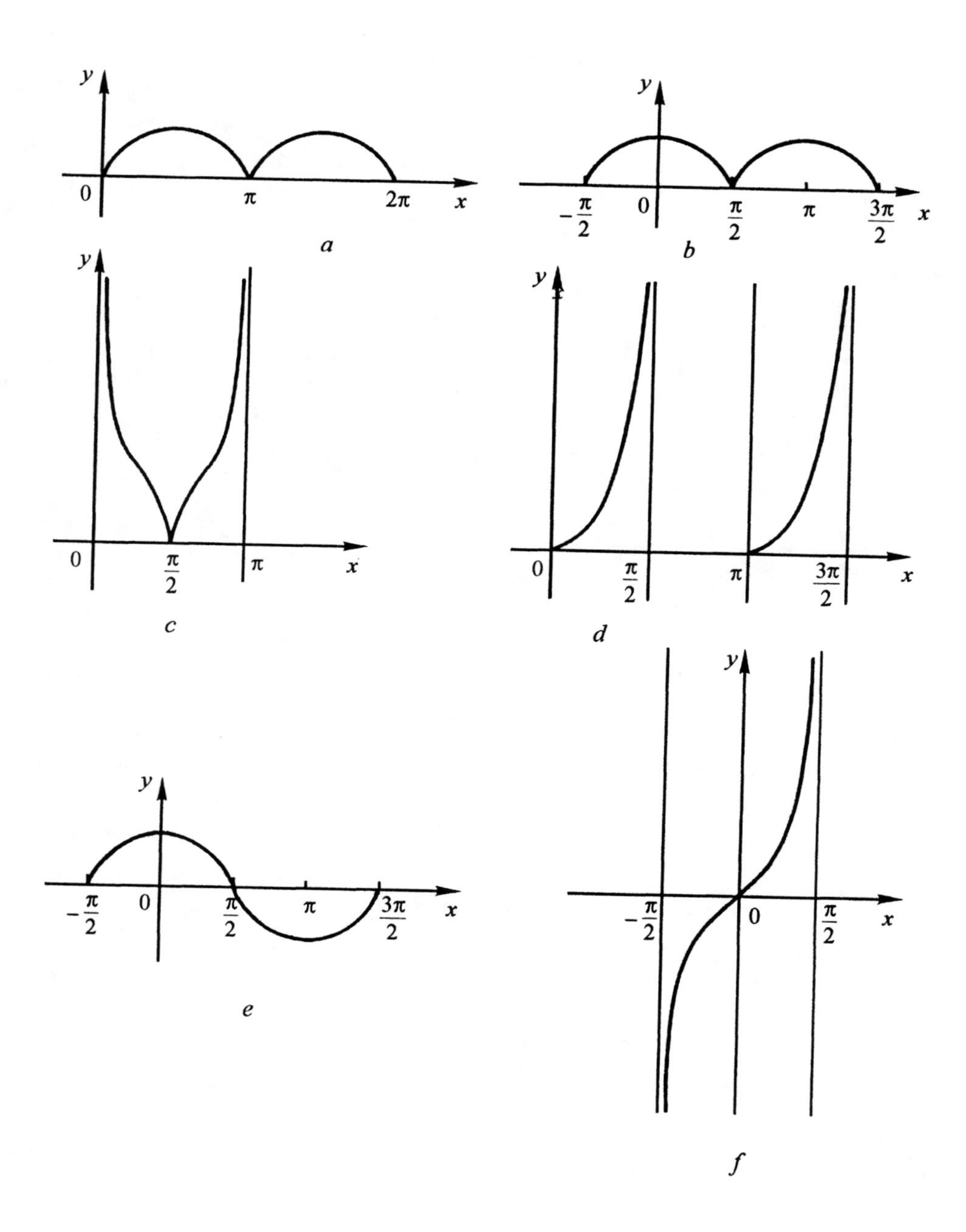

Figure 5.85. The graphs of functions: (*a*) $y = \sin^{4/3}x$; (*b*) $y = \cos^{2/3}x$; (*c*) $y = \cot^{4/3}x$; (*d*) $y = \tan^{7/6}x$; (*e*) $y = \cos^{3/5}x$; (*f*) $y = \tan^{1/3}x$.

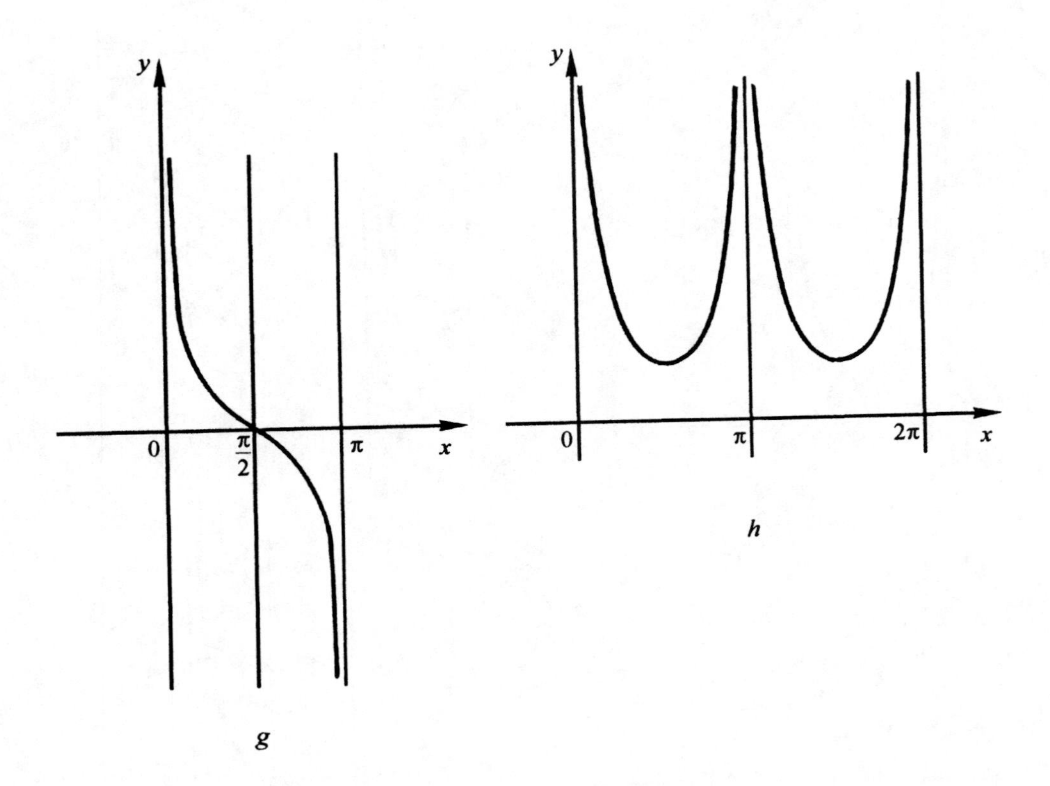

Figure 5.85. The graphs of functions: (*g*) $y = \cot^{7/5}x$; (*h*) $y = \sin^{-2/3}x$.

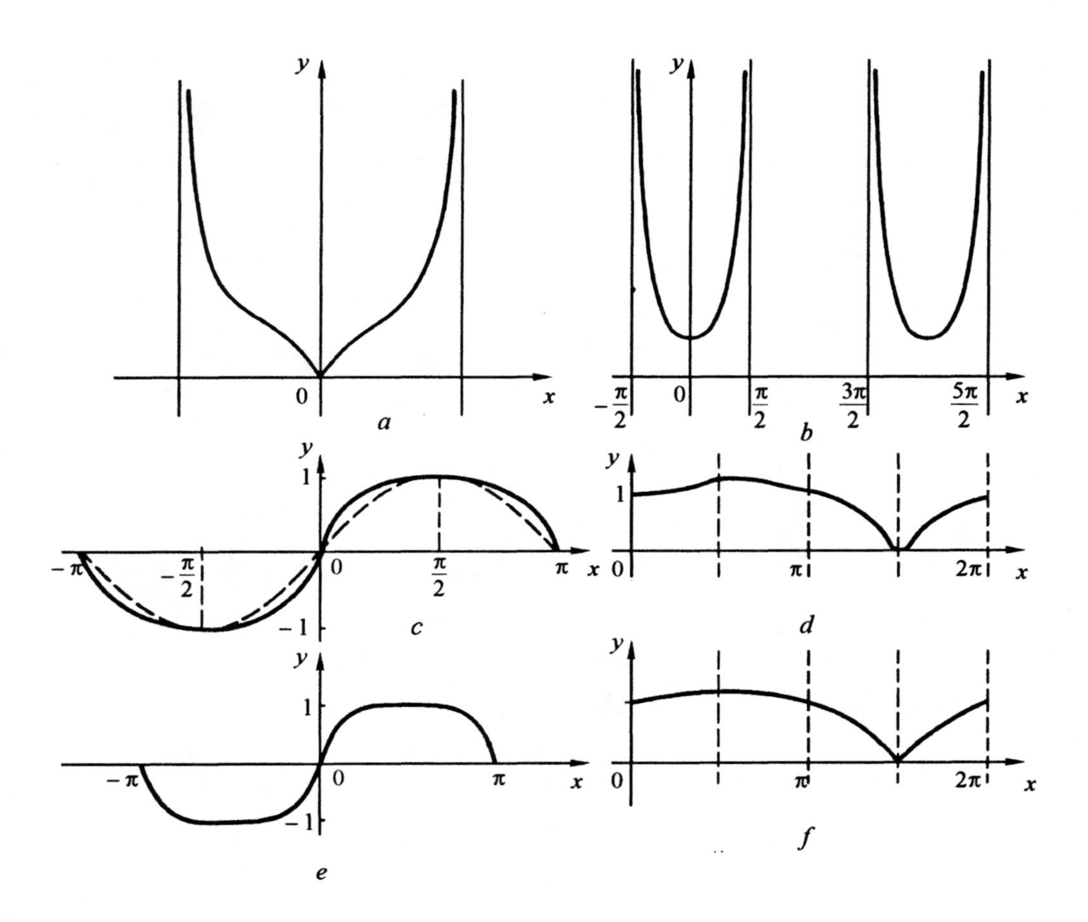

Figure 5.86. The graphs of functions: (*a*) $y = \sin^{3/4}x$; (*b*) $y = \cos^{1/2}x$; (*c*) $y = \sin^{1/5}x$; (*d*) $y = \sqrt{1 + \sin^3 x}$; (*e*) $y = \sin^{1/3}x$; (*f*) $y = \sqrt[4]{1 + \sin x}$.

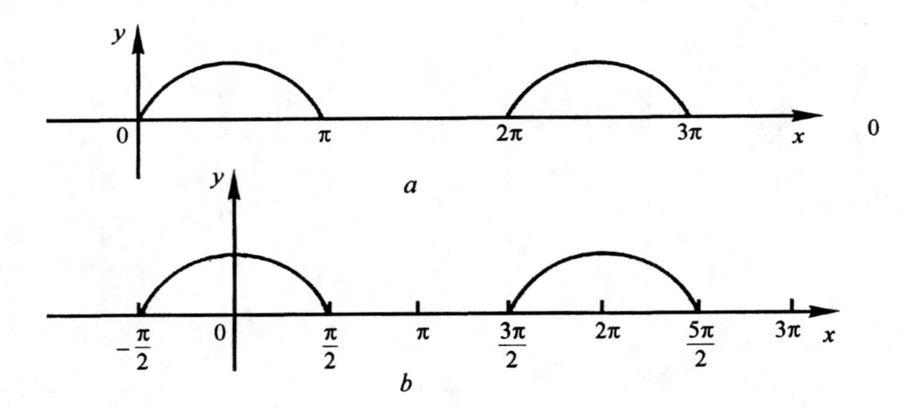

Figure 5.87. The graphs of functions: (*a*) $y = \sin^{3/4}x$; (*b*) $y = \cos^{1/2}x$.

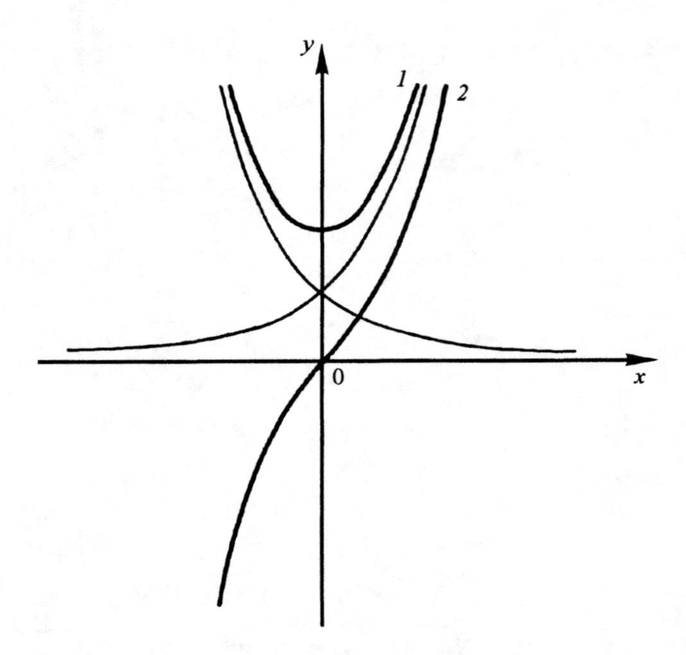

Figure 5.88. *1*. Hyperbolic cosine function $y = \cosh x$, *2*. hyperbolic sine function $y = \sinh x$.

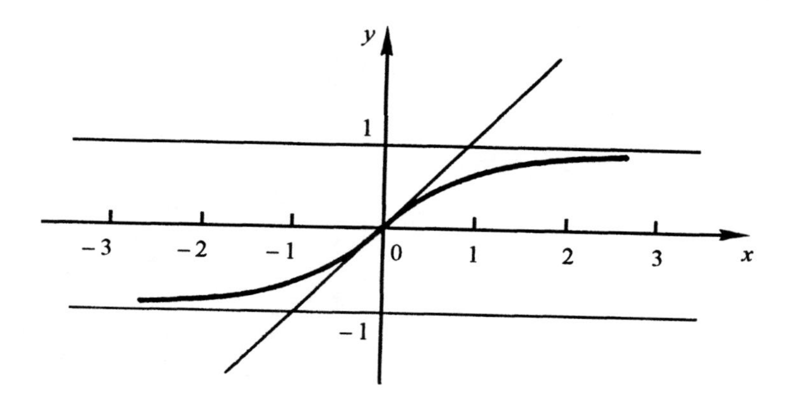

Figure 5.89. Hyperbolic tangent function $y = \tanh x$.

$$\operatorname{sech} x = \frac{1}{\cosh x} = \frac{2}{e^x + e^{-x}}.$$

Hyperbolic cosecant function, cosech x, is the function

$$\operatorname{cosech} x = \frac{1}{\sinh x} = \frac{1}{e^x - e^{-x}}.$$

5.3.2. Inverse hyperbolic functions

The inverse hyperbolic functions $y = \operatorname{Asinh} x$, $y = \operatorname{Acosh} x$, $y = \operatorname{Atanh} x$ and $y = \operatorname{Acoth} x$ are the functions inverse to $x = \sinh y$, $x = \cosh y$, $x = \tanh y$, and $x = \coth y$ respectively. These functions can also be expressed in terms of the logarithmic function:

$$\operatorname{Asinh} x = \ln (x + \sqrt{x^2 + 1}), \quad \operatorname{Acosh} x = \pm \ln (x + \sqrt{x^2 - 1}), \quad (x \geq 1),$$

$$\operatorname{Atanh} h = \frac{1}{2} \ln \frac{1 + x}{1 - x}, \quad |x| < 1, \quad \operatorname{Acoth} = \frac{1}{2} \ln \frac{x + 1}{x - 1}, \quad |x| > 1.$$

The plots of the inverse hyperbolic functions can be obtained by the symmetry transformation of the plots of the corresponding hyperbolic function with respect to the bisectrix of the first quadrant.

The plot of the function $\operatorname{Asinh} x$ is shown in Figure 5.91. The domain of the function is the set $(-\infty; +\infty)$. The range of the function is $(-\infty; +\infty)$. The function is odd. The point $(0; 0)$ is the center of symmetry and also an inflection point. The function does not have asymptotes.

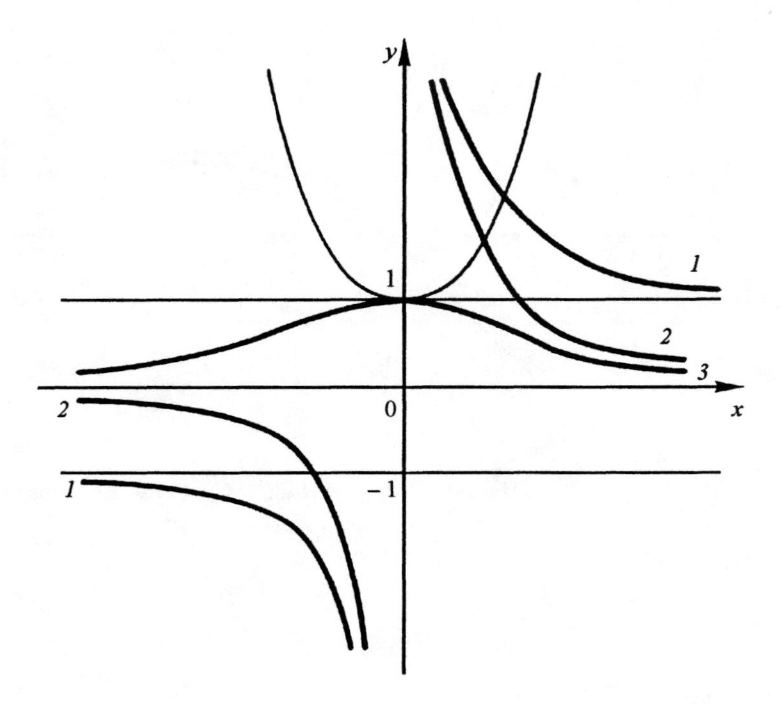

Figure 5.90. *1*. Hyperbolic cotangent function $y = \coth x$, *2*. Hyperbolic cosecant function $y = \operatorname{cosech} x$, *3*. Hyperbolic secant function $y = \operatorname{sech} x$.

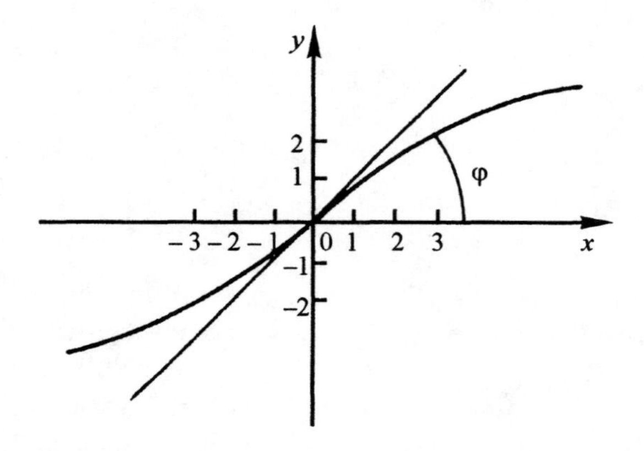

Figure 5.91. Inverse hyperbolic function function $y = \operatorname{Asinh} x$.

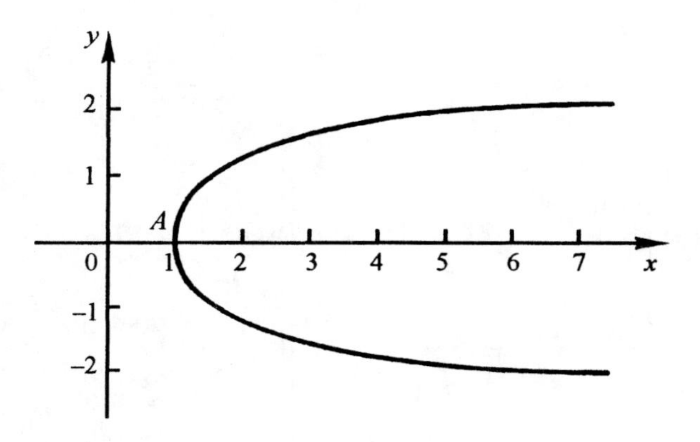

Figure 5.92. The function $y = \text{Acosh}\, x$.

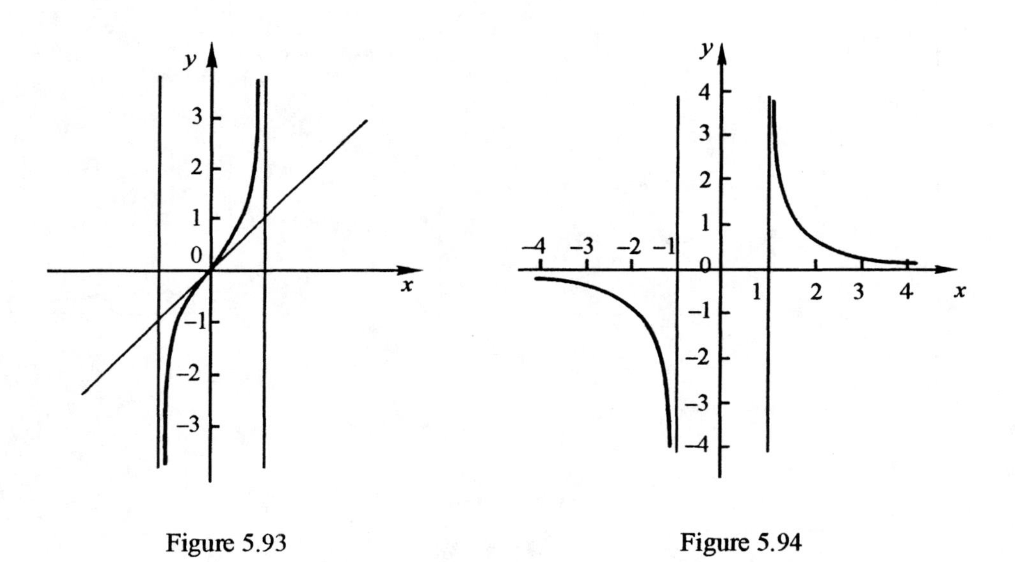

Figure 5.93

Figure 5.94

Figure 5.93. The function $y = \text{Atanh}\, x$.

Figure 5.94. The function $y = \text{Acoth}\, x$.

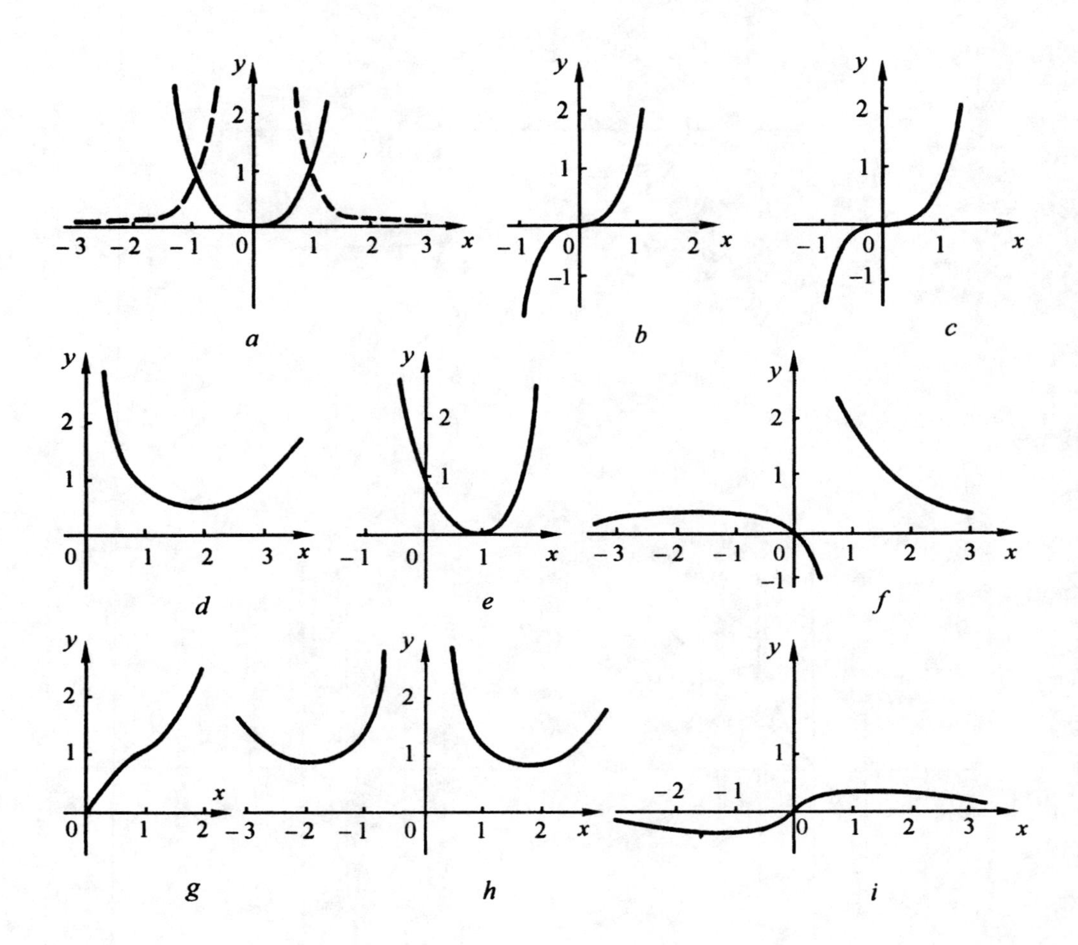

Figure 5.95. The graphs of functions: *(a)* $y = \sinh^2 x$, $y = \dfrac{1}{\sinh^2 x}$; *(b)* $y = \sinh^3 x$; *(c)* $y = x^2 \sinh x$; *(d)* $y = \dfrac{\sinh x}{x^2}$; *(e)* $y = (\sinh x - 1)^2$; *(f)* $y = \dfrac{x}{\sinh x - 1}$; *(g)* $y = \sinh\sqrt{x}$; *(h)* $y = \dfrac{\cosh x}{x^2}$; *(i)* $y = \dfrac{x}{\cosh x + 1}$.

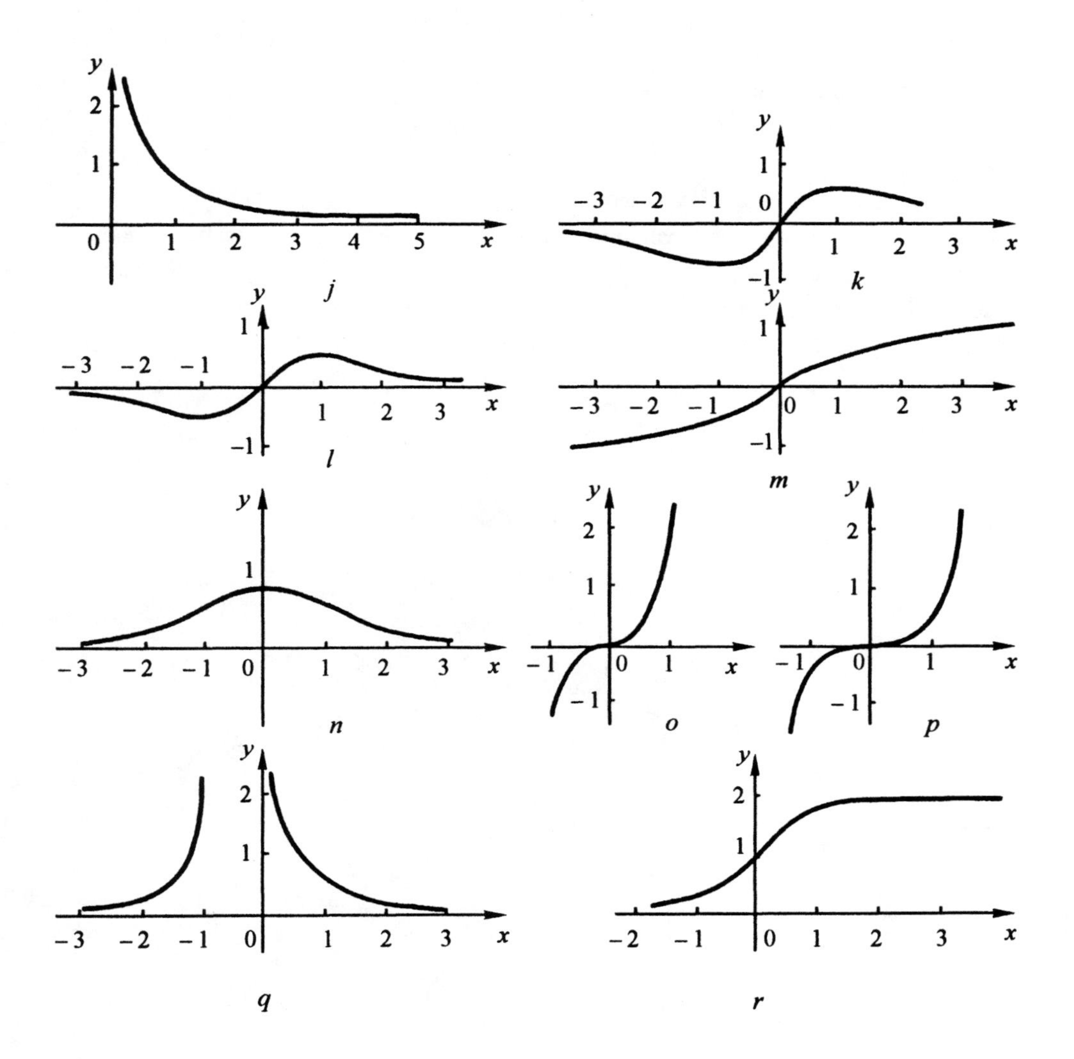

Figure 5.95. The graphs of functions: (*j*) $y = \dfrac{x}{\cosh^2 x - 1}$; (*k*) $y = \dfrac{\sinh x}{\cosh^2 x}$;
(*l*) $y = \dfrac{x}{\cosh^2 x}$; (*m*) $y = \tanh \dfrac{x}{2}$; (*n*) $y = 1 - \tanh^2 \dfrac{x}{2}$; (*o*) $y = \sinh^3 x \cosh x$;
(*p*) $y = \sinh x(\cosh x - 1)$; (*q*) $y = \dfrac{\cosh x}{\sinh x\,(\sinh x + 1)}$; (*r*) $y = 1 + \tanh x$.

The domain of the function $y = \text{Acosh}\, x$ (Figure 5.92) is $[1; +\infty)$, its range is $(-\infty; +\infty)$. The curve is symmetric with respect to the x-axis. The curve is tangent to the vertical line $x = 1$ at the point $A(1; 0)$, and then the absolute value of the function increases.

The domain of the function $y = \text{Atanh}\, x$ (Figure 5.93) is $[-1; 1)$, the range is $(-\infty; +\infty)$. The function is odd. The point $(0; 0)$ is a center of symmetry and an inflection point. Vertical asymptotes are given by $x = \pm 1$.

The domain of the function $y = \text{Acoth}\, x$ (Figure 5.94) is $(-\infty; -1) \cup (1; +\infty)$. The range of the function is $(-\infty; +\infty)$. The function is odd. On the set $(-\infty; -1)$, the function is monotone decreasing, with values ranging from 0 to $-\infty$. On the set $(0; +\infty)$, the function monotonically decreases from $+\infty$ to 0. Vertical asymptotes are $x = \pm 1$; a horizontal asymptote is $y = 0$. Graphs of hyperbolic function are shown in Figure 5.95.

Chapter 6

Plots of parametrically defined curves

6.1. A study of parametrically defined functions

Let a function be defined parametrically,

$$\begin{cases} x = \varphi(t), \\ y = \psi(t), \end{cases} \qquad (1)$$

where t is a parameter, $t_0 \le t \le t_1$. In this case, a study and plotting of the graph of the function is similar to the case where the function is given by $y = f(x)$. First, the graphs of the functions $x = \varphi(t)$ and $y = \psi(t)$ are plotted in the coordinates Otx and Oty, respectively. Then use the plots of the functions $x = \varphi(x)$ and $y = \psi(x)$ to study the function $y = y(x)$ along the scheme given in Section 1.2.2. Let us note some properties of the graph of function (1):

- It is symmetric with respect to the y-axis if the replacement of t with $-t$ leaves the y-coordinate unchanged and the x-coordinate changes sign.
- It is symmetric with respect to the x-axis, if the replacement of t with $-t$ leaves the x-coordinate unchanged and the y-coordinate becomes $-y$.
- The period of the function is determined from the periods of the functions $x = \varphi(t)$ and $y = \psi(t)$.
- To find the points where the graph intersects the y-axis, it is necessary to find t such that $x = 0$, i.e. to solve the equation $\varphi(t) = 0$ and find the corresponding value of $y = \psi(t)$.
- To find the points where the graph intersects the x-axis, it is necessary to find t such that $y = 0$, i.e. to solve the equation $\psi(t) = 0$ and to find the corresponding value of $x = \varphi(t)$.
- It is sometimes useful to find the points of intersection of the graph and bisectors of the quadrants, $y = x$ and $y = -x$. To do this, we solve the equations $\psi(t) = \varphi(t)$ and $\psi(t) = -\varphi(t)$. The values of $x = \varphi(t)$ and $y = \psi(t)$ give the needed coordinates.
- Vertical asymptotes can be found if there is t such that y becomes infinite whereas x remains finite; horizontal asymptotes are found similarly.
- Oblique asymptotes $y = kx + b$, if they exist, are found from the formulas:

$$k = \lim_{x \to \pm\infty} \frac{y}{x}, \quad b = \lim_{x \to \pm\infty} (y - kx).$$

- Points of self intersection of the curve, if they exist, can be found from the condition that the coordinates are the same for different values of t:

$$x_s = \varphi(t_1) = \varphi(t_2) \quad \text{and} \quad y_s = \psi(t_1) = \psi(t_2).$$

Example 6.1.1. Plot the graph of the function $x = 1 - t, y = 1 - t^2$.

We reduce the plotting and studying of the parametrically defined function to that of an explicitly defined function $y = y(x)$.

Plot the graph of the function $x = 1 - t$ in the coordinate system Otx (Figure 6.1 a), and the graph of the function $y = 1 - x^2$ in the coordinates Oty (Figure 6.1 b). Then study the function $y = y(x)$. Find that its domain is $(-\infty; +\infty)$, and its range is $(-\infty; 1)$. Since $y(-t) = y(t)$, we have that $t = 0$, i.e. $x = 1$ is a symmetry axis. The limit values of the functions are $\lim_{x \to \pm\infty} y = -\infty$. To find the point of intersection of the curve and the y-axis, solve the equation $x = 0$, i.e. $1 - t = 0$. This means that $t = 1$. For this t, the value of the function $y = t^2 - 1$ is 0; hence the graph of the function $y = y(x)$ intersects the y -axis at the point $O(0;0)$.

To find the point of intersection with the x-axis, solve the equation $y \equiv 1 - t^2 = 0$. We find that $t = \pm 1$; hence $x_1 = 0, x_2 = 2$. The graph intersects the x-axis at the points $O(0;0)$, $C(2;0)$.

To find where the graph intersects the bisector $y = x$, we solve the equation $1 - t^2 = 1 - t$. This gives $t_1 = 0, t_2 = 1$. Thus the points of intersection of the graph and the bisector $y = x$ are $D(1;1)$, $O(0;0)$.

Considering points of intersection of the graph and the bisector $y = -x$ we solve the equation $1 - t^2 = -(1 - t)$ or $t^2 + t - 2 = 0$. Whence we find that $t_1 = -2$, $t_2 = 1$. Thus the points of intersection of the graph and the bisector $y = -x$ are $O(0;0)$ and $E(3; -3)$. The graph does not have asymptotes.

Now we plot the graph of the function $y = y(x)$. Draw the straight line $y = 1$ (the graph lies below it), and find the points at which the graph of the function $y = y(x)$ intersects the coordinate axes and the bisectors of the coordinate angles, the points O, C, D, E (Figure 6.1 b).

Graphs of other parametric curves are shown in Figures 6.2 – 6.10.

Remark 6.1.1. It is sometimes possible (in most simple cases) to plot the graph of a parametrically defined funtion by directly plotting points of the graph. Let us look at an example. Suppose we need to plot the function $x = t^2, y = t^3$. Let us calculate the values:

t	0	1	2	3	–1	–2	–3
x	0	1	4	9	1	4	9
y	0	1	8	27	–1	–8	–27

and plot the needed curve (Figure 6.11).

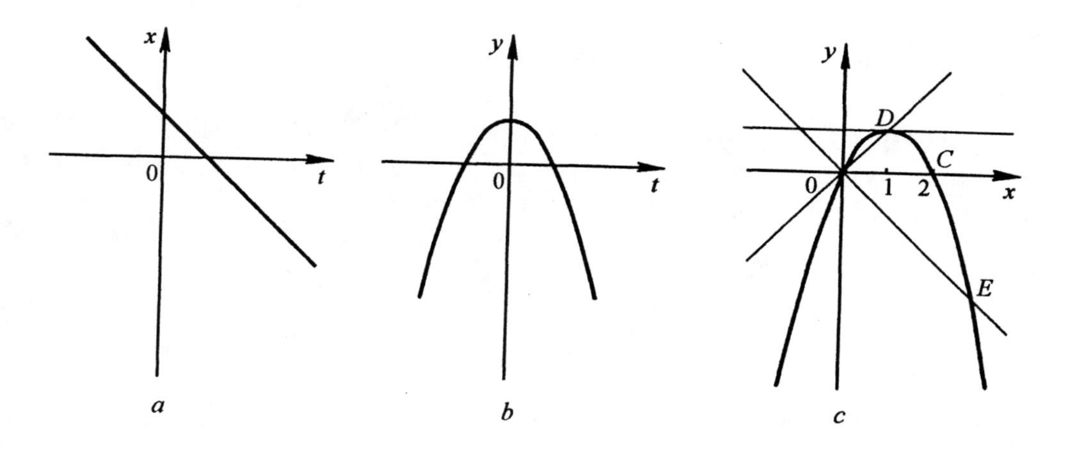

Figure 6.1. The graphs of functions: (*a*) $x = 1 - t$, (*b*) $y = 1 - t^2$, (*c*) $x = 1 - t$, $y = 1 - t^2$.

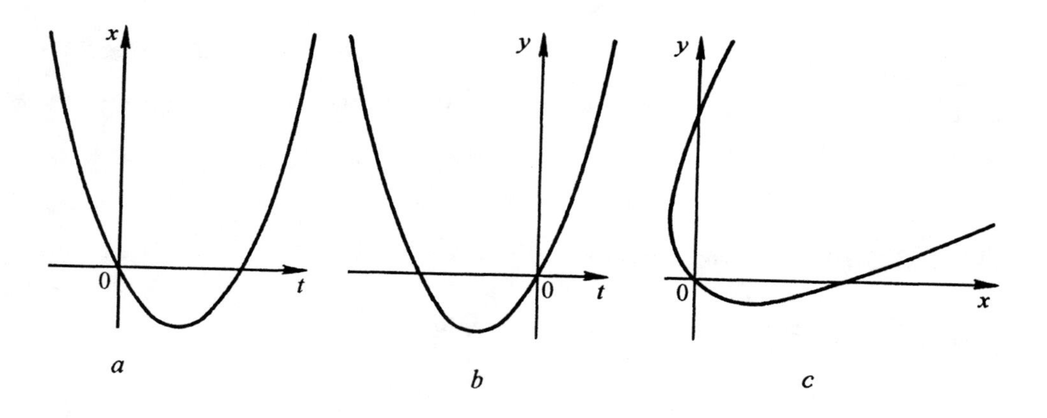

Figure 6.2. The graphs of functions: (*a*) $x = t^2 - 2t$, (*b*) $y = t^2 - 2t$, (*c*) $x = t^2 - 2t$, $y = t^2 + 2t$.

6.2. The method of eliminating the parameter t

If one of the functions $\varphi(t)$ or $\psi(t)$ (see formula (1) in Section 6.1) has an inverse, $t = \varphi^{-1}(x)$ or $t = \psi^{-1}(t)$, then the parameter t can be eliminated from equations (1). It is clear that not all parametrically defined functions admit such an elimination.

Consider some examples of functions where the parameter can be eliminated (Figures 6.12 and 6.13).

Example 6.2.1. Plot the graph of the function $x = \frac{a}{\cos t}$, $y = b\tan t$, $-\pi/2 < t < \pi/2$, $\pi/2 < t < 3/2\pi$. We have

$$\left.\begin{aligned} \frac{x}{a} &= \frac{1}{\cos t} \\ \frac{y}{b} &= \tan t \end{aligned}\right\} \Rightarrow \frac{x^2}{a^2} - \frac{y^2}{b^2} = \frac{1}{\cos^2 t} - \tan^2 t \Rightarrow \frac{x^2}{a^2} - \frac{y^2}{b^2} = 1,$$

hence, it is a hyperbola (Figure 6.14).

In the same way we find that the graph of the function $x = x_0 + \frac{a}{\cos t}$, $y = y_0 + b\tan t$, where $-\pi/2 < t < \pi/2$, $\pi/2 < t < 3/2\pi$, is the hyperbola

$$\frac{(x - x_0)^2}{a^2} - \frac{(y - y_0)^2}{b^2} = 1.$$

(Figure 6.15).

The graph of the curve $x = a\cos t + d\sin t$, $y = b\sin t + c\cos t$, where $ab \neq cd$, $t_1 < t < t_2$, is shown in Figure 6.16.

6.3. Methods for finding points of intersection of the curve $x = \varphi(t)$, $y = \psi(t)$ and the straight line $y = kx$

Let us go back to the plotting of the curve

$$\begin{aligned} x &= a\cos t + d\sin t \quad (ab \neq cd), \\ y &= b\sin t + c\cos t \quad (t_1 < t < t_2). \end{aligned}$$

We have

$$\left.\begin{aligned} y &= kx \\ y &= b\sin t + c\cos t \\ x &= a\cos t + d\sin t \end{aligned}\right\} \Rightarrow b\sin t + c\cos t = k(a\cos t + d\sin t),$$

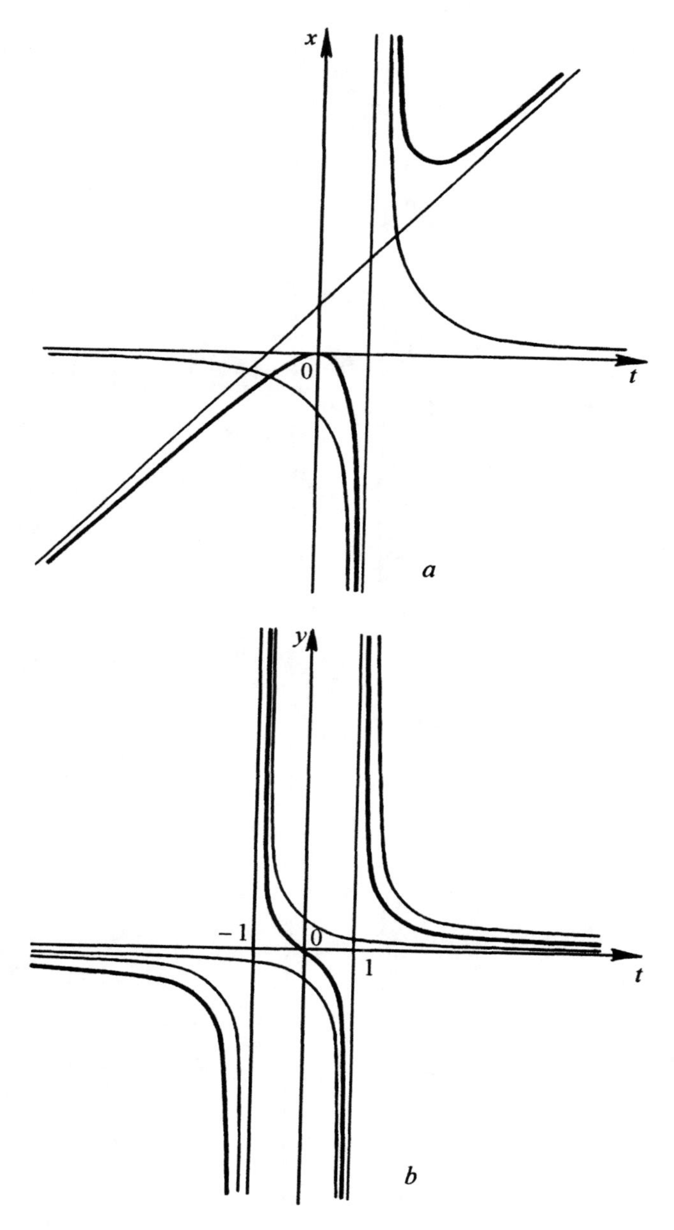

Figure 6.3. The graphs of functions: (*a*) $x = t^2/(t-1)$, (*b*) $y = t/(t^2-1)$.

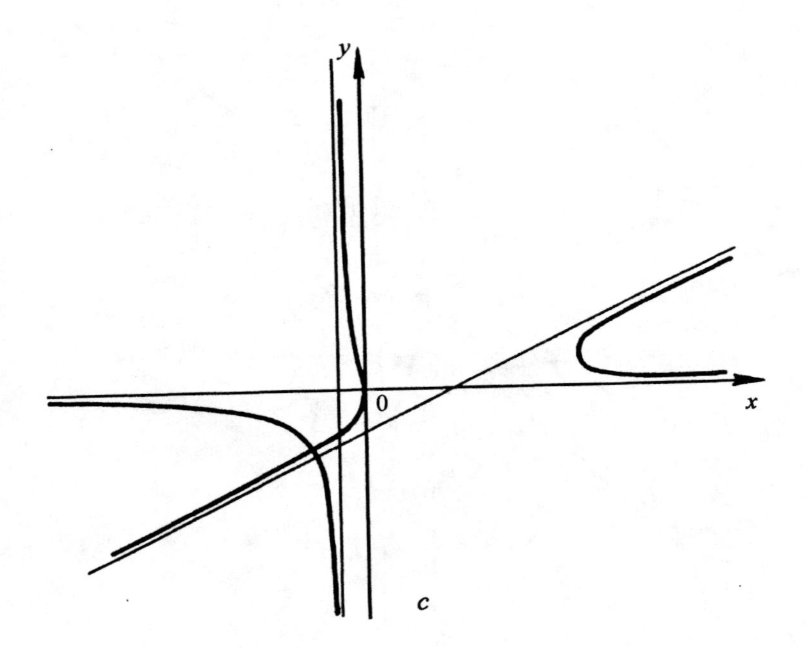

Figure 6.3. The graphs of functions: $(c)\ x = t^2/(t-1),\ y = t/(t^2-1)$.

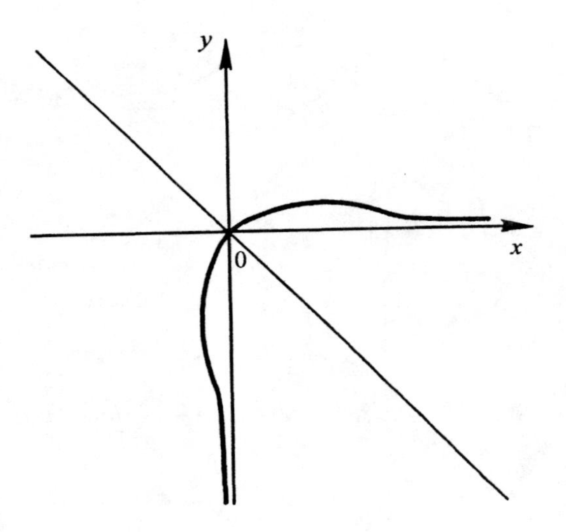

Figure 6.4. The graph of the function $x = te^t,\ y = te^{-t}$.

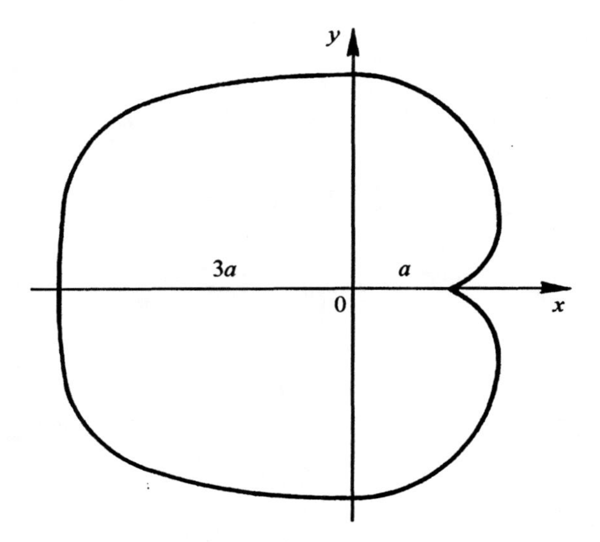

Figure 6.5. The cardioid $x = 2a \cos t - a \cos 2t$, $y = 2a \sin t - a \sin 2t$.

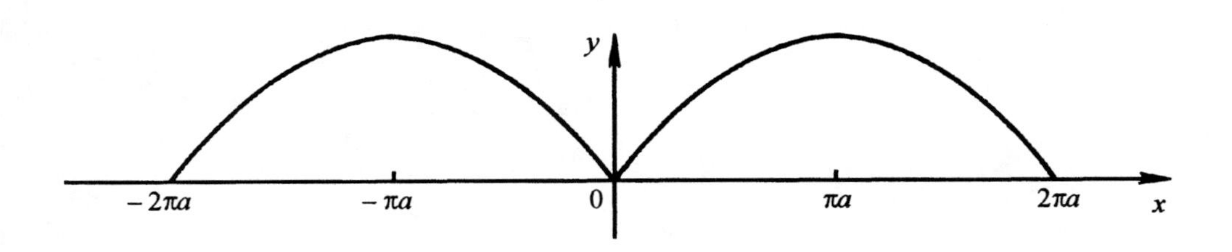

Figure 6.6. The cycloid $x = a(t - \sin t)$, $y = a(1 - \cos t)$.

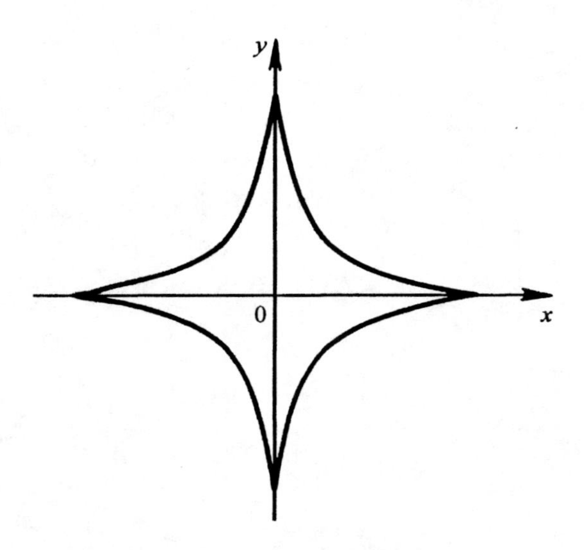

Figure 6.7. The astroid $x = a\cos^3 t, y = a\sin^3 t$.

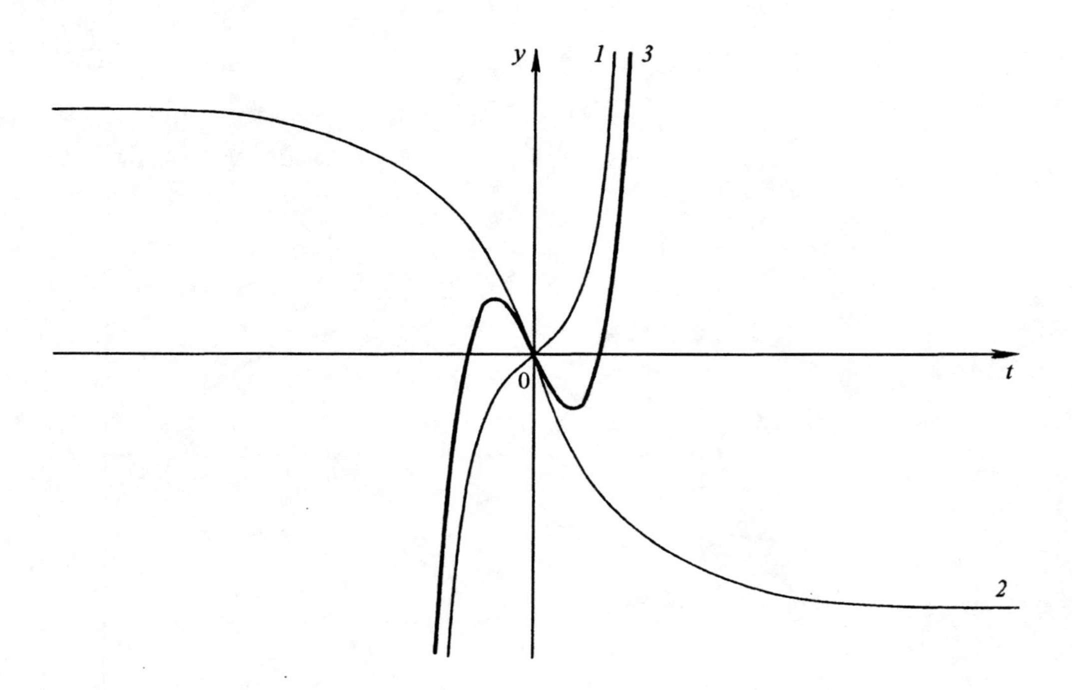

Figure 6.8. The graphs of curves: *(1)* $y = t^3$, *(2)* $y = -6\arctan t$, *(3)* $y = t^3 - 6\arctan t$.

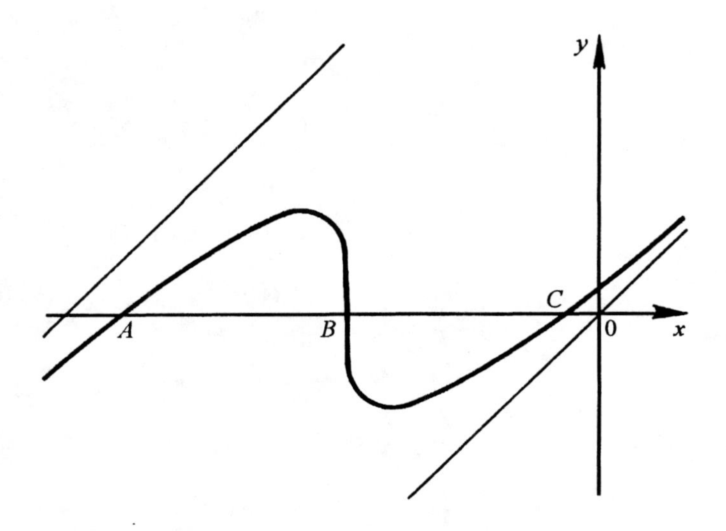

Figure 6.9. The graph of the function $x = t^3 - 3\pi$, $y = t^3 - 6 \arctan t$.

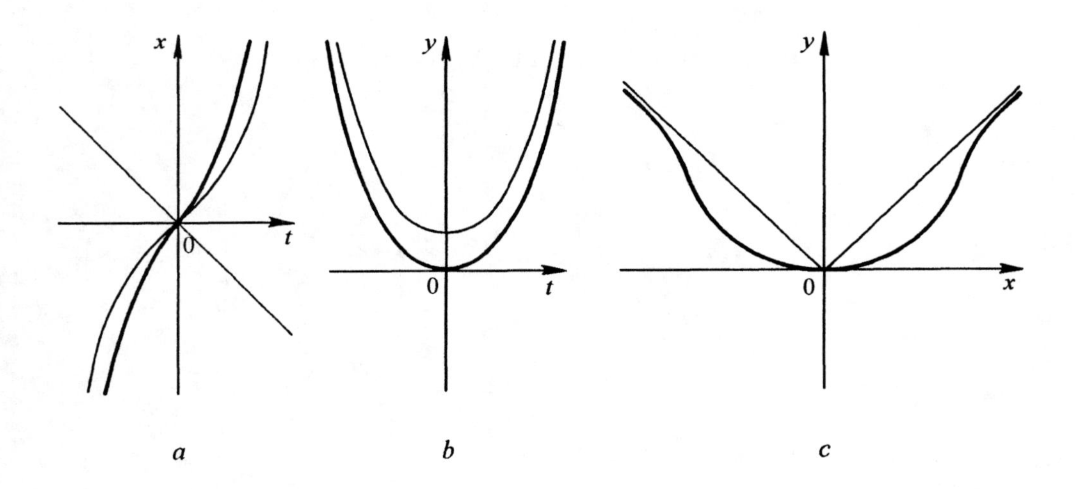

Figure 6.10. The graphs of functions: (*a*) $x = a(\sinh t - t)$; (*b*) $y = a(\cosh t - 1)$; (*c*) $x = a(\sinh t - t)$, $y = a(\cosh t - 1)$, $a > 0$.

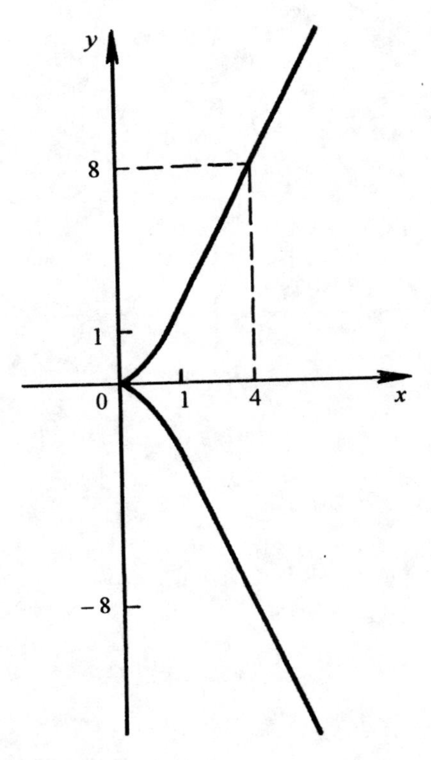

Figure 6.11. The graph of the function $x = t^2, y = t^3$.

whence

$$\tan t = \frac{ka - c}{b - kd} \Rightarrow t = \arctan\frac{ka - c}{b - kd}.$$

Then

$$\sin \arctan \frac{ka - c}{b - kd} = \pm \frac{ka - c}{\sqrt{(b - kd)^2 + (ka - c)^2}},$$

$$\cos \arctan \frac{ka - c}{b - kd} = \pm \frac{b - kd}{\sqrt{(b - kd)^2 + (ka - c)^2}}.$$

So,

$$x = \pm \frac{ab - cd}{\sqrt{(b - kd)^2 + (ka - c)^2}}, \quad y = \pm \frac{k(ab - cd)}{\sqrt{(b - kd)^2 + (ka - c)^2}}.$$

As k takes the values 0, 1, -1, ∞, etc., we get

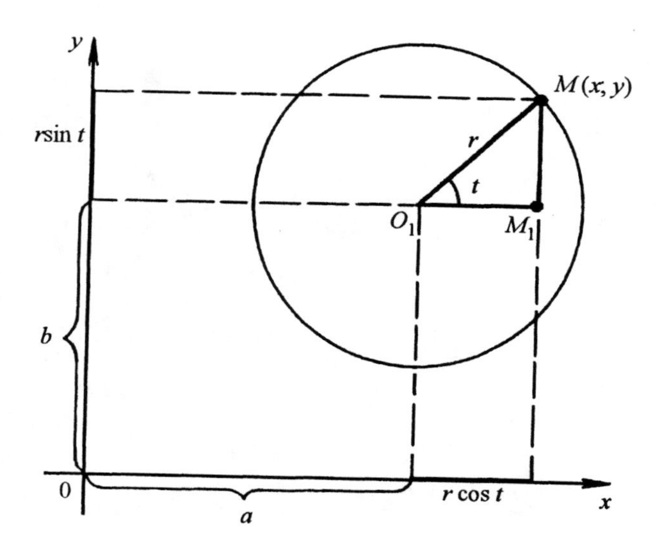

Figure 6.12. The circle $(x-a)^2+(y-b)^2=R^2$.

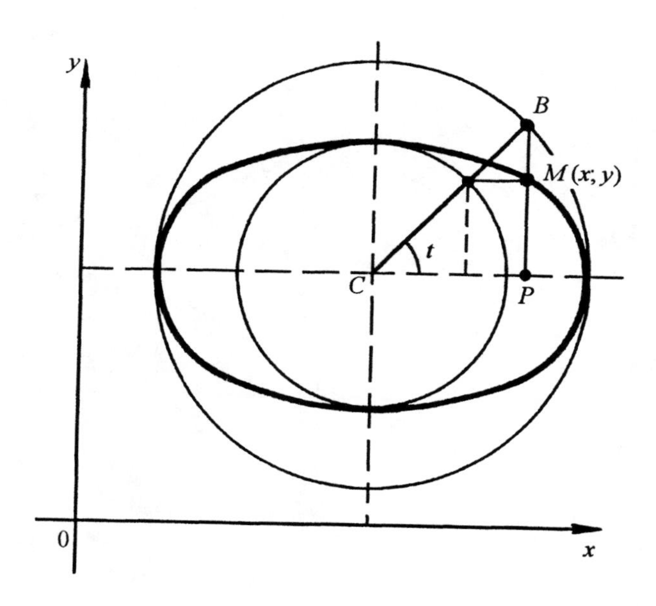

Figure 6.13. The ellipse: $\dfrac{(x-x_0)^2}{a^2}+\dfrac{(y-y_0)^2}{b^2}=1$.

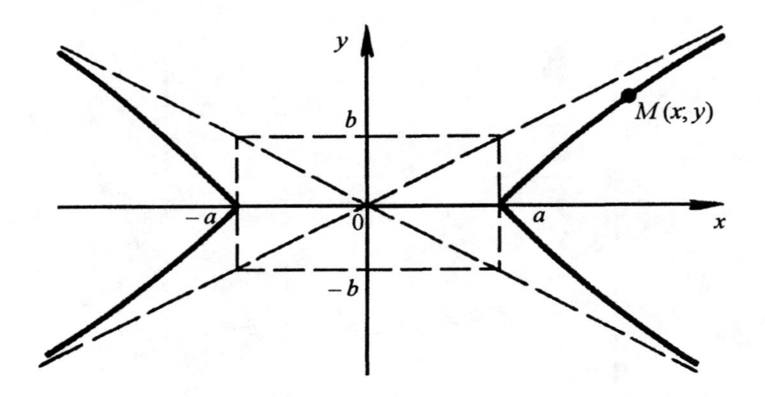

Figure 6.14. The hyperbola $\frac{x^2}{a^2} - \frac{y^2}{b^2} = 1$.

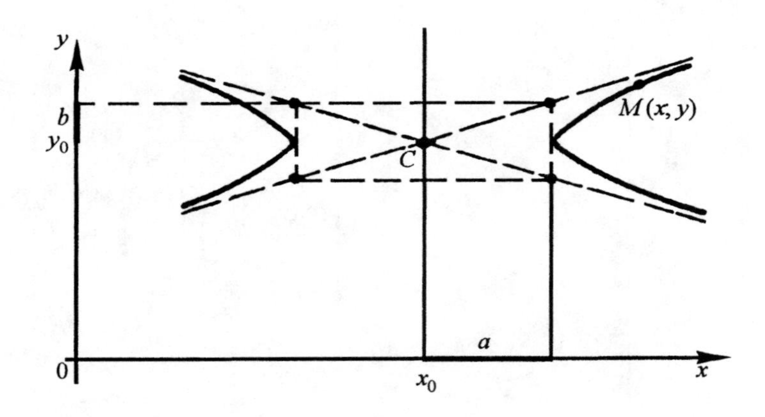

Figure 6.15. The hyperbola $\frac{(x - x_0)}{a^2} - \frac{(y - y_0)^2}{b^2} = 1$.

$$x=\pm\frac{ab-cd}{\sqrt{b^2+c^2}},\quad y=0;\quad y=x=\pm\frac{ab-cd}{\sqrt{(b-d)^2+(a-c)^2}};$$

$$y=-x;\quad x=\pm\frac{ab-cd}{\sqrt{(b+d)^2+(a+c)^2}};$$

$$x=0,\quad y=\frac{ab-cd}{\sqrt{a^2+d^2}}.$$

These values of x are used to plot the graph, see Figure 6.16.

Example 6.3.1. Plot the graph of the cissoid of Diocles:

$$x=\frac{at^2}{1+t^2},$$

$$y=\frac{at^3}{1+t^2}\qquad (y^2=\frac{x^3}{a-x}).$$

We have:

$$y=kx$$

$$y=\frac{at^3}{1+t^2}\ \Rightarrow\ \frac{at^3}{1+t^2}=k\frac{at^2}{1+t^2}\ \Rightarrow\ t=k,\ t=0$$

$$x=\frac{at^2}{1+t^2}$$

Hence, $x=\frac{ak^2}{1+k^2}, y=\frac{ak^3}{1+k^2}$. As k takes the values 0, ± 1, ± 2, ± 3, etc., we get

$$x=0,\quad y=0,$$

$$x=\frac{a}{2},\quad y=\pm\frac{a}{2},$$

$$x=\frac{4}{5}a,\quad y=\pm\frac{27}{10}a,$$

and so on. Now we use these values of x and y to plot the graph (Figure 6.17).

Example 6.3.2. Plot the graph of the folium of Descartes:

$$x=\frac{3at}{1+t^3},$$

$$y=\frac{3at^2}{1+t^3}\qquad (x^3+y^3-3axy=0).$$

We have:

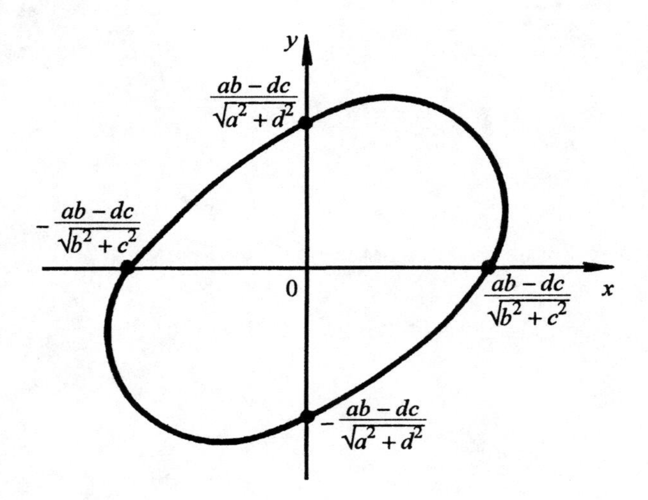

Figure 6.16. The graph of the function $x = a\cos t + d\sin t$, $y = b\sin t + c\cos t$ $(ab \neq cd)$.

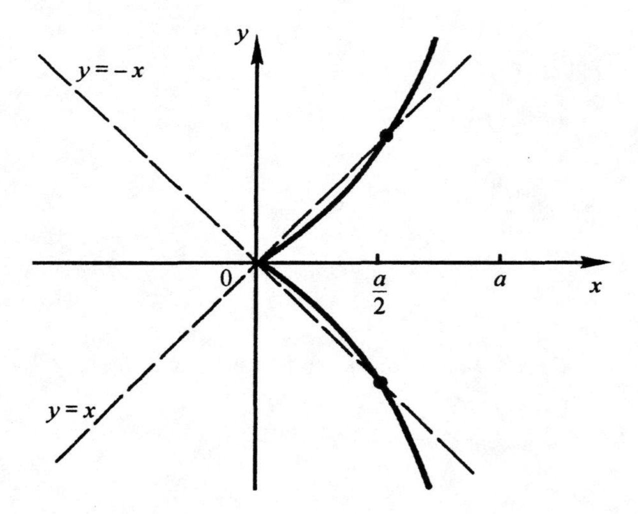

Figure 6.17. The cissoid of Diocles $x = \dfrac{at^2}{1+t^2}, y = \dfrac{at^3}{1+t^2}$.

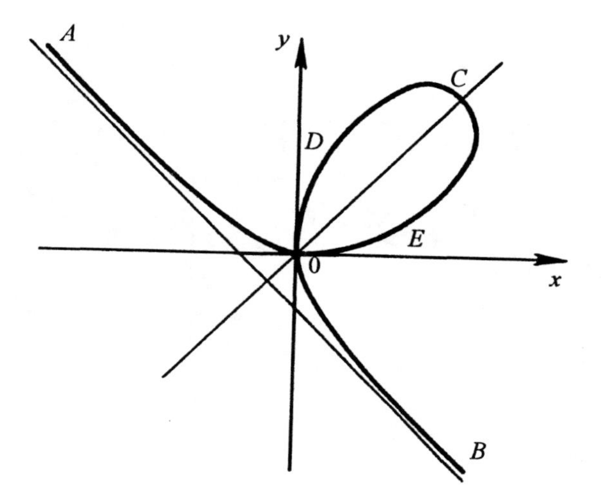

Figure 6.18. The folium of Descartes $x = \dfrac{3t}{1+t^3}, y = \dfrac{3t^2}{1+t^3}$.

$$y = kx$$

$$y = \frac{3at^2}{1+t^3} \quad \Rightarrow \quad \frac{3at^2}{1+t^3} = k\frac{3at}{1+t^3} \quad \Rightarrow \quad t = k, \quad t = 0$$

$$x = \frac{3at}{1+t^3}.$$

Hence,

$$x = \frac{3ak}{1+k^3}, \qquad y = \frac{3ak^2}{1+k^3} \qquad (x + y + a \neq 0 \Leftrightarrow t \neq -1).$$

Now let k range over 0, 1/2, 1, 2, −1/2, −2, etc. We get:

$$x = 0, \ y = 0, \ x = \frac{4}{3}a, \ y = \frac{2}{3}a, \ x = \frac{3}{2}a, \ y = \frac{3}{2}a,$$

$$x = \frac{2}{3}a, \ y = \frac{4}{3}a, \ x = -\frac{12}{7}a, \ y = \frac{6}{7}a, \ y = -\frac{12}{7}a,$$

and so on. By using these values of x and y we plot the curve (Figure 6.18).

Graphs of parametrically defined functions are shown in Figures 6.19 and 6.20.

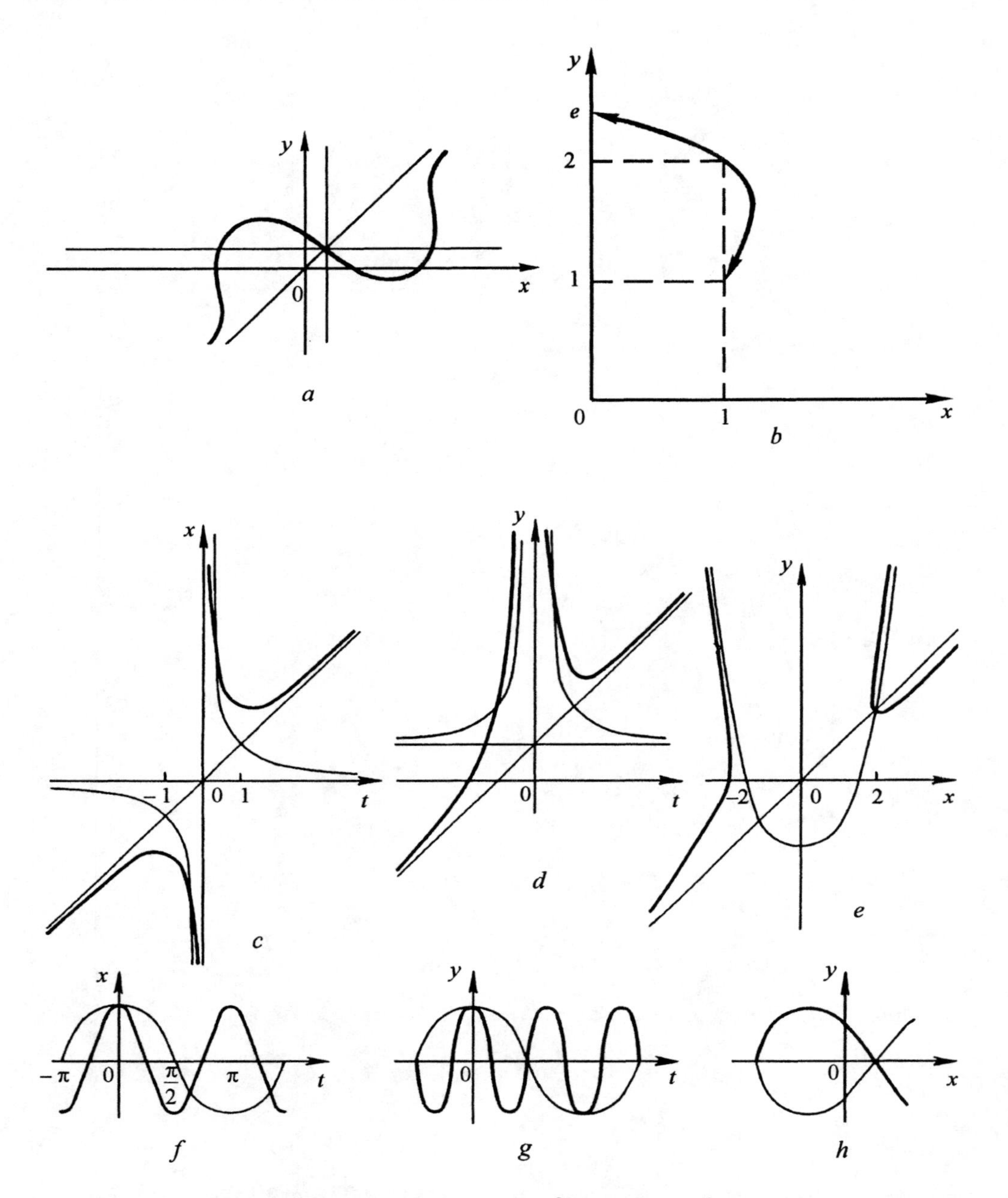

Figure 6.19. The graphs of functions: (*a*) $x = t^3 + 3t + 1, y = t^3 - 3t + 1$; (*b*) $x = t^{1/(t+1)}$, $y = (t+1)^{1/t}$; (*c*) $x = \dfrac{t^2+1}{t}$; (*d*) $y = \dfrac{t^3+1}{t^2}$; (*e*) $x = t + 1/t$, $y = t + 1/t^2$; (*f*) $x = a\cos 2t$; (*g*) $y = a\cos 3t$; (*h*) $x = a\cos 2t, y = a\cos 3t, a > 0$.

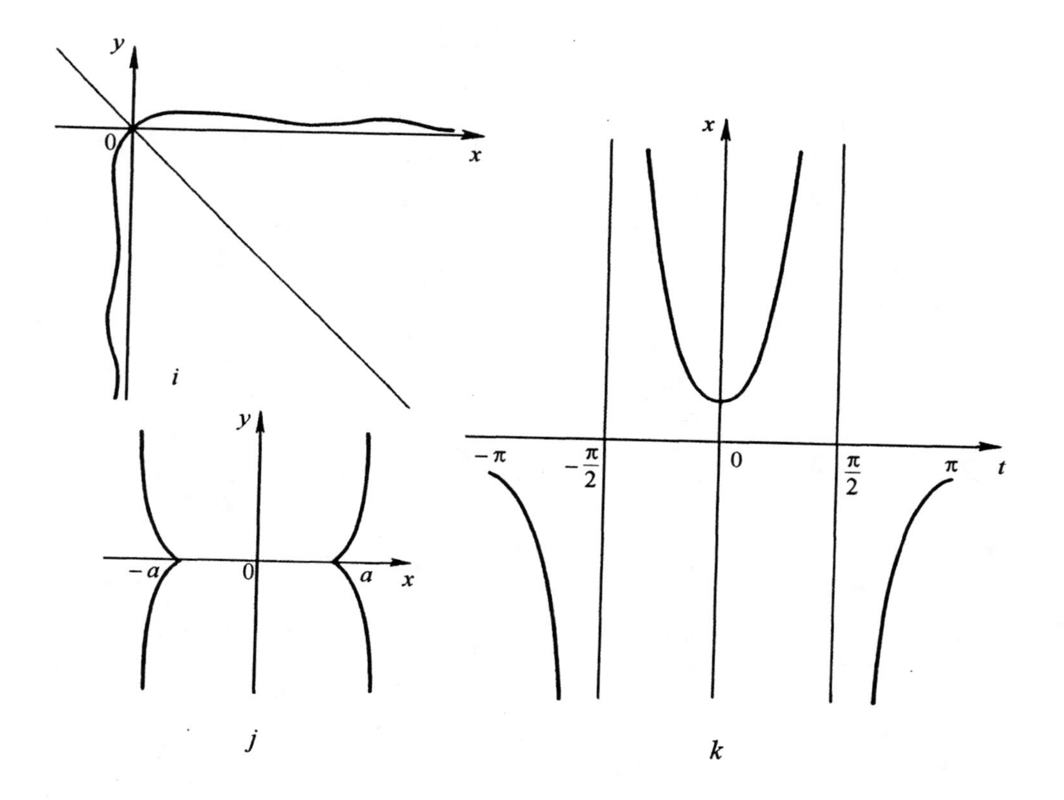

Figure 6.19. The graphs of functions: (i) $x = t \ln t, y = \ln t/t$; (j) $x = a/\cos^3 t$; (k) $x = a/\cos^3 t, y = a \tan^3 t$.

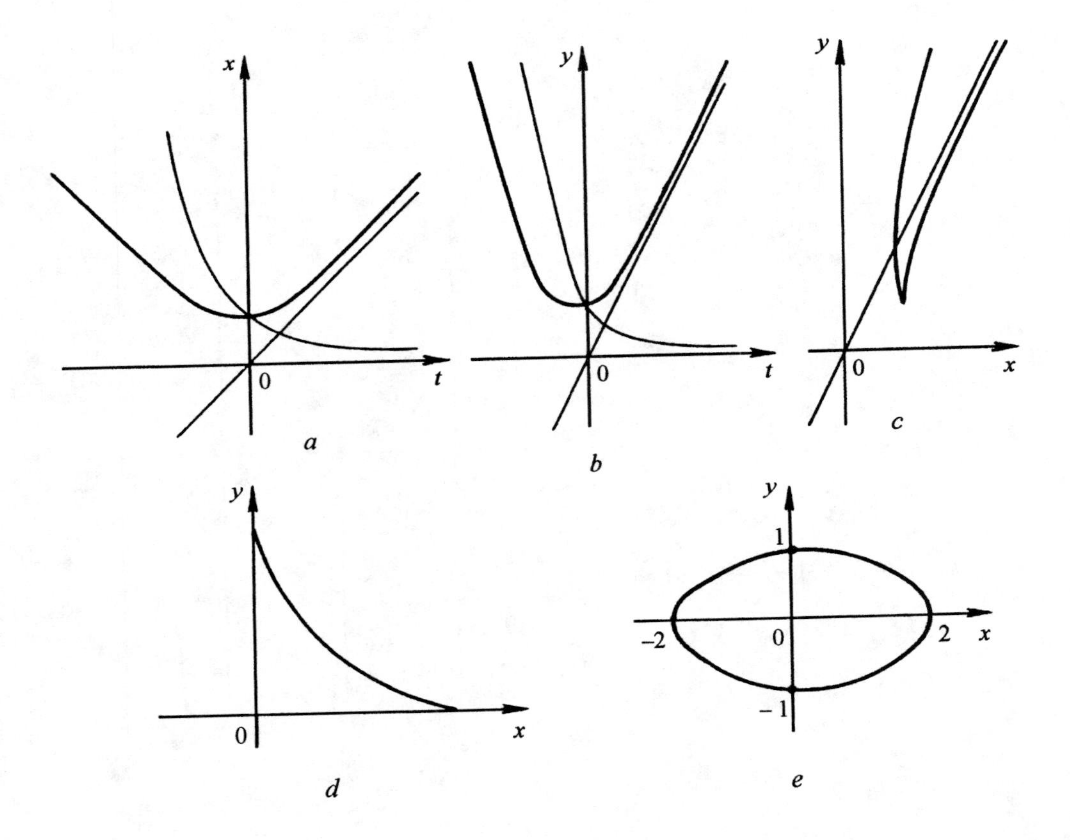

Figure 6.20. The graphs of functions: (*a*) $x = t + e^{-t}$; (*b*) $y = 2t + e^{-2t}$; (*c*) $x = t + e^{-t}, y = 2t + e^{-2t}$; (*d*) $x = \cos^4 t, y = \sin^4 t$; (*e*) $x = 2\cos t, y = \sin t$.

Chapter 7

Plots of graphs in polar coordinates

7.1. Functions in polar coordinates

A general form of a function defined in polar coordinates is $\rho = f(\varphi)$, or $F(\rho,\varphi) = 0$ in the implicit form.

The function $\rho = f(\varphi)$ can also be studied in polar coordinates by comparing it with a function in the Cartesian system, $y = f(x)$, which is obtained if, in the original function, ρ is replaced with y and φ with x. It is natural then to study the behavior of the function $\rho = f(\varphi)$ by using the scheme used for a function $y = f(x)$, see Chapter 2, Section 2.

Consider certain properties of the graph of a function $\rho = f(\varphi)$ (comparing it with the graph of the function $y = f(x)$).

The domain $a \le x \le b$ of the function $y = f(x)$ corresponds to the domain $a \le \varphi \le b$ of the function $\rho = f(\varphi)$. Singular points $x_1, x_2, \ldots$ of the function $y = f(x)$ correspond to singular points $\varphi_1 = x_1, \varphi_2 = x_2, \ldots$ of the function $\rho = f(\varphi)$.

Let $y = f(x)$ be an even function. Because $f(x) = f(-x)$, two points $A(x; y)$ and $B(-x; y)$ of the curve $y = f(x)$ correspond to two points $A_1(\rho,\varphi)$ and $B_1(\rho,\pi - \varphi)$ of the curve $\rho = f(\varphi)$ (Figure 7.1); points $A(x; -y)$ and $B(-x; -y)$ correspond to the points $A_1(\rho; 2\pi - \varphi)$ and $B_1(\rho; \pi + \varphi)$ (Figure 7.2).

Let $y = f(x)$ be an odd function. Then two points $A(x; y)$ and $B(-x; -y)$, which are symmetric with respect to the origin in the Cartesian coordinate system, correspond to the points $A_1(\rho; \varphi)$ and $B_1(\rho; \pi + \varphi)$ (Figure 7.3), whereas the points $A(-x; y)$ and $B(x; -y)$ in the Cartesian system correspond to the points $A_1(\rho; \pi/2 + \varphi)$ and $B_1(\rho; \frac{3}{2}\pi + \varphi)$ in polar coordinates (Figure 7.4).

If a curve $y = f(x)$ is symmetric with respect to the x-axis for $x > 0$, then points $A(x; y)$ and $B(x; -y)$ of this curve in Cartesian coordinates will correspond to the points $A_1(\rho; \varphi)$ and $B_1(\rho; 2\pi - \varphi)$ of the curve $\rho = f(\varphi)$ in the polar coordinates (Figure 7.5).

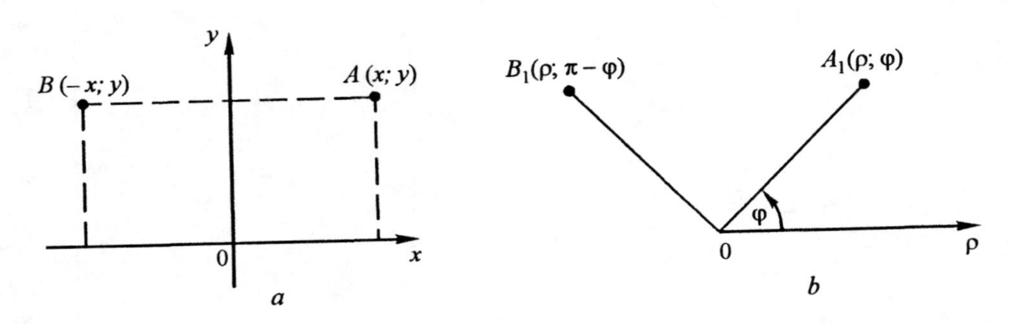

Figure 7.1. Coordinates of points $A(x; y)$, $B(-x; y)$ in Cartesian (*a*) and polar (*b*) coordinates.

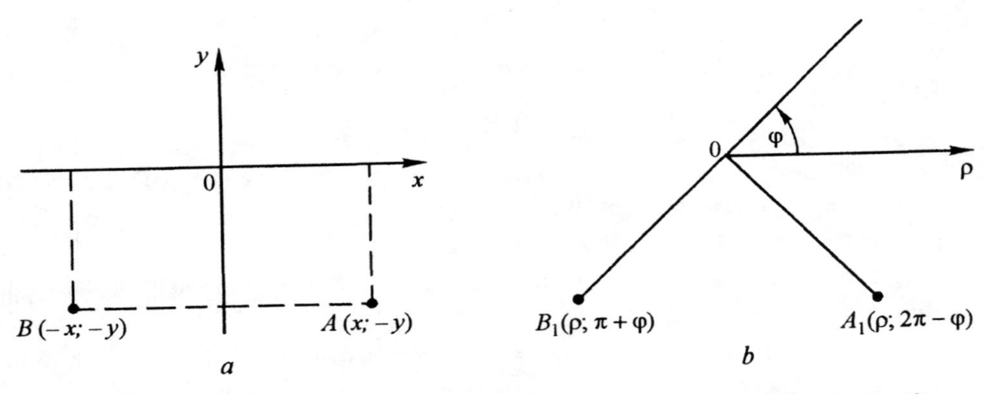

Figure 7.2. Coordinates of points $A(x; -y)$, $B(-x; -y)$ in Cartesian (*a*) and polar (*b*) coordinates.

If a curve $y = f(x)$ is symmetric with respect to the *x*-axis for $x < 0$, then points $A(-x; y)$ and $B(-x; -y)$ of this curve in the Cartesian coordinates correspond to the points $A_1(\rho; \frac{\pi}{2} + \varphi)$ and $B_1(\rho; \frac{3}{2}\pi - \varphi)$ in polar coordinates (Figure 7.6).

Periods of the functions $y = f(x)$ and $\rho = f(\varphi)$ are the same. This means that it will suffice to plot the graph of the function $\rho = f(\varphi)$ within the angle equal to the period of the function and then continue the plot of the graph by rotating it by the period.

If the function $y = f(x)$ is bounded $(M < f(x) < N)$, then its graph is located between the straight lines $y = M$ and $y = N$. For the corresponding function $\rho = f(\varphi)$, the same inequality holds, $M < f(\varphi) < N$, i.e. the graph of the function $\rho = f(\varphi)$ lies inside the annulus with the outer and inner radii equal to N and M, respectively.

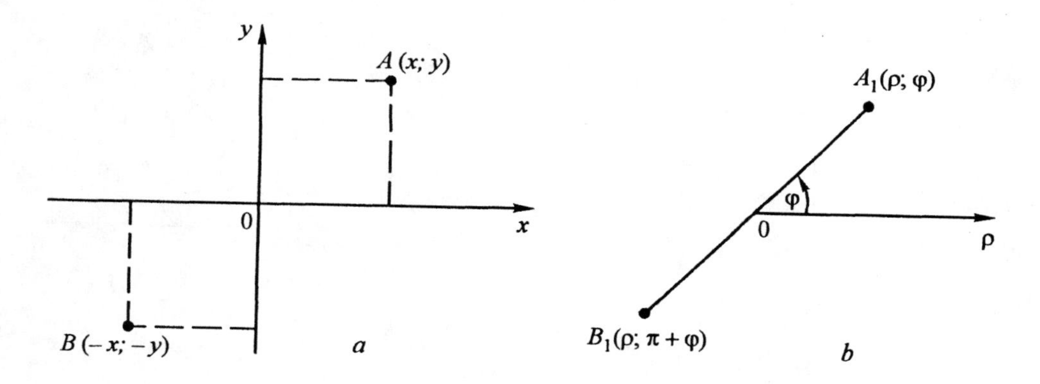

Figure 7.3. Coordinates of points $A(x; y)$, $B(-x; -y)$ in Cartesian (a) and polar (b) coordinates.

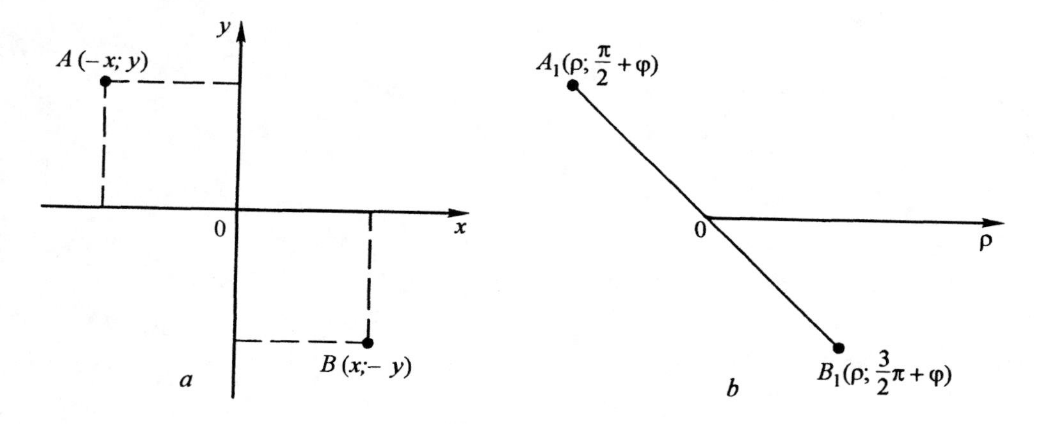

Figure 7.4. Coordinates of points $A(-x; y)$, $B(x; -y)$ in Cartesian (a) and polar (b) coordinates.

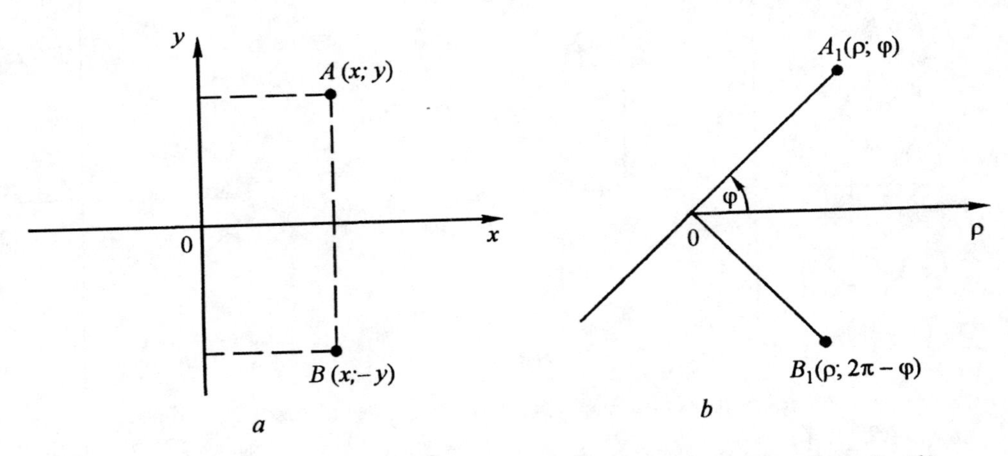

Figure 7.5. Coordinates of points $A(x; y)$, $B(x; -y)$ in Cartesian (a) and polar (b) coordinates.

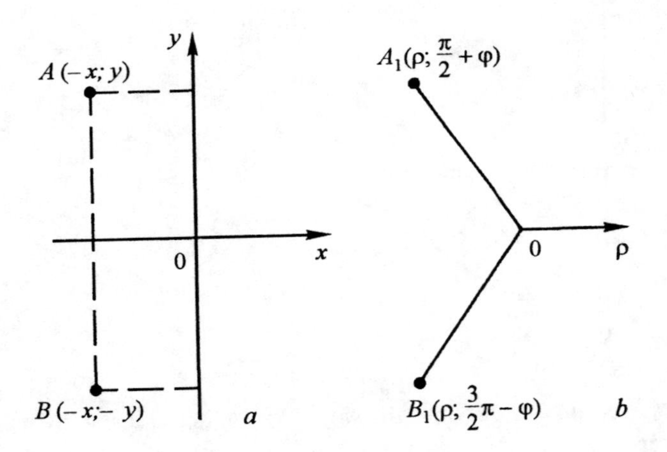

Figure 7.6. Coordinates of points $A(-x; y)$, $B(-x; -y)$ in Cartesian (a) and polar (b) coordinates.

If the function $y = f(x)$ has an extremum at $x = x_0$, then the function $\rho = f(\varphi)$ will have an extremum at $\varphi = x_0$. If the function $y = f(x)$ is decreasing in some interval, then the radius in the graph $\rho = f(\varphi)$ decreases if going clockwise and increases in the other case.

A horizontal asymptote $y = c$ of the curve $y = f(x)$ in Cartesian coordinates becomes an asymptotic circle $\rho = c$ in polar coordinates. In particular, if $c = 0$, then the circle degenerates into a point.

A vertical asymptote given in Cartesian coordinates by $x = b$ becomes, in polar coordinates, a ray given by $\varphi = b$. In particular, if $b = 0$, the asymptote $x = 0$ goes into the polar axis; if $b = \pi/2 + 2k\pi$, k is an integer, then the asymptote $x = b$ becomes the vertical ray $\varphi = \pi/2 + 2k\pi$.

An oblique asymptote $y = ax + b$ of the curve $y = f(x)$ in Cartesian coordinates will be the Archimedean spiral $\rho = a\varphi + b$ in the polar coordinates. In particular, an asymptote $y = ax$ of the curve $y = f(x)$ becomes the Archimedean spiral $\rho = a\,\varphi$.

Remark 7.1.1. To plot the graph of a function $\rho = f(\varphi)$ for φ corresponding to x such that $f(x) < 0$, it is sufficient to plot the graph $y = |f(x)|$, use this graph to plot the curve in the polar coordinates, and rotate the curve around the pole by π. We thus obtain a curve that corresponds to negative values of the function $\rho = f(\varphi)$. From this we conclude that to plot the curve $\rho = f(\varphi)$, we first plot it for φ such that $f(x) > 0$ and then for φ such that $f(x) < 0$ for the corresponding x.

7.2. Plots of graphs in polar coordinates

In Section 7.1 we looked in detail at polar coordinates and their connection with Cartesian coordinates. Results of this section were used in an essential way for constructing plots of functions in polar coordinates.

A scheme for plotting the graph of a function $\rho = f(\varphi)$ (by using properties of the function in the Cartesian coordinates) is as follows:

1. For the function $\rho = f(\varphi)$, consider the corresponding function $y = f(x)$.
2. Make an analysis of the function $\rho = f(\varphi)$ by comparing it with the corresponding function $y = f(x)$ as described in Section 2.1.
3. Use the graph of the function $y = f(x)$ to plot the graph of $\rho = f(\varphi)$.

It is sometimes possible in the most simple cases to draw the graph by plotting its points.

As an example, let us plot the graph of the function $\rho = a\varphi$, $a > 0$ (the *Archimedean spiral*). Write a table for $\varphi > 0$:

φ	0	$\frac{\pi}{4}$	$\frac{\pi}{2}$	$\frac{3}{4}\pi$	π	$\frac{5}{4}\pi$	$\frac{3}{2}\pi$	$\frac{7}{4}\pi$	2π	$\frac{9}{4}\pi$	$\frac{5}{2}\pi$	$\frac{11}{4}\pi$	3π
ρ	0	$0.8a$	$1.6a$	$2.5a$	$3.1a$	$3.9a$	$4.7a$	$5.5a$	$6.3a$	$7.1a$	$7.9a$	$8.7a$	$9.5a$

(the values in the table are approximate). By plotting these points in the coordinate plane and connecting them with a smooth line, we get a plot of the function for $\varphi > 0$.

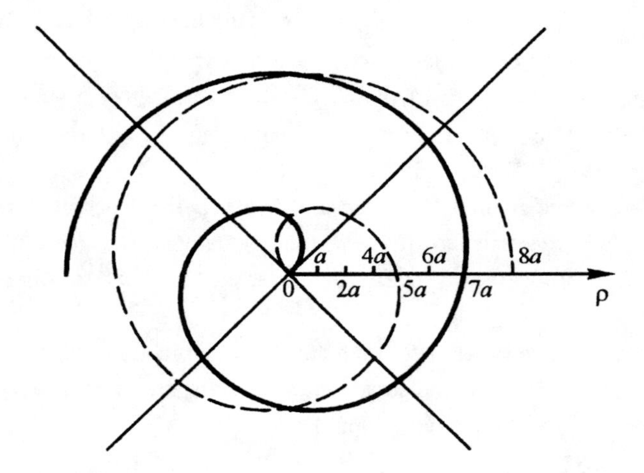

Figure 7.7. The Archimedean spiral $\rho = a\varphi$. Solid line for $\varphi > 0$, dotted — for $\varphi < 0$.

It is clear from the expression $\rho = a\varphi$ that the domain of the function is $(-\infty; +\infty)$. As $0 < \varphi < +\infty$, we have that $0 < \rho < +\infty$. If $-\infty < \varphi < 0$, then $-\infty < \rho < 0$. This means that the Archimedean spiral consists of two branches that are symmetric with respect to the straight line perpendicular to the polar axis (Figure 7.7).

Remark 7.2.1. The equation $\rho = a\varphi + b$ is an Archimedean spiral. Indeed, rotating the polar axis by the angle $\alpha = -b/a$, we get the Archimedean spiral $\rho = a\,\varphi$.

7.2.1. Plots of the curves $\rho = a$, $\rho = a\sin\varphi$, $\rho = a\cos\varphi$, $\rho = a\cos(\varphi - \varphi_0)$

For plotting these curves, it will be convenient to use a simple method for reduction of equations of curves to an equation of a circle in Cartesian coordinates.

1. By using the formula $\rho = \sqrt{x^2 + y^2}$, write the equation $\rho = a$ in the form $x^2 + y^2 = a^2$. This shows that $\rho = a$ is a circle of radius a with center at the origin (Figure 7.8).
2. Writing the curve $\rho = a\sin\varphi$ in the Cartesian coordinates we get $x^2 + (y - a/2)^2 = (a/2)^2$. This is a circle centered at $(0; a/2)$ of radius $a/2$ (Figure 7.9).
3. Write the curve $\rho = a\cos\varphi$ in Cartesian coordinates: $(x - a/2)^2 + y^2 = (a/2)^2$. This is a circle of radius $a/2$ with center at the point $(a/2; 0)$ (Figure 7.10).
4. Write the curve $\rho = a\cos(\varphi - \varphi_0)$ in Cartesian coordinates:

$$(x - \frac{a}{2}\cos\varphi_0)^2 + (y - \frac{a}{2}\sin\varphi_0)^2 = \left(\frac{a}{2}\right)^2.$$

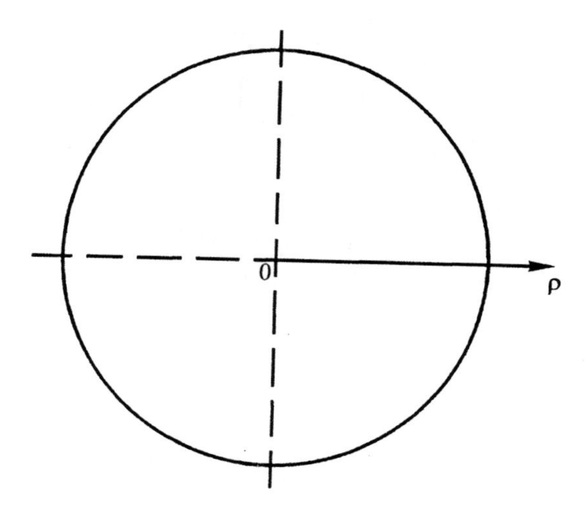

Figure 7.8. The circle $\rho = a$.

This is a circle of radius $a/2$ with center at the point $(\frac{a}{2}\cos\varphi_0; \frac{a}{2}\sin\varphi_0)$, where φ_0 is the angle between the polar axis and the ray passing through the origin and the center of the circle (Figure 7.11).

Note that the only curves that can be reduced to circles in Cartesian coordinates are those that have the equation in the form $\rho = M\cos\varphi + N\sin\varphi$, where M, N are constants. Such a circle will have a radius $\sqrt{M^2 + N^2}/2$ and its center at $(M/2; N/2)$.

Plotting curves $\rho = af(\varphi)$ by finding points of intersection of the graph and concentric circles

Let $\rho = af(\varphi)$ be a curve in polar coordinates. Then

$$\sqrt{x^2 + y^2} = af(\varphi) \Rightarrow x^2 + y^2 = \Phi^2(\varphi), \tag{1}$$

where $\Phi(\varphi) = af(\varphi)$. If $\varphi = \text{const}$, then we get the circle $x^2 + y^2 = \Phi^2$.

Let $\varphi \neq \text{const}$. To plot the graph of the function $\rho = af(\varphi)$ in such a case, one can use points of intersection of the curves $\rho = f(\varphi)$ with concentric circles. Let us take a closer look at this method.

Consider the equations $\rho = af(\varphi)$, $x^2 + y^2 = \Phi^2(\varphi)$.

Note that, for fixed values of φ_i from the domain of the function $f(\varphi)$, points of the curve $\rho = f(\varphi)$ lie on the circle (1). Hence, for $\varphi = \varphi_i$, equation (1) gives a family of concentric circles of radii $\rho = \Phi(\varphi_i)$ intersecting the curve $\rho = af(\varphi)$ at points M_i; the angles φ_i are angles between ρ_i and the polar axis (Figure 7.12). It is convenient to use this method in the cases where the

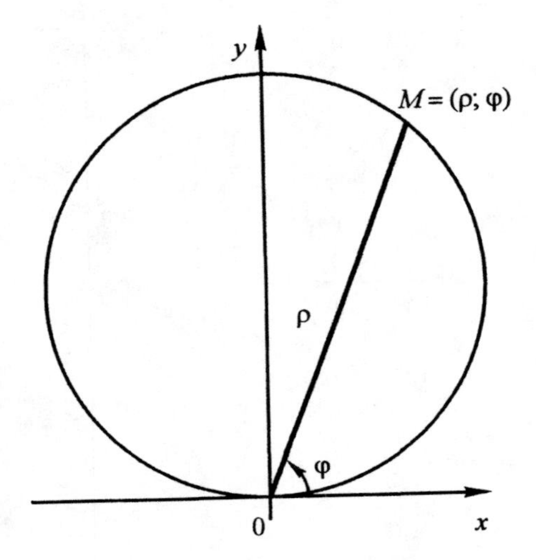

Figure 7.9. The circle $\rho = a \sin \varphi$.

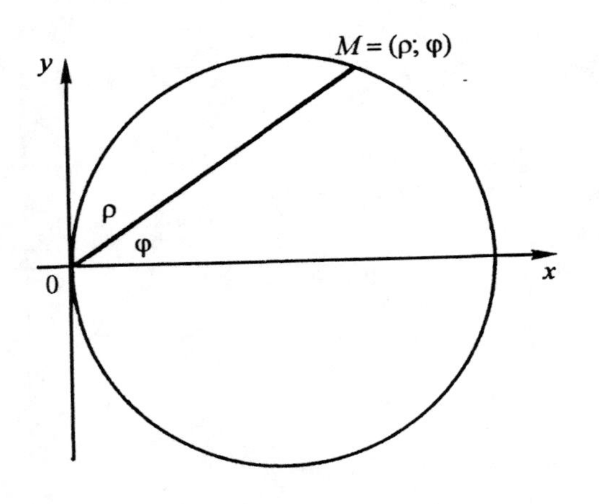

Figure 7.10. The circle $\rho = a \cos \varphi$.

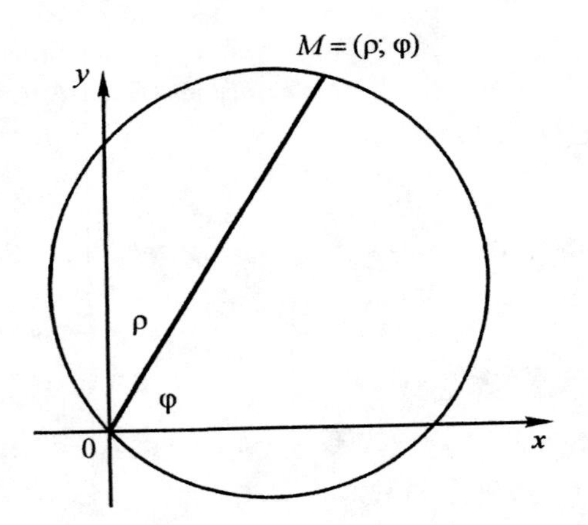

Figure 7.11. The circle $\rho = a \cos (\varphi - \varphi_0)$.

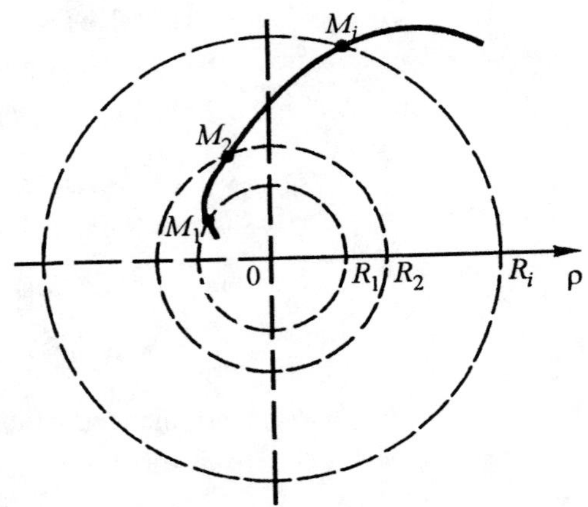

Figure 7.12. The graph of the curve $\rho = af(\varphi)$.

equation of the curve $\rho = af(\varphi)$ cannot be reduced to an equation of a circle in Cartesian coordinates.

7.2.2. Plots of the graphs of curves $\rho = a\sin n\varphi$, $\rho = a\cos n\varphi$ ($n \in \mathbf{N}, n > 1$)

Graphs of the curves $\rho = a\sin n\varphi$, $\rho = a\cos n\varphi$ are leaf "roses" (see Chapter 3, part II). In this section we consider these curves in generalized polar coordinates which differ from the usual polar coordinates in that ρ is allowed to be negative; in this case the point is located from the pole at the distance equal to the absolute value of ρ and is drawn in the direction opposite to the ray with the angle φ.

Let us use the method of finding points of intersection of curves with concentric circles.

1. To plot the graph of the curve $\rho = a\sin n\varphi$, subdivide the circle into n parts. The number of points in which the curve $\rho = a\sin n\varphi$ intersects the circle of radius r ($0 < r < a$) equals $2n$. The greatest value is $\rho = a$, the least — $\rho = 0$. To find the angle at which $\rho = a$, it is necessary to solve the equation $\sin n\varphi = 1$, and to find the ranges of the angle φ, one needs to solve the inequality $\sin n\varphi \geq 0$. The first point we start with is located at the intersection of the circle $\rho = a$ and the ray coming out at the angle found from the equation $\sin n\varphi = 1$. Hence, each leaf of the curve lies within the angle that can be found from the inequality $\sin n\varphi \geq 0$, and intersects the circle of radius r ($0 < r < a$) in two points.

Example 7.2.1. Plot the graph of the curve $\rho = 3\sin 2\varphi$.

Use the method of finding points of intersection of the curve with concentric circles of radii 3/2, 3. First find the angle of the rays that pass through points of intersection of the curve $\rho = 3\sin 2\varphi$ and the circle $\rho = 3$:

$$\rho = 3\sin 2\varphi, \quad \rho = 3.$$

We get $\varphi = \pi/4 + k\pi$, where $k \in \mathbf{Z}$.

Making k range over the values 0, 1, we get $\pi/4$, $\pi/4 + \pi$. Note that for $k \geq 2$, the angles repeat, i.e. $\varphi = \pi/4 + 2\pi$ coincides with the angle $\pi/4$, and if $k < 0$, say for example $k = -1$, then $\varphi = \pi/4 - \pi$, i. e. the angle is $\pi/4 + \pi$. Hence k will assume only the values 0, 1, 2, ...

Now we find the angles of rays that pass through points of intersection of the curve $\rho = 3\sin 2\varphi$ and the circle $\rho = 3/2$:

$$\rho = 3\sin 2\varphi, \quad \rho = 3/2.$$

We have $\varphi = (-1)^n\pi/12 + n\pi/2$. Make n take the values 0, 1, 2, 3, and get the corresponding values of φ: $\pi/12$, $-\pi/12 + \pi/2$, $\pi/12 + \pi$, $-\pi/12 + 3/2\pi$.

The curve $\rho = 3\sin 2\varphi$ intersects circles of radius r ($0 < r < 3$) in four points. It is clear that $\rho_{max} = 3$, $\rho_{min} = 0$. The range of the angle φ can be found from the inequality $\sin 2\varphi \geq 0$. This gives $2k\pi \leq 2\varphi \leq \pi + 2k\pi$, or $k\pi \leq \varphi \leq \pi/2 + k\pi$.

Hence, for $k = 0$ and 1, we have $0 \leq \varphi \leq \pi/2$, $\pi \leq \varphi \leq 3/2\pi$, correspondingly (Figure 7.13).

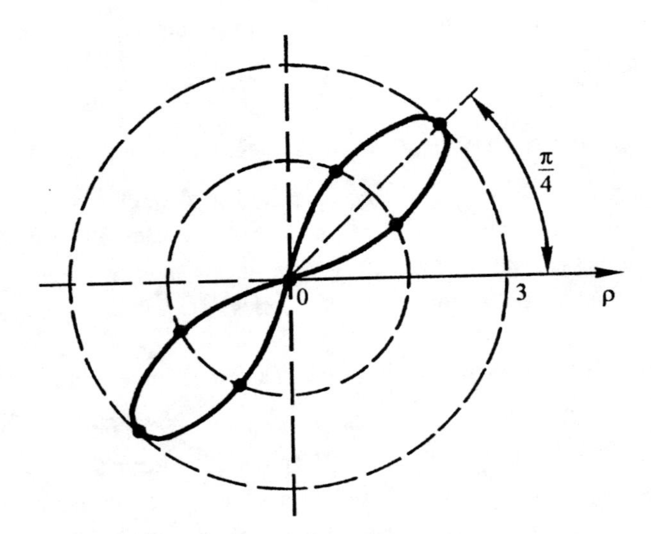

Figure 7.13. The curve $\rho = 3\sin 2x$.

Remark 7.2.2. Let us plot the graph of the same curve $\rho = 3\sin 2\varphi$ in the generalized polar coordinates by using the method of comparing it with the graph of the corresponding function in Cartesian coordinates.

First we plot the graph of the function $y = 3\sin 2x$ (see Figure 4.32), stretched by 3 along the y-axis.

Study the function $\rho = 3\sin 2\varphi$ by comparing it with the function $y = 3\sin 2x$. The function $y = 3\sin 2x$ is defined for all x, hence the function $\rho = 3\sin 2x$ is also defined for all φ. The function $y = 3\sin 2x$ is odd, i.e. the curve $\rho = 3\sin 2\varphi$ is symmetric with respect to the pole. Because the function $y = 3\sin 2x$ is periodic with the period π, the period of the function $\rho = 3\sin 2\varphi$ is also π . Since the function $y = 3\sin 2x$ is bounded ($|3\sin 2x| \le 3$), the function $\rho = 3\sin 2\varphi$ is also bounded ($|3\sin 2\varphi| \le 3$). The function $y = 3\sin 2x$ attains a maximum on $[0; \pi]$ at $x = \pi/4$ ($y = 3$) and a minimum at $x = 3/4\pi$ ($y_{min} = -3$). At $x = 0, \pi/2, \pi$, the function $y = 3\sin 2x$ equals zero. The corresponding function $\rho = 3\sin 2\varphi$ has extremums at $\varphi = \pi/4$ and $\varphi = 3/4\pi$ that equal 3 and –3.

The function $y = 3\sin 2x$ is increasing for $x \in (0; \frac{\pi}{4}) \cup (\frac{3}{4}\pi; \pi)$, and decreasing for $x \in (\frac{\pi}{4}; \frac{\pi}{2}) \cup (\frac{\pi}{2}; \frac{3}{4}\pi)$, and hence the function $\rho = 3\sin 2\varphi$ will be increasing on $(0; \frac{\pi}{4}) \cup (\frac{3}{4}\pi; \pi)$ and decreasing on $(\frac{\pi}{4}; \frac{\pi}{2}) \cup (\frac{\pi}{2}; \frac{3}{4}\pi)$.

The curve $y = 3\sin 2x$ does not have asymptotes, and hence there are no asymptotes for the curve $\rho = 3\sin 2\varphi$. Thus the curve $\rho = 3\sin 2\varphi$ is located inside a disc centered at the pole and having radius 3.

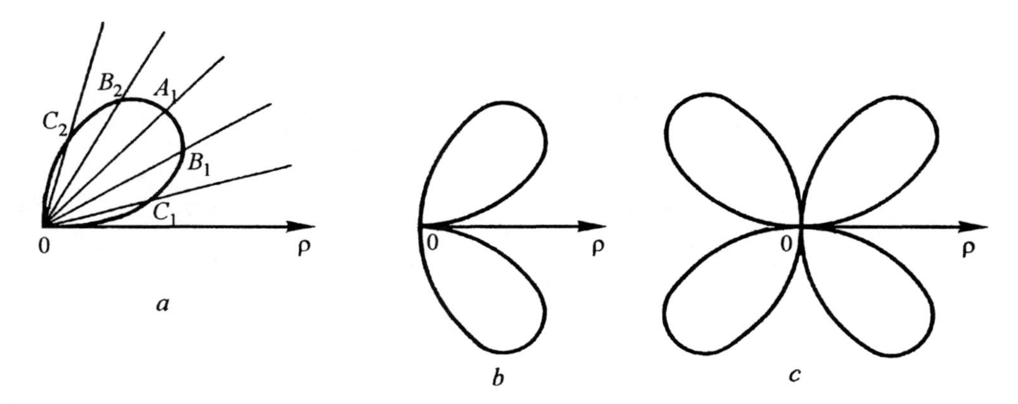

Figure 7.14. Construction of the graph of the function $\rho = 3\sin 2\varphi$ (four-leaf "rose").

Using symmetry of the curve $\rho = 3\sin 2\varphi$ with respect to the pole and its periodicity, we plot the curve $\rho = 3\sin 2\varphi$ for $0 \le \varphi \le \pi/2$. The plotting is carried out as follows: mark the points of the maximum, $A_1(3;\ \pi/4)$, and the minimums, $A_2(0;\ 0)$, $A_3(0;\ \pi/2)$, where $\rho = 0$; after that we plot the points $B_1(\rho;\ \pi/6)$, $B_2(\rho,\ 3/8\ \pi)$. Note that the value of ρ for points B_1 , B_2 is found by taking the y value of the graph of the function $y = 3\sin 2x$ at $x = \pi/8$ and $x = 3/8\pi$. In the same way we construct points C_1 and C_2. Plotting a curve that passes through these points we obtain the graph of the function $\rho = 3\sin 2\varphi$ for $0 \le \varphi \le \pi/2$ (Figure 7.14a). Using the fact that the curve is symmetric with respect to the pole, we plot the curve $\rho = 3\sin 2\varphi$ for $-\pi/2 \le \varphi \le 0$ (Figure 7.14b). Finally, rotating the graph about the pole by π we obtain the graph of the function $\rho = 3\sin 2\varphi$ (Figure 7.14c).

Example 7.2.2. Plot the graph of the function $\rho = a\cos 2\varphi$ $(a = 3)$.

We will plot this curve using both methods. At first, we apply the method of finding intersections of the curve $\rho = 3\cos 2\varphi$ with concentric circles. Finding the angles of rays that pass through points of intersection of the curve and circles of radii $3/2$ and 3, we get:

$$\rho = 3\cos 2\varphi, \quad \rho = 3; \quad \rho = 3\cos 2\varphi, \quad \rho = 3/2.$$

We get $\varphi = k\pi$, $\varphi = \pm\pi/6 + k\pi$.

For $k = 0$, $k = 1$, we get from $\varphi = k\pi$ that $\varphi = 0, \varphi = \pi$, and from $\varphi = \pm\pi/6 + k\pi$ that $\varphi = \pm\pi/6$, $\varphi = \pm\pi/6 + \pi$. We also find that $\rho_{max} = 3$ is attained if $\varphi = 0$ or $\varphi = \pi$.

Determine bounds for the range of the angle φ: $\cos 2\varphi \ge 0 \Rightarrow\ -\pi/4 + k\pi \le \varphi \le \pi/4 + k\pi$; whence, setting $k = 0$, $k = 1$, we get

$$-\pi/4 \le \varphi \le \pi/4, \qquad -\pi/4 + \pi \le \varphi \le \pi/4 + \pi.$$

Now use this data to plot the graph of the curve $\rho = 3\cos 2\varphi$ (Figure 7.15).

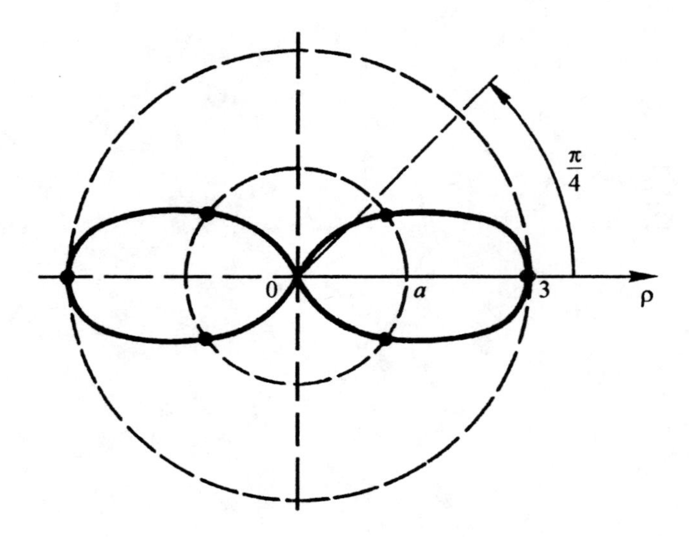

Figure 7.15. The graph of the function $\rho = 3\cos 2\varphi$.

Using Remark 7.2.2 as well as the formula $\cos 2\varphi = \sin(\pi/2 + 2\varphi)$, we see that the graph of the curve $\rho = 3\cos 2\varphi = 3\sin(\pi/2 + 2\varphi)$ can easily be obtained from the graph of the curve $\rho = 3\cos 2\varphi$ by rotating it at the angle $\pi/4$ (Figure 7.16).

Example 7.2.3. Plot the graph of the curve $\rho = a\sin 3\varphi$.

Find the angles of the rays that pass through points of intersection of the curve $r = a\sin 3\varphi$ and the circle $\rho = a$:

$$\begin{matrix}\rho = a\sin 3\varphi \\ \rho = a\end{matrix} \Rightarrow \varphi = \frac{\pi}{6} + \frac{2k\pi}{3}.$$

Ranging k over the values 0, 1, 2, we get three values of φ: $\pi/6, \pi/6 + 2/3\pi, \pi/6 + 4/3\pi$. So there are three points of intersection of the curve $\rho = a\sin 3\varphi$ and the circle $\rho = a$.

In the same way we find the angles of the rays that pass through points of intersection of the curve $\rho = a\sin 3\varphi$ and the circle $\rho = a/2$:

$$\begin{matrix}\rho = a\sin 3\varphi \\ \rho = a/2\end{matrix} \Rightarrow \varphi = (-1)^n \frac{\pi}{18} + \frac{n\pi}{3}.$$

As n takes the values 0, 1, 2, 3, 4, 5, the corresponding angles become $\pi/18$, $\pi/18 + \pi/3$, $\pi/18 + 2/3\pi$, $-\pi/18 + \pi, \pi/18 + 4/3\pi$, $-\pi/18 + 5/3\pi$. The curve intersects the circle of radius $\rho = a/2$ in six points. Determine the range of the angle φ: $\sin 3\varphi \geq 0 \Rightarrow 2/3\pi k \leq \varphi \leq \pi/3 + 2/3\pi k$.

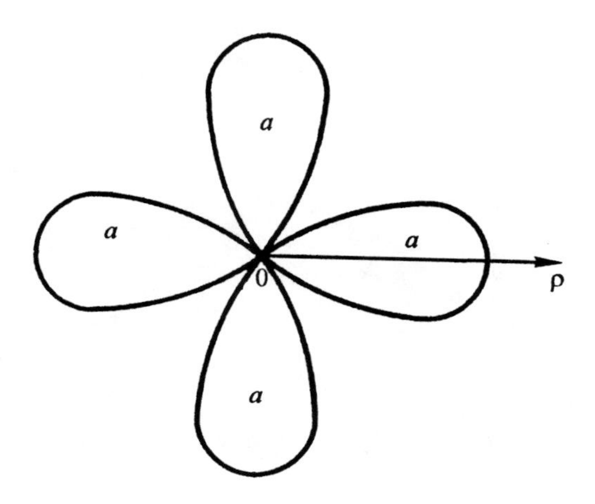

Figure 7.16. The graph of the curve $\rho = a \cos 2\varphi$ (four-leaf "rose").

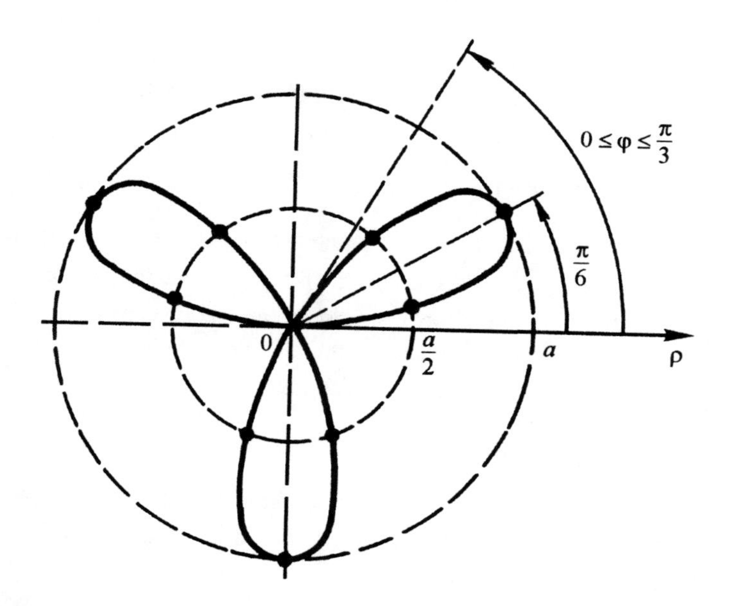

Figure 7.17. The graph of the curve $\rho = a \sin 3\varphi$.

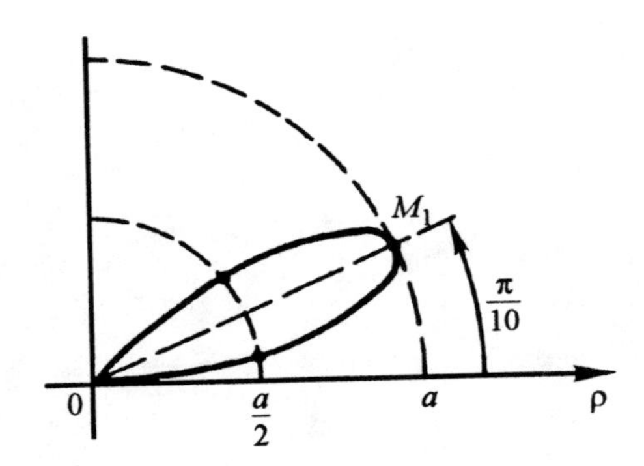

Figure 7.18. Construction of a leaf of the curve $\rho = a \sin 5\varphi$.

For k equal 0, 1, 2, the corresponding values satisfy $0 \le \varphi \le \pi/3$, $2/3\pi \le \varphi \le \pi/3 + 2/3\pi$, $4/3\pi \le \varphi \le \pi/3 + 4/3\pi$. Now use this data to plot the curve $\rho = a \sin 3\varphi$ (Figure 7.17).

Example 7.2.4. Plot the graph of the curve $\rho = a \sin 5\varphi$.

Let us use a simplified procedure. Find the starting angle, $\sin 5\varphi = 1 \Rightarrow \varphi = \pi/10$.

Determine the ranges of the angle for the sector containing the first leaf: $\sin 5\varphi \ge 0 \Rightarrow 0 \le \varphi \le \pi/5$. Now find the angles of the rays passing through points of intersection of the first leaf of the curve $\rho = a \sin 5\varphi$ and the circle $\rho = a/2$:

$$\sin 5\varphi = 1/2 \quad \Rightarrow \quad \varphi_1 = \pi/30, \quad \varphi_2 = \pi/6.$$

Hence, the first leaf can be plotted by drawing the angle of $\varphi = 2/5\pi$ from the point M_1 counterclockwise. Then subdivide the circle into five parts. We thus obtain vertices of the remaining leafs. Drawing rays at the angle of $\varphi = \pi/10$, counting clockwise and counterclockwise from the obtained points, we find sectors containing the rest of the leaves (Figure 7.18). Plotting a leaf in each sector, we get the graph of the curve $\rho = a \sin 5\varphi$ (Figure 7.19).

Example 7.2.5. Plot the graph of the curve $\rho = a \cos 6\varphi$ by using the simplified procedure (see Example 7.2.4, Figure 7.20). Mark the angles $\varphi = 2/6\pi = \pi/3$, each counting from the point M counterclockwise, which subdivides the circle into six parts. We thus obtain vertices of the leafs. Starting at these points, draw angles of $\varphi = \pi/12$ clockwise and counterclockwise. This gives sectors that contain the rest of the leaves (Figure 7.21).

7.2.3. Plots of the graphs of curves $\rho = a(1 \pm \sin \varphi)$, $\rho = a(1 \pm \cos \varphi)$

To plot the graphs of these curves, we use the method of finding intersections of the graph of the curve and corresponding concentric circles [3] (Figure 7.22).

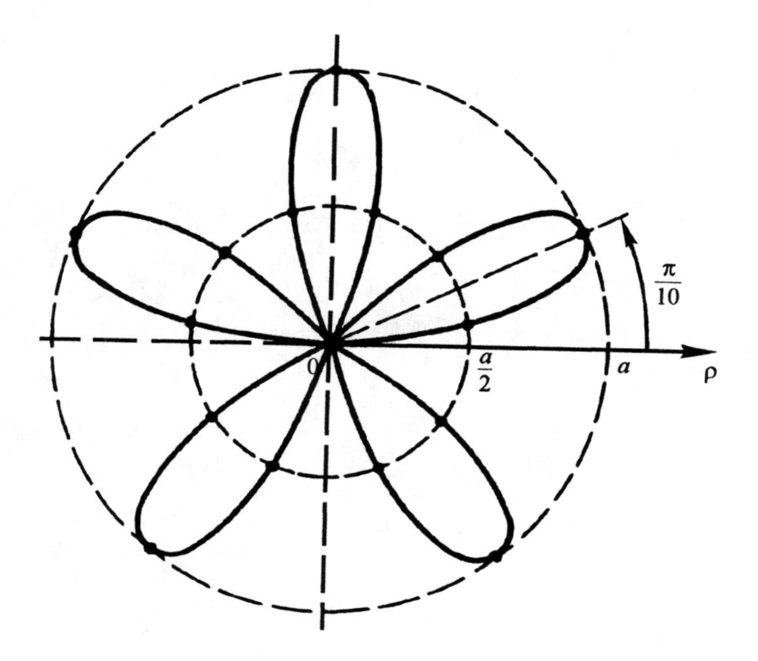

Figure 7.19. The graph of the curve $\rho = a \sin 5\varphi$.

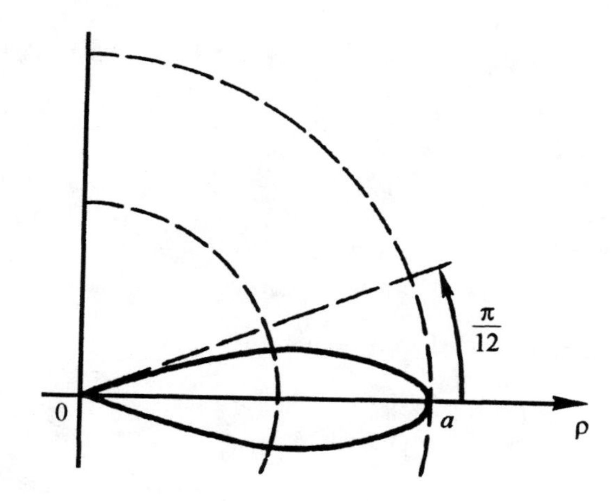

Figure 7.20. The first leaf of the curve $\rho = a \cos 6\varphi$.

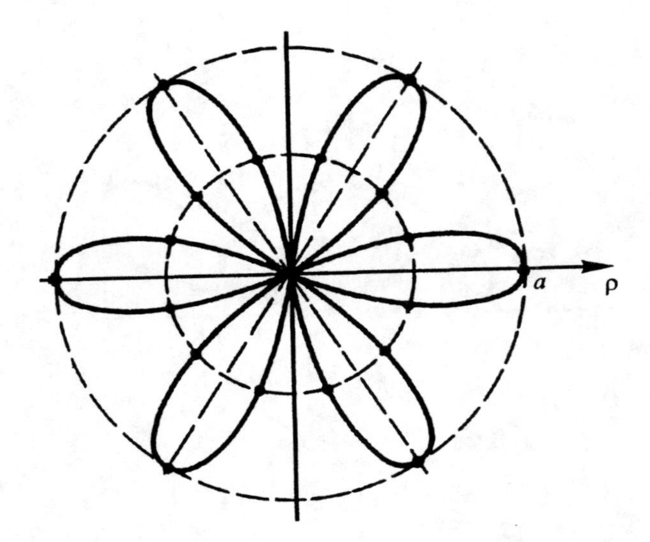

Figure 7.21. The curve $\rho = a\cos 6\varphi$.

Curve	$\rho_{max} = 2a$	$\rho_{min} = 0$
$\rho = a(1 + \sin\varphi)$	$\sin\varphi = 1, \varphi = \pi/2$	$\sin\varphi = -1, \varphi = -\pi/2$
$\rho = a(1 - \sin\varphi)$	$\sin\varphi = -1, \varphi = -\pi/2$	$\sin\varphi = 1, \varphi = \pi/2$
$\rho = a(1 + \cos\varphi)$	$\cos\varphi = 1, \varphi = 2\pi$	$\cos\varphi = -1, \varphi = \pi$
$\rho = a(1 - \cos\varphi)$	$\cos\varphi = -1, \varphi = \pi$	$\cos\varphi = 1, \varphi = 0$

$r = a/2$		$r = a$		$r = 3/2a$	
ρ	φ	ρ	φ	ρ	φ
0	p	$-p/6$	$\pi/6 + \pi$	$p/6$	$-\pi/6 + \pi$
0	π	$\pi/6$	$-\pi/6 + \pi$	$-\pi/6$	$\pi/6 + \pi$
$-\pi/3 + \pi$	$\pi/3 + \pi$	$-\pi/2$	$\pi/2$	$-\pi/3$	$\pi/3$
$-\pi/3$	$\pi/3$	$-\pi/2$	$\pi/2$	$-\pi/2 + \pi$	$\pi/2 + \pi$

Use this data to plot the heart-shaped curves.

7.2.4. Plots of graphs of curves $\rho = a(1 \pm \sin n\varphi)$, $\rho = a(1 \pm \cos n\varphi)$ ($n \in \mathbf{N}, n > 1$)

Each of these curves has n maximal values of ρ (Figures 7.23 –7.34).

Curve	ρ_{max}	φ	$k = n$
$\rho = a(1 + \sin n\varphi)$	$\sin n\varphi = 1$	$\varphi = \pi/(2n) + 2k\pi/n$	$\varphi = \pi/(2n) + 2\pi$
$\rho = a(1 - \sin n\varphi)$	$\sin n\varphi = -1$	$\varphi = -\pi/(2n) + 2k\pi/n$	$\varphi = -\pi/(2n) + 2\pi$
$\rho = a(1 + \cos n\varphi)$	$\cos n\varphi = 1$	$\varphi = 2\pi k/n$	$\varphi = 2\pi$
$\rho = a(1 - \cos n\varphi)$	$\cos n\varphi = -1$	$\varphi = \pi/n + 2k\pi/n$	$\varphi = \pi/n + 2\pi$

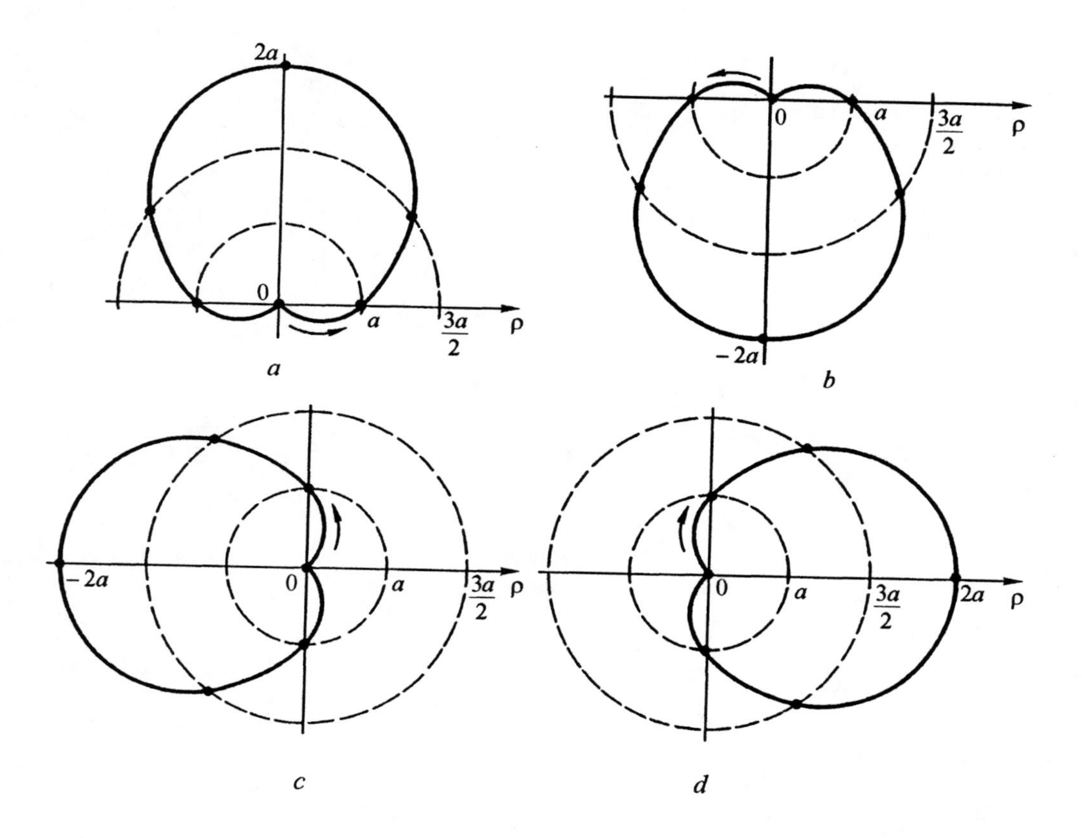

Figure 7.22. The curves (*a*) $\rho = a(1 + \sin\varphi)$; (*b*) $\rho = a(1 - \sin\varphi)$; (*c*) $\rho = a(1 - \cos\varphi)$; (*d*) $\rho = a(1 + \cos\varphi)$.

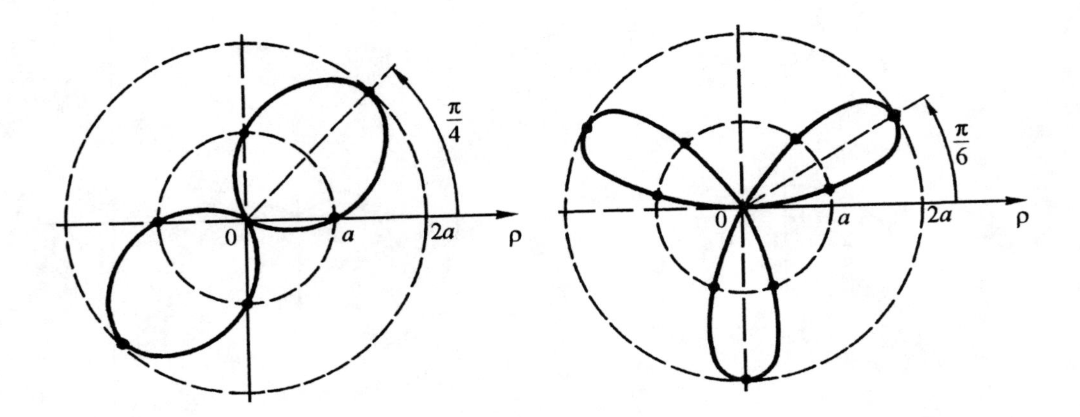

Figure 7.23. The curve $\rho = a(1 + \sin 2\varphi)$.

Figure 7.24. The curve $\rho = a(1 + \sin 3\varphi)$.

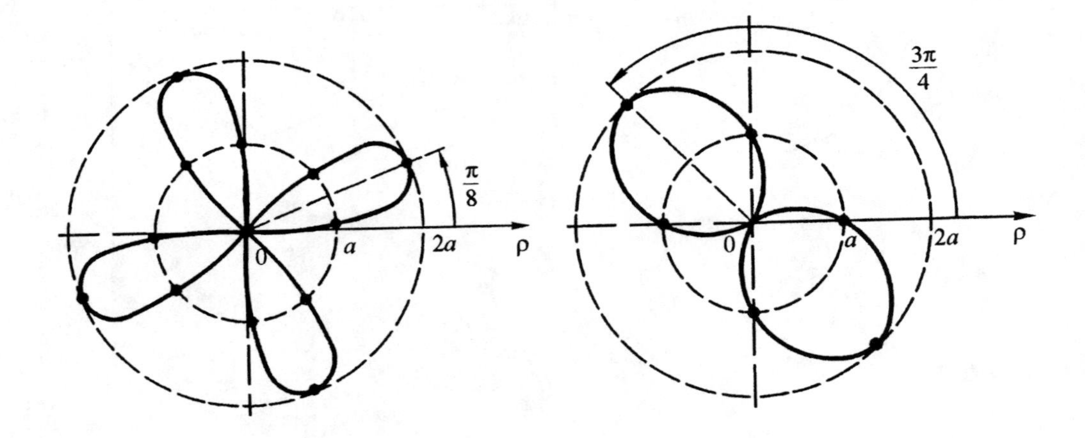

Figure 7.25. The curve $\rho = a(1 + \sin 4\varphi)$.

Figure 7.26. The curve $\rho = a(1 - \sin 2\varphi)$.

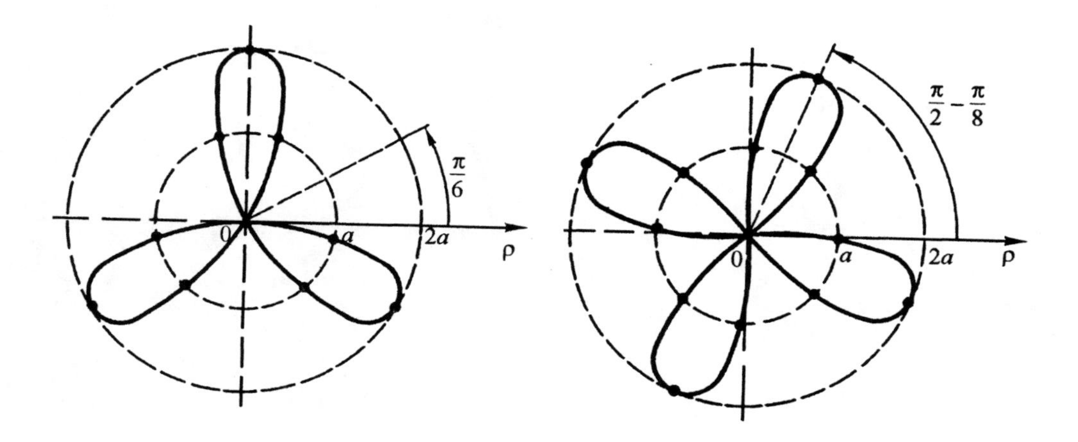

Figure 7.27. The curve $\rho = a(1 - \sin 3\varphi)$.

Figure 7.28. The curve $\rho = a(1 - \sin 4\varphi)$.

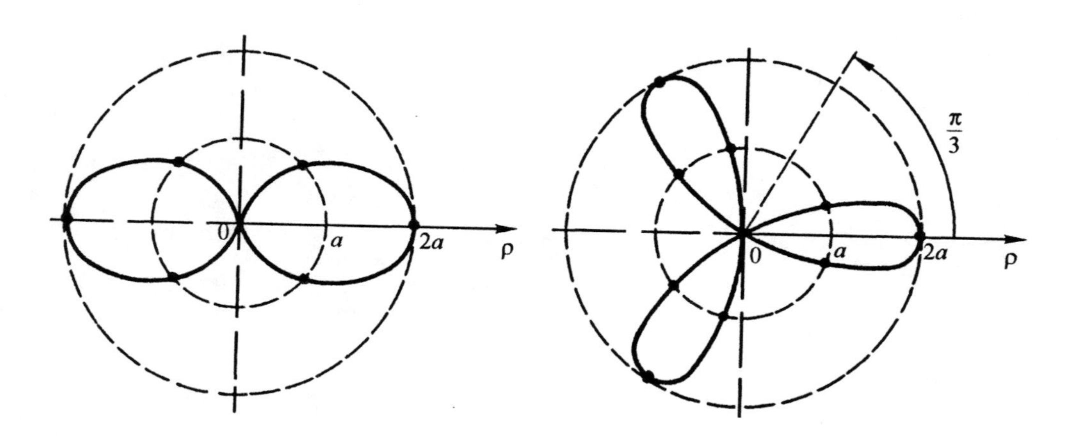

Figure 7.29. The curve $\rho = a(1 + \cos 2\varphi)$.

Figure 7.30. The curve $\rho = a(1 + \cos 3\varphi)$.

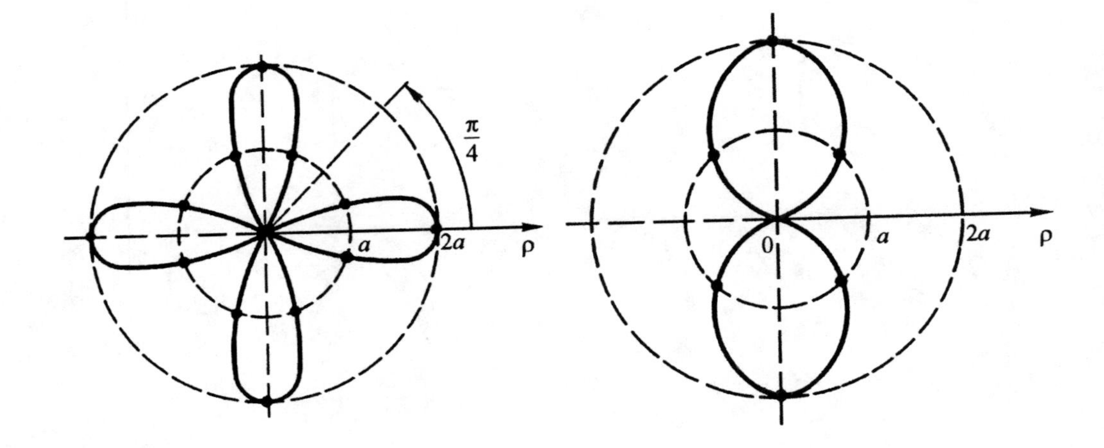

Figure 7.31. The curve
$\rho = a(1 + \cos 4\varphi)$.

Figure 7.32. The curve
$\rho = a(1 - \cos 2\varphi)$.

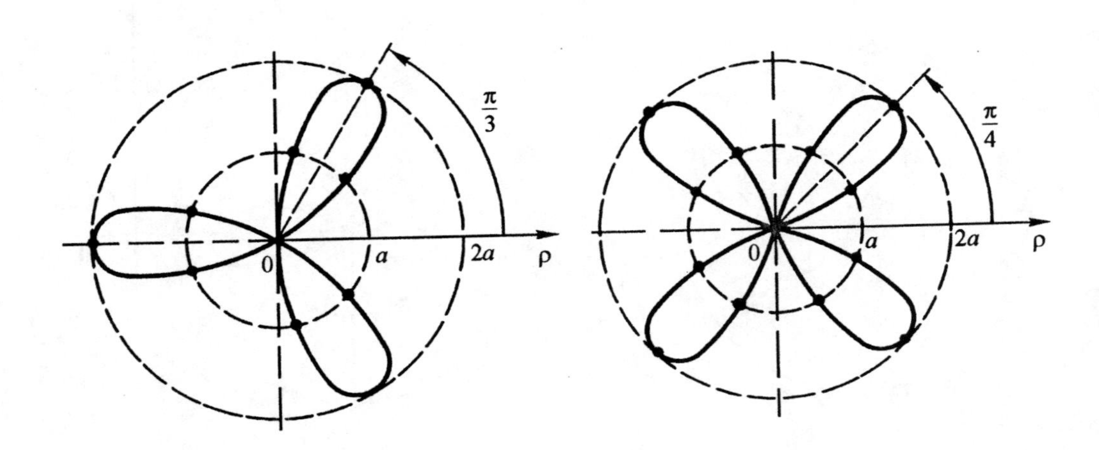

Figure 7.33. The curve
$\rho = a(1 - \cos 3\varphi)$.

Figure 7.34. The curve
$\rho = a(1 - \cos 4\varphi)$.

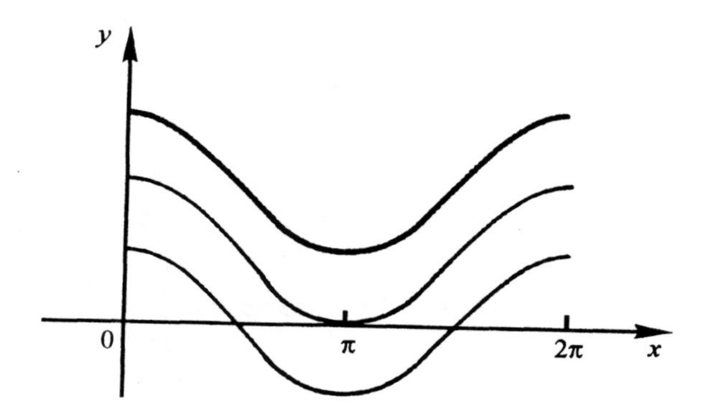

Figure 7.35. The curve $y = b + a\cos x$, $a = 2$, $b = 1$.

Each of the curves $\rho = a(1 \pm \sin n\varphi)$, $\rho = a(1 \pm \cos n\varphi)$ intersects circles of radius r, $0 < r < 2a$, in $2n$ points. The curves are plotted by using the method described above.

Example 7.2.6. Plot the graph of the function $\rho = b + a\cos\varphi$ $(0 < a \le b,\ \rho \ge 0)$.

Let us plot the graph of the function $y = b + a\cos x$ (Figure 7.35).

First, study the function $\rho = b + a\cos\varphi$. Since the domain of the function is $(-\infty; +\infty)$, the function $\rho = b + a\cos\varphi$ is defined for φ such that $\rho \ge 0$, i.e. $|\varphi| \le \alpha$, where $\alpha = \arccos(-b/a)$.

For $\varphi = \pm\alpha$, the graph of the function $\rho = b + a\cos\varphi$ intersects the polar axis and the half line $(-\infty; 0)$. Because the function $y = b + a\cos x$ is bounded for $|x| \le \alpha$, $-b \le y \le b + a$, the function $\rho = b + a\cos\varphi$ is also bounded for $|\varphi| \le \alpha$, $0 \le \rho \le b + a$.

The function $y = b + a\cos x$ is periodic with the period 2π. The function $\rho = b + a\cos\varphi$ is thus also periodic with the period 2π. When plotting the curve $\rho = b + a\cos\varphi$, its periodicity is not used, since we need to plot the graph only for $\rho \ge 0$. On the line segment $|x| \le \alpha$, the function $y = b + a\cos x$ reaches maximum at the point $x = 0$ with $y_{max} = a + b$. The function is increasing on $(-\alpha; 0)$, and is decreasing on $(0; \alpha)$. Hence, the function $\rho = b + a\cos\varphi$ has a maximum at $\varphi = 0$. Going clockwise with φ ranging over the interval $(\alpha; 0]$, the value of ρ decreases, and it increases on the interval $(0; \alpha]$. Because the graph of the function $y = b + a\cos x$ intersects the x-axis in points $(-\alpha; 0)$ and $(\alpha; 0)$, the graph of the function $\rho = b + a\cos\varphi$ intersects the polar axis in the points $(-\alpha; 0)$ and $(\alpha; 0)$. The graph of the function $y = b + a\cos x$ intersects the y-axis in the point $(\pi/2; a + b)$; the graph of the function $\rho = b + a\cos\varphi$ intersects the polar axis in the point $(0; \arccos(-b/a))$ (Figure 7.36).

Example 7.2.7. Plot the graph of the function $\rho = a(1 + b\cos\varphi)$ $(a > 0,\ b > 1)$ (the Pascal snail).

Let us plot the graph of the function $y = a(1 + b\cos x)$ (Figure 7.37a).

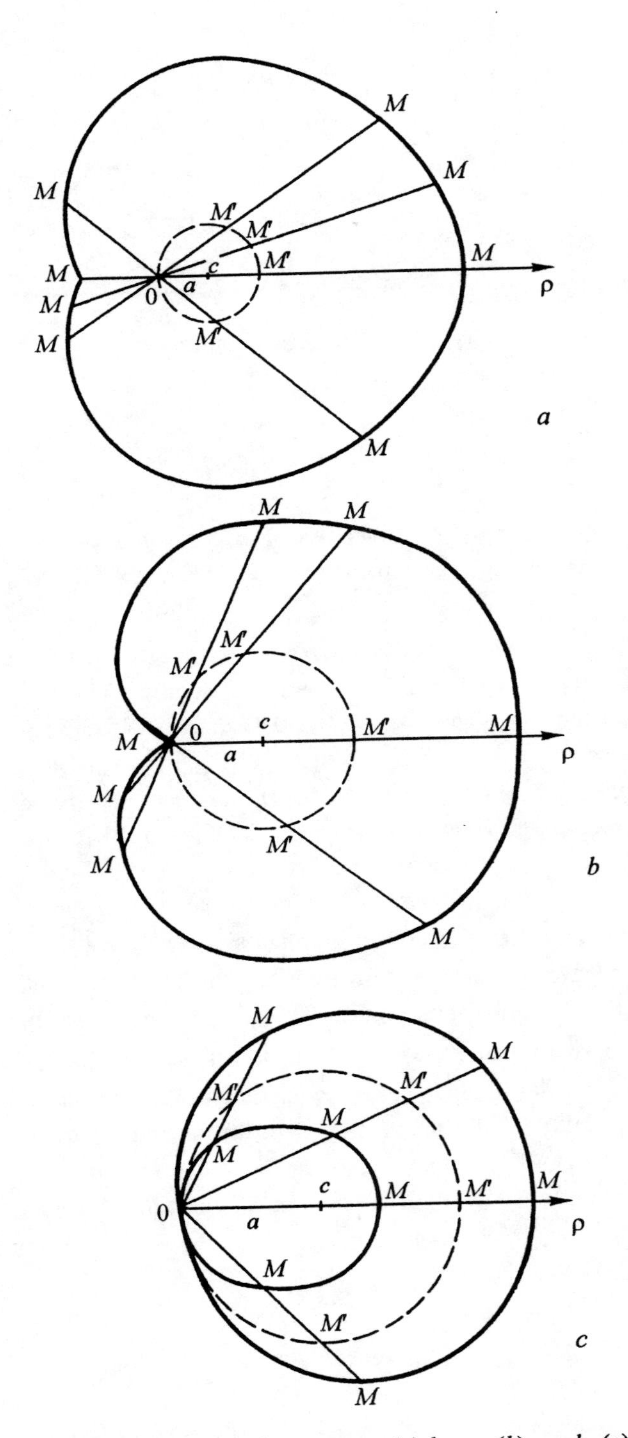

Figure 7.36. The curve $\rho = b + a \cos \varphi$: (*a*) $b > a$, (*b*) $a = b$, (*c*) $b < a$.

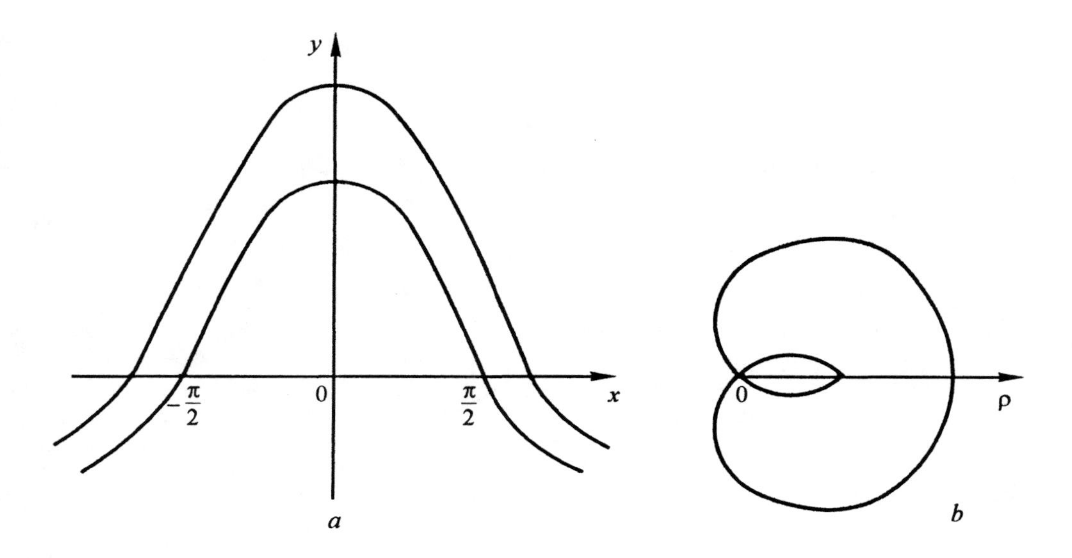

Figure 7.37. The curve $y = a(1 + b\cos x)$: (*a*) $\rho = a(1 + b\cos\varphi)$, (*b*) $a = 1, b = 2$.

Study the function $\rho = a(1 + b\cos\varphi)$. The function is defined for all φ; its graph is symmetric with respect to the polar axis. The function is periodic with the period 2π. The principal interval is $[-\pi; \pi]$; at the endpoints the function takes the values $\rho = a(1 - b)$, $\rho = a(1 + b)$. For $\varphi \in [-\pi; \pi]$, the function is bounded, $a(1 - b) \le \rho \le a(1 + b)$. The function attains an extremum at the point $\varphi = 0$ ($\rho_{max} = a(1 + b)$). As φ changes clockwise in the interval $[-\pi; 0]$ ($[0; \pi]$), the value of ρ decreases (increases). The graph of the function $\rho = a(1 + b\cos\varphi)$ intersects the polar axis in the points $(0; -2/3\pi)$ and $(0; 2/3\pi)$.

7.2.5. Plots of the graphs of curves $\rho = a\tan\varphi$, $\rho = a\cot\varphi$.

We will use two methods to plot the graphs — the method of finding points of intersection of the graph and concentric circles, and the method of comparing the graph with the graph of the corresponding function in Cartesian coordinates.

1. Find angles of the rays passing through points of intersection of the curves $\rho = a\tan\varphi$, $\rho = a\cot\varphi$ and the circles of radii $a/\sqrt{3}$, a, $a\sqrt{3}$:

Curve	$r = a/\sqrt{3}$		$r = a$		$r = a\sqrt{3}$	
$\rho = a\tan\varphi$	$\pi/6$	$\pi/6 + \pi$	$\pi/4$	$\pi/4 + \pi$	$\pi/6$	$\pi/6 + \pi$
$\rho = a\cot\varphi$	$\pi/3$	$\pi/3 + \pi$	$\pi/4$	$\pi/4 + \pi$	$\pi/6$	$\pi/6 + \pi$.

It is clear that each curve intersects the corresponding circle in two points. It is easy to see that the curves intersect any circle of radius r, $0 < r < +\infty$, in two points. Determine bounds for the angle φ:

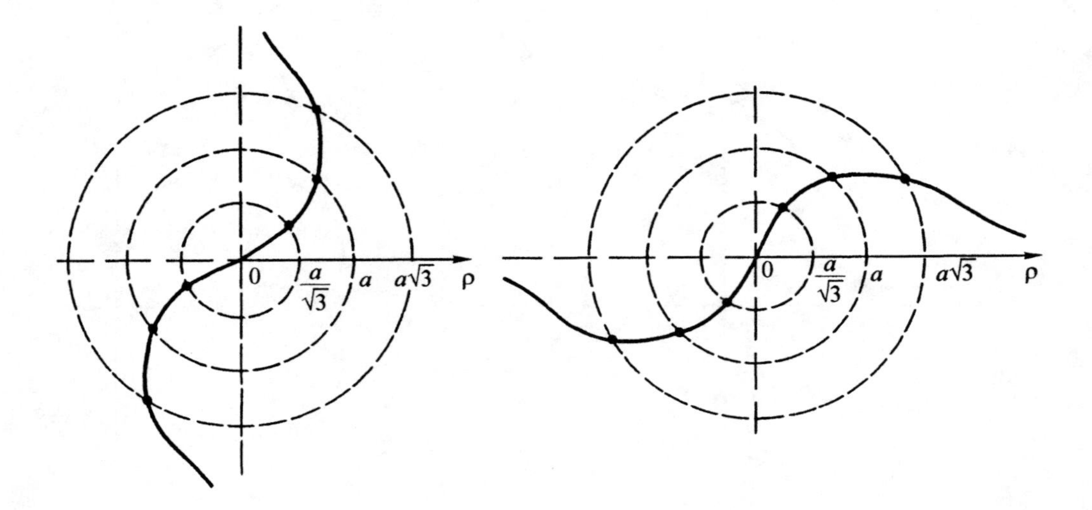

Figure 7.38. The curve $\rho = a \tan \varphi$.

Figure 7.39. The curve $\rho = a \cot \varphi$.

$$\rho \geq 0 \Rightarrow \begin{cases} \tan \varphi \geq 0 \\ \cot \varphi \geq 0 \end{cases} \Rightarrow \begin{matrix} \pi \leq \varphi < \frac{3}{2}\pi, \\ \pi < \varphi \leq \frac{3}{2}\pi \end{matrix}$$

(Figures 7.38 and 7.39).

2. In extended polar coordinates, plot the graphs of the curves $\rho = a \tan \varphi$ and $\rho = a \cot \varphi$ by comparing them with the graph of the corresponding function in Cartesian coordinates (Figures 7.40 and 7.41).

Plot the graph of the function $y = a \tan x$ (see Figure 3.24), stretched along the y-axis by a. Study the function $\rho = a \tan \varphi$. The domain of the function is the set of all φ save for $\varphi = \frac{\pi}{2}(2n+1)$, $n \in \mathbf{N}$. The range of the function is $(-\infty; \infty)$.

Since the function $y = a \tan x$ is odd, the graph of the curve $\rho = a \tan \varphi$ is symmetric with respect to the pole. The function $\rho = a \tan \varphi$ is periodic with the period π.

Because the function $\rho = a \tan \varphi$ is symmetric and periodic, we plot the curve for $\varphi \in [0; \pi/2)$. Since the function $y = a \tan x$ is increasing for $x \in (-\pi/2; \pi/2)$, the value of ρ decreases when φ changes clockwise. The graph of the function $\rho = a \tan \varphi$ passes through the point $(0; 0)$.

Vertical asymptotes of the graph of the function $y = a \tan x$ are given by $x = \frac{\pi}{2}(2n+1)$, $n \in \mathbf{N}$. The corresponding asymptotes of the graph of the function $\rho = a \tan \varphi$ are

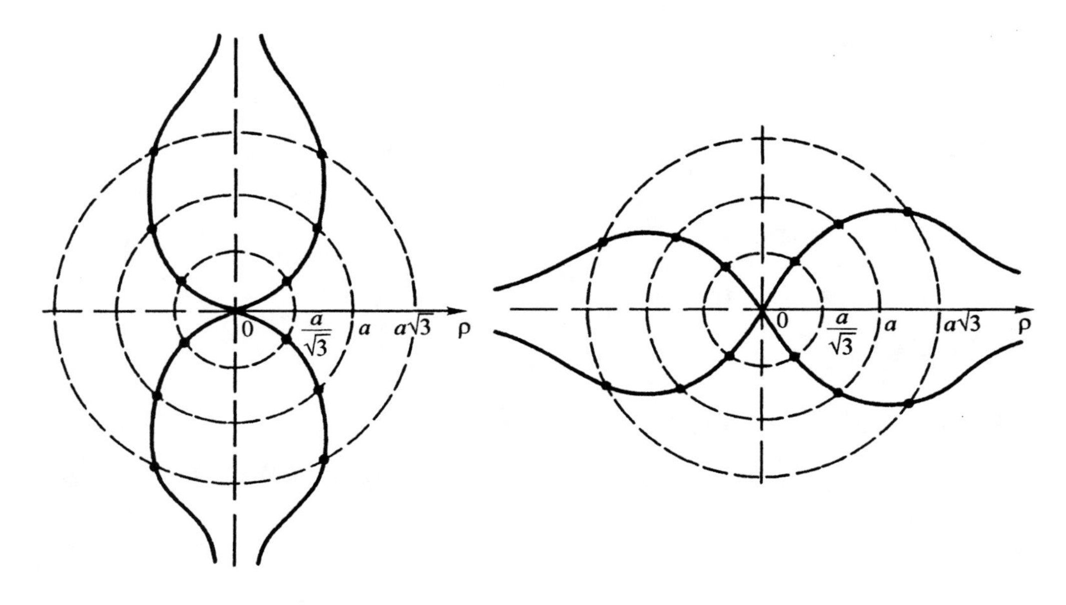

Figure 7.40 Figure 7.41

Figure 7.40. The curve $\rho = a \tan \varphi$ (in extended polar coordinates).

Figure 7.41. The curve $\rho = a \cot \varphi$ (in extended polar coordinates).

$\varphi = \frac{\pi}{2}(2n+1)$. The are no oblique asymptotes for the function $y = a \tan x$; the same is true for the graph of the function $\rho = a \tan \varphi$.

Rotating the graph of the function $\rho = a \tan \varphi$, $0 \le \varphi \le \pi/2$, by the angle π, we obtain the graph of the function $\rho = a \tan \varphi$ for $\pi/2 \le \varphi \le \pi$.

7.2.6. Plots of the graphs of curves $\rho = a\tan n\varphi$, $\rho = a\cot n\varphi$ ($n \in \mathbf{N}, n > 1$)

Let us use the method of finding points of intersection of the graph and concentric circles.

Let us check that the curves $\rho = a \tan n\varphi$, $\rho = a \cot n\varphi$ intersect any circle of radius r, $0 < r < +\infty$. Indeed, let $\rho = am$. Then

$$\left.\begin{array}{l}\tan n\varphi = m\\ \cot n\varphi = m\end{array}\right\} \Rightarrow \left\{\begin{array}{l}\varphi = \frac{1}{n}\arctan m + \frac{k\pi}{n}\\ \varphi = \frac{1}{n}\operatorname{arccot} m + \frac{k\pi}{n}\end{array}\right\} \Rightarrow \left\{\begin{array}{l}\varphi = \frac{1}{n}\arctan m + 2\pi\\ \varphi = \frac{1}{n}\operatorname{arccot} m + 2\pi \quad (k = 2n)\end{array}\right.$$

Note that the curves $\rho = a \tan n\varphi$, $\rho = a \cot n\varphi$ have n asymptotes that form the angles with the polar axis found from the relations:

$$\left.\begin{array}{l}\tan n\varphi = \infty \\ \cot n\varphi = \infty\end{array}\right\} \Rightarrow \left\{\begin{array}{l}\varphi = \dfrac{\pi}{2} + 2\pi, \\ \varphi = \pi.\end{array}\right.$$

Find the bounds for the angle φ:

$$\rho \geq 0 \Rightarrow \left\{\begin{array}{l}\tan n\varphi \geq 0 \\ \cot n\varphi \geq 0\end{array}\right. \Rightarrow \left\{\begin{array}{l}\dfrac{2k\pi}{n} \leq \varphi < \dfrac{\pi}{2n} + \dfrac{2k\pi}{n}; \quad \dfrac{\pi}{n} + \dfrac{2k\pi}{n} \leq \varphi < \dfrac{3\pi}{2n} + \dfrac{2k\pi}{n}, \\ \dfrac{2k\pi}{n} < \varphi \leq \dfrac{\pi}{2n} + \dfrac{2k\pi}{n}; \quad \dfrac{\pi}{n} + \dfrac{2k\pi}{n} < \varphi \leq \dfrac{3\pi}{2n} + \dfrac{2k\pi}{n}\end{array}\right.$$

(see Figure 7.42).

7.2.7. Plots of the graphs of curves $\rho = a(1 \pm \tan \varphi)$, $\rho = a(1 \pm \cot \varphi)$

Let us make a table of values of angles of the rays that pass through points of intersection of the curves and circles of radii a, $2a$, as well as a table of values of bounds for the angle φ:

Curve	$r = a$		$r = 2a$	
$\rho = a(1 + \tan \varphi)$	0	π	$\pi/4$	$\pi/4 + \pi$
$\rho = a(1 - \tan \varphi)$	0	π	$-\pi/4$	$-\pi/4 + \pi$
$\rho = a(1 + \cot \varphi)$	$\pi/2$	$\pi/2 + \pi$	$\pi/4$	$\pi/4 + \pi$
$\rho = a(1 - \cot \varphi)$	$\pi/2$	$\pi/2 + \pi$	$-\pi/4$	$-\pi/4 + \pi$

Curves	$\rho \geq 0$	φ	
$\rho = a(1 + \tan \varphi)$	$\tan \varphi \geq -1$	$-\frac{\pi}{4} \leq \varphi < \frac{\pi}{2}$	$\frac{3}{4}\pi \leq \varphi < \frac{3}{2}\pi$
$\rho = a(1 - \tan \varphi)$	$\tan \varphi \leq 1$	$-\frac{\pi}{2} \leq \varphi < \frac{\pi}{4}$	$\frac{\pi}{2} \leq \varphi < \frac{5}{4}\pi$
$\rho = a(1 + \cot \varphi)$	$\cot \varphi \geq -1$	$0 < \varphi \leq \frac{3}{4}\pi$	$\pi < \varphi \leq \frac{7}{4}\pi$
$\rho = a(1 - \cot \varphi)$	$\cot \varphi \leq 1$	$\frac{\pi}{4} \leq \varphi < \pi$	$\frac{5}{4}\pi \leq \varphi < 2\pi$

(see Figure 7.43).

7.2.8. Plots of the graphs of curves $\rho = a(1 \pm \tan n\varphi)$, $\rho = a(1 \pm \cot n\varphi)$ $(n \in \mathrm{N}, n > 1)$

The curves are plotted by using points of intersection of the curves and concentric circles. Let the circles have radius a, $2a$,.... Solve the trigonometric equations $1 \pm \tan n\varphi = 1$, $1 \pm \cot n\varphi = 1$ if, for example, $r = a$. We find that the curves $\rho = a(1 \pm \tan n\varphi)$, $\rho = a(1 \pm \cot n\varphi)$ intersect these circles in $2n$ points. The graphs have n asymptotes that make angles with the polar axis found from the equations $\tan n\varphi = \pm\infty$ and $\cot n\varphi = \pm\infty$, which yields $\varphi = \pm\pi/2$ and $\varphi = \pi$, respectively (Figures 7.44 and 7.45).

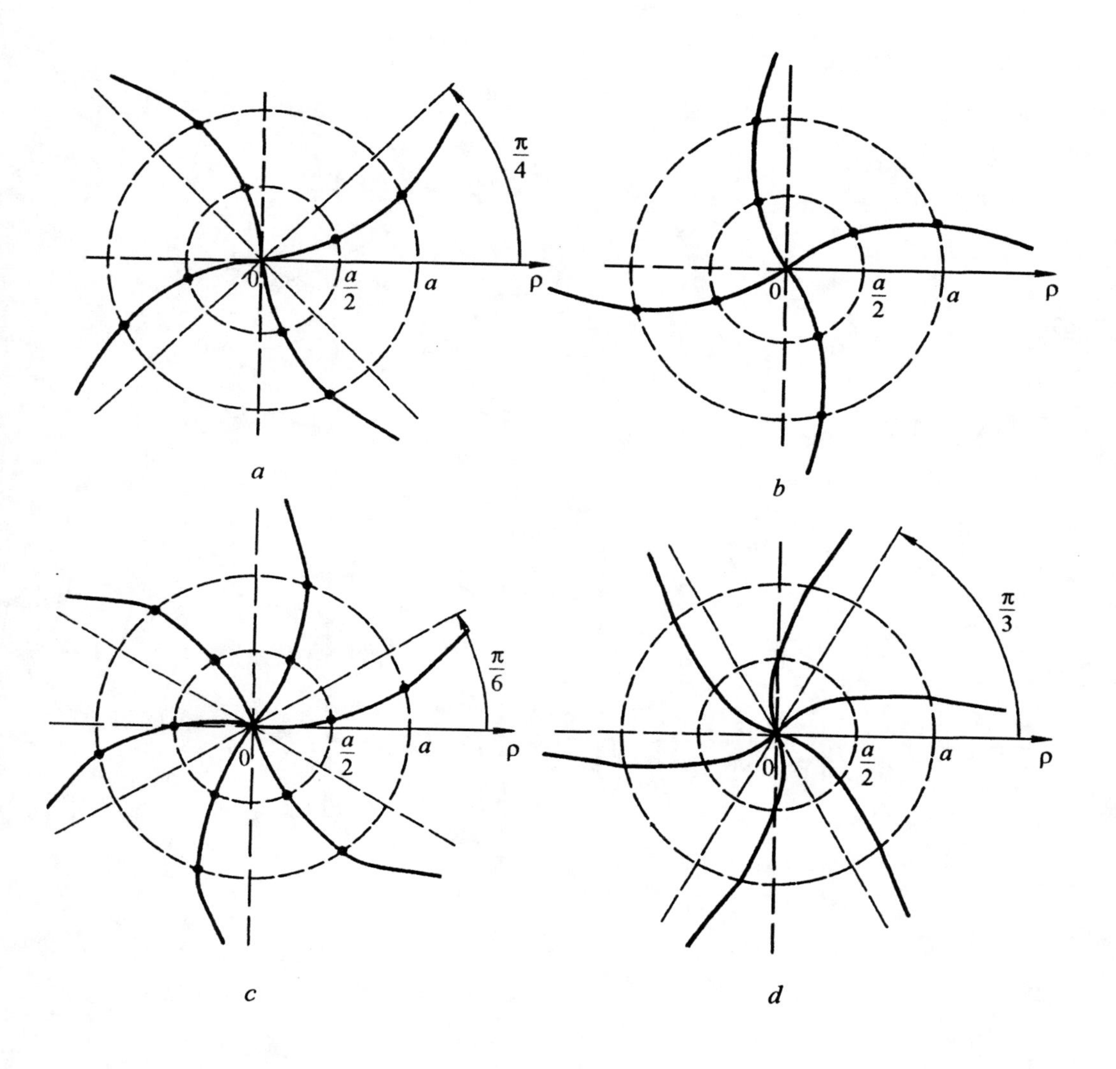

Figure 7.42. The graphs of functions: (*a*) $\rho = a \tan 2\varphi$; (*b*) $\rho = a \cot 2\varphi$; (*c*) $\rho = a \tan 3\varphi$; (*d*) $\rho = a \cot 3\varphi$.

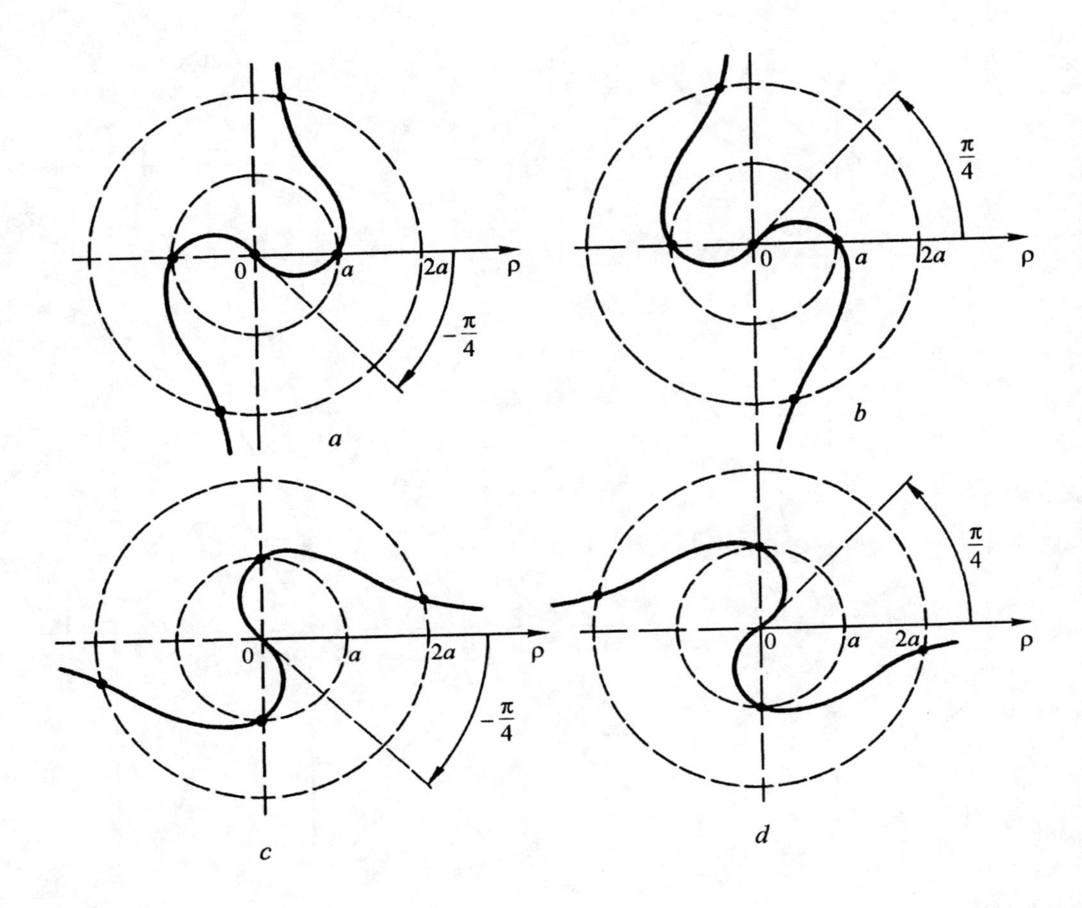

Figure 7.43. The graphs of functions: (*a*) $\rho = a(1 + \tan\varphi)$; (*b*) $\rho = a(1 - \tan\varphi)$; (*c*) $\rho = a(1 + \cot\varphi)$; (*d*) $\rho = a(1 - \cot\varphi)$.

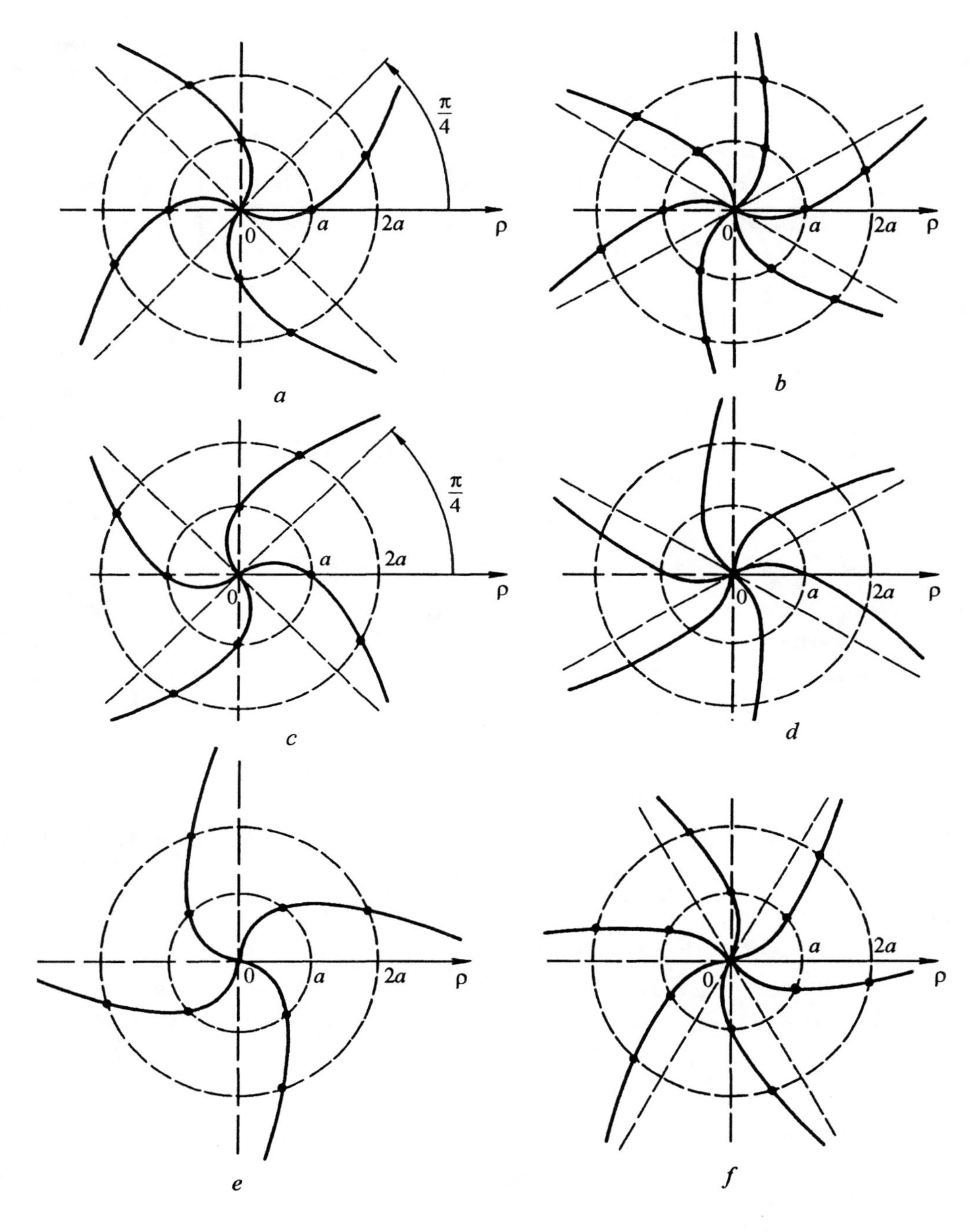

Figure 7.44. The graphs of functions: (*a*) $\rho = a(1 + \tan 2\varphi)$; (*b*) $\rho = a(1 + \tan 3\varphi)$; (*c*) $\rho = a(1 - \tan 2\varphi)$; (*d*) $\rho = a(1 - \tan 3\varphi)$; (*e*) $\rho = a(1 + \cot 2\varphi)$; (*f*) $\rho = a(1 + \cot 3\varphi)$.

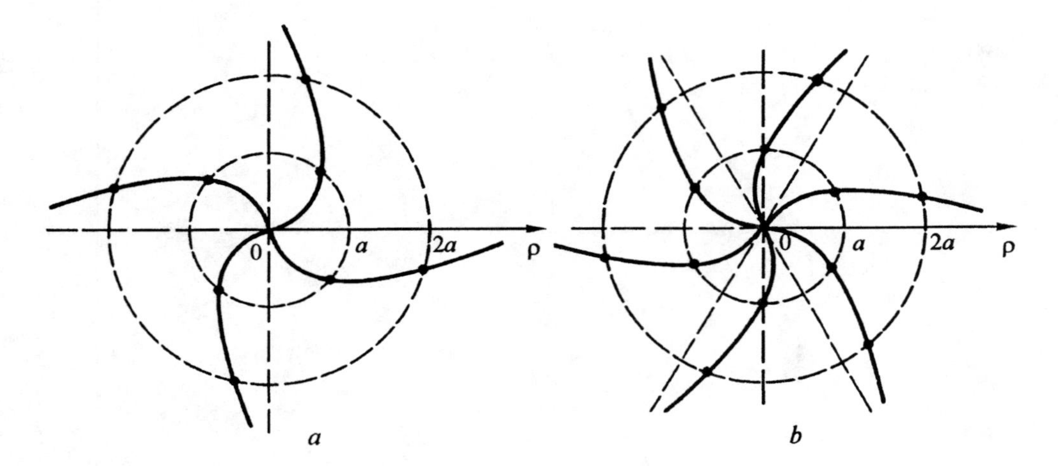

Figure 7.45. The graphs of functions: (*a*) $\rho = a(1 - \cot 2\varphi)$;
(*b*) $\rho = a(1 - \cot 3\varphi)$.

7.2.9. Plots of the graphs of curves $\rho = b \pm a\sin \varphi$, $\rho = b \pm a\cos \varphi$ $(b \neq a)$

We plot these curves by determining the angles φ such that the value of ρ achieves a maximum and minimum, as well as angles of the rays passing through points of intersection of the curves and concentric circles of radius r for $\rho_{min} < r < \rho_{max}$. Also calculate bounds for the angle φ so that $\rho \geq 0$. The graphs are shown in Figures 7.46 and 7.47.

7.2.10. Plots of the graphs of curves $\rho = b \pm a\sin n\varphi$, $\rho = b \pm a\cos n\varphi$ $(n \in \mathbf{N}, b \neq a, n > 1)$

Solve the trigonometric equations $\sin n\varphi = +1$, $\sin n\varphi = -1$, $\cos n\varphi = +1$, $\cos n\varphi = -1$. These equations have n solutions with $\rho_{max} = b + a$. If $b < a$, $\rho_{min} = 0$, and if $b > a$, $\rho_{min} = b - a$.

The curves $\rho = b \pm a\sin n\varphi$, $\rho = b \pm a\cos n\varphi$ intersect a concentric circle of radius r in $2n$ points if $b < a$ and $0 < r < \rho_{max}$, and if $b > a$ and $\rho_{min} < r < \rho_{max}$. The graphs of the curves $\rho = b \pm a\sin n\varphi$, $\rho = b \pm a\cos n\varphi$ are n-leaf curves if $b < a$, and star-shaped curves if $b > a$ (Figures 7.48 – 7.52).

7.2.11.Plots of the graphs of curves $\rho = b \pm a \tan n\varphi$, $\rho = b \pm a \cot n\varphi$ $(b \neq a, n \in \mathbf{N}, n \geq 1)$

The analysis needed for plotting these curves is the same as for the curves $\rho = b \pm a \tan n\varphi$, $\rho = b \pm \cot n\varphi$. The graphs of the curves are shown in Figures 7.53 – 7.56.

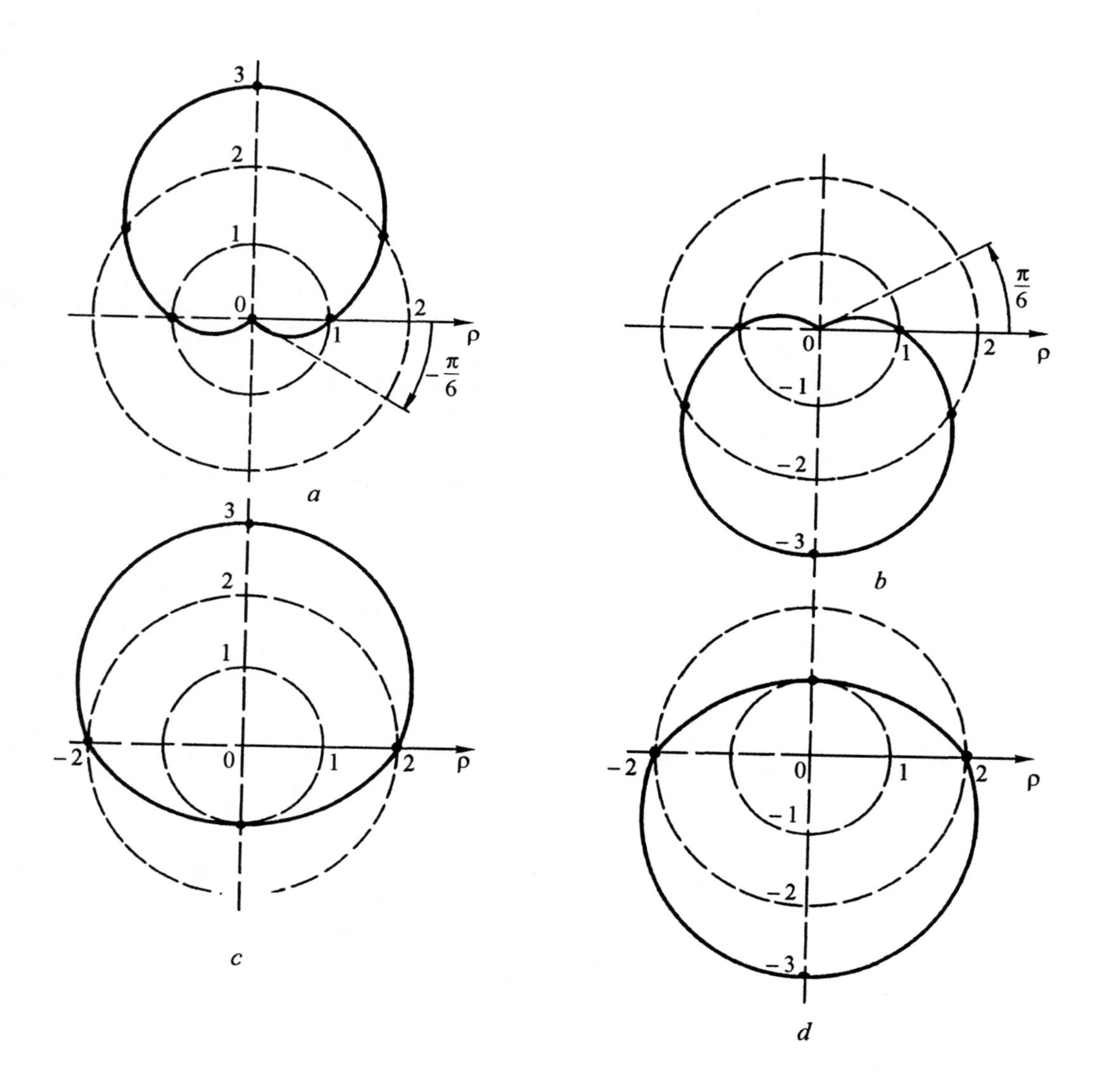

Figure 7.46. The graphs of functions: (*a*) $\rho = 1 + 2\sin\varphi$; (*b*) $\rho = 1 - 2\sin\varphi$; (*c*) $\rho = 2 + \sin\varphi$; (*d*) $\rho = 2 - \sin\varphi$.

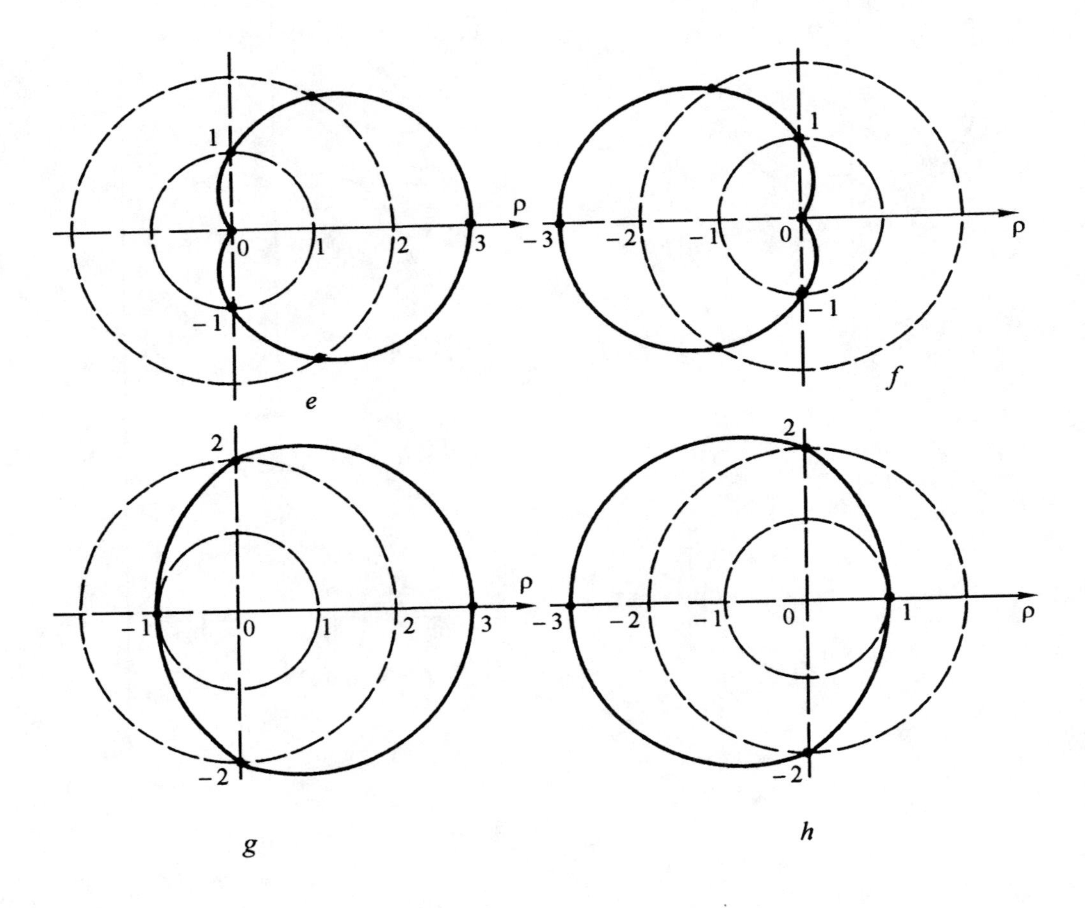

Figure 7.46. The graphs of functions: (*e*) $\rho = 1 + 2\cos\varphi$; (*f*) $\rho = 1 - 2\cos\varphi$; (*g*) $\rho = 2 + \cos\varphi$; (*h*) $\rho = 2 - \cos\varphi$.

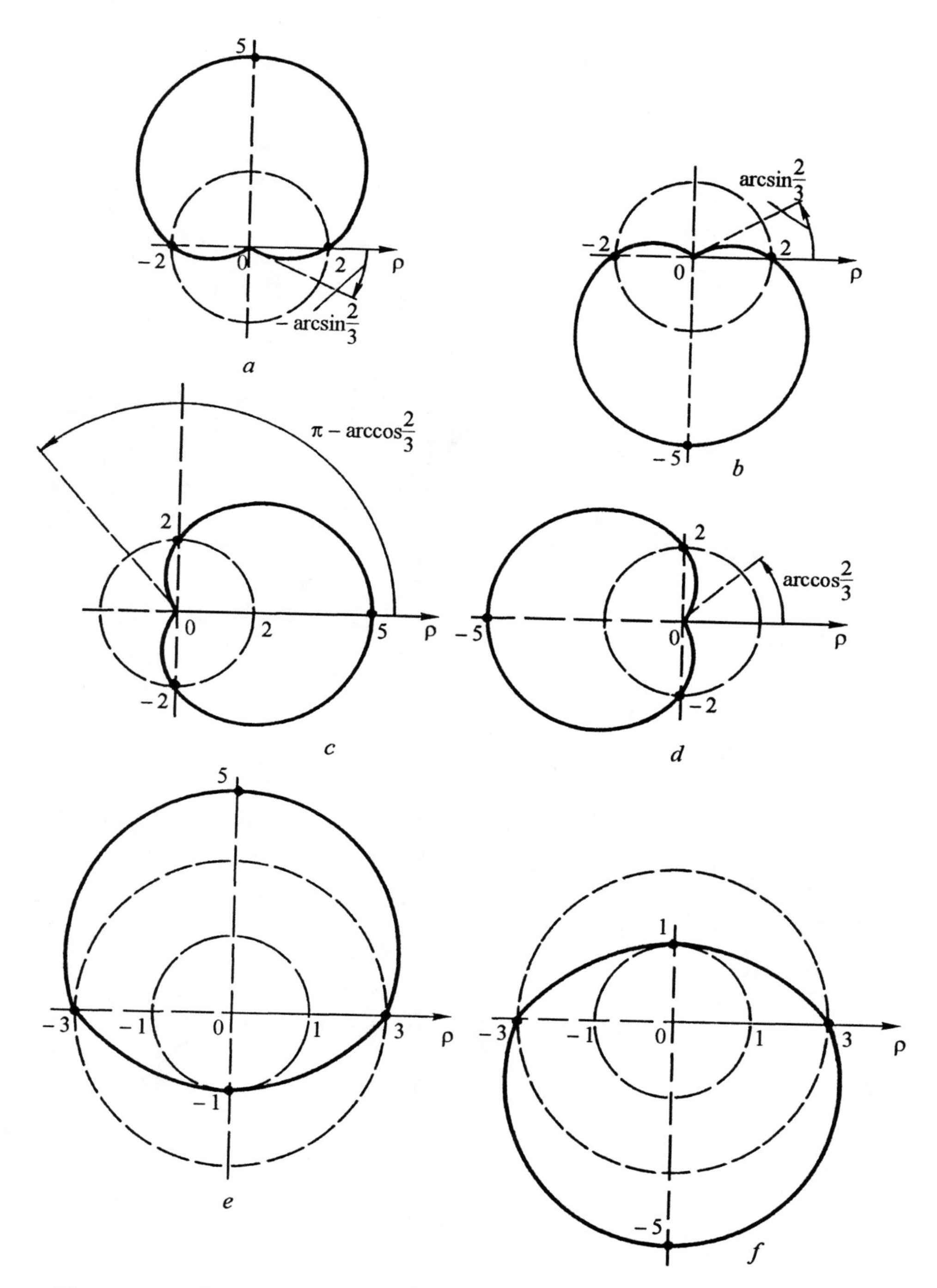

Figure 7.47. The graphs of functions: (*a*) $\rho = 2 + 3\sin\varphi$; (*b*) $\rho = 2 - 3\sin\varphi$; (*c*) $\rho = 2 + 3\cos\varphi$; (*d*) $\rho = 2 - 3\cos\varphi$; (*e*) $\rho = 3 + 2\sin\varphi$; (*f*) $\rho = 3 - 2\sin\varphi$.

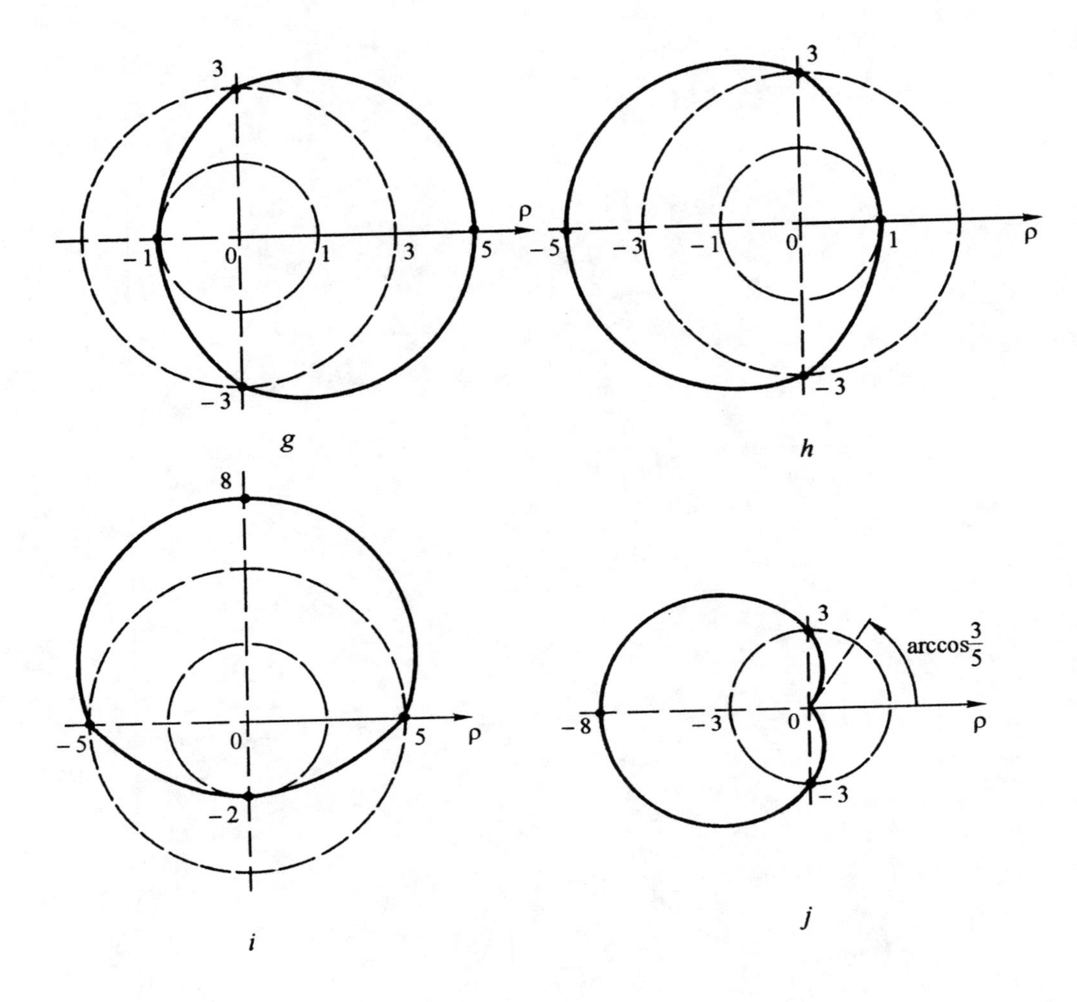

Figure 7.47. The graphs of functions: (*g*) $\rho = 3 + 2\cos\varphi$; (*h*) $\rho = 3 - 2\cos\varphi$; (*i*) $\rho = 5 + 3\sin\varphi$; (*j*) $\rho = 3 - 5\sin\varphi$.

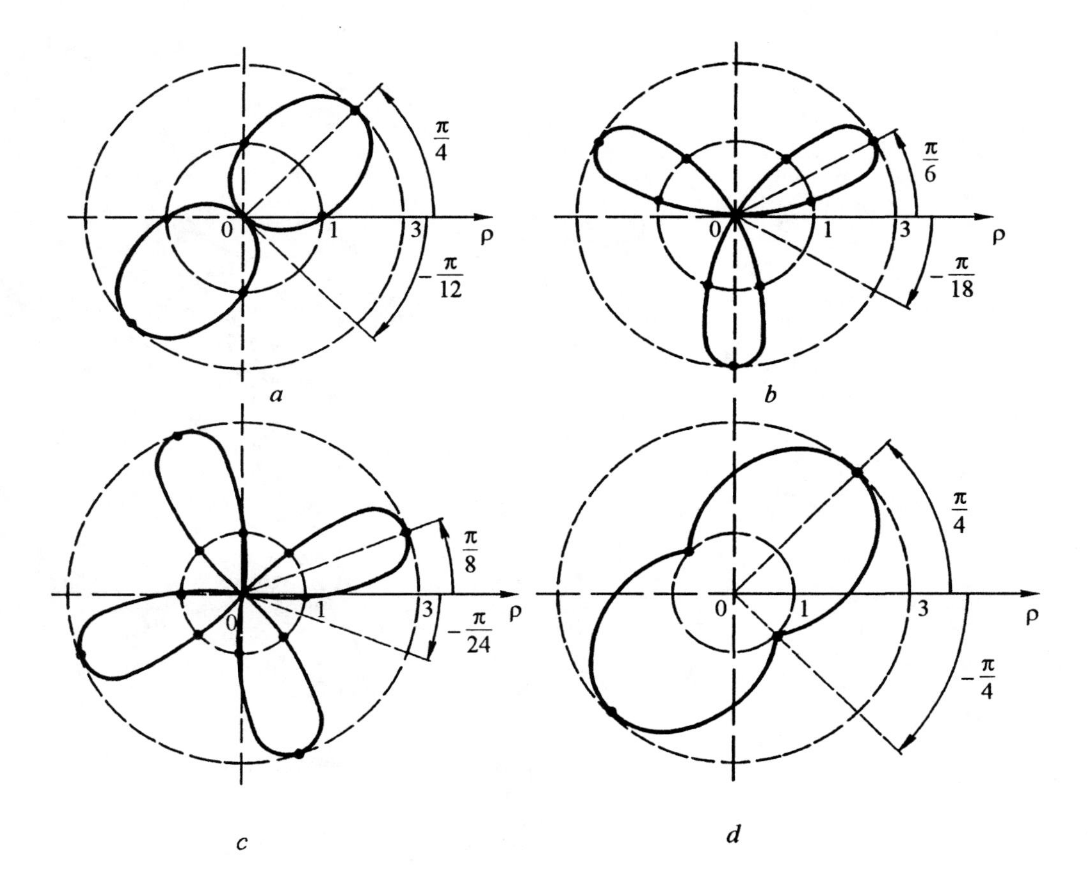

Figure 7.48. The graphs of functions: (*a*) $\rho = 1 + 2\sin 2\varphi$; (*b*) $\rho = 1 + 2\sin 3\varphi)$; (*c*) $\rho = 1 + 2\sin 4\varphi$; (*d*) $\rho = 2 + \sin 2\varphi$.

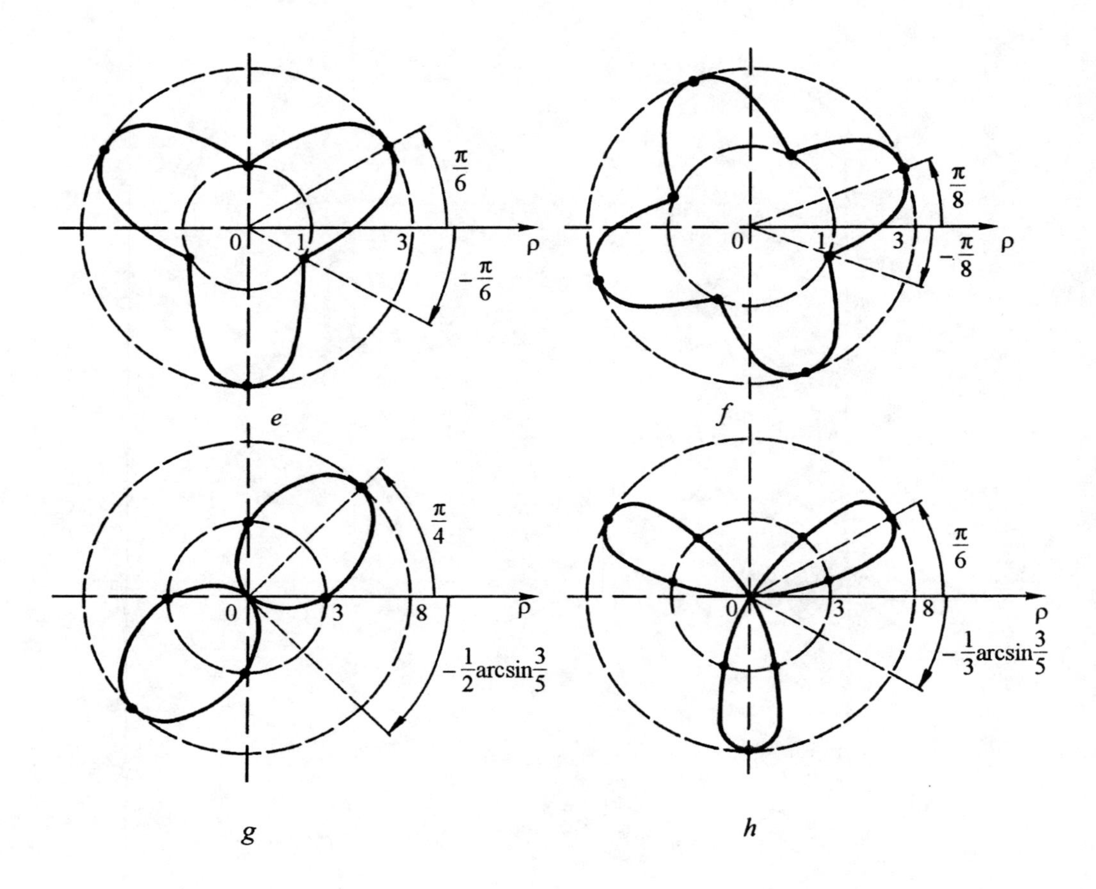

Figure 7.48. The graphs of functions: (*e*) $\rho = 2 + \sin 3\varphi$; (*f*) $\rho = 2 + \sin 4\varphi$; (*g*) $\rho = 3 + 5\sin 2\varphi$; (*h*) $\rho = 3 + 5\sin 3\varphi$.

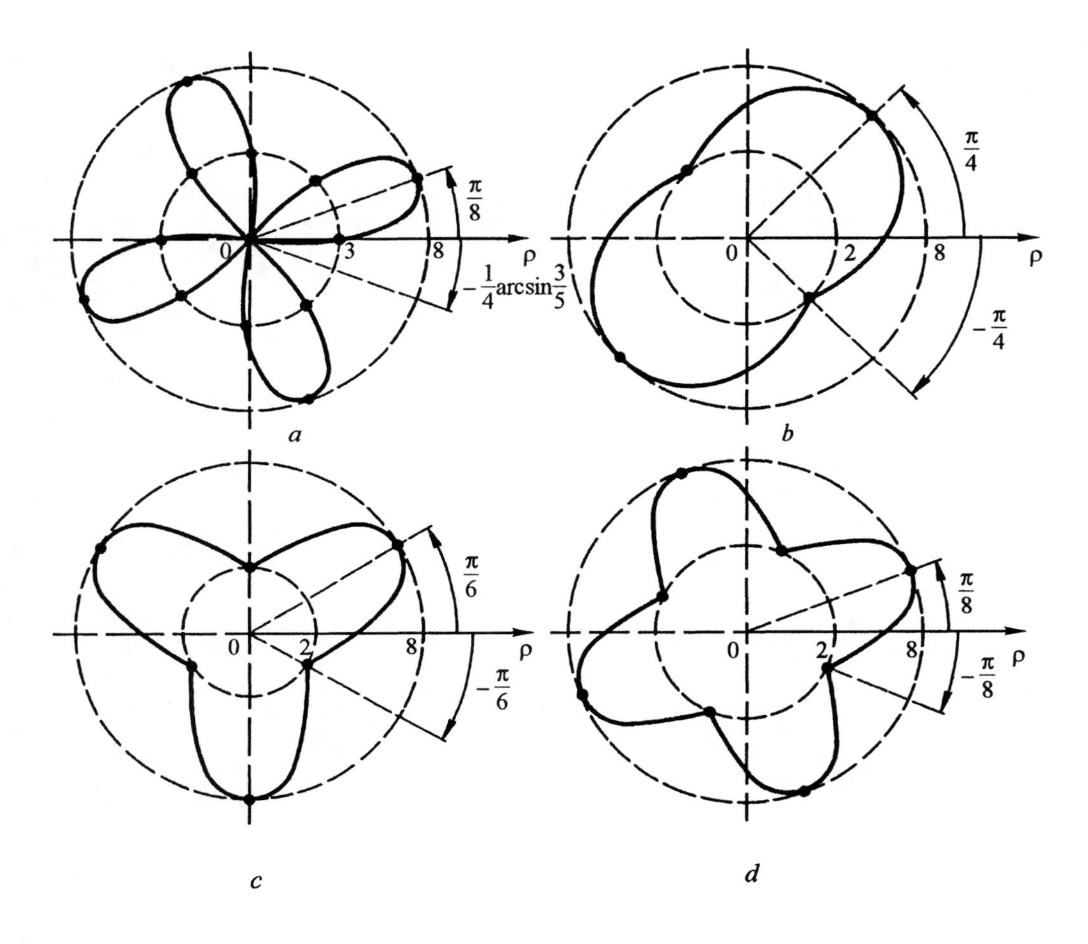

Figure 7.49. The graphs of functions: (*a*) $\rho = 3 + 5\sin 4\varphi$; (*b*) $\rho = 5 + 3\sin 2\varphi$; (*c*) $\rho = 5 + 3\sin 3\varphi$; (*d*) $\rho = 5 + 3\sin 4\varphi$.

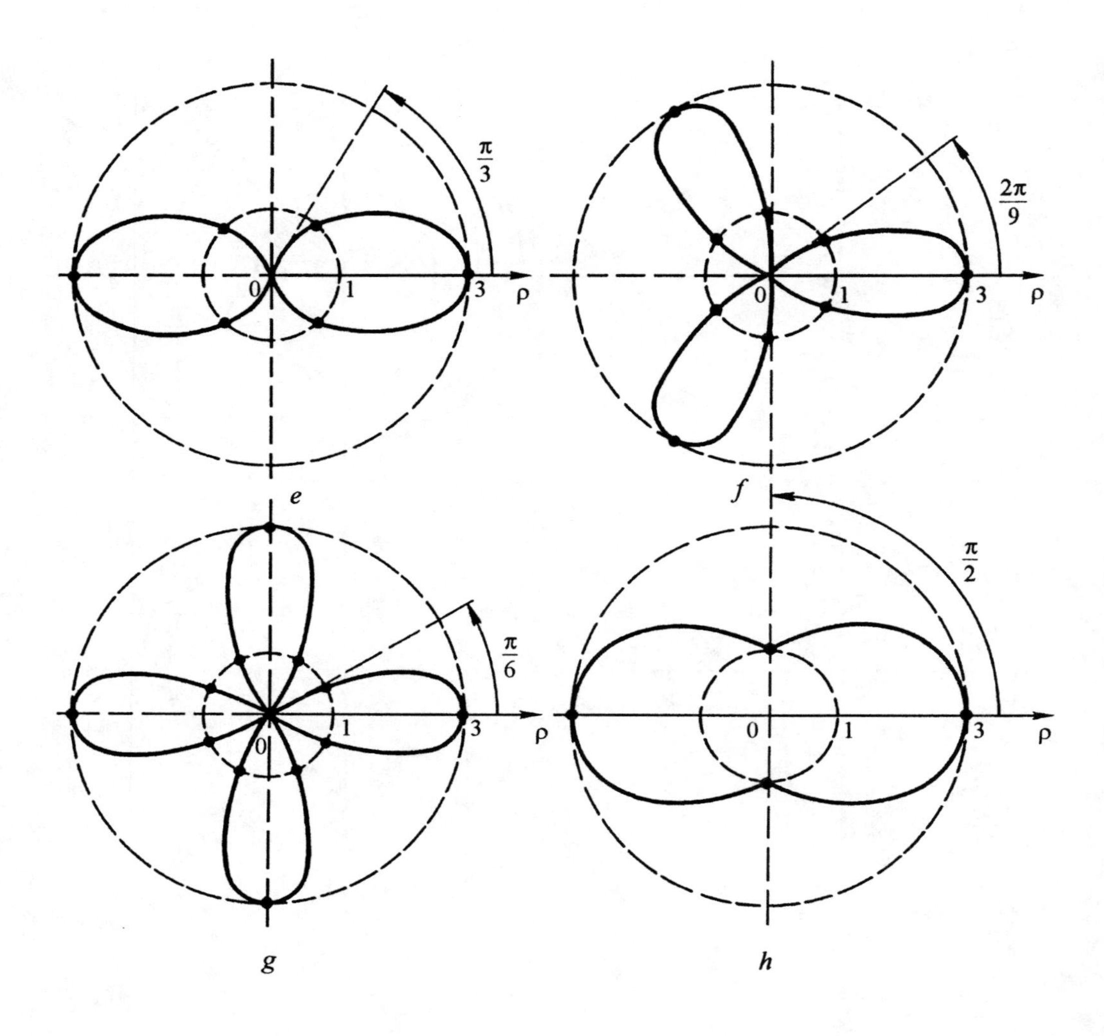

Figure 7.49. The graphs of functions: (*e*) $\rho = 1 + 2\cos 2\varphi$; (*f*) $\rho = 1 + 2\cos 3\varphi$; (*g*) $\rho = 1 + 2\cos 4\varphi$; (*h*) $\rho = 2 + \cos 2\varphi$.

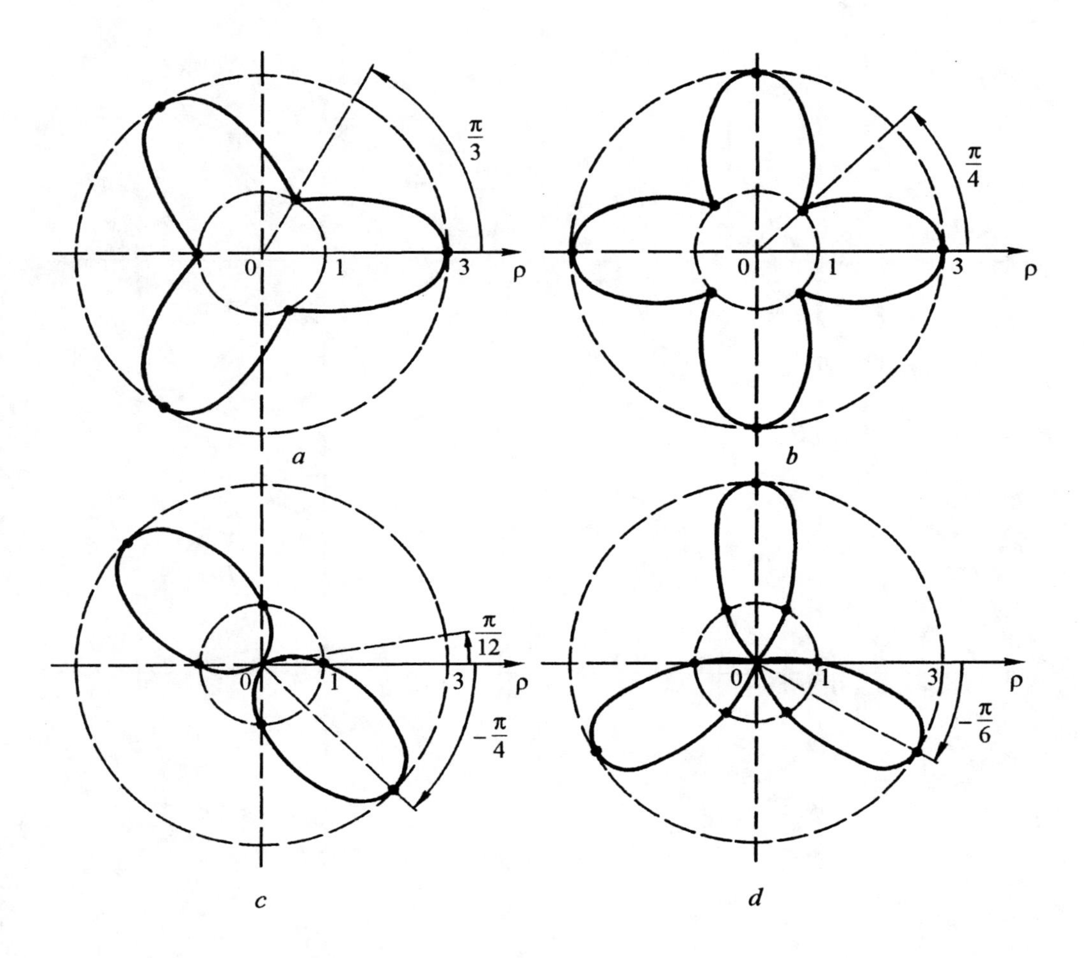

Figure 7.50. The graphs of functions: (*a*) $\rho = 2 + \cos 3\varphi$; (*b*) $\rho = 2 + \cos 4\varphi$; (*c*) $\rho = 1 - 2\sin 2\varphi$; (*d*) $\rho = 1 - \sin 3\varphi$.

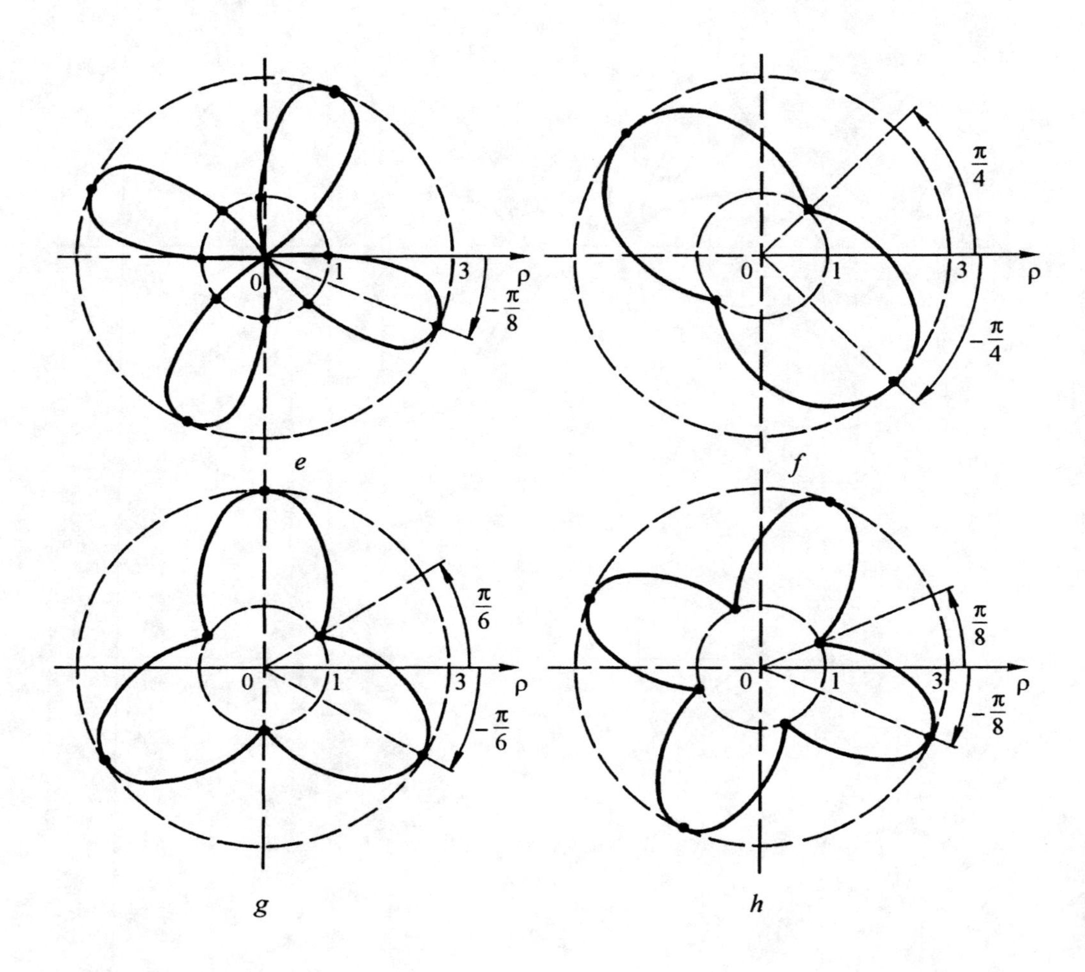

Figure 7.50. The graphs of functions: (*e*) $\rho = 1 - \sin 3\varphi$; (*f*) $\rho = 2 - \sin 2\varphi$; (*g*) $\rho = 2 - \sin 3\varphi$; (*h*) $\rho = 2 - \sin 4\varphi$.

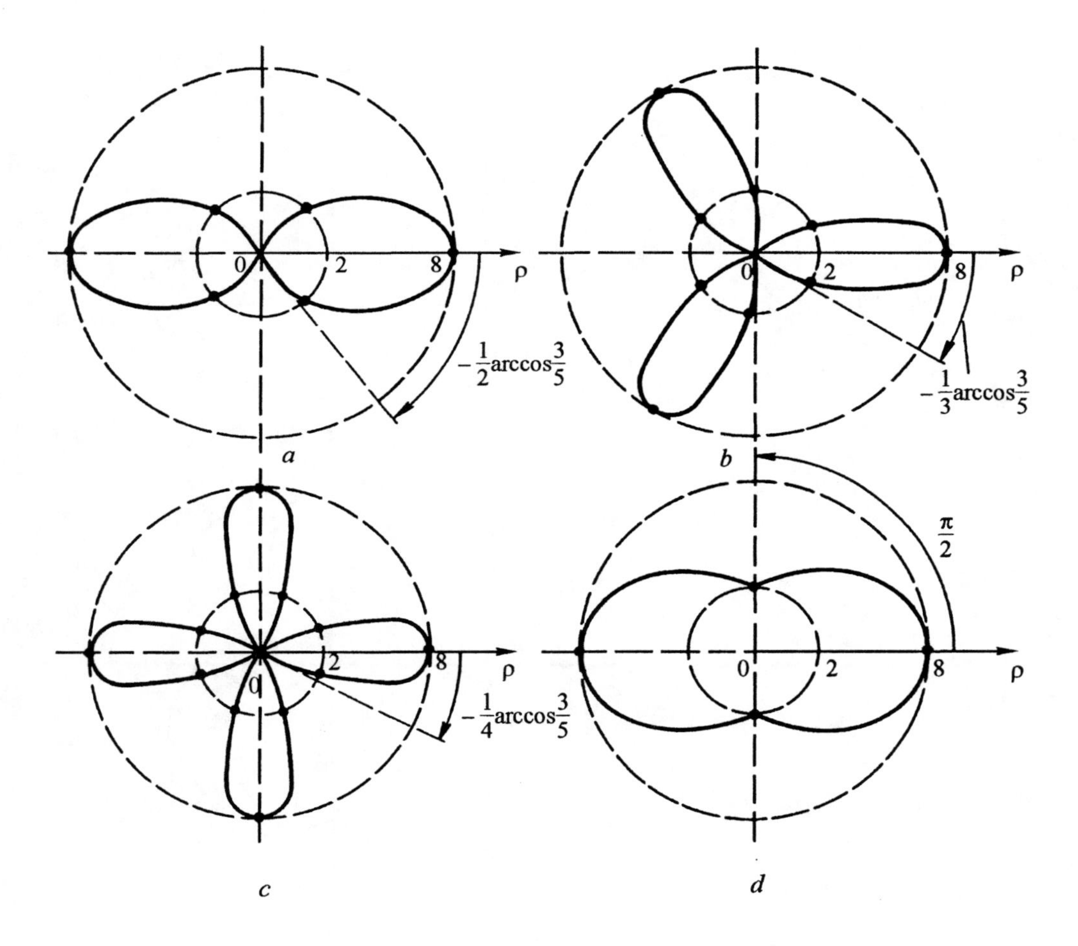

Figure 7.51. The graphs of functions: (*a*) $\rho = 3 + 5\cos 2\varphi$; (*b*) $\rho = 3 + 5\cos 3\varphi$; (*c*) $\rho = 5 + 3\cos 4\varphi$; (*d*) $\rho = 5 + 3\cos 2\varphi$.

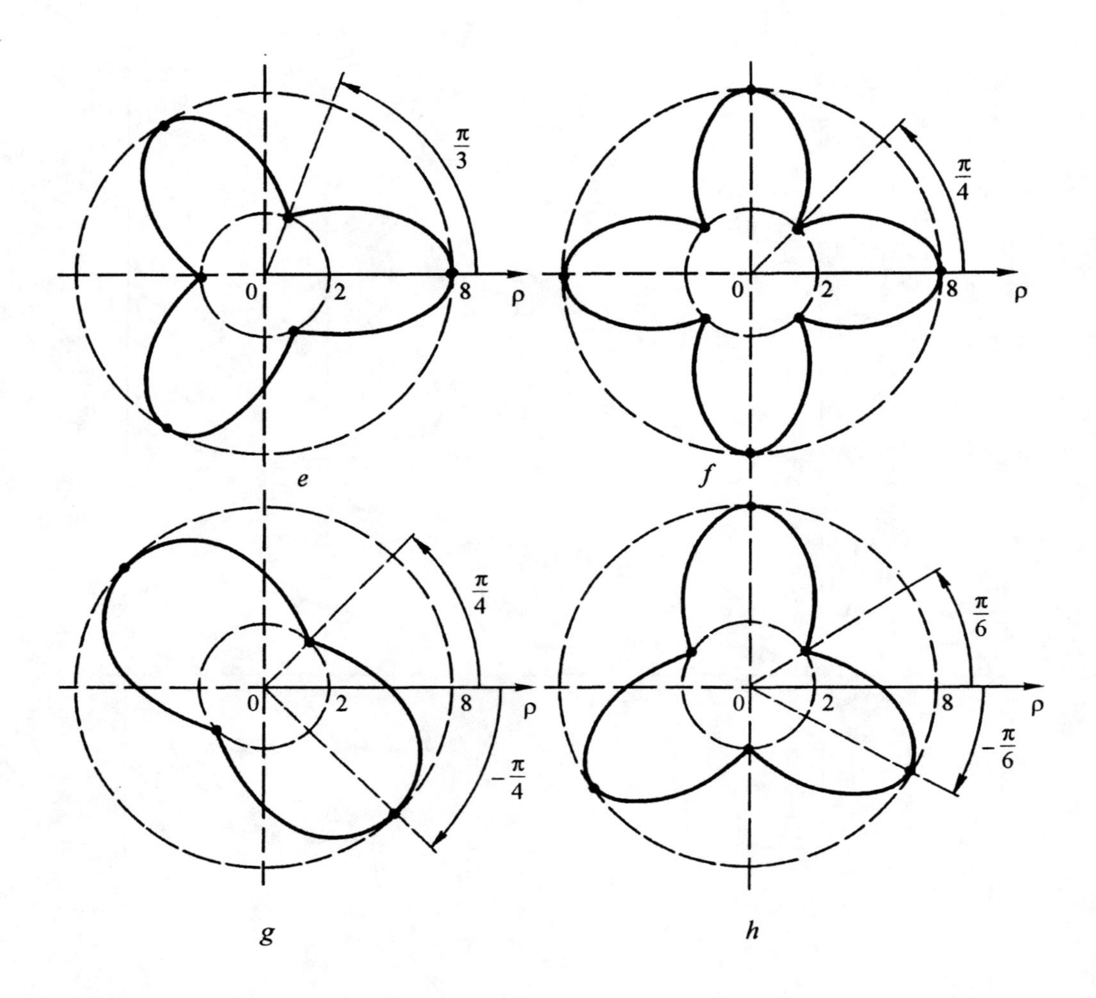

Figure 7.51. The graphs of functions: (*e*) $\rho = 5 + 3\cos 3\varphi$; (*f*) $\rho = 5 + 3\cos 4\varphi$; (*g*) $\rho = 5 - 3\sin 2\varphi$; (*h*) $\rho = 5 - 3\sin 3\varphi$.

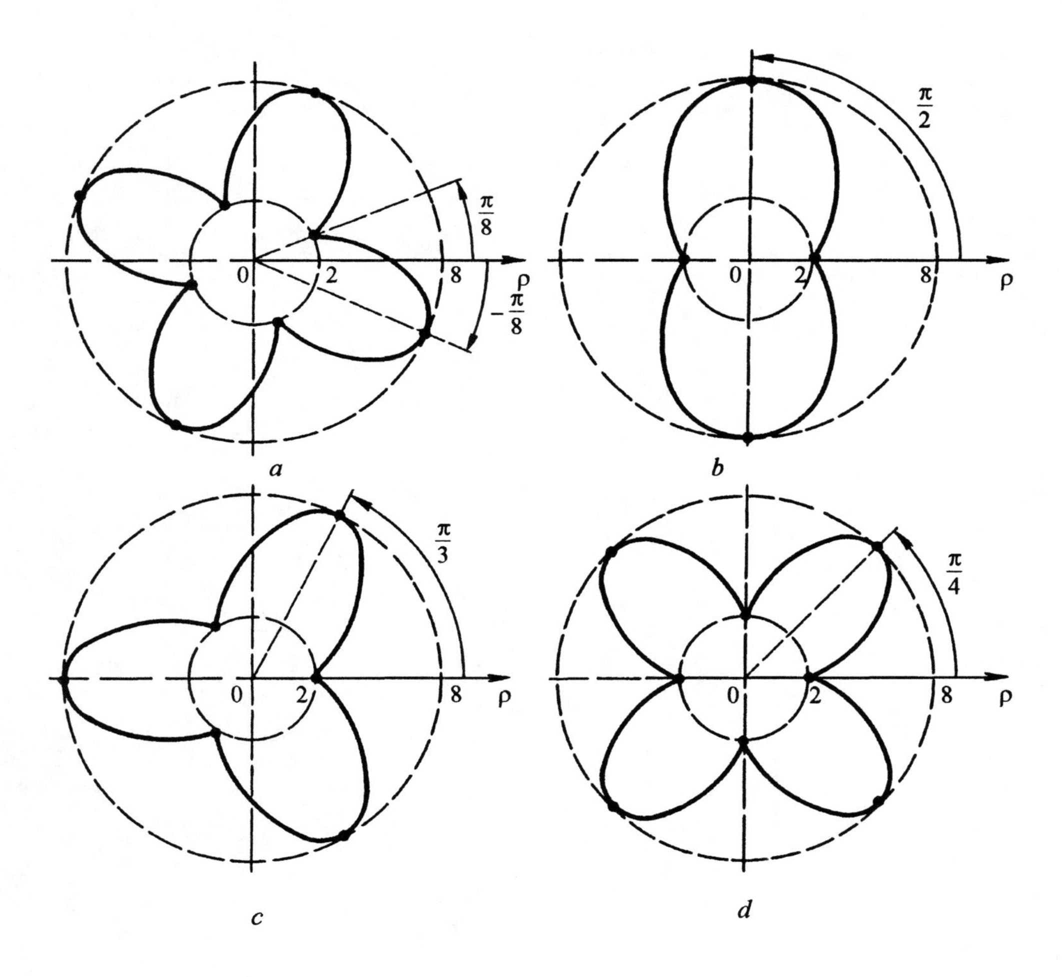

Figure 7.52. The graphs of functions: (*a*) $\rho = 5 - 3\sin 4\varphi$; (*b*) $\rho = 5 - 3\cos 2\varphi$; (*c*) $\rho = 5 - 3\cos 3\varphi$; (*d*) $\rho = 5 - 3\cos 4\varphi$.

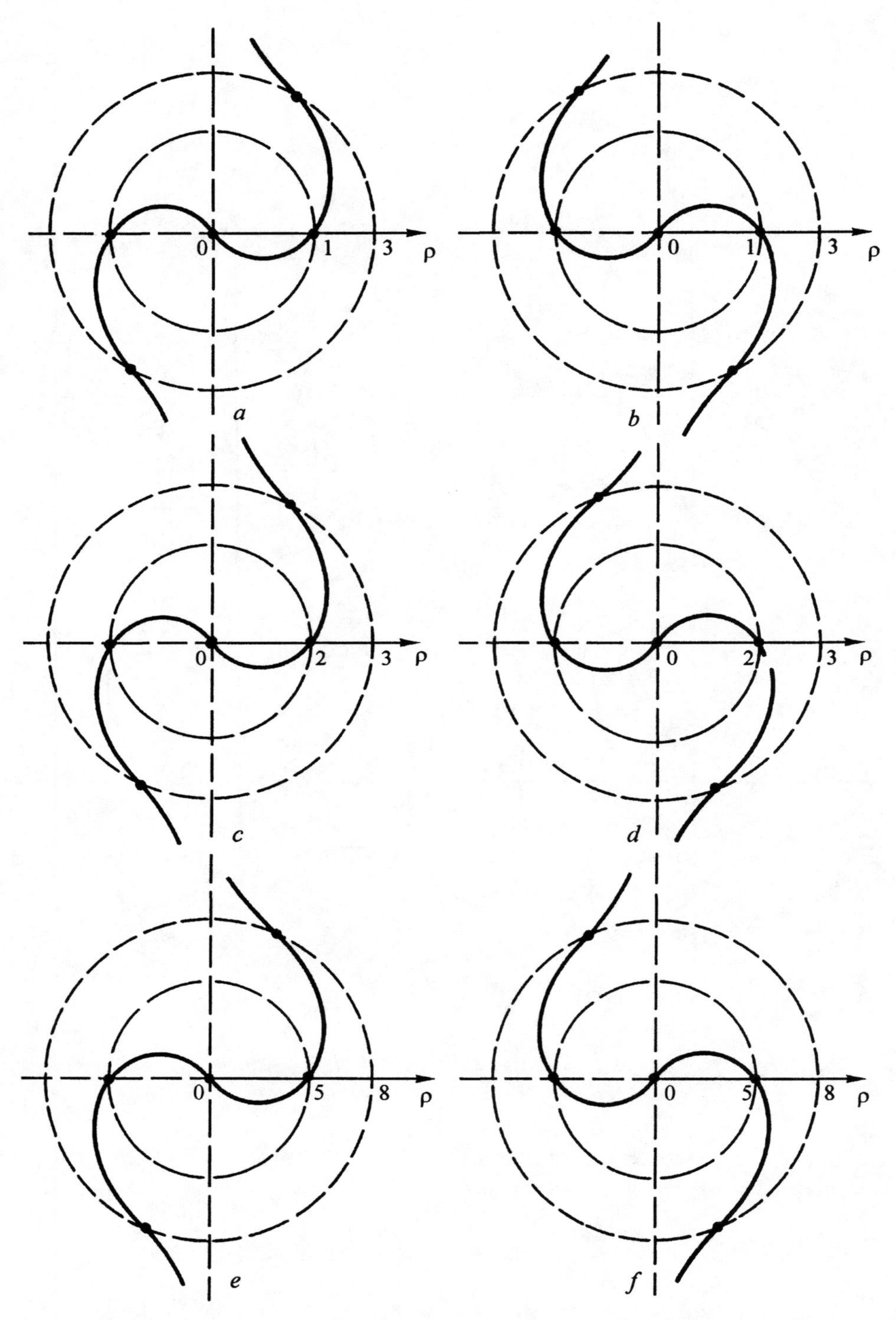

Figure 7.53. The graphs of functions: (*a*) $\rho = 1 + 2\tan\varphi$; (*b*) $\rho = 1 - 2\tan\varphi$; (*c*) $\rho = 2 + \tan\varphi$; (*d*) $\rho = 2 - \tan\varphi$; (*e*) $\rho = 5 + 3\tan\varphi$; (*f*) $\rho = 5 - 3\tan\varphi$.

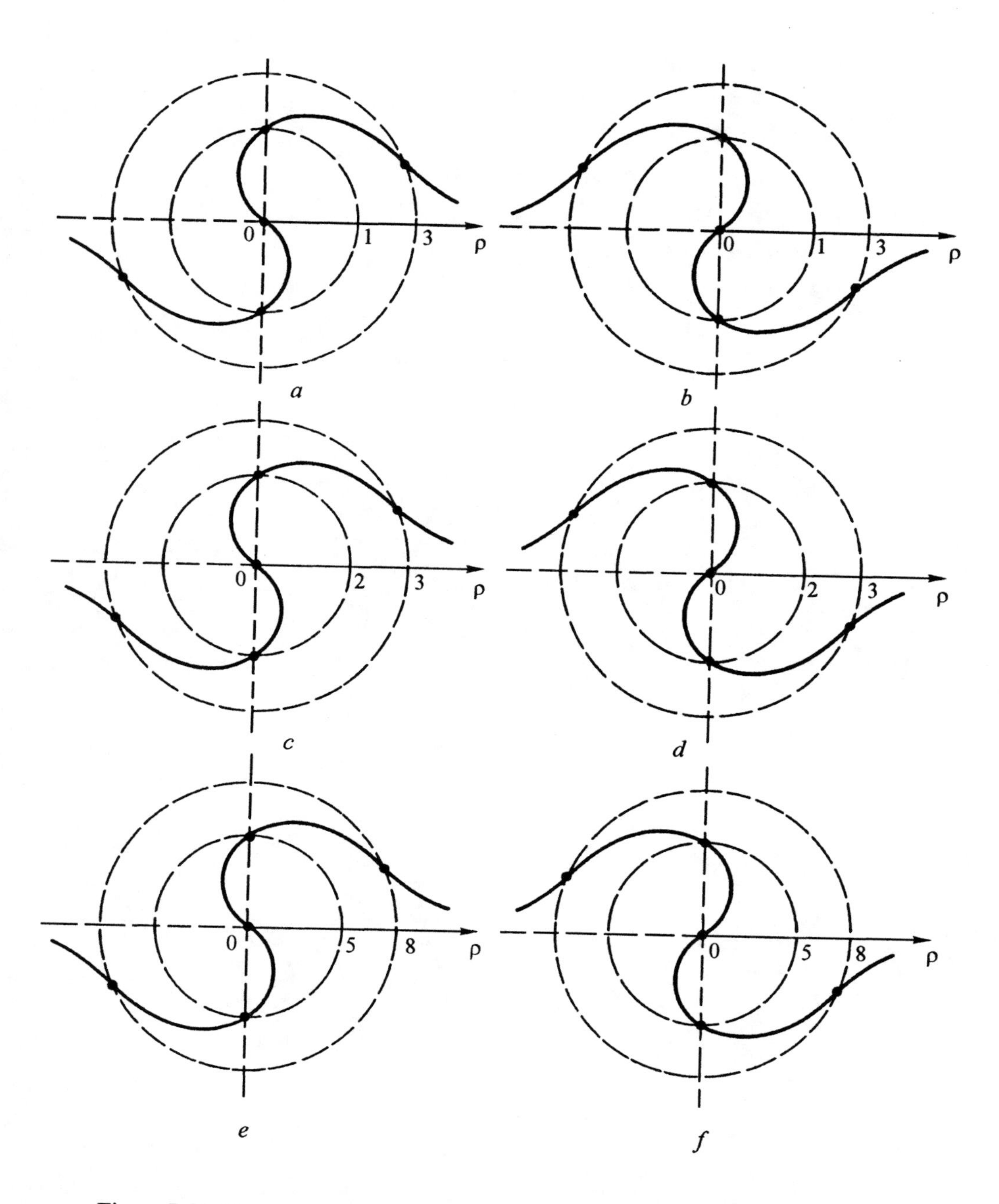

Figure 7.54. The graphs of functions: (*a*) $\rho = 1 + 2\cot\varphi$; (*b*) $\rho = 1 - 2\cot\varphi$; (*c*) $\rho = 2 + \cot\varphi$; (*d*) $\rho = 2 - \cot\varphi$; (*e*) $\rho = 5 + 3\cot\varphi$; (*f*) $\rho = 5 - 3\cot\varphi$.

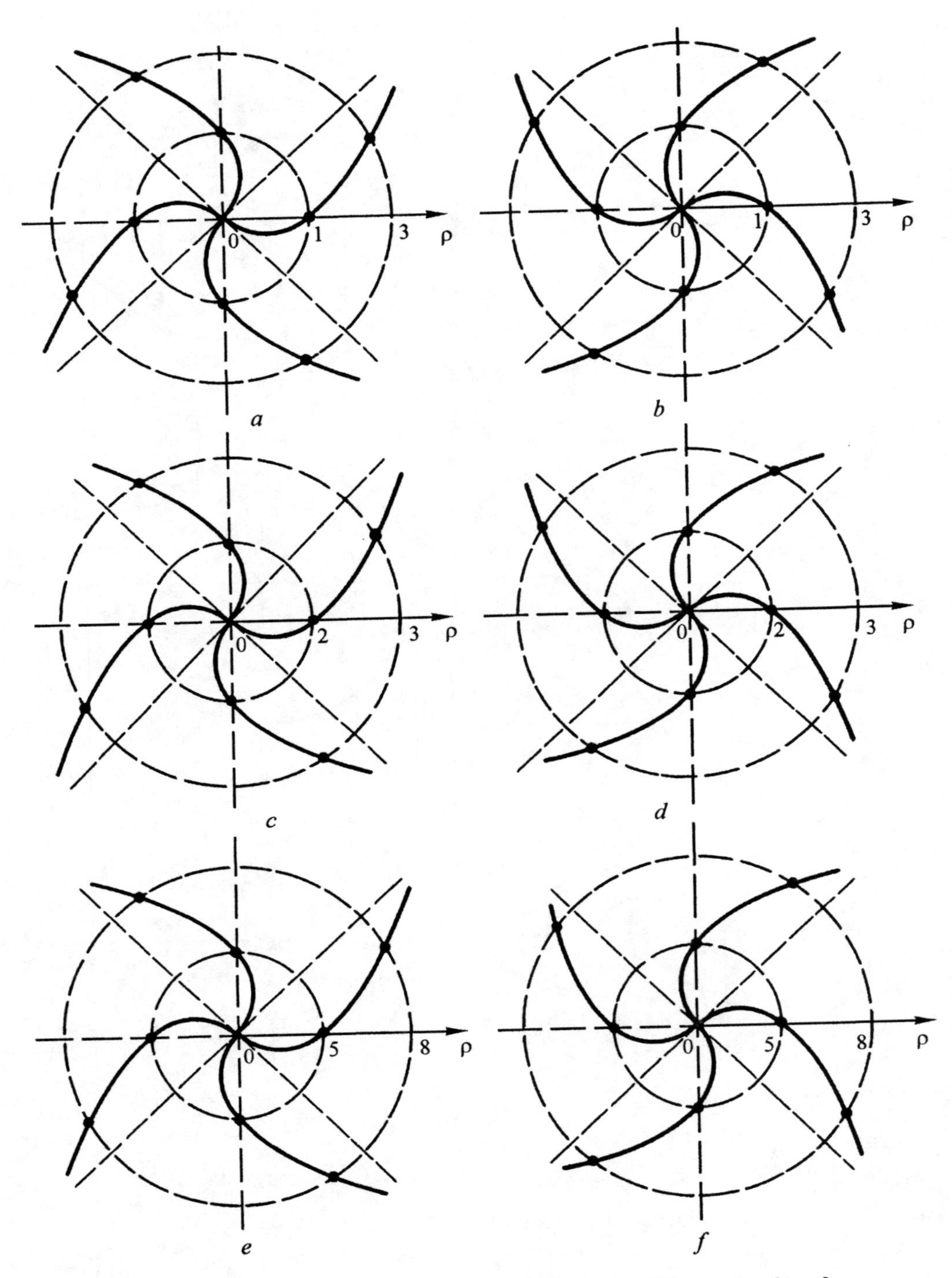

Figure 7.55. The graphs of functions: (*a*) $\rho = 1 + 2\tan 2\varphi$; (*b*) $\rho = 1 - 2\tan 2\varphi$; (*c*) $\rho = 2 + \tan 2\varphi$; (*d*) $\rho = 2 - \tan 2\varphi$; (*e*) $\rho = 5 + 3\tan 2\varphi$; (*f*) $\rho = 5 - 3\tan 2\varphi$.

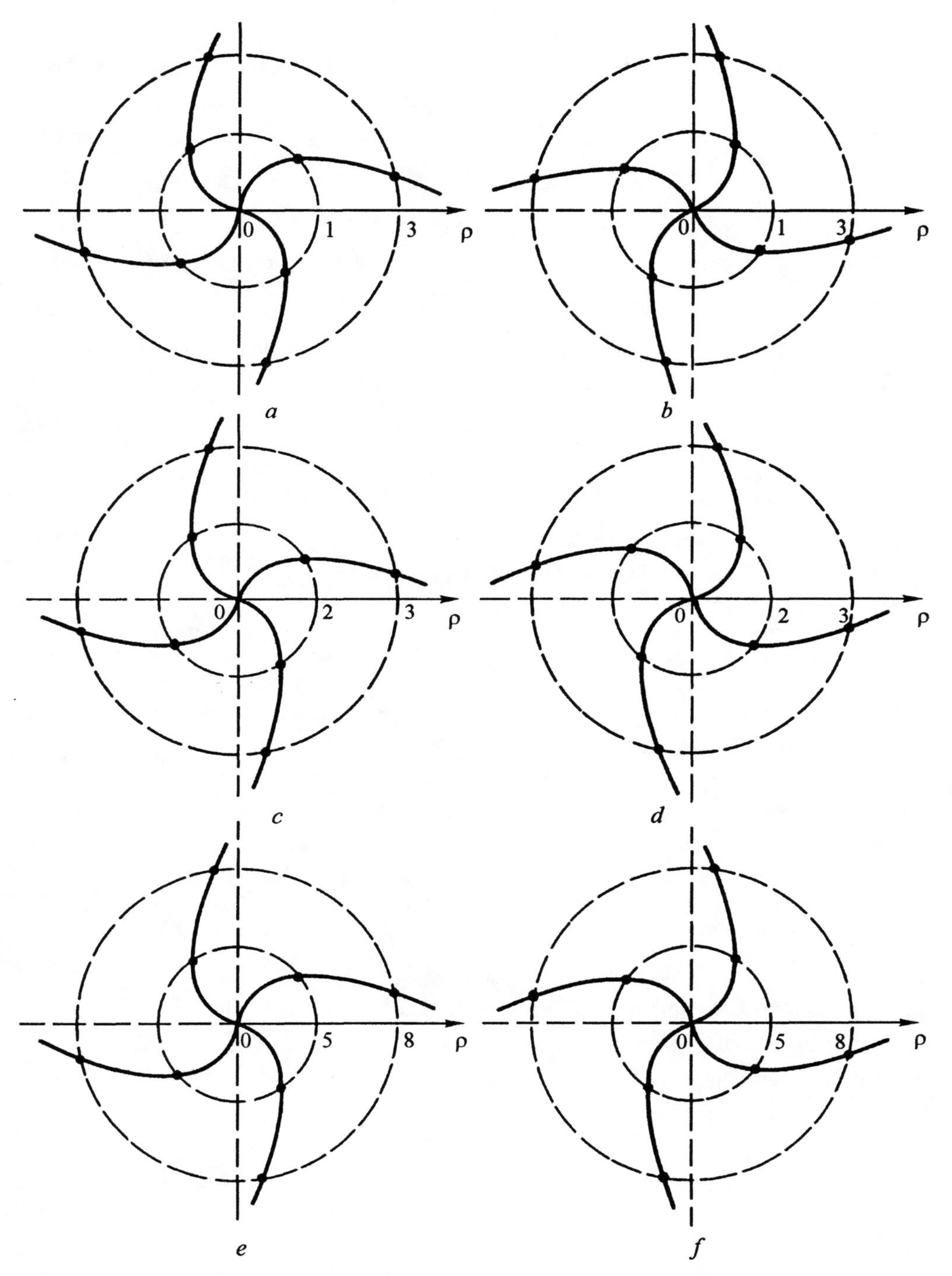

Figure 7.56. The graphs of functions: (*a*) $\rho = 1 + 2\cot 2\varphi$; (*b*) $\rho = 1 - 2\cot 2\varphi$; (*c*) $\rho = 2 + \cot 2\varphi$; (*d*) $\rho = 2 - \cot 2\varphi$; (*e*) $\rho = 5 + 3\cot 2\varphi$; (*f*) $\rho = 5 - 3\cot 2\varphi$.

7.2.12. Plots of the graphs of curves $\rho = a \sin^2 n\varphi$, $\rho = a \cos^2 n\varphi$ ($n \in \mathbf{N}$)

By solving the equations $\sin^2 n\varphi = 1$, $\cos^2 n\varphi = 1$, we find that ρ achieves its maximum $2n$ times with $\rho_{max} = a$. These curves and a circle of radius r, $0 \le r < \rho_{max}$, intersect $4n$ times.

Finding values of the angles such that $\rho = \rho_{max}$ and the angles at which the curves intersect a circle, say of radius $a/2$, we plot the graphs of the curves (Figure 7.57).

7.2.13. Plots of the graphs of curves $\rho = a \sin^3 n\varphi$, $\rho = a \cos^3 n\varphi$ ($n \in \mathbf{N}$)

Solving the trigonometric equations $\sin^3 n\varphi = 1$, $\cos^3 n\varphi = 1$, or the equations $(\sin n\varphi - 1)(\sin^2 n\varphi + \sin n\varphi + 1) = 0$ and $(\cos n\varphi - 1)(\cos^2 n\varphi + \cos n\varphi + 1) = 0$, we see that for the curves $\rho = a \sin^3 n\varphi$ and $\rho = a \cos^3 n\varphi$ there are n values of φ such that $\rho_{max} = a$.

The curves $\rho = a \sin^3 n\varphi$, $\rho = a \cos^3 n\varphi$ intersect a circle of radius r, $0 < r < \rho_{max}$, in $2n$ points (Figure 7.58).

The plots of curves $\rho = a \sin^m n\varphi$, $\rho = a \cos^m n\varphi$, $n \in \mathbf{N}$, $m \in \mathbf{N}$, $m > 3$, can be constructed finding φ such that $\rho = \rho_{max}$ by solving the equations $\sin^m n\varphi = 1$, $\cos^m n\varphi = 1$, and finding points of intersection of these curves with a circle of radius r, $0 < r < \rho_{max}$.

7.2.14. Plots of the graphs of curves $\rho = a \tan^2 n\varphi$, $\rho = a \cot^2 n\varphi$ ($n \in \mathbf{N}$, $n > 1$)

The curves $\rho = a \tan^2 n\varphi$, $\rho = a \cot^2 n\varphi$ intersect a circle of radius r, $0 < r < +\infty$, in $4n$ points. The curves have $4n$ asymptotes coming from the pole.

To plot these curves, we find points of intersection of the curves with circles of radii $a, \sqrt{3}\,a$, then construct asymptotes, and draw lines through the found points of intersection approaching the asymptotes (Figure 7.59).

7.2.15. Plots of curves $\rho = a\tan^3 n\varphi$, $\rho = a\cot^3 n\varphi$ ($n \in \mathbf{N}$, $n > 1$)

The curves $\rho = a \tan^3 n\varphi$, $\rho = a \cot^3 n\varphi$ intersect a circle of radius r, $0 < r < \infty$, in $2n$ points. These curves have $2n$ asymptotes coming from the pole.

Finding points of intersection of the curves with circles of radii a, $a\sqrt{3}$, and drawing the asymptotes, we plot the graph by making it pass through the points of intersection and letting it tend to the asymptotes (Figure 7.60).

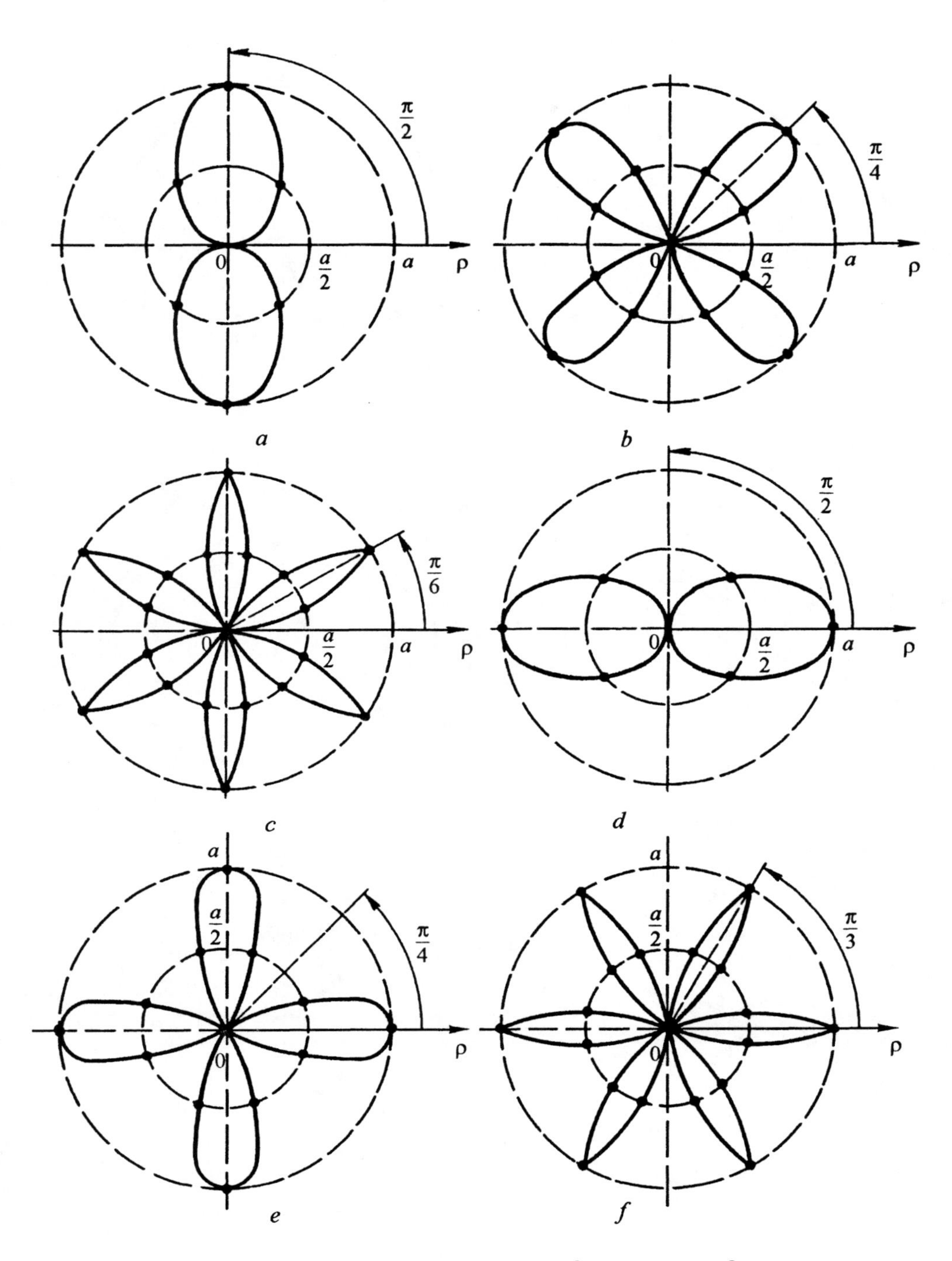

Figure 7.57. The graphs of functions: (*a*) $\rho = a \sin^2\varphi$; (*b*) $\rho = a \sin^2 2\varphi$; (*c*) $\rho = a \sin^2 3\varphi$; (*d*) $\rho = a \cos^2\varphi$; (*e*) $\rho = a \cos^2 2\varphi$; (*f*) $\rho = a \cos^2 3\varphi$.

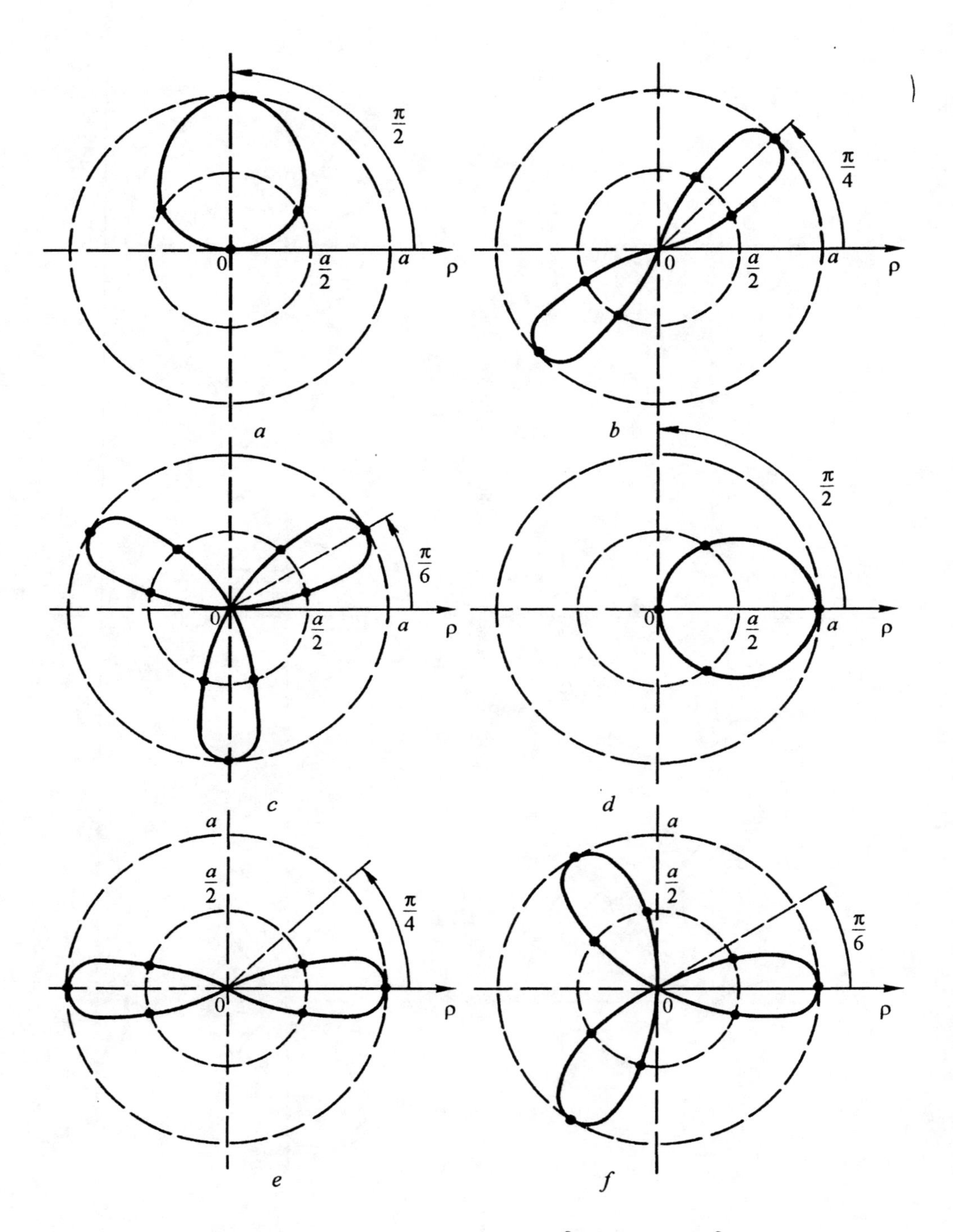

Figure 7.58. The graphs of functions: (*a*) $\rho = a \sin^3\varphi$; (*b*) $\rho = a \sin^3 2\varphi$; (*c*) $\rho = a \sin^3 3\varphi$; (*d*) $\rho = a \cos^3\varphi$; (*e*) $\rho = a \cos^3 2\varphi$; (*f*) $\rho = a \cos^3 3\varphi$.

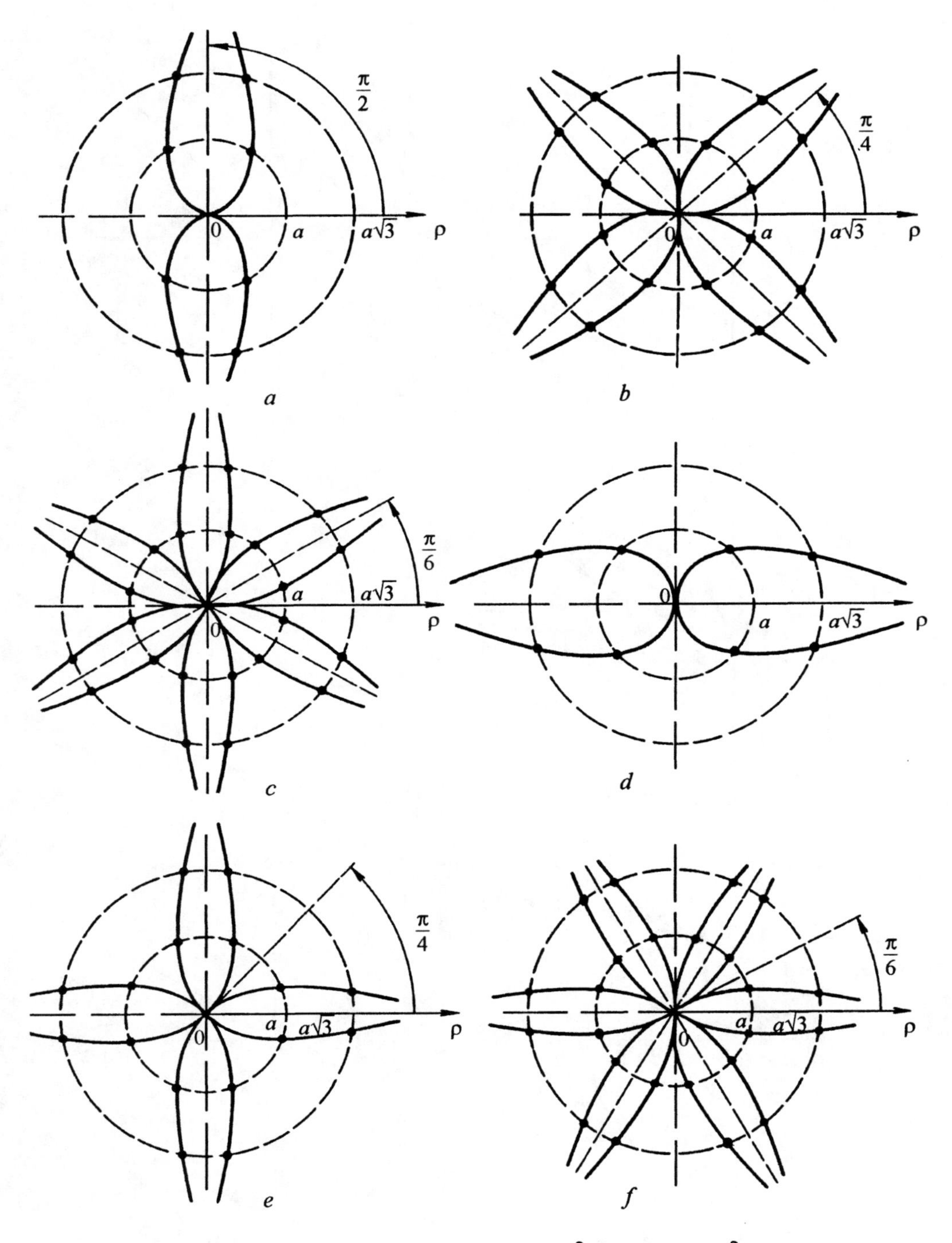

Figure 7.59. The graphs of functions: (*a*) $\rho = a\tan^2\varphi$; (*b*) $\rho = a\tan^2 2\varphi$; (*c*) $\rho = a\tan^2 3\varphi$; (*d*) $\rho = \cot^2\varphi$; (*e*) $\rho = \cot^2 2\varphi$; (*f*) $\rho = \cot^2 3\varphi$.

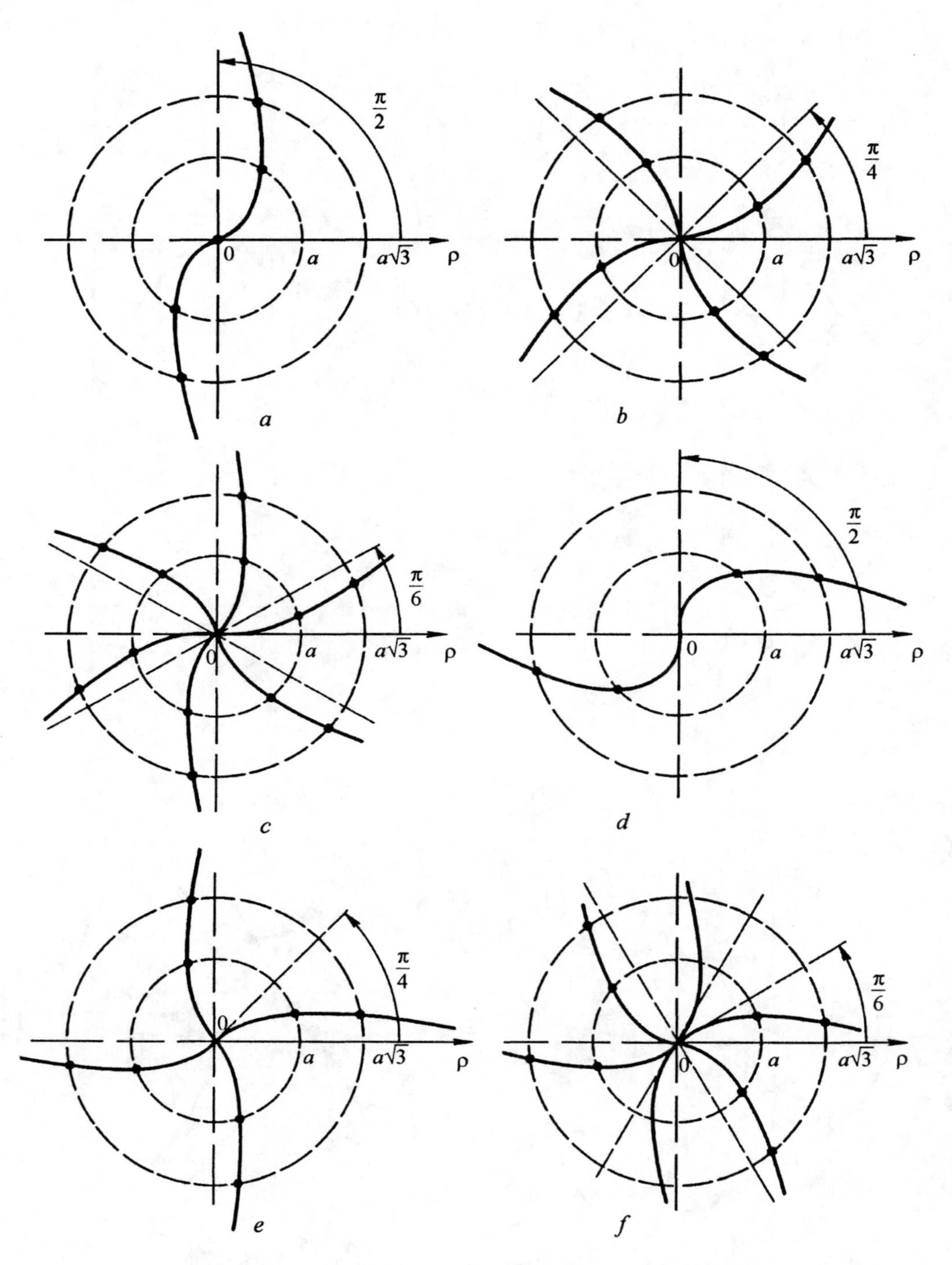

Figure 7.60. The graphs of functions: (*a*) $\rho = a\tan^3\varphi$; (*b*) $\rho = a\tan^3 2\varphi$; (*c*) $\rho = a\tan^3 3\varphi$; (*d*) $\rho = a\cot^3\varphi$; (*e*) $\rho = a\cot^3 2\varphi$; (*f*) $\rho = a\cot^3\varphi$.

Plots of curves $\rho = a\sin\frac{p}{q}\varphi$, $\rho = a\cos\frac{p}{q}\varphi$, $\rho = a\tan\frac{p}{q}\varphi$, $\rho = a\cot\frac{p}{q}\varphi$, $\rho = a(1\pm\sin\frac{p}{q}\varphi)$, $\rho = a(1\pm\cos\frac{p}{q}\varphi)$, $\rho = a(1\pm\tan\frac{p}{q}\varphi)$, $\rho = a(1\pm\cot\frac{p}{q}\varphi)$, $\rho = a\sin^n\frac{p}{q}\varphi$, $\rho = a\cos^n\frac{p}{q}\varphi$, $\rho = a\tan^n\frac{p}{q}\varphi$, $\rho = a\cot^n\frac{p}{q}\varphi$ $(p, q, n \in \mathrm{N})$

The curves are plotted by finding intersections of the curve and concentric circles. Note that if $p<q$, then the plots of the curves $\rho = a\sin\frac{p}{q}\varphi$, $\rho = a\cos\frac{p}{q}\varphi$, $\rho = a(1\pm\sin\frac{p}{q}\varphi)$, $\rho = a(1\pm\cos\frac{p}{q}\varphi)$, $\rho = a\sin^n\frac{p}{q}\varphi$, $\rho = a\cos^n\frac{p}{q}\varphi$ are certain closed heart-shaped curves; if $p>q$, then they are leaf curves (Figures 7.61 – 7.95).

7.2.16. Plots of curves $\rho = a/\sin^k n\varphi$, $\rho = a/\cos^k n\varphi$ $(n, k \in \mathrm{N})$

If $n = 1$, $k = 1$, we have a corresponding straight line $\rho = a/\sin\varphi$ that is parallel to the axis Ox and lies at the distance a (Figure 7.96). The curve $\rho = a/\cos\varphi$ is parallel to Oy and lies at the distance a (Figure 7.97).

The curves $\rho = a/\sin^k n\varphi$, $\rho = a/\cos^k n\varphi$ have $\rho_{\min} = a$.

For $k = 2m$, the graphs of these curves consist of two lines that are symmetric with respect to the x-axis (y-axis). If $k = 2m+1$, then they consist of a single line located inside the angle $(0<\varphi<\pi)$ $(-\pi<\varphi<\pi/2)$ for $n = 1$. Indeed, for $k = 2m$, we have that

$$\left.\begin{aligned}\rho &= \frac{a}{\sin^{2m} n\varphi}\\ \rho &= \frac{a}{\cos^{2m} n\varphi}\end{aligned}\right\}\Rightarrow$$

$$\Rightarrow\begin{cases}(\sin^2 n\varphi - \sqrt[2m]{a/\rho}\,(\sin^{2m-2} n\varphi + \dots + \sqrt[2m]{(a/\rho)^{2m-2}}\,) = 0\\ (\cos^2 n\varphi - \sqrt[2m]{a/\rho}\,(\cos^{2m-2} n\varphi + \dots + \sqrt[2m]{(a/\rho)^{2m-2}}\,) = 0\end{cases}\Rightarrow$$

$$\Rightarrow\begin{cases}\sin n\varphi = \pm\sqrt{\sqrt[2m]{a/\rho}}\,,\\ \cos n\varphi = \pm\sqrt{\sqrt[2m]{a/\rho}}\,.\end{cases}$$

The expression $\sin n\varphi = \pm\sqrt[2m]{a/\rho}$ defines four values of the angles of rays that pass through points of intersection of the curves $\rho = a/\sin^{2m} n\varphi$ and concentric circles of radii $\rho \ge a$; these angles are such that points of intersection are symmetric with respect to the x-axis (Figure 7.98). Expressions $\cos n\varphi = \pm\sqrt[2m]{a/\rho}$ are analyzed (Figure 7.99) in a similar way. If k is odd, $k = 2m+1$, considerations are similar (Figures 7.100 and 7.101). The graphs of curves $\rho = a/\sin^k n\varphi$, $\rho = a/\cos^k n\varphi$ are shown in Figures 7.102 – 7.104.

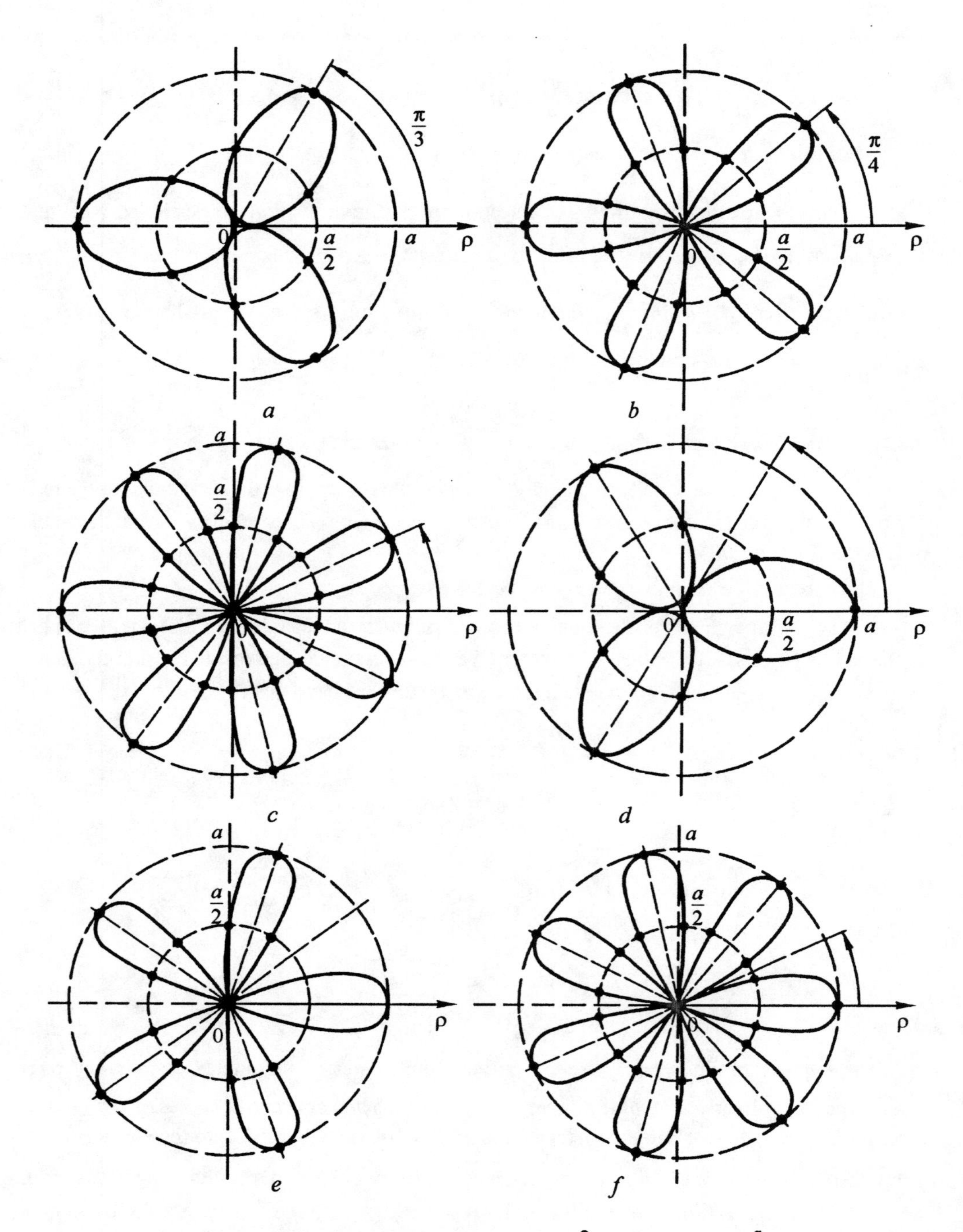

Figure 7.61. The graphs of functions: (*a*) $\rho = a \sin \frac{3}{2} \varphi$; (*b*) $\rho = a \sin \frac{5}{2} \varphi$; (*c*) $\rho = a \sin \frac{7}{2} \varphi$; (*d*) $\rho = a \cos \frac{3}{2} \varphi$; (*e*) $\rho = a \cos \frac{5}{2} \varphi$; (*f*) $\rho = a \cos \frac{7}{2} \varphi$.

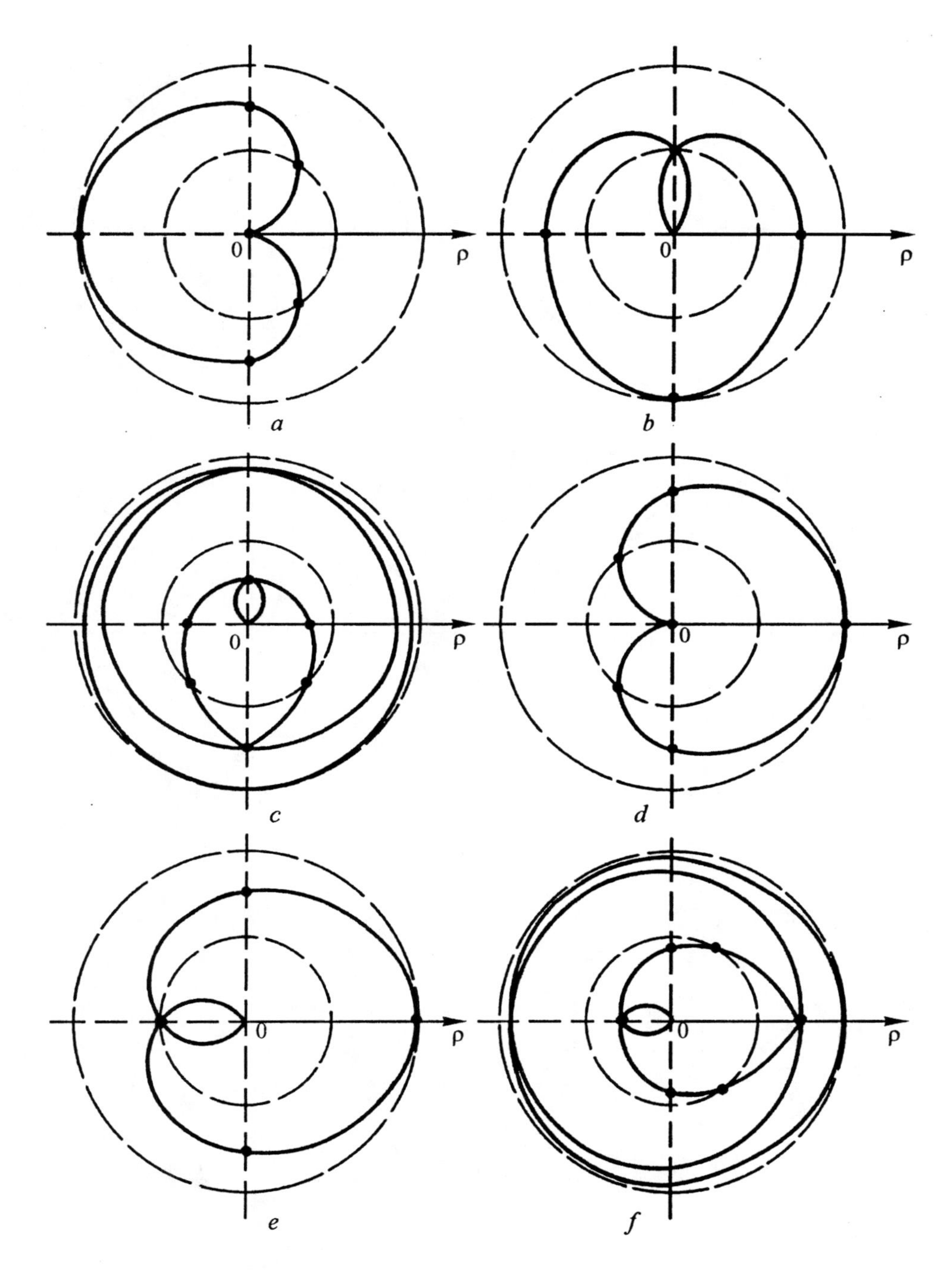

Figure 7.62. The graphs of functions: (*a*) $\rho = a \sin \frac{\varphi}{2}$; (*b*) $\rho = a \sin \frac{\varphi}{3}$; (*c*) $\rho = a \sin \frac{\varphi}{4}$; (*d*) $\rho = a \cos \frac{\varphi}{2}$; (*e*) $\rho = a \cos \frac{\varphi}{3}$; (*f*) $\rho = a \cos \frac{\varphi}{4}$.

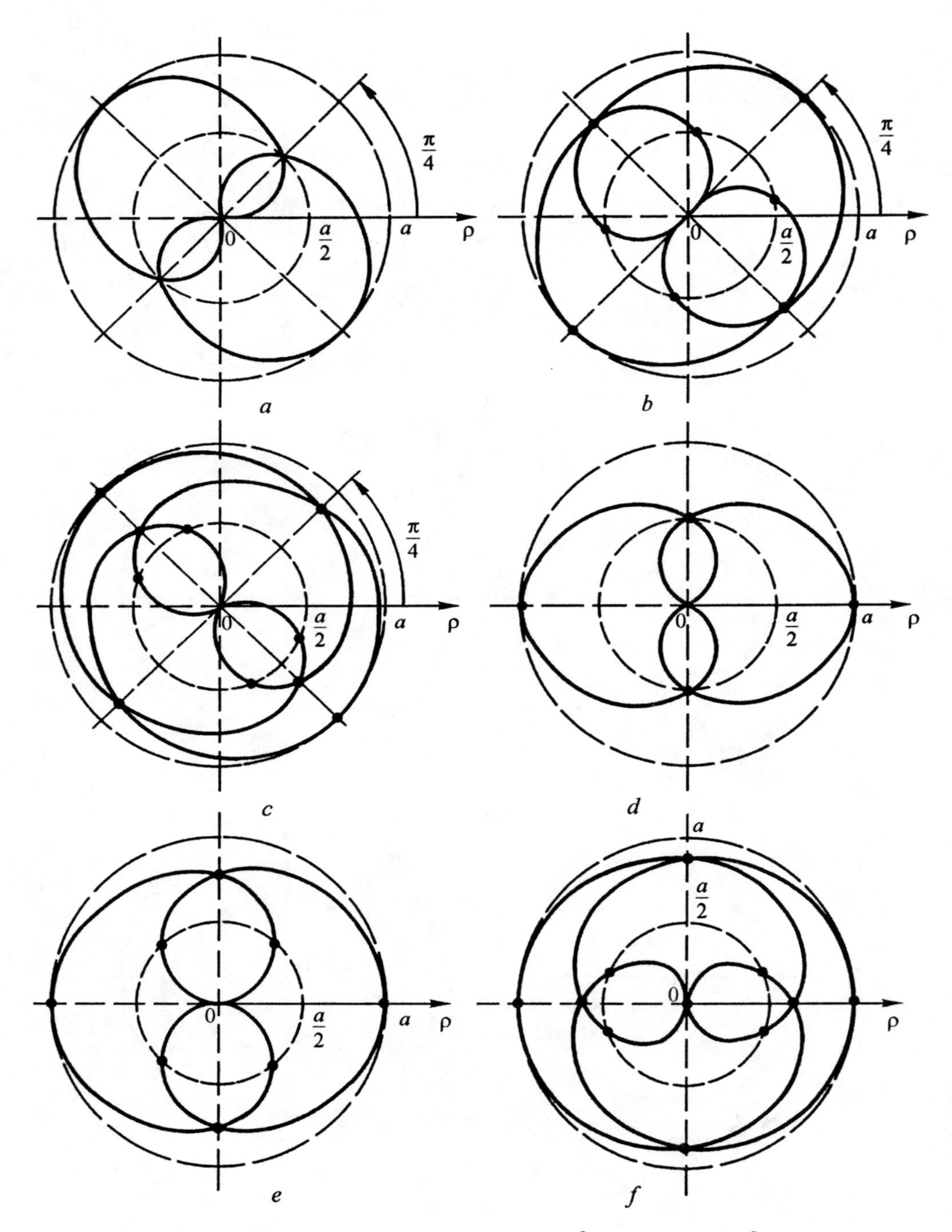

Figure 7.63. The graphs of functions: (*a*) $\rho = a \sin \frac{2}{3} \varphi$; (*b*) $\rho = a \sin \frac{2}{5} \varphi$; (*c*) $\rho = a \sin \frac{2}{7} \varphi$; (*d*) $\rho = a \cos \frac{2}{3} \varphi$; (*e*) $\rho = a \cos \frac{2}{5} \varphi$; (*f*) $\rho = a \cos \frac{2}{7} \varphi$.

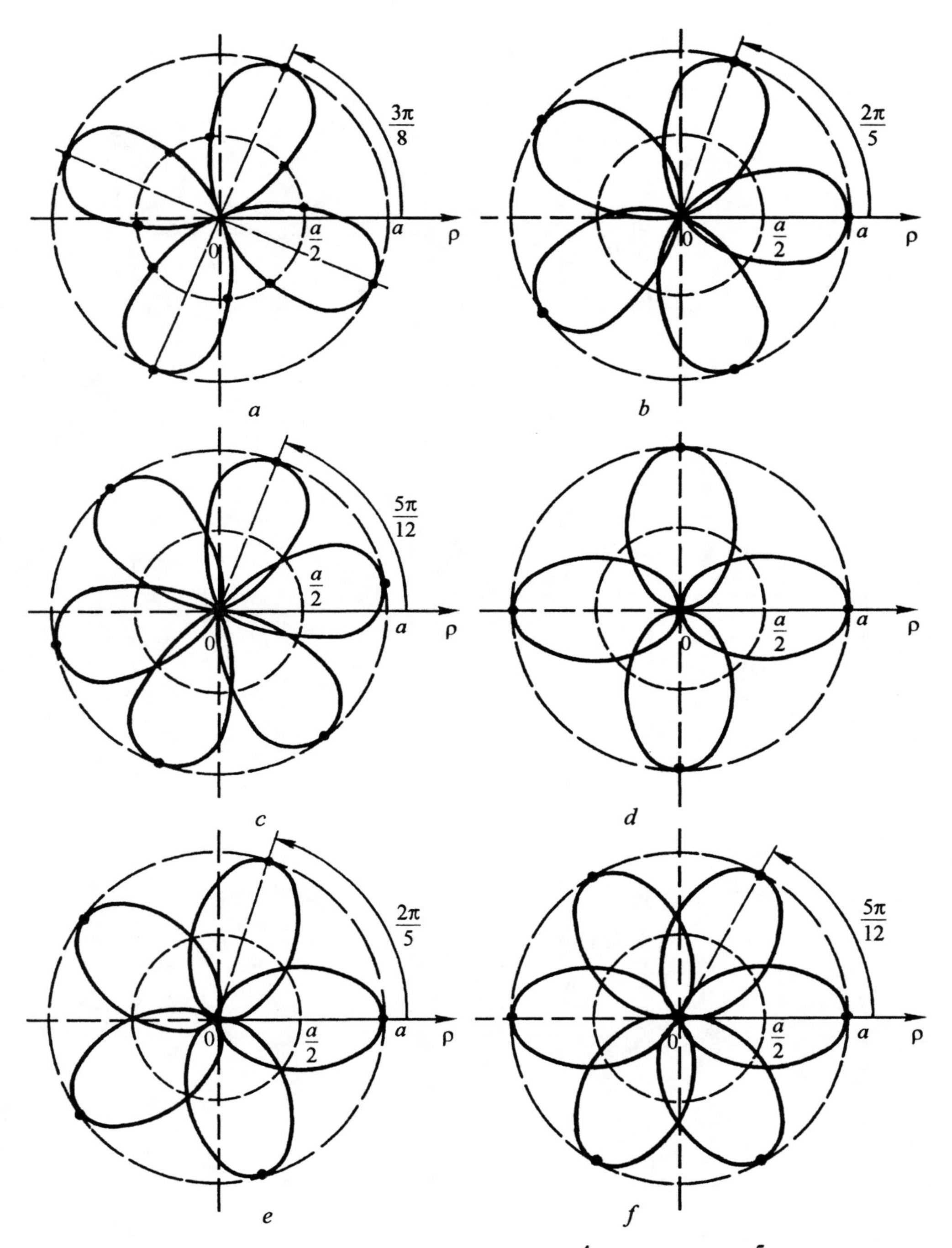

Figure 7.64. The graphs of functions: (*a*) $\rho = a\sin\frac{4}{3}\varphi$; (*b*) $\rho = a\sin\frac{5}{4}\varphi$; (*c*) $\rho = a\sin\frac{6}{5}\varphi$; (*d*) $\rho = a\cos\frac{4}{3}\varphi$; (*e*) $\rho = a\cos\frac{5}{4}\varphi$; (*f*) $\rho = a\cos\frac{6}{5}\varphi$.

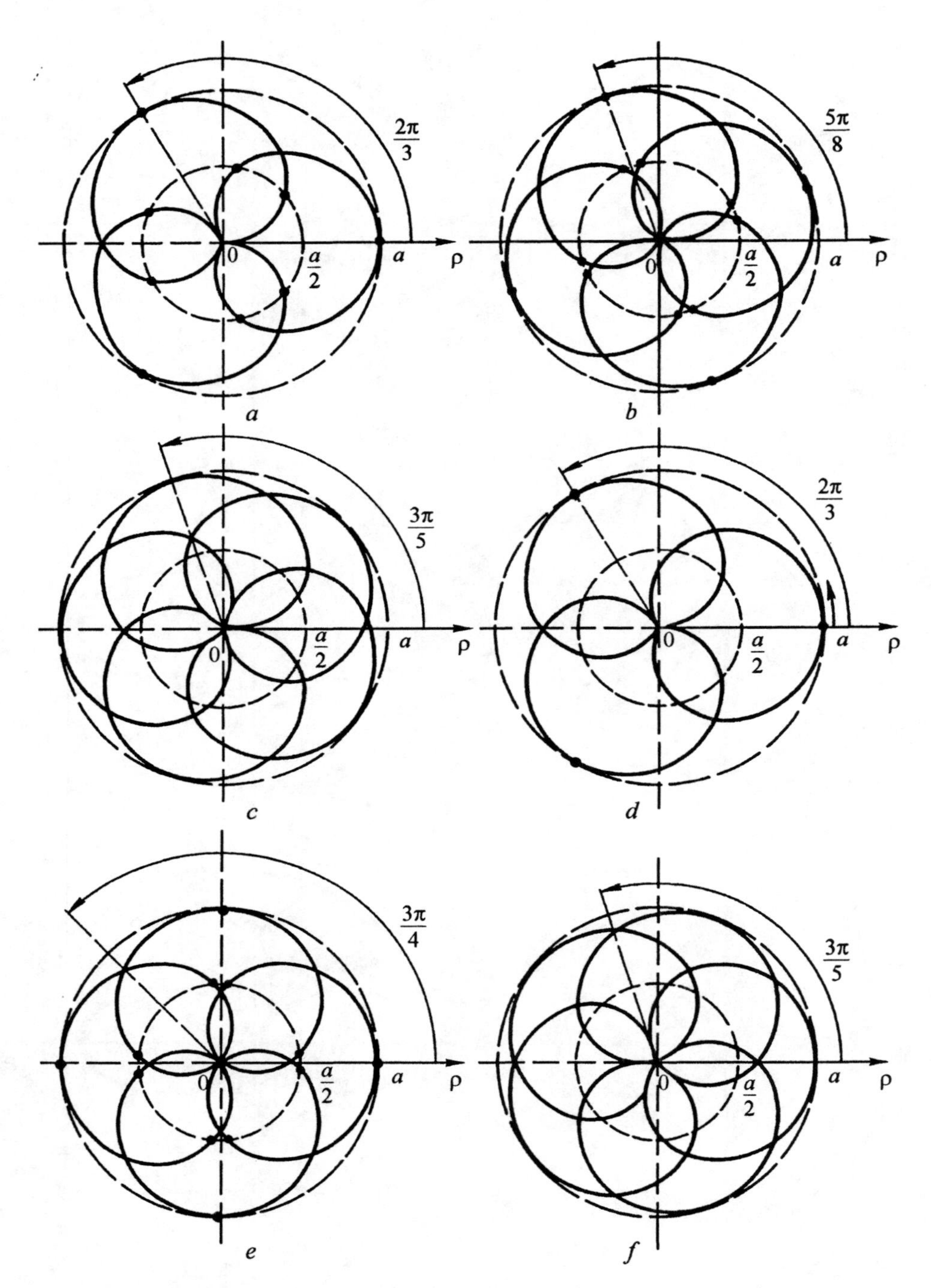

Figure 7.65. The graphs of functions: (*a*) $\rho = a \sin \frac{3}{4} \varphi$; (*b*) $\rho = a \sin \frac{4}{5} \varphi$; (*c*) $\rho = a \sin \frac{5}{6} \varphi$; (*d*) $\rho = a \cos \frac{3}{4} \varphi$; (*e*) $\rho = a \cos \frac{4}{5} \varphi$; (*f*) $\rho = a \cos \frac{5}{6} \varphi$.

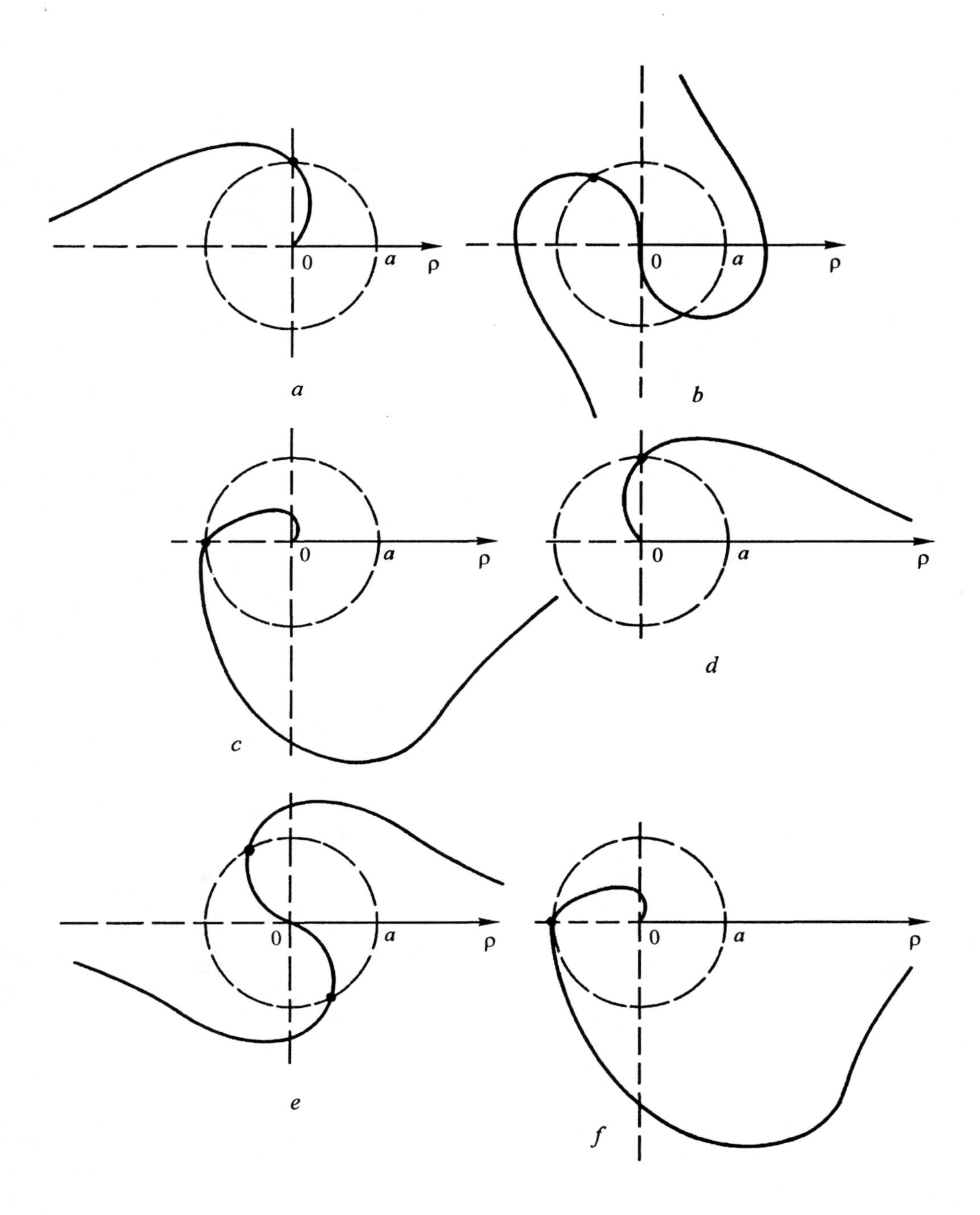

Figure 7.66. The graphs of functions: (*a*) $\rho = a \tan \frac{\varphi}{2}$; (*b*) $\rho = a \tan \frac{\varphi}{3}$; (*c*) $\rho = a \tan \frac{\varphi}{4}$; (*d*) $\rho = a \cot \frac{\varphi}{2}$; (*e*) $\rho = a \cot \frac{\varphi}{3}$; (*f*) $\rho = a \cot \frac{\varphi}{4}$.

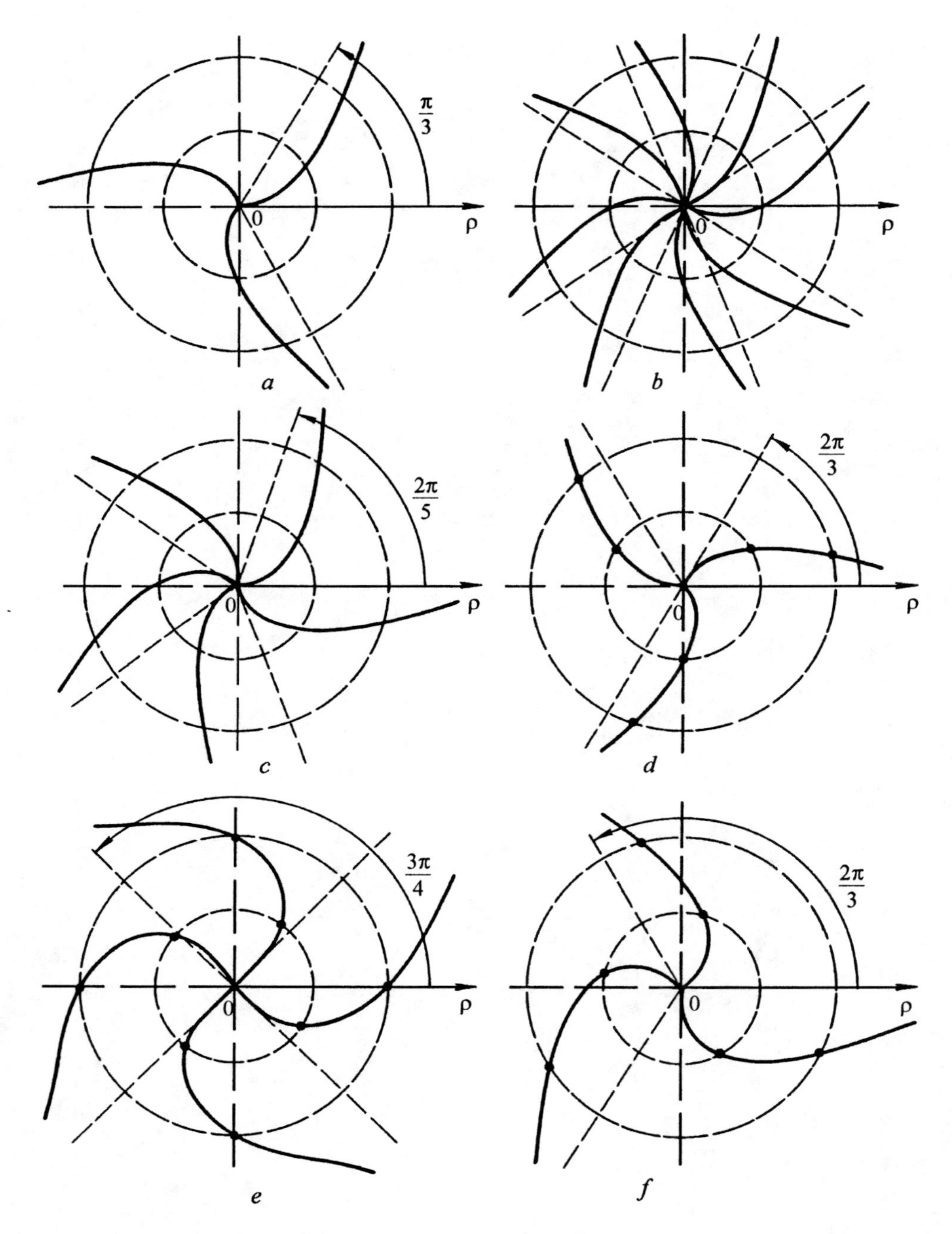

Figure 7.67. The graphs of functions: (*a*) $\rho = a \tan \frac{3}{2} \varphi$; (*b*) $\rho = a \tan \frac{4}{3} \varphi$; (*c*) $\rho = a \tan \frac{5}{4} \varphi$; (*d*) $\rho = a \cot \frac{3}{2} \varphi$; (*e*) $\rho = a \tan \frac{2}{3} \varphi$; (*f*) $\rho = a \tan \frac{3}{4} \varphi$.

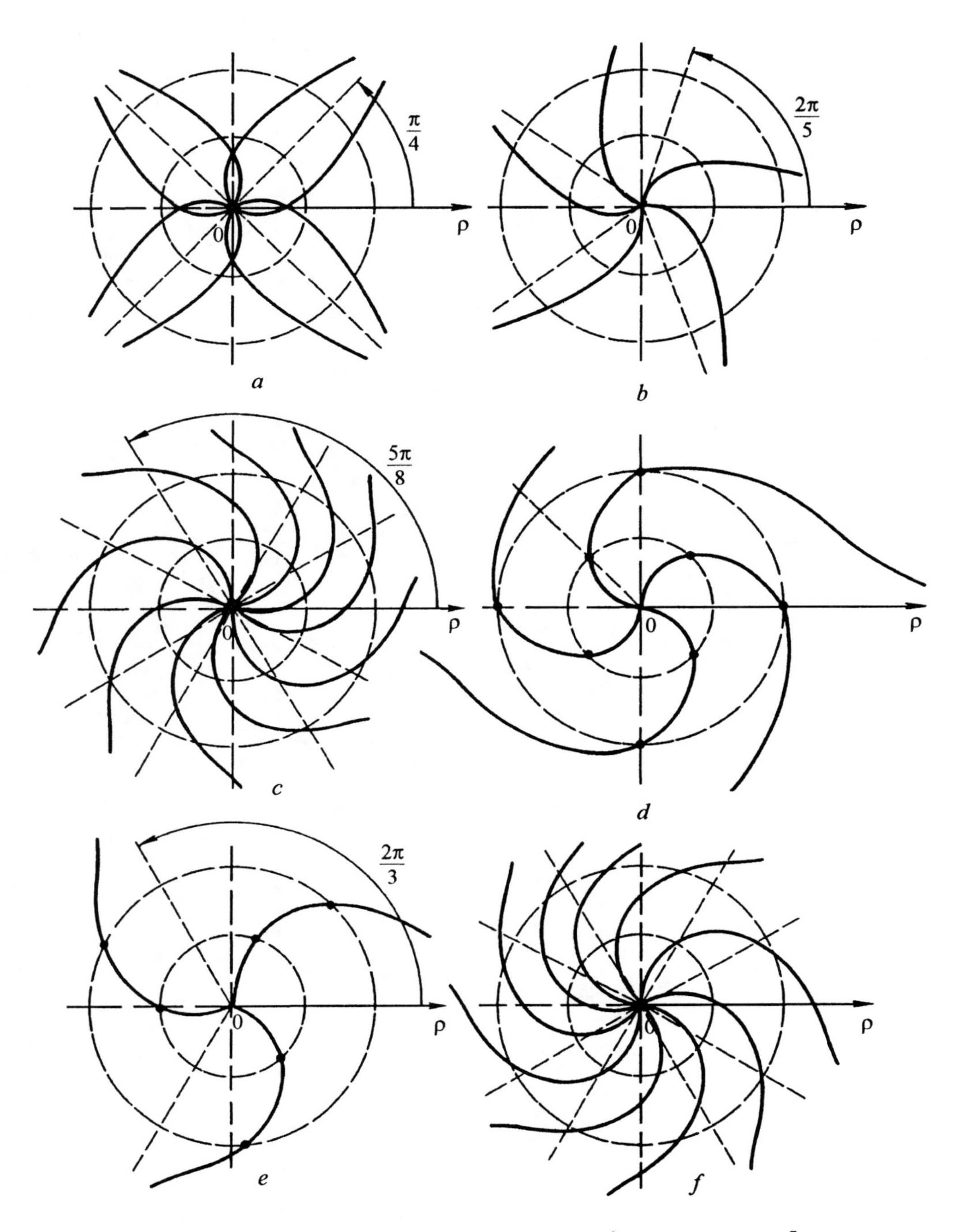

Figure 7.68. The graphs of functions: (*a*) $\rho = a\cot\frac{4}{3}\varphi$; (*b*) $\rho = a\cot\frac{5}{4}\varphi$; (*c*) $\rho = a\tan\frac{4}{5}\varphi$; (*d*) $\rho = a\cot\frac{2}{3}\varphi$; (*e*) $\rho = a\cot\frac{3}{4}\varphi$; (*f*) $\rho = a\cot\frac{4}{5}\varphi$.

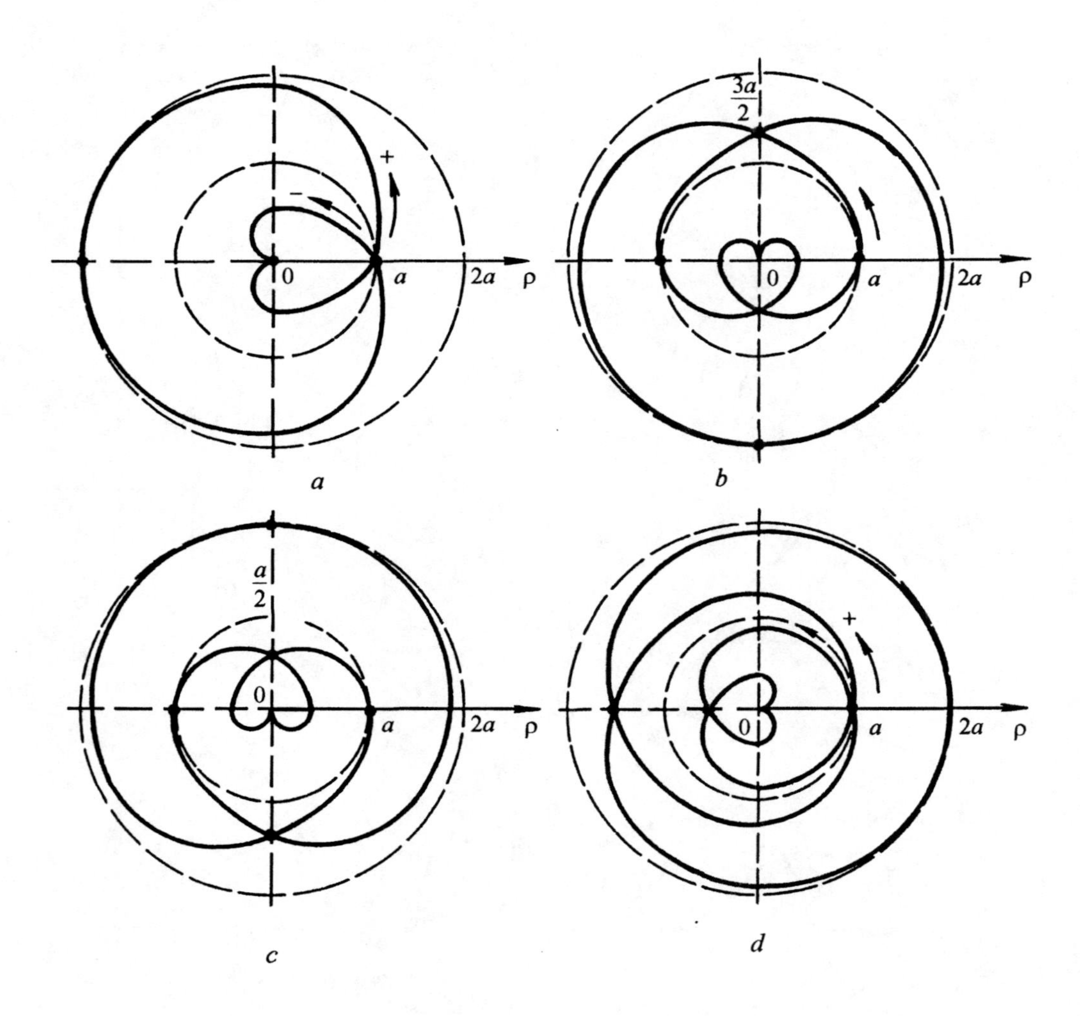

Figure 7.69. The graphs of functions: (*a*) $\rho = a(1 \pm \sin\frac{\varphi}{2})$; (*b*) $\rho = a(1 + \sin\frac{\varphi}{3})$; (*c*) $\rho = a(1 - \sin\frac{\varphi}{2})$; (*d*) $\rho = a(1 \pm \sin\frac{\varphi}{4})$.

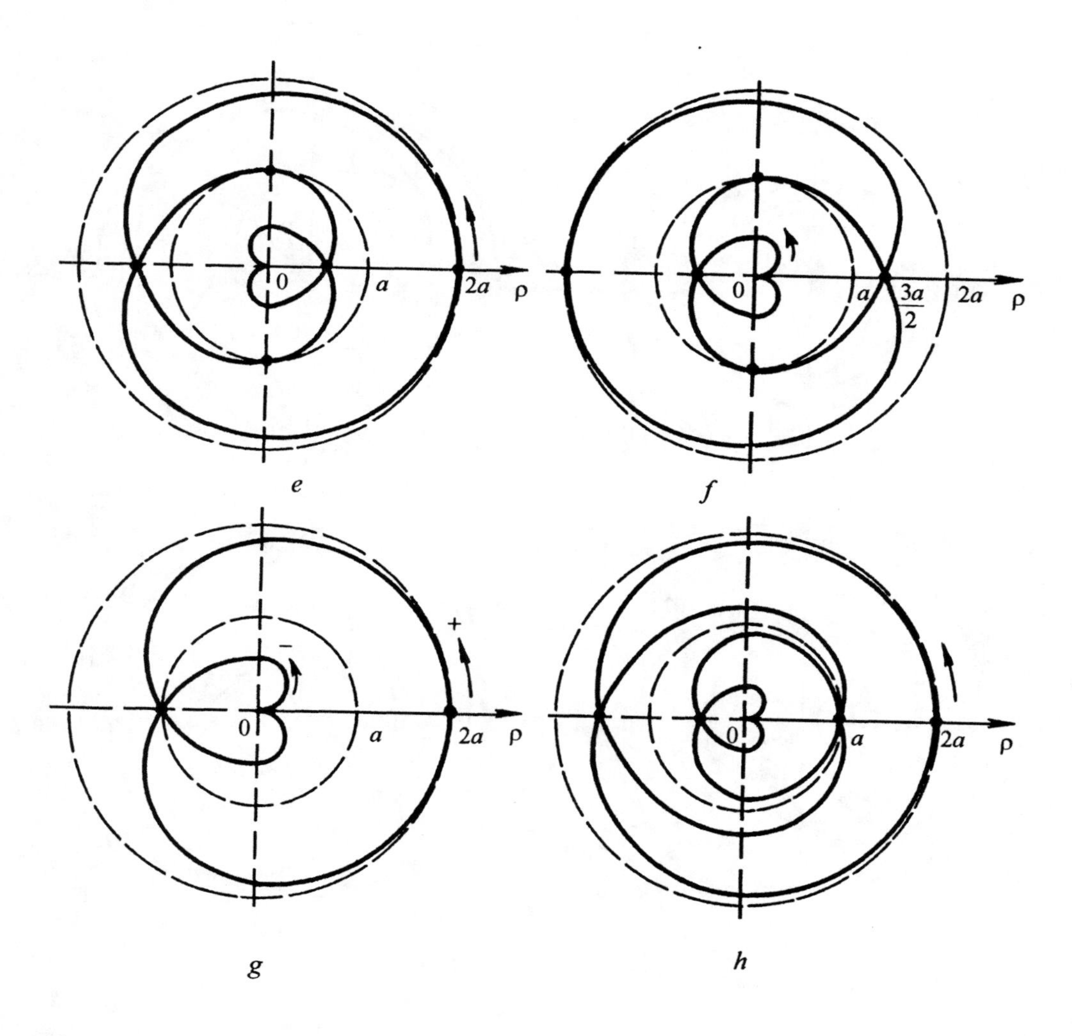

Figure 7.69. The graphs of functions: (*e*) $\rho = a(1 + \cos\frac{\varphi}{3})$; (*f*) $\rho = a(1 - \cos\frac{\varphi}{3})$; (*g*) $\rho = a(1 \pm \cos\frac{\varphi}{2})$; (*h*) $\rho = a(1 + \cos\frac{\varphi}{4})$.

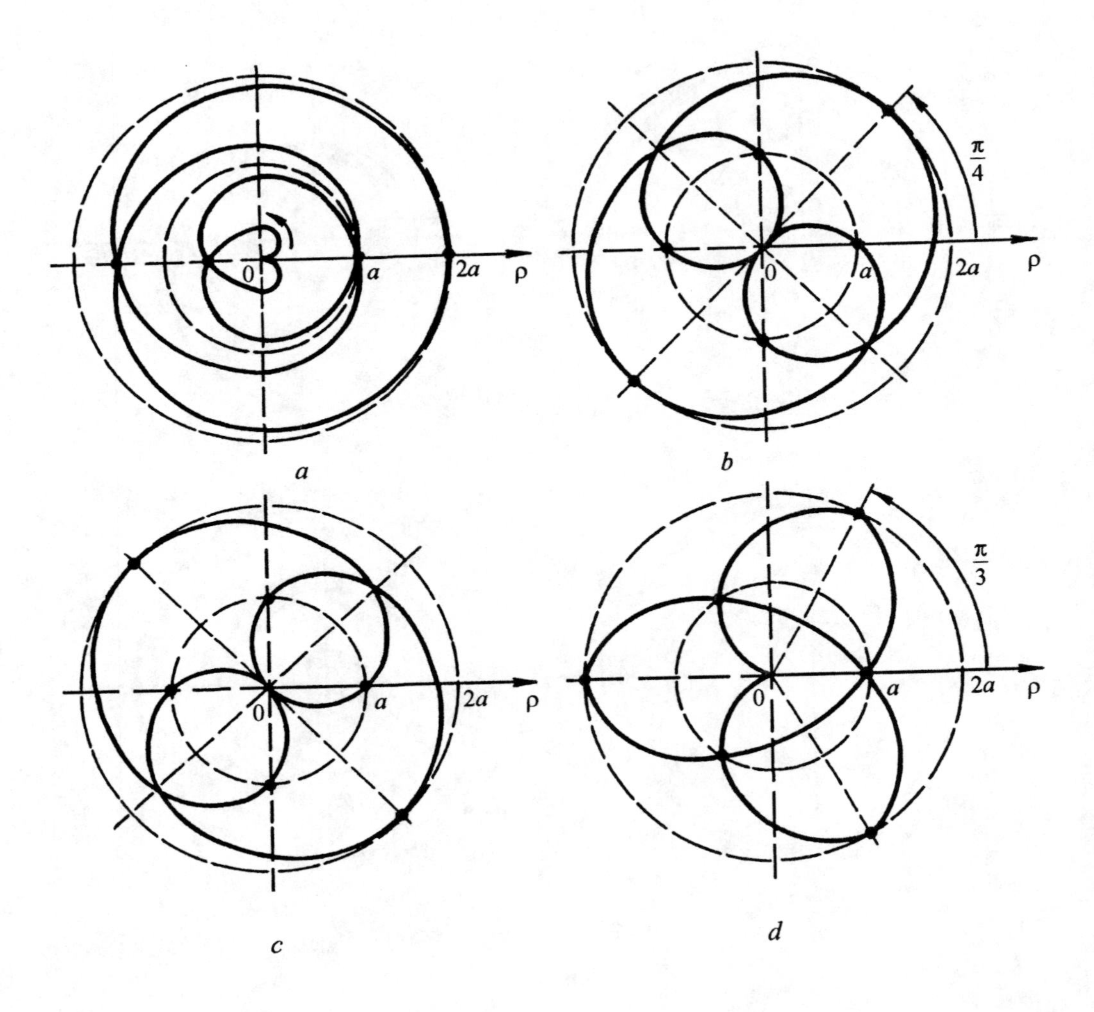

Figure 7.70. The graphs of functions: (*a*) $\rho = a(1 - \cos\frac{\varphi}{4})$;

(*b*) $\rho = a(1 - \sin\frac{2}{3}\varphi)$; (*c*) $\rho = a(1 + \sin\frac{2}{3}\varphi)$; (*d*) $\rho = a(1 - \sin\frac{3}{2}\varphi)$.

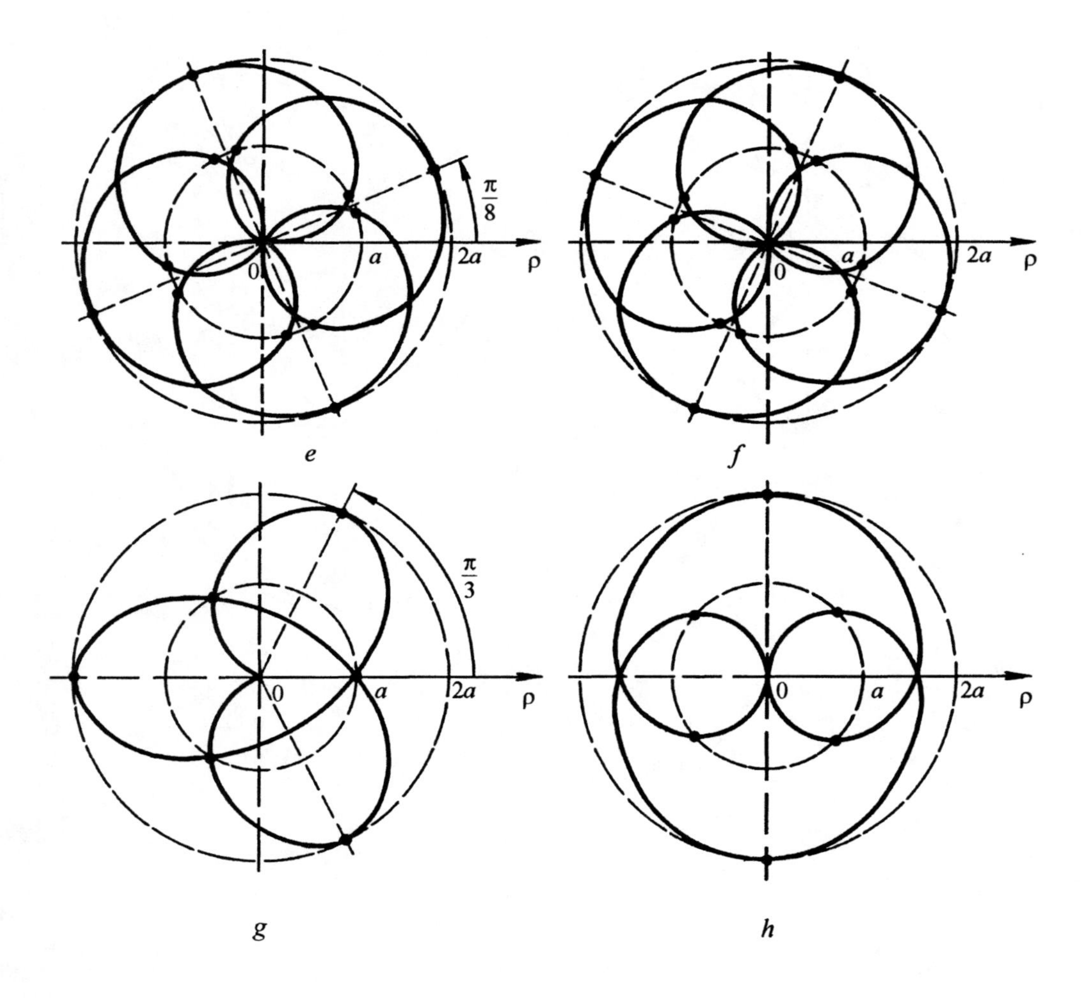

Figure 7.70. The graphs of functions: (*e*) $\rho = a(1 - \sin\frac{4}{3}\varphi)$;

(*f*) $\rho = a(1 + \sin\frac{4}{3}\varphi)$; (*g*) $\rho = a(1 + \sin\frac{3}{2}\varphi)$; (*h*) $\rho = a(1 - \cos\frac{2}{3}\varphi)$.

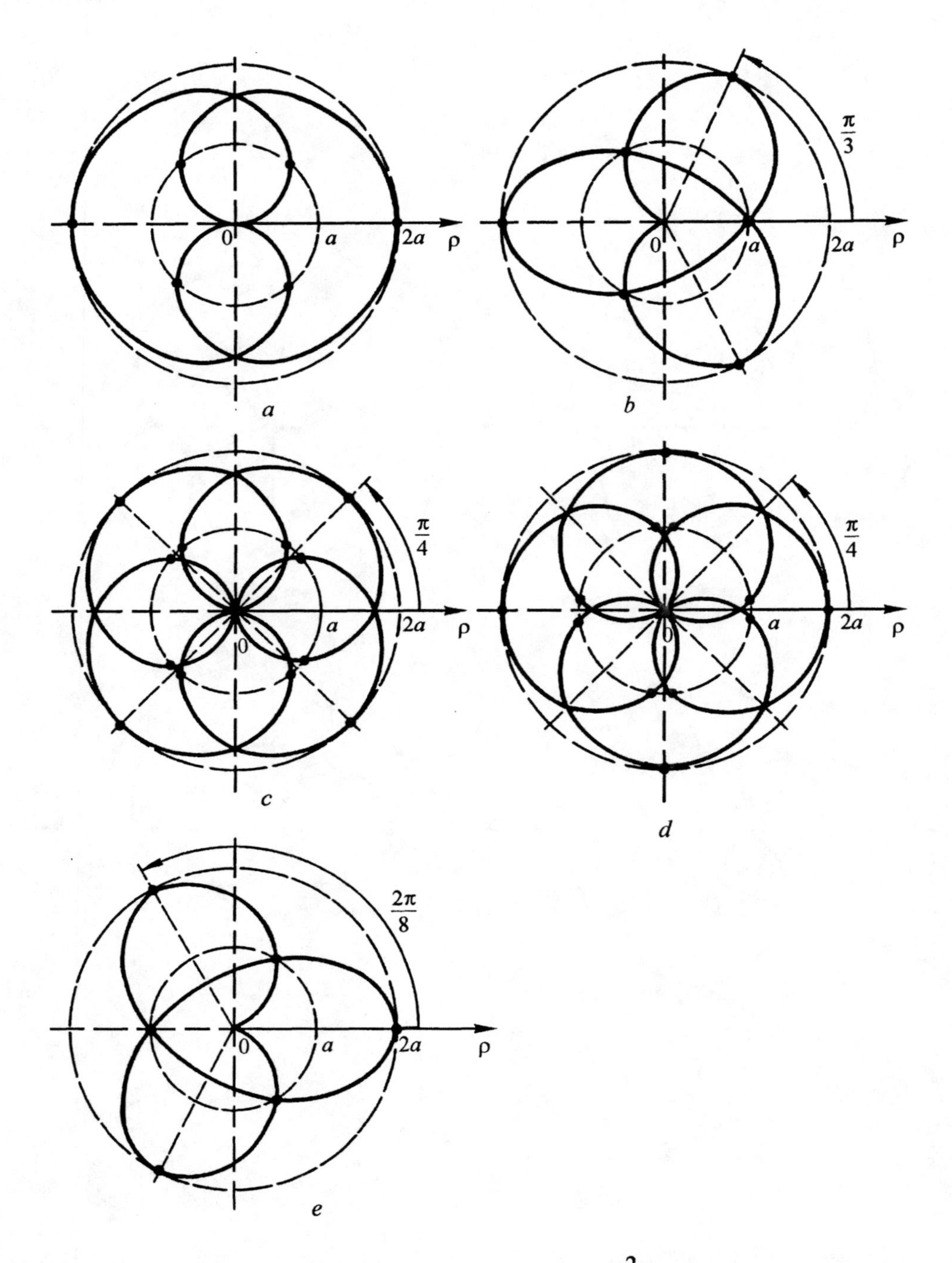

Figure 7.71. The graphs of functions: (*a*) $\rho = a(1 + \cos \frac{2}{3} \varphi)$;

(*b*) $\rho = a(1 - \cos \frac{3}{2} \varphi)$; (*c*) $\rho = a(1 - \cos \frac{4}{3} \varphi)$; (*d*) $\rho = a(1 + \cos \frac{4}{3} \varphi)$;

(*e*) $\rho = a(1 + \cos \frac{3}{2} \varphi)$.

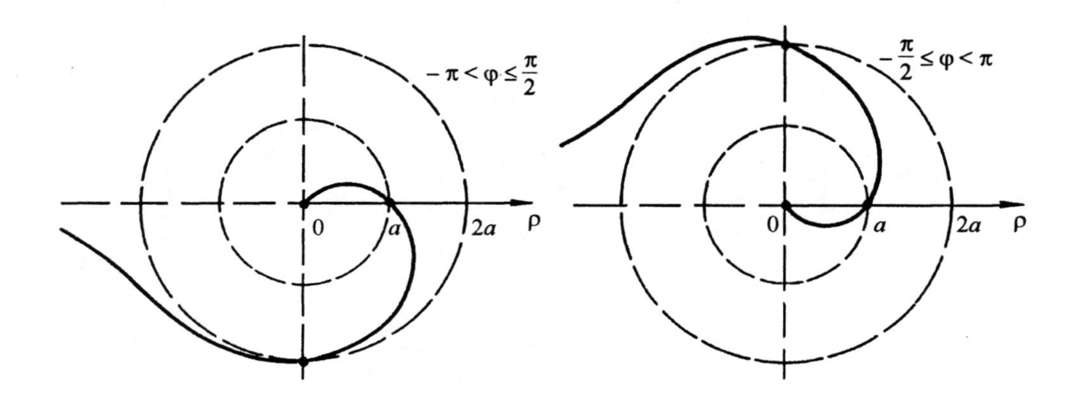

Figure 7.72. The graph of the function $\rho = a(1 - \tan\frac{\varphi}{2})$.

Figure 7.73. The graph of the function $\rho = a(1 + \tan\frac{\varphi}{2})$.

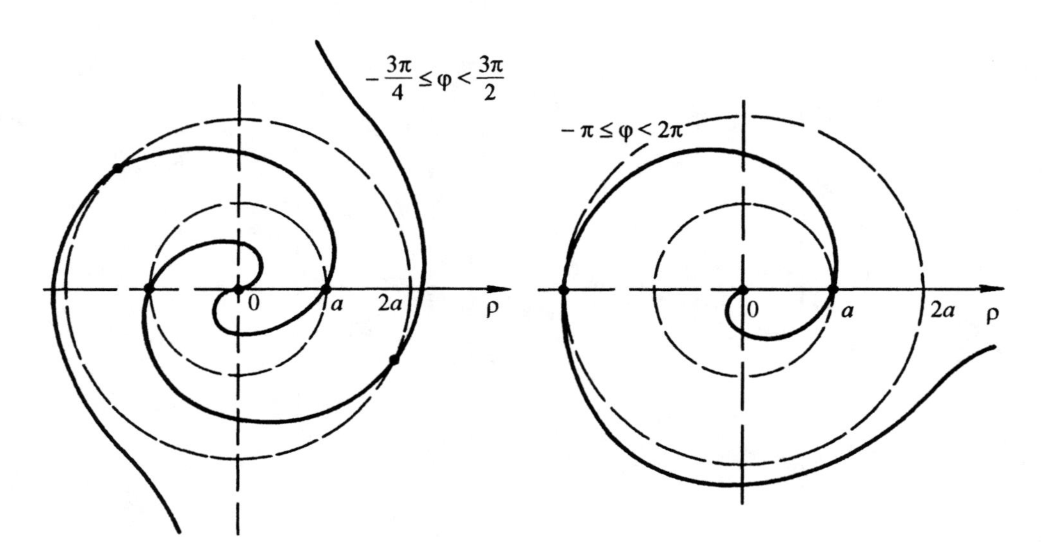

Figure 7.74. The graph of the function $\rho = a(1 + \tan\frac{\varphi}{3})$.

Figure 7.75. The graph of the function $\rho = a(1 + \tan\frac{\varphi}{4})$.

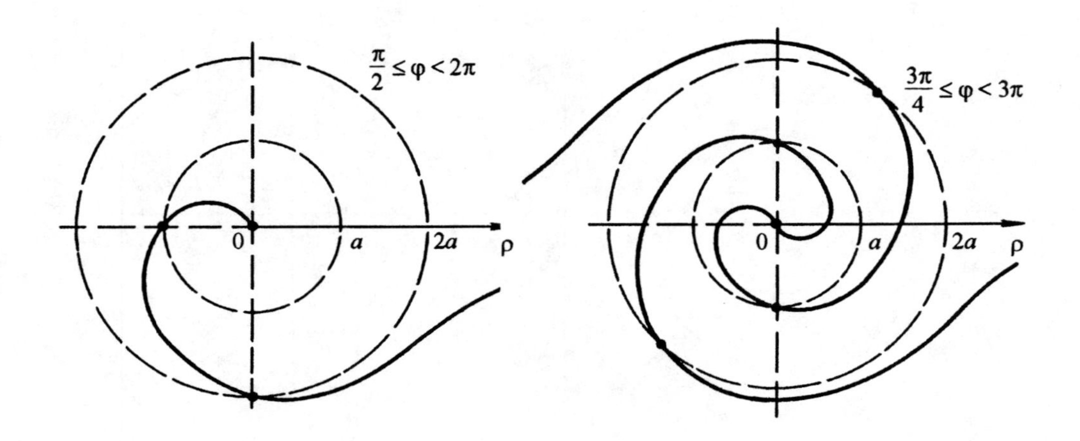

Figure 7.76. The graph of the function $\rho = a(1 - \cot\frac{\varphi}{2})$.

Figure 7.77. The graph of the function $\rho = a(1 - \cot\frac{\varphi}{3})$.

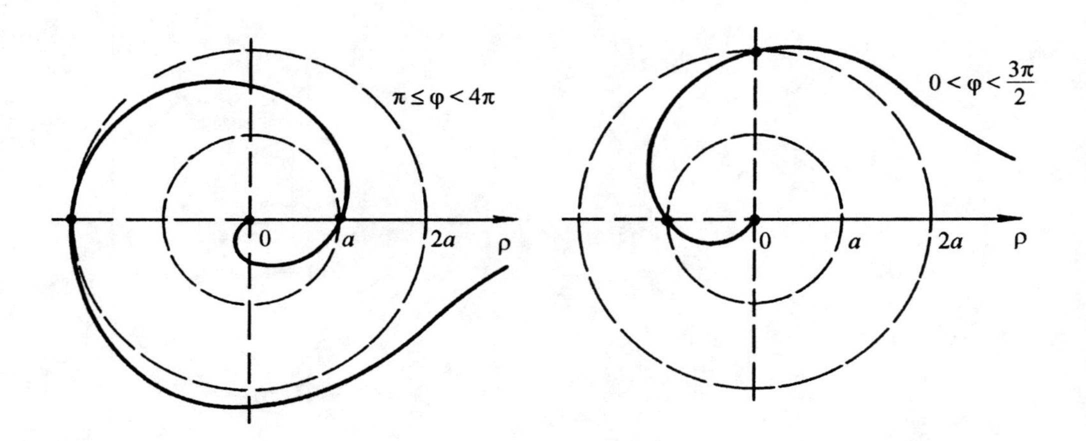

Figure 7.78. The graph of the function $\rho = a(1 - \cot\frac{\varphi}{4})$.

Figure 7.79. The graph of the function $\rho = a(1 + \cot\frac{\varphi}{2})$.

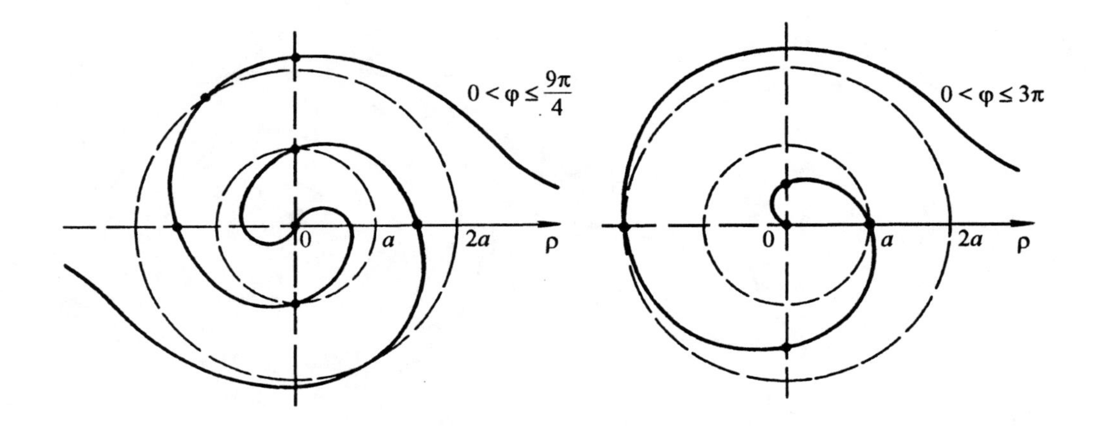

Figure 7.80. The graph of the function $\rho = a(1 + \cot \frac{\varphi}{3})$.

Figure 7.81. The graph of the function $\rho = a(1 + \cot \frac{\varphi}{4})$.

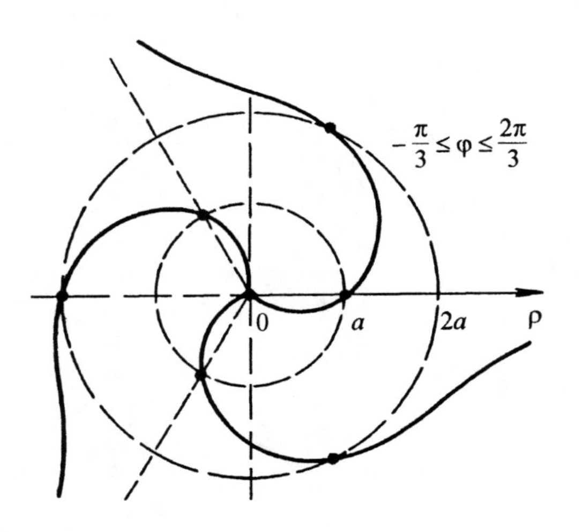

Figure 7.82. The graph of the function $\rho = a(1 + \tan \frac{3}{4} \varphi)$.

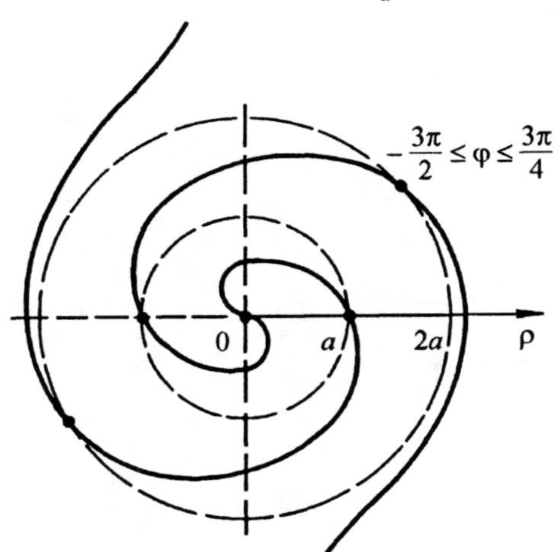

Figure 7.83. The graph of the function $\rho = a(1 - \tan \frac{\varphi}{3})$.

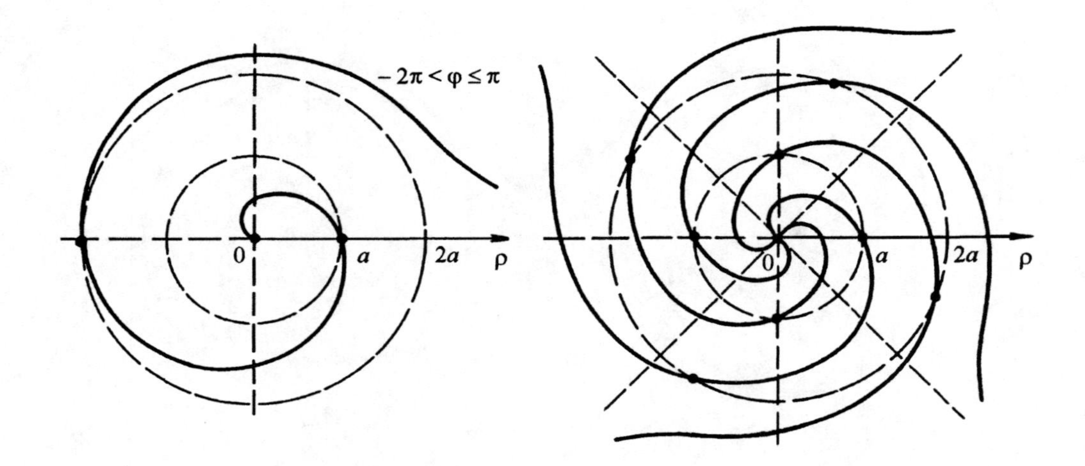

Figure 7.84. The graph of function $\rho = a(1 - \tan\frac{\varphi}{4})$.

Figure 7.85. The graph of the function $\rho = a(1 - \tan\frac{2}{5}\varphi)$.

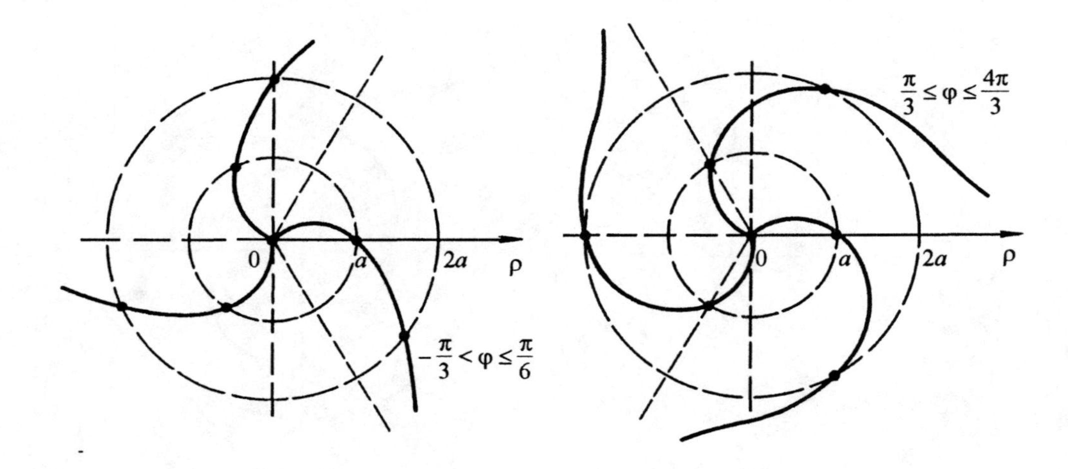

Figure 7.86. The graph of the function $\rho = a(1 - \tan\frac{3}{2}\varphi)$.

Figure 7.87. The graph of the function $\rho = a(1 + \cot\frac{3}{4}\varphi)$.

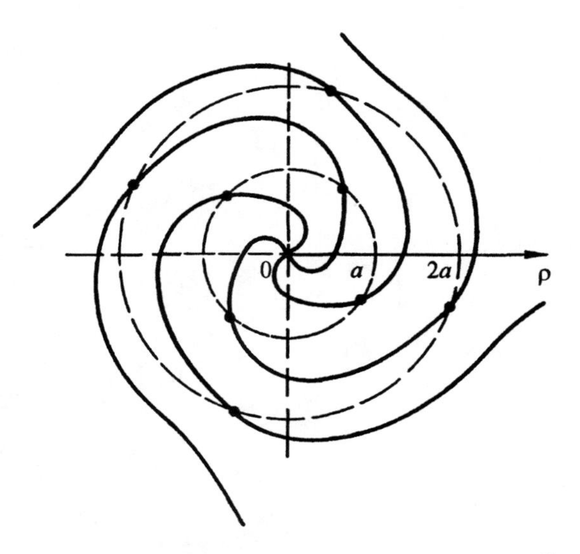

Figure 7.88. The graph of the function $\rho = a(1 - \cot \frac{2}{5} \varphi)$.

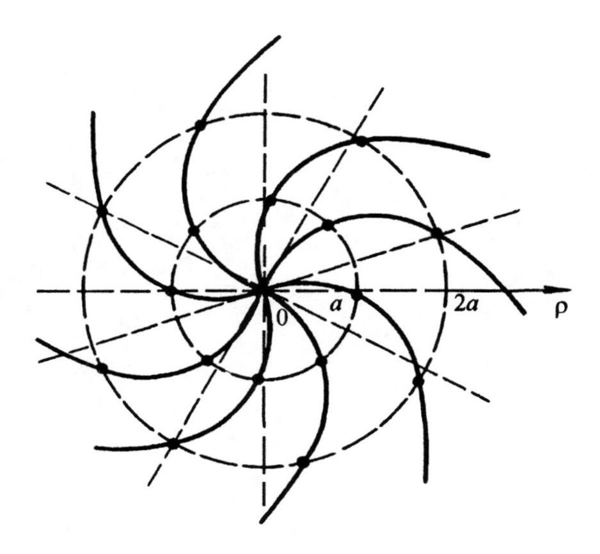

Figure 7.89. The graph of the function $\rho = a(1 + \cot \frac{4}{3} \varphi)$.

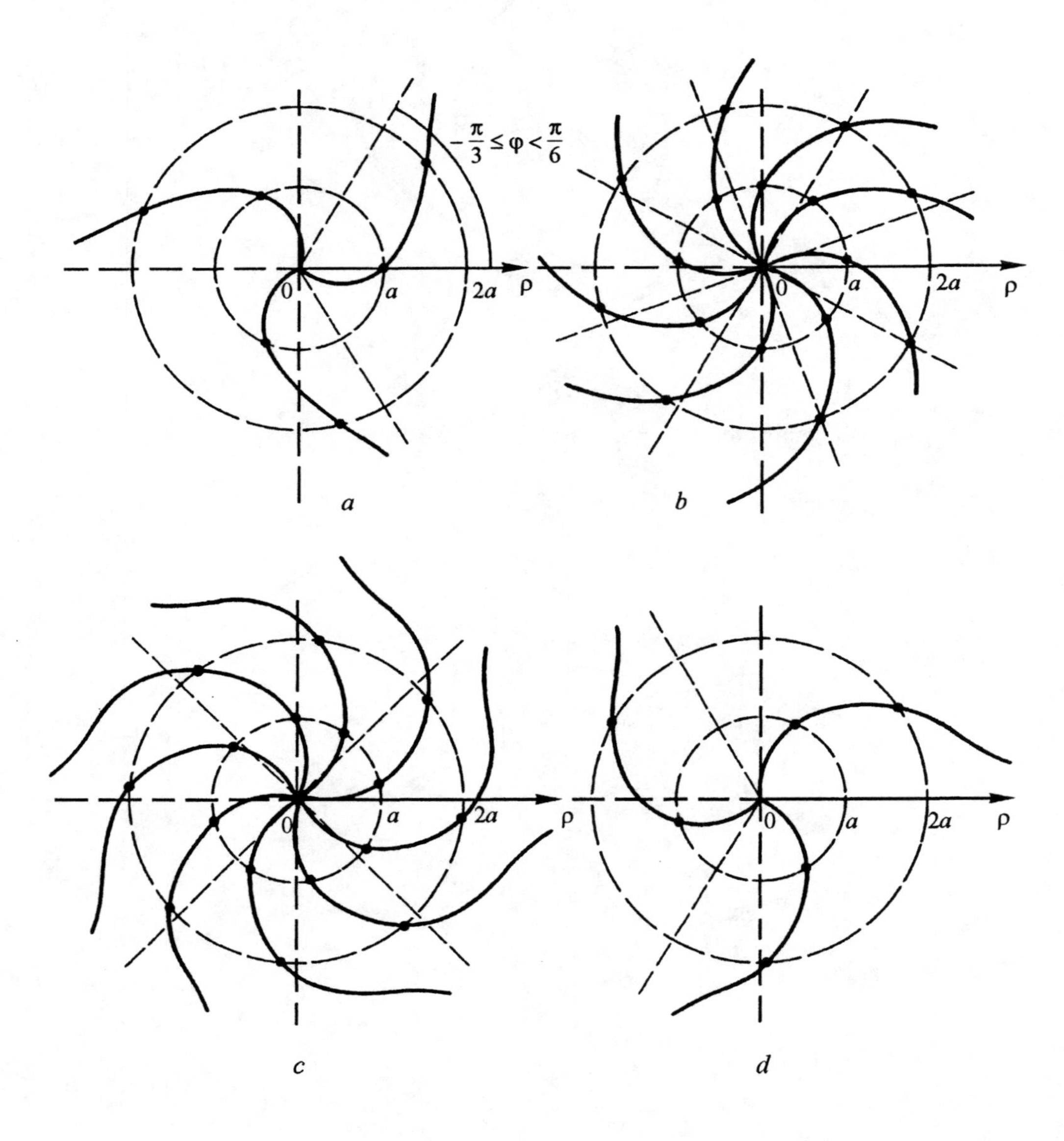

Figure 7.90. The graph of functions: (*a*) $\rho = a(1 + \tan\frac{3}{2}\varphi)$;
(*b*) $\rho = a(1 - \tan\frac{4}{3}\varphi)$; (*c*) $\rho = a(1 - \cot\frac{4}{3}\varphi)$; (*d*) $\rho = a(1 + \cot\frac{3}{2}\varphi)$.

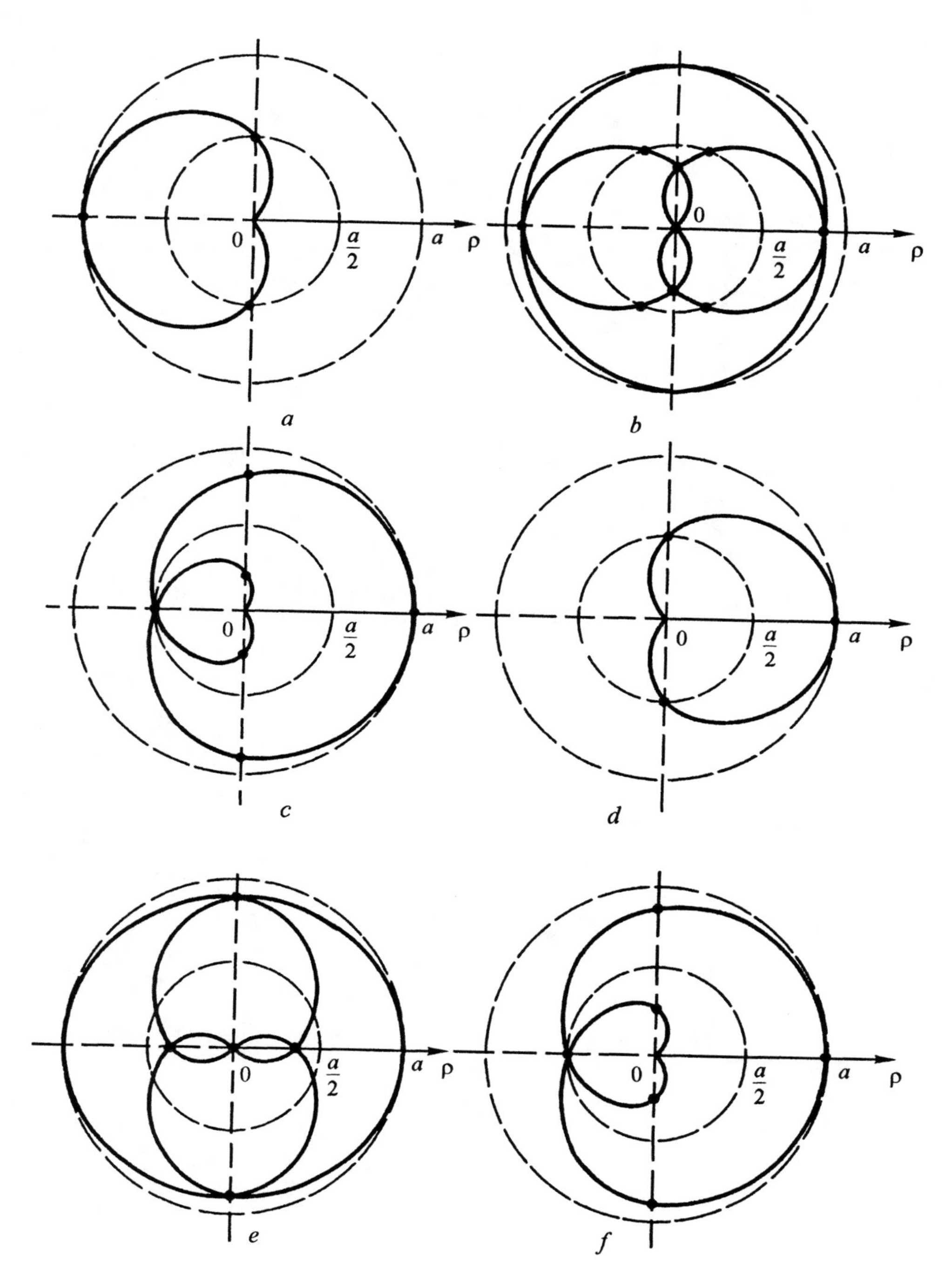

Figure 7.91. The graphs of functions: (*a*) $\rho = a\sin^2\frac{\varphi}{2}$; (*b*) $\rho = a\sin^2\frac{\varphi}{3}$; (*c*) $\rho = a\sin^2\frac{\varphi}{4}$; (*d*) $\rho = a\cos^2\frac{\varphi}{2}$; (*e*) $\rho = a\cos^2\frac{\varphi}{3}$; (*f*) $\rho = a\cos^2\frac{\varphi}{4}$.

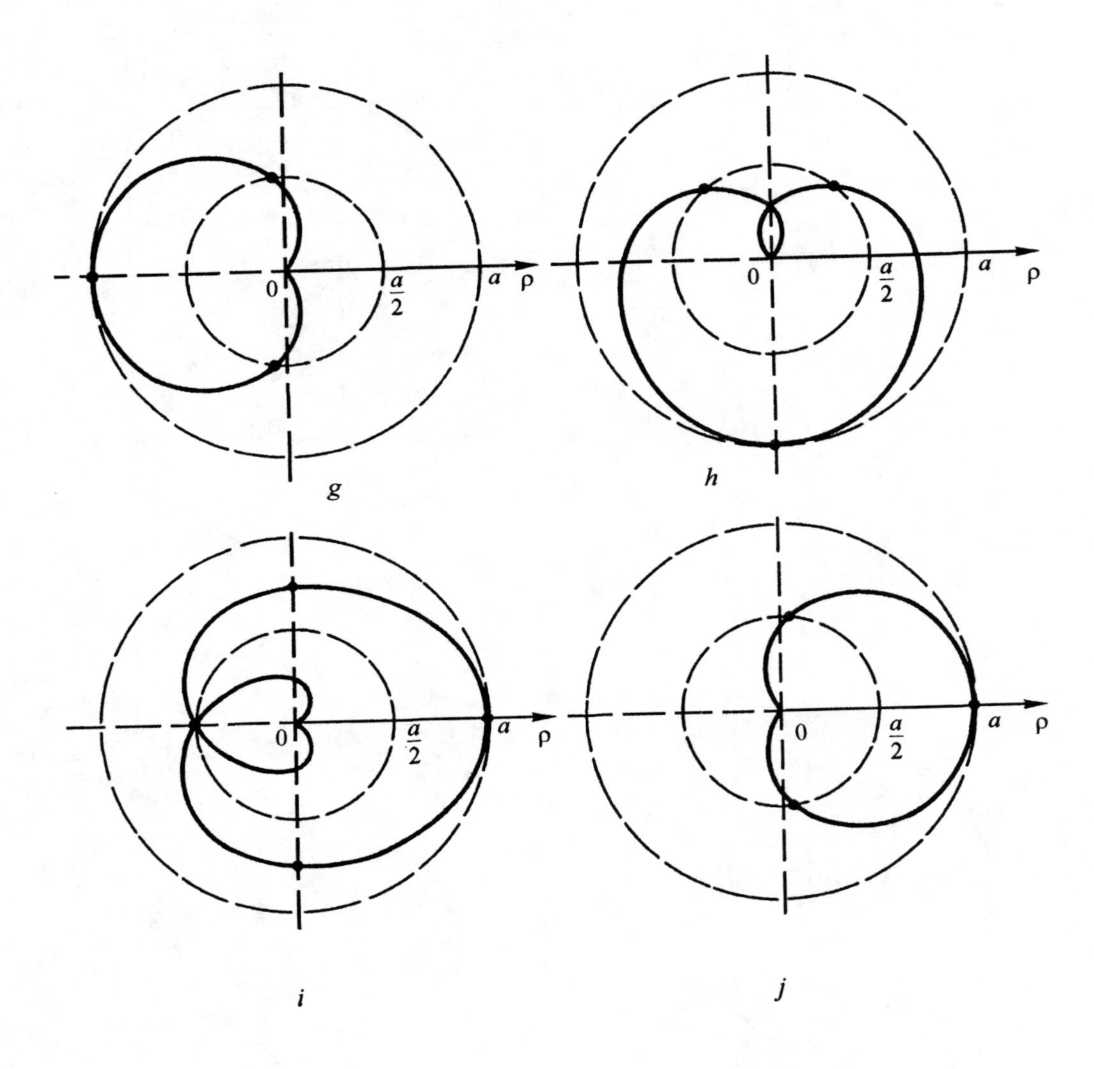

Figure 7.91. The graphs of functions: (*g*) $\rho = a\sin^3\frac{\varphi}{2}$; (*h*) $\rho = a\sin^3\frac{\varphi}{3}$; (*i*) $\rho = a\sin^3\frac{\varphi}{4}$; (*j*) $\rho = a\cos^3\frac{\varphi}{2}$.

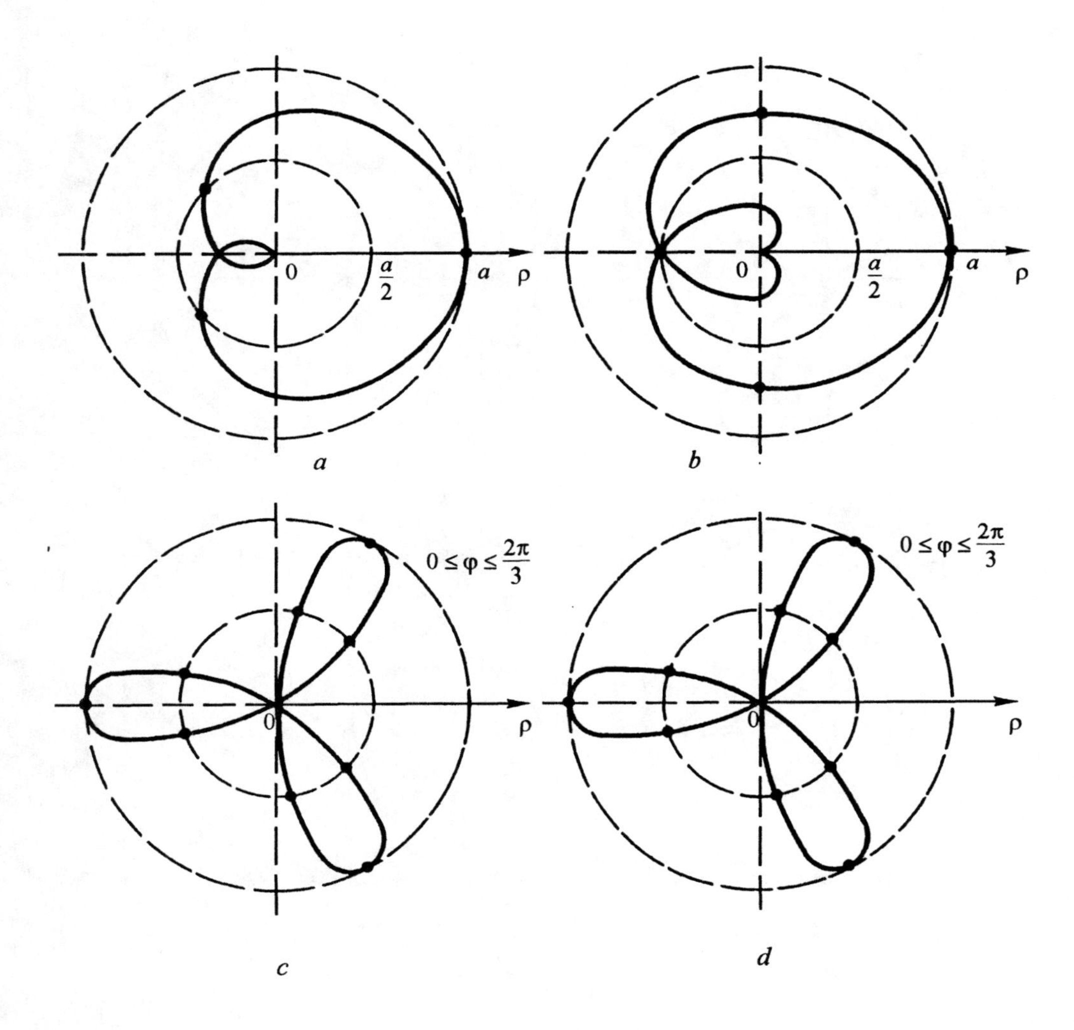

Figure 7.92. The graphs of functions: (*a*) $\rho = a\cos^3\frac{\varphi}{3}$; (*b*) $\rho = a\cos^3\frac{\varphi}{4}$;

(*c*) $\rho = a\sin^2\frac{3}{2}\varphi$; (*d*) $\rho = a\sin^3\frac{3}{2}\varphi$.

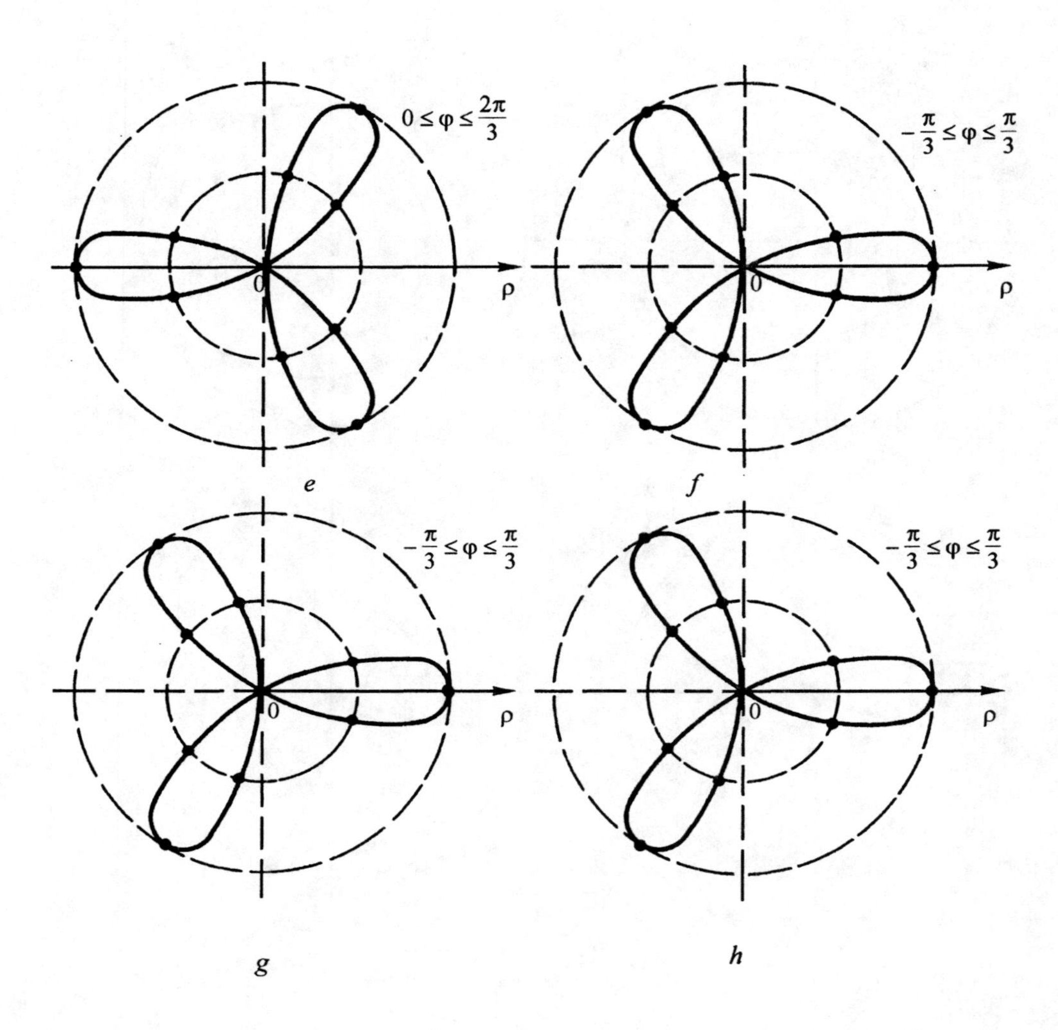

Figure 7.92. The graphs of functions: (*e*) $\rho = a \sin^4 \frac{3}{2} \varphi$; (*f*) $\rho = a \cos^2 \frac{3}{2} \varphi$; (*g*) $\rho = a \cos^3 \frac{3}{2} \varphi$; (*h*) $\rho = a \cos^4 \frac{3}{2} \varphi$.

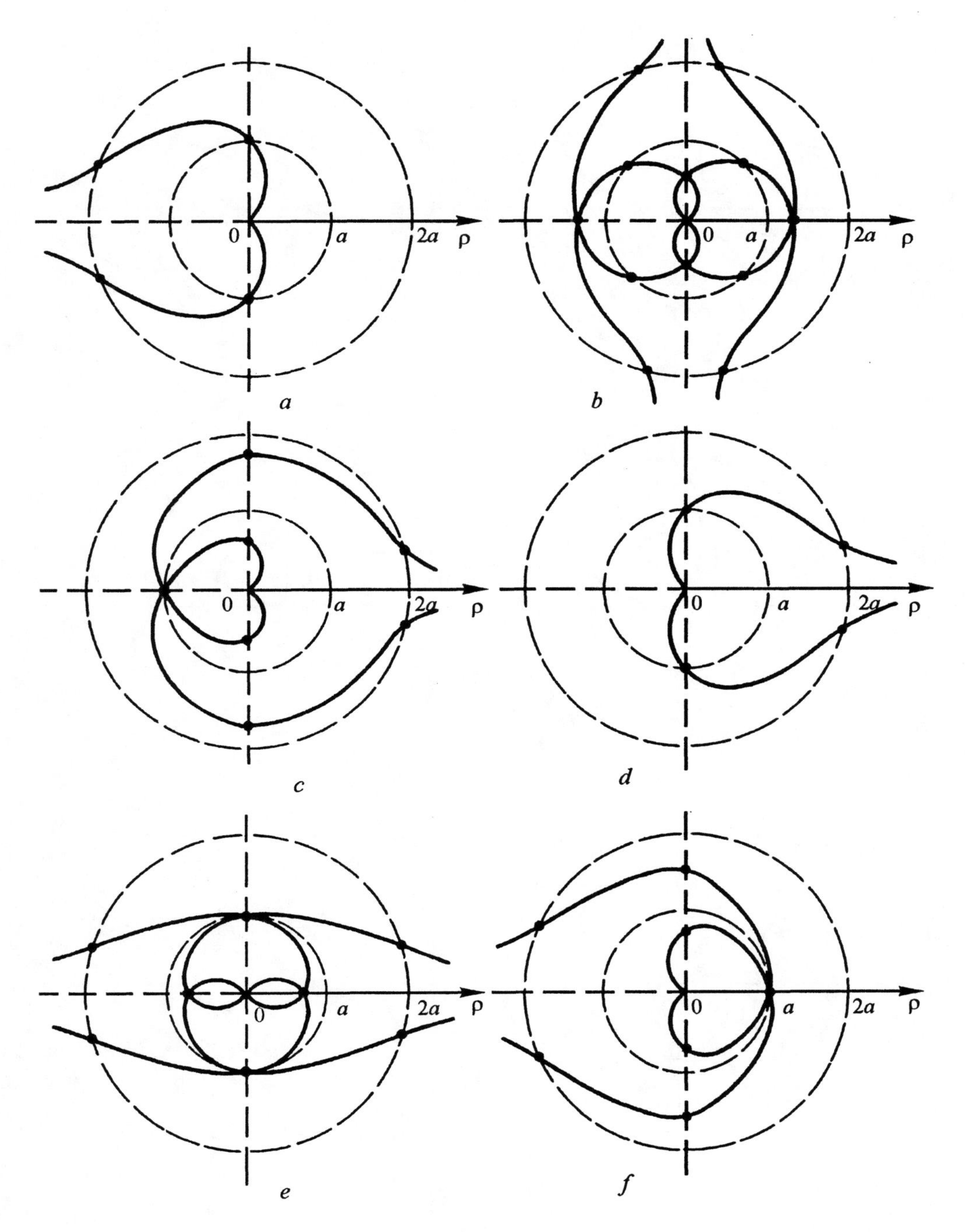

Figure 7.93. The graphs of functions: (*a*) $\rho = a\tan^2\frac{\varphi}{2}$; (*b*) $\rho = a\tan^2\frac{\varphi}{3}$; (*c*) $\rho = a\tan^2\frac{\varphi}{4}$; (*d*) $\rho = a\cot^2\frac{\varphi}{2}$; (*e*) $\rho = a\cot^2\frac{\varphi}{3}$; (*f*) $\rho = a\cot^2\frac{\varphi}{4}$.

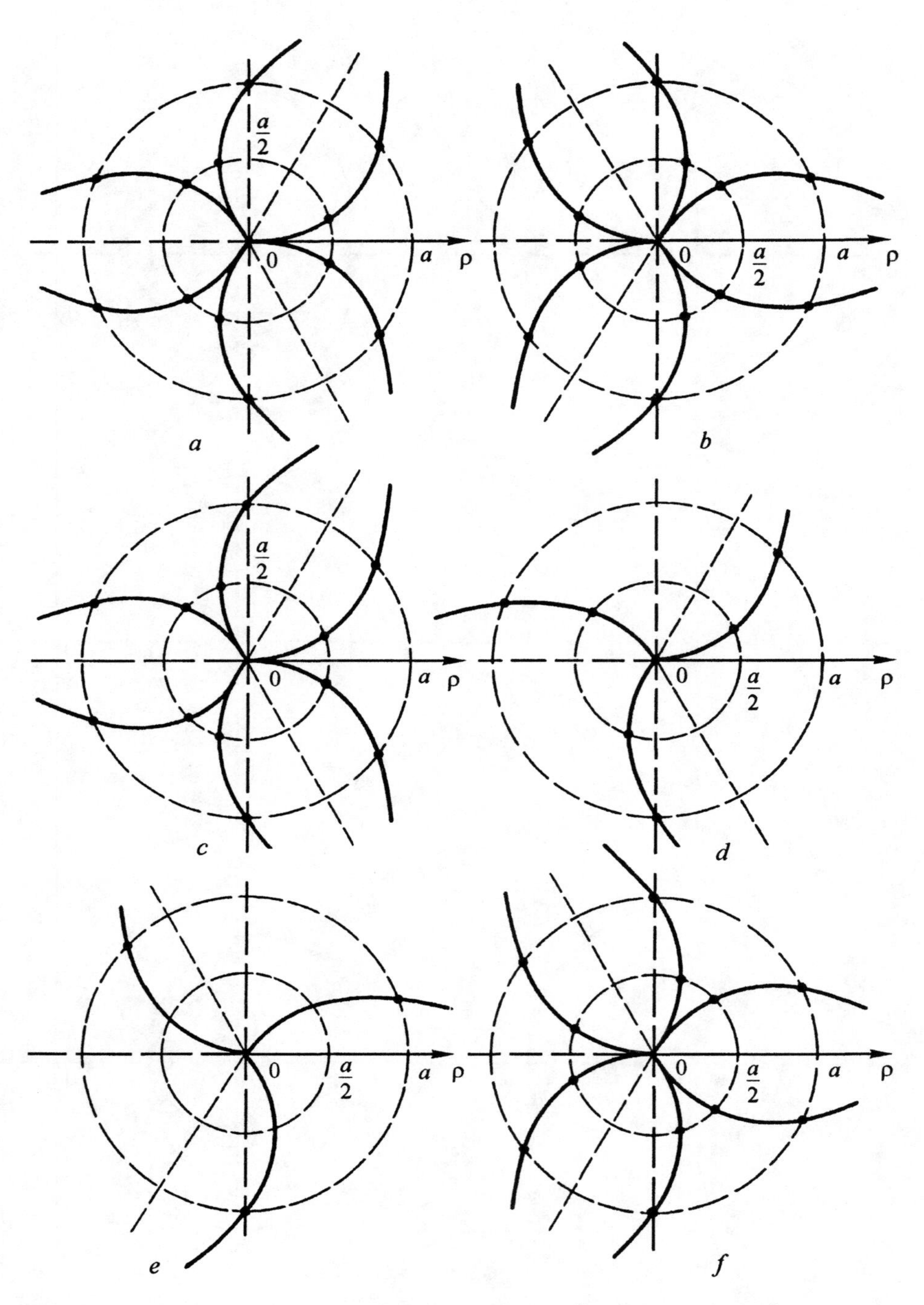

Figure 7.94. The graphs of functions: (*a*) $\rho = a\tan^4\frac{3}{2}\varphi$; (*b*) $\rho = a\cot^2\frac{3}{2}\varphi$; (*c*) $\rho = a\tan^2\frac{3}{2}\varphi$; (*d*) $\rho = a\tan^3\frac{3}{2}\varphi$; (*e*) $\rho = a\cot^3\frac{3}{2}\varphi$; (*f*) $\rho = a\cot^4\frac{3}{2}\varphi$.

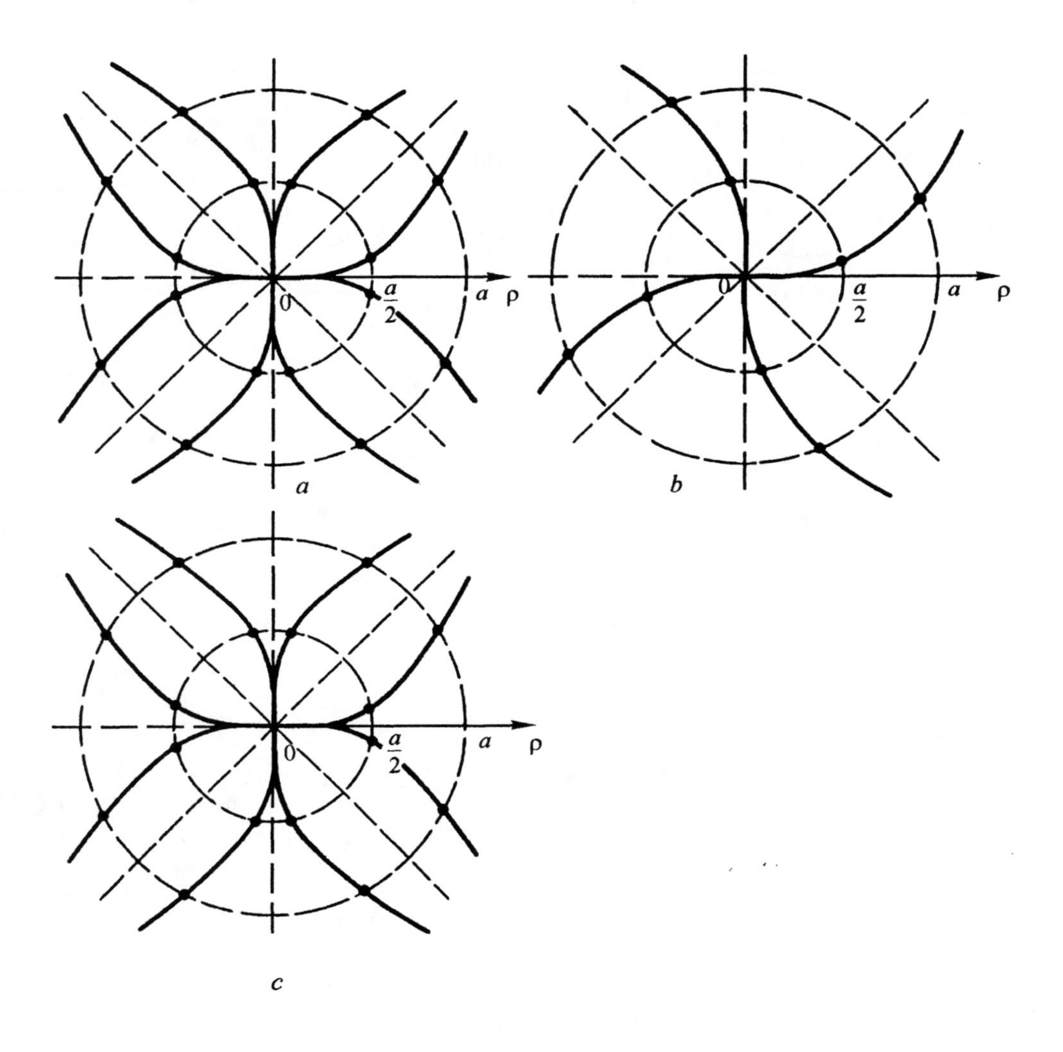

Figure 7.95. The graphs of functions: (*a*) $\rho = a\tan^2\frac{2}{3}\varphi$; (*b*) $\rho = a\tan^3\frac{2}{3}\varphi$; (*c*) $\rho = a\tan^4\frac{2}{3}\varphi$.

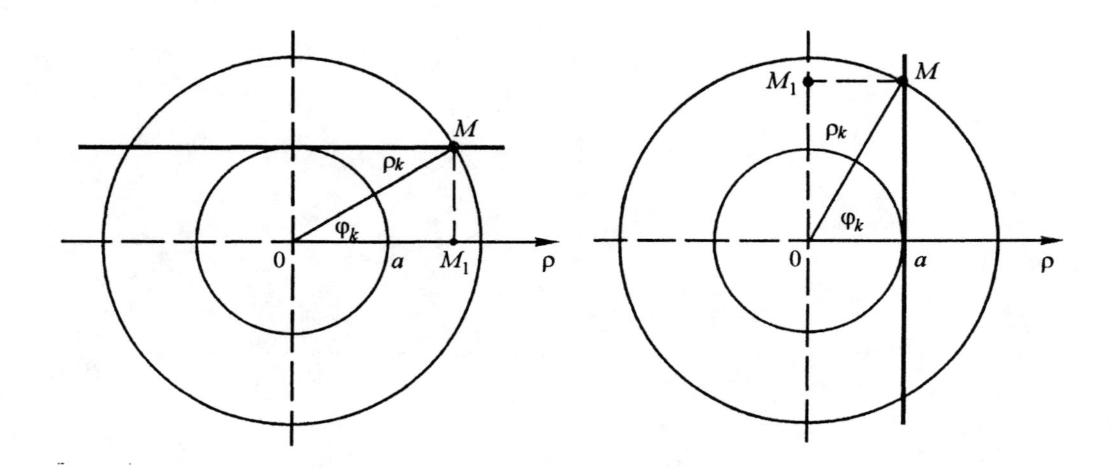

Figure 7.96. The graph of the function $\rho = \dfrac{a}{\sin \varphi}$.

Figure 7.97. The graph of the function $\rho = \dfrac{a}{\cos \varphi}$.

7.3. Transformations of graphs in polar coordinates

Plotting the graph of a function $\rho = mf(\varphi + \alpha) + b$ in polar coordinates using the graph of the function $\rho = f(\varphi)$ can be done similarly as in Cartesian coordinates. The construction is carried out using simple geometric transformations according to the following properties of the function $\rho = f(\varphi)$:

1. The graphs of the functions $\rho = -f(\varphi)$ and $\rho = f(\varphi)$ are symmetric with respect to the pole.
2. The graph of the function $\rho = f(-\varphi)$ is symmetric to the graph of the function $\rho = f(\varphi)$ with respect to the polar axis.
3. The graph of the function $\rho = mf(\varphi)$, $m > 0$, is obtained by stretching or expanding the graph of the function $\rho = f(\varphi)$ along the polar axis.
4. The graph of the function $\rho = f(\varphi + \alpha)$ is a graph obtained by rotating the graph of the function $\rho = f(\varphi)$ by the angle α.
5. The graph of the function $\rho = f(\varphi) + b$ is a graph obtained from the graph of the function $\rho = f(\varphi)$ by a parallel translation by b along the polar axis.

Example 7.3.1. Plot the graph of the function $\rho = \dfrac{a}{\varphi - \pi/4}$.

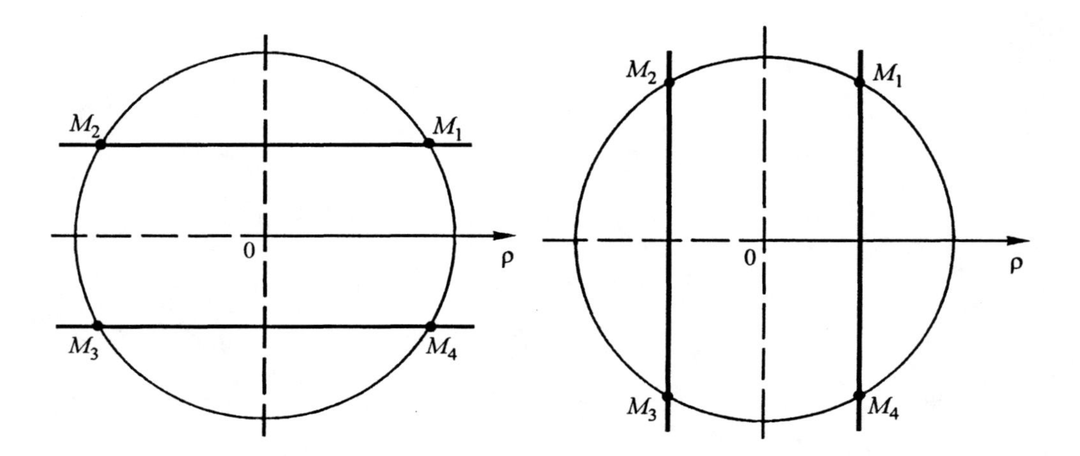

Figure 7.98. The graph of the function $\rho = \dfrac{a}{\sin^{2m} n\varphi}$.

Figure 7.99. The graph of the function $\rho = \dfrac{a}{\cos^{2m} n\varphi}$.

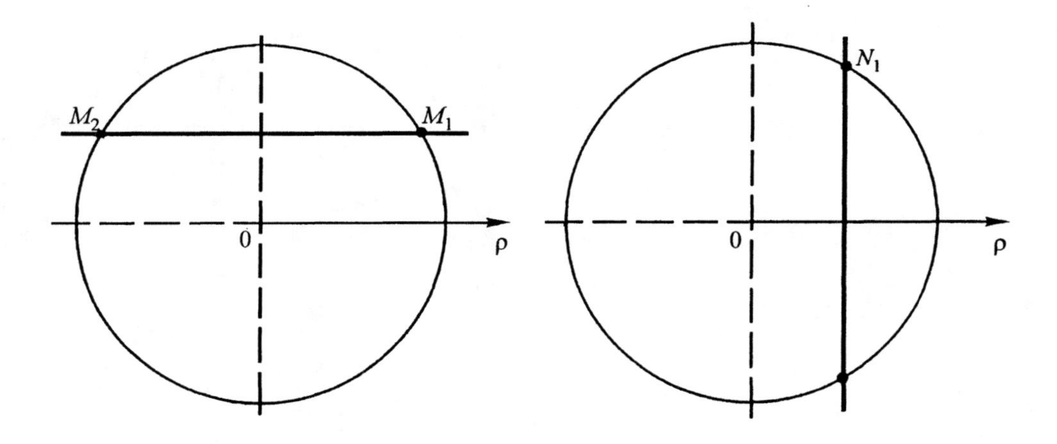

Figure 7.100. The graph of the function $\sin n\varphi = \sqrt[2k+1]{\dfrac{a}{\rho}}$.

Figure 7.101. The graph of the function $\cos n\varphi = \sqrt[2k+1]{\dfrac{a}{\rho}}$.

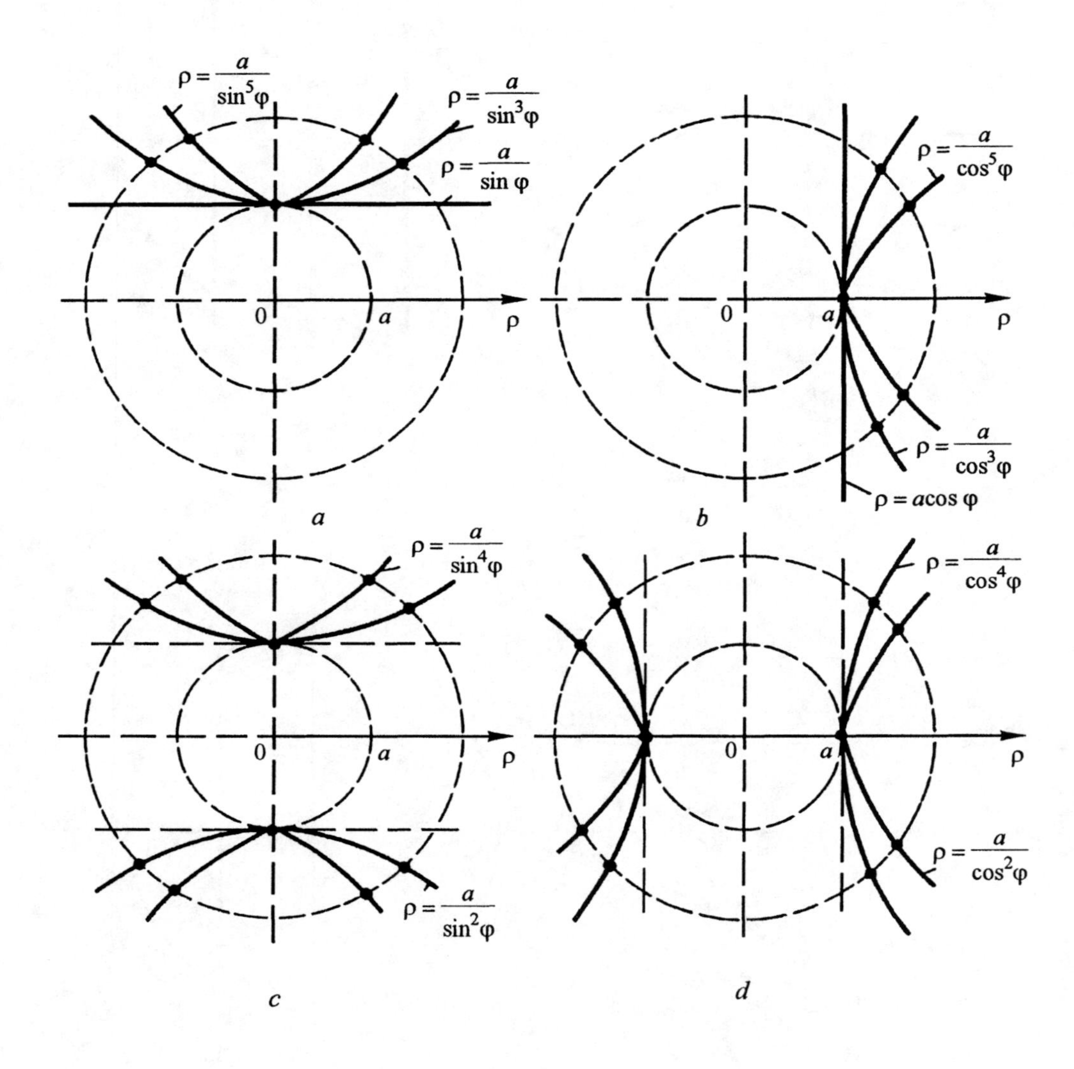

Figure 7.102. The graphs of functions: (*a*) $\rho = \frac{a}{\sin^{2k+1}\varphi}$; (*b*) $\rho = \frac{a}{\cos^{2k+1}\varphi}$; (*c*) $\rho = \frac{a}{\sin^{2k}\varphi}$; (*d*) $\rho = \frac{a}{\cos^{2k}\varphi}$.

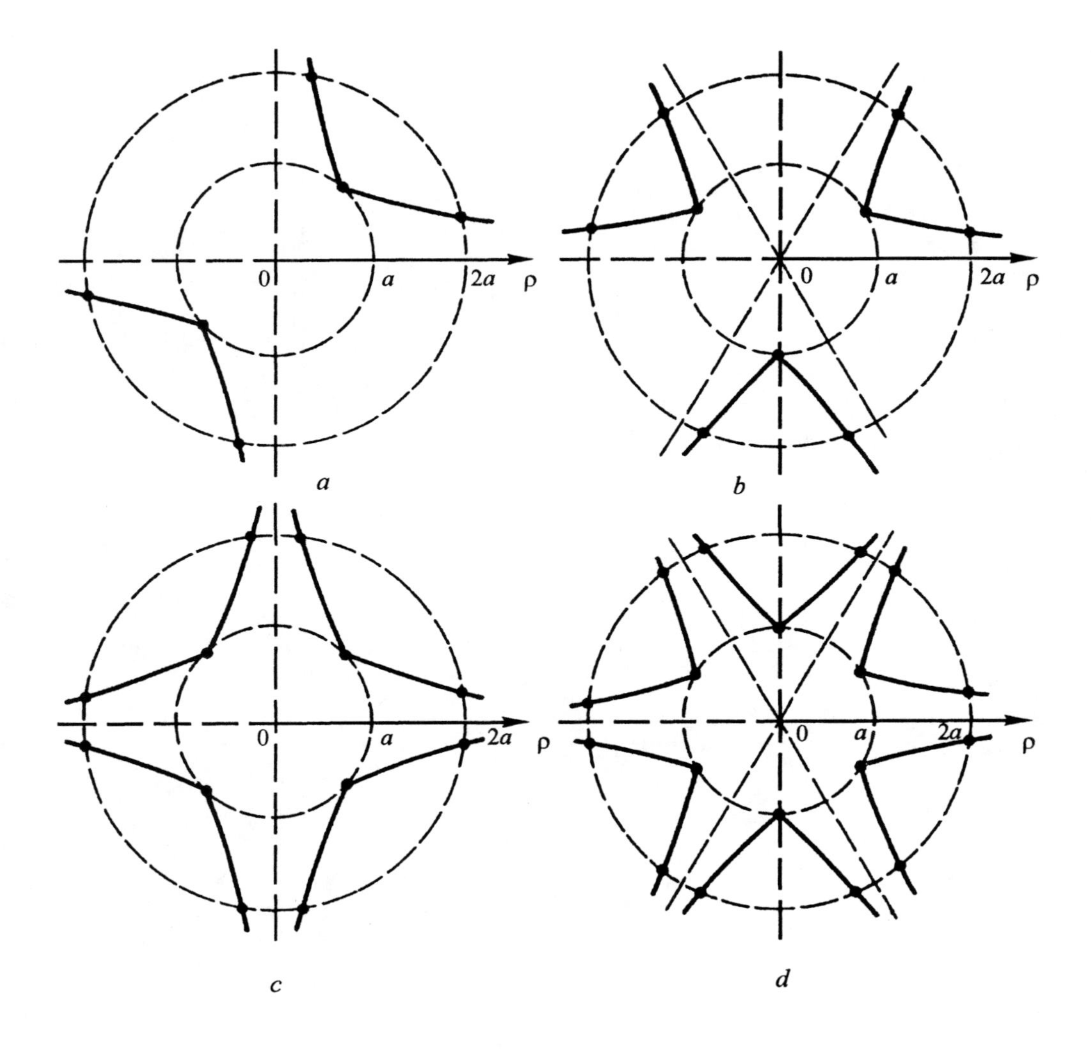

Figure 7.103. The graphs of functions: (*a*) $\rho = \dfrac{a}{\sin 2\varphi}$; (*b*) $\rho = \dfrac{a}{\sin 3\varphi}$; (*c*) $\rho = \dfrac{a}{\sin^2 2\varphi}$; (*d*) $\rho = \dfrac{a}{\sin^2 3\varphi}$.

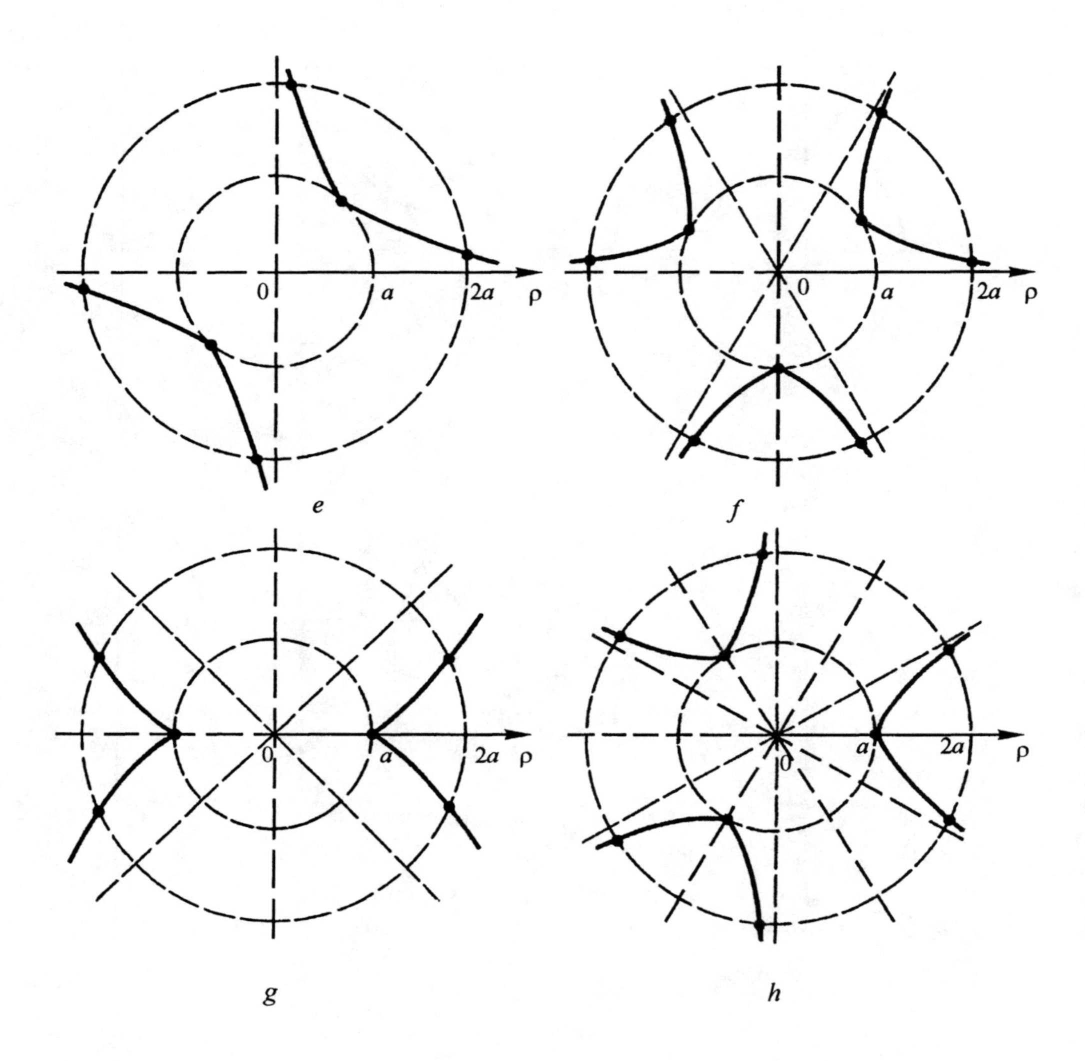

Figure 7.103. The graphs of functions: (*e*) $\rho = \dfrac{a}{\sin^3 2\varphi}$; (*f*) $\rho = \dfrac{a}{\sin^3 3\varphi}$;

(*g*) $\rho = \dfrac{a}{\cos 2\varphi}$; (*h*) $\rho = \dfrac{a}{\cos 3\varphi}$.

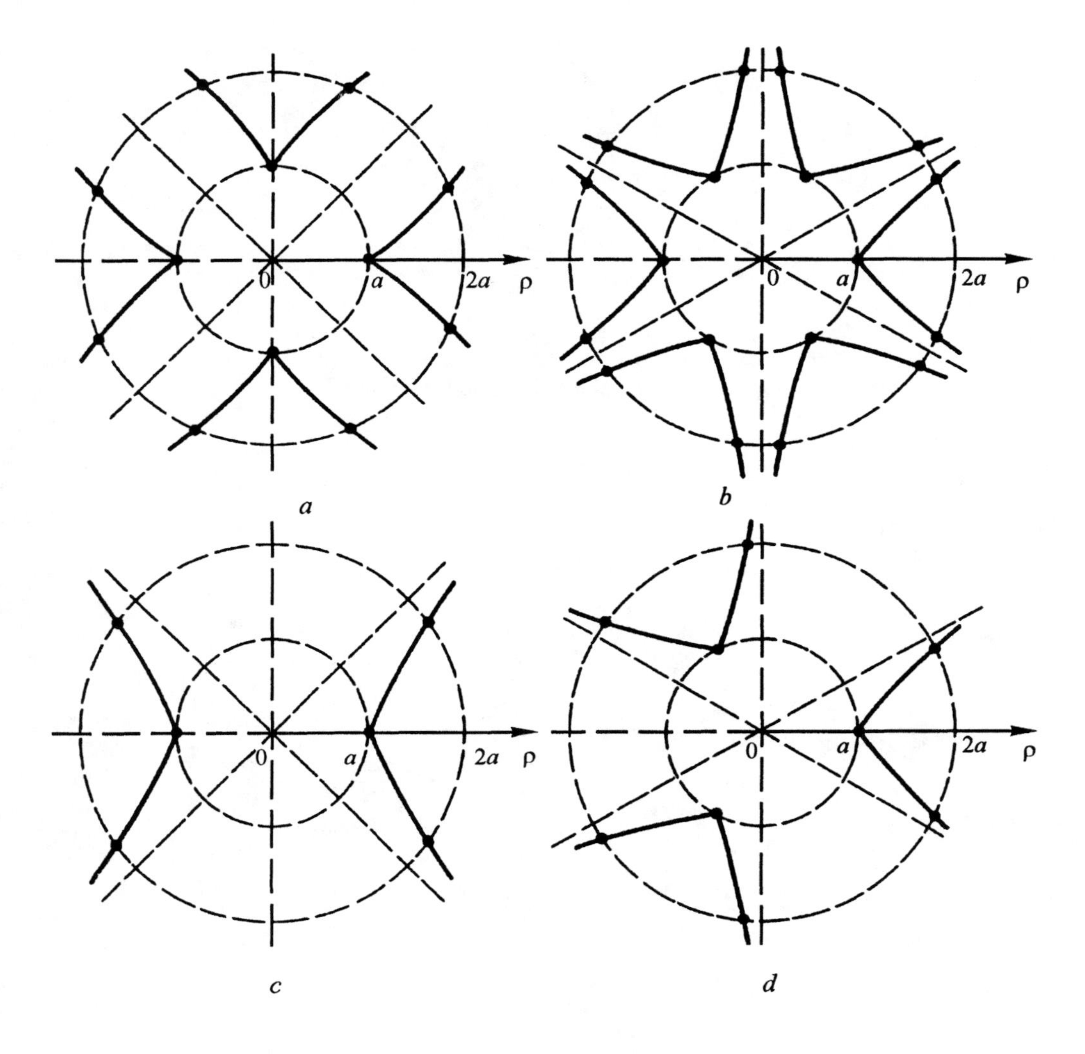

Figure 7.104. The graphs of functions: (*a*) $\rho = \dfrac{a}{\cos^2 2\varphi}$; (*b*) $\rho = \dfrac{a}{\cos^2 4\varphi}$;

(*c*) $\rho = \dfrac{a}{\cos^3 3\varphi}$; (*d*) $\rho = \dfrac{a}{\cos^3 3\varphi}$.

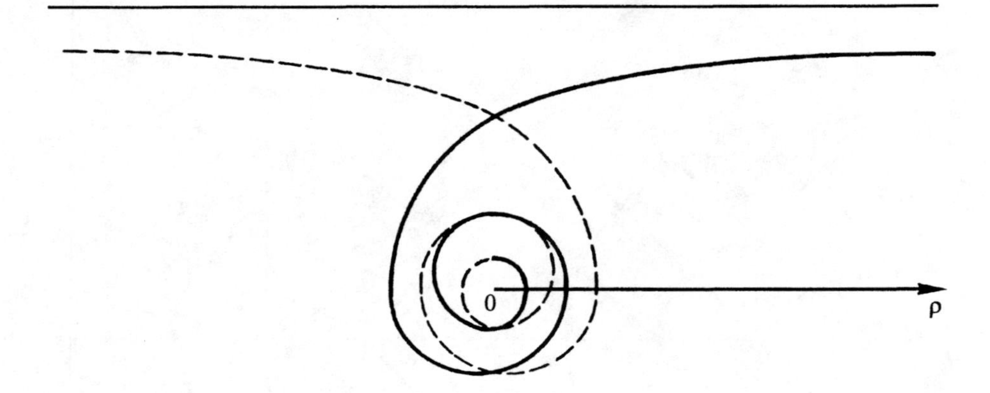

Figure 7.105. The graph of the function $\rho = \frac{a}{\varphi}$.

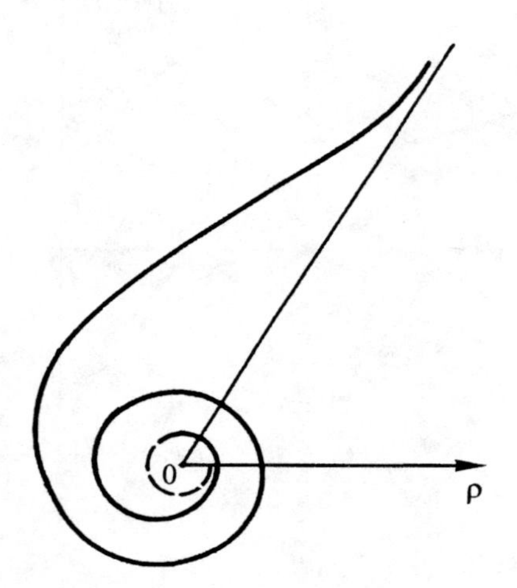

Figure 7.106. The graph of the function $\rho = \frac{a}{\varphi - \pi/4}$.

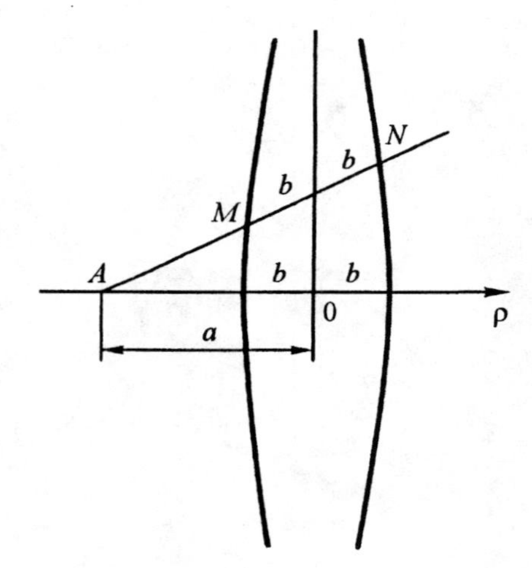

Figure 7.107. The curve $\rho = \frac{a}{\cos\varphi} + b, a = 2, b = 1$ (the conchoid).

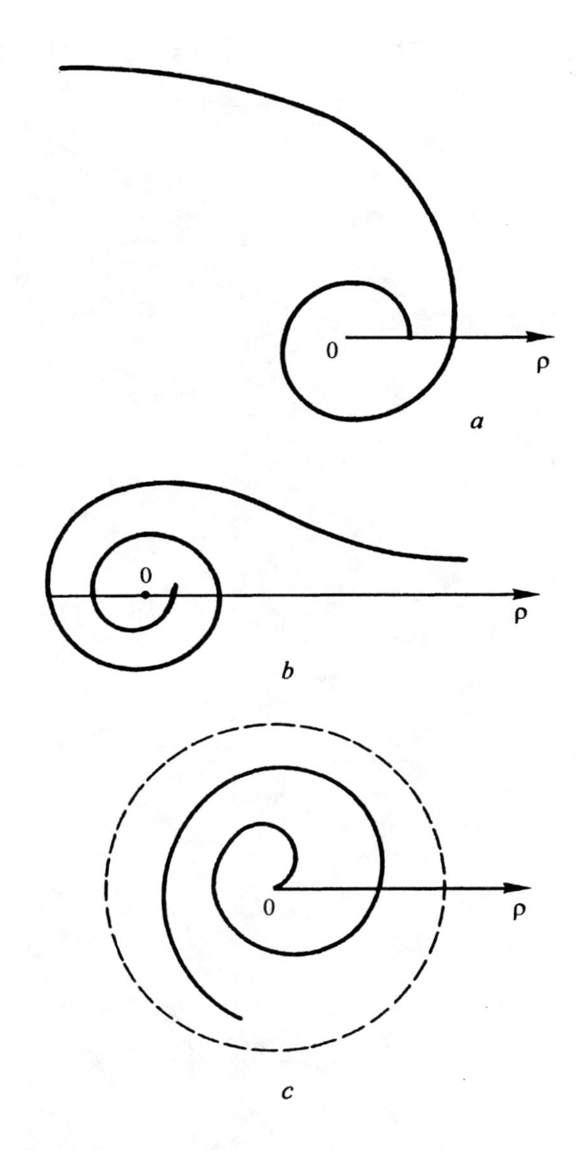

Figure 7.108. The graphs of functions: (*a*) $\rho = e^{a\varphi}$, $a > 0$ (logarithmic spiral); (*b*) $\rho = \sqrt{\dfrac{\pi}{\varphi}}$; (*c*) $\rho = \dfrac{\varphi}{\varphi + 1}$, $0 < \varphi < +\infty$.

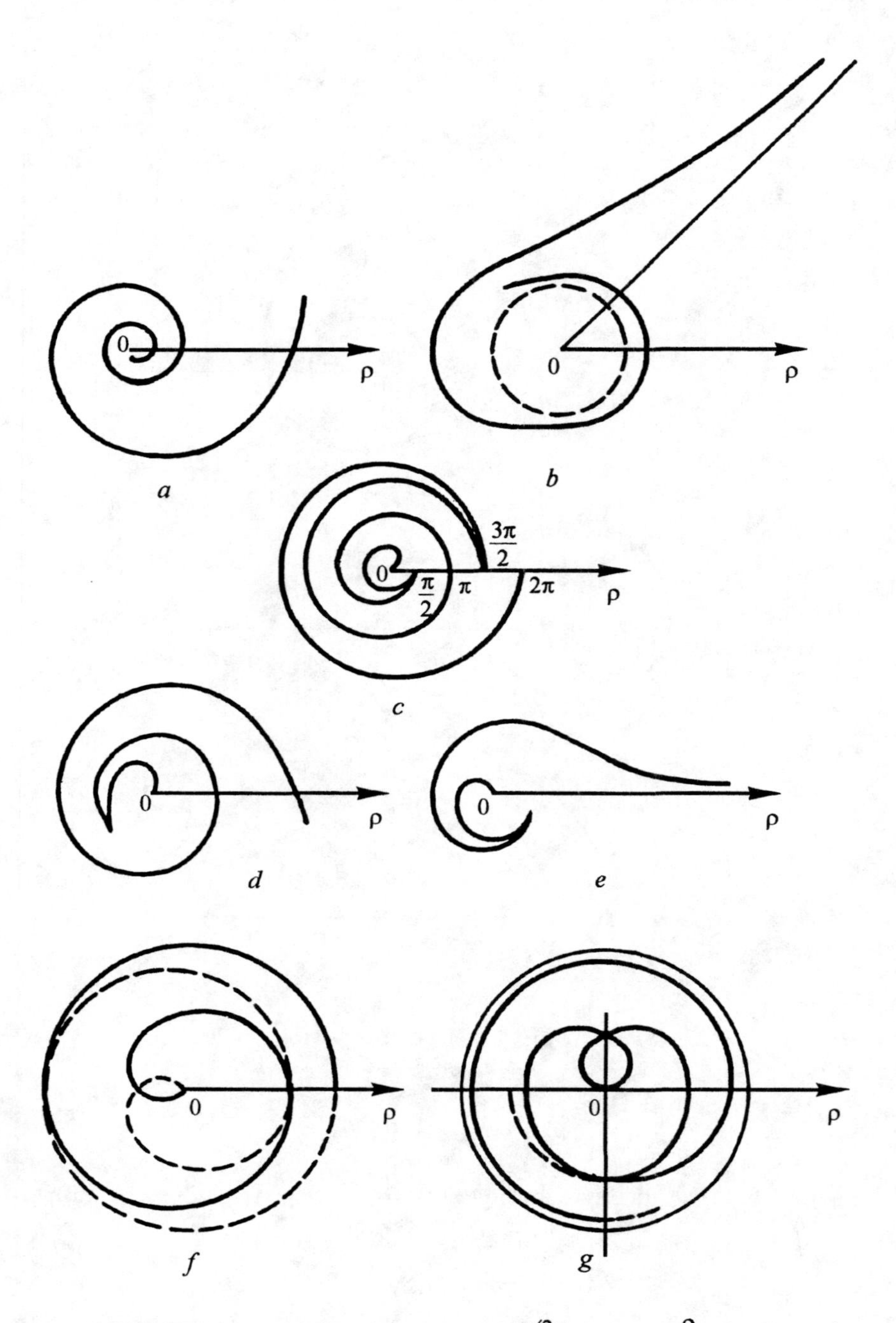

Figure 7.109. The graphs of functions: (*a*) $\rho = 2^{\varphi/2\pi}$; (*b*) $\varphi = \dfrac{\rho}{\rho - 1}$; (*c*) $\varphi = 2\pi \sin \rho$; (*d*) $\varphi = 4\rho - \rho^2$; (*e*) $\varphi = \dfrac{12\rho}{1 + \rho^2}$; (*f*) $\rho^2 + \varphi^2 = 100$; (*g*) $\rho = \dfrac{2}{\pi} \arctan \dfrac{\varphi}{\pi}$.

First we plot the curve $\rho = a/\varphi$ for $a > 0$ (the hyperbolic spiral) (Figure 7.105). The pole is an asymptotic point of the curve.

A straight line parallel to the polar axis and located at the distance a is an asymptote for the curve, where the curve approaches the asymptote counterclockwise as φ increases. As φ increases from 0 to $+\infty$, a point in the spiral runs over an asymptotic portion of the curve and, crossing the polar axis, approaches the pole, making an infinite number of windings about it with the distance between the windings rapidly tending to zero.

The graph of the function $\rho = \dfrac{a}{\varphi - \pi/4}$ is obtained from the graph of the function $\rho = a/\varphi$ by rotating it by the angle $\pi/4$ counterclockwise (Figure 7.106).

Example 7.3.2. Plot the graph of the function $\rho = a/\cos\varphi \pm b$ (the conchoid).

Plot the curve $\rho = a/\cos\varphi$ (see Figure 7.97). By using property 5), we obtain the plot of the conchoid. Note that, for different values of a and b, the conchoid has a different appearance (Figure 7.107). Graphs of some functions in polar coordinates are given in Figures 7.108 and 7.109.

Chapter 8

Plots of implicitly defined functions

8.1. Implicitly defined functions

Consider a function defined implicitly, $F(x, y) = 0$.

Depending on whether the function $F(x, y)$ is algebraic or transcendental, the curve defined by the equation $F(x, y) = 0$ is called *algebraic* or *transcendental.*

If the function $F(x, y)$ can be decomposed into factors $\varphi_1(x, y)$, $\varphi_2(x, y)$, ..., $\varphi_n(x, y)$, then this equation corresponds to the family of curves $\varphi_1(x, y) = 0$, $\varphi_2(x, y) = 0$, ..., $\varphi_n(x, y) = 0$.

Properties of graphs of the function $F(x, y) = 0$

If the equation $F(x, y) = 0$ does not change when x is replaced with $-x$, then the curve is symmetric with respect to the y-axis. If the equation $F(x, y) = 0$ remains unchanged when y is replaced with $-y$, then the graph of the function is symmetric with respect to the x-axis. If replacing x and y with $-x$ and $-y$ does not change the equation $F(x, y) = 0$, then the curve is symmetric with respect to the origin. In the case where the equation $F(x, y) = 0$ is not changed when the variables x and y are interchanged, the graph is symmetric with respect to the bisector $y = x$.

The graph of the curve $F(x + a, y) = 0$ is obtained from the graph $F(x, y) = 0$ by a parallel translation of the latter along the x-axis by $|a|$ in the direction opposite to the sign of a. The graph of the curve $F(x, y + b) = 0$ is obtained by a parallel translation of the graph of the curve $F(x, y) = 0$ along the y-axis by $|b|$ in the direction opposite to the sign of b. The graph of the curve $F(x/p, y) = 0$ is obtained by stretching (contracting) the graph of the curve $F(x, y) = 0$ along the x-axis by p. The graph of the curve $F(x, y/q) = 0$ is obtained by stretching (contracting) the graph of the function $F(x, y) = 0$ along the y-axis by q.

By applying these transformations to the graph of the curve $F(x, y) = 0$, one obtains the graph of the curve $F(x/p + a, y/q + b) = 0$.

Points of intersection of the curve $F(x, y) = 0$ and the axes

Points where the curve $F(x, y) = 0$ intersects the x-axis are solutions of the system $F(x, 0) = 0$, $F(x, y) = 0$; points of intersection of the curve and the y-axis are given by the system $F(0, y) = 0$, $F(x, y) = 0$.

Note that in order to obtain a better plot of the curve $F(x, y) = 0$, it is sometimes useful to calculate other points not lying on the axes. These auxiliary points could be found as points of intersection of the curve and straight lines $y = kx$ for different values of k.

If the equation of the curve $F(x, y) = 0$ can be written in the form

$$\varphi_1(x, y)\, \varphi_2(x, y) + \psi_1(x, y)\, \psi_2(x, y) = 0,$$

then coordinates of the points satisfying any of the following systems,

$$\begin{cases} \varphi_1 = 0 \\ \psi_1 = 0 \end{cases}, \quad \begin{cases} \varphi_1 = 0 \\ \psi_2 = 0 \end{cases}, \quad \begin{cases} \varphi_2 = 0 \\ \psi_1 = 0 \end{cases}, \quad \begin{cases} \varphi_2 = 0 \\ \psi_2 = 0 \end{cases}$$

will also satisfy the equation of the curve and, hence, lie on it.

Asymptotes of the curve $F(x, y) = 0$

To find horizontal asymptotes of the curve given by $F(x, y) = 0$, set the coefficient in the term with the highest power of x in the equation equal to zero. If this coefficient is a constant, then there are no horizontal asymptotes.

To find vertical asymptotes of the curve $F(x, y) = 0$, set the coefficient in the term containing the highest power of y equal to zero.

To find an equation for an oblique asymptote of the curve $F(x, y) = 0$, replace y in the equation of the curve with $kx + b$, set the coefficients in two terms containing the highest power of x to zero, and solve the obtained system with respect to k and b.

Example 8.1.1. The curve $x^2y^2 + y^4 - 16x^2 - 0$ has two horizontal asymptotes.

Indeed, let us rewrite the equation as follows: $(y^2 - 16)x^2 + y^4 = 0$. Set the coefficient at x^2 to zero. We get $y^2 - 16 = 0$, whence $y - 4 = 0$ and $y + 4 = 0$ define two horizontal asymptotes of the given curve.

Example 8.1.2. The curve $3y^2 - xy^2 - 5x^2y = 0$ has one vertical asymptote.

Indeed, write the equation in the form: $(3 - x)y^2 - 5x^2y = 0$. Whence, $x = 0$ is a vertical asymptote. This curve has also a horizontal asymptote $y = 0$.

Example 8.1.3. The curve $x^3 + y^3 - 3x^2 = 0$ has an oblique asymptote. Indeed, replace y with $kx + b$:

$$x^3 + (kx + b)^3 - 3x^2 = 0,$$
$$(1 + k^3)x^3 + 3(bk^2 - 1)x^2 - 3bkx - 3x^2 = 0.$$

Hence, $k = -1$, $b = 1$. Thus, $y = -x + 1$ is an oblique asymptote of the curve.

8.2. Plotting the graphs of implicitly defined functions

8.2.1. Plotting the graphs by transforming an implicitly defined function into an explicit function

Example 8.2.1. Study and plot the function $x^4 + 2y^2 - x^2 - y^4 = 0$.

To determine the domain of the implicitly defined function, solve the equation with respect to y. We get $y = \pm\sqrt{1 + \sqrt{x^4 - x^2 + 1}}$, $y = \pm\sqrt{1 - \sqrt{x^4 - x^2 + 1}}$, i.e. we get two branches of the curve. In the domain of each function, we must have that $x^4 - x^2 + 1 \geq 0$, or $(x^2 - 1)^2 + x^2 \geq 0$, which holds true for all x.

For points in the second branch, we must have that $\sqrt{x^4 - x^2 + 1} \leq 1$, or $x^2 - 1 \leq 0$, which implies that $|x| \leq 1$. Hence the domain of the first function is $(-\infty; +\infty)$, and for the second function, the domain is $[-1;1]$.

The curve is symmetric with respect to the coordinate axes. No horizontal or vertical asymptotes exist, since the coefficients at the highest powers of x and y are constants. Let us find oblique asymptotes. We have $x^4 - (kx + b)^4 - x^2 + 2(kx + x)^2 = 0$. Set the coefficients at x^4 and x^3 equal to zero, $k^4 - 1 = 0$, $4k^{3b} = 0$, whence $k = \pm 1$, $b = 0$. Thus the straight lines $y = x$, $y = -x$ are oblique asymptotes for the curve.

Let us find a few additional points of the curve. To do this, we solve the system $x^4 + 2y^2 - x^2 - y^4 = 0$, $y = kx$ for different values of k (Figure 8.1):

$$x_{1,2} = 0, \quad x^2 = \frac{1 - 2k^2}{1 - k^4}.$$

Example 8.2.2. Plot the graph of the function $y^2 = x^3 + 1$.

Solve the equation with respect to y: $y = \pm\sqrt{x^3 + 1}$. The function is defined for $x \geq -1$. The curve is symmetric with respect to the x-axis. The curve does not have asymptotes.

Let us write the function $y = \sqrt{x^3 + 1}$ in terms of an auxiliary variable t: $y = \sqrt{t}$, $t = x^3 + 1$. Using graphs of these functions, we plot the graph of the upper branch of the function under consideration, and then reflecting it symmetrically about the x-axis, obtain the lower branch of the graph (Figure 8.2).

Graphs of other functions are shown in Figures 8.3 – 8.6.

Remark 8.2.1. This method cannot always be used to plot graphs of implicitly defined functions, since one needs to solve an equation, possibly of a high degree. As it is well known,

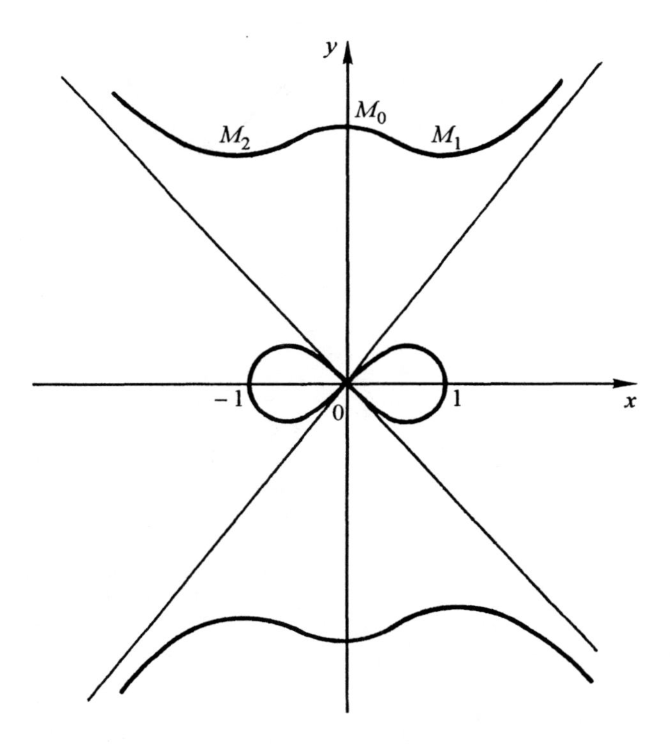

Figure 8.1. The graph of the function $x^4 + 2y^2 - x^2 - y^4 = 0$.

a solution in radicals exists, i.e. one can solve the equation explicitly, only for an equation of a degree not greater that four.

8.2.2. Plotting the graphs by using the method of "linear relation"

Let a curve be given by an equation $F(x, y) = 0$. Write the function F as $F(x, y) = f(x, y) + \varphi(x, y) = 0$. If now we plot the curves $f(x, y) = 0$, $\varphi(x, y) = 0$, then points of intersection of these two curves will lie on the graph of the curve $F(x, y) = 0$. If we can write the function F as

$$F(x, y) = f_1(x, y) f_2(x, y) f_3(x, y) + \lambda(\varphi_1(x, y)\, \varphi_2(x, y)\, \varphi_3(x, y), \quad (1)$$

then the more various representations of the function we get, the more points of the graph can be found. For example, in the case of (1), we can find solutions of the following systems:

$$\overset{\text{I}}{\begin{cases} f_1 = 0 \\ \varphi_1 = 0 \end{cases}}, \quad \overset{\text{II}}{\begin{cases} f_1 = 0 \\ \varphi_2 = 0 \end{cases}}, \quad \overset{\text{III}}{\begin{cases} f_1 = 0 \\ \varphi_3 = 0 \end{cases}}, \quad \overset{\text{IV}}{\begin{cases} f_2 = 0 \\ \varphi_1 = 0 \end{cases}}, \quad \overset{\text{V}}{\begin{cases} f_2 = 0 \\ \varphi_2 = 0 \end{cases}}, \quad \overset{\text{VI}}{\begin{cases} f_2 = 0 \\ \varphi_3 = 0 \end{cases}},$$

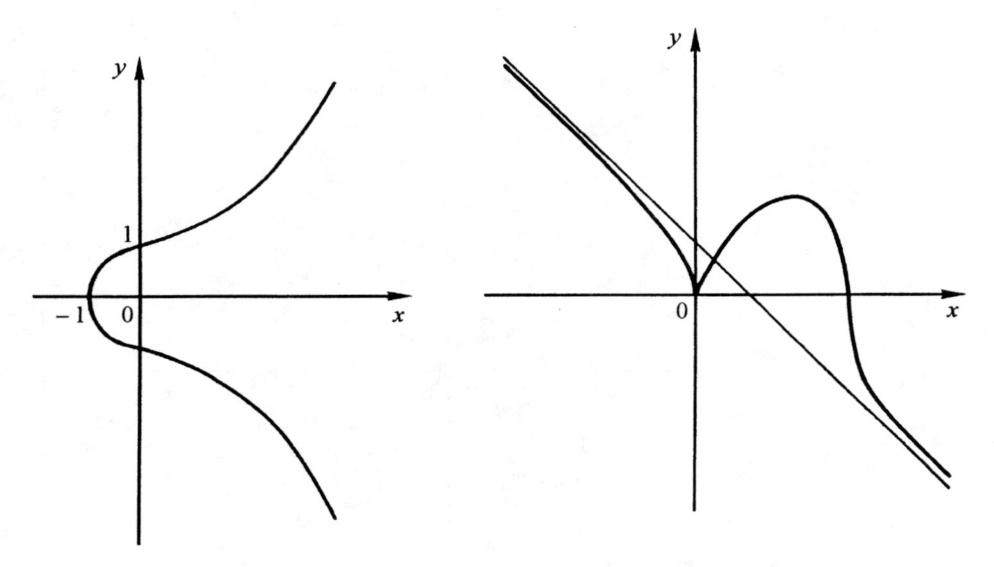

Figure 8.2 Figure 8.3

Figure 8.2. The graph of the function $y^2 = x^3 + 1$.

Figure 8.3. The graph of the function $y^3 = 6x^2 - x^3$.

$$\begin{matrix} \text{VII} & \text{VIII} & \text{IX} \\ \begin{cases} f_3 = 0 \\ \varphi_1 = 0 \end{cases}, & \begin{cases} f_3 = 0 \\ \varphi_2 = 0 \end{cases}, & \begin{cases} f_3 = 0 \\ \varphi_3 = 0 \end{cases}. \end{matrix}$$

Solutions of these systems give nine groups of points of intersection. It can happen that some of the functions f or φ are linear.

Example 8.2.3. Plot the graph of the curve $F(x, y) = x^4 + y^4 - 3x^2 - 3x^2 - 5y^2 + 6 = 0$.

Note that the function is symmetric with respect to the coordinate axes. Decompose the function into two summands: $x^4 + y^4 - 3x^2 - 5y^2 + 6 = (x^2 - 1)(x^2 - 2) + (y^2 - 1)(y^2 - 4) = 0$ or $(x + 1)(x - 1)(x + \sqrt{2})(x - \sqrt{2}) + (y + 1)(y - 1)(y + 2)(y - 2) = 0$. We thus obtain $4 \cdot 4 = 16$ points of intersection of the auxiliary lines $x \pm 1 = 0$, $x \pm \sqrt{2} = 0$ with the lines $y \pm 1 = 0$, $y \pm 2 = 0$ (Figure 8.7).

This function can also be written as $F(x, y) = x^2(x^2 - 3) + (y^2 - 3)(y^2 - 2) = 0$ or

$$xx(x + \sqrt{3})(x - \sqrt{3}) + (y + \sqrt{3})(y - \sqrt{3})(y + \sqrt{2})(y - \sqrt{2}) = 0.$$

Note that the straight lines defined by the second summand have two points in common with the straight line $x^2 = 0$, i.e. with the y-axis, and hence the curve under consideration is tangent at these points to the straight lines $(y \pm \sqrt{3}) = 0$ and $(y \pm \sqrt{2}) = 0$.

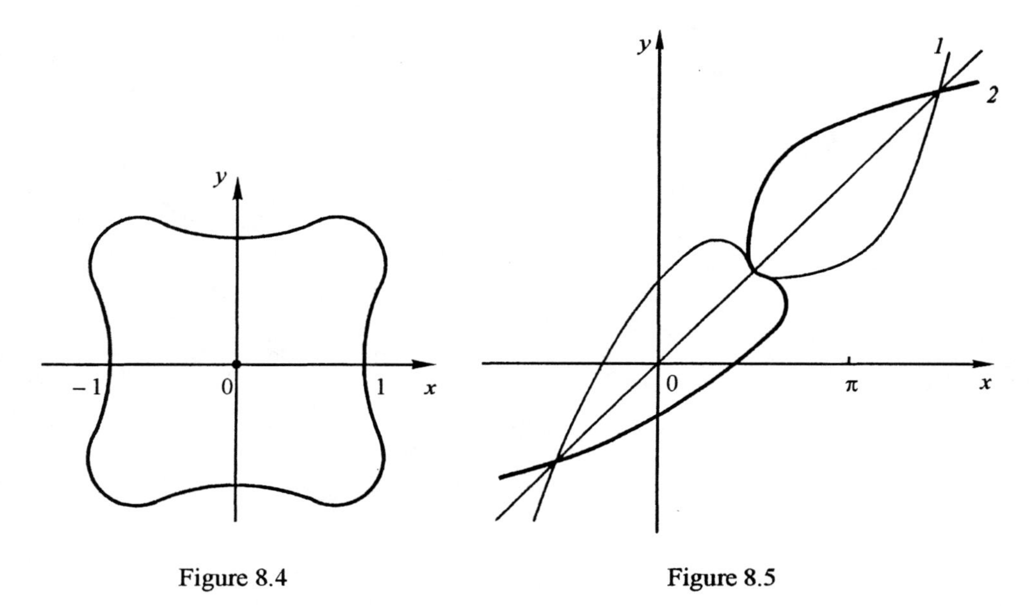

Figure 8.4 Figure 8.5

Figure 8.4. The graph of the function $x^4 + y^4 = x^2 + y^2$.

Figure 8.5. The graphs of functions: (1) $y = x + \cos x$; (2) $y + \cos y - x = 0$.

The second decomposition of the function yields 12 more points of the curve $F(x, y) = 0$. One can also determine extreme points. Rewrite the equation $F(x, y) = 0$ as

$$F(x, y) = x^4 - 3x^2 + 9/4 + y^4 - 5y^2 + 15/4 = 0 \text{ or } (x^2 - 3/2)^2 + y^4 - 5y^2 + 15/4 = 0.$$

Here one summand is a square of a certain expression of x, the other is a biquadratic function. Write $(x^2 - 3/2)^2 = 0$, $y^4 - 5y^2 + 15/4 = 0$. The latter equation has four roots, and thus can be decomposed into the product of four terms of degree one corresponding to the graphs of four straight lines that are parallel to the x-axis. At the same time, points of intersection of the auxiliary lines are points at which they are tangent to the curve $F(x, y) = 0$. Thus the x-coordinates of points having the greatest and smallest y-coordinate are found from the equation $x^2 - 3/2 = 0$, $x = \pm\sqrt{3}/2$. The corresponding y-coordinates are found from equation (2). In the same way, one can find tangents that are parallel to the y-axis, $y^2 - 5/2 = 0$, $y = \pm\sqrt{5/2}$.

8.2.3. Plotting of graphs by using the method of "cells"

Consider a function $F(x, y) = 0$. To each pair of values $x = a$, $y = b$ there corresponds a point $M(a; b)$ in the coordinate plane at which the function $F(x, y)$ can take positive or negative values (complex values correspond to the case where the curve of the function has no points in the corresponding region). Points that satisfy the condition $F(x, y) = 0$ define a certain curve and separate points in the coordinate plane at which the function $F(x, y)$ takes positive and negative

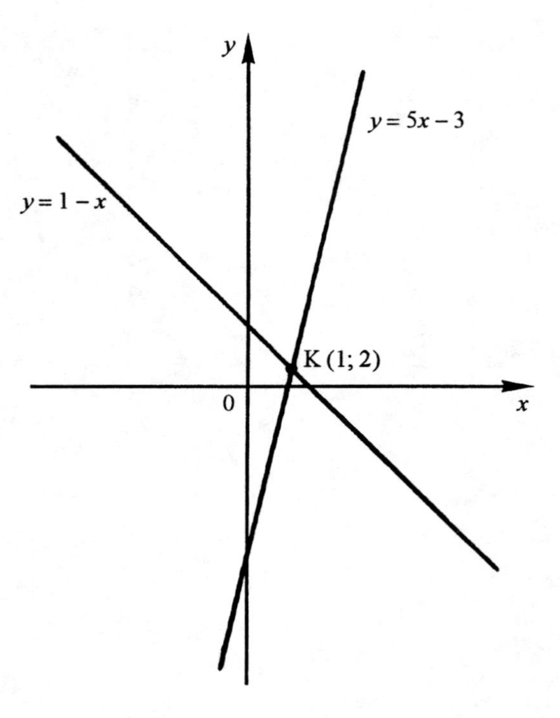

Figure 8.6. The graph of the function $5x^2 + 4xy - y^2 - 8x - 2y + 3 = 0$.

values. If now we consider the function $F(x, y) = f_1(x, y)f_2(x, y)f_3(x, y)... - \lambda\varphi_1(x, y)\varphi_2(x, y)\varphi_3(x, y)$ then, for every pair of values x and y, each one of the functions f and φ assumes certain values positive or negative.

The values of x and y such that F becomes zero can be found from the equality $f_1f_2f_3... = \varphi_1\varphi_2\varphi_3...$. It is clear that this equality holds only if the signs of the left-hand side and the right-hand side are the same, or $f_1f_2f_3...\varphi_1\varphi_2\varphi_3... > 0$, i.e. the product of all auxiliary functions f and φ is positive. This condition is necessary.

Thus, by plotting the curve defined by $F(x, y) = 0$ in the plane, we at the same time subdivide it into the region of positive and negative values of the function. By doing this for all the functions f and φ, we obtain a subdivision of the coordinate plane into regions ("cells") such that the sign of the function is plus in some cells and minus in the rest of them. It is clear that the curve $F = 0$ cannot have points in the cells where the product is negative. We thus exclude regions in the coordinate plane where the curve cannot have points.

If the function F has another decomposition, i.e. if F can also be expressed as

$$F = \Phi^1\Phi^2\Phi^3... - \lambda\Theta^1\Theta^2\Theta^3...,$$

then we get another set of cells where the curve $F = 0$ cannot have points.

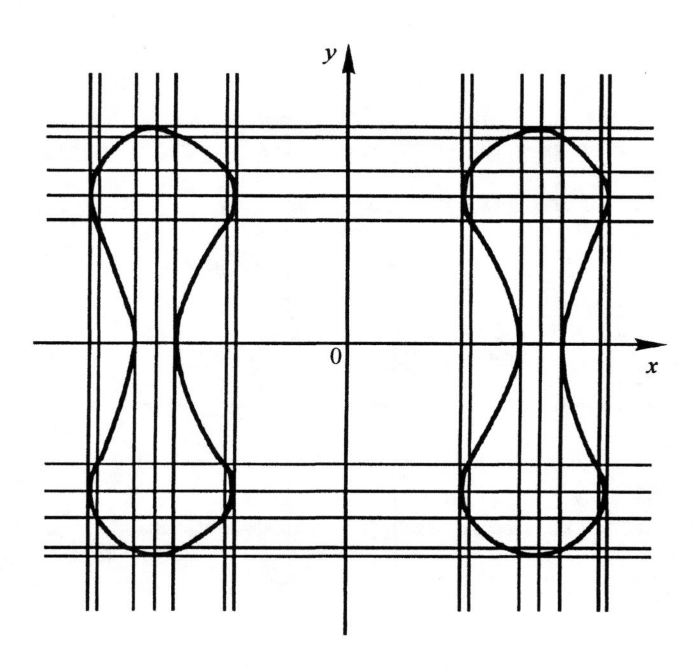

Figure 8.7. The graph of the function $x^4 - 3x^2 + y^4 - 5y^2 + 6 = 0$.

Example 8.2.4. Plot the graph of the curve $F(x, y) = x^4 - 2ay^3 - 3a^2y^2 - 2a^2x^2 + a^4 = 0$.

We have

$$x^4 - 2a^2x^2 = 2ay^3 + 3a^2y^2 - a^4 \Rightarrow$$

$$\Rightarrow xx(x + a\sqrt{2})(x - a\sqrt{2}) = 2a(y + a)(y + a)(y + a/2),$$

or $x^2(x + 1.41a)(x - 1.41a) = 2a(y + a)^2(y - a/2)$. Note that the terms having even power do not influence the sign of the product, since they always have sign " + ", and as far as the subdivision into cells is concerned, we have only the product

$$(x + 1.41a)(x - 1.41a)(y - a/2) > 0.$$

Plot the graph of the straight line $f_2 = x + 1.41a = 0$ (Figure 8.8). Let us determine the direction where the function f_1 is positive; the function f_2 will be negative in the opposite direction. To do this, it will suffice to determine the sign of f_2 in any point of the coordinate plane not lying on the line $f_2 = 0$ itself. The most simple way is to determine this at the point $O(0;0)$; we have $f_2(0) = + 1.41a$, i.e. the points lying on the same side of the line as the origin will correspond to positive values of f_2. We mark this area with the sign "+", and the other with the sign "–".

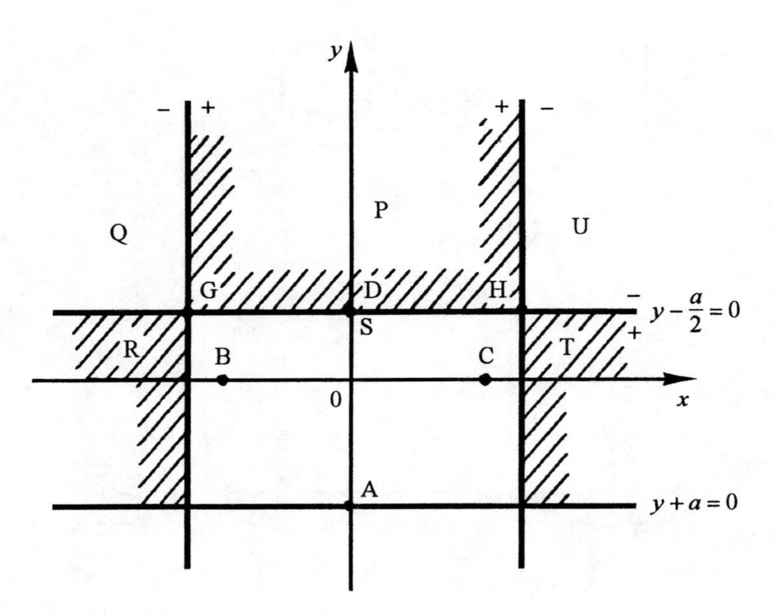

Figure 8.8. The graph of the function $x^4 - 2a^2x^2 = 2ay^3 + 3a^2y^2 - a^4$.

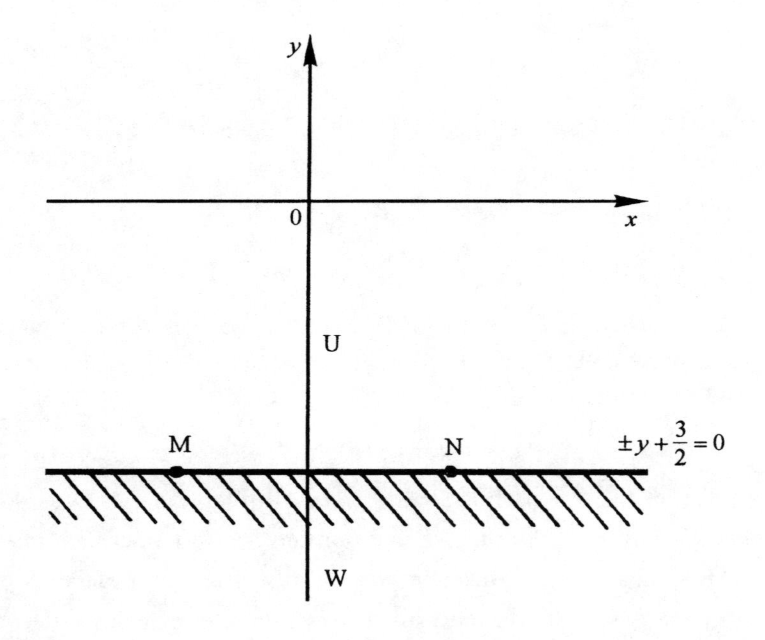

Figure 8.9. The graph of the function $(x + a)^2(x - a)^2 = 2ay^2(y + 3/2a)$.

Do the same thing with other factors in the product. We thus obtain 6 cells denoted by P, Q, R, S, T, U.

In the cell P, the function f_2 is positive and f_3 is negative; the function φ_1 is positive, and, hence, the sign for the sell P is negative, i.e. there are no points in the cell P belonging to the curve $F = 0$. We shade this cell. The neighboring cells Q, U, and S have the sign opposite to the sign of the cell P. This follows from the fact that one of the factors in the product changes its sign. Using this we can shade cells in the chessboard pattern.

By taking functions f and φ in pairs, we can determine a number of points of the curve $F(x, y) = 0$ as points of intersection of the curves f_1, f_2, f_3 and φ_1^2, φ_3. This gives six points, five of which are tangent points and one point that is singular. Indeed, the point of intersection of the curves f_1^2 and φ_1^2 (point A in Figure 8.8) is a point of intersection of two double lines; at the point of intersection of f_2^2 and φ_2 (point D), the curve $F(x, y) = 0$ and straight line $\varphi_3 = 0$ are tangent to each other. In the same way one can find points E and F of the curve $F(x, y) = 0$ at which the curve is tangent to the lines $f_3 = 0$ and $f_1 = 0$, and intersects the line $\varphi_3 = 0$. Points G and H are the points where the straight lines $f_2 = 0, f_1 = 0$ intersect the straight line $\varphi_3 = 0$.

Additional information for plotting the curve $F = 0$ can be obtained if another decomposition of the function F into products is used. For example, $x^4 - 2a^2x^2 + a^4 = 2ay^3 + + 3ay^2 \Rightarrow (x + a)^2(x - a)^2 = 2ay^2(y + 3/2a), f_1 f_2 f_3 f_4 = 2a\varphi_1\varphi_2\varphi_4$, where $f_1 = f_2, f_3 = f_4, \varphi_1 = \varphi_2$.

For points of the curve $F = 0$ to be in a cell, we must have $\varphi_3 > 0$ (Figure 8.9).

Plot the line $\varphi_3 = 0$. We get two cells V and W. For points in the first cell, the values of the function φ_3 are positive; in the cell W this function takes negative values. Thus we obtained additional points of the curve $F = 0$. The graph cannot lie below the straight line $\varphi_3 = 0$.

It follows from $f_1^2 f_3^2 = 2a\varphi_1^2\varphi_3$ that the curve $F = 0$ has two more singular points (double points) B and C (which are points at which the double lines $f_1^2 = 0$ and $f_3^2 = 0$ intersect the straight line $\varphi_1^2 = 0$), and points M and N at which the curve $F = 0$ is tangent to the straight line $\varphi_3 = 0$ and intersects the double straight lines (Figure 8.10).

Example 8.2.5. Plot the graph of the curve $F(x, y) = x^4y - 2x^2y - y^2 + 3y - 1 = 0$.

The curve is symmetric with respect to the y-axis. Let us write the equation as

$$x^2y(x^2 - 2) = y^2 - 3y + 1, xxy(x + \sqrt{2})(x - \sqrt{2}) = (y - 2.62)(y - 0.38),$$

i.e. we have $f_1 f_2 f_3 f_4 f_5 = \varphi_1\varphi_2$.

Among the auxiliary lines five are significant. Those are $f_3 = y = 0$, $f_4 = x + \sqrt{2} = 0$, $f_5 = x - \sqrt{2} = 0$, $\varphi_1 = y - 2.62 = 0$, $\varphi_3 = y - 0.38 = 0$. All of them are straight lines. Plot them in the coordinate system with the corresponding signs (Figure 8.10). For cell P, where $f_3 > 0$, $f_4 > 0, f_5 < 0$, $\varphi_1 > 0$, $\varphi_2 > 0$, determine the sign of $f_3 f_4 f_5 - \varphi_1\varphi_2$. It is negative, hence the graph has no points in the cell P. We shade the cell.

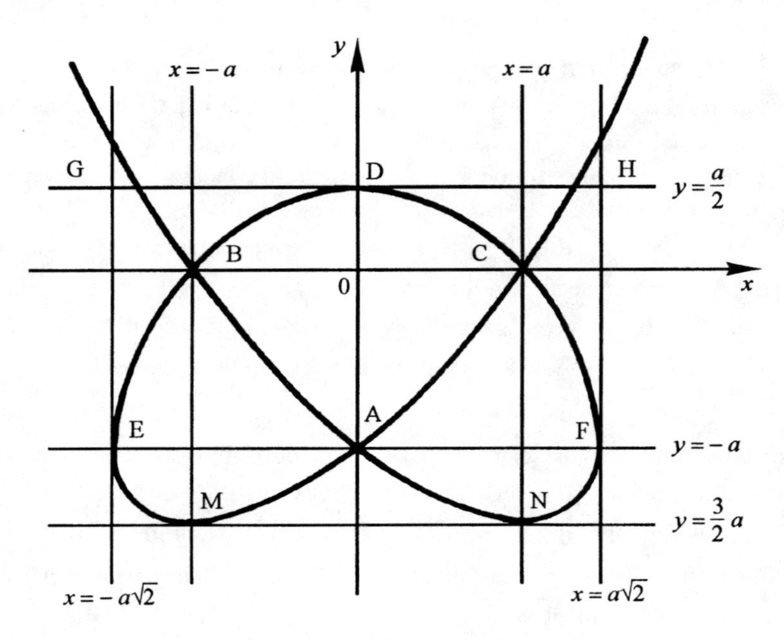

Figure 8.10. The graph of the function $x^4 - 2ay^3 - 3a62y^2 - 2a^2x^2 + a^4 = 0$.

Decomposing the function we find eight points of intersection of auxiliary lines, two of which are tangent points: point A is a point of intersection of $\varphi_1 = 0$ and $f_1^2 = 0$, at point B, $\varphi_2 = 0$ intersects $f_1^2 = 0$. The other points are: $C = (-\sqrt{2};2.62)$, $D(\sqrt{2};+2.62)$, $E(-\sqrt{2};0.38)$, $F(\sqrt{2};0.38)$, $G(-\sqrt{2};0)$, $H(\sqrt{2};0)$.

It is clear from the equation of the curve that if $y = \infty, x = \infty$, i. e. the axis Ox is an asymptote. It is also clear from the form of the factored equation that the curve has no points in the region below the line $y = 0$.

Consider one more decomposition of F: $y(x-1)^2(x+1)^2 = (y-1)^2$, $f_1 f_2 f_3 f_4 f_5 = \varphi_1\varphi_2$, where $f_2 = f_3, f_4 = f_5, \varphi_1 = \varphi_2$. The only significant line is $f = y = 0$, which agrees with the earlier conclusion that the curve $F = 0$ has no points below the x-axis.

The last term shows that the curve $F = 0$ has two singular points $I(-1;1)$, $K(1;1)$, the points at which the double lines $f_2^2 = 0$ and $f_4^2 = 0$ intersect the straight line $\varphi_1^2 = 0$.

Example 8.2.6. Plot the graph of the curve $F(x, y) = (x^2 + y^2 - 1)(x^2 - y^2) - 2xy = 0$.

Let us write the equation of the curve as follows: $(x^2 + y^2 - 1)(x + y)(x - y) = 2xy \Rightarrow f_1^2 f_3 f_4 = 2\varphi_1\varphi_2$, where $f_1 = f_2 = x^2 + y^2 - 1 = 0$ is the equation of a circle that intersects each of the lines $\varphi_1 = x = 0$ and $\varphi_2 = y = 0$ in two points (Figure 8.11). This intersection of the straight

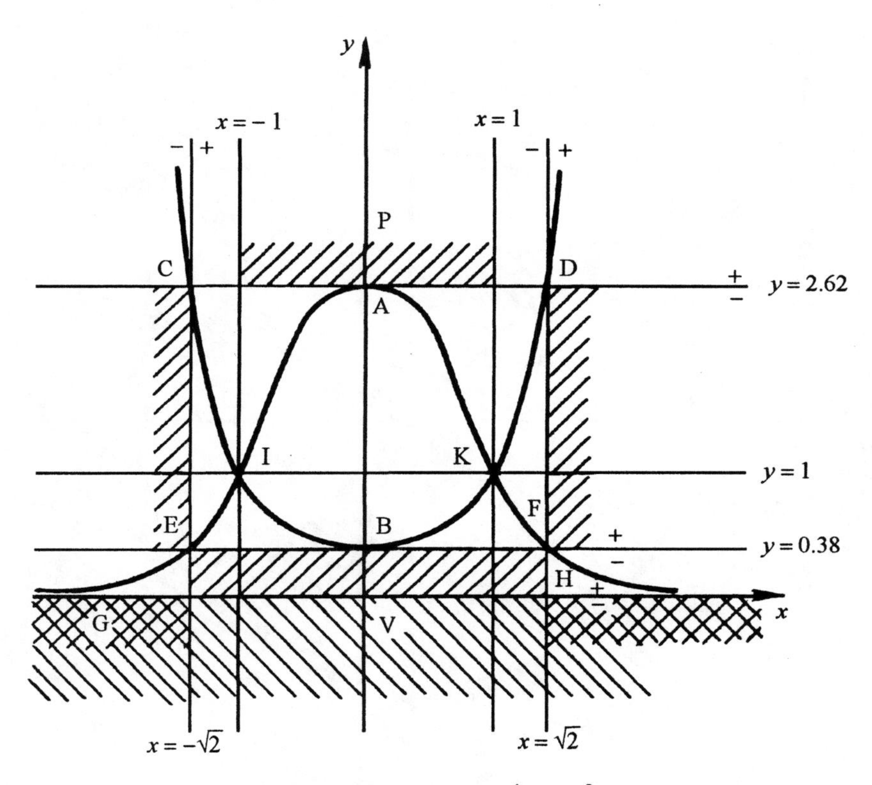

Figure 8.11. The graph of the function $x^4y - 2x^2y + 3y - 1 = 0$.

line $\varphi_1 = 0$ and circle $f_1^2 = 0$ yields two points $A(0;1)$ and $B(0; -1)$, and the straight line φ_2 and this circle yields points $C(-1; 0)$ and $D(1; 0)$. The bisectors $f_3 = x + y = 0$ and $f_4 = x - y = 0$ intersect the axes $\varphi_1 = x = 0$ and $\varphi_2 = y = 0$ in the point $O(0; 0)$.

Calculating the signs we draw the cells. We have four significant lines, and it is possible for the curve $F = 0$ to have points in any of the cells where $f_3 f_4 f_1 \varphi_2 > 0$.

Drawing the straight lines, we obtain eight cells (Figure 8.12) which are separated with radial straight lines. To determine the sign of the product in an arbitrary cell, it is sufficient to find its sign in any point of the cell. Take, for example, point $M(1; 2)$. We get $f_3(1; 2) = 3$, $f_4(1; 2) = -1$, $\varphi_1(1; 2) = 1$, $\varphi_2(1; 2) = 2$. The product is negative; thus the cell containing point M should be shaded. We then shade every other cell.

This function can also be written in another way:

$$(x^2 + y^2 - 1)(x^2 + y^2 - 1)(x^2 - y^2) - 2xy = 0,$$

or
$$(x^2 + y^2)(x^2 - y^2)(x^2 + y^2 - 2) - (y \pm (1 + \sqrt{2})x) = 0,$$

or $(x^2 + y^2)(x + y)(x - y)(x^2 + y^2 - 2) - (y - 0.41x)(y + 2.4x) = 0$, or $\Phi_1\Phi_2\Phi_3\Phi_4 - \psi_1\psi_2 = 0$.

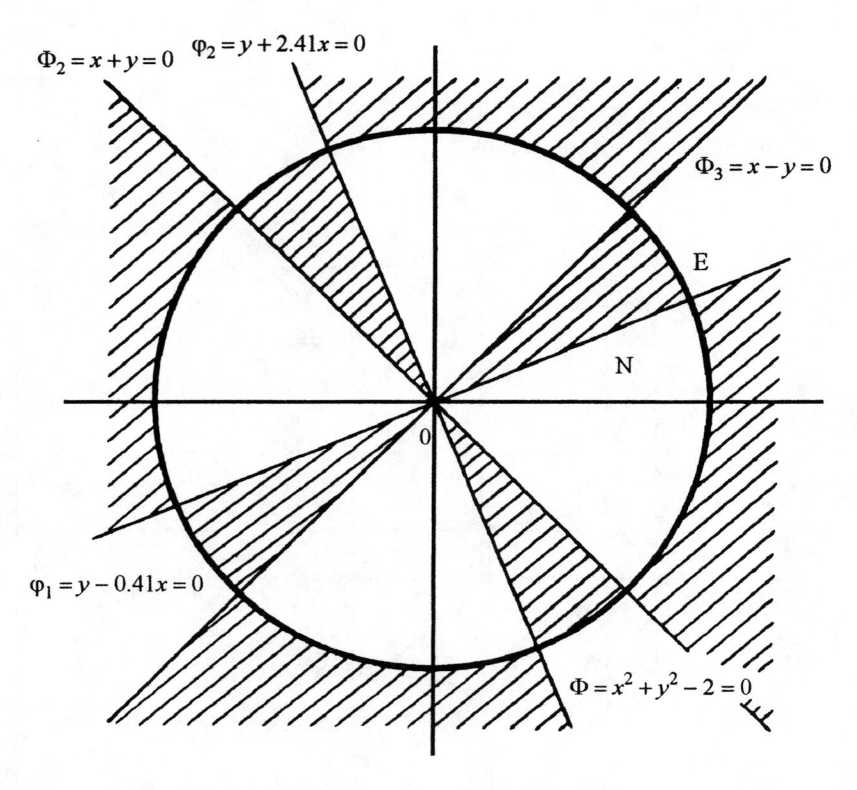

Figure 8.12. Plotting the function $(x^2 + y^2 - 1)^2(x^2 - y^2) = 2xy$.

In this decomposition, the factor $\Phi_1 = (x^2 + y^2)$ gives only the point $O(0; 0)$; the factors $\Phi_2 = x + y = 0$, $\Phi_3 = x - y = 0$ are bisectors; $\Phi_4 = x^2 + y^2 - 2 = 0$ is a circle; and $\psi_1 = = y - 0.41x = 0$, $\psi_2 = y + 2.41x = 0$ are equations of two radial straight lines. Significant lines are all but $\Phi_1 = 0$. To determine the sign of the product $\Phi_2\Phi_3\Phi_4\psi_1\psi_2$, take the point $N(1.5; 0.2)$. We get $\Phi_2(N) = 1.7$, $\Phi_3(N) = 1.3$, $\Phi_4(N) = 0.29$, $\psi_1(N) = -0.415$, $\psi_2(N) = 2.811$. Hence the sign of the product is "–", and the cell containing this point should be shaded.

From the initial decomposition of the equation, we find 10 points of the curve $F(x, y) = 0$ with only four points being new. These are points E and G formed by intersection of the circle $\Phi_4 = 0$ and the line $\psi_1 = 0$, and points F and H, in which this circle intersects the line $\psi_2 = 0$.

It is clear that $\Phi_1 = \Phi_2 = \Phi_3 = 0$ at a point common to the lines $\psi_1 = 0$ and $\psi_2 = 0$. The curve $F = 0$ has a point that coincides with three points; in other words, the curve has an inflection point relative to the straight lines $\psi_1 = 0$ and $\psi_2 = 0$ at the origin $O(0; 0)$.

It is easy to find asymptotes. Divide the initial equation by x and set $x = \infty$. We get $(1 + k^2)^2(1 - k^2) = 0$, whence $k_1 = 1$, $k_2 = -1$. Hence the asymptotes are $y = x$, $y = -x$ (Figure 8.13).

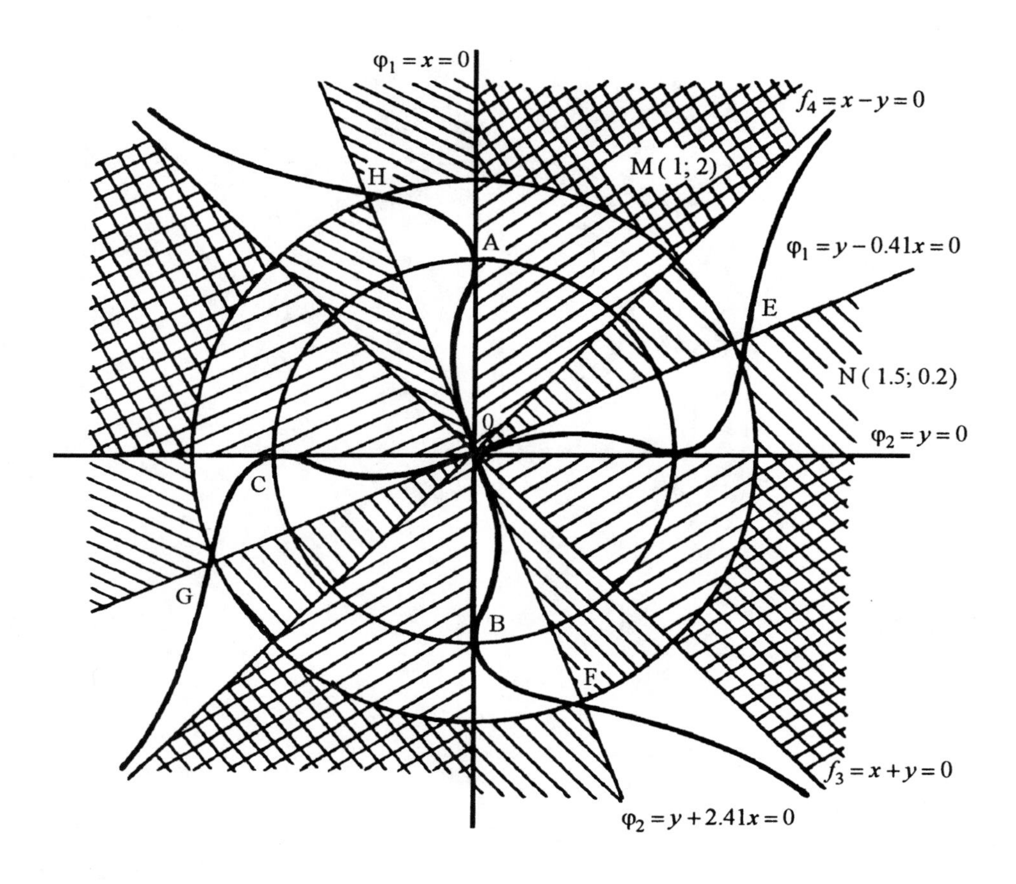

Figure 8.13. The graph of the function $(x^2 + y^2 - 1)(x^2 - y^2) - 2xy = 0$.

Example 8.2.7. Plot the graph of the curve $F(x, y) = x^4 + x^2y^2 - 6x^2y + y^2 = 0$.

Let us show by this example how the auxiliary factor λ can be used. Note that the curve passes through the origin and is symmetric with respect to the y-axis.

Write the equation as

$$x^2(x^2 + y^2 - 6y) + y^2 = 0 \quad \Rightarrow \quad x^2[x^2 + (y-3)^2 - 9] + y^2 = 0, \qquad (3)$$

or else $f_1 f_2 f_3 + \varphi_1\varphi_2 = 0$, where $f_1 = f_2$, $\varphi_1 = \varphi_2$.

It is apparent from (3) that the curve will have real points only if the value of the first term is negative, $[x^2 + (y-3)^2 - 9] < 0$. Since $x^2 + (y-3)^2 - 9 = 0$ is an equation of a circle with its center at the point (0; 3), this inequality shows that all points of the curve under consideration lie inside the circle (Figure 8.14).

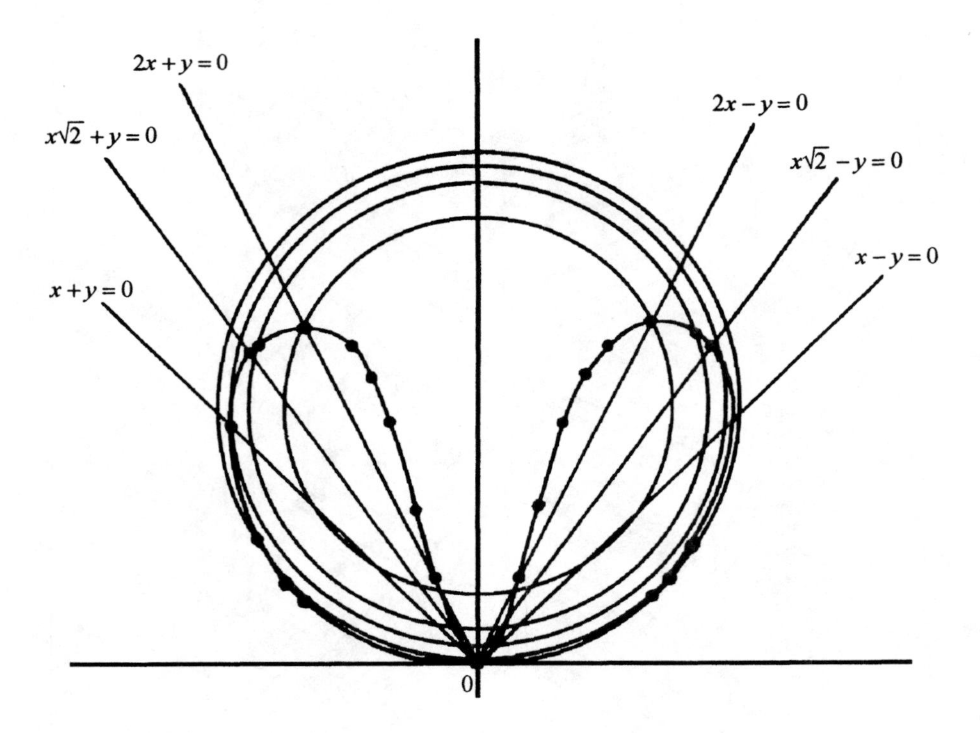

Figure 8.14. The graph of the function $x^4 + x^2y^2 - 6x^2y + y^2 = 0$.

The initial equation can be replaced by the system of equations that hold for all λ: $x^2 + (y-3)^2 - 9 = -\lambda^2, \lambda^2x^2 = y^2$, or $f = x^2 + (y-3)^2 - 9 + \lambda^2 = 0$, $\varphi = \lambda^2x^2 - y^2 = 0$.

Using the functions f and φ, write the equation $x^2[x^2 + (y-3)^2 - 9 + \lambda^2] = \lambda^2x^2 - y^2$. This is an equation of a curve that intersects the considered curve in points at which the circle of radius $\sqrt{9-\lambda^2}$ intersects the curve $\lambda^2x^2 - y^2 = 0$. Let λ take the values: $\lambda^2 = 1, \lambda^2 = 2, \lambda^2 = 4$. Then we have:

$$x^2[x^2 + (y-3)^2 - 8] = x^2 - y^2 = (x+y)(x-y),$$
$$x^2[x^2 + (y-3)^2 - 7] = 2x^2 - y^2 = (x\sqrt{2} + y)(x\sqrt{2} - y),$$
$$x^2[x^2 + (y-3)^2 - 5] = 4x^2 - y^2 = (2x + y)(2x - y).$$

By using the first equation we determine points of intersection of the circle with its center at $C(0; 3)$ of radius $\sqrt{8}$ and the straight lines $x + y = 0$, $x - y = 0$, the points A, B and A_1, B_1. In the same way, by using the second equation, we find points D, E, D_1, E_1 formed by intersection of the circle of radius $\sqrt{7}$ and the straight lines $x\sqrt{2} + y = 0$, $x\sqrt{2} - y = 0$. The points G, H, G_1, H_1 are found as points of intersection of the circle of radius $\sqrt{5}$ and the straight lines $2x + y = 0$, $2x - y = 0$.

Finally, note that the initial equation of the curve implies that the curve $F = 0$ intersecting the line $f_1^2 = x^2 = 0$ in the point $O(0; 0)$ is tangent to the straight line $\varphi_1 = 0$. At the same time, this point is a singular point of the curve $F = 0$, since it is a point of intersection of the lines $f_1^2 = 0$, $\varphi_1^2 = 0$.

At the end of this section, we outline a scheme useful for plotting implicitly defined functions.

1. Is it possible to write the function in an explicit form?
2. Is there a symmetry with respect to the coordinate axes?
3. Is there a symmetry with respect to bisectors of the quadrants?
4. Does the curve pass through the origin?
5. Are there points of the curve at infinity?
6. Are there areas in the plane that do not contain points of the curve?
7. Determine, if possible, where the curve intersects the coordinate axes and bisectors of the quadrants.
8. Are there asymptotes parallel to the coordinate axes?
9. Are there other asymptotes?
10. Do the asymptotes intersect the curve in finite points?
11. Does the curve have tangents parallel to the coordinate axes?
12. Does the curve have singular points?
13. Determine the direction of tangents at singular points.
14. Determine the direction of tangents in other characteristic points.
15. Are there inflection points?
16. Write a linear relation.
17. Find points of the curve as points of intersection of auxiliary lines determined by factors.
18. By using the rule of signs determine the cells where the curve has no points.

Examples of graphs of implicitly defined functions are shown in Figures 8.15 – 8.18.

8.3. Curves defined by second-degree algebraic equation

Consider curves defined by a second-degree equation in more detail. A general equation of degree two in two variables has the form

$$Ax^2 + 2Bxy + Cy^2 + 2Dx + 2Ey + F = 0, \qquad (1)$$

where A, B, C, D, E, and F are given numbers such that at least one of the coefficients A, B, or C is different from zero.

If $A = C$ and $B = 0$, this becomes an equation of a circle. For example, consider the equation $x^2 + y^2 - 4x + 4y + 2 = 0$. Write this equation as $(x - 2)^2 - 4 + (y + 2)^2 - 4 + 2 = 0$ or else $(x - 2)^2 + (y + 2)^2 = 6$. This defines an equation of a circle with its center at the point $(2; -2)$ and radius $\sqrt{6}$.

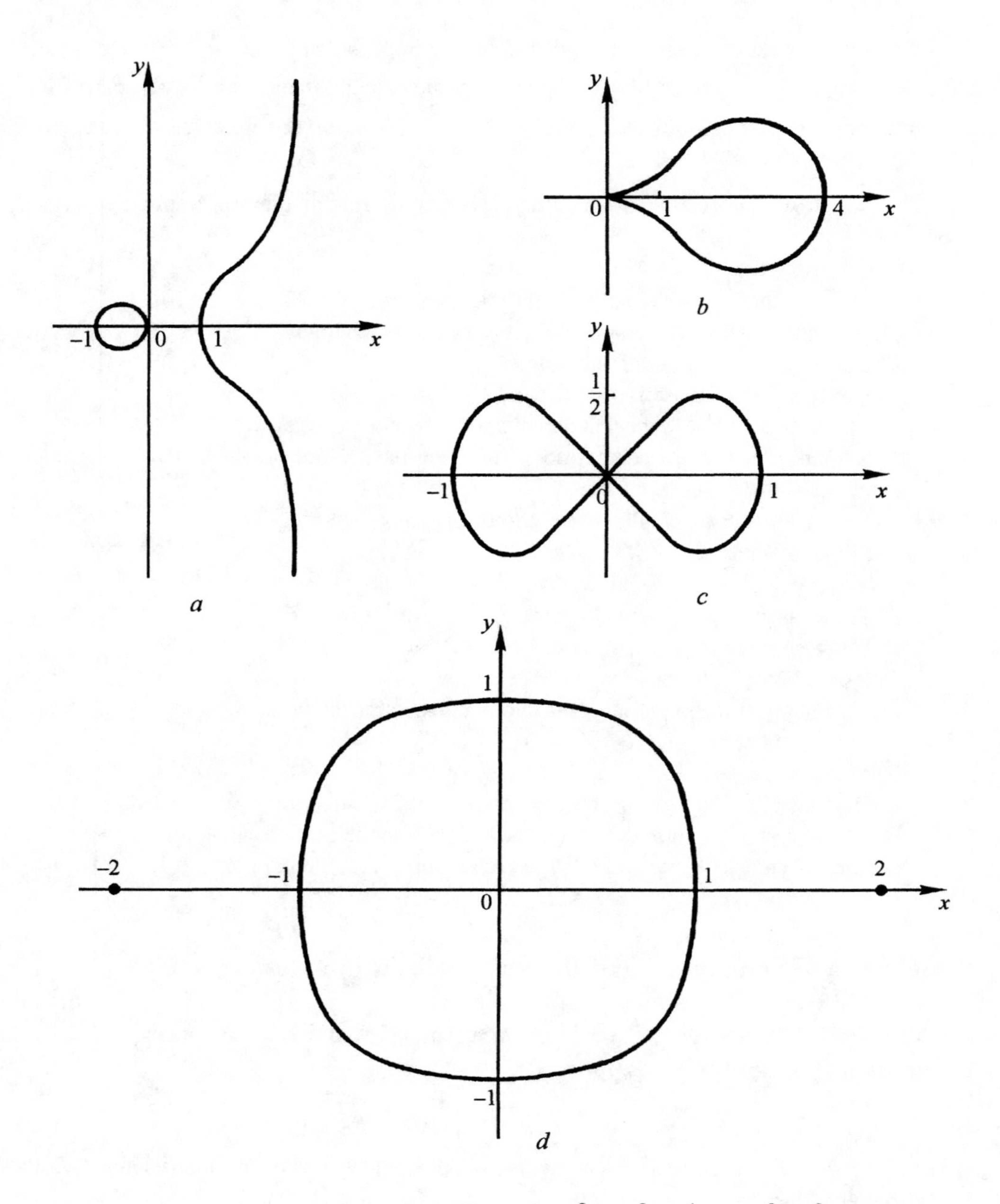

Figure 8.15. The graphs of functions: *(a)* $9y^2 = 4x^3 - x^4$; *(b)* $y^2 = x^3 - x$; *(c)* $y^2 = x^2 - x^4$; *(d)* $16y^2 = (x^2 - 4)^2(1 - x^2)$.

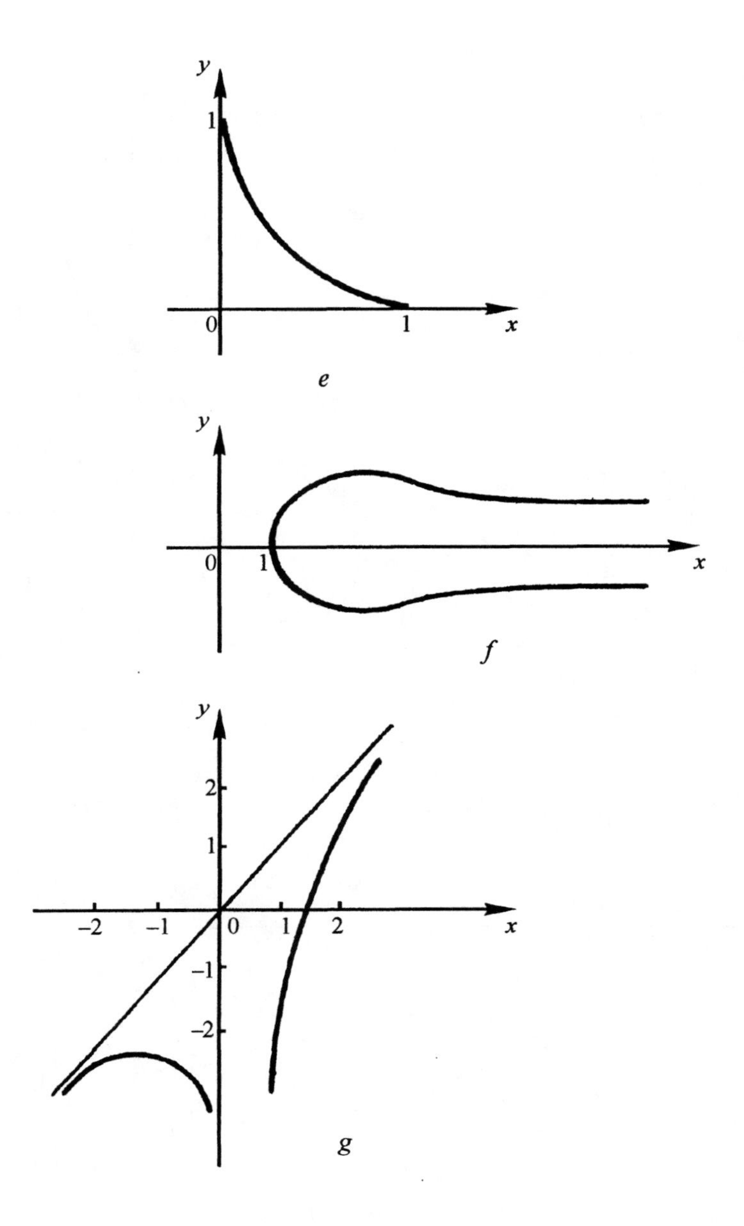

Figure 8.15. The graphs of functions: (*e*) $\sqrt{x} + \sqrt{y} = 1$; (*f*) $x^2y^2 = 4(x - 1)$; (*g*) $(y - x)x^4 + 8 = 0$.

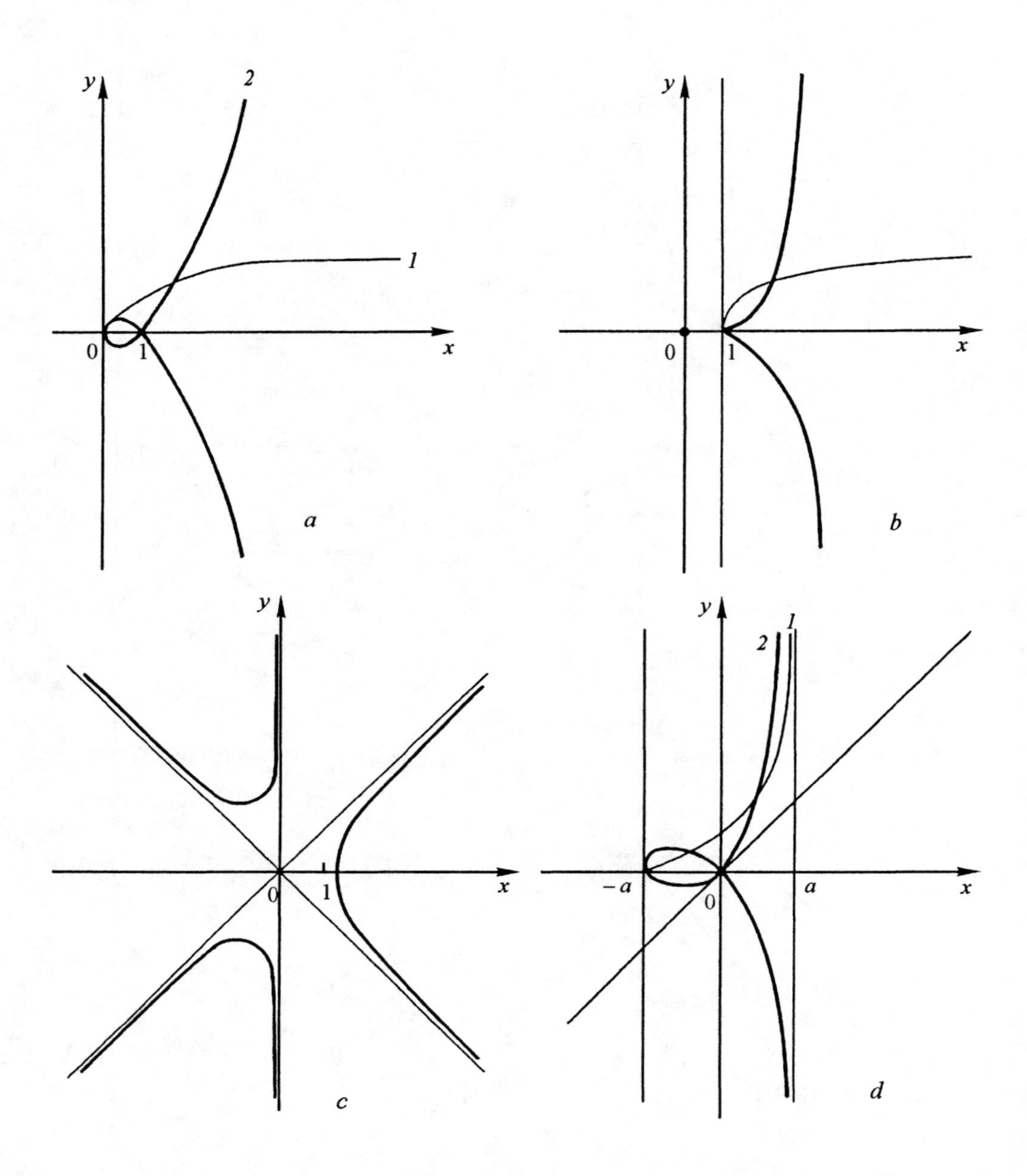

Figure 8.16. The graphs of functions: (*a*) *1*. $y = \sqrt{x}$, *2*. $y^2 = x(x - 1)^2$; (*b*) $y^2 = x^2(x - 1)$ (O is an isolated point); (*c*) $3xy^2 = x^3 - 2$; (*d*) *1*. $y = \frac{a + x}{a - x}$, *2*. $y = x^2 \frac{a + x}{a - x}$, $a = 2$.

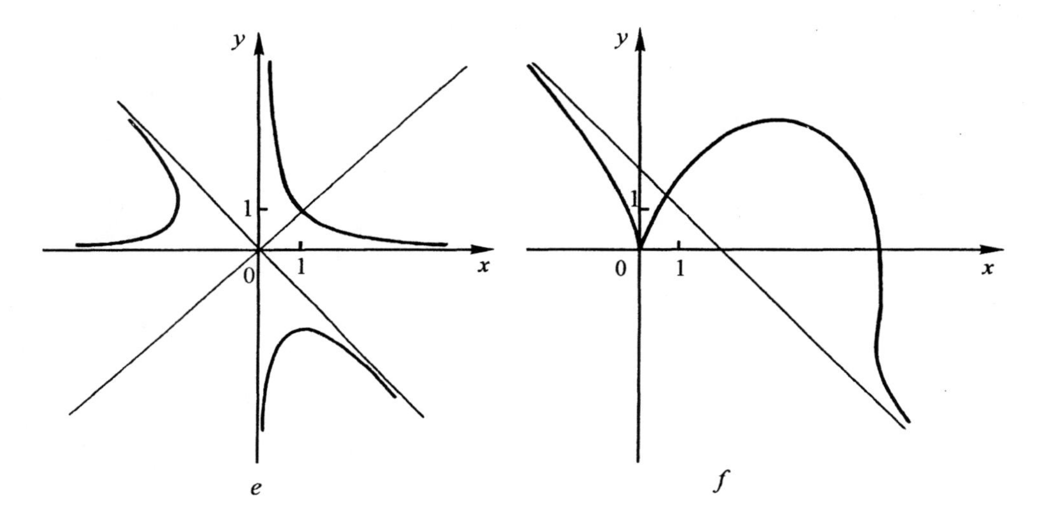

Figure 8.16. The graphs of functions: (*e*) $yx^2 + xy^2 = 2$; (*f*) $y^3 = 6x^2 - x^3$.

In a general case, by using parallel translation and rotation, equation (1) can be reduced to the most simple form resulting in a canonical equation of an ellipse, hyperbola, parabola, or a pair of equations of the first degree.

Indeed, let us use the parallel translation defined by $x = x' + a$, $y = y' + b$, where x', y' are new coordinates a and b are coordinates of the new origin. Substituting these x and y into (1), we obtain:

$$Ax'^2 + 2Bx'y' + Cy'^2 + 2(Aa + Bb + D)x' +$$
$$+ 2(Ba + Cb + E)y' + Aa^2 + 2Bab + Cb^2 + 2Da + 2Eb + D = 0. \quad (2)$$

Choose a and b such that the coefficients of x' and y' become zero,

$$Aa + Bb + D = 0, \quad Ba + Cb + E = 0. \quad (3)$$

If the determinant of system (3) is not equal to zero, $AC - B^2 \neq 0$, then equation (2) becomes:

$$Ax'^2 + 2Bx'y' + Cy' + (Aa^2 + 2Bab + Cb^2 + 2Da + 2Eb + D) = 0. \quad (4)$$

Denote $Aa^2 + 2Bab + Cb^2 + 2Da + 2Eb + D = k$. Then equation (4) can be rewritten as

$$Ax'^2 + 2Bx'y' + Cy'^2 + k = 0. \quad (5)$$

Studying equation (5), there are two cases to be considered.

Case 1: $k = 0$. Equation (5) becomes

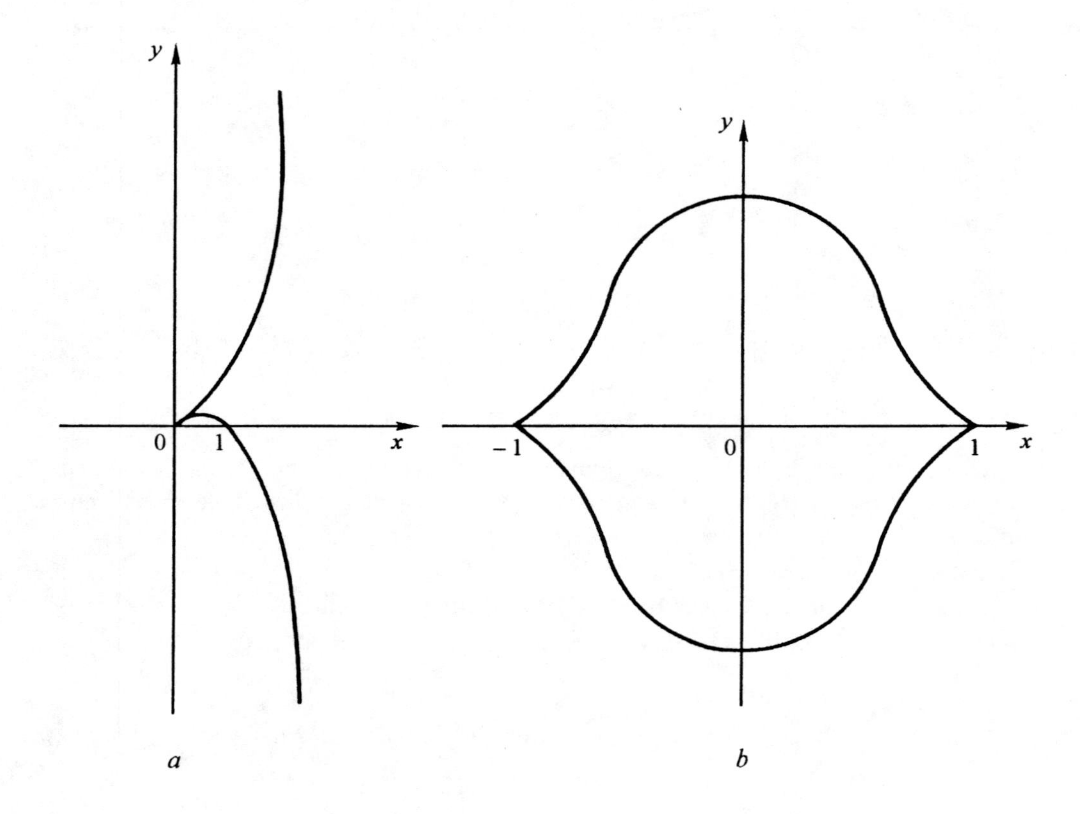

Figure 8.17. The graphs of functions: (*a*) $(y - x^2)^2 = x^5$; (*b*) $y^2 = (1 - x^2)^3$.

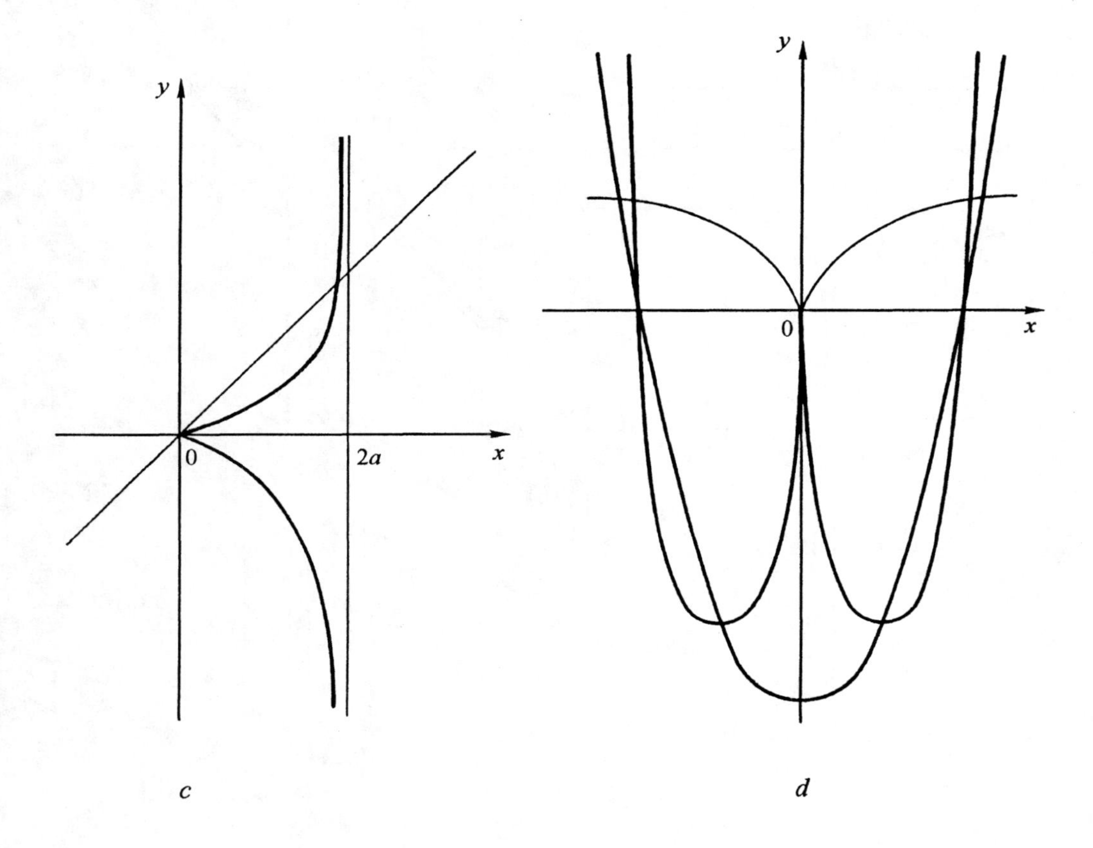

Figure 8.17. The graphs of functions: (*c*) $y^2(2a - x) = x^3$, $a > 0$;
(*d*) $y^3 = x^2(x^2 - 4)^3$.

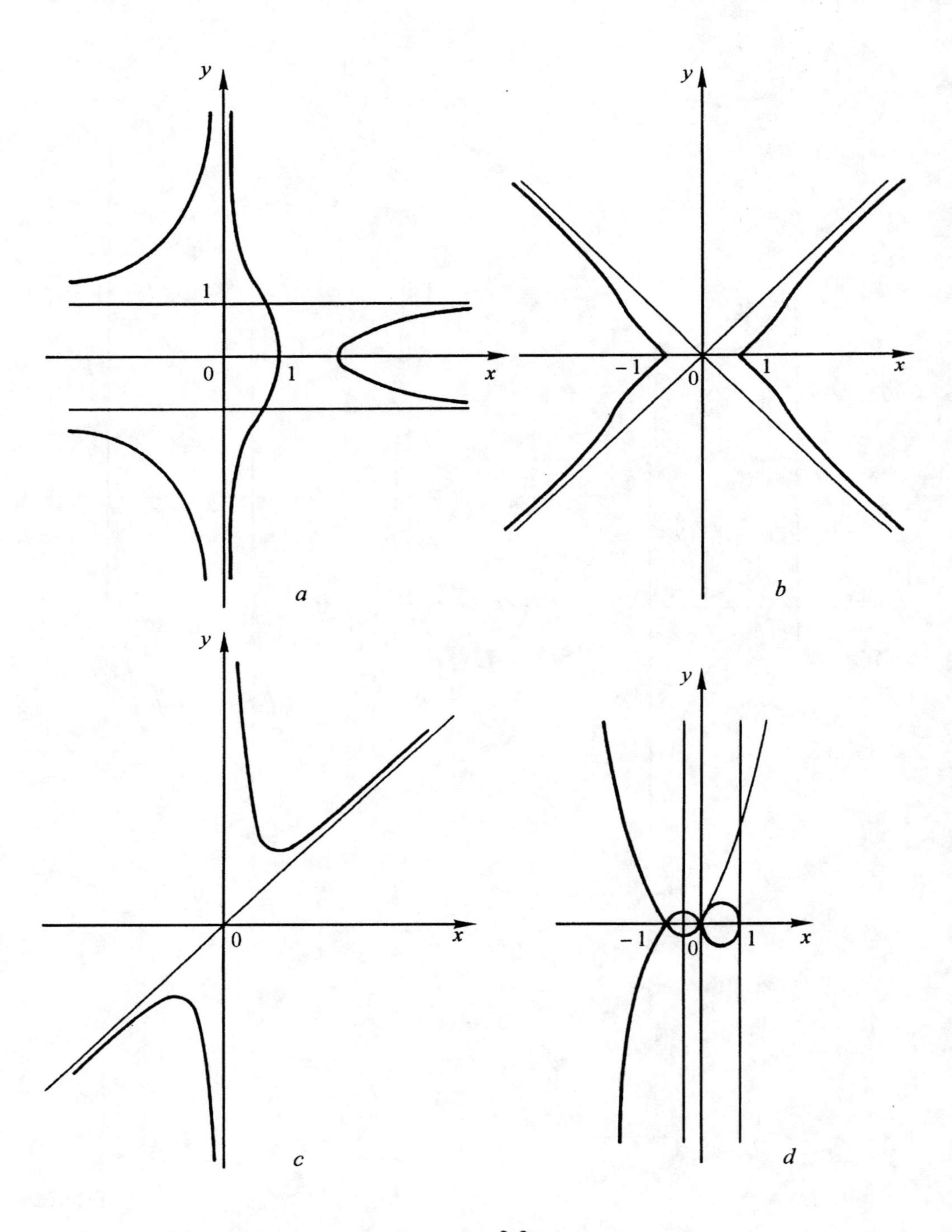

Figure 8.18. The graphs of functions: (*a*) $x^2y^2 = (x-1)(x-2)$;
(*b*) $y^2x^4 = (x^2-1)^3$; (*c*) $x^3(y-x) = x^3+1$; (*d*) $y^2 = x^2(x+1)^2(1-x)$.

$$Ax'^2 + 2Bx'y' + Cy'^2 = 0. \qquad (6)$$

Then: a) if $A = C = 0$, $B \neq 0$ (since $AC - B^2 \neq 0$) equation (5) takes the form $2Bx'y' = 0$. Whence $x' = 0$, $y' = 0$. Hence, we have two straight lines (the new coordinate axes) intersecting at the point (0;0). b) If one of the coefficients A or C is different from zero, say $A \neq 0$, then, by multiplying equation (6) by A, we get $A^2x'^2 + 2ABx'y' + ACy'^2 = 0$. Let us write it as

$$(Ax' + By')^2 + (AC + B^2)y'^2 = 0. \qquad (7)$$

If $AC - B^2 > 0$, equation (7) becomes the point (0;0); if $AC - B^2 < 0$, the equation gives two intersecting straight lines, $Ax' + By' = \pm\sqrt{AC - B^2}\, y'$. Hence, if $k = 0$, equation (5) degenerates into two straight lines or a point*.

Case 2: $k \neq 0$. To study equation (5), make a rotation of the coordinates $Ox'y'$ by an angle α. Make a change of coordinates by the formulas:

$$x' = \bar{x}' \cos\alpha - \bar{y}' \sin\alpha, \quad y' = \bar{x}' \sin\alpha + \bar{y}' \cos\alpha,$$

$$A_1\bar{x}'^2 + 2B_1\bar{x}'\bar{y}' + C_1\bar{y}'^2 + k = 0, \qquad (8)$$

where $A_1 = A\cos^2\alpha + 2B\cos\alpha\sin\alpha + C\sin^2\alpha$, $B_1 = (C - A)\sin\alpha\cos\alpha + B(\cos^2\alpha - \sin^2\alpha)$, $C_1 = A\sin^2\alpha - 2B\cos\alpha\sin\alpha + C\cos^2\alpha$..

Choose α in such a way that $B_1 = 0$, $(C - A)\sin\alpha\cos\alpha + B(\cos^2\alpha - \sin^2\alpha) = 0$, or $\frac{C - A}{2}\sin 2\alpha + B\cos 2\alpha = 0$. Whence,

$$\tan 2\alpha = -\frac{2B}{C - A}. \qquad (9)$$

Equation (9) has an infinite number of solutions but there are only four values that are essentially different. Indeed, if α_0 is one of the angles satisfying (9), then it follows that $2\alpha = 2\alpha_0 + \pi k$ (k is an integer). Hence, $\alpha = \alpha_0 + \frac{\pi}{2}k$, i.e. $\alpha = \alpha_0$ or $\alpha = \alpha_0 + \frac{\pi}{2}$, or $\alpha = \alpha_0 + \pi$, or $\alpha = \alpha_0 + \frac{3}{2}\pi$. All other values of α give the same position of the new coordinate system. If $A = C$, then the value $\frac{2B}{A - C}$ is unbounded, and $2\alpha = \pm\pi/2$, i.e. $\alpha = \pm\pi/4$. If $A = C$ and $B = 0$, the expression $2B/(A - C)$ is not defined and (4) gives an equation of a circle.

With condition (9), equation (8) becomes $A_1x'^2 + B_1y'^2 + k = 0$, where A_1 and C_1 are determined from the system of equations $A_1 + C_1 = A + C$, $A_1C_1 + B_1^2 = AC - B^2$ (invariants of a second-degree curve).

---

* A circle or ellipse could be real or imaginary.

Let us consider the curve defined by (1) in the case where $AC - B^2 = 0$. Suppose that A and C are not zero, $A \neq 0$, $C \neq 0$. Multiplying equation (1) by A we get:

$$A^2x^2 + 2ABxy + ACy^2 + 2ADx + 2AEy + AF = 0.$$

Since $AC - B^2 = 0$, the equation can be written as:

$$(Ax + By + \lambda)^2 + 2(AD - \lambda A)x + 2(AE - \lambda B)y + (AF - \lambda^2) = 0, \tag{10}$$

where λ is an arbitrary parameter.

Choose a new coordinate system such that the straight line $Ax + By + \lambda = 0$ is the x-axis, and the straight line $2(AD - \lambda A)x + 2(AC - \lambda B)y + AF - \lambda^2 = 0$ is the y-axis. Since these lines are orthogonal, $2(AD - \lambda A)A + 2(AE - \lambda B)B = 0$, whence $\lambda = \dfrac{A^2D - ABE}{A^2 + B^2}$. Substituting λ into this equation we get:

$$(Ax + By + \lambda)^2 + m(Bx - Ay) + n = 0, \tag{11}$$

where m and n are determined from coefficients of equation (1).

Let $m = 0$. Then equation (11) becomes:

$$(Ax + By + \lambda)^2 + n = 0. \tag{12}$$

If $n > 0$, equation (12) defines an imaginary set of points; if $n = 0$, the equation degenerates into the straight line $Ax + By + \lambda = 0$; and if $n < 0$, i.e. $n = -l^2$, we get two parallel straight lines $Ax + By + \lambda - l = 0, Ax + By + \lambda + l = 0$.

Let $m \neq 0$, $m/n = p$. Then equation (11) takes the form:

$$(Ax + By + \lambda)^2 + m(Bx - Ay + p) = 0. \tag{13}$$

Choose a new coordinate system as follows: for the x'-axis take the straight line $Ax + By + \lambda = 0$; for the y'-axis — $Bx - Ay + p = 0$. Relative to the new coordinates, formulas that define the transformation become

$$x' = \pm\frac{Bx - Ay + p}{\sqrt{A^2 + B^2}}, \quad y' = \pm\frac{Ax + By + \lambda}{\sqrt{A^2 + B^2}},$$

and equation (13) assumes the form

$$(A^2 + B^2)y'^2 + m(A^2 + B^2)^{1/2}x' = 0,$$

or $y'^2 = \pm\dfrac{m}{\sqrt{A^2 + B^2}}x'$, thus obtaining an equation of a parabola with the sign of $\sqrt{A^2 + B^2}$ coinciding with the sign of m.

8.4. Curves defined by a third-degree polynomial

It is convenient to classify curves of degree three following Newton (see, for example, [4]). By using elementary transformations, a general equation of degree three,

$$Ax^3 + 3Bx^2y + 3Cxy^2 + Dy^3 + 3Ex^2 + 6Fxy + 3Gy^2 + Hx + Ty + K = 0,$$

can be reduced into one of the following canonical forms:

a) $xy^2 + ey = ax^3 + bx^2 + cx + d$;

b) $xy = ax^3 + bx^2 + cx + d$;

c) $y^2 = ax^3 + bx^2 + cx + d$;

d) $y = ax^3 + bx^2 + cx + d$.

Next, the following equation of degree four or three is considered:

$$ax^4 + bx^3 + cx^3 + d{\cdot}x + \frac{1}{4}e^2 = 0 \text{ or } ax^3 + bx^2 + cx + d = 0.$$

This equation is called *characteristic.*

Depending on various relations between roots of the characteristic equation, Newton has divided all curves of degree three into seven classes.

Curves that can be reduced to canonical form a) are split into four classes: *hyperbolic hyperbola* ($a > 0$), *defective hyperbola* ($a < 0$), *parabolic hyperbola* ($a = 0, b \neq 0$), and *conic section hyperbolisms* ($a = b = 0$).

Curves that can be reduced to canonical forms b), c), and d) contain, correspondingly, a single class: *trident* (or *hyperbola parabolism*), *prolate parabola*, or *cubical parabola.*

Let us consider the dependence between roots of the characteristic equation and the form of the curve of degree three.

8.4.1. Hyperbolic hyperbola

A hyperbolic hyperbola defined by the equation $xy^2 + ey = ax^3 + bx^2 + cx + d$ ($a > 0$) has three usual asymptotes: $x = 0, y = \pm\sqrt{a}\ (x + b/2a)$.

They intersect at the points $(0; b/2\sqrt{a}), 0; -b/2\sqrt{a}), -b/2a; 0)$. These points form vertices of the asymptotic triangle.

In its turn, a hyperbolic hyperbola is subdivided into four types: with no diameters ($e \neq 0$), with one diameter ($e = 0,\ b^2 \neq 4ac$), with three diameters ($e = 0,\ b^2 = 4ac$), and hyperbolic hyperbola with asymptotes intersecting in one point ($b = 0$).

Hyperbolic hyperbola with no diameters

The characteristic equation for a hyperbolic hyperbola with no diameters is

$$ax^4 + bx^3 + cx^2 + dx + \frac{1}{4}e^2 = 0$$

and has four roots: x_1, x_2, x_3, x_4.

Let us consider nine possible cases.

1. All roots of the characteristic equation are real, distinct, and all of them have the same sign.

Suppose that $0 < x_1 < x_2 < x_3 < x_4$. In the interval $[x_2; x_3]$, where y assumes only finite values, the curve has a closed form (an oval). This oval lies entirely inside the asymptotic triangle. On the intervals $(x_1; x_2)$, $(x_3; x_4)$, the value of y takes imaginary values. Save for the oval, the curve has three unbounded hyperbolic branches. The curve has no double points. It intersects the asymptotes at a finite distance in the points

$$(0; \frac{d}{e}), \quad \left(\frac{4ad \pm 2be\sqrt{a}}{b^2 - 4ac \mp 4ae\sqrt{a}}; \quad \frac{8a^2d + b^3 - 4abc}{2a(b^2 - 4ac \mp 4ae\sqrt{a})}\right)$$

and intersects the x-axis (see Figure 8.19).

2. All roots of the characteristic equation are real and distinct; two of them are positive, and two are negative.

Let $x_1 < x_2 < 0 < x_3 < x_4$. The curve consists of three unbounded branches, one of which is serpentine and approaches the asymptote $x = 0$ from two different sides; the other two branches are hyperbolic. The branches do not intersect each other, but all of them intersect the x-axis. The serpentine branch intersects all three asymptotes at a finite distance (Figure 8.20).

3. All roots of the characteristic equation are real, two of them coincide, and the multiple root is greater or less than the two distinct roots and has opposite sign.

Let $x_1 = x_2 < 0 < x_3 < x_4$. The variable y takes real values if $x \le x_3$ and $x \ge x_4$. If $x_1 = x_2 = x$, then the curve has a node. The curve consists of three unbounded branches, two of which intersect in the node (the x-coordinate of the point equals the double root of the characteristic equation). The node lies outside of the asymptotic triangle. The curve intersects the x-axis in three points, and intersects all asymptotes at a finite distance (Figure 8.21).

4. All conditions are as in item **3** but signs of all roots of the characteristic equations are the same.

Let $0 < x_1 < x_2 < x_3 = x_4$. The variable y takes real values if $x \le x_1$ and $x \ge x_2$. The curve has three branches with a loop, since it takes only finite values in the interval $[x_2; x_3]$. The node is located inside the asymptotic triangle (Figure 8.22).

5. Conditions are the same as in **4** but the multiple root is greater than one and less than another root of multiplicity one.

Let $0 < x_1 < x_2 = x_3 < x_4$. The variable y does not take real values in the intervals $(x_1; x_2)$ and $(x_2; x_4)$, hence the point $(x_2; -e/2x_2)$ is isolated. It is located inside the asymptotic triangle. Except for the isolated point, the curve has three unbounded branches on the intervals $(-\infty; 0)$, $(-\infty; x_1]$, and $[x_4; +\infty)$, approaches the asymptote $x = 0$ from different sides, and intersects all three asymptotes at a finite distance (Figure 8.23).

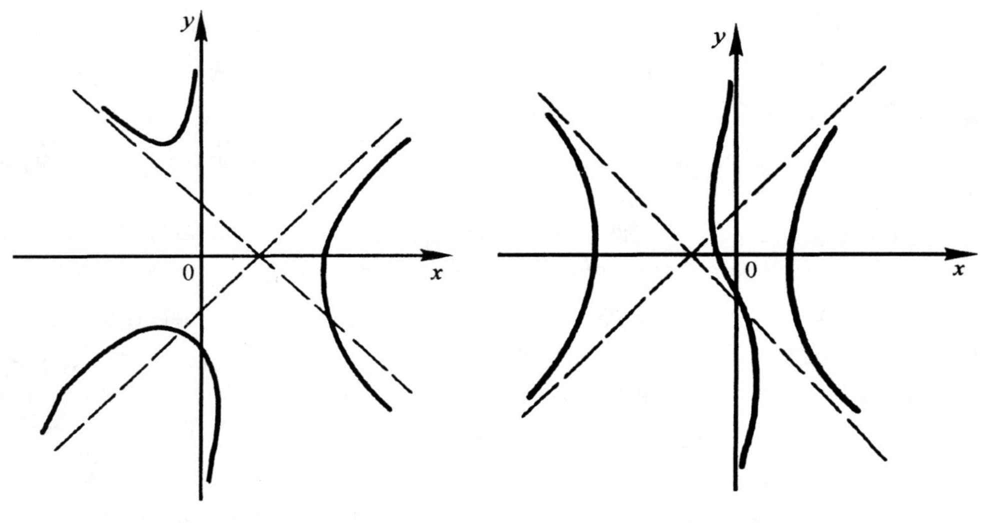

Figure 8.19 Figure 8.20

Figure 8.19. The hyperbolic hyperbola, $xy^2 + ey = ax^3 + bx^2 + cx + d$, $0 < x_1 < x_2 < x_3 < x_4$.

Figure 8.20. The hyperbolic hyperbola, $x_1 < x_2 < 0 < x_3 < x_4$.

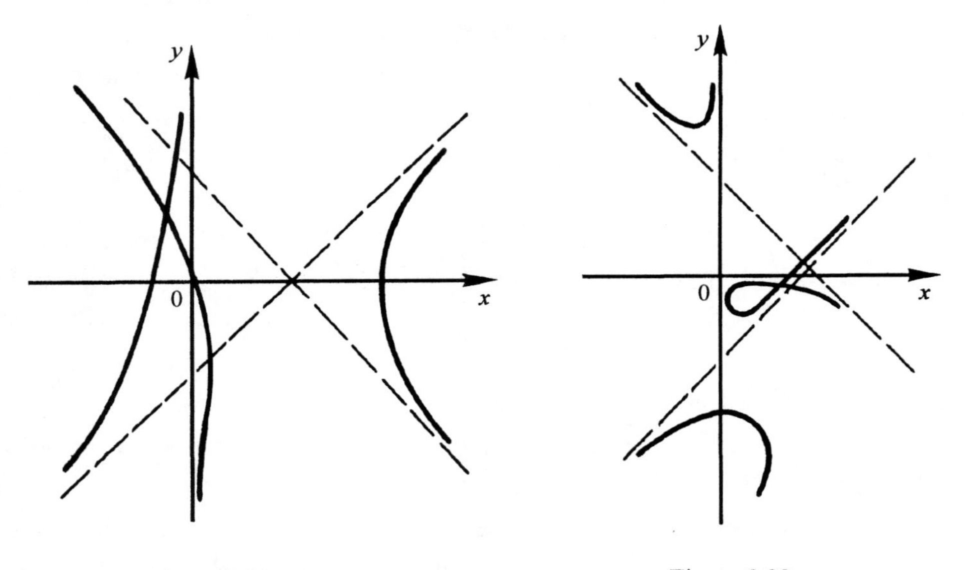

Figure 8.21 Figure 8.22

Figure 8.21. The hyperbolic hyperbola, $x_1 = x_2 < 0 < x_3 < x_4$.

Figure 8.22. The hyperbolic hyperbola, $0 < x_1 < x_2 < x_3 = x_4$.

6. All roots of the characteristic equation are real; three of them coincide.

Let $x_1 = x_2 = x_3 < x_4$. The curve has three branches with a turning point of genus $(x_1; -e/2x_1)$ that lies inside the asymptotic triangle (Figure 8.24).

7. Two roots of the characteristic polynomial are complex conjugate; the other two are real and distinct.

Let $x_1 < x_2$, $x_{3,4} = \alpha \pm \beta i$. The variable y takes real values for $x \le x_1$ and $x \ge x_2$. The curve has three branches of hyperbolic type (Figure 8.25).

8. Two roots of the characteristic equation are complex conjugate; the other two are real and equal, $x_1 = x_2$, $x_{3,4} = \alpha + i\beta$.

The curve has three unbounded branches of hyperbolic type. Two of them intersect in the node $(x_1; -e/2x_1)$ that lies outside of the asymptotic triangle (Figure 8.26).

9. All roots of the characteristic equation are complex, pairwise complex conjugate.

The curve has three branches: two hyperbolic and one serpentine (Figure 8.27).

Hyperbolic hyperbola ($e = 0$, $b^2 \ne 4ac$, $b \ne 0$) with one diameter $y = 0$

The characteristic equation is given by $ax^4 + bx^3 + cx^2 + dx = 0$.

Let us look at possible cases.

1. All roots of the characteristic equation are real, distinct, and have the same sign.

Let $x_1 = 0 < x_2 < x_3 < x_4$. The variable y takes real values on the intervals $[-\infty; 0)$, $[x_2; x_3]$, $[x_4; +\infty)$. The curve has an oval on the interval $[x_1; x_2]$ lying inside the asymptotic triangle (Figure 8.28).

2. All roots of the characteristic equation are real, there are roots that have opposite signs, and $|x_1| > x_3 + x_4$ if $x_1 < x_2 = 0 < x_3 < x_4$.

The variable y takes real values on intervals $(-\infty; x_1]$, $(0; x_3]$, $[x_4; +\infty)$. The curve has three separate branches. Two of them are of hyperbolic type, and the third one has conchoid type; it is located on the side of the asymptote $x = 0$ and approaches it from both sides starting from the base of the asymptotic triangle (Figure 8.29).

3. Conditions are as in item **2** but $|x_1| < x_3 + x_4$.

The conchoid type branch approaches the asymptote $x = 0$ starting from the vertex of the asymptotic triangle (Figure 8.30).

4. Two roots of the characteristic equation are equal and less than the other distinct roots, moreover, the sign of the multiple root is opposite to the sign of the other one. Also $|x_1| > 1/2x_4$ if $x_1 = x_2 < x_3 = 0 < x_4$.

The variable y takes real values on the intervals $(-\infty; 0)$ and $[x_4; +\infty)$. The curve has hyperbolic branches, two of which intersect in the node $(x_1; 0)$ that lies outside of the asymptotic triangle on the side of its vertex (Figure 8.31).

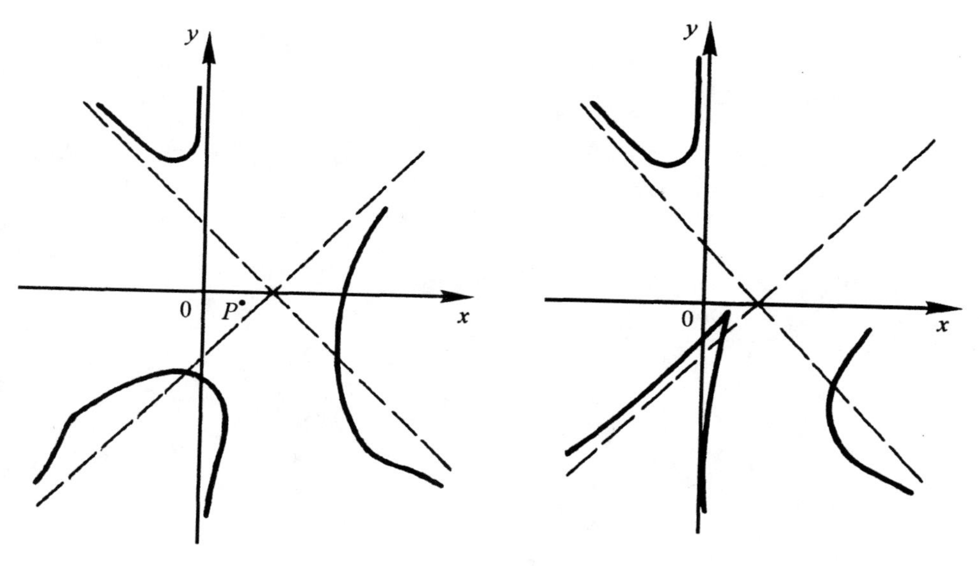

Figure 8.23 Figure 8.24

Figure 8.23. The hyperbolic hyperbola, $0 < x_1 < x_2 = x_3 < x_4$.

Figure 8.24. The hyperbolic hyperbola, $x_1 = x_2 = x_3 < x_4$.

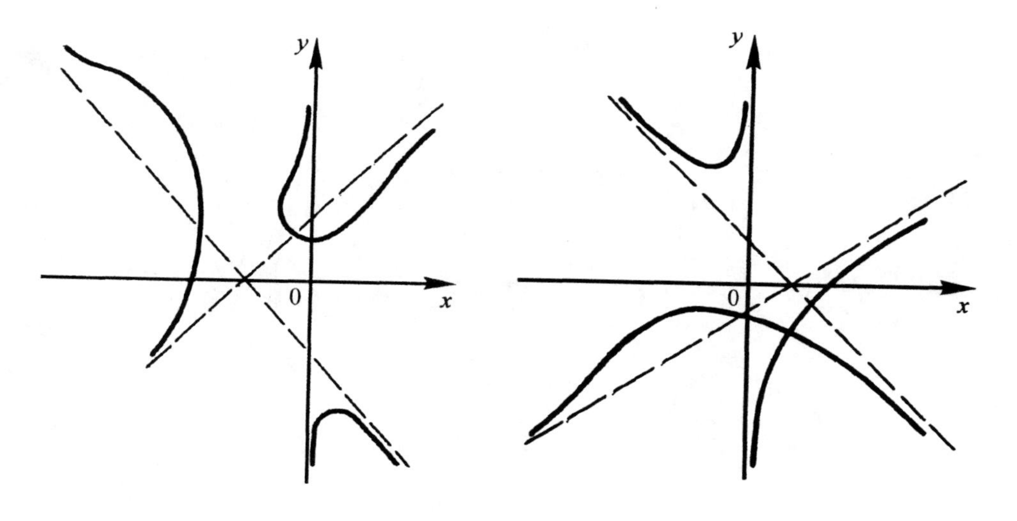

Figure 8.25 Figure 8.26

Figure 8.25. The hyperbolic hyperbola, $x_1 < x_2$, $x_{3,4} = \alpha \pm \beta i$.

Figure 8.26. The hyperbolic hyperbola, $x_1 < x_2$, $x_{3,4} = \alpha \pm i\beta$.

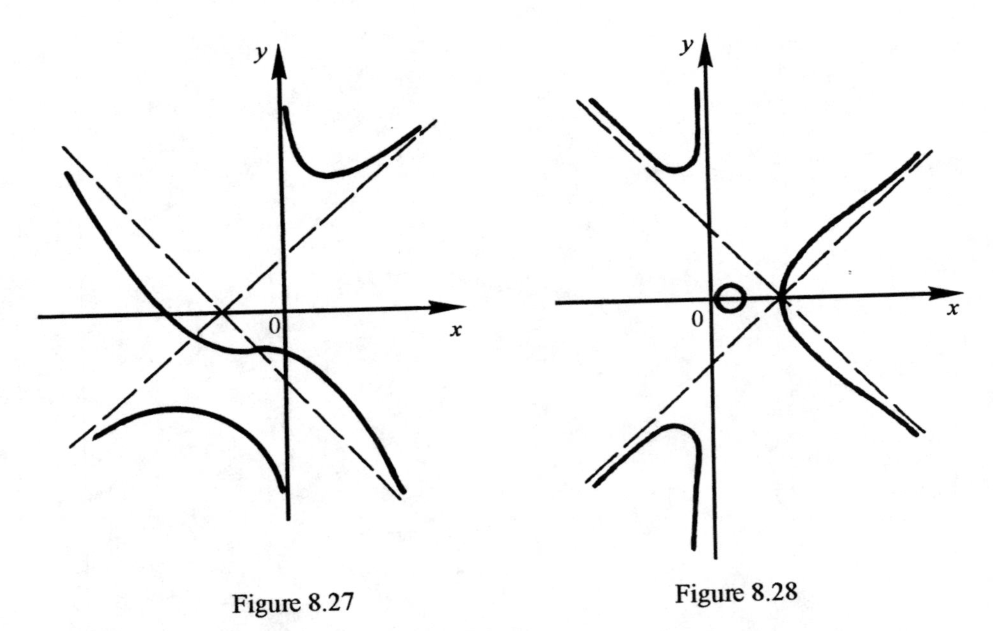

Figure 8.27 Figure 8.28

Figure 8.27. The hyperbolic hyperbola, all roots are complex, pairwise complex conjugate.

Figure 8.28. The hyperbolic hyperbola with one diameter, $x_1 = 0 < x_2 < x_3 < x_4$.

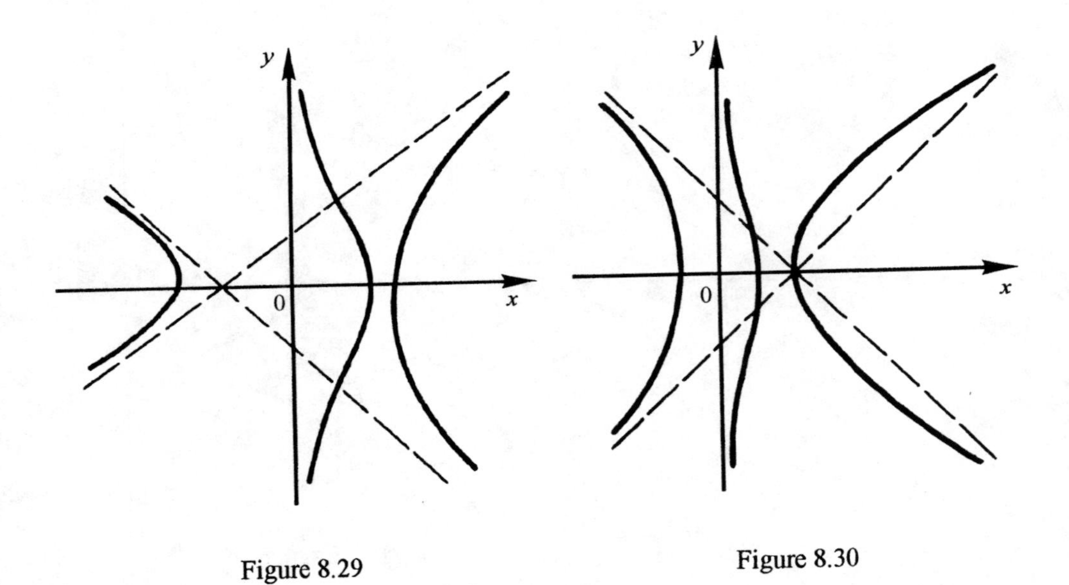

Figure 8.29 Figure 8.30

Figure 8.29. The hyperbolic hyperbola, $|x_1| > x_3 + x_4$, $x_1 < x_2 = 0 < x_3 < x_4$.

Figure 8.30. The hyperbolic hyperbola, $|x_1| < x_3 + x_4$.

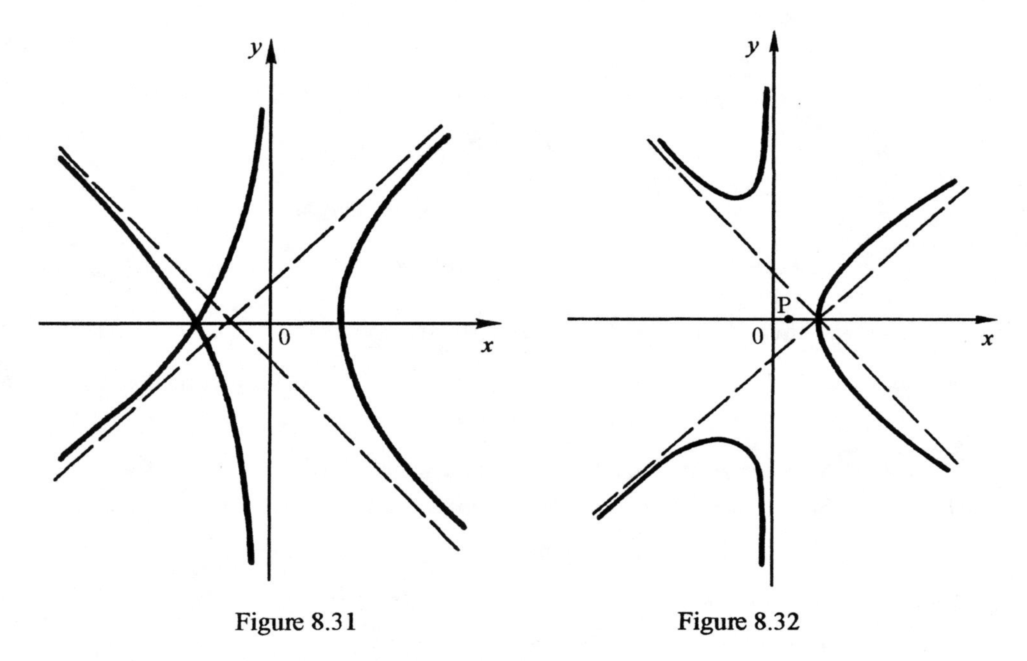

Figure 8.31. The hyperbolic hyperbola with one diameter, $x_1 = x_2 < x_3 = 0 < x_4$, $|x_1| > 1/2x_4$, at $|x_1| < 1/2x_4$.

Figure 8.32. The hyperbolic hyperbola with one diameter, $x_1 = 0 < x_2 = x_3 < x_4$.

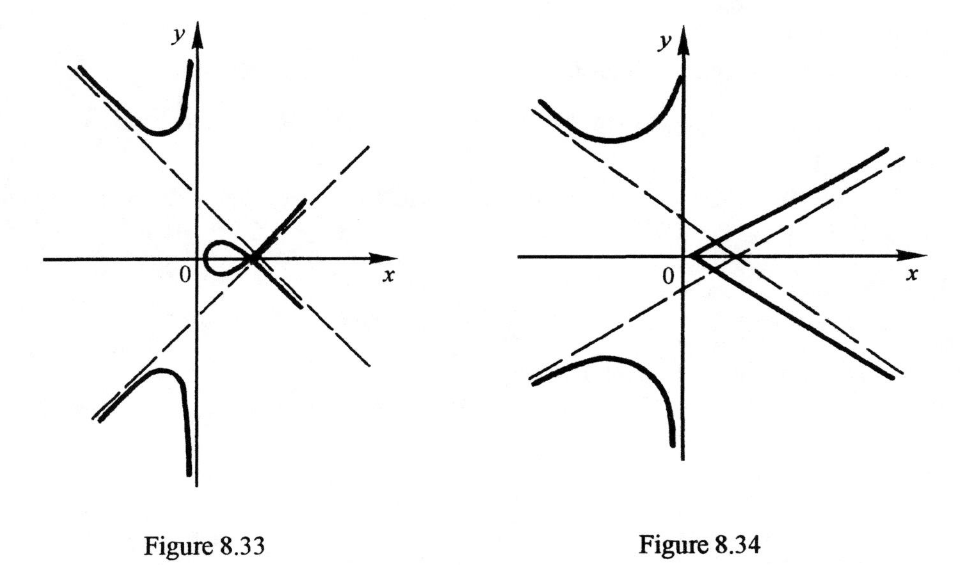

Figure 8.33

Figure 8.34

Figure 8.33. The hyperbolic hyperbola with one diameter, $x_1 = 0 < x_2 < x_3 = x_4$.

Figure 8.34. The hyperbolic hyperbola with one diameter, all roots are real, three of them are equal.

5. Conditions are the same but $|x_1| < 1/2x_4$.
The node lies outside of the asymptotic triangle on the side of its base (Figure 8.32).

6. The multiple root is less than one and greater than the second distinct root, and the signs of all roots are the same.

Let $x_1 = 0 < x_2 = x_3 < x_4$. The curve has three unbounded branches and the isolated point $(x_2; 0)$ lying inside the asymptotic triangle. There are two possible cases: if $x_2 < \frac{1}{2}x_4$, then all three branches are inscribed into the angles formed by the asymptotes; if $\frac{1}{2}x_4 < x_2 < x_4$, then the branch that passes through the point $(x_4; 0)$ is circumscribed about the corresponding angle formed by the asymptotes (Figure 8.33).

7. The multiple root is greater than two distinct roots, and all roots have the same sign.

Let $x_1 = 0 < x_2 < x_3 = x_4$. The curve has three unbounded branches. One of them forms a loop with the node at $(x_3; 0)$ located inside the asymptote triangle (Figure 8.34).

8. All roots of the characteristic equation are real and three of them coincide.
The curve has a turning point of the first kind at $(x_1; 0)$ (Figure 8.35).

9. The characteristic equation has two real roots, 0 and $r \neq 0$, and two complex conjugate roots, $\alpha \pm i\beta$, and α and r have the same sign, and $|r| \leq 2|\alpha|$.

The curve has three separate branches with no singularities, one of which is circumscribed about a vertex of the asymptotic triangle (Figure 8.36).

10. Conditions are the same but $|r| > 2|\alpha|$.
In this case the mentioned branch (item **9**) is inscribed in the angle formed by the asymptotes.

11. Conditions are the same as in item **9** but r and α have opposite signs, $|r| < 2|\alpha|$, $r(r - 4\alpha) > 4\beta^2$.

The same branch approaches the asymptotes from the outside of the main side of the asymptotic triangle (Figure 8.37).

12. Conditions are the same as in item **11** but $r(r - 4\alpha) < 4\beta^2$.
The same branch approaches the asymptote from the inside (Figure 8.38).

13. r and α have opposite signs, and $|r| > 2|\alpha|$.
All three branches are inscribed in angles formed by asymptotes.

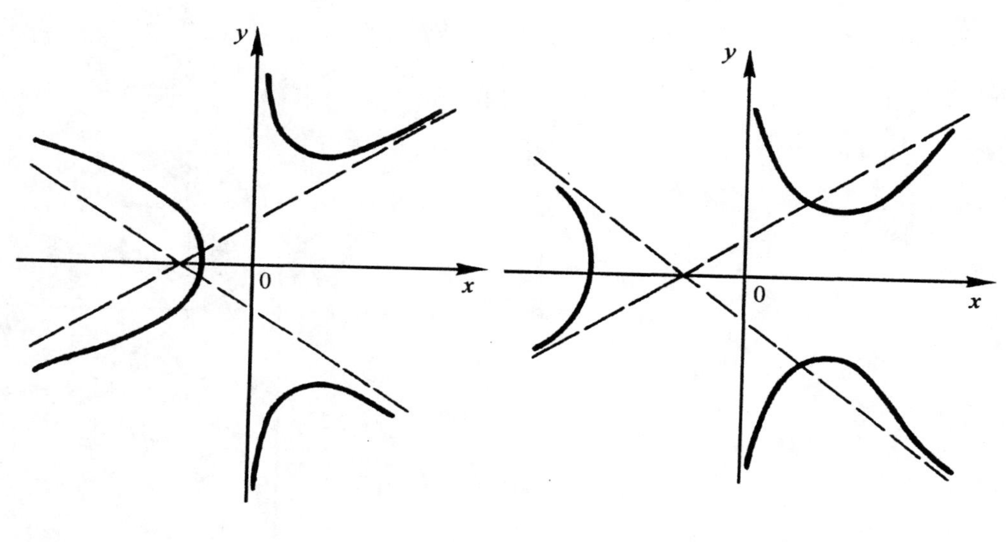

Figure 8.35

Figure 8.36

Figure 8.35. The hyperbolic hyperbola with one diameter, $r \neq 0$, $\alpha \pm i\beta$, $|r| \leq 2|\alpha|$.

Figure 8.36. The hyperbolic hyperbola with one diameter, $|r| > 2|\alpha|$.

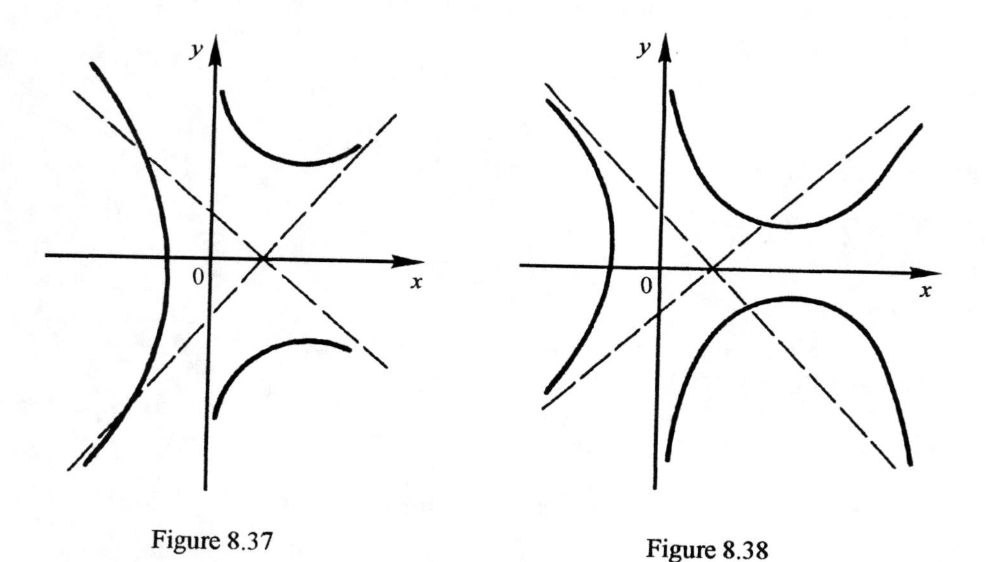

Figure 8.37

Figure 8.38

Figure 8.37. The hyperbolic hyperbola with one diameter, $|r| < 2|\alpha|$, $r(r - 4\alpha) > 4\beta^2$.

Figure 8.38. The hyperbolic hyperbola with one diameter, $r(r - 4\alpha) < 4\beta^2$.

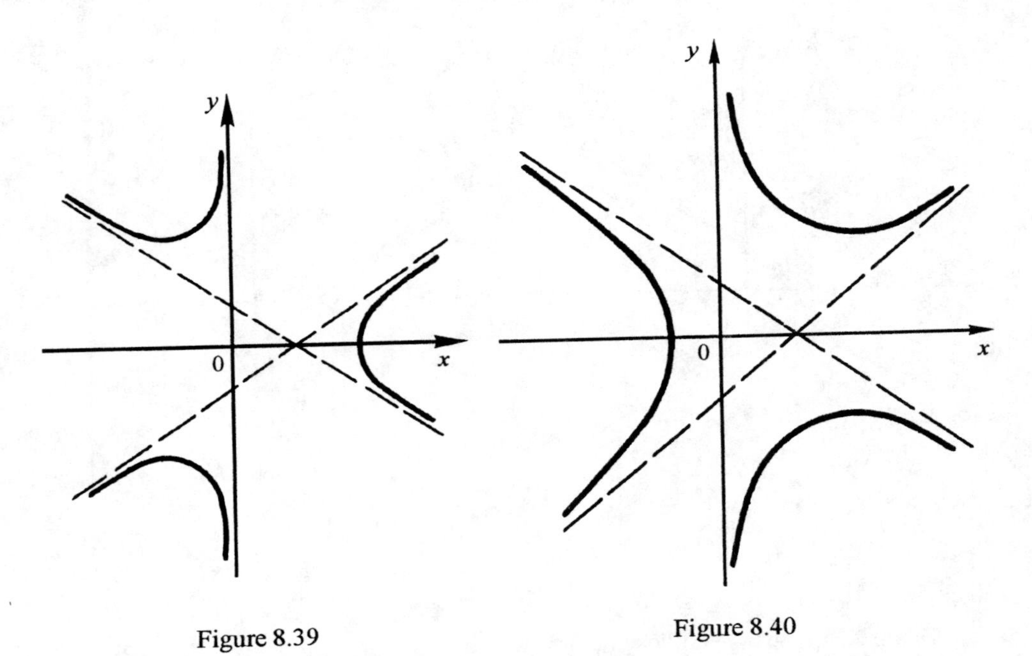

Figure 8.39 Figure 8.40

Figure 8.39. The hyperbolic hyperbola with three diameters (three branches are inscribed in angles vertical to angles of asymptotic triangle).

Figure 8.40. The hyperbolic hyperbola with three diameters (three branches are inscribed in angles of asymptotic triangle and lying outside of it).

Hyperbolic hyperbola with three diameters ($e = 0, b^2 = 4ac$)*

Consider two cases.

1. The curve consists of three branches inscribed in angles vertical to the angles of the asymptotic triangle (Figure 8.39).

2. Three branches of the curve are inscribed in angles of the asymptotic triangle and lie outside of it (Figure 8.40).

Note that a hyperbolic hyperbola with three diameters has three inflection points at infinity.

Hyperbolic hyperbola with asymptotes intersecting in one point ($b = 0$)

These curves are obtained from the previously considered curves in the case where the asymptotic triangle degenerates into a point. In such a case there are curves with no diameter

* If the chords of a curve of degree three parallel to one of its asymptotes have middle points lying on one straight line, then this line is called the *diameter* of the curve.

($e \neq 0$), curves with one diameter ($e = 0$, $c \neq 0$), and curves with three diameters ($e = c = 0$) (Figures 8.41 – 8.44).

8.4.2. Defective hyperbola

A defective hyperbola is given by the equation $xy^2 + ey = ax^3 + bx^2 + cx + d$ ($a < 0$) and has one real asymptote $x = 0$ and two imaginary asymptotes.

Defective hyperbola with no diameters ($e \neq 0$)

The curve has the asymptote $x = 0$ and intersects it in the point $(0; d/e)$. The curve has a point at infinity that is not an inflection point. There are six possible cases.

1. All roots of the characteristic equation $ax^4 + bx^3 + cx^2 + dx + e/4 = 0$ are distinct.

Let $x_1 < x_2 < x_3 < 0 < x_4$. Note that if $a < 0$, it can happen that all roots do not have the same sign. It is clear that y takes real values for $x_1 \leq x \leq x_2$, $x_3 \leq x \leq x_4$. The curve has an oval on $[x_1;x_2]$ and a serpentine on the interval $[x_3; x_4]$ (Figure 8.45).

2. All roots of the characteristic equation are real, and two of them coincide. The multiple root is greater or less than two other roots.

Let $x_1 < 0 < x_2 < x_3 < 0 < x_4$. The variable y takes real values for $x_1 \leq x \leq x_2$ and $x = x_3$. The curve has a serpentine branch and an isolated point at $(x_3; -e/2x_3)$ (Figure 8.46).

3. Same conditions as in item **2** but the multiple root is less than one and greater than the other root at the same time.

Let $x_1 < x_2 = x_3 < 0 < x_4$. The variable y takes real values on $[x_1; x_4]$. The curve has one branch with a loop and a node at the point $(x_2; -e/2x_2)$ (Figure 8.47).

4. Three real roots of the characteristic equation coincide.

Let $x_1 = x_2 = x_3 < 0 < x_4$. The curve has one branch on $[x_1; x_4]$ with a turning point at $(x_1; -e/2x_1)$ (Figure 8.48).

5, 6. Two roots of the characteristic equation are real, and two roots are complex conjugate.

Let $x_1 < x_2$, $x_{3,4} = \alpha \pm \beta i$. The curve has one serpentine branch on $[x_1; x_2]$ which could or not pass through the origin (Figures 8.49 and 8.50).

Defective hyperbola with one diameter ($e = 0$)

The curve has the asymptote $x = 0$ with which it does not intersect at a finite distance. The curve has an inflection point at infinity. There are seven possible cases.

1. All roots of the characteristic equation $ax^4 + bx^3 + cx^2 + dx = 0$ are real, distinct, and have the same sign (Figure 8.51).

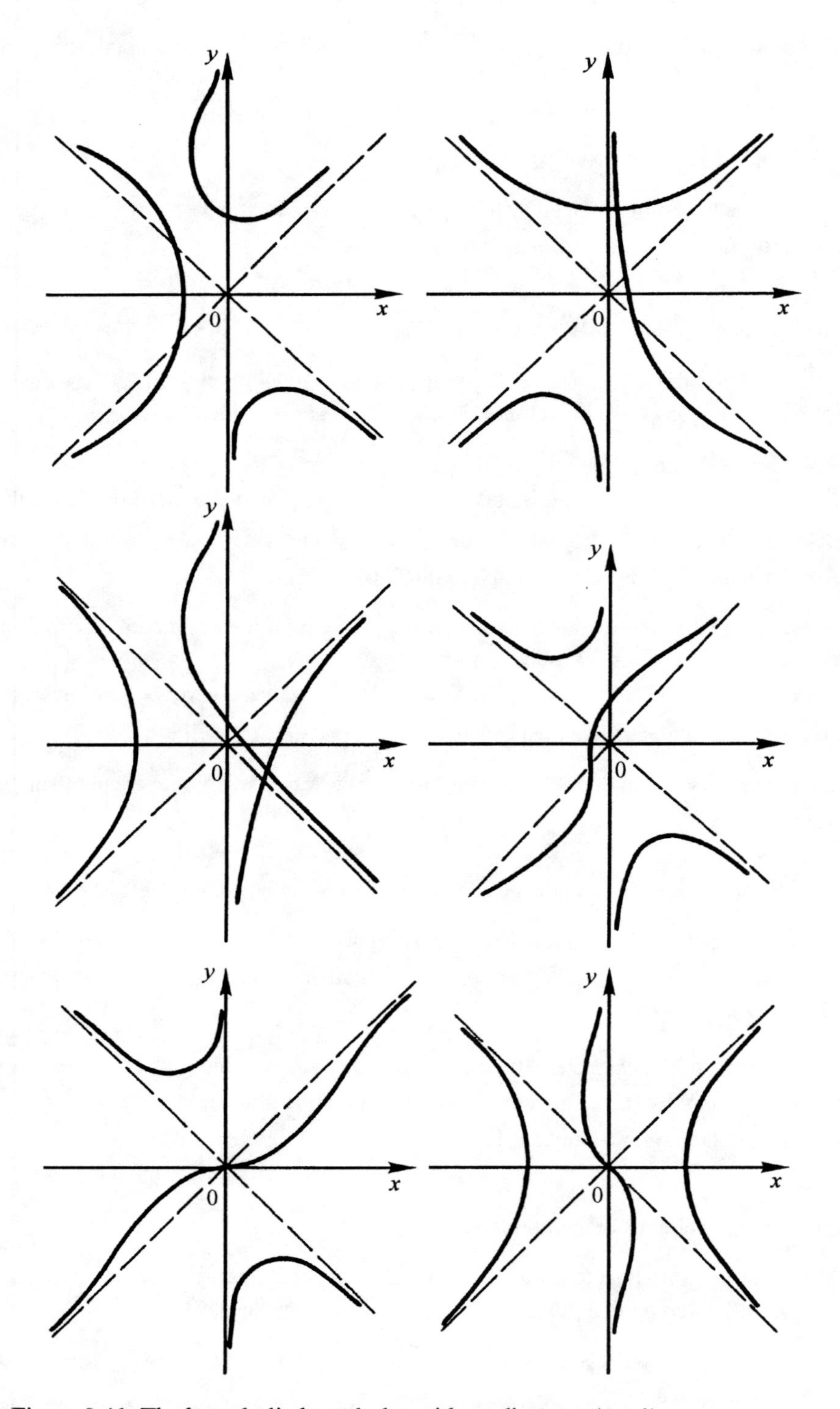

Figure 8.41. The hyperbolic hyperbolas with no diameter ($c \neq 0$).

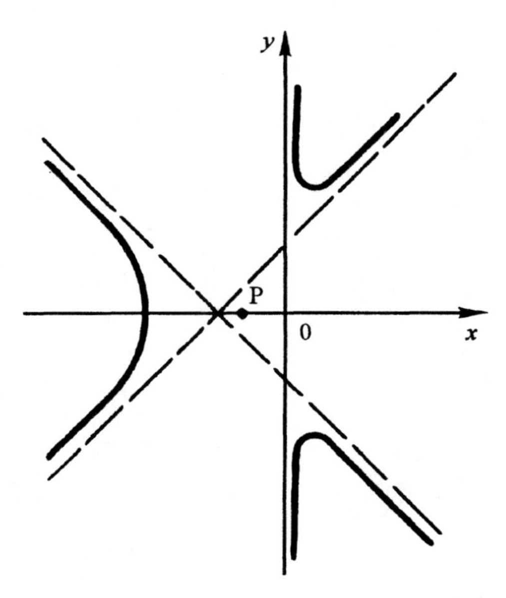

Figure 8.42. The hyperbolic hyperbola with one diameter ($e = 0$, $c \neq 0$)
P is an isolated point.

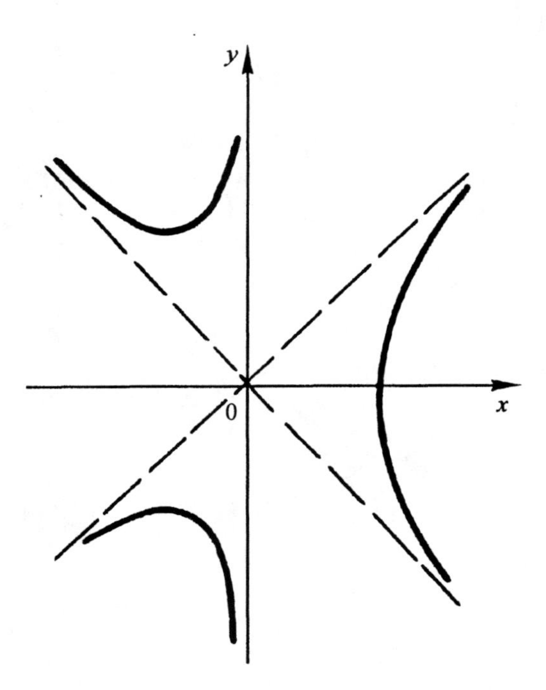

Figure 8.43. The hyperbolic hyperbola with one diameter ($e = 0$, $c \neq 0$).

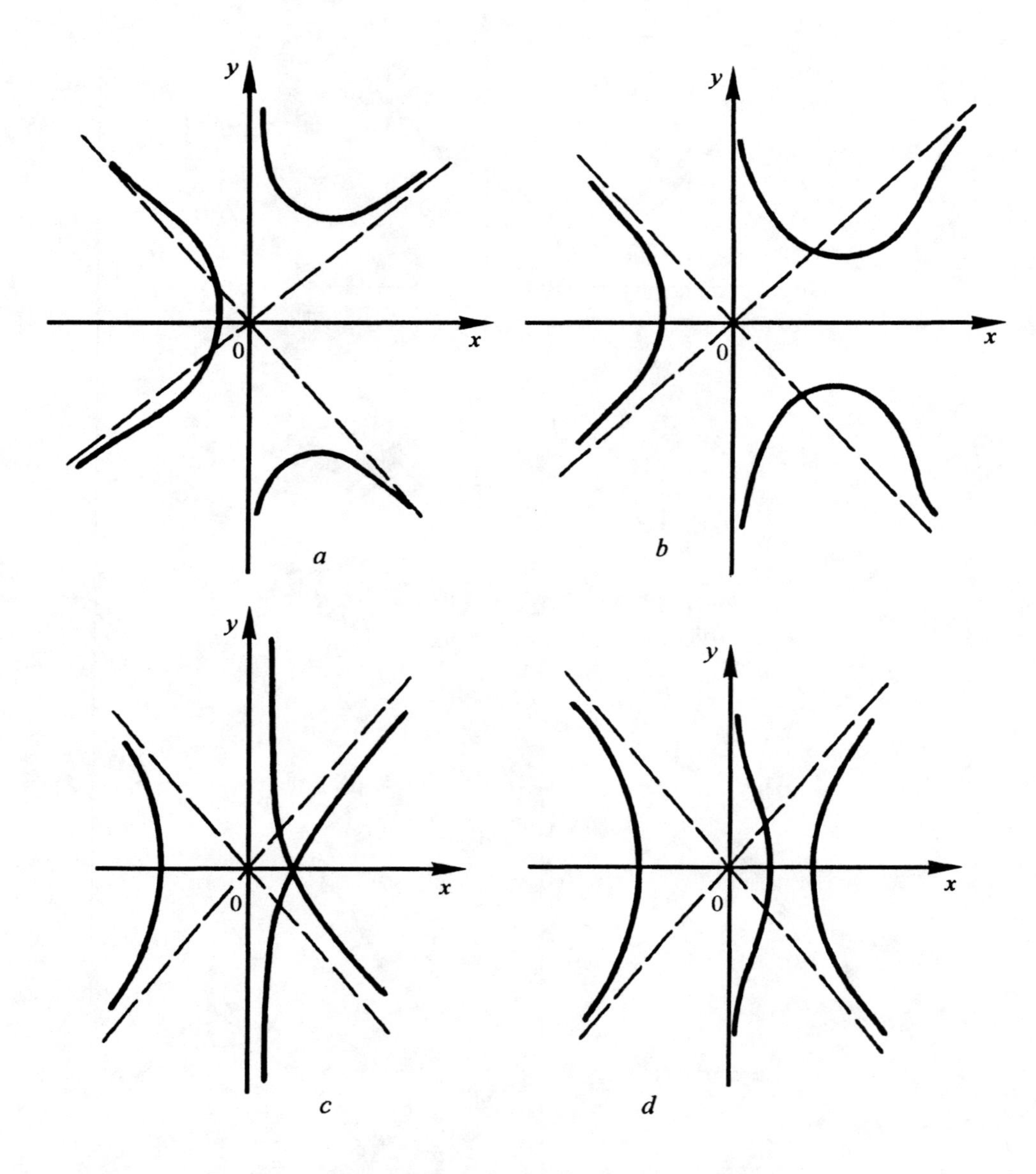

Figure 8.44. The hyperbolic hyperbolas with three diameters ($e = c = 0$).

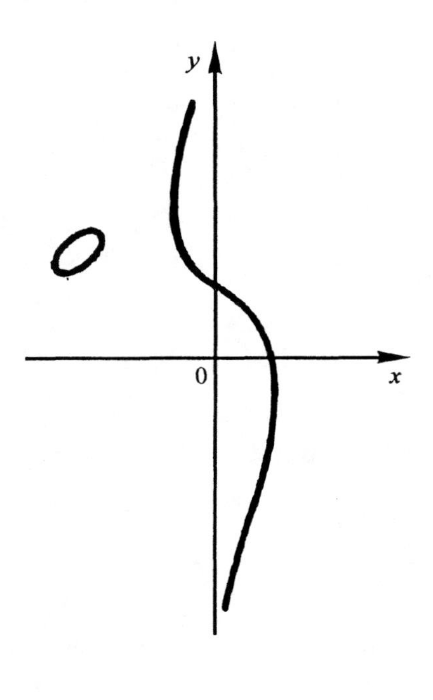

Figure 8.45. The defective hyperbola ($x_1 < x_2 < x_3 < 0 < x_4$).

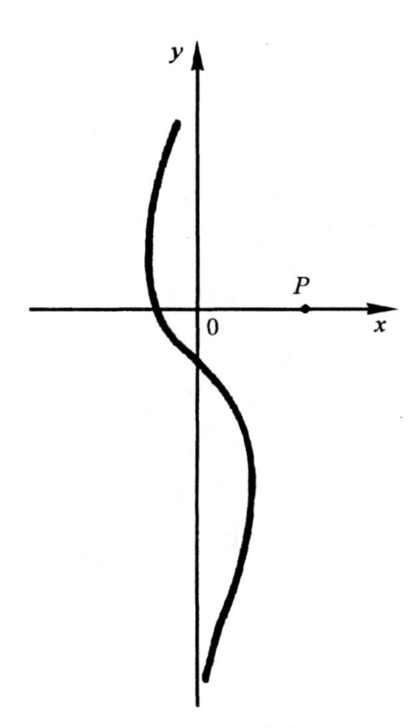

Figure 8.46. The defective hyperbola ($x_1 < 0 < x_2 < x_3 <$ $< 0 < x_4$); P is an isolated point.

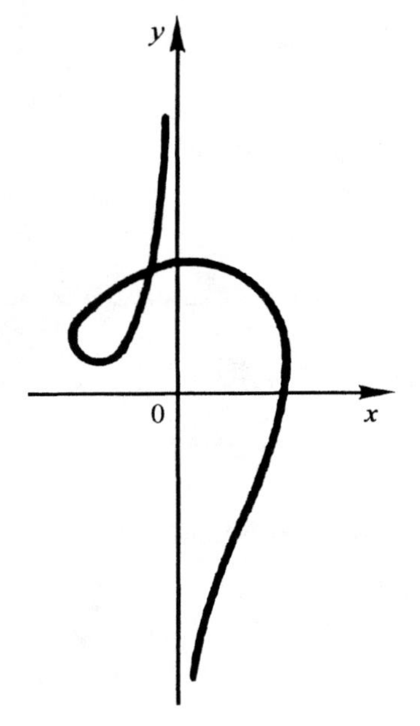

Figure 8.47. The defective hyperbola ($x_1 < x_2 = x_3 < 0 < x_4$).

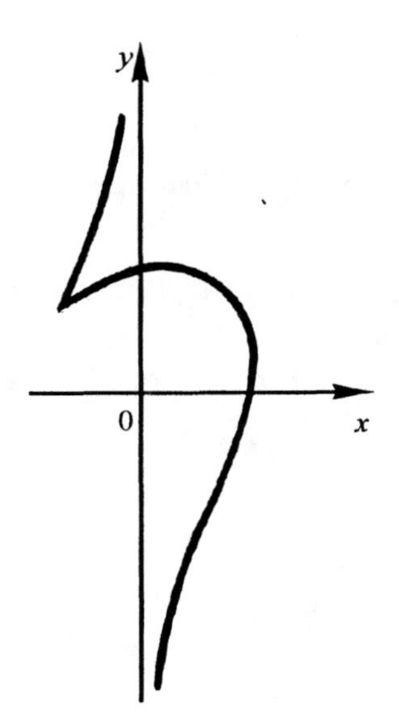

Figure 8.48. The defective hyperbola ($x_1 = x_2 = x_3 < 0 < x_4$).

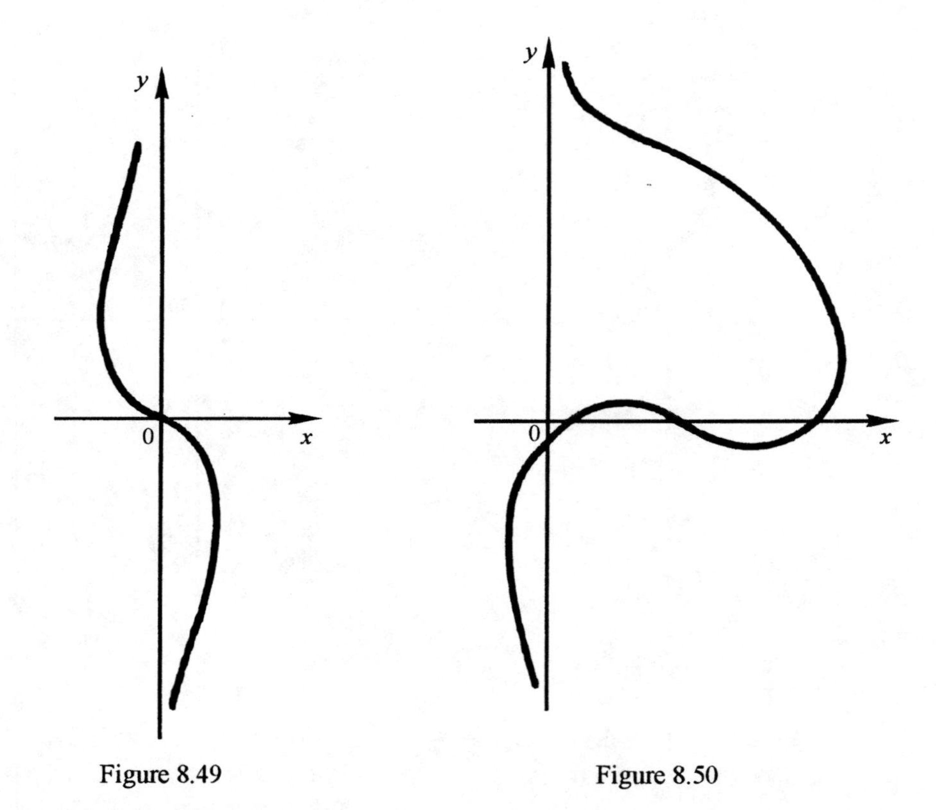

Figure 8.49 Figure 8.50

Figure 8.49. The defective hyperbola ($x_1 < x_2$, $x_{3,4} = \alpha + \beta i$).

Figure 8.50. The defective hyperbola ($x_1 < x_2$, $x_{3,4} = \alpha - \beta i$).

2. All roots of the characteristic equation are real, distinct, but there are roots with opposite signs (Figure 8.52).

3. The root with multiplicity two is greater or less than two simple roots, and all the roots have the same sign.

Let $x_1 = 0 < x_2 < x_3 = x_4$. The curve has one branch on $(0; x_2)$ and an isolated point at $(x_2; -\frac{e}{2x_3})$. This branch approaches the asymptote $x = 0$ from one side (Figure 8.53).

4. The same conditions but there are roots having opposite signs (Figure 8.54).

5. The root of multiplicity two is greater than one simple root and greater than the other.

Let $x_1 = 0 < x_2 = x_3 < x_4$. The curve has one branch with a loop and a node at $(x_2; -e/2x_2)$ (Figure 8.55).

6. Three roots of the equation coincide. The curve has a turning point of the first kind (Figure 8.56).

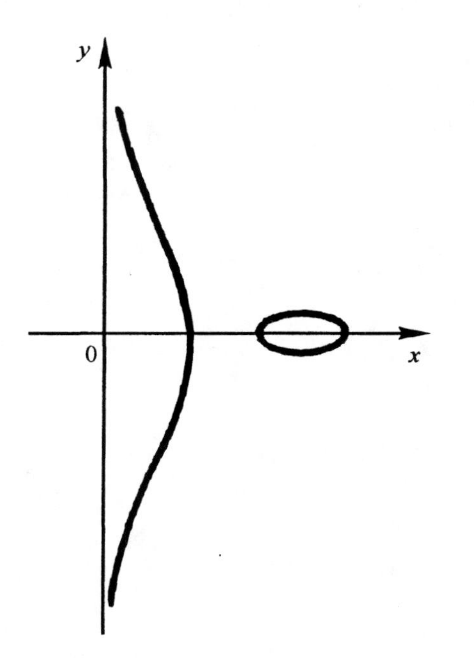

Figure 8.51. The defective hyperbola with one diameter (the roots are real, distinct, and have the same sign).

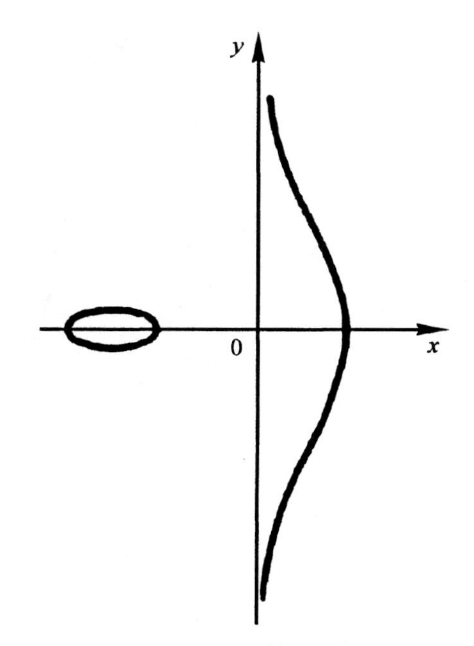

Figure 8.52. The defective hyperbola with one diameter (the roots are real, distinct, some have the same sign).

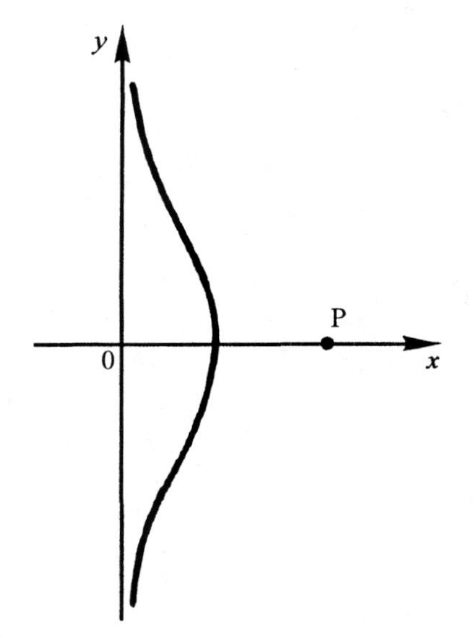

Figure 8.53. The defective hyperbola with one diameter ($x_1 = 0 < x_2 < x_3 < 0 = x_4$), P is an isolated point.

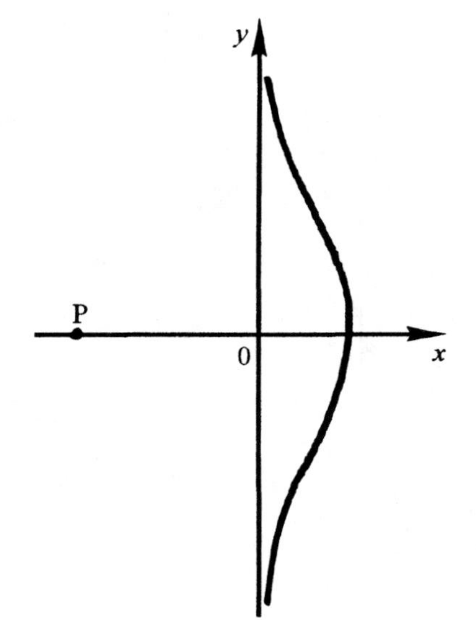

Figure 8.54. The defective hyperbola with one diameter (the roots have opposite signs).

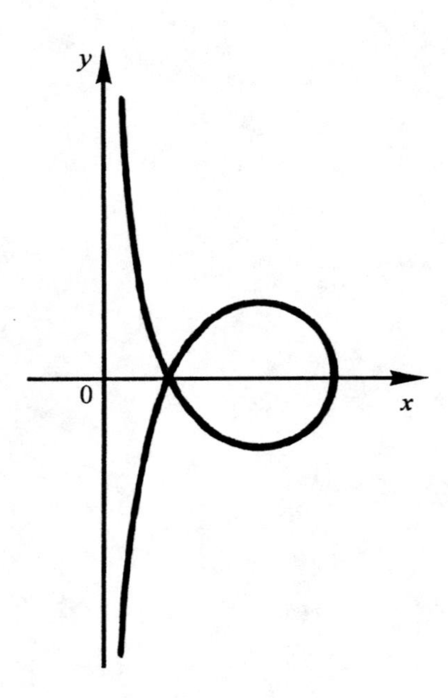

Figure 8.55. The defective hyperbola with one diameter
($x_1 = 0 < x_2 = x_3 < 0 = x_4$).

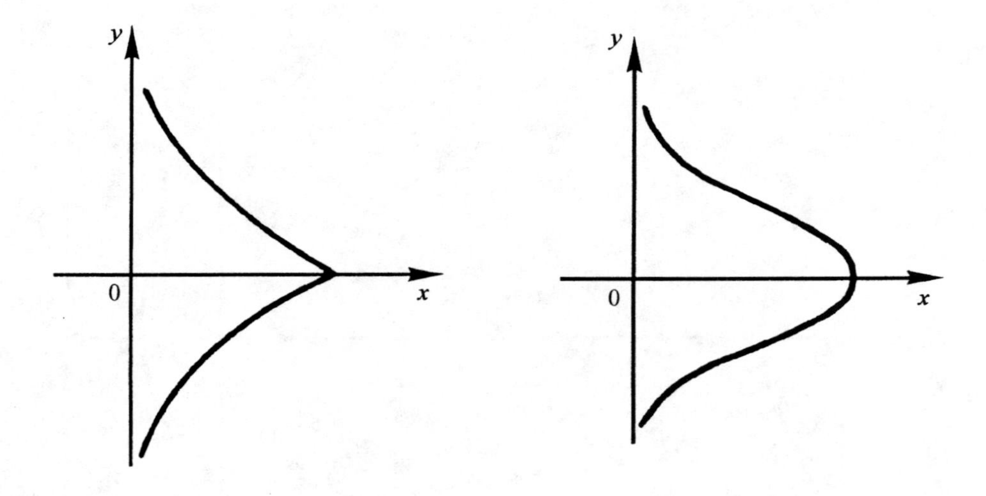

Figure 8.56 Figure 8.57

Figure 8.56. The defective hyperbola with one diameter (the roots coincide).

Figure 8.57. The defective hyperbola with one diameter (the roots are complex conjugate).

7. The characteristic equation has two complex conjugate roots.
The curve has one branch (Figure 8.57).

8.4.3. Parabolic hyperbola

A parabolic hyperbola is given by the equation $xy^2 + ey = bx^2 + cx + d$ $(b \neq 0)$. The curve has asymptotes: the straight line $x = 0$ and parabola $y^2 = bx + c$.

Parabolic hyperbola with no diameters ($e \neq 0$)

The curve intersects the asymptote $x = 0$ in the point $(0; d/e)$ and the asymptote $y^2 = bx + c$ in the point $\left(\frac{d^2 - ce^2}{be^2}; \frac{d}{e}\right)$. The point at infinity is not an inflection point.

There are seven possible cases.

1. All roots of the characteristic equation $bx^3 + cx^2 + dx + \frac{1}{4}e^2 = 0$ are real, distinct, and have the same sign (opposite to the sign of b).
The curve has two unbounded hyperbolic-parabolic branches and an oval located inside the asymptotic triangle (Figure 8.58).

2. All roots of the characteristic equation are real, distinct, and there are roots of opposite signs.
The curve has two unbounded branches, serpentine and parabolic (Figure 8.59).

3. If two smaller roots of the characteristic polynomial are equal and roots have the same sign, then the oval is merged with one of the unbounded branches and makes a loop with a saddle point (Figure 8.60).

4. If all roots are equal, then the saddle point becomes a turning point (Figure 8.61).

5. If the multiple root of the characteristic equation is greater than the simple root, and all the roots have the same sign, then the oval becomes an isolated point (Figure 8.62).

6. If the multiple root has the sign opposite to the sign of the simple root, then the curve has two unbounded branches that intersect in a node (Figure 8.63).

7. If the characteristic equation has two complex conjugate roots, then the curve has two branches of hyperbolic-parabolic type with no singularities (Figure 8.64).

Parabolic hyperbola with one diameter ($e = 0$)

The curve is defined by the equation $xy^2 = bx^2 + cx + d$. Curves of this type do not intersect with asymptotes at a finite distance, and have an inflection point at infinity.

Let us consider the following cases.

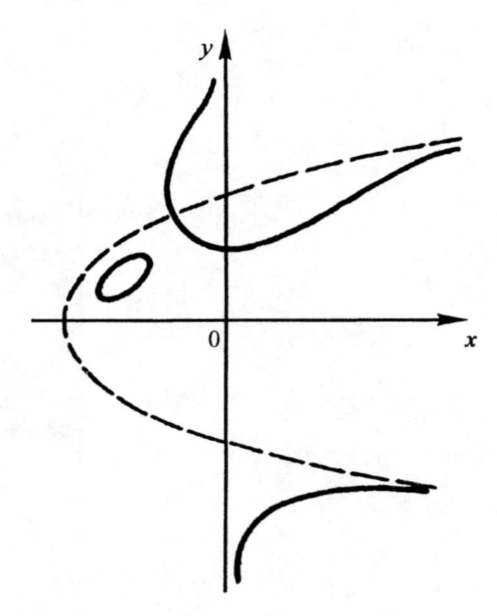

Figure 8.58. The parabolic hyperbola with no diameter (the roots are real, distinct, and have the same sign).

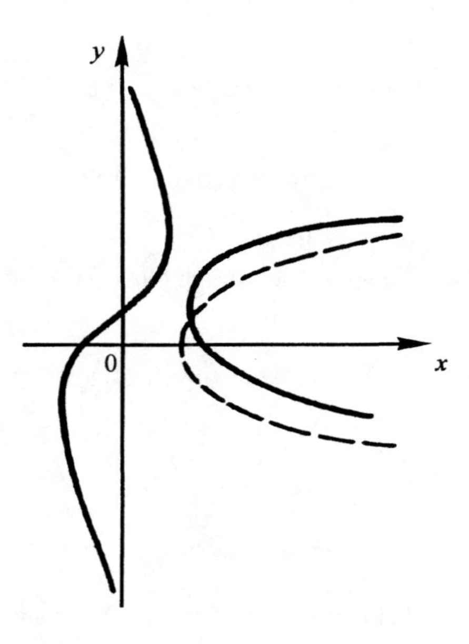

Figure 8.59. The parabolic hyperbola with no diameter (the roots are real, distinct; some have opposite signs).

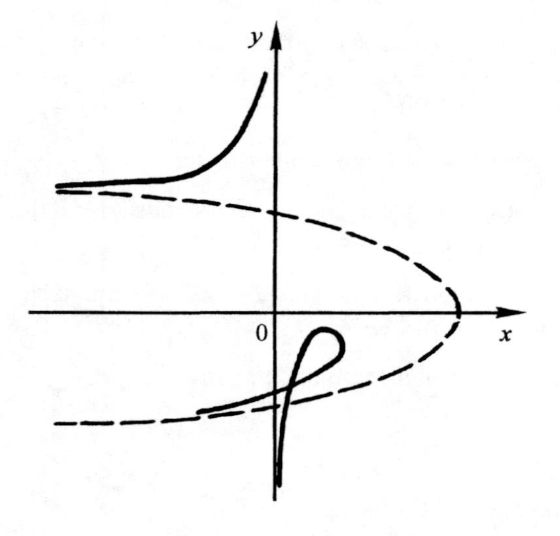

Figure 8.60. The parabolic hyperbola with no diameter (the roots have the same sign; two smallest roots are equal).

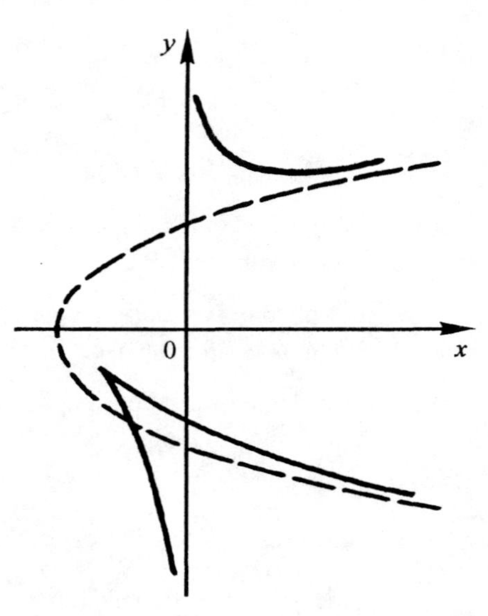

Figure 8.61. The parabolic hyperbola with no diameter (three roots are equal).

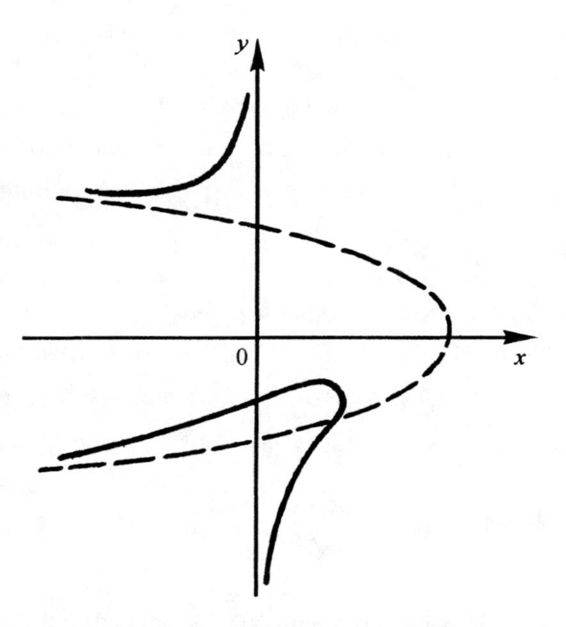

Figure 8.62. The parabolic hyperbola with no diameter (the roots are real, distinct, and have the same sign).

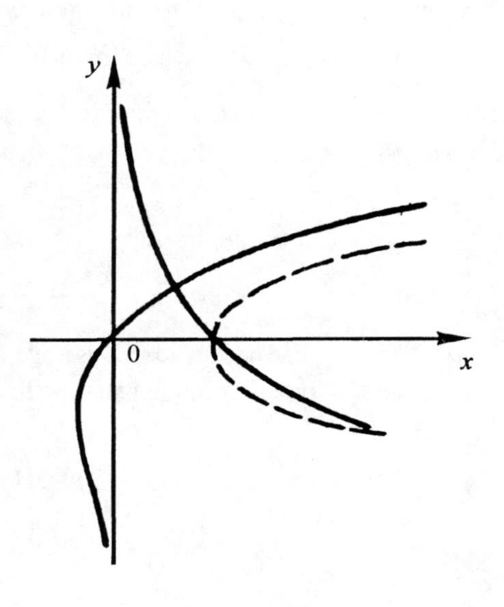

Figure 8.63. The parabolic hyperbola with no diameter (the roots have the same sign; the multiple root is greater than the simple root).

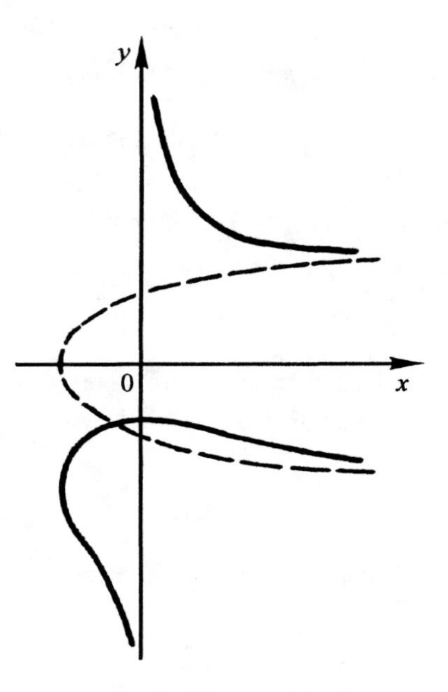

Figure 8.64. The parabolic hyperbola with no diameter (the roots are complex conjugate).

1. All roots of the characteristic equation $bx^3 + cx^2 + dx = 0$ are real, distinct, and have signs that coincide with the sign of b.

Let $b > 0$ and $x_1 = 0 < x_2 < x_3$. The condition $bx(x - x_2)(x - x_2) \geq 0$ defines the domain of the curve, $(0; x_2] \cup [x_3; +\infty)$. The curve has two unbounded branches: a conchoid branch, which approaches the asymptote $x = 0$ from one side; and a parabolic one. Both branches are located on one side of the asymptote $x = 0$ (Figure 8.65).

2. Nonzero roots of the characteristic equation are real and have opposite signs.

Let $b > 0$ and $x_1 < x_2 = 0 < x_3$. The condition $b(x - x_1(x - x_3) \geq 0$ defines the domain of the function, $[x_1; 0) \cup [x_3; +\infty)$. The curve has a behavior that is similar to the previous type of curve, but both branches are located on both sides of the asymptote $x = 0$ (Figure 8.66).

3. Two roots are equal and have the same sign as b.

Let $b > 0$ and $x_1 = x_2 > x_3 = 0$. The condition $bx(x - x_1)^2 \geq 0$ defines the domain of the curve, $(0; +\infty)$. The curve has two branches of hyperbolic-parabolic type intersecting in a node at $(x_1; 0)$ (Figure 8.67).

4. The characteristic equation has two complex conjugate roots $\alpha \pm i\beta$.

The curve has branches of hyperbolic-parabolic type with no singularities (Figure 8.68).

5. All roots are real and distinct; the sign of b is opposite to the sign of the other two roots.

Let $b > 0$ and $x_1 < x_2 < x_3 = 0$. The condition $bx(x - x_1)(x - x_2) \geq 0$ defines the domain of the curve, $[x_1; x_2] \cup (0; +\infty)$. The curve has an oval and two unbounded branches (Figure 8.69).

6. Two roots are equal and the sign of the multiple root is opposite to the sign of b.

Let $x_1 = x_2 > x_3 = 0, b > 0$. The curve has two unbounded branches defined on $(-\infty; 0)$ and an isolated point $(x_1; 0)$ (Figure 8.70).

8.4.4. Hyperbolisms of conic sections

The conic section hyperbolism, $xy^2 + ey = cx + d$ $(a = b = 0)$, can be considered as a special case of a parabolic hyperbola that has a parabolic asymptote split into two parallel straight lines, $y = \pm\sqrt{c}$.

In a general case, *hyperbolism of a curve* $F(x, y) = 0$ is the curve $F(x, \frac{xy}{\alpha}) = 0$, where $\alpha = \text{const} \neq 0$. For example, for $\alpha = 1$, the curves

$$xy^2 = -\frac{b^2}{a^2}x + 2p, \quad xy^2 = 2p, \quad xy^2 = \frac{b^2}{a^2}x + 2p$$

are hyperbolisms of the corresponding ellipse, parabola, and hyperbola given by the equations:

$$y^2 = 2px - \frac{b^2}{a^2}x^2, \quad y^2 = 2px, \quad xy^2 = \frac{b^2}{a^2}x + 2p.$$

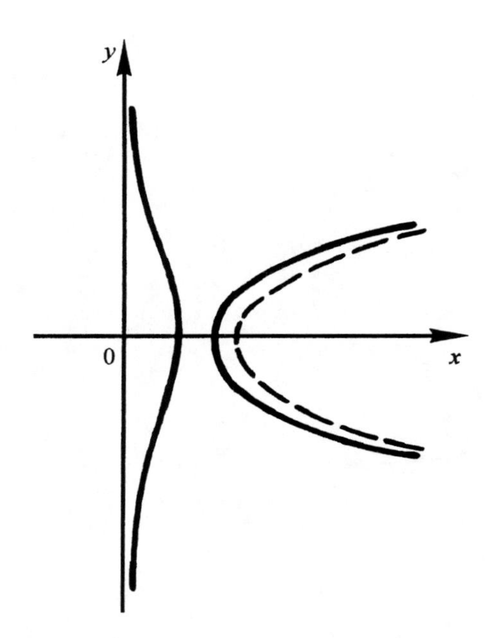

Figure 8.65. The parabolic hyperbola with one diameter ($b > 0, x_1 = 0 < x_2 < x_3$).

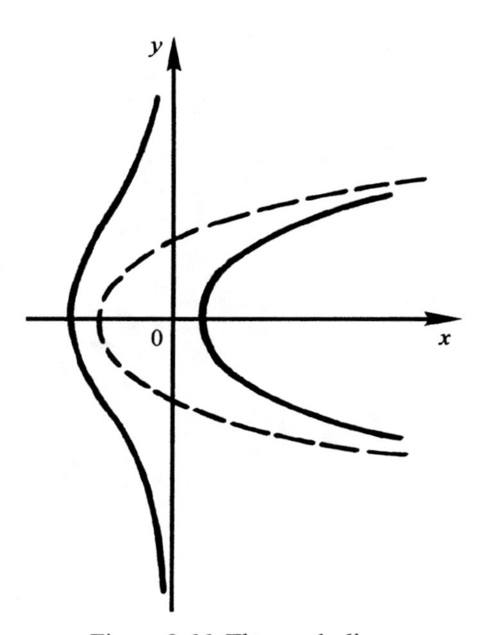

Figure 8.66. The parabolic hyperbola with one diameter ($b > 0, x_1 < x_2 = 0 < x_3$).

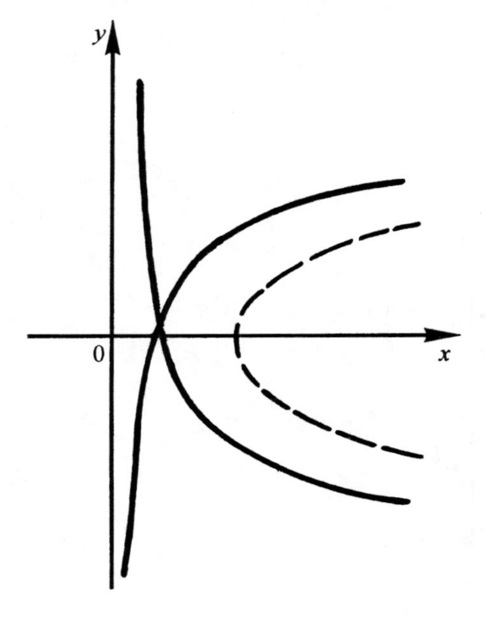

Figure 8.67. The parabolic hyperbola with one diameter ($b > 0, x_1 = x_2 > x_3 = 0$).

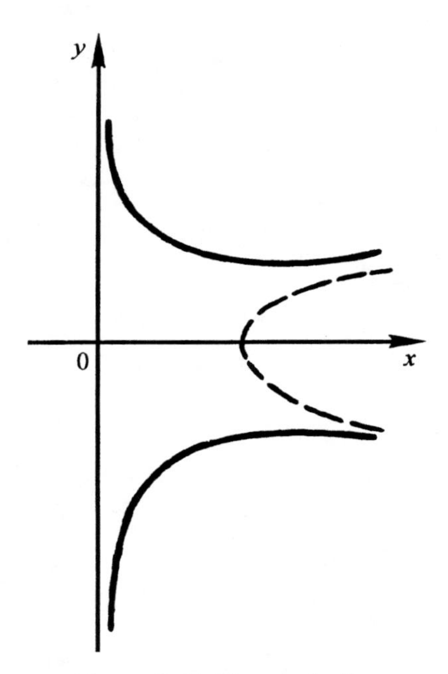

Figure 8.68. The parabolic hyperbola with one diameter (two roots are complex conjugate).

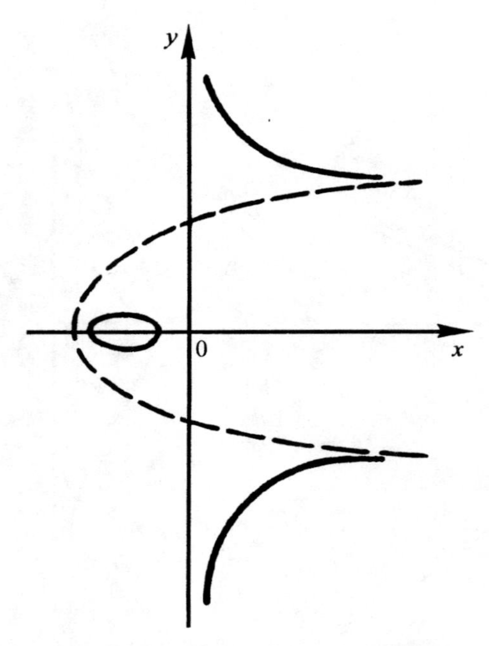

Figure 8.69. The parabolic hyperbola with one diameter
($b > 0, x_1 < x_2 < x_3 = 0$).

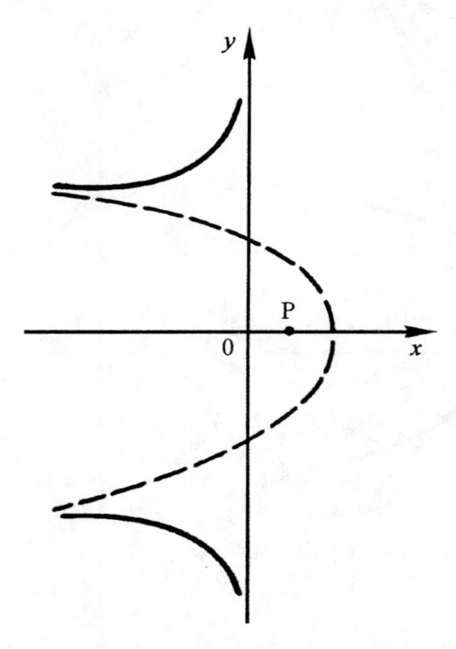

Figure 8.70. The parabolic hyperbola with one diameter
($b > 0, x_1 = x_2 > x_3 = 0$).

Hyperbolisms of conic sections have three asymptotes: $x = 0$, $y = \pm\sqrt{c}$, and a double point at infinity. Let x_1, x_2 be roots of the characteristic equation $cx^2 + dx + \frac{1}{4}e^2 = 0$.

Hyperbola hyperbolism ($c > 0$)

The point at infinity is a node. All three asymptotes are real.

1. $e \neq 0$, $d^2 > ce^2$. Roots of the characteristic equation are real, distinct, and have the same sign. Let $0 < x_1 < x_2$. In this case, we have that $|y| < \sqrt{c}$ on $[x_2; +\infty)$, and $|y| > \sqrt{c}$ on $(-\infty; x_1]$. The curve has one branch that lies between the asymptotes $y = \pm\sqrt{c}$ on $[x_2; +\infty)$, and has two branches that approach the asymptote $x = 0$ from different sides. One of these branches is defined on $(-\infty; 0]$, the other — on $(-\infty; x_1]$. Both of them approach the asymptote $y = \pm\sqrt{c}$ from the outside (Figure 8.71).

2. $e \neq 0$, $d \neq 0$, $d^2 < ce^2$. Roots of the characteristic equation are complex. The curve has a serpentine branch lying between the asymptotes $y = \pm\sqrt{c}$, and two hyperbolic branches lying outside the strip between the asymptotes (Figure 8.72).

3. $e \neq 0$, $d = 0$. The curve has a shape similar to the one in item 2, but its center is located at the point $O(0;0)$ (Figure 8.73).

4. $e = 0$, $d \neq 0$. In this case, $x_2 = 0$, and the equation of the curve can be written as $xy^2 = cx - cx_1$.

The curve has three branches: one is located on $[x_1; +\infty)$ between the asymptotes $y = \pm\sqrt{c}$ if $x_1 > 0$; the other two branches lie outside of the strip bounded by these asymptotes. The branches tend to the asymptote $x = 0$ from one side. This curve has a diameter $y = 0$ (Figure 8.74).

If $e \neq 0$, $d^2 = ce^2$, the hyperbola hyperbolism breaks down into the conic section $xy + x\sqrt{c} + e = 0$ and the straight line $y = \sqrt{c}$.

Ellipse hyperbolism $c < 0$

The double point at infinity is isolated. The curve has one real asymptote $x = 0$ and two imaginary asymptotes $y = \pm\sqrt{c}$.

1. $e \neq 0$, $d \neq 0$. Roots of the characteristic equation are real, distinct, and have opposite signs, $x_1 < 0 < x_2$ (Figure 8.75). The curve has one serpentine branch on $[x_1; x_2]$ without singularities.

2. $e \neq 0$, $d = 0$. Two roots have equal absolute values and opposite signs. The curve has one serpentine branch centered at $O(0; 0)$ (Figure 8.76).

3. $e = 0$, $d \neq 0$. One of roots of the characteristic equation equals zero. The curve has one conchoid branch on $(0; x_1]$ if $x_1 > 0$ (Figure 8.77).

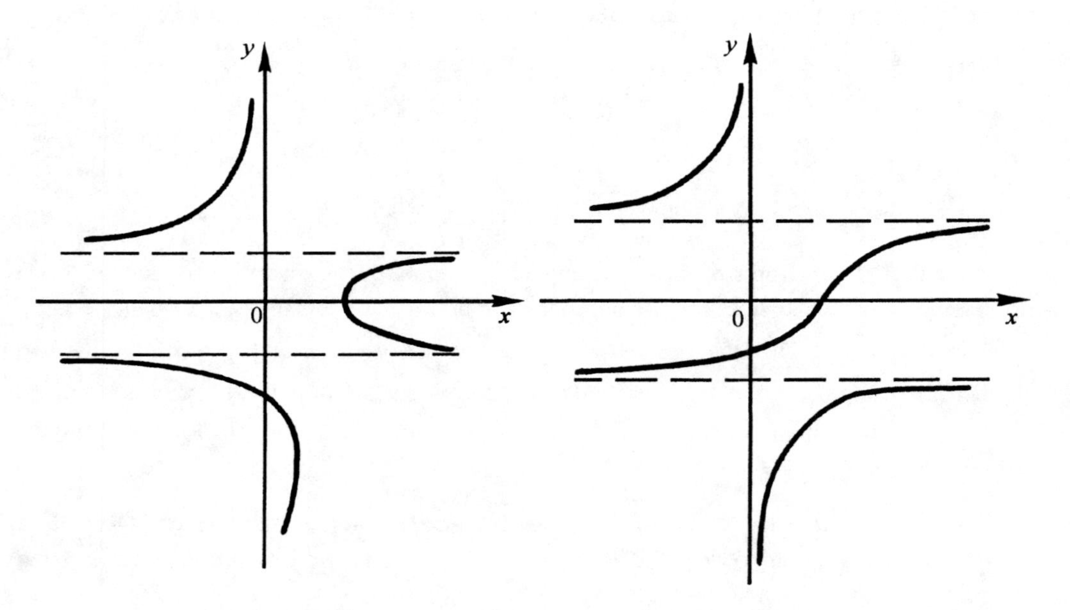

Figure 8.71. The hyperbola hyperbolism ($e \neq 0$, $d^2 > ce^2$, $0 < x_1 < x_2$).

Figure 8.72. The hyperbola hyperbolism ($e \neq 0$, $d \neq 0$, $d^2 < ce^2$, the roots are complex).

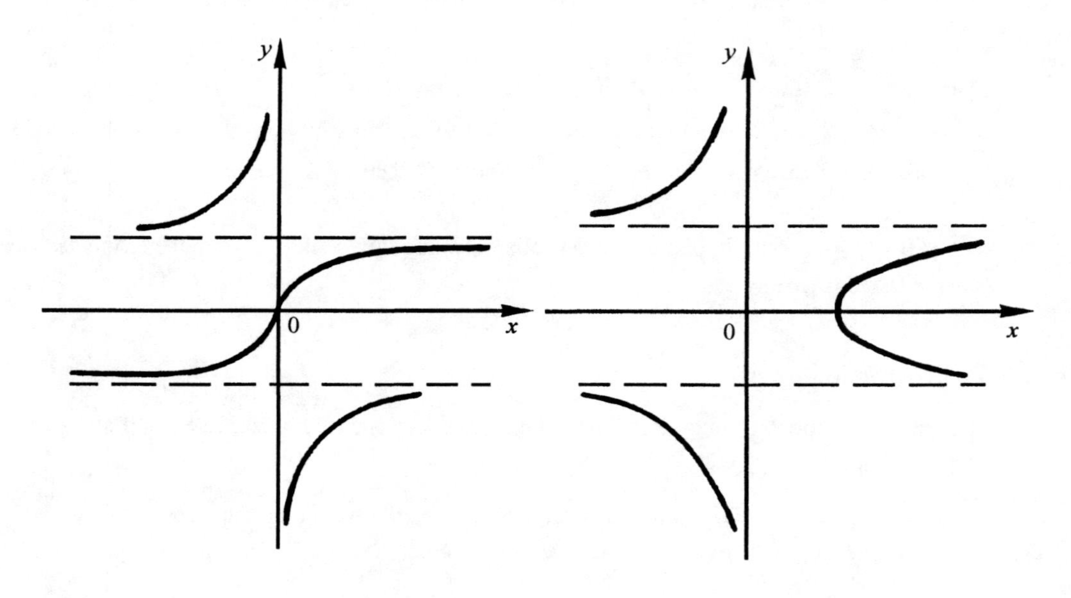

Figure 8.73. The hyperbola hyperbolism ($e \neq 0$, $d = 0$).

Figure 8.74. The hyperbola hyperbolism ($e = 0$, $d \neq 0$).

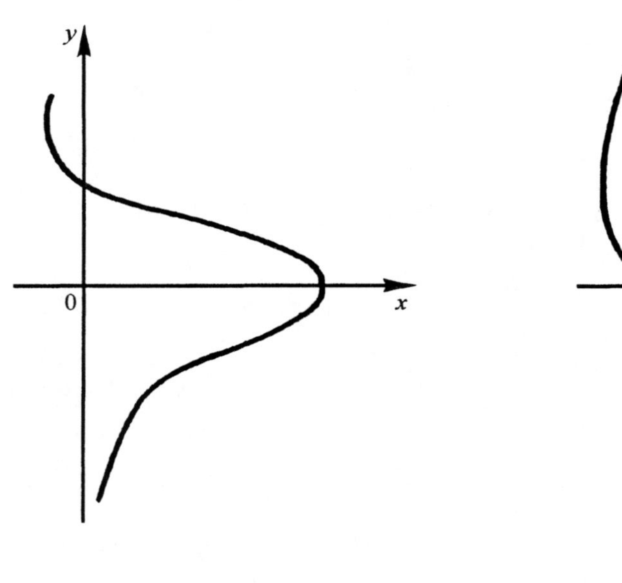

Figure 8.75. The ellipse hyperbolism ($e \neq 0$, $d \neq 0$, the roots are real and have opposite signs).

Figure 8.76. The ellipse hyperbolism ($e \neq 0$, $d = 0$, two roots have opposite signs and equal absolute values).

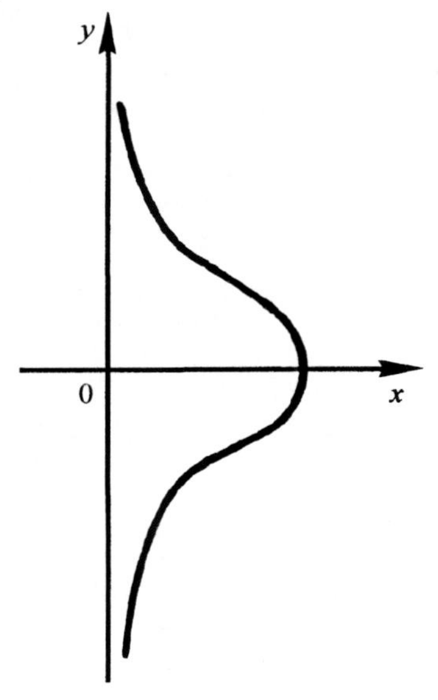

Figure 8.77. The ellipse hyperbolism ($e = 0$, $d \neq 0$, one of the roots posite signs and equals zero).

Parabola hyperbolism ($c = 0$)

Curves of this type have a turning point at infinity, and three real asymptotes, two of which coincide and make the tangent $y = 0$ at the turning point.

Parabola hyperbolism consists of two hyperbolic branches located in adjacent angles formed by the asymptotes.

1) $e \neq 0$. The curve has no diameter (Figure 8.78).

2) $e = 0$. The curve has the diameter $y = 0$ (Figure 8.79).

8.4.5. Trident (Hyperbola parabolism)

The trident $xy = ax^3 + bx^2 + cx + d$ has two asymptotes, one of which is the straight line $x = 0$, and the other one is the parabola $y = ax^2 + bx + c$. The characteristic equation for the trident is $ax^3 + bx^2 + cx + d = 0$.

The trident consists of two hyperbolic-parabolic branches (Figure 8.80). The shape of the curve somewhat changes in the case where roots of the characteristic equation are real and distinct (Figure 8.81).

8.4.6. Prolate parabola

The prolate parabola $y^2 = ax^3 + bx^2 + cx + d$ has an inflection point at infinity and the asymptotic curve of degree three, $y^2 = ax^3$.

1. Roots of the characteristic equation $ax^3 + bx^2 + cx + d = 0$ are real and distinct.

Let $0 < x_1 < x_2 < x_3$. The curve has an oval on $[x_1; x_2]$ and an unbounded branch on $[x_3; +\infty)$ (Figure 8.82).

2. Two roots of the characteristic equation are equal; the multiple root is greater than the simple root.

Let $x_1 < x_2 = x_3$. The curve has one branch on $[x_1; +\infty)$ with a loop and a node at $(x_2; 0)$ (Figure 8.83).

3. The multiple root is less than the simple root.

Let $x_1 = x_2 < x_3$. The curve has one branch on $[x_3; +\infty)$ and an isolated point at $(x_1; 0)$ (Figure 8.84).

4. All three roots are equal. The curve has a turning point (Figure 8.85). This is a semicubic parabola $y^2 = a(x - x_1)^3$.

5. Two of the roots are complex conjugate.

The curve has one branch (Figure 8.86).

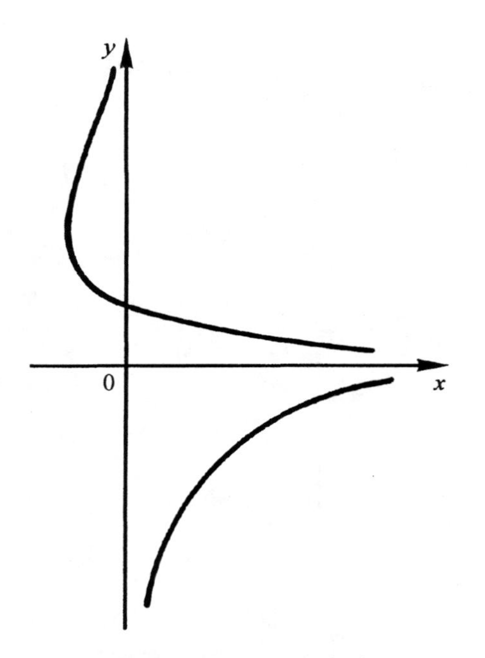

Figure 8.78. The parabola hyperbolism ($e \neq 0$, no diameter).

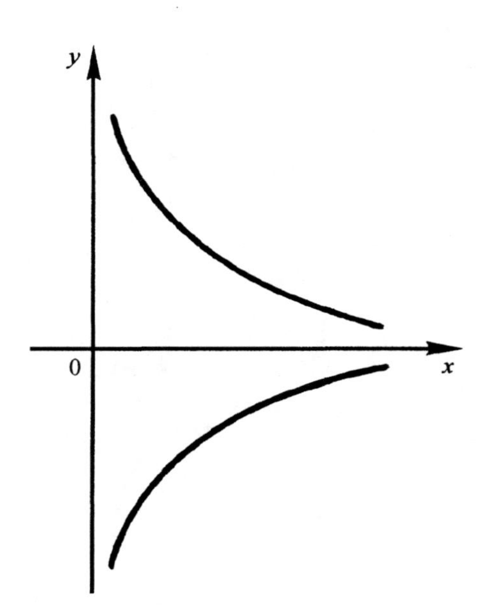

Figure 8.79. The parabola hyperbolism ($e = 0$, the curve has a diameter).

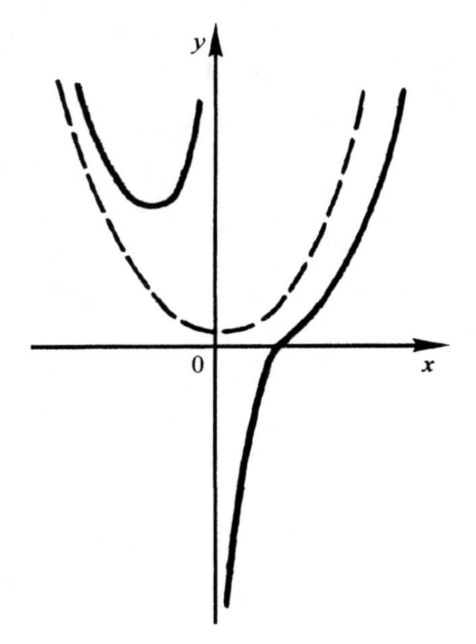

Figure 8.80. The hyperbola parabolism (the curve with two parabolic branches).

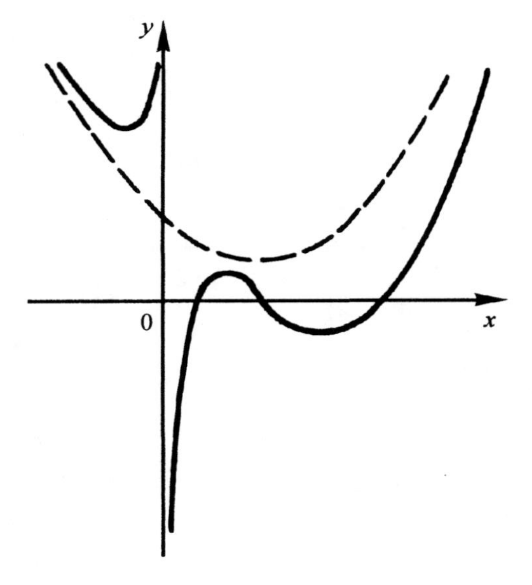

Figure 8.81. The hyperbola parabolism (the roots are real and distinct).

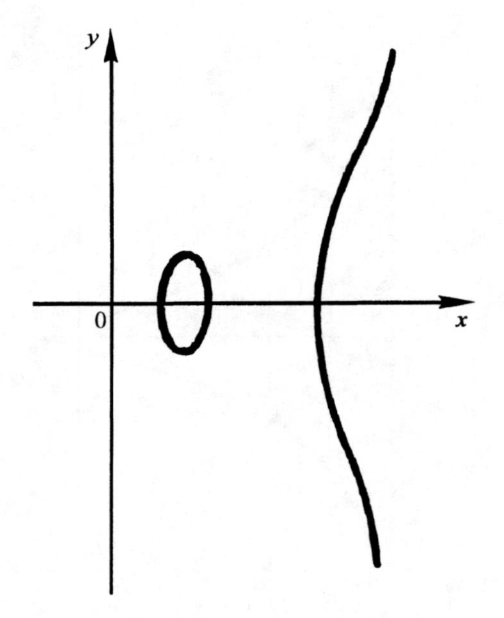

Figure 8.82. The prolate parabola ($0 < x_1 < x_2 < x_3$).

Figure 8.83. The prolate parabola ($x_1 < x_2 = x_3$).

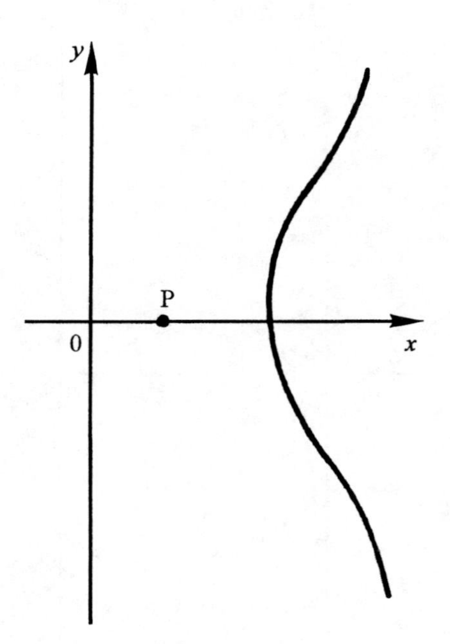

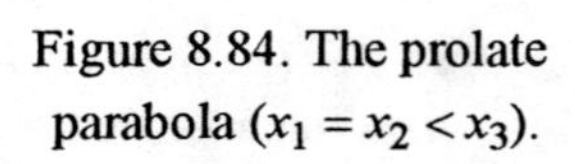
Figure 8.84. The prolate parabola ($x_1 = x_2 < x_3$).

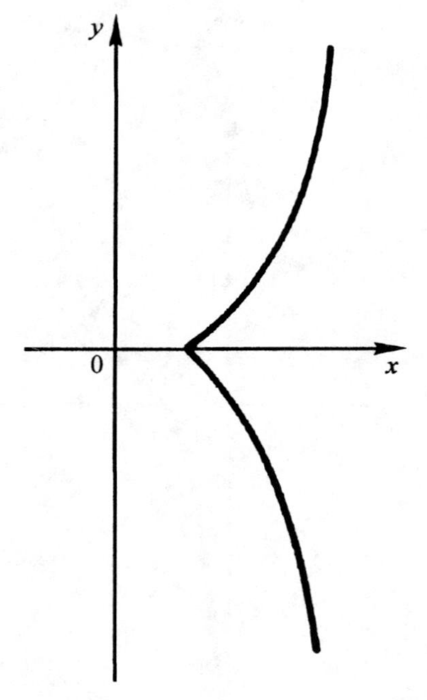

Figure 8.85. The prolate parabola (all roots are equal).

8.4.7. Cubic parabola

Cubic parabola $y = ax^3 + bx^2 + cx + d$ has neither asymptotes, singular points, singular points at a finite distance, nor a turning point at infinity (Figure 8.87).

8.5 Graphs of implicitly defined functions containing the absolute value sign

Example 8.5.1. Plot the locus $|x| + |y| = 1$.

The expression is symmetric with respect to the coordinate axes, hence it will suffice to plot the curve in the first quadrant, i.e. for $x \geq 0, y \geq 0$.

If $|x| = x$ and $|y| = y$, the expression becomes $x + y = 1$, or $y = 1 - x$. The set of points we are looking for contains only the line segment of this straight line lying in the first quadrant. The graph in the quadrants $II - IV$ is constructed by the symmetry (Figure 8.88).

Graphs of the curves $|x| - |y| = 1$ and $|y| - |x| = 1$ are shown in Figure 8.89.

Example 8.5.2. Plot the graph $||x| - |y|| = 1$.

This equation is equivalent to two equations: $|x| - |y| = 1$ and $|x| - |y| = -1$. Hence the graph is the union of two graphs (Figure 8.90).

Example 8.5.3. Plot the set of points satisfying $|y - 2| = |x + 1|$.

Choose a new coordinate system, $y^1 = y - 2$, $x^1 = x + 1$. Then the equation becomes $|x^1| = |y^1|$.

Plotting the graph $|y^1| = |x^1|$ in the coordinate system $O_1x^1y^1$, where $O_1(-1;2)$, we get the needed graph in the coordinate system Oxy (Figure 8.91).

Example 8.5.4. Plot the set of points satisfying $y + |y| = x + |x|$.

The equation is neither even nor odd with respect to the variables. Consider this equation for every quadrant separately. In the first quadrant, the equation becomes $y = x$.

Hence the graph in the first quadrant is the bisector.

In the second quadrant, $|y| = y$, $|x| = -x$, hence $y = 0$, and this is the x-half-axis.

In the third quadrant, the equation becomes $y - y = x - x$. This is an identity, which means that any point in the third quadrant has coordinates satisfying the equation.

In the forth quadrant, the equation is $x = 0$, the y-half-axis (Figure 8.92).

Examples of graphs of analytic expressions containing the absolute value sign are shown in Figures 8.93 – 8.95.

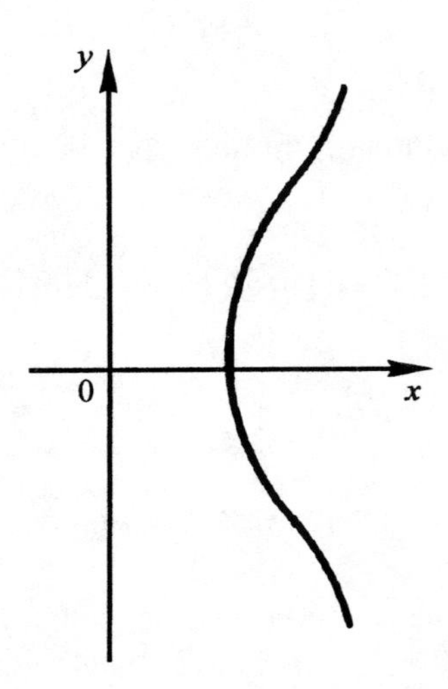

Figure 8.86. The prolate parabola
(two roots are complex conjugate).

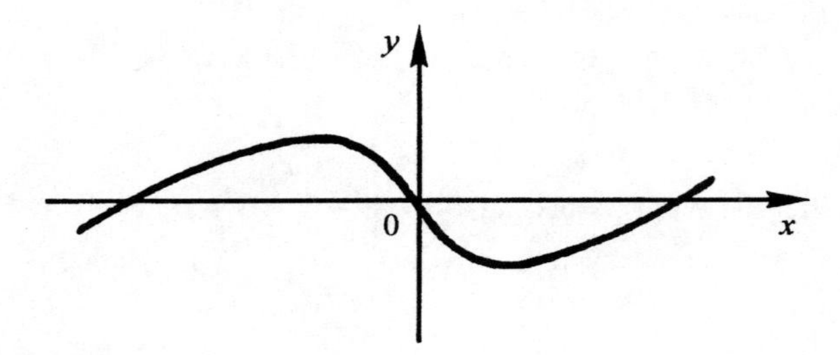

Figure 8.87. Cubic parabola.

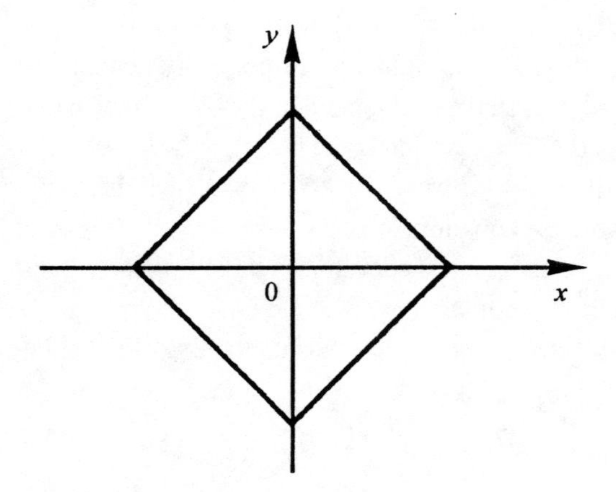

Figure 8.88. The graph of the function $|x| + |y| = 1$.

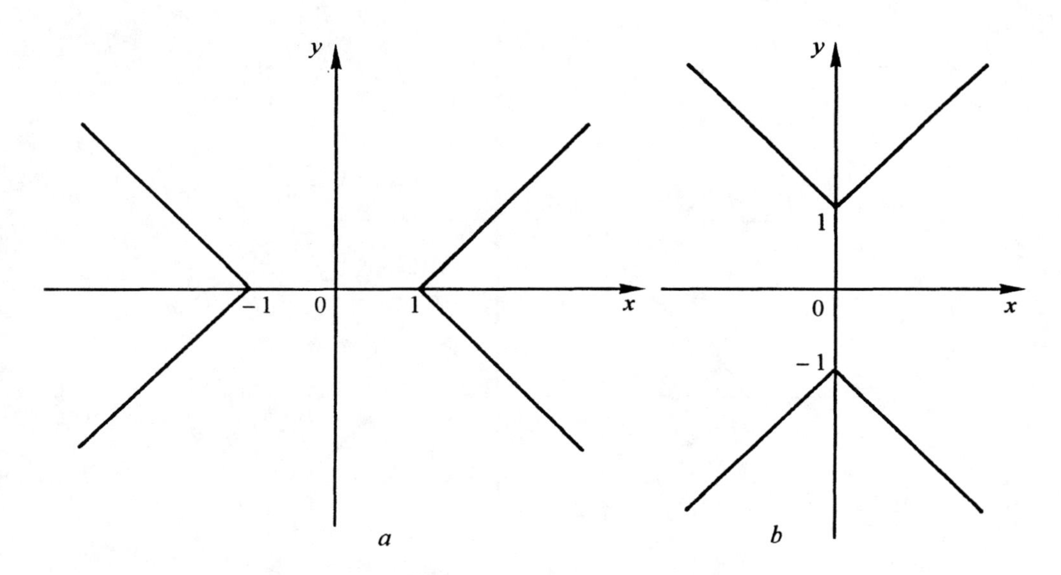

Figure 8.89. The graph of functions: (*a*) $|x| - |y| = 1$, (*b*) $|y| - |x| = 1$.

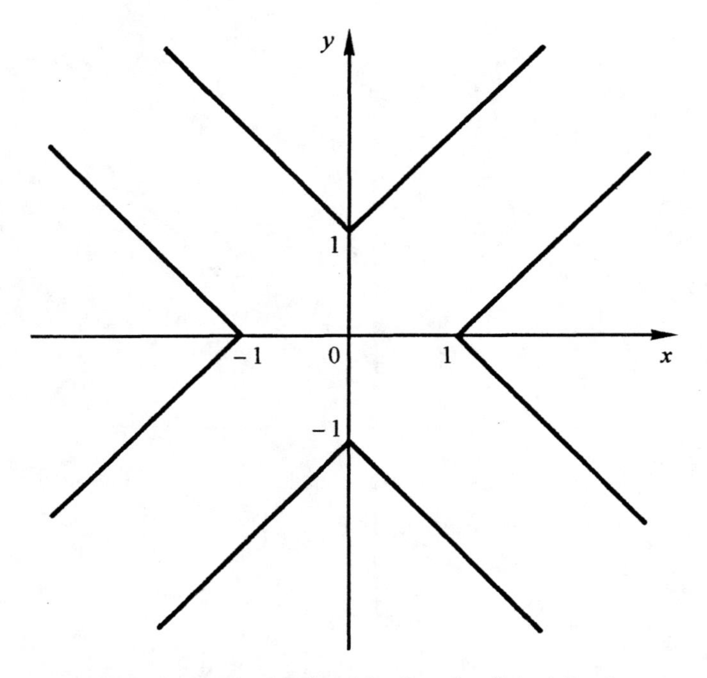

Figure 8.90. The graph of the function $||x| - |y|| = 1$.

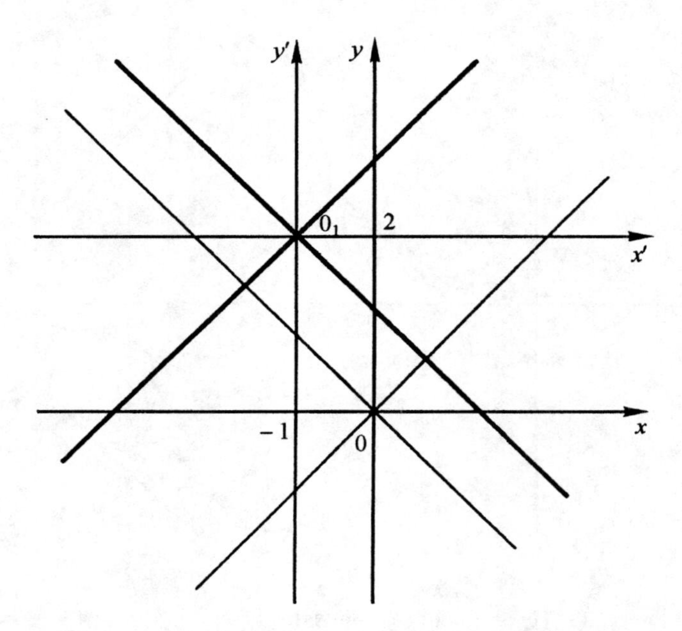

Figure 8.91. The graph of the function $|y-2|=|x+1|$.

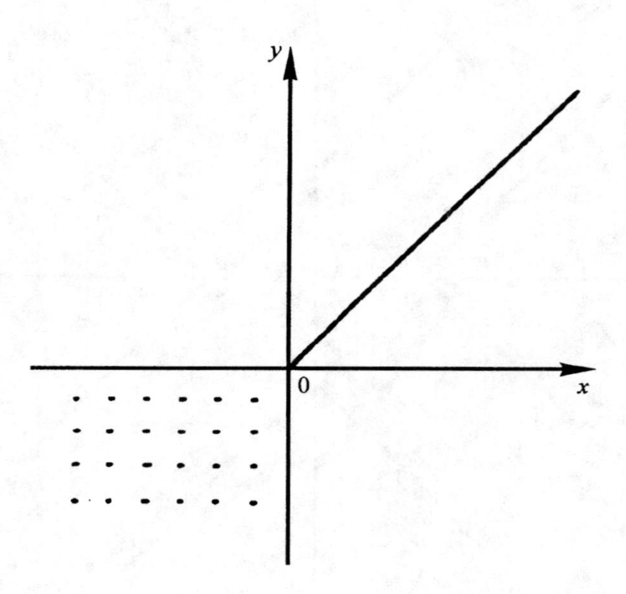

Figure 8.92. The graph of the function $y+|y|=x+|x|$.

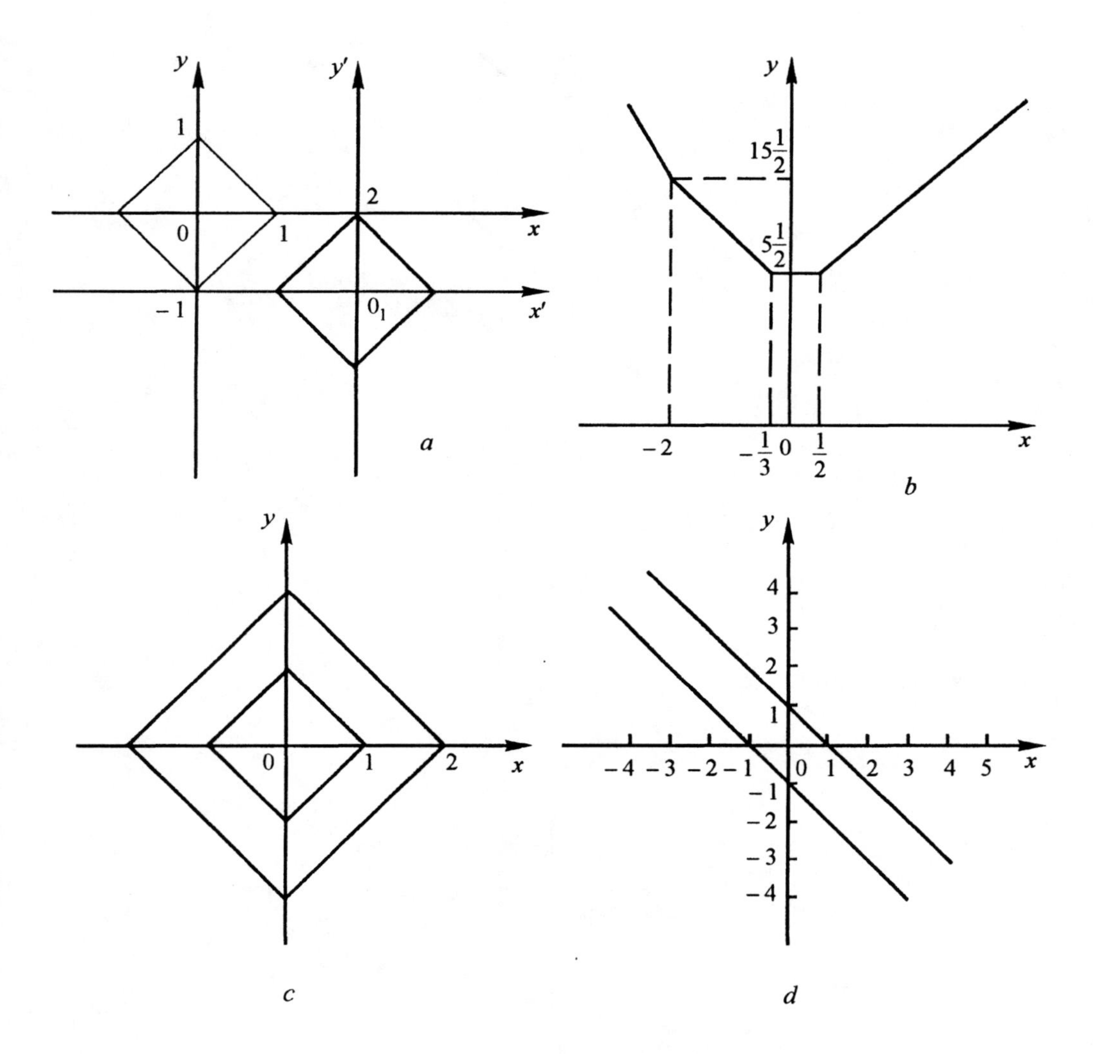

Figure 8.93. The graphs of the functions: (*a*) $|x-2|+|y|=1$;
(*b*) $y=|x+2|+|13x+1|-|x-1/2|-3x+2|$; (*c*) $||x|+|y|-3/2|=1/2$;
(*d*) $|y+x|=1$.

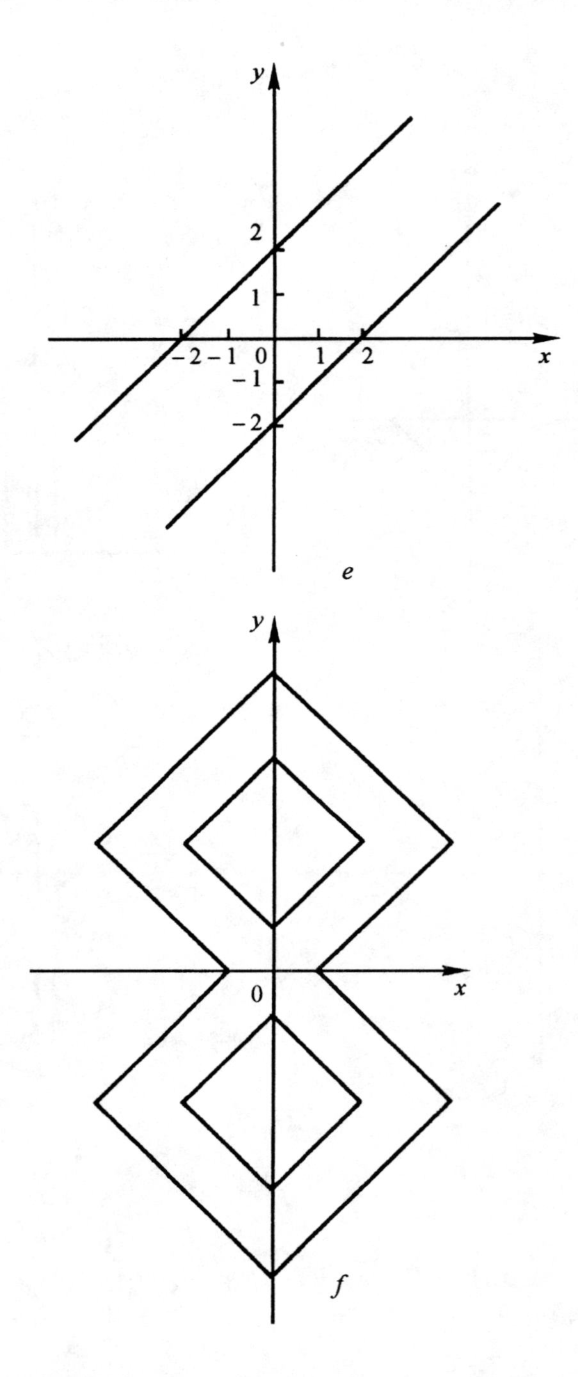

Figure 8.93. The graphs of the functions: (*e*) $|x - y| = 2$; (*f*) $||x| + |y - 3| - 3| = 1$.

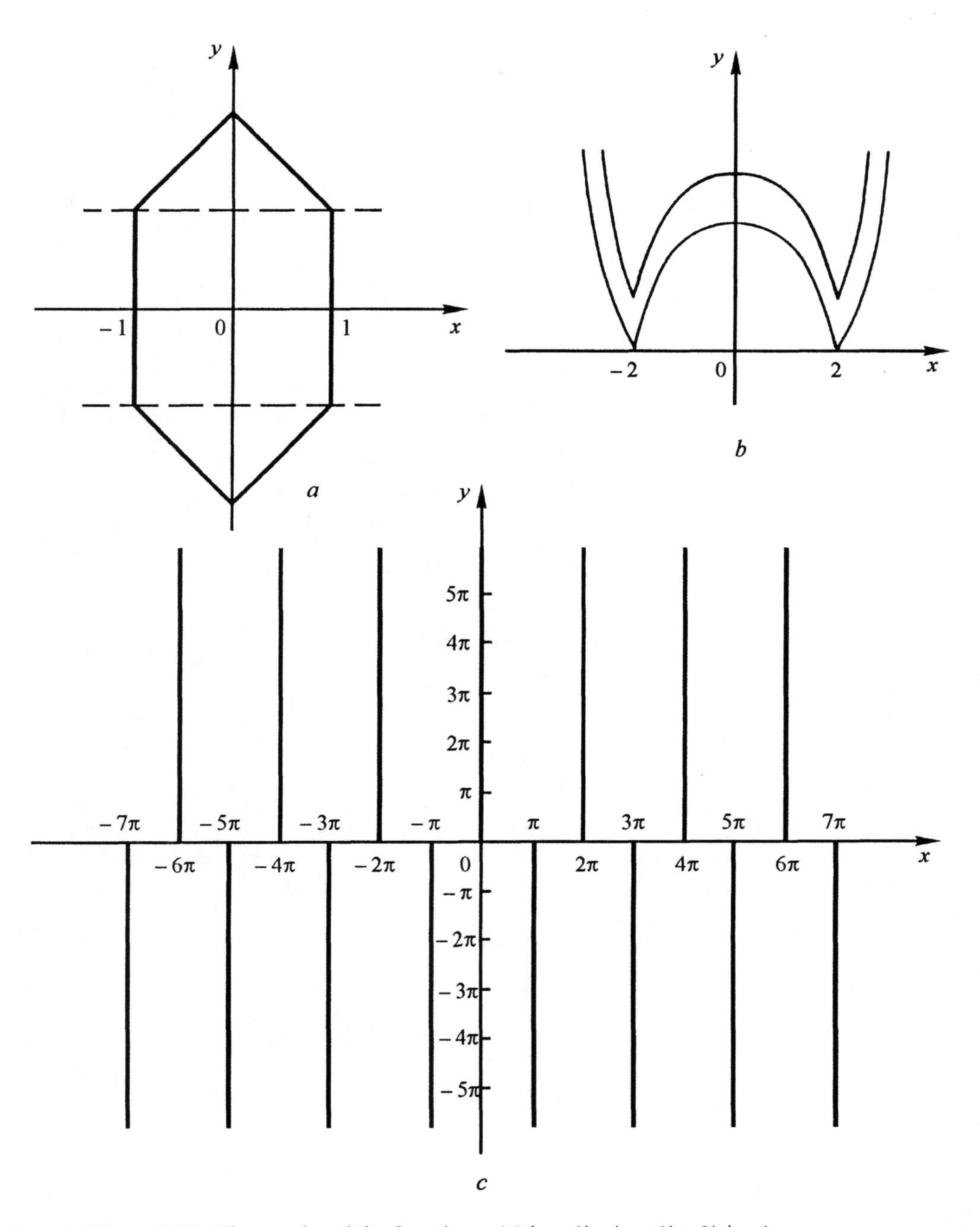

Figure 8.94. The graphs of the functions: (*a*) $|y-1|+|y+1|+2|x|=4$; (*b*) *1*. $y=|x^2-4|$, *2*. $|y-1|=1+|x^2-4|$; (*c*) $y=|y|\cos x$.

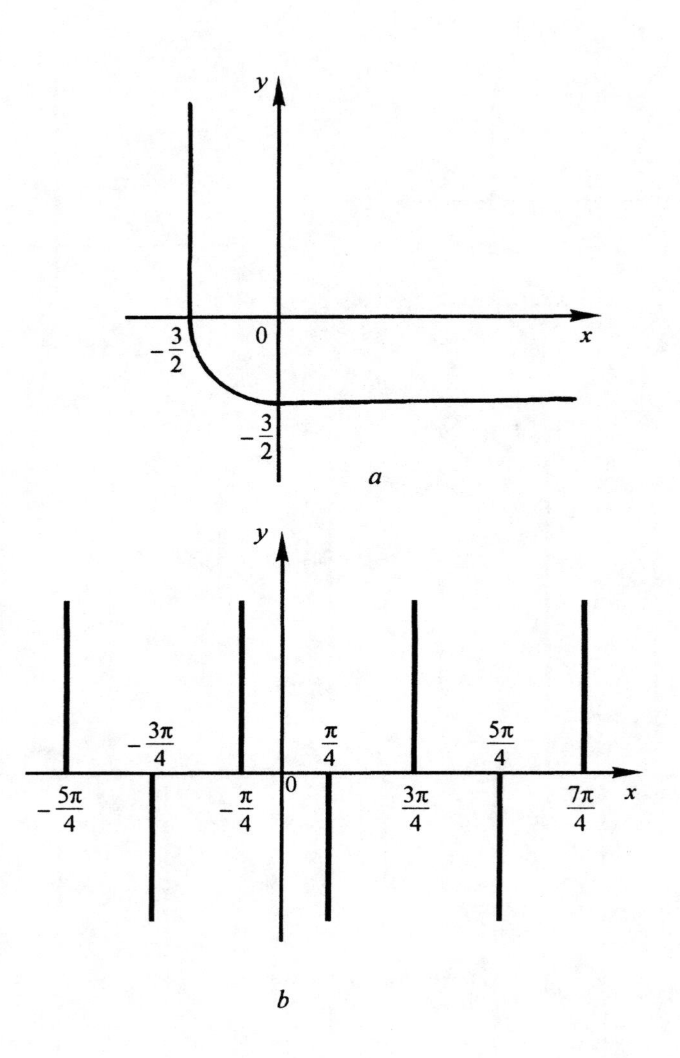

Figure 8.95. The graphs of the functions: (*a*) $(x-|x|)^2)+(y-|y|)^2=9$; (*b*) $y+y\tan x=0$.

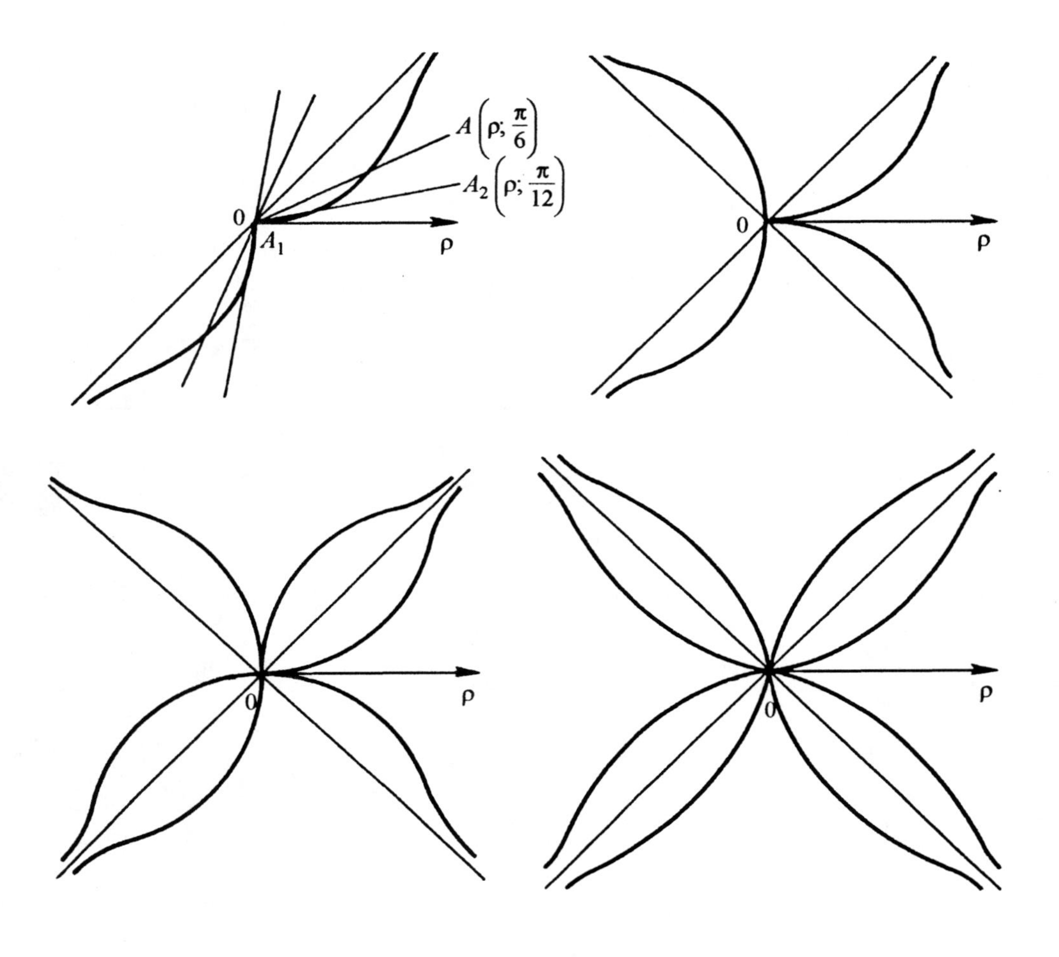

Figure 8.96. The graph of the functions $\rho = \pm \tan 2\varphi$.

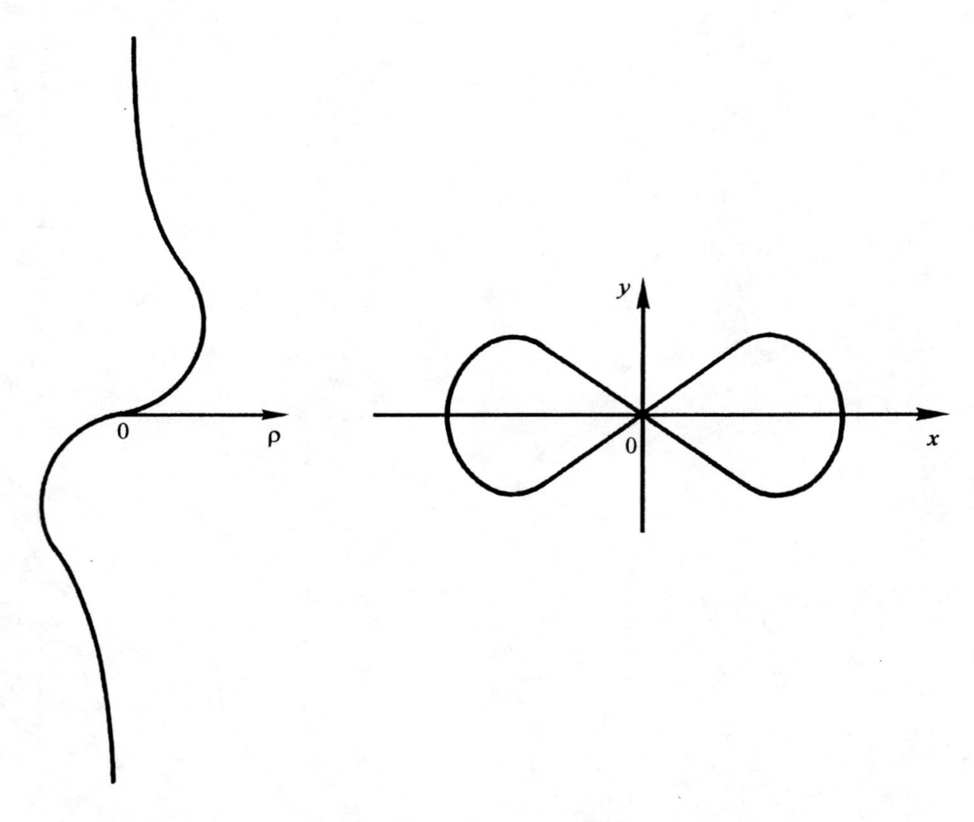

Figure 8.97 Figure 8.98

Figure 8.97. The graph of functions: $\rho = \pm\sqrt{\tan\varphi}$, $(x^2+y^2)x = a^2y$.

Figure 8.98. The graph of functions $(x^2+y^2)^2 = a^2(x^2-y^2)$ or $\rho = \pm\sqrt{\cos 2\varphi}$ (Bernoulli lemniscate).

8.6. Examples of implicitly defined functions

for which it is convenient to plot the graphs in polar coordinates

Example 8.6.1. Plot the graph of the function $(x^2+y^2)^3 = 4a^2x^2y^2$.

Pass to the polar coordinates, $x = \rho\cos\varphi$, $y = \rho\sin\varphi$. The equation becomes $\rho^2 = a^2\sin^2 2\varphi$, whence $\rho = \pm a\sin 2\varphi$. The graph of the two functions is a four-leaf rose (see Figure 7.14).

Example 8.6.2. Plot the graph of the function $(x^2+y^2)(x^2-y^2) = 4x^2y^2$.

The graph is symmetric with respect to the bisectors $y = x, y = -x$, and hence it is symmetric with respect to the origin. Thus we plot the graph only in the first quadrant. The graph in the other quadrants is obtained by symmetry with respect to the axes.

Use the polar coordinates, $x = \rho\cos\varphi$, $y = \rho\sin\varphi$. The equation becomes $\rho^2 = \tan^2 2\varphi$, whence $\rho = \pm\tan 2\varphi$ (Figure 8.96).

Example 8.6.3. Plot the graph of the function $(x^2+y^2)x = a^2y$.

Using polar coordinates, we get $\rho^2 = a^2\tan\varphi$, whence $\rho = \pm\sqrt{\tan\varphi}$ (Figure 8.97). Graphs of similar functions are shown in Figure 8.98.

Remark 8.6.1. For a more thorough study and plotting of implicitly defined functions, a better way is to use methods of calculus (see Part II). However, in many cases as we saw before, graphs of implicitly defined functions $F(x, y) = 0$ can be plotted without the use of calculus.

Chapter 9

Plots of more complicated functions

9.1. Functions defined by several analytical expressions

Let us look at examples where the function is defined by different analytical expressions on different sets from its domain (see Figures 9.1 – 9.5).

9.2. Function $y = \operatorname{sign} f(x)$, $y = f(\operatorname{sign} x)$

The graph of the function

$$y = \operatorname{sign} x = \begin{cases} 1, & x > 0, \\ 0, & x = 0, \\ -1, & x < 0 \end{cases}$$

consists of two half-lines and the point (0; 0) (Figure 9.6). The points (0; 1) and (0; −1) do not belong to the graph. The domain of the function $y = \operatorname{sign} f(x)$ coincides with the domain of the function $y = f(x)$. It follows from the definition of the function $\operatorname{sign} x$ that

$$y = \operatorname{sign} f(x) = \begin{cases} 1, & f(x) > 0, \\ 0, & f(x) = 0, \\ -1, & f(x) < 0. \end{cases}$$

Using this, we get the following scheme for constructing the graph of the function $\operatorname{sign} f(x)$: plot the straight line $y = 1$ on the intervals where the graph of the function $f(x)$ is in the upper half-plane, plot the line $y = -1$ for x such that the graph of $f(x)$ is in the lower half-plane; the points in which $f(x)$ intersects the x-axis remain unchanged (Figure 9.7).

Example 9.2.1. Plot the graph of the function $y = \operatorname{sign} \tan x$.

The domain of the function consists of all x such that $x \neq \pi/2 + \pi n$, $n \in \mathbf{Z}$. The function is periodic with the period $\omega = \pi$; it is bounded, since $|\operatorname{sign} (\tan x)| \leq 1$. Because the function is

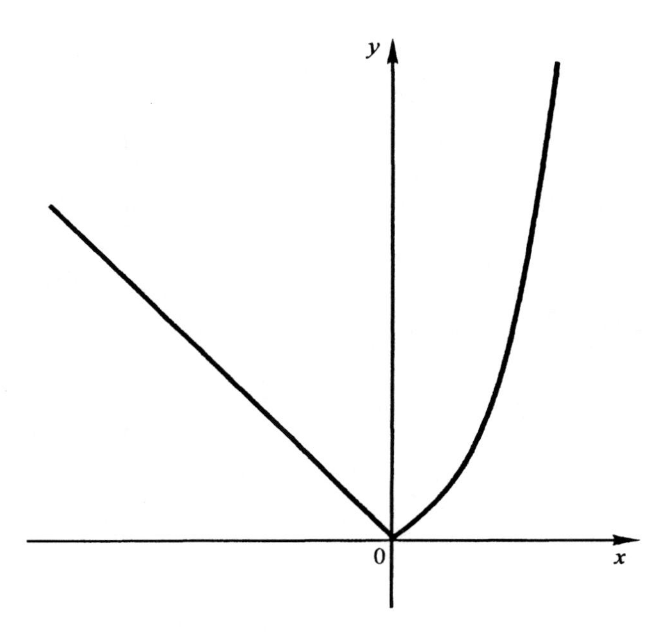

Figure 9.1. The graph of the function $y = \begin{cases} x^2, & x \geq 0, \\ -x, & x < 0. \end{cases}$.

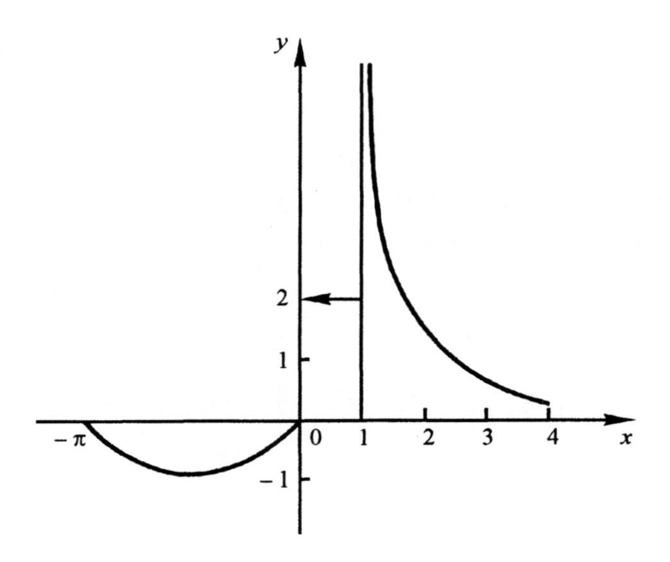

Figure 9.2. The graph of the function $y = \begin{cases} \sin x, & -\pi \leq x \leq 0, \\ 2, & 0 < x \leq 1, \\ \dfrac{1}{x-1}, & 1 < x \leq 4. \end{cases}$.

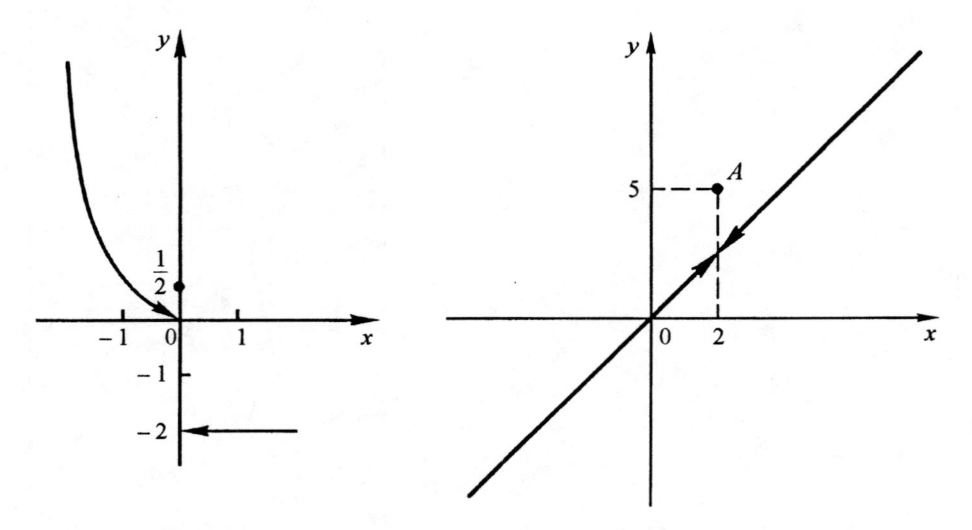

Figure 9.3. Figure 9.4.

Figure 9.3. The graph of the function $y = \begin{cases} -2, & x > 0, \\ \frac{1}{2}, & x = 0, \\ -x^3, & x < 0. \end{cases}$

Figure 9.4. The graph of the function $y = \begin{cases} x, & x \neq 2, \\ 5, & x = 2. \end{cases}$

bounded, it has no vertical asymptotes, and since it is periodic, there are neither horizontal nor oblique asymptotes (Figure 9.8).

Graphs of functions $y = \operatorname{sign} f(x)$ are shown in Figures 9.9 – 9.10.

Plotting the graph of $y = f(\operatorname{sign} x)$ we use that

$$y = f(\operatorname{sign} x) = \begin{cases} f(1), & x > 0, \\ f(0), & x = 0, \\ f(-1), & x < 0, \end{cases}$$

if $-1,0,1$ are in the domain of the function $f(x)$ (Figure 9.11).

Example 9.2.2. Plot the graph of the function $y = 1 + (\operatorname{sign} x)^2$.

Write the function in the form

$$y = 1 + (\operatorname{sign} x)^2 = \begin{cases} 2, & x \neq 0, \\ 1, & x = 0. \end{cases}$$

The graph is shown in Figure 9.12.

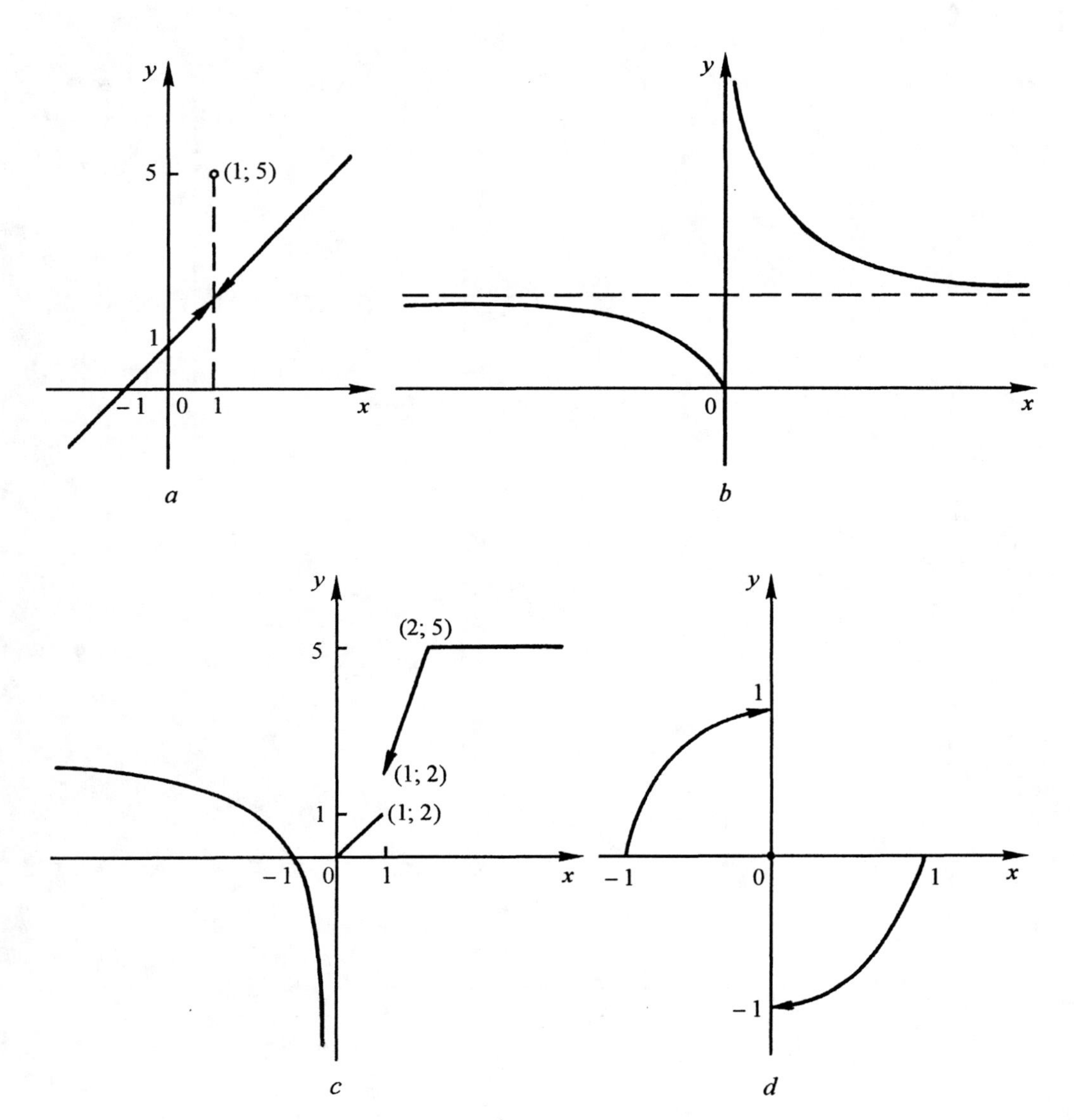

Figure 9.5. The graph of functions: (*a*) $y = \begin{cases} \dfrac{x^2 - 1}{x - 1}, & x \neq 1, \\ 5, & x = 1; \end{cases}$

(*b*) $y = \begin{cases} e^{1/x}, & x \neq 0, \\ 0, & x = 0; \end{cases}$ (*c*) $y = \begin{cases} \log x, & x < 0, \\ x, & 0 \leq x \leq 1, \\ x^2 + 1, & 1 < x \leq 2, \\ 5, & 2 < x < +\infty; \end{cases}$

(*d*) $y = \begin{cases} -x^2 + 1, & -1 \leq x < 0, \\ 0, & x = 0, \\ x^2 - 1, & 0 < x < 1; \end{cases}$

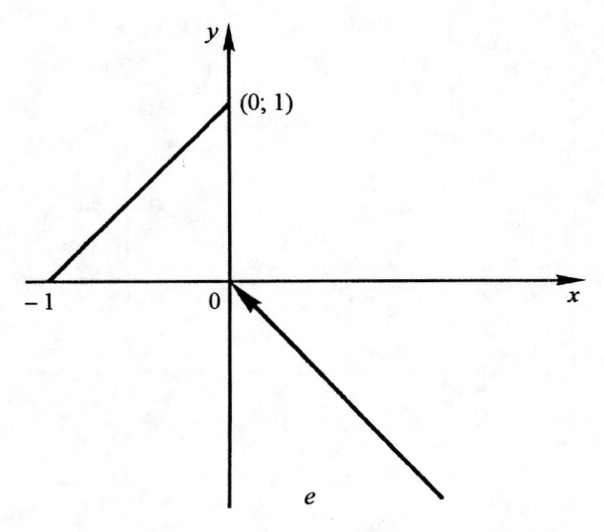

Figure 9.5. The graph of functions: (e) $y = \begin{cases} x+1, & -1 < x < 0, \\ -x, & 0 < x < 1. \end{cases}$

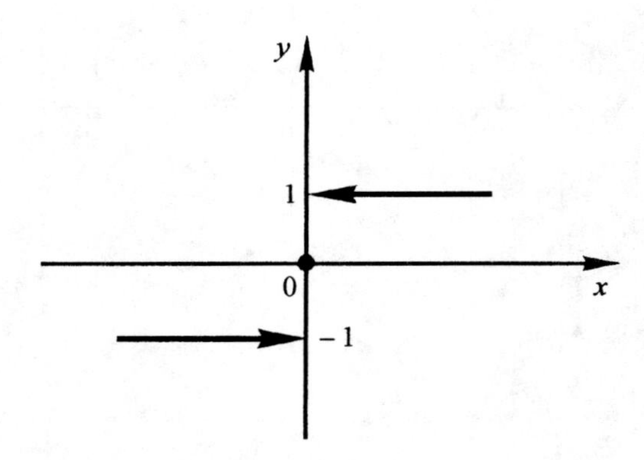

Figure 9.6. The graph of function $y = \text{sign}\, x = \begin{cases} 1, & x > 0, \\ 0, & x = 0, \\ -1, & x < 0 \end{cases}$

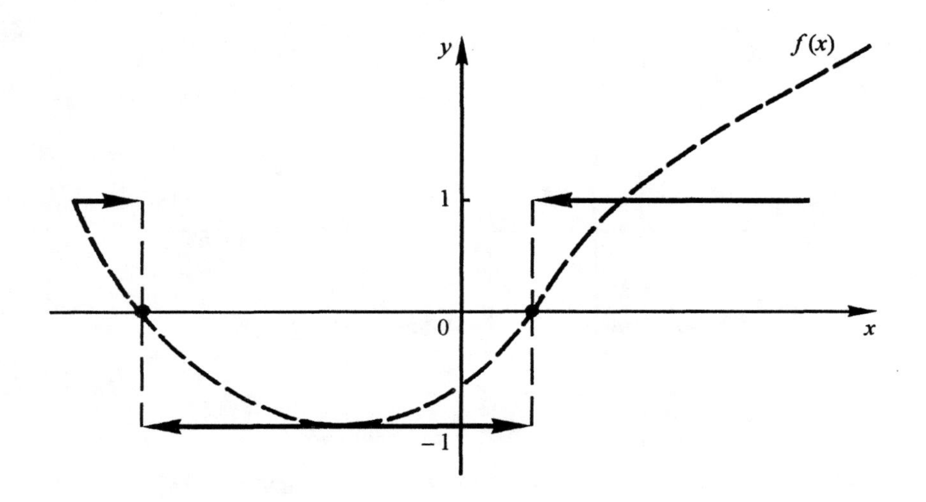

Figure 9.7. The graph of the function $y = \text{sign}\,(f(x))$.

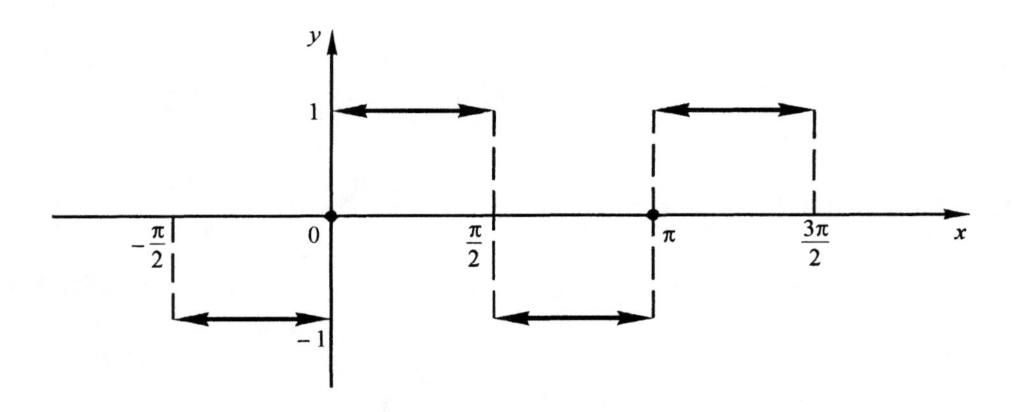

Figure 9.8. The graph of the function $y = \text{sign}\,(\tan x)$.

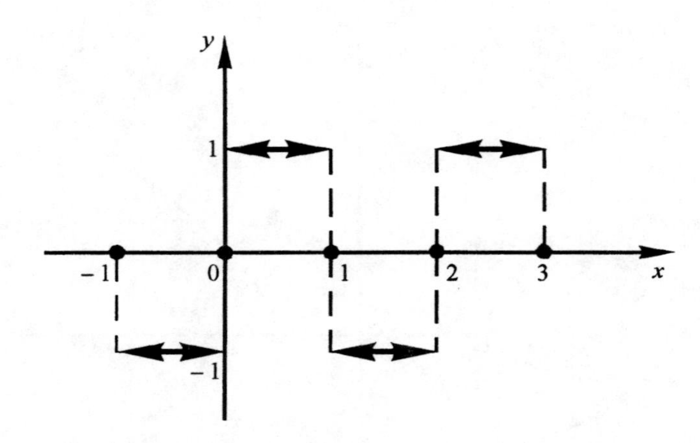

Figure 9.9. The graphs of the function $y = \text{sign}(\sin \pi x)$.

Example 9.2.3. The graph of the function $y = \text{sign}\, x(1 - \text{sign}^2 x)$ clearly coincides with the graph $y = 0$ (the whole x-axis), since $y = \text{sign}\, x(1 - \text{sign}^2 x) = 0$ for all x.

Example 9.2.4. The graph of the function $y = x \,\text{sign}\, x$ is clearly the same as the graph of the function $y = |x|$, since

$$y = x \,\text{sign}\, x = \begin{cases} x, & x > 0, \\ 0, & x = 0, \\ -x, & x < 0. \end{cases}$$

9.3. Plots of functions defined by a recurrence relation

Let us plot the graph of a function $y = f(x)$, $x \in \mathbf{R}$, such that $f(x + 1) = 2f(x)$ and $f(x) = x(1 - x)$ for $0 \le x \le 1$.

For a natural n we have: $f(x + n) = 2f(x + n - 1) = 2^2 f(x + n - 2) = \ldots = 2^n(f(x)$, i.e. $f(x + n) = 2^n f(x)$, $n \in \mathbf{Z}$. Because $f(x) = x(1 - x)$ for $0 \le x \le 1$, by setting $t = x + n$ we get $f(t) = 2^n(t - n)(n + 1 - t)$, $n \le t \le n + 1$, or $f(x) = 2^n(x - n)(n + 1 - x)$, $n \le x \le n + 1$, $n \in \mathbf{Z}$. Whence,

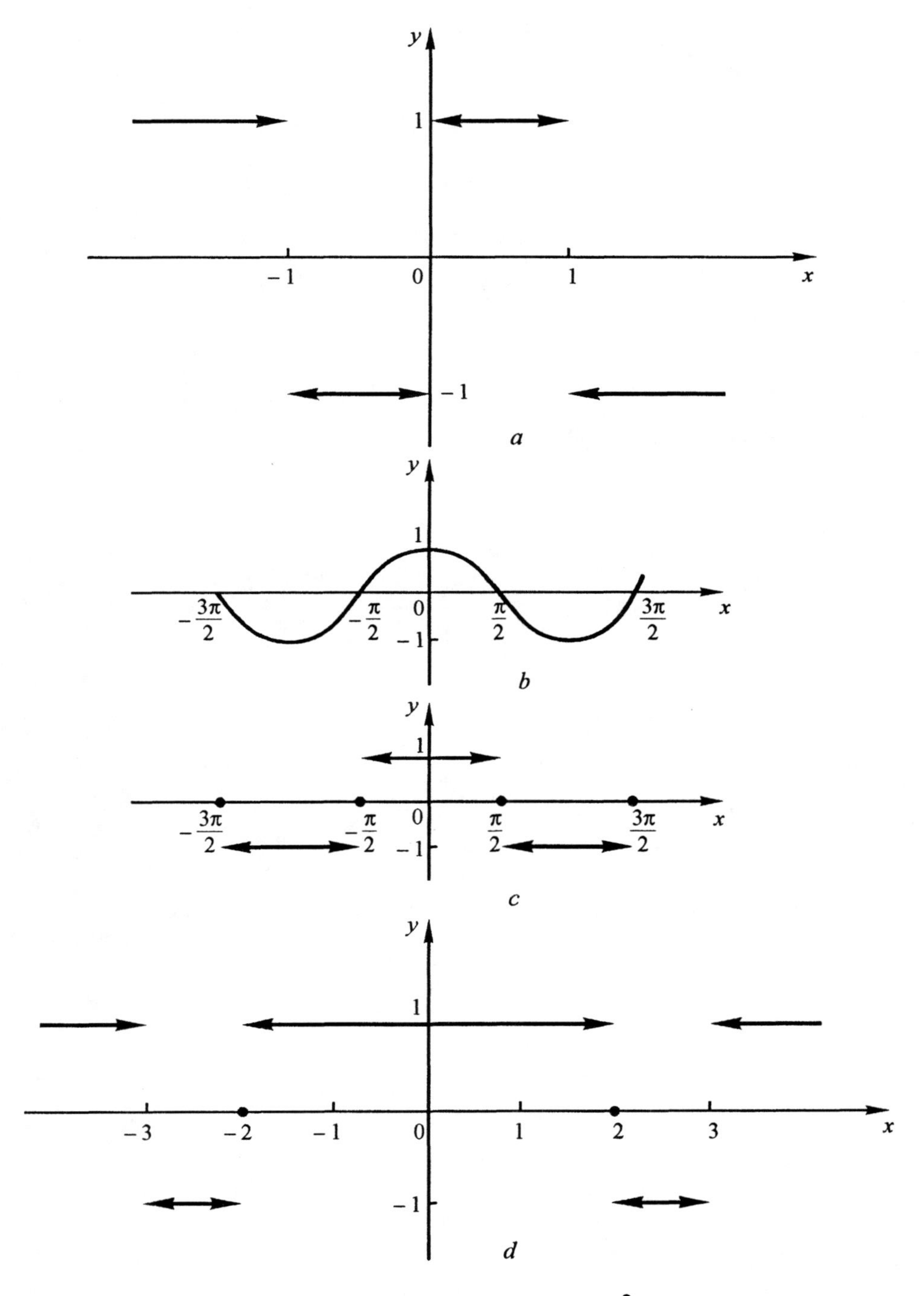

Figure 9.10. The graph of functions: (*a*) $y = \text{sign}\,(x(1 - x^2))$;
(*b*), (*c*) $y = \text{sign}\,(\cos x)$; (*d*) $y = \text{sign}\left(\dfrac{x^2 - 4}{x^2 - 9}\right)$.

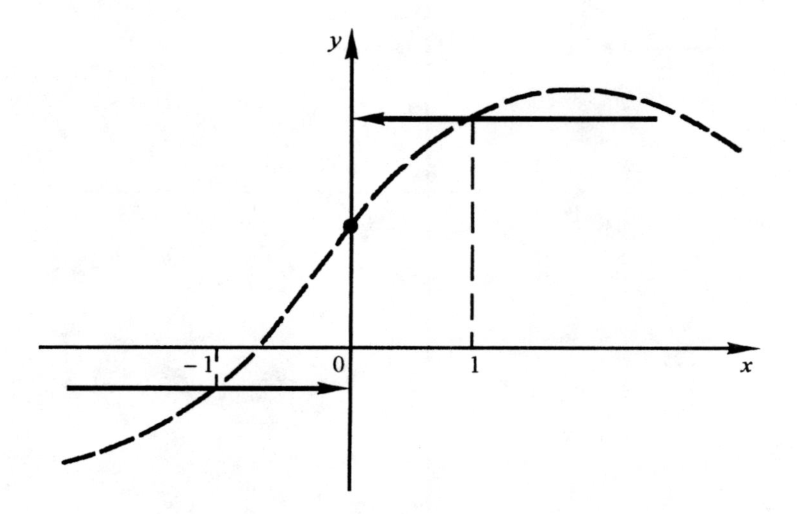

Figure 9.11. The graph of the function $y = f(\operatorname{sign} x)$.

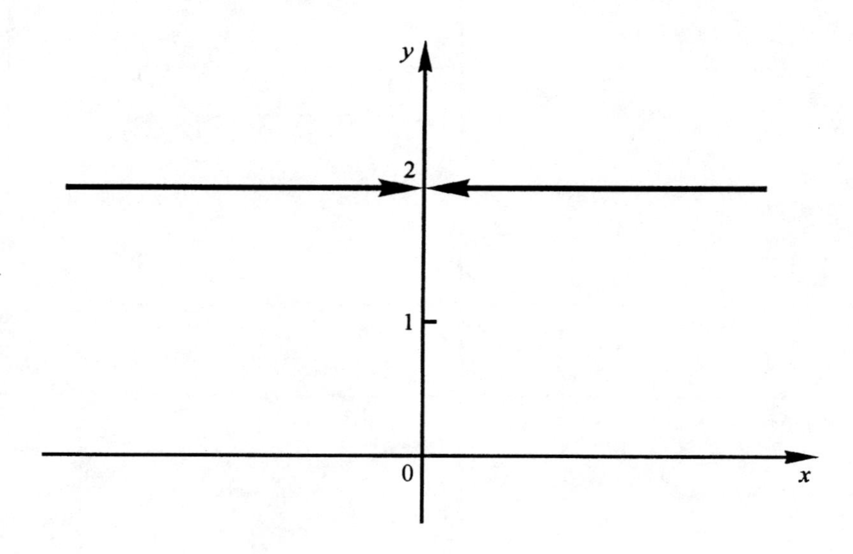

Figure 9.12. The graph of the function $y = 1 + (\operatorname{sign} x^2)$.

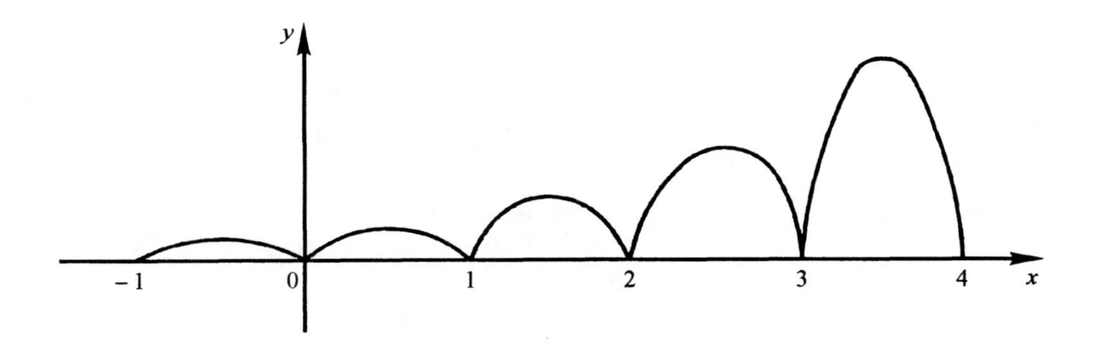

Figure 9.13. The curve $y = f(x)$, $x \in \mathbf{R}$, if $f(x+1) = 2f(x)$, and $f(x) = x(1-x)$, $0 < x < 1$.

$$f(x) = \begin{cases} \quad . \; . \; . \\ 2^{-1}(x+1)(-x), & -1 \le x \le 0, \\ x(1-x), & 0 \le x \le 1, \\ 2(x-1)(2-x), & 1 \le x \le 2, \\ 2^2(x-2)(3-x), & 2 \le x \le 3, \\ 2^3(x-3(4-x), & 3 \le x \le 4, \\ \quad . \; . \; . \end{cases}$$

The graph is shown in Figure 9.13.

Let us now plot the graph of a function $y = f(x)$, $x \in \mathbf{R}$, if $f(x+\pi) = f(x) + \sin x$, $f(x) = 0$ for $0 \le x \le \pi$. We have:

$$\begin{gathered} . \; . \; . \\ f(x+\pi) = f(x) + \sin x, \\ f(x+2\pi) = f(x+\pi) + \sin(x+\pi) = f(x), \\ f(x+3\pi) = f(x+2\pi) + \sin(x+2\pi) = f(x) + \sin x, \\ f(x+4\pi) = f(x+3\pi) + \sin(x+3\pi) = f(x), \\ . \; . \; . \end{gathered} \tag{1}$$

$$f(x+n\pi) = \begin{cases} f(x), & n \text{ is even}, \\ f(x) + \sin x, & n \text{ is odd}. \end{cases}$$

Replacing x with $x - \pi$ in the equality $f(x) = f(x+\pi) - \sin x$, we get $f(x-\pi) = f(x) + \sin x$. Now, as before, we get equality (1) for integer negative values of n. Let $x + n = t$. Then, if $0 \le x \le \pi$, we have that $n\pi \le t \le (n+1)\pi$, $n \in \mathbf{Z}$, and $f(x) = 0$. Thus,

$$f(x) = \begin{cases} 0, & n \text{ is even}, \\ \sin x, & n \text{ is odd}. \end{cases}$$

Now we plot the graph of the function (Figure 9.14).

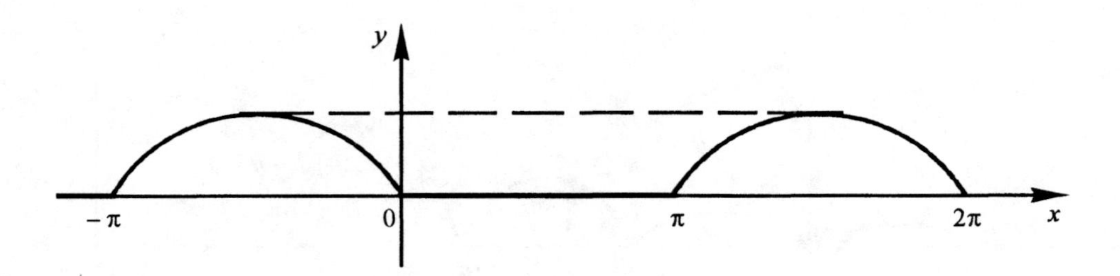

Figure 9.14. The curve $y = f(x)$, $x \in \mathbf{R}$, if $f(x+\pi) = f(x) + \sin x$, and $f(x) = 0$, $0 \le x \le \pi$.

9.4. Plots of functions defined in terms of limits

Let us consider a function defined as the limit of a sequence of functions that have the same domain.

Example 9.4.1. The function $f(x) = \lim\limits_{n\to\infty} \dfrac{x^n}{1+x^n}$ is not defined at $x = -1$. For $x \in (-1; 1)$ and $n \to \infty$, we have that $x^n \to 0$, $1 + x^n \to 1$. If $x = 1$, $x^n = 1$, $1 + x^n = 2$. For $x \in (-\infty; -1) \cup (1; +\infty)$, we have that $x^n \to \infty$ if $n \to \infty$.

Then $\dfrac{x^n}{1+x^n} = \dfrac{1}{1/x^n + 1} \to 1$ as $n \to \infty$. Thus the function $f(x)$ can be written in the form

$$f(x) = \begin{cases} 0, & |x| < 1, \\ \dfrac{1}{2}, & x = 1, \\ 1, & |x| > 1. \end{cases}$$

The graph of the function is shown in Figure 9.15.

Example 9.4.2. The function

$$f(x) = \lim_{n\to\infty} \frac{x + x^2 e^{nx}}{1 + e^{nx}}$$

can be written as

$$f(x) = \lim_{n \to \infty} \frac{x + x^2 e^{nx}}{1 + e^{nx}} = \begin{cases} x^2, & x > 0, \\ 0, & x = 0, \\ x, & x < 0. \end{cases}$$

The graph of the function is shown in Figure 9.16.

Graphs of functions defined as limits are shown in Figure 9.17.

Consider examples of functions defined in terms of the extremum of some functions.

Example 9.4.3. The points $(1; y)$ lie on the graph of $F(x, y) = \min(x, y) = 1$ for all $y \geq 1$. The points $(x; 1)$ lie on this graph for all $x \geq 1$ (Figure 9.18).

Example 9.4.4. The points $(1; y)$, for all $y \leq 1$, and the points $(x; 1)$, for all $x \leq 1$, lie on the graph of the function $F(x, y) = \max(x, y) = 1$ (Figure 9.19).

Example 9.4.5. The following points belong to the graph of the function $F(x, y) =$ $= \max(|x|, |y|) = 1$: $(1; |y|)$ for all $|y| \leq 1$, $(-1; |y|)$ for $|y| \leq 1$, $(|x|; 1)$ for $|x| \leq 1$, $(|x|; -1)$ for $|x| \leq 1$ (Figure 9.20).

9.5. Plots of functions $y = [f(x)]$

To plot the graph of a function $y = [f(x)]$, properties of the functions $y = f(x)$ and the integer function $y = [x]$ (Figure 9.21) are used in an essential way.

The domain of the function $y = [f(x)]$ coincides with the domain of the function $f(x)$. If the function $y = f(x)$ is even, then the function $y = [f(x)]$ is even. If the function $f(x)$ is periodic, the function $y = [f(x)]$ is also periodic.

Let the graph of a function $y = f(x)$ be given. The construction of the graph of $y = [f(x)]$ is carried out as follows:

Draw the straight lines $y = n$, $n \in Z$, and consider one of the strips bounded by $y = n$, $y = n + 1$.

Points of intersection of the straight lines $y = n$, $y = n + 1$ and the graph of the function $y = f(x)$ belong to the graph of $y = [f(x)]$, since their y-coordinates are integers; other points of the graph of $y = [f(x)]$ are obtained by projecting the part of the graph of $y = f(x)$ which belongs to this strip onto the line $y = n$, since the y-coordinate of any point M of this portion of the graph equals y_0 with $n \leq y_0 < n + 1$, i.e. $[y_0] = n$.

For every other strip containing points of the graph of $y = f(x)$, constructions are carried out in a similar way (Figure 9.22).

Examples of graphs of functions $y = [f(x)]$ are shown in Figures 9.23 – 9.25.

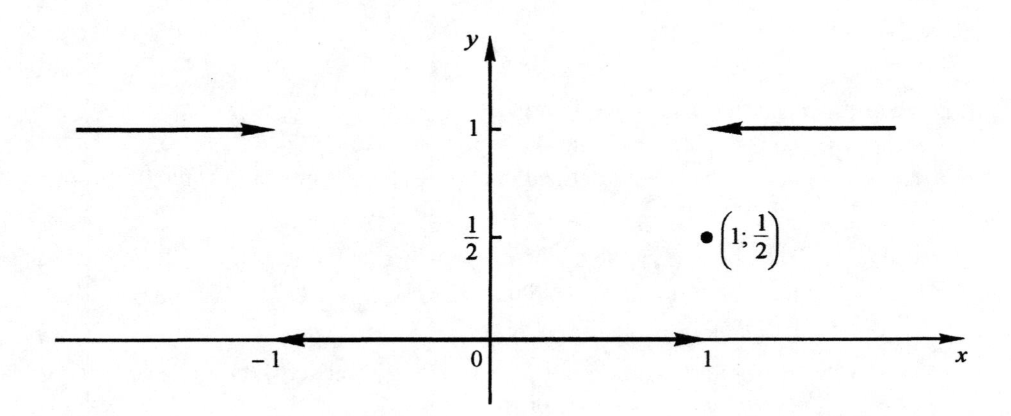

Figure 9.15. The graph of the function $f(x) = \lim\limits_{n\to\infty} \dfrac{x^n}{1+x^n}$.

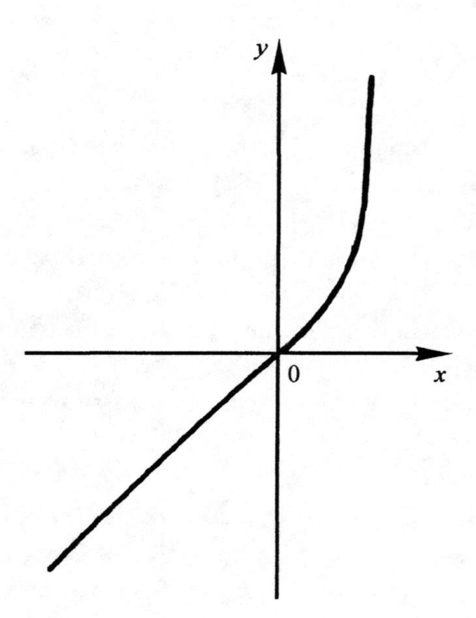

Figure 9.16. The graph of the function $f(x) = \begin{cases} x^2, & x > 0, \\ 0, & x = 0, \\ x, & x < 0. \end{cases}$

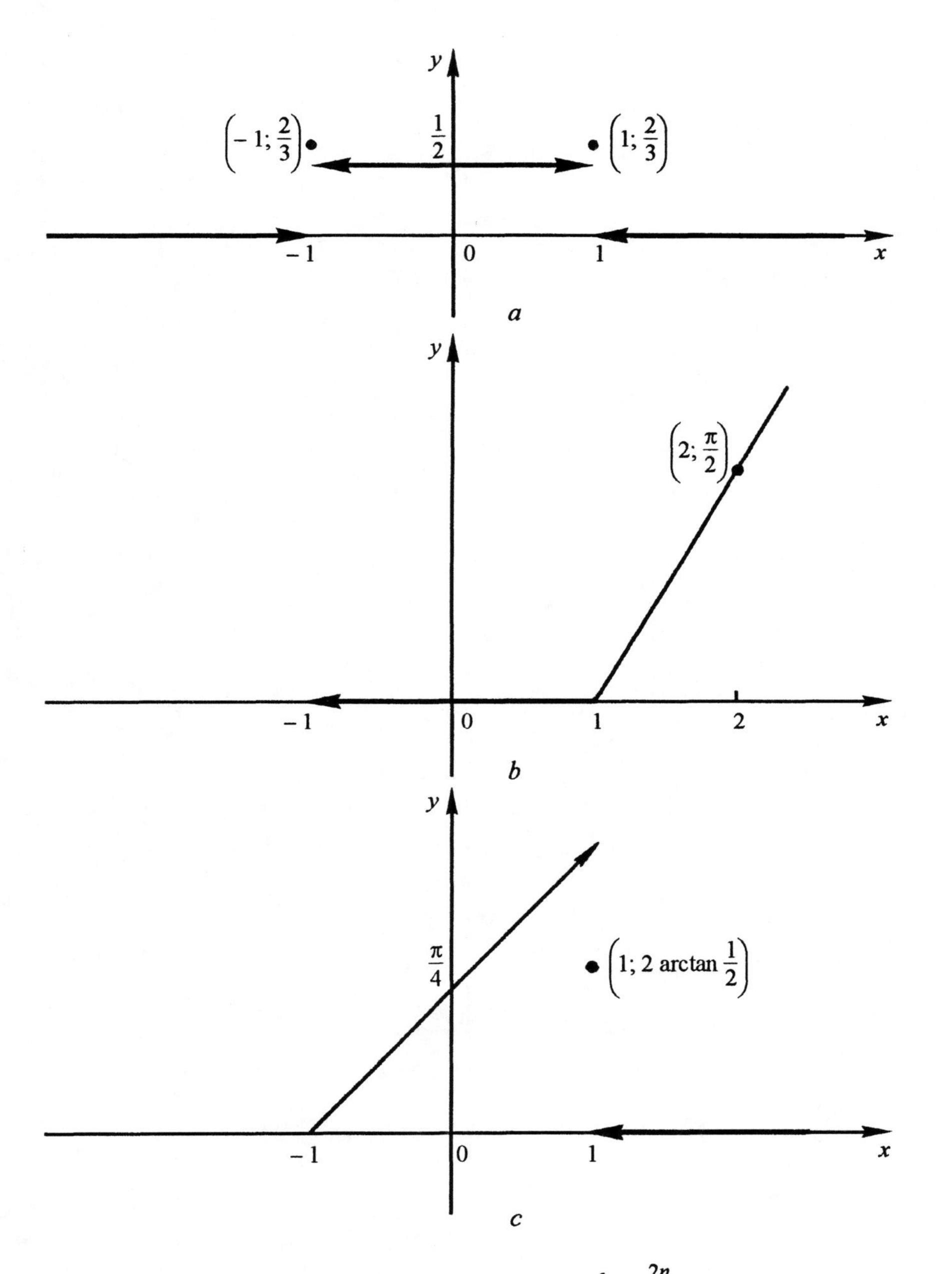

Figure 9.17. The graph of functions: (*a*) $f(x) = \lim\limits_{n\to\infty} \dfrac{1+x^{2n}}{2+x^{4n}}$;

(*b*) $f(x) = \lim\limits_{n\to\infty} (x-1)\arctan x^n$; (*c*) $y = \lim\limits_{n\to\infty} (x+1)\arctan \dfrac{1}{x^{2n}+1}$.

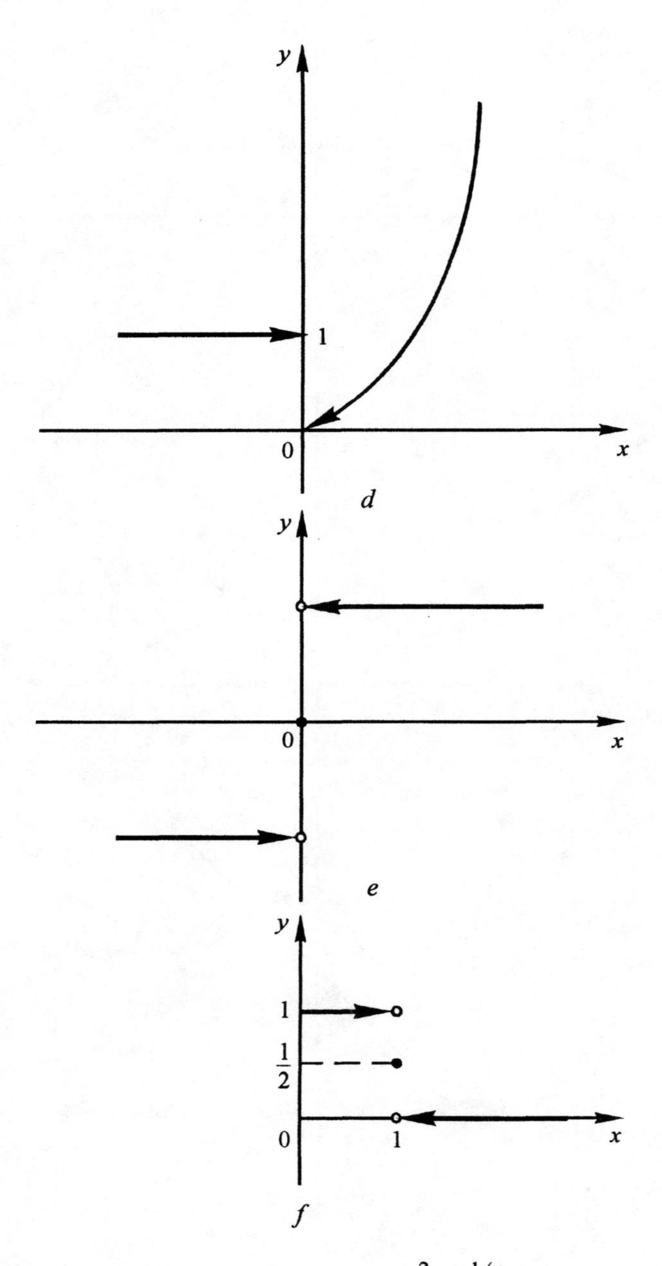

Figure 9.17. The graph of functions: (*d*) $y = \lim\limits_{n\to\infty} \dfrac{x^2 + 2^{1/x}}{1 + 2^{1/x}}$;

(*e*) $y = \lim\limits_{n\to\infty} \dfrac{n^x - n^{-x}}{n^x + n^{-x}}$; (*f*) $y = \lim\limits_{n\to\infty} \dfrac{1}{1 + x^n}$, $x \geq 0$.

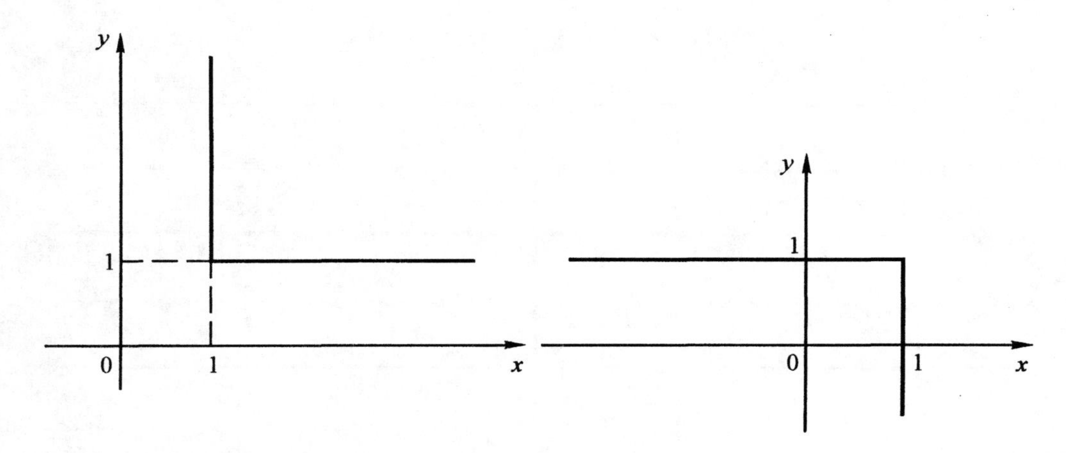

Figure 9.18 Figure 9.19

Figure 9.18. The graph of the function min $(x, y) = 1$.

Figure 9.19. The graph of the function max $(x, y) = 1$.

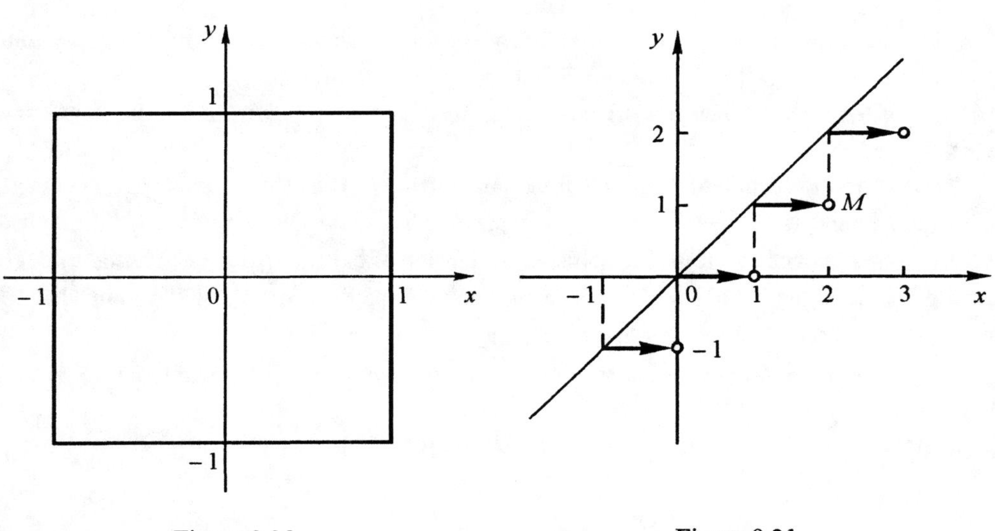

Figure 9.20 Figure 9.21

Figure 9.20. The graph of the function max $(|x|, |y|) = 1$.

Figure 9.21. The graph of the function $y = [x]$.

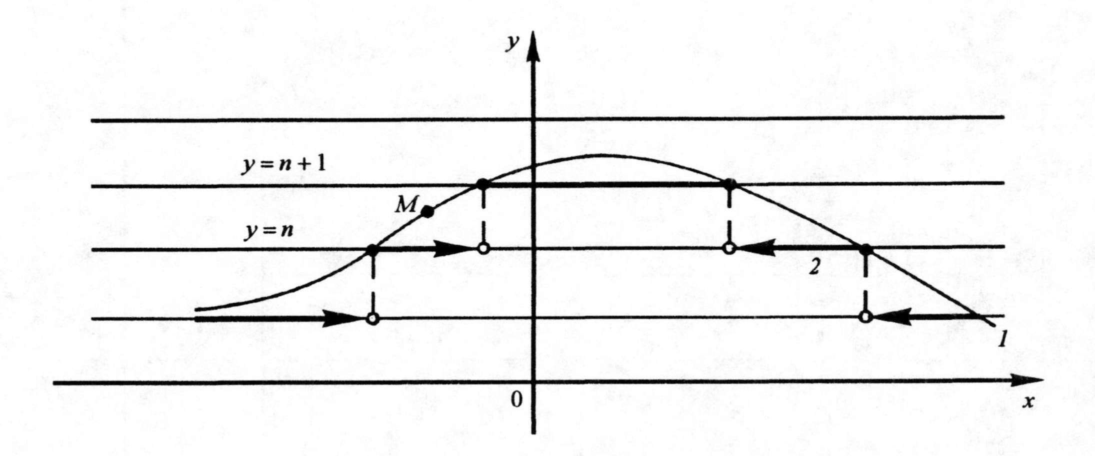

Figure 9.22. The graphs of functions: 1. $y = f(x)$, 2. $y = [f(x)]$.

9.6. Plots of functions $y = f([x])$

For the graph of a function $y = f(x)$ already constructed, a procedure for plotting the graph of the function $y = f([x])$ is the following (Figure 9.26).

Draw the straight lines $x = n$, $n \in \mathbf{Z}$, and consider the strip formed by the lines $x = n$, $x = n + 1$.

Points of intersection of the graph of the function $y = f(x)$ and these straight lines belong to the graph of $y = f([x])$, since their x-coordinates are integers. Other points of the graph of $y = f([x])$ are obtained by projecting points of the portion of the graph of $y = f(x)$ lying inside the strip onto the straight line $y = f(n)$, since any point N of this portion of the graph has the x-coordinate x_0 such that $n \leq x_0 < n + 1$, i.e. $[x_0] = n$.

For every other strip containing points of the graph of $y = f(x)$, construction is carried out in the same way (Figure 9.26).

Graphs of functions $y = f([x])$ are shown in Figures 9.27 – 9.29.

9.7. Plots of functions $y = \{f(x)\}$

To plot the graph of $y = \{f(x)\}$, the fractional function $y = \{x\}$ (Figure 9.30) and its properties (see Section 1.1) as well as properties of the function $y = f(x)$ are used. Since $\{f(x)\} = f(x) - [f(x)]$, plotting the graph of $y = \{f(x)\}$ can be reduced to plotting the graphs of $f(x)$ and $[f(x)]$.

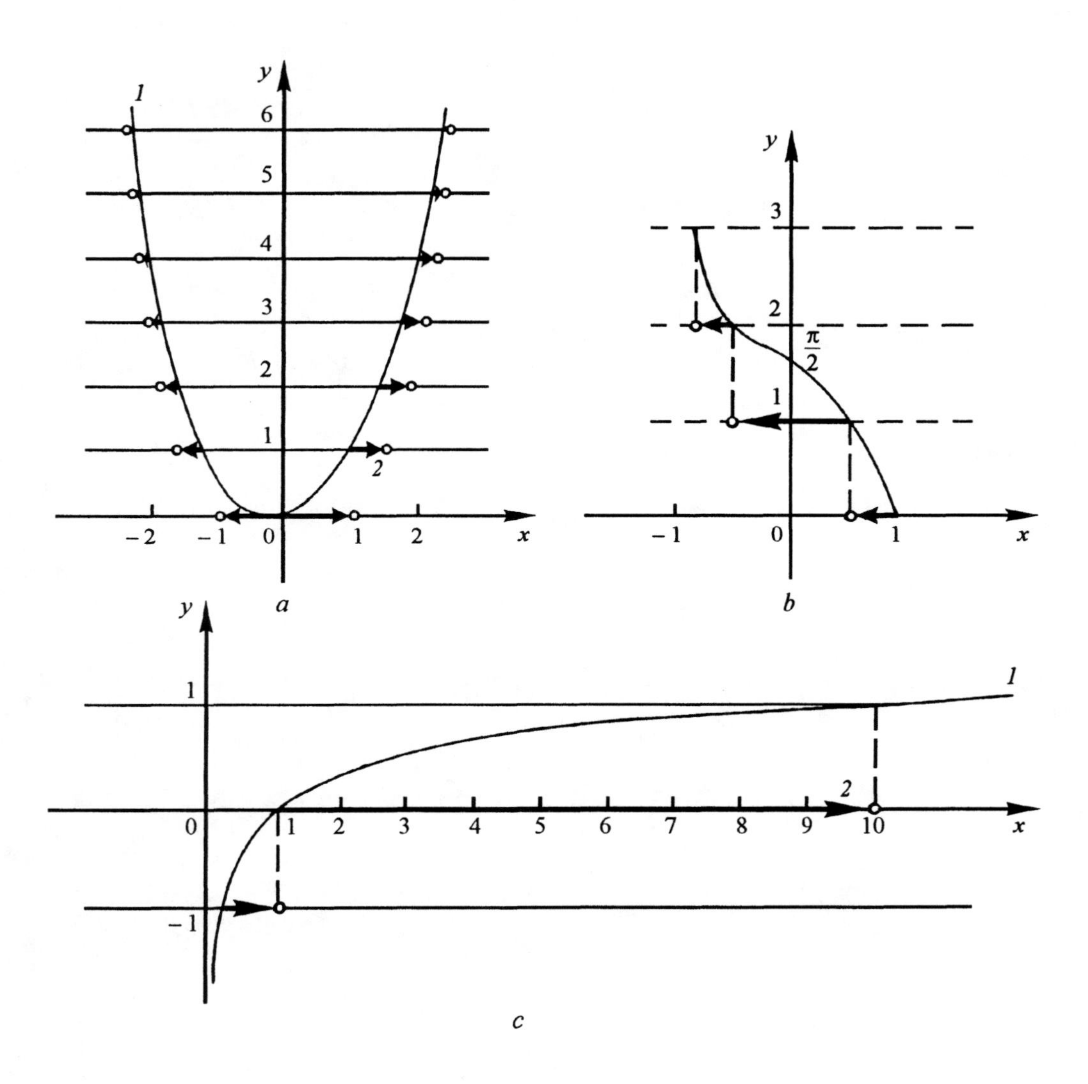

Figure 9.23. The graphs of functions: (*a*) *1.* $y = x^2$, *2.* $y = [x^2]$; (*b*) $y = [\arccos x]$; (*c*) *1.* $y = \lg x$, *2.* $y = [\log x]$.

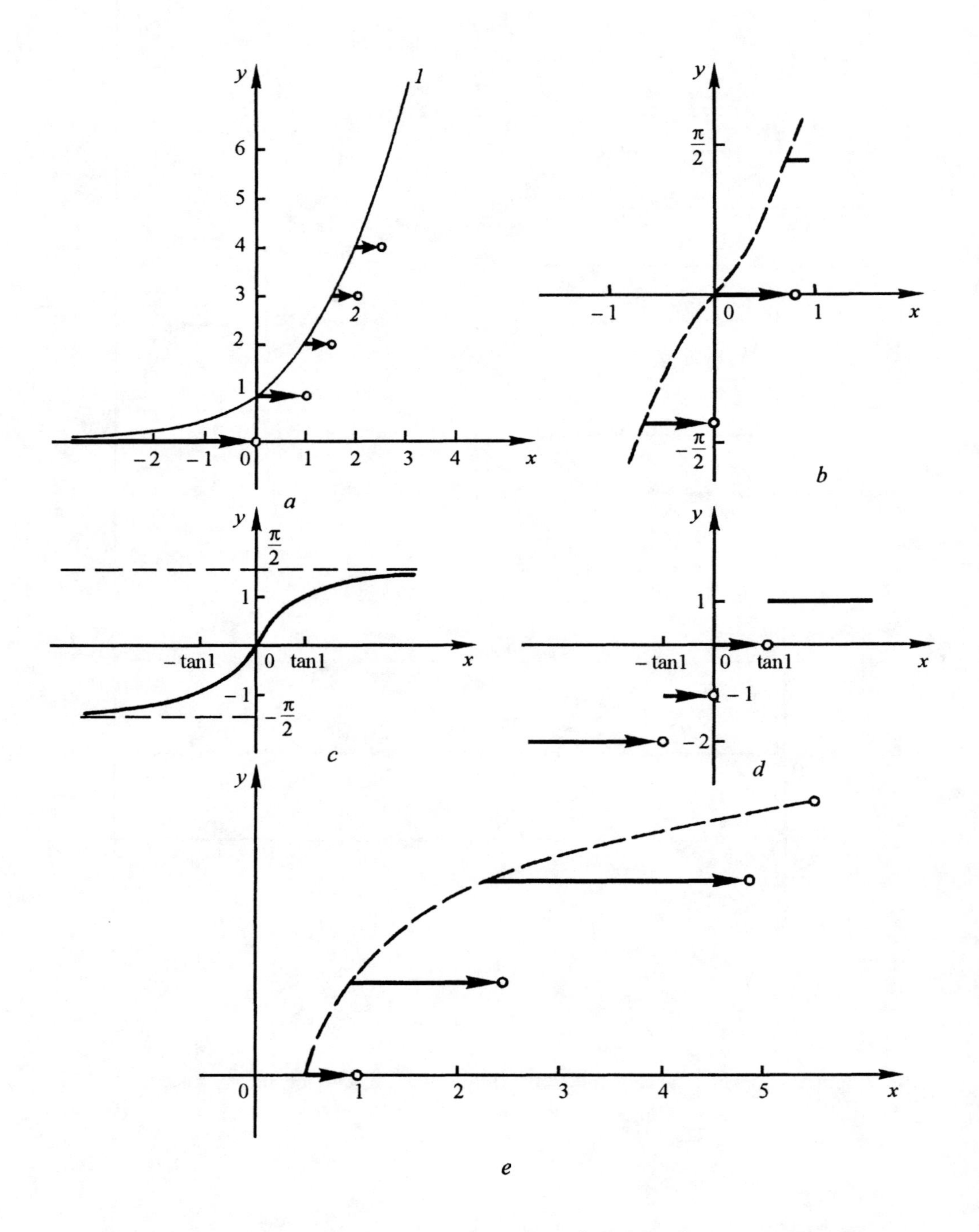

Figure 9.24. The graphs of functions: (*a*) *1*. $y = 2^x$, *2*. $y = [2^x]$; (*b*) $y = [\arcsin x]$; (*c*), (*d*) $y = [\arctan x]$; (*e*) $y = [\sqrt{2x-1}$.

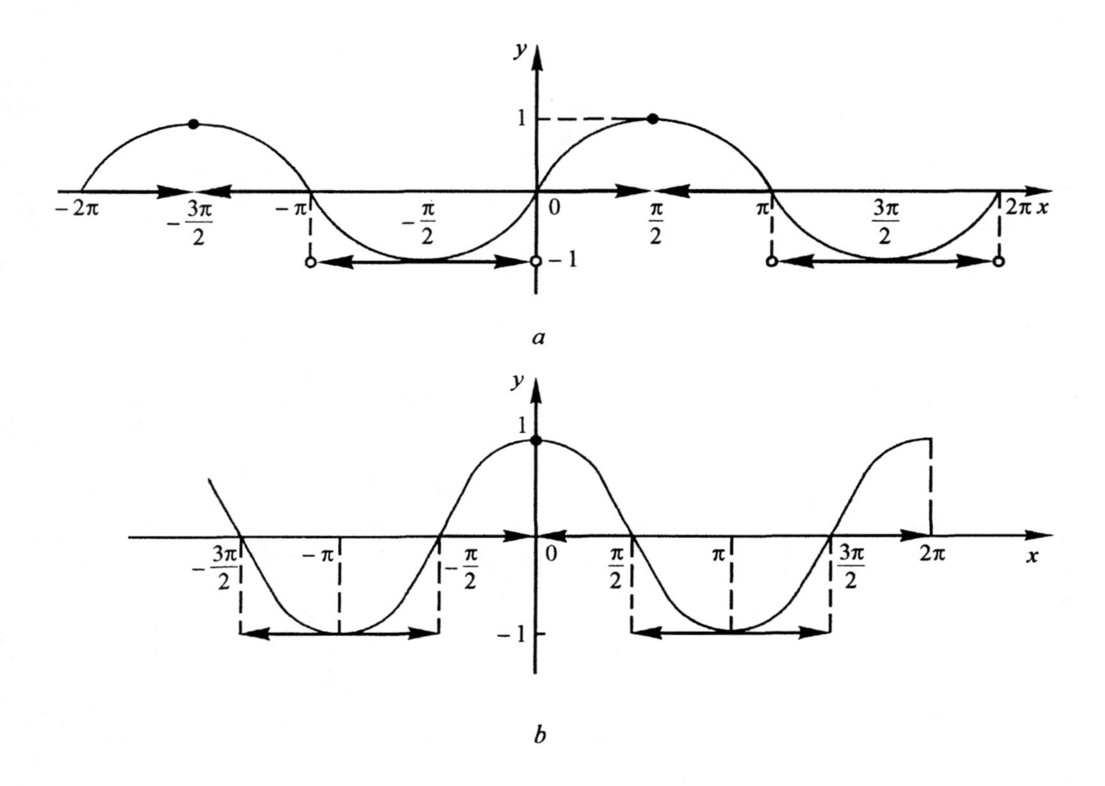

Figure 9.25. The graphs of functions: (*a*) $y = [\sin x]$; (*b*) $y = [\cos x]$.

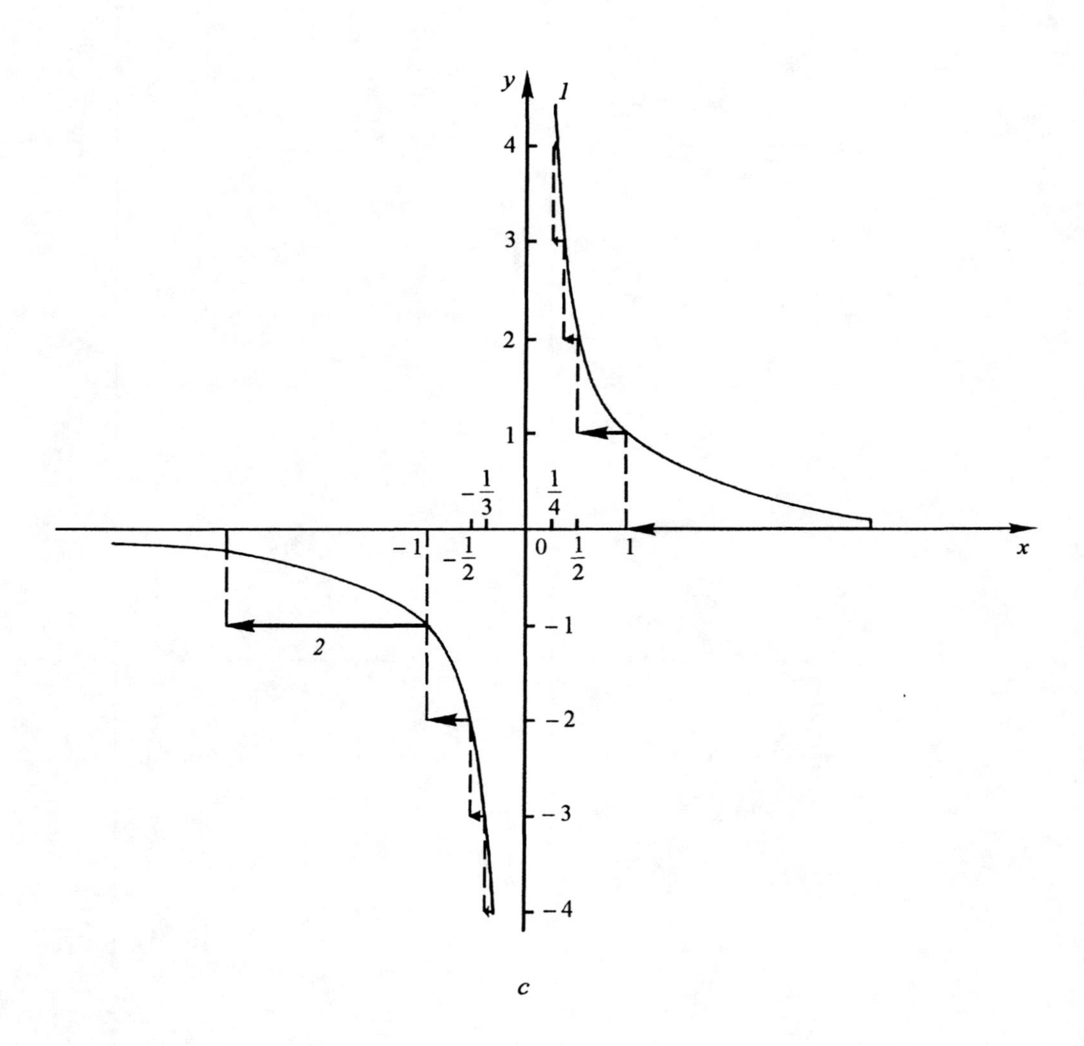

Figure 9.25. The graphs of functions: (*c*) *1*. $y=\frac{1}{x}$, *2*. $y=\left[\frac{1}{x}\right]$.

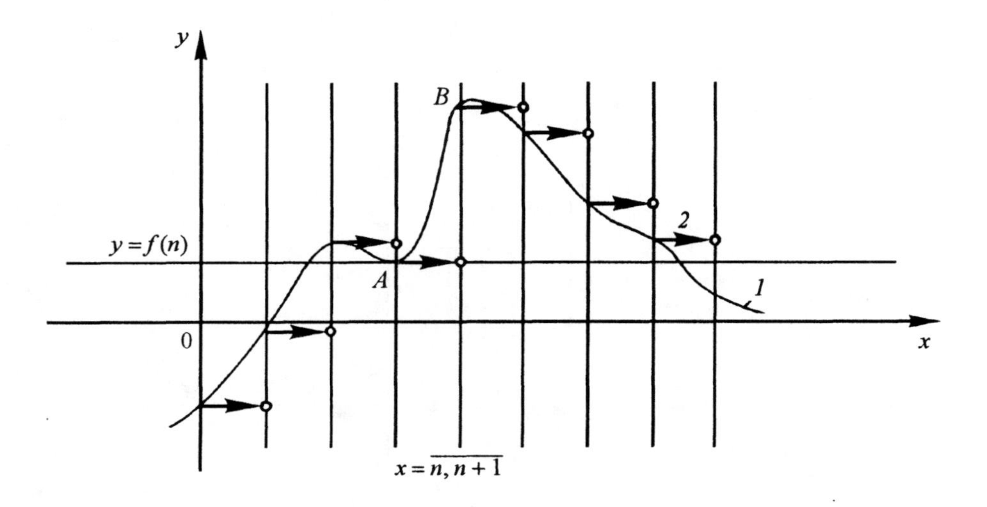

Figure 9.26. The graphs of functions: *1.* $y = f(x)$, *2.* $y = f([x])$.

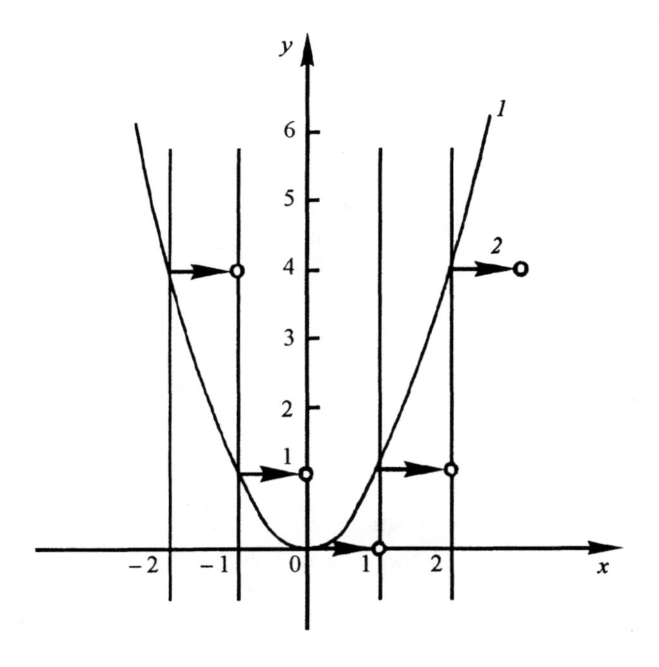

Figure 9.27. The graphs of functions: *1.* $y = x^2$, *2.* $y = [x]^2$.

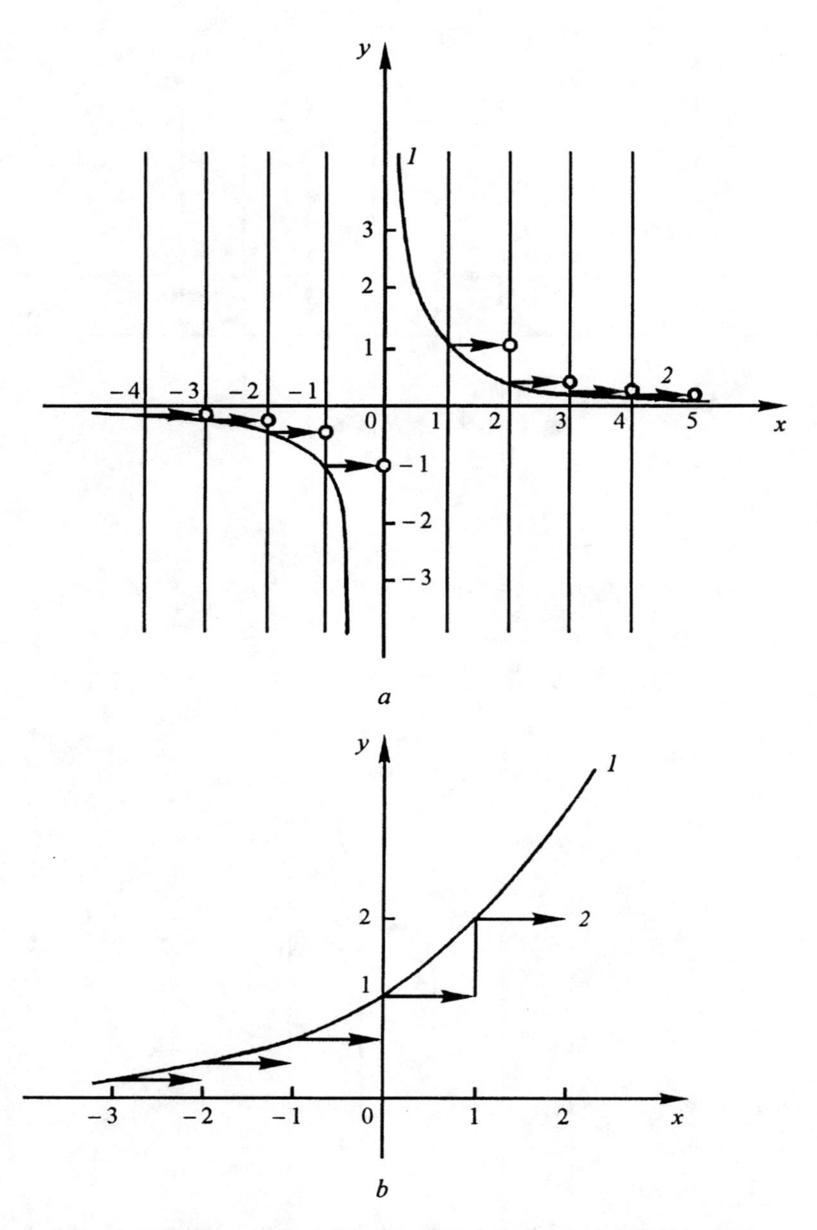

Figure 9.28. The graphs of functions: (*a*) *1*. $y = 1/x$, *2*. $y = 1/[x]$;
(*b*) *1*. $y = 2^x$, *2*. $y = 2^{[x]}$.

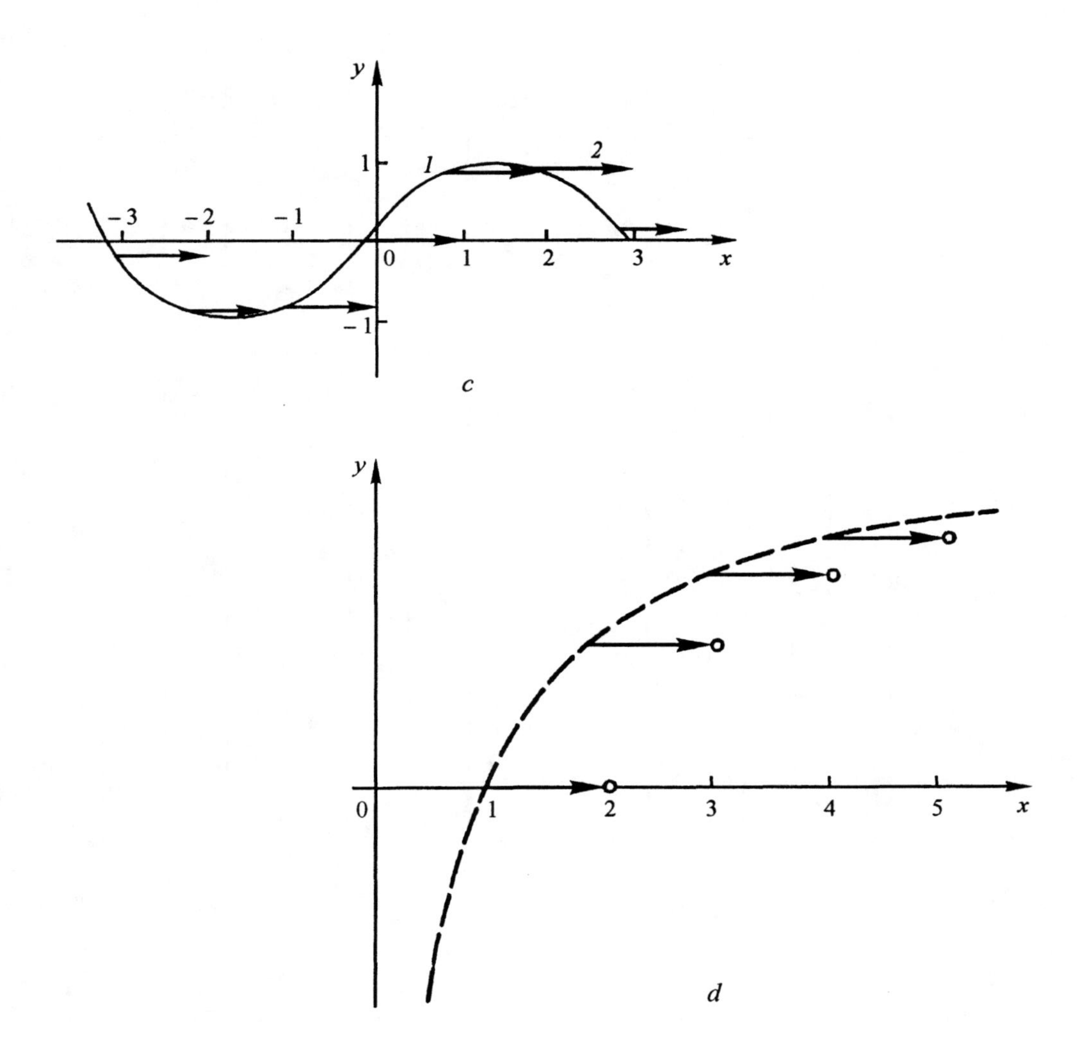

Figure 9.28. The graphs of functions: (*c*) *1*. $y = \sin x$, *2*. $y = \sin [x]$;
(*d*) $y = \log_3 [x]$.

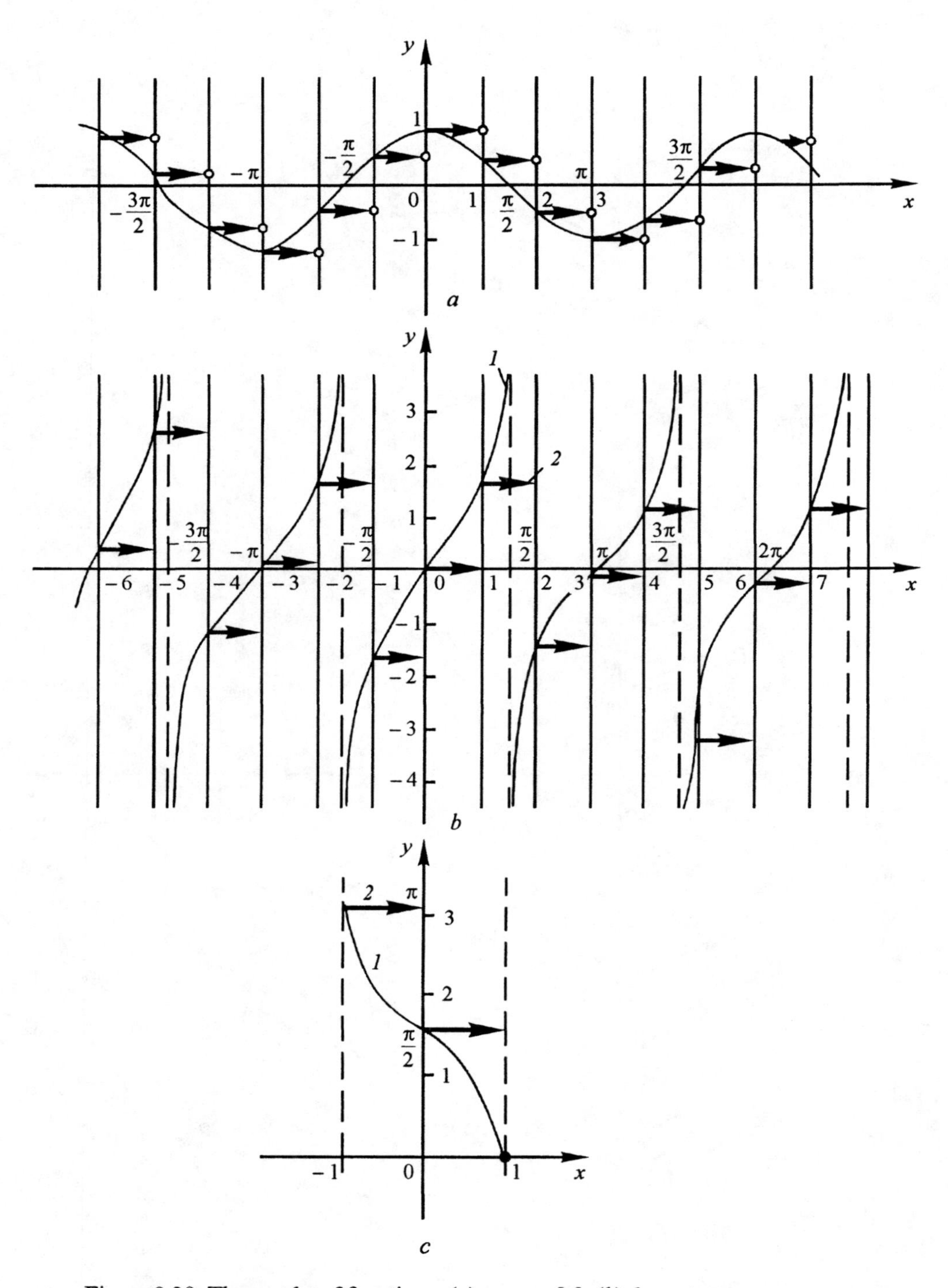

Figure 9.29. The graphs of functions: (*a*) $y = \cos [x]$; (*b*) *1.* $y = \tan x$, *2.* $y = [\tan x]$;(*c*) *1.* $y = \arccos x$, *2.* $y = \arccos [x]$.

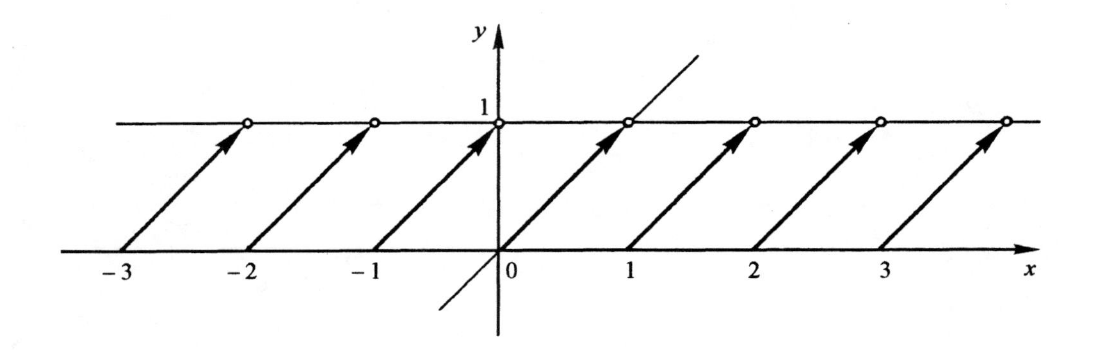

Figure 9.30. The graph of the function $y = \{x\}$.

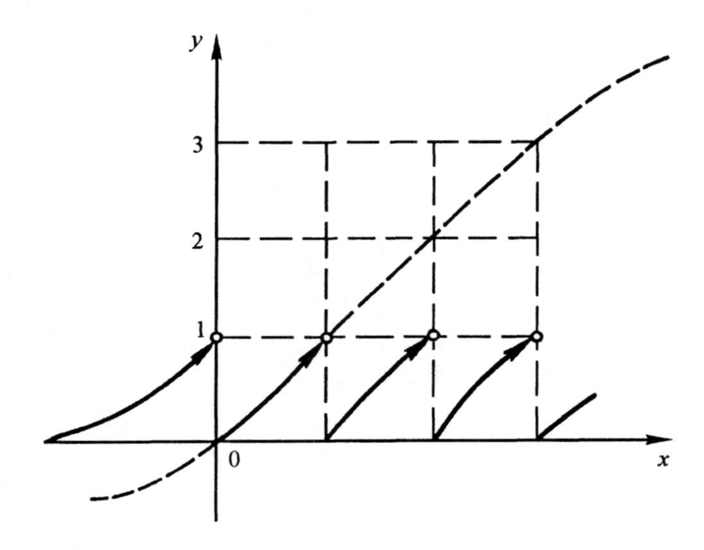

Figure 9.31. The graph of the function $y = \{f(x)\}$.

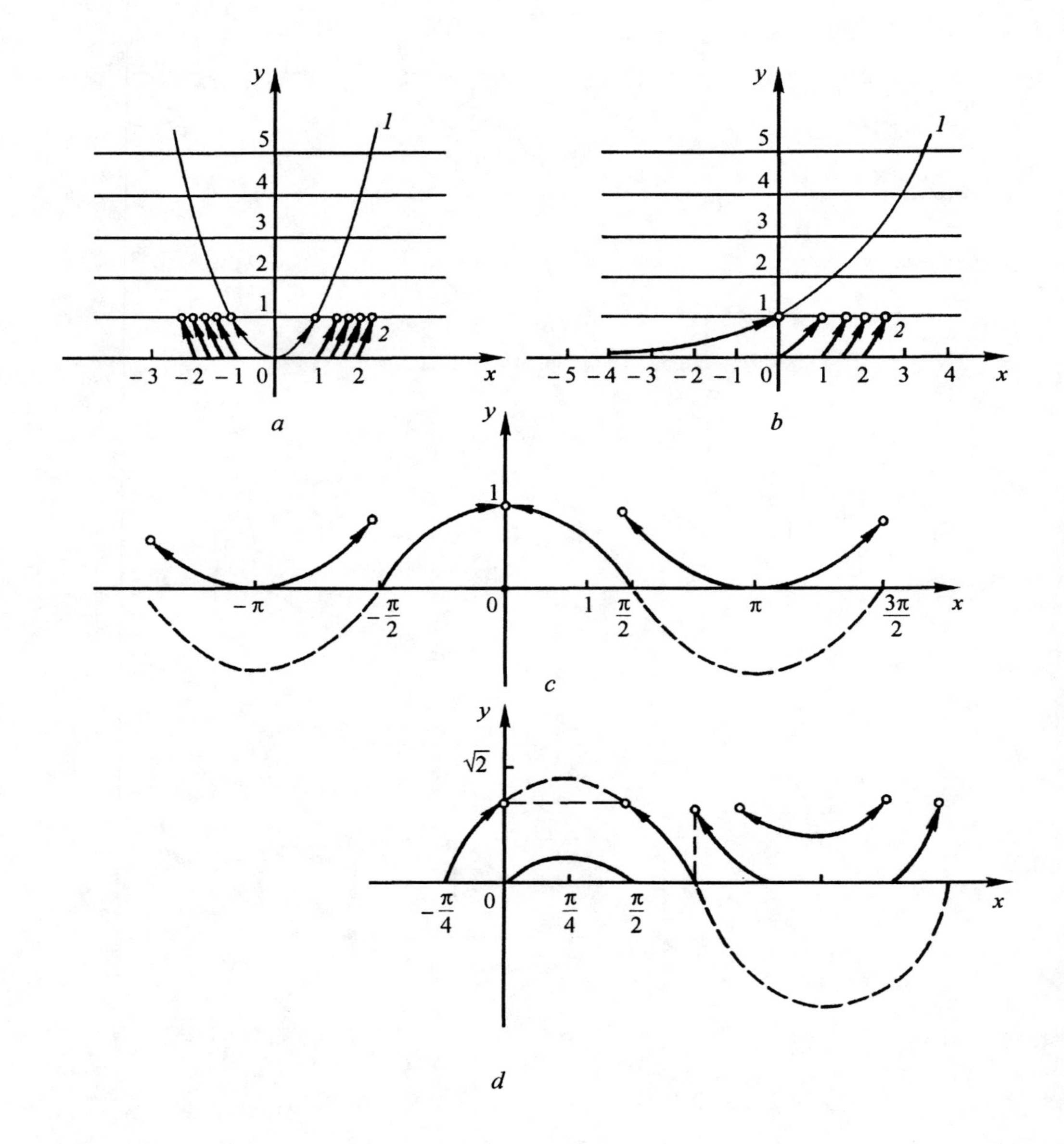

Figure 9.32. The graphs of functions: (*a*) *1.* $y = x^2$, *2.* $y = \{x^2\}$; (*b*) *1.* $y = a^x$, *2.* $y = \{a^x\}$ $(a = 2)$; (*c*) $y = \{\cos x\}$; (*d*) $y = \{\sin x + \cos x\}$.

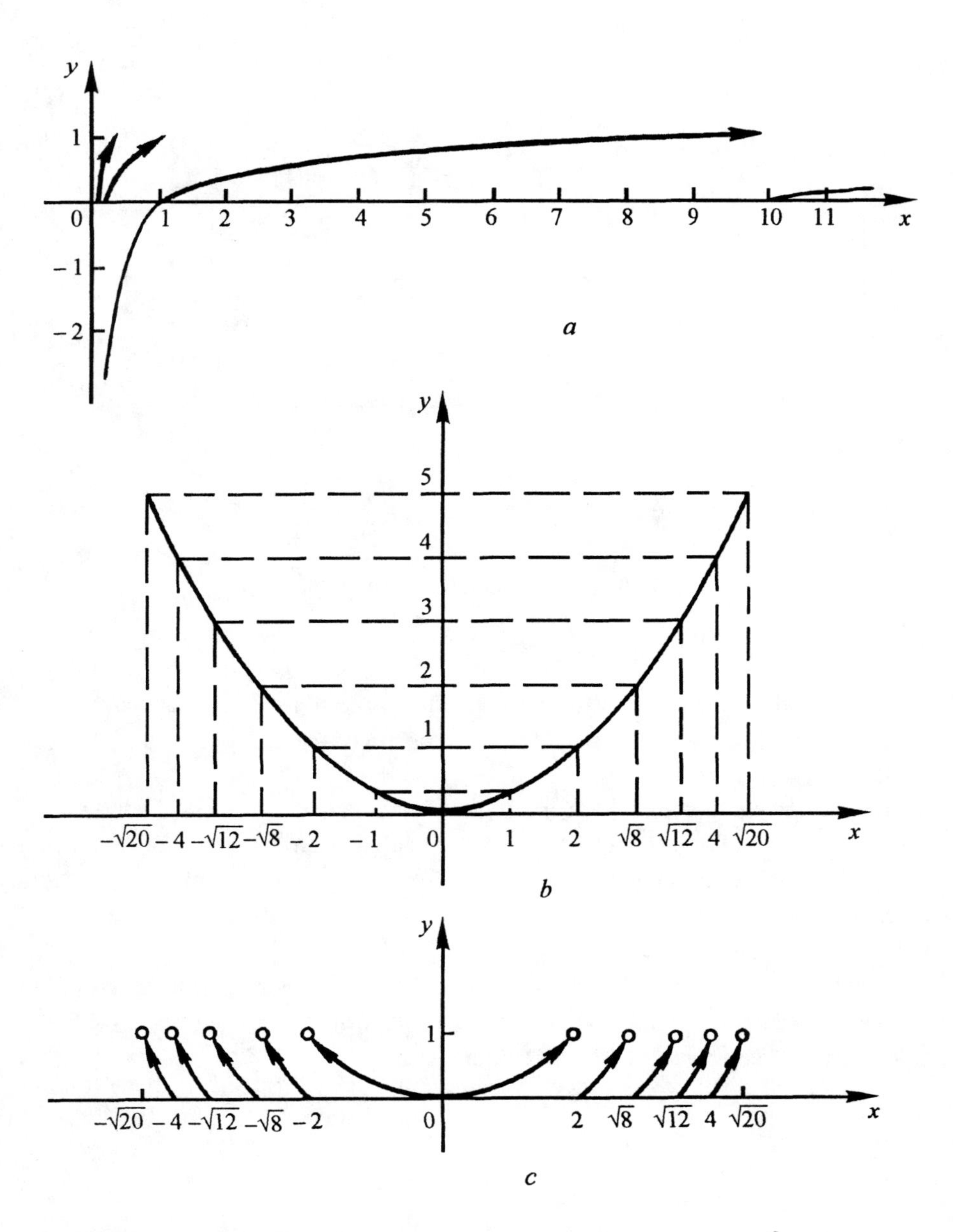

Figure 9.33. The graphs of functions: *(a)* $y = \{\tan x\}$; *(b, c)* $y = \{x^2/4\}$.

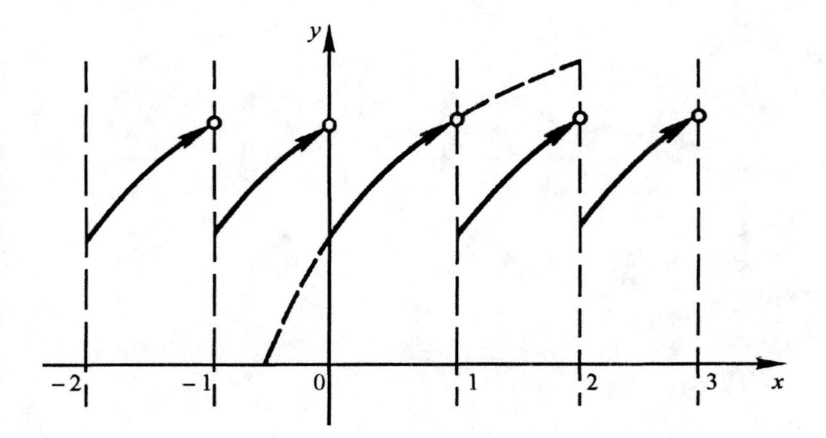

Figure 9.34. The graph of the function $y = f(\{x\})$.

Consider the simplest way to construct the graph of $y = \{f(x)\}$. Suppose the graph of $y = f(x)$ has already been constructed. The construction of the graph of $y = \{f(x)\}$ is carried out as follows:

a) Draw the straight lines $y = n$, $n \in \mathbf{Z}$.

b) Draw straight lines parallel to the y-axis that pass through points of intersection of the graph of $y = f(x)$ and the straight lines $y = n$. The fractional parts of the function $y = f(x)$ lie inside the generated rectangles.

Portions of the graph of $y = f(x)$ that lie inside the rectangles in the upper part of the plane are moved towards the x-axis by n. The portions lying inside the rectangles in the lower part of the plane are moved upwards by $1 + |n|$ (Figure 9.31).

Graphs of functions of the type $y = \{f(x)\}$ are shown in Figures 9.32 and 9.33.

9.8. Plots of functions $y = f(\{x\})$

The function $y = f(\{x\})$ is periodic with the period $\omega = 1$, and being restricted to $[0; 1)$, it is given by $f(\{x\}) = f(x)$. This shows how to plot the graph of $y = f(\{x\})$.

a) Plot the function $y = f(x)$ for $x \in [0; 1)$.

b) Continue this graph by periodicity, since $y = f(\{x\})$ is periodic (Figure 9.34).

The graphs of functions $y = f(\{x\})$ are shown in Figure 9.35.

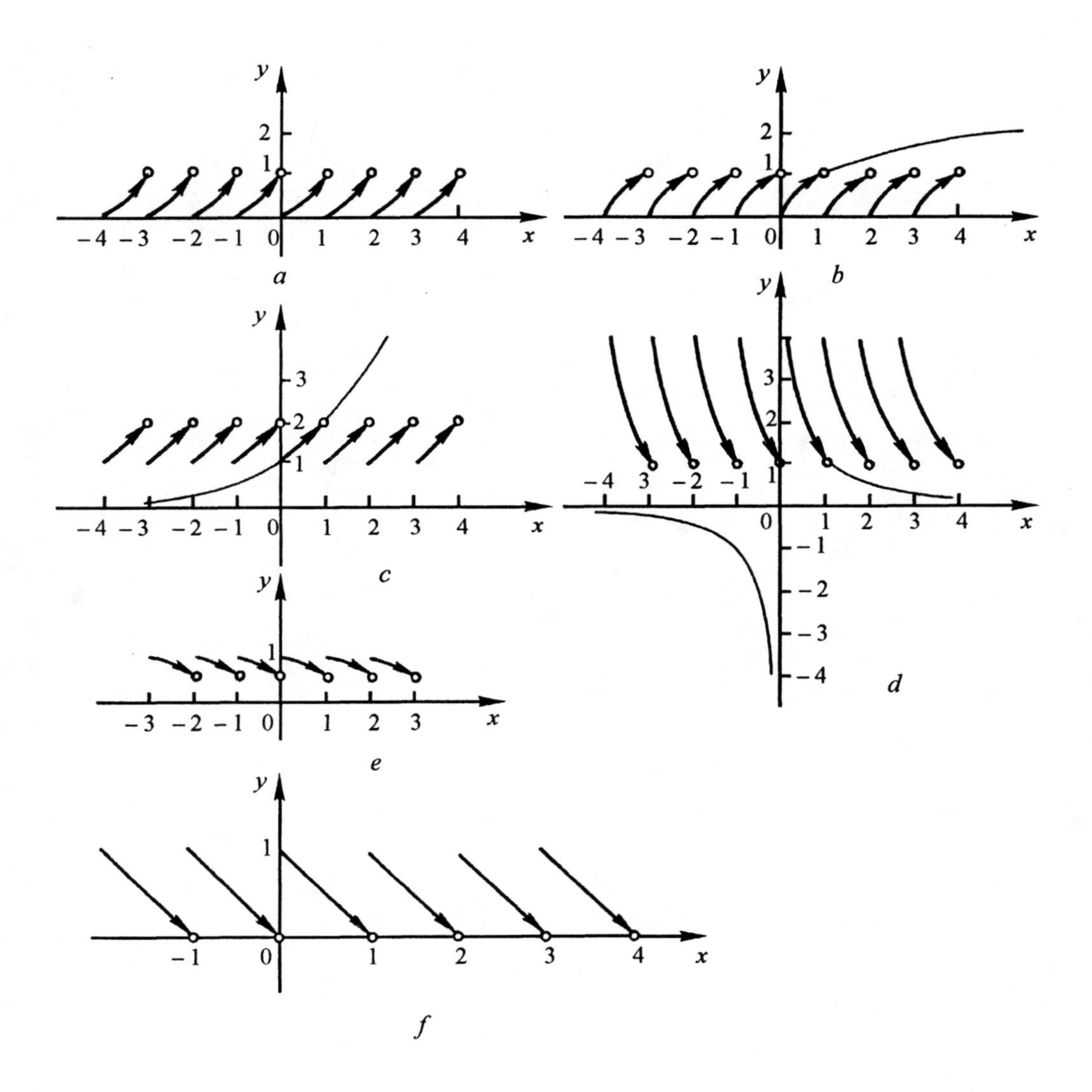

Figure 9.35. The graphs of functions: (*a*) $y = \{x\}^2$; (*b*) $y = \{x\}$; (*c*) $y = a^{\{x\}}$, $a = 2$; (*d*) $y = \dfrac{1}{\{x\}}$; (*e*) $y = \cos\{x\}$; (*f*) $y = \dfrac{1 - \{x\}}{1 + \{x\}}$.

Part II

Plots of graphs with the use of calculus. Graphs of special functions

Chapter 1

Derivatives and differentials. Their applications to plotting of graphs

1.1. Derivative of a function of one variable. Properties

Let a function $y = f(x)$ be defined in a neighborhood of a point x_0 and x be a point in this neighborhood, $x \neq x_0$. Consider the ratio $\frac{f(x) - f(x_0)}{x - x_0}$. If the limit of this ratio exists for $x \to x_0$, then it is called *derivative* of the function $f(x)$ at the point x_0, $f'(x_0) = \lim_{x \to x_0} \frac{f(x) - f(x_0)}{x - x_0}$.

Denote $\Delta x = x - x_0$. Then

$$f'(x_0) = \lim_{x \to x_0} \frac{f(x_0 + \Delta x) - f(x_0)}{\Delta x} = \lim_{\Delta x \to 0} \frac{\Delta f(x)}{\Delta x}. \tag{1}$$

Definition 1.1.1. For a function $f(x)$ and a point x_0, the limit of the ratio of the increment of the function at the point x_0, Δy, and the increment of the variable, Δx, as $\Delta x \to 0$, is called the *derivative* of the function $f(x)$ at the point x_0.

The derivative is denoted by: y', y'_x, $f'(x)$ (Lagrange); dy/dx, df/dx (Leibniz); Dy, Df (Cauchy).

The value of the derivative at $x = a$ is denoted by $f'(a)$ or $y'|_{x=a}$. In every point at which limit (1) exists, the derivative dy/dx is a measure of the rate of change of y with respect to x. The process of finding the derivative is called *differentiation* of this function.

Geometric interpretation of differentiation

The derivative $f'(x)$ of a function $f(x)$ at a point x is the slope of the tangent to the curve $y = f(x)$ at the point x, $f'(x) = \tan\alpha$, where α is the angle formed by the x-axis and the tangent to the curve at the point x counted from the x-axis counterclockwise (Figure 1.1).

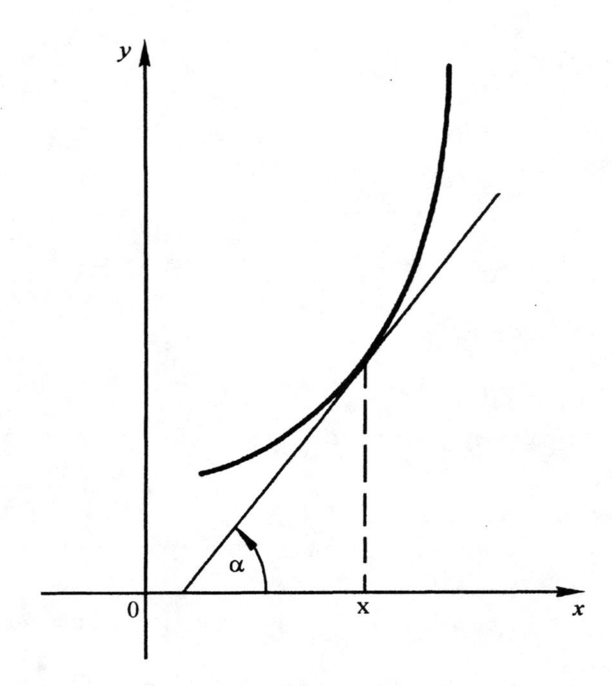

Figure 1.1. The geometric interpretation of the derivative of a function, $f'(x) = \tan\alpha$.

If a function $y = f(x)$ has a derivative at a point $x = x_0$, i.e. limit (1) exists, then it is said that the function is *differentiable* at the point x_0. If a function is differentiable at every point of an interval $[a; b]$ or $(a; b)$, then it is said to be differentiable on the interval $[a; b]$ or $(a; b)$, correspondingly.

If a function $f(x)$ is differentiable at a point $x = x_0$, then it is continuous at this point.

If the derivative does not exist for a given $x = x_1$, then either the function does not have a tangent at this point or the tangent makes the angle with the x-axis of 90°. For example, the derivative of the function $f(x) = \sqrt{x}, f'(x) = \frac{1}{2\sqrt{x}}$, does not exist at the point $x = 0$, since it takes the infinite value. Limit (1) does not exist for the function $f(x) = x\sin\frac{1}{x}$ at the point $x = 0$.

If limit (1) does not exist at a point $x = 0$ but the left-hand and right-hand limits exist, then the limits are called *left-hand* and *right-hand* derivatives, respectively. Geometric interpretation of the left- and right-hand derivatives is the following: $f'(x-0) = \tan\alpha_1$, $f'(x+0) = \tan\alpha_2$ (Figure 1.2). The curve has a kink at the point $x = a$. For example, the function $f(x) = |x|$, at the point $x = 0$, has the following left- and right-hand derivatives: $f'(0-0) = -1, f'(0+0) = 1$.

The derivative of the derivative of a function $f(x)$ is called the *second-order derivative*.

The *derivative of order n* of a function $f(x)$ is the derivative of the derivative of order $n-1$. Derivatives of higher order are denoted by $, \dots y^{(n)}, \dots$, or $f''(x), f'''(x), f^{IV}(x), \dots,$ $f^{(n)}(x)$, or $d^2f/dx^2, d^3f/dx^3, d^4f/dx^4, \dots, d^nf/dx^n, \dots$.

To find the derivative of a higher order, it is necessary to know all the derivatives of the lower order.

The derivative of the product of two functions is found from Leibniz's formula:

$$(uv)^{(n)} = uv^{(n)} + \frac{n}{1}u'v^{(n-1)} + \frac{n(n-1)}{2}u''v^{(n-2)} + \dots$$

$$+ \frac{n(n-1)\dots(n-m+1)}{m!}u^{(m)}v^{(n-m)} + \dots + u^{(n)}v,$$

or $(un)^{(n)} = \sum_{m=0}^{n} C_n^m u^{(m)} v^{(n-m)}$.

1.2. Differential of a function of one variable

Let $y = f(x)$ be differentiable on an interval X, i.e.

$$f'(x) = \lim_{\Delta x \to 0} \frac{\Delta y}{\Delta x}, \quad \Delta y = f'(x)\Delta x + \alpha\Delta x, \tag{2}$$

or $\Delta y = f'(x)\Delta x + o(\Delta x)$, where $\alpha \to 0$ for $\Delta x \to 0$, $o(\Delta x)$ is an infinitely small with respect to Δx variable of order 1. Then $f'(x)\Delta x$ is the principal part of the increment of the function; it is linear with respect to Δx and is called the *differential* of the function and is denoted by $dy = df(x) = f'(x)dx$, since $\Delta x = dx$.

A function $y = f(x)$ has the differential at a point x if and only if it is differentiable at this point.

The differential of a function is characterized by two main properties: it is a linear homogeneous function of the increment Δx, and it differs from the increment of the function by a value that is infinitely small with respect to Δx for $\Delta x \to 0$.

Geometric interpretation of the differential

The differential of a function $f(x)$, for given values of x and Δx, is the increment of the y-coordinate of the tangent to the curve $y = f(x)$ at the point x (Figure 1.3).

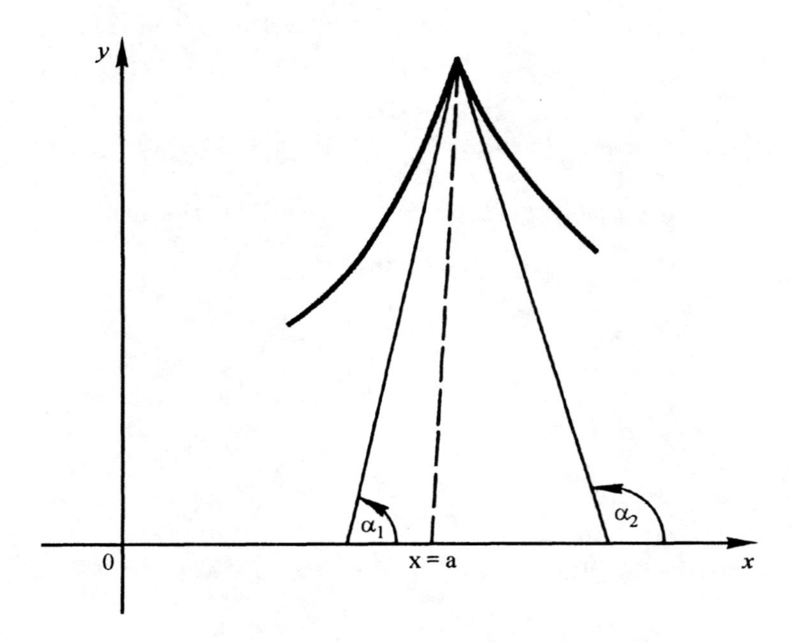

Figure 1.2. The geometric interpretation of the left- and right-hand derivatives of a function.

1.3. Fundamental theorems of calculus

Fermat's theorem. If a function $y = f(x)$ is defined on some interval, reaches a maximum or minimum at an interior point $x = c$ (i.e. $f(c) > f(x)$ or $f(c) < f(x)$), and has a finite derivative at the point c, then this derivative equals zero, $f'(c) = 0$.

Geometric interpretation of the theorem. The tangent to the graph of the function at points A and B is parallel to the x-axis (Figure 1.4).

Rolle's theorem. Let a function $y = f(x)$ be defined and continuous on an interval $[a; b]$, and have a finite derivative at least in the open interval $(a; b)$, and suppose that $f(a) = f(b)$. Then there exists at least one point c between a and b $(a < c < b)$ such that the derivative of the function at this point equals zero, $f'(c) = 0$.

Geometric interpretation of the theorem. If the y-coordinates of the curve $y = f(x)$ are equal at the endpoints, then there exists a point at which the tangent is parallel to the x-axis (Figure 1.5).

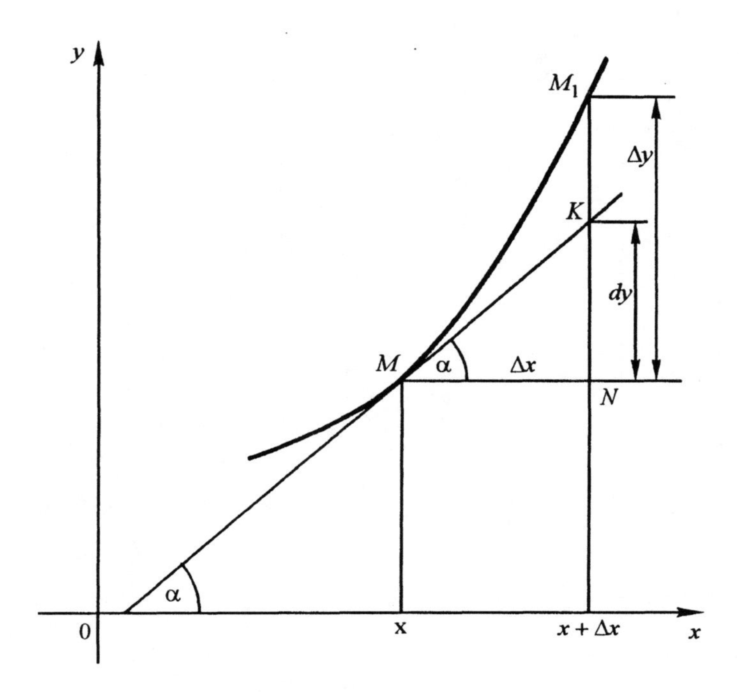

Figure 1.3. The geometric interpretation of the differential of a function.

Lagrange's theorem (mean value theorem). If a function $y = f(x)$ is defined and continuous on an interval $[a; b]$ and is differentiable at least in the open interval $(a; b)$, then there exists at least one point c in the interval $(a; b)$, $(a < c < b)$ such that

$$\frac{f(b) - f(a)}{b - a} = f'(c). \tag{3}$$

Geometric interpretation of the theorem. There is always at least one point M on the arc AB of the curve $y = f(x)$ such that the tangent at this point is parallel to the chord AB (Figure 1.6). Formula (3) is called *Lagrange's formula* or the *formula of finite increments.* It can be rewritten as

$$\Delta f(x_0) = f'(c)\Delta x, \quad \text{where} \quad c = x_0 + \theta\Delta x, \quad 0 < \theta < 1,$$

if formula (3) is applied to the interval $[x_0; x_0 + \Delta x]$, $\Delta x > 0$. This formula gives an exact value of the increment of the function for any finite increment Δx of the argument.

Cauchy theorem (generalized mean value theorem). Let functions $f(x)$ and $g(x)$ be continuous on an interval $[a; b]$, differentiable at least in the open interval $(a; b)$, $g'(x) \neq 0$ on $(a; b)$. Then there exists a point c in $(a; b)$ such that

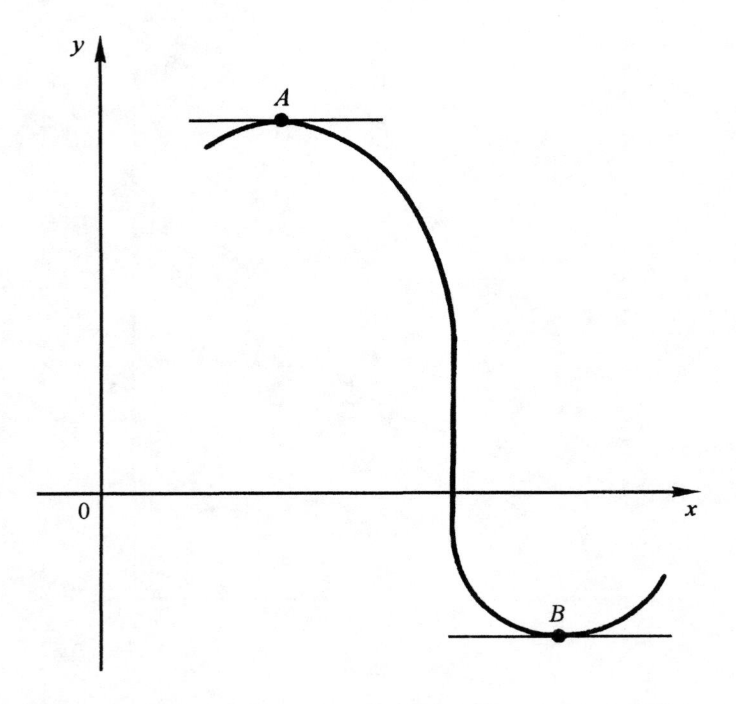

Figure 1.4. The geometric interpretation of Fermat's theorem.

$$\frac{f(b)-f(a)}{g(b)-g(a)}=\frac{f'(c)}{g'(c)}. \tag{4}$$

Formula (4) is called the Cauchy formula.

Geometric interpretation of the theorem. If a curve is defined parametrically (Figure 1.6), $x = g(t)$, $y = f(t)$, where the point A corresponds to the value of the parameter $t = a$, and the point B to the value $t = b$, then the slope of the tangent to the curve at the point c is

$$\tan\alpha=\frac{f(b)-f(a)}{g(b)-g(a)}=\frac{f'(c)}{g'(c)}.$$

Taylor's formula. If a function $f(x)$ is continuous on an interval $[a; b]$ and has continuous derivatives up to and including an order $n + 1$, then the following equality holds (*Taylor's formula*):

$$f(b)=f(a)+\frac{f'(a)}{1!}(b-a)+\frac{f''(a)}{2!}(b-a)^2+\ldots$$

$$+\ldots+\frac{f^{(n)}(a)}{n!}(b-a)^n+\frac{f^{(n+1)}(\xi)}{(n+1)!}(b-a)^{n+1}, \tag{5}$$

where $\xi = a + \theta(b - a)$, $0 < \theta < 1$.

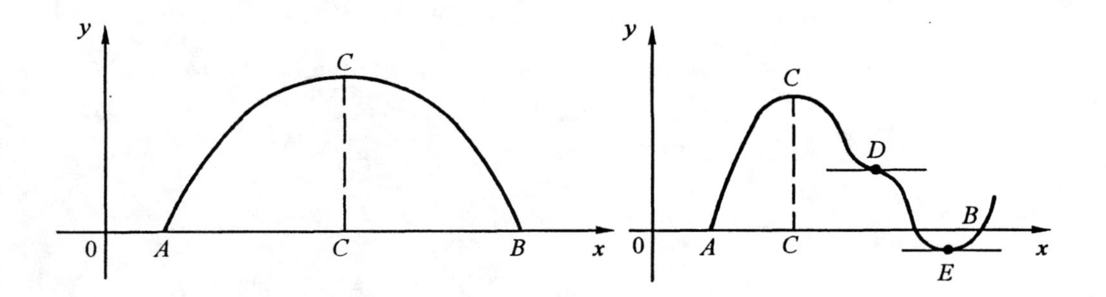

Figure 1.5. The geometric interpretation of Rolle's theorem.

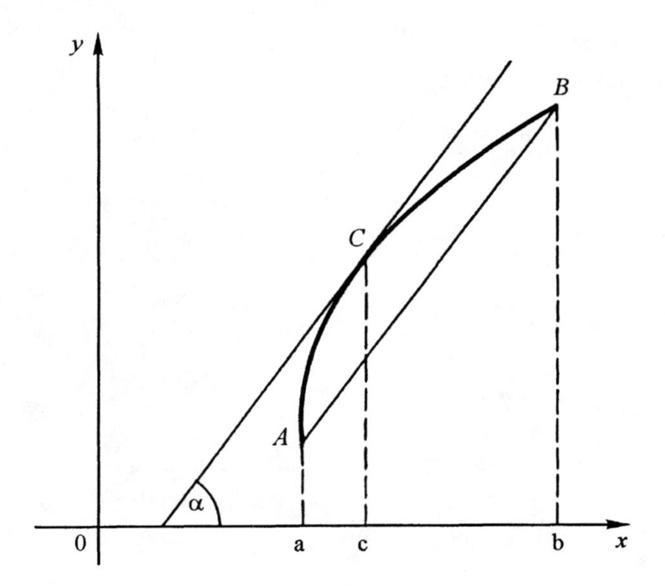

Figure 1.6. The geometric interpretation of Lagrange's theorem.

The remainder term $R_n = \frac{f^{(n)}(\xi)}{(n+1)!}(b-a)^{n+1}$ is called the *Lagrange form* of the remainder.

If a is a constant and b varies, then b is replaced with x and the formula is written as

$$f(x) = f(a) + \frac{f'(a)}{1!}(x-a) + \ldots + \frac{f^{(n)}(a)}{n!}(x-a)^n + \frac{f^{(n+1)}(\xi)}{(n+1)!}(x-a)^{n+1}.$$

For $a = 0$, *Maclaurent's formula* is obtained:

$$f(x) = f(0) + \frac{f'(0)}{1!}x + \ldots + \frac{f^{(n)}(0)}{n!}x^n + \frac{f^{(n+1)}(\xi)}{(n+1)!}x^{n+1}. \quad (6)$$

The remainder in the *Cauchy form* is written as

$$R_n = \frac{f^{(n+1)}(\xi)}{n!}(x-\xi)^n(x-a),$$

or, for $\xi = a + \theta(x-a)$,

$$R_n = \frac{f^{(n+1)}[a+\theta(x-a)]}{n!}(1-\theta)^n(x-a)^{n+1}.$$

The remainder in the Schlömilch–Roch form is

$$R_n = \frac{f^{(n+1)}(\xi)}{n!p}(x-\xi)^{n+1-p}(x-a)^p,$$

where $p > 0$.

1.4. Study of functions by using derivatives

1.4.1. Condition for constancy

Theorem 1. Let a function $f(x)$ be continuous on an interval $[a; b]$ and have derivative $f'(x)$ at least in the interval $(a; b)$. The function $f(x)$ is a constant on the interval $[a; b]$ if and only if $f'(x) = 0$ in all points of the interval $(a; b)$.

1.4.2. Condition for monotonicity

Theorem 2. For a function $f(x)$ which has a finite or infinite derivative to be increasing (decreasing) in every point of the interval $[a; b]$, it is necessary and sufficient that $f'(x) \geq 0$ ($f'(x) \leq 0$), $f(x) \not\equiv 0$ on any subinterval of $[a; b]$.

Geometric interpretation of the theorem. If a function $f(x)$ is increasing on an interval $[a; b]$, then a tangent to the curve $y = f(x)$ at every point of this interval makes an acute angle with the

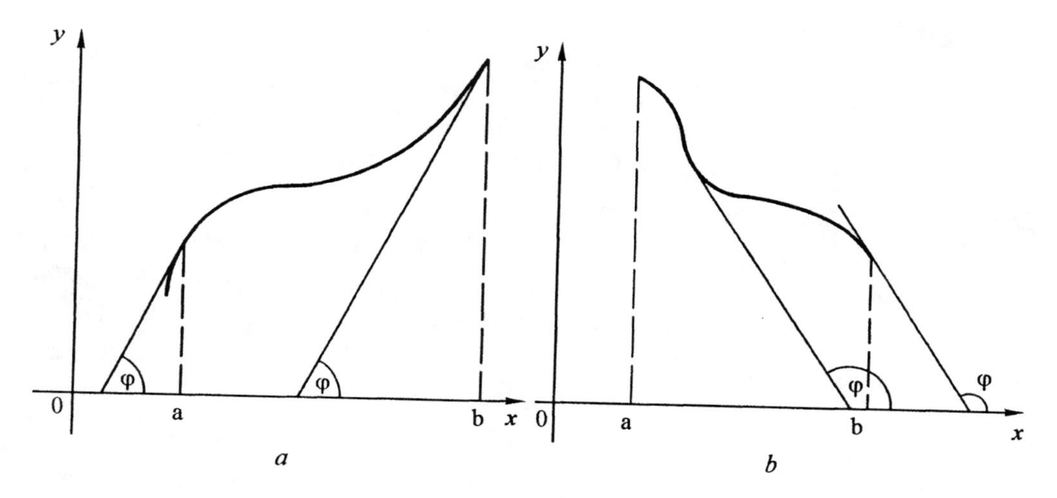

Figure 1.7. (*a*) Increasing function; (*b*) decreasing function.

x-axis, φ, or (in some points) the tangent can be horizontal; the slope of this tangent is nonnegative ($f'(x) = \tan \varphi \geq 0$) (Figure 1.7).

If a function is decreasing on an interval $[a; b]$, then the angle formed by the tangent and the x-axis is obtuse or (in some points) the tangent is horizontal; the slope of the tangent is not positive.

Hence the sign of the derivative determines whether the function is increasing or decreasing.

Rule. To find intervals on which the function is strictly increasing (decreasing), it is necessary to solve the inequality $f'(x) > 0$ ($f'(x) < 0$), and also determine where the points at which the derivative becomes zero are located, and whether or not they form a subinterval of the interval on which the function is considered.

1.4.3. Maxima and minima of a function

Let a function $f(x)$ be defined in a neighborhood of a point $x = x_0$. The function $f(x)$ has a maximum at the point $x = x_0$ if there is a δ-neighborhood of the point x_0 such that $f(x) < f(x_0)$ for all x, $x_0 - \delta < x < x_0 + \delta$, i.e. the value of the function at this point is greater than the value of the function at points sufficiently close to x_0.

A function $f(x)$ has a minimum at a point $x = x_0$ if there exists a δ-neighborhood of the point x_0 such that $f(x_0) < f(x)$ for all x such that $x_0 - \delta < x < x_0 + \delta$, i.e. the value of the function at x_0 is less than the value of the function at all points that are sufficiently close to x_0.

If the function has a maximum or minimum at a point, then it is said that the function reaches an *extremum* at this point, and the value of the function is called *extremal*.

If a function is defined on an closed interval, it can assume a maximum or minimum only at an interior point of this interval. The maximum (minimum) is not necessarily the greatest (least) value of the function in its domain. There could exist points outside of the neighborhood of the point at which the function assumes values that are greater (less) than the value at the point. A function can also have several maximum (minimum) points in its domain.

A necessary condition for an extremum. If a function $y = f(x)$ has an extremum at a point $x = x_0$, then, at this point, its derivative equals zero or does not exist. Points at which the derivative equals zero or does not exist, but at which the function is continuous, are called *critical points*. So, a function could have an extremum at a critical point, but not every critical point is a point of extremum. For example, the function $y = x^3$ has a critical point at $x = 0$, since the derivative of the function, $y' = 3x^2$, equals zero at this point. However, the function does not reach either maximum or minimum at this point (Figure 1.40c). The derivative of the function $y = \sqrt[3]{x}$ is infinite at $x = 0$, i.e. the point $x = 0$ is critical, but it is not a point of extremum. Because $y < 0$ for $x < 0$, and $y > 0$ for $x > 0$, $y = 0$ at $x = 0$ (see Figure 4.52).

Sufficient conditions for an extremum. If a function $f(x)$ is continuous on $(a; b)$ which contains a critical point x_0, differentiable in all points of this interval with a possible exception of the point x_0, and if the derivative changes its sign passing through this point, then the function $f(x)$ has an extremum at $x = x_0$. In such a case, if the derivative changes its sign from plus to minus, the function reaches a maximum at this point; if the sign of the derivative changes from minus to plus, the point $x = x_0$ is a point of minimum. If the derivative does not change sign passing through a critical point, then this is not a point of extremum.

A procedure for finding extremums of a function (by using the first derivative)

1) Find the first derivative $f'(x)$.

2) Find critical points of the function by solving the equation $f'(x) = 0$ for real roots, and points at which the derivative does not exist.

Suppose the critical points are $x_1, x_2, \ldots, x_n$ and contained in the interval $(a; b)$. Plot all the points in the interval $(a; b)$, $a < x_1 < x_2 < \ldots < x_n < b$. Determine the sign of the derivative at any point of every interval $(a; x_1)$, $(x_1; x_2)$, . . . , $(x_k; x_k + 1)$ . . . $(x_n; b)$.

Consider the signs of the derivative $f'(x)$ in neighboring intervals, going from the first interval to the last one. If the derivative $f'(x)$ changes sign from "+" to "–", then the function has a maximum at this point. If the derivative changes from "–" to "+", then this point is a point of minimum. If the sign of the derivative in two neighboring intervals is the same, then the critical point is not a point of extremum.

One calculates values of the function at points of extremum (extremal values). Graphs of functions that have extremums are shown in a neighborhood of a critical point in Figure 1.8.

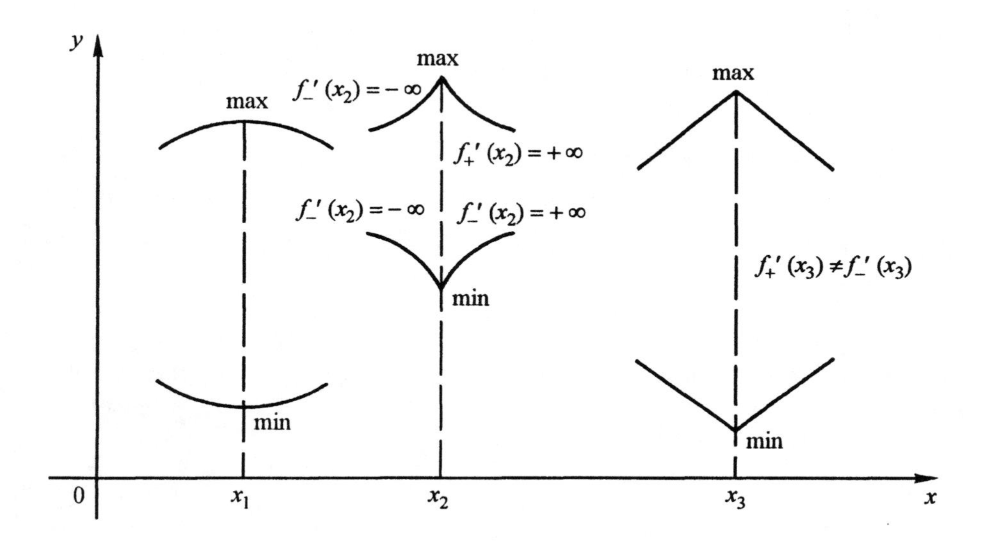

Figure 1.8. Extremums of functions.

1.4.4. Determining extremums of a function by using the second derivative

If the derivative of a function $f(x)$ equals zero at $x = x_1$ and the second derivative of the function does not vanish at this point, then the function $f(x)$ has a maximum at $x = x_1$ if $f''(x) < 0$, and a minimum if $f''(x) > 0$.

1.4.5. Determining extremums of a function by using Taylor's formula

If $f'(x_0) = 0$ at some point $x = x_0$, and $f''(x_0) = 0$, then one can use higher-order derivatives to investigate whether the function achieves an extremum at this point. In such a case the following lemma about preservation of sign is used: if a function is continuous and not zero at $x = x_0$, then it is not zero in a neighborhood of the point x_0 and has there the same sign as $f(x_0)$.

Assume that a function $f(x)$ has derivatives up to and including an order n at a point x_0, $f'(x_0) = f''(x_0) = \ldots = f^{(n-1)}(x_0)$, and $f^{(n)}(x_0) \neq 0$, and $f^{(n)}$ is continuous at $x = x_0$. Then, if n is odd, the function $f(x)$ does not have an extremum at the point x_0. If n is even, then the function has a maximum if $f^{(n)} < 0$, and a minimum if $f^{(n)} > 0$. This gives a method for studying the possibility of having an extremum of a function in terms of higher-order derivative.

Example 1.4.1. Determine extremums of the function $f(x) = 5x^4$.

Find the derivatives: $f'(x) = 20x^3, f''(x) = 60x^2, f'''(x) = 120x, f^{IV}(x) = 120$.

At the critical point $x = 0$, we have that $f'(0) = f''(0) = f'''(0) = 0$, $f^{IV}(0) \neq 0$ and f^{IV} is positive at $x = 0$. Hence the function has a minimum at $x = 0$.

Example 1.4.2. Determine extremuma of the function $f(x) = 2x^5 + 1$.

Find the derivatives: $f'(x) = 10x^4$, $f''(x) = 40x^3$, $f'''(x) = 120x^2$, $f^{IV}(x) = 240x$, $f^{V}(x) = 240 \neq 0$. Hence the function does not have an extremum at the point $x = 0$.

1.4.6. The maximal and minimal value of a function on an interval

Let $f(x)$ be a function defined and continuous on an interval $[a; b]$. Then there are always points at which the function reaches a maximal and minimal value. These points are either critical points or the endpoints of the interval.

Rule. In order to find the maximal and minimal value of the function defined on an interval, it is necessary: to determine critical points, and to calculate the value of the function at the critical points and at the endpoints of the interval; among the determined values, to choose the largest and the smallest value of the function on the interval $[a; b]$.

An important particular case should be noted. If a point x_1 is a unique point of extremum of the function $f(x)$ on the interval $[a; b]$, then $f(x_1)$ will be the maximal value of the function if $f(x_1)$ is a maximum, and the minimal value of $f(x)$ if $f(x_1)$ is a minimum.

1.4.7. Convexity and concavity of a curve. Inflection points

Let $f(x)$ be a differentiable function. The curve $y = f(x)$ is called *convex* (*concave*) at a point x_0 if there exists a neighborhood of the point x_0 such that the graph of the curve over this neighborhood is located below (above) the tangent to the curve at the point $M(x_0; f(x_0))$. If a curve is convex (concave) at all points of an interval, then it is called convex (concave) on this interval (Figure 1.9).

A sufficient condition for the curve $y = f(x)$ to be convex (concave). If $f''(x) < 0$ ($f''(x) > 0$) at all points of the interval $(a; b)$, then the curve $y = f(x)$ will be convex (concave) on this interval.

An *inflection point* of a continuous curve is a point at which the curve changes from being convex to concave or vice versa. At this point the tangent intersects the curve (Figure 1.10).

Sufficient condition for a given point to be an inflection point. Let a curve be defined by an equation $y = f(x)$. If $f''(a) = 0$ or $f''(a)$ does not exist and $f''(x)$ changes sign as x passes through $x = a$, then the point of the curve with the x-coordinate equal to a is an inflection point.

Rule. In order to find inflection points and study convexity and concavity of the curve, calculate the second derivative $f''(x)$, solve the equation $f''(x) = 0$, and determine the values of x such that $f''(x)$ does not exist. Then consider the sign of $f''(x)$ in neighborhoods of every root of the equation $f''(x) = 0$ and the points where $f''(x)$ does not exist. Let $x = x_1$ be one of such

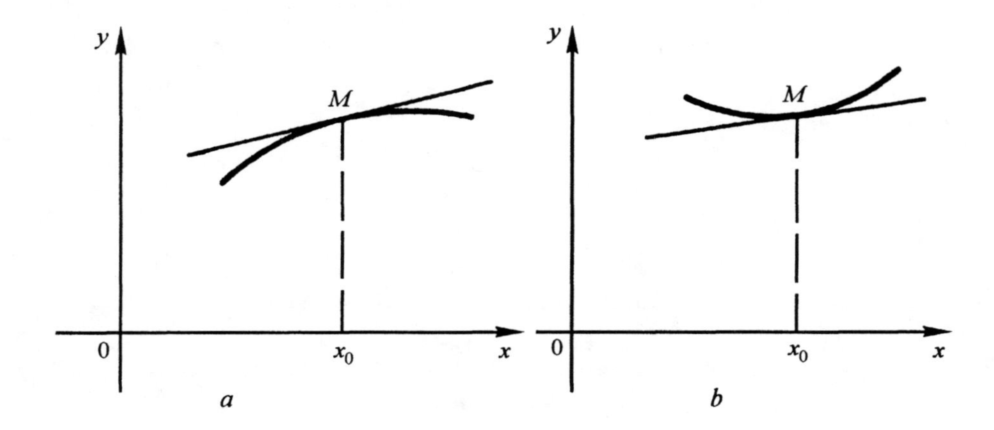

Figure 1.9. (*a*) Convex curve; (*b*) concave curve.

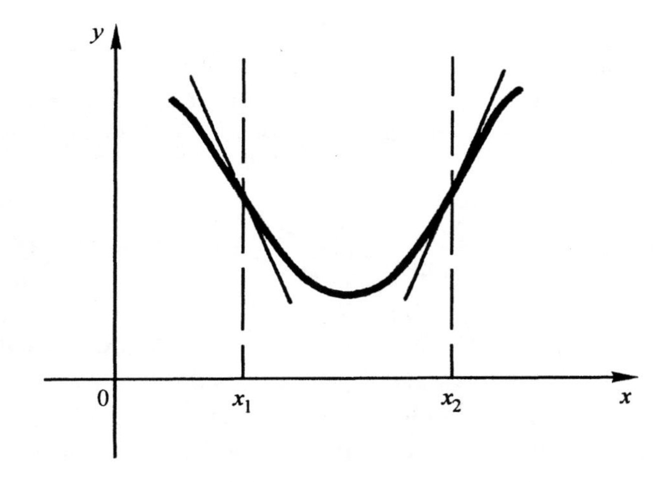

Figure 1.10. Inflection points of a curve.

points. Then look at the sign of $f''(x)$ for $x < x_1$ and for $x > x_1$. If these signs are different, then $M(x_1; f(x_1))$ is an inflection point. If the signs are the same, then the point $M(x_1; f(x_1))$ is not an inflection point. At the points where $f''(x) > 0$, the curve is concave; at the points where $f''(x) < 0$, the curve is convex.

1.5. Graph plotting with the use of derivatives

A study of the function with the use of derivatives and plotting its graph is carried out along the following scheme:

1. Find the domain of the function (see Part I, Section 2.2) Determine points of discontinuity and intervals on which the function is continuous.
2. Find out whether the function is even, odd, or periodic. Locate points in which the curve intersects the axes. Find intervals on which the function does not change sign (see Part I, Section 2.2).
3. Find points of extremum and the values of the function at these points. Determine intervals on which the function is monotone (see Part II, Section 1.4).
4. Find inflection points and determine where the function is convex or concave (see Part II, Section 1.4).
5. Find asymptotes of the graph of the function (Part I, Section 2.2).
6. Plot the graph of the function (for better precision, calculate the value of the function at certain points). For example Figure 1.11.

It is better to plot the graph and study properties of the function at the same time. Sometimes it is more convenient to draw the asymptotes, points of extremums, and inflection points.

If the function turns out to be even or odd, it is important to use these properties.

1.6. Plotting the graph of $f'(x)$ and $f''(x)$ from the graph of $f(x)$

It is sometimes important, for example in the theory of machine tools and transfer devices, where one needs to construct the acceleration chart from a velocity chart, to plot the graphs of the first and second derivatives of a function from the graph of the function.

Let a differentiable function $y = f(x)$ be given. The first and second derivatives of the function are, in their turn, functions of x. Thus their graphs can be constructed by using general procedures for constructing graphs. However, if the graph of the function $f(x)$ is given, then the construction of the graphs of $f'(x)$, $f''(x)$ can sometimes be greatly simplified. There the following considerations should be used. a) At the points where the function $f(x)$ has an extremum, $f'(x) = 0$. b) At inflection points, the function $f'(x)$ has an extremum (maximum if the derivative changes at this point from an increasing to decreasing function, and minimum if the derivative from decreasing becomes an increasing function); as x increases ranging in the intervals where the function $f(x)$ is convex, the derivative $f'(x)$ is decreasing, on intervals where

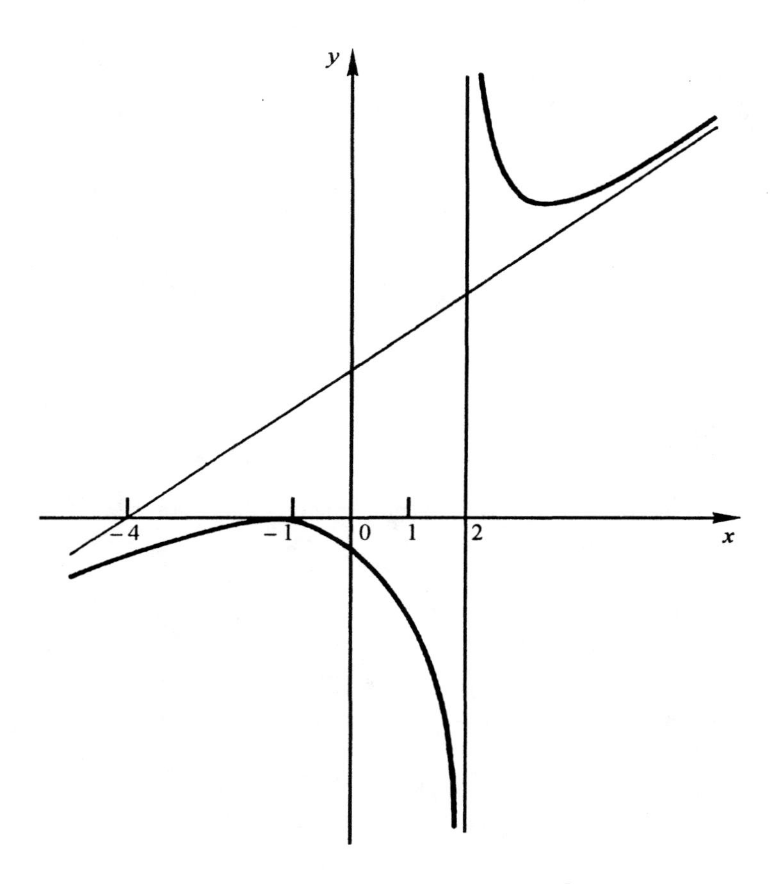

Figure 1.11. The curve $y = \dfrac{(x+1)^2}{x-2}$.

$f(x)$ is concave, $f'(x)$ is increasing. c) Inflection points of the derivative $f'(x)$ are found as usual. Oblique asymptotes of the function $f(x)$ correspond to horizontal asymptotes of the derivative $f'(x)$.

Remark 1.6.1. The same reasoning is used for plotting the graph of the second derivative $f''(x)$ on the base of the graph of the first derivative $f'(x)$.

Example 1.6.1. Plot the graph of the function $y = (2x-1)/(x-1)^2$ and its derivatives y', y''.

Domain of the function is $(-\infty; 1) \cup (1; +\infty)$. The straight line $x = 1$ is a vertical asymptote. The point $(0; -1)$ is a point of minimum. The curve has an inflection point at $(-1/2; -8/9)$. The curve passes through the points $(0; -1)$ and $(1/3; 0)$. The curve is convex for $x \in (-\infty; -1/2)$, and concave on $x \in (-1/2; 1) \cup (1; \infty)$.

Domain of the derivative $y' = -2x/(x-1)^3$ is $(-\infty; 1) \cup (1; +\infty)$. The curve passes through the point (0; 0), and reaches a minimum at $(-1/2; -8/27)$. The derivative $y' < 0$ for $x \in (1; +\infty)$, on this interval, the function is convex.

The graph of $y = 2(2x+1)/(x-1)^4$ passes through the points $(-1/2; 0)$ and (0; 2). The curve is concave for $x \in (-\infty; 1) \cup (1; +\infty)$. The graphs of the functions $y = \dfrac{2x-1}{(x-1)^2}$, $y' = \dfrac{-2x}{(x-1)^3}$, $y'' = \dfrac{2(2x+1)}{(x-1)^4}$ are shown in Figure 1.12.

1.7. Graphical differentiation

Analytical methods of differentiation and integration of functions cannot always be used in practice. For example, if a function is defined by a table or graph, then it is impossible to find a derivative or take an integral of such a function. Sometimes, even if a function is given analytically, it is also not possible to carry out differentiation or integration explicitly. For example, elliptic integrals cannot be expressed in terms of elementary functions, etc.

Finding a derivative of a function $f(x)$ graphically is reduced to calculating the slope of the curve $y = f(x)$ at a given point by virtue of the formula $f'(x) = \tan\alpha$, where α is the angle formed by the tangent and the x-axis.

The graphical differentiation is done as follows. Plot the graph of the function $y = f(x)$ (if the function is defined by a table). Plot the tangent to the curve at the point, and calculate the slope.

Note that if the scale of the axes is different, then the slope of the curve is calculated by the formula

$$f'(x) = \frac{dy}{dx} = \frac{r_x}{r_y}\frac{y_2 - y_1}{x_2 - x_1},$$

where r_x and r_y are scales of the x and y axes, $(x_1; y_1)$, $(x_2; y_2)$ are arbitrary points.

The graph should be plotted very carefully, with a template. It is good to use plotting paper. The graph of the function $y = f(x)$ should be plotted in such a way that the angle of the tangent at the point x lies between 35° and 55°. This minimizes the error of finding the value of the derivative. This can be achieved by choosing appropriate scales r_x, r_y.

To plot the tangent to the curve, one needs to use a transparent triangle, which is slowly rotated about the point on the convex side of the curve until a right position of the tangent is found.

Example 1.7.1. Graphically find the derivative of a function at the point $x = 0.28$. The function is given by the following table.

x	0.0	0.1	0.2	0.3	0.4	0.5	0.6
$f(x)$	–0.0200	–0.0272	–0.0284	–0.0176	–0.0112	–0.0640	–0.1468

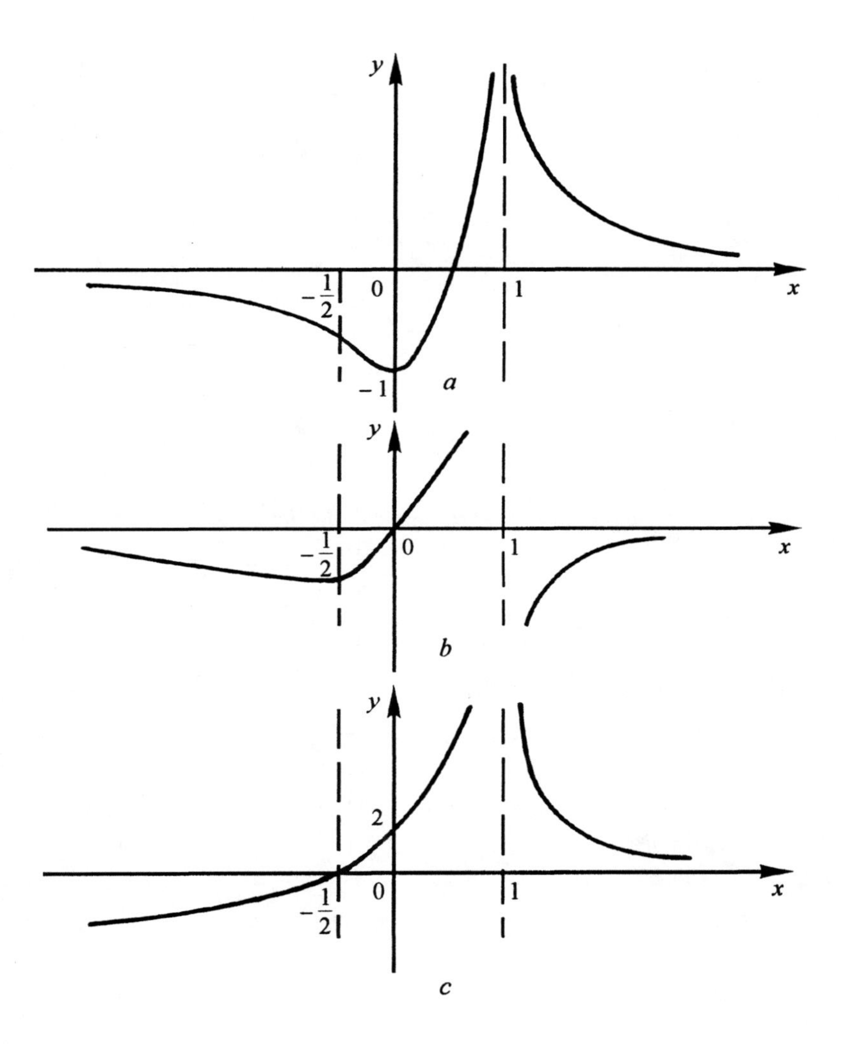

Figure 1.12. The curves (*a*) $y = f(x)$; (*b*) $y = f'(x)$; (*c*) $y = f''(x)$.

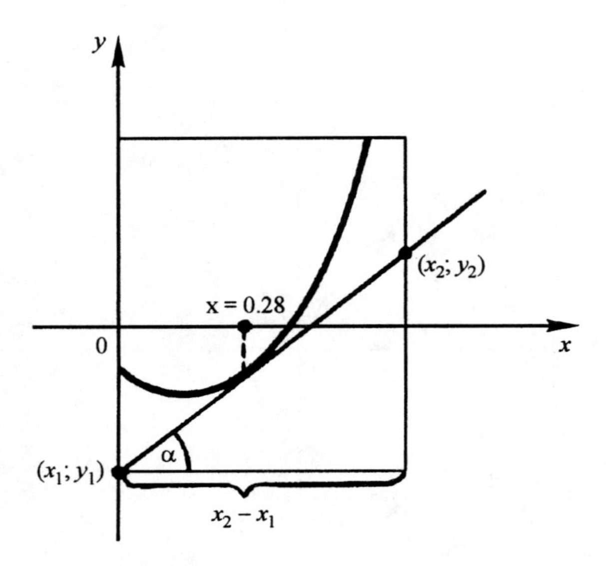

Figure 1.13. Graphical differentiation.

Take the scale of the y-axis to be five times the scale of the x-axis, i.e. $r_x/r_y = 1/5 = 0.2$.

The segments $y_2 - y_1$ and $x_2 - x_1$ measure directly from the graph (Figure 1.13). Then using the formula $f'(x) = \frac{r_x}{r_y}\frac{y_2 - y_1}{x_2 - x_1}$ find that $f'(x) = 0.2\,\frac{23.1}{29.7} = 0.156$.

The function $f'(x)$ is dimensionless. The function considered above, which is defined by the table, has the form $f(x) = x^3 - 0.082x - 0.020$. Its derivative is $f'(x) = 3x^2 - 0.082$, and hence $f'(x)|_{x = 0.28} = 0.1532$.

1.8. Graphical integration

Let a function be given by its graph. We find an integral of it graphically.

Consider the integral

$$F(x) = \int_0^x f(x)\, dx.$$

Using the known trapezium rule we write

$$F(x_k) = \int_a^{a+kh} f(x)\, dx = h\left(\frac{y_0 + y_1}{2} + \frac{y_1 + y_2}{2} + \ldots + \frac{y_{k-1} + y_k}{2}\right),$$

$$F(x_{k+1}) = \int_a^{a+(k+1)h} f(x)\,dx = h\left(\frac{y_0 + y_1}{2} + \frac{y_1 + y_2}{2} + \ldots + \frac{y_{k-1} + y_k}{2} + \frac{y_k + y_{k+1}}{2}\right),$$

or

$$F(x_{k+1}) = F(x_k) + h\frac{y_k + y_{k+1}}{2}, \tag{1}$$

where $x_k = a + kh$, $k \in \mathbf{N}$.

Hence, formula (1) gives an expression for the value $F(x_k + h)$ in terms of the previous value $F(x_k)$; the value of h could be different at each step.

Subdivide the interval of integration with points $x_0 = a, x_1, x_2, \ldots, x_n = b$, and plot the points with the y-coordinates $y_k = f(x_k)$ and the middle points of these segments, $y_{k+1/2}$. If each of the integration intervals is sufficiently small, we can take $\frac{y_k + y_{k+1}}{2} = y_{k+1/2}$, and then get that $F(x_{k+1} = f(x_k) + hy_{k+1/2}$.

Project the endpoints with the y-coordinate equal to $y_{1/2}, y_{3/2}, y_{5/2}, \ldots$ onto the y-axis, and join the obtained points $A_1, A_2, A_3, \ldots$ with the point $P(-1; 0)$.

Draw a ray through the point $M_0(x_0; 0)$ parallel to the line segment PA_1, and find the point of intersection with the line segment at x_1, M_1. In the same way find the line segments M_1M_2, $M_2M_3, \ldots$, which are parallel to the corresponding rays. The points of intersection will be points of the curve under construction, $F(x)$ (Figure 1.14), since

$$\overline{x_1M_1} = hy_{1/2} = F(x_1),$$

$$\overline{x_2M_2} = F(x_1) + hy_{3/2} = F(x_2),$$

$$\overline{x_3M_3} = F(x_2) + hy_{5/2} = F(x_3),$$

$$\ldots .$$

The polygon $M_0M_1M_2M_3\ldots$ is an approximation of the graph of the first integral

$$F(x) = \int_{x_0}^{x} f(x)\,dx.$$

For points that are not nodes, this construction does not give a good approximation; however, if we are interested in the value of the function at a point x, then this point could be taken as a node, and thus drawing a curve through the points M_0, M_1, M_2 gives a graph of the function and more precise results.

On intervals where the integrand $f(x)$ assumes negative values, the values of the first integral at the corresponding points decrease.

In Figure 1.14, we have $x_0 = 0.2, f(x) = \cos x$, hence

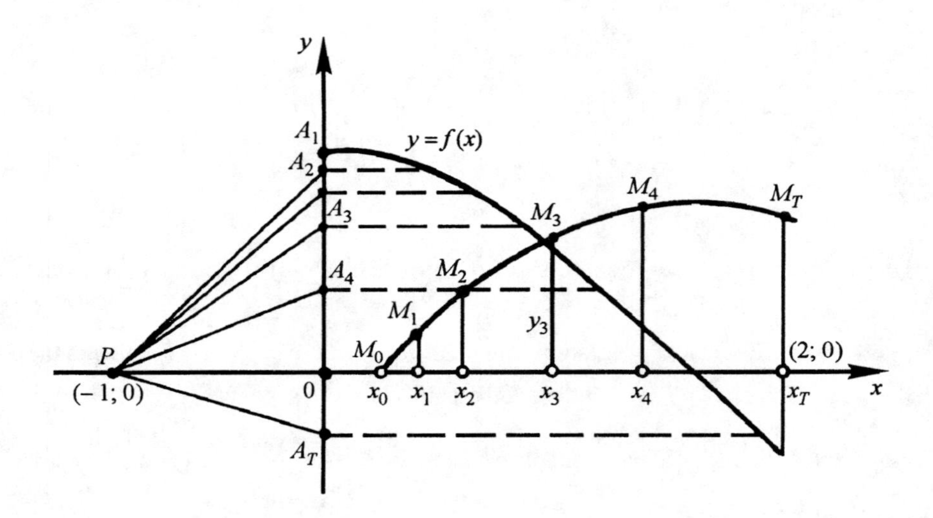

Figure 1.14. Graphical integration.

$$F(x) = \int_{0.2}^{x} \cos x, dx = \sin x - \sin 0.2 = \sin x - 0.1987.$$

The precision of the graphical method of integration can be improved by increasing the number of points and the scale of the plot.

1.8.1. The rule of Bernoulli–L'Hospital

Sometimes in doing an investigation of a function for plotting its graph, one encounters indefiniteness of the form $0/0$, ∞/∞, $0\cdot\infty$, $\infty - \infty$, 0^0, 1^∞, or ∞^0. In many instances the calculation is done by using the rules of Bernoulli-L'Hospital.

Indefiniteness of the type 0/0. Let $f(x)$ and $\varphi(x)$ be functions defined on an interval $(a; b)$ which is finite or infinite. Making sense out of the indefiniteness of the type $0/0$ means to find the limit $\lim\limits_{a \to a+0} \dfrac{f(x)}{\varphi(x)}$ if $\lim\limits_{x \to a+0} f(x) = \lim\limits_{x \to a+0} \varphi(x) = 0$ and $\varphi(x) \neq 0$ for all $x \in (a; c)$, where $a < c < b$, or to find the limit $\lim\limits_{x \to b-0} \dfrac{f(x)}{\varphi(x)}$ if $\lim\limits_{x \to b-0} f(x) = \lim\limits_{x \to b-0} \varphi(x) = 0$ and $\varphi(x) \neq 0$ for all $x \in (c; b)$, where $a < c < b$.

It is indeed an "indefiniteness," since the limit could exist or not exist, and even if it exists, its value is not clear.

Theorem 1 (the first rule of Bernoulli–L'Hospital). Let functions $f(x)$ and $\varphi(x)$ be differentiable in an interval $(a; b)$. If $\lim\limits_{x\to a+0} f(x) = \lim\limits_{x\to a+0} f(x) = 0$, $\varphi(x) \neq 0$ for all $x \in (a; c_1)$, $a < c_1 < b$, and a finite or not limit $\lim\limits_{x\to a+0}\dfrac{f'(x)}{\varphi'(x)}$ exists, then $\lim\limits_{x\to a+0}\dfrac{f(x)}{\varphi(x)} = \lim\limits_{x\to a+0}\dfrac{f'(x)}{\varphi'(x)}$.

If $\lim\limits_{x\to b-0} f(x) = \lim\limits_{x\to b-0} \varphi(x) = 0$, $\varphi'(x) \neq 0$ for all $x \in (c_2; b)$, $a < c_2 < b$, and a finite or infinite limit $\lim\limits_{x\to b-0}\dfrac{f'(x)}{\varphi'(x)}$ exists, then $\lim\limits_{x\to b-0}\dfrac{f(x)}{\varphi(x)} = \lim\limits_{x\to b-0}\dfrac{f'(x)}{\varphi'(x)}$.

Corollary. Let functions $f(x)$ and $\varphi(x)$ be differentiable in intervals $(x_0 - \delta; x_0)$ and $(x_0; x_0 + \delta)$, where $\delta > 0$. If $\lim\limits_{x\to x_0} f(x) = \lim\limits_{x\to x_0} \varphi(x) = 0$, $\varphi'(x) \neq 0$ for all $x \in (x_0 - \delta; x_0) \cup \cup (x_0; x_0 + \delta)$, and a finite or infinite limit $\lim\limits_{x\to x_0}\dfrac{f'(x)}{\varphi'(x)}$ exists, then $\lim\limits_{x\to x_0}\dfrac{f(x)}{\varphi(x)} = \lim\limits_{x\to x_0}\dfrac{f'(x)}{\varphi(x)}$.

Indefiniteness of the type ∞/∞

Theorem 2 (the second rule of Bernoulli–L'Hospital). Let functions $f(x)$, $\varphi(x)$ be differentiable in an interval $(a; b)$ which could be finite or infinite. If $\lim\limits_{x\to a+0} f(x) = \infty$, $\lim\limits_{x\to a+0} \varphi(x) = \infty$, $\varphi(x) \neq 0$ for all $x \in (a; c_1)$, $a < c_1 < b$, and a finite or infinite limit $\lim\limits_{x\to a+0}\dfrac{f'(x)}{\varphi'(x)}$ exists, then $\lim\limits_{x\to a+0}\dfrac{f(x)}{\varphi(x)} = \lim\limits_{x\to a+0}\dfrac{f'(x)}{\varphi'(x)}$.

If $\lim\limits_{x\to b-0} f(x) = \lim\limits_{x\to b-0} \varphi(x) = 0$, $\varphi(x) \neq 0$ for all $x \in (c_2; b)$, $a < c_2 < b$, and a finite or infinite limit $\lim\limits_{x\to b-0}\dfrac{f'(x)}{\varphi'(x)} = A$ exists, then $\lim\limits_{x\to b-0}\dfrac{f(x)}{\varphi(x)} = \lim\limits_{x\to b-0}\dfrac{f'(x)}{\varphi'(x)}$.

Corollary. Let functions $f(x)$, $\varphi(x)$ be differentiable in intervals $(x_0 - \delta; x_0)$ and $(x_0; x_0 + \delta)$, where $\delta > 0$. If $\lim\limits_{x\to x_0} f(x) = \infty$, $\lim\limits_{x\to x_0} \varphi(x) = \infty$, $\varphi(x) \neq 0$ for all $x \in (x_0 - \delta; x_0) \cup (x_0; x_0 + \delta)$, and a finite or infinite limit $\lim\limits_{x\to x_0}\dfrac{f'(x)}{\varphi'(x)}$ exists, then $\lim\limits_{x\to x_0}\dfrac{f(x)}{\varphi(x)} = \lim\limits_{x\to x_0}\dfrac{f'(x)}{\varphi'(x)}$.

This corollary is used in applications most often.

Indefiniteness of the type $0\cdot\infty$. By writing $f(x)\varphi(x) = f(x)/1/\varphi(x)$ an indefiniteness of this type is reduced to an indefiniteness of the type $0/0$.

Indefiniteness of the type 0^0, 1^∞, ∞^0. By using the equality $f(x)^{\varphi(x)} = e^{\varphi(x)\ln f(x)}$ and the fact that the exponential function is continuous, we find that $\lim\limits_{x \to a+0} f(x)^{\varphi(x)} =$
$= \exp(\lim\limits_{x \to a+0} \varphi(x)\ln f(x))$, or $\lim\limits_{x \to b-0} f(x)^{\varphi(x)} = \exp(\lim\limits_{x \to b-0} \varphi(x)\ln f(x))$.

Hence, indefiniteness of the considered type can be reduced to an indefiniteness of the type $0\cdot\infty$ which, in its turn, can be regarded as $0/0$.

Remark 1.8.1. If the ratio of the derivatives $\dfrac{f'(x)}{\varphi'(x)}$ is again an indefiniteness, then the rule of Bernoulli-L'Hospital is applied again until the limit is calculated or shown that it does not exist.

Example 1.8.1. Find $\lim\limits_{x \to +\infty} \dfrac{\ln x}{x^n}$.

We have

$$\lim_{x \to +\infty} \frac{\ln x}{x^n} = \lim_{x \to +\infty} \frac{1/x}{nx^{n-1}} = \lim_{x \to +\infty} \frac{1}{nx^n} = 0,$$

for $n > 0$. As $x \to +\infty$, the logarithmic function increases much slower than does the power function with an arbitrary exponent ($n > 0$).

As $x \to +\infty$, the exponential function a^x increases much faster than does any power function ($n > 0$).

1.9. Approximations for solutions of equations

Consider an equation $f(x) = 0$. If this equation is algebraic and its degree is not greater than 4, then, in general, there are formulas for finding roots of this equation. For an equation of a degree higher than 4, no such formulas exist.

Consider several methods for finding approximate solutions of an equation $f(x) = 0$.

1.9.1. The chord method

Let $f(x) = 0$ be a given equation, where $f(x)$ is a continuous, two times differentiable function on $[a; b]$.

Find an interval $[x_1; x_2]$ in $[a; b]$ such that the function is monotone on this interval and assumes values with opposite signs at the endpoints. Assume that $f(x_1) < 0$ and $f(x_2) > 0$ (Figure 1.15). Since the function $y = f(x)$ is continuous on $[x_1; x_2]$, there is a point between x_1 and x_2 at which its graph intersects the x-axis.

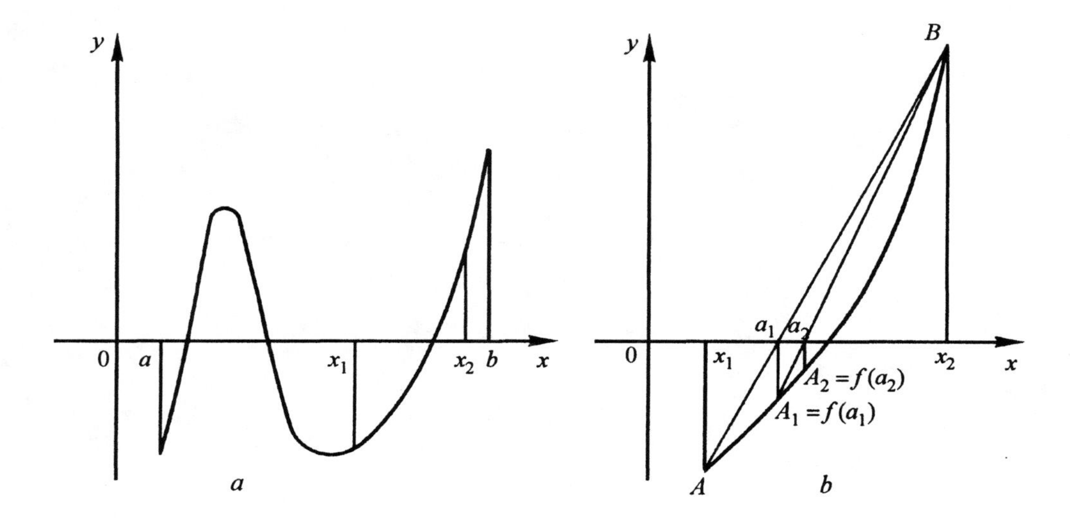

Figure 1.15. The chord method: (*a*) $f(x_1) < 0$; (*b*) $f(x_2) > 0$.

Draw a chord that connects the points on the curve $y = f(x)$ that have x-coordinates equal to x_1 and x_2. The x-coordinate of the point of intersection of the chord and the x-axis will be an approximation to the root of the equation $f(x) = 0$ (Figure 1.15*b*),

$$a_1 = \frac{x_1 f(x_2) - x_2 f(x_1)}{f(x_2) - f(x_1)}.$$

To find a better approximation to the root, evaluate $f(a_1)$. If $f(a_1) < 0$, then do the same thing and use the last formula for the interval $[x_1; a_1]$.

By repeating this process, we will be getting better approximations to the root, $a_2, a_3, \ldots$.

1.9.2. The tangent method (Newton's method)

Let again $f(x_1) < 0$, $f(x_2) > 0$, and assume that the first derivative of the function does not change sign on $[x_1; x_2]$. Then there is one root of the equation $f(x) = 0$ in the interval $(x_1; x_2)$. Suppose that the second derivative also does not change its sign, i.e. the function is either convex or concave on $[x_1; x_2]$.

Let us draw a tangent at the point B (Figure 1.16). The point a_1, the point at which the tangent intersects the x-axis, will be an approximation to the root, $a_1 = x_2 - f(x_2)/f'(x_2)$. By drawing a tangent at the point $B_1(a_1; f(a_1))$, we, in a similar way, get a better approximation to the root, a_2. By iterating this process several times, one can compute an approximation to the root with an arbitrary precision.

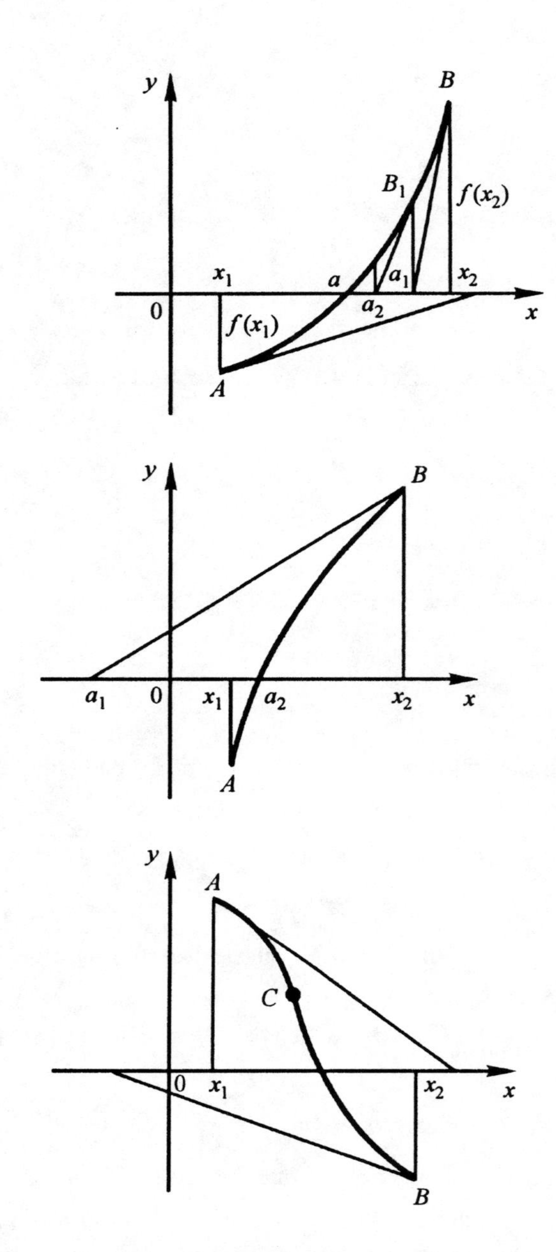

Figure 1.16. The tangent method.

Note that the tangent should be constructed at the endpoint at which the sign of the values of the function and its second derivative is the same. The same holds if $f'(x) < 0$. If the tangent is constructed at the left endpoint, then a_1 is computed by the formula $a_1 = x_1 - f(x_1)/f'(x_2)$.

If there is an inflection point C in the interval $(x_1; x_2)$, then the tangent method could give an approximation that lies outside of the interval $(x_1; x_2)$.

In applications, a method which is a combination of both methods is used, that is the chord method and tangent method are used simultaneously. Applying the chord and tangent methods in the interval $[x_1; x_2]$ we get two points, a_1 and $\bar{a}_1$, which lie on both sides of the root a, since $f(a_1)$ and $f(\bar{a}_1)$ have opposite signs. Now the chord and tangent methods are used in the interval $(a_1; \bar{a}_1)$, obtaining two more points, a_2 and $\bar{a}_2$, which lie still closer to the root. Continue this process until the difference between the found approximations gets within a given accuracy.

Example 1.9.1. Find a real root of the equation $f(x) = x^3 + 3x - 2x^2 - 5 = 0$ with the precision of 10^{-4}.

1. First we make sure that this equation has only one root. This follows, since $f'(x) = 3x^2 + 3 - 4x > 0$. Because $f(1) = -3 < 0$ and $f(2) = 1 > 0$, we conclude that this equation has one positive root lying in the interval $(1; 2)$.

By using the chord method we get $a_1 = \dfrac{1 \cdot 1 - 2 \cdot (-3)}{1 - (-3)} = \dfrac{7}{4} = 1.75$. But $f(a_1) = f(1.75) =$ $= -0.5156 < 0$ and $f(2) = 1 > 0$. Hence, $1.75 < a_2 < 2$. Use the chord method once again but in the interval $[1.75; 2]$, $a_2 = \dfrac{1.751 - 2(-0.5156)}{2 - (-0.5156)} = 1.8350$. Check that $f(1.835) = -0.05059 < 0$, hence $1.835 < a_3 < 2$. Decrease the interval. Indeed, $f(1.9) = 0.339 > 0$. So, $1.835 < a_3 < 1.9$.

Now use the chord method in the interval $[1.835; 1.9]$, and get $a_3 = 1.8434$. We find that $a_4 = 1.8437$, $a_5 = 1.8438$. Then $f(1.8437) < 0$ and $f(1.8438) > 0$, and the value $a_5 = 1.8438$ can be taken as an approximate value of the root.

2. The tangent method gives an approximation to the root much faster. Let us use this method.

Since $f(2) = 1 > 0$ and $f''(x) = 6x - 4 > 0$ on $[1; 2]$, and the first derivative $f'(x)$ does not change sign on $[1; 2]$, the tangent method can be used.

We have $f'(x) = 3x^2 - 4x + 3$, $f'(2) = 7$. Hence,

$$a_1 = 2 - \frac{1}{7} = 1.857,$$

$$a_2 = 1.857 - \frac{f(1.857)}{f'(1.857)} = 1.857 - \frac{0.0779}{5.9275} = 1.8439,$$

$$a_3 = 1.8439 - \frac{f(1.8439)}{f'(1.8439)} = 1.8438,$$

hence a_3 is already within the necessary precision.

Chapter 2

Graphs of all types of functions

The investigation of the function is carried out along the scheme described in Part II, Section 1.5. In some cases, additional study will be necessary. The investigation will sometimes be simplified or the order changed.

2.1. Graphs of functions $y = f(x)$ in Cartesian coordinates

Example 2.1.1. Plot the graph of the function $y = 6x/(1 + x^2)$.

The domain of the function is $(-\infty; +\infty)$. The function is odd, hence its graph is symmetric with respect to the origin. From this point on we will consider the function with nonnegative x.

It follows from the relations

$$\lim_{x \to +\infty} y = \lim_{x \to +\infty} \frac{6x}{1 + x^2} = 0, \quad \lim_{x \to -\infty} y = -0$$

that y approaches zero as $x \to \pm 0$ remaining positive (negative), i.e. $y = 0$ for $x = 0$. Since the denominator is always positive, $y > 0$ if $x > 0$. By symmetry, $y < 0$ if $x < 0$.

Check if there are asymptotes:

$$k = \lim_{x \to \pm\infty} \frac{f(x)}{x} = \lim_{x \to \pm\infty} \frac{6}{1 + x^2} = 0,$$

$$b = \lim_{x \to \pm\infty} [f(x) - kx] = 0.$$

The x-axis is a horizontal asymptote.

Let us find extremums and intervals where the function is monotone. Find the derivatives:

$$y' = 6\frac{1 - x^2}{(1 + x^2)^2}, \quad y'' = 12\frac{x(x - \sqrt{3})(x + \sqrt{3})}{(1 + x^2)^3}.$$

Now, $y' = 0$ for $x = -1$ and $x = 1$. Since y' is defined everywhere, there are no other critical points. Check if the second sufficient condition for existence of an extremum holds at $x = 1$: $y'(1) < 0$. Hence, the function reaches a maximum at $x = 1$ ($y_{max} = 3$). By symmetry, $x = -1$ is a point of minimum ($y_{min} = -3$). Since $y' > 0$ for $\frac{1-x^2}{(1+x^2)^2} > 0$, i.e. for $|x| < 1$, the function is increasing on $(-1; 1)$ and decreasing on $(-\infty; -1) \cup (1; +\infty)$.

Let us find inflection points and intervals on which the function is convex and concave. We have $y'' =$ for $x = -\sqrt{3}$, $x = 0$, $x = \sqrt{3}$. Because of the symmetry, it is enough to check the sufficient condition only for the two latter points: $y''(0-0) > 0$, $y''(0+0) < 0$, $y''(\sqrt{3}-0) < 0$, $y''(\sqrt{3}+0) > 0$. Thus the points $(0; 0)$, $(\sqrt{3}; \frac{3\sqrt{3}}{2})$, $(-\sqrt{3}; -\frac{3\sqrt{3}}{2})$ are inflection points.

On the interval $(-\sqrt{3}; 0) \cup (\sqrt{3}; +\infty)$ the curve is concave; it is convex on the interval $(-\infty; -\sqrt{3}) \cup (0; \sqrt{3})$ (Figure 2.3).

Example 2.1.2. Plot the graph of the irrational function $y = \frac{x-3}{\sqrt{4+x^2}}$.

Domain of the function is $(-\infty; +\infty)$. The function is continuous at all points of $(-\infty; +\infty)$. The graph intersects the coordinate axes in the points $(3; 0)$, $(0; -3/2)$.

The first-order derivative is $y' = \frac{3x+4}{\sqrt{(4+x^2)^3}}$. To determine critical points, solve the equation $y' = 0$, i.e. $\frac{3x+4}{\sqrt{(4+x^2)^3}} = 0$, whence $x = -4/3$. For $x < -4/3$, $y' < 0$, hence the function is decreasing; for $x > -4/3$, $y' > 0$, and the function is increasing. Thus the function has a minimum at $x = -4/3$, $y_{min} = -1.8$.

Computing the second-order derivative we have:

$$y'' = \frac{3\sqrt{(4+x^2)^3} - (3x+4)\cdot\frac{3}{4}(4+x^2)^{1/2}\cdot 2x}{(4+x^2)^3} =$$

$$= \frac{3(4+x^2) - 3x(3x+4)}{\sqrt{(4+x^2)^5}} = -6\frac{x^2+2x-2}{\sqrt{(4+x^2)^5}} =$$

$$= -6\frac{(x+1)^2-3}{\sqrt{(4+x^2)^5}} = -6\frac{(x+1-\sqrt{3})(x+1+\sqrt{3})}{\sqrt{(4+x^2)^5}},$$

$y'' = 0$ for $x_1 = -(1+\sqrt{3})$ and $x_2 = \sqrt{3} - 1$.

For $x < -(1+\sqrt{3})$, $y'' < 0$, and for $-(1+\sqrt{3}) < x < \sqrt{3} - 1$, $y'' > 0$, and for $x > \sqrt{3} - 1$, $y'' < 0$. Hence the graph of the function is convex for $x < -(1+\sqrt{3})$ and $x > 1 - \sqrt{3}$, and concave for $-(1+\sqrt{3}) < x < \sqrt{3} - 1$.

Inflection points are $P_1(-2.73; -1.7)$, $P_2(0.73; -1.6)$. Finding asymptotes we have:

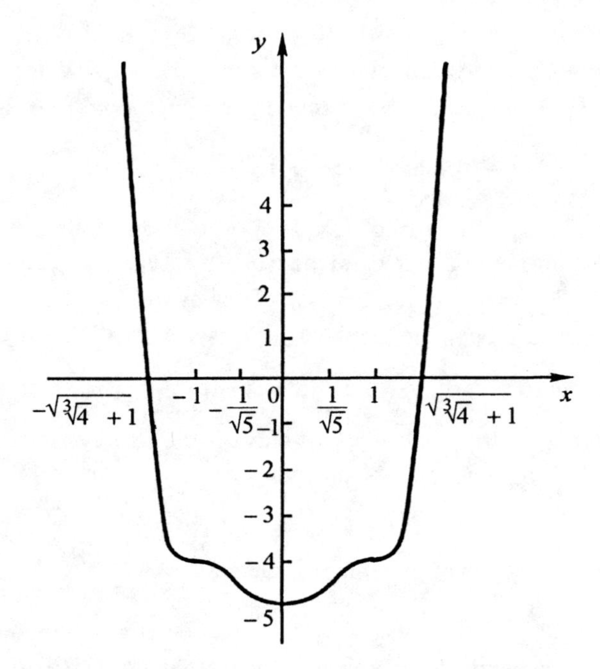

Figure 2.1. The graph of the function $y = x^6 - 3x^4 + 3x^2 - 5$.

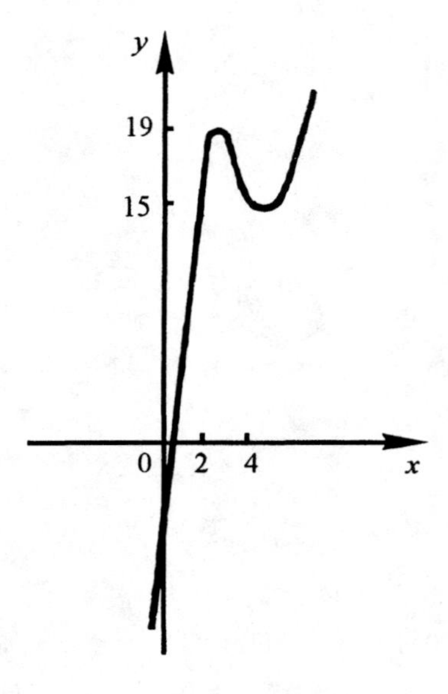

Figure 2.2. The graph of the function $y = x^3 - 6x^2 + 24x - 1$.

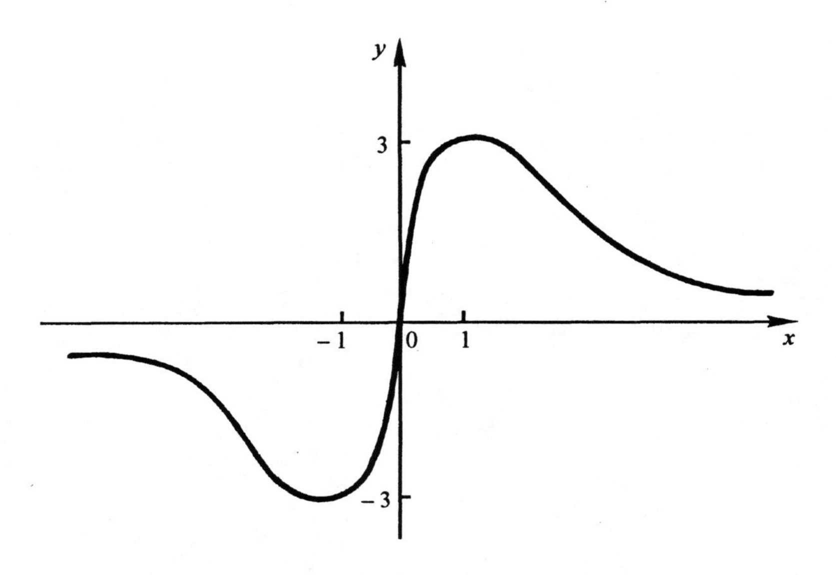

Figure 2.3. The graph of the function $y = \dfrac{6x}{1+x^2}$.

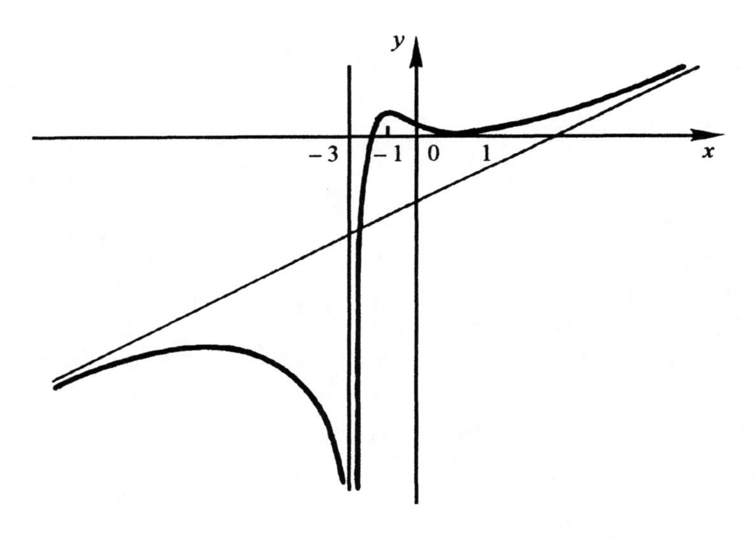

Figure 2.4. The graph of the function $y = \dfrac{x^3 - 3x + 2}{2(x+3)^2}$.

$$\lim_{x\to\pm\infty} y = \lim_{x\to\pm\infty} \frac{x-3}{\sqrt{4+x^2}} = \pm 1.$$

Horizontal asymptotes are $y = 1$ and $y = -1$. There are no vertical or oblique asymptotes (Figure 2.5).

Example 2.1.3. Plot the graph of the function $y = x^2 e^{1/x}$.

Domain of the function is $(-\infty; 0) \cup (0; +\infty)$. The function is continuous on each of the intervals $(-\infty; 0)$ and $(0; +\infty)$. The point $x = 0$ is a point of discontinuity. The graph does not intersect the coordinate axes.

A vertical asymptote is the straight line $x = 0$, since

$$\lim_{x\to 0+0} x^2 e^{1/x} = +\infty, \qquad \lim_{x\to 0-0} x^2 e^{1/x} = 0.$$

The graph does not have oblique asymptotes.

The first-order derivative of the function is $y' = 2xe^{1/x} - x^2 e^{1/x}\frac{1}{x^2} = e^{1/x}(2x-1)$. Solving the equation $y' = 0$, i.e. $e^{1/x}(2x-1) = 0$, we get $x = 1/2$, which is a critical point. For $x > 1/2$, $y' > 0$, and for $x < 1/2, y' < 0$. Thus $x = 1/2$ is a point of minimum ($y_{min} = e^2/4 \approx 1.87$).

The second-order derivative is given by

$$y'' = 2x^{1/x} - \frac{2}{x}e^{1/x}\frac{1}{x^2} = \frac{e^{1/x}(2x^2 - 2x + 1)}{x^2} > 0,$$

$$y'' = 0 \quad \Rightarrow \quad 2x^2 - 2x + 1 = 0.$$

The last equation has no real roots; hence there are no inflection points. The graph is shown in Figure 2.9.

Example 2.1.4. Plot the graph of the function $f(x) = \sqrt{1 - e^{-x^2}}$.

Domain of the function is $(-\infty; +\infty)$. The function is even; hence the graph symmetric with respect to the y-axis, $f(0) = 0$. The function is continuous on $(-\infty; +\infty)$.

Since $\lim_{x\to\infty} f(x) = 1, y = 1$ is a horizontal asymptote.

Let us now look at the monotonicity of the function and find extremums. We have:

$$f'(x) = \frac{xe^{-x^2}}{\sqrt{1-e^{-x^2}}} = \frac{xe^{-x^2}}{|x|\sqrt{1+0(1)}}.$$

The derivative does not exist at $x = 0$. There are no other critical points. If $x < 0, f'(x) < 0$, and $f'(x) > 0$ if $x > 0$. Thus the function has a minimum at the point $x = 0$ ($f_{min} = f(0) = 0$). The function is decreasing for $x < 0$ and increasing for $x > 0$.

Calculating the second-order derivative we have:

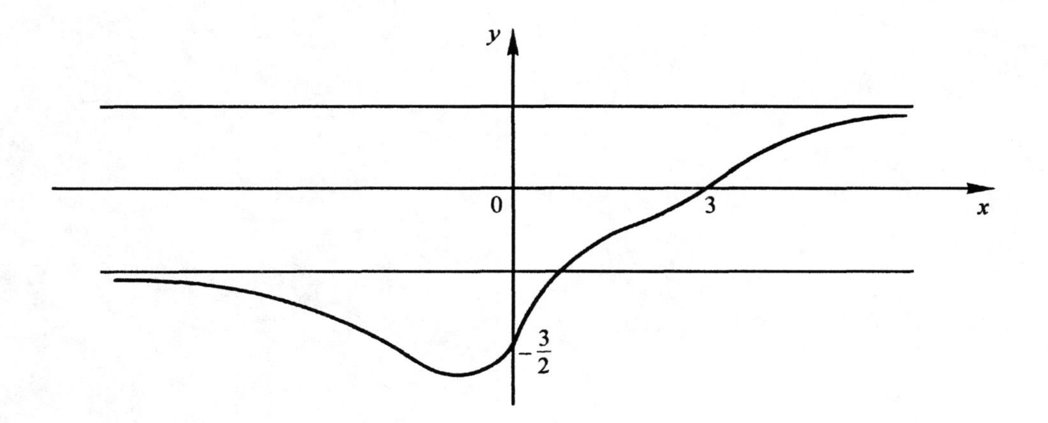

Figure 2.5. The graph of the function $y = \dfrac{x-3}{\sqrt{4+x^2}}$.

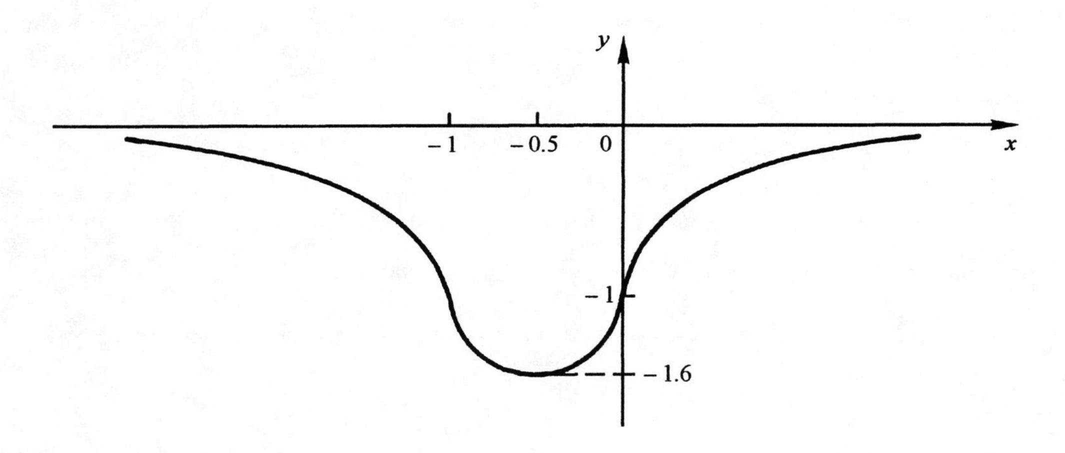

Figure 2.6. The graph of the function $y = \sqrt[3]{x} - \sqrt[3]{x+1}$.

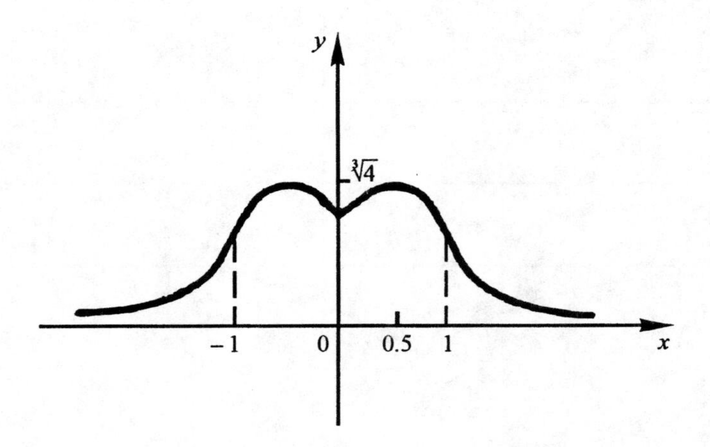

Figure 2.7. The graph of the function $y = \sqrt[3]{x^2} - \sqrt[3]{x^2 - 1}$.

$$f'' = \frac{(x^2 - 1) + (1 + 2x^2)e^{-x^2}}{(1 - e^{-x^2})^{3/2}}.$$

The second-order derivative $f''(x)$ does not exist at $x = 0$. It is negative for all $x \neq 0, f''(x) < 0$. The curve is convex for $x \in (-\infty; 0) \cup (0; +\infty)$. There are no inflection points. The graph of the curve is shown in Figure 2.11.

Example 2.1.5. Plot the graph of the function $y = 2^{\sqrt{x^2+1} - \sqrt{x^2-1}}$.

The domain of the function is $(-\infty; -1) \cup (1; +\infty)$. The function is continuous on its domain, even, and takes positive values. Since $\lim\limits_{x \to \pm\infty} y = \lim\limits_{x \to \pm\infty} 2^{\sqrt{x^2+1} - \sqrt{x^2-1}} = 1$, $y = 1$ is a horizontal asymptote. The graph has no other asymptotes.

Calculating the first-order derivative we have:

$$y' = 2^{\sqrt{x^1+1} - \sqrt{x^2-1}} \ln 2 \left(\frac{x}{\sqrt{x^2+1}} - \frac{x}{\sqrt{x^2-1}} \right) =$$

$$= xy \ln 2 \left(\frac{\sqrt{x^2-1} - \sqrt{x^2+1}}{x^2 - 1} \right).$$

Critical points are $x = \pm 1$. For $x < -1$, $y' > 0$, and the function is increasing. For $x > 1$, $y' < 0$, the function is decreasing.

We have:

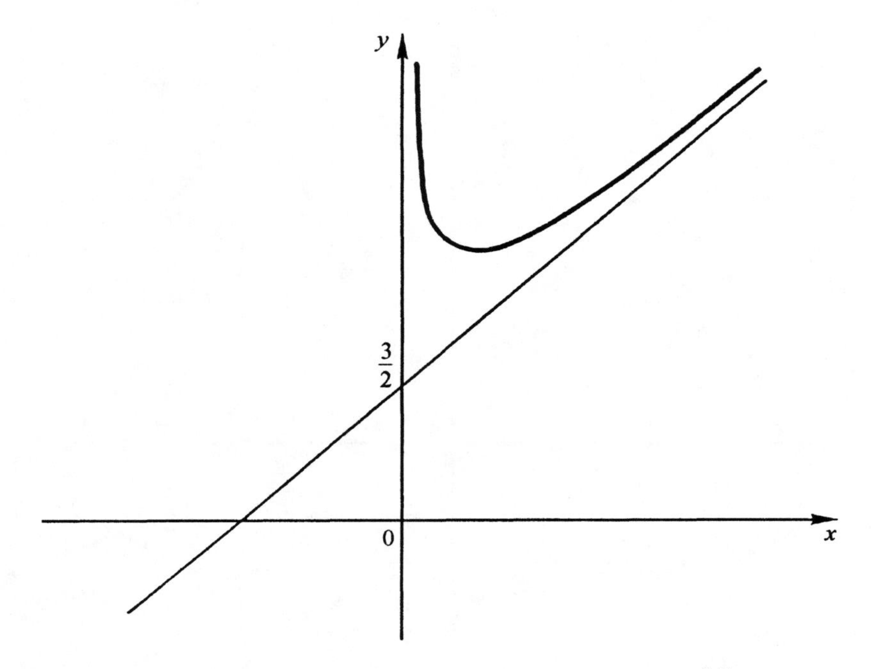

Figure 2.8. The graph of the function $y = \frac{|1+x|^{3/2}}{\sqrt{x}}$.

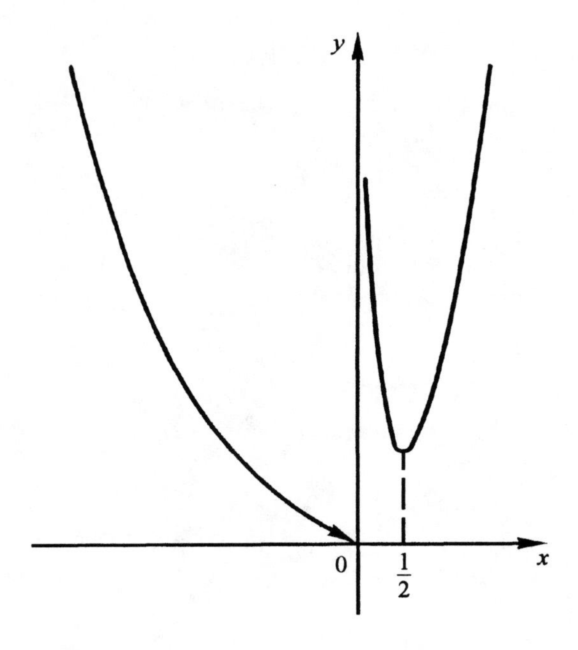

Figure 2.9. The graph of the function $y = x^2 e^{1/x}$.

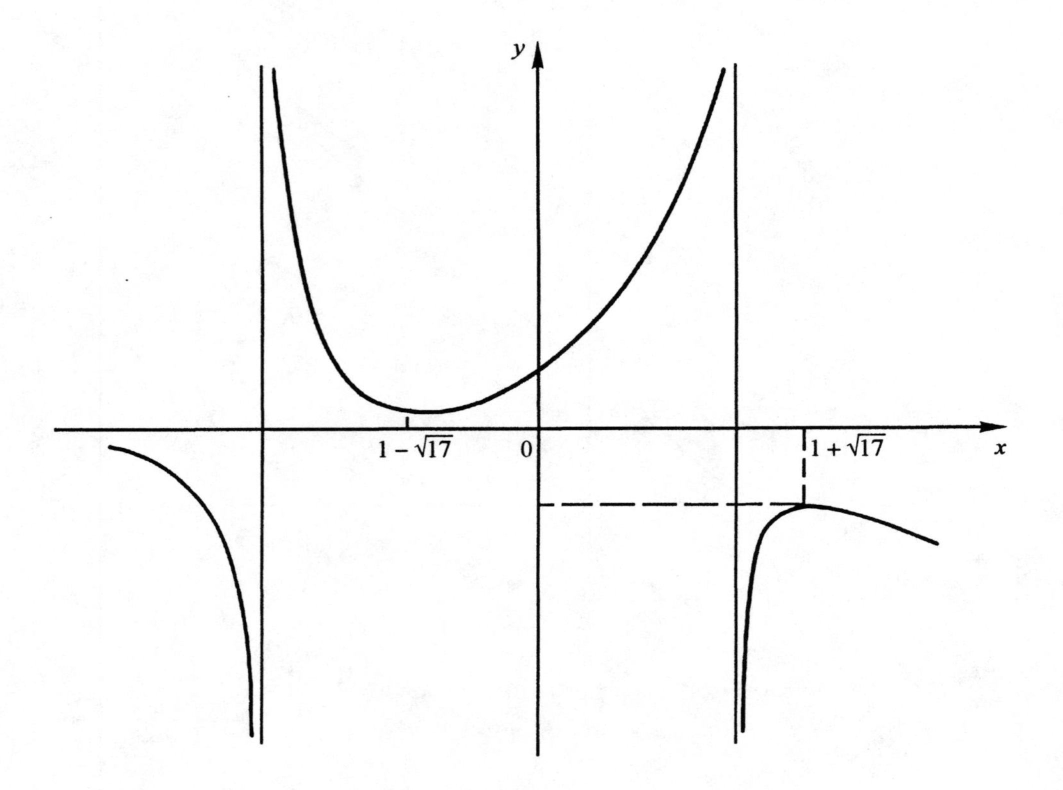

Figure 2.10. The graph of the function $y = \dfrac{e^x}{16 - x^2}$.

$$\lim_{x \to -1-0} y = \lim_{x \to -1-0} 2^{\sqrt{x^2+1} - \sqrt{x^2-1}} = 2^{\sqrt{2}}.$$

$$\lim_{x \to 1+0} y = \lim_{x \to 1+0} 2^{\sqrt{x^2+1} - \sqrt{x^2-1}} = 2^{\sqrt{2}}.$$

The function has a left-hand maximum at the point $x = -1$, and a right-hand maximum at $x = 1$.

The second-order derivative is

$$y'' = y \ln 2 \left[\frac{1}{\sqrt{x^2+1}} - \frac{1}{\sqrt{x^2-1}} + x^2 \ln \left(\frac{1}{\sqrt{x^2+1}} - \frac{1}{\sqrt{x^2-1}} \right)^2 + \right.$$

$$\left. + x^2 \left(\frac{1}{\sqrt{(x^2-1)^3}} - \frac{1}{\sqrt{(x^2+1)^3}} \right) \right].$$

It is clear that

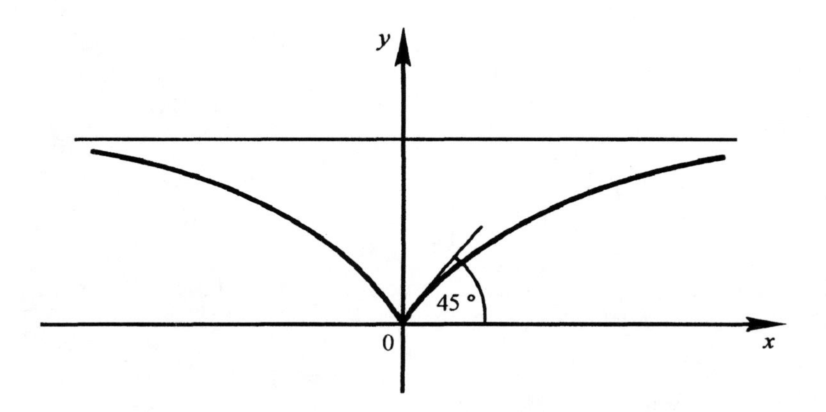

Figure 2.11. The graph of the function $y = \sqrt{1 - e^{-x^2}}$.

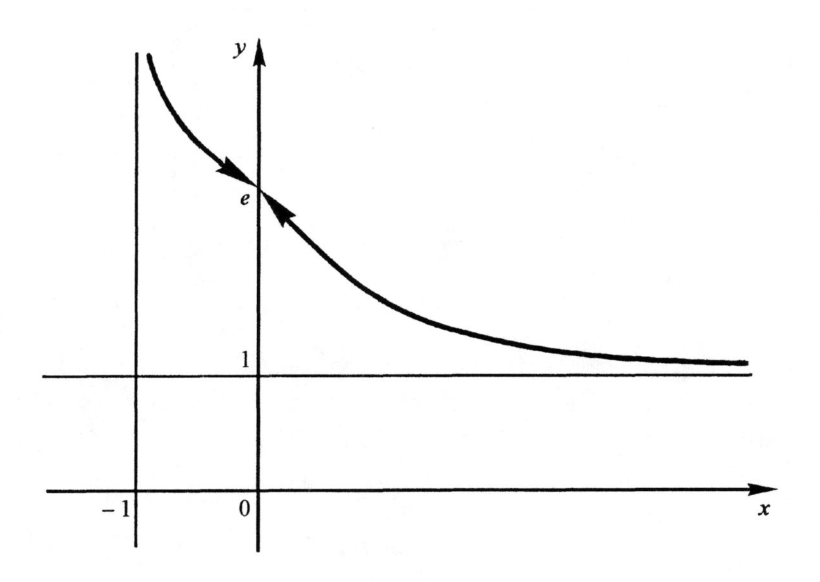

Figure 2.12. The graph of the function $y = (1 + x)^{1/x}$.

$$\left[\frac{1}{\sqrt{x^2+1}}-\frac{1}{\sqrt{x^2-1}}+x^2\ln 2\left(\frac{1}{\sqrt{x^2+1}}-\frac{1}{\sqrt{x^2-1}}\right)^2+x^2\left(\frac{1}{\sqrt{(x^2-1)^3}}-\frac{1}{\sqrt{(x^2+1)^3}}\right)\right]>$$

$$\left[\frac{1}{\sqrt{x^2+1}}-\frac{1}{\sqrt{(x^2-1)}}+\frac{x^2}{\sqrt{(x^2-1)^3}}-\frac{x^2}{\sqrt{(x^2+1)^3}}\right],$$

i.e.

$$y''>y\left[\frac{1}{\sqrt{(x^2+1)^3}}+\frac{1}{\sqrt{(x^2-1)^3}}\right]\ln 2>0$$

in the domain of the function. The curve is concave and has no inflection points. The graph of the function is shown in Figure 2.14.

Example 2.1.6. Plot the graph of the function $y=\log_2(1-x^2)$.

Domain of the function is $(-1;\,1)$. The straight lines $x=1, x=-1$ are vertical asymptotes. The function is even; its graph is symmetric with respect to the y-axis. The graph passes through the point $(0;\,0)$. If $-1<x<1, y<0$.

Let us find the points of extremum. We have $y'=-2x/(1-x^2)\ln 2$. The points $x=0,1,-1$ are critical; however the points $x=\pm 1$ do not belong to the domain. Hence we need to study the behavior of the derivative as x passes through the point $x=0$. We have $y'(0-0)>0$ and $y'(0+0)<0$. Thus $x=0$ is a point of maximum with $y_{max}=f(0)=0$. The function is increasing on $(-1;\,0)$ and decreasing on $(0;\,1)$.

Let us find inflection points. We have

$$y''=-\frac{2}{\ln 2}\,\frac{(1-x^2)-2x\cdot x}{(1-x^2)^2}=-\frac{2(1+x^2)}{(1-x^2)^2\ln 2}.$$

Thus $y''<0$ for all $x\in(-1;\,1)$. There are no inflection points; the graph is convex. It is shown in Figure 2.15.

Example 2.1.7. Plot the graph of the function $y=\sin^4 x+\cos^4 x$.

The function is defined and continuous on $(-\infty;\,+\infty)$. The function is even; thus its graph is symmetric with respect to the y-axis. The function is periodic with the period $\omega=\pi/2$. Indeed,

$$\sin^4 x+\cos^4 x=(\sin^2 x+\cos^2 x)^2-2\sin^2 x\cos^2 x=$$
$$=1-\frac{1}{2}\sin^2 2x=1-\frac{1}{4}(1-\cos 4x)=\frac{3}{4}+\frac{1}{4}\sin\left(4x+\frac{\pi}{2}\right),$$

whence $\omega=2\pi/4=\pi/2$. It is thus necessary to plot the graph only for $x\in[0;\,\pi/2]$.

The graph lies above the x-axis, since $y=\sin^4 x+\cos^4 x>0$ for all $x\in[0;\,\pi/2]$ and passes through the point $(0;\,1)$.

The first-order derivative is

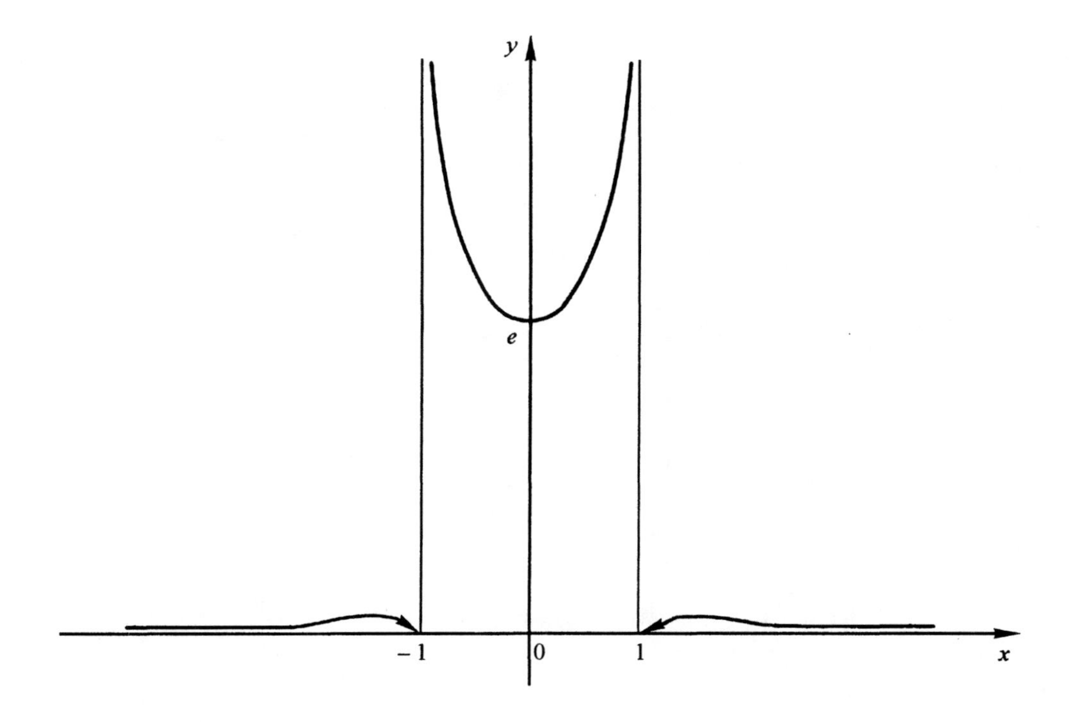

Figure 2.13. The graph of the function $y = \dfrac{e^{1/(1-x^2)}}{1+x^2}$.

$$y' = 4\sin^3 x \cos x - 4\cos^3 x \sin x = -4\sin x\cos x\,(\cos^2 x - \sin^2 x) =$$
$$= -2\sin 2x \cos 2x = -\sin 4x.$$

Critical points are found from the equation $y' = -\sin 4x = 0$, whence $x = k\pi/4$, $k \in \mathbf{Z}$.

Consider $x = k\pi$ for $k = 0,1,2$. The behavior of the function in critical points is found by considering the second-order derivative, $y'' = -4\cos 4x$. We have: $y''(0) = -4 < 0$, and the function has a maximum at $x = 0$ ($y_{max} = 1$); $y''(\pi/4) = 4 > 0$ and $x = \pi/4$ is a point of minimum ($y_{min} = 1/2$); $y''(\pi/2) = -4 < 0$, thus $x = \pi/2$ is a point of maximum with $y_{max} = 1$.

Inflection points are found from the equation $y'' = -4\cos 4x = 0$, whence $x = \pi/8(2k+1)$, $k \in \mathbf{Z}$, moreover, $y''(\pi/8 - 0) < 0$, $y''(\pi/8 + 0) > 0$, $y''(3/8\pi - 0) > 0$, $y''(3/8\pi + 0) < 0$.

The graph of the function is convex on $(0; \pi/8)$, concave on $(\pi/8; 3/8\pi)$, and convex on $(3/8\pi; \pi/2)$. The graph of the function is shown in Figure 2.20.

Example 2.1.8. Plot the graph of the function $y = \cos x - \cos^2 x$.

The function is defined and continuous on $(-\infty; +\infty)$. It is periodic with the period $\omega = 2\pi$; thus it suffices to plot the graph on an interval of length 2π. Let us plot the graph on

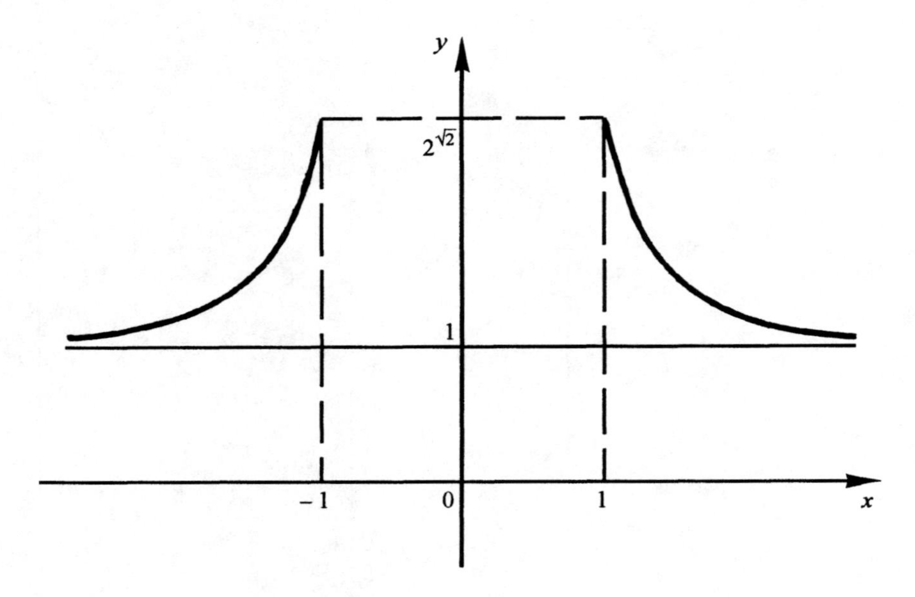

Figure 2.14. The graph of the function $y = 2^{\sqrt{x^2+1} - \sqrt{x^2-1}}$.

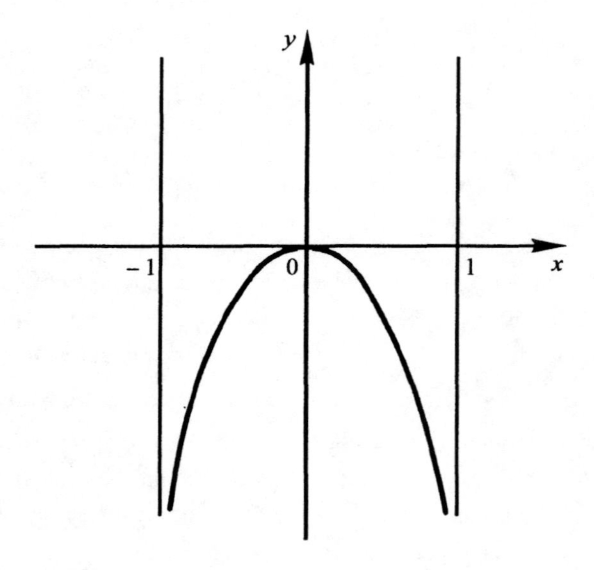

Figure 2.15. The graph of the function $y = \log_2(1 - x^2)$.

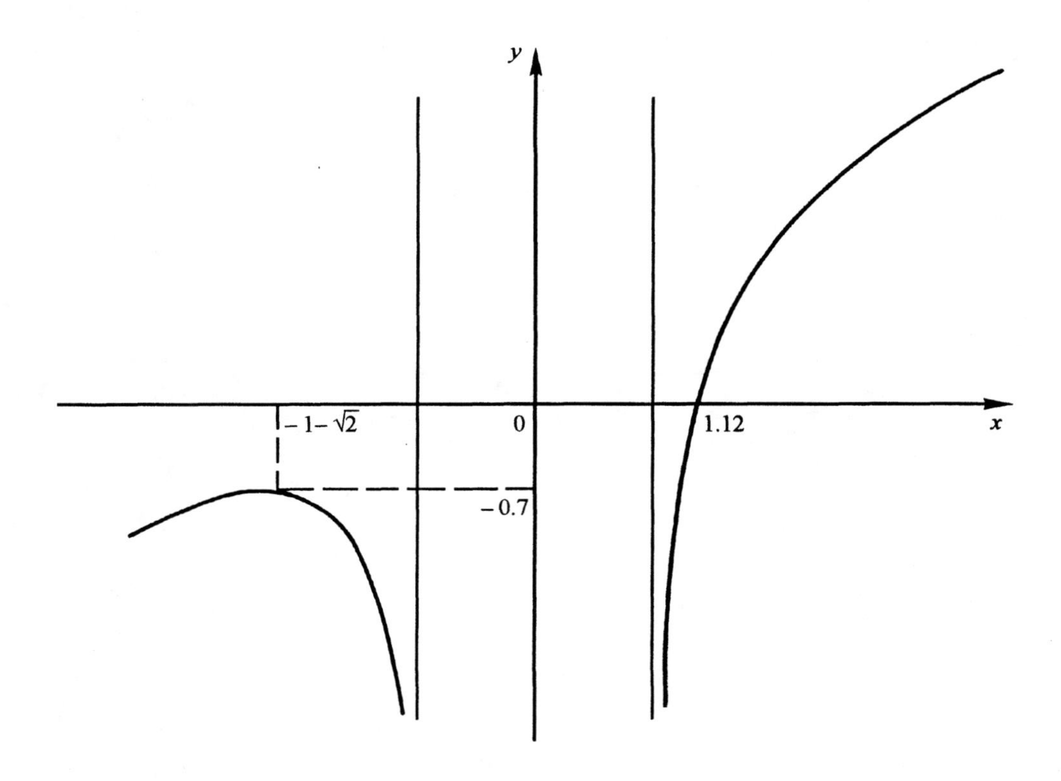

Figure 2.16. The graph of the function $y = \ln(x^2 - 1) + x$.

$[-\pi; \pi]$. The function is even; hence we can study its behavior on $[0; \pi]$. At the endpoints of this interval we have that $y(0) = 0$, $y(\pi) = -2$. By symmetry, $y(-\pi) = -2$.

Find zeros of the function: $y = 0 \Rightarrow \cos x(1 - \cos x) = 0$. Thus on $[0; \pi]$, $\cos x = 0$ at $x = \pi/2$, and $\cos x = 1$ at $x = \pi/2$. By symmetry, $x = -\pi/2$ is also a zero of the function.

Because $y(\pi/3) = 1/4 > 0$, $y(\pi) = -2 < 0$, the function is positive on $(0; \pi/2)$ and negative on $(\pi/2; \pi)$. By symmetry we also have that $y > 0$ for $x \in (-\pi/2; 0)$ and $y < 0$ for $x \in (-\pi; \pi/2)$.

Periodic functions do not have oblique or horizontal asymptotes. There are no vertical asymptotes, since the function is defined on $(-\infty; +\infty)$.

Calculating the derivatives we have $y' = 2\sin x(\cos x - \frac{1}{2})$, $y'' = 4\cos^2 x - \cos x - 2$. Thus $y' = 0$ if $\sin x = 0$ or $\cos x = 1/2$, i.e. for $x = 0$, $x = \pi/3$, $x = \pi$. We also have that $y''(0) = 1 > 0$, i.e. $x = 0$ is a minimum of the function with $y_{min} = y(0) = 0$; $y''(\pi/3) = -3/2 < 0$, hence $x = \pi/3$ is a point of maximum with $y_{max} = y(\pi/3) = 1/4$; $y''(\pi) = 3 > 0$, and $x = \pi$ is a point of minimum ($Y_{min} = y(\pi) = -2$).

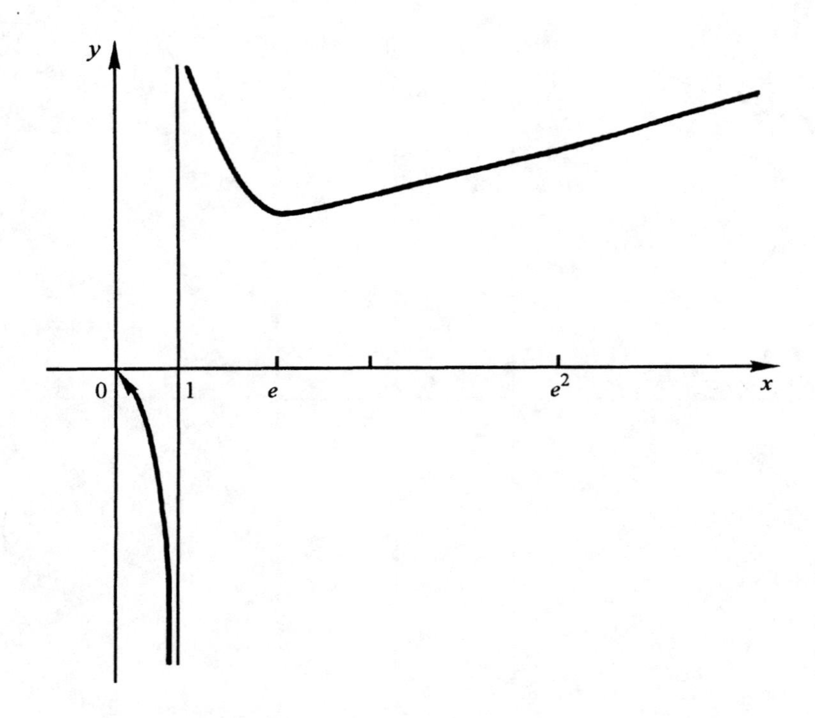

Figure 2.17. The graph of the function $y = \frac{x}{\ln x}$.

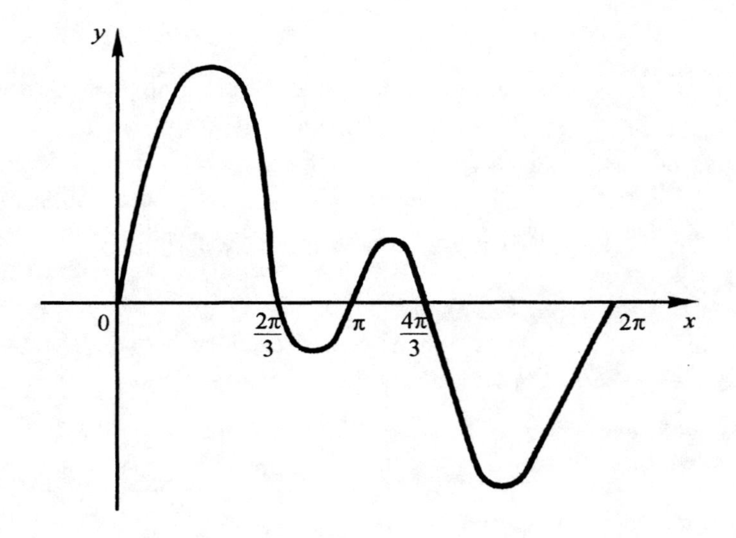

Figure 2.18. The graph of the function $y = \sin x + \sin 2x$.

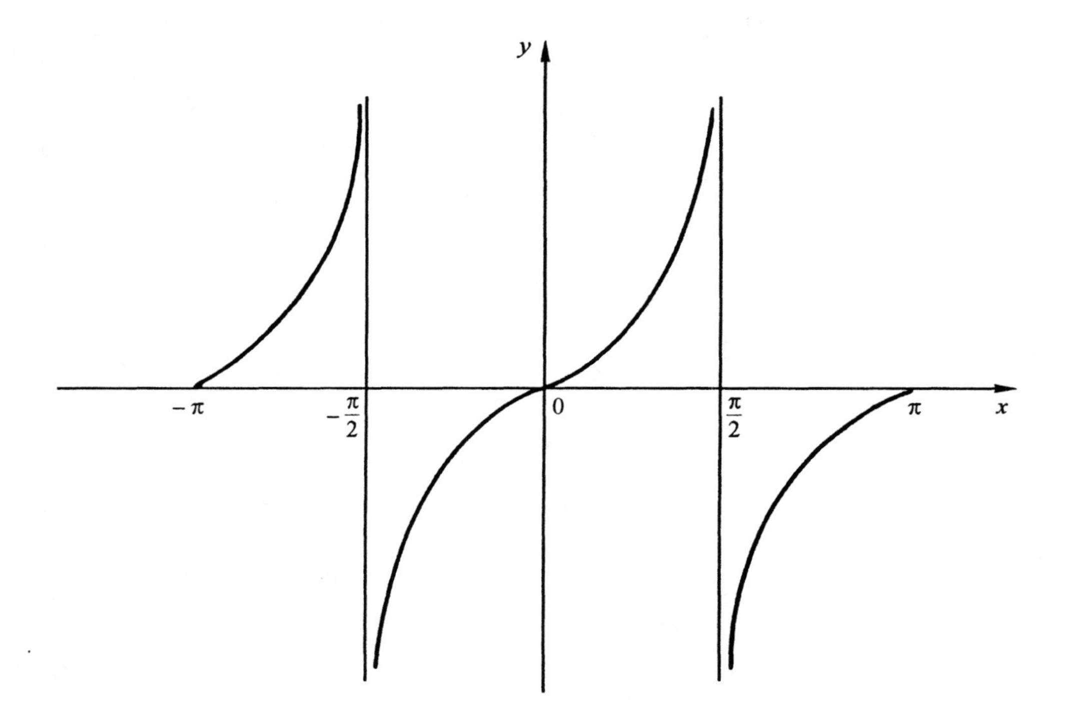

Figure 2.19. The graph of the function $y = \tan x + \sin x$.

Due to symmetry, the function also has a minimum at $x = -\pi$ and a maximum at $x = -\pi/3$.

Consider the equation $y'' = 0$, i.e. $4\cos^2 x - \cos x - 2 = 0$. Approximate roots of this equation are $x_1 = 0.57$, $x_2 = 2.20$, and the corresponding values of the function are $y(x_1) = 0.13$, $y(x_2) = -0.95$. Write the expression for the second-order derivative, $y'' = 4(\cos x - x_1)(\cos x - \cos x_2)$, and apply the sufficient condition for existence of an inflection point. We have that the points $(2.20; -0.95)$, $(0.57; 0.13)$, $(-2.20; -0.95)$, $(-0.57; 0.13)$ are inflection points of the graph. The graph of the function is shown in Figure 2.21.

Example 2.1.9. Plot the graph of the function $f(x) = \arcsin \frac{1 - x^2}{1 + x^2}$.

The domain of the function is found from the condition $-1 \le \frac{1 - x^2}{1 + x^2} \le 1$, i.e. $x \in (-\infty; +\infty)$. The function is even and the y-axis is a symmetry axis of the graph. Find the

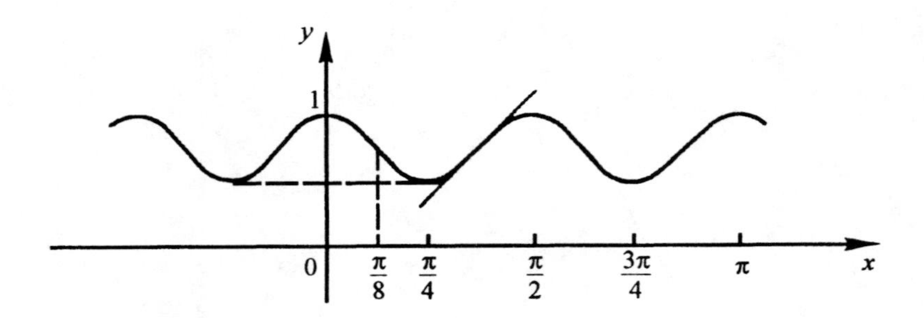

Figure 2.20. The graph of the function $y = \sin^4 x + \cos^4 x$.

points at which the graph intersects the axes; $y = 0$ if $x = \pm 1$, $y = \pi/2 \approx 1.57$ if $x = 0$. We also have that $f(x) > 0$ for $-1 < x < 1$, and $f(x) < 0$ for $x < -1$ or $x > 1$.

Let us find the asymptotes. We have:

$$k = \lim_{x \to \infty} \frac{\arcsin \dfrac{1 - x^2}{1 + x^2}}{x} = 0, \quad b = \lim_{x \to \infty} \arcsin \frac{1 - x^2}{1 + x^2} = -\frac{\pi}{2}.$$

Thus $y = -\pi/2$ is a horizontal asymptote. The graph of the function is located under the asymptote.

The expression for the first-order derivative is $f'(x) = -\dfrac{x}{|x|} \dfrac{2}{(1 + x^2)}$. If $x < 0, f'(x) < 0$, and the function is monotone decreasing on the interval $(0; +\infty)$. For $x < 0, f'(x) > 0$, and the function is monotone decreasing on $(-\infty; 0)$. The derivative does not exist at the point $x = 0$. Let us look at the behavior of the derivative in a vicinity of this point. We have:

$$\lim_{x \to -0} \left(-\frac{x}{|x|} \frac{2}{1 + x^2} \right) = -2 \lim_{x \to -0} \frac{x}{|x|} = 2,$$

$$\lim_{x \to +0} \left(-\frac{x}{|x|} \frac{2}{1 + x^2} \right) = -2.$$

Hence, the first-order derivative has a discontinuity of the first kind at the point $x = 0$; the derivative changes sign from "+" to "–" at this point. Since the function is continuous at this point, it has a maximum with $y_{max} = \pi/2$.

The second-order derivative is $y'' = \dfrac{4x}{(1 + x^2)^2}$, and $y'' > 0$ for $x \in (0; +\infty)$, thus the graph is concave, and $y'' < 0$ for $x \in (-\infty; 0)$, and the graph is convex on this interval.

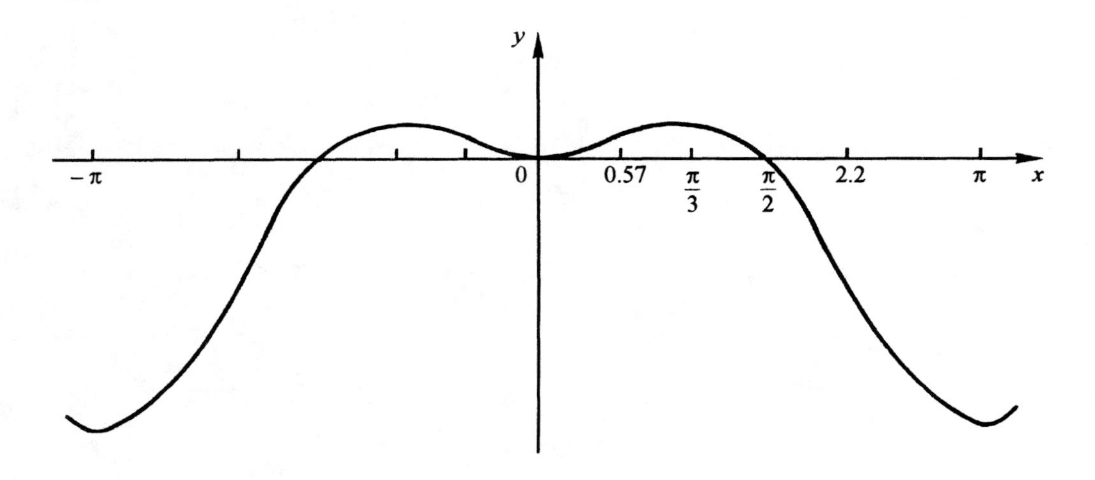

Figure 2.21. The graph of the function $y = \cos x - \cos^2 x$.

The graph has no inflection points. It is shown in Figure 2.22.

Graphs of various kinds of functions in the Cartesian coordinates are given in Figures 2.1 – 2.39.

2.2. Plots of parametrically defined functions

In the first part of the book, parametrically defined functions $x = \varphi(t)$, $y = \psi(t)$ were studied and their graphs were plotted by using elementary methods. Now we extend the discussed methods for studying and plotting such functions.

2.2.1. A study of parametrically defined functions by using derivatives

Symmetry. If the function $\varphi(t)$ is even and $\psi(t)$ is odd, then the curve is symmetric with respect to the x-axis. If the function $\varphi(t)$ is odd and $\psi(t)$ is even, then the curve is symmetric with respect to the y-axis. If both functions $\varphi(t)$ and $\psi(t)$ are odd, then the function is symmetric with respect to the origin.

Note that these are sufficient conditions for the symmetry, but they are not necessary.

Points of intersection with the coordinate axes. By solving the equation $\psi(t) = 0$, we find values of t that correspond to points of intersection of the curve with the x-axis.

Values of t that correspond to points of intersection of the curve with the y-axis are found by solving the equation $\varphi(t) = 0$.

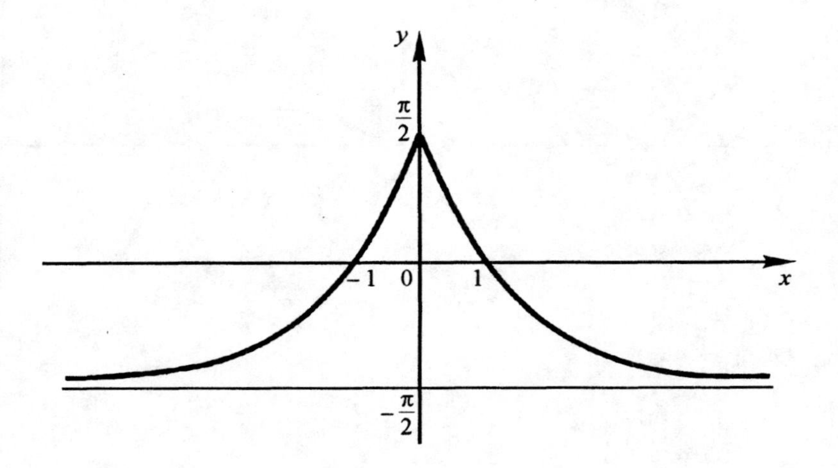

Figure 2.22. The graph of the function $y = \arcsin\frac{1-x^2}{1+x^2}$.

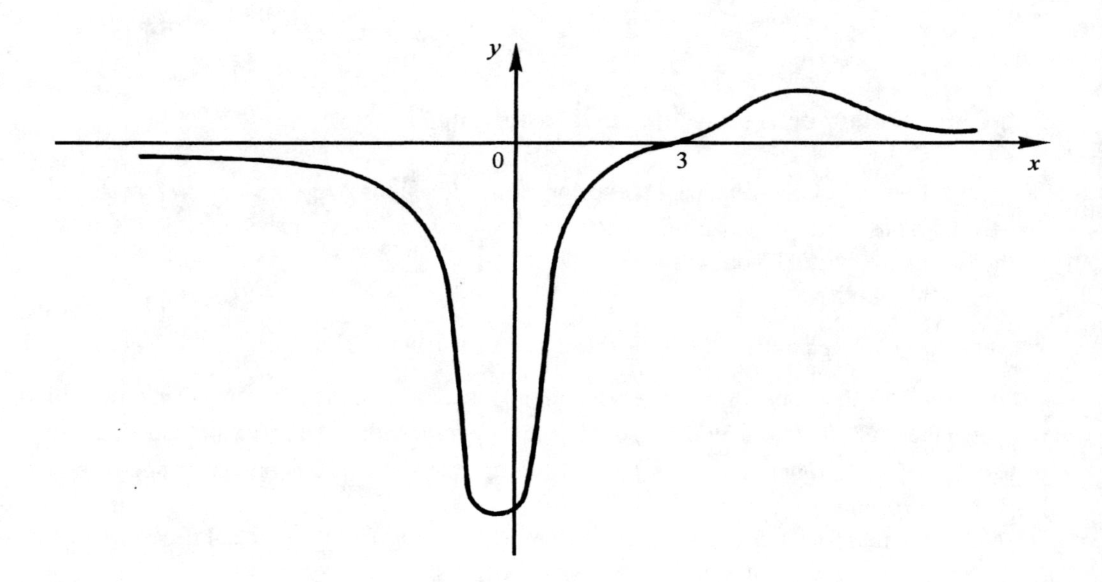

Figure 2.23. The graph of the function $y = \arctan\frac{x-3}{x^2+4}$.

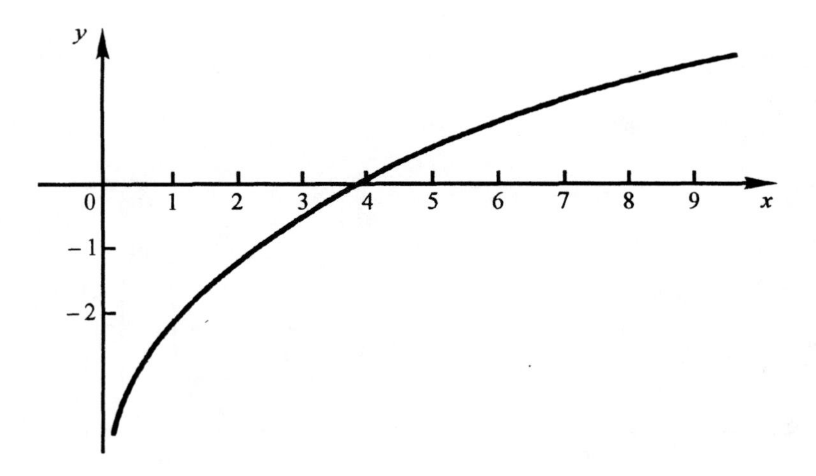

Figure 2.24. The graph of the function $y = \ln x - \arctan x$.

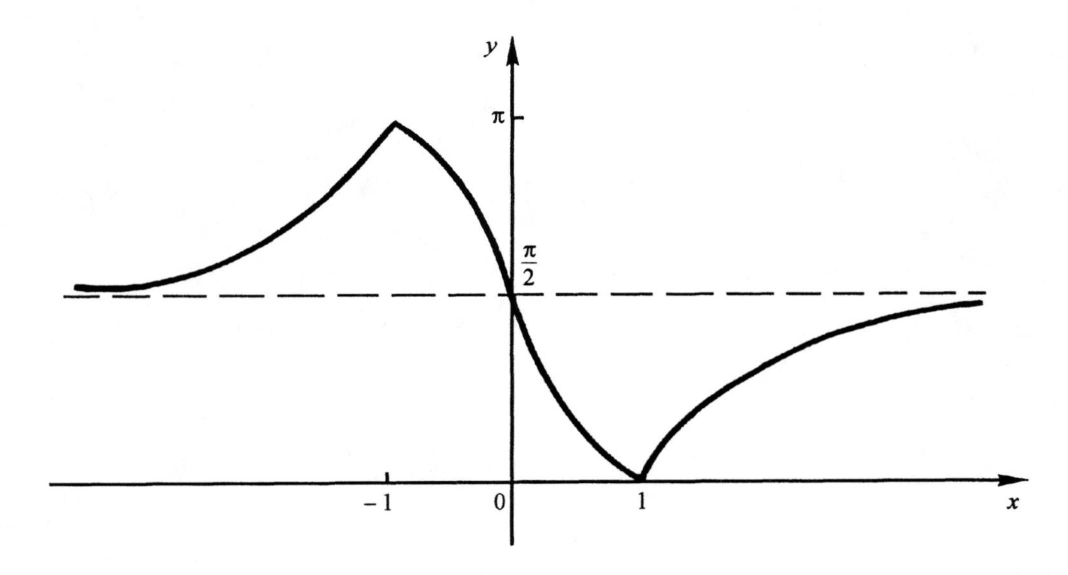

Figure 2.25. The graph of the function $y = \arccos \dfrac{2x}{1 + x^2}$.

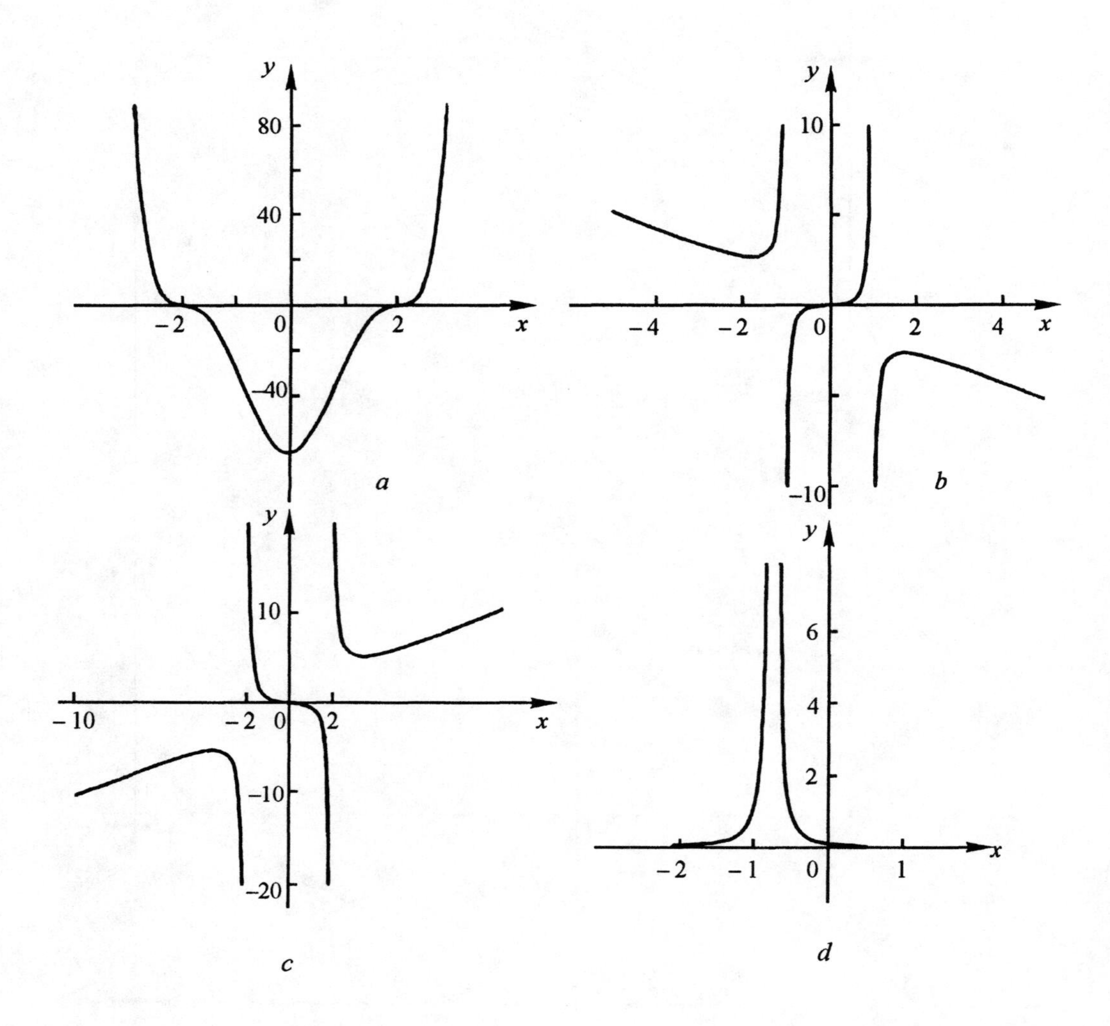

Figure 2.26. The graphs of functions: (*a*) $y = (x^2 - 4)^3$; (*b*) $y = \dfrac{x^3}{1 - x^2}$;

(*c*) $y = \dfrac{x^3}{x^2 - 4}$; (*d*) $y = \dfrac{1}{(3 + 4x)^2}$.

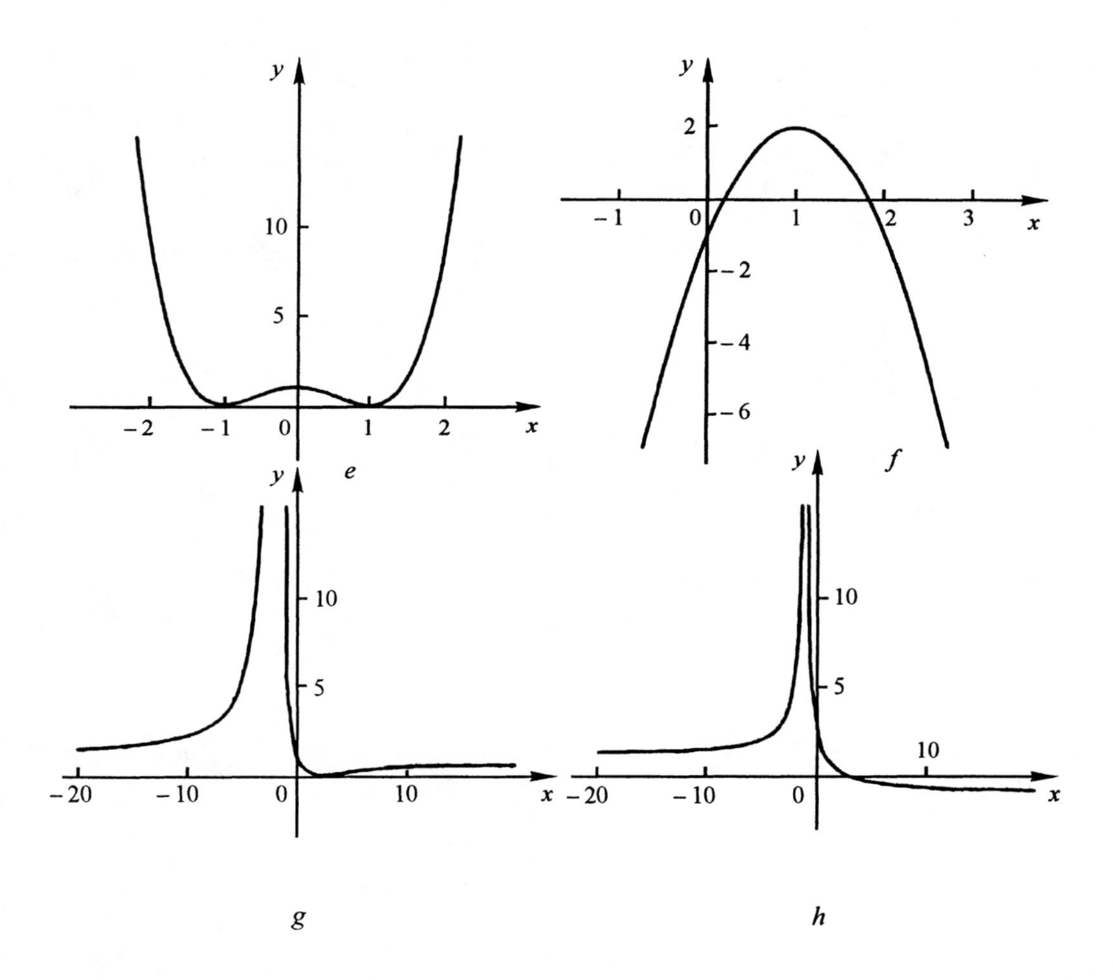

Figure 2.26. The graphs of functions: (*e*) $y = (x^2 - 1)^2$; (*f*) $y = 2 - 3(x - 1)^2$;
(*g*) $y = \left(\dfrac{x-2}{x+2}\right)^2$; (*h*) $y = \dfrac{3-x}{|x+1|}$.

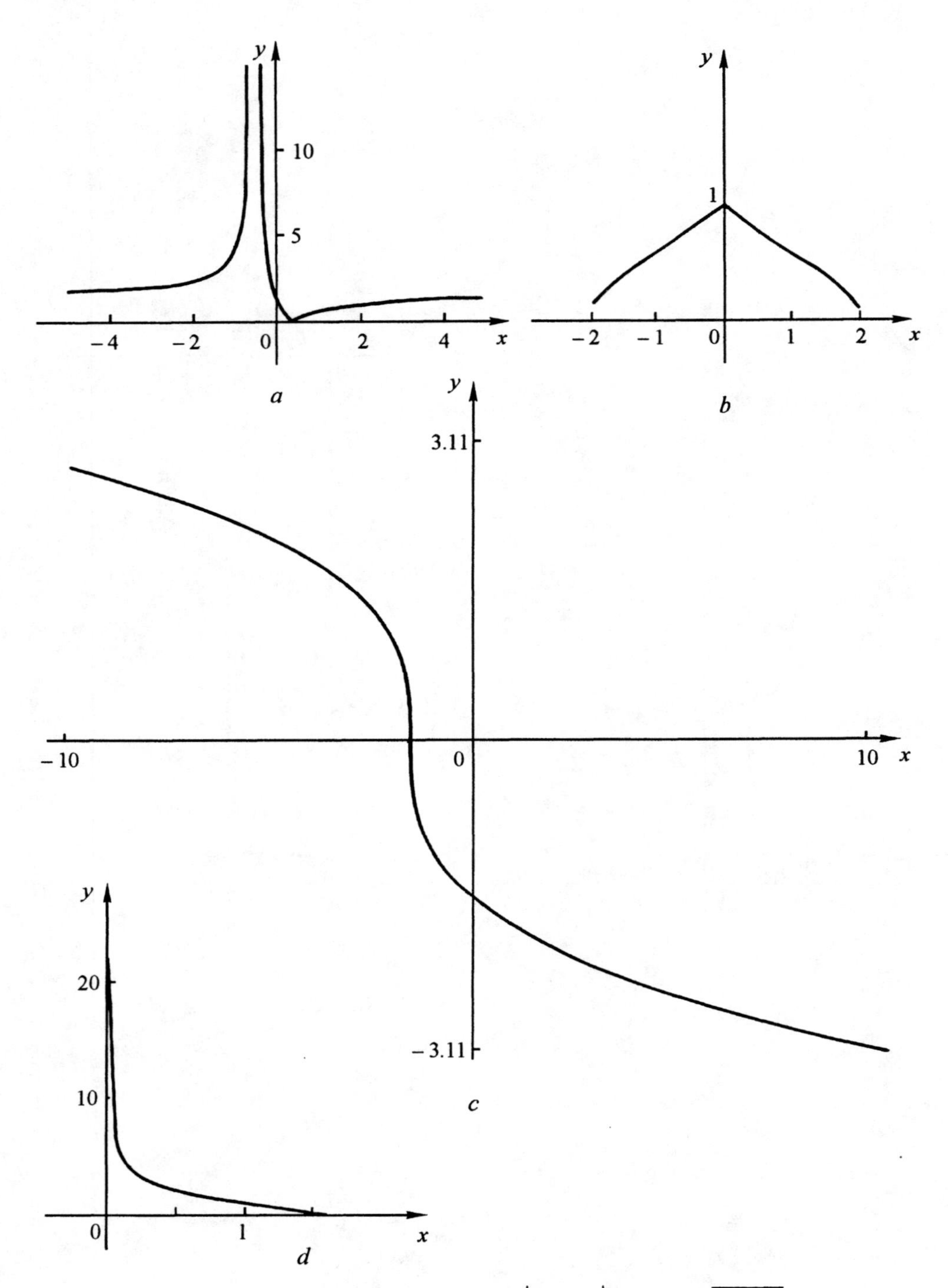

Figure 2.27. The graphs of functions: (*a*) $y=\left|\dfrac{1-3x}{2x+1}\right|$; (*b*) $y=\sqrt{\dfrac{2-|x|}{2+|x|}}$;
(*c*) $y=\sqrt[3]{-3-2x}$; (*d*) $y=\sqrt{\dfrac{3}{x}-2}$.

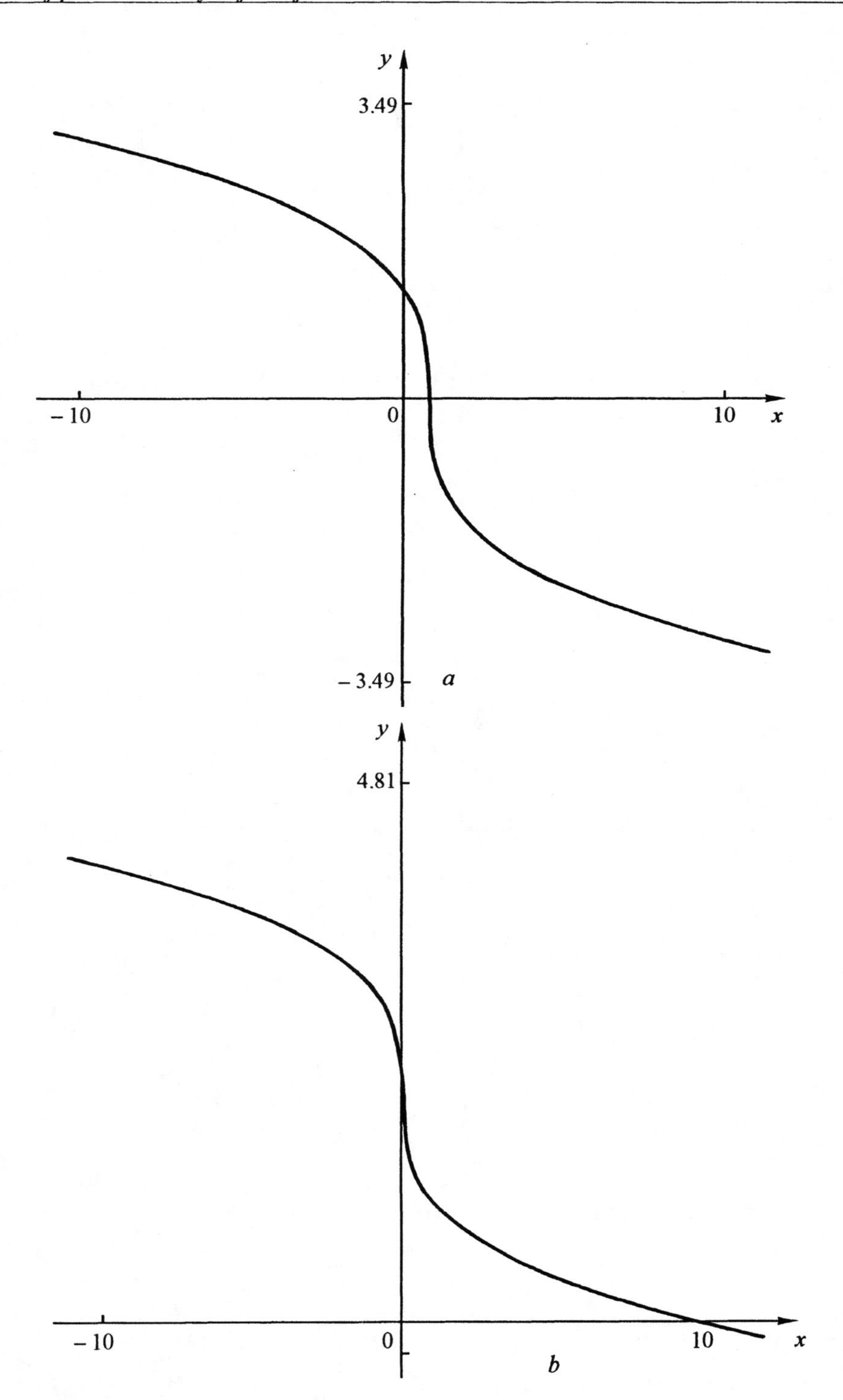

Figure 2.28. The graphs of functions: (*a*) $y = \sqrt[3]{2 - 3x}$; (*b*) $y = 2 - \sqrt[3]{x}$.

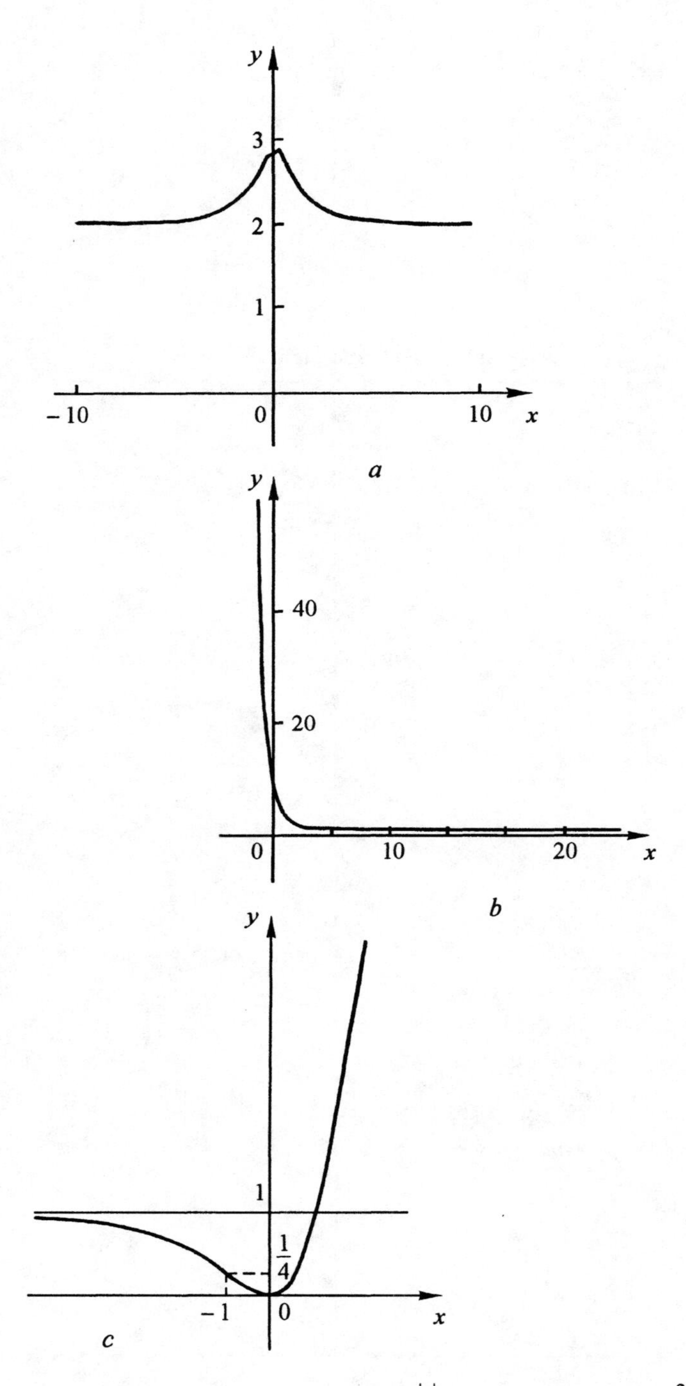

Figure 2.29. The graphs of functions: (*a*) $y = \left(\frac{1}{2}\right)^{|x|} + 2$; (*b*) $y = \left[\left(\frac{1}{2}\right)^{x} + 1\right]^{3}$; (*c*) $y = (1 - 2^{x})^{2}$.

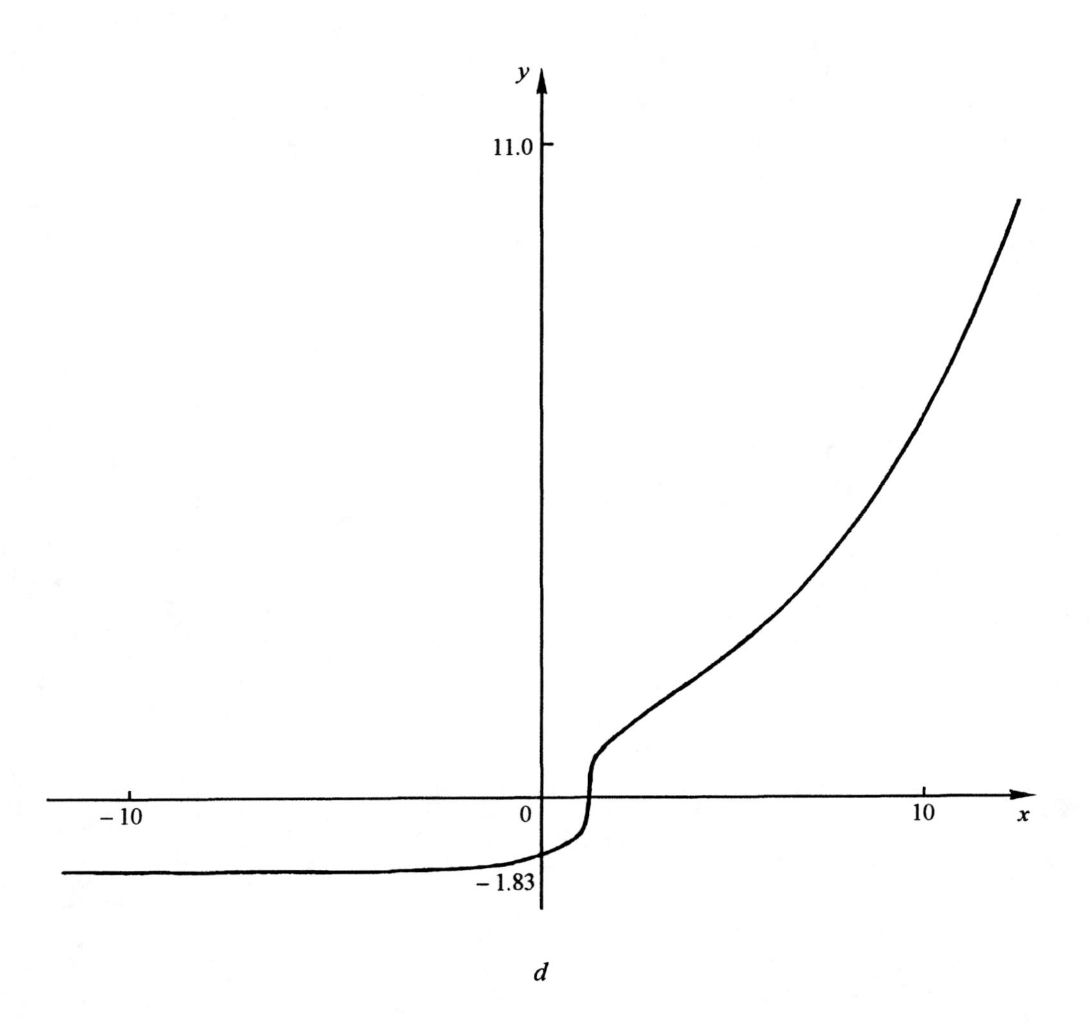

Figure 2.29. The graphs of functions: (*d*) $y = \sqrt[3]{2^x - 2}$.

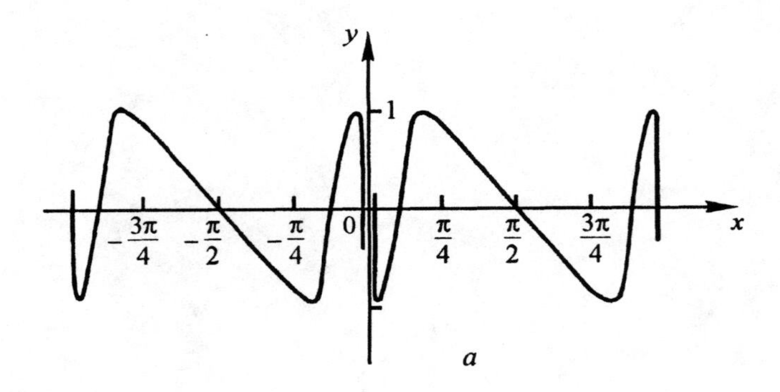

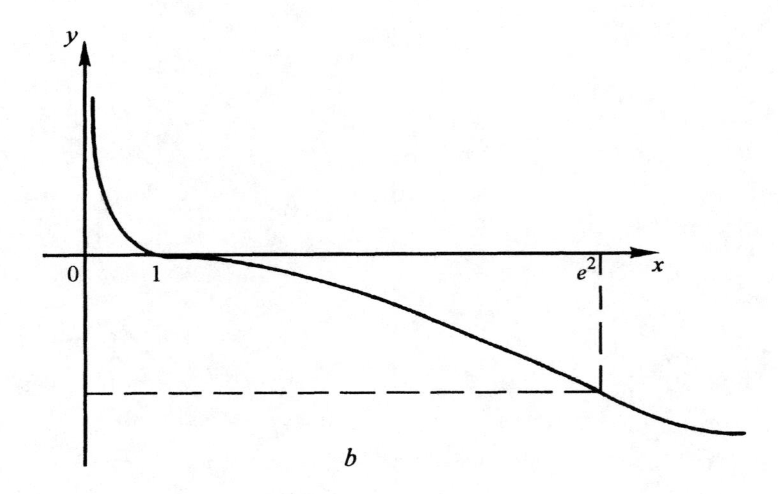

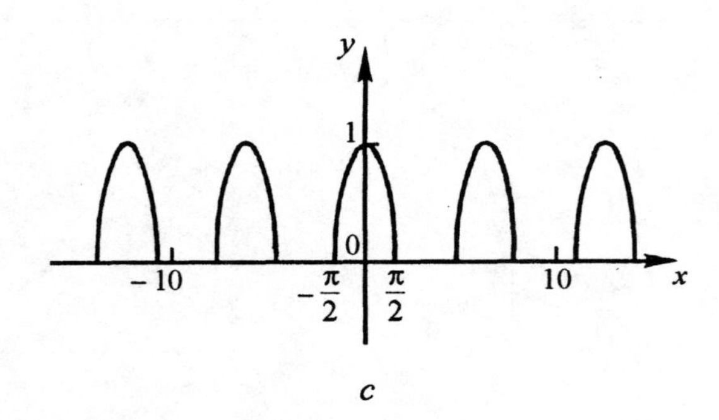

Figure 2.30. The graphs of functions: (*a*) $y = \sin(\cot x)$; (*b*) $y = \log^3_{1/2} x$; (*c*) $y = \sqrt{\cos x}$.

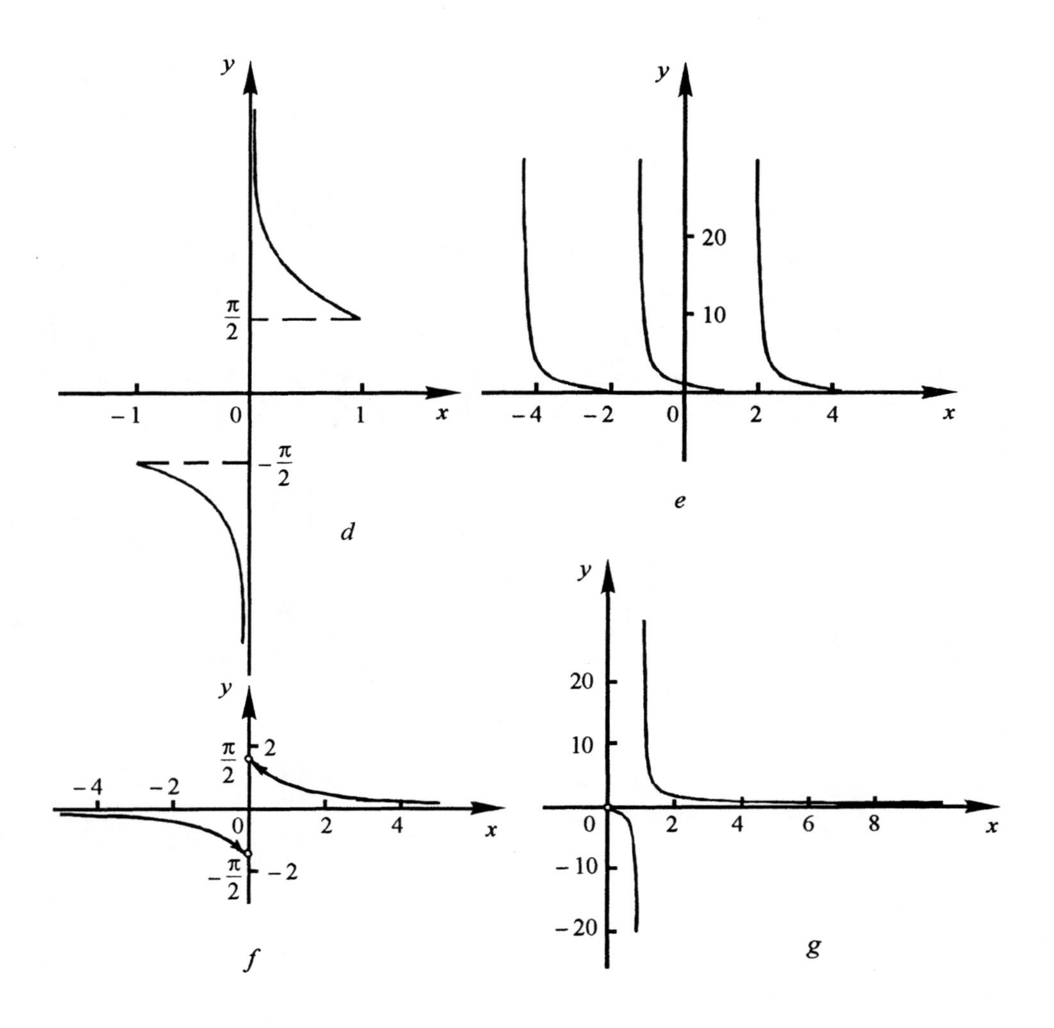

Figure 2.30. The graphs of functions: (*d*) $y = \dfrac{1}{\arcsin x}$; (*e*) $y = \left(\dfrac{1}{3}\right)^{\tan x}$; (*f*) $y = \log_x 3$; (*g*) $y = \arctan\dfrac{1}{x}$.

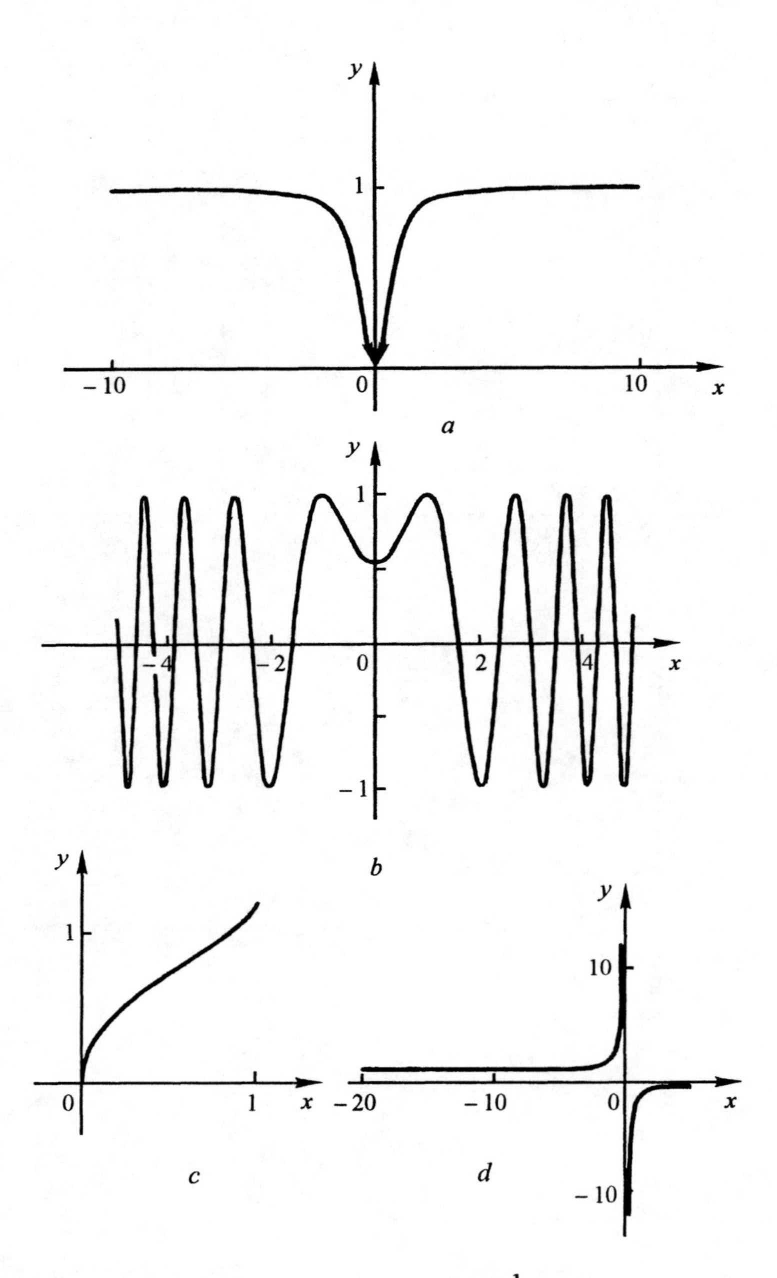

Figure 2.31. The graphs of functions: (*a*) $y = x^2 \sin^2 \frac{1}{x}$; (*b*) $y = \cos(1 - x^2)$;

(*c*) $y = \sqrt{\arcsin x}$; (*d*) $y = \frac{1}{1 - 2^x}$.

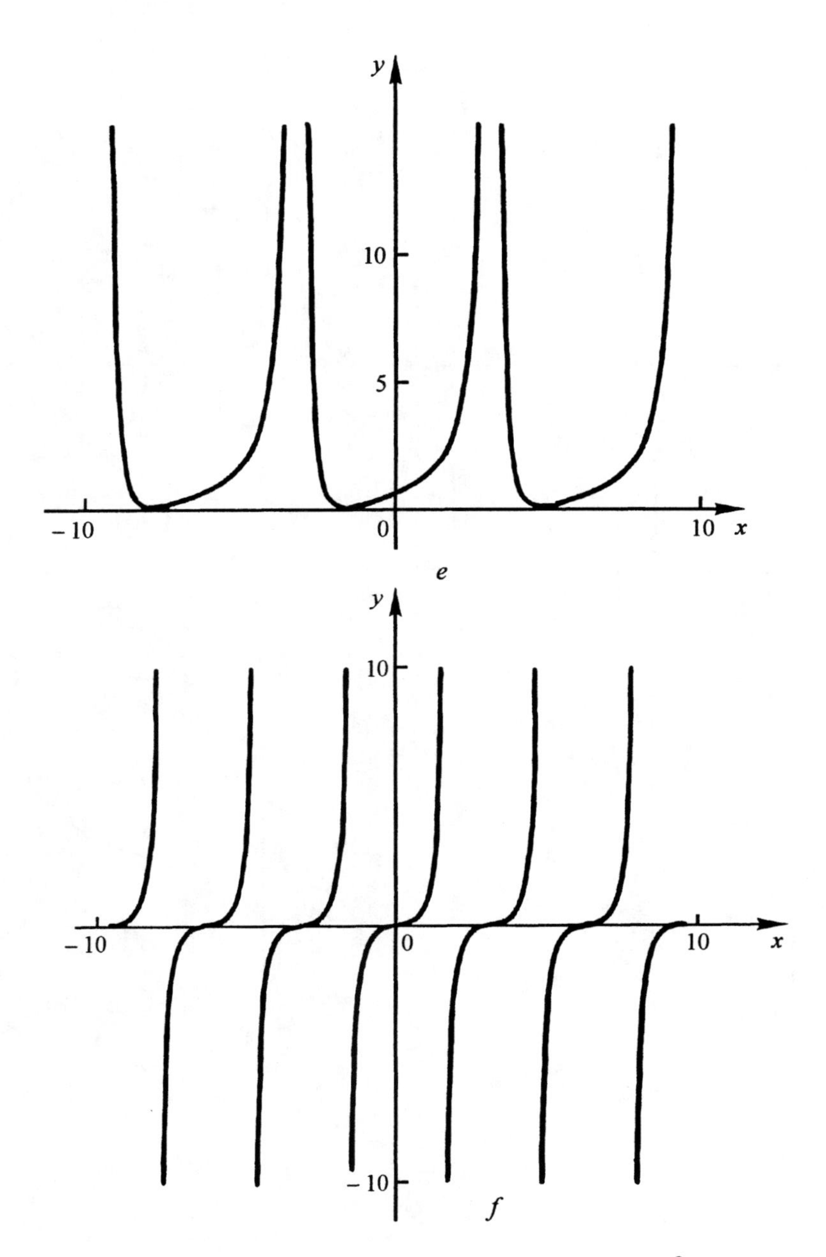

Figure 2.31. The graphs of functions: (*e*) $y = \dfrac{1+\sin x}{1+\cos x}$; (*f*) $y = \dfrac{\sin^3 x}{\cos x}$.

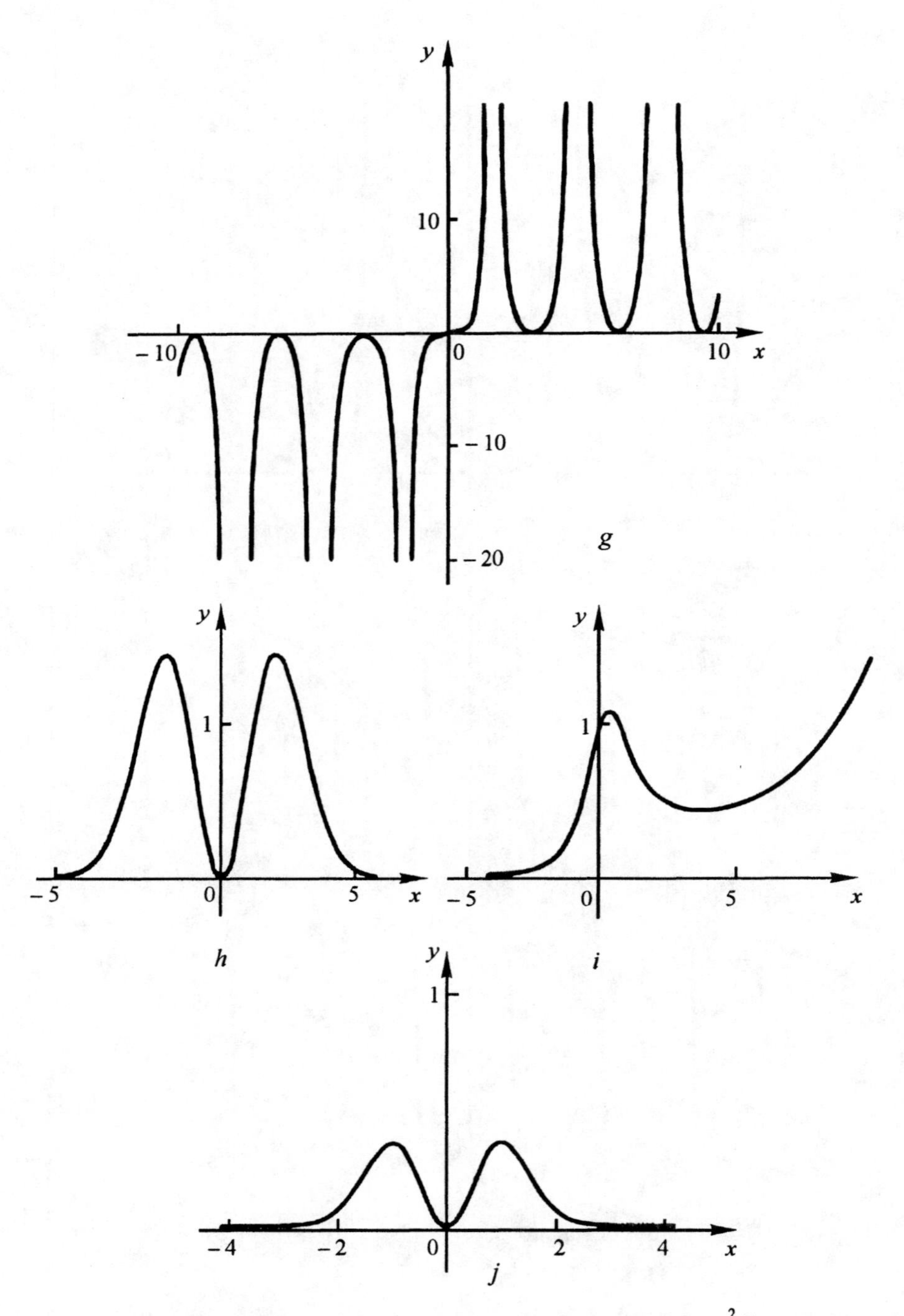

Figure 2.31. The graphs of functions: (g) $y = x\tan^2 x$; (h) $y = x^2 e^{-x^2/4}$;
(i) $y = \dfrac{e^{x/2}}{1+x^2}$; ($j$) $y = x^2 e^{-x^2}$.

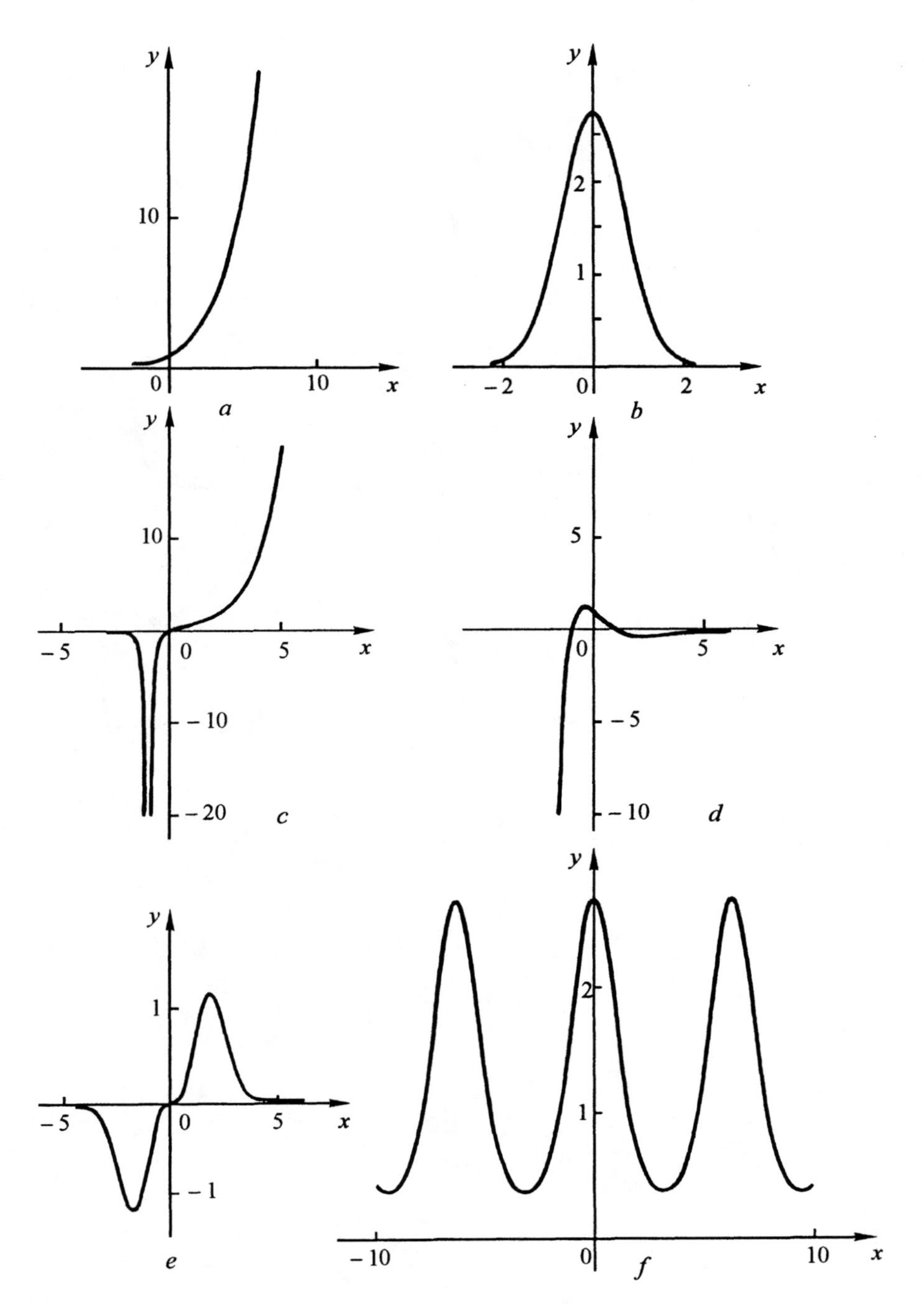

Figure 2.32. The graphs of functions: (*a*) $y = \dfrac{e^x}{\sqrt{1+e^x}}$; (*b*) $y = e^{1-x^2}$;

(*c*) $y = \dfrac{xe^x}{(1+x)^2}$; (*d*) $y = (1-x^2)e^{-x}$; (*e*) $y = x^3 e^{x^2/2}$; (*f*) $y = e^{\cos x}$.

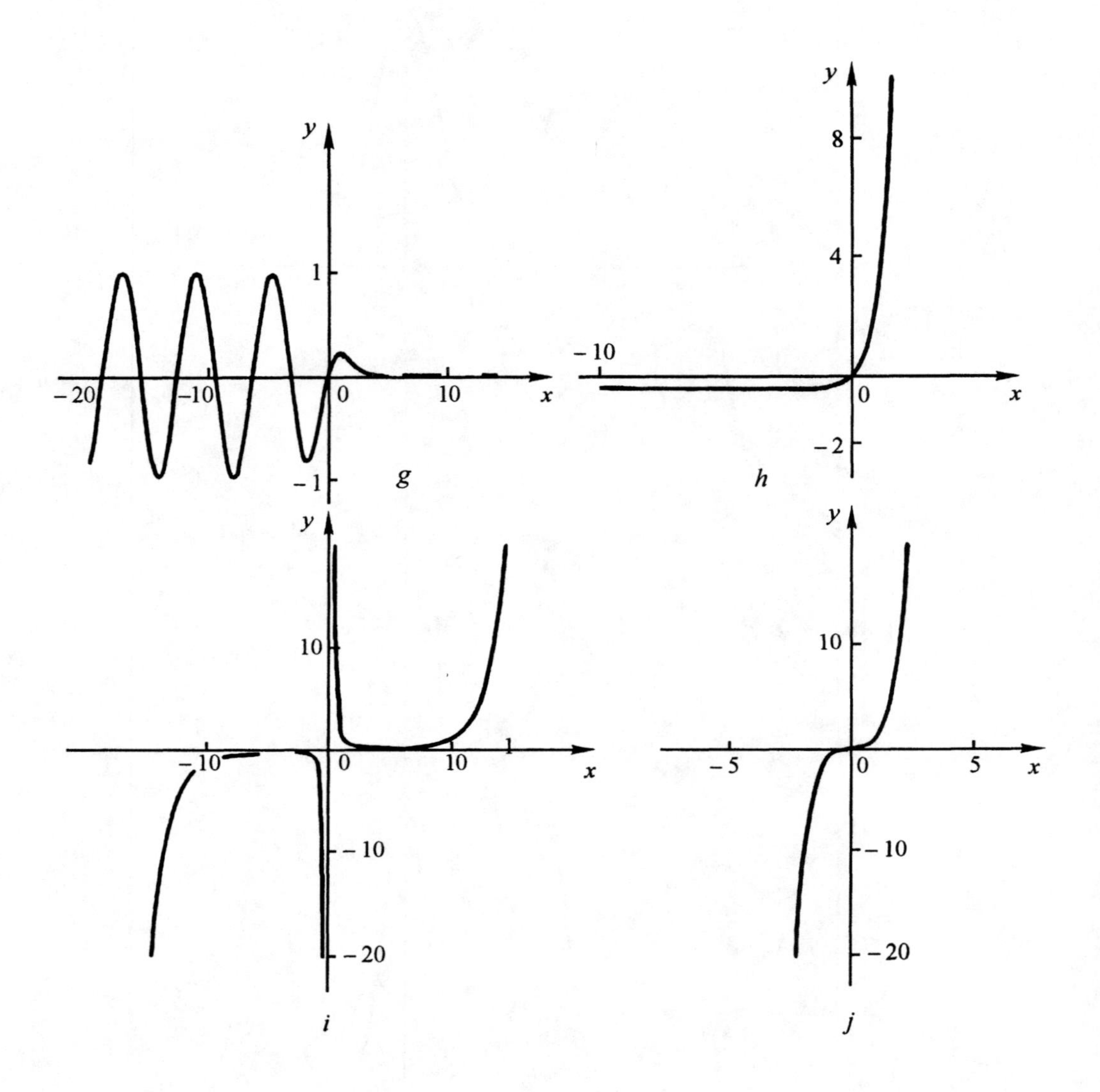

Figure 2.32. The graphs of functions: (*g*) $y = \dfrac{\sin x}{1 + e^x}$; (*h*) $y = e^x \sinh x$;

(*i*) $y = \dfrac{\sinh x}{x^4}$; (*j*) $y = \dfrac{\sinh^3 x}{\cosh x}$.

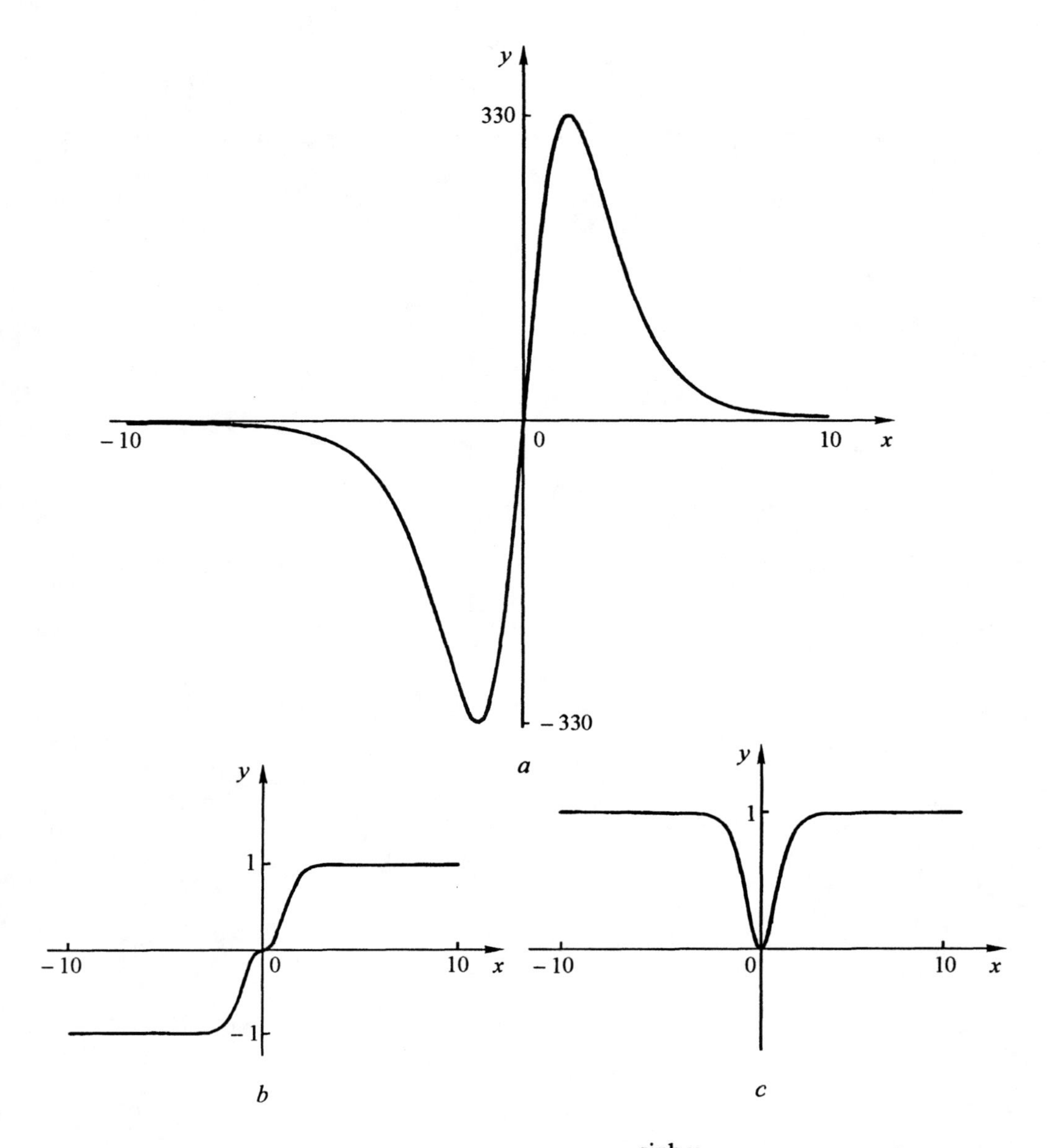

Figure 2.33. The graphs of functions: (*a*) $y = \dfrac{\sinh x}{(\cosh x + 1)\cosh x}$; (*b*) $y = \tanh^3 x$; (*c*) $y = \tanh^2 x$.

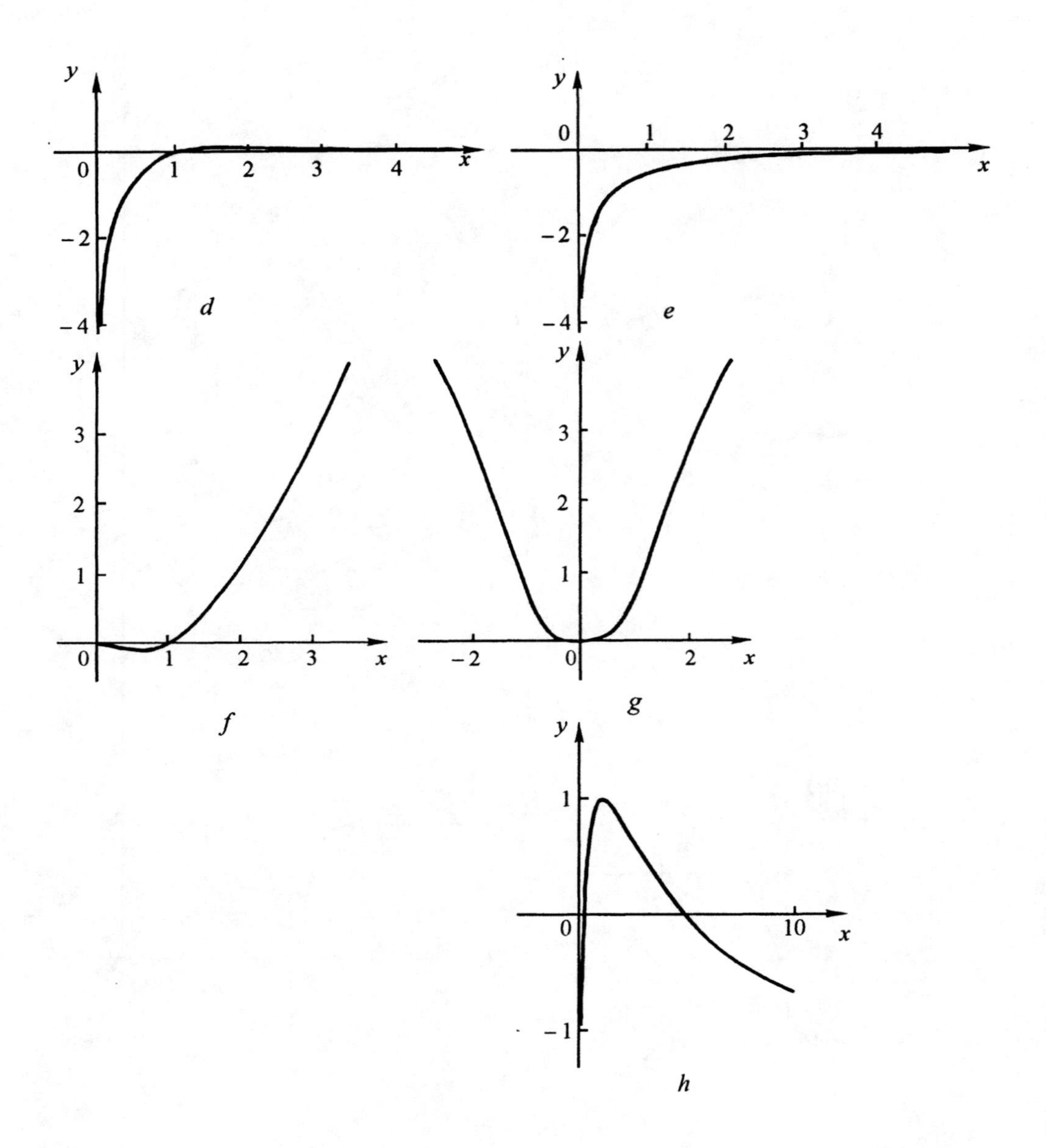

Figure 2.33. The graphs of functions: (*d*) $y = \dfrac{\ln x}{1+x^3}$; (*e*) $y = \dfrac{\ln x}{1-x^2}$; (*f*) $y = \dfrac{x^3 \ln x}{1+x^2}$;

(*g*) $y = \ln(1+x^4)$; (*h*) $y = \cos \ln x$.

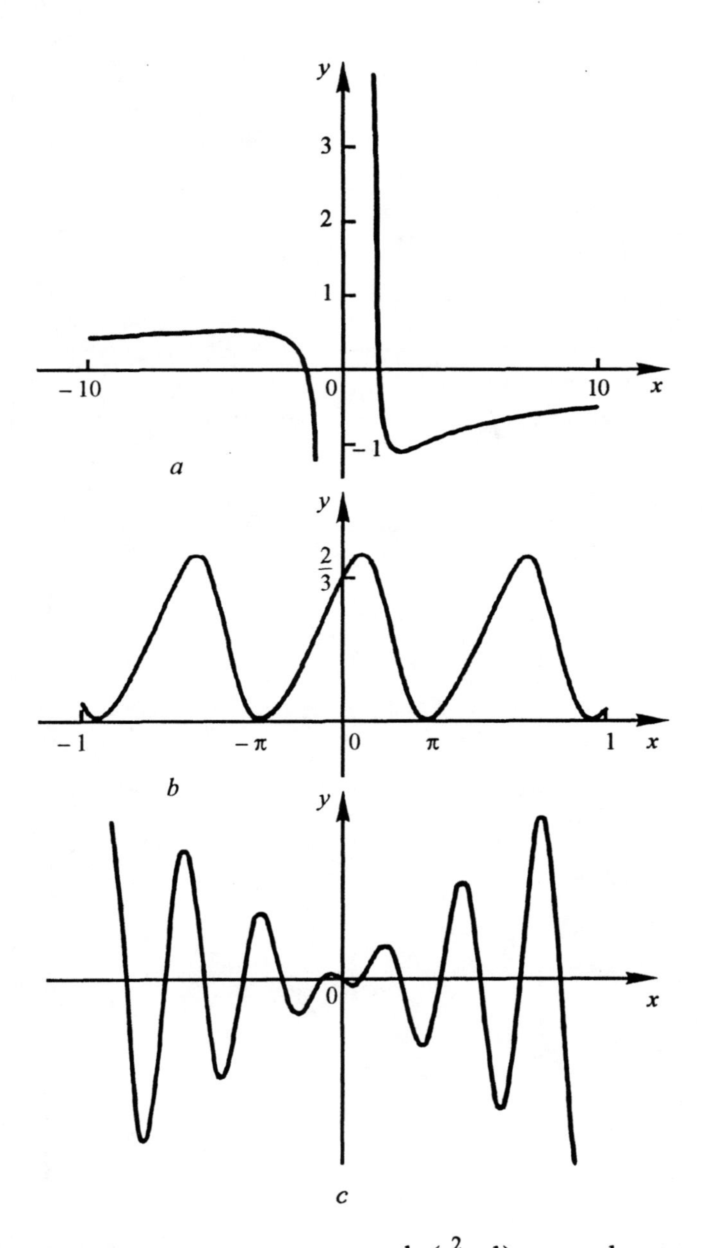

Figure 2.34. The graphs of functions: (*a*) $y = \dfrac{\ln(x^2 - 1)}{1 - x}$; (*b*) $y = \dfrac{1 + \cos x}{3 - \sin x}$;

(*c*) $y = -x \cos x$.

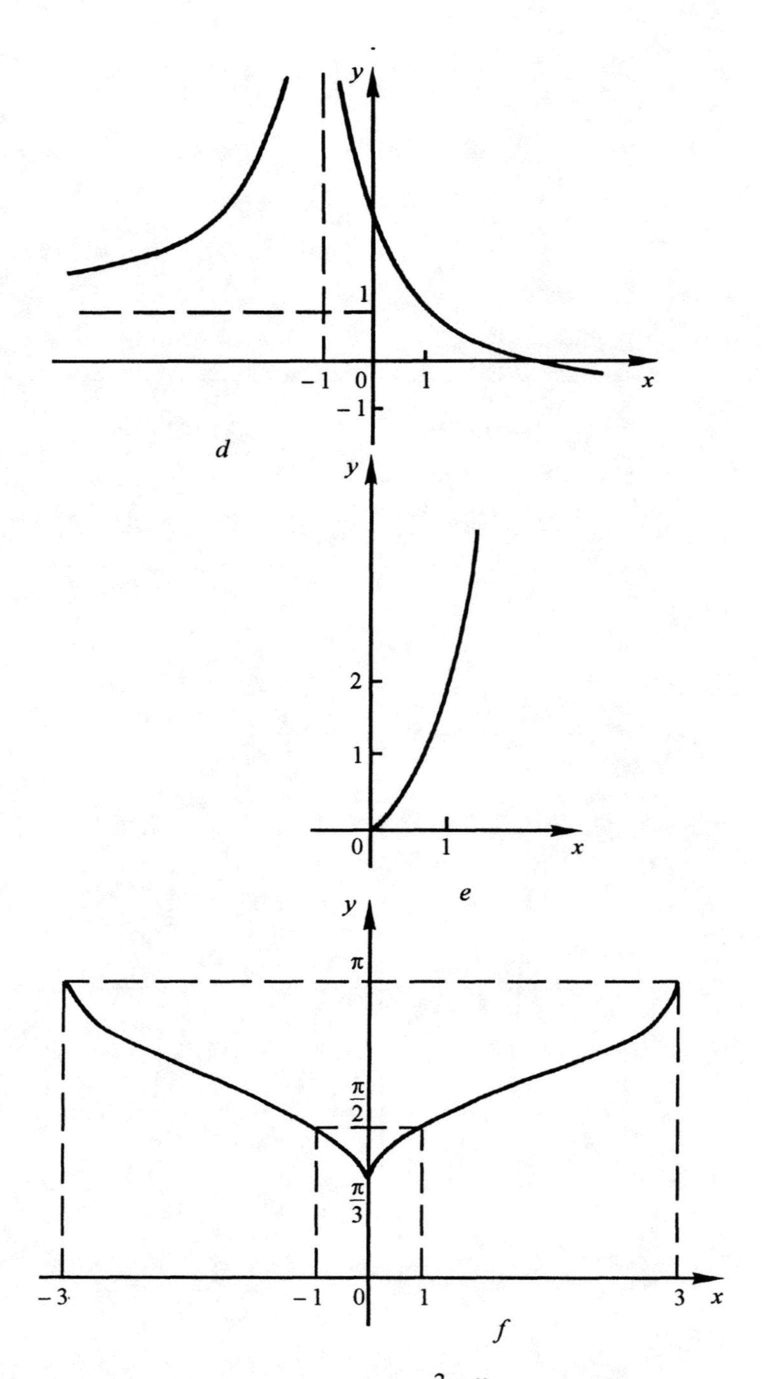

Figure 2.34. The graphs of functions: (*d*) $y = \dfrac{3-x}{|1+x|}$; (*e*) $y = x\sqrt{x} + |x|$;

(*f*) $y = \arccos\left(\dfrac{1-|x|}{2}\right)$.

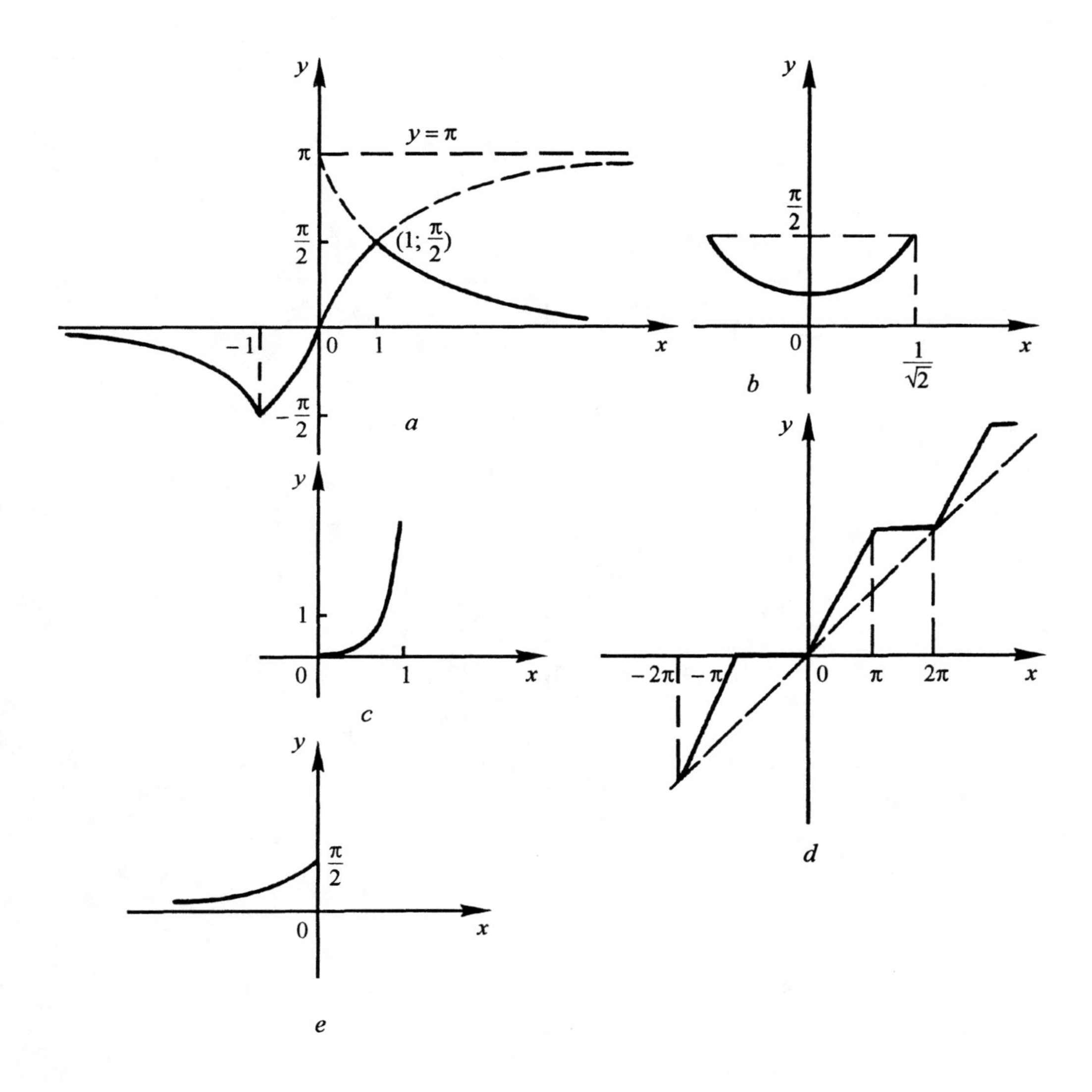

Figure 2.35. The graphs of functions: (a) $y = \arcsin \dfrac{2x}{1+x^2}$; ($b$) $y = \arcsin\left(\dfrac{1}{2} + x^2\right)$; ($c$) $y = \dfrac{\arcsin x}{\sqrt{1-x^2}}$; ($d$) $y = x + \arccos(\cos x)$; (e) $y = \arcsin\left(\dfrac{4}{3}\right)^x$.

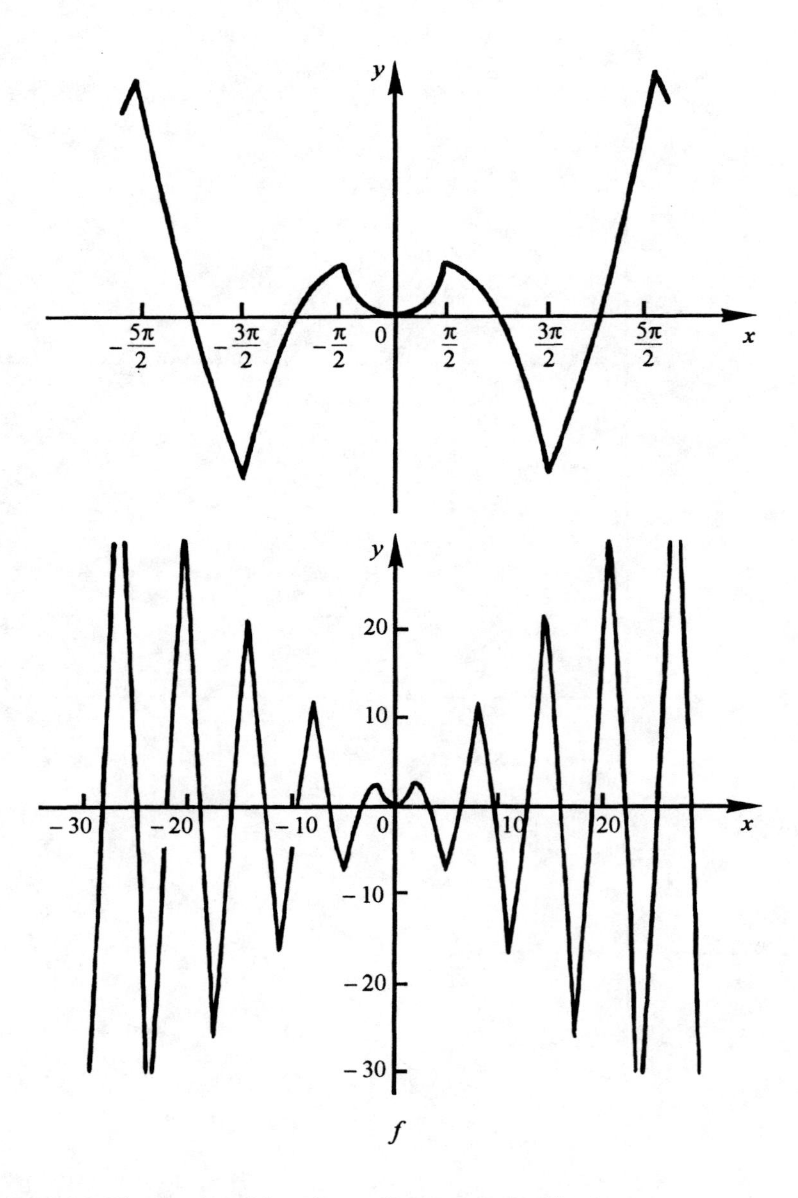

Figure 2.35. The graphs of functions: (*f*) $y = x \arcsin(\sin x)$.

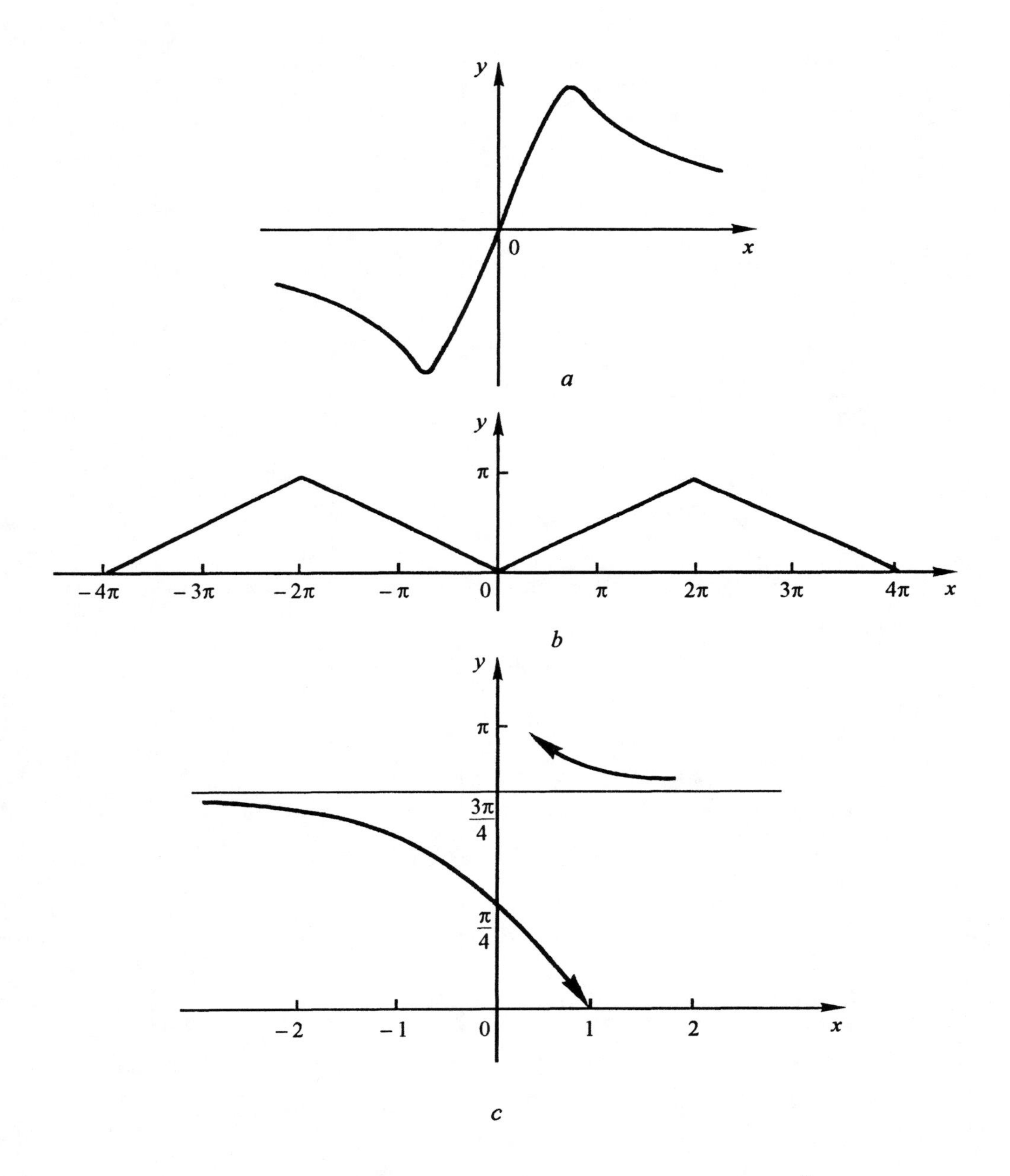

Figure 2.36. The graphs of functions: (*a*) $y = \arcsin \frac{3x}{1+x^2}$; (*b*) $y = \arccos(\cos \frac{x}{2})$; (*c*) $y = \arctan \frac{x+1}{1-x}$.

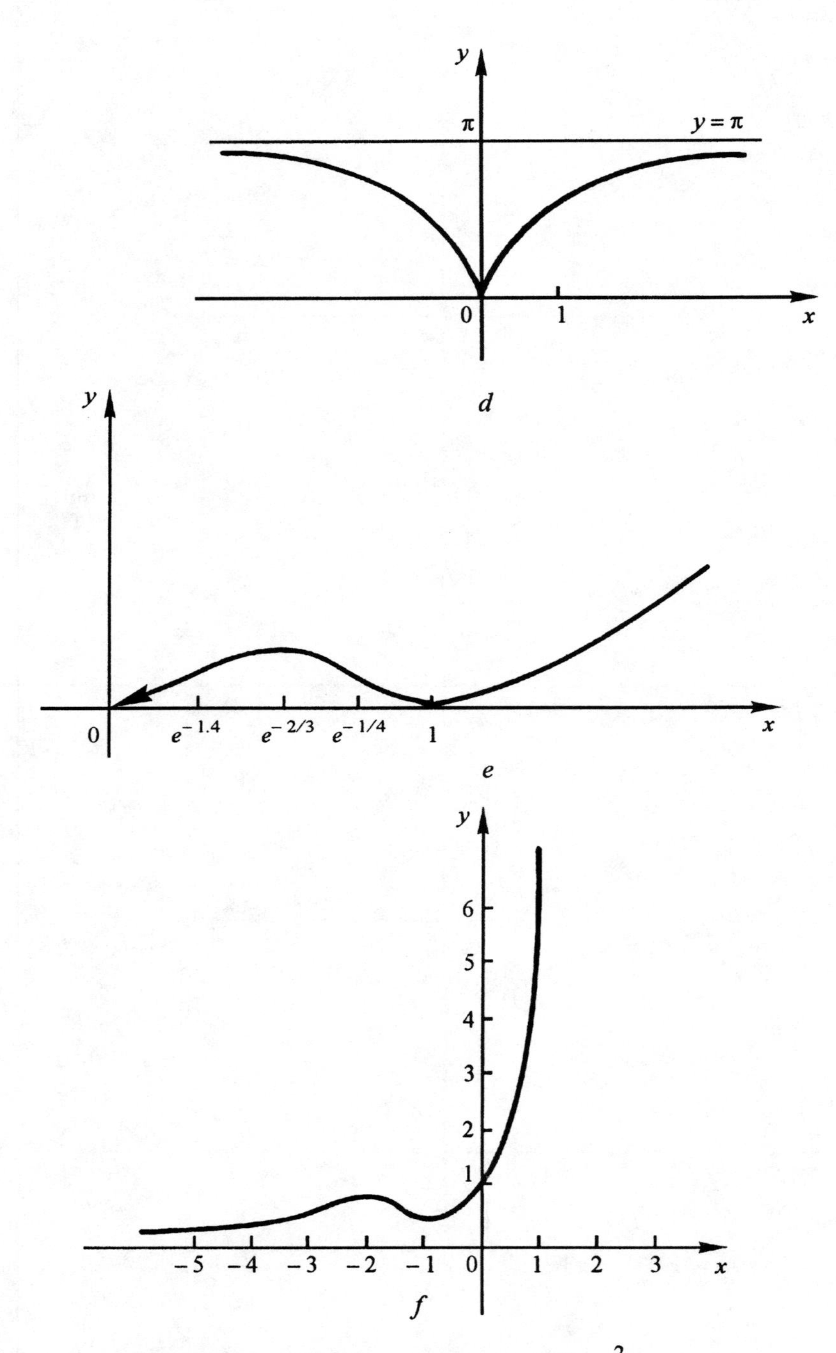

Figure 2.36. The graphs of functions: (*d*) $y = \arccos \dfrac{1-x^2}{1+x^2}$; (*e*) $y = x^3 \ln^2 x$; (*f*) $y = (1+x^2)e^x$.

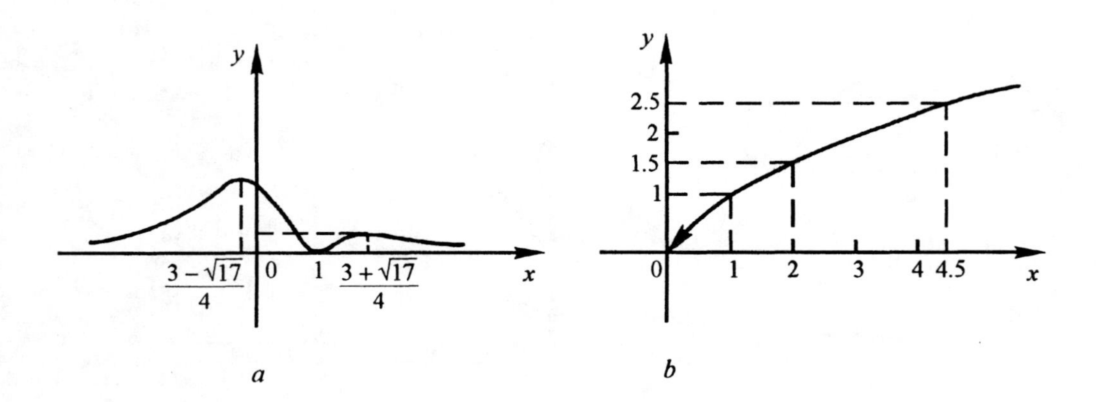

Figure 2.37. The graphs of functions: (*a*) $y = \frac{\sqrt[3]{(x-1)^2}}{x^2+1}$; (*b*) $y = x^2 \ln(x+2)$.

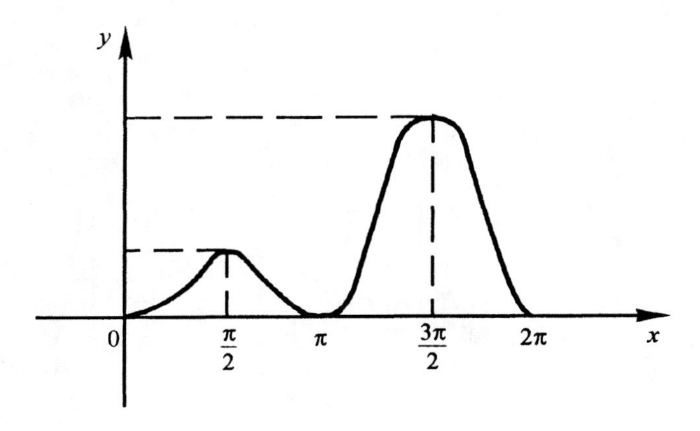

Figure 2.38. The graph of the function $y = \frac{\sin^2 x}{2 + \sin x}$.

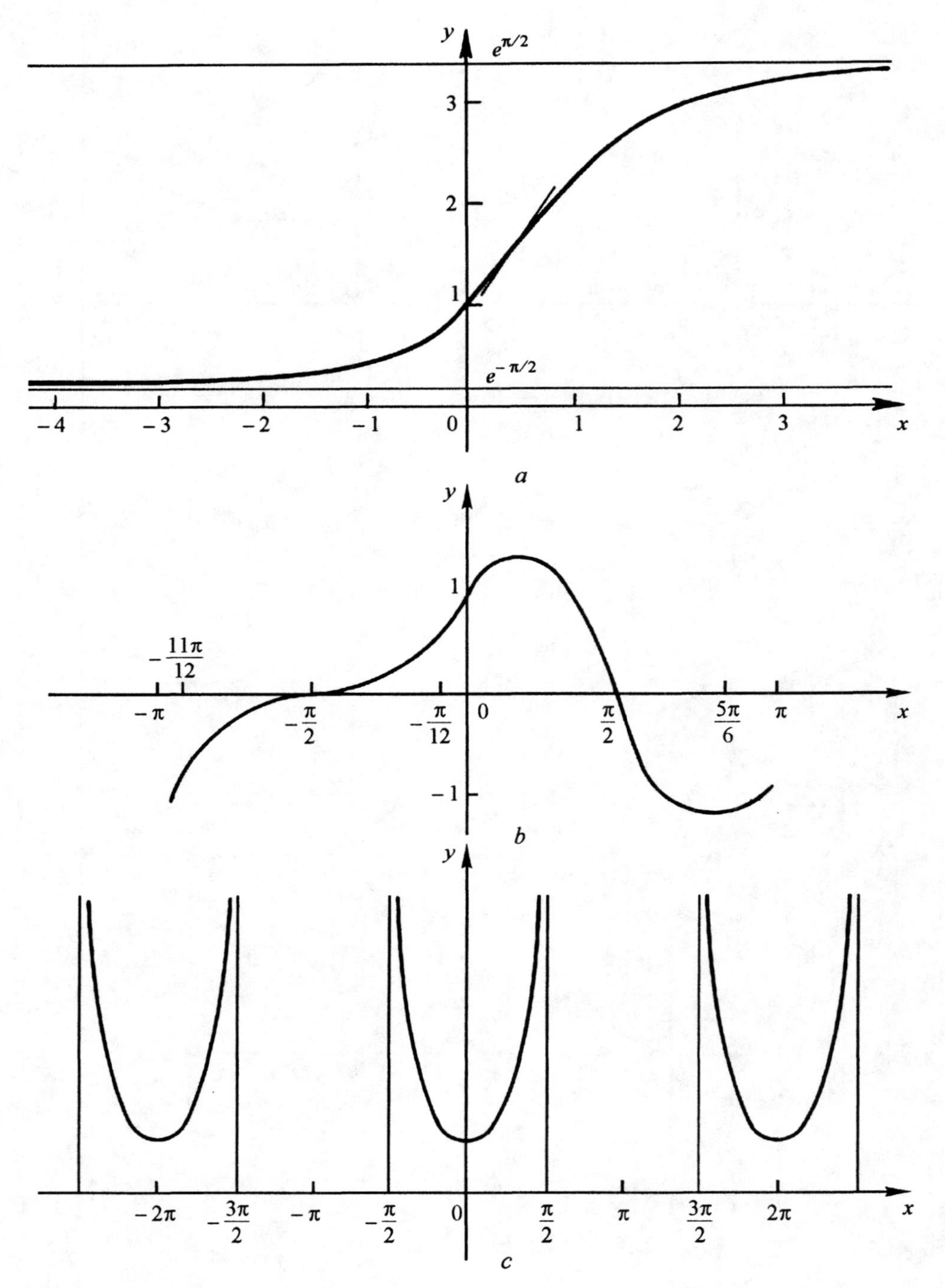

Figure 2.39. The graphs of functions: (*a*) $y = e^{\arctan x}$; (*b*) $y = \frac{1}{2}\sin 2x + \cos x$; (*c*) $y = \cos x - \ln \cos x$.

Singular points. A point $(x_0; y_0)$ of a curve $\begin{cases} x = \varphi(t) \\ y = \psi(t) \end{cases}$ is called *singular* if $x(t_0) = x_0$, $y(t_0) = y_0, x'(t_0) = 0, y'(t_0) = 0$ for some $t = t_0$. Other points such that at least one of the derivatives $\varphi(t)$ or $\psi(t)$ is not zero are called *nonsingular*. At any nonsingular point $(x_0; y_0)$, an equation of the tangent is given by

$$y - y_0 = \frac{\psi'(t_0)}{\varphi'(t_0)}(x - x_0), \quad \text{if} \quad \varphi'(t_0) \neq 0$$

$$x - x_0 = \frac{\varphi'(t_0)}{\psi'(t_0)}(y - y_0), \quad \text{if} \quad \psi'(t_0) \neq 0.$$

Determining points at which the tangent is parallel to an axis. A sufficient condition for a tangent to the curve to be parallel to the x-axis, as well known, is the condition that $y'_x = 0$, i.e. the values of t that correspond to such points are found from the equation $y'_t = 0$, if $x'_t \neq 0$; the values of t such that the tangent is parallel to the y-axis are found from the condition $x'_t = 0$, if $y'_t \neq 0$.

Inflection points. Inflection points are determined from the equation $x'_t y''_t - y'_t x''_t = 0$, if $x'_t \neq 0$.

Determining a double point. A point $(x_0; y_0)$ is a *double point* if it is a point of intersection of two branches of the curve that have different tangents. This implies that there exist two different values of the parameter t, t_1 and t_2, such that $x_1 = \varphi(t_1) = \varphi(t_2)$, $y_1 = \psi(t_1) = \psi(t_2)$. Solving this system of equations, we find values of the tangents to the curve that correspond to the found values of t, i.e. $\psi'(t_1)/\varphi'(t_1)$, $\psi'(t_2)/\varphi'(t_2)$. If these values are different, then the curve has a double point.

Determining a turning point. A *turning point* is determined from the equations $\varphi'(t) = 0$ and $\psi'(t) = 0, x'''_t y''_t - y'''_t x''_t \neq 0$, i.e. this is a point at which the slope of the tangent is undefined.

Asymptotes of the curve. The curve has a vertical asymptote $x = a$, if there exists t_1 (finite or not) such that $\lim\limits_{t \to t_1} \varphi(t) = a$, $\lim\limits_{t \to t_1} \psi(t) = \infty$, or $\lim\limits_{t \to t+1+0} \varphi(t) = a$, $\lim\limits_{t \to t_1+0} \psi(t) = \infty$, or $\lim\limits_{t \to t+1-0} \varphi(t) = a$, $\lim\limits_{t \to t_1-0} \psi(t) = \infty$. The curve has a horizontal asymptote $y = b$, if there exists t_1 (finite or not) such that $\lim\limits_{t \to t_1-0} \varphi(t) = \infty$, $\lim\limits_{t \to t_1-0} \psi(t) = b$, or $\lim\limits_{t \to t_1+0} \varphi(t) = \infty$, $\lim\limits_{t \to t_1+0} \psi(t) = b$, or $\lim\limits_{t \to t_1} \varphi(t) = \infty$, $\lim\limits_{t \to t_1} \psi(t) = b$.

To find an oblique asymptote $y = kx + b$, $k \neq 0$, it is necessary to find such a value t_1 (finite or not) so that $x = \infty$, $y = \infty$. If such a value of t is found, then k and b are determined from the formulas: $k = \lim\limits_{t \to t_1} \frac{\varphi(t)}{\psi(t)}$, $b = \lim\limits_{t \to t_1} [\psi(t) - k\varphi(t)]$, if both limits exist.

2.2.2. Graphs of parametrically defined functions

Example 2.2.1. Plot the graph of the function $x = \frac{t^2}{t^2+1}$, $y = \frac{t(1-t^2)}{t^2+1}$. The domain of the function is [0; 1). Indeed, the value of $x = \frac{t^2}{t^2+1}$ will be positive and less than one. Hence the curve is located between the straight lines $x = 0$ and $x = 1$.

Range of the function is $(-\infty; +\infty)$. Since $x(t)$ is even, and $y(t)$ is odd, the curve is symmetric with respect to the x-axis.

Let us find points in which the curve intersects the axes: if $t = 0$, $x = 0$ and $y = 0$, hence the curve passes through the point (0; 0); if $t = 1$, we have that $y = 0$, and the curve passes through the point (1/2; 0).

Calculating the derivatives we have:

$$x'_t = \frac{2t}{(t^2+1)^2}, \quad y'_t = \frac{1-4t-t^4}{(t^2+1)^2}, \quad y'_x = \frac{1-4t-t^4}{2t}.$$

Two values $t = \pm\sqrt{-2+\sqrt{5}}$ determine points A and B at which the tangent is parallel to the x-axis. We also have that, for $0 < x < 1/2$, $y > 0$, $d^2y/dx^2 < 0$, which means that the curve is convex, and $d^2y/dx^2 > 0$ for $1/2 < x < 1$, $y > 0$, and the curve is concave.

Let us check if the curve has a double point. We have:

$$\frac{t_1^2}{t_1^2+1} = \frac{t_2^2}{t_2^2+1}, \quad \frac{t_1(1-t_1^2)}{t_1^2+1} = \frac{t_2(1-t_2^2)}{t_2^2+1}.$$

Solving this system we find two different values $t_1 = 1$ and $t_2 = -1$ for which the values of x and y are the same. Hence, the point (1/2; 0) is a double point of the curve. Let us calculate the slopes k_1 and k_2 of the tangents to corresponding branches that intersect at this point:

$$k_1 = \frac{y'(t_1)}{x'(t_1)} = -2, \quad k_2 = \frac{y'(t_2)}{x'(t_2)} = 2.$$

Consider the limits

$$\lim_{t \to \infty} x(t) = \lim_{t \to \infty} \frac{t^2}{t^2+1} = 1, \quad \lim_{t \to \infty} y(t) = \lim_{t \to \infty} \frac{t(1-t^2)}{t^2+1} = \infty.$$

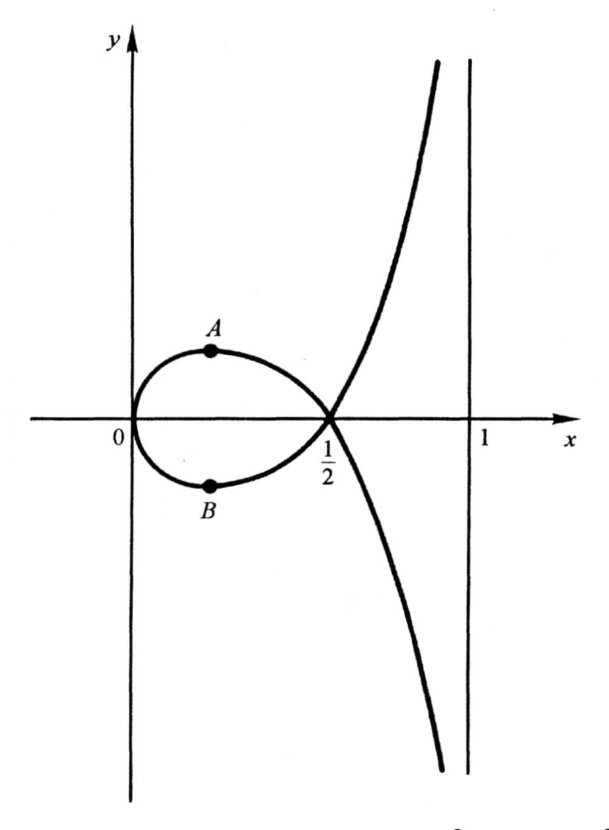

Figure 2.40. The graph of the function $x = \frac{t^2}{t^2+1}, y = \frac{t(1-t^2)}{t^2+1}$.

The curve has only the vertical and no oblique asymptotes. The graph of the curve is shown in Figure 2.40.

Example 2.2.2. Plot the graph of the function $x = 2t - t^2, y = 3t - t^3$.

The domain of the function is $(-\infty; 1]$, its range is $(-\infty; +\infty)$. The functions $x(t)$ and $y(t)$ are continuous for $-\infty < t < +\infty$.

Let us find the first-order derivative: $\frac{dy}{dx} = \frac{3}{2}\frac{1-t^2}{1-t}$. For $t_1 = -1$ $(x_1 = -3)$, $\frac{dy}{dx} = 0$, for $t_2 = 1$ $(x_2 = 1)$, $\frac{dy}{dx} = 3$. Thus the function is decreasing on $(-\infty < x < -3)$ $(-1 < t < 1)$, since $dy/dx < 0$, and increasing on $(-3 < x < 1)$, since $dy/dx > 0$. The function has a minimum, $y_{min} = -2$, for $x = -3$, and a maximum, $y_{max} = 2$, for $x = 1$.

Calculating the second-order derivative we have:

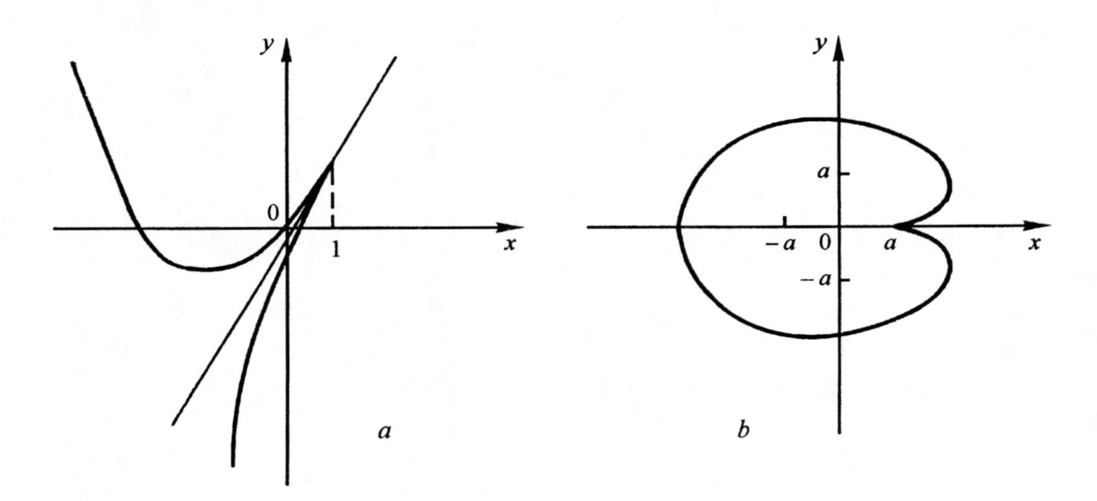

Figure 2.41. The graphs of functions: (*a*) $x=2t-t^2$, $y=3t-t^3$; (*b*) $x=2a\cos t-a\cos 2t, y=2a\sin t-a\sin 2t$.

$$\frac{d^2y}{dx^2}=\frac{3}{4}\frac{(1-t^2)^2}{(1-t)^3}.$$

For $-\infty<x<1, y''_{xx}>0$ and the curve is concave, and if x decreases from 1 to $-\infty$, the curve is convex. The point (1; 2) is an inflection point.

Since $\lim\limits_{t\to\pm\infty} x(t)=-\infty$, $\lim\limits_{t\to\pm\infty} y(t)=\mp\infty$, $\lim\limits_{t\to\pm\infty}\frac{y(t)}{x(t)}=\pm\infty$, the curve does not have asymptotes. Its graph is shown in Figure 2.41.

Example 2.2.3. Plot the graph of the function $x=\frac{a}{\cos^3 t}, y=a\tan^3 t\ (a>0)$.

The domain of the function is $(-\infty; a]\cup[a; +\infty)$. The range of the function is $(-\infty; +\infty)$. The function $x(t)$ is even, $y(t)$ is odd; thus the graph of the function $y=y(x)$ is symmetric with respect to the x-axis. For $0<t<\pi/2$, the graph of the function is symmetric with respect to the y-axis.

Let us calculate the first-order derivative of the function y: $dy/dx=\sin t$, $dy/dx>0$ if $0<t<\pi/2$. Thus the function increases on $0<t<\pi/2$ with $x\to+\infty, y\to+\infty$.

The second-order derivative is $\frac{d^2y}{dx^2}=\frac{1}{3a}\frac{\cos^5 t}{\sin t}, \frac{d^2y}{dx^2}>0$ for $0<t<\pi/2$, and the curve $y(x)$ is concave for $0<t<\pi/2$.

To check if there are turning points we have: $x'(0)=0, y'(0)=0$, or

$$x'(t) = -3a\cos^{-4}t\sin t = 0,\ y'(t) = 3a\tan^2 t\frac{1}{\cos^2 t} = 0,$$

whence $t = 0,\pi$, and also $x''_t\, t'_t - y''_t x'_t \neq 0$ at these points. Hence $(a;\,0)$ and $(-a;\,0)$ are turning points.

Let us find asymptotes:

$$\lim_{t \to \pi/2 - 0} x(t) = \lim_{t \to \pi/2 - 0} \frac{a}{\cos^3 t} = +\infty,$$

$$\lim_{t \to \pi/2 - 0} y(t) = \lim_{t \to \pi/2 - 0} a\tan^3 t = +\infty.$$

There are no vertical asymptotes. Let us check if there are oblique asymptotes:

$$k = \lim_{t \to \pi/2 - 0} \frac{y(t)}{x(t)} = \lim_{t \to \pi/2 - 0} \left(a\tan^3 t \div \frac{a}{\cos^3 t}\right) = 1,$$

$$b = \lim_{t \to \pi/2 - 0} [y(t) - x(t)] = a \lim_{t \to \pi/2 - 0} \left(\tan^3 t - \frac{1}{\cos^3 t}\right) = -\infty.$$

This shows that the graph does not have oblique asymptotes.

Example 2.2.4. Plot the graph of the function $x = \dfrac{t^2}{t-1},\ y = \dfrac{t}{t^2-1}$.

The domain of the function is $(-\infty;\,0) \cup [4;\,+\infty)$. The first-order derivative is

$$\frac{dy}{dx} = \frac{y'_t}{x'_t} = -\frac{t^2+1}{t(t-2)(t+1)^2}.$$

If $-\infty < x \le 0$ (for $0 \le t < 1$), then $y'_x > 0$, and the function is increasing. If $1 < t \le 2$, then $4 \le x < +\infty$, $2/3 \le y < +\infty$, $y'_x > 0$, and the function is also increasing. For $2 \le t < +\infty$, we have that $0 < x \le 4$, $0 < y \le 2/3$, $y'_x < 0$, and the function is decreasing. The function is also decreasing for $-\infty < t < -1$ ($-\infty < x < -1/2$ and $-\infty < y < 0$), since $y'_x < 0$.

The second-order derivative is given by

$$\frac{d^2y}{dx^2} = \frac{2(t-1)^3(t^3+3t+1)}{t^3(t-2)^3(t+1)^3}.$$

For $t_1 \approx -3.22$ ($x(t_1) \approx -0.17$) and $t_2 = 1$ ($y(t_2) \approx 0.37$), $y'' = 0$.

For $-1 < t < t_1$, $y''_{xx} > 0$, and for $t_1 < t < 0$, $y''_{xx} < 0$. This means that the point $P(-0.17;\,0.37)$ is an inflection point of the graph of $y = y(x)$. For $0 \le t < 1$ ($-\infty < x \le 0$), $y''_{xx} > 0$, and the curve is concave. For $1 < t \le 2$ ($4 \le x < +\infty$), $y''_{xx} < 0$, and the curve is concave. For $2 \le t < +\infty$ ($0 < x \le 4$), $y''_{xx} > 0$, and the curve is concave.

Since

$$\lim_{t\to\pm\infty} x(t) = \pm\infty, \quad \lim_{t\to\pm\infty} y(t) = 0,$$

$$\lim_{t\to-1+0} x(t) = \frac{1}{2}, \quad \lim_{t\to-1-0} x(t) = \frac{1}{2},$$

$$\lim_{t\to-1+0} y(t) = +\infty, \quad \lim_{t\to-1-0} y(t) = -\infty,$$

we see that $y = 0$ is a horizontal asymptote for the graph of the function $y = t(x)$, and $x = 1/2$ is its vertical asymptote.

Let us check if there exist oblique asymptotes:

$$k = \lim_{t\to 1} \frac{y}{x} = \lim_{t\to 1} \frac{\dfrac{t}{t^2-1}}{\dfrac{t^2}{t-1}} = \frac{1}{2}, \qquad b = \lim_{t\to 1}\left(\frac{t}{t^2-1} - \frac{1}{2}\frac{t^2}{t-1}\right) = -\frac{3}{4};$$

hence $y = x/2 - 3/4$ is an oblique asymptote for the graph (see Part I, Figure 6.3*c*).

Example 2.2.5. Plot the graph of the function $x = t + e^{-t}$, $y = 2t + e^{-2t}$.

The domain of the function is $[1; +\infty)$; its range is $[1; +\infty)$.

We have that $x(t) = 1 - e^{-t}$, $x(t) = 0$ if $t = 0$; $x'(t) = e^{t}$, and $x'(t) > 0$ for all t. We also have $y(t) = 2(1 - e^{-2t})$, $y(t) = 0$ if $t = 0$, and $y'(t) = 4e^{-2t}$; hence $y'(t) > 0$ for all t. We see that $x_{\min} = 1$, $y_{\min} = 0$ for $t = 0$, and $dy/dx = 2(1 + e^{-t})$.

For $-\infty < t < 0$ $(1 < x < +\infty,\ 1 < y < +\infty)$, $y'_x > 0$, and the function is increasing. For $0 < t < +\infty$ $(1 < x < +\infty,\ 1 < y < +\infty)$, $y'_x > 0$, and the function is increasing. The function $y = y(x)$ attains a minimum $(y_{\min} = 1)$ at the point $x_{\min} = 1$.

Calculate the second-order derivative, $d^2/dx^2 = -2(e^{-t} - 1)$. For $t < 0, y''_{xx} > 0$, and the curve is concave; for $t > 0$, $y''_{xx} < 0$, and the curve is convex. The straight line $y = 2x$ is an oblique asymptote for the graph of the function as $t \to +\infty$. The graph of the curve is shown in Figure 6.20 *c*.

2.3. Graphs of implicitly defined functions

In the first part of the book, implicitly defined functions $F(x, y) = 0$ were studied and their graphs were plotted by using elementary methods. Now we extend the discussed methods for studying and plotting such functions.

Singular points of the curve $F(x, y) = 0$. A point $M(x_0; y_0)$ is called a *singular point* if its coordinates satisfy the following system:

$$F(x_0; y_0) = 0, \quad F_x'(x_0; y_0) = 0, \quad F_y'(x_0; y_0) = 0.$$

The slope of the tangent to the curve, $k = -\dfrac{F_x'}{F_y'}$, is not defined at a singular point.

Suppose that the second-order derivatives F_{xx}'', F_{xy}'', F_{yy}'' are not all zero at a singular point $M(x_0; y_0)$. The point $M(x_0; y_0)$ is a double point of the curve, and the curve, in a vicinity of the critical point, is described by the quantity

$$\Delta = F_{xy}^2(x_0; y_0) - F_{xx}(x_0; y_0)\, F_{yy}(x_0; y_0)\,.$$

If $\Delta > 0$, then the curve has a node (Figure 2.42).

If $\Delta < 0$, then $M(x_0; y_0)$ is an isolated point (Figure 2.43).

If $\Delta = 0$, the point $M(x_0; y_0)$ could be a turning point of the first kind (Figure 2.44) or of the second kind (Figure 2.45), or an isolated point, or a point at which the curve is tangent to itself (Figure 2.46).

If the origin is a double point of an algebraic curve, then the reduced equation of such a curve should not contain the variables with powers zero and one.

Note that equations of tangents at a singular point that coincides with the origin could be found by equating the expression containing terms with the lowest powers of the variables to zero. If the singular point does not coincide with the origin, then equations of tangents can be obtained by translating the singular point into the origin and doing the same as above.

Points of extremums of the curve $F(x, y) = 0$. To find critical points of the curve defined by an equation $F(x, y) = 0$, note that the slope of the tangent at an arbitrary point of the curve is given by $k = -F_x/F_y$.

To find points at which the tangent is parallel to the x-axis, we need to solve the system $F_x(x, y) = 0$, $F(x, y) = 0$. Let x_1, y_1 be a solution of this system, and $F_y(x_1, y_1) \neq 0$. The point $M(x_1; y_1)$ is a point of maximum, y_{max}, if $F_{xx}(x_1; y_1)F_y(x_1, y_1) > 0$, and it is a point of minimum, y_{min}, if $F_{xx}(x_1, y_1)F_y(x_1, y_1) < 0$.

To find points at which the tangent is parallel to the y-axis, it is necessary to solve the system $F_y(x, y) = 0$, $F(x, y) = 0$. Let x_1, y_1 be a solution of this system, and $F_x(x_1, y_1) \neq 0$. Then M is a point of maximum, x_{max}, if $F_{yy}(x_1, y_1) > 0$, and a point of minimum, x_{min}, if $F_{yy}(x_1, y_1)F_x(x_1, y_1) < 0$.

Inflection points of the curve $F(x, y) = 0$. If the equation $F(x, y) = 0$ cannot be solved explicitly with respect to y, then it is difficult to find a general expression for inflection points. Sometimes one can use the following fact: inflection points of an algebraic curve $F(x, y) = 0$ must be points of intersection of the curve with its Hessian, the curve given by the equation $F_{xx}F_y^2 - 2F_{xy}F_xF_y + F_{yy}F_x^2 = 0$.

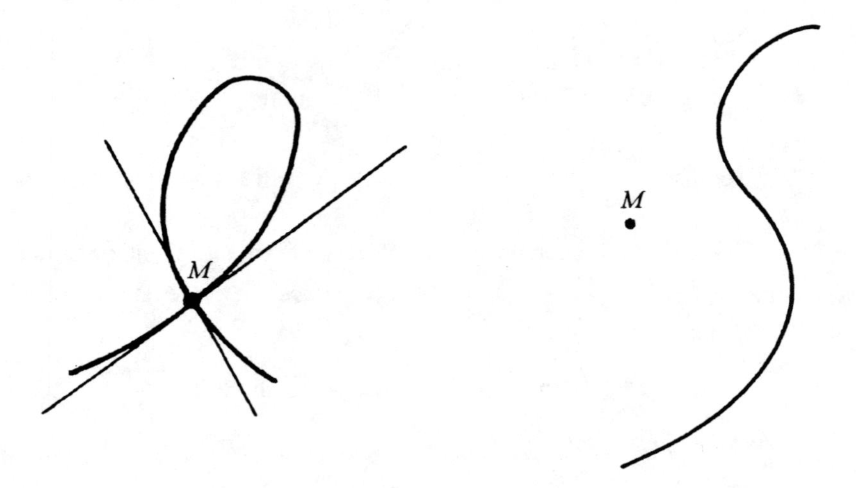

Figure 2.42 Figure 2.43

Figure 2.42. Singular point (node) M of the curve $F(x, y) = 0$.

Figure 2.43. Singular (isolated) point M of the curve $F(x, y) = 0$.

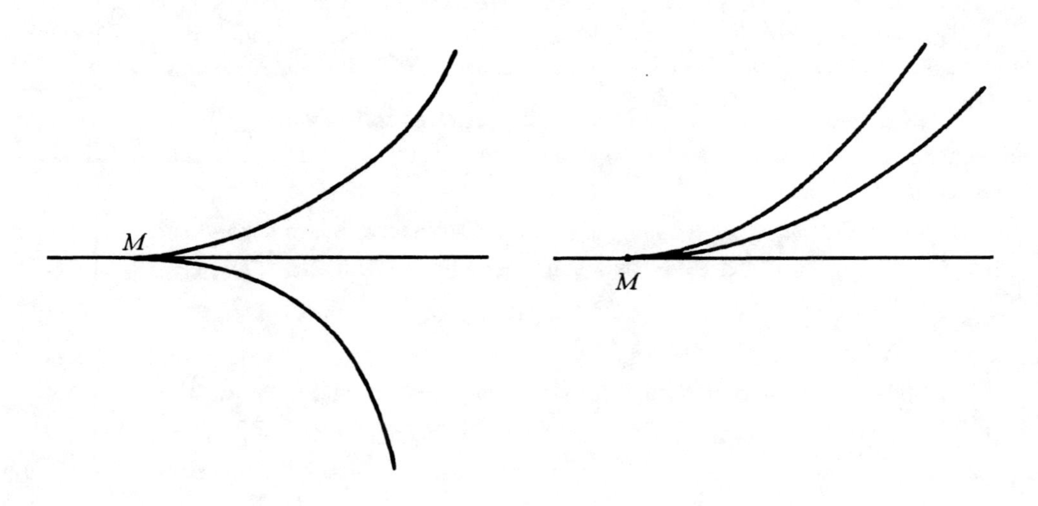

Figure 2.44 Figure 2.45

Figure 2.44. M is a turning point of the first kind of the curve $F(x, y) = 0$.

Figure 2.45. M is a turning point of the second kind of the curve $F(x, y) = 0$.

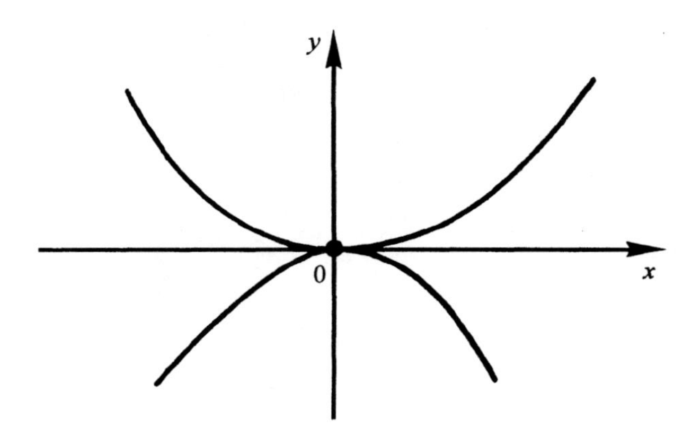

Figure 2.46. M is a point of self-tangency of the curve $F(x, y) = 0$.

Remark 2.3.1. To study and plot the graph of an implicitly defined function, it is sometimes more convenient to write the equations of this curve in a parametric form.

Example 2.3.1. Plot the curve defined by $x^4 - y^4 - x^2 + 2y^2 = 0$.

The curve is symmetric with respect to the coordinate axes. The critical points are found from the system: $F_x = 4x^3 - 2x = 0$, $F_y = -4y^3 + 4y = 0$. Coordinates of points at which the tangent is parallel to the x axis are found by solving the following system: $F(x, y) = x^4 - y^4 - x^2 + 2y = 0$, $F_x(x, y) = 4x^3 - 2x = 0$. We get the points $(0; \pm\sqrt{2})$, $\left(\pm\frac{\sqrt{2}}{2}; \pm\sqrt{\frac{2 \pm \sqrt{3}}{2}}\right)$, $(0; 0)$. Now, $F_{xx} = 12x^2 - 2$, $F_{yy} = -12y^2 + 4$. The point $M_0(0; \sqrt{2})$ is a point of maximum, y_{max}, since $(F_{xx}F_y)|_{x=0,\, y=\sqrt{2}} > 0$. The points $M_{1,2}\left(\pm\frac{\sqrt{2}}{2}; \sqrt{\frac{2+\sqrt{3}}{2}}\right)$ will be points of minimum, y_{min}, since $(F_{xx}F_y)|_{x=\pm\sqrt{2}/2,\, y=\sqrt{2+\sqrt{3}}/2} < 0$.

Coordinates of points at which the tangent is parallel to the y-axis are found by solving the system $F_y = -4y^3 + 4y = 0$, $F = x^4 - y^4 - x^2 + 2y^2 = 0$. We get the points $(\pm 1; 0)$, $(0; 0)$. At the point $A_1(1; 0)$, x reaches a maximum, x_{max}, since $(F_{yy}F_x)|_{x=1,\, y=0} > 0$, and reaches a minimum at $A_2(-1; 0)$, since $(F_{yy}F_x)|_{x=-1,\, y=0} < 0$.

Coordinates of singular points satisfy the system: $F_x = 4x^3 - 2x^2 = 0$, $F_y = -4y^3 + 4y = 0$, $F = x^4 - y^4 - x^2 + 2y^2 = 0$. The point $(0; 0)$ is the only singular point of the curve. Because $\Delta = F_{xy}^2 - F_{xx}F_{yy} > 0$, the point $(0; 0)$ is a node. Solving the equation $x^2 - 2y^2 = 0$, we get the equations of two tangents: $x - \sqrt{2}\,y = 0$, $x + \sqrt{2}\,y = 0$.

Other information about this curve can be obtained by using the methods discussed in Part I.

Example 2.3.2. Plot the curve given by the equation $x^2y^2 = x^3 - y^3$.

Write the curve in a parametric form. Let $y = tx$. Then $x^2t^2 = x^3 - t^3x^3$, whence $x = 1/t^2 - t, y = 1/t - t^2$, and $t \neq 0$.

The functions $x(t) = 1/t^2 - t, y(t) = 1/t - t^2$ are defined on the set $(-\infty; 0) \cup (0; +\infty)$; the function $y = y(x)$ is defined on $(-\infty; +\infty)$.

The curve is symmetric with respect to the straight line $x + y = 0$. Indeed, let $x - y = m$, $x + y = n$. Then it follows from the equation of the curve that $(n^2 - m^2)^2 = 12n^2m + 4m^3$. It follows from this equality that the graph is symmetric with respect the straight line $n = 0$.

Finding the first-order derivative we get

$$\frac{dy}{dx} = \frac{t(1 + 2t^3)}{2 + t^3}, \quad t \neq 0.$$

The derivative does not exist if $t = -\sqrt[3]{2} \approx -1.26$ ($x \approx 2.38$, $y \approx -2.38$), and equals zero, $dy/dx = 0$, if $t = -\sqrt[3]{1/2} \approx -0.79$ ($x \approx 2.38, y \approx -1.89$). For $-\infty < t < -\sqrt[3]{2}$, $\frac{dy}{dx} < 0$ and the function is decreasing; for $-\sqrt[3]{2} < t < -\sqrt[3]{1/2}$, the function is increasing, since $\frac{dy}{dx} > 0$; for $-\sqrt[3]{1/2} < t < 0$, $\frac{dy}{dx} < 0$ and the function decreases; for $0 < t < +\infty$, $\frac{dy}{dx} > 0$ and the function is increasing.

The second-order derivative is

$$\frac{d^2y}{dx^2} = -\frac{2t^3(t^6 + 7t^3 + 1)}{(2 + t^3)^3}, \quad t \neq 0.$$

If $t = -\frac{\sqrt[3]{7 + 3\sqrt{5}}}{2} \approx -1.90$ ($x \approx 2.18$, $y \approx -4.14$) and $t = -\frac{\sqrt[3]{7 - 3\sqrt{5}}}{2} \approx -0.53$ ($x \approx 4.14$, $y \approx -2.18$), we have that $\frac{d^2y}{dx^2} = 0$. The second-order derivative is negative, $\frac{d^2y}{dx^2} < 0$, if $-\infty < t < -\frac{\sqrt[3]{7 - 3\sqrt{5}}}{2}$ or $-\sqrt[3]{2} < t < -\frac{\sqrt[3]{7 - 3\sqrt{5}}}{2}$, and the curve is concave; for $-\frac{\sqrt[3]{7 + 3\sqrt{5}}}{2} < t < -\sqrt[3]{2}$ or $0 < t < +\infty$, the second-order derivative, $\frac{d^2y}{dx^2}$, is positive, and the curve is convex. The graph of the curve is shown in Figure 2.47.

Example 2.3.3. Plot the curve given by the equation $\cosh^2 x - \sinh^2 y - 1 = 0$.

The curve is symmetric with respect to the coordinate axes. For $x > 0, y > 0$, the equation of the curve can be written as $\sinh x = \cosh y$, whence $x = \ln(\cosh y + \sqrt{1 + \cosh^2 y})$.

Finding asymptotes we have:

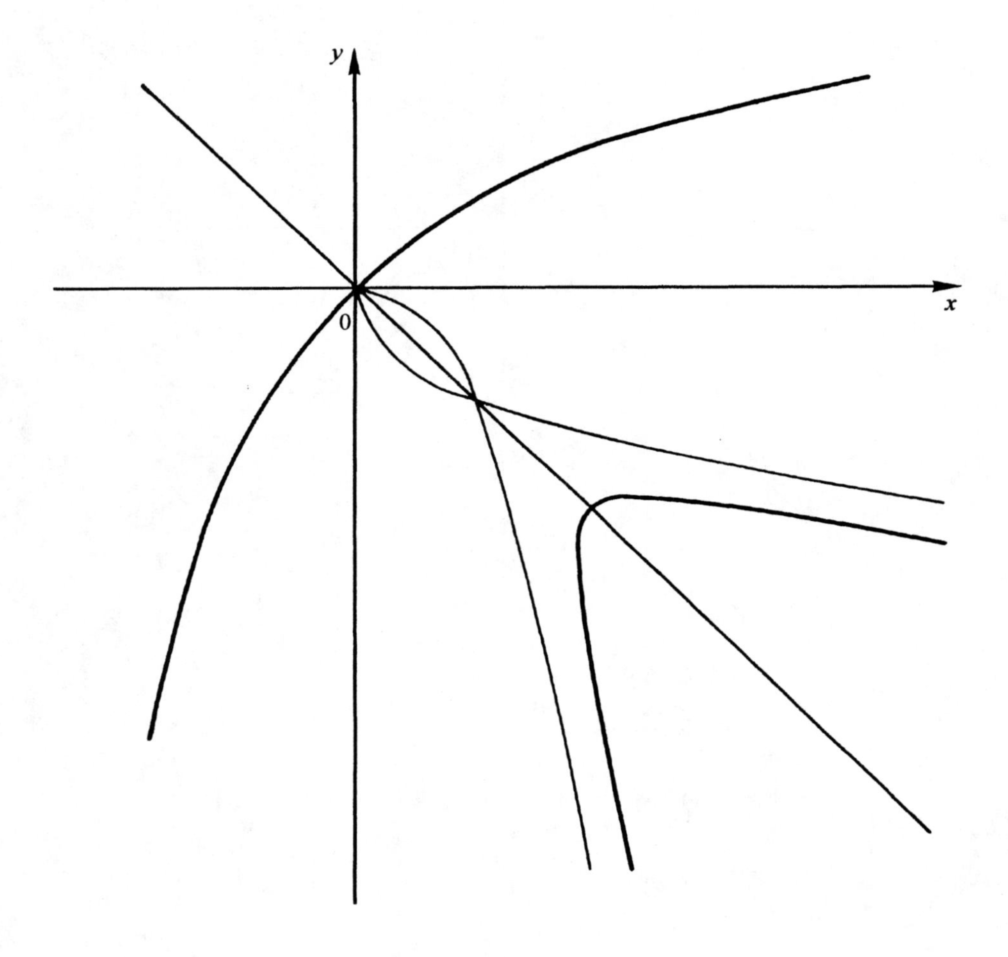

Figure 2.47. The curve $x^2y^2 = x^3 - y^3$.

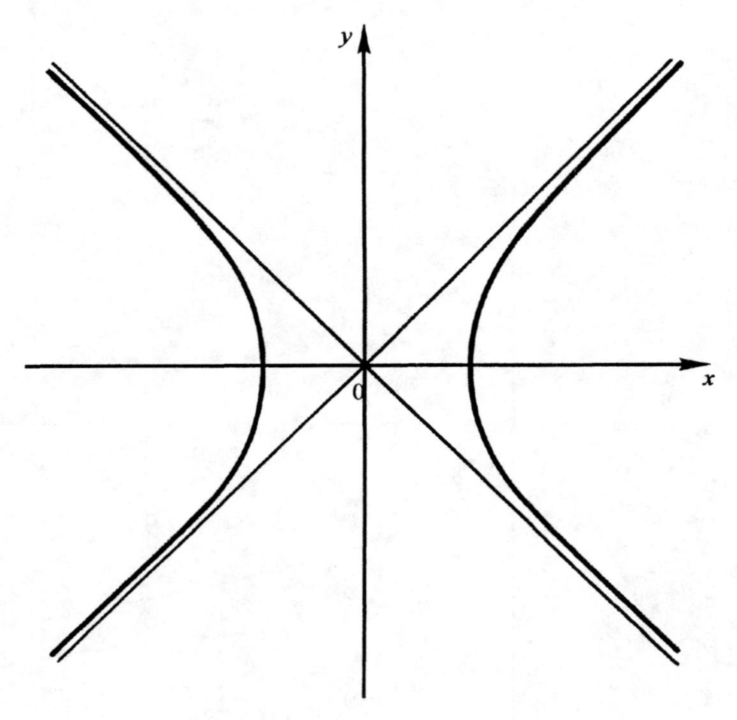

Figure 2.48. The curve $\cosh^2 x - \cosh^{2y} - 1 = 0$.

$$k = \lim_{y \to +\infty} \frac{x(y)}{y} = \lim_{y \to +\infty} \frac{\ln(\cosh y + \sqrt{1 + \cosh^2 y})}{y} = 1,$$

$$b = \lim_{y \to +\infty} [x(y) - y] = \lim_{t \to +\infty} [\ln(\cosh y + \sqrt{1 + \cosh^2 y}) - y] = 0.$$

Thus $y = \pm x$ are oblique asymptotes.

The first-order derivative is

$$\frac{dy}{dx} = \frac{\sinh y}{\sqrt{1 + \cosh^2 y}},$$

and $\frac{dx}{dy} > 0$ if $y > 0$ and thus the function increases, and $\frac{dx}{dy} < 0$ for $y < 0$, and the function decreases. The function reaches a minimum at ($y = 0$; $x = \ln(1 + \sqrt{2})$).

The second-order derivative is given by

$$\frac{d^2x}{dy^2} = \frac{2\cosh y}{(1 + \cosh^2 y)^{3/2}},$$

whence $\frac{d^2x}{dy^2} > 0$ if $y > 0$ (Figure 2.48).

2.4. Graphs of functions in polar coordinates

In the first part of the book, graphs were studied and plotted by using elementary methods. Now using methods of calculus we extend the discussed methods for studying and plotting such graphs in polar coordinates.

Consider a function of the form $\rho = f(\varphi)$, where $-\infty < \varphi < +\infty$. We will plot negative values for ρ at the angle $\varphi + \pi$, i.e. opposite to the direction defined by the angle φ.

If $\rho = f(\varphi)$ is bounded, i.e. $|\rho| \le a$, then the entire curve is located inside the disk of radius a. If $f(-\varphi) = f(\varphi)$, then the curve is symmetric with respect to the pole. In the case where $f(\pi - \varphi) = f(\varphi)$, the curve is symmetric with respect to the straight line which is perpendicular to the polar axis at the pole. If $f(\pi + \varphi) = f(\varphi)$, then the curve is symmetric with respect to the pole.

Extremums of the function $\rho = f(\varphi)$ are found as usual by solving the equation $f'(\varphi) = 0$. For a root of this equation, $\varphi = \varphi_1, f(\varphi_1) \in \rho_{max}$ if $f'(\varphi_1)$ changes its sign from "+" to "–" as φ passes through the value φ_1, and $f(\varphi_1) \in \rho_{min}$ if $f'(\varphi)$ changes sign from "–" to "+" at this point. The tangent at points ρ_{max} and ρ_{min} is perpendicular to the radius vector to this point.

To find points on the curve $\rho = f(\varphi)$ at which the tangent is perpendicular to the polar axis, one needs to make the substitution $\mu = k\pi + \pi/2 - \varphi$ in the formula $\tan\mu = \rho/\rho'$ (see Figure 2.49). Solutions φ of this equation give the needed points.

If the curve has inflection points, then the corresponding φ are roots of the equation

$$\rho^2 + 2\rho'^2 - \rho'\rho'' = 0$$

having odd multiplicity.

The equation of an asymptote in polar coordinates is given by $\rho = p/\cos(\varphi - \gamma)$, where γ is the angle between the polar axis and the perpendicular dropped onto the asymptote from the pole and p is the length of this perpendicular (Figure 2.50).

To determine γ from a given equation, find $\varphi = \varphi_1$ such that $\rho = \infty$. If such φ can be found, the curve has an unbounded branch. Find $p = \lim\limits_{\varphi \to \varphi_1} \dfrac{\varphi^2}{|\varphi'|}$. If this limit exists, then the unbounded branch has an asymptote. Otherwise the unbounded branch of the curve is parabolic. If $p > 0$, the asymptote is located to the right of the radius vector going to infinity, and if $p < 0$, then the asymptote lies to the left of this radius vector.

Example 2.4.1. Plot the graph of the function $\rho = a/\cos^2\varphi$.

The curve has unbounded branches, is symmetric with respect to the polar axis, and perpendicular to it at the pole. The function is periodic with the period $\omega = 2\pi$.

Let us find extremums of the function: $\rho' = 2a\sin\varphi/\cos^3\varphi$, whence $\varphi_1 = 0, \varphi_2 = \pi$. We have $\rho'(0-0) < 0$, $\rho'(0+0) > 0$, and φ_1 is a point of minimum with $\rho_{min} = a$. At the point $\varphi = \pi$, we have $\rho'(\pi - 0) < 0$ and $\rho'(\pi + 0) > 0$, and $\rho_{min} = a$.

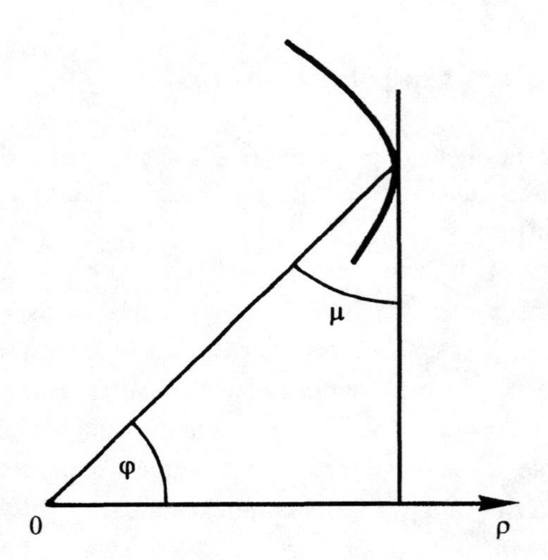

Figure 2.49. The curve $\tan \mu = \rho/\rho'$.

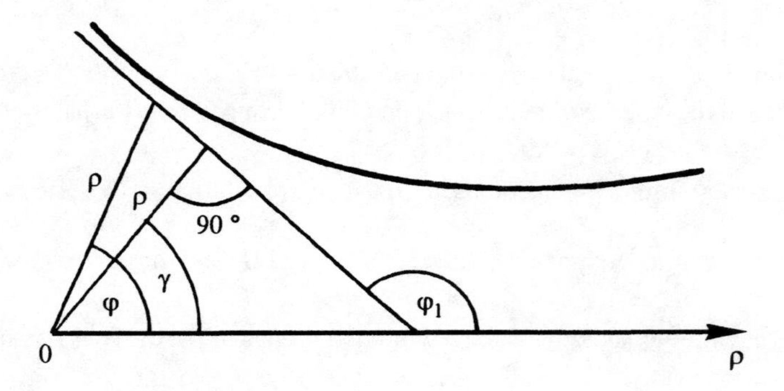

Figure 2.50. Asymptote to the curve $\rho = \rho(\varphi)$, $\rho = p/\cos(\varphi - \gamma)$.

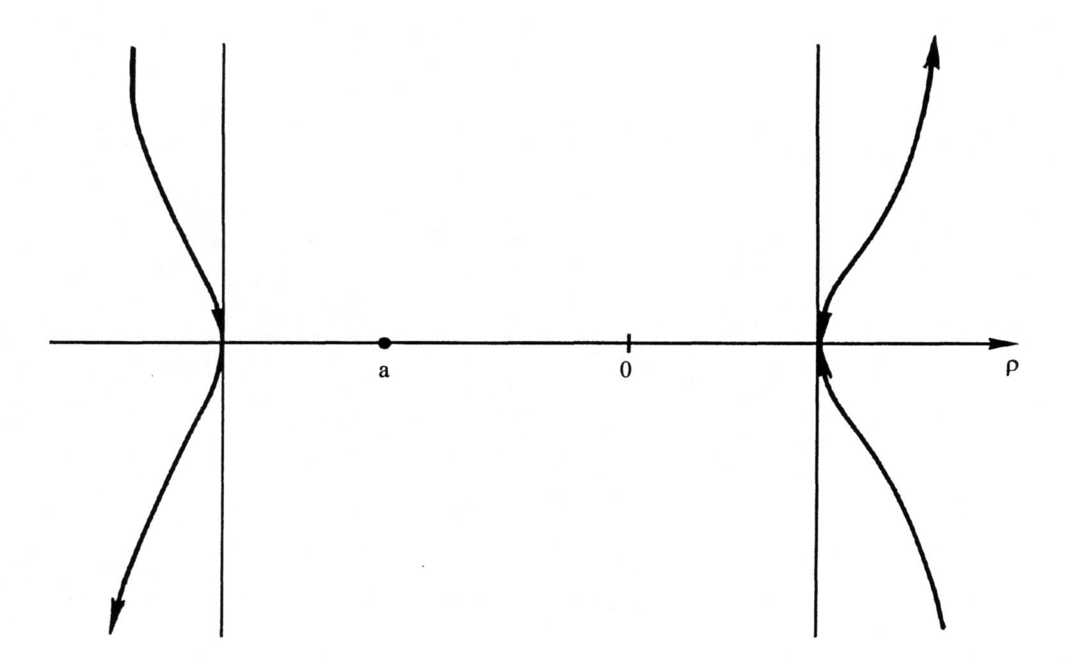

Figure 2.51. The curve $\rho = a/\cos^2\varphi$.

Let us find points at which the tangent to the curve is parallel to the polar axis. Set $\mu = k\pi - \varphi$ in the formula $\tan\mu = \rho/\rho'$. We get $\tan(k\pi - \varphi) = 1/2\cot\varphi$, whence $\tan^2\varphi = -1/2$, and the equation has no real roots.

To find points at which the tangent is perpendicular to the polar axis, set $\mu = k\pi + \frac{\pi}{2} - \varphi$ in the equation $\tan\mu = \rho/\rho'$. We have $\tan(k\pi + \pi/2 - \varphi) = 1/2\cot\varphi$, whence $\varphi_1 = 0$, $\varphi_2 = \pi$. Thus the points at which the tangent is perpendicular to the polar axis are $(a; 0)$ and (a,π).

Solving the equation $\rho^2 + 2\rho'^2 - \rho\rho'' = 0$ we find inflection points. We get $\cos\varphi = \pm\sqrt{2/3}$, i.e. there are four values of φ in $[0; 2\pi]$ that give inflection points.

To find asymptotes, we solve $\rho' = 0$ getting $\varphi = \pi/2$, but since $p = \lim\limits_{\varphi\to\pi/2} \dfrac{a}{\sin 2\varphi} = \infty$, the curve has no asymptotes. The plot of the curve is given in Figure 2.51.

Example 2.4.2. Plot the graph of the function $\varphi = \arccos\dfrac{\rho - 1}{\rho^2}$.

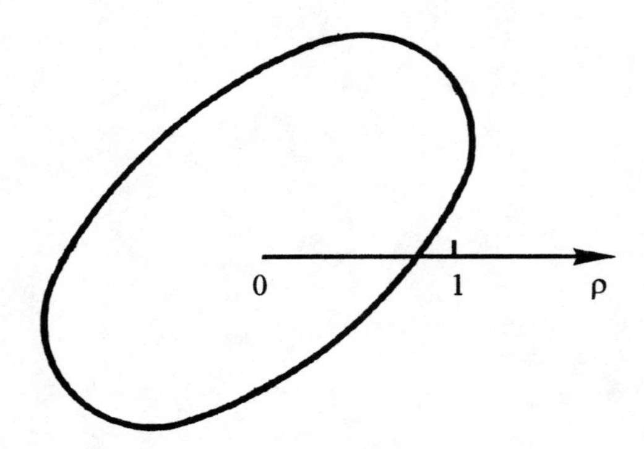

Figure 2.52. The function $\varphi = \arccos \dfrac{\rho - 1}{\rho^2}$.

The domain of the function is determined from $\left|\dfrac{\rho - 1}{\rho^2}\right| \le 1$, whence $-\rho^2 \le \rho - 1 \le \rho^2$ or $\dfrac{-1+\sqrt{5}}{2} \le \rho < +\infty$. We have

$$\lim_{\rho \to \frac{-1+\sqrt{5}}{2}+0} \varphi(\rho) = \lim_{\rho \to \frac{-1+\sqrt{5}}{2}+0} \arccos \frac{\rho - 1}{\rho^2} = \pi,$$

and $\lim\limits_{\rho \to +\infty} \varphi(\rho) = \lim\limits_{\rho \to +\infty} \arccos \dfrac{\rho - 1}{\rho^2} = \pi/2$, $\varphi = \pi/2$.

Because $\rho - 1 \neq \rho^2$, the function $\varphi(\rho)$ has no zeros and takes positive values.

The first-order derivative is

$$\frac{d\varphi}{d\rho} = \frac{\rho - 2}{\sqrt{1 - \left(\frac{1}{\rho} - \frac{1}{\rho^2}\right)^2}},$$

hence $\dfrac{d\varphi}{d\rho} = 0$ for $\rho = 2$, and $\dfrac{d\varphi}{d\rho} < 0$ if $\dfrac{-1+\sqrt{5}}{2} < \rho < 2$, and $\dfrac{d\varphi}{d\rho} > 0$ if $\rho > 2$. Thus the function has a minimum at $\rho = 2$ with $\varphi = \arccos 1/4$. At the point $\rho = \dfrac{-1+\sqrt{5}}{2}$, the derivative does not exist, and the function takes the maximal value π. The graph is shown in Figure 2.52.

Chapter 3

Graphs of certain remarkable curves

3.1. Algebraic curves

To an infinite set of equations in two variables and a given coordinate system there corresponds an infinite set of different kinds of curves. Since the characteristics of a curve are determined by properties of the corresponding equation, it is natural to classify curves on the basis of the nature of the corresponding equation that can be either algebraic or transcendental. It should be noted, however, that the nature of the curve does not depend only on the nature of the corresponding equation but also depends on the coordinate system in which the curve is considered. The same curve could be defined by an algebraic equation in one coordinate system and by a transcendental equation in another system. For example, the equation of a circle with its center at the origin is $x^2 + y^2 = R^2$ with respect to the Cartesian coordinates, and is clearly algebraic. This equation remains algebraic in the polar system, if the center is placed at the pole (the equation is $\rho = R$). However, if the center is located at some other point, $(\rho_1; \varphi_1)$, then the equation is $\rho^2 + \rho_1^2 - 2\rho\rho_1 \cos(\varphi - \varphi_1) = R^2$, and thus becomes transcendental.

There is no such drawback if only Cartesian coordinate systems are considered. Thus it is natural to split curves into algebraic or transcendental depending on whether the corresponding equation is algebraic or transcendental in a Cartesian coordinate system.

Algebraic curves, in their turn, are subdivided into curves of different degrees. The degree of a curve is determined by the highest power of the defining equation.

An *algebraic curve of degree n* is a curve the defining equation of which can be written, after elimination of fractions and radicals, in the form

$$A_0x^n + A_1x^{n-1}y + A_2x^{n-2}y^2 + \ldots + A_ny^n + B_0x^{n-1} + B_1x^{n-2}y + B_2x^{n-3}y^2 + \ldots +$$
$$+ B_{n-1}y^{n-1} + \ldots + C_0x^{n-2} + C_1x^{n-3} + \ldots + C_{n-2}y^{n-2} + \ldots + L_0x + L_1y + M = 0.$$

The number of terms in the equation equals $(n+1)(n+2)/2$. In particular cases some of the coefficients may equal zero.

If the left-hand side of the equation can be factored, with factors being $f_1(x, y), f_2(x, y), \ldots$, then this equation corresponds to the union of curves defined by the equations $f_1(x, y) = 0$, $f_2(x, y) = 0, \ldots$. In such a case the curve is called *decomposable* or *reducible*.

In particular, if the left-hand side of the equation that defines the curve, $f(x, y)$, is a homogeneous function of degree n, then the curve decomposes into the set of straight lines. Indeed, since the function is homogeneous, setting $y = \alpha x$ we obtain $f(x, y) = x^n f(1,\alpha)$, and if $\alpha_1, \alpha_2, \ldots, \alpha_n$ are roots of the equation $f(1,\alpha) = 0$, then

$$f(x, y) = A_0(\alpha - \alpha_1)(\alpha - \alpha_2)\ldots(\alpha - \alpha_n)x^n =$$
$$A_0\left(\frac{y}{x} - \alpha_1\right)\left(\frac{y}{x} - \alpha_2\right)\ldots\left(\frac{y}{x} - \alpha_n\right)x^n = A_0(y - \alpha_1 x)(y - \alpha_2 x)\ldots(y - \alpha_n x).$$

By equating each factor to zero, we obtain a set of n curves (some of which could be imaginary).

The following are some general theorems about algebraic curves.

1. The degree of an algebraic curve does not depend on the location of the curve with respect to the coordinate system.
2. Two irreducible algebraic curves that have degrees m and n, correspondingly, intersect in not more than mn points.
3. An algebraic curve of degree n is determined by its $n(n + 3)/2$ points.
4. Every curve of degree n that passes through $n(n + 3)/2 - 1$ points also passes through another $(n - 1)(n - 2)/2$ points the location of which depends on the position of the initial points.
5. **Newton's theorem:** If two parallel lines are drawn through two given points O and O_1, and if these lines intersect a curve of degree n in points $P_1, P_2, \ldots, P_n$ and $Q_1, Q_2, \ldots, Q_n$, respectively, then the ratio

$$\frac{OP_1 OP_2 \ldots OP_n}{O_1Q_1 O_1Q_2 \ldots O_1Q_n}$$

is a constant and does not depend on the direction of the parallel lines.

6. **Cotes' theorem:** If on every line that passes through a fixed point O, a point P is found such that $n/OP = 1/OP_1 + 1/OP_2 + \ldots + 1/OP_n$, where $P_1, P_2, \ldots, P_n$ are points in which the straight line intersects the curve, then the set of all such points P form a straight line.
7. **Chasles' theorem:** For a family of parallel tangents to a curve, the center of midpoints of distances to the points of tangency does not depend on the direction of the tangents.

Remark 3.1.1. Besides the classification of algebraic curves by the degree of the curve, there are other classifications based on the notions of class, genus, etc.

Before looking at curves of degree n, let us consider some forms of the equation of a straight line (the curve of degree one) (Figure 3.1): $y = kx + b$, where $k = \tan\alpha$, $b = OB$; $x = ly + a$, where $l = \tan\beta$, $a = OA$; $x/a + y/b = 1$, where $a = OA$, $b = OB$, $Ax + By + C = 0$; $x\cos\gamma_1 + y\cos\gamma_2 - p = 0$, $x\cos\gamma_1 + y\sin\gamma_1 - p = 0$, where γ_1 is the angle between the

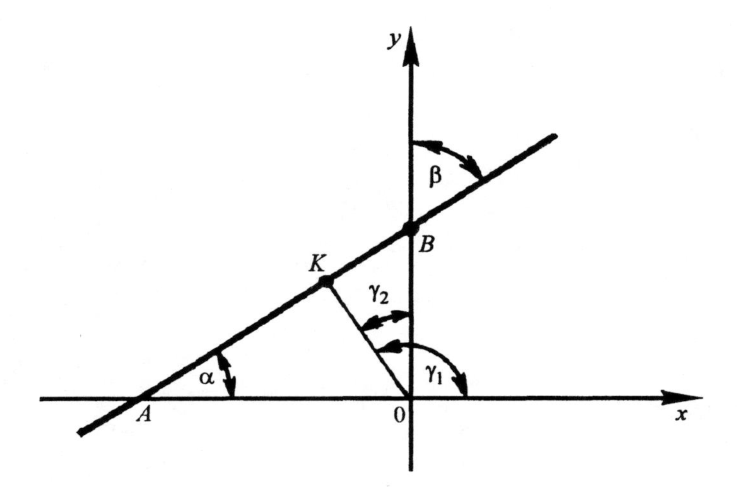

Figure 3.1. Equations of a curve of degree one.

perpendicular dropped to the line from the origin and the x-axis, γ_2 is the angle between this perpendicular and the y-axis, $p = OK$,

$$p = \frac{C}{\pm\sqrt{A^2 + B^2}}, \quad \cos\gamma_1 = \frac{A}{\pm\sqrt{A^2 + B^2}}, \quad \cos\gamma_2 = \sin\gamma_1 = \frac{B}{\pm\sqrt{A^2 + B^2}}.$$

Equations of a straight line in polar coordinates are: $\rho = p/\cos(\varphi - \gamma_1)$, where $p = OC$, $\gamma_1 = \angle AOC$, or $\rho = a\sin\alpha/\sin(\alpha - \varphi)$, where $a = OA$, α is the angle between the straight line and the polar axis (Figure 3.2).

3.2. Curves of degree two

3.2.1. The circle

The circle with center at a point $M(a; b)$ and radius R is the locus of points in the plane such that their distance to a point M is a constant R. The equation of the circle is

$$(x - a)^2 + (y - b)^2 = R^2, \tag{1}$$

where $M = M(a; b)$ is the center and R is the radius (Figure 3.3). The equation of the circle with its center at the origin is $x^2 + y^2 = R^2$. Equation (1) can be written as $x^2 + y^2 + px + qy + m = 0$, where $p = -2a$, $q = -2b$, $m = a^2 + b^2 - R^2$. In such a general case, the equation of the circle in polar coordinates will be $\rho^2 - 2r\rho\cos(\varphi - \alpha) + r^2 - R^2 = 0$, where r is the distance between the

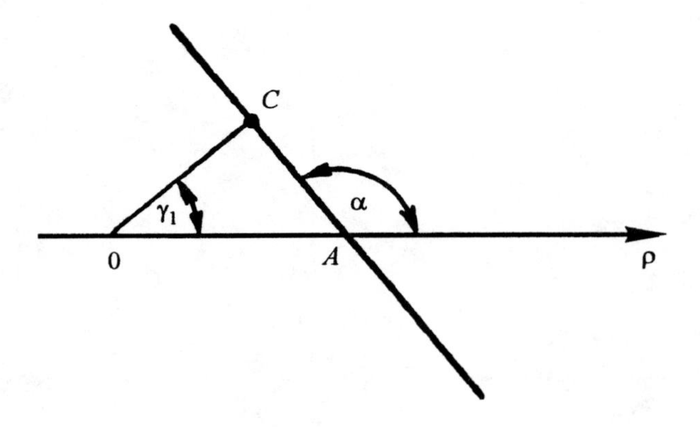

Figure 3.2. Equations of a straight line in polar coordinates.

center of the circle and the pole and α is the angle between the radius vector OM and the polar axis. The equation of the circle centered at the pole ($r = 0$), in polar coordinates, will be $\rho = R$.

Certain circles that have a simple form of the equation due to its location are shown in Figure 3.4.

Circle I has its center in the y-axis. Its equation is $x^2 + y^2 + qy + m = 0$ and has no term with x in power one.

Circle II has its center in the x-axis. Its equation is $x^2 + y^2 + px + m = 0$ and does not contain x in power one.

Circle IV passes through the origin, and the equation $x^2 + y^2 + px + qy = 0$ does not contain the free term.

Circle V has its center in the y-axis and passes through the origin. Its equation $x^2 + y^2 + px = 0$ has neither the free term nor y in power one.

Circle III has its center in the x-axis and passes through the origin. Its equation $x^2 + y^2 + qy = 0$ does not contain the free term or the term with the first power of y.

Let M be a point not lying on a circle. Draw a straight line through the point M intersecting the circle. We get two points K and L. The line segments MK and ML will be parts of a chord if the point M lies inside the circle (Figure 3.5), and line segments of the secant of the circle if the point M lies outside the circle (Figure 3.6). The value of the product $MK \cdot ML$, taken with "+" or "–" sign depending on whether the point M lies outside or inside the circle, is called the *degree* of the point M with respect to the circle. In all cases, the degree of the point M with respect to the circle equals $d^2 - R^2$, where d is the distance between the point M and the center of the circle, and R is its radius.

The degree of a point that lies on the circle is taken to equal zero. If a point lies outside of a circle, then its degree equals the square of the length of the tangent to the circle drawn from this point (Figure 3.7).

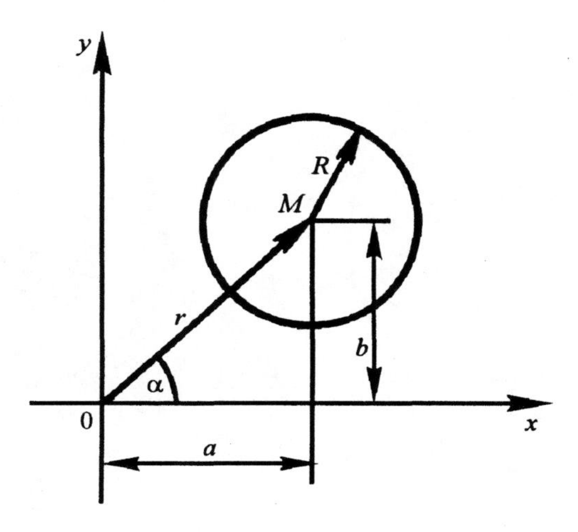

Figure 3.3. The graph of the curve $(x-a)^2+(y-b)^2=R^2$.

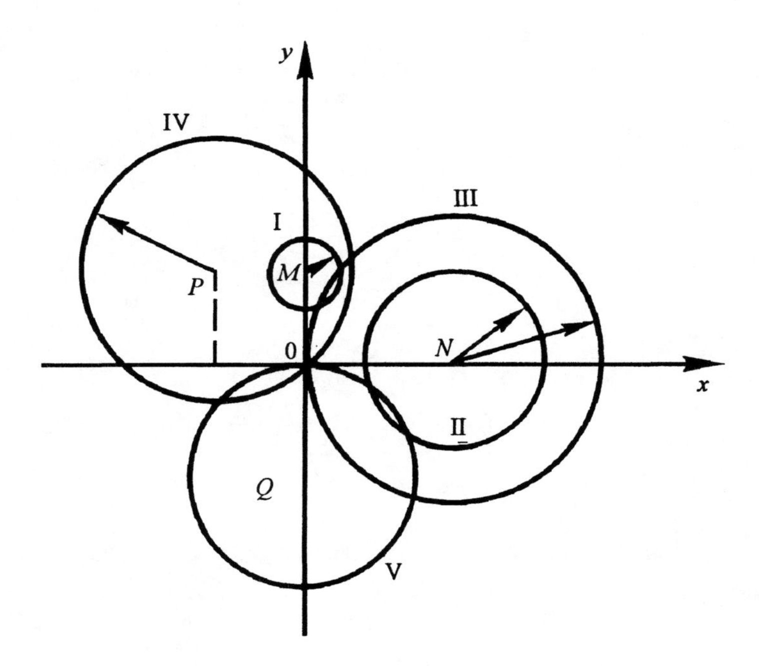

Figure 3.4. The circle: *I* $x^2+y^2+qy+m=0$; *II* $x^2+y^2+px+m=0$; *III* $x^2+y^2+px=0$; *IV* $x^2+y^2+px+qy=0$; *V* $x^2+y^2+qy=0$.

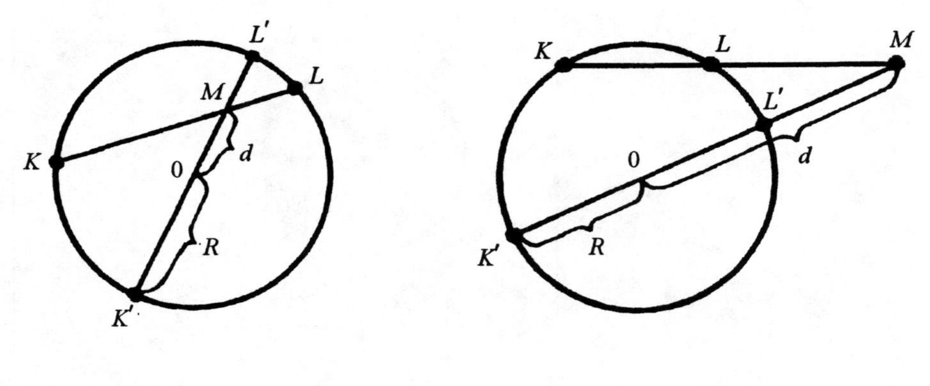

Figure 3.5 Figure 3.6

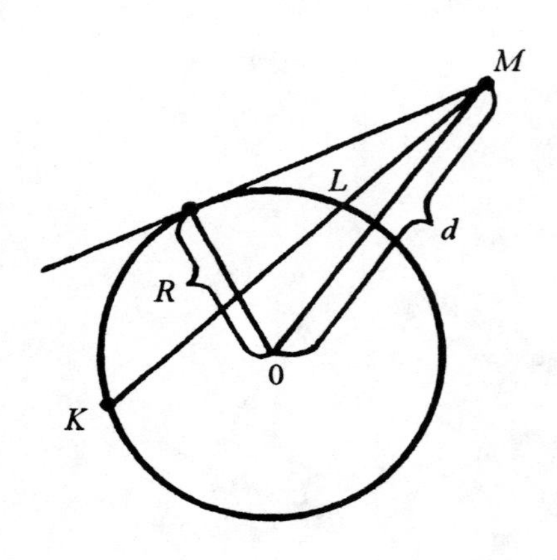

Figure 3.7

Figure 3.5 – 3.7. The degree of a point.

Let a circle have the corresponding equation given by (1), and M have coordinates x_0, y_0. Then the degree of this point with respect to the circle is equal to $(x_0 - a)^2 + (y_0 - b)^2 - R^2$.

3.2.2. The ellipse

The *ellipse* is the locus of points for which the sum of their distances from two fixed points, called foci, is a constant ($2a$) (Figure 3.8a). For each of these distances (focal radius vector), we have $r_1 = MF_1 = a - \varepsilon x$, $r_2 = MF_2 = a + \varepsilon x$, $r_1 + r_2 = 2a$, AB is the major axis ($2a$), CD is the minor axis ($2b$), A, B, C, D are vertices, O is the center, F_1, F_2 — foci (the points that lie on the

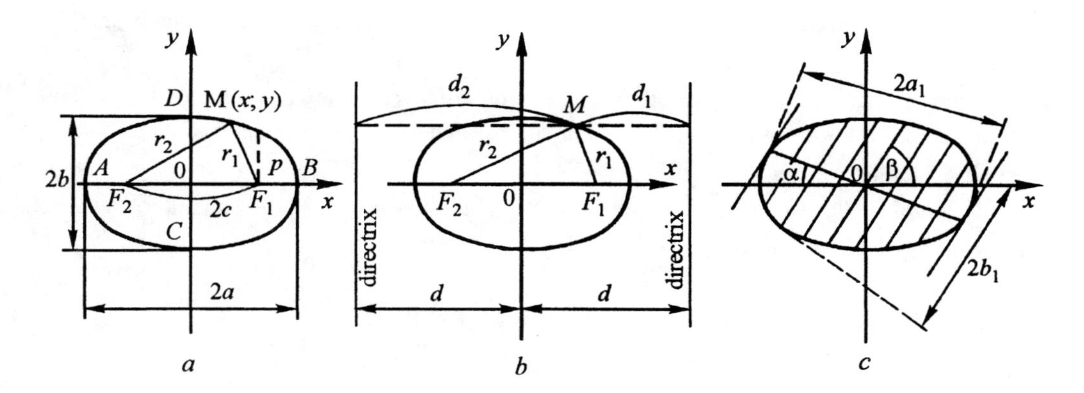

Figure 3.8. The ellipse (definitions).

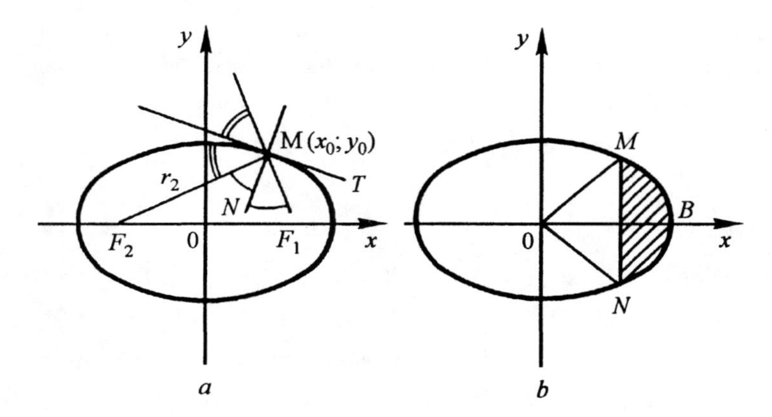

Figure 3.9. The circle $\frac{x^2}{a^2}+\frac{y^2}{b^2}=1$.

major axis on both sides of the center at the distance $c=\sqrt{a^2-b^2}$), $\varepsilon = c/a$ is the ($\varepsilon < 1$) eccentricity, p is the focal parameter (half of the chord passing through a focus and parallel to the minor axis), $p = b^2/a$.

The canonical equation of an ellipse (if the coordinate axes coincide with the axes of the ellipse) is $x^2/a^2 + y^2/b^2 = 1$.

The equation of an ellipse with the axes parallel to the coordinate axes and the center at a point $O_1(\alpha; \beta)$ is $(x-\alpha)^2/a^2 + (y-\beta)^2/b^2 = 1$.

The parametric equation of an ellipse is $x = a\cos t$, $y = b\sin t$ $(0 \le t \le 2\pi)$. In polar coordinates, an ellipse is given by the equation $\rho = p/(1+\varepsilon\cos\varphi)$, if the pole coincides with a focus of the ellipse and the polar axis is directed towards the closest vertex.

The directrixes are straight lines parallel to the minor axis and located from it at the distance $d = a/\varepsilon$ (Figure 3.8*b*). For any point $M(x; y)$ of the ellipse, $r_1/d_1 = r_2/d_2 = \varepsilon$.

Diameters are chords that pass through the center of an ellipse and are divided by it in half. The locus of the middle points of chords that are parallel to a given diameter of an ellipse is a diameter conjugate to the given one (Figure 3.8*c*).

If k_1 and k_2 are slopes of conjugate diameters, then $k_1k_2 = -b^2/a^2$. A tangent to an ellipse at a point $M(x_1; x_2)$ has the equation

$$xx_1/a^2 + yy_1/b^2 = 1.$$

The straight line given by the equation $Ax + By + C = 0$ is a tangent to the ellipse if $Aa^2 + B^2b^2 - C^2 = 0$. The radius of curvature at a point $M(x_0; y_0)$ (Figure 3.9*a*) is given by

$$R = a^2b^2\left(\frac{x_1^2}{a^4} + \frac{y_1^2}{b^4}\right)^{3/2} = \frac{(r_1r_2)^{3/2}}{ab} = \frac{p}{\sin^3\varphi},$$

where φ is the angle between the tangent and the radius vector to the point of tangency.

The area of the ellipse is $S = \pi ab$. The area of the sector S_{BOM} equals $ab/2\arccos x/a$. The area of the segment MBN is given by $ab\arccos x/a - xy$ (Figure 3.9*b*).

3.2.3. The hyperbola

The *hyperbola* is a locus of points the difference of distances from which to two given points (foci) is a constant ($2a$). The points satisfying $r_1 - r_2 = 2a$ are located on the same branch of the parabola (left branch, Figure 3.10). The points for which $r_2 - r_1 = 2a$ lie on the right branch. Each distance is expressed by the formula $r_1 = \pm(\varepsilon x - a)$, $r_2 = \pm(\varepsilon x + a)$, where "+" is taken for the right branch, and "–" for the left one. We also have $r_2 - r_1 = \pm 2a$.

Let AB be the transverse axis ($2a$), A, B — vertices, O — center, F_1, F_2 — foci lying on the transverse axis on two sides of the center at the distance c (which is greater than a), CD — the conjugate axis ($2b = 2\sqrt{c^2 - a^2}$), p is the focal parameter, $p = b^2/a$, $\varepsilon = c/a > 1$ — the eccentricity.

The canonical equation of the hyperbola is $x^2/a^2 - y^2/b^2 = 1$, if the transverse axis is coincident with the x-axis.

The parametric equation of the hyperbola is given by $a = a\cosh t$, $y = b\sinh t$, or else $x = a\sec t$, $y = b\tan t$.

The equation of the hyperbola in polar coordinates is $\rho = p/(1 + \varepsilon\cos\varphi)$, if the pole is placed at a focus and the polar axis is directed towards the closest vertex. This equation gives only one branch of the hyperbola.

The directrixes are straight lines located at the distance $d = a/\varepsilon$ from the center and perpendicular to the transverse axis (Figure 3.10).

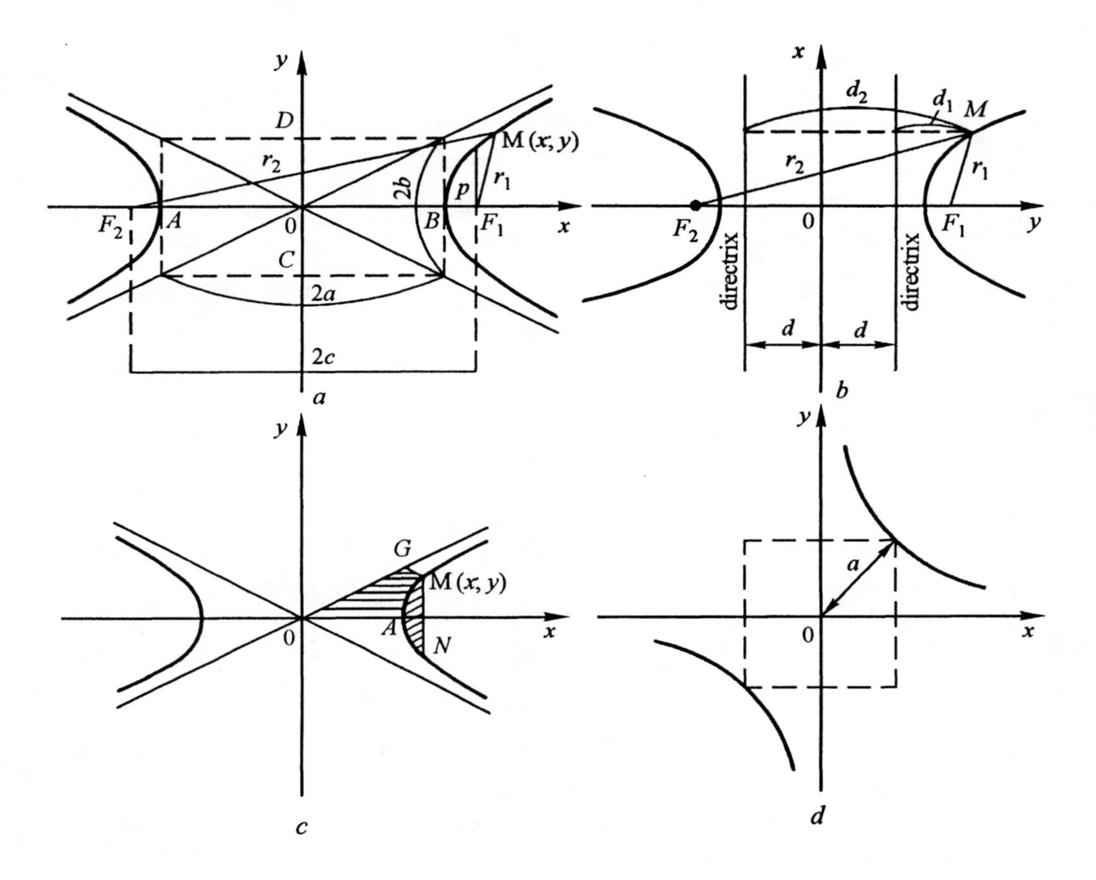

Figure 3.10. The hyperbola (definition).

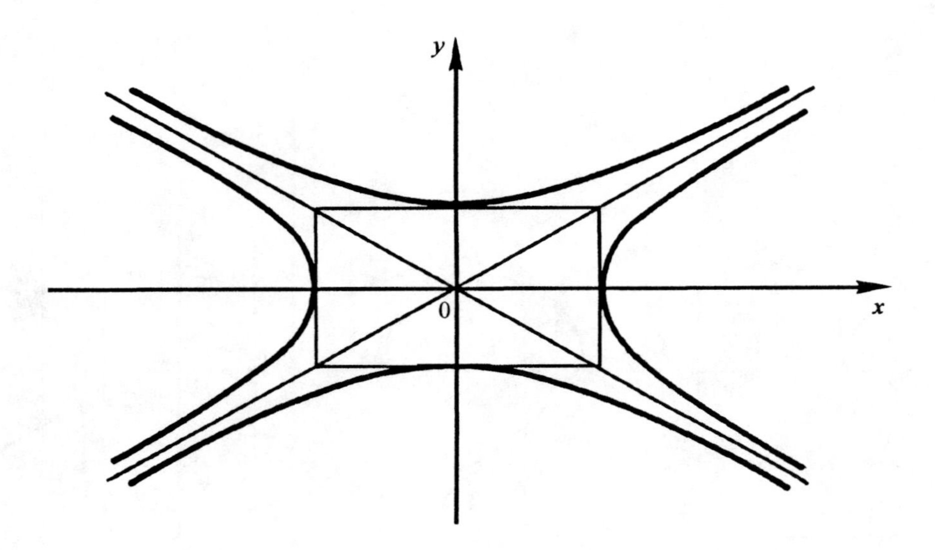

Figure 3.11. Conjugate hyperbolas.

The tangent to the hyperbola at a point $M(x_1; y_1)$ is given by the equation $\frac{xx_1}{a^2} - \frac{yy_1}{b^2} = 1$. The straight line $Ax + By + C = 0$ is a tangent to the hyperbola, if $A^2a^2 - B^2b^2 = c^2$.

Equations of asymptotes are $y = \pm\frac{b}{a}x$. The conjugate hyperbolas $x^2/a^2 - y^2/b^2 = 1$ and $y^2/b^2 - x^2/a^2 = 1$ share the same asymptotes (Figure 3.11). The transverse axis of one of the parabolas is the conjugate axis of the other, and vice versa.

Diameters are chords of the hyperbola and the conjugate hyperbola, if they pass through the common center and are divided by it in half. Two diameters having slopes k_1 and k_2 are called conjugate if $k_1\, k_2 = b^2/a^2$.

The radius of curvature at a point $M(x_1; y_1)$ of the hyperbola is calculated by

$$R = a^2b^2\left(\frac{x_1^2}{a^4} + \frac{y_1^2}{b^4}\right)^{3/2} = \frac{(r_1r_2)^{3/2}}{ab} = \frac{p}{\sin^3\varphi},$$

where φ is the angle between the tangent and the radius vector to the tangency point. Calculated at the vertices A and B, $R = p = b^2/a$ (see Figure 3.10a).

The area of the segment AMN (see Figure 3.10c) is given by $xy - ab\ \ln(x/a + y/b) = = xy - ab\ \text{Arch}\,\frac{x}{a}$.

An isosceles hyperbola is a hyperbola that has equal axes, $a = b$ (see Figure 3.10*d*). The equation of an isosceles hyperbola is $x^2 - y^2 = a^2$. Asymptotes of such a hyperbola are mutually orthogonal. If the axes are chosen to coincide with the asymptotes, then the equation of the isosceles hyperbola is $xy = a^2/2$.

3.2.4. The parabola

The *parabola* is a locus of points $M(x, y)$ that are equidistant from a given point (focus) and a given straight line (directrix), $MF = MK = x + p/2$, where MF is the focal radius vector of the point of the parabola, Ox is the axis of the parabola, O is the vertex, F — focus (the point that is located at the distance $p/2$ from the vertex), N_1N_2 is the directrix (the straight line that lies at the distance $p/2$ on the other side of the focus), and p is the focal parameter (Figure 3.12*a*). The eccentricity of the parabola equals one. The canonical equation of the parabola is $y^2 = 2px$, if Ox coincides with the axis of the parabola and its vertex is located to the left of the origin.

In the polar coordinates, the equation of the parabola is $\rho = \dfrac{p}{1 + \varepsilon \cos \varphi}$, if the pole coincides with the focus and the polar axis is directed from the focus to the vertex of the parabola.

If the axis of the parabola is vertical, then its equation is $y = a x^2 + bx + c$, where $p = 1/2|a|$ (Figure 3.12*b*). If $a > 0$, then the parabola has its vertex on top, and if $a < 0$, — on the bottom. The coordinates of the vertex are $x_0 = - b/2a$, $y_0 = 4ac - b^2/4a$.

The diameter is a straight line parallel to the axis of the hyperbola. The diameter divides in half the chords that are parallel to the tangent constructed at the end of the diameter. The equation of the diameter is $y = p/k$, where k is the slope of these chords (Figure 3.12*c*).

The tangent to the parabola at a point $M(x_1; y_1)$ is given by the equation $yy_1 = p(x + x_1)$ (Figure 3.12e). The straight line defined by the equation $y = kx + b$ is a tangent to the parabola if $p = 2bk$. The radius of curvature of the parabola at a point $M(x_1; y_1)$ is

$$R = (p + 2x_1)^{3/2}\sqrt{p} = p/\sin^3\varphi.$$

At the vertex O the radius of curvature equals $R = p$.

The area of the segment MON of the parabola equals 2/3 the area of $PQNM$ (Figure 3.12*d*).

Consider other parametric equations of the ellipse, hyperbola, and parabola.

Construct an ellipse and hyperbola having the same semiaxes a and b (Figure 3.13). Let M be an arbitrary point of the ellipse. Extend the radius vector OM until it intersects, at a point K', the tangent at the vertex common to the ellipse and hyperbola, and draw a straight line passing through the point K' and parallel to the axis Ox until it intersects the closest branch of the hyperbola at a point K. If this construction is carried out starting with a point K lying on the hyperbola, then denote the point of intersection of the straight line and the closest portion of the ellipse by M.

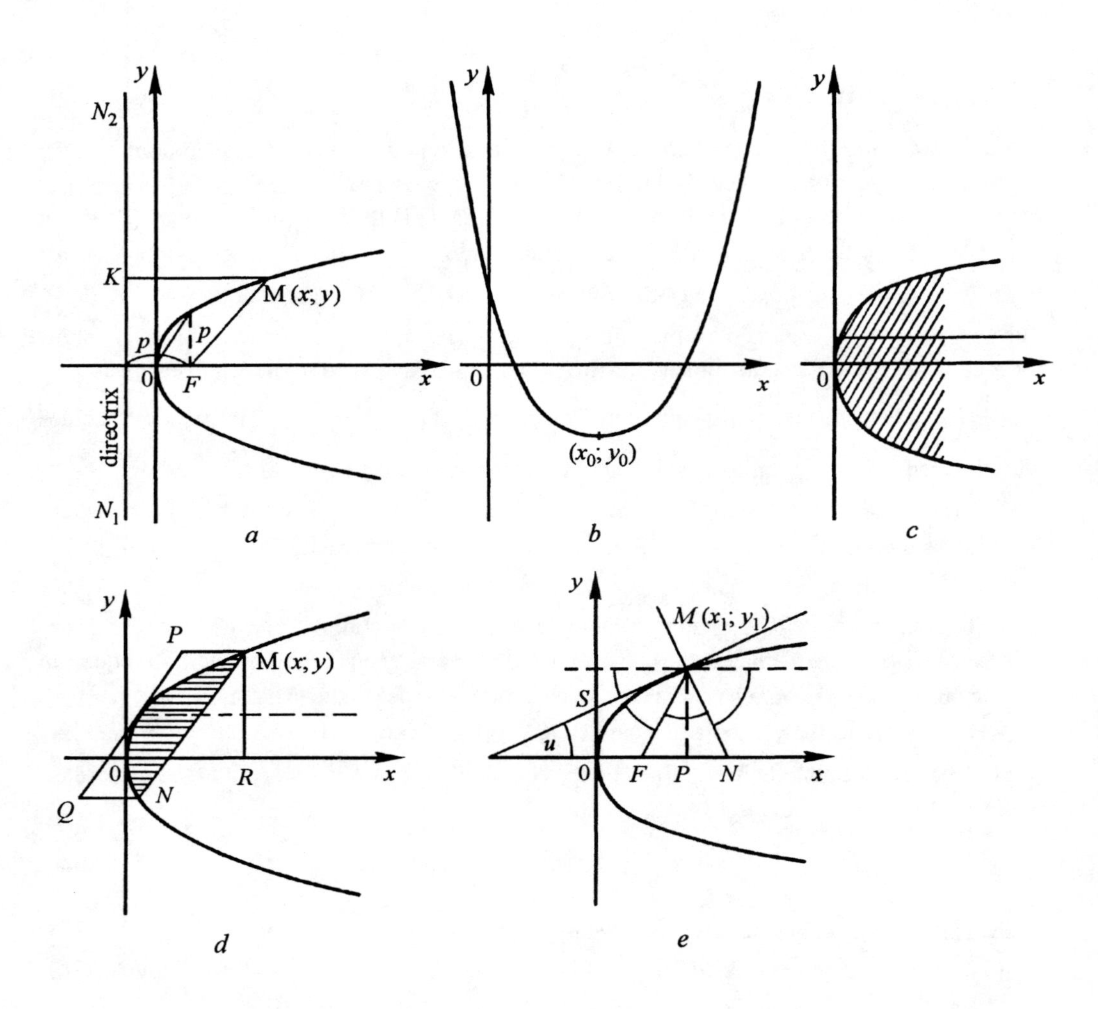

Figure 3.12. The parabola.

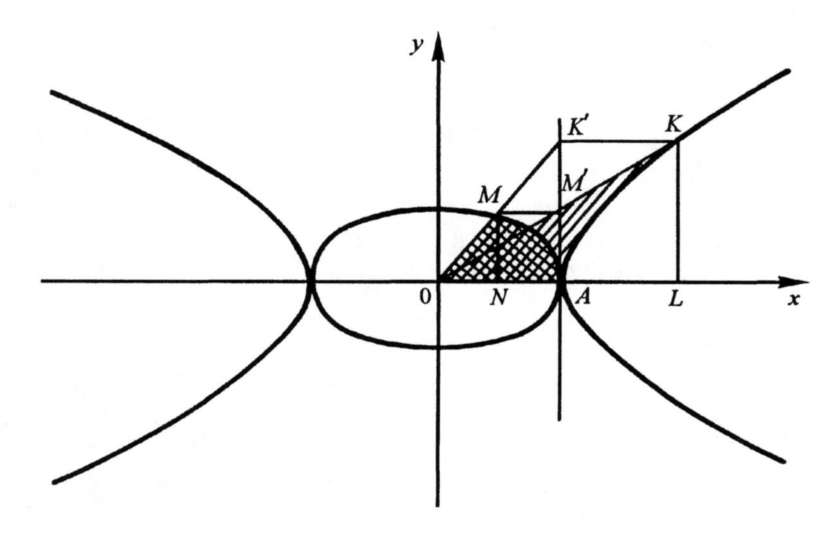

Figure 3.13. The ellipse and hyperbola with the same semiaxes.

Let 2 times the area of MOA/ab be equal to φ, and let 2 times the area of KOA/ab be equal to θ. Then $\tan\frac{\varphi}{2} = \tanh\frac{\theta}{2}$. Denote this common value by τ. Then we have the following parametric equations: for the ellipse $x = a/\cosh\theta, y = b\tanh\theta$ $(-\infty < \theta < +\infty)$, or

$$x = a\frac{1 = \tau^2}{1+\tau^2}, \quad y = b\frac{2\tau}{1+\tau^2}, \quad -\infty < \tau < +\infty,$$

and for the hyperbola

$$x = a\frac{1+\tau^2}{1-\tau^2}, \quad y = \frac{2\tau}{1+\tau^2}, \quad -\infty < \tau < +\infty.$$

The parametric equation for the parabola can be written as

$$x = \tau^2, \quad y = \sqrt{2p}\,\tau, \quad -\infty < \tau < +\infty.$$

3.3. Curves of degree three

In the first part of the book, certain properties of curves of degree three were considered. This allowed us to see the behavior of the curve before plotting the graph. We continue studying the properties of curves of degree three.

1. Theorem of Maclaurin. The points at which three tangents constructed at the points of intersection of a curve of degree three with a line intersect the curve lie on a straight line.

2. A straight line that pass through two inflection points of a curve of degree three also passes through the third inflection point of this curve.

3. If a curve of degree two is constructed passing through four points of the curve of degree three, and there are two more points of intersection of these curves, then the straight line that passes through these two points intersects the curve of degree three in a point which lies on all curves of degree two that pass through the initial four points.

This statement implies the following assertions.

3′. If a curve of degree three has three inflection points, then they lie on the same straight line.

4. Any curve of degree three without a double point has at least one inflection point. The number of inflection points of a curve of degree three cannot exceed nine with more than three real points.

5. A curve of degree three cannot have more than one double point.

6. Theorem of Salmon. Four tangents to a curve of degree three with no double points constructed from any point of this curve have the double ratio which does not depend on the choice of the point.

Consider some curves of degree three.

3.3.1. Folium of Descartes

The *folium of Descartes* (Figure 3.14) is a curve of degree three given by the equation $x^3 + y^3 - 3axy = 0$; in the Cartesian coordinates, its parametric equation is $x = \frac{3at}{1+t^3}$, $y = \frac{3at^2}{1+t^3}$. In the polar coordinates, with O being the pole and Ox the polar axis, the curve is given by the following equation:

$$\rho = \frac{3a\cos\varphi\sin\varphi}{\cos^3\varphi + \sin^3\varphi}.$$

Coordinates x and y enter into the equation symmetrically, which implies that the curve is symmetric with respect to the bisector $y = x$. Analyzing singular points (see Section 3.2) we see that the origin (0; 0) is a node. To get equations for tangents, set the expression of the least degree equal to zero, $3axy = 0$. Whence, $x = 0$, $y = 0$ are tangents at the node. They coincide with the coordinate axes; thus the curve intersects itself at a straight angle.

In the first quadrant, the curve makes a loop that intersects the line $y = x$ at the point $A(\frac{3}{2}a; \frac{3}{2}a)$. The points of the loop at which tangents are parallel to the coordinate axes are

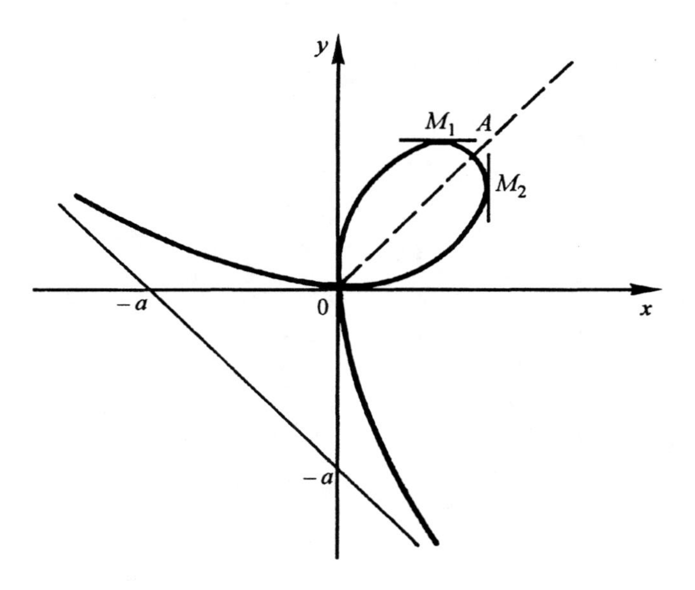

Figure 3.14. The folium of Descartes $x^3 + y^3 - 3axy = 0$.

$M_1(a\sqrt[3]{2}\,;\, a\sqrt[3]{4}\,)$, $M_2(a\sqrt[3]{4}\,;\, a\sqrt[3]{2}\,)$; $OA = \dfrac{3a}{\sqrt{2}}$. The radius of curvature at the origin equals $R = 3a/2$.

Let us find an asymptote $y = kx + b$. Replace y with $kx + b$ in the equation of the curve, and set the coefficients of the two highest — degree terms equal to zero. We get $1 + k^2 = 0$, $k^2 b - 3ak = 0$, whence $k = -1$, $b = -a$. So the folium of Descartes has the asymptote $y = -x - a$ or $x + y + a = 0$. The area of the loop is

$$S_1 = \frac{9a^2}{2}\int_0^{\infty} \frac{t^2\,dt}{(1+t^3)^2} = \frac{3a^2}{2}.$$

The area bounded by the curve and asymptote equals $S_2 = \frac{3}{2}a^2$.

Let us now plot the folium of Descartes. Take the symmetry axis of the curve to be the x-axis. Then the equation becomes

$$y^2 = -\frac{1}{3}\,\frac{x - \dfrac{3a}{\sqrt{2}}}{x + \dfrac{a}{\sqrt{2}}}\,x^2.$$

Suppose we have a circle of radius r centered at the point $M(r/2; 0)$. Draw the straight line $x = -h$, and let Q be an arbitrary point of this circle. Drawing the straight lines QA and

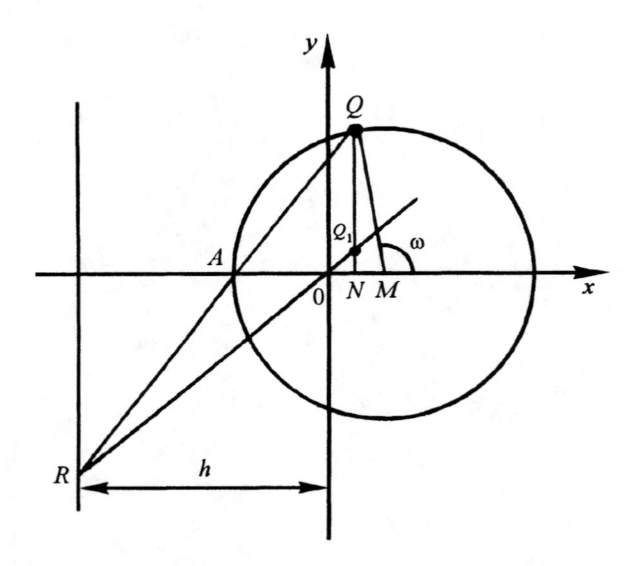

Figure 3.15. Construction of a folium of Descartes.

$QN \perp Ox$ (Figure 3.15), which passes through the point of intersection of the straight line QA and $x = -h$, R, we plot the straight line RO until it intersects the straight line QN in the point Q_1. The locus of points Q_1 is a folium of Descartes.

The folium of Descartes is a curve that is defined by an equation of the form $(x + y)(x^2 - 2kxy + y^2) = 2(1 + k)axy$. If $k = 1/2$, we get the equation of a folium of Descartes; if $k = 0$, we get the equation of a strophoid, $(x + y)(x^2 + y^2) = 2axy$, see Figure 3.16. This curve has the same asymptote as the folium of Descartes, and the same node at $O(0; 0)$.

The equation of the strophoid in polar coordinates is $\rho = \frac{a}{\sqrt{2}} \frac{\sin 2\varphi}{\sin(\pi/4 + \varphi)}$; its parametric equation is

$$x = \frac{2at}{(1 + t)(1 + t^2)}, \; y = \frac{2at^2}{(1 + t)(1 + t^2)} \quad (-\infty < t < +\infty, \; t = \tan \angle MOX).$$

Note. The first time in the history of mathematics that the curve later to be called the folium of Descartes was mentioned was in a letter of Descartes to Fermat (in 1638). Descartes defines it as a curve for which the sum of volumes of the cubes with edges coinciding with the x- and y-coordinates at every point equals the volume of a parallelepiped that has edges equal to the x- and y-coordinate and a certain constant.

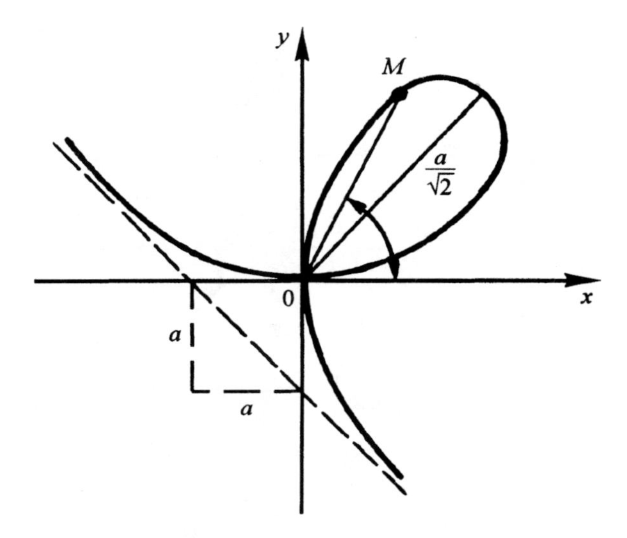

Figure 3.16. The strophoid.

3.3.2. Cissoid of Diocles

The *cissoid of Diocles* is a locus of points M such that $OM = PQ$, where P is an arbitrary point of an initial circle with diameter a (Figure 3.17).

There is a simple way to construct the cissoid of Diocles. Take a circle with diameter $OA = 2a$ and construct a tangent to it, AQ. Draw a ray passing through the point O, OQ, and mark the line segment $OM = QP$. The point M thus constructed is a point of the cissoid.

The equation of the cissoid in the Cartesian coordinates is given by $y^2 = x^3/(a - x)$. In the polar coordinates (O is the pole, Ox is the polar axis), the equation is $\rho = 2a\sin^2\varphi/\cos\varphi$. The equation in the parametric form is $x = a/(t^2 + 1)$, $y = a/t(t^2 + 1)$.

The curve is symmetric with respect to the x-axis and has two unbounded branches. The origin is a turning point of the first kind. The asymptote is given by $x = 2a$. The area bounded by the curve and the asymptote equals $S = 3\pi a^2$.

The volume of the solid of revolution obtained by rotating the strip formed by the cissoid and the asymptote about the y-axis is given by $V = 5{\cdot}2\pi^2a^3$.

If a perpendicular is dropped from the vertex of the parabola to the tangents, then the locus of the feet of these perpendiculars will be a cissoid of Diocles, $y^2 = -x^3/(2p + x)$. Thus the cissoid of Diocles is a locus of points which are symmetric to the vertex of the parabola with respect to the tangents.

Note that the locus of points symmetric to the origin with respect to the tangents to a parabola can be regarded as the trajectory of another parabola rolling along the given one. This gives another method (kinematic) for constructing a cissoid, tracing the trajectory of the vertex of a parabola that rolls on another parabola without slipping.

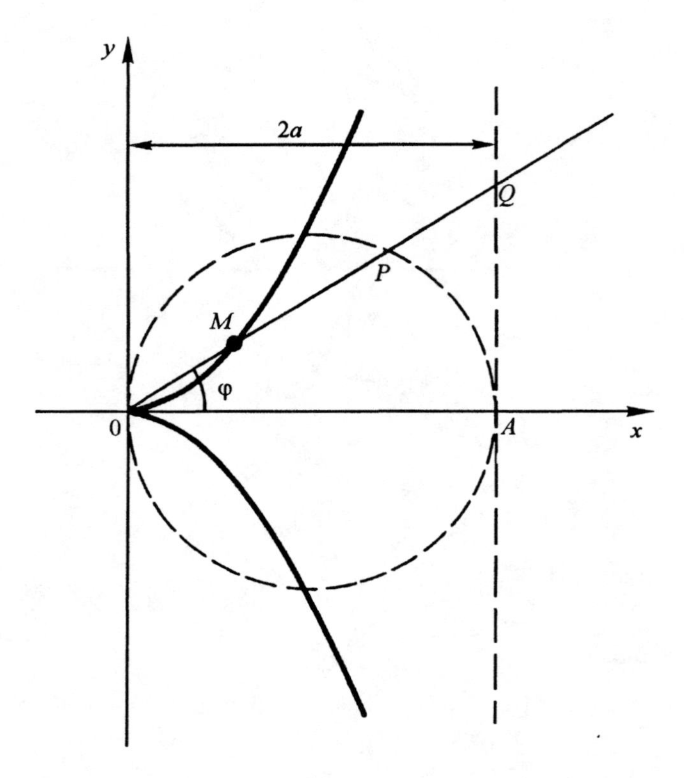

Figure 3.17. The cissoid of Diocles, $y^2 = \dfrac{x^3}{a - x}$.

Let x_c be the x-coordinate of the center of mass of the portion of the plane bounded by a cissoid and its asymptote. Then, by using the theorem of Gulden, we find the relation $x_c/(2a - x_c) = 5/1$, i.e. the center of mass of the portion of the plane bounded by a cissoid and its asymptote divides the line segment on the x-axis lying between the curve and its asymptote into two parts in the ratio of one to five. This implies that the volume of the solid of revolution of a cissoid about its asymptote equals the volume of a torus obtained by revolving the generating circle.

The length of the arc of a cissoid bounded by the origin and a point with the coordinate x is given by

$$l = a\int_0^x \frac{1}{2a - 1}\sqrt{\frac{8a - 3x}{2a - x}}\,dx = a(u - 2) + \frac{a\sqrt{3}}{2}\ln\frac{(u - \sqrt{3})(2 + \sqrt{3})}{(u + \sqrt{3})(2 - \sqrt{3})},$$

where $u = \sqrt{(4a - 3x)/(a - x)}$.

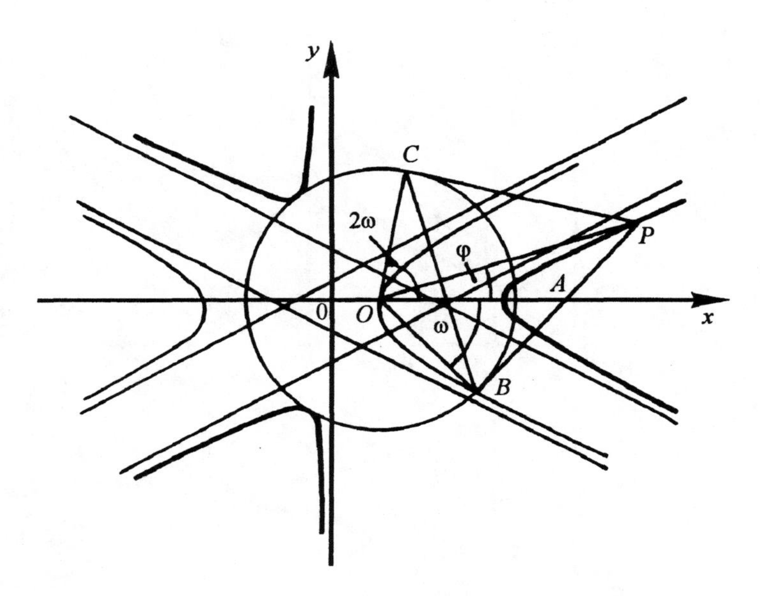

Figure 3.18. The trisectrix of Longchamps, $\rho = \dfrac{a}{\cos 3\varphi}$.

Note. The cissoid has been known since the third century B.C., when there was a search for solution of the famous Dalosian problem of doubling the cube, i.e. determining the edge of the cube that would have a volume twice as large as that of the given cube.

Remark. One of the cissoids is a second-degree curve called the *trisectrix of Longchamps* (Figure 3.18). It is defined as a locus of points P which are points of intersection of tangents to the circle constructed at the endpoints B and C of arcs $AB = \omega, AC = 2\omega$ for a fixed point A and arbitrary central angle ω. The equation of the trisectrix in polar coordinates is

$$\rho = \frac{a}{\cos 3\varphi},$$

whereas in the Cartesian coordinate, it is

$$(x + \frac{a}{3})(y^2 - \frac{1}{3}x^2) + \frac{4}{9}ax^2 = 0.$$

Note that, by applying the inversion transformation, the trisectrix gives rise to the tree folium "rose'' $\rho = a\cos 3\varphi$.

The equation $(x - \frac{p}{2})y^2 - 2px^2 = 0$ is an equation of the curve known as the *mixed cubic* (Figure 3.19). Note that the parabola $y^2 = 2px$ is a curvilinear asymptote of the mixed cubic. Longchamps has called this curve "mixed cubic," since its branch has, on one hand, hyperbolic behavior, and on the other hand, parabolic behavior.

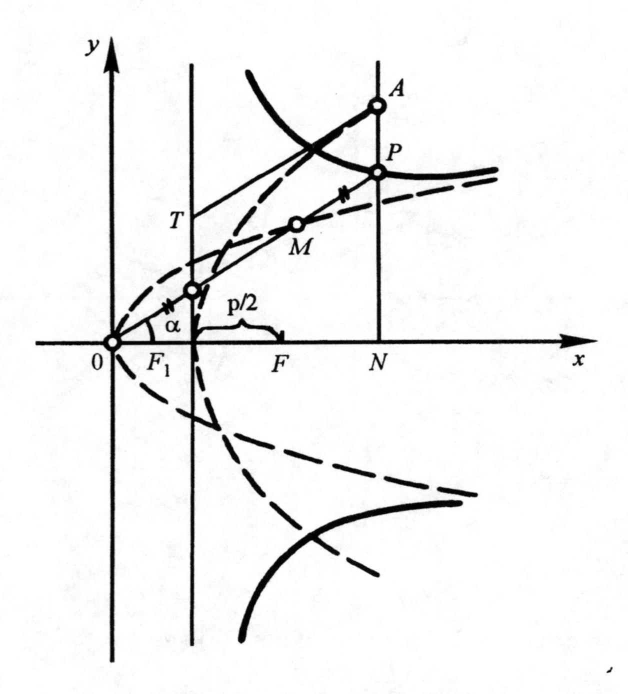

Figure 3.19. The mixed cubic.

3.3.3. Strophoid

The *strophoid* is a locus of points M_1 and M_2 that lie on arbitrary rays passing through a point A and such that $PM_1 = PM_2 = OP$, where P is an arbitrary point on the y-axis (Figure 3.20).

The equation of a strophoid in Cartesian coordinates is

$$y^2 = x^2\left(\frac{a+x}{a-x}\right);$$

in parametric form, the curve is given by

$$x = a\frac{t^2-1}{t^2+1}, \quad y = at\frac{t^2-1}{t^2+1} \quad (-\infty < t < +\infty);$$

in polar coordinates, in which O is the pole and Ox is the pole axis, the equation is

$$\rho = -a\frac{\cos 2\varphi}{\cos \varphi}.$$

The curve is symmetric with respect to the x-axis. The origin $(0; 0)$ is a node of the curve with the tangents $y = \pm x$. The asymptote is given by $x = a$.

Let us list the main properties of a strophoid.

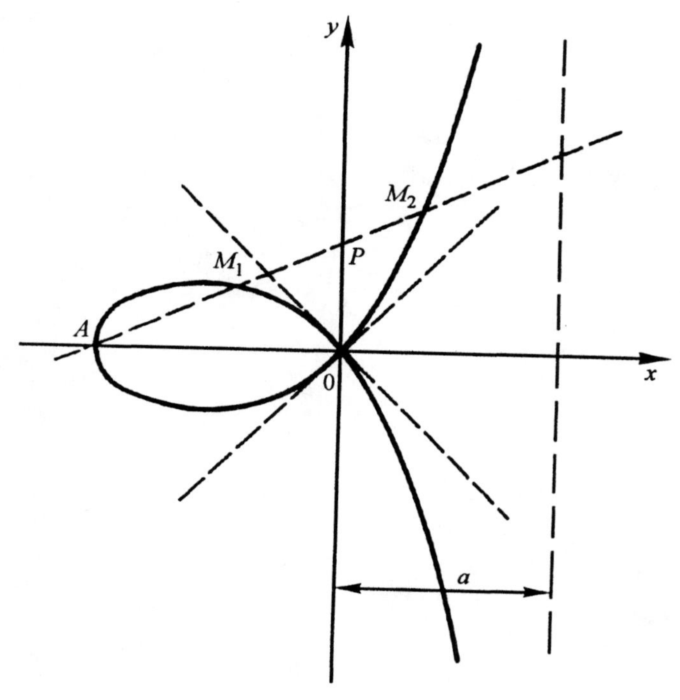

Figure 3.20. The strophoid $y^2 = x^2\left(\frac{a+x}{a-x}\right)$.

1. The strophoid is a pedal of a parabola* with respect to the point of intersection of the axis of the parabola and its directrix.

2. The locus of points which are intersections of two tangents constructed at conjugate points M_1 and M_2 of a strophoid is a cissoid of Diocles.

3. Inversion of a strophoid gives the same strophoid, if the inversion pole coincides with the point A and the inversion degree equals a^2.

4. If two tangents to a strophoid are constructed from its arbitrary point M, with P and Q being the points of tangency, then the points M, P, and Q lie on a circle that passes through the origin (Figure 3.21).

The area of the loop is $S = 2a^2 - \pi a^2/2$. The area bounded by the curve and the asymptote is $S_2 = 2a^2 + \pi a^2/2$. The radius of curvature at the node is $R = a\sqrt{2}$.

---

* The pedal of a given curve K with respect to a point in the plane is another curve which is the locus of feet of perpendiculars dropped from the point to the tangents to the given curve.

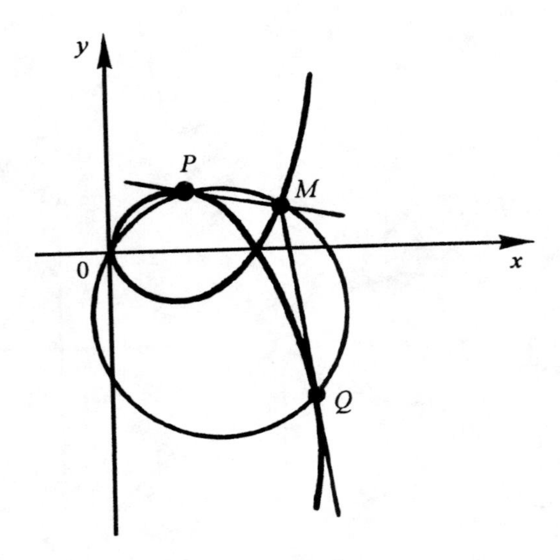

Figure 3.21. A property of strophoid.

The strophoid that has been considered is a particular case of the so-called *slant strophoid.* Let us look at a method for constructing a slant strophoid.

Consider an angle with the vertex at O. Take an arbitrary point A on one of its sides, and draw an arbitrary ray through it intersecting the other side of the angle at a point P. Then the points M_1, M_2, which are constructed such that $PM_2 = PM_1 = OP$, belong to a slant strophoid.

The equation of a slant strophoid in polar coordinates is $\rho = OA \dfrac{\sin\alpha \pm \sin\varphi}{\sin(\alpha+\varphi)}$. If $\alpha = \pi/2$, we get the equation of the so-called straight strophoid we considered earlier.

Note. Torricelli was the first to study the strophoid, in 1645. Besides numerous applications in geometry, the strophoid is also used to solve certain problems in optics, descriptive geometry, etc.

3.3.4. Ophiurida

The *ophiurida* is a curve that can be considered as a pedal of a parabola with respect to an arbitrary point of the tangent to the parabola at the vertex (Figure 3.22). Its equation is $x(x^2+y^2) = y(cy-bx)$. If we set $b=0$ in the equation of the ophiurida, then it becomes the equation of a cissoid. Ophiurida has a node at the origin with the tangents $y=0$ and $y=b/cx$. The straight line $x=c$ is an asymptote of the ophiurida. Note that the straight line $x=c$ is a directrix of the parabola.

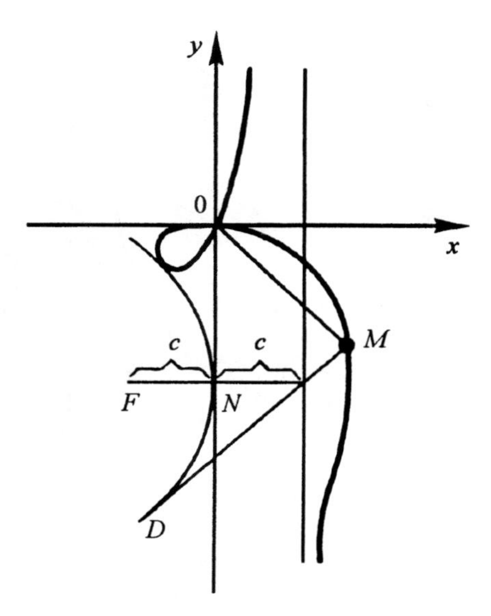

Figure 3.22. The ophiurida $x(x^2 + y^2) = y(cy - bx)$.

3.3.5. Trisectrix of Maclaurin

The *trisectrix of Maclaurin* is a curve defined as a pedal of a parabola with respect to a point on its axis such that the distance between its point and the directrix equals the distance between the directrix and the focus (Figure 3.23).

Its equation in Cartesian coordinates is $(x - 3a)(x^2 + y^2) - 4a^3 = 0$, and, in polar coordinates, $\rho = a/\cos\varphi/3$.

This implies that the trisectrix of Maclaurin can be obtained by inversion of the rose $\rho = a\cos\varphi/3$ with respect to the pole.

3.3.6. Cubic of Tschirnhausen

The *cubic of Tschirnhausen* or *trisectrix of Catalan* is a curve that has a parabola as a pedal (Figure 3.24). If the parabola is taken to be $y^2 = 2px + p^2$, then the equation of the cubic of Tschirnhausen in Cartesian coordinates becomes $2(2p + x)^3 = 27p(x^2 + y^2)$; in polar coordinates it is $\rho\cos^3\pi - \varphi/3 = p/2$. The equation of the cubic of Tschinhausen in parametric form is

$$x = p\sin\frac{3\omega}{2}/2\sin^3\frac{\omega}{2}, \quad y = -p\cos\frac{3\omega}{2}/2\sin^3\frac{\omega}{2}.$$

The curve is symmetric with respect to the x-axis and intersects it in the points $(-p/2; 0)$ and $(4p; 0)$. The latter point is a node with tangents that make a 60° angle.

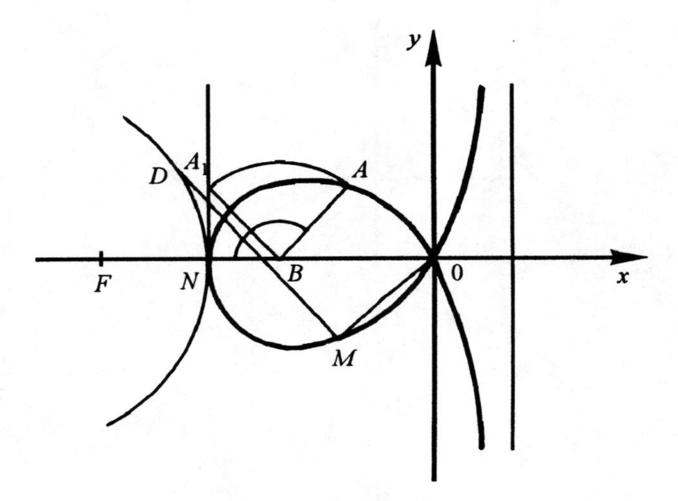

Figure 3.23. The trisectrix of Maclaurin $(x-3a)(x^2+y^2+4a^3)=0$.

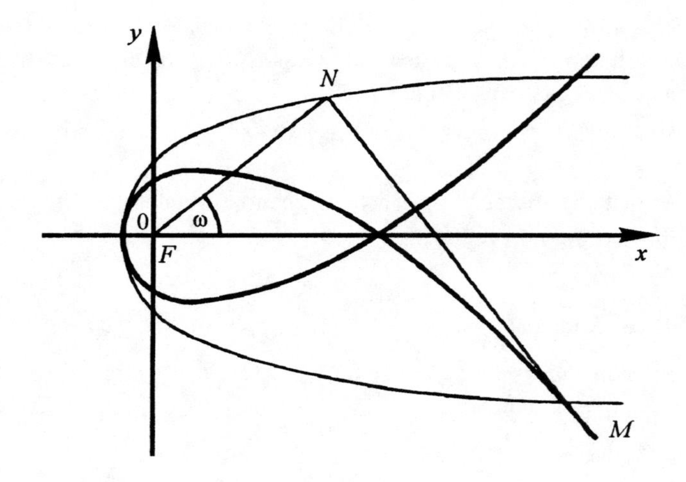

Figure 3.24. Cubic of Tschirnhausen $2(2p+x)^3=27p(x^2+y^2)$.

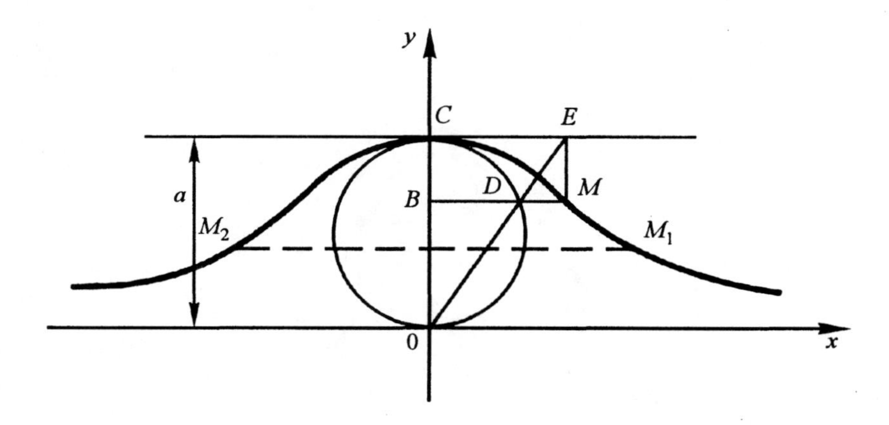

Figure 3.25. The versiera of Agnesi.

We remark that the cubic of Tschirnhausen is an envelope of rays of light reflected from a parabola if they are incident at a straight angle to the axis of the parabola. This envelope is called the catacaustic of the parabola.

3.3.7. Versiera of Agnesi

Construct a circle with the diameter being the line segment $OC = a$ (Figure 3.25). Extend the semichord BD to the point M determined from the ratio $BM \div BD = OC \div OB$. The locus of such points M is a *versiera*.

The equation of the curve in Cartesian coordinates is $y = a^3/(a^2 + x^2)$. The curve is symmetric with respect to the y-axis. The x-axis is an asymptote of the curve. At the point C, the curve and the circle have a common tangent parallel to the x-axis.

The function reaches maximum at the point $C(0; a)$; the radius of curvature at this point is $R = a/2$. Inflection points are $M_1(a/\sqrt{3};\ 3/4a)$, $M_2(-a/\sqrt{3};\ 3/4a)$. The slopes of the curve at inflection points are $\tan\varphi = \pm 3\sqrt{3}/8$. The area of the portion of the plane lying between the asymptote and the curve is $S = \pi a^2$. The volume of the solid obtained by revolving the curve about its asymptote, V, is twice the volume of the solid obtained by revolving the initial circle about the same axis, $V = \pi^2 a^3/2$, $V_1 = \pi^2 a^3/4$.

Remark. The pseudoversiera is a curve obtained from a versiera by doubling all its y-coordinates. Its equation is $y = 2a^3/(a^2 + x^2)$.

A generalization of a versiera is a locus of points. It is constructed in the same way as the versiera, however, the x-axis is not a tangent to the generating circle, but translated parallel to itself at a certain distance. Such a curve is know as anguine of Newton.

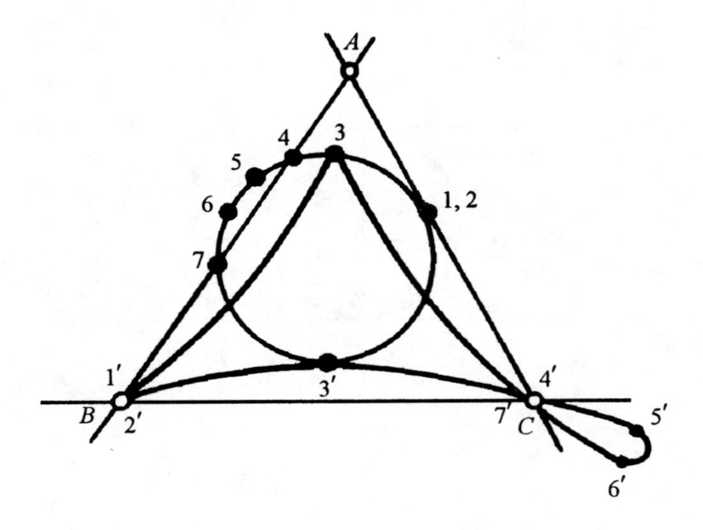

Figure 3.26. Singular points of a curve of degree four.

A study of the versiera is connected with the work of Maria Agnesi (1718 – 1799). Sometimes this curve is called *witch of Agnesi*.

3.3.8. Semicubic parabola (parabola of Neil)

The equation of the curve is given by $a^2x^3 - y^2 = 0$, $a > 0$; its parametric equation is $x = t^2$, $y = at^3$, $(-\infty < t < +\infty)$. The curvature of the curve at a point x is $K = 6a/\sqrt{x} \cdot \cdot (4 + 9a^2x)^{3/2}$. The length of the arc starting at the origin and ending at a point M with the x-coordinate equal to x is $l = ((4 + 9a^2x)^{3/2} - 8)/27a^2$.

3.4. Curves of degree four

It is convenient to classify curves of degree four according to the number and nature of singular points of the curve. We have the following cases: 1) curves with no singular points; 2) curves with one singular point; 3) elliptic curves, which are the curves with two double points; 4) rational curves, which are the curves with three double points or one triple point. Let us look at some of these curves.

Rational curves of degree four can be obtained by using a quadratic transformation of curves of degree two. If the transformed curve of degree two, for example a circle, intersects the coordinate triangle in two points, then the corresponding curve of degree four will have two tangents at the opposite vertex, and, hence, this point is a node. If these two points coincide, then these two tangents also coincide and the curve has a turning point at the vertex of the triangle.

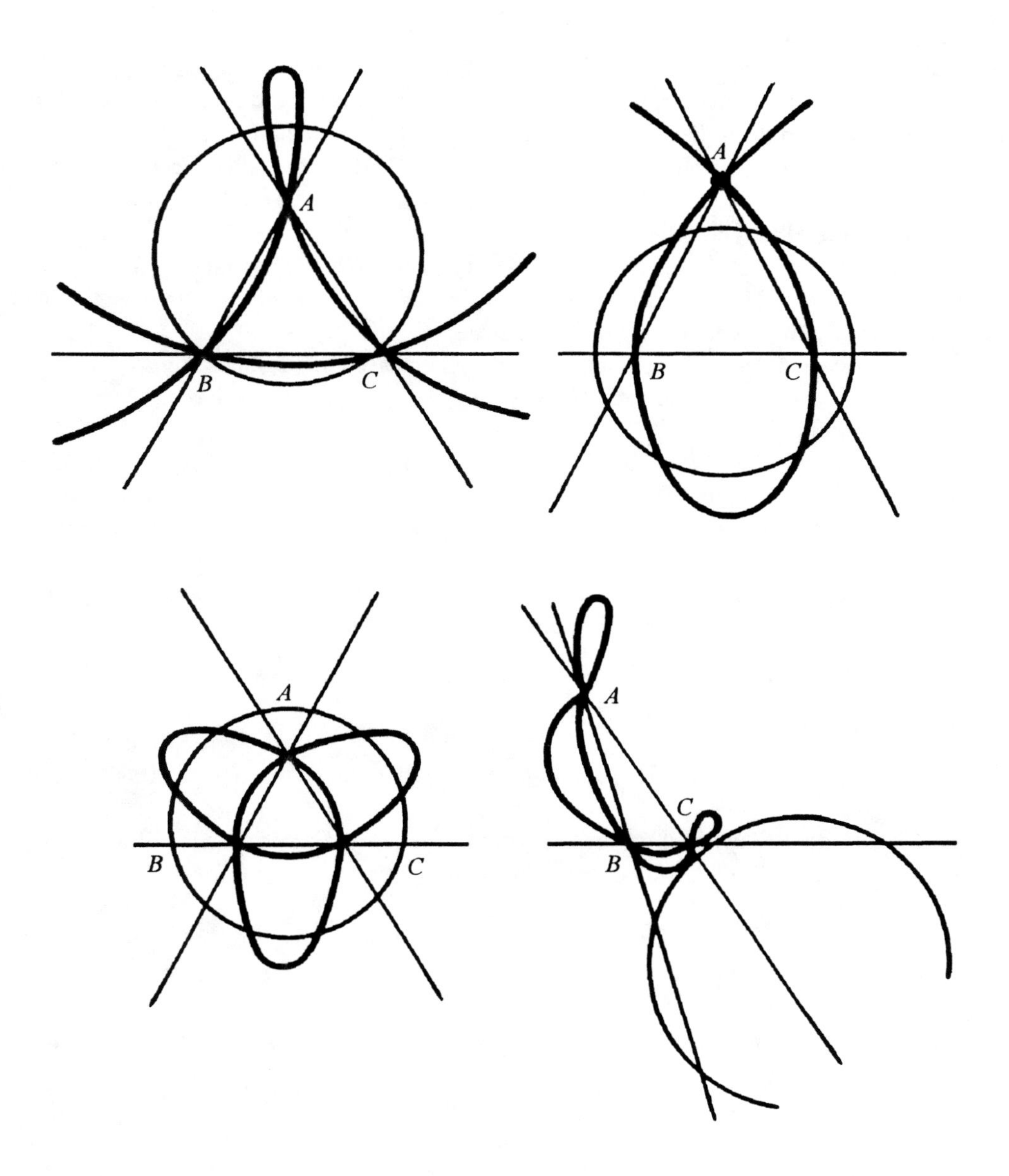

Figure 3.27. Various curves of degree four.

If the curve of degree two intersects a side of the triangle in imaginary points, the opposite vertex is an isolated point of the curve of degree four (Figure 3.26).

It follows that if the transformed curve of degree four intersects the coordinate triangle in six points, then the curve of degree four has three nodes. If these six points are imaginary, then there will be three isolated points; if they pairwise coincide, then the curve has three turning points. Singular forms of the curve of degree four depend on whether the curve of degree two intersects the sides of the coordinate triangle or their continuation (Figure 3.27).

Note that the curve of degree four obtained as the result of a quadratic transformation of a curve of degree two will have points at infinity only in the case where the initial curve intersects the circle circumscribed about the coordinate triangle.

Elliptic curves of degree four, similar to rational curves, could be obtained via a quadratic transformation of curves of degree three that have no double points and pass through two vertices of the coordinate triangle.

Bicircular quartics are characterized by the fact that they have two double points coinciding with cyclic points of the plane. These curves can be obtained in several ways, in particular, by using inversion of a circular curve of degree three; the foci of this curve will correspond to the foci of the curve of degree four.

If a bicircular curve is at the same time rational, then it has only one double point located at a finite distance. The general form of the equation of such curves is

$$(x^2+y^2)^2+(dx+ey)(x^2+y^2)+ax^2+bxy+cy^2=0.$$

The equation of the tangent at the double point is $ax^2+bxy+cy^2=0$. Depending on the sign of the discriminant b^2-4ac, the double point of the curve will be an isolated point, turning point, or a node.

Bicircular rational curves of degree four could be obtained by an inversion of a curve of degree two if the center of inversion does not lie on this curve.

Inversion of an ellipse is given in Figure 3.28*a*. Here the center of inversion is located at the center of the ellipse, which is an isolated point of the curve. Inversions of a parabola and hyperbola are shown in Figures 3.28*b*,c and 3.28*d*, correspondingly, with the center of inversion located at the focus of the hyperbola.

It should be remarked that bicircular rational curves of degree four can be obtained as pedals of an ellipse or hyperbola.

Remarkable curves of degree four include the limacon of Pascal, the cardioid, the elliptic and hyperbolic lemniscate of Booth, the lemniscate of Bernoulli, and others. General type bicircular curves of degree four (not necessarily rational) contain the following remarkable curves: ovals of Descartes, Cassinian ovals, curves of Perseus, and others.

3.4.1. Conchoid

A *conchoid* of a given curve is the curve obtained by increasing or decreasing the radius vector at every point of the curve by a constant length l. If the equation of the curve is given in polar coordinates by $\rho=f(\varphi)$, then the equation of its conchoid is $\rho=f(\varphi)\pm l$.

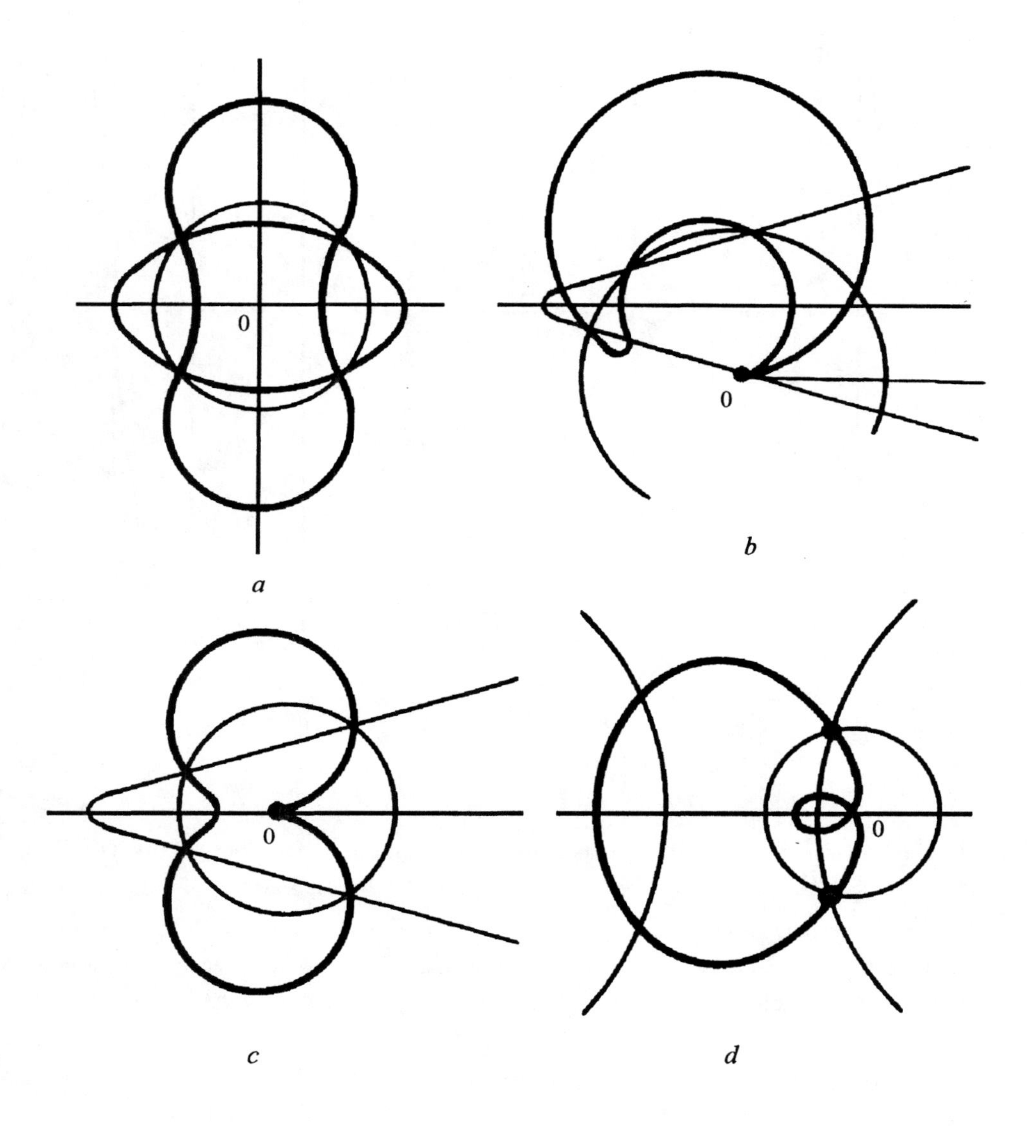

Figure 3.28. Inversion of (*a*) the ellipse, (*b*), (*c*) parabola, (*d*) hyperbola.

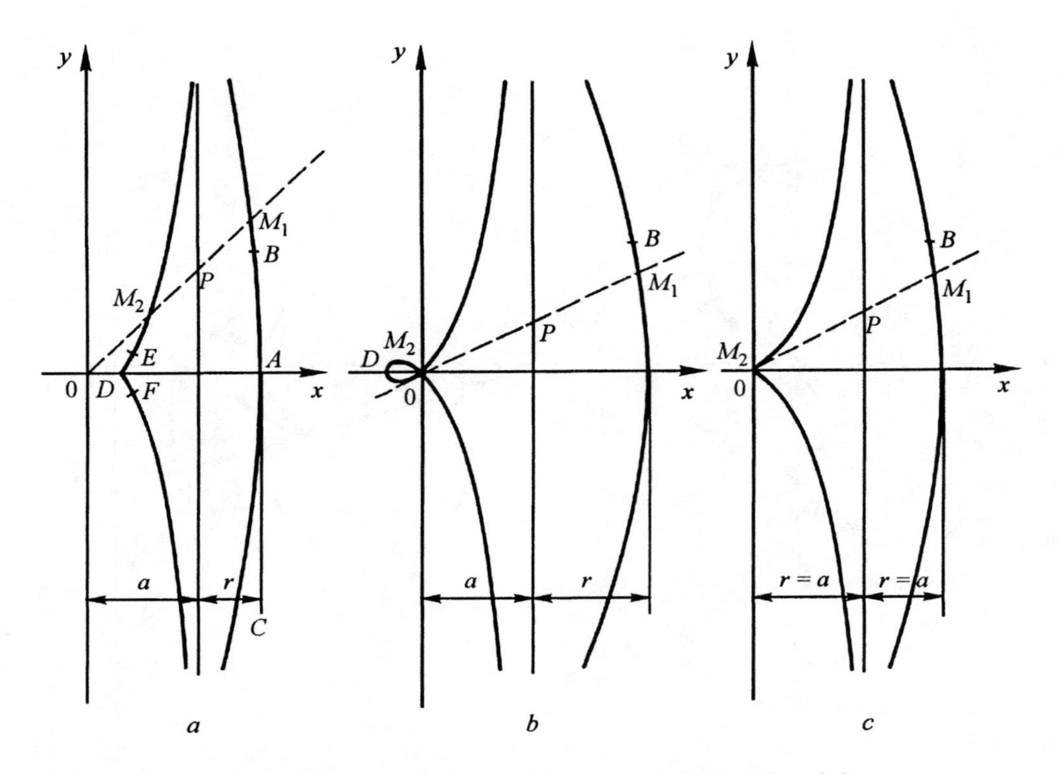

Figure 3.29. The conchoid of Nicomedes $(x-a)^2(x^2+y^2)-l^2x^2=0$.

Conchoid of Nicomedes. This is a conchoid of a straight line, i.e. the locus of points M such that $OM = OP \pm l$ ("+" is taken for the outer branch, and "–" for the inner, see Figure 3.29). Its equation in Cartesian coordinates is given by $(x-a)^2(x^2+y^2)-l^2x^2=0$. The equation in parametric form is $x = a + l\cos\varphi$, $y = a\tan\varphi + l\sin\varphi$, and in polar coordinates, $\rho = a/\cos\varphi \pm l$.

For the outer branch, the straight line $x = a$ is an asymptote, and the vertex is located at the point $A(a + l; 0)$; the curve has two inflection points at B and C. The area of the portion of the plane lying between the curve and the asymptote is unbounded, $S = \infty$. For the inner branch, $x = a$ is an asymptote, $D(a - l; 0)$ is a vertex, the point $O(0; 0)$ is a double point with the behavior of the curve at this point depending on the parameters a and l: a) if $l < a$, the point is isolated (Figure 3.29a), and the curve has two inflection points E and F; b) if $l > a$, the point is a node (Figure 3.29b), and the curve has a maximum and minimum for $x = a - \sqrt[3]{ab^2}$; c) if $l = a$, then the point is a turning point (Figure 3.29c).

The inflection points of all conchoids with the same value of the parameter a but different l lie on the parabola of Neil, $y^3 = 2ax^2$.

Indeed, points of inflection of a curve in polar coordinates are such that $\rho^2 + 2\rho'^2 - \rho\rho'' = 0$. In this case, $\rho = 2a\cos^2\varphi/\sin^3\varphi$, or, in Cartesian coordinates, $y^3 = 2ax^2$.

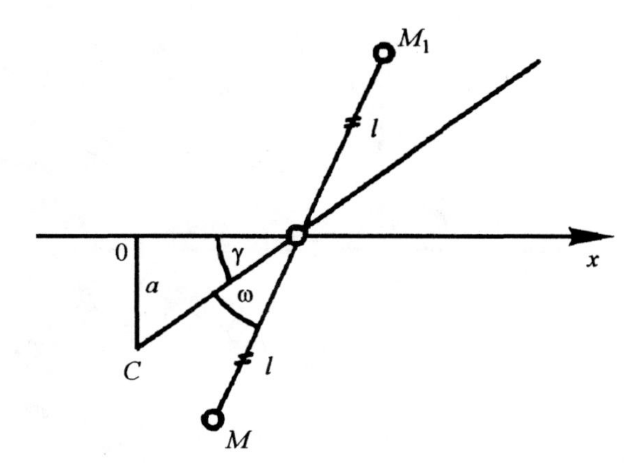

Figure 3.30. A scheme for constructing an arbitrary point of an oblique conchoid.

The area bounded by the straight line that passes through the pole of the conchoid and perpendicular to its baseline, arcs of the upper and lower branches of the curve, and a radius vector OM_1 which corresponds to an angle φ, is computed by the formula:

$$S = \frac{1}{2}\int_{\pi/2}^{\varphi}\left[\left(\frac{a}{\sin\varphi}+l\right)^2 - \left(\frac{a}{\sin\varphi}+l\right)^2\right]d\varphi = 2al\ln\tan\frac{\varphi}{2}.$$

This shows, in particular, that the entire area bounded by branches of a conchoid is infinitely large.

The radius of curvature at the point $O(0; 0)$ is given by $R = l\sqrt{l^2 - a^2}/2a$.

Remark. The basis of a conchoid of Nicomedes is a straight line. In the general case, the basis could be any curve. If a segment of a fixed length is drawn from the radius vector, which joins the origin and a point on the base curve, not along the radius vector but at a certain angle to it, say ω, then we come to a generalization of the notion of a conchoid. For example, let the base line, as in the case of the conchoid of Nicomedes, be a straight line. Take the foot of the perpendicular dropped from the pole C to the base line to be the origin, and the length of this perpendicular to be a. Then the parametric equations of this conchoid will be $x = a\cot\gamma - l\cos(\omega+\gamma)$, $y = -l\sin(\omega+\gamma)$.

The angle γ (Figure 3.30) plays the role of the parameter, and the constant angle is ω for one branch of the curve and $-\omega+\pi$ for the other.

By eliminating the parameter γ from these equations, we get an equation of the so-called *oblique conchoid*, $[xy\cos\omega - (y^2 + ay - l^2)\sin\omega]^2 = (x\sin\omega + y\cos\omega)(l^2 - y^2)$. In particular, we get the usual conchoid of Nicomedes by setting $\omega = 0$ and $\omega = \pi$. Depending on the value of the angle ω, the oblique conchoid could have various forms which sometimes could be fairly complicated (Figure 3.31).

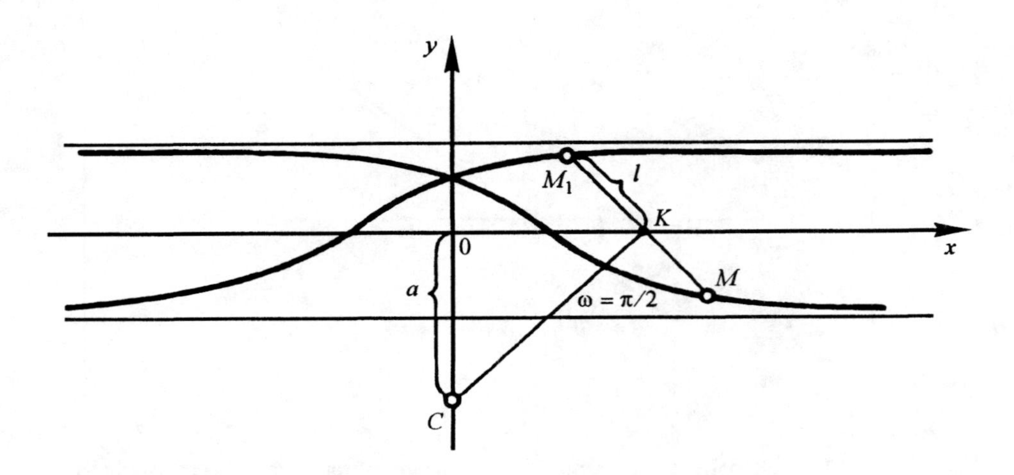

Figure 3.31. The orthogonal conchoid.

Note. It is believed that the first who studied the conchoid was Nicomedes (third century B.C.), who applied it to the problem of trisection of the angle. This curve is also interesting from the point of view of the history of mathematics, for it was the curve used by Descartes to demonstrate his method for construction of tangents and normal lines to curves.

3.4.2. The limacon of Pascal

The *limacon of Pascal* can be defined as a conchoid of a circle, $OM = OP \pm l$ (the pole lies on the circle), see Figure 3.32.

The equation of the curve in the Cartesian coordinate system is

$$(x^2 + y^2 - ax)^2 = l^2(x^2 + y^2),$$

its parametric equation is $x = a\cos^2\varphi + l\cos\varphi$, $y = a\sin\varphi\cos\varphi + l\sin\varphi$. In polar coordinates ($O$ is the pole, Ox is the polar axis), the equation is $\rho = a\cos\varphi + l$ (l is the diameter of the circle). The vertices are A, $B(a \pm l; 0)$. The form of the curve depends on the values of a and l (Figure 3.32).

The curve has four extremums if $a > l$, and there are two extremums if $a \le l$, C, D, $F(\cos\varphi = \dfrac{-l \pm \sqrt{l^2 + 8a^2}}{4a})$.

The inflection points G, $H(\cos\varphi = -\dfrac{2a^2 + l^2}{3al})$ if $a < l < 2a$. The origin is a double point; it is isolated if $a < l$, a node if $a > l$, and a turning point if $a = l$.

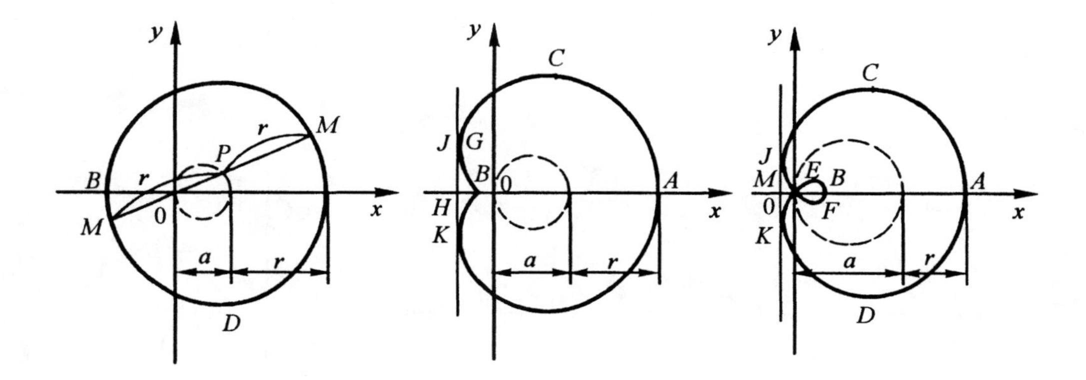

Figure 3.32. Limacon of Pascal $(x^2 + y^2 - ax)^2 = l^2(x^2 + y^2)$.

The area enclosed by having the polar radius make a complete revolution is $S = \left(\frac{a^2}{2} + l^2\right)\pi$. If the curve has a loop, then

$$S_1 = \left(\frac{a^2}{2} + l^2\right)\varphi_1 + \frac{3}{2}\,l\sqrt{a^2 - l^2} \quad \left(\varphi_1 = \arccos\left(-\frac{l}{a}\right)\right);$$

$$S_2 = \left(\frac{a^2}{2} + l^2\right)\varphi_2 - \frac{3}{2}\,l\sqrt{a^2 - l^2} \quad \left(\varphi_1 = \arccos\left(\frac{l}{a}\right)\right);$$

where S_1 is the area bounded by the outer loop, and S_2 is the area bounded by the inner one.

Note that the limacon of Pascal can also be regarded as a pedal of a circle with respect to an arbitrarily chosen point in the plane.

A limacon of Pascal is a trisectrix, a curve that allows one to trisect the angle. The limacon of Pascal has many applications in engineering (for example, it is used in the device for lifting and lowering the semaphore arm signal, for drawing the profile of the eccentric that makes harmonic oscillations, etc.).

3.4.3. Cycloidal curves

One of the kinematic ways to construct a curve is the following. One curve rolls on the other curve without slipping. Here a point on the first curve traces a certain new line. Among the lines constructed in such a way there are curves that are trajectories of a fixed point on a circle that rolls on another circle without slipping. The curves thus obtained are called *cycloidal curves*.

Cycloidal curves can be algebraic or transcendental. Some important types of such curves are algebraic. We will look at some examples of these curves.

Epicycloid. An *epicycloid* is a curve traced by a point on a circle that rolls on the circumference of another circle without slipping (Figure 3.33).

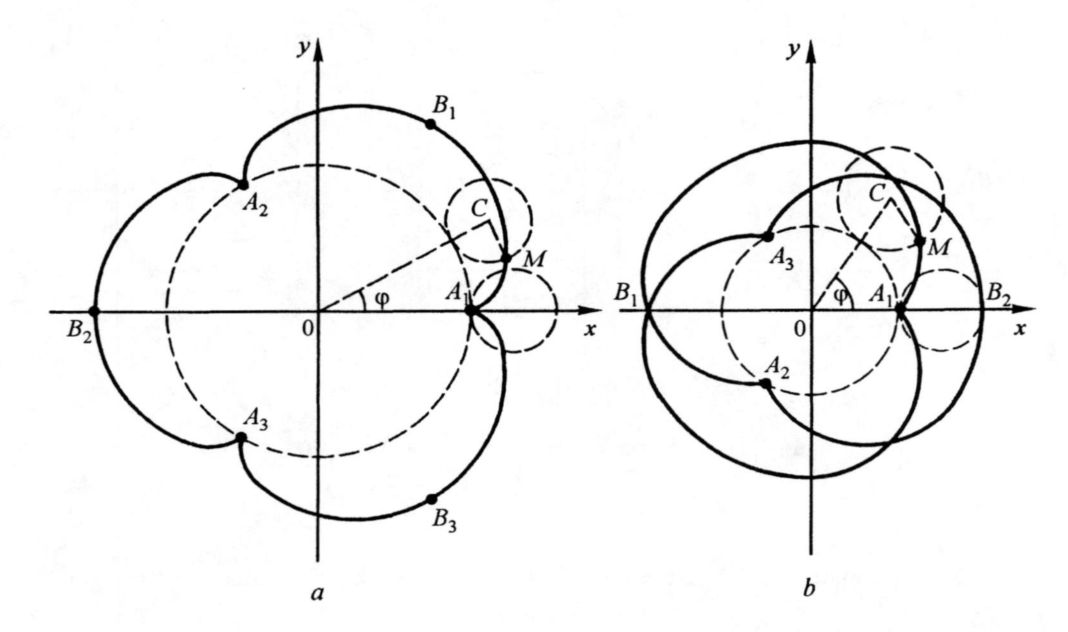

Figure 3.33. The epicycloid $x = (A + a)\cos\varphi - a\cos\left(\frac{A+a}{a}\varphi\right)$, $y = (A + a)\sin\varphi - a\sin\left(\frac{A+a}{a}\varphi\right)$. (*a*) $m = 3$; (*b*) $m = 3/2$.

The parametric equation of the curve has the form:

$$x = (A + a)\cos\varphi - a\cos\left(\frac{A+a}{a}\varphi\right),\quad y = (A + a)\sin\varphi - a\sin\left(\frac{A+a}{a}\varphi\right),$$

where A is the radius of the fixed circle, a is the radius of the moving circle, $\varphi = \angle COx$. The form of the curve depends on the ratio $A/a = m$. In particular, if $m = 1$, the curve is a cardioid.

If m is an integer, the epicycloid consists of m branches that enclose the fixed circle. The turning points are $A_1, A_2, \ldots, A_m$ ($\rho = A$, $\varphi = 2k\pi/m$ ($k = 0, 1, \ldots, m-1$). The vertices are located at the points $B_1, B_2, \ldots, B_m$ ($\rho = A + 2a$, $\varphi = 2\pi/m(k + 1/2)$). If m is a fraction, then the moving point M, having traced a finite number of branches, returns in the initial point. If m is irrational, the number of branches is infinite, and the point M never returns in the initial position.

The length of one branch is $L_{A_1B_1A_2} = 8(A + a)/m$. If m is an integer, the length of the entire curve equals $L = 8(A + a)$.

The area of the sector $A_1B_1A_2A_1$ is $S = \pi a^2\left(\frac{3A+2a}{A}\right)$. The radius of curvature is $R = \frac{4a(A+a)}{2a+A}\sin\frac{A\varphi}{2a}$, and at the vertices it is $R_B = \frac{4a(A+a)}{2a+A}$. The natural equation of the regular epicycloid is given by $R^2/\tilde{a}^2 + (s - \tilde{b})^2/\tilde{b}^2 = 1$, $0 \le s \le \tilde{b}$, where $\tilde{a} = 4a(A \pm a)/(A \pm 2a)$;

$\tilde{b} = 4a(A \pm a)A$, R is the radius of curvature, s is the length of the arc that starts at one of the initial points. The latter equation describes the following kinematic property of the epicycloid: if the arc of a regular epicycloid rolls on a straight line without slipping, then the center of curvature of the tangent point moves along an ellipse that has its center at the point of the straight line on which the vertex of the epicycloid rolls.

If $m < 0$ but $|m| > 1$, then the generating disk encloses the fixed one, and the curve obtained lies outside of the fixed disk. Such types of cycloids are called *pericycloids.*

Let us mention a few properties of epicycloids: 1) the epicycloid with a rational m is an algebraic curve; 2) the tangent at an arbitrary point of the epicycloid passes through a point of the generating disk that is diametrically opposite to the point at which it touches the fixed disk; the normal line passes through the point of contact; the evolute of the epicycloid is an epicycloid similar to the given one with the ratio equal to $k = 1/(1 + 2m)$ and rotated relative to the given epicycloid by the angle $m\pi$; 4) the epicycloid is a catacaustic of the circle; 5) the pedal of the epicycloid relative to the center of the fixed circle is a "rose''.

Hypocycloids. A *hypocycloid* is a curve described by a point on a circle that rolls on the interior of another circle without slipping (Figure 3.34).

The equation of a hypocycloid, the coordinates of the vertices and turning points, formulas for the length of arc, area, and radius of curvature are the same if "+" is replaced with "–". The number of turning points is the same as for the epicycloid.

Extended and shortened hypocycloids. These curves, sometimes called *epitrochoids and hypotrochoids* are traced by a point located on the circumference or inside a circle rolling on, correspondingly, the exterior or interior of another circle without slipping (Figures 3.35, 3.36).

Parametric equations are of the form:

$$x = (A + a)\cos\varphi - \lambda a\cos\left(\frac{A+a}{a}\varphi\right),\ y = (A + a)\sin\varphi - \lambda a\sin\left(\frac{A+a}{a}\varphi\right),$$

where A is the radius of the fixed circle, a is the radius of the moving circle, the sign "+" in the expression for the hypocycloid is replaced with "–", $\lambda = CM$ ($\lambda > 1$ for the extended hypocycloid, $\lambda < 1$ for the shortened one). If $A = a$ (λ is arbitrary), the hypocycloid becomes an ellipse with semiaxes $a(1 + \lambda)$ and $a(1 - \lambda)$. In the case where $A = a$, we get a limacon of Pascal, $x = a(2\cos\varphi - \lambda\cos 2\varphi)$, $y = a(2\sin\varphi - \lambda\sin 2\varphi)$.

Remark. The considered types of epicycloid and hypocycloid are particular cases of the cycloidal curves.

In the general case, the point describing the curve is located not on the circumference of the generating disk but at a certain distance h from the center. Curves thus obtained are called *trochoids*.

If $h > a$, the trochoid is called *extended,* if $h < a$, the trochoid is *shortened,* if $h = a$, the trochoid becomes an epicycloid or hypocycloid.

If we set $A = 2a$ in the equation of a hypocycloid, then, as we mentioned, the obtained curve will be an ellipse. This leads to the use of hypocycloids in designing elliptic compasses, etc.

Cardioid. A *cardioid* can be defined to be an epicycloid with equal diameters of the fixed and moving circles ($m = 1$), see Figure 3.37.

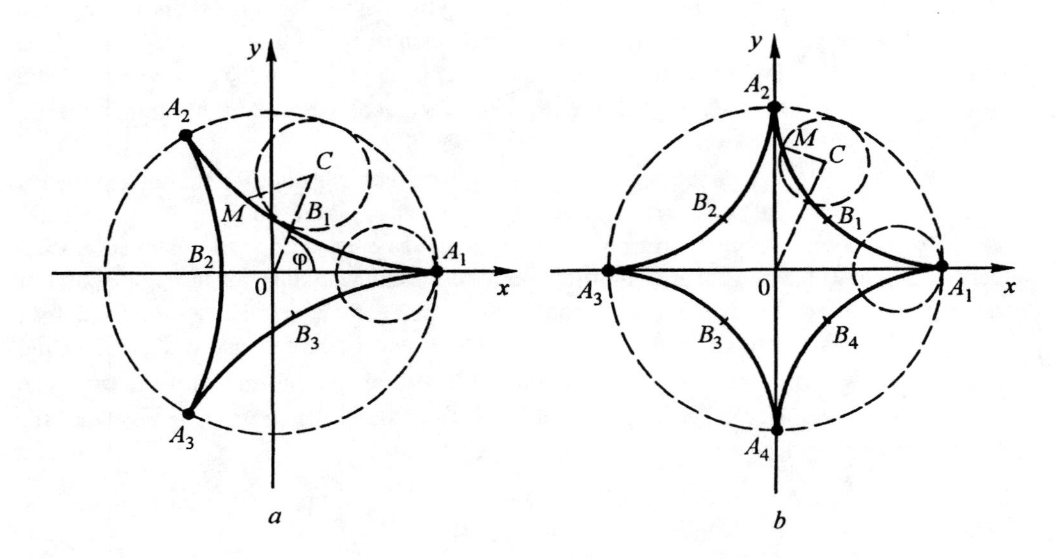

Figure 3.34. The hypocycloid. (a) $m = 3$; (b) $m = 4$.

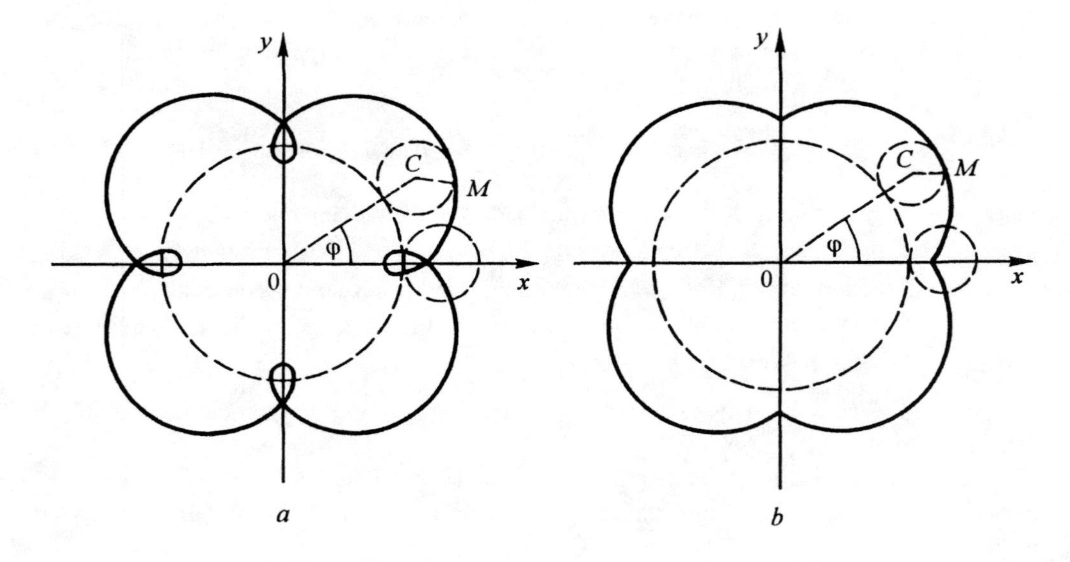

Figure 3.35. The epitrochoid: (a) $\lambda > 1$, $a > 0$; (b) $\lambda < 1$, $a > 0$.

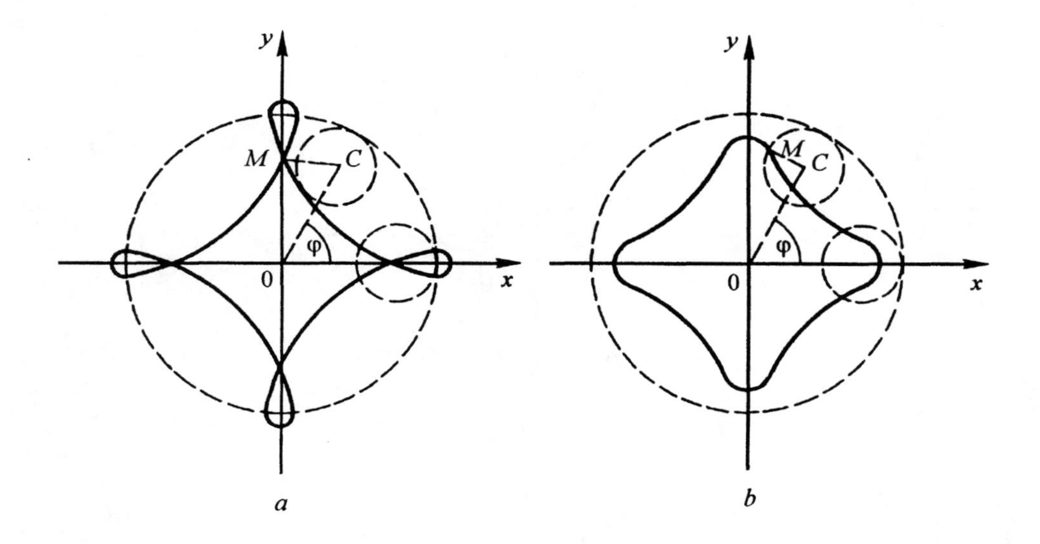

Figure 3.36. The epitrochoid: (*a*) $\lambda > 1, a > 0$; (*b*) $\lambda < 1, a < 0$.

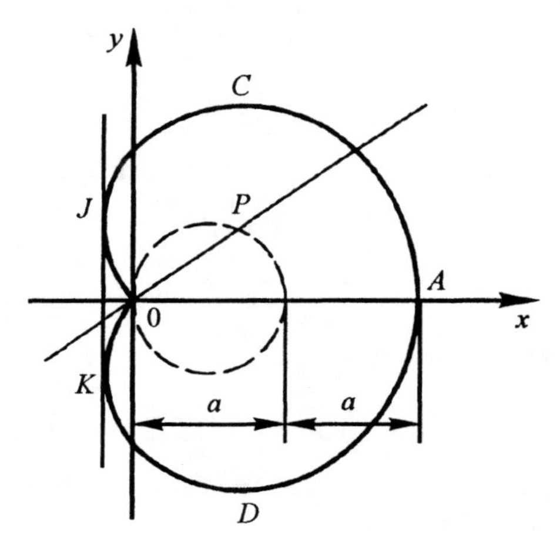

Figure 3.37. The cardioid $x = a\cos\varphi(1 + \cos\varphi), y = a\sin\varphi(1 + \cos\varphi)$.

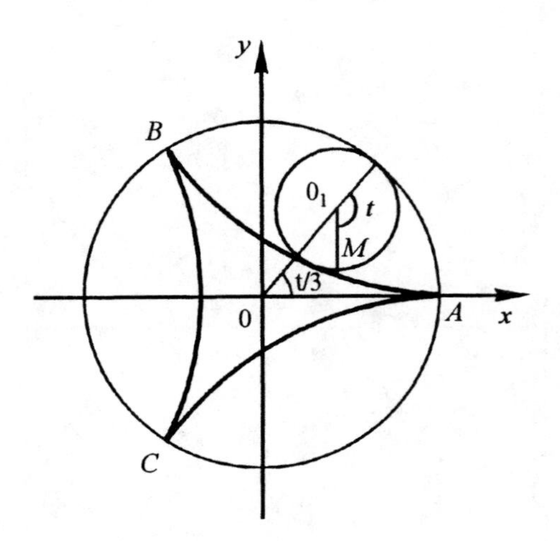

Figure 3.38. The Steiner curve $x = 2a\cos t/3 + a\cos 2t/3$, $y = 2a\sin t/3 - a\sin 2t/3$.

The equation of the cardioid in Cartesian coordinates is $(x^2+y^2)^2 - 2ax(x^2+y^2) = a^2y^2$, $a > 0$; in parametric form it is given by $x = a\cos t(1 + \cos t)$, $y = a\sin t(1 + \cos t)$, $0 \le t \le 2\pi$; and in the polar coordinates, it is $\rho = a(1 + \cos\varphi)$.

The origin is a turning point, the vertex lies in $A(2a; 0)$, the points of maximum and minimum ($\cos\varphi = 1/2$) are at C, D, $\left(\frac{3}{4}a; \pm\frac{3\sqrt{3}}{4}a\right)$. The area of the cardioid is $S = \frac{3}{2}\pi a^2$; the length of the whole cardioid, $L = 8a$.

Since the cardioid is a particular case of the epicycloid ($m = 1$), it possesses the same properties as the epicycloid. In particular, the properties are: a) tangents to the cardioid constructed at the endpoints of a chord that passes through the pole are mutually perpendicular, also note that the locus of points of intersection of these tangents is the circle $x^2 + y^2 = 9a^2$; b) the volume of the solid obtained by revolution of the cardioid about its axis equals $\frac{64}{3}\pi a^3$; c) the area of the surface of this body of revolution is computed by the formula $\frac{128}{5}\pi a^2$; d) the cardioid is a conchoid of a disk, and it is also a pedal of a circle with respect to a point that belongs to this circle; e) inversion of a cardioid relative to the turning point gives a parabola.

Steiner curve. A *Steiner curve* can be defined as a hypocycloid such that the radius of the moving circle is three times smaller than the radius of the fixed circle, i.e. $m = 1/3$ (Figure 3.38).

The parametric equation of the curve is $x = 2a\cos\frac{t}{3} + a\cos\frac{2t}{3}$, $y = 2a\sin\frac{t}{3} - a\sin\frac{2t}{3}$, where t is the angle of rotation of the generating circle. In Cartesian coordinates, the equation of the curve is $(x^2+y^2)^2 + 8ax(3y^2 - x^2) + 18a^2(x + y^2) - 27a^4 = 0$.

Note an interesting property of the Stein curve: The point of intersection of mutually perpendicular tangents to a Steiner curve lies on the circle inscribed in this curve.

The radius of curvature at an arbitrary point is given by $K = 8a\sin\frac{t}{2}$. The length of the arc counting from the initial point A to some point M is $l = \frac{16}{3}R\sin^2\frac{t}{4}$, the corresponding length of one branch equals $16/3a$, and the length of the entire curve is $L = 16a$. The area bounded by the curve is $S = 2\pi a^2$.

The evolute of the Steiner curve is also a Steiner curve similar to the given one with ratio $k = 3$ and rotated by the angle $\pi/3$ relative to the initial curve.

A Steiner curve is also interesting because of its pedals. The equation of a pedal of the Steiner curve with respect to the point $S(b; 0)$ has the form $(x^2 + y^2)^2 + [(b + 3a)y^2 + (b - a)x^2]x = 0$, or $\rho = 4a\cos^3\varphi - (b + 3a)\cos\varphi$, whence we see the following.

1) If $b = 0$, we get a tree folium rose, the pedal of the Steiner curve with respect to the origin (Figure 3.39).

2) If $b = a$, we get $\rho = -4a\cos\varphi\sin^2\varphi$, a *straight double folium*, which is a pedal of the Steiner curve with respect to the point S (Figure 3.40).

3) If $b = -a$, then $\rho = 2a\cos\varphi\cos 2\varphi$, which gives a straight trifolium. It is a pedal of the Steiner curve with respect to the point S_1 (Figure 3.41).

4) If $b = -3a$, then $\rho = 4a\cos^3\varphi$, which is a *simple folium* and a pedal of the Steiner curve with respect to the point S_2 (Figure 3.42).

Astroid. An *astroid* is a particular case of the hypocycloid for $m = \frac{1}{4}$ (Figure 3.43).

An astroid thus is the trajectory traced by a point of a disk of radius R that rolls on the interior of another fixed disk with the radius four times the radius of the first disk. The equation of the curve is $(x^2 + y^2 - R^2)^3 + 27R^2x^2y^2 = 0$, which is an algebraic curve of degree six. The parametric equation is given by $x = R\cos^3\frac{t}{4}$, $y = R\sin^2\frac{t}{4}$. Eliminating t from these equations, we get an extended equation of the astroid, $x^{2/3} + y^{2/3} = R^{2/3}$.

The radius of curvature at an arbitrary point is given by the formula $R_k = \frac{3}{2}R\sin\frac{t}{2}$. The length of the asteroid counting from the point A to a point M is given by $l = \frac{3}{2}R\sin^2\frac{t}{4}$. The length of one branch of the asteroid equals $\frac{3}{2}R$, the length of the entire curve is $L = 6R$.

An evolute of an astroid is also an astroid homothetic to the initial one with the ratio coefficient $k = 2$ and rotated at the angle of $\pi/4$.

The area bounded by the whole astroid equals $\frac{3}{8}\pi R^2$. The volume of the body of revolution is $\frac{32}{105}\pi R^3$. Its surface area equals $\frac{12}{5}\pi R^2$.

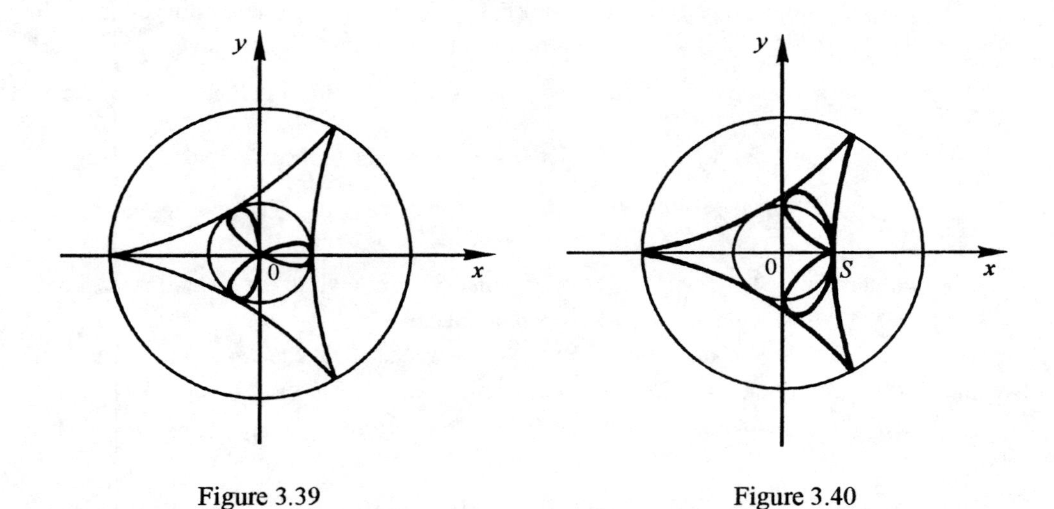

Figure 3.39 Figure 3.40

Figure 3.39. The pedal of a Stein curve with respect to the origin.

Figure 3.40. The straight double folium.

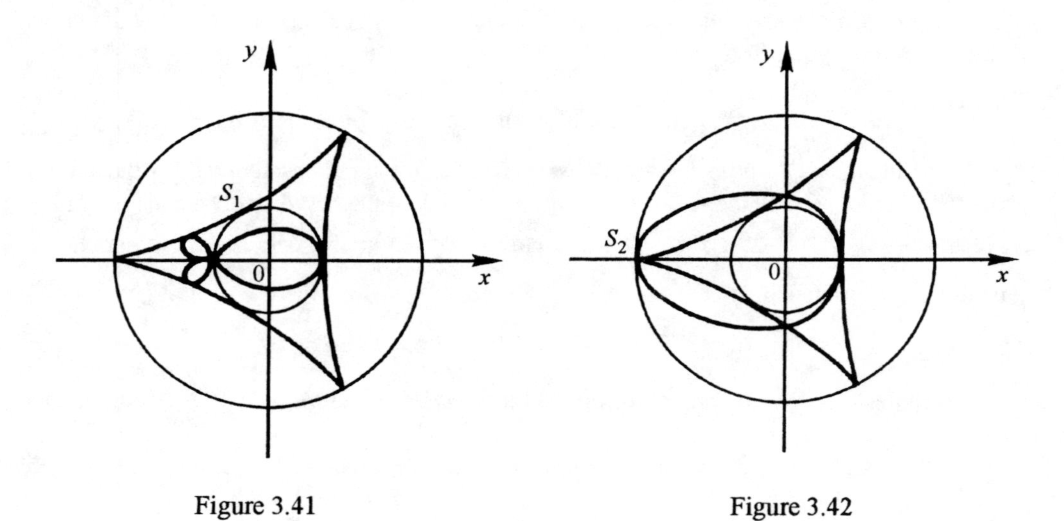

Figure 3.41 Figure 3.42

Figure 3.41. The straight trifolium.

Figure 3.42. The simple folium.

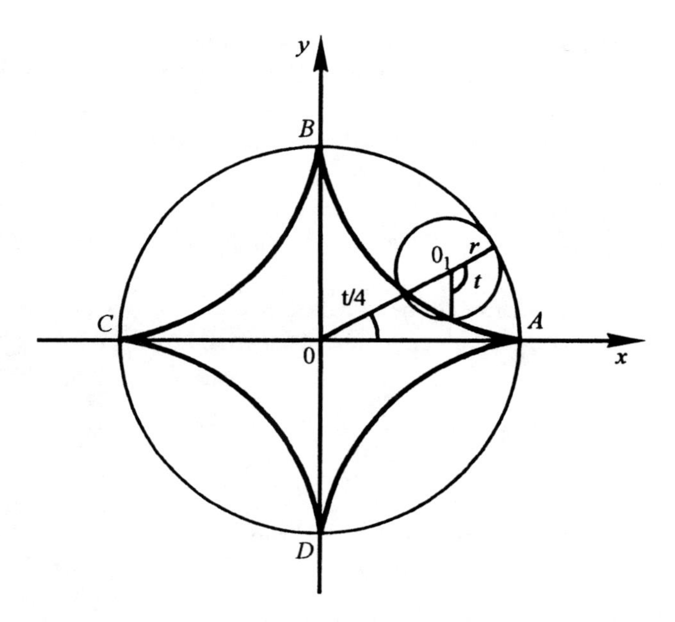

Figure 3.43. The astroid (definition).

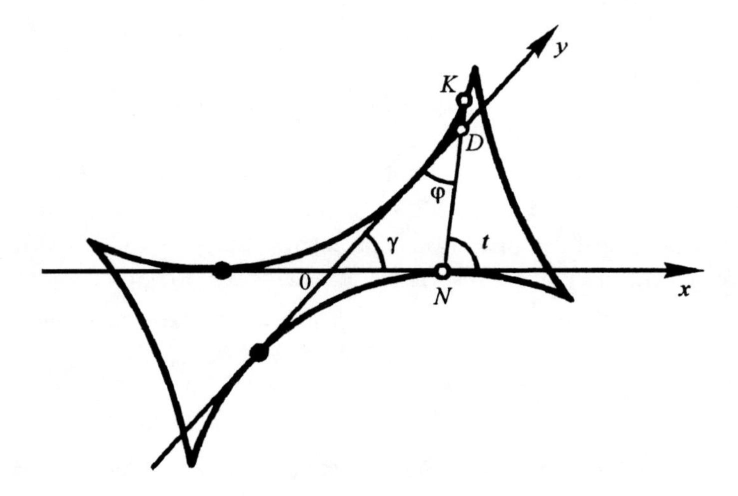

Figure 3.44. The slant astroid.

The following is a list of some properties of the astroid.

1. An astroid is an envelope of a line segment of constant length the endpoints of which slide on two mutually perpendicular lines.
2. An astroid is a locus of feet of perpendiculars dropped from a vertex of a rectangle to its diagonal.
3. Each tangent intersects the astroid in two points such that the tangents to the astroid at these points intersect in a point that lies on a circle circumscribed about the astroid.

A generalization of this astroid is the so-called *slant astroid*, which is an envelope of a line segment ND of a constant length R the endpoints of which slide on two straight lines that intersect at an arbitrary angle γ (Figure 3.44). The parametric equation of a slant astroid is given by $x=\frac{R}{\sin^2\gamma}\cos t\sin^2(t-\gamma)$, $y=\frac{R}{\sin^2\gamma}\cos(t-\gamma)\sin^2 t$. If $\gamma=\pi/2$, these equations define the usual (straight) astroid.

Oval of Descartes. The *oval of Descartes* is a locus of points whose sum of distances to two fixed points, called foci, multiplied by given numbers is a constant, $mMF+nMF_1=c$ (Figure 3.45). An equation defining ovals of Descartes in polar coordinates is $(n^2-m^2)\rho^2+2\rho(mc-$ $-n^2d\cos\varphi)+n^2d^2-c^2=0$. In the Cartesian coordinate system, the equation is $m\sqrt{x^2+y^2}+n\sqrt{(x-d)^2+y^2}=c$, or $(x^2+y^2-2rx)^2-l^2(x^2+y^2)-k=0$, where r,l,k are certain constants connected via certain relations with the parameters n,m,d.

Chasles has shown that, besides the foci F, F_1, ovals of Descartes have the third focus. Thus ovals of Descartes become loci of points whose sum of distances to three fixed points multiplied by given numbers equals zero (Figure 3.46).

An analysis of ovals of Descartes shows that each oval consists of two closed curves, one of which encloses the other.

Note that if $k=0$, the oval in this case becomes a limacon of Pascal, the node of which coincides with one of the foci of the oval. If $k>0$, the oval consists of two closed lines, one of which lies inside the loop of the limacon and the other lies outside of it.

As k increases, the inner oval shrinks to the second focus F_1. The outer oval, having a heart shape for values of k close to zero, becomes a convex oval. In the case where $k<0$, the inner oval is located inside the limacon but outside of its loop and, having the shape of a chipped half moon for small absolute values of k, becomes a convex oval as the absolute value of k increases, and then shrinks to a point that coincides with the third focus of the oval.

Cassinian ovals. A *Cassinian curve* (often called a *Cassinian oval* although actually it is not always an oval) is a locus of points M such that the product $MF_1{\cdot}MF_2$ of distances from the endpoints of the line segment $F_1F_2=2c$ equals the square of a given number, $MF_1{\cdot}MF_2=a^2$ (Figure 3.47). The points F_1 and F_2 are called *foci*, the straight line F_1F_2 is the axis of the Cassinian curve, and the middle point O of the line segment F_1F_2 is the *center*.

The curve is defined in the Cartesian coordinate system by the equation

$$(x^2+y^2)^2-2c^2(x^2-y^2)=a^4-c^4$$

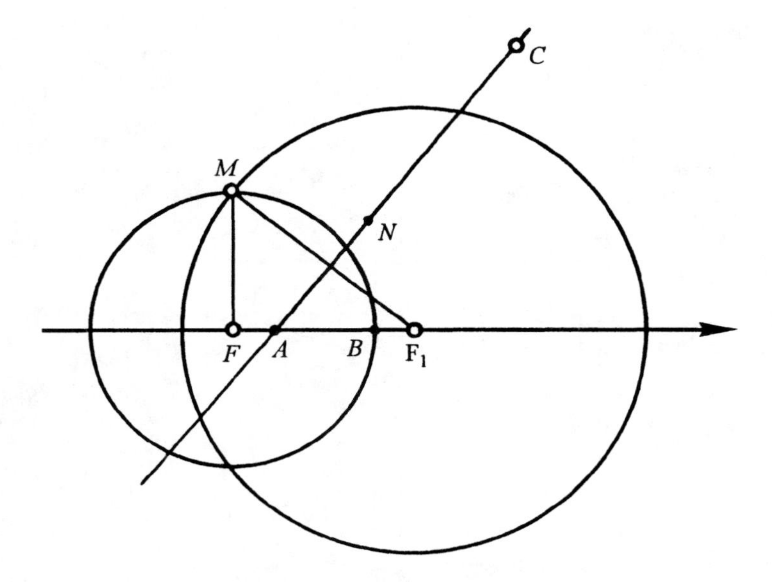

Figure 3.45. The oval of Descartes (definition).

Figure 3.46. The ovals of Descartes (in the case of existence of the third focus).

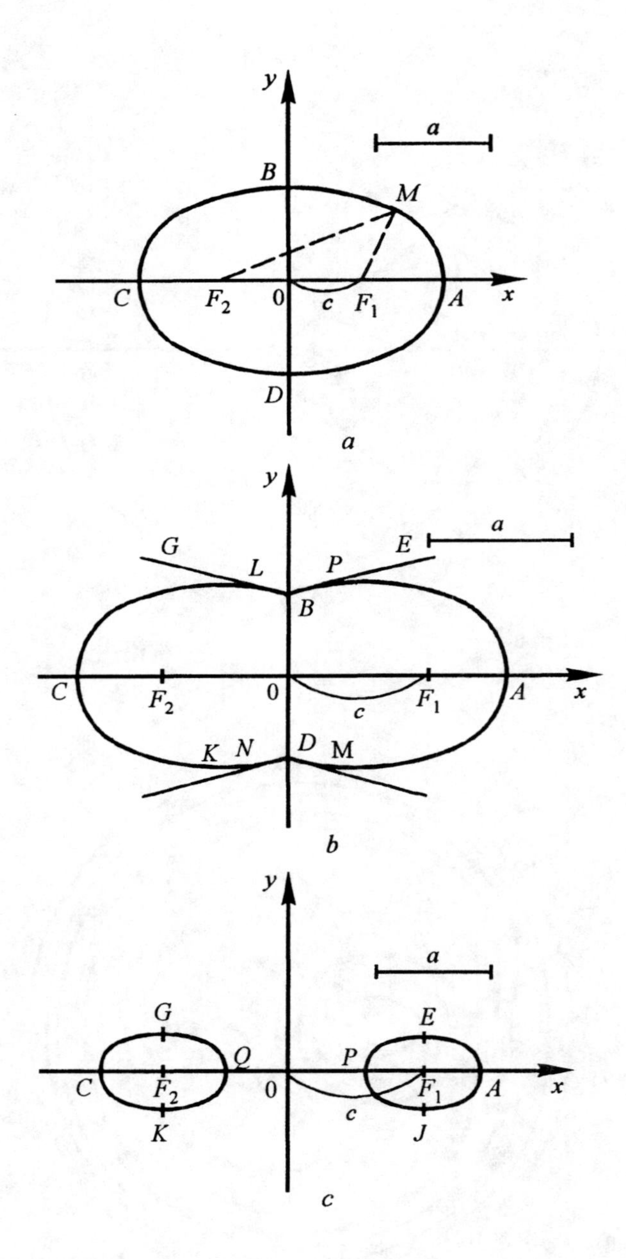

Figure 3.47. The Cassinian curves: (*a*) $a > c\sqrt{2}$; (*b*) $c < a < c\sqrt{2}$; (*c*) $a < c$.

(O is the origin, F_1F_2 lies on the x-axis). In the polar coordinates, the curve is given by

$$\rho^4 - 2c^2\rho^2\cos 2\varphi + c^4 - a^4 = 0, \text{ or } \rho^2 = c^2\cos 2\varphi \pm \pm\sqrt{a^4 - c^4\sin^2 2\varphi},$$

or else $\rho^2 = c^2\cos 2\varphi \pm \sqrt{c^4\cos^2 2\varphi + (a^4 - c^4)}$. Both signs "+" and "–" are taken if $a < c$, and only " + " otherwise.

The Cassinian curve is symmetric with respect to the axes Ox and Oy, and hence, with respect to the origin O. The shape of the curve depends on the ratio of a and c.

1) If $a > c\sqrt{2}$, then we have an elliptical oval (Figure 3.47*a*) which intersects the x-axis in the points $A,C(\pm\sqrt{a^2+c^2};\, 0)$, and the y-axis in the points $B,D(0;\, \pm\sqrt{a^2-c^2})$. If $a = c\sqrt{2}$, then the shape of the curve is the same; the points of intersection with the axes are $A,C(\pm c\sqrt{3};\, 0)$, and $B,C(0;\, \pm c)$. The curvature at the points B and D equals zero.

2) If $c < a < c\sqrt{2}$, we get an oval (Figure 3.47*b*). It intersects the axes in the same points as in the case 1); the curve has maximums and minimums at the points $B,D(0;\, \pm c)$, $E,G,K,I\left(\pm\frac{\sqrt{4c^4 - a^4}}{2c};\, \pm\frac{a^2}{2c}\right)$. The curve has four inflection points which are P,L,M,N $\left(\pm\sqrt{\frac{m-n}{2}};\, \pm\sqrt{\frac{m+n}{2}}\right)$, where $n = \frac{a^4 - c^4}{3c^2}$, $m = \sqrt{\frac{a^4 - c^4}{3}}$.

3) If $a = c$, the curve is a lemniscate (see below).

4) If $a < c$, we get two ovals (Figure 3.47*c*). The curve intersects the x-axis in points $A,C(\pm\sqrt{a^2+c^2};\, 0)$ and $P,Q(\pm\sqrt{c^2-a^2};\, 0)$; maximums and minimums are reached at the points $E,G,K,I\left(\pm\frac{\sqrt{4c^4 - a^4}}{2c};\, \pm\frac{a^2}{2c}\right)$. The radius of curvature is given by $R = \frac{2a^2\rho^3}{c^4 - a^4 + 3\rho^4}$, where ρ is the radius vector.

Note. Cassinian curves were conceived by astronomer D. Cassini in the seventeenth century in an attempt to describe the orbit of the Earth, which was not, in his opinion, an ellipse. Cassinian ovals can be observed by looking at an aragonite or silicon plate in a polarized light.

Flowing in two infinitely long conducting straight lines, the electric current induces the field in the plane perpendicular to the conductors such that its equipotential lines form Cassinian ovals.

Lemniscate of Bernoulli. The *lemniscate of Bernoulli* is a locus of points such that the product of distances to the endpoints of a given line segment F_1F_2 equals c^2, i.e. it is a special case of the Cassinian oval for $a = c$ (Figure 3.48).

The straight line F_1F_2 is the axis of the lemniscate. The equation of a lemniscate in Cartesian coordinates is $(x^2+y^2)^2 = 2c^2(x^2 - y^2)$, where O is the midpoint of the line segment F_1F_2 and the axis Ox is directed along F_2F_1. The equation in polar coordinates (with the pole at O and polar axis Ox) is $\rho^2 = 2c^2\cos 2\varphi$, where the angle φ ranges over the intervals $(-\pi/4;\, \pi/4)$ and $(3/4\pi;\, 5/4\pi)$. The parametric equation in rational form is

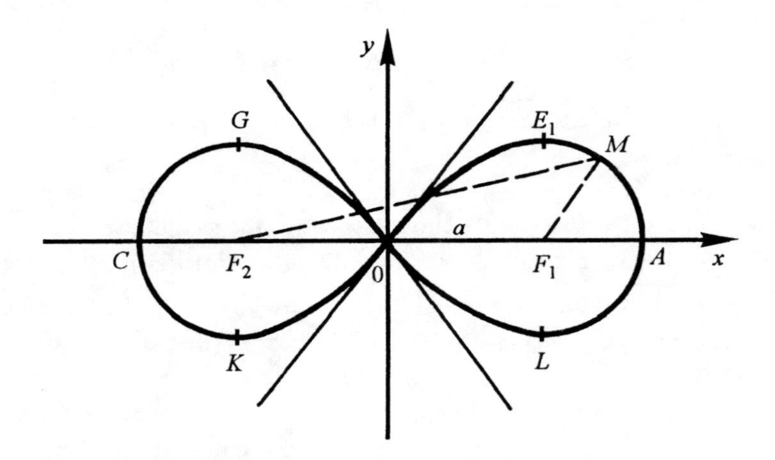

Figure 3.48. The lemniscate of Bernoulli.

$$x = c\sqrt{2}\frac{t+t^3}{1+t^4}, \quad y = c\sqrt{2}\frac{t-t^3}{1+t^4} \quad (-\infty < t < +\infty),$$

with the relation between t and φ being $t^2 = \tan(\pi/4 - \varphi)$.

The lemniscate has two symmetry axes: the straight line F_1F_2 (Ox) and the axis $Oy \perp Ox$. The point O is a node; the tangents are given by $y = \pm x$. This point is also an inflection point.

The curve intersects the axis Ox in points $A, C(\pm a\sqrt{2};\ 0)$ and achieves maximums and minimums at the points $E, G, K, L(\pm a\sqrt{3}/2; \pm a/2)$ with the polar angle at these points being $\varphi = \pm\pi/6$. The radius of curvature is $R = 2c^2/3\rho$, or $R = \rho/3\cos 3\varphi$. The area of each loop is given by $2S(\pi/4) = c^2$.

If perpendiculars are dropped from the center O of an equilateral hyperbola that has vertices at the points A_1 and A_2 to tangents to the curve, then the locus of feet of these perpendiculars will be a lemniscate that has the same vertices.

Let us recall some remarkable properties of the lemniscate of Bernoulli.

1) The angle μ formed by the tangent at an arbitrary point of the lemniscate and the radius vector equals $2\varphi + \pi/2$.

2) Tangents to the lemniscate drawn from the endpoints of a chord that passes through the pole are parallel. (It can be seen from Figure 3.49 that the angle between the perpendicular dropped to the tangent to the curve passing through the pole and the polar axis is three times the polar angle of the tangency point. Thus the lemniscate can be used for trisection of the angle.)

3) The locus of projections of centers of curvature of a lemniscate to the corresponding radius vectors is also a lemniscate with the equation given by $\rho^2 = \frac{4a^2}{9}\cos 2\varphi$.

4) Inversion of the lemniscate $\rho^2 = 2a^2\cos 2\varphi$ gives an equilateral hyperbola $\rho^{-2} = 2a^2\cos 2\varphi$.

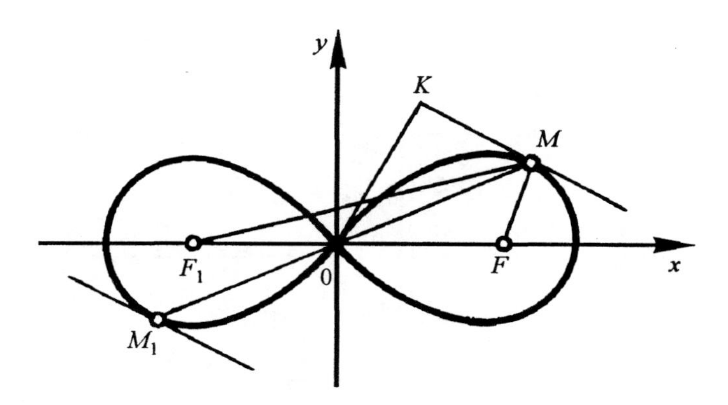

Figure 3.49. One of properties of the lemniscate of Bernoulli.

5) The pedal of the lemniscate is the sine spiral $\rho^{2/3} = (\sqrt{2}a)^{2/3}\cos\frac{2}{3}\varphi$ (the lemniscate is a pedal of an equilateral hyperbola).

6) The lemniscate is a locus of points symmetric to the center of an equilateral hyperbola with respect to its tangents.

7) Kinematically, the lemniscate can be obtained as the trajectory of the midpoint of the largest member of the hinged antiparallelogram, the opposite member of which is fixed.

8) The length of the line segment of the bisector of the angle between focal radius vectors is equal to the distance between the center of the lemniscate and the point of intersection of this bisector and the axis of the lemniscate.

9) The area of the sector formed by the axis of the lemniscate and the radius vector that corresponds to an angle φ is given by

$$S_\varphi = \frac{1}{2}\int_0^\varphi \rho^2 d\varphi = a^2\int_0^\varphi \cos 2\varphi \, d\varphi = \frac{a^2}{2}\sin 2\varphi,$$

and $S_{\pi/4} = \frac{a^2}{2} = \frac{1}{4}(a\sqrt{2}\,)^2$ for $\varphi = \pi/4$, i.e. the area bounded by the lemniscate equals the area of a square with side $a\sqrt{2}$.

Note that the quadrature of the lemniscate calculated by di Fagnano de Toschi (1750) was a remarkable event in mathematics at the time, since it proved that the statement of Tschirnhausen about the impossibility of finding the quadrature of curves containing several foliums was erroneous.

10) The perpendicular dropped from a focus of the lemniscate to the radius vector to any of its points divides the area of the corresponding sector in half.

11) The length of the arc cut by points with $\varphi_1 = 0$ and $\varphi_2 = \varphi$ equals

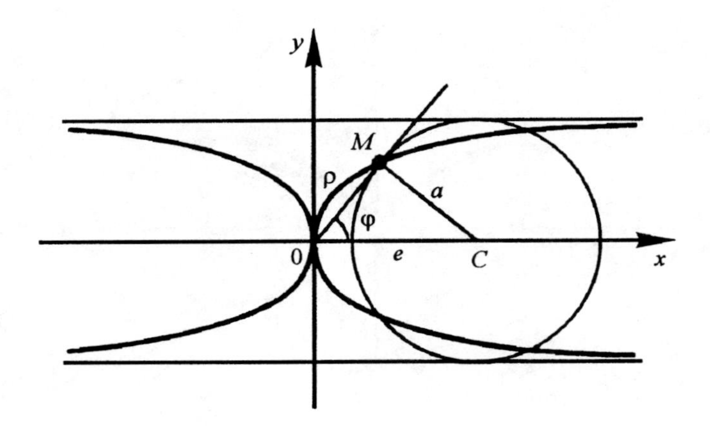

Figure 3.50. The kappa curve $\rho = a\cot\varphi$.

$$l = a\int_0^{\varphi}\frac{d\varphi}{\sqrt{\cos 2\varphi}} = a\int_0^{\varphi}\frac{d\varphi}{\sqrt{1-2\sin^2\varphi}}\,.$$

Hence the arc length of the lemniscate is determined from the incomplete elliptic integral of the first kind. Write the expression for l as

$$l = \frac{a}{\sqrt{2}}\int_0^{\pi/2}\frac{d\theta}{\sqrt{1-\frac{1}{2}\sin^2\theta}} = \frac{a}{\sqrt{2}}F\left(\frac{1}{\sqrt{2}},\theta\right).$$

The length of the quarter of the lemniscate is given by the complete elliptic integral of the first kind.

Note. The equation of a lemniscate appears for the first time in 1694, in works of J. Bernoulli. Lemniscates have broad applications, particularly in engineering. For example, one is used as a transient curve on small-radius rail roundings (for railways in mountain areas), etc.

Kappa curve. The *Kappa curve* is a locus of points of tangency of a circle of radius a and the tangents drawn to the circle from the origin with the center of the circle moving along the x-axis (Figure 3.50). The equation of the kappa curve in polar coordinates is given by $\rho = \cot\varphi$, and its equation in Cartesian coordinates is $(x^2+y^2)y^2 = a^2x^2$.

Note that the kappa curve belongs to the family of curves that have the equations $\rho = a\cot k\varphi$ and are called *node curves.* All these curves have a node at the origin (0; 0) and asymptotes parallel to the coordinate axes. In particular, the node curves also include the strophoid, with $k = 1/2$, and the wind mill curve, for which $k = 2$ (Figure 3.51).

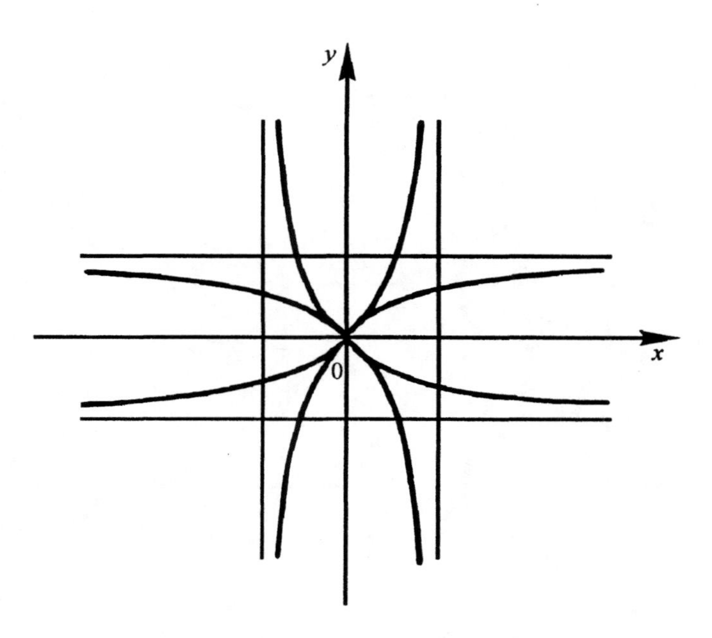

Figure 3.51. The wind mill curve $\rho = \cot 2\varphi$.

The kappa curve is an algebraic curve of degree four. It is symmetric with respect to the coordinate axes. The straight lines $y = \pm a$ are asymptotes of the kappa curve.

Note. The quadrature of the kappa curve was first calculated by Huygens. The study of kappa curves is related in the history of mathematics to a solution of the problem of Sluze (1662), which was to find a curve such that the line segment MC, which is perpendicular to the radius vector OM, has the same length (see Figure 3.51).

The curves of Perseus. The *curves of Perseus* or *spiric curves* are curves obtained by intersecting a torus with a plane parallel to its axis. These curves are defined by the equation

$$(x^2 + y^2 + p^2 + d^2 - R^2)^2 = 4d^2(x^2 + p^2),$$

where R is the radius of the torus-generating circle, d is the distance from the center of the generating circle to the origin, p is a parameter for the cutting plane, and $y_1 = p$. The coordinate axes are axes of symmetry for the curves of Perseus (Figure 3.52).

The Cassinian ovals, lemniscate of Bernoulli, and lemniscate of Booth are all curves of Perseus.

The equation of the lemniscate of Booth is

$$(x^2 + y^2)^2 - (2m^2 + c)x^2 + (2m^2 - x)y^2 = 0.$$

The form of the curve depends upon the relation between the parameters m and c.

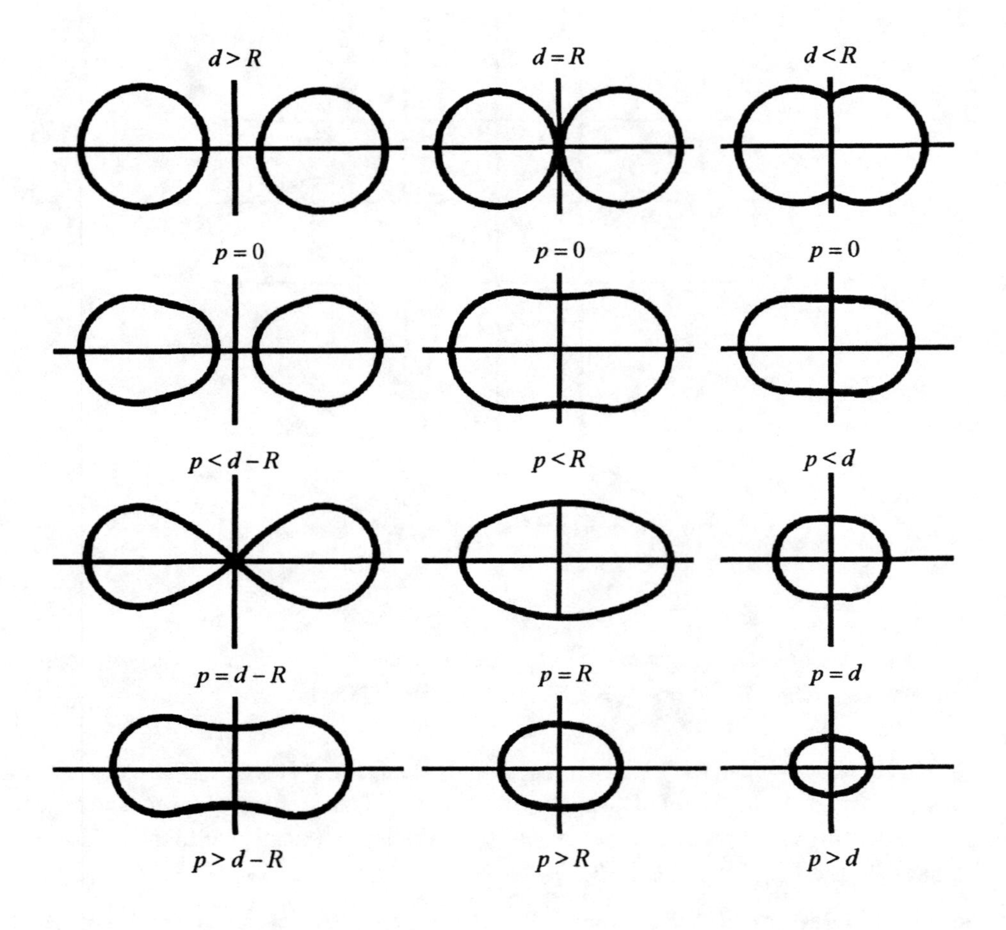

Figure 3.52. The curves of Perseus.

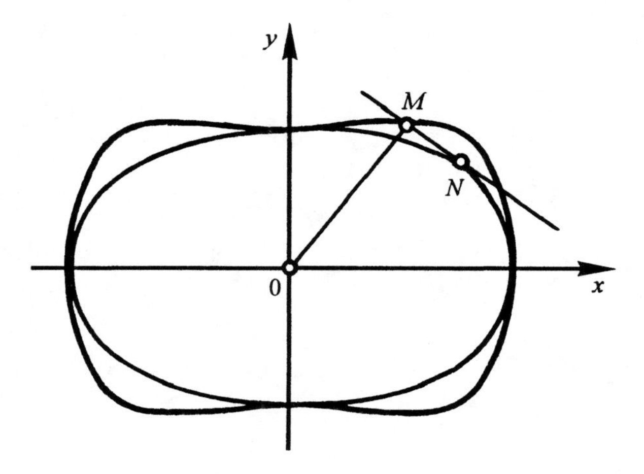

Figure 3.53. The elliptic lemniscate of Booth.

If $c > 2m^2$, then the equation of the lemniscate of Booth is

$$(x^2 + y^2)^2 = a^2x^2 + b^2y^2, \tag{1}$$

where $a^2 = 2m^2 + c$, $b^2 = c - 2m^2$. If $c < 2m^2$, then setting $2m^2 + c = a^2$ and $2m^2 - c = b^2$, we get an equation of the lemniscate of Booth in the form

$$(x^2 + y^2)^2 = a^2x^2 - b^2y^2. \tag{2}$$

In the first case, the lemniscate of Booth is a pedal of an ellipse with respect to its center; in the second case, it is a pedal of a hyperbola relatively to the center.

Because of this, one considers the elliptic lemniscate of Booth given by equation (1) and the hyperbolic one defined by (2). For the former curve, the point (0; 0) is an isolated point (Figure 3.53); for the latter, $O(0; 0)$ is a node with tangents given by $y = \pm \frac{a}{b} x$ (Figure 3.54).

Note that if $c = 2m^2$, then the lemniscate of Booth degenerates into a pair of circles, $x^2 + y^2 \pm 2mx = 0$; if $c = 0$, we obtain a lemniscate of Bernoulli.

The lemniscate of Booth is an orthogonal projection of the line of intersection of the paraboloid $x^2 + y^2 = cz$ and the cone $a^2x^2 + b^2y^2 = c^2z^2$ to the plane Oxy.

If the equation of the lemniscate of Booth is written in polar coordinates, $\rho^2 = a^2c^2\varphi \pm b^2\sin^2\varphi$, then it becomes clear that the curve can be obtained by inverting the second-degree curve $a^2x^2 \pm b^2y^2 = k^4$ with the center at the origin and the degree of inversion equal to k^4.

The area of the elliptic lemniscate of Booth is calculated as

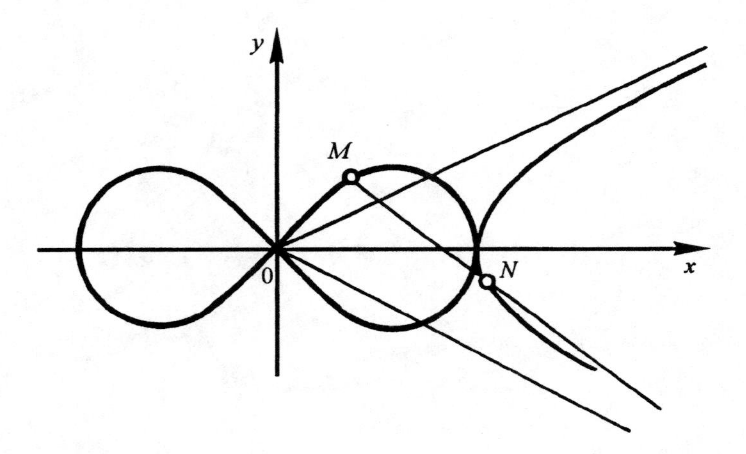

Figure 3.54. The hyperbolic lemniscate of Booth.

$$S_1 = 2\int_0^{\pi/2} ((a^2\cos^2\varphi + b^2\sin^2\varphi)\, d\varphi = \frac{\pi}{2}(a^2 + b^2).$$

The area of the hyperbolic lemniscate of Booth is given by

$$S_2 = 2\int_0^{\arctan\frac{a}{b}} (a^2\cos^2\varphi - b^2\sin^2\varphi)\, d\varphi = \frac{a^2 - b^2}{2}\arctan\frac{a}{b} + \frac{ab}{2}.$$

The arc length of the lemniscate of Booth is expressed in therms of elliptic integrals.

Sinusoidal spirals. *Sinusoidal spirals* are curves that are given in polar coordinates by one of the equations $\rho^m = a^m\sin m\varphi$ or $\rho^m = a^m\cos m\varphi$. If m is rational, these curves are algebraic with the degree dependent on the value of m: for $m = 1$, $\rho = a\cos\varphi$ is a circle; for $m = -1$, $\rho = a/\cos\varphi$ is a straight line; for $m = 2$, $\rho^2 = a^2\cos 2\varphi$ is a lemniscate of Bernoulli; for $m = -2$, $\rho^2 = a^2/\cos 2\varphi$ is an equilateral hyperbola; for $m = 1/2$, $\rho = a\cos^2\varphi/2$ is a cardioid; for $m = -1/2$, $\rho = 2a/(1 + \cos\varphi)$ is a parabola, etc.

Let us take a closer look at some properties of sinusoidal spirals. If the value of the index m is positive, then the curve passes through the pole and lies entirely inside the circle of radius a. If m is negative, then the radius vector can become arbitrarily large, and in this case the curve has infinite branches and no longer passes through the pole. Since the equation $\rho^m = a^m\cos m\varphi$ does not change if φ is replaced with $-\varphi$, the spirals of this type are symmetric with respect to the polar axis.

Sinusoidal spirals that have the index $m = p/q$, where p and q are mutually prime, have p symmetry axes that pass through the pole.

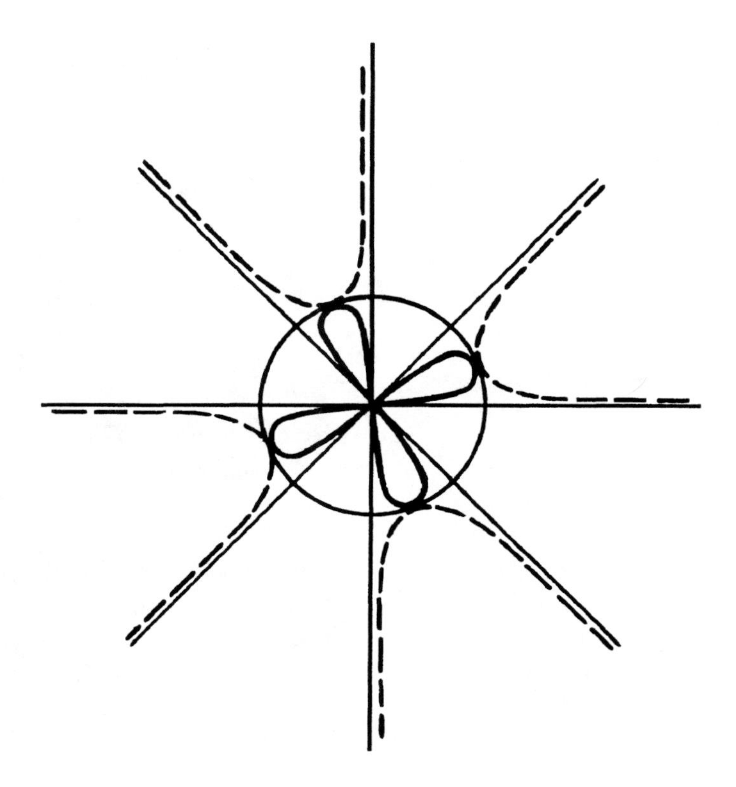

Figure 3.55. The sinusoidal spiral.

For a positive integer m, the function is periodic with period $2\pi/m$. As the polar angle ranges from 0 to 2π, the radius vector passes through the maximum m times, and hence the curve has m foliums, each of which lies inside the angle π/m. Such a spiral has a multiple point at the pole.

The sinusoidal spiral with a positive rational $m = p/q$ also has p foliums which may intersect; in such a case the curve will have multiple points in addition to a multiple point at the pole.

Sinusoidal spirals with an integer negative index have m infinite branches. Each of these branches touches the basis circle and asymptotically approaches the rays the angle between which equals π/m.

The sinusoidal spirals with index $m = 4$ (solid line) and $m = -4$ (dashed line) are shown in Figure 3.55.

The angle μ formed by the tangent at an arbitrary point and the radius vector to the point of tangency equals $m\varphi + \pi/2$, where φ is the polar angle of the point of tangency. For spirals $\rho^m = a^m \sin m\varphi$, the corresponding value is $\mu = m\varphi$.

The radius of curvature at an arbitrary point is calculated by $R = \rho/(m + 1)\cos m\varphi$.

The inversion of a sinusoidal spiral with respect to the pole again yields a sinusoidal spiral with the sign of the index opposite to the sign of the index of the initial spiral.

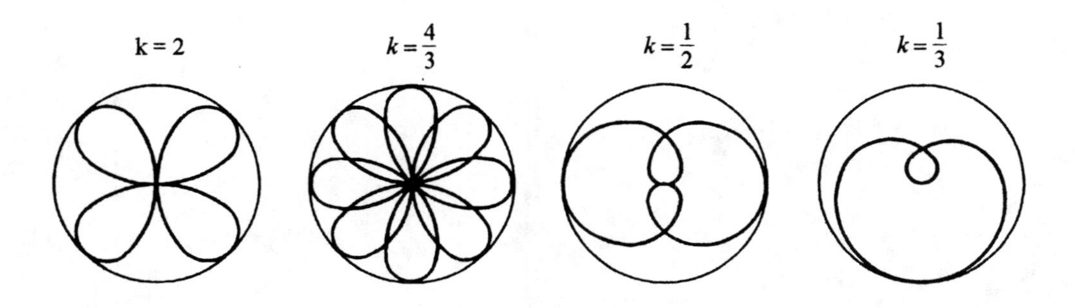

Figure 3.56. The curves of Guido Grandi (rose curves) $\rho = a\sin k\varphi$, $\rho = a\cos k\varphi$.

Rose curves. *Rose curves* or *curves of Guido Grandi* are curves having the equation in polar coordinates of the form $\rho = a\sin k\varphi$ or $\rho = a\cos k\varphi$, where a and k are some constants (we take them to be positive). If the modulus k is a rational number, m/n, then the rose curve is algebraic of even degree. In Cartesian coordinates, equations of the rose curves are

$$a^n\left[\binom{m}{1}x^{m-1}y-\binom{m}{3}x^{m-3}y^3+\binom{m}{5}x^{m-5}y^5-\ldots\right]=$$

$$=\binom{n}{1}(x^2+y^2)^{\frac{m+1}{2}}[a^2-(x^2+y^2)]^{\frac{n-1}{2}}-\binom{n}{3}(x^2+y^2)^{\frac{m-1}{2}}[a^2-(x^2+y^2)]^{\frac{n-3}{2}}+\ldots$$

A study of the form of rose curves. Since the right-hand side does not exceed a, we see that the curve lies entirely inside the circle of radius a. The function $\sin k\varphi$ ($\cos k\varphi$) is periodic; hence the rose curve consists of congruent foliums which are symmetric relative to the largest radii. The number of these foliums depends on the number k. 1) If k is an odd integer, then the rose curve has k foliums, and $2k$ foliums in the case of an even integer. 2) If k is a rational number, m/n ($n > 1$), then the curve has m foliums if both numbers m and n are odd, and $2m$ foliums if one of them is even; the foliums in this case partially overlap. 3) If k is irrational, then the rose curve has an infinite number of partially overlapping foliums (Figure 3.56).

Rose curves are cycloidal curves. They are trochoids, more exactly, hypotrochoids if $k < 1$, and epitrochoids if $k > 1$. The area of one rose curve is given by

$$S=\frac{1}{2}\int_0^{\frac{\pi}{k}}\rho^2\,d\varphi=\frac{a^2}{2}\int_0^{\frac{\pi}{k}}\sin^2 k\varphi\,d\varphi=\frac{\pi a^2}{4k}.$$

The arc length of the rose curve is given in terms of an elliptic integral of the second kind.

It is an interesting fact that the four-folium rose is a locus of vertices of straight angles formed by tangents to an astroid.

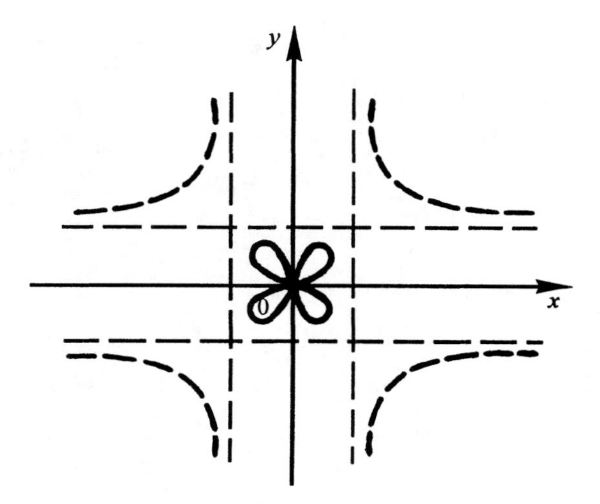

Figure 3.57. The cross curve $\frac{a^2}{x^2} + \frac{a^2}{y^2} = 1$.

A four-folium rose curve can be obtained by inverting the curve $a^2/x^2 + a^2/y^2 = 1$ (or $\rho = 2a/\sin 2\varphi$ in polar coordinates) with respect to the origin. This curve is called a *cross curve* (Figure 3.57).

Note. The rose curve was first studied by Guido Grandi in 1728. Ghabenight has found equations for many foliums, flowers. For example the equation of a "leaf of dock" curve is $\rho = 4(1 + \cos 3\varphi) + 4\sin^2 3\varphi$; a "leaf of ivy" curve is

$$\rho = 3(1 + \cos^2\varphi) + 2\cos\varphi + \sin^2\varphi - 2\sin^2 3\varphi \cos^4\varphi/2.$$

"Spikelet" curves are given by the equation $\rho = a/\sin k\varphi$ or $\rho = a/\cos k\varphi$. These curves can be obtained from rose curves by inverting the rose curve $\rho = a\sin k\varphi$ or $\rho = a\cos k\varphi$ with respect to the circle of radius a.

If k is an odd integer, then the curves $\rho = a/\sin k\varphi$ consist of k congruent hyperbolic branches. In the case where k is an even integer, the number of branches equals $2k$.

Examples of the "spikelet" curves include the trisectrix of Longchamps, $\rho = a/\cos 3\varphi$ and the trisectrix of Maclaurin, $\rho = a/\cos \varphi/3$.

Slide curves. Kinematically, *slide curves* are obtained as follows. Let two curves l_1 and l_2 be given in a plane P_1 and a triangle ABC in a plane P_2. Let the plane P_2 slide on the plane P_1 in such a way that two vertices of the triangle ABC move along the lines l_1 and l_2 in the plane P_1 (Figure 3.58). Then the third vertex of the triangle, C, describes in the plane P_1 a certain curve (the slide curve), the shape of which depends on the form and mutual location of the curves l_2 and l_2. For example, one can consider sliding on two mutually perpendicular straight lines, on a straight line and a circle, on two circles, etc.

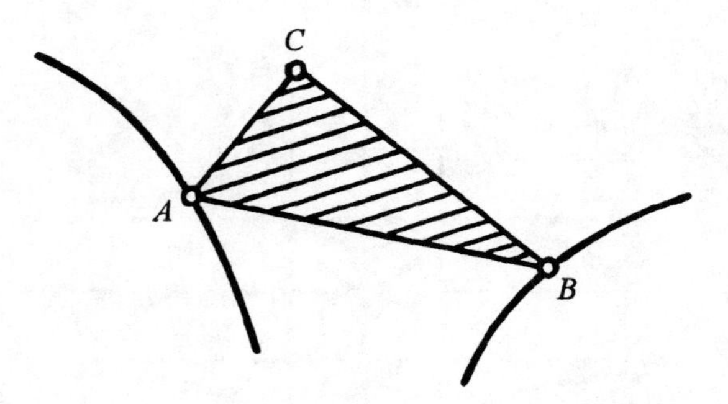

Figure 3.58. The construction of slide curves.

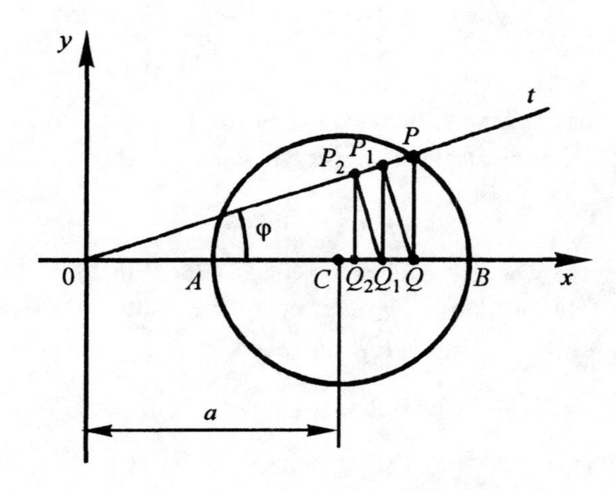

Figure 3.59. The construction of the oval of Munger.

Ovals of Munger. Suppose we have a circle with its center at C and radius r. Take a point inside the disk and draw a ray t through the point O. It intersects the circle in a point P. Project the point P orthogonally to the diameter and get a point Q. The point Q is again projected to the ray t, and denote the projection by P_1. Continuing this process, we obtain points on t, P_1, P_2, ..., P_n. The locus of points P_n formed by moving the ray t is called an *oval of Munger* (Figure 3.59).

The equation of the curve in Cartesian coordinates is

$$(x^2 + y^2)^{2n+1} - 2dx^{n+1}(x^2 + y^2)^n + (d^2 - r^2)\, x^{4n} = 0.$$

Ovals of Munger are algebraic curves of degree $2(2n + 1)$. The point $O(0; 0)$ is a point of multiplicity $4n$ with tangents that coincide with the coordinate axis. Besides the origin, the curves

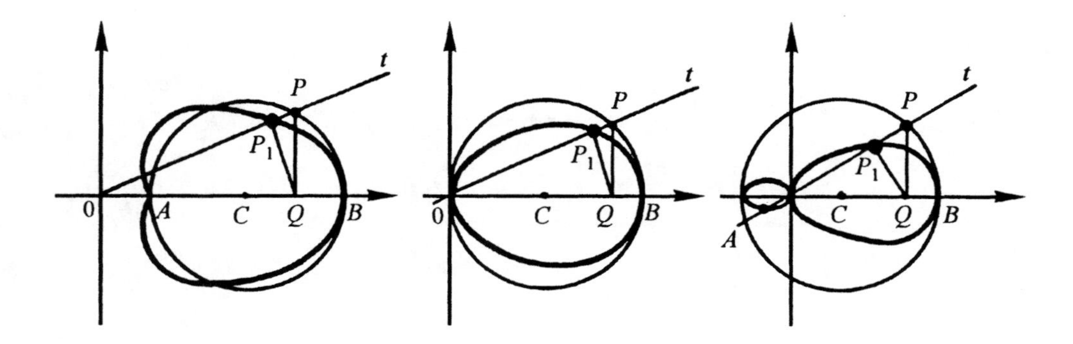

Figure 3.60. Shapes of the ovals of Munger.

intersect the x-axis in the points with the x-coordinates $d \pm r$, i.e. in the points A and B (d is the length of the line segment OC). It is immediately apparent from the equation that ovals of Munger are symmetric with respect to the x-axis.

Setting $n = 0$ in the equation, we get a circle. If $n = 1$, the equation becomes

$$(x^2 + y^2)^3 - 2dx^3(x^2 + y^2) + (d^2 - r^2)\, x^4 = 0,$$

and this is a curve of degree 6. Depending on whether the point O lies outside, inside, or on the circle, ovals of Munger have different shapes (Figure 3.60).

The polar equation that defines the oval of Munger has the form

$$\rho_n^2 - 2d\rho_n \cos^{2n+1}\varphi + (d^2 - r^2) \cos^{4n}\varphi = 0.$$

The case of interest is where $d = 0$. The curve is symmetric with respect to the y-axis and consists of two identical ovals. Munger called the curve the *double line*. Equations of the double line in polar coordinates are $\rho = r \cos^2\varphi$; in the Cartesian coordinates, they are given by $(x^2 + y^2)^3 - r^2x^4 = 0$. If the polar equation of the double line is written in the form $\rho = r/2\cos 2\varphi + r/2$, then it is clear that the double line is a conchoid of a rose curve.

Curves of Lamé. Curves of Lamé are defined by the equation $(x/a)^m + (y/b)^m = 1$, where the exponent m may be any rational number, the constants a and b are positive numbers. Let us consider various cases.

1. Curves of the type

$$\left(\frac{x}{a}\right)^m + \left(\frac{y}{b}\right)^n = 1,$$

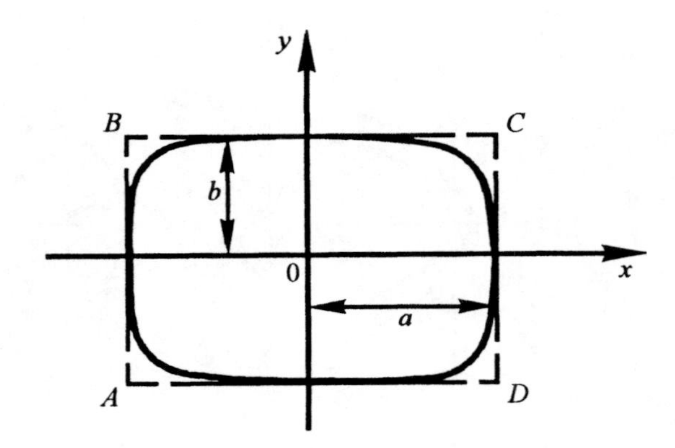

Figure 3.61. One of the forms of Lamé curves $\left(\frac{x}{a}\right)^m + \left(\frac{y}{b}\right)^n = 1.$

where m, n are positive integers. It is clear that if $m = n = 1$, the resulting curve is a straight line that has intercepts a and b; if $m = 1$, $n = 2$ or $m = 2$, $n = 1$, the curve is a parabola; if $m = 2$, $n = 2$, we get an ellipse.

If m and n are even, $m > 2$, $n > 2$, then we get an oval with semiaxes a and b (Figure 3.61). Different shapes of Lamé curves are shown in Figure 3.62. (*a*) m is even and n is odd, $m \geq 2$, $n \geq 3$; (*b*) the same but $m < n$; (*c*), (*d*) n is even, m is odd, $n \geq 2$, $m \geq 3$, $m > n$; (*e*) $m = n$, m, n are odd, $m \geq 3$, $n \geq 3$, $m > n$; (*f*) the same but $m < n$; (*g*) the same but $m \leq n$.

2. Curves of the type

$$\left(\frac{x}{a}\right)^m - \left(\frac{y}{b}\right)^n = 1.$$

If $m = n = 1$, the curve is a straight line with intercepts a and b. If $m = 2$, $n = 1$ or $m = 1$, $n = 2$, the curves are parabolas; if $m = 2$, $n = 2$, the curves are hyperbolas.

For different cases, shapes of Lamé curves are shown in Figure 3.63: (*a*) m and n are even, $m > 2, n > 2, m \leq n$; (*b*) $m > n$; (*c*) m is even, n is odd, $m \geq 2, n \geq 3, m > n$; (*d*) $m < n$; (*e*) m is odd, n is even, $m \geq 3, n \geq 2, m < n$; (*f*) $m > n$; (*g*) m and n are odd, $m \geq 3, n \geq 3, m = n$; (*h*) m and n are odd, $m \geq 3, n \geq 3, m > n$; (*i*) $m < n$.

3. Let us consider Lamé curves in the case where m is a positive rational, $m = p/q$.

Let $\frac{p}{q} > 1$. If p is even, q is odd, the curve is symmetric with respect to the coordinate axes (Figure 3.64a). If p is even, q is odd, the curve is located in the first quadrant and consists of three arcs, one of which passes through the points $(a; 0)$ and $(0; b)$, and two other tend to infinity asymptotically approaching the straight line $y = \frac{b}{a}x$. The tangent to the curve at the point $(a; 0)$

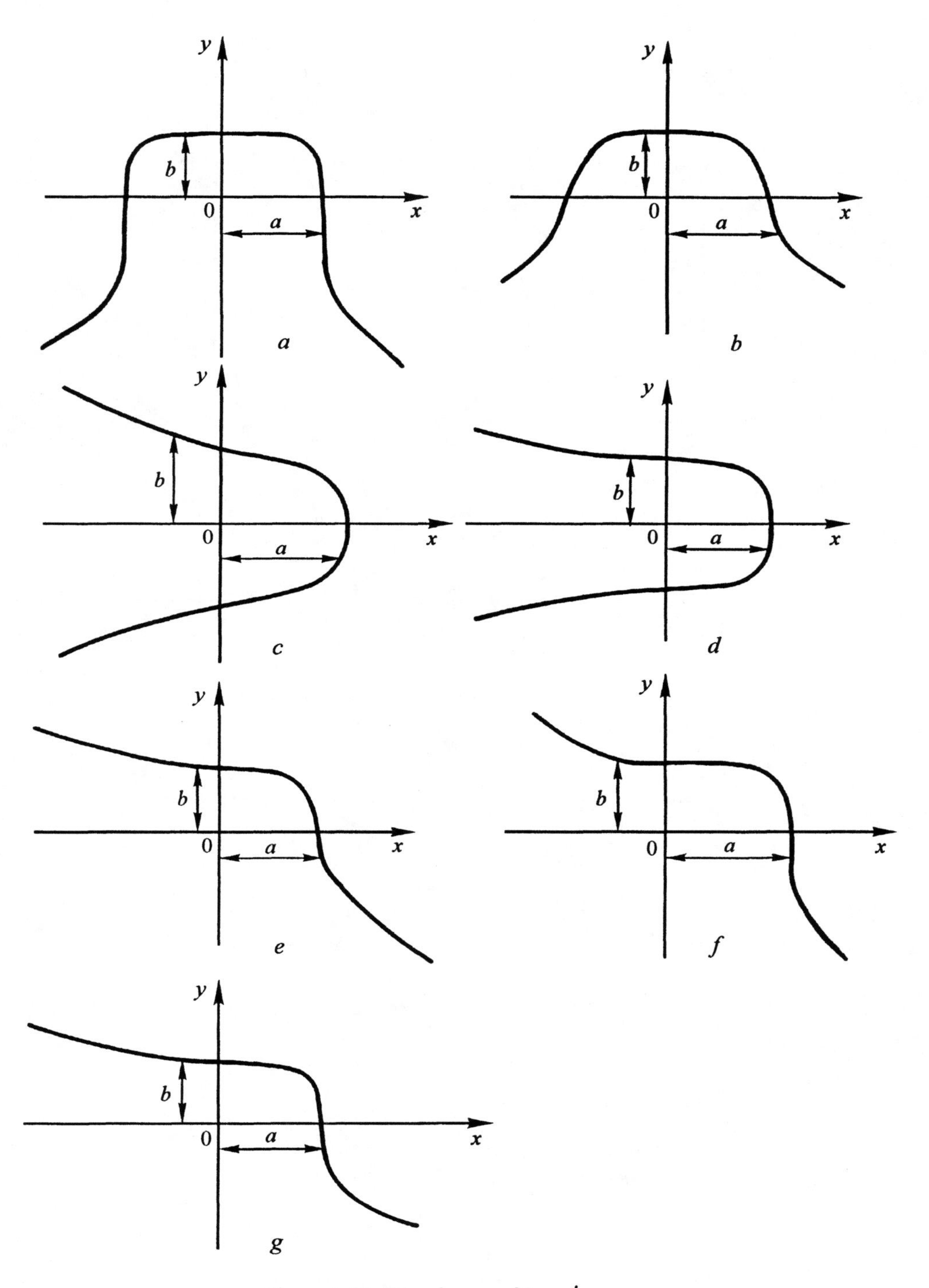

Figure 3.62. The shapes of Lamé curves.

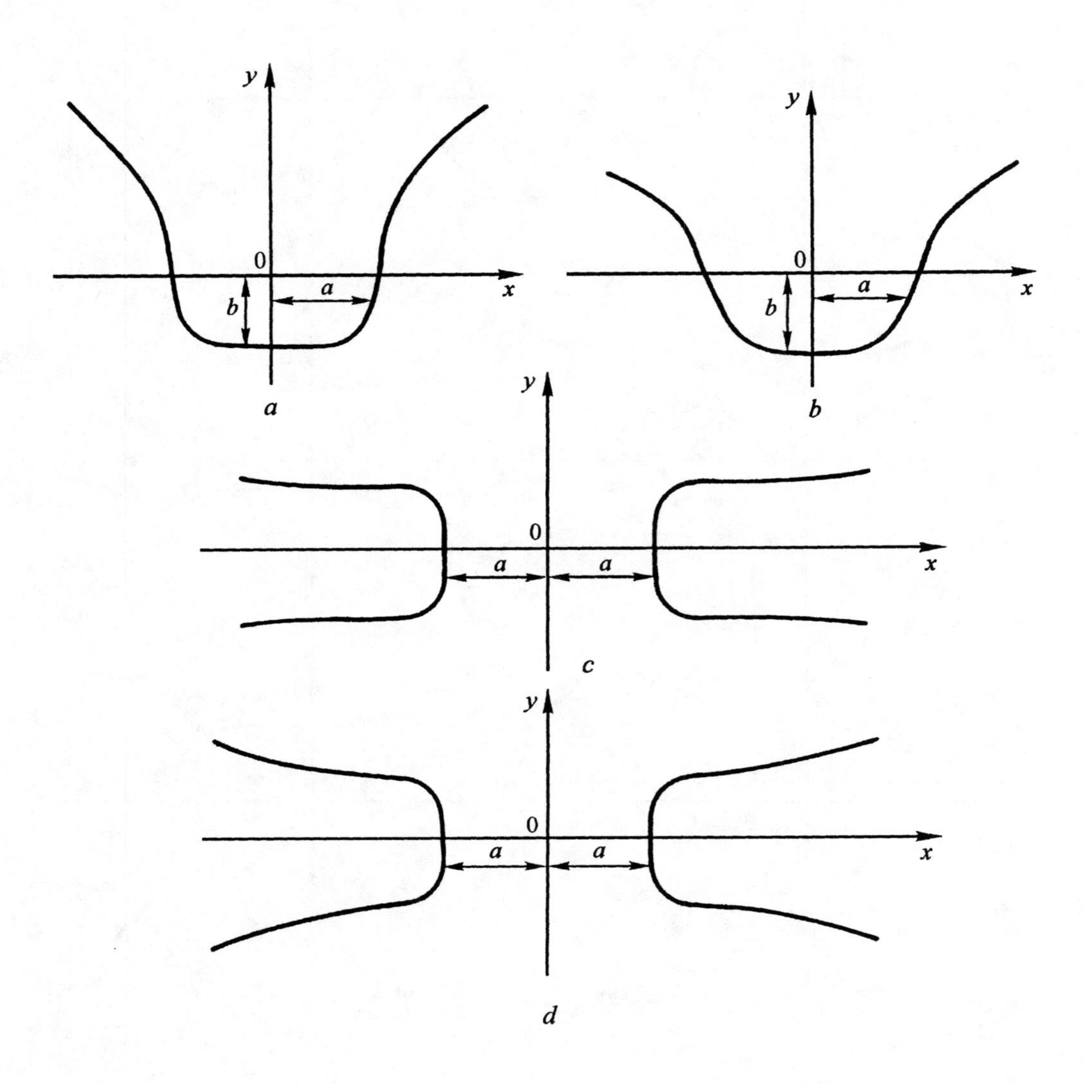

Figure 3.63. The shapes of the curve $\left(\frac{x}{a}\right)^m - \left(\frac{y}{b}\right)^n = 1$.

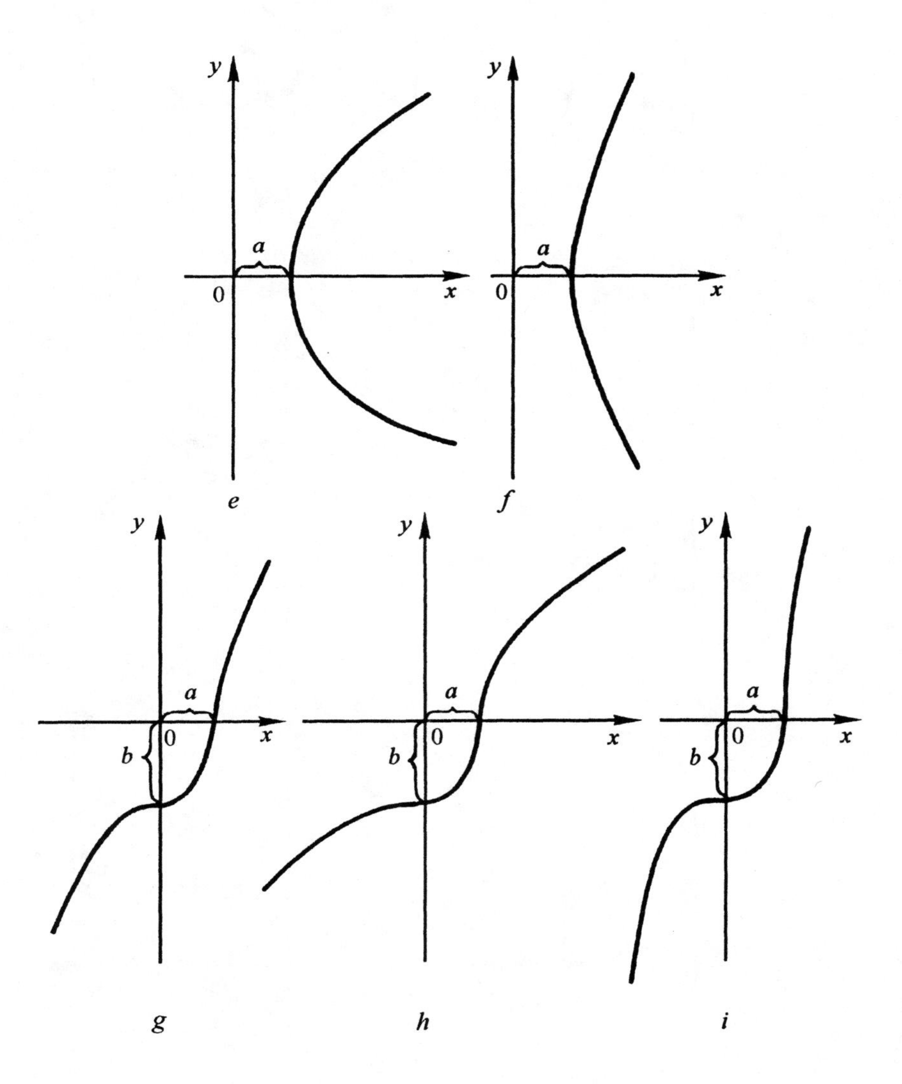

Figure 3.63. The shapes of the curve $\left(\frac{x}{a}\right)^m - \left(\frac{y}{b}\right)^n = 1$.

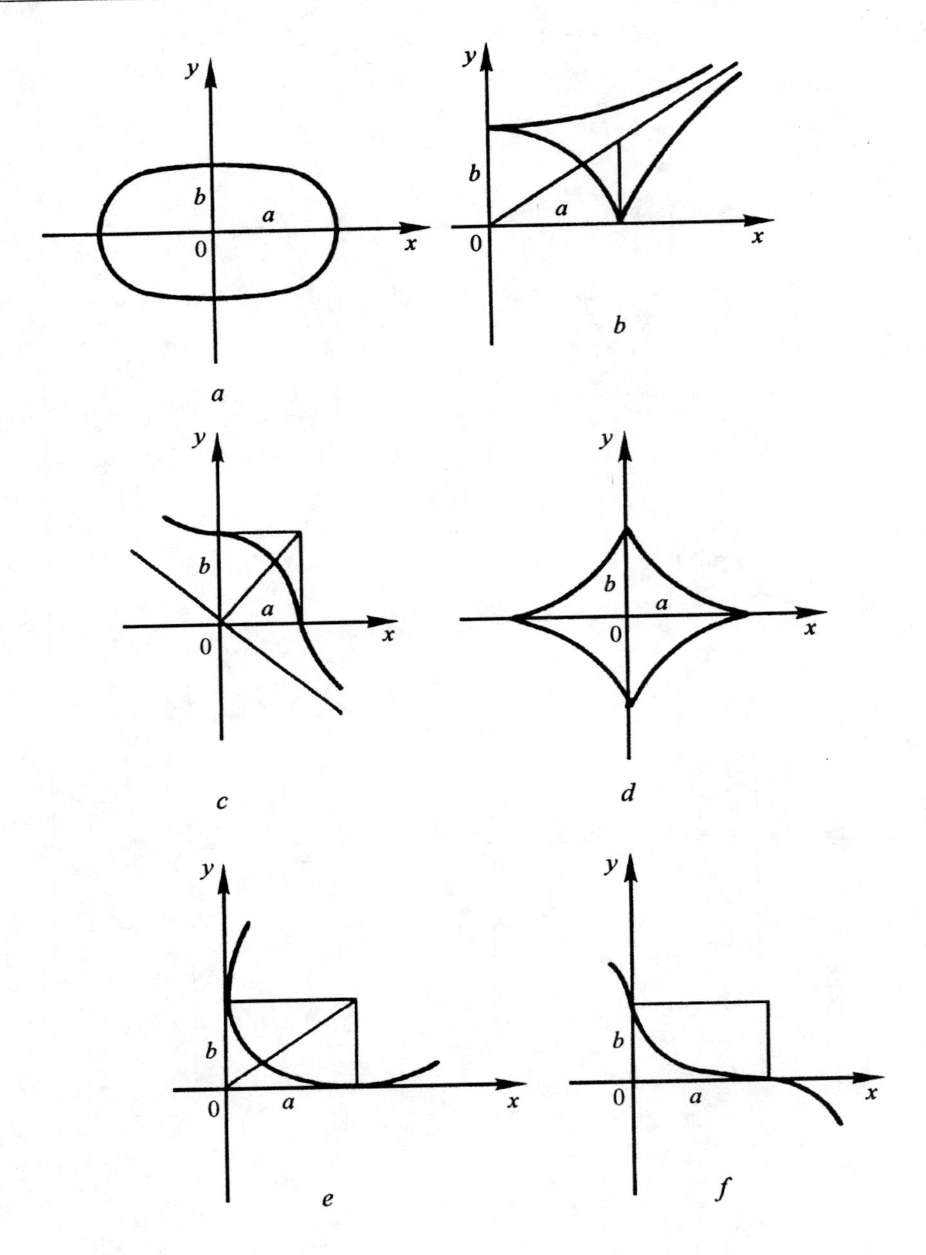

Figure 3.64. The curves of Lamé for rational m.

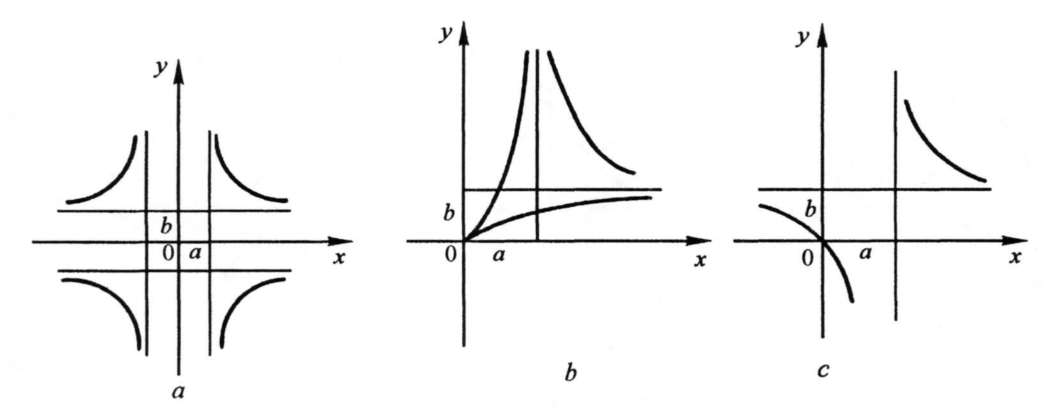

Figure 3.65. The curves of Lamé for $n = m = -\frac{p}{q}$.

is parallel to the y-axis, and at the point (0; b) to the x-axis (Figure 3.64b). If p and q are odd, the curve is located in the first and fourth quadrants and has the asymptote $y = -\frac{b}{a}x$; the tangents at the points (a; 0) and (0; b) are parallel to the corresponding coordinate axes (Figure 3.64c).

Let $p/q < 1$. If p is even and q is odd, then the curve is symmetric with respect to the coordinate axes. At the points ($\pm a$; 0) and (0; $\pm b$), the curve has a cusp (Figure 3.64d). If p is odd, q even, the curve is located in the first quadrant and resembles a parabola. At the points (a; 0) and (0; b), the curve is tangent to the corresponding coordinate axes (Figure 3.64e). If both p and q are odd, then the curve is located in the first, second, and fourth quadrants and tends to infinity, and at the points (a; 0), (0; b) the tangents coincide with the coordinate axes (Figure 3.64f).

Let $m = -p/q$. If p is even and q odd, the curve is symmetric with respect to the coordinate axes and consists of four branches which asymptotically approach the straight lines $x = \pm a$, $y = \pm b$; the point O(0; 0) is an isolated point (Figure 3.65a). If p is odd and q is even, then the curve is located in the first quadrant, consists of three branches asymptotically tending to the straight line $x = a$, $y = b$, and the point (0,0) is a peak with the tangent $y = \frac{b}{a}x$ (Figure 3.65b). The curve has two branches of hyperbolic type, one of which passes through the origin. The straight lines $x = a$, $y = b$ are asymptotes (Figure 3.65c).

Note that the curves of Lamé with $m = p/q > 0$ will be algebraic of degree pq, if $m = -p/q < 0$, then the degree of the curve equals $2pq$.

Parabolic and hyperbolic curves. These curves are defined by the equation $y = ax^m$, where a is a constant which is assumed to be positive and m is a rational number. If $m > 0$, then the curves are parabolic; if $m < 0$, the curve is hyperbolic. Nail's parabola $y = ax^{3/2}$, $a > 0$, is an example of such a curve.

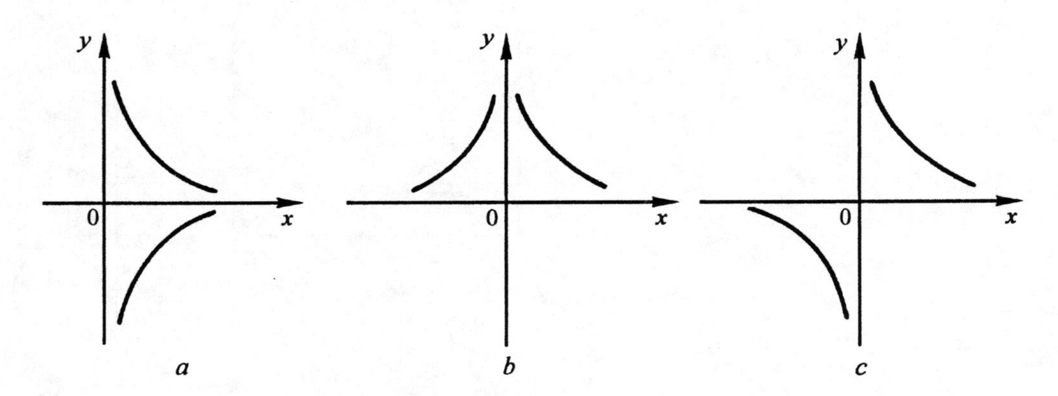

Figure 3.66. The hyperbolic curves $y = ax^{-m}$, $m = \frac{p}{q} > 0$.

Hyperbolic curves are generalizations of the hyperbolas $xy = a$. All three cases of the curve depending on the exponent $m = p/q$ are shown in Figure 3.66: (*a*) p is odd, q is even; (*b*) p is even, q is arbitrary; (*c*) p and q are odd.

Parabolic and hyperbolic curves belong to the class of polytropic curves.

Let us consider a few examples of remarkable algebraic curves.

1) The curve $x^4 = (x^2 - y^2)y$. Its parametric equation is $x = t - t^3$, $y = t^2 - t^4$, $-\infty < t < +\infty$, where $t = \tan MOx$ (Figure 3.67).

2) The curve $x^5 + y^5 = 5ax^2y^2$. The parametric equation of the curve is $x = 5at^4/(1 + t^5)$, $y = 5at^5/(1 + t^5)$, $-\infty < t < +\infty$, where $t = \tan MOx$ (Figure 3.68).

3) The curve $(y - x^2)^2 = x^5$ (Figure 3.69*a*). The parametric equation of the curve is $x = (t - 1)^2$, $y = t(t - 1)^4$, $-\infty < t < +\infty$.

4) The bicorn $y = \dfrac{a^2 - x^2}{2a \pm \sqrt{a^2 - x^2}}$ (Figure 3.69*b*). Its parametric equation is given by

$$x = a\cos\theta,\ y = a\frac{\sin^2\theta}{2 + \sin\theta},\ 0 \le \theta \le 2\pi, \text{ or by } x = a\frac{1 - t^2}{1 + t^2},\ y = \frac{2t^2}{(1 + t^2)(1 + t + t^2)},\ -\infty < t < +\infty.$$

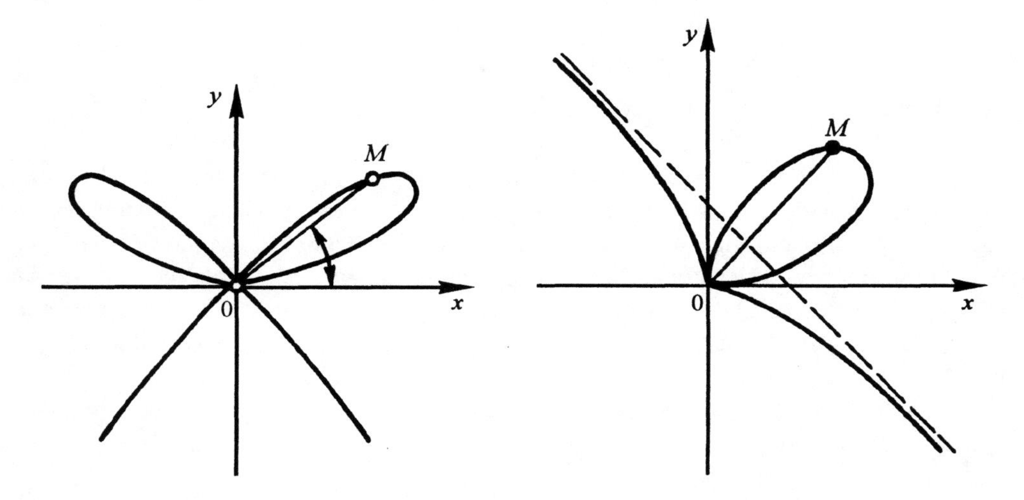

Figure 3.67. The curve $x^4 = (x^2 - y^2)y$. Figure 3.68. The curve $x^5 + y^5 = 5ax^2y^2$.

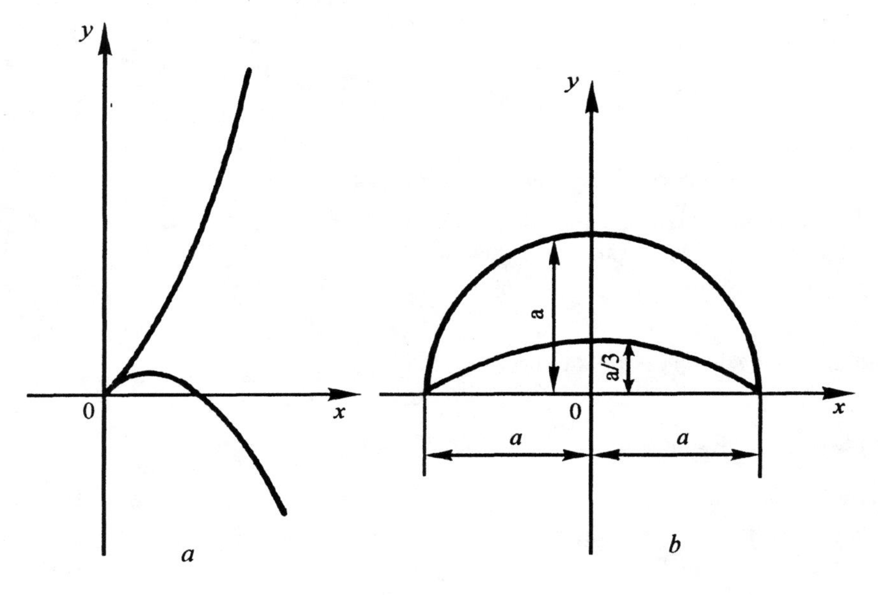

Figure 3.69. The graphs of functions: (*a*) $(y - x^2)^2 = x^5$; (*b*) $y = \dfrac{a^2 - x^2}{2a \pm \sqrt{a^2 - x^2}}$.

3.5. Transcendental curves

A *transcendental* curve is one such that its equation in Cartesian coordinates is not algebraic (in other coordinate systems it could be algebraic). In some cases, transcendental curves are considered as infinite-degree algebraic curves, since the number of some points of interest, e.g. points of intersection with a straight line, inflection points, singular points, etc., could be infinite. Moreover, transcendental curves can also have special points that algebraic curves do not have. For example, a *vanishing point*, is the point such that a circle of sufficiently small radius intersects this curve in only one point (for example, the curve $y = x\ln x$ has a vanishing point at the origin); a *corner point* is a point at which two branches of the curve end, and each branch has its tangent at the point; an *asymptotic point* is a point to which the curve approaches making an infinite number of turns about this point.

It should be noted that the transcendental curve may have a dotted branch consisting of a finite number of isolated points.

For well known transcendental curves, as in the case of algebraic curves, the slope y' of the tangent at every point of the curve is a root of an algebraic equation with the coefficients that are polynomials in x and y, i.e. a differential equation for transcendental curves has the form

$$\sum_{k=0}^{n} f_k(x, y)\, y'^{\,n-k} = 0,$$

where $f_0, f_1, \ldots, f_n$ are polynomials without common factors.

D. Loria has combined all algebraic and almost all known transcendental curves in a single group, the so-called *panalgebraic curves*, that differ in the degree, equal to the number n, i.e. the degree of the differential equation, and the rank μ, equal to the highest degree of polynomials $f_0, f_1, \ldots, f_n$.

Let us look at some important transcendental curves.

3.5.1. Spiral of Archimedes

The *spiral of Archimedes* is a curve traced by a point moving with a constant velocity v along a ray that rotates about the origin O with constant angular velocity ω (Figure 3.70).

Kinematic terms in this definition can be replaced by the condition that the distance $\rho = OM_1$ be proportional to the angle φ of the turn of the ray. To a rotation at a given angle from any position of the ray there corresponds the same increment of ρ. In particular, a complete turn corresponds to the same displacement $MM_1 = a$ (Figure 3.70). The value of a is called the *step* of the spiral of Archimedes. To a given step a there correspond two spirals of Archimedes that differ in the direction of rotation. If rotation is performed counterclockwise, the curve is called a *right spiral* (Figure 3.70, thick line); if the ray rotates clockwise, the curve is called a *left spiral* (Figure 3.70, dotted line). The right and left spirals with the same step can be considered as two

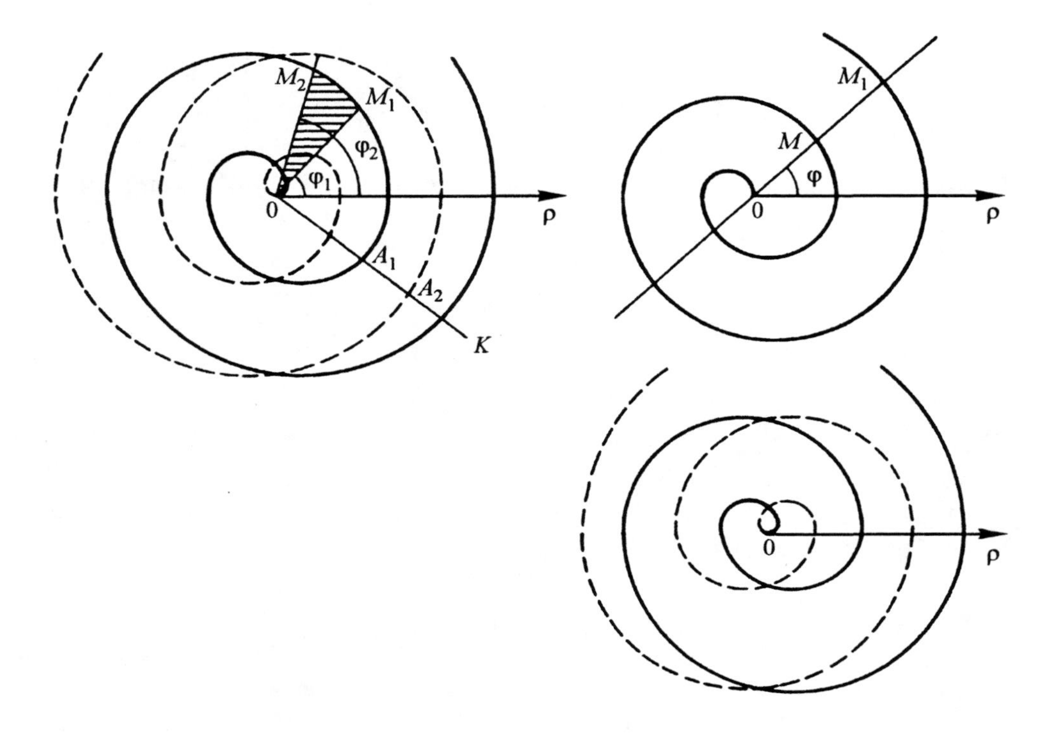

Figure 3.70. The spiral of Archimedes $\rho = k\varphi$.

branches of the curve traced by the point M that runs on the whole ray and passes through the point O.

The polar equation of the curve is $\rho = k\varphi$, where k is the parameter of the spiral, $k = y/\omega$; $k = a/2\pi$ is the displacement of the point M along the ray rotated by one radian.

The right branch corresponds to positive values of φ, the left branch to negative values of φ. The area S_n bounded by the nth branch of the spiral and the line segment OA_n is calculated by the formula

$$S_n = \frac{n^3 - (n-1)^3}{3}\pi a^2 = \frac{n^3 - (n-1)^3}{3n^2} S_n',$$

where S_n' is the area of the disc of radius OA_n.

The area of the nth annulus of the spiral of Archimedes is

$$S_n = S_{n+1} - S_n = 6nS_1,$$

where $S_1 = \frac{1}{3}\pi a^2$ is the area of the zero annulus. The length of the arc OM is calculated by

$$l = \frac{k}{2}\,[\varphi\sqrt{\varphi^2+1} + \ln(\varphi + \sqrt{\varphi^2+1}\,)] =$$

$$= \frac{1}{2}\left[\frac{\rho\sqrt{\rho^2+k^2}}{k} + k\ln\frac{\rho+\sqrt{\rho^2+k^2}}{k}\right] = \frac{1}{2}k\,[\tan\alpha\cdot\sec\alpha + \ln(\tan\alpha + \sec\alpha)],$$

$2l/k\varphi^2 \to 0$ for large φ, where α is the acute angle between the tangent MT and the polar radius OM.

The radius of curvature

$$R = \frac{(\rho^2+k^2)^{3/2}}{\rho^2+2k^2} = k\frac{(\varphi^2+1)^{3/2}}{\varphi^2+2} = k\frac{(\tan^2\alpha+1)^{3/2}}{\sec^2\alpha+1}.$$

At the point (0; 0), we have that $R_0 = k/2$.

Note that the spiral of Archimedes is an orthogonal projection, on the plane Oxy, of the line of intersection of the helical surface $z = k\arctan y/x$ and the surface of the cone $x^2 + y^2 = a^2z^2$.

A generalization of the spiral of Archimedes is the curve $\rho = k\varphi + l$, which is a conchoid of the spiral of Archimedes; it is called a *neoide*.

Note. Discovery of the spiral of Archimedes is attributed to Conon of Samos (according to Pappus); however, a detailed study was carried out by Archimedes. Rectification of the spiral of Archimedes was carried out by Cavalieri and others in the seventeenth century.

The spiral of Archimedes is widely used in engineering, in particular in cam boxes, spinning frames, etc.

Since the equation of the spiral of Archimedes connects the radius vector and the angle by a linear relation, the constructed spiral of Archimedes can be used to subdivide any angle into any number of equal parts.

3.5.2. Algebraic spirals

An *algebraic spiral* is a line the equation of which in polar coordinates is algebraic with respect to ρ and φ, $f(\rho,\varphi) = 0$.

Algebraic spirals can be described stereometrically in the same way: they are orthogonal projections, on the plane Oxy, of the line of intersection of the helical surface $z = k\arctan\frac{y}{z}$ and the surface of revolution $f(z,\sqrt{x^2+y^2}) = 0$, where f is an algebraic function. The most simple example is the spiral of Archimedes.

Hyperbolic spiral $\rho = a/\varphi$. The curve consists of two branches that are located symmetrically with respect to the line perpendicular to the polar axis at the pole (Figure 3.71).

The straight line $y = a$ is an asymptote of each branch and the origin O is an asymptotic point. The area of the sector M_1OM_2 is calculated as

$$S = \frac{a^2}{2}\left(\frac{1}{\varphi_1} - \frac{1}{\varphi_2}\right), \qquad \lim_{\varphi_2\to\infty} S = \frac{a^2}{2\varphi}.$$

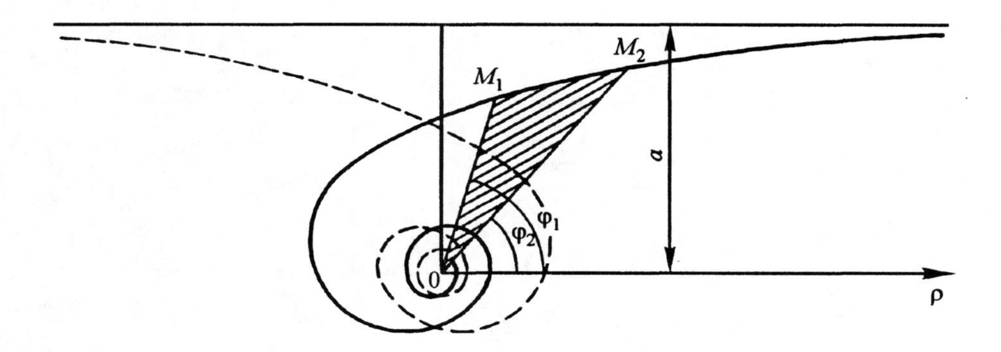

Figure 3.71. The hyperbolic spiral $\rho = a/\varphi$.

The radius of curvature, $r = \dfrac{a}{\varphi}\left(\dfrac{\sqrt{1+\varphi^2}}{\varphi}\right)^3$.

Comparing the equation of the spiral of Archimedes and hyperbolic spiral, note that each one can be obtained from the other by inversion with respect to the pole.

The length of the arc of the hyperbolic spiral between points $M_1(\rho_1;\ \varphi_1)$ and $M_2(\rho_2;\ \varphi_2)$ is found from the formula

$$L = \int_{\varphi_1}^{\varphi_2} \sqrt{\rho^2 + \rho'^2}\, d\varphi = a \int_{\varphi_1}^{\varphi_2} \frac{\sqrt{1+\varphi^2}}{\varphi^2}\, d\varphi =$$

$$= a\left[-\frac{\sqrt{1+\varphi^2}}{\varphi} + \ln\left(\varphi + \sqrt{1+\varphi^2}\,\right)\right]_{\varphi_1}^{\varphi_2}.$$

Conchoid of the hyperbolic spiral. This curve is a generalization of the hyperbolic spiral. Its equation is $\rho = a/\varphi + l,\ l \neq 0$ (Figure 3.72). The asymptote is given by $y = a$; inflection points are found from the equation $\varphi(l\varphi + a)^2 - 2al = 0$.

If $\varphi = 0$, the left-hand side is negative and increases as φ increases, assuming the value zero only once. If φ is negative, it is always negative. Thus the curve has only one inflection point at $\varphi = \varphi_1$, where $0 < \varphi_1 < 1$.

As φ increases from 0 to $+\infty$, we have that $\rho \to l$, i.e. the curve winding counterclockwise asymptotically approaches the circle $\rho = l$. As φ varies from 0 to $-\infty$, ρ starts at $-\infty$, becomes zero at $\varphi = -l/a$, and begins to be marked in the opposite direction. Thus the second branch of the curve has a self-intersection, and then asymptotically approach the same circle but from the inside.

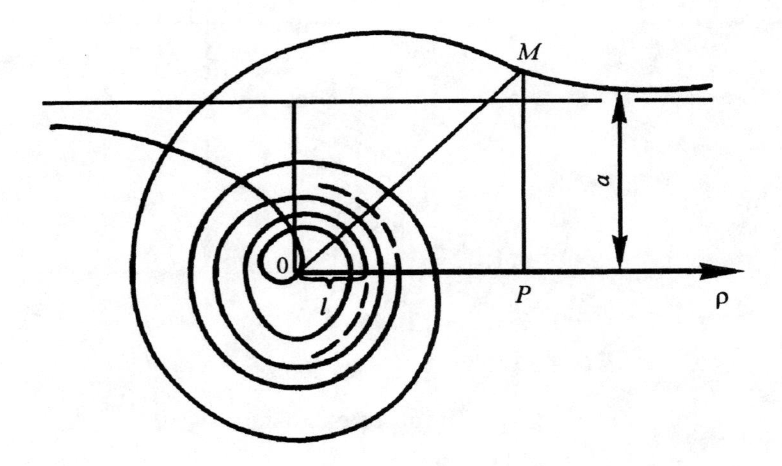

Figure 3.72. The conchoid of hyperbolic spiral, $\rho = \frac{a}{\varphi} + l$.

Spiral of Galileo $\rho = a\varphi^2 - l, l \geq 0$. This curve is shown in Figure 3.73. The general form of this curve is $\rho = a_1\varphi^2 + a_2\varphi + a_3$. By a rotation of the polar axis, this equation can be brought into the form $\rho = a\varphi^2 - l$.

The curve is symmetric with respect to the polar axis and has a double point at the origin with the tangents making the angles of $\pm\sqrt{l/a}$ with the polar axis. The curve has an infinite number of double points located in the polar axis and given by the equation $\rho = ak^2\pi^2 - l$, $k \in \mathbf{N}$. If $l = 0$, we have the spiral $\rho = a\varphi^2$.

Note. The spiral of Galileo has been known since the seventeenth century in connection with determining the shape of the trajectory of a particle that makes a free fall on the earth near the equator (without initial velocity).

The spiral $\rho = a/\varphi^2$. This curve can be easily obtained from the spiral of Galileo by the inversion $\rho\rho_1 = a^2$. The curve has two infinite branches; O is an asymptotic double point (Figure 3.74).

Fermat's spiral $\rho = a\sqrt{\varphi}$. The curve has central symmetry and consists of two branches, one of which corresponds to positive values of ρ and the other to negative. Both branches start at the pole, which is an inflection point. Both branches tending to infinity make an infinite number of windings about the pole (Figure 3.75).

This spiral has a special property that the distance between the windings decreases to zero as the distance to the pole increases (for the spiral of Archimedes, this distance remains constant;

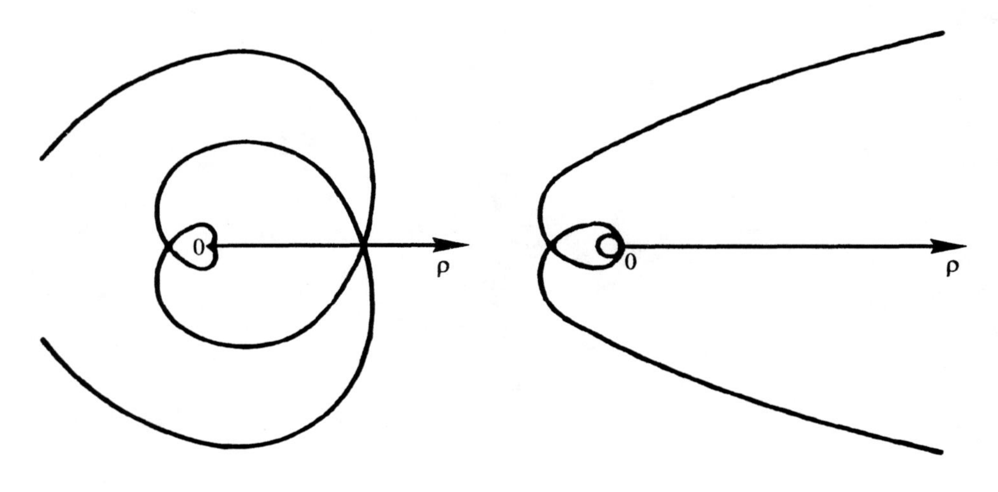

Figure 3.73 Figure 3.74

Figure 3.73. The spiral of Galileo, $\rho = a\varphi^2 - l$ for $l = 0$.

Figure 3.74. The spiral $\rho = \dfrac{a}{\varphi^2}$.

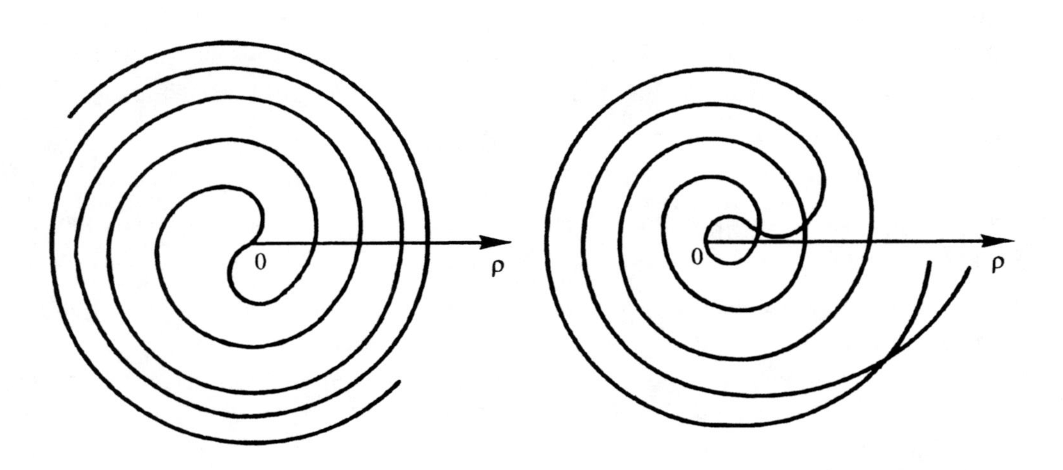

Figure 3.75 Figure 3.76

Figure 3.75. Fermat's spiral $\rho = a\sqrt{\varphi}$.

Figure 3.76. The parabolic spiral $\rho = a\sqrt{\varphi} + l$.

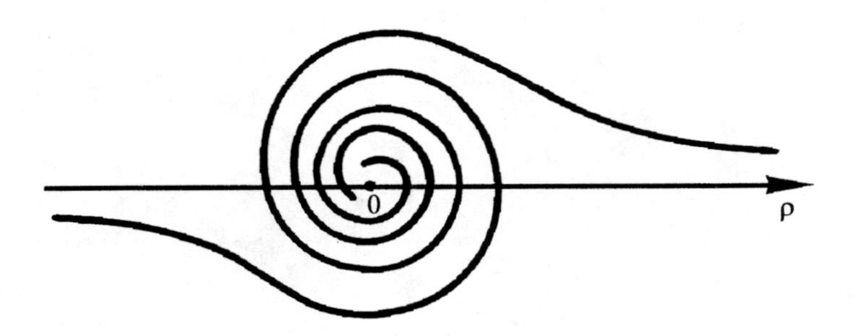

Figure 3.77. The lituus $\rho = \frac{a}{\sqrt{\varphi}}$.

it increases for the logarithmic spiral). Indeed, the distance between two nearest windings of the curve is $h = a(\sqrt{\varphi + 2\pi} - \sqrt{\varphi})$. Hence $\lim\limits_{\varphi \to \infty} l = 0$.

Parabolic spiral $\rho = a\sqrt{\varphi} + l, l > 0$. This spiral is a conchoid of Fermat's spiral. The curve has two branches: $\rho = a\sqrt{\varphi} + l$, $\rho = -a\sqrt{\varphi} + l$. The second branch makes with itself and with the first branch an infinite number of double points. The curve has one inflection point (Figure 3.76).

Lituus $\rho = a/\sqrt{\varphi}$. The curve has two branches, each of which asymptotically approaches the pole. The polar axis is an asymptote (Figure 3.77). The inflection points are $(1/2; a\sqrt{2})$, $(-1/2; -a\sqrt{2})$. This curve is a locus of the endpoints of subnormals of Fermat's spiral.

Logarithmic spiral. A *logarithmic spiral* is a curve that intersects all rays coming out from the point O at the same angle α (Figure 3.78). The equation of the curve in polar coordinates is $\rho = \rho_0 e^{k\varphi}$, or $\rho = a^{\varphi}$, $-\infty < \varphi < +\infty$, where k is a parameter ($k = \ln q/2\pi$), q is the growth coefficient, and a is an arbitrary positive number.

It follows from the equation $\rho = a^{\varphi}$ that, if $\varphi \to +\infty$, ρ tends to infinity and the spiral unwinds counterclockwise, and if $\varphi \to -\infty$, ρ tends to zero, the curve winds clockwise, making an infinite number of windings about the pole (for $a > 1$) and approaching it as an asymptotic point (Figure 3.78). If $a < 1$, then the curve winds about the pole counterclockwise.

A particular property of the logarithmic spiral is that the angle between the tangent and the radius vector to an arbitrary point depends only on the parameter a and, hence, a constant for each logarithmic spiral.

Because of this property, the logarithmic spiral is also called an *equiangular spiral.* Save for the logarithmic spiral, only a circle has this property; it intersects its radius vectors at a straight angle. The radius of curvature is calculated as $R = \rho\sqrt{1 + \ln^2 a}$, which shows that the radius of curvature is proportional to the length of the radius vector to the point.

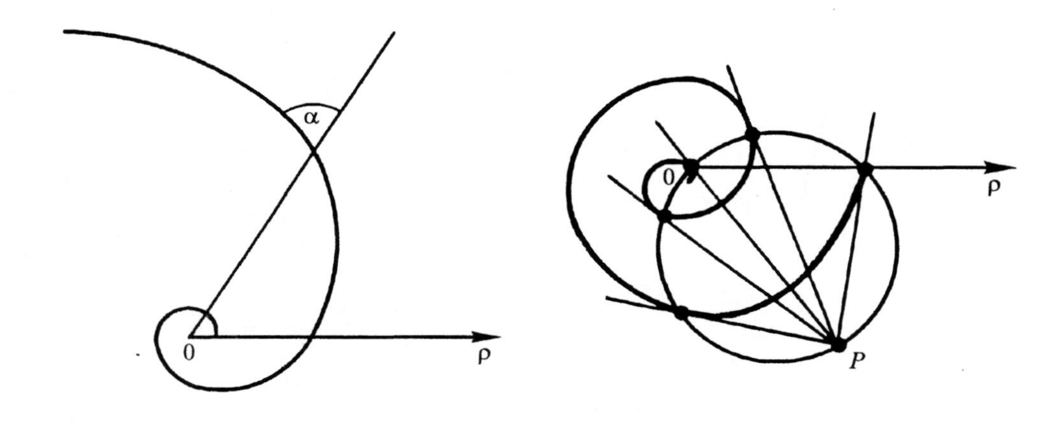

Figure 3.78

Figure 3.79

Figure 3.78. The logarithmic spiral $\rho = \rho_0 e^{k\varphi}$.

Figure 3.79. One of properties of the logarithmic spiral.

The length of the arc of the logarithmic spiral bounded by points $(\rho_1;\ \varphi_1)$ and $(\rho_2;\ \varphi_2)$ equals $l = \frac{\sqrt{1+\ln^2 a}}{\ln a}(\rho_2 - \rho_1)$. The length of the arc from the pole to an arbitrary point is $L = \rho\sqrt{1+\ln^2 a}/\ln a$. The natural equation of the logarithmic spiral has the form $S = kR$, where $k = 1/\ln a$. Thus the length of the arc from the pole to a point is proportional to the radius of curvature at the endpoint of this arc.

The curve has the property that its nature does not change under various transformations. For example, a pedal, evolute, and catacaustic of a logarithmic spiral is again a logarithmic spiral.

The locus of the points of tangency of tangents to a logarithmic spiral that pass through a point P is a circle passing through the point P and the pole of the spiral (Figure 3.79).

A *loxodromic curve*, i.e. the curve lying on a surface and intersecting all meridians at the same angle, is projected by a stereographic projection to a logarithmic spiral.

Note. The logarithmic spiral is mentioned for the first time in a letter of Descartes to Mersenne (1638), in which Descartes defines a new spiral as a curve such that the ratio of the length of the arc and the radius vector is a constant.

Applications of the logarithmic spiral in engineering are based on the property of intersecting the radius vectors at a constant angle. Logarithmic spirals are used in rotary blades of various cutting tools, in gear wheels with variable transmission ratio, etc. Many objects in nature have the form of a logarithmic spiral: a shell, a sunflower, etc.

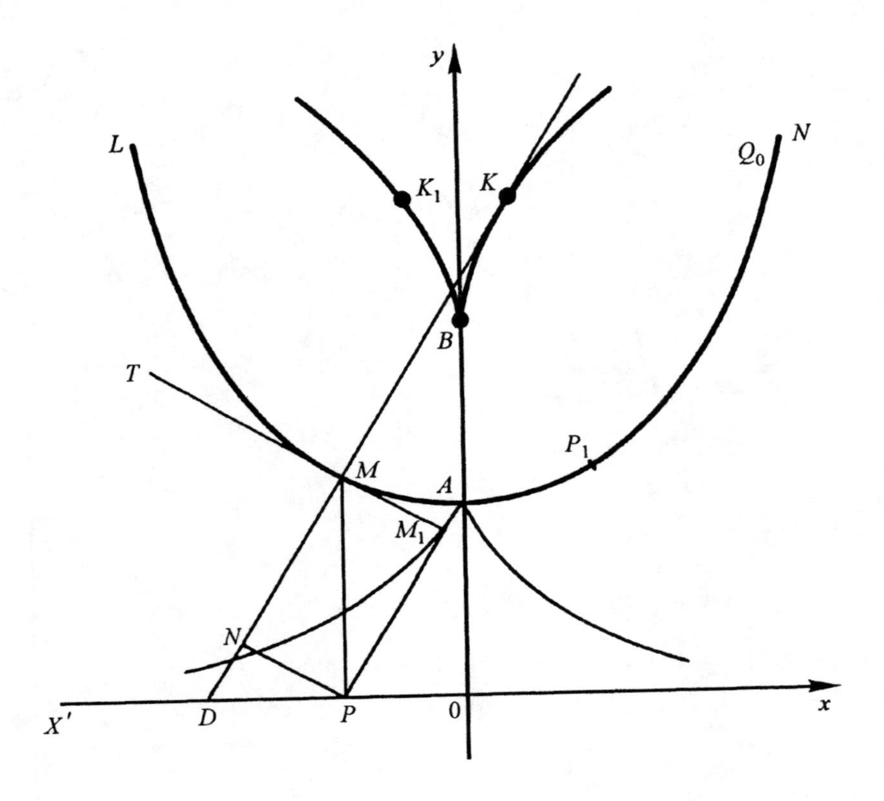

Figure 3.80. The catenary.

3.5.3. Catenary

The *catenary* is the form that a homogeneous nonstretchable string assumes when suspended by its ends (Figure 3.80). Depending on the position of the ends and the length of the string, the shape of the whipping can be different.

The lowest point of the catenary is called the *vertex*.

If the y-axis is directed upward and the vertex is taken to be the origin, then the equation of the catenary is $y = \frac{a}{2}(x^{x/a} + e^{-x/a}) - a$, where a is a parameter of the catenary. If the origin O lies at the distance a below the vertex, then we get a simple equation, $y = a\cosh x/a$.

The straight line XX' (Figure 3.80) that is parallel to the tangent at the vertex and located below the vertex at the distance a is called the *directrix* of the catenary.

The length of arc AM, s, of the catenary equals the projection MM_1 of the y-coordinate PM to the tangent MT,

$$s = \overset{\cup}{AM} = MM_1 = \frac{a}{2}\left(e^{x/a} - e^{-x/a}\right) = a\,\text{sh}\,\frac{x}{a}\,.$$

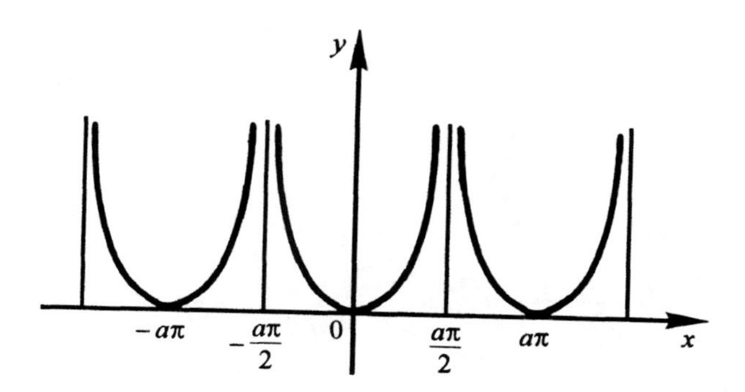

Figure 3.81. The even resistance catenary.

The following relation takes place: $s^2 + a^2 = y^2$, where $y = PM$. The projection MH of the y-coordinate MP of the catenary to the normal MD has a constant length, $MH = OA = a$.

The radius of curvature can be calculated by

$$R = MD = \frac{a}{4}\left(e^{x/a} + e^{-x/a}\right) = a\,\mathrm{ch}^2\frac{x}{a} = \frac{y^2}{a}.$$

Note that the length of the normal line at an arbitrary point of the catenary is

$$N = y\sqrt{1 + y'^2} = a\,\mathrm{ch}^2\frac{x}{a} = \frac{y^2}{a}.$$

This shows that the radius of curvature of a catenary coincides with the length of the normal at the point. This property can be used to easily find the center of curvature of the catenary at any point. The natural equation of the catenary has the form $R = s^2/a + a$. It has the following kinematic meaning: if the catenary rolls on a straight line without slipping, then the center of curvature at the contact point describes a parabola.

The area of the curvilinear trapezoid $OAMP$ equals the area of a rectangle with sides a, s, so that $S = as = a^2 \sinh x/a$. If a straight angle moves in the plane in such a way that its sides slide on a catenary, then the product of lengths of arcs between the vertex of the catenary and the contact point is a constant. The sum of curvatures of a catenary at the points at which the tangents are perpendicular is a constant for each catenary.

An interesting variant of the catenary is an *even-resistance catenary*, $e^{y/a}\left|\cos\frac{x}{a}\right| = 1$. This is the form taken by a flexible heavy string that has a variable cross section, the area of which is proportional to the strain force in the string at the cross section.

The curve passes through the origin. Since the function is aperiodic, the curve consists of an infinite number of identical branches symmetric with respect to the straight lines $x = a\pi k$, where k is an integer (Figure 3.81).

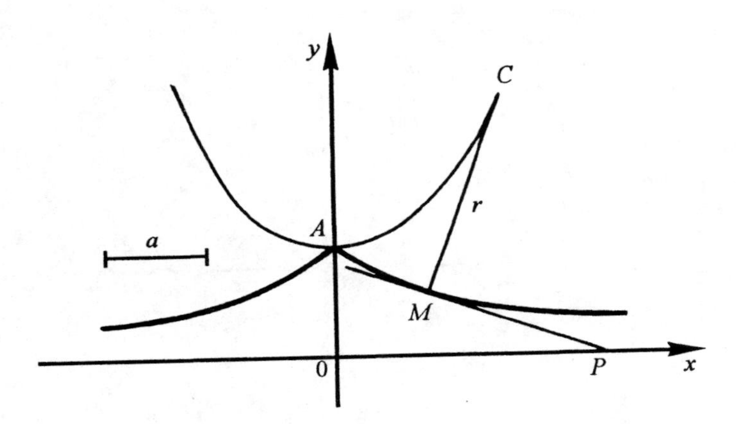

Figure 3.82. The tractrix $x = \arctan\frac{a}{x} \pm \sqrt{a^2 - y^2}$.

Note that the slope of the tangent to this curve is proportional to the x-coordinate of the point of tangency. This implies that the straight lines $x = (2k+1)a\pi/2$ are asymptotes. The length of the even-resistance catenary equals $l = a \ln \tan(\pi/4 + x/2a)$. The radius of curvature is $R = \dfrac{a}{\cos\frac{x}{a}}$. The natural equation of the curve is $R = \frac{a}{2}(e^{s/a} + e^{-s/a})$.

If an even-resistance catenary rolls without slipping on a straight line, the locus of its centers of curvature corresponding to the points of contact is the usual catenary.

Remark. The questions concerning the form of a suspended curve were subjects of studies by Galileo (1638), and later by Huygens, Leibniz, and others.

3.5.4. Tractrix

A *tractrix* is a curve such that the length of the line segment between the point of tangency, M, and the point P of intersection of this line with a given line is a constant (Figure 3.82). If a mass point M is attached to one end of a nonstretchable string of length a, and the second end moves along the x-axis, then the trajectory of the point M is a tractrix. The point that is the most distant from the guiding line is called the *vertex* of the tractrix.

The equation of the tractrix is

$$x = a \operatorname{Arch}\frac{a}{y} \pm \sqrt{a^2 - y^2} = \left(a \ln\frac{a \pm \sqrt{a^2 - y^2}}{y} \mp \sqrt{a^2 - y^2}\right).$$

The parametric equation is given by $x = a\cos t + a\ln\tan t/2$, $y = a\sin t$. The natural equation is $s = a\ln\sqrt{R^2/a^2 + 1}$. The x-axis is an asymptote of the curve. The point $A(a;\ 0)$ is a turning point (with vertical tangent). The curve is symmetric with respect to the coordinate axes.

The area bounded by the curve and the asymptote is determined by the formula

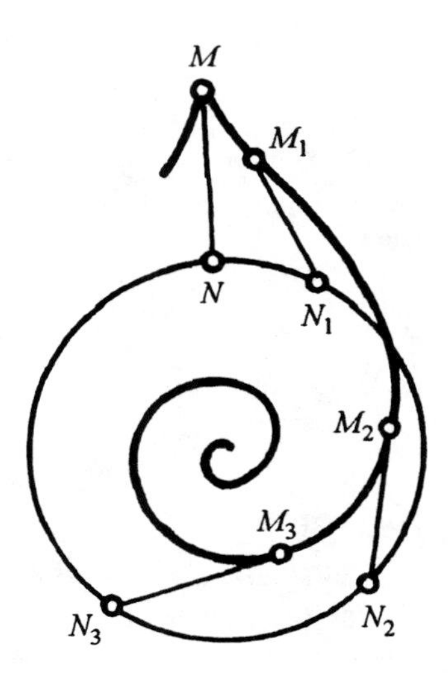

Figure 3.83. The tractrix of a circle.

$$S = \int_{-\infty}^{\infty} y\,dx = \int_{0}^{\pi} a \sin\varphi \,\frac{a \cos^2\varphi}{\sin\varphi}\,d\varphi = \frac{\pi a^2}{2}.$$

The volume of the body of revolution obtained by revolving the tractrix about its asymptote is calculated by

$$V = \pi \int_{-\infty}^{\infty} y^2\,dx = \pi \int_{0}^{\pi} a^2 \sin^2\varphi \,\frac{a \cos^2\varphi}{\sin\varphi}\,d\varphi = \frac{2}{3}\pi a^3,$$

the surface area of the body of revolution is

$$P = 4\pi \int_{\pi/2}^{\pi} a \sin\varphi \,\frac{a \cos\varphi}{\sin\varphi}\,d\varphi = 4\pi a^2.$$

The radius of curvature of the curve at an arbitrary point is given by $R = a\cot t$. The length of the arc from the point (0; 0) to an arbitrary point that corresponds to an arbitrary value of the parameter t is $l = -a\ln \sin t$.

The tractrix of the circle is defined by the equation

$$\varphi = -\arcsin\frac{\rho}{2a} + \frac{\sqrt{4a^2 - \rho^2}}{\rho},$$

where a is the radius of the circle. The curve consists of two branches symmetric with respect to the polar axis (Figure 3.83). The center of the circle is an asymptotic point of the curve.

Polar tractrix. A *polar tractrix* is a curve such that the length of the polar tangent is the same for any point of the curve.

The equation of the polar tractrix is

$$\varphi = -\arccos\frac{\rho}{a} + \frac{\sqrt{a^2-\rho^2}}{\rho},$$

or

$$\varphi = -\arcsin\frac{\rho}{a} + \frac{\sqrt{a^2-\rho^2}}{\rho}.$$

Note that the tractrix is, at the same time, a tractrix of the circle. The polar tractrix is a pedal of a hyperbolic spiral with respect to the pole.

Note. The tractrix is mentioned for the first time in connection with Perro (seventeenth century). Later, the curve was studied by Leibniz, Huygens, and others.

The tractrix attracted attention in relation to finding the curve described by a heavy mass point M fixed at one end of a string while the other end of the string moves on a certain straight line. Later the straight line was replaced by an arbitrary curve (equitangential curve). An *equitangential curve* for an arbitrary curve is the locus of the endpoint of a line segment of constant length drawn from the point of tangency along the tangent to the given curve.

The tractrix has played an important role in mathematics in connection with development of the theory of non-Euclidean geometries. The elliptic geometry of Riemann and the hyperbolic geometry of Lobachevsky can be realized on surfaces of constant curvature. Such surfaces can be obtained by revolving, about the x-axis, a curve that has the property that the product of the radius of curvature, R, and the normal N at the point is a constant, $RN = C$.

If $C > 0$, the corresponding surface has a constant positive curvature (for example, a sphere), and can be used to realize the geometry of Riemann. If $C < 0$, the corresponding surface has negative curvature, and can serve for a realization of the geometry of Lobachevsky. The tractrix satisfies the equality $RN = \text{const}$.

Note that the surface obtained by revolving a tractrix has negative curvature and is called a *pseudosphere*.

3.5.5. Quadratrix of Dinostratus

The *quadratrix of Dinostratus* can easily be obtained in the following way. Let a radius $OA = a$ make uniform revolutions with angular velocity $\pi/2T$ about the center, the straight line perpendicular to AD, OB, and uniformly move from the point A to the point D with the velocity a/T. The point of intersection, M, will describe a quadratrix (Figure 3.84).

Parametric equation of the quadratrix is

$$x = \frac{2a}{\pi}t, \quad y = \frac{2a}{\pi}t\cot t,$$

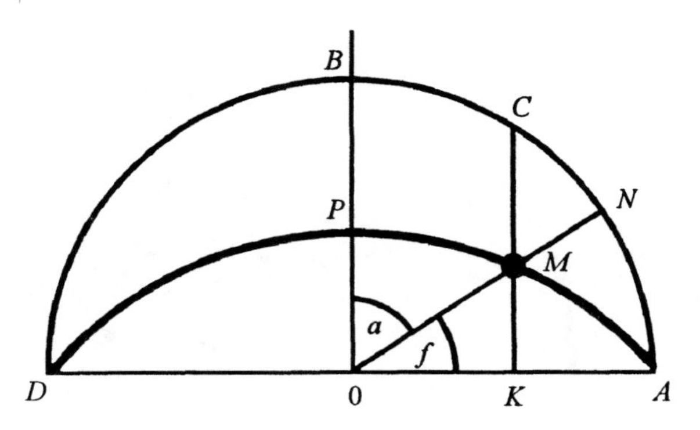

Figure 3.84. The quadratrix of Dinostratus.

where $t = \frac{\pi}{2}\left(1 - \frac{\tau}{T}\right) = \angle NOB$ and τ is the time counted from the start of the motion. In the Cartesian coordinate system, the equation takes the form $y = x \cot \frac{\pi x}{2a}$. The polar equation of the quadratrix is $\rho = a(\pi - 2\varphi)/\pi\cos\varphi$. Since

$$\lim_{x \to 0} y = \lim_{x \to 0} x \cot \frac{\pi x}{2a} = \frac{2a}{\pi},$$

the initial y-coordinate of the curve is $2a/\pi$.

The curve intersects the x-axis in an infinite number of points, $x = \pm a, \pm 3a, \pm 5a, \ldots$. The straight lines $x = \pm 2a, \pm 4a, \pm 6a, \ldots$ are asymptotes. Inflection points of the curve are the points in which it intersects the straight lines $y = 2a/\pi$ (Figure 3.85).

The area bounded by the curve over the interval $-a \le x \le a$ equals $S = 4a^2 \ln 2/\pi$.

There is a stereometric method for obtaining the quadratrix: if a helical surface is intersected by a plane that passes through a generating line, then the projection of the resulting curve to the plane perpendicular to the axis of the helical surface is the quadratrix of Dinostratus.

Note. The quadratrix of Dinostratus is one of the curves known from ancient times when there were attempts to solve the problems of partitioning the angle at a given ratio and quadrature of a circle. The curve was first constructed by Hippias of Elis (420 BC) in connection with solution of the first problem. The second problem was solved using this curve by Dinostratus (fourth century BC). Later it appeared that the quadratrix of Dinostratus is not the unique curve that can be used for quadrature of a circle. Other curves can be used to solve this problem: quadratics of Tschirnhausen $y = a \sin \frac{\pi x}{2a}$, the curve of Otsanama $x = 2a \sin^2 \frac{y}{2a}$ and the cochleoid $\rho = a \frac{\sin\varphi}{\varphi}$.

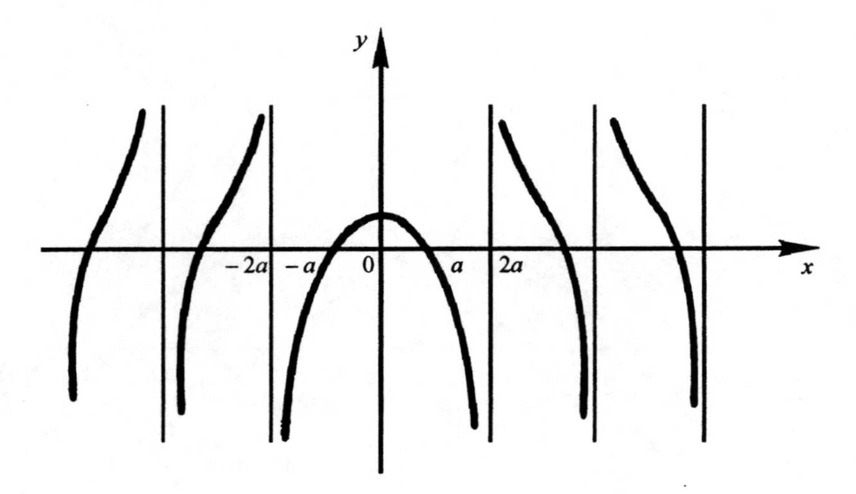

Figure 3.85. Inflection points of the quadratrix of Dinostratus.

3.5.6. Cochleoid

The *cochleoid* is a curve that belongs to the family of quadratrixes. The equation of the curve is $\rho = a\dfrac{\sin\varphi}{\varphi}$. As φ tends to infinity, the length of the radius vector tends to zero, and, since values of the sine function oscillate, the values of ρ are also oscillating, become zero at $\varphi = 0, \pi, 2\pi, \ldots$, and attain relative maximums for values of φ that are roots of the equation $\tan\varphi - \varphi = 0$ (Figure 3.86).

The cochleoid has the following property: Every straight line passing through the pole of the cochleoid intersects the curve in an infinite number of points at which the tangents pass through the same point A. The locus of such a point is a circle with its center at the pole of the cochleoid and radius OA.

A cochleoid can be obtained by projecting a helical curve to the plane that is perpendicular to the axis passing through any point of the helical curve.

3.5.7. Exponential curves

Exponential curves are curves of the form $y = a^x$ (see Part I of the book). Consider examples of power-exponential curves.

1. $y = x^x$. The curve consists of a single branch. The point $(1/e; (1/e)^{1/e})$ is a vertex of the curve; the vanishing point is located at (0; 1) (Figure 3.87).

2. $x^y = y^x$. The curve consists of the straight line $y = x$ and a branch located in the first quadrant. Asymptotes of the branch are $y = 1$ and $x = 1$. The branches intersect in the point $(e; e)$ (Figure 3.88).

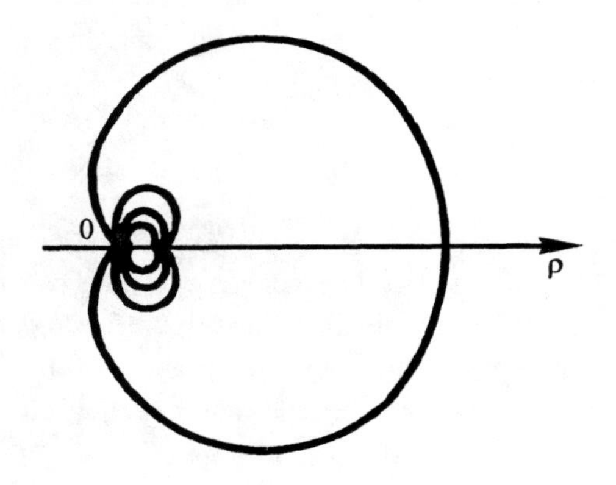

Figure 3.86. The cochleoid.

Figure 3.87. The curve $y = x^x$.

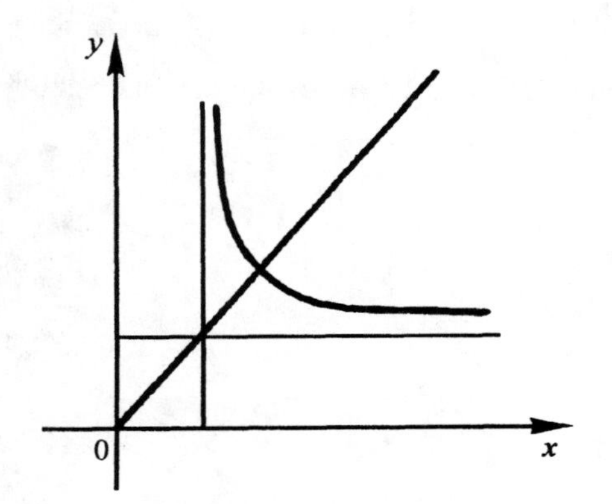

Figure 3.88. The curve $x^y = y^x$.

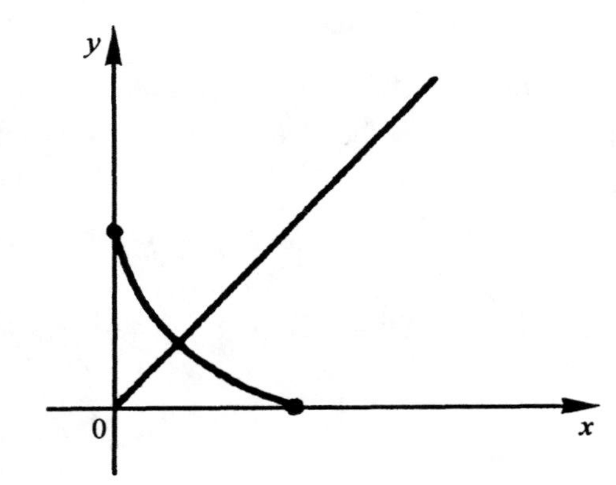

Figure 3.89. The curve $x^x = y^y$.

3. $x^x = y^y$. The curve consists of the straight line $y = x$ and a bounded branch that has endpoints at (0; 1) and (1; 0). The branch intersects the straight line in the point $(1/e;\ 1/e)$ at a right angle (Figure 3.89).

3.5.8. Cycloid

A *cycloid* is the curve traced by a point fixed in the plane of a disk that rolls without slipping on a straight line. If the point that traces a cycloid lies inside the disk, then the cycloid is called *curtaite* (Figure 3.90*c*); if it lies outside of the disk, then it is called *prolate* (Figure 3.90*b*); if the point is located on the circumference, it is called usual or simply cycloid (Figure 3.90*a*).

The equation of the cycloid in parametric form is $x = a(t - \sin t)$, $y = a(1 - \cos t)$, where a is the radius of the disk, $t = \angle MC_1B$. In Cartesian coordinates, the cycloid is given by

$$x + \sqrt{y(2a - y)} = a \arccos \frac{a - y}{a}.$$

The curve is periodic with the period (the basis of the cycloid) $OO_1 = 2\pi a$. The turning points are $O, O_1, O_2, \dots (2\pi ka;\ 0)$ and the vertices lie at $A_1, A_2, \dots [(2k + 1)\pi a;\ 2a]$. Curtaite and prolate cycloids do not have turning points.

The length of OM is given by $L = 8a\sin^2 t/4$; the length of one arc of the cycloid is calculated by $L_{OA_1O_1} = 8a$; the area bounded by the arc OA_1O_1O is $S = 3\pi a^2$. The radius of curvature is $R = 4a\sin t/2$, and $R_A = 4a$ at the vertices. The volume of the solid of revolution obtained by revolving an arc of the cycloid about the x-axis is

$$V = \pi \int_0^{2\pi a} y^2\, dx = 5\pi^2 a^2.$$

The surface area of this body is given by

$$P = 2pi \int_0^{2\pi a} y\, dx = \frac{64}{3} \pi a^2.$$

The volume and surface area of the body of revolution obtained by revolving the arc of the cycloid about the y-axis equal $6\pi^3 a^2$, $32/3\pi a^2$, respectively. The volume and surface area of the body of revolution in the case where the cycloid is revolving about its symmetry axis are given by $\frac{\pi a^3}{6}(9\pi^2 - 16)$ and $8\pi\left(\pi - \frac{4}{3}\right)a^2$, respectively.

An evolute of a cycloid is again a cycloid (Figure 3.90, dashed line). For curtaite and prolate cycloids, the equation in the parametric form is given by $x = a(t - \lambda\sin t)$, $y = a(1 - \lambda\cos t)$, where a is the radius of the disk, $t = \angle MC_1P$, $\lambda a = C_1M$ (for prolate cycloid, $\lambda > 1$, for curtaite, $\lambda < 1$).

The curves are periodic with the period $OO_1 = 2\pi a$; maximums are achieved at $A_1, A_2, \dots,$ $[(2k + 1)\pi a;\ (1 + \lambda)a]$, minimums at $B_0, B_1, \dots, [(2k\pi a;\ (1 - \lambda)a]$. For the prolate cycloid, the

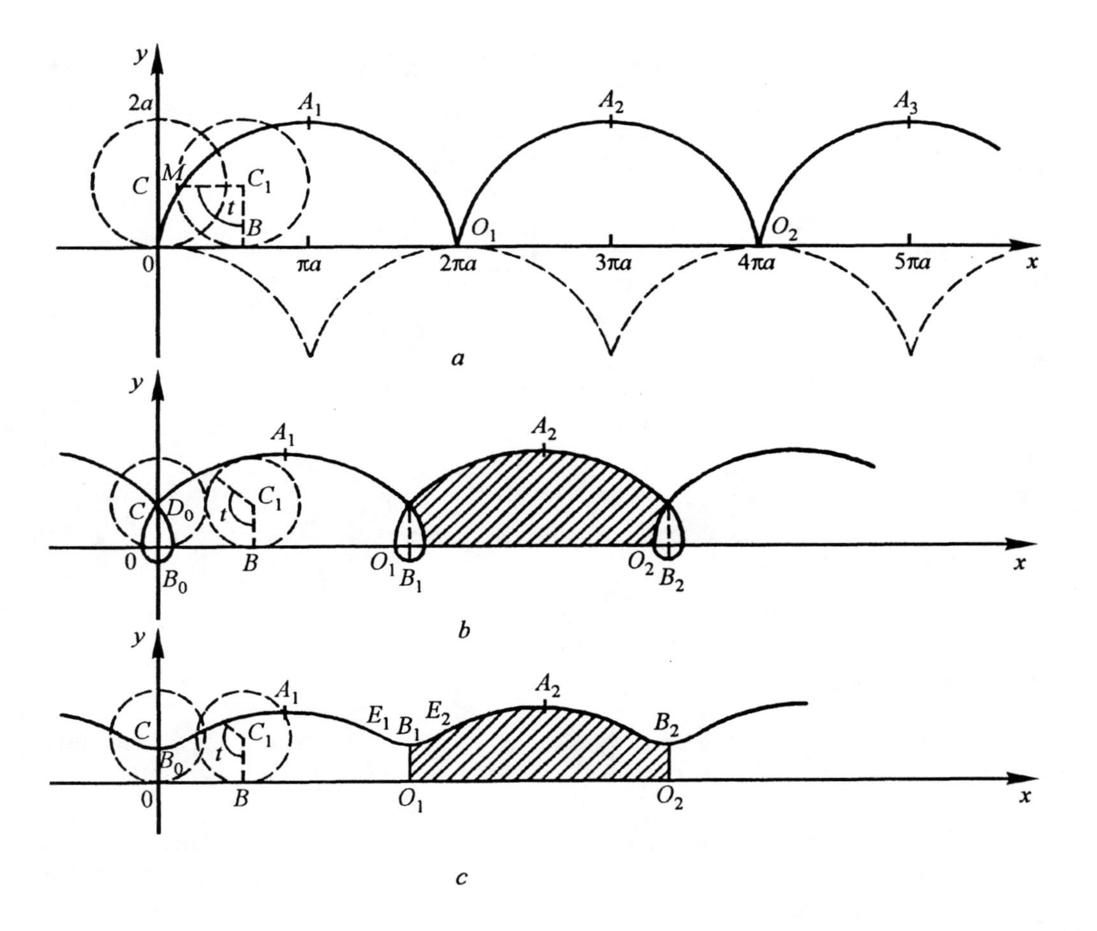

Figure 3.90. The cycloid $x = a(t - \sin t)$, $y = a(1 - \cos t)$.

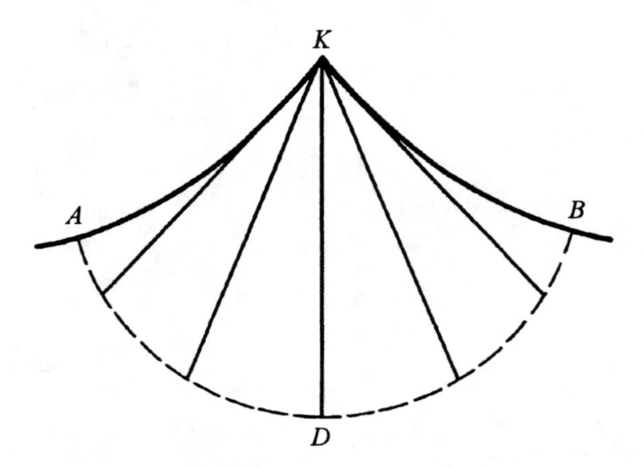

Figure 3.91. The tautochrone property of the cycloid.

nodes are at the points $D_0, D_1, D_2, \ldots$ $[2k\pi;\ a(1-\sqrt{\lambda^2-t_0^2})]$, where t_0 is the least positive root of the equation $t=\lambda\sin t$. Inflection points of the curtaite cycloid are the points $E_1, E_2, \ldots,$ $a(\arccos\lambda-\lambda\sqrt{1-\lambda^2});\ a(1-\lambda^2)]$. The length of one cycle is $L=a\int_0^{2\pi}\sqrt{1+\lambda^2-2\lambda\cos t}\,dt$. The area of the shaded part in Figure 3.90 is $S=\pi a^2(2+\lambda^2)$.

The radius of curvature of the curve is $R=\dfrac{(1+\lambda^2-2\lambda\cos t)^{3/2}}{\lambda(\cos t-\lambda)}$ at the points of maximum, $R_a=-a(1+\lambda)^2/\lambda$, and $R_B=a(1-\lambda)^2/\lambda$ at the points of minimum.

Curtaite and prolate cycloids are sometimes called *trochoids*. All cycloids (the usual, curtaite, and prolate) can be defined as parallel projections of a helical line to a surface perpendicular to the axis of the line. Depending on the value of the angle between the direction of the projection and the generating line (less, equal, or greater than the angle of ascent, μ, which is the angle between the tangent to the helical line and the z-axis), we get an ordinary, prolate, or curtaite cycloid.

The natural equation of one arc of the ordinary cycloid is given by $R^2+(s-4a)^2=(4a)^2$, $0<s<2$, where R is the radius of curvature.

Ordinary cycloid has the following kinematic property: If an ordinary cycloid rolls without slipping on a straight line, then the center of curvature of the point of contact describes a circle. The radius of this circle is four times the radius of the initial circle, and its center coincides with the point at which the trajectory of the vertex of the cycloid intersects the straight line.

The cycloid has also the following tautochrone property: a mass point that moves on an ordinary cycloid ADB (Figure 3.91) under the action of the force of gravity reaches the lowest position D in time $t=\pi\sqrt{a/g}$, where a is the radius of the circle and g is gravitational acceleration.

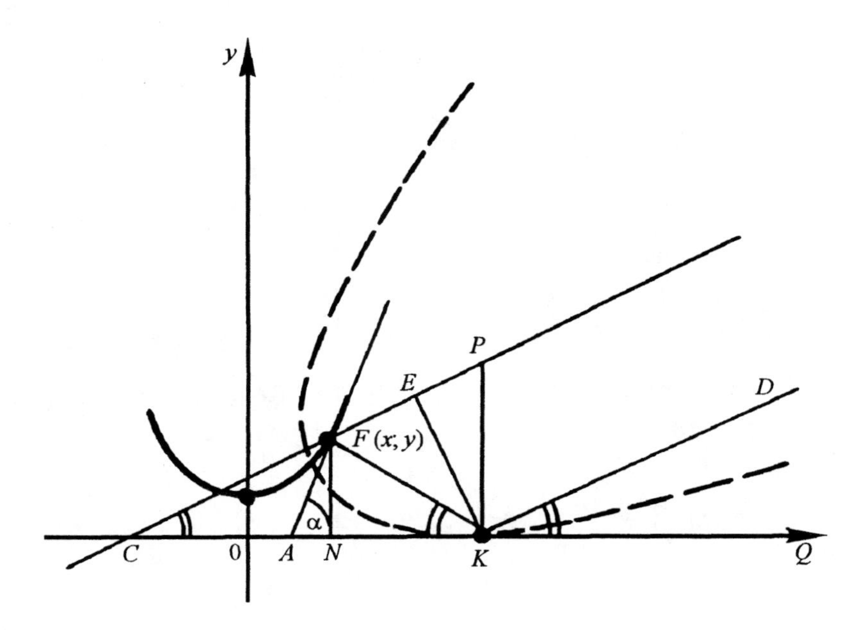

Figure 3.92. The trajectory of the focus of a parabola that rolls on a straight line.

This time does not depend on the initial position of the point. Thus the period of oscillation of a pendulum ($T = 4t$) does not depend on its amplitude. The string of the cycloidal pendulum is attached at the initial point K of another cycloid AKB, which is an evolute of the cycloid ADB.

The brachistichrone of a point that moves, under the action of the force of gravity in a medium with negligible resistance, from a given point A to a point B not lying on the same vertical, is an ordinary cycloid.

Note. The cycloid was studied for the first time by Galileo, and later by Torricelli, Roberval, Descartes, Fermat, and others.

In 1673 Huygens discovered the isochrone property of a cycloidal pendulum and saw it as one of the laws of nature.

Discovered in 1696 by J. Bernoulli, the property of cycloid to be a brachistochrone has had a great impact. Actually, this was a basis for the main ideas in the calculus of variations.

3.5.9. Curves of Sturm

If an ellipse, hyperbola, or parabola rolls without slipping on a straight line, then a focus traces the curve called the *curve of Sturm*. In the case where a parabola rolls on the x-axis, the equation of the curve described by the focus is $y = \frac{p}{2} \cosh \frac{2}{p} x$, where p is the parameter of the parabola. Hence this trajectory is a catenary (Figure 3.92). The form of the curve traced by a focus of a hyperbola that rolls on a straight line is close to the form of a prolate cycloid.

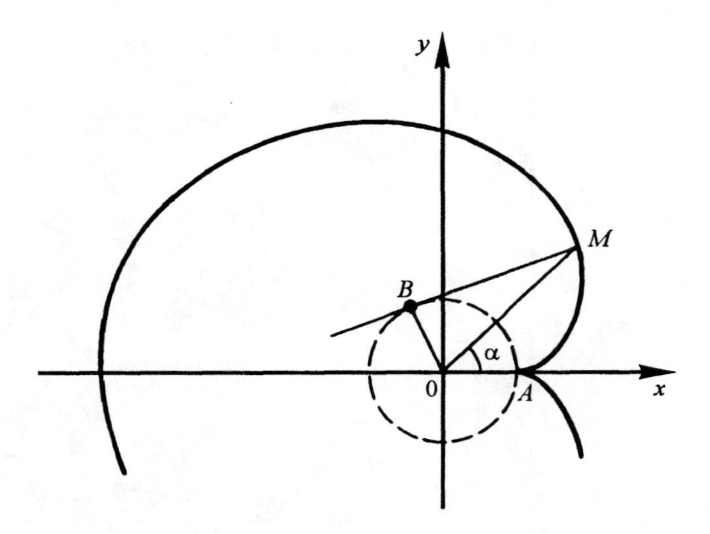

Figure 3.93. The involute of a circle,
$x = k(\cos\alpha + \alpha\sin\alpha, y = k(\sin\alpha - \alpha\cos\alpha)$.

The differential equation of the trajectory generated by the center of an ellipse that rolls on a straight line is

$$dx = \frac{y^2\,dy}{\sqrt{(a^2 - y^2)(y^2 - b^2)}}.$$

Trajectories of points in the plane of an ellipse rolling on another ellipse are called *epiellipsids*.
Trajectories of a point in the plane of the rolling curve, which changes its position relative to the curve, generate a *curve of Manheim*.

If the rolling curve has the natural equation $R = f(s)$, then the corresponding curve of Manheim is obtained as follows. The straight line on which the curve rolls is taken to be the x-axis; the length of the arc of the curve is being measured starting from the origin at the initial position of the curve. Then s will be the x-coordinate of the point of the curve of Manheim; R is the corresponding y-coordinate. Hence, the curve of Manheim will be given by the equation $y = f(x)$. For example, the curve of Manheim of the catenary given by the equation $R = s^2/a + a$ will be the parabola defined by the equation $y = x^2/a + a$; for the logarithmic spiral $R = ks$, the corresponding curve of Manheim is $y = kx$.

3.5.10. Involute of a circle

The *involute of a circle* is the curve described by the end of a string that unwinds from a circle ($\breve{AB} = BM$) (Figure 3.93). The same circle gives rise to an infinite number of involutes (which correspond to different initial positions of the point). The foot of the perpendicular dropped from the center O to the tangent to the involute, MT, generates a spiral of Archimedes.

The polar equation of the involute of a circle is

$$\varphi = \frac{\sqrt{\rho^2 - k^2}}{k} - \arccos\frac{k}{\varphi},$$

where O is the center of the given circle, the polar axis Ox is directed towards the initial radius OA, and k is radius of the circle.

The parametric equation of the involute of a circle is given by

$$x = k(\cos\alpha + \alpha\sin\alpha), \quad k(\sin\alpha - \alpha\cos\alpha), \quad \alpha = \angle MOA \ (-\infty < \lambda < +\infty).$$

The natural equation of the curve is $R^2 = 2ks$. The curve consists of two branches, which are symmetric with respect to the x-axis for $-\infty < \lambda \le 0$, $0 \le \alpha < +\infty$. The two branches intersect in the point $A(k; 0)$. The point A is a turning point.

The tangent to the involute at an arbitrary point is parallel to the radius that joins the center of the circle and the point of tangency. The length of the arc $\overset{\smile}{AM}$ is $l = \frac{k\alpha^2}{2}$. The radius of curvature is given by $R(\alpha) = k|\alpha|$. All centers of curvature lie on a circle with its center at the point O and of radius k. If an involute of a circle rolls on a straight line, then the locus of its centers of curvature corresponding to the point of contact makes a parabola. (This is a kinematic property of the involute.)

The area of the sector OMA is $S_c = \frac{1}{6}R^2\alpha^3$ or $S_c = \frac{1}{3}SR_k$, i.e. the area of the sector OMA of the involute is one third of the product of the arc length and the radius of curvature at the endpoint of this arc.

The pedal of the involute of a circle with respect to the origin is a spiral of Archimedes.

Stereometrically, the involute of a circle is given as the line of intersection of the surface of tangents to a helical curve and a plane perpendicular to the axis of the curve.

Note. Since normals to the involute are tangents to the evolute, the involute of a curve can be constructed by a kinematic method.

There exist many various relations between the evolute and the corresponding involute, which follow from definitions of the curves (the *evolute* of a curve is a locus of the centers of curvature, and the curve itself is an involute with respect to its evolute). For example, the differential of the arc of an evolute equals the radius of curvature of the involute; an extremum of the radius of curvature of an involute corresponds to a turning point of the evolute; the points of the involute at which the curvature equals zero correspond to the point at infinity for the evolute and the corresponding unbounded branch with the asymptote being a normal to the involute at the point of zero curvature.

The involute of an involute is the *second involute* of the curve. For a circle, for instance, the natural equation of the second involute is $R^3 = \frac{9}{2}ks^2$. The curve given by this equation is called the *spiral of Sturm* . Actually, this is a curve such that the radius of curvature at any of its points equals the length of the radius vector to this point.

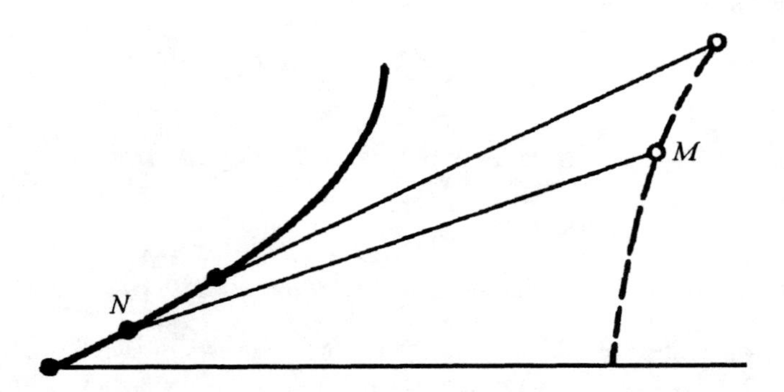

Figure 3.94. The construction of a pursuit curve.

3.5.11. Pursuit curve

Suppose we have a point $M_1(x_1; y_1)$ in a curve and a point $N(x_2; y_2)$ on a plane (Figure 3.94). Let a point M uniformly move on the given curve. We need to find a curve traced by the point N moving uniformly such that at any time the tangent to the curve at the point N passes through the point M.

This is a formulation of the problem that defines the pursuit curve. The name of the curve comes from the fact that the curve is the trajectory of the point N which "pursues" the point M. The pursuit curve is the most effective trajectory, since the point N moving along the pursuit curve is always directed towards the point M along the tangent, i.e. the shortest path joining the points N and M at any instant.

Let, at an initial time, the point M be located in the x-axis at the distance a from the origin, and then move along the straight line $\xi = a$ in the positive direction. The equation of the pursuit curve, in this simple case, will be

$$y = \frac{ak}{2}\left[\frac{\left(1-\frac{x}{a}\right)^{1+1/k}}{1+k} + \frac{\left(1-\frac{x}{a}\right)^{1+1/k}}{1-k}\right] + \frac{ak}{k^2-1},$$

where $k \neq 1$. If $k = 1$, the equation of the pursuit curve is

$$y = \frac{a}{2}\left[\frac{\left(1-\frac{x}{a}\right)^{2}}{2} - \ln\left(1-\frac{x}{a}\right)\right] - \frac{a}{4}.$$

The curve has the following main properties. The tangent to the curve at the origin coincides with the x-axis. The derivative y' is positive on $(0; a)$, thus the curve lies above the x-axis. As $x \to a$, $y' \to +\infty$, i.e. the tangent to the curve is perpendicular to the x-axis. If $x = a$ and $k > 1$,

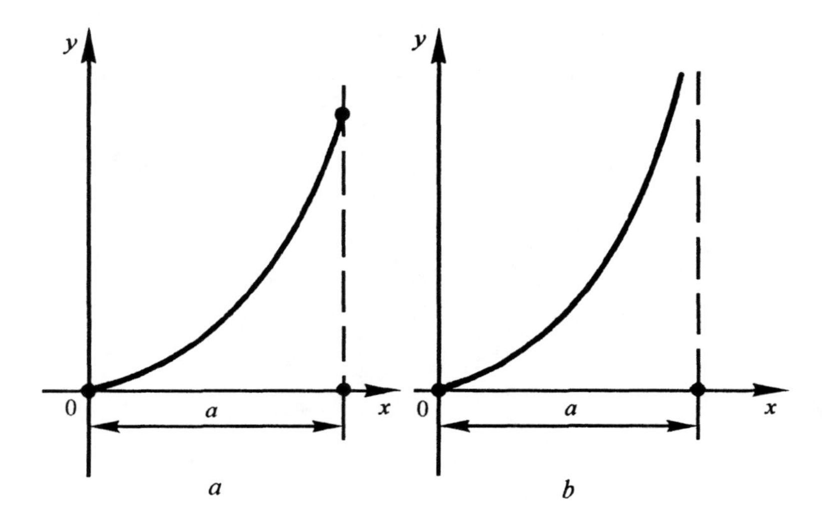

Figure 3.95. The shapes of a pursuit curve for $k > 1$ and $k \leq 1$.

we have that $y = ak/(k^2 - 1)$. If $k < 1$ or $k = 1$ and $x \to a$, $y \to +\infty$. This shows that, for $k > 1$, the pursuit curve intersects the straight line $\xi = a$, and the point N will catch up with the point M (Figure 3.95*a*). If $k \leq 1$, the pursuit curve asymptotically approaches the straight line $x = a$, and hence will not be able to catch up with the point M in a finite time (Figure 3.95*b*).

Note that if k is rational, $k \neq 1$, the pursuit curve is algebraic, and if k is integer, the curve is of parabolic type (for $k = 2$, the pursuit curve is a cubic of Tschirnhausen). If $k = 1$ or k is irrational, the pursuit curve is transcendental.

Note. The problem about the pursuit curve was considered for the first time by Leonardo da Vinci, and solved by Bouguer in 1732.

3.5.12. Curves of Ribaucour

A property of the tractrix is that the product of the radius of curvature, R, at any point and the normal N at this point is a constant, $RN = \text{const}$. Curves that satisfy the equation $R/N = \text{const}$ are called *curves of Ribaucour*.

The differential equation that these curves satisfy is $y'' = \frac{n}{y}(1 + y'^2)$, where $1/n = k$. The parametric equation of a curve of Ribaucour (in the case where $h = 1/k$, h is an arbitrary number) is

$$x = (m + 1)C \int_0^t \sin^{m+1} t \, dt, \quad y = C \sin^{m+1} t,$$

where $m = -(n+1)/n$, C is an arbitrary constant. Examples of curves of Ribaucour are: the circle $x = -C\cos\varphi, y = C\sin\varphi$ $(m = 0)$; the cycloid $x = C/2(2t - \sin 2t)$, $y = C/2(1 - \cos 2t)$ $(m = 1)$; the catenary $y = \frac{C}{2}\cosh\frac{x}{C}$ $(m = -2)$; and the parabola $x^2 = 2C(y - C)$ $(m = -3)$.

In the general case there are four groups of curves of Ribaucour that correspond to these examples:

1) higher degree circles, if $(m+1)$ is a positive odd number;
2) curves of cycloid type, if $(m+1)$ is a positive even number;
3) curves of catenary type, if $(m+1)$ is a negative odd number;
4) higher degree parabolas, if $(m+1)$ is a negative even number.

The arc length of a curve of Ribaucour is calculated by

$$l = (m+1)C\int_0^t \sin^m t\, dt\,.$$

The radius of curvature is given by $R = -(m+1)C\sin^m t$. The natural equation of the curve is

$$s = -\frac{1}{m}\int \frac{dR}{\sqrt{\left[\frac{R}{(m+1)C}\right]^{-2/m} - 1}}\,.$$

Remark that curves of Ribaucour and sinusoidal spirals belong to the same class of curves which satisfy the natural equation

$$s = \int \frac{dR}{\sqrt{(R/b)^m - 1}}\,.$$

Such curves in this class are called *curves of Cesaro*.

Curves of Cesaro are defined to be the curves such that the radius of curvature at an arbitrary point P is proportional to the corresponding segment of the normal bounded by the point P polar with respect to a circle.

Note. The problem of finding curves satisfying the equation $R/N = \text{const}$ was posed by J. Bernoulli (1716) and solved by Leibniz. A more detailed study of these curves was carried out by Ribaucour (1880).

3.5.13. Cornu's spiral

Cornu's spiral is a curve such that the radius of curvature is inversely proportional to the length of the arc, $R = a^2/s$ is the natural equation of the curve. The parametric equation of Cornu's spiral is

$$x = a\sqrt{\pi}\int_0^t \cos\frac{\pi\tau^2}{2}\,dt, \quad y = a\sqrt{\pi}\int_0^t \sin\frac{\pi\tau^2}{2}\,dt,$$

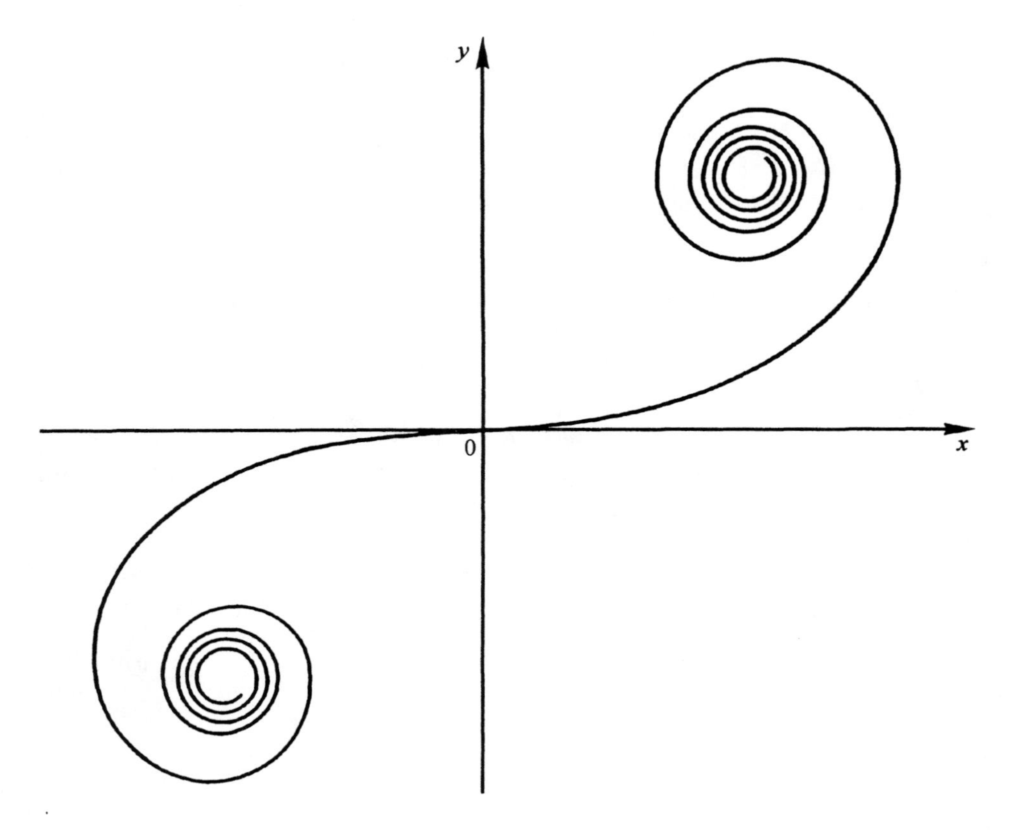

Figure 3.96. The Cornu's curve $x = a\sqrt{\pi}\int_0^t \cos\frac{\pi\tau^2}{2}\,dt,\ y = a\sqrt{\pi}\int_0^t \sin\frac{\pi\tau^2}{2}\,dt.$

where $t = s/a\sqrt{\pi}$, $s = OM$ (Figure 3.96).

The Cornu's curve is symmetric with respect to the origin.

The curve has two asymptotic points $A(a\sqrt{\pi}/2;\ a\sqrt{\pi}/2)$ and $B(-a\sqrt{\pi}/2;\ -a\sqrt{\pi}/2)$. The origin is an inflection point of the curve.

The curvature at the point $O(0;\ 0)$ equals zero. As the length of the arc increases, the curvature increases proportionally, and the curve turns about the point A (B) making an infinite number of windings.

Note that the natural equation of the Cornu's curve (as well as of the logarithmic spiral) is a special case of the equation $R = as^m$. Curves defined by this equation are called *pseudohelixes*. Examples of pseudohelixes are the logarithmic spiral ($m = 1$), Cornu's curve ($m = -1$) and involute of a circle ($m = 1/2$).

Pseudohelixes have the following properties: The evolute of a pseudohelix is a pseudohelix. For example, the natural equation of the evolute of a Cornu's curve is $R_1 = -\frac{1}{a}s_1^3$.

Let α be the angle between the tangent to the Cornu's curve and the x axis. The relation between the angle α and the arc length of the Cornu's curve has the form $\lambda = \frac{1}{2a}s^2$. Krause studied the curves that satisfy the general equation $\alpha = f(s)$. Among those, there are curves that have a very specific shape. They satisfy the equation $\alpha = k\sin s$, or the natural equation

$$R = k \sec s. \qquad (1)$$

Curves that are defined by the natural equation of type (1) are called *patterned curves* (Figure 3.97).

3.5.14. Radial curves

Suppose a point moves on a curve that has a variable value and direction of curvature.

Fix a point on the line, and draw, from this point, line segments with length and direction equal to the value and direction of the curvature. The locus of the endpoints of these line segments is called a *radial curve*.

Let $s = \psi(R)$ be the natural equation of the curve, φ be the angle between the polar axis and the radius vector ρ to an arbitrary point P of the radial curve. The polar equation of the radial curve is $\varphi = \int \frac{\psi'(\rho)}{\rho}\, d\rho$, the direction of the polar axis coincides with the direction of the radius of curvature at the initial point. For curves given by a natural equation of the form $s = \int \frac{\lambda\, dR}{\sqrt{(R/a)^{\mu} - 1}}$, the equation of the radial curve is

$$\rho = \frac{a}{\left(\cos^{2\mu} \frac{\mu\varphi}{2\lambda}\right)}.$$

Special cases are the following: For a cycloid, the radial curve is the circle $\rho = a\cos\varphi$ ($\mu = -2$, $\lambda = 1$); for a catenary of uniform resistance, it is the straight line $\rho = a/\cos\varphi$ ($\mu = 2$, $\lambda = 1$); for the tractrix $R = a\sqrt{e^{2s/a} - 1}$, it is the kappa curve $\rho = a\tan\varphi$.

A curve that is a solution of the inverse problem of finding the curve from its radial curve is called *antiradial* with respect to the given curve.

The natural equation of the antiradial curve is $s = \int Rf'(R)\, dR$, where $\varphi = f(\rho)$ is the given curve.

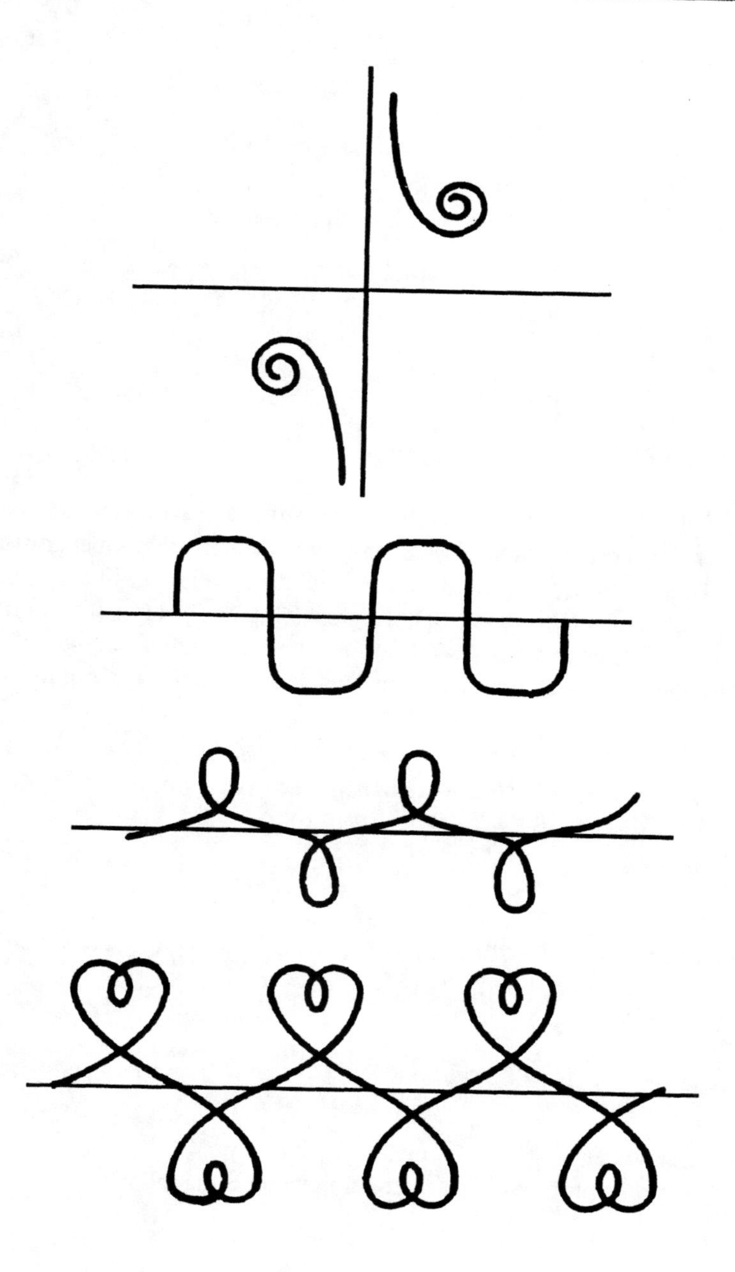

Figure 3.97. The patterned curves.

Chapter 4

Graphs of special functions

To solve and investigate many scientific and engineering problems, one needs to study special functions. The most important classes of special functions are: elementary transcendental functions (exponential, logarithmic, trigonometric, and hyperbolic); the integral power function; the gamma-function (and related functions); the probability integral function; Fresnel integrals; Legendre functions; elliptic functions; Mathieu functions; orthogonal polynomials; and others.

Elementary transcendental functions were considered in the first part of the book.

4.1. Factorial function $n!$. Bernoulli numbers and polynomials.

Euler numbers and polynomials. Binomial coefficients. Pochhammer polynomials

The factorial function $n!$ This function is defined only for nonnegative integers, and takes positive integer values:

$$n! = 1\cdot 2\cdot 3\cdot \ldots \cdot (n-1)n = \prod_{k=1}^{n} k, \qquad n = 1, 2, 3, \ldots$$

This definition is supplemented by defining $0! = 1$.

The most important property of the factorial function is the following:

$$(n+1)! = (n+1)n!, \qquad n = 0, 1, 2, \ldots,$$
$$(n+m)! = n!(n+1)(n+2)\ldots(n+m) = n!(n+1)_m, \qquad n, m = 0, 1, 2, \ldots,$$

where $(n+1)_m$ is the Pochhammer symbol,

$$(a)_0 = 1,\ (a)_m = a(a+1)\ldots(a+m-1)_m, \qquad m = 1, 2, 3, \ldots .$$

For the ratio of two factorials, we have $n!/(n-m)! = n(n-1)(n-2)...(n-m+1)$ or $\frac{n!}{(n-m)!} = \sum_{k=0}^{m} S_m^{(k)} n^k$, where $S_m^{(k)}$ is the Stirling number of the first kind. The Stirling formula is

$$n! = \sqrt{2\pi n}\, n^n e^{-n}\left[1 + \frac{1}{12n} + \frac{1}{288n^2} - \frac{139}{51840n^3} - ...\right] \text{ at } n \to \infty.$$

We have

$$\ln(n!) = n \ln n - n + \frac{1}{2}\ln 2\pi n + \frac{1}{12n} - \frac{1}{360n^3} + ..., \quad n \to \infty.$$

The *double factorial* is defined by

$$n!! = \begin{cases} 1, & n = -1, 0, \\ n(n-2)(n-4)...5\cdot 3\cdot 1, & n = 1, 3, 5, ..., \\ n(n-2)(n-4)...6\cdot 4\cdot 2, & n = 2, 4, 6, ... \end{cases}$$

The *triple factorial* is

$$n!!! = \begin{cases} 1, & n = -2, -1, 0, \\ n(n-3)(n-6)...7\cdot 4\cdot 1, & n = 1, 4, 7, ..., \\ n(n-3)(n-6)...8\cdot 5\cdot 2, & n = 2, 5, 8, ..., \\ n(n-3)(n-6)...9\cdot 6\cdot 3, & n = 3, 6, 9, ... \end{cases}$$

The following Wallis formula holds:

$$\frac{(n-1)!!}{n!!} = \frac{n!}{2^n[(\frac{n}{2})!]^2} = \frac{2}{\pi}\int_0^{\pi/2} \sin^n x dx, \quad n = 0, 2, 4, ...$$

The graph of the factorial function $n!$ (with the logarithmic scale used for the y-coordinate) is shown in Figure 4.1.

The *Bernoulli numbers* B_n are defined for all nonnegative integers n by the formula

$$\frac{e^x}{e^x - 1} = \sum_{n=0}^{\infty} B_n \frac{x^n}{n!},$$

and the *Bernoulli polynomial* $B_n(x)$ is defined by

$$\frac{ze^{zx}}{e^z - 1} = \sum_{n=0}^{\infty} B_n(x)\frac{z^n}{n!}, \quad |z| < 2\pi .$$

The expansion for the Bernoulli numbers is

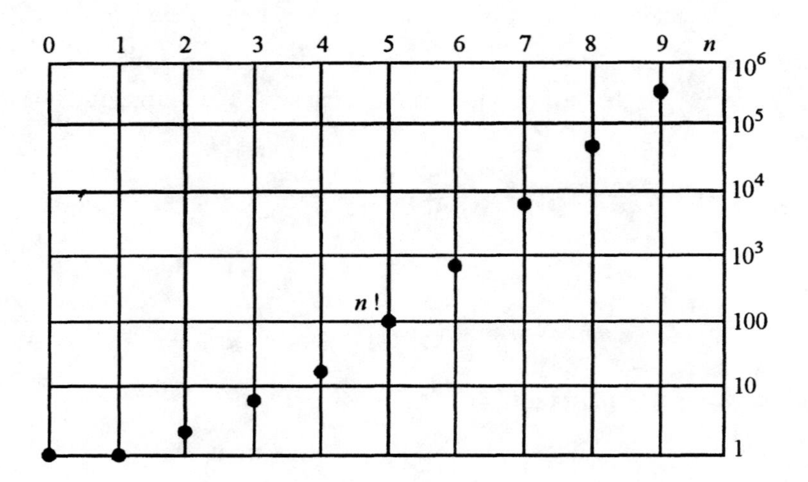

Figure 4.1. The factorial function $n!$.

$$B_{2n} = \frac{2(-1)^{n+1}(2n)!}{(2\pi)^{2n}} \sum_{k=1}^{\infty} \frac{1}{k^{2n}}, \quad n = 1, 2, 3, \ldots,$$

$$B_{2n+1} = 0, \quad n = 1, 2, 3, \ldots .$$

Thus all Bernoulli numbers with odd indices, save for B_1, equal zero. All B_n with the index n divisible by 4, except for B_0, are negative (Figure 4.2).

The integral representation of the Bernoulli numbers is the following:

$$B_{2n} = (-1)^{n+1}\pi \int_0^{\infty} \frac{t^{2n}dt}{\operatorname{sh}^2(\pi t)},$$

$$B_{2n} = \frac{(-1)^{n+1}\pi}{1 - 2^{1-2n}} \int_0^{\infty} \frac{t^{2n}dt}{\operatorname{ch}^2(\pi t)}.$$

There is a recurrence formula,

$$B_n = -n! \sum_{k=0}^{n-1} \frac{B_k}{k!(n-1-k)!}, \quad n = 2, 3, 4, \ldots ,$$

$B_0 = 1$. This formula easily follows from a more concise formula,

$$\sum_{k=0}^{n} \binom{n}{k} B_k = 0, \quad n = 2, 3, 4, \ldots ,$$

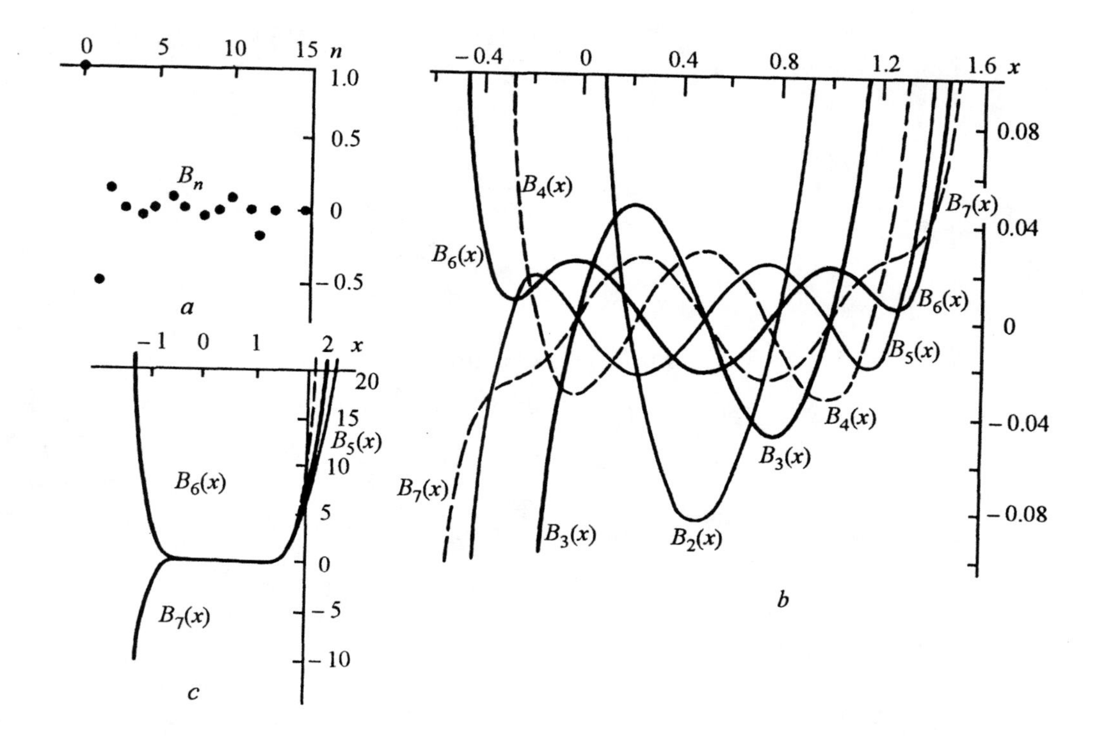

Figure 4.2. (a) The Bernoulli numbers B_n; (b), (c) the Bernoulli polynomials B_n.

where $\binom{n}{k}$ is the binomial coefficient. The approximate formula

$$B_n \approx (-1)^{\frac{n+1}{2}} \frac{2n!}{(2\pi)^n} [1 + \frac{1}{2^n}], \quad n = 6, 8, 10, \dots,$$

follows from

$$B_n = (-1)^{\frac{n+2}{2}} \frac{2n!}{(2\pi)^n} [1 + \frac{1}{2^n} + \frac{1}{3^n} + \dots], \quad n = 2, 4, 6, \dots .$$

The values of the Bernoulli numbers are

$$B_0 = 1, B_1 = -\frac{1}{2}, B_2 = \frac{1}{6}, B_3 = 0, B_4 = -\frac{1}{30}, B_6 = \frac{1}{42}, B_8 = -\frac{1}{30}, \dots .$$

Bernoulli polynomials admit the following integral representation:

$$B_{2n}(x) = (-1)^{n+1}(2n) \int_0^\infty \frac{\cos(2\pi x) - e^{-2\pi t}}{\cosh(2\pi t) - \cos(2\pi x)} t^{2n-1}\, dt,$$

$$0 < \operatorname{Re} x < 1, \quad n = 1, 2, 3, \ldots,$$

$$B_{2n+1}(x) = (-1)^{n+1}(2n+1) \int_0^\infty \frac{\sin(2\pi x)\, t^{2n}\, dt}{\cosh(2\pi t) - \cos(2\pi x)},$$

$$0 < \operatorname{Re} x < 1, \quad n = 0, 1, 2, \ldots .$$

The Bernoulli numbers can also be expressed in terms of the zeta function:

$$B_{2n} = (-1)^{n+1}(2\pi)^{-2n} \cdot 2 \cdot (2n)!\, \zeta(2n), \quad n = 0, 1, 2, \ldots,$$

$$B_{2n} = -2n\, \zeta[-(2n-1)], \quad n = 1, 2, 3, \ldots .$$

Note that the Bernoulli numbers are values of the Bernoulli polynomials at $x = 0$, $B_n = B_n(0)$, $n = 0,1,2,\ldots$.

The Bernoulli numbers and polynomials of degree m are defined by

$$\alpha_1 \ldots \alpha_m z^m \left[(e^{\alpha_1 z} - 1)\ldots(e^{\alpha_m z} - 1)\right]^{-1} = \sum_{n=0}^\infty B_n^{(m)}(\alpha_1 \ldots \alpha_m) \frac{z^n}{n!}, \quad |z| < \frac{2\pi}{|\alpha_l|},$$

$$\alpha_1 \ldots \alpha_m z^m \left[(e^{\alpha_1 z} - 1)\ldots(e^{\alpha_m z} - 1)\right]^{-1} e^{xz} = \sum_{n=0}^\infty B_n^{(m)}(x|\alpha_1 \ldots \alpha_m) \frac{z^n}{n!}, \quad |z| < \frac{2\pi}{|\alpha_l|},$$

where $m \in \mathbf{N}$, $\alpha_1, \ldots, \alpha_m$ are some parameters, $|\alpha_l| = \max[|\alpha_1|, \ldots, |\alpha_m|]$.

Euler numbers E_n. The *Euler numbers* are defined by the formula

$$\frac{1}{\cosh x} = \frac{2e^x}{e^{2x} + 1} = \sum_{n=0}^\infty E_n \frac{x^n}{n!}, \quad |x| < \frac{\pi}{2}.$$

They admit the following integral representation

$$E_n = (-1)^{n/2} \left(\frac{2}{\pi}\right)^{n+1} \int_0^\infty \frac{t^n}{\cosh}\, dt \qquad (n = 0, 2, 4, \ldots).$$

The recurrence formula for the Euler numbers has the form

$$\sum_{k=0}^n \binom{2n}{2k} E_{2k} = 0 \qquad (n = 1, 2, 3, \ldots)$$

or

$$E_n = (-n)! \sum_{k=0}^n \frac{E_{2k}}{(n-2k)!(2k)!}, \qquad (n = 2, 4, 6, \ldots)$$

where $E_0 = 1$. The Euler numbers E_n are defined for all nonnegative integers n, and they themselves are integers. The Euler numbers with odd index are all zero and, for even values of the index, they are positive or negative depending on particular values of n (Figure 4.3).

The formula for approximate values of the Euler numbers,

$$E_n \approx (-1)^{n/2} 2n! \left(\frac{2}{\pi}\right)^{n+1}, \qquad n = 4, 6, 8, \dots,$$

is obtained from the formula

$$E_n \approx (-1)^{n/2} 2n! \left(\frac{2}{\pi}\right)^{n+1} [1 - \frac{1}{3^{n+1}} + \frac{1}{5^{n+1}} - \dots], \qquad (n = 0, 2, 4, \dots).$$

The values of first several Euler numbers are the following: $E_0 = 1$, $E_1 = 0$, $E_2 = -1$, $E_4 = 5$, $E_6 = -61$, $E_8 = 1385$, ...

The *Euler polynomials* $E_n(x)$ are defined by

$$\frac{2e^{xz}}{e^z + 1} = \sum_{n=0}^{\infty} E_n(x) \frac{z^n}{n!}, \qquad |z| < \pi .$$

For odd m,

$$E_n(x + 1) + E_n(x) = 2x^n,$$

$$E_n(mx) = m^n \sum_{k=0}^{m-1} (-1)^k E_n(x + \frac{k}{m}),$$

and for even m,

$$E_n(mx) = -2m^n(n + 1)^{-1} \sum_{k=0}^{m-1} (-1)^k B_{n+1}(x + \frac{k}{m}).$$

Integral representations for the Euler numbers are the following:

$$E_{2n}(x) = (-1)^n \cdot 4 \int_0^{\infty} \frac{t^{2n} \sin(\pi x) \cosh(\pi t)}{\cosh(2\pi t) - \cos(2\pi x)} dt, \qquad n = 0, 1, 2, \dots,$$

$$(0 < \operatorname{Re} x < 1),$$

$$E_{2n+1}(x) = (-1)^{n+1} \cdot 4 \int_0^{\infty} \frac{\cos(\pi x) \sinh(\pi t)}{\cosh(2\pi t) - \cos(2\pi x)} dt, \qquad n = 0, 1, 2, \dots$$

$$(0 < \operatorname{Re} x < 1).$$

Note that the Euler numbers are related to the values of the Euler polynomials at $x = 1/2$, $E_n = 2^n E_n(1/2)$, $n = 0,1,2,\dots$.

The Euler numbers and polynomials of higher degree (degree m) are defined by

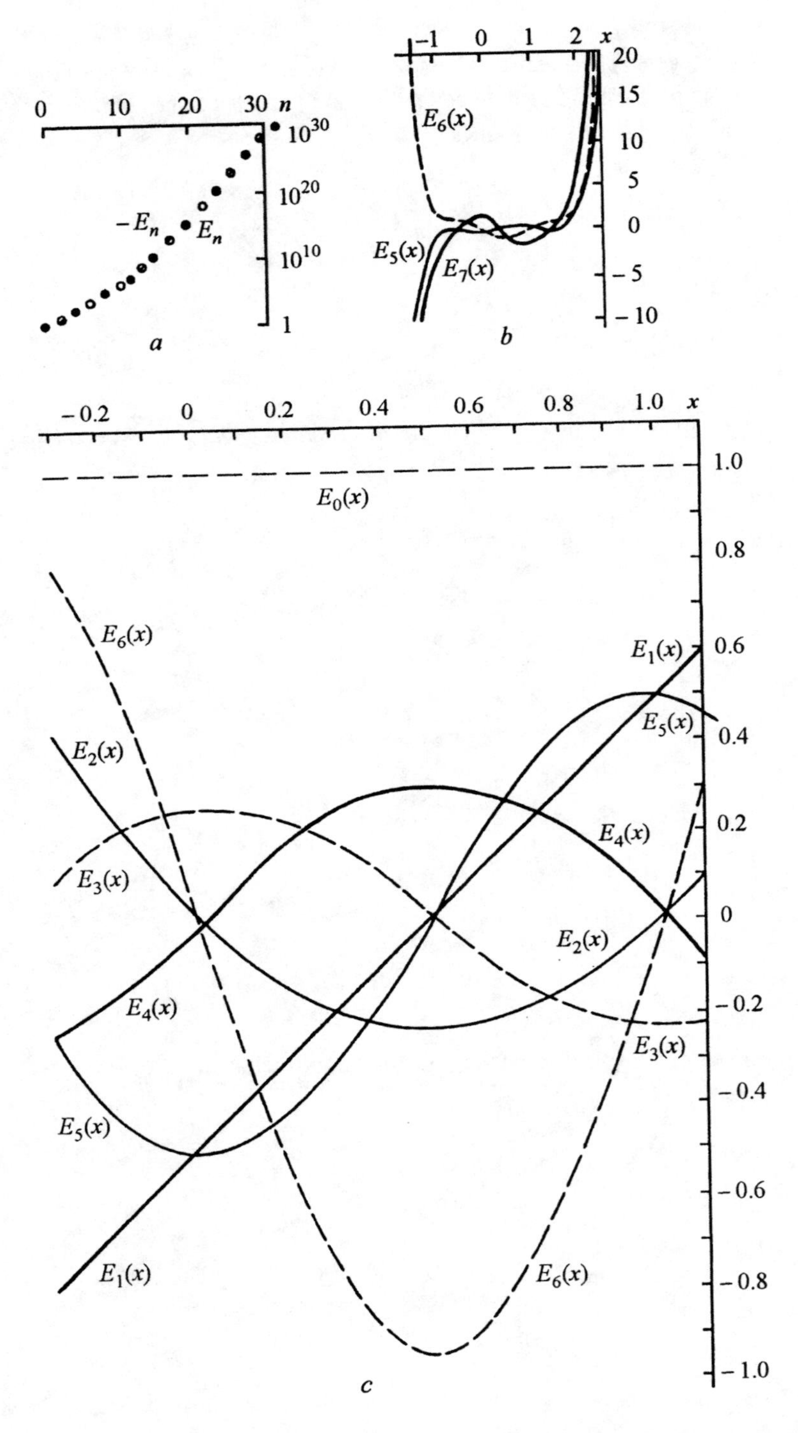

Figure 4.3. (*a*) The Euler numbers E_n; (*b*), (*c*) the Euler polynomials $E_n(x)$.

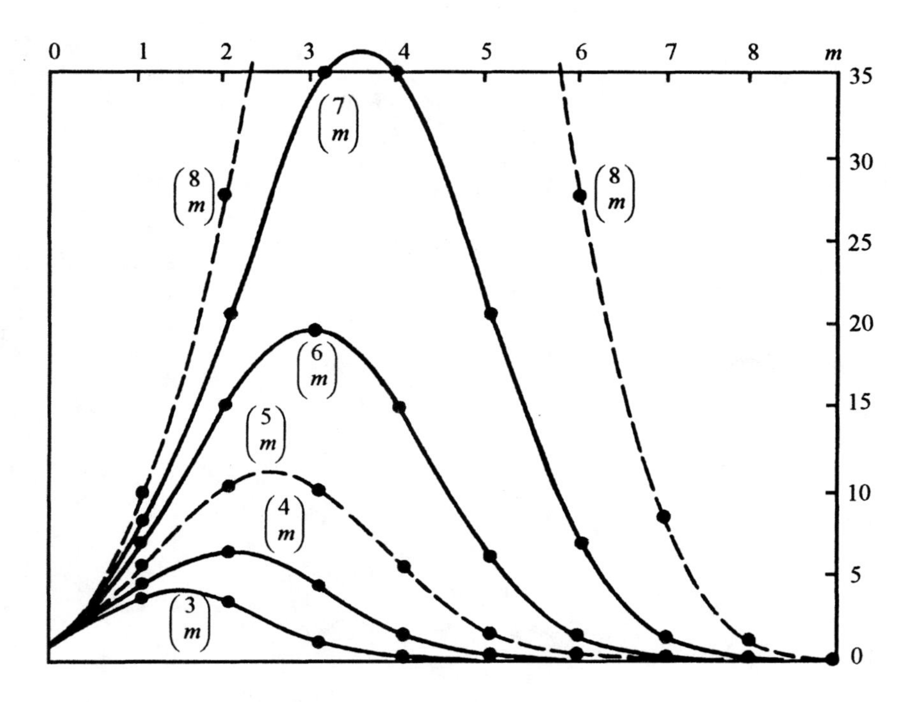

Figure 4.4. The binomial coefficients $\binom{\nu}{m}$.

$$2^m e^{z(\alpha_1 + \ldots + \alpha_m)}[(e^{2\alpha_1 z} + 1)\ldots(e^{2\alpha_m z} + 1)]^{-1} =$$

$$= [\cosh(\alpha_1 z)\ldots\cosh(\alpha_m z)]^{-1} = \sum_{n=0}^{\infty} E_n^{(m)}(\alpha_1\ldots\alpha_m)\frac{z^n}{n!},$$

$$2^m e^{xz}[(e^{\alpha_1 z} + 1)\ldots(e^{\alpha_m z} + 1)]^{-1} = \sum_{n=0}^{\infty} E_n^{(m)}(x|\alpha_1\ldots\alpha_m)\frac{z^n}{n!},$$

where the first series is convergent for $|z| < \pi/2|\alpha_i|^{-1}$, and the second series converges for $|z| < \pi|\alpha_i|^{-1}$, where $|\alpha_i| = \max\{|\alpha_1|, \ldots, |\alpha_m|\}$, $m \in \mathbf{N}$, and $\alpha_1, \alpha_2, \ldots, \alpha_m$ are arbitrary parameters.

Binomial coefficients. The *binomial coefficient* is defined to be the finite product

$$\binom{\nu}{m} = \left(\frac{\nu - m + 1}{1}\right)\left(\frac{\nu - m + 2}{2}\right)\left(\frac{\nu - m + 3}{3}\right)\ldots\binom{\nu}{m} = \prod_{k=1}^{m}\left(1 + \frac{\nu - m}{k}\right),$$

or

$$\binom{\nu}{m} = \frac{(\nu - m + 1)_m}{m!},$$

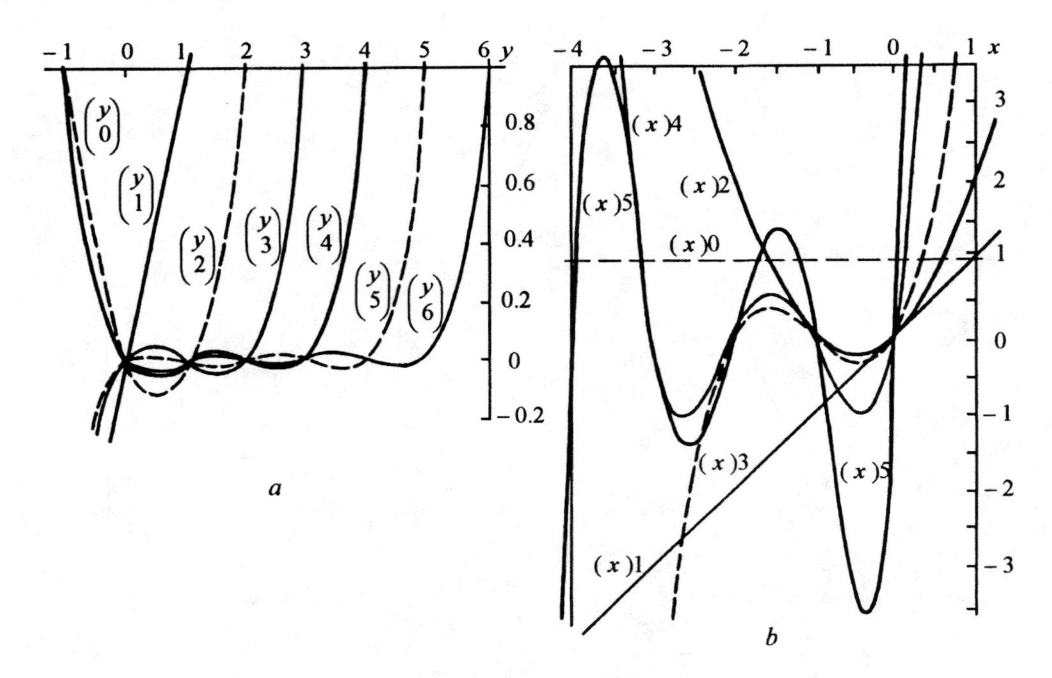

Figure 4.5. Particular cases of binomial coefficients (*a*). The functions of Pochhammer $(x)_n$ (*b*).

or else

$$\binom{\nu}{m} = \frac{\Gamma(\nu+1)}{m!\Gamma(\nu-m+1)} = \frac{\Gamma(\nu+1)}{\Gamma(m+1)\,\Gamma(\nu-m+1)}.$$

By definition, $\binom{\nu}{0} = 1$.

For positive integer ν, the binomial coefficient is denoted by C_n^m.

The lower number m of the binomial coefficient $\binom{\nu}{m}$ is always a nonnegative integer; the upper ν may be arbitrary. The binomial coefficient $\binom{n}{m}$ is positive if $m = 0,1,2,\ldots$, and equals zero if $m = n+1, n+2, n+3, \ldots$.

For a given n, $\binom{n}{m}$ is maximal for $m = n/2$, if n is even, and for $m = (n \pm 1)/2$, if n is odd (Figure 4.4).

If ν is not a positive integer or zero, then the binomial coefficient $\binom{\nu}{m}$ could be positive or negative, integer or not. The binomial coefficient equals zero if $\nu = 0,1,2,\ldots,(m-1)$, and its value is close to zero for $\nu \in (-\frac{1}{2}; m-\frac{1}{2})$ (Figure 4.5*a*).

Sometimes the binomial coefficients are defined as coefficients of the series

$$(1+x)^{\nu} = \sum_{m=0}^{\infty} \binom{\nu}{m} x^m \qquad (|x| < 1).$$

In the case where the upper index of the binomial coefficient is a positive integer, the binomial coefficient is expressed in terms factorials:

$$\binom{n}{m} = \frac{n!}{m!(n-m)!} \qquad (n \geq m).$$

The values of the binomial coefficients can be recurrently determined from Pascal's triangle:

$$\begin{array}{ccccccccccccccc}
 & & & & & & & 1 & & & & & & & \\
 & & & & & & 1 & & 1 & & & & & & \\
 & & & & & 1 & & 2 & & 1 & & & & & \\
 & & & & 1 & & 3 & & 3 & & 1 & & & & \\
 & & & 1 & & 4 & & 6 & & 4 & & 1 & & & \\
 & & 1 & & 5 & & 10 & & 10 & & 5 & & 1 & & \\
 & 1 & & 6 & & 15 & & 20 & & 15 & & 6 & & 1 & \\
1 & & 7 & & 21 & & 35 & & 35 & & 21 & & 7 & & 1
\end{array}$$

Each coefficient is the sum of the two left and right upper coefficients. The outmost left and right coefficients equal 1 for all n, $C_n^0 = C_n^n = 1$. The nth row consists of the coefficients C_n^0, C_n^1, C_n^2, ..., C_n^n. The following binomial formula holds for all real a, b, and $n \in \mathbf{N}$:

$$(a+b)^n = \sum_{k=0}^{n} C_n^k a^{n-k} b^k = C_n^0 a^n b^0 + C_n^1 a^{n-1} b^1 + \ldots + C_n^n a^0 b^n .$$

Polynomial coefficient. The *polynomial coefficient* is the function $C_n(k_1, k_2, ..., k_r)$ or

$$\binom{n}{k_1, k_2, ..., k_r} = C_n(k_1, k_2, ..., k_r) = \frac{n!}{k_1!\, k_2! \ldots! \, k_r!},$$

defined for all natural n and all collections of nonnegative integers $\{k_1, k_2, ..., k_r\}$ such that $\sum_{i=1}^{r} k_i = n$.

The polynomial formula is the following:

$$(a_1 + a_2 + \ldots + a_r)^n = \sum_{k_1 + k_2 + \ldots + k_r = n} C_n(k_1, k_2, ..., k_r)\, a_1^{k_1} a_2^{k_2} \ldots a_r^{k_r},$$

where all $a_r \in \mathbf{R}$, $n \in \mathbf{N}$. The summation is taken over the collection of all nonnegative integers $(k_1, k_2, ..., k_r)$ such that $\sum_{i=1}^{r} k_i = n$.

Pochhammer polynomials $(x)_n$. These polynomials are defined by

$$(x)_n = x(x+1)(x+2)\ldots(x+n-1) = \prod_{k=0}^{n-1}(x+k) \quad (n \in \mathbf{N})$$

with the condition that $(x)_0 = 1$ (Figure 4.5*b*).

One can also define the Pochhammer polynomials by using the factorial function and the binomial coefficients as $(x)_n = n!\binom{x+n-1}{n}$, $n = 0,1,2,\ldots$. Another definition is by the formula

$$(1-t)^{-x} = e^{-x\ln(1-t)} = \sum_{n=0}^{\infty}(x)_n \frac{t^n}{n!}.$$

The Pochhammer polynomial is written as a power series as follows:

$$(x)_n = (-1)^n \sum_{k=1}^{n} S_n^k (-x)^k = \sum_{k=0}^{n} |S_n^{(k)}|\, x^k,$$

where $S_n^{(k)}$ are Stirling numbers of the first kind. We remark that the Stirling numbers of the first kind satisfy the formula

$$\sum_{k=1}^{n} S_n^{(k)} = 0, \quad n = 2, 3, 4, \ldots, \quad \sum_{k=1}^{n} |S_n^{(k)}| = n!, \quad n \in \mathbf{N}.$$

The asymptotic formula for the Pochhammer polynomials is

$$(x)_n \sim \frac{n^{x-1}n!}{\Gamma(x)}\left[1 + \frac{x(x-1)}{2n} + \frac{x(x-1)(x-2)(3x-1)}{24n^2} + \ldots\right] \text{ for } n \to \infty.$$

The following formulas are also of interest:

$$\left(\frac{1}{2}\right)_n \sim \frac{n!}{\sqrt{\pi n}}, \quad n \to \infty,$$

$$\frac{d(x)_n}{dx} = (x)_n \sum_{k=0}^{n-1} \frac{1}{x+k} = (x)_n\,[\psi(n+x) - \psi(x)],$$

where ψ is the digamma function.

4.2. Gamma function and related functions

4.2.1. Gamma function $\Gamma(x)$

Following Euler, the gamma function $\Gamma(x)$ can be defined as

$$\Gamma(x) = \int_0^{\infty} e^{-t} t^{x-1} dt \qquad (x > 0),$$

following Gauss, as

$$\Gamma(x) = \lim_{n \to \infty} \frac{n! n^{x-1}}{x(x+1)(x+2) \ldots (x+n-1)}$$

(for $x \in \mathbf{R}$, $x \neq 0, -1, -2, \ldots$), and following Weierstrass, as

$$\frac{1}{\Gamma(x)} = x e^{\gamma x} \prod_{n=1}^{\infty} [(1 + \frac{x}{n}) e^{-x/n}],$$

where $\gamma = \lim_{m \to \infty} \left(\sum_{n=1}^{m} \frac{1}{n} - \ln m \right) = 0.5772156649\ldots$ is Euler's constant.

The gamma function can also be defined as a solution of the functional equation $\Gamma(z+1) = z\Gamma(z)$, $\Gamma(1) = 1$. The function, considered as a function of $z = x + iy$, is a meromorphic function with poles at the points $z = -n$, $n = 0, 1, 2, \ldots$. For real positive values of the argument $z = x$, the values of the function are real positive and satisfy the inequality $(\Gamma(x))^2 < \Gamma(x)\Gamma(x)''$, which means that the function is logarithmically convex.

The gamma function is also known as the *Euler integral of the second kind.* Sometimes it is denoted by $x!$ or $\pi(x)$ (Figure 4.6). The reciprocal function $1/\Gamma(x)$ is continuous and has zeros at $x = 0, -1, -2, \ldots$.

For real values of $z = x > 2$, the gamma function rapidly increases, and for large x, it behaves as $e^{x \ln (x/e)}/x$.

The main relations for the gamma function are the following:

$$\Gamma(2x) = (2\pi)^{-1/2} 2^{-1/2} \Gamma(x)\Gamma(x + \frac{1}{3}) = \frac{4^x}{2\sqrt{\pi}} \Gamma(x)\Gamma(\frac{1}{2} + x),$$

$$\Gamma(n) = (n-1)!, \quad n = 1, 2, 3, \ldots,$$

$$\Gamma(n + \frac{1}{2}) = (2n-1)!!\sqrt{\pi}\, 2^{-n}, \quad n = 0, 1, 2, \ldots,$$

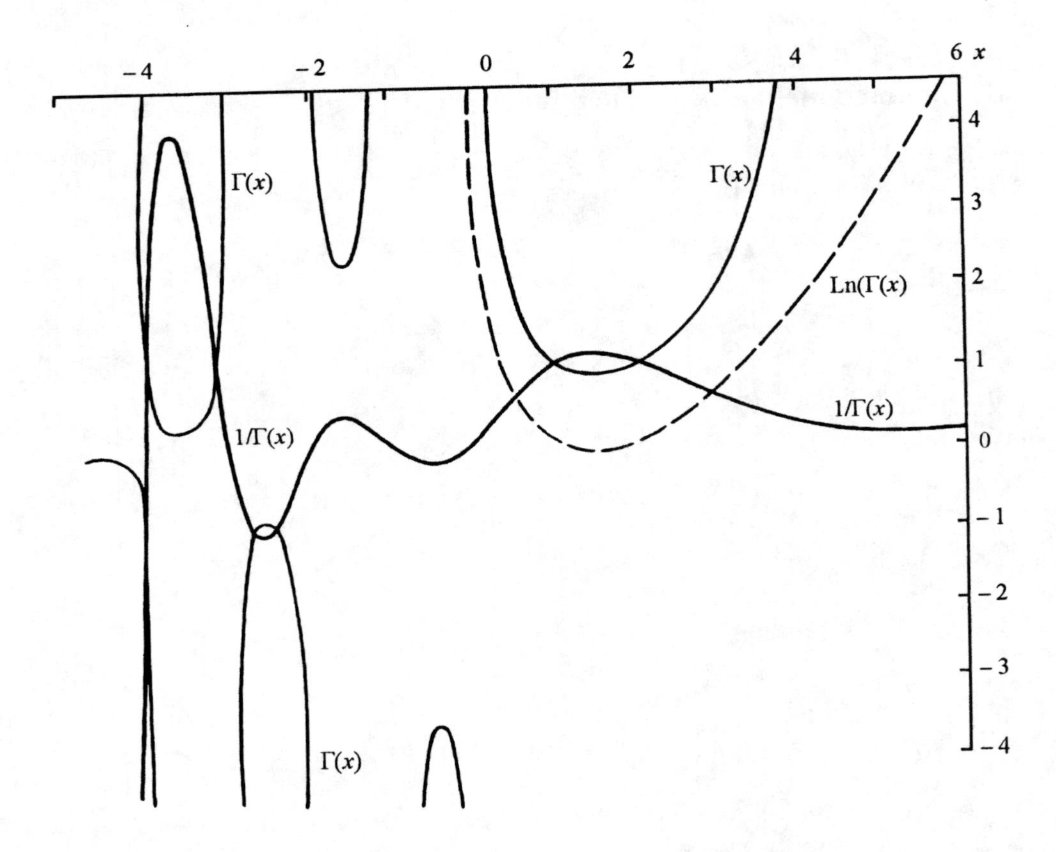

Figure 4.6. The gamma function $\Gamma(x)$.

$$\Gamma(n+\frac{1}{4}) = (4n-3)!!!!\,\Gamma\left(\frac{1}{4}\right)4^{-n}, \quad n = 0, 1, 2, \ldots,$$

$$\Gamma(-x) = \frac{-\pi \operatorname{cosec} \pi x}{x\Gamma(x)}, \quad \frac{\Gamma(x-n)}{\Gamma(x)} = \frac{(-1)^n}{(1-x)_n} \quad (n = 1, 2, 3, \ldots).$$

The function satisfies the complement formulas: $\Gamma(z)\Gamma(1-z) = -z\Gamma(-z)\Gamma(z) = \pi/\sin \pi z$, $\Gamma(1+z)\Gamma(1-z) = \pi z/\sin \pi z$, $\Gamma(1/2+z)\Gamma(1/2-z) = \pi/\cos \pi z$. The following reduction formulas also hold: $\Gamma(z+n) = z(z+1)(z+2)\ldots(z+n-1)\Gamma(z)$, $\Gamma(z-n) = \Gamma(z)/(z-1)(z-2)\ldots$ $\ldots(z-n)$. The function takes the following values: $\Gamma(1/2) = \sqrt{\pi}$, $\Gamma(-1/2) = -2\sqrt{\pi}$, $\Gamma(3/2) = \sqrt{\pi}/2$, $\Gamma(1/4) = 3.625609\ldots$, $\Gamma(1/3) = 2.678938\ldots$ (see Figure 4.6).

The logarithm of the gamma function can be expressed in terms of the integral as follows:

$$\ln(\Gamma(1+x)) = \int_0^1 \left(\frac{t^x - 1}{t-1} - x\right)\frac{dt}{\ln t} \quad (x > -1),$$

or

$$\frac{d\Gamma(z)}{dz} = \int_0^\infty x^{z-1} e^{-x} \ln x \, dx,$$

$$\Gamma^{(k)}(z) = \int_0^\infty x^{z-1} e^{-x} (\ln x)^k \, dx.$$

4.2.2. Beta function $B(x, y)$ (Euler integral of the first kind)

This function is defined by

$$B(x, y) = \int_0^1 t^{x-1}(1-t)^{y-1} dt, \quad x > 0, \ y > 0,$$

or

$$B(x, y) = \int_0^\infty (t^{x-1} + t^{y-1})(1+t)^{-x-y} dt, \quad x > 0, \ y > 0,$$

$$B(x, y) = \int_0^\infty t^{x-1}(1+t)^{-x-y} dt.$$

The relationship between the gamma and beta functions is the following:

$$B(x, y) = \frac{\Gamma(x)\Gamma(y)}{\Gamma(x+y)} = B(y,x).$$

The function $B(x, y)$ can also be represented as the product

$$B(x, y) = \prod_{k=0}^{\infty} \frac{(k+1)(x+y+k)}{(x+k)(y+k)},$$

$$B(x,-x) = \begin{cases} \dfrac{(-1)^x}{x}, & x \in \mathbf{Z}, \\ 0, & x \notin \mathbf{Z}. \end{cases}$$

If the arguments of the beta function are positive integers, then

$$\frac{1}{B(x, n)} = x \binom{n+x-1}{n-1}.$$

The main relations that the beta function satisfies are:

$$B(x, y+1) = \frac{y}{x} B(x+1, y) = \frac{y}{x+y} B(x, y),$$

$$B(x, y)\, B(x+y, z) = B(y, z)\, B(y+z, x) = B(z, x)\, B(x+z, y).$$

The asymptotic formulas for the beta function are:

$$\Gamma(az+b) \sim \sqrt{2\pi}\, e^{-az}(az)^{az+b-1/2} \qquad (|\arg z| < \pi,\ a > 0),$$

$$\ln\Gamma(z) \sim (z-\frac{1}{2})\ln z - z + \frac{1}{2}\ln(2\pi) + \sum_{m=1}^{\infty} \frac{B_{2m}}{2m(2m-1)z^{2m-1}} \qquad (z \to \infty,\quad |\arg z| < \pi),$$

where B_{2m} are the Bernoulli numbers.

4.2.3. Psi function (digamma function $\psi(x)$)

The *psi function* is defined to be the logarithmic derivative of the gamma function,

$$\psi(x) = \frac{d\ln\Gamma(x)}{dx} = \frac{\Gamma'(x)}{\Gamma(x)} \quad \text{or} \quad \ln\Gamma(x) = \int_1^x \psi(t)dt.$$

Another definition of the psi function is

$$\psi(x) = \lim_{n\to\infty} [\ln n - \sum_{k=0}^{n} \frac{1}{k+x}].$$

The psi function admits the following integral representation:

$$\psi(x) = -\gamma + \int_0^1 \frac{1-t^{x-1}}{1-t}\,dt, \qquad x > 0,$$

where γ is Euler's constant, or

$$\psi(x) = \int_0^{\infty} \left(\frac{e^{-t}}{t} - \frac{e^{-xt}}{1-e^{-t}}\right) dt = \int_0^{\infty} \frac{e^{-t}-(1+t)^{-x}}{t}\,dt, \qquad x > 0,$$

$$\psi(x) = \ln x - \frac{1}{2x} - 2\int_0^{\infty} \frac{tdt}{(t^2+x^2)(e^{2\pi t}-1)}, \qquad x > 0.$$

The function $\psi(x)$ has discontinuities at $x = 0, -1, -2, \ldots$ (Figure 4.7). If $x \to \infty$, the function approaches infinity as the logarithmic function.

The derivative $d\psi(x)/dx$ is always positive (Figure 4.8).

Some functional relations satisfied by the psi function are the following:

$$\psi(x) = \psi(1+x) - \frac{1}{x}, \quad \psi(n+1) = 1 + \frac{1}{2} + \frac{1}{3} + \ldots + \frac{1}{n} - \gamma,$$

$$\psi(x+n) = \frac{1}{x} + \frac{1}{x+1} + \ldots + \frac{1}{x+n-1} + \psi(x), \qquad n \in \mathbf{N},$$

$$\psi(x) - \psi(1-x) = -\pi \cot \pi z \qquad \text{(symmetry formula)}$$

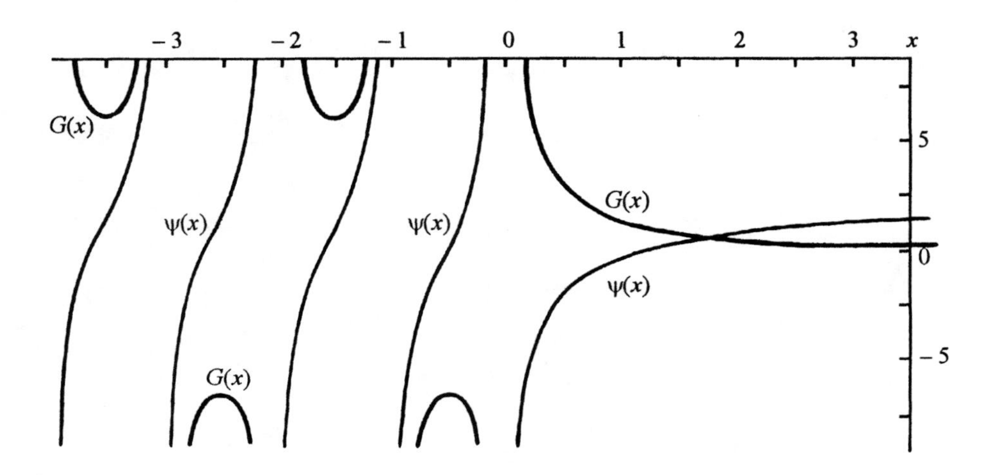

Figure 4.7. The psi function $\psi(x)$, the function $G(x)$.

$$\psi(2x) = \frac{1}{2}\psi(x) + \frac{1}{2}\psi(x + \frac{1}{2}) + \ln 2 \quad \text{(duplication formula)}.$$

Values of the psi function at certain points are the following:

$$\psi(1) = -\gamma, \ \psi'(1) = \frac{\pi^2}{6}, \ \psi\left(\frac{3}{2}\right) = 2 - \gamma - \ln 4, \ \psi'\left(\frac{3}{2}\right) = \frac{\pi^2}{2} - 4,$$

$$\psi(2) = 1 - \gamma, \ \psi'(2) = \frac{\pi^2}{6} - 1, \ \psi(3) = \frac{3}{2} - \gamma, \ \psi'(3) = \frac{\pi^2}{6} - \frac{5}{4}.$$

The *polygamma function* is defined by subsequent differentiation of the psi function, $\psi^{(n)}(x) = d^n\psi(x)/dx^n$, $n \in \mathbf{N}$. Sometimes, $\psi'(x)$, $\psi''(x)$, $\psi^{(3)}(x)$, $\psi^{(4)}(x)$ are called *trigamma, tetragamma, pentagamma*, and *hexagamma* functions. The polygamma functions take only positive values for odd n, and if n is even, the functions take both positive and negative values. The polygamma functions have discontinuities at $x = 0, -1, -2, \ldots$ (Figure 4.6 – 4.8).

The polygamma functions also can be defined as the integral

$$\psi^{(n)}(x) = (-1)^n \int_0^\infty \frac{t^n e^{-xt}}{t-1}\, dt, \quad n \in \mathbf{N}.$$

They satisfy the following recurrence formula:

$$\psi^{(n)}(x+1) = \psi^{(n)}(x) - n!/(-x)^{n+1}, \quad n \in \mathbf{N}.$$

For large values of x, we have

$$\psi^{(n)}(x) \approx -(n-1)!/(1/2 - x)^n.$$

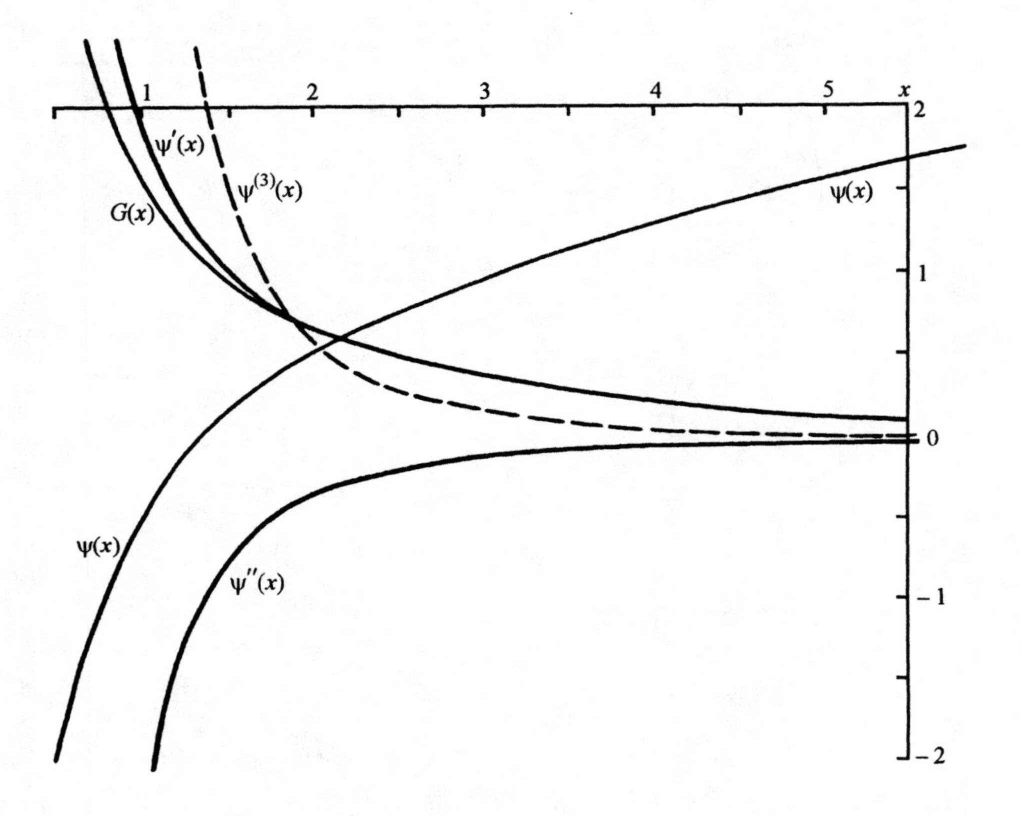

Figure 4.8. The psi function and its derivatives $\psi(x)$, $\psi'(x)$, $\psi''(x)$, $\psi^{(3)}(x)$.

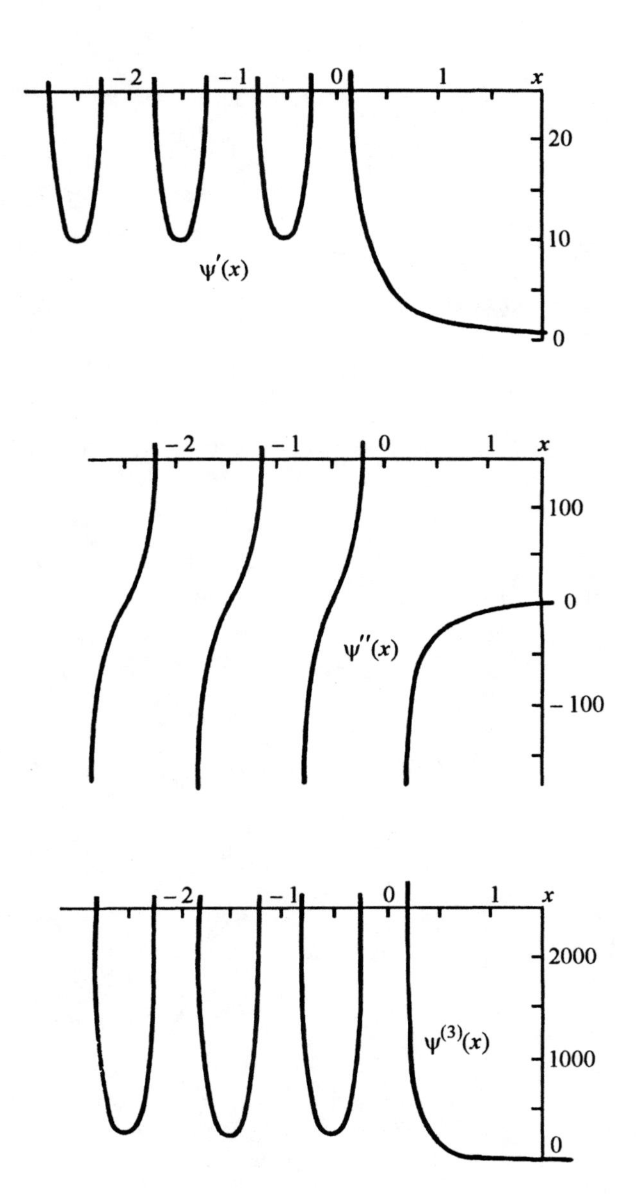

Figure 4.8. The psi function and its derivatives $\psi(x)$, $\psi'(x)$, $\psi''(x)$, $\psi^{(3)}(x)$.

The following power series representation holds:

$$\psi^{(n)}(x) = n! \sum_{k=0}^{\infty} (-\frac{1}{k+x})^{n+1}, \quad n \in \mathbf{N}.$$

4.2.4. The function $G(x)$ (Bateman function)

The function $G(x)$ is defined by the equality

$$G(x) = \psi\left(\frac{x+1}{2}\right) - \psi\left(\frac{x}{2}\right)$$

or

$$G(x) = 2\int_0^1 \frac{t^{x-1}}{1+t}\,dt = 2\int_0^{\infty} \frac{e^{-xt}}{1+e^{-t}}\,dt$$

(Figures 4.4, 4.5). Relations that $G(x)$ satisfies are the following:

$$G(1+x) = \frac{2}{x} - G(x), \quad G(1-x) = 2\pi \operatorname{cosec} \pi x - G(x),$$

$$G(x) = \sum_{k=0}^{\infty} \frac{2}{(2k+x)(2k+x+1)} = \sum_{k=0}^{\infty} \frac{k!}{2^k (x)_{k+1}},$$

$$G(mx) = -\frac{2}{m} \sum_{k=0}^{m-1} (-1)^k \psi(x + \frac{k}{m}), \quad m \text{ — even},$$

$$G(mx) = \frac{1}{m} \sum_{k=0}^{m-1} (-1)^k G(x + \frac{k}{m}), \quad m \text{ — odd}.$$

4.2.5. Incomplete gamma functions

The main *incomplete gamma functions* are $\gamma(\nu; x)$, $P(\nu, x)$, $Q(\nu, x)$. They are sometimes denoted by $\Gamma_x(\nu)$, $\gamma(\nu; x)/\Gamma(\nu)$, $\Gamma(\nu; x)/\Gamma(\nu)$. The notations $E_\nu(x)$ and $K_\nu(x)$ are sometimes used to denote the functions $x^{\nu-1}\Gamma(1-\nu; x)$.

The functions $\gamma(\nu; x)$ and $\Gamma(\nu; x)$ are defined by

$$\gamma(\nu; x) = \int_0^x t^{\nu-1} e^{-t}\,dt \qquad (x \geq 0, \ \nu > 0),$$

$$\Gamma(\nu; x) = \int_x^{\infty} t^{\nu-1} e^{-t}\,dt \qquad (x > 0).$$

The integrals

$$\Gamma(\nu; x) = \frac{x^{\nu} e^{-x}}{\Gamma(1-\nu)} \int_0^{\infty} \frac{t^{-\nu} e^{-t}}{t+x} dt \qquad (x > 0, \ \nu < 1),$$

$$\gamma^*(\nu; x) = \frac{1}{\Gamma(\nu)} \int_0^1 t^{\nu-1} e^{-xt} dt \qquad (\nu > 0)$$

can be used to define the functions for a larger set of values of the argument and parameter. Here $\gamma^*(\nu; x)$ is the incomplete gamma function, which is single-valued and does not have singularities in the finite part of the complex plane,

$$\gamma^*(a; x) = x^{-a} P(a: x) = \frac{x^{-a}}{\Gamma(a)} \gamma(a; x).$$

For negative values of the argument, the functions $\gamma(\nu; x)$ and $\Gamma(\nu; x)$ may have complex values and be many valued. Let $x \geq 0$. The function $\gamma(\nu; x)$, which is defined everywhere but $\nu = 0, -1, -2, \ldots$, is monotonically increasing. For $\nu > 0$, $\gamma(\nu; x) = 0$ at $x = 0$, and is positive elsewhere. The graphs of the functions $\gamma(\nu; x)$, $\Gamma(\nu; x)$ are shown in Figures 4.9.

If the parameter is a positive integer, then the incomplete gamma functions are greatly simplified. For example, $\Gamma(1; x) = e^{-x}$, $\gamma(1; x) = 1 - e^{-x}$, $\gamma^*(1; x) = \dfrac{1 - e^{-x}}{x}$.

For a negative integer parameter, $\gamma^*(-n; x) = x^n$, $n = 0, 1, 2, \ldots$.

The incomplete gamma functions satisfy the following formulas:

$$\gamma(\nu; x) + \Gamma(\nu; x) = \Gamma(\nu), \quad \gamma(\nu + 1; x) = \nu\gamma(\nu; x) - x^{\nu} e^{-x},$$

$$\Gamma(\nu + 1; x) = \nu\Gamma(\nu; x) + x^{\nu} e^{-x},$$

$$\gamma^*(\nu + 1; x) = \frac{\gamma^*(\nu; x)}{x} - \frac{e^{-x}}{x\Gamma(1+\nu)}, \qquad x \neq 0,$$

$$\gamma(\nu + n; x) = (\nu)_n \left[\gamma(\nu; x) - x^{\nu} e^{-x} \sum_{k=0}^{n-1} \frac{x^k}{(\nu)_{k+1}}\right],$$

$$\frac{d}{dx}\gamma(\nu; x) = -\frac{d}{dx}\Gamma(\nu; x) = x^{\nu-1} e^{-x} = \gamma(\nu; x) - (\nu - 1)\gamma(\nu - 1; x),$$

$$\frac{d^r}{dx^r}[x^{\nu} e^{x} \gamma^*(\nu; x)] = x^{\nu-r} e^{x} \gamma^*(\nu - r; x),$$

where r can take only integer nonzero values.

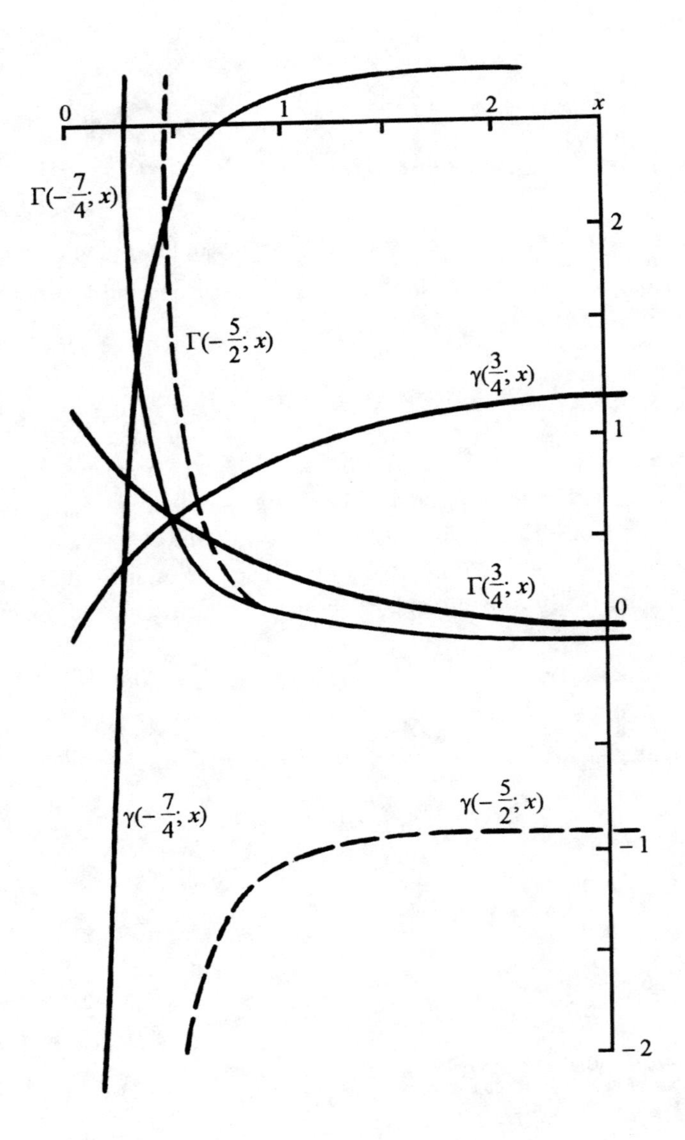

Figure 4.9. The functions $\gamma(\nu; x)$, $\Gamma(\nu; x)$.

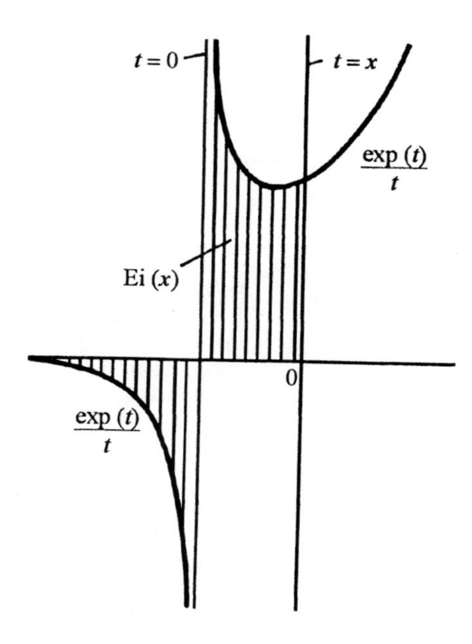

Figure 4.10. The integral exponential function Ei(x).

4.3. Integral exponential and related functions

Integral exponential function. The *integral exponential function* is defined in terms of the principal value of the integral

$$\mathrm{Ei}(x) = -\,\mathrm{v.p.}\int_{-x}^{\infty} \frac{e^{-t}}{t}\,dt = \mathrm{v.p.}\int_{-\infty}^{x} \frac{e^{t}}{t}\,dt \qquad (x > 0),$$

see Figure 4.10.

If the path of integration does pass through the origin and does not intersect the negative part of the real axis, then

$$\mathrm{Ei}(z) = \int_{x}^{\infty} \frac{e^{-t}}{t}\,dt \qquad (|\arg z| < \pi)\,.$$

The *integer integral function* is defined by $\mathrm{Ein}(x) = \gamma + \ln(|x|) - \mathrm{Ei}(x)$, where γ is the Euler's constant. The integral exponential function is also denoted by $\mathrm{Ei}^{*}(x)$, $\mathrm{E}^{*}(x)$, $\overline{\mathrm{Ei}(x)}$, etc. The integral exponential function can be expressed as the integral

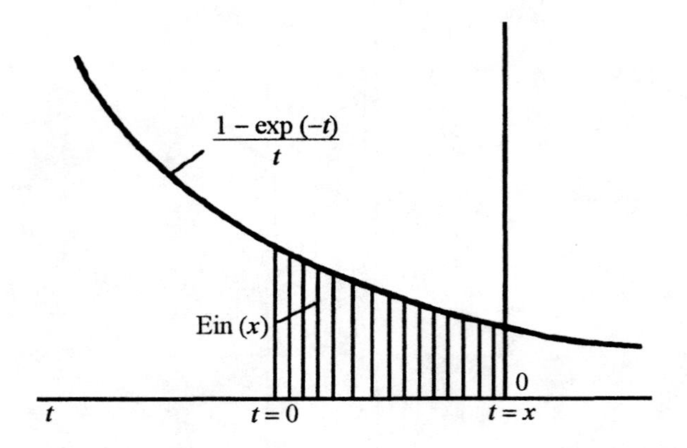

Figure 4.11. The integral exponential function Ein(x).

$$\mathrm{Ein}(x) = \int_0^x \frac{1 - e^{-t}}{t}\, dt$$

(see Figure 4.11).

The related functions are introduced via

$$\mathrm{li}(x) = \mathrm{v.p.} \int_0^x \frac{dt}{\ln t} = \mathrm{Ei}(\ln x) \qquad (x > 1),$$

$$E_n(z) = \int_1^\infty e^{-zt} \frac{dt}{t^n} \qquad (n = 0, 1, 2, \ldots, \mathrm{Re}\, z > 0),$$

$$\alpha_n(z) = \int_1^\infty t^n e^{-zt} dt \qquad (n = 0, 1, 2, \ldots, \mathrm{Re}\, z > 0),$$

$$\beta_n(z) = \int_{-1}^1 t^n e^{-zt} dt \qquad (n = 0, 1, 2, \ldots)$$

(see Figures 4.12, 4.13).

Special cases are

$$E_n(0) = \frac{1}{n-1}, \quad n = 2, 3, 4, \ldots, \quad E_0(x) = \frac{e^{-x}}{x},$$

$$\alpha_0(x) = \frac{e^{-x}}{x}, \quad \beta_0(x) = \frac{2}{x} \sinh x .$$

Asymptotic formulas are

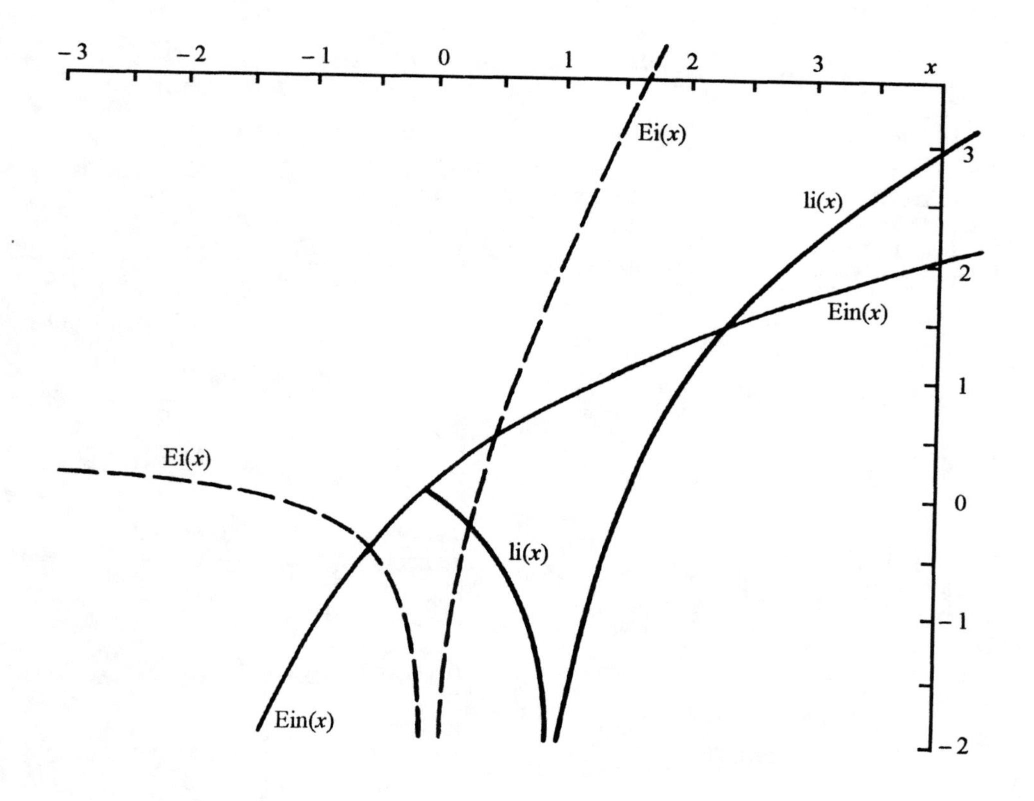

Figure 4.12. The integral exponential functions.

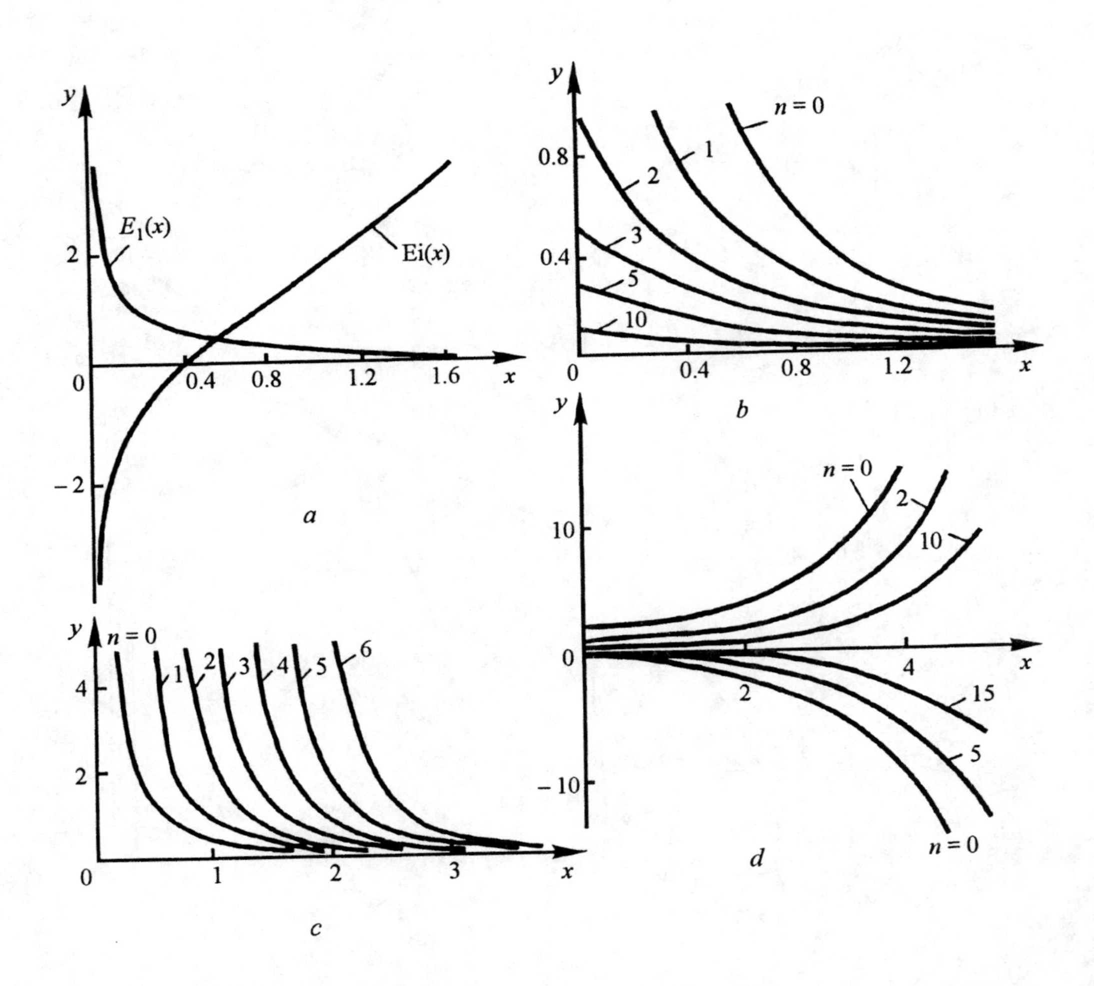

Figure 4.13. The graphs of functions: (a) $y = \mathrm{Ei}(x)$, $y = E_1(x)$; (b) $y = E_n(x)$, $n = 0,1,2,3,5,10$; (c) $y = \alpha_n(x)$, $n = 0,1,2,3,4,5,6$; (d) $y = \beta_n(x)$, $n = 0,1,2,5,10,15$.

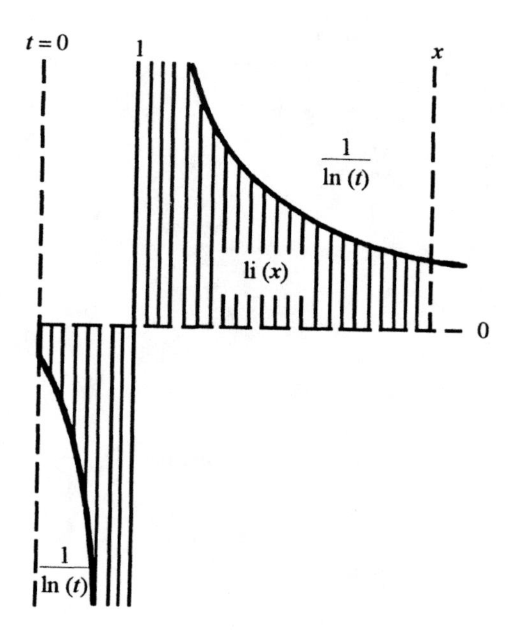

Figure 4.14. The logarithmic integral function li(x).

$$E_n(x) \sim \frac{e^{-x}}{x} [1 - \frac{n}{x} + \frac{n(n+1)}{x^2} - \ldots - \frac{(n)_k}{(-x)^k} + \ldots] \qquad (x \to \infty),$$

$$E_n(x) \sim \frac{e^{-x}}{x+n} [1 + \frac{n}{x+n} + \frac{n(n-2x)}{(x+n)^4} + \ldots] \qquad (x > 0),$$

n is sufficiently large.

The *alpha* and *beta exponential functions* can be expressed in terms of the incomplete gamma function by

$$\alpha_n(x) = x^{-n-1}\Gamma(n+1, x), \quad \beta_n(x) = x^{-n-1}[\Gamma(n+1, -x) - \Gamma(n+1, x)].$$

The *logarithmic integral function* (Figure 4.14) is defined by the integral

$$\mathrm{li}(x) = x \int_0^1 \frac{dt}{\ln x + \ln t}.$$

The *dilogarithm* or *Spencer's integral* is the function

$$\mathrm{diln}(x) = \int_1^x \frac{\ln t}{t-1} dt = \int_0^x \frac{\ln t}{t-1} dt - \frac{\pi^2}{6}$$

(see Figure 4.15, 4.16).

The *trilogarithm* is defined by

$$\operatorname{triln}(x) = \int_1^x \frac{\operatorname{diln} t}{t-1}\,dt = -\sum_{k=1}^{\infty} \frac{(1-x)^k}{k^3},$$

the *polylogarithm* is

$$\operatorname{poln}_\nu(x) = -\sum_{k=1}^{\infty} \frac{(1-x)^k}{k^\nu}.$$

4.4. Integral sine and cosine functions

The *integral sine function* is given by the formula

$$\operatorname{Si}(x) = \int_0^x \frac{\sin t}{t}\,dt = \frac{\pi}{2} - \int_x^{\infty} \frac{\sin t}{t}\,dt.$$

The *integral cosine function* is defined by

$$\operatorname{Ci}(x) = -\int_x^{\infty} \frac{\cos t}{t}\,dt = \gamma + \ln|x| + \int_0^x \frac{\cos t - 1}{t}\,dt \qquad (|\arg x| < \pi).$$

The entire function $\int_0^x \frac{1-\cos t}{t}\,dt$ is sometimes taken as the main function. It is denoted by $\operatorname{Cin}(x) = -\operatorname{Ci}(x) + \ln|x| + \gamma$ (see Figure 4.17).

For hyperbolic integral sine and cosine functions, the notations Shi (or Sih) and Chi (see Figure 4.18) are used,

$$\operatorname{Shi}(x) = \int_0^x \frac{\sinh t}{t}\,dt = \frac{\operatorname{Ei}(x) - \operatorname{Ei}(-x)}{2},$$

$$\operatorname{Chi}(x) = \gamma + \ln|x| + \int_0^x \frac{\cosh t - 1}{t}\,dt \qquad (|\arg x| < \pi),$$

or

$$\operatorname{Chi}(x) = \int_0^x \frac{\cosh t}{t}\,dt = \frac{\operatorname{Ei}(x) + \operatorname{Ei}(-x)}{2}.$$

Similarly,

$$\operatorname{Chin}(x) = \int_0^x \frac{\cosh t - 1}{t}\,dt = -\frac{\operatorname{Ein}(x) + \operatorname{Ein}(-x)}{2}.$$

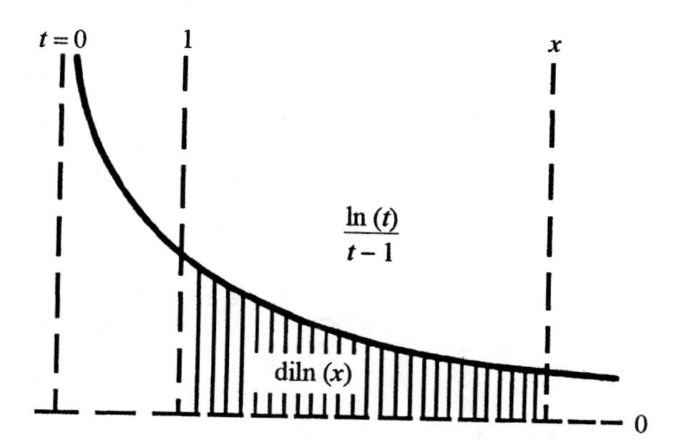

Figure 4.15. The dilogarithm diln(x) (definition).

The functions Ci, Chi, Chin are even; Shi, Si are odd. The power series expansions are the following:

$$\mathrm{Si}(x) = \sum_{k=0}^{\infty} \frac{(-1)^k x^{2k+1}}{(2k+1)(2k+1)!},$$

$$\mathrm{Ci}(x) = C + \ln|x| + \sum_{k=1}^{\infty} \frac{(-1)^k x^{2k}}{2k(2k)!},$$

$$\mathrm{Shi}(x) = \sum_{k=0}^{\infty} \frac{x^{2k+1}}{(2k+1)(2k+1)!},$$

$$\mathrm{Chi}(x) = C + \ln|x| + \sum_{k=1}^{\infty} \frac{x^{2k}}{2k(2k)!},$$

where C is Euler's constant.

The integral representation is:

$$\mathrm{Si}(x) = -\int_0^{\pi/2} e^{-x\cos t}\cos(x\sin t)\,dt,$$

where $\mathrm{Si}(x) = \mathrm{Si}(x) - \pi/2$, $\mathrm{Ci}(x) + E_1 x = \int_0^{\pi/2} e^{-x\cos t}\sin(x\sin t)\,dt.$

Particular values are:

$$\mathrm{Si}(\pm 2\pi n) = \mp \pi/2 \mp \mathrm{fi}(2\pi n), \quad \mathrm{Si}(\pm(2n-1)\pi) = \pm\pi/2 \pm \mathrm{fi}((2n-1)\pi),$$

$$\mathrm{Ci}(\pm(2n-1/2)\pi) = \mp \mathrm{gi}((2n-1/2)\pi), \quad \mathrm{Ci}(\pm(2n-3/2)\pi) = \pm \mathrm{gi}((2n-3/2)\pi) \quad (n \in \mathbf{N}),$$

where the functions fi and gi are auxiliary integral functions.

Auxiliary integral sine and cosine functions. These functions are defined in terms of definite integrals:

$$\mathrm{fi}(x) = \int_0^\infty \frac{\sin t}{t+x}dt = \int_0^\infty \frac{e^{-xt}}{t^2+1}dt; \quad \mathrm{gi}(x) = \int_0^\infty \frac{\cos t}{t+x}dt,$$

see Figure 4.19.

The relationship between the functions fi(x), gi(x) and the integral sine and cosine functions is given by

$$\mathrm{fi}(x) = \sin x\,\mathrm{Ci}(x) + \cos x[\pi/2 - \mathrm{Si}(x)],$$

$$\mathrm{gi}(x) = \sin x[\pi/2 - \mathrm{Si}(x)] - \cos x\,\mathrm{Ci}(x).$$

Asymptotic formulas for the auxiliary functions are

$$\mathrm{fi}(x) \sim \frac{1}{x}\left(1 - \frac{2!}{x^2} + \frac{4!}{x^4} - \frac{6!}{x^6} + \ldots\right) \quad \text{as} \quad x \to \infty,$$

$$\mathrm{gi}(x) \sim \frac{1}{x^2}\left(1 - \frac{3!}{x^2} + \frac{5!}{x^4} - \frac{7!}{x^6} + \ldots\right) \quad \text{as} \quad x \to \infty,$$

The following formulas also hold:

$$\frac{d}{dx}\mathrm{fi}(x) = -\,\mathrm{gi}(x), \quad \frac{d}{dx}\mathrm{gi}(x) = \mathrm{fi}(x) - \frac{1}{x}.$$

Integral arc tangent function. The function is defined by

$$\mathrm{Ti}_2(x) = \int_0^x \frac{\arctan t}{t}dt,$$

see Figure 4.20.

The main relations the function satisfies are:

$$\mathrm{Ti}_2(x) - \mathrm{Ti}_2\left(\frac{1}{x}\right) = \frac{\pi}{2}\mathrm{sgn}(x)\ln|x|, \quad \mathrm{Im}\,x = 0,$$

$$\lim_{x\to\infty} \mathrm{Ti}_2(x) = \frac{\pi}{2}\ln x,$$

$$\lim_{x\to-\infty} \mathrm{Ti}_2(x) = -\frac{\pi}{2}\ln(-x).$$

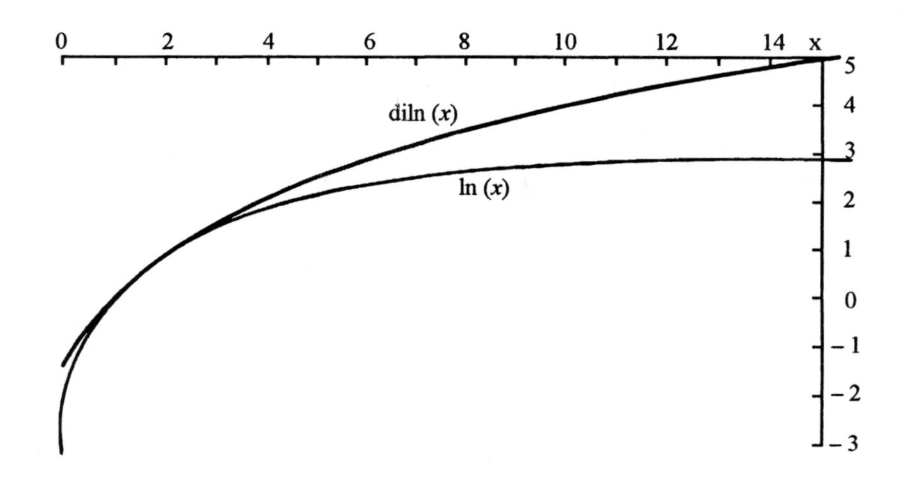

Figure 4.16. The behavior of diln(x).

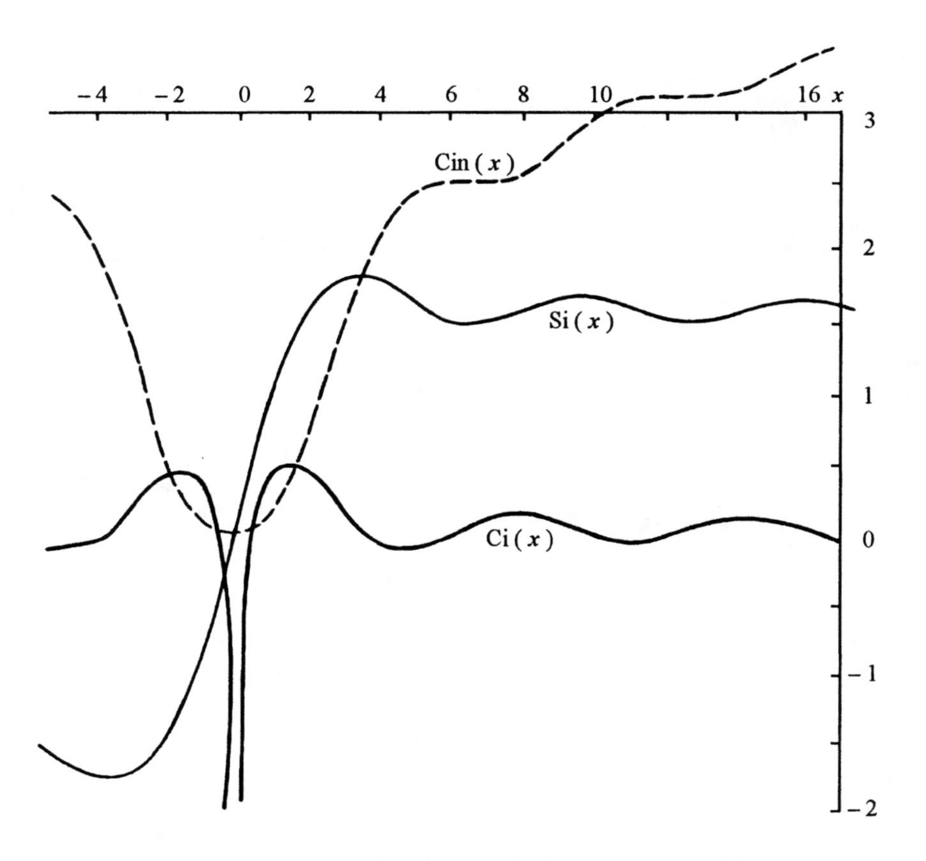

Figure 4.17. The integral sine function Si(x), the integral cosine functions Ci(x) and Cin(x).

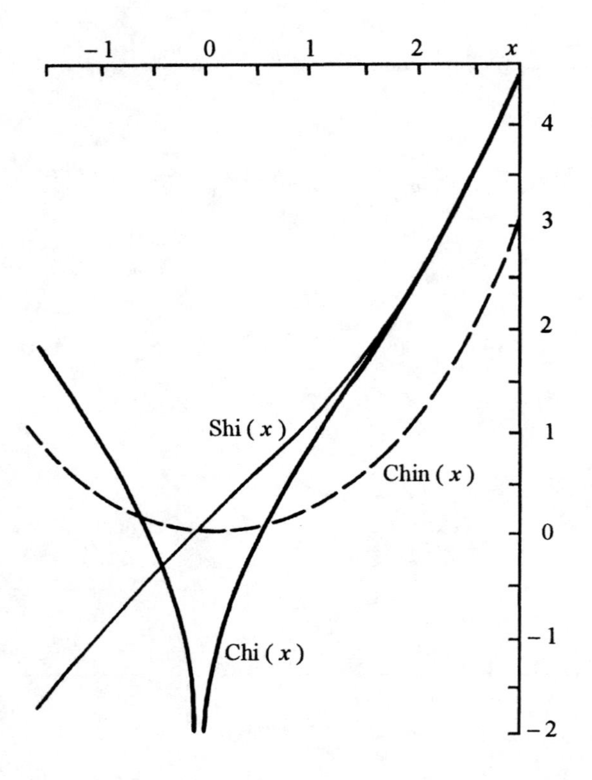

Figure 4.18. The integral hyperbolic sine function, Shi(x), cosine functions, Chi(x), Chin(x).

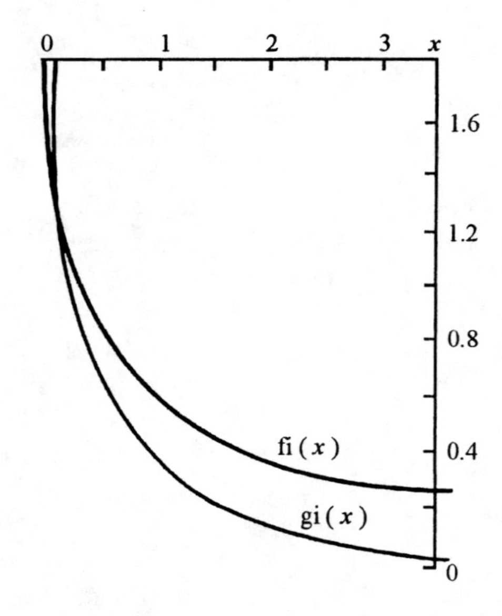

Figure 4.19. The functions fi(x), gi(x).

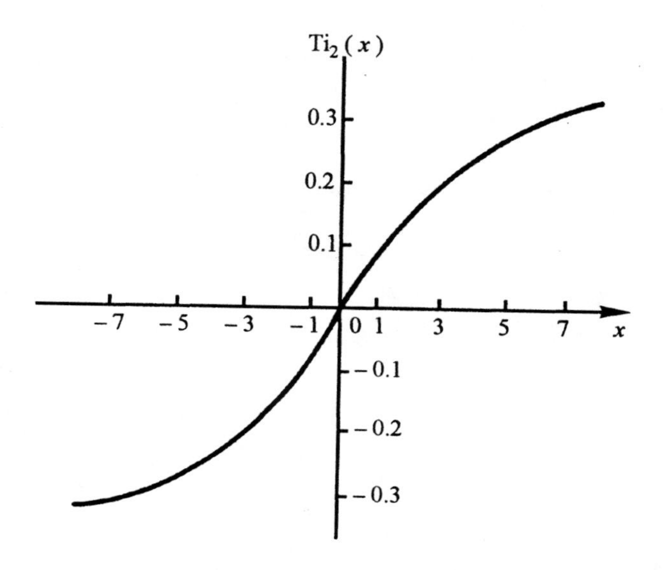

Figure 4.20. The integral arc tangent, $\mathrm{Ti}_2(x)$.

We have $\mathrm{Ti}_2(\tan\frac{\pi}{2}) = \mathrm{Ti}_2(2-\sqrt{3})$. The power series representation for the integral arc tangent function is

$$\mathrm{Ti}_2(x) = \sum_{k=0}^{\infty} \frac{(-1)^k x^{k+1}}{(2k+1)^2}, \quad |x| \le 1,$$

$$\mathrm{Ti}_2(x) = \frac{\pi}{2} \ln|x| + \sum_{k=0}^{\infty} \frac{(-1)^k}{(2k+1)^2 x^{2k+1}}, \quad |x| \ge 1.$$

4.5. Fresnel integrals

The *Fresnel integral sine* and *cosine functions* are defined by the formulas

$$\mathrm{S}(x) = \sqrt{\frac{2}{\pi}} \int_0^x \sin t^2 dt = \frac{\operatorname{sgn} x}{\sqrt{2\pi}} \int_0^{x^2} \frac{\sin t}{\sqrt{t}}\, dt,$$

$$\mathrm{C}(x) = \sqrt{\frac{2}{\pi}} \int_0^x \cos t^2 dt = \frac{\operatorname{sgn} x}{\sqrt{2\pi}} \int_0^{x^2} \frac{\cos t}{\sqrt{t}}\, dt.$$

The following functions are also called *Fresnel integrals*,

$$S(x) = \int_0^x \sin\frac{\pi}{2}t^2 dt, \quad C(x) = \int_0^x \cos\frac{\pi}{2}t^2 dt,$$

see Figure 4.21.

The Fresnel integral sine and cosine functions are odd. The power series expansions for these functions are

$$S(x) = x^3\sqrt{\frac{2}{\pi}}\sum_{k=0}^{\infty}\frac{(-x^4)^k}{(2k+1)!(4k+3)},$$

$$C(x) = x\sqrt{\frac{2}{\pi}}\sum_{k=0}^{\infty}\frac{(-x^4)^k}{(2k)!(4k+1)},$$

$$C(x) \to 1/2, \quad S(x) \to 1/2, \quad \text{as} \quad x \to \infty.$$

Generalized Fresnel integrals are correspondingly defined by

$$S(x;\nu) = \int_x^{\infty} t^{\nu-1}\sin t\, dt \qquad (x \ge 0, \nu < 1),$$

$$C(x;\nu) = \int_x^{\infty} t^{\nu-1}\cos t\, dt \qquad (x \ge 0, \nu < 1).$$

In the case where $\nu = 0$, they are reduced to the integral sine and cosine functions, $S(x; 0) = \pi/2 - Si(x)$, $C(x; 0) = -Ci(x)$. The following formulas give a relationship between the generalized and usual Fresnel integrals:

$$S(x;\tfrac{1}{2}) = \sqrt{2\pi}\,(\tfrac{1}{2} - S(\sqrt{x})),\ C(x;\tfrac{1}{2}) = \sqrt{2\pi}\,(\tfrac{1}{2} - C(\sqrt{x})).$$

The following recurrence relations hold:

$$S(x;\nu) = -\frac{C(x;\nu+1)}{\nu} - \frac{x^{\nu}\sin x}{\nu},$$

$$C(x;\nu) = -\frac{C(x;\nu+1)}{\nu} - \frac{x^{\nu}\cos x}{\nu}.$$

Special cases of the functions are: $C(0;\nu) = \Gamma(\nu)\cos\frac{\nu\pi}{2}$, $S(0;\nu) = \Gamma(\nu)\sin\frac{\nu\pi}{2}$.

The functions admit the power series expansions:

$$S(x;\nu) = S(0;\nu) - x^{1+\nu}\sum_{k=0}^{\infty}\frac{(-x^2)^k}{(2k+1)!(2k+1+\nu)},$$

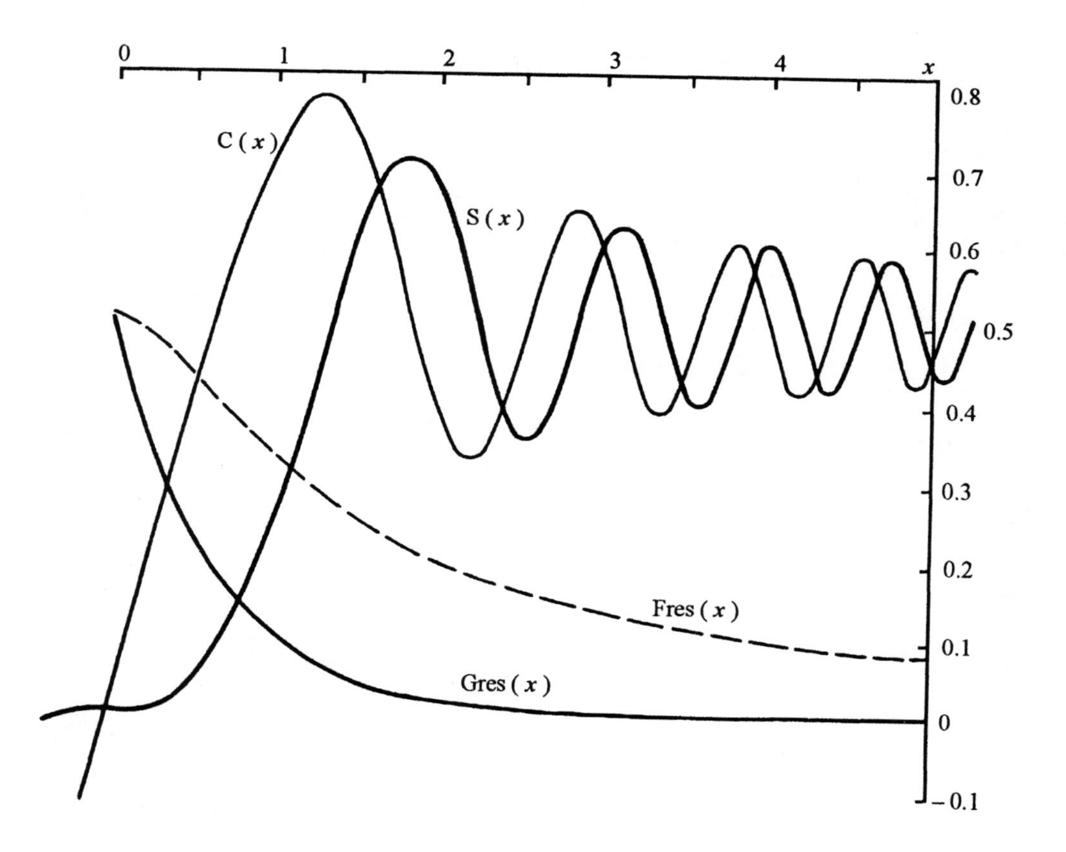

Figure 4.21. The functions S(x), C(x), Gres(x), Fres(x).

$$C(x;\nu) = C(0;\nu) - x^{\nu} \sum_{k=0}^{\infty} \frac{(-x^2)^k}{(2k)!(2k+\nu)}.$$

Auxiliary Fresnel integral sine and *cosine functions* are defined by the integrals

$$\text{Gres}\,(x) = \sqrt{\frac{2}{\pi}} \int_0^{\infty} e^{-2xt} \sin t^2 \, dt \qquad (x \geq 0),$$

$$\text{Fres}\,(x) = \sqrt{\frac{2}{\pi}} \int_0^{\infty} e^{-2xt} \cos t^2 \, dt \qquad (x \geq 0),$$

see Figure 4.21.

4.6. Probability integral and related functions

The *probability integral function* is the function

$$\text{erf}(x) = \frac{2}{\sqrt{\pi}} \int_0^x e^{-t^2} dt,$$

or

$$\text{erf}(x) = \frac{\text{sgn}\, x}{\sqrt{\pi}} \int_0^{x^2} \frac{e^{-t}}{\sqrt{t}} \, dt, \quad \text{erf}(x) = \frac{2}{\pi} \int_0^{\infty} \frac{e^{-t^2} \sin 2xt}{t} \, dt,$$

$$\text{erf}(x) = \frac{2x}{\sqrt{\pi}} \int_0^1 e^{-x^2 t^2} \, dt.$$

Sometimes these functions are defined by the expressions

$$\Phi(x) = \frac{1}{\sqrt{2\pi}} \int_{-\infty}^{x} e^{-t^2/2} \, dt,$$

$$\Phi_0(x) = \frac{1}{\sqrt{2\pi}} \int_0^x e^{-t^2/2} \, dt = \frac{1}{2} \text{erf}\left(\frac{x}{\sqrt{2}}\right),$$

$$\Phi^{-}(x) = \sqrt{\frac{2}{\pi}} \int_0^x e^{-t^2/2} \, dt = 2\Phi_0(z).$$

The *complementary function* erfc(x) (*error function*) is given by

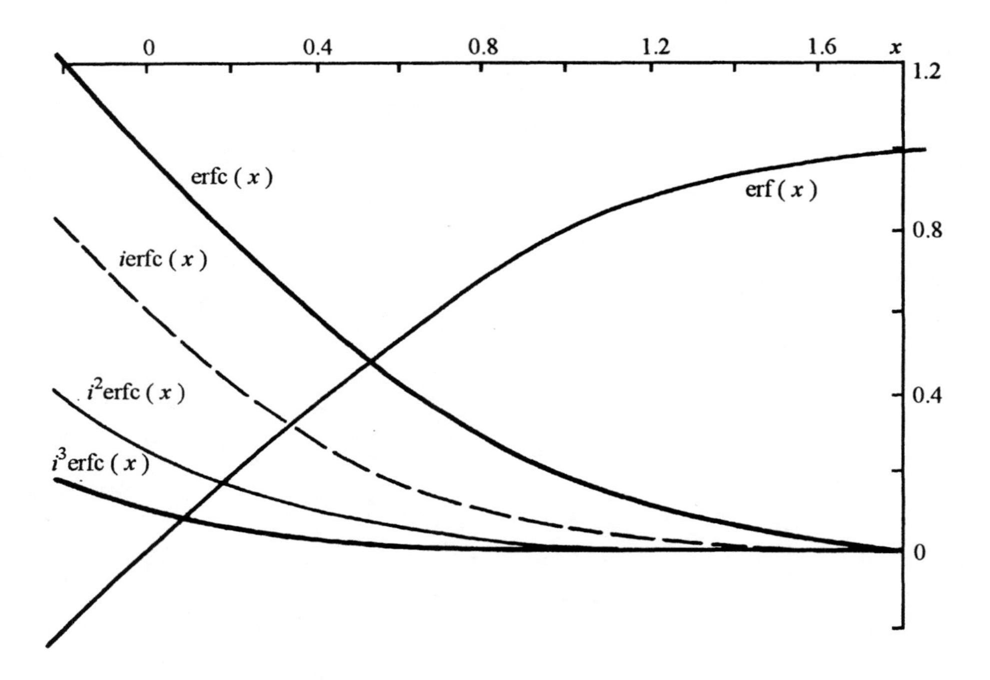

Figure 4.22. The functions erf(x) and erfc(x).

$$\operatorname{erfc}(x) = \frac{2}{\sqrt{\pi}} \int_x^{\infty} e^{-t^2}\, dt = 1 - \operatorname{erf}(x),$$

see Figure 4.22.

The function erf(x) is odd, the function erfc(x) satisfies the identity erfc$(-x) = 2 - $erfc($x$).

The *generalized complementary function* is defined by

$$i^n\operatorname{erfc}(x) = \frac{2}{\sqrt{\pi}} \int_x^{\infty} \frac{(t-x)^n}{n!} e^{-t^2}\, dt, \quad n = 2, 3, 4, \ldots,$$

whence

$$i\operatorname{erfc}(x) = i^1\operatorname{erfc}(x) = \frac{2}{\sqrt{\pi}} \int_x^{\infty} (t-x)\, e^{-t^2}\, dt = \int_x^{\infty} \operatorname{erfc}(t)\, dt,$$

$$i^0\operatorname{erfc}(x) = \frac{2}{\sqrt{\pi}} \int_x^{\infty} e^{-t^2}\, dt = \operatorname{erfc}(x),$$

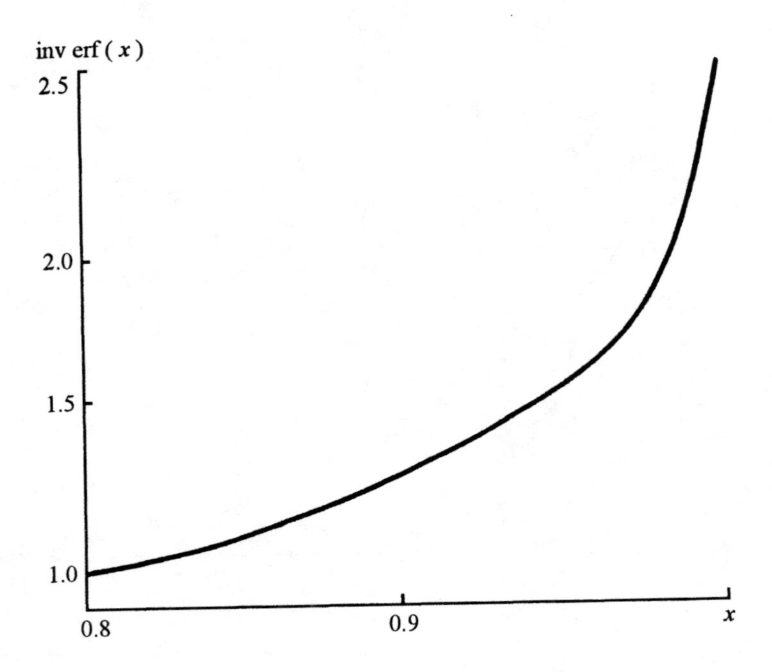

Figure 4.23. The function inv erf(x).

$$i^{-1}\text{erfc}\,(x) = \frac{2}{\sqrt{\pi}}\; e^{-x^2} .$$

The *function* inv erf(x) is the function inverse to the probability integral, $y = \text{inv erf}(x)$ if $x = \text{erf}(y)$.

The *function* inv erfc(x) is the function inverse to the function erfc(x), $y = \text{inv erfc}(x)$ if $x = \text{erfc}(y)$. These functions have the following properties: $\text{inv erfc}(x) = \text{inv erf}(1 - x)$, $\text{inv erf}(0) = 0$, $\text{inv erf}(1) = \infty$, $\text{inv erfc}(0) = \infty$, $\text{inv erfc}(1) = 0$ (Figure 4.23).

Important for applications are the functions of the form $e^x\text{erfc}(\sqrt{x}\,)$, in particular, $\sqrt{\pi x}\; e^x\text{erfc}(\sqrt{x}\,)$, $e^{x^2}\text{erfc}(x)$ and others (Figures 4.24, 4.25). The function $e^x\text{erfc}(\sqrt{x}\,)$ can be defined as one of the integrals:

$$e^x\text{erfc}(\sqrt{x}\,) = \frac{1}{\sqrt{\pi}} \int\limits_x^\infty \frac{e^{x-t}dt}{\sqrt{t}} = \frac{2}{\sqrt{\pi}} \int\limits_{\sqrt{\pi}}^\infty e^{x-t^2} dt,$$

$$e^x\text{erfc}(\sqrt{x}\,) = \frac{2}{\sqrt{\pi}} \int\limits_0^\infty e^{-t^2 - 2t\sqrt{x}} dt,$$

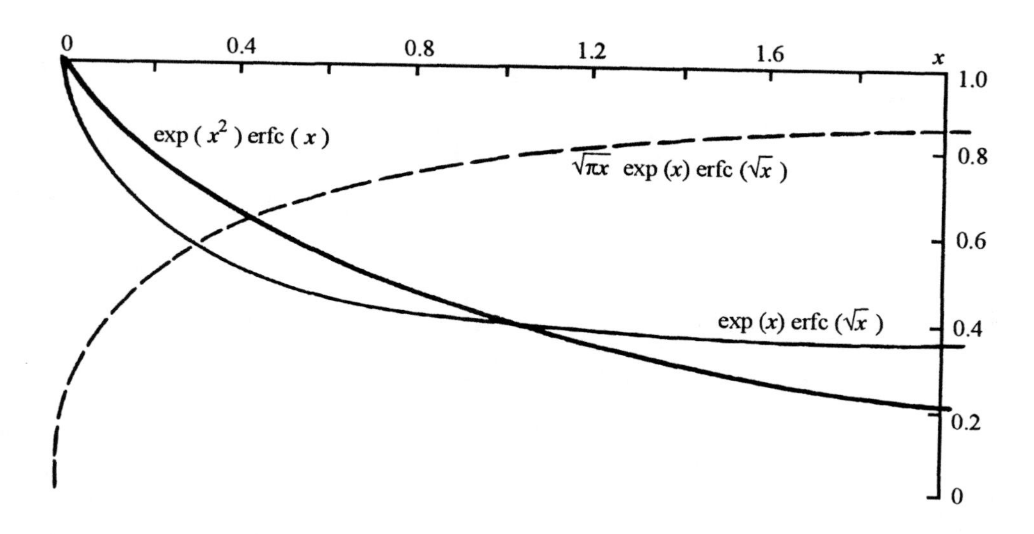

Figure 4.24. The functions $\sqrt{\pi x}\, e^{x}\operatorname{erfc}(\sqrt{x}\,)$, $e^{x^2}\operatorname{erfc}(\sqrt{x}\,)$, $e^{x}\operatorname{erfc}(\sqrt{x}\,)$.

$$e^{x}\operatorname{erfc}(\sqrt{x}\,) = \frac{1}{\sqrt{\pi}} \int_0^\infty \frac{e^{-t}dt}{\sqrt{t+x}} = \frac{2\sqrt{x}}{\pi} \int_0^\infty \frac{e^{-t^2}dt}{t+x},$$

$$e^{x}\operatorname{erfc}(\sqrt{x}\,) = \sqrt{\frac{x}{\pi}} \int_0^\infty \frac{e^{-xt}dt}{\sqrt{t+1}} = \frac{1}{\pi} \int_0^\infty \frac{e^{-xt}dt}{(t+1)\sqrt{t}} .$$

The following power series expansion and asymptotic formulas hold:

$$e^{x}\operatorname{erfc}(\pm\sqrt{x}\,) = \sum_{k=0}^{\infty} \frac{(\mp\sqrt{x}\,)^k}{\Gamma(1+\frac{k}{2})},$$

$$\sqrt{\pi x}\; e^{x}\operatorname{erfc}(\sqrt{x}\,) \sim 1 - \frac{1}{2x} + \frac{3}{4x^2} - \ldots + \frac{(2k-1)!!}{(-2x)^k} + \ldots \quad \text{as} \quad x \to \infty .$$

Similar formulas exist for the functions of the form $e^{x}\operatorname{erf}(\sqrt{x}\,)$. Here the following relations are useful:

$$e^{x}\operatorname{erf}(\sqrt{x}\,) = e^{x} - e^{x}\operatorname{erfc}(\sqrt{x}\,), \quad e^{x}\operatorname{erf}(\sqrt{x}\,) = \frac{e^{x}\operatorname{erfc}(-\sqrt{x}\,) - e^{x}\operatorname{erfc}(\sqrt{x}\,)}{2} .$$

The *Dawson integral,* daw(x), is a particular case (for $n = 2$) of the functions defined by the integral

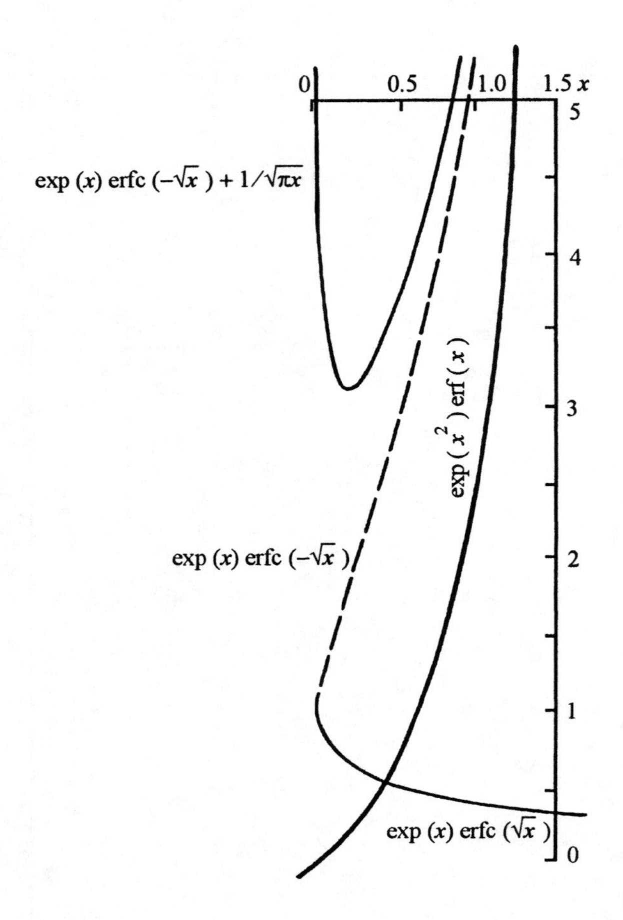

Figure 4.25. The behavior of the functions erf(x), erfc(x).

$$\int_0^x e^{t^n - x^n} dt \qquad (n = 2, 3, 4, \ldots),$$

$$\operatorname{daw}(x) = \int_0^x e^{t^2 - x^2} dt = \frac{\operatorname{sgn} x}{2} \int_0^{x^2} \frac{e^{t - x^2}}{\sqrt{t}} dt .$$

The function daw(x) is connected with the function erf by the relation daw(x) = $= \frac{-i\sqrt{\pi}}{2} e^{-x^2}$erf($ix$). In the literature, one also uses the notation $D(x) = \sqrt{2x}\,\operatorname{daw}(\sqrt{x/2}\,)$ (Figure 4.26).

The Dawson integral is an odd function, daw(0) = 0, daw(∞) = 0.

The power series expansion and asymptotic formulas for the function are the following:

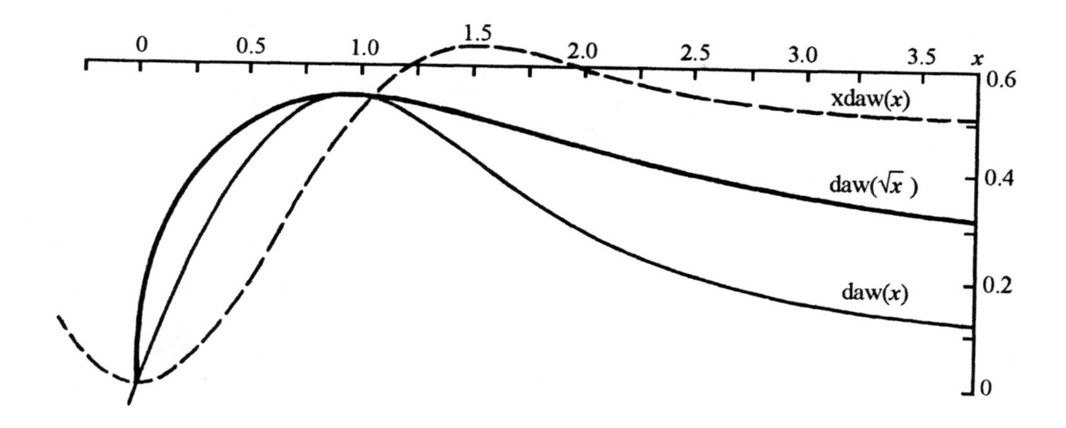

Figure 4.26. The functions daw(x), daw($\sqrt{x}$), xdaw(x).

$$\operatorname{daw}(x) = x \sum_{k=0}^{\infty} \frac{(-2x^2)^k}{(2k+1)!!},$$

$$\operatorname{daw}(x) \sim \frac{1}{2x} + \frac{1}{4x^3} + \ldots + \frac{(2k-1)!!}{(2x^2)^k 2x} + \ldots \quad \text{as} \quad x \to \infty .$$

4.7. Bessel functions

The *Bessel functions* (cylinder functions) are defined as solutions of Bessel's equation

$$z^2 \frac{d^2u}{dz^2} + z \frac{du}{dz} + (z^2 - \nu^2)\, u = 0.$$

Solutions of this equation are Bessel functions of the first kind, $J_\nu(z)$, $Y_\nu(z)$, second kind (sometimes called *Weber functions* or *Neumann function*), denoted by $N_\nu(z)$, and third kind, $H_\nu^{(1)}(z)$ and $H_\nu^{(2)}(z)$, called sometimes *Hankel functions*. Here z is a complex variable. We will consider Bessel functions mainly as functions of the real variable x. The order of the function, ν, may be real or complex. All Bessel functions are analytic functions of z in the complex plane with the cut on the negative part of the real axis. If $\nu = \pm n$, $n \in \mathbf{N}$, the function $J_\nu(x)$ does not have singularities and is an entire function of z.

The function $J_\nu(z)$, $Re\,\nu \geq 0$, is bounded if $arg\, z$ is bounded, $|J_\nu(x)| \leq 1$ for $\nu \geq 0$, and $|J_\nu(x)| \leq 1/\sqrt{2}$ for $\nu \geq 1$.

The functions J_ν and $J_{-\nu}$ are linearly independent for any ν; the functions $H_\nu^{(1)}, H_\nu^{(2)}$ are also linearly independent for any ν.

Let us discuss the functions $J_n(x)$ for $n \in \mathbf{N}$ in more detail. All functions $J_n(x)$, save for $J_0(x)$, have zero at $x = 0$ (Figure 4.27). As n increases, the function $J_n(x)$ gets closer to the x-axis. Each maximum or minimum of the function $J_0(x)$ is a zero of the function $J_1(x)$; however this is not true for the functions of greater order. For $n \geq 1$, each relative maximum or minimum of the function $J_n(x)$ corresponds to the point of intersection of the graphs of the functions $J_{n-1}(x)$ and $J_{n+1}(x)$. All zeros of the functions $J_n(x), J_{n-1}(x), J_{n+1}(x)$ have the same magnitude but opposite signs.

The power series expansion is

$$J_\nu(x) = \left(\frac{x}{2}\right)^\nu \sum_{k=0}^{\infty} \frac{(-x^2/4)^k}{k!\Gamma(1+k+\nu)}.$$

The Neumann series is

$$f(x) = \left(\frac{2}{x}\right)^\nu \sum_{k=0}^{\infty} (\nu+k) b_k J_{\nu+k}(x),$$

where ν is any integer other than a negative integer,

$$b_k = \sum_{i=0}^{m} \frac{4^i \Gamma(m+i+\nu)}{(m-i)!} a_{2i} \quad (2m = k = 0, 2, 4, \dots),$$

if k is even, and

$$b_k = 2\sum_{i=0}^{m} \frac{4^i \Gamma(m+i+1+\nu)}{(m-i)!} a_{2i+1} \quad (2m+1 = k = 1, 3, 5, \dots),$$

if k is odd.

The *Neumann function* for noninteger ν is defined in terms of the Bessel functions,

$$Y_\nu(x) = \frac{\cos \nu\pi J_\nu(x) - J_{-\nu}(x)}{\sin \nu\pi} \quad (\nu \in \mathbf{Z}).$$

For $\nu = n \in \mathbf{Z}$, we have

$$Y_n(x) = \lim_{\nu \to n} [\cot\nu\pi J_\nu(x) - \operatorname{cosec}\nu\pi\, J_{-\nu}(x)].$$

For an arbitrary ν (integer or not), the linear combination $C_1 J_\nu(x) + C_2 Y_\nu(x)$ gives a general solution of Bessel's equation. Similarly, the linear combination $C_1 J_\nu(2\sqrt{x}) + C_2 Y_\nu(2\sqrt{x})$

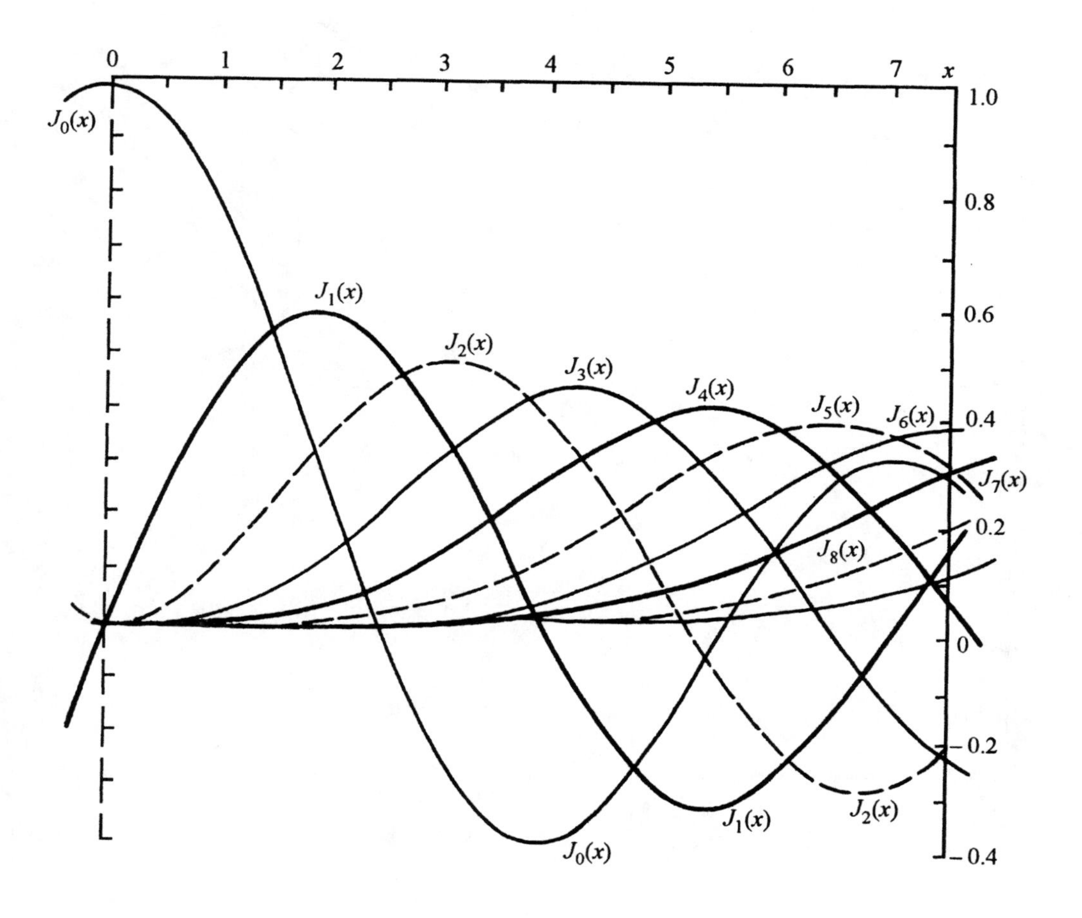

Figure 4.27. The Bessel functions $J_n(x)$, $n = 0,1,2,3,4,5,6,7,8,9$.

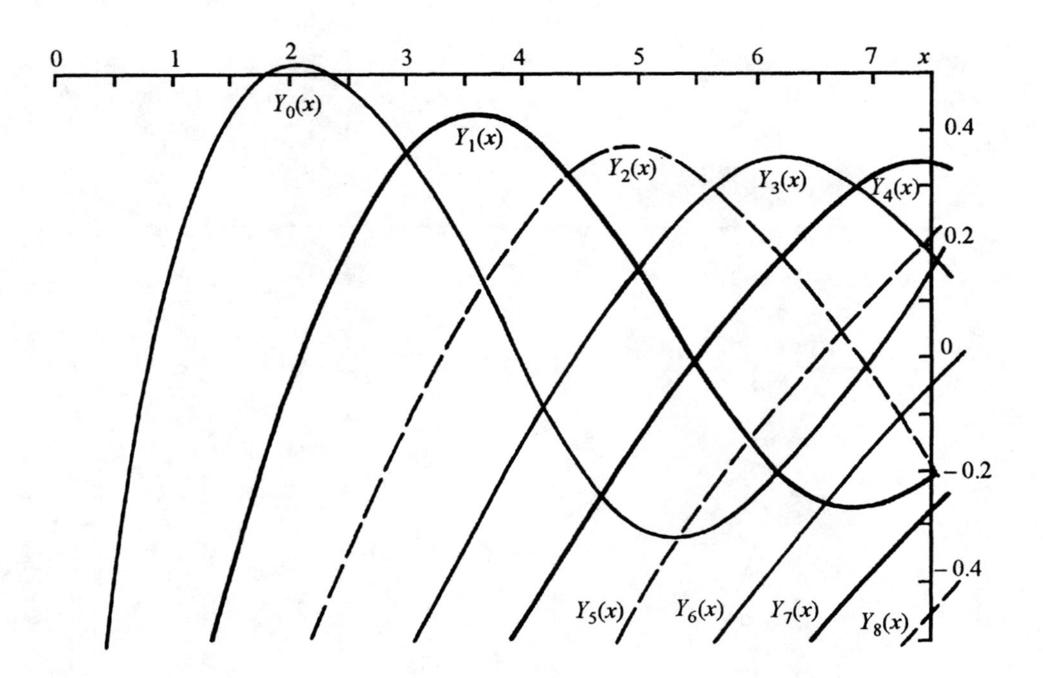

Figure 4.28. The functions $Y_\nu(x)$.

satisfies the Bessel–Clifford equation $x\dfrac{d^2u}{dx^2} + (1+\nu)\dfrac{du}{dx} + u = 0$ for all ν. The coefficients C_1, C_2 are arbitrary constants. The following asymptotic formulas take place:

$$Y_\nu(x) \sim \sqrt{\frac{2}{\pi x}}\ \sin\left(x - \frac{\nu\pi}{2} - \frac{\pi}{4}\right) \quad (x \to \infty),$$

$$Y_\nu(x) \sim -\sqrt{-\frac{2}{\pi x}}\ \left(\frac{-ex}{2\nu}\right)^\nu \cos\nu\pi \quad (\nu \to -\infty, x = \text{const} > 0),$$

$$Y_\nu(x) \sim -\sqrt{\frac{2}{\pi x}}\ \left(\frac{2\nu}{ex}\right)^\nu \quad (\nu \to \infty, x = \text{const}).$$

(see Figures 4.28, 4.29).

The *Bessel functions of the third kind* (*Hankel functions*) are defined by the formulas

$$H_\nu^{(1)}(z) = J_\nu(z) + iY_\nu(z) = i\operatorname{cosec}\nu\pi[e^{-\nu\pi i}J_\nu(z) - J_{-\nu}(z)],$$

$$H_\nu^{(2)}(z) = J_\nu(z) - iY_\nu(z) = i\operatorname{cosec}\nu\pi[J_{-\nu}(z) - e^{\nu\pi i}J_\nu(z)].$$

The *modified Bessel functions* are solutions of the differential equation

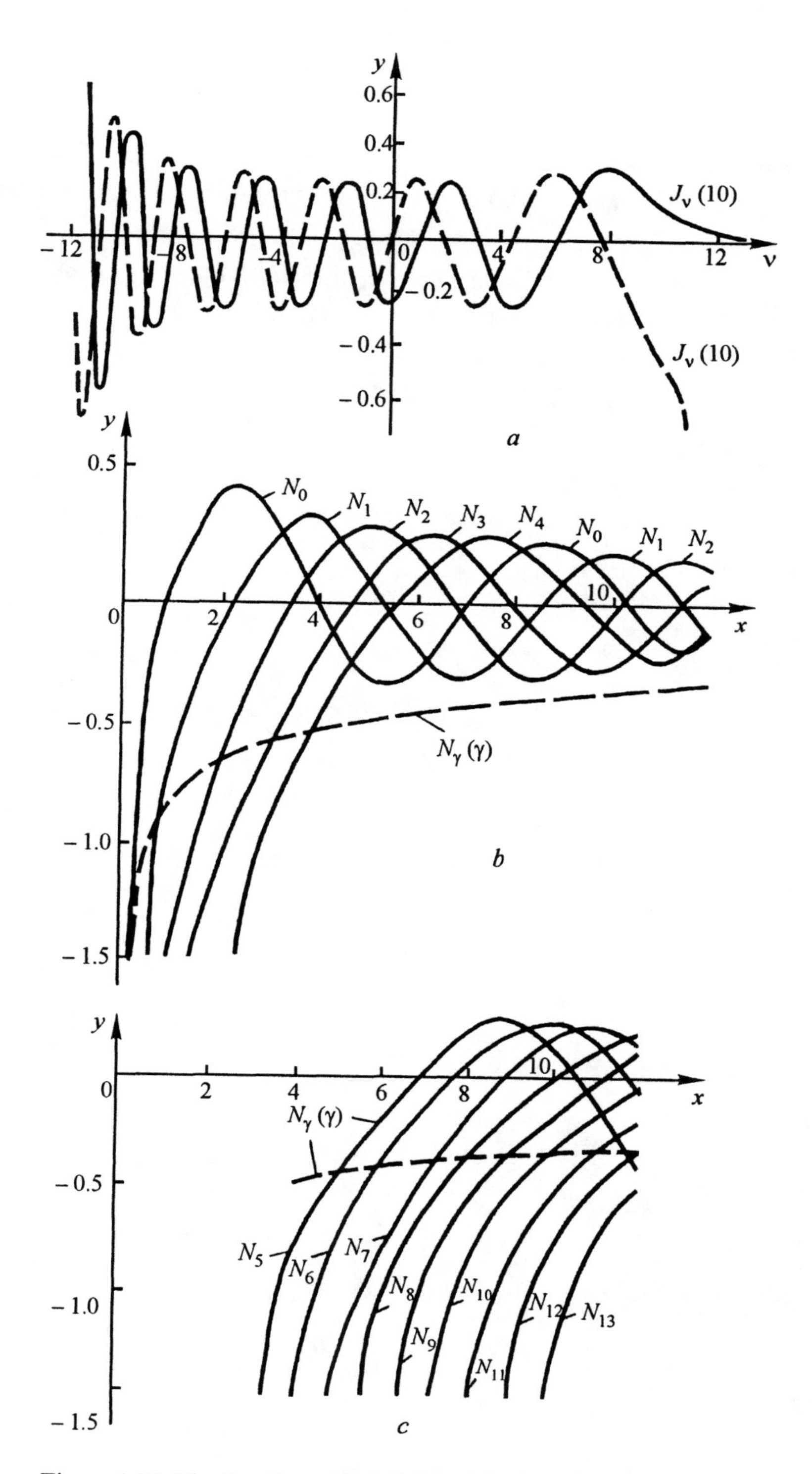

Figure 4.29. The functions (*a*) $J_ν(10)$ and $Y_ν(10)$; (*b*) and (*c*) $N_ν(x)$.

$$z^2\frac{d^2u}{dz^2}+z\frac{du}{dz}-(z^2-\nu^2)u=0,$$

obtained by replacing z by iz in Bessel's equation. Solutions of this equation are the *modified Bessel functions of the first kind,* $I_\nu(z)$, $I_{-\nu}(z)$, as well as the *modified Bessel functions of the third kind,* $K_\nu(z)$ (sometimes they are called *Basset functions, Bessel functions of the second kind of imaginary argument, Macdonald functions,* or *modified Hankel functions*).

The functions $I_{\pm}(z)$ and $K_\nu(z)$ are regular on the complex plane cut along the negative part of the real axis. For $\nu=\pm n$, $J_n(z)$ is an entire function of z.

If Re $\nu \ge 0$, then $I_\nu(z)$ is bounded in a neighborhood of zero; the functions $I_\nu(z)$ and $I_{-\nu}(z)$ are linearly independent if ν is not an integer.

The function $K_\nu(z)$ exponentially tends to zero as $|z|\to 0$ in the sector $|arg\ z|\le\pi/2$. The functions $I_\nu(z)$ and $K_\nu(z)$ are linearly independent for any ν. If $\nu>-1$, the functions $I_\nu(z)$ and $K_\nu(z)$ are real and positive for $z>0$.

The *functions $I_\nu(z)$ and $K_\nu(z)$ are defined by*

$$I_\nu(z)=\begin{cases} e^{-\pi\nu i/2}J_\nu(ze^{\pi i/2}) & (-\pi<\arg z\le\pi/2),\\ e^{\frac{3}{2}\pi i\nu}J_\nu(ze^{3\pi i/2}) & (\pi/2<\arg z\le\pi),\end{cases}$$

$$K_\nu(z)=\frac{\pi}{2}\frac{I_{-\nu}(z)-I_\nu(z)}{\sin\nu\pi}\quad(\nu\ne n).$$

If ν is an integer or zero, then the right-hand side in the last formula is replaced with the limit

$$K_n=\frac{\pi}{2}\lim_{\nu\to n}\left[\frac{I_{-\nu}(z)-I_\nu(z)}{\sin\nu\pi}\right].$$

There are symmetry relations:

$$I_{-n}(z)=I_n(z),\quad K_{-\nu}(z)=K_\nu(z),\quad I_n(-z)=(-1)^nI_n(z),$$

$$I_\nu(z)\sim\left(\frac{z}{2}\right)^\nu/\Gamma(\nu+1)\quad(\nu\ne-1,-2,\ldots);$$

and $K_0(z)\sim-\ln z$, $z\to 0$, $K_\nu(z)\sim\frac{1}{2}\Gamma(\nu)\left(\frac{z}{2}\right)^{-\nu}$ (Re $\nu>0$), see Figures 4.30, 4.31.

The functions have the following power series expansions:

$$I_\nu(z)=\left(\frac{z}{2}\right)^\nu\sum_{k=0}^{\infty}\frac{\left(\frac{z^2}{4}\right)^k}{k!\Gamma(\nu+k+1)},$$

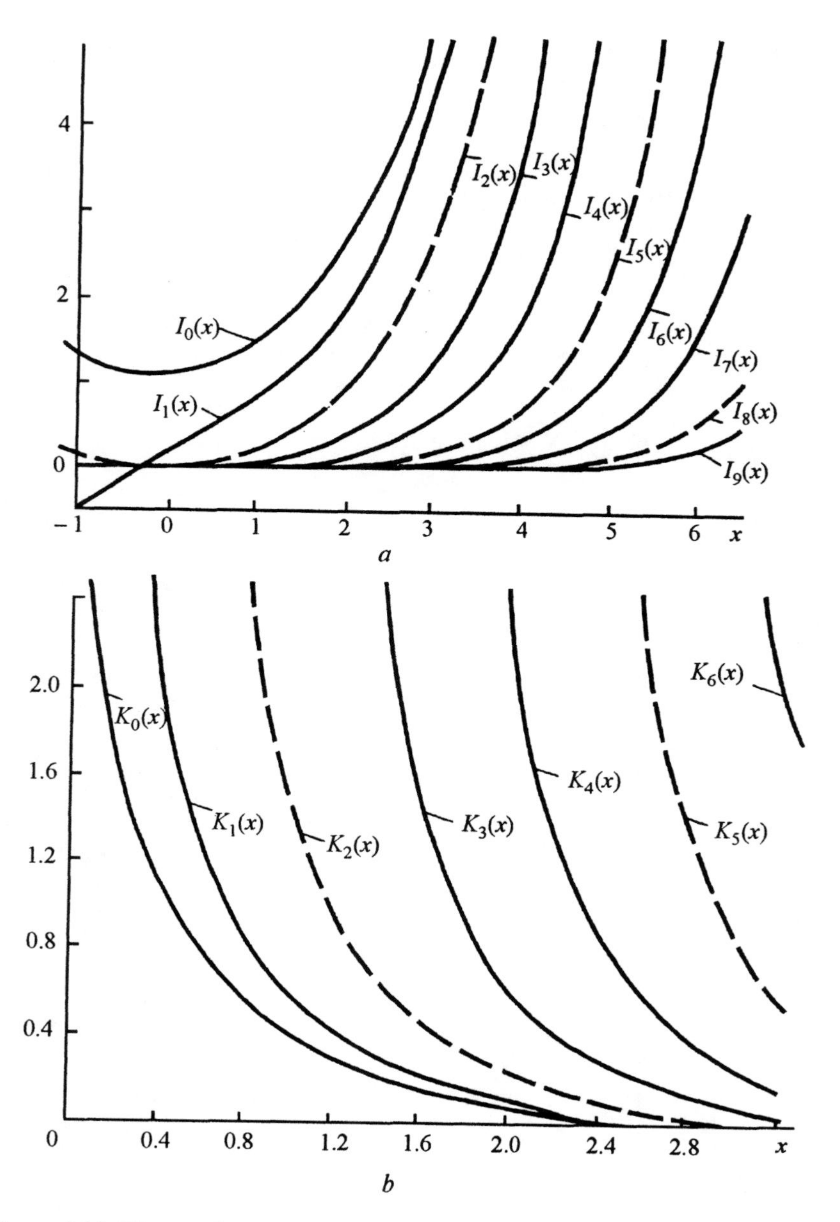

Figure 4.30. The graphs of functions: (*a*) $I_n(x)$, $n = 0,1,2,3,4,5,6,7,8,9$; (*b*) $K_n(x)$, $n = 0,1,2,3,4,5,6$.

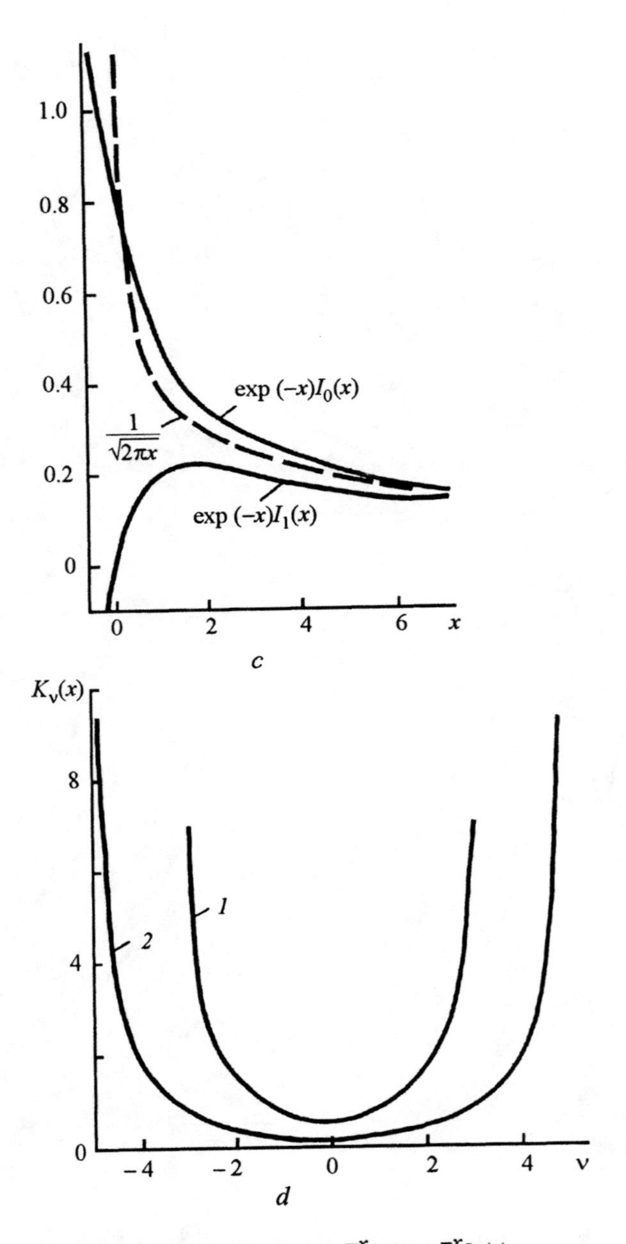

Figure 4.30. The graphs of functions: (*c*) $e^{-x}I_0(x)$, $e^{-x}I_1(x)$; (*d*) 1. $K_\nu(1)$, 2. $K_\nu(2)$.

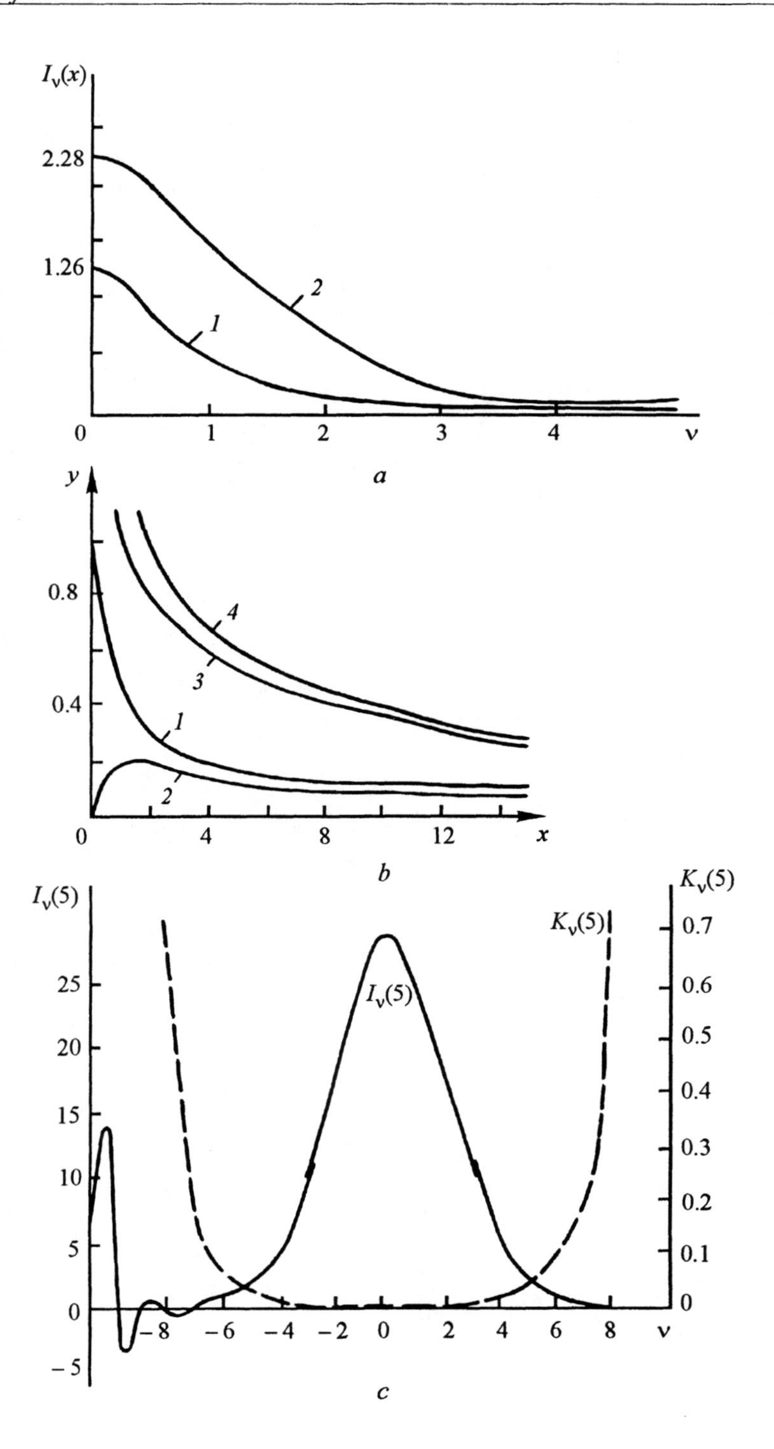

Figure 4.31. The graphs of functions: (*a*) *1.* $I_\nu(1)$, *2.* $I_\nu(2)$; (*b*) *1.* $e^{-x}I_0(x)$, *2.* $e^{-x}I_1(x)$, *3.* $e^xK_0(x)$, *4.* $e^xK_1(x)$; (*c*) $I_\nu(5)$, $K_\nu(5)$.

$$K_\nu(z) = \frac{\Gamma(\nu)}{2}\left(\frac{z}{2}\right)^{-\nu}\sum_{k=0}^{\infty}\frac{\left(\frac{z^2}{4}\right)^k}{k!(1-\nu)_k} + \frac{\Gamma(-\nu)}{2}\left(\frac{z}{2}\right)^{\nu}\sum_{k=0}^{\infty}\frac{\left(\frac{z^2}{4}\right)^k}{k!(1+\nu)_k} \quad (\nu \neq 0, \pm 1, \pm 2, \ldots),$$

$$I_0(z) = \sum_{k=0}^{\infty}\frac{\left(\frac{z^2}{4}\right)^k}{(k!)^2},$$

$$I_1(z) = \frac{z}{2}\sum_{k=0}^{\infty}\frac{\left(\frac{z^2}{4}\right)^k}{k!(k+1)!}.$$

Integral representations for the functions are given by

$$I_0(z) = \frac{1}{\pi}\int_0^\pi e^{\pm z\cos t}dt = \frac{1}{\pi}\int_0^\pi \cosh(z\cos t)dt = \frac{2}{\pi}\int_0^1 \frac{\cosh ztdt}{\sqrt{1-t^2}},$$

$$K_0(z) = -\frac{1}{\pi}\int_0^\pi e^{\pm z\cos t}(\gamma + \ln(2z\sin^2 t))dt,$$

$$I_1(z) = \frac{1}{\pi z}\int_0^{2z}\frac{(z-t)e^{z-1}}{\sqrt{2zt-t^2}}dt = \frac{2z}{\pi}\int_0^1 \sqrt{1-t^2}\cosh zt\,dt,$$

$$I_n(z) = \frac{1}{\pi}\int_0^\pi e^{z\cos t}\cos(nt)dt \quad (n \in \mathbf{Z}),$$

$$I_n(z) = \frac{z^n}{(2n-1)!!\pi}\int_0^\pi \sin^{2n} t e^{\pm z\cos t}dt \quad (n = 0, 1, 2, \ldots),$$

$$I_\nu(z) = \frac{1}{\pi}\int_0^\pi e^{z\cos t}\cos\nu t dt - \frac{\sin\nu\pi}{\pi}\int_0^\infty e^{z\cosh t - \nu t}dt \quad (|\arg z| < \pi/2),$$

$$K_\nu(z) = \frac{\sqrt{\pi}\left(\frac{z}{2}\right)^\nu}{\Gamma(\nu+\frac{1}{2})}\int_0^\infty e^{-z\cosh t}\sinh^{2\nu} t\,dt = \frac{\sqrt{\pi}\left(\frac{z}{2}\right)^\nu}{\Gamma(\nu+\frac{1}{2})}\int_1^\infty e^{-zt}(t^2-1)^{\nu-\frac{1}{2}}dt$$

$$\left(\mathrm{Re}\,\nu > -\frac{1}{2}, |\arg z| < \frac{\pi}{2}\right),$$

$$K_\nu(z) = \int_0^\infty e^{-z\cosh t} \cosh \nu t dt \quad (|\arg z| < \pi/2).$$

Note that the functions $I_0(x)$, $I_1(x)$ can be represented as fractional derivatives of certain expressions. For example,

$$I_0(x) = e^{-x} \frac{d^{-1/2}}{dx^{-1/2}} \left(\frac{e^{2x}}{\sqrt{\pi x}} \right) = \frac{e^x}{\sqrt{2}} \frac{d^{1/2}}{dx^{1/2}} \operatorname{erf}(\sqrt{2x}),$$

$$I_1(x) = \frac{xe^x}{2} \frac{d^{1/2}}{dx^{1/2}} \left(\frac{e^{-2x}}{\sqrt{\pi x^3}} \right) = \frac{e^x}{x} \frac{d^{-1/2}}{dx^{-1/2}} \left[\frac{\operatorname{erf}(\sqrt{2x})}{\sqrt{8}} - \sqrt{\frac{x}{\pi}}\, e^{-2x} \right].$$

The *Kelvin functions* are the functions $\operatorname{ber}_\nu(x)$, $\operatorname{bei}_\nu(x)$, $\operatorname{ker}_\nu(x)$, $\operatorname{kei}_\nu(x)$ defined by

$$\operatorname{ber}_\nu(x) + i \operatorname{bei}_\nu(x) = J_\nu(xe^{3\pi i/4}) = e^{\nu\pi i} J_\nu(xe^{-\pi i/4} = e^{\nu\pi i/2} I_\nu(xe^{\pi i/4}) = e^{3\nu\pi i/2} I_\nu(xe^{-3/4\pi i}),$$

$$\operatorname{ker}_\nu(x) + i \operatorname{kei}_\nu(x) = e^{-\nu\pi i/2} K_\nu(xe^{\pi i/4}) = \frac{1}{2} \pi i H_\nu^{(1)}(xe^{3/4\pi i}) = -\frac{1}{2} \pi i e^{-\nu\pi i} H_\nu^{(2)}(xe^{-\pi i/4}).$$

In the case where $\nu = 0$, the Kelvin function is written without an index.

The Kelvin functions (sometimes called *Thompson functions*) are real and imaginary parts of solutions of the differential equation

$$x^2 \frac{d^2u}{dx^2} + x \frac{du}{dx} - (ix^2 + \nu^2)u = 0.$$

Here ν is real, and x is real nonnegative. The functions $u = \operatorname{ber}_\nu x + i\operatorname{bei}_\nu x$, $u = \operatorname{ber}_{-\nu} x + i\operatorname{bei}_{-\nu} x$, $u = \operatorname{ker}_\nu x + i\operatorname{kei}_\nu x$, actually make a fundamental system of solutions of this equation.

All Kelvin functions take real values for $x >> 0$. The functions $\operatorname{ber}_\nu x$ and $\operatorname{bei}_\nu x$ are bounded in a neighborhood of the origin, change sign, and exponentially increase for $x >> 1$. The functions $\operatorname{ker}_\nu x$ and $\operatorname{kei}_\nu x$, $\nu > 0$, are unbounded in a neighborhood of zero, change sign and exponentially decrease for $x >> 1$.

The functions take the following values: $\operatorname{ber}(0) = 1$, $\operatorname{ker}_{\pm 2}(0) = \frac{1}{2}$, $\operatorname{kei}(0) = \frac{\pi}{4}$, $\operatorname{ker}_\nu(\infty) = 0$, $\operatorname{kei}_\nu(\infty) = 0$ (Figure 4.32).

The behavior of the functions

$$A = \sqrt{2\pi x}\, e^{-x/2} \operatorname{ber}(x), \qquad B = \sqrt{2\pi x}\, e^{-x/\sqrt{2}} \operatorname{bei}(x),$$
$$C = \sqrt{2x/\pi}\, e^{x/\sqrt{2}} \operatorname{ker}(x), \qquad D = \sqrt{2x/\pi}\, e^{x/\sqrt{2}} \operatorname{kei}(x)$$

for large values of the argument is shown in Figure 4.33. In some cases, the functions

$$\operatorname{her}_\nu(x) = \frac{2}{\pi} \operatorname{kei}_\nu(x), \qquad \operatorname{hei}_\nu(x) = -\frac{2}{\pi} \operatorname{ker}_\nu(x)$$

are used instead of the functions $\operatorname{ker}_\nu(x)$ and $\operatorname{kei}_\nu((x)$.

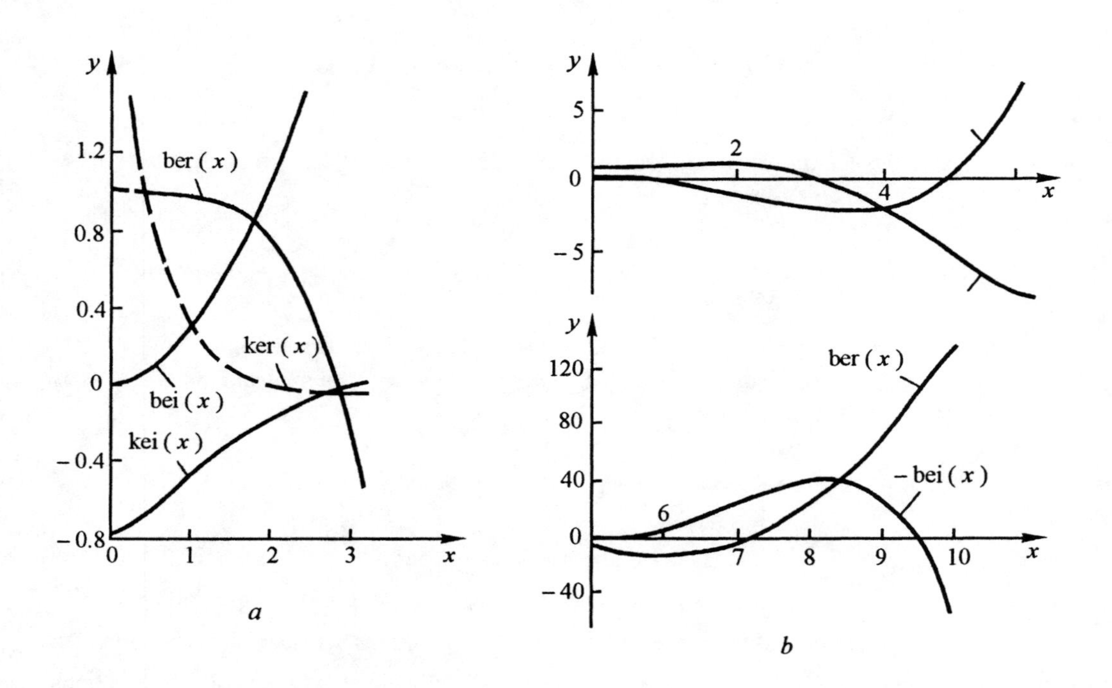

Figure 4.32. The graphs of functions: (*a*) ber(*x*), bei(*x*), ker(*x*), kei(*x*); (*b*) ber(*x*), – bei(*x*).

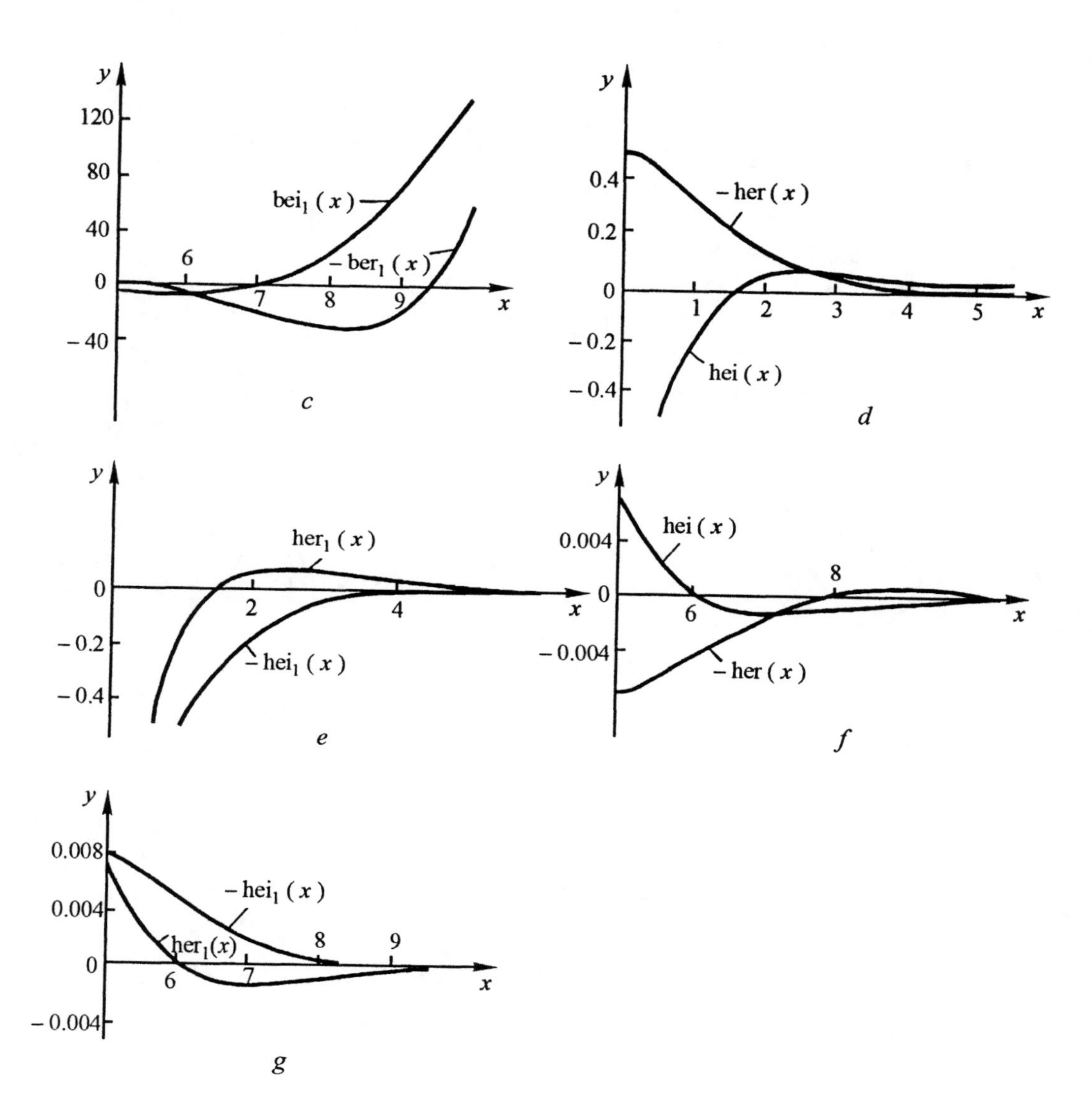

Figure 4.32. The graphs of functions: (*c*) $bei_1(x)$, $-ber_1(x)$; (*d*) $-her(x)$, $hei(x)$; (*e*) $hei(x)$, $-her(x)$; (*f*) $her_1(x)$, $-hei_1(x)$; (*g*) $-hei_1(x)$, $her_1(x)$.

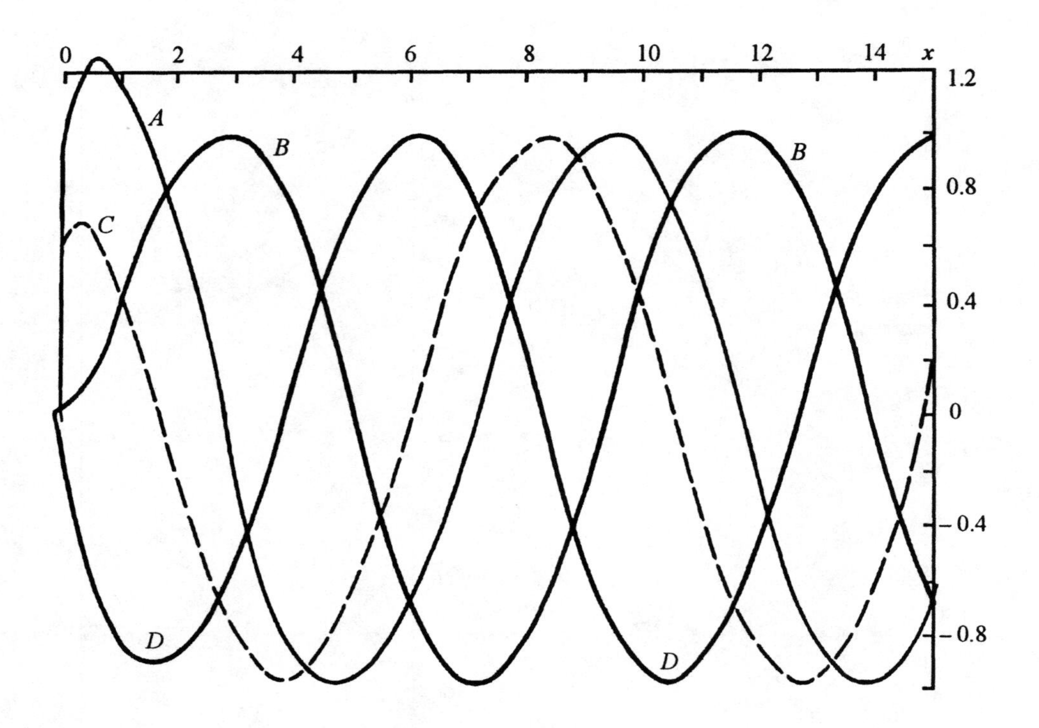

Figure 4.33. The functions A, B, C, D.

The moduli M_ν, N_ν and phases Θ_ν, Φ_ν of the Kelvin functions are defined by the formulas

$$M_\nu = \sqrt{\mathrm{ber}_\nu^2(x) + \mathrm{bei}_\nu^2(x)}\,, \qquad \Theta_\nu(x) = \arctan\frac{\mathrm{bei}_\nu(x)}{\mathrm{ber}_\nu(x)}\,,$$

$$\mathrm{ber}_\nu(x) = M_\nu \cos\Theta_\nu, \qquad \mathrm{bei}_\nu(x) = M_\nu \sin\Theta_\nu\,,$$

$$N_\nu = \sqrt{\mathrm{ker}_\nu^2(x) + \mathrm{kei}_\nu^2(x)}\,, \qquad \Phi_\nu(x) = \arctan\frac{\mathrm{kei}_\nu(x)}{\mathrm{ker}_\nu(x)}\,,$$

$$\mathrm{ker}_\nu(x) = N_\nu \cos\Phi_\nu, \qquad \mathrm{kei}_\nu(x) = N_\nu \sin\Phi_\nu\,.$$

The phases of the Kelvin functions are shown in Figure 4.34.

The Kelvin functions of order zero, ker and kei, satisfy the corresponding formulas:

$$\mathrm{ker}(x) = -\frac{x^2}{8}\int_0^\infty \mathrm{Ci}\left(\frac{1}{t}\right) e^{-x^2 t/4} dt = \int_0^\infty \frac{t^3 J_0(xt)}{1+t^4}\,dt = \frac{x}{4}\int_0^\infty \ln\,(1+t^4)\,J_1(xt)dt,$$

$$\mathrm{kei}(x) = \frac{x^2}{8}\int_0^\infty \left[\mathrm{Si}\left(\frac{1}{t}\right) - \frac{\pi}{2}\right] e^{-x^2 t/4} dt = -\int_0^\infty \frac{t\,J_0(xt)}{1+t^4}\,dt = -\frac{x}{2}\int_0^\infty \mathrm{arccot}\, t^2\, J_1(xt)dt,$$

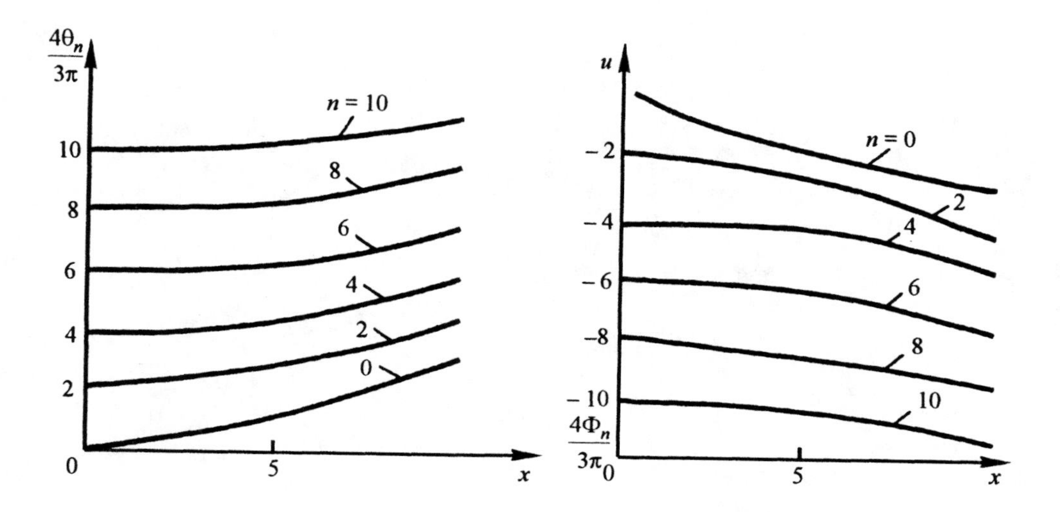

Figure 4.34. The phases of the Kelvin functions.

$$\mathrm{ber}_{-\nu}(x) = \cos\nu\pi\,\mathrm{ber}_\nu(x) + \sin\nu\pi\,\mathrm{bei}_\nu(x) + \frac{2}{\pi}\sin\nu\pi\,\mathrm{ker}_\nu(x),$$

$$\mathrm{bei}_{-\nu}(x) = -\sin\nu\pi\,\mathrm{ber}_\nu(x) + \cos\nu\pi\,\mathrm{bei}_\nu(x) + \frac{2}{\pi}\sin\nu\pi\,\mathrm{kei}_\nu(x),$$

$$\mathrm{ker}_{-\nu}(x) = \cos\nu\pi\,\mathrm{ker}_\nu(x) - \sin\nu\pi\,\mathrm{kei}_\nu(x),$$

$$\mathrm{kei}_{-\nu}(x) = \sin\nu\pi\,\mathrm{ker}_\nu(x) + \cos\nu\pi\,\mathrm{kei}_\nu(x).$$

The last four formulas can be written as a single formula, $f_{-n}(x) = (-1)^n f_n(x)$, where $f =$ ber, bei, ker, kei, $n \in \mathbf{Z}$.

The function has the power series expansion:

$$\mathrm{ber}_\nu(x) = \left(\frac{x}{2}\right)^\nu \sum_{k=0}^{\infty} \frac{\cos\left(\frac{3}{4}\nu + \frac{k}{2}\right)\pi}{k!\,\Gamma(\nu + k + 1)} \left(\frac{x^2}{4}\right)^k,$$

$$\mathrm{bei}_\nu(x) = \left(\frac{x}{2}\right)^\nu \sum_{k=0}^{\infty} \frac{\sin\left(\frac{3}{4}\nu + \frac{k}{2}\right)\pi}{k!\,\Gamma(\nu + k + 1)} \left(\frac{x^2}{4}\right)^k,$$

$$\mathrm{ker}_n(x) = \frac{1}{2}\left(\frac{x}{2}\right)^{-n} \sum_{k=0}^{n-1} \cos\left(\frac{3}{4}n + \frac{k}{2}\right)\pi\,\frac{(n-k-1)!}{k!}\left(\frac{x^2}{4}\right)^k - \ln\frac{x}{2}\,\mathrm{ber}_n(x) +$$

$$+ \frac{1}{4}\pi\,\mathrm{bei}_n(x) + \frac{1}{2}\left(\frac{x}{2}\right)^n \sum_{k=0}^{\infty} \cos\left(\frac{3}{4}n + \frac{k}{2}\right)\pi\,\frac{\psi(k+1) + \psi(n+k+1)}{k!\,(n+k)!}\left(\frac{x^2}{4}\right)^k,$$

$$\mathrm{kei}_n(x) = -\frac{1}{2}\left(\frac{x}{2}\right)^{-n} \sum_{k=0}^{n-1} \sin\left(\frac{3}{4}n + \frac{k}{2}\right)\pi \frac{(n-k-1)!}{k!}\left(\frac{x^2}{4}\right)^k - \ln\frac{x}{2}\,\mathrm{bei}_n(x) -$$

$$-\frac{1}{4}\pi\,\mathrm{ber}_n(x) + \frac{1}{2}\left(\frac{x}{2}\right)^n \sum_{k=0}^{\infty} \sin\left(\frac{3}{4}n + \frac{k}{2}\right)\pi \frac{\psi(k+1) + \psi(n+k+1)}{k!\,(n+k)!}\left(\frac{x^2}{4}\right)^k.$$

and the asymptotic formulas:

$$\mathrm{ber}_\nu(x) \approx \frac{1}{\sqrt{2\pi x}}\, e^{x/\sqrt{2}} \cos(x/\sqrt{2} + \nu\pi/2 - \pi/8),$$

$$\mathrm{bei}_\nu(x) \approx \frac{1}{\sqrt{2\pi x}}\, e^{x/\sqrt{2}} \sin(x/\sqrt{2} + \nu\pi/2 - \pi/8),$$

$$\mathrm{ker}_\nu(x) \approx \sqrt{\frac{\pi}{2x}}\, e^{-x/\sqrt{2}} \cos(x/\sqrt{2} + \nu\pi/2 + \pi/8),$$

$$\mathrm{kei}_\nu(x) \approx \sqrt{\frac{\pi}{2x}}\, e^{-x/\sqrt{2}} \sin(x/\sqrt{2} + \nu\pi/2 + \pi/8),$$

$$\left[\mathrm{ber}_\nu^2(x) + \mathrm{bei}_\nu^2(x)\right]\left[\mathrm{ker}_\nu^2(x) + \mathrm{kei}_\nu^2(x)\right] \approx \frac{1}{4x^2}\ \mathrm{ber}_\nu(x)\,\mathrm{ker}_\nu(x) + \mathrm{bei}_\nu(x)\,\mathrm{kei}_\nu(x) =$$

$$= \mathrm{ber}_\nu(x)\,\mathrm{kei}_\nu(x) - \mathrm{bei}_\nu(x)\,\mathrm{ker}_\nu(x) \approx \frac{1}{x\sqrt{8}}.$$

Bessel functions of fractional order. The most widely used Bessel functions of fractional order are spherical Bessel functions, modified spherical Bessel functions, and Airy functions.

The *spherical Bessel functions* are particular solutions of the differential equation

$$z^2 \frac{d^2u}{dz^2} + 2z\frac{du}{dz} + [z^2 - n(n+1)]u = 0 \qquad (n \in \mathbf{Z}).$$

The *spherical Bessel functions of the first*, *second*, and *third kind* are, correspondingly, the functions $j_n(z) = \sqrt{\frac{\pi}{2z}}\, J_{n+1/2}(z)$, $y_n(z) = \sqrt{\frac{\pi}{2z}}\, Y_{n+1/2}(z)$, and $h_n^{(1)}(z) = j_n(z) + i\,y_n(z) = \sqrt{\frac{\pi}{2z}} H_{n+1/2}^{(1)}(z)$, $h_n^{(2)}(z) = j_n(z) - iy_n(z) = \sqrt{\frac{\pi}{2z}}\, H_{n+1/2}^{(1)}(z)$.

Spherical Bessel functions are bounded everywhere and have many zeros.

Values of the functions at particular points are the following:

$$j_0(z) = \sin z/z,\quad j_1(z) = \sin z/z^2 - \cos z/z,$$

$$y_0(z) = -j_{-1}(z) = -\cos z/z,\quad y_1(z) = j_{-2}(z) = -\cos z/z^2 - \sin z/z,$$

$$j_n(z)\, y_{n-1}(z) - j_{n-1}(z)\, y_n(z) = z^{-2},$$

$$j_{n+1}(z)\, y_{n-1}(z) - j_{n-1}(z)\, y_{n+1}(z) = (2n+1)z^{-3},\quad j_n(z)e^{m\pi i} = e^{mn\pi i} j_n(z),$$

$$y_n(ze^{m\pi i}) = (-1)^m e^{mn\pi i} y_n(z), \quad h_n^{(1)}(ze^{(2m+1)\pi i}) = (-1)^n h_n^{(2)}(z),$$

$$h_n^{(2)}(ze^{(2m+1)\pi i}) = (-1)^n h_n^{(1)}(z), \quad h_n^{(r)}(ze^{2m\pi i}) = h_n^{(r)}(z).$$

where $r = 1,2$; $m,n = 0,1,2,\ldots$ (Figure 4.35).

The *modified spherical Bessel functions* are partial solutions of the differential equation

$$z^2 \frac{d^2u}{dz^2} + 2z\frac{du}{dz} - [z^2 + n(n+1)]u + 0 \quad (n \in \mathbb{Z}).$$

The modified spherical Bessel functions of the first, second, and third kind are, respectively,

$$i_n(z) = \sqrt{\frac{\pi}{2z}}\, I_{n+1/2}(z) = \begin{cases} e^{-n\pi i/2} j_n(ze^{\pi i/2}) & (-\pi < \arg z \le \pi/2), \\ e^{3n\pi i/2} j_n(ze^{-3\pi i/2}) & (\pi/2 < \arg z \le \pi), \end{cases}$$

$$\tilde{i}_n(z) = \sqrt{\frac{\pi}{2z}}\, I_{n-1/2}(z) = \begin{cases} e^{3(n+1)\pi i/2} y_n(ze^{\pi i/2}) & (-\pi < \arg z \le \pi/2), \\ e^{-(n+1)\pi i/2} y_n(ze^{-3\pi i/2}) & (\pi/2 < \arg z \le \pi), \end{cases}$$

$$k_n(z) = \sqrt{\frac{\pi}{2z}}\, K_{n+1/2}(z) = \frac{\pi}{2}(-1)^{n+1}\sqrt{\frac{\pi}{2z}}\left[I_{n+1/2}(z) - I_{-n-1/2}(z)\right].$$

The modified spherical Bessel functions that have integer order are bounded at the origin, and exponentially increase for $x >> 1$. The functions of the second kind having integer order are unbounded at the origin, and exponentially increase for $x >> 1$. The functions of the third kind are also unbounded at the origin, and exponentially decrease for $x >> 1$ (Figure 4.36).

The functions satisfy the following formulas:

$$i_n(z) = z^n \left(\frac{1}{z}\frac{d}{dz}\right)^n \frac{\sinh z}{z}.$$

$$\tilde{i}_n(z) = z^n \left(\frac{1}{z}\frac{d}{dz}\right)^n \frac{\cosh z}{z} \qquad (n = 0, 1, 2, \ldots) \text{ Relay's formulas.}$$

Special cases are given by

$$i_0(z) = \sinh z/z, \qquad i_1(z) = -\sinh z/z^2 + \cosh z/z,$$

$$\tilde{i}_0(z) = \cosh z/z, \qquad \tilde{i}_1(z) = \sinh z/z - (\cosh z)/z^2,$$

$$k_0(z) = \frac{\pi}{2z}e^{-z}, \qquad k_1(z) = \frac{\pi}{2z}e^{-z}(1 + 1/z).$$

The functions have the power series expansions:

$$i_n(z) = \sum_{k=0}^{\infty} \frac{z^{2k+n}}{(2k!)(2n+2k+1)!!},$$

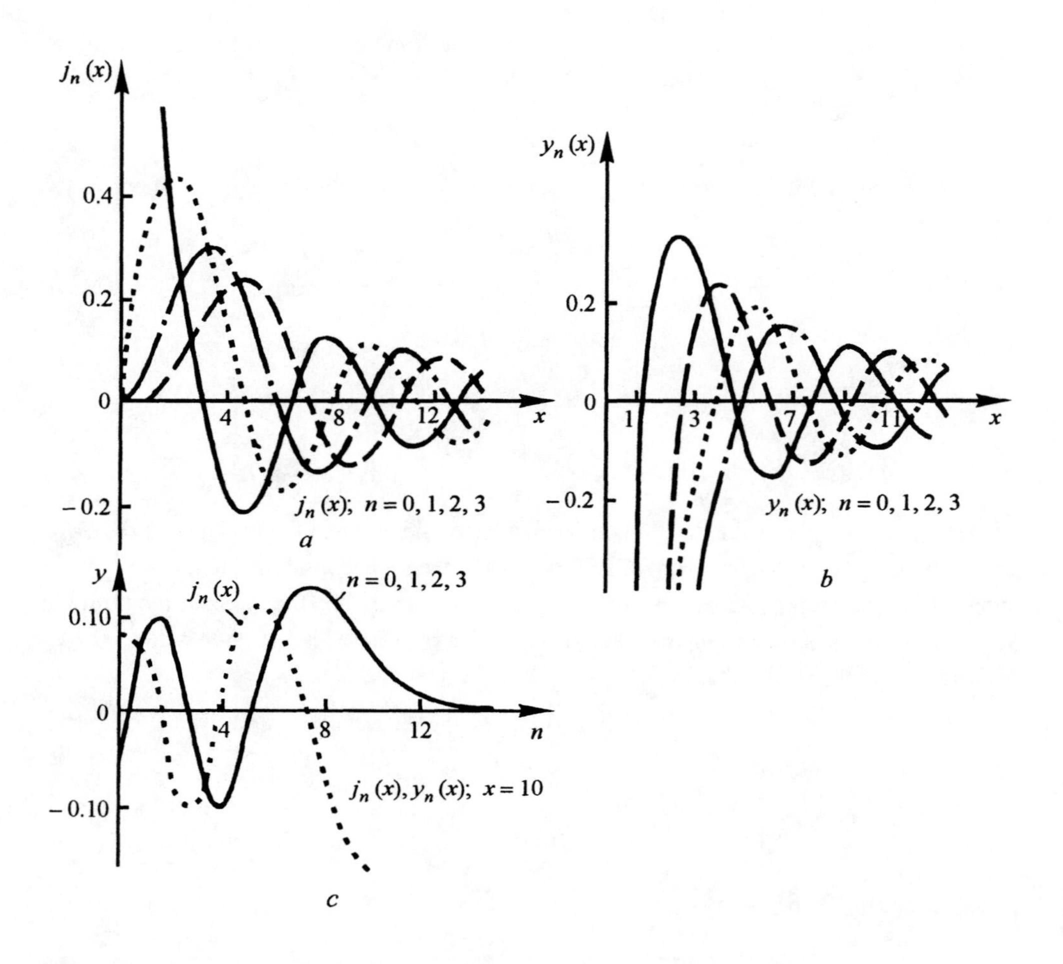

Figure 4.35. The graphs of functions: (*a*) $J_n(x)$, $n = 0,1,2,3$; (*b*) $V_n(x)$, $n = 0,1,2,3$; (*c*) $J_n(x)$, $V_n(x)$, $x = 10$.

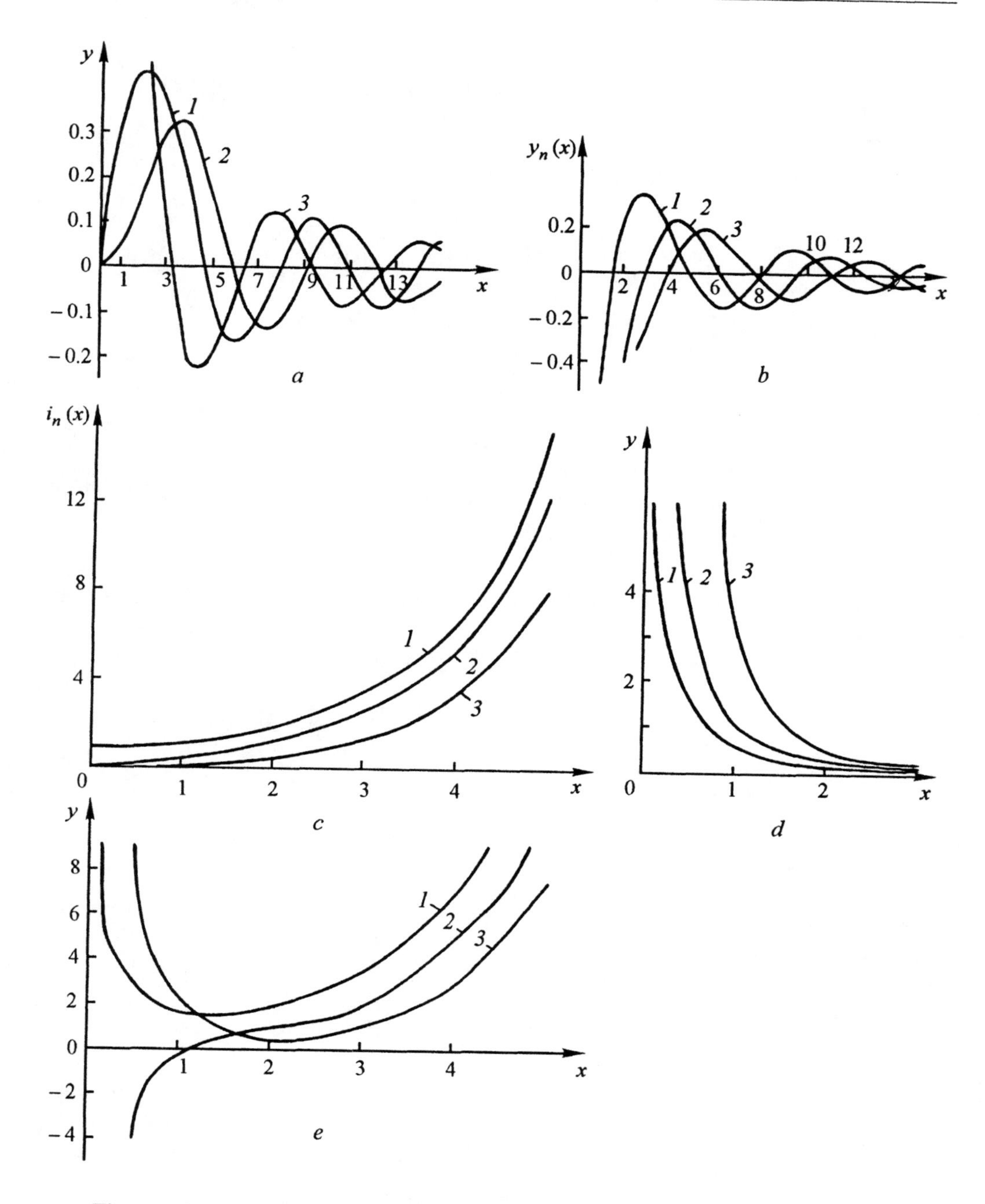

Figure 4.36. The spherical Bessel functions: (*a*) of the first kind *1.* $J_0(x)$, *2.* $J_1(x)$, *3.* $J_2(x)$; (*b*) of the second kind *1.* $Y_0(x)$, *2.* $Y_1(x)$, *3.* $Y_2(x)$; (*c*) modified spherical Bessel functions of the first kind *1.* $i_0(x)$, *2.* $i_1(x)$, *3.* $i_2(x)$; (*d*) modified spherical Bessel functions of the second kind *1.* $\widetilde{i}_0(x)$, *2.* $\widetilde{i}_1(x)$, *3.* $\widetilde{i}_2(x)$; (*e*) modified spherical Bessel functions of the third kind *1.* $K_0(x)$, *2.* $K_1(x)$, *3.* $K_2(x)$.

$$\widetilde{i}_n(z) = \sum_{k=0}^{n-1} \frac{(-1)^{n+k}(2n-2k-1)!!}{(2k)!!} z^{2k-n+1} + (-1)^n \sum_{k=n}^{\infty} \frac{(2k+1-2n)!!}{(2k)!!} z^{k-n-1} .$$

The following doubling formula holds:

$$k_n(2z) = \sqrt{\frac{2}{\pi}}\, n! z^{n+1} \sum_{k=0}^{n} \frac{(-1)^k(2n-2k+1)}{k!(2n-k+1)!} k_{n-k}^2(z) .$$

Riccati–Bessel functions are particular solutions of the equation

$$z^2 \frac{d^2u}{dz^2} + [z^2 - n(n+1)]u = 0, n \in \mathbf{Z}.$$

These functions are denoted by $zj_n(z)$, zy_n, $zh_n^{(1)}(z)$, $zh_n^{(2)}(z)$.

All properties of these functions follow from properties of spherical Bessel functions. We thus have $zj_0(z) = \sin z$, $zj_1(z) = \frac{\sin z}{z} - \cos z$, $zy_0(z) = -\cos z$, $zy_1(z) = -\sin z - \cos z$.

Airy functions. Pairs of linearly independent solutions of the differential equation $\frac{d^2u}{dz^2} - zu = 0$ are the functions of the form

$$\mathrm{Ai}(z),\ \mathrm{Bi}(z);\ \mathrm{Ai}(z),\ \mathrm{Ai}(ze^{2\pi i/3});\ \mathrm{Ai}(z),\ \mathrm{Ai}(ze^{-2\pi i/3}) .$$

The integral

$$\mathrm{Ai}(x) = \frac{1}{\pi}\int_0^{\infty} \cos\left(xt + \frac{t^3}{3}\right) dt$$

defines the function Ai(x) for all values of the argument; the function Bi(z) is defined for all values of the argument as the sum of the integrals

$$\mathrm{Bi}(x) = \frac{1}{\pi}\int_0^{\infty} e^{xt - t^3/3} dt + \frac{1}{\pi}\int_0^{\infty} \sin\left(xt + \frac{t^3}{3}\right) dt .$$

The Airy functions Ai(x), Bi(x) are bounded at the origin. For $x >> 1$, the function Ai(x) exponentially decreases; for $-x >> 1$, the function changes sign and the absolute value of the function decreases. For $x >> 1$, the function Bi(x) is monotonic exponentially increasing; for $-x >> 1$, the function changes sign and its absolute value decreases.

For $x \geq 0$, both functions Ai(x) and Bi(x) are positive; as $x \to \infty$, the function Bi(x) tends to infinity very fast, whereas the function Ai(x) approaches zero (Figure 4.37, 4.38).

The Airy functions can be expressed in terms of Bessel functions of fractional order:

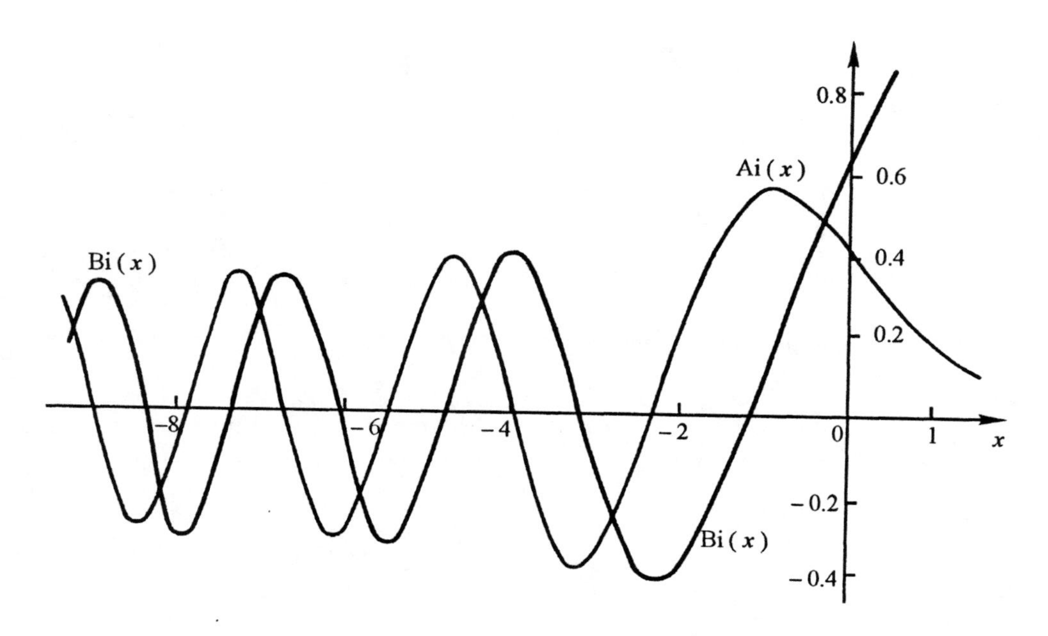

Figure 4.37. The Airy functions Ai(x), Bi(x).

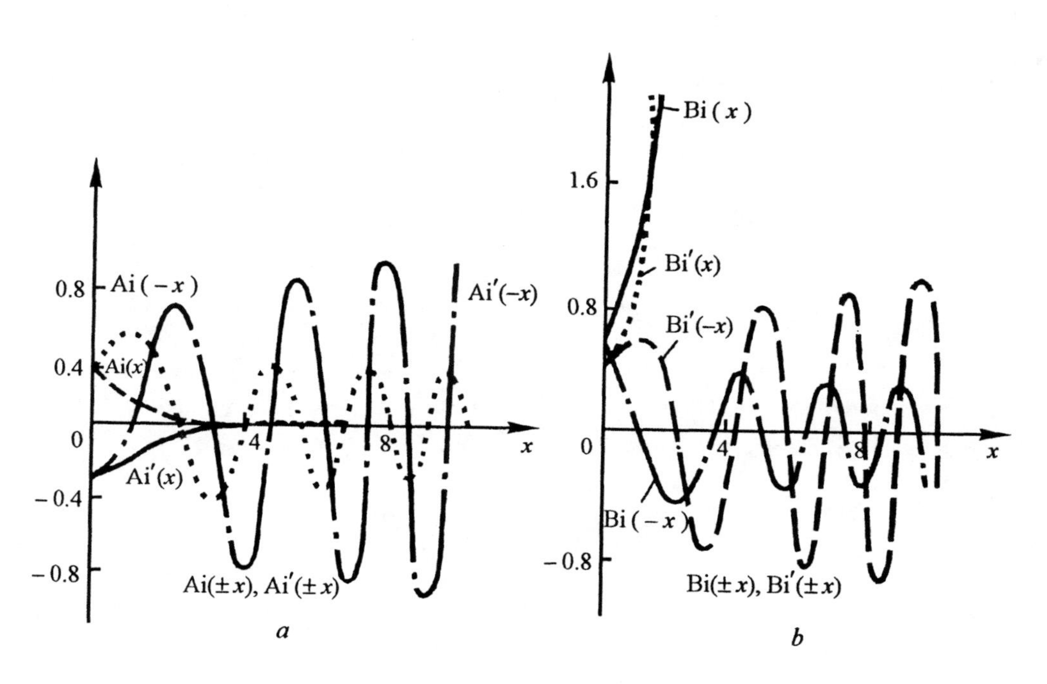

Figure 4.38. The graphs of functions: (*a*) Ai($\pm x$), Ai'($\pm x$); (*b*) Bi($\pm x$), Bi'($\pm x$).

$$\mathrm{Ai}(x) = \begin{cases} \dfrac{\sqrt{x}}{3}\,[I_{-1/3}(X) - I_{1/3}(X)] = \dfrac{1}{\pi}\sqrt{\dfrac{x}{3}}\,K_{1/3}(X), & x > 0 \\ \dfrac{\sqrt{-x}}{3}\,[J_{1/3}(X) + J_{1/3}(X)] = \sqrt{\dfrac{-x}{3}}\,[Y_{-1/3}(X) - Y_{1/3}(X)], & x < 0, \end{cases}$$

$$\mathrm{Bi}(x) = \begin{cases} \dfrac{\sqrt{x}}{3}\,[I_{-1/3}(X) + I_{1/3}(X)], & x > 0 \\ \sqrt{\dfrac{-x}{3}}\,[J_{-1/3}(X) - J_{1/3}(X)] = -\dfrac{\sqrt{-x}}{3}\,[Y_{-1/3}(X) + Y_{1/3}(X)], & x < 0. \end{cases}$$

Auxiliary Airy functions are defined by

$$f(x) = \mathrm{fai}(x) = \frac{3^{2/3}\Gamma(2/3)}{2}\left[\frac{\mathrm{Bi}(x)}{\sqrt{3}} + \mathrm{Ai}(x)\right] = \begin{cases} \dfrac{\Gamma(2/3)\sqrt{x}}{3^{1/3}}\,I_{-1/3}(X), & x > 0, \\ \dfrac{\Gamma(2/3)\sqrt{-x}}{3^{1/3}}\,J_{-1/3}(X), & x < 0, \end{cases}$$

$$g(x) = \mathrm{gai}(x) = \frac{3^{1/3}\Gamma(1/3)}{2}\left[\frac{\mathrm{Bi}(x)}{\sqrt{3}} - \mathrm{Ai}(x)\right] = \begin{cases} \dfrac{\Gamma(1/3)\sqrt{x}}{3^{2/3}}\,I_{1/3}(X), & x > 0, \\ \dfrac{\Gamma(1/3)\sqrt{-x}}{3^{2/3}}\,J_{1/3}(X), & x < 0. \end{cases}$$

In these formulas, we used the notation $X = 2/3(|x|)^{3/2}$. The graphs are shown in Figure 4.39.

For large positive x, the approximation formulas for these functions are $\mathrm{Ai}(x) \approx \dfrac{e^{-2x^{3/2}/3}}{2\sqrt{\pi}\sqrt{x}}$,

$\mathrm{Bi}(x) \approx \dfrac{e^{2x^{3/2}/3}}{\sqrt{\pi}\sqrt{x}}$.

For negative x, we have the following formulas:

$$\mathrm{Ai}(x)\mathrm{Bi}(x) \approx 1/(2\pi\sqrt{x}\,),\quad \mathrm{Ai}^2(x) + \mathrm{Bi}^2(x) \approx 1/(\pi\sqrt{-x}\,)$$

$$\mathrm{Ai}(-x) \approx \frac{1}{\sqrt{\pi}}\,x^{-1/4}\sin\left(\frac{2}{3}x^{3/2} + \frac{\pi}{4}\right),\quad \mathrm{Bi}(-x) \approx \frac{1}{\sqrt{\pi}}\,x^{-1/4}\cos\left(\frac{2}{3}x^{3/2} + \frac{\pi}{4}\right).$$

The Anger and Weber functions. The *Anger function* $J_\nu(z)$ and the *Weber function* $E_\nu(z)$ are obtained from the integral representation of the Bessel function,

$$J_\nu(z) \pm iE_\nu(z) = \frac{1}{\pi}\int_0^\pi e^{\pm i(\nu t - zt)}dt,$$

whence,

$$J_\nu(z) = \frac{1}{\pi}\int_0^\pi \cos(\nu t - z\sin t)dt,\quad E_\nu(z) = \frac{1}{\pi}\int_0^\pi \sin(\nu t - z\sin t)dt.$$

Sometimes the Weber function $E_\nu(z)$ is denoted by $Q_\nu(z)$ and called the *Lommel–Weber function*.

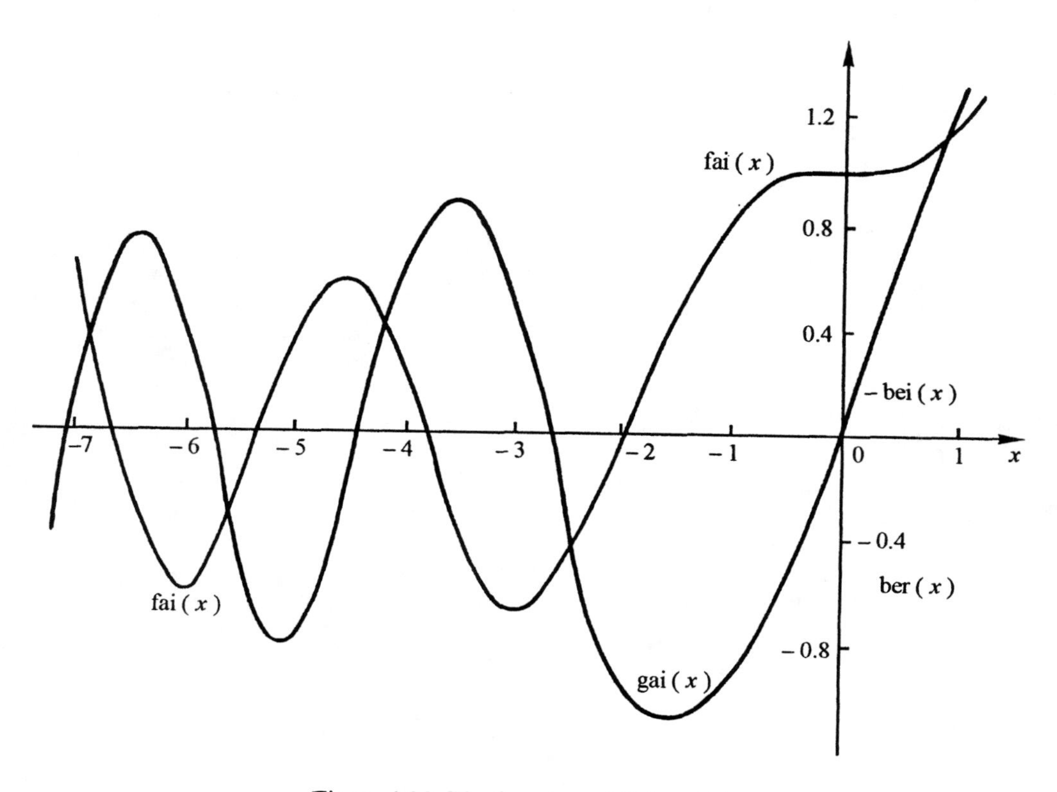

Figure 4.39. The functions fai(x), gai(x).

For an integer $\nu = n$, the Anger function coincides with the Bessel function $J_n(z) = J_n(z)$ (Figure 4.40).

The Anger and Weber functions are solutions of the nonhomogeneous Bessel's equations:

$$z^2 \frac{d^2u}{dz^2} + z\frac{du}{dz} + (z^2 - \nu^2)\, u = \begin{cases} \dfrac{1}{\pi}(z - \nu)\sin \nu\pi, \\ -\dfrac{1}{\pi}[(z + \nu) + (z - \nu)\cos \nu\pi]. \end{cases}$$

If ν is not an integer, the functions $J_\nu(z)$ and $E_n(z)$ satisfy the relations

$$\sin \nu\pi J_\nu(z) = \cos \nu\pi E_\nu(z) - E_{-\nu}(z),\ \sin E_\nu(z) = J_{-\nu}(z) - \cos \nu\pi J_\nu(z).$$

The recurrence relations for these functions are:

$$J_{\nu-1}(z) + J_{\nu+1}(z) = 2\nu z^{-1} J_\nu(z) - 2(\pi z)^{-1} \sin \nu\pi,$$

$$E_{\nu-1}(z) + E_{\nu+1}(z) = 2\nu z^{-1} E_\nu(z) - 2(\pi z)^{-1}(1 - \cos \nu\pi).$$

The functions $J_\nu(z)$ and $E_\nu(z)$ have the following power series expansion:

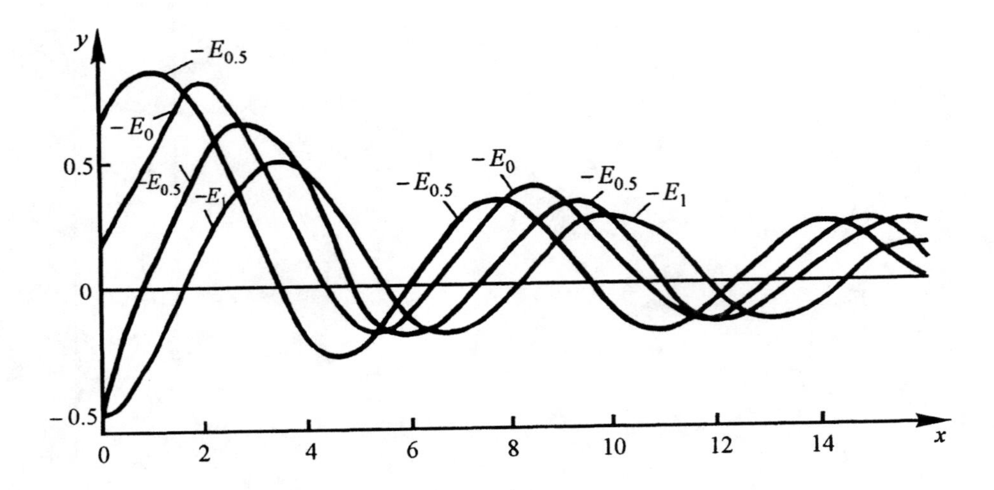

Figure 4.40. The Weber functions $E_\nu(x)$: $E_{-.5}, -E_0, -E_{.5}, -E_1$.

$$J_\nu(z) = g(z)\cos\frac{\nu\pi}{2} + f(z)\sin\frac{\nu\pi}{2}, \text{ where}$$

$$g(z) = \sum_{k=0}^{\infty} \frac{(-1)^k (z/2)^{2k}}{\Gamma(k+1+\nu/2)\Gamma(k+1-\nu/2)},$$

$$f(z) = \sum_{k=0}^{\infty} \frac{(-1)^k (z/2)^{2k+1}}{\Gamma(k+3/2+\nu/2)\Gamma(k+3/2-\nu/2)},$$

$$E_\nu(z) = g(z)\sin\frac{\nu\pi}{2} - f(z)\cos\frac{\nu\pi}{2}.$$

Struve functions. The *Struve functions* are the functions

$$H_\nu(z) = \frac{2}{\sqrt{\pi}} \frac{(z/2)^\nu}{\Gamma(\nu+1/2)} \int_0^{\pi/2} \sin(z\cos t)\sin^{2\nu} t\,dt =$$

$$= \frac{2}{\sqrt{\pi}} \frac{(z/2)^\nu}{\Gamma(\nu+1/2)} \int_0^1 (1-t^2)^{\nu-1/2} \sin xt\, dt,$$

obtained from the integral representation of Bessel functions for Re $\nu > -1/2$. They also satisfy the following nonhomogeneous Bessel's equation,

$$z^2 \frac{d^2u}{dz^2} + z\frac{du}{dz} + (z^2 - \nu^2)\, u = \frac{4(z/2)^{\nu+1}}{\sqrt{\pi}\ \Gamma(\nu + \frac{1}{2})}.$$

There is a relation between the Struve functions of integer order, $\nu = n$, and Weber functions. In particular, we have that $H_0(z) = -E_0(z)$, $H_1(z) = -E_1(z) + 2/\pi$. Similarly to the Bessel functions, $H_\nu(z)$ is real for negative x only if its order is an integer. This is an oscillating function, with the amplitude decreasing for increasing x. The maximums of the amplitude are not symmetric (Figures 4.41 – 4.43).

The Struve functions of order $\nu = n + 1/2$ (n is an integer) are elementary functions; for example, $\sqrt{\frac{\pi}{2}}\, H_{1/2}(z) = \frac{1 - \cos z}{\sqrt{z}}$.

There are relations between the Struve functions and other special functions:

$$H_\nu(z) - Y_\nu(z) = \frac{2(z/2)^\nu}{\sqrt{\pi}\ \Gamma(\nu + 1/2)} \int_0^\infty (1 + t^2)^{\nu - 1/2} e^{-zt}\, dt \ (\mathrm{Re}\,\nu > -1/2),$$

$$H_{-n-1/2}(z) = Y_{-n-1/2}(z) = (-1)^n J_{n+1/2}(z) =$$

$$= \sqrt{\frac{2z}{\pi}}\ y_{-n}(z) = (-1)^n \sqrt{\frac{2x}{\pi}}\ j_n(z) \quad (n = 0, 1, 2, \dots).$$

The main formulas for the functions are:

$$H_{n+1/2}(z) = \frac{2n-1}{z} H_{n-1/2}(z) - H_{n-3/2}(z) + \sqrt{\frac{2}{\pi z}}\ \frac{(z/2)^n}{n!} \quad (n = 0, 1, 2, \dots),$$

$$H_{n+1}(z) + H_{n-1}(z) = \frac{2n}{z} H_n(z) \begin{cases} + \dfrac{2z^n}{\pi(2n+1)!!}, & n = -1, 0, 1, 2, \dots, \\ \dfrac{2(-2n-3)!!}{\pi(-x)^{-n}}, & n = -1, -2, -3, \dots, \end{cases}$$

$$H_n(-z) = (-1)^{n+1} H_n(z) \quad (n \in \mathbb{Z}), \quad H_0(z) = \frac{4}{\pi} \sum_{k=0}^{\infty} \frac{J_{2k+1}(z)}{2k+1},$$

$$H_1(z) = -\frac{2}{\pi}\left[1 - J_0(z) + 2\sum_{k=0}^{\infty} \frac{J_{2k}(z)}{4k^2 - 1}\right],$$

$$H_\nu(z) = \frac{2}{\pi}\ \frac{(z/2)^{\nu+1}}{\Gamma(\nu + 3/2)} \left[1 + \sum_{k=0}^{\infty} \frac{(-z^2)^k}{(2k+1)!!} \prod_{i=1}^{k} (2\nu + 2i + 1)^{-1}\right].$$

The *modified Struve functions* $L_\nu(z)$ are expressed in terms of the Struve functions as follows:

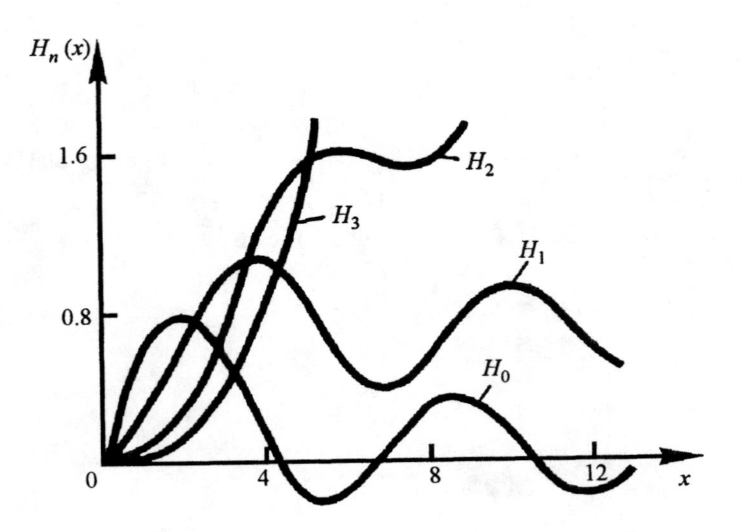

Figure 4.41. The Struve functions $H_n(x)$, $n = 0,1,2,3$.

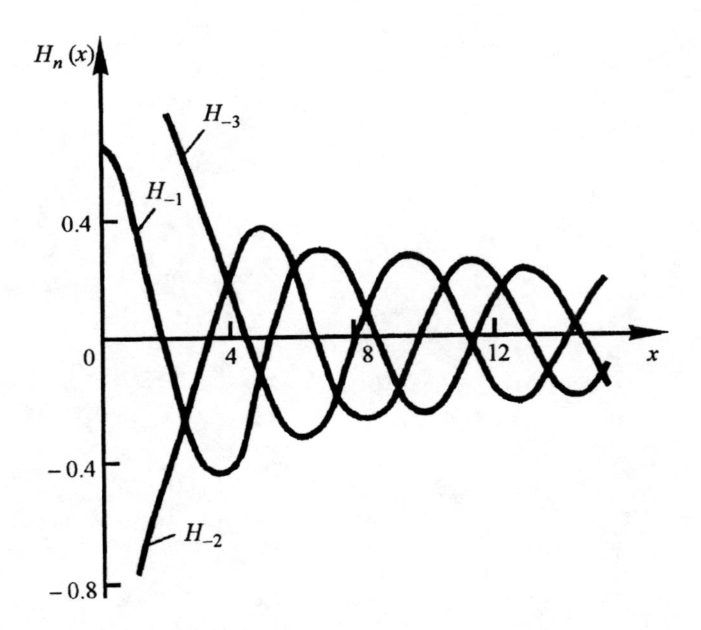

Figure 4.42. The function $H_n(x)$, $n = -1, -2, -3$.

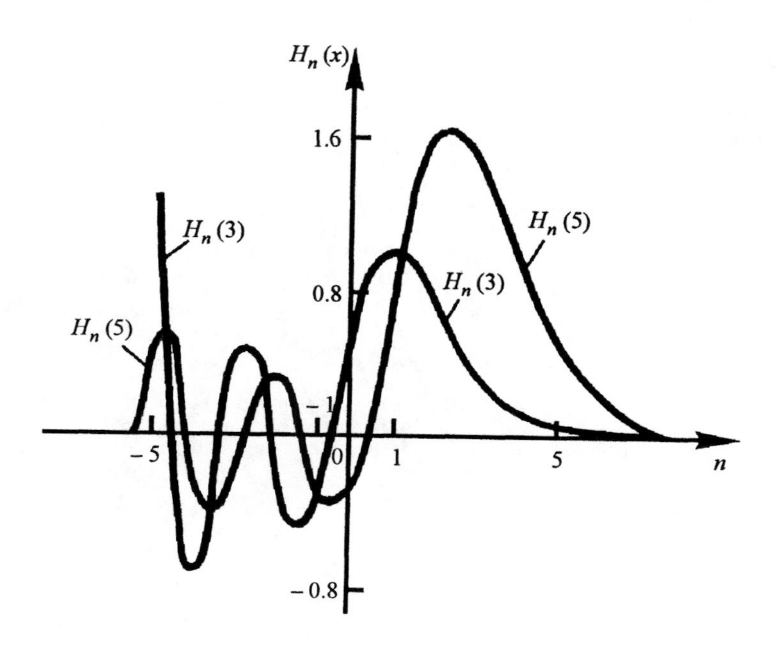

Figure 4.43. The function $H_n(x)$, $x = 3,5$.

$$L_\nu(z) = -ie^{-i\nu\pi/2} H_\nu(iz),$$

$$L_\nu(z) = \frac{2(z/2)^\nu}{\sqrt{\pi}\,\Gamma(\nu + 1/2)} \int_0^{\pi/2} \sinh (z \cos t) \sin^{2\nu} t \, dt =$$

$$= \frac{2(z/2)^\nu}{\sqrt{\pi}\,\Gamma(\nu + 1/2)} \int_0^1 (1 - t^2)^{\nu - 1/2} \sinh zt \, dt \qquad \left(\operatorname{Re} \nu > -\frac{1}{2}\right).$$

The function $L_\nu(z)$ is sometimes called the *hyperbolic Struve function*. It is related to the hyperbolic Bessel function via the formula:

$$L_\nu(x) - I_{-\nu}(x) = \frac{-2(x/2)^\nu}{\sqrt{\pi}\,\Gamma(\nu + 1/2)} \int_0^\infty (1 + t^2)^{\nu - 1/2} \sin xt \, dt.$$

The function $L_\nu(x)$ is defined (as a real-valued function) for negative x only if ν is integer. The function $L_n(x)$ is odd, if n is even, see Figures 4.44, 4.45.

The functions satisfy the recurrence relations:

$$L_{\nu - 1}(z) - L_{\nu + 1}(z) = \frac{2\nu}{z} L_\nu + \frac{(z/2)^\nu}{\sqrt{\pi}\,\Gamma(\nu + 3/2)},$$

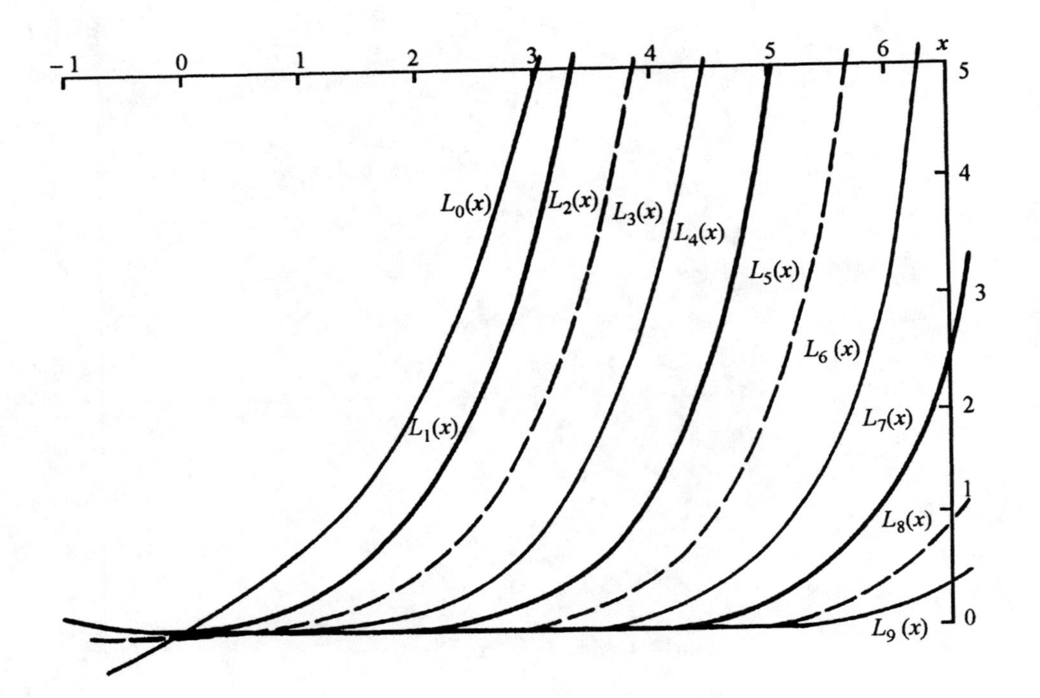

Figure 4.44. The modified Struve functions $L_n(x)$, $n = 0,1,2,3,4,5,6,7,8,9$.

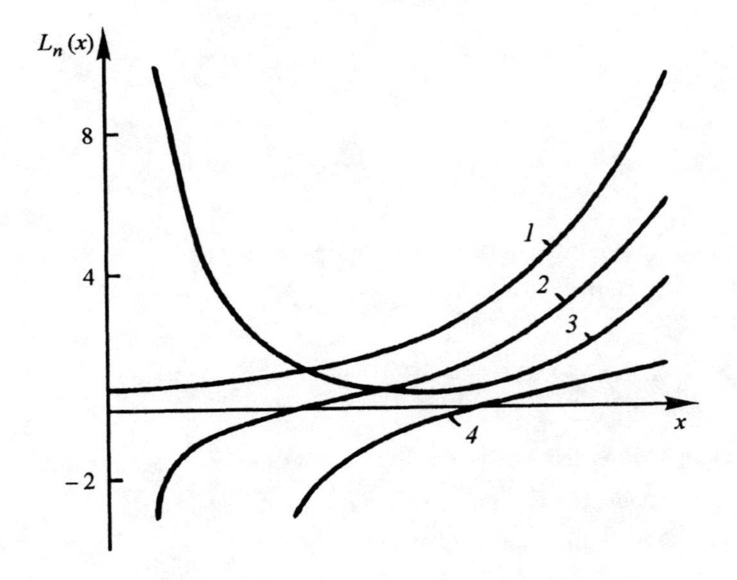

Figure 4.45. The modified Struve functions $L_n(x)$, $n = -1, -2, -3, -4$.

$$L_{\nu-1}(z)+L_{\nu+1}(z)=2L'_{\nu}-\frac{(z/2)^{\nu}}{\sqrt{\pi}\,\Gamma(\nu+3/2)}.$$

The asymptotic formula for large $|z|$ is

$$L_{\nu}(z)-I_{-\nu}(z)\sim\frac{1}{\pi}\sum_{k=0}^{\infty}\frac{(-1)^{k+1}\Gamma(k+1/2)}{\Gamma(\nu+1/2-k)\,(z/2)^{2k-\nu+1}}\qquad(|\arg z|<\frac{\pi}{2}).$$

The Coulomb wave functions. The *Coulomb wave functions* F_L, G_L are particular solutions of the differential equation

$$\frac{d^2u}{d\rho^2}+\left[\frac{1-2\eta}{\rho}-\frac{L(L+1)}{\rho^2}\right]u=0\quad(\rho>0,\quad -\infty<\eta<+\infty),$$

where L is a nonnegative integer. The equation has one regular singularity with indices $L+1$ and $-L$ at $\rho=0$, and an irregular singularity at $\rho=\infty$.

The functions F_L and G_L can be defined in terms of the integral

$$F_L+iG_L=\frac{ie^{-i\rho}\,\rho^{-L}}{(2L+1)!C_L(\eta)}\int_0^{\infty}e^{-t}t^{L-i\eta}(t+2i\rho)^{L+i\eta}dt,$$

where

$$C_L(\eta)=\frac{2^Le^{-\pi\eta/2}|\Gamma(L+1+i\eta)|}{\Gamma(2L+2)}.$$

The main formulas for the functions are:

$$F_L(\eta,\rho)=C_L(\eta)\,\rho^{L+1}\,\Phi_L(\eta,\rho),$$

where

$$\Phi_L(\eta,\rho)=\sum_{k=L+1}^{\infty}A_k^L(\eta)\,\rho^{k-L-1},\quad A_{L+1}^L=1,\quad A_{L+2}^L=\frac{\eta}{L+1},$$

$$(k+L)(k-L-1)A_k^L=2\eta A_{k-1}^L-A_{k-2}^L\qquad(k>L+2),$$

$$F'_L=\frac{d}{d\rho}F_L(\eta,\rho)=C_L(\eta)\,\rho^L\,\Phi_L^*(\eta,\rho),$$

$$\Phi_L^*(\eta,\rho)=\sum_{k=L+1}^{\infty}kA_k^L(\eta)\,\rho^{k-L-1},$$

$$F_L(\eta,\rho)=C_L(\eta)\frac{(2L+1)!}{(2\eta)^{2L+1}}\rho^{-L}\sum_{k=2L+1}^{\infty}b_kt^{k/2}I_k(2\sqrt{t})\qquad(t=2\eta\rho,\quad\eta>0).$$

If $L > 0$, $\rho = 0$, we have:

$$F_L = 0, \quad F_L' = 0, \quad G_L = \infty, \quad G_L' = -\infty,$$

if $L = 0$, $\rho = 0$,

$$F_0 = 0, \quad F_0' = G_0(\eta), \quad G_0 = \frac{1}{C_0(\eta)}, \quad G_0' = -\infty,$$

if $L = 0$, $\eta = 0$,

$$F_0 = \sin\rho, \quad F_0' = \cos\rho, \quad G_0 = \cos\rho, \quad G_0' = -\sin\rho,$$

and, as $L \to \infty$,

$$C_L(\eta) \sim \frac{2_L L!}{(2L+1)!} e^{-\pi\eta/2}$$

(Figure 4.46).

4.8. Hypergeometric functions

The hypergeometric equation has the form

$$z(1-z)\frac{d^2u}{dz^2} + [c - (a+b+1)z]\frac{du}{dz} - ab \cdot u = 0,$$

where a,b,c are parameters that may be complex. The function

$$F(a, b; c; z) = {}_2F_1(a, b; c; z) = \sum_{n=0}^{\infty} \frac{(a)_n (b)_n}{(c)_n} \frac{z^n}{n!} \quad (1)$$

is one of solutions of the equation, and it is regular at $z = 0$ if $c \neq 0, -1, -2, ..., n \in \mathbf{N}$.

The function $F(a, b; c; z)$ is called the *hypergeometric function* or *Gauss hypergeometric function*.

If none of the quantities a, b, $c-a$, or $c-b$ is a negative integer, then the function $F(a, b; c; z)$ is defined only for real values of the argument in the region $-\infty < x < 1$. If $c > a + b$, then $x = 1$ can be included in the domain. If $c = -n$, where $n = 0,1,2,...$, then the function

$$z^{n+1} \sum_{k=0}^{\infty} \frac{(a+n+1)_k (b+n+1)_k}{(n+2)_k} \frac{z^k}{k!} = z^{n+1} {}_2F_1(a+n+1, b+n+1, n+2, z)$$

is also a solution of the hypergeometric equation.

Series (1) defines the Gauss hypergeometric function $F(a, b; c; x)$ for $|x| < 1$. The transformation

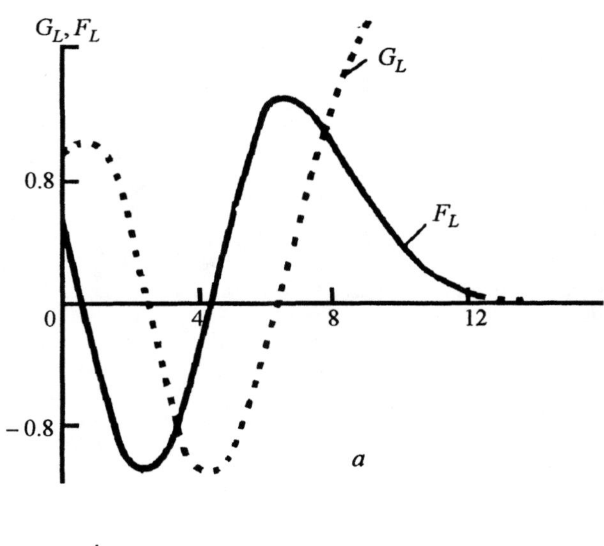

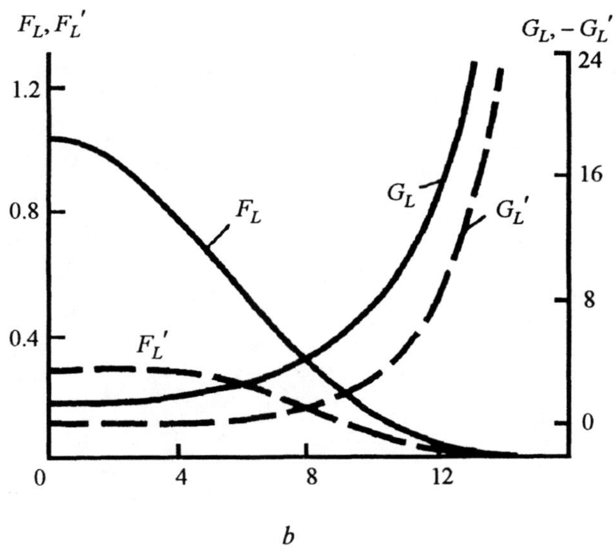

Figure 4.46. The Coulomb wave functions: (*a*) $F_L(\eta, \rho)$, $G_L(\eta, \rho)$, $\eta = 1$, $\rho = 10$; (*b*) F_L, F_L', G_L, G_L', $\eta = 10$, $\rho = 20$.

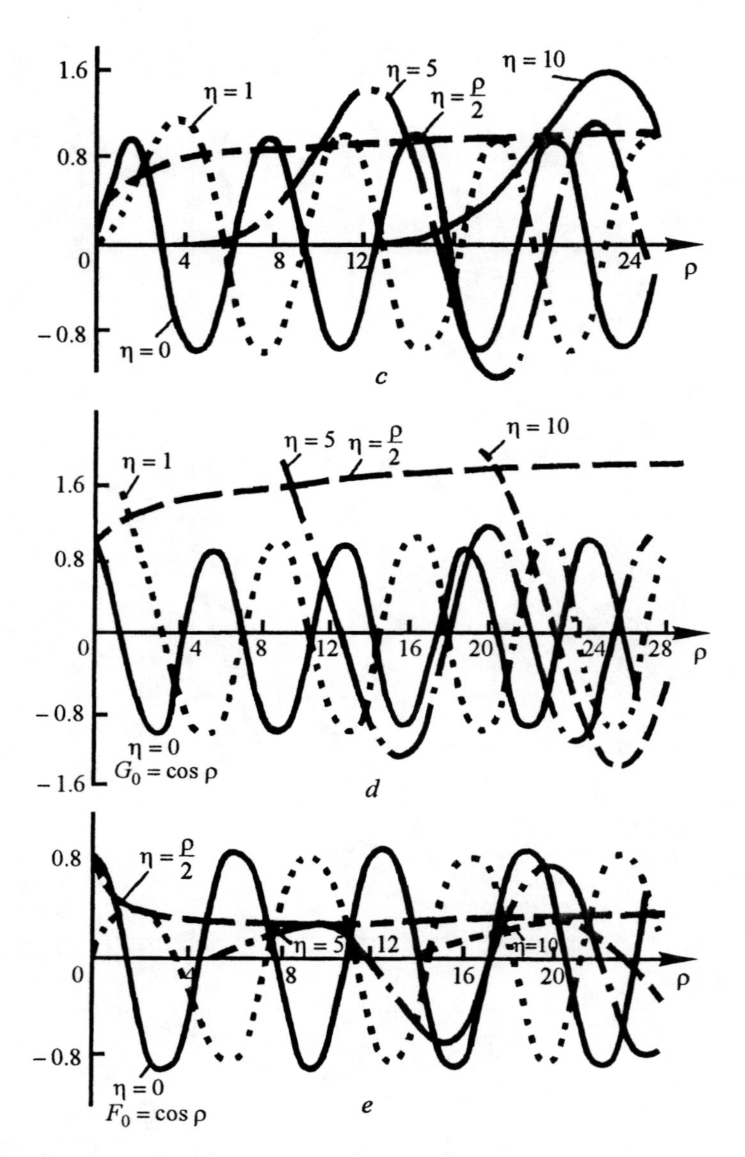

Figure 4.46. The Coulomb wave functions: (*c*) $F_0(\eta, \rho)$, $\eta = 0,1,5,10$; (*d*) $G_{0L}(\eta, \rho)$, $\eta = 0,1,5,10,\rho/2$; (*e*) $F_0'(\eta, \rho)$, $\eta = 0,1,5,10,\rho/2$.

$$F(a, b; c; x) = \frac{F\left(a, c-b; c; \frac{x}{x-1}\right)}{(1-x)^a} = \frac{F\left(c-a, b; c; \frac{x}{x-1}\right)}{(1-x)^b}$$

continues the function $F(a, b; c; x)$ to the interval $-\infty < x < 1/2$. In some cases, the Gauss hypergeometric function is defined by the integrals:

$$\frac{\Gamma(c)}{\Gamma(b)\Gamma(c-b)} \int_0^1 \frac{t^{b-1}(1-t)^{c-b-1}}{(1-xt)^a} dt \qquad (c > b > 0, \ x < 1),$$

$$\frac{\Gamma(c)}{\Gamma(b)\Gamma(c-b)} \int_1^\infty \frac{t^{a-c}(t-1)^{c-b-1}}{(t-x)^a} dt \qquad (1 + a > c > b, \ x < 1),$$

$$\frac{\Gamma(c)}{\Gamma(b)\Gamma(c-b)} \int_0^\infty \frac{t^{b-1}(1+t)^{a-c}}{(1+x-xt)^a} dt \qquad (c > b > 0, \ x < 1).$$

Special cases are:

$$F(a, b; c; 0) = 1,$$

$$F(a, b; c; 1) = \frac{\Gamma(c)\,\Gamma(c-a-b)}{\Gamma(c-a)\,\Gamma(c-b)} \qquad (c > a + b),$$

$$F(a, b; c; -\infty) = \begin{cases} 0, & a < 0, \ b < 0, \\ 1, & a \text{ (or } b) = 0, \\ \infty, & a \text{ (or } b) > 0, \end{cases} \qquad (b \text{ (or } a) < 0),$$

$$F(a, b; c; -\infty) = \frac{\sqrt{\pi}\,\Gamma(1+a-b)}{2^a\,\Gamma(1+\frac{a}{2}-b)\,\Gamma\left(\frac{1+a}{2}\right)} \qquad (a - b \neq -1, -2, -3, \ldots),$$

$$F\left(a, b; \frac{1+a+b}{2}; \frac{1}{2}\right) = \frac{\sqrt{\pi}\,\Gamma\left(\frac{1+a+b}{2}\right)}{\Gamma\left(\frac{1+a}{2}\right)\Gamma\left(\frac{1+b}{2}\right)} \qquad (a + b \neq -1, -3, -5, \ldots),$$

$$F\left(a, 1-a; c; \frac{1}{2}\right) = \frac{2\sqrt{\pi}\,\Gamma(c)}{2^c\,\Gamma\left(\frac{a+c}{2}\right)\Gamma\left(\frac{1+c-a}{2}\right)} \qquad (c \neq 0, -1, -2, \ldots),$$

$$F(1, 1; 2; z) = -\frac{1}{z}\ln(1-z), \quad F\left(\frac{1}{2}, 1; \frac{3}{2}; z^2\right) = \frac{1}{2z}\ln\left(\frac{1+z}{1-z}\right),$$

$$F\left(\frac{1}{2}, 1; \frac{3}{2}; -z^2\right) = \frac{\arctan z}{z},$$

$$F\left(\frac{1}{2}, \frac{1}{2}; \frac{3}{2}; z^2\right) = (1-z^2)^{1/2} F(1, 1; \frac{3}{2}; z^2) = \frac{\arcsin z}{z},$$

$$F(a, b; b; z) = (1-z)^{-a}, \quad F(-a, a; \frac{1}{2}; \sin^2 z) = \cos(2az).$$

Degenerate (confluent) hypergeometric functions. The *Kummer function* $M(a; c; z)$ (sometimes denoted by $\Phi(a; c; z)$) is one of solutions of the confluent hypergeometric equation

$$a\frac{d^2u}{dz^2}+(c-z)\frac{du}{dz}-au=0,$$

where a,c are parameters. If $c\neq 0,-1,-2,\ldots$, the function $M(a;c;z)$ has the form

$$M(a;c;z)=\sum_{n=0}^{\infty}\frac{(a)_n}{(c)_n}\frac{z^n}{n!}.$$

If a and c are fixed, then the function $M(a;c;z)$ is an entire function of z.

For real a and c and real values of the variable $z=x$, the function is real-valued (Figures 4.47 – 4.50).

The function $M(a;c;z)$ can be obtained from the Gauss hypergeometric function by passing to the limit,

$$M(a;c;z)=\lim_{b\to\infty}F\left(a,b;c;\frac{z}{b}\right).$$

The function $M(a;c;z)$ has a number of integral representations that can be taken as a definition of the function. For example,

$$M(a;c;x)=\frac{\Gamma(c)x^{1-c}}{\Gamma(c-a)\Gamma(a)}\int_0^x\frac{t^{a-1}e^t}{(x-t)^{1+a-c}}\,dt\qquad(0<a<c),$$

$$M(a;c;z)=\frac{\Gamma(c)}{\Gamma(c-a)\Gamma(a)}\int_0^1\frac{t^{a-1}e^{zt}}{(1-t)^{1+a-c}}\,dt\qquad(\operatorname{Re}c>\operatorname{Re}a>0),$$

$$M(a;c;x)=\frac{2^{1-c}e^{x/2}\Gamma(c)}{\Gamma(c-a)\Gamma(a)}\int_{-1}^1\frac{(t+1)^{a-1}e^{xt/2}}{(1-t)^{1+a-c}}\,dt\qquad(0<a<c).$$

Main formulas for the function are:

$$M(a;a;x)=e^x,\qquad M(a;1+a;x)=\frac{a}{(-x)^a}\gamma(a;-x),$$

$$M(a;2a;x)=\Gamma\left(\frac{1}{2}+a\right)e^{x/2}\left(\frac{x}{4}\right)^{\frac{1}{2}-a}I_{a-1/2}\left(\frac{x}{2}\right),$$

$$M(1;c;x)=1+x^{1-c}e^x\gamma(c;x),$$

$$\lim_{a\to\infty}M(a;c+1;\frac{-x}{a})=\frac{\Gamma(1+c)}{x^{c/2}}J_c(2\sqrt{x})\quad(x\geq 0),$$

$$\lim_{a\to\infty}M(a;c+1;\frac{x}{a})=\frac{\Gamma(1+c)}{x^{c/2}}I_c(2\sqrt{x})\quad(x\geq 0),$$

$$M(a;c;-x)=e^{-x}M(c-a;c;x),$$

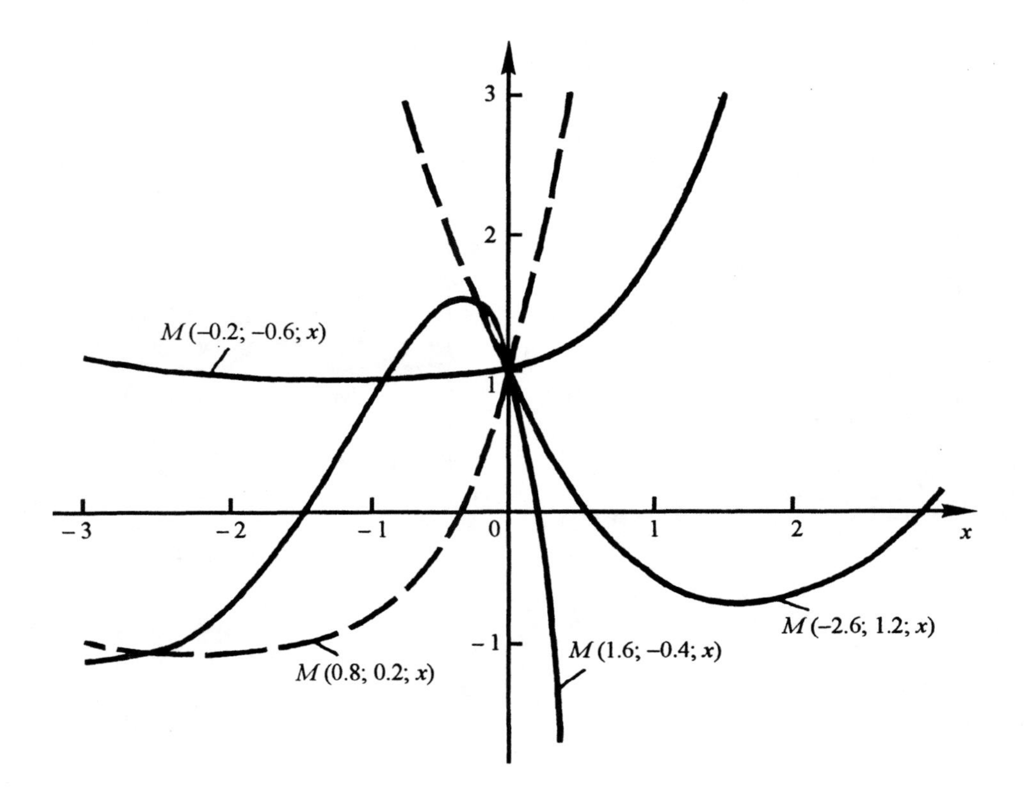

Figure 4.47. The Kummer function $M(a; c; x)$.

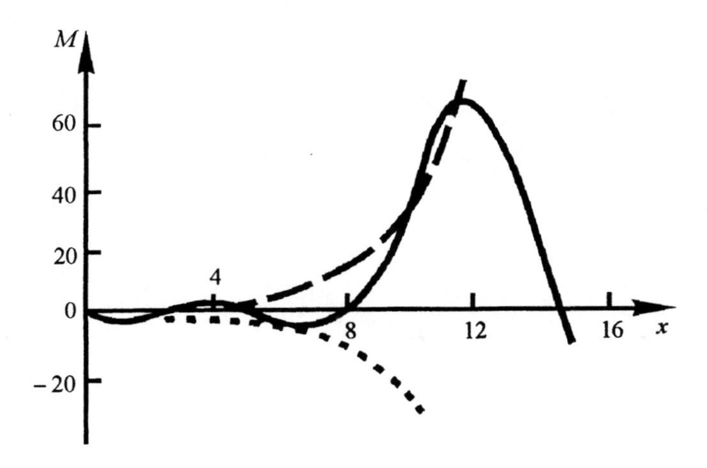

Figure 4.48. The function $M(-4.5; 1; x)$.

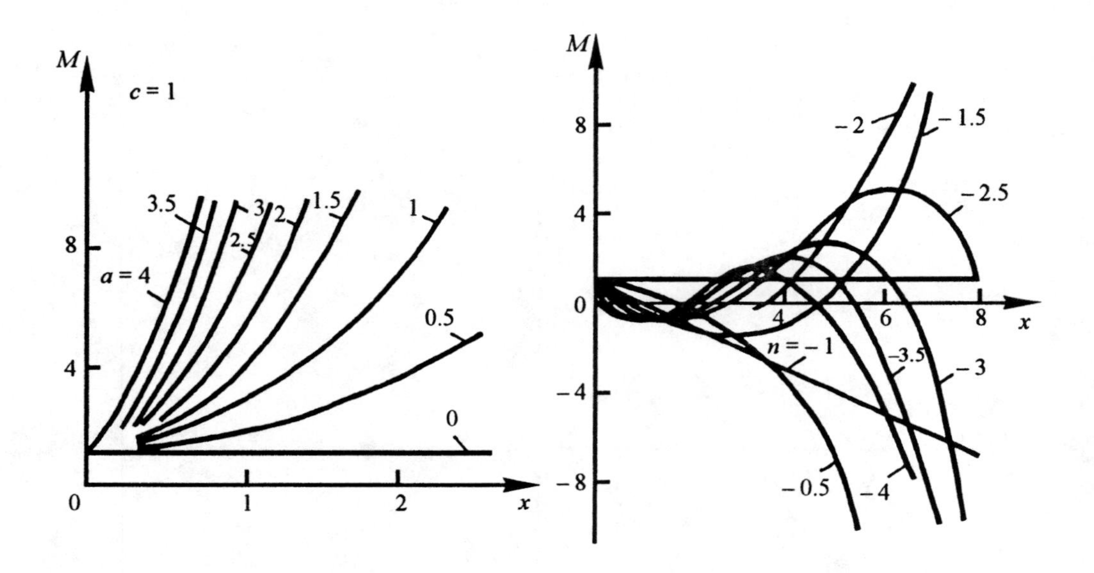

Figure 4.49. The function $M(a;\ 1;\ x)$.

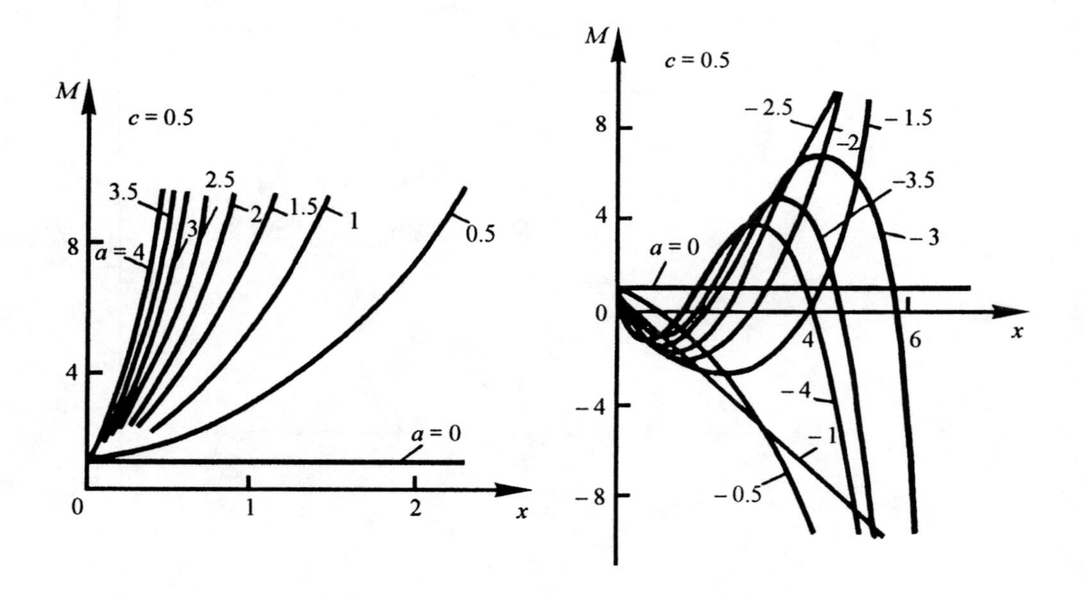

Figure 4.50. The function $M(a;\ 0.5;\ x)$.

$$M(a;\,c;\,0) = 1, \qquad M(a;\,c;\,x) \approx 1 + \frac{ax}{c},$$

$$M(a;\,c;\,x) \approx \frac{\Gamma(c)}{\sqrt{\pi}}\left(\frac{cx}{2} - ax\right)^{\frac{1-2c}{4}} e^{x/2} \cos\left(\frac{\pi}{4} - \frac{c\pi}{2} + \sqrt{2cx - 4ax}\right).$$

We also have that, for large negative a,

$$M(a;\,c;\,x) \approx \Gamma(c)\left(ax - \frac{cx}{2}\right)^{\frac{1-c}{2}} e^{x/2}\, I_{c-1}(\sqrt{4ax - 2cax}\,),$$

for large positive a,

$$M(a;\,c;\,x) \approx \left(1 - \frac{x}{c}\right)^{-a},$$

for large positive or negative c,

$$M(a;\,c;\,c) \approx \frac{\Gamma(c)}{\Gamma(c-a)}\,(-x)^{a},$$

for large negative x and $c - a \neq 0, -1, -2, \ldots$

$$M(a;\,c;\,x) \approx \frac{\Gamma(c)}{\Gamma(a)}\, x^{a-c}\, e^{x}.$$

Tricomi function U(a; c; z). A linearly independent solution of the confluent hypergeometric equation is the Tricomi function U(a; c; z) (it is sometimes denoted by $\Psi(a;\,c;\,z)$).

The *Tricomi function* is defined as the limit of the Gauss hypergeometric function,

$$\mathrm{U}(a;\,c;\,z) = z^{-a} \lim_{\gamma \to \infty} F\left(a,\, 1 + a - c;\, \gamma;\, 1 - \frac{\gamma}{z}\right),$$

or in terms of the integral

$$\mathrm{U}(a;\,c;\,z) = \frac{1}{\Gamma(a)} \int_0^\infty \frac{t^{a-1}\, e^{-zt}\, dt}{(1+t)^{1+a-c}} \qquad (\operatorname{Re} c > \operatorname{Re} a > 0),$$

or else in terms of the Kummer function,

$$\mathrm{U}(a;\,c;\,z) = \frac{\Gamma(1-c)}{\Gamma(1+a-c)} M(a;\,c;\,z) + \frac{\Gamma(c-1)}{\Gamma(a)\, z^{c-1}} M(1 + a - c;\, 2 - c;\, z),$$

if c is not an integer.

A more general definition of the Tricomi function is given by

$$\mathrm{U}(a;\,n+1;\,z) = \frac{(n-1)!}{\Gamma(a)\, z^n} \sum_{k=0}^{n-1} \frac{(a-n)_k\, z^k}{(1-n)_k\, k!} - \frac{(-1)^n}{n!\,\Gamma(a-n)} \times$$

$$\times \left\{ \ln z \cdot M(a; n+1; z) + \sum_{k=0}^{\infty} \frac{(a)_k z^k}{(1+n)_k k!} [\Psi(a+k) - \Psi(1+k) - \Psi(1+n+k)] \right\} \quad (n = 0, 1, 2, \ldots).$$

For negative values of the argument, we have

$$\mathrm{U}(a; c; -z) = \frac{\Gamma(1-c)}{\Gamma(b)} M(a; c; -z) - \frac{(-1)^c \Gamma(c-1)}{\Gamma(a) z^{c-1}} M(b; 2-c; -z) \quad (z > 0),$$

where $b = 1 + a - c$, $(-1)^c = \cos \pi c + i \sin \pi c$.

For fixed a and c, the function $\mathrm{U}(a; c; z)$ is a many-valued analytic function of z with a unique finite branching point at $z = 0$. For real positive $z = x > 0$, the function has a real branch.

Whittaker functions $M_{k,\mu}(z)$, $W_{k,\mu}(z)$. The *Whittaker functions* are solutions of the equation

$$\frac{d^2u}{dz^2} + \left[-\frac{1}{4} + \frac{k}{z} + \frac{\frac{1}{4} - \mu^2}{z^2} \right] u = 0,$$

$$M_{k,\mu}(z) = e^{-z/2} z^{1/2+\mu} M\left(\frac{1}{2} + \mu - k; 1 + 2\mu; z\right),$$

$$M_{k,\mu}(z) = e^{-z/2} z^{1/2+\mu} \mathrm{U}\left(\frac{1}{2} + \mu - k; 1 + 2\mu; z\right),$$

$$(-\pi < \arg z \le \pi, \quad k = c/2 - a, \quad \mu = c/2 - 1).$$

For such a choice of the parameters, the Whittaker functions look fairly symmetric,

$$M_{k,\mu}(-z) = (-1)^{1/2+\mu} M_{-k,\mu}(z) = (-\sin \mu\pi + i \cos \mu\pi) M_{-k,\mu}(z),$$

$$W_{k,\mu}(z) = W_{k-\mu}(z),$$

$$W_{k,\mu}(z) = \frac{\Gamma(-2\mu)}{\Gamma\left(\frac{1}{2} - \mu - k\right)} M_{k,\mu}(z) + \frac{\Gamma(2\mu)}{\Gamma\left(\frac{1}{2} + \mu - k\right)} M_{k,-\mu}(z).$$

Functions of the paraboloid of revolution are solutions of the degenerate hypergeometric equation

$$\frac{d^2u}{d\xi^2} + \frac{1}{\xi} \frac{du}{d\xi} + (4\xi^2 - \frac{p^2}{\xi^2} - 4\tau) u = 0,$$

where $\tau, \xi \in \mathbf{R}$.

Parabolical cylinder functions $D_\nu(z)$. Solutions of the differential Weber equation

$$\frac{d^2u}{dz^2} + (\nu + \frac{1}{2} - \frac{z^2}{4}) u = 0$$

are called *parabolical cylinder functions*, $D_\nu(z)$, (sometimes *Weber–Hermite functions*, or *Weber functions*).

The function $D_\nu(z)$ can be defined as one of the integrals:

$$D_\nu(z) = \sqrt{\frac{2}{\pi}}\, e^{z^2/4} \int_0^\infty e^{-t^2/2}\, t^\nu \cos\left(zt - \frac{\nu\pi}{2}\right) dt \qquad (\operatorname{Re}\nu > -1),$$

$$D_\nu(z) = \frac{2^{\nu/2}}{\Gamma(-\nu/2)}\, e^{-z^2/4} \int_0^\infty e^{-tz^2/2}\, t^{-1-\nu/2} (1+t)^{(\nu-1)/2}\, dt \qquad \left(\operatorname{Re}\nu < 0, \quad |\arg z| \le \frac{\pi}{4}\right),$$

$$D_\nu(z) = \frac{2^{(\nu-1)/2}\, z}{\Gamma((1-\nu)/2)}\, e^{-z^2/4} \int_0^\infty e^{-tz^2/4}\, t^{(1+\nu)/2} (1+t)^{\nu/2}\, dt \qquad \left(\operatorname{Re}\nu < 1, \quad |\arg z| \le \frac{\pi}{4}\right),$$

$$D_\nu(z) = \frac{e^{-z^2/4}}{\Gamma(-\nu)} \int_0^\infty e^{-zt - t^2/2}\, t^{-\nu-1}\, dt \qquad (\operatorname{Re}\nu < 0).$$

The parabolical cylinder functions are entire functions of z. If $\nu = n$ is a nonnegative integer, then $e^{z^2/4} D_n(z) = \frac{1}{\sqrt{2^n}} H\left(\frac{z}{\sqrt{2}}\right)$ is a polynomial (here $H_n(z)$ is a Hermite polynomial). If ν is not an integer, then $D_\nu(z)$ and $D_\nu(-z)$ are linearly independent. The functions $D_\nu(z)$ and $D_{-\nu-1}(iz)$ are also linearly independent.

If ν and z are real, then the value of $D_\nu(x)$ is real. For large ν, the function $D_\nu(x)$ has zeros, maximums, and minimums in the interval $-\infty < x < +\infty$ (Figure 4.51). For even (odd) nonnegative ν, the function $D_\nu(x)$ is even (odd).

Special cases are:

$$D_0(x) = e^{-x^2/4}; \qquad D_1(x) = x e^{-x^2/4},$$

$$D_{-1}(x) = \sqrt{\frac{\pi}{2}}\, e^{x^2/4} \operatorname{erfc}\left(\frac{x}{\sqrt{2}}\right),$$

$$D_{-n-1}(x) = \frac{(-1)^n}{n!} \sqrt{\frac{\pi}{2}}\, e^{-x^2/4} \frac{d^n}{dx^n}\left[e^{x^2/2} \operatorname{erfc}\left(\frac{x}{\sqrt{2}}\right)\right] \qquad (n = 0, 1, 2, \ldots)$$

or

$$D_{-n-1}(x) = \sqrt{\frac{\pi}{2}}\, 2^{n/2} e^{x^2/4}\, i^n \operatorname{erfc}\left(\frac{x}{\sqrt{2}}\right) \qquad n = 0, 1, 2, \ldots,$$

$$D_{1/2}(x) = \sqrt{\frac{x}{2\pi}}\, K_{1/4}\left(\frac{x^2}{4}\right), \qquad x > 0$$

$$D_\nu(z) = 2^{(\nu-1)/2} e^{-z^2/4}\, z\, \mathrm{U}\left(\frac{1-\nu}{2}; \frac{3}{2}; \frac{z^2}{2}\right),$$

$$D_\nu(z) = 2^{(\nu+1/2)/2} z^{-1/2}\, W_{(\nu+1/2)/2,\,-1/4}\left(\frac{z^2}{2}\right).$$

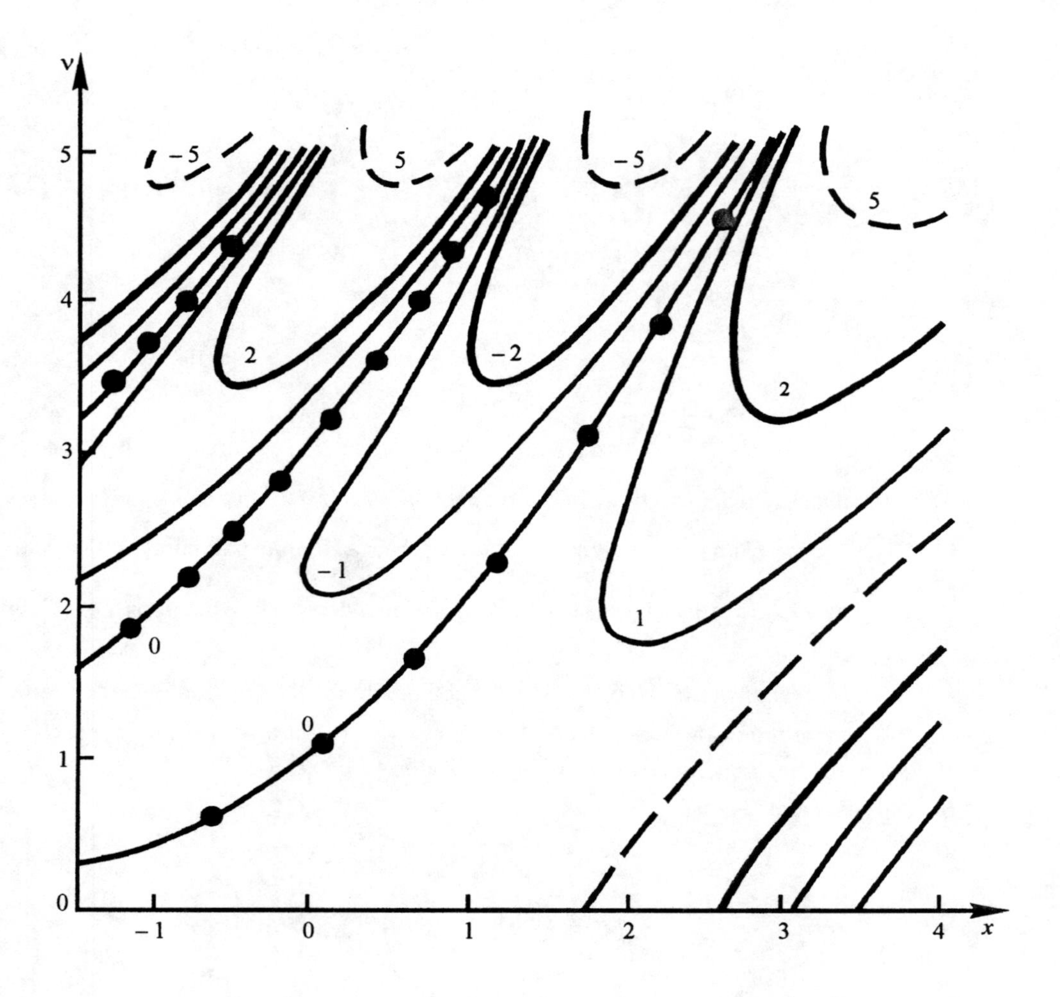

Figure 4.51. The parabolical cylinder functions $D_\nu(x)$.

4.9. Orthogonal polynomials

4.9.1. General properties of orthogonal polynomials

Let an interval $[a; b]$ be given. A system of polynomials $p_n(x)$ (n is degree of the polynomial) is called *orthogonal* on $[a; b]$ with a weight function $\omega(x) \geq 0$, $x \in [a; b]$, if

$$\int_a^b \omega(x)\, p_n(x)\, p_m(x) dx = h_n \delta_{mn},$$

where δ_{mn} is the Kronecker δ-symbol,

$$\delta_{mn} = \begin{cases} 0, & m \neq n, \\ 1, & m = n \quad (n, m = 0, 1, 2, \ldots), \end{cases}$$

$$p_n(x) = k_n x^n + k_{n-1} x^{n-1} k_{n-2} x^{n-2} + \ldots + k_0 \ (n = 0, 1, \ldots).$$

The differential equation satisfied by the polynomials is $g_2(x)p_n''(x) + g_1(x)p_n'(x) + a_n(x) = 0$, where the functions $g_i(x)$, $i = 1,2$, do not depend on n, and a_n are constants depending only on n. The polynomials also satisfy the recurrence formula, $p_{n+1} = (a_n + xb_n)p_n - c_n p_{n-1}$, where

$$b_n = \frac{k_{n+1}}{k_n}, \quad a_n = b_n\left(\frac{k'_{n+1}}{k_{n+1}} - \frac{k'_n}{k_n}\right), \quad c_n = \frac{k_{n+1} k_{n-1} h_n}{h_{n-1} k_n^2}.$$

The following Rodriguez formula holds:

$$p_n = \frac{1}{e_n \omega(x)} \frac{d^n}{dx^n} \{\omega(x) [g(x)]^n\},$$

where $g(x)$ is a polynomial independent of n.

The system $\left\{\frac{dp_n(x)}{dx}\right\}$ is also a system of orthogonal polynomials.

Let $f^2(x)$ be a Lebesgue integrable function on $[a; b]$, and assume that a sequence $\{P_n(x)\}$ is summable. The function $\varphi_n(x) = \sum_{k=0}^{n} b_k p_n$ is called the nth order approximation of the function $f(x)$. The accuracy of such an approximation is given by the value of the integral

$$\int_a^b \left[f(x) - \sum_{k=0}^{n} b_k p_k(x) \right]^2 \omega(x) dx.$$

If this integral achieves a minimal value, then φ_n is called the best quadratic estimate of the function $f(x)$.

Among all n order approximations of the function $f(x)$, the best in the sense of square means is an approximation such that $b_k = a_k$, $k = 0,1,2,...$, where

$$a_k = \frac{1}{h_k}\int_a^b f(x)p_n(x)\omega\,(x)dx.$$

The approximation $\varphi(x) = \sum_{k=0}^{\infty} a_k$ is called the Fourier expansion of the function $f(x)$ on $[a; b]$ with respect to the system of orthogonal polynomials $\{p_n(x)\}$. If

$$\varphi_n(x) = \sum_{k=0}^{n} a_k\, p_n(x),$$

then the difference $\varphi(x) - \varphi_n(x)$ has at least $(n-1)$ zeros on $[a; b]$. Zeros of orthogonal polynomials are mutually simple and lie in $[a; b]$.

4.9.2. Jacobi polynomials $P_n^{(\alpha,\beta)}(x)$ (Figure 4.52)

The differential equation for the Jacobi polynomials $P_n^{(\alpha,\beta)}(x)$ has the form

$$(1-x^2)\,y'' + [(\beta-\alpha) - (\lambda+1)x]\,y' + n(n+\lambda)\,y = 0,$$

where $\alpha > -1$, $\beta > -1$, $\lambda = \alpha + \beta + 1$. The Rodrigues formula has the form

$$2^n n!\; P_n^{(\alpha,\,\beta)}(x) = (-1)^n(1-x)^{-\alpha}(1+x)^{-\beta}\frac{d^n}{dx^n}\left[(1-x)^{\alpha+n}\,(1+x)^{\beta+n}\right].$$

The recurrence formulas for Jacobi polynomials are:

$$\begin{aligned}
&2(n+1)(n+\alpha+\beta+1)(2n+\alpha+\beta)P_{n+1}^{(\alpha,\,\beta)}(x) = (2n+\alpha+\beta+1)\times\\
&\times\left[(2n+\alpha+\beta)(2n+\alpha+\beta+2)x+\alpha^2-\beta^2\right]P_n^{(\alpha,\,\beta)}(x) -\\
&-2(n+\alpha)(n+\beta)(2n+\alpha+\beta+2)\,P_{n-1}^{(\alpha,\,\beta)}(x),\\
&(1-x)\,P_n^{(\alpha+1,\,\beta)}(x) + (1+x)\,P_n^{(\alpha,\,\beta+1)}(x) = 2P_n^{(\alpha,\,\beta)}(x),\\
&P_n^{(\alpha,\,\beta-1)}(x) - P_n^{(\alpha-1,\,\beta)}(x) = P_{n-1}^{(\alpha,\,\beta)}(x).
\end{aligned}$$

Special cases are:

$$P_0^{(\alpha,\,\beta)}(x) = 1,\quad P_1^{(\alpha,\,\beta)}(x) = \frac{1}{2}\left(\alpha-\beta+\frac{1}{2}\right)+\frac{1}{2}(\lambda+1)x,$$

$$\left| P_n^{(\alpha,\beta)}(x) \right| \le \begin{cases} \binom{n+q}{n}, & q = \max(\alpha,\beta) \ge -\frac{1}{2} \\ \left| P_n^{(\alpha,\beta)}(\tilde{x}) \right| \approx \sqrt{\frac{1}{n}}, & q < -\frac{1}{2}, \end{cases}$$

where $\tilde{x}$ is the point of maximum closest to $\frac{\beta-\alpha}{\alpha+\beta+1}$. We also have that

$$\lim_{n\to\infty} \left[\frac{1}{n^\alpha} P_n^{(\alpha,\beta)}\left(\cos\frac{x}{n}\right) \right] = \lim_{n\to\infty} \frac{1}{n^\alpha} P_n^{(\alpha,\beta)}\left(1 - \frac{x^2}{2n^2}\right) = \left(\frac{2}{x}\right)^\alpha J_\alpha(x),$$

$$P_n^{(\alpha,\beta)}(1) = \binom{n+\alpha}{n} = \frac{(\alpha+1)_n}{n!}; \quad P_n^{(\alpha,\beta)}(-1) = \frac{(-1)^n(\beta+1)_n}{n!},$$

$$P_n^{(\alpha,\beta)}(-x) = (-1)^n P_n^{(\beta,\alpha)}(x).$$

Jacobi polynomials are expressed in terms of the hypergeometric function,

$$P_n^{(\alpha,\beta)}(x) = \binom{n+\alpha}{n} F\left(-n, n+\alpha+\beta+1; \alpha+1; \frac{1-x}{2}\right) =$$

$$= (-1)^n \binom{n+\beta}{n} F\left(-n, n+\alpha+\beta+1; \beta+1; \frac{1+x}{2}\right) =$$

$$= \binom{n+\alpha}{n} \left(\frac{x+1}{2}\right)^n F\left(-n, -n-\beta; \alpha+1; \frac{x-1}{x+1}\right) =$$

$$= \binom{n+\beta}{n} \left(\frac{x-1}{2}\right)^n F\left(-n, -n-\alpha; \beta+1; \frac{x+1}{x-1}\right).$$

These formulas give a differentiation formula,

$$2^m \frac{d^m}{dx^m}\left[P_n^{(\alpha,\beta)}(x)\right] = (n+\alpha+\beta+1)_m P_{n-m}^{(\alpha+m,\beta+m)}(x) \quad (m = 1, 2, \dots, n).$$

Orthogonality relations have the form:

$$\int_{-1}^{1} (1-x)^\alpha (1+x)^\beta P_n^{(\alpha,\beta)}(x) P_m^{(\alpha,\beta)}(x) = \frac{2^\lambda \Gamma(n+\alpha+1)\Gamma(n+\beta+1)}{(2n+\lambda)\Gamma(n+\lambda)\, n!} \delta_{mn}.$$

The following power series expansions hold:

$$P_n^{(\alpha,\beta)}(x) = 2^{-n} \sum_{k=0}^{n} \binom{n+\alpha}{k} \binom{n+\beta}{n-k} (x-1)^{n-k}(x+1)^k;$$

$$P_n^{(\alpha,\beta)} = \sum_{r=0}^{n} k_{r,n}^{(\alpha,\beta)} x^r,$$

where

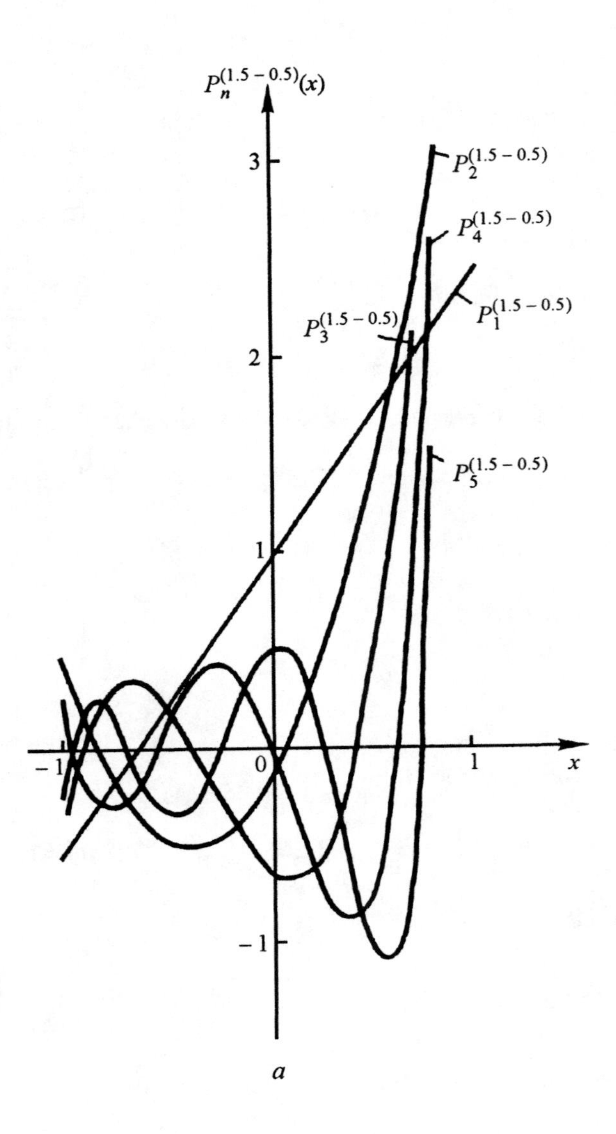

Figure 4.52. The Jacobi polynomials $P_n^{(\alpha,\beta)}(x)$: (*a*) $\alpha = 1.5$, $\beta = -.5$, $n = 1,2,3,4,5$.

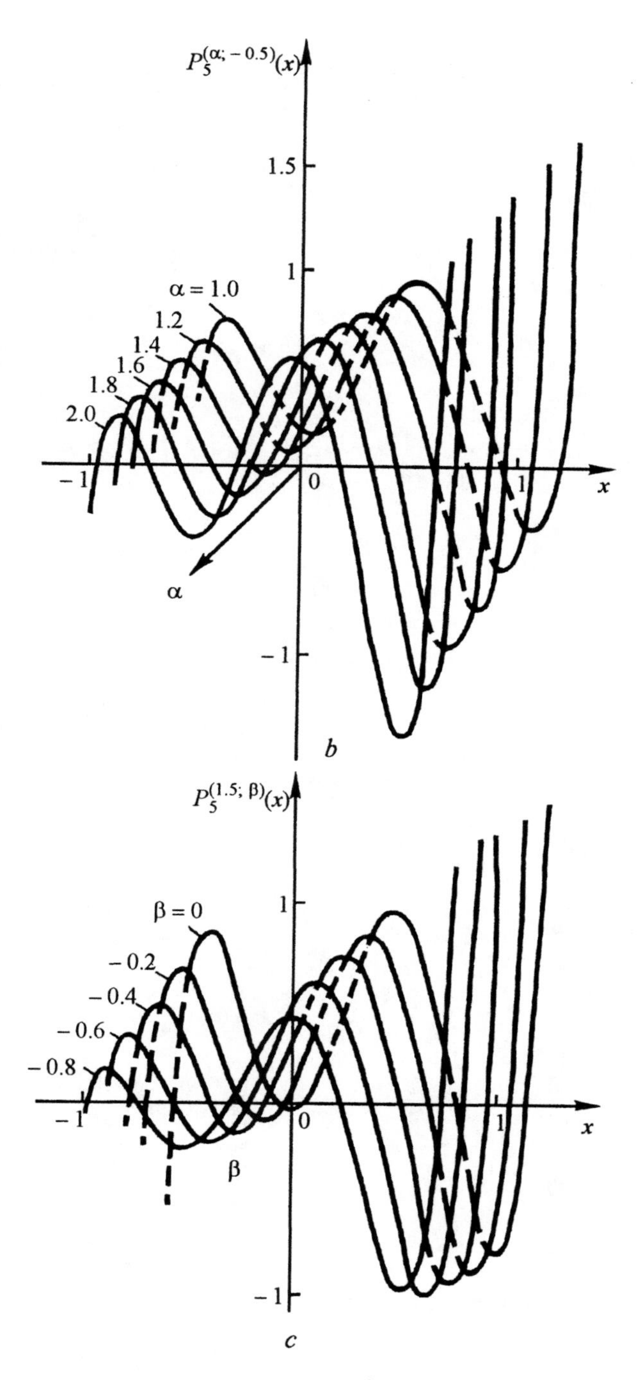

Figure 4.52. The Jacobi polynomials $P_n^{(\alpha,\beta)}(x)$: (*b*) $\alpha = 1.0,1.2,1.4,1.6,1.8,2$, $\beta = -.5$, $n = 5$; (*c*) $\alpha = -.5$; $\beta = 0, -.2,.4, -.6, -.8$, $n = 5$.

$$k_{r,n}^{(\alpha,\beta)} = \frac{(-1)^r(\alpha+1)_n(-n)_r(n+\lambda)_r}{r!(\alpha+1)_r 2^r n!} F\left(r-n, n+\lambda+r; \alpha+1+r; \frac{1}{2}\right) =$$

$$= \frac{(-1)^n(\beta+1)_n(-n)_r(n+\lambda)_r}{r!(\beta+1)_r 2^r n!} F\left(r-n, n+\lambda+r; \beta+1+r; \frac{1}{2}\right).$$

The following are decomposition formulas for some functions with respect to Jacobi polynomials:

$$x^m = \sum_{n=0}^{m} \alpha_n P_n^{(\alpha,\beta)}(x),$$

$$\alpha_n = \frac{2^n m!\,\Gamma(n+\lambda)}{(m-n)!\,\Gamma(2n+\lambda)} F(n-m, \alpha+n+1; 2n+\lambda+1; 2),$$

$$(1-x)^m = \sum_{n=0}^{m} c_n P_n^{(\alpha,\beta)}(x),$$

$$c_n = \frac{(-1)^n 2^m m!\,\Gamma(\alpha+m+1)(2n+\lambda)\,\Gamma(n+\lambda)}{(m-n)!\,\Gamma(\alpha+n+1)\,\Gamma(m+n+\lambda+1)}.$$

Translated Jacobi polynomials are:

$$R_n^{(\alpha,\beta)}(x) = P_n^{(\alpha,\beta)}(2x-1), \quad P_n^{(\alpha,\beta)}(x) = R_n^{(\alpha,\beta)}\left(\frac{1+x}{2}\right),$$

$$R_n^{(\alpha,\beta)}(x) = \frac{(\alpha+1)_n}{n!} F(-n, n+\lambda; \alpha+1-x) =$$

$$= (-1)^n \frac{(\beta+1)_n}{n!} F(-n, n+\lambda; \beta+1; x),$$

$$R_n^{(\alpha,\beta)}(1-x) = (-1)^n R_n^{(\beta,\alpha)}(x),$$

$$R_n^{(\alpha,\beta)}(1) = \frac{(\alpha+1)_n}{n!}, \quad R_n^{(\alpha,\beta)}(0) = (-1)^n \frac{(\beta+1)_n}{n!},$$

$$R_n^{(\alpha,\beta)}(x) = \frac{(-1)^n}{n!}(1-x)^{-\alpha} x^{-\beta} \frac{d^n}{dx^n}\left[(1-x)^{\alpha+n} x^{\beta+n}\right].$$

4.9.3. Jacobi functions of the second kind

The Jacobi functions of the second kind, $Q_n^{(\alpha,\beta)}(x)$, satisfy the same equation as the Jacobi polynomials $P_n^{(\alpha,\beta)}$. The Jacobi functions of the second kind are not polynomials; however, they satisfy the same differentiation formula except for the case $n = 0$. If $\mathrm{Re}\,(\alpha+\beta) > -n-1$, the functions $Q_n^{(\alpha,\beta)}(x)$ vanish at infinity.

There are many identities that relate Jacobi polynomials and Jacobi functions of the second kind. For example,

$$Q_n^{(\alpha,\beta)}(x) = -\frac{\pi}{2\sin\alpha\pi} P_n^{(\alpha,\beta)}(x) + 2^{(\alpha+\beta-1)}\frac{\Gamma(\alpha)\Gamma(n+\beta+1)}{\Gamma(n+\alpha+\beta+1)} \times$$

$$\times (x-1)^{-\alpha}(x+1)^{-\beta} F\left(n+1, -n-\alpha-\beta; 1-\alpha; \frac{1-x}{2}\right),$$

$$Q_n^{(\alpha,\beta)}(x) = \frac{1}{2(x-1)^{\alpha}(x+1)^{\beta}} \int_{-1}^{1} \frac{(1-t)^{\alpha}(1+t)^{\beta}}{x-t} P_n^{(\alpha,\beta)}(t)\, dt.$$

The integral relation is valid for all points of the complex plane cut along the interval $[-1; 1]$. Here the function $Q_n^{(\alpha,\beta)}(x)$ approaches different values that depend on whether the point x tends to the point ξ in the cut from the upper part of the plane, $(\xi + i0)$, or the lower, $(\xi - i0)$.

The polynomials

$$q_n^{(\alpha,\beta)}(x) = \int_{-1}^{1} \frac{(1-t)^{\alpha}(1+t)^{\beta}}{t-x}\left[P_n^{(\alpha,\beta)}(t) - P_n^{(\alpha,\beta)}(x)\right] dt$$

are polynomials *associated with Jacobi polynomials* $P_n^{(\alpha,\beta)}(x)$.

4.9.4. Gegenbauer polynomials (ultraspherical polynomials)

These polynomials, denoted by $C^{\alpha}{}_n(x)$ or by $G_n^{\alpha}(x)$, are a special case of the Jacobi polynomials,

$$C_n^{\alpha+1/2}(x) = \frac{(2\alpha+1)_n}{(\alpha+1)_n} P_n^{(\alpha,\alpha)}(x) \quad \left(\alpha > -\frac{1}{2}\right).$$

They satisfy the differential equation $(1-x^2)y'' - (2\alpha+1)xy' + n(n+2\alpha)y = 0$. The orthogonality condition takes the form

$$\int_{-1}^{1} (1-x^2)\, x^{\alpha-1/2}\, C_n^{\alpha}(x)\, C_m^{\alpha}(x)\, dx = \frac{\pi 2^{1-2\alpha}\Gamma(n+2\pi)}{n!\,(n+\alpha)\,[\Gamma(\alpha)]^2}\,\delta_{mn}.$$

The recurrence formula and Rodrigues formula are, correspondingly,

$$(n+1)C_{n+1}^{\alpha}(x) = 2(\alpha+n)x\, C_n^{\alpha}(x) - (2\alpha+n-1)\, C_{n-1}^{\alpha}(x),$$

$$C_0^{\alpha}(x) = 1, \quad C_1^{\alpha}(x) = 2\alpha x,$$

and

$$C_n^\alpha(x) = (-1)^n (1-x^2)^{\frac{1}{2}-\alpha} \frac{\Gamma(\alpha+\frac{1}{2})\,\Gamma(n+2\alpha)}{2^n\, n!\,\Gamma(\alpha+n+\frac{1}{2})} \frac{d^n}{dx^n}\left[(1-x^2)^{\alpha+n-\frac{1}{2}}\right].$$

Special cases are:

$$C_n^\alpha(1) = \frac{(2\alpha)_n}{n!},\quad C_0^0(1) = 1,\quad C_n^0(1) = \frac{2}{n}\quad (n = 1, 2, \ldots),$$

$$C_n^0(x) = \lim_{\alpha\to\infty} \alpha^{-1}\, C_n^\alpha(x) = 2\,\frac{(n-1)!}{\left(\frac{1}{2}\right)_n} P_n^{(-\frac{1}{2},-\frac{1}{2})}(x),$$

$$C_n^\alpha(-x) = (-1)^n\, C_n^\alpha(x),$$

$$C_n^\alpha(\cos t) = \sum_{m=0}^{n} \frac{(\alpha)_m\,(\alpha)_{n-m)}}{m!\,(n-m)!} \cos(n-2m)t,$$

$$C_n^\alpha(0) = \begin{cases} 0, & \text{if } n \text{ is odd,} \\ (-1)^m\,(\alpha)_m/m!, & \text{if } n \text{ is even.} \end{cases}$$

The Gegenbauer polynomials are expressed in terms of the Gauss hypergeometric function as follows:

$$C_n^\alpha(x) = \frac{(2\alpha)_n}{n!} F\left(-n, n+2\alpha; \alpha+\frac{1}{2}; \frac{1-x}{2}\right) =$$

$$= \frac{(-1)^n}{n!}(2\alpha)_n F\left(-n, n+2\alpha; \alpha+\frac{1}{2}; \frac{1+x}{2}\right) =$$

$$= 2^n \frac{(\alpha)_n}{n!}(x-1)^n F\left(-n, -n-\alpha+\frac{1}{2}; -2n-2\alpha+1; \frac{2}{1-x}\right) =$$

$$= \frac{(2\alpha)_n}{n!}\left(\frac{1+x}{2}\right)^n F\left(-n, -n-\alpha+\frac{1}{2}; \frac{x-1}{x+1}\right).$$

Integral representations for the polynomials ($\alpha > 0$) are given by:

$$C_n^\alpha(x) = \frac{2^{1-2\alpha}\,\Gamma(2\alpha+n)}{n!\,\Gamma^2(\alpha)} \int_0^\pi (x+\sqrt{x^2-1}\cos t)^n \sin^{2\alpha-1}t\, dt,$$

$$C_n^\alpha(\cos\varphi) = \frac{2^\alpha\,\Gamma(\alpha+\frac{1}{2})\,(2\alpha)_n}{\sqrt{\pi}\; n!\,\Gamma(\alpha)} \sin^{1-2\alpha}\varphi \int_0^\varphi \frac{\cos(n+\alpha)t\, dt}{(\cos t-\cos\varphi)^{1-\alpha}}.$$

There is a relationship between the Gegenbauer polynomials and Legendre functions,

$$C_n^{\alpha}(x) = \Gamma\left(\alpha + \frac{1}{2}\right)\frac{(2\alpha)_n}{n!}\left(\frac{x^2-1}{4}\right)^{\frac{1-2\alpha}{4}} P_{n+\alpha-1/2}^{1/2-\alpha}(x)\,.$$

The graphs are shown in Figures 4.53, 4.54.

4.9.5. Legendre polynomials (spherical polynomials) $P_n(x)$

Legendre polynomials are a particular case of Jacobi polynomials ($\alpha = \beta = 0$) and Gegenbauer polynomials ($\alpha = 1/2$). Legendre polynomials are solutions of the differential equation $(1 - x^2)y'' - 2xy' + n(n + 1)y = 0$. Orthogonality relations are

$$\int_{-1}^{1} P_n(x)\, P_m(x)\, dx = \frac{2}{2n+1}\,\delta_{mn}\,.$$

The following recurrence formula holds:

$$(n+1)\, P_{n+1}(x) = (2n+1)\, x\, P_n(x) - nP_{n-1}(x),$$

$$P_0(x) = 1,\ P_1(x) = x,\ P_2(x) = \frac{3}{2}x^2 - \frac{1}{2},$$

$$P_3(x) = \frac{5x^3 - 3x}{2},\ P_4(x) = \frac{35x^4 - 30x^2 + 3}{8},\ P_5(x) = \frac{63x^5 - 70x^3 + 15x}{8}.$$

The Rodrigues formula is

$$P_n(x) = \frac{1}{2^n\, n!}\frac{d^n (x^2-1)^n}{dx^n}\,.$$

The main relations between Legendre polynomials are the following:

$$P_n(x) = F\left(-n, n+1; 1; \frac{1-x}{2}\right),$$

$$P_n(-x) = (-1)^n P_n(x),\quad P_n(\pm 1) = (\pm 1)^n,$$

$$P_{2m}(0) = (-1)^m g_m,\quad P_{2m+1}(0) = 0,$$

$$P_{2m}^1(0) = 0,\quad P_{2m+1}^1(0) = (-1)^m (2m+1)\, g_m,$$

$$g_m = \frac{(1/2)_m}{m!} = 2^{-2m}\binom{2m}{n}.$$

For $x \le 1$,

$$P_n(\cos t) = P_n(x)\ \ (t = \arccos x),\quad |\, P_n(x)\,| \le 1.$$

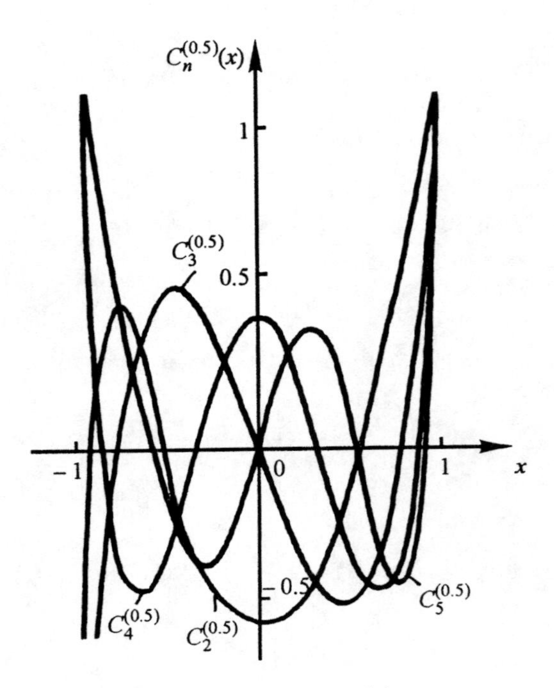

Figure 4.53. The Gegenbauer polynomials $C_n^{(\alpha)}(x)$, $a = .5$, $n = 2,3,4,5$.

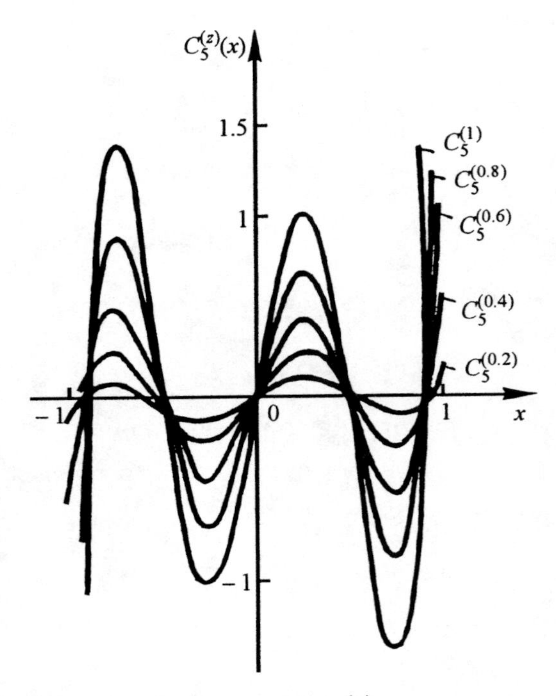

Figure 4.54. The Gegenbauer polynomials $C_n^{(\alpha)}(x)$, $a = .2,.4,.6,.8,1$, $n = 5$.

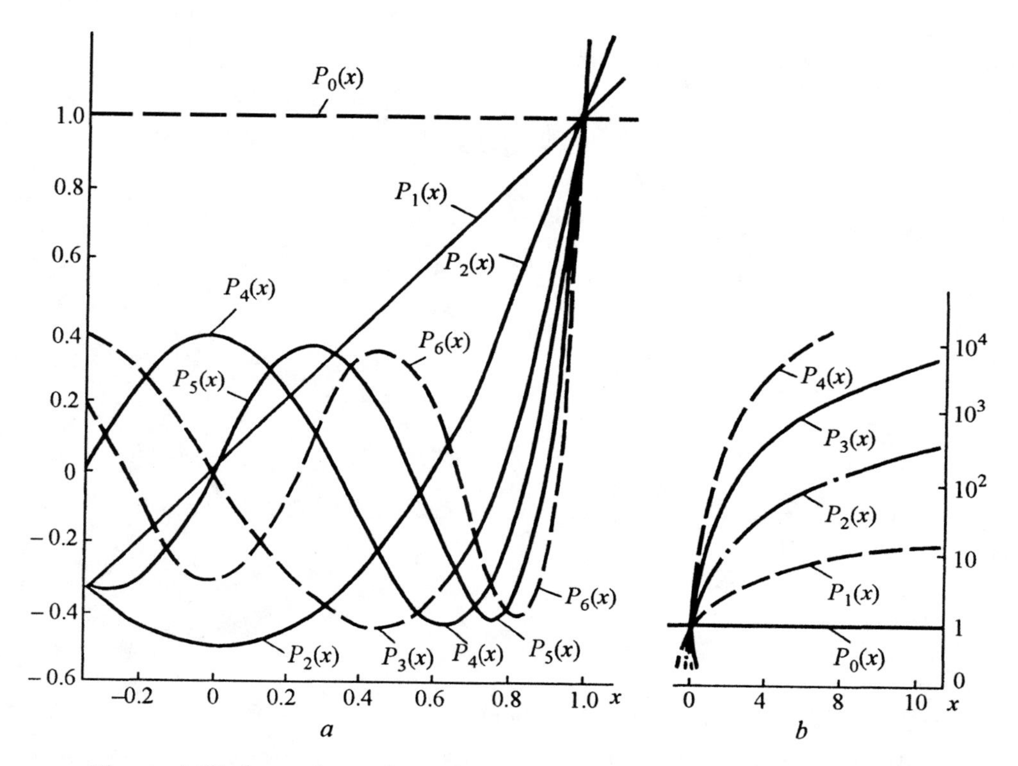

Figure 4.55. Legendre polynomials: (*a*) $P_n(x)$, $n = 0,1,2,3,4,5,6$; (*b*) $P_n(x)$, $x = 0,2,4,6,8,10,12$.

The polynomials P_n satisfy the inequalities: $1 = P_0(x) < P_1(x) < P_2(x) < \ldots$ for $x > 1$, and $1 = P_0(x) < -P_1(x) < P_2(x) < -P_3(x) < \ldots$ for $x < -1$ (Figure 4.55).

It is convenient in some cases to define the Legendre polynomials via the following generating function (actually, such a generating function exists for each previously considered polynomial)

$$\frac{1}{\sqrt{1 - 2xt + t^2}} = \begin{cases} \sum\limits_{n+0}^{\infty} P_n(x)\, t^n & (|t| < 1, \; |x| \leq 1), \\ \sum\limits_{n=0}^{\infty} P_n(x)\, t^{-n-1} & (|t| > 1, \; |x| \leq 1). \end{cases}$$

The power series

$$P_n(x) = \sum_k (-1)^{\frac{k}{2}} \frac{(2n-k-1)!!}{k!!\,(n-k)!} x^{n-k} \quad \left(k = 0, 2, 4, \ldots, n - \frac{1}{2} \pm \frac{1}{2}\right)$$

shows that the Legendre polynomial contains $(n+2)/2$ terms if n is even, and $(n+1)/2$ terms if n is odd. An equivalent expression, for $|x| < 1$, is

$$P_n(\cos t) = \frac{1}{2^n} \sum_{k=0}^{n} \frac{(2k-1)!!\,(2n-2k-1)!!}{k!\,(n-k)!} \cos(n-2k)\,t.$$

The polynomials admit the following integral representation:

$$P_n(x) = \frac{1}{\pi} \int_0^{\pi} (x + \sqrt{x^2-1}\cos t)^n \, dt \quad (x > 1),$$

$$P_n(\cos t) = \frac{\sqrt{2}}{\pi} \int_0^{\pi} \frac{\sin(n+\frac{1}{2})\,\varphi}{\cos t - \cos\varphi} \, d\varphi.$$

The Hilbert transformation with respect to a Legendre polynomial gives rise to a Legendre function of the second kind, $Q_n(\tau)$:

$$\frac{1}{\pi} \int_{-1}^{1} \frac{P_n(t)\,dt}{t-\tau} = -\frac{2}{\pi} Q_n(\tau).$$

The function $Q_n(x)$ satisfies the same equation as the polynomials $P_n(x)$. Note that $Q_n(x)$ is unbounded at $x = 1$ and $x = -1$. The function $Q_n(x)$ is a particular case of the Legendre function of the second kind, $Q_\nu(x)$, for $\nu = 0,1,2,3,\ldots$. But even for an integer ν, the functions $Q_\nu(x)$ are not polynomials. For example, $Q_0 = \arctan x$, $|x| < 1$, $Q_1(x) = x\,\mathrm{arcth}\,x - 1$, $|x| < 1$, $Q_2(x) = \frac{3x^2-1}{2}\,\mathrm{arcth}\,x - \frac{3x}{2}$, $|x| < 1$ (Figure 4.56).

4.9.6. Laguerre polynomials $L_n(x)$

Laguerre polynomials satisfy the differential equation $xy'' + (1-x)y' + ny = 0$. The orthogonality relations for the Laguerre polynomials are

$$\int_0^{\infty} e^{-x} L_n(x)\, L_m(x) dx = \delta_{mn}\,.$$

The Rodrigues formula is:

$$L_n(x) = \frac{e^x}{n!} \frac{d^n (x^n e^{-x})}{dx^n} \quad (n = 0, 1, 2, \ldots).$$

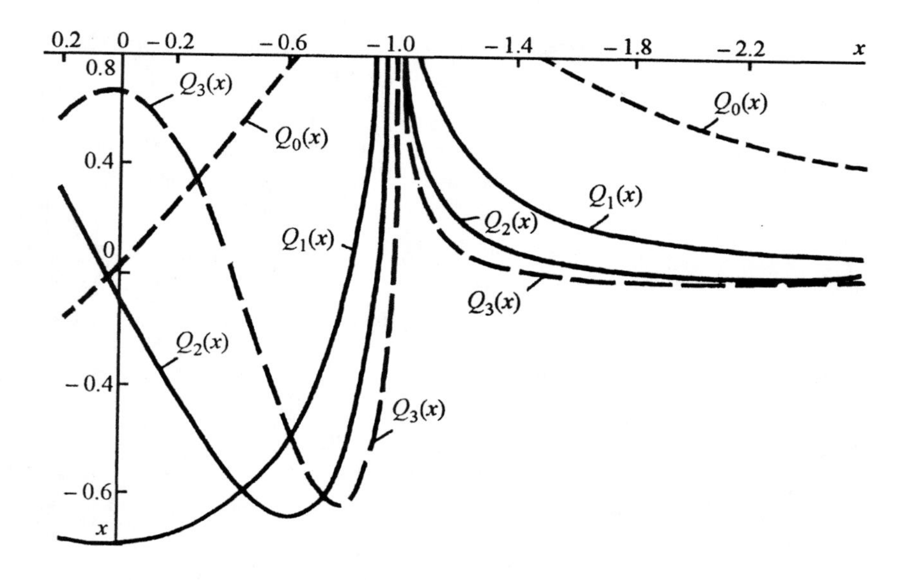

Figure 4.56. The Legendre function of the second kind, $Q_n(x)$.

The Laguerre polynomial $L_n(x)$ is defined for all values of the argument x and all nonnegative integers n. For odd n, the Laguerre polynomials range from $-\infty$ to $+\infty$, and for even n, the polynomials are bounded, $-e^{x/2} \leq L_n(x) \leq e^{x/2}$, $x \geq 0$ (Figures 4.57, 4.58).

Laguerre polynomials can be defined as a particular case of the Jacobi polynomials:

$$L_n(x) = \lim_{\beta \to \infty} P_n^{(0,\,\beta)}\left(1 - \frac{2x}{\beta}\right),$$

or by using the generating function

$$\frac{1}{1-z}\, e^{-xt/(1-t)} = \sum_{n=0}^{\infty} L_n(x)\, t^n \quad (|\,t\,| < 1).$$

The main formulas for Laguerre polynomials are:

$$L_n(x) = \frac{e^x}{n!} \int_0^{\infty} t^n e^{-t} J_0\left(2\sqrt{tx}\right) dt,$$

$$L_\nu(x) = M(-\nu;\, 1;\, x) = \sum_{k=0}^{\infty} \frac{(-\nu)_k\, x^k}{(k!)^2}, \quad L_0(x) = 1, \quad L_1(x) = -x + 1,$$

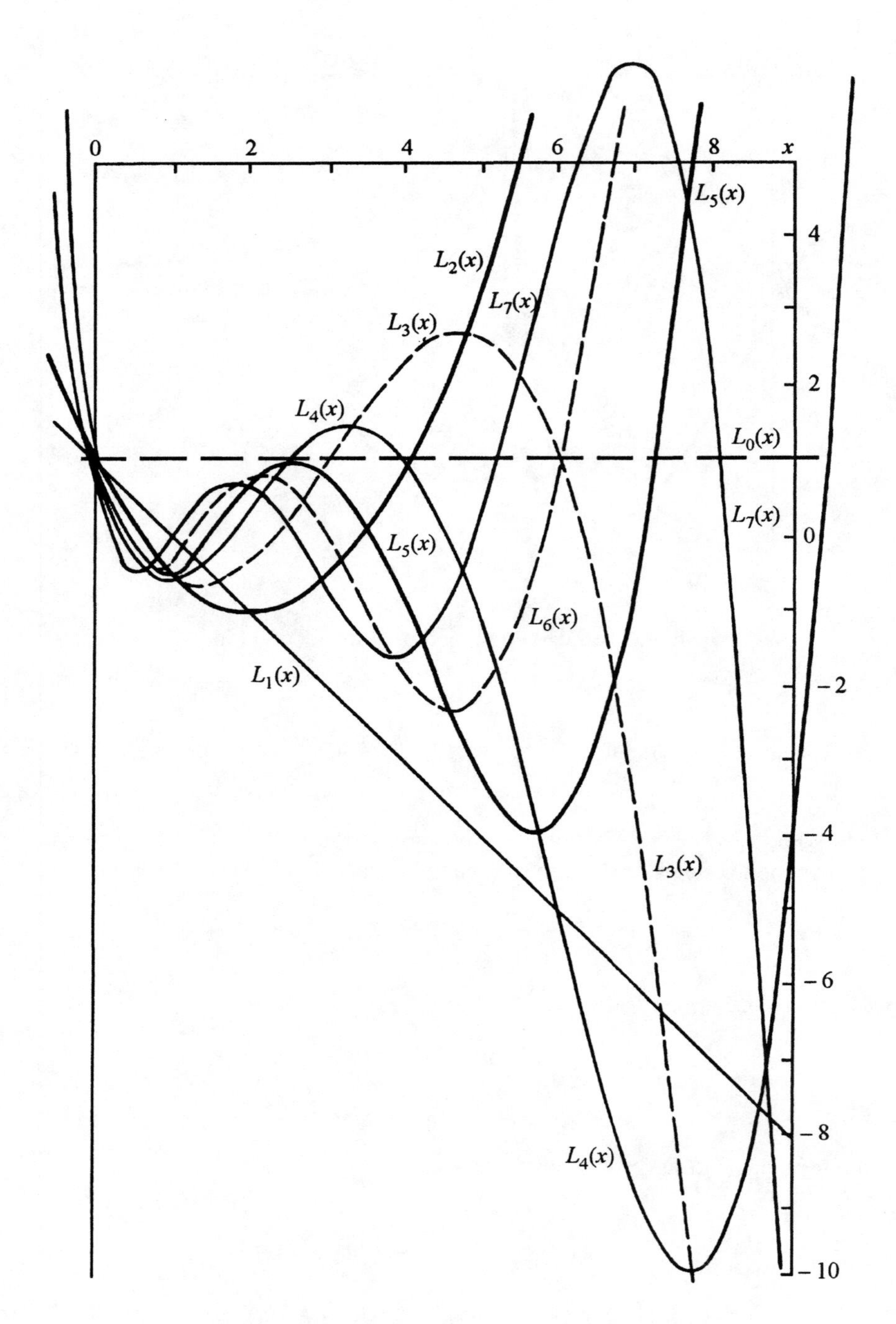

Figure 4.57. The Laguerre polynomials $L_n(x)$.

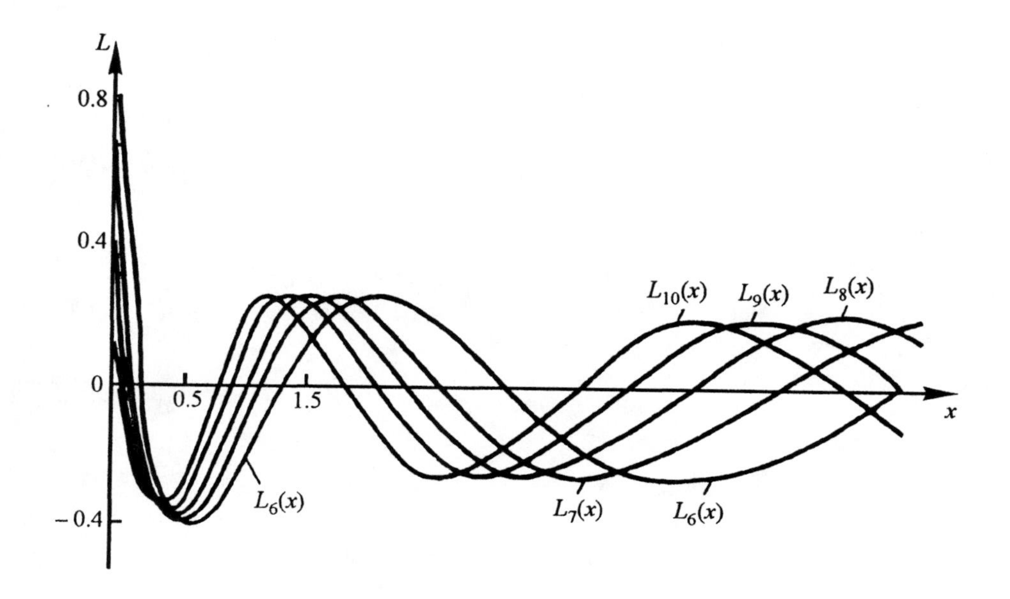

Figure 4.58. The Legendre functions $L_n(x)$, $n = 6,7,\ldots,10$.

$$L_2(x) = \frac{x^2}{2} - 2x + 1, \quad L_3 = -\frac{x^3}{6} + \frac{3x^2}{2} - 3x + 1.$$

The recurrence formula takes the form

$$L_n(x) = \frac{2n - 1 - x}{n} L_{n-1}(x) - \frac{n-1}{n} L_{n-2}(x), \quad n = 2, 3, 4, \ldots$$

The power series expansion for Laguerre polynomials is the following:

$$L_n(x) = \sum_{k=0}^{n} \binom{n}{n-k} \frac{(-x)^k}{k!}.$$

The Laguerre polynomials are a particular case of the so-called *generalized Laguerre polynomials* L_n^α. The polynomials $L_n^\alpha(x)$ verify the equation $xy'' + (\alpha + 1 - x)y' + ny = 0$. The main relations for L_n^α are the following:

$$L_n^0(x) = L_n(x),$$

$$\frac{1}{(1-t)^{\alpha+1}} e^{-xt/(1-t)} = \sum_{n=0}^{\infty} L_n^\alpha(x)\, t^n \quad (|t| < 1),$$

$$L_n^\alpha(x) = \frac{e^x}{n!\,x^\alpha}\frac{d^n}{dx^n}\left[x^{n+\alpha}\,e^{-x}\right],\ L_0^\alpha(x) = 1,\ L_1^\alpha(x) = \alpha + 1 - x,$$

$$L_n^\alpha(x) = \lim_{\beta\to\infty} P_n^{(\alpha,\beta)}\left(1 - \frac{2x}{\beta}\right),\ L_n^\alpha(x) = \frac{e^x}{n!x^{\alpha/2}}\int_0^\infty t^{(2n+\alpha)/2} J_\alpha(2\sqrt{tx}\ )\,dt,$$

$$L_n^\alpha(x) = \sum_{k=0}^{n}\binom{n+\alpha}{n-k}\frac{(-x)^k}{k!},\ L_n^\alpha(x) = \binom{n+\alpha}{n} M(-n;\alpha+1;x),$$

$$\int_0^\infty t^\alpha\,e^{-t}\,L_n^\alpha(t)\,L_m^\alpha(t)\,dt = \begin{cases}0, & n\neq m,\\ 1, & n = m.\end{cases}$$

We remark that the expression

$$(n+1)!\,e^{-\frac{x}{2}}x^k\frac{d^{2k+1}}{dx^{2k+1}}L_{n+1}(x)$$

defines the Laguerre functions.

4.9.7. Hermite polynomials $H_n(x)$

Hermite polynomials are solutions of the differential equation $y'' - 2xy + 2ny = 0$, where $n = 0,1,2,...$, and can be expressed by

$$H_n(x) = (-1)^n\,e^{x^2}\frac{d^n}{dx^n}\left(e^{-x^2}\right).$$

These polynomials can also be defined via a generating function or by the integral representation:

$$H_n(x) = \frac{2^{n+1}}{\sqrt{\pi}}\,e^{x^2}\int_0^\infty t^n\,e^{-t^2}\cos\left(2xt - \frac{n\pi}{2}\right)dt.$$

The orthogonality relations for Hermite polynomials are

$$\int_{-\infty}^{\infty} e^{-x^2} H_n(x)\,H_m(x)\,dx = 2^n\,n!\,\sqrt{\pi}\ \ \delta_{mn}\,.$$

The first several polynomials are: $H_0(x) = 1$, $H_1(x) = 2x$, $H_2(x) = 4x^2 - 2$, $H_3(x) = 8x^3 - 12x$, $H_4(x) = 16x^4 - 48x^2 + 12$.

For a Hermite polynomial we have that $H_{n(}-x) = (-1)^n H_n(x)$, whence it follows that, depending on n, the Hermite polynomial is an even or odd function.

The power series expansion is the following:

$$H_n(x) = n! \sum_{k=0}^{N} \frac{(-1)^k (2x)^{n-2k}}{k!\,(n-2k)!},$$

where $N = [n/2]$ or $N = (n-1)/2$ depending on whether n is even or odd. The Hermite polynomials have precisely n zeros (Figures 4.59, 4.60).

The Hermite polynomials can be expressed in terms of the Laguerre and Gegenbauer polynomials as follows:

$$H_{2k}(x) = (-1)^k 2^{2k} m!\, L_k^{-\frac{1}{2}}(x^2),$$

$$H_n(x) = \lim_{\alpha \to \infty} \frac{n!}{\alpha^{n/2}} C_n^{\alpha}\left(\frac{x}{\sqrt{\alpha}}\right),$$

$$H_{2k+1}(x) = (-1)^k 2^{2k+1} m!\, L_k^{\frac{1}{2}}(x^2).$$

The main relations are:

$$H_{n+1}(x) - 2x\,H_n(x) + 2n\,H_{n-1}(x) = 0,$$

$$\sum_{k=0}^{n} \frac{H_k(x)\,H_k(y)}{2^k\,k!} = \frac{H_{n+1}(x)\,H_n(y) - H_n(x)\,H_{n+1}(y)}{2^{n+1}\,n!\,(x-y)},$$

$$H_{2k}(0) = (-1)^k \frac{(2k)!}{k!}, \quad H_{2k+1}(0) = 0,$$

$$H_n(x) = \sqrt{2^n}\, e^{x^2/2} D_n(\sqrt{2}\,x) = 2^n \psi\left(-\frac{n}{2}; \frac{1}{2}; x^2\right),$$

$$\lim_{m \to \infty}\left[\frac{(-1)^m \sqrt{m}}{2^{2m}\,m!} H_{2m}\left(\frac{x}{2\sqrt{m}}\right)\right] = \frac{\cos x}{\sqrt{\pi}},$$

$$\lim_{m \to \infty}\left[\frac{(-1)^m}{2^{2m}\,m!} H_{2m+1}\left(\frac{x}{2\sqrt{m}}\right)\right] = \frac{2\sin x}{\sqrt{\pi}},$$

$$\int_{-\infty}^{\infty} e^{-t^2} t^n H_n(xt)\,dt = \sqrt{\pi}\, n!\, P_n(x),$$

$$\frac{dH_n(x)}{dx} = 2nH_{n-1}(x)\,.$$

For large values of n, we have:

$$H_n(x) = \begin{cases} (-2)^{n/2}(n-1)!!\, e^{x^2/2} \cos(x\sqrt{2n+1}\,), & n \to \infty, \\ (-2)^{(n-1)/2} \sqrt{\frac{2}{\pi}}\, n!!\, e^{x^2/2} \sin(x\sqrt{2n+1}\,), & n \to \infty, \end{cases}$$

where the first case is for even n; the second is for odd n.

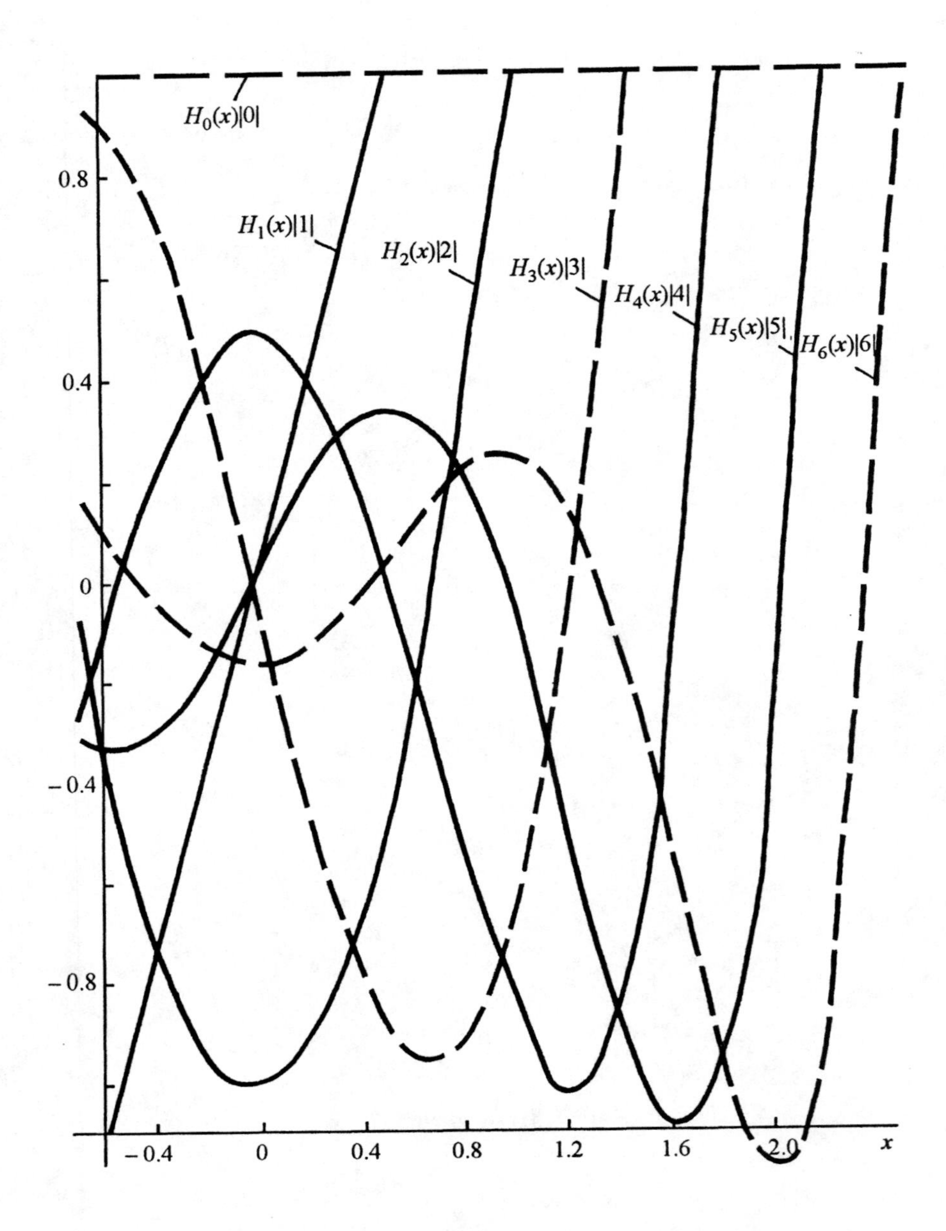

Figure 4.59. The Hermite polynomials $\frac{H_n(x)}{n!}$.

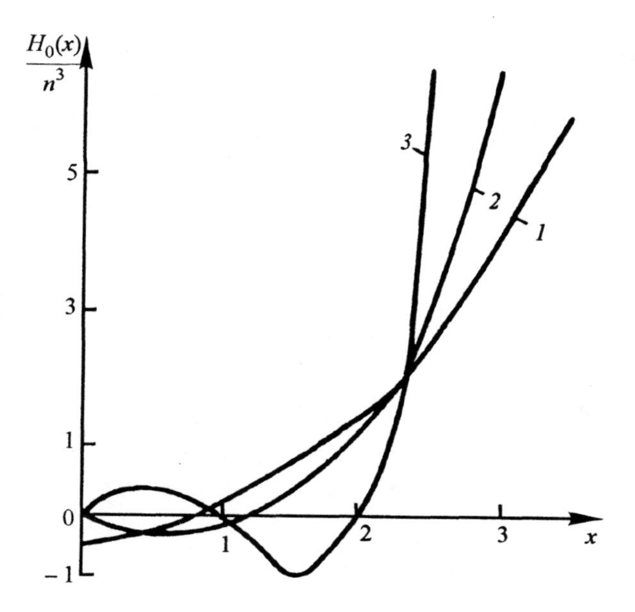

Figure 4.60. The Hermite polynomials $\frac{H_n(x)}{n^3}$. 1. $n = 2$, 2. $n = 3$, 3. $n = 5$.

For large values of x, we have:

$$H_n(x) \approx \left(2x - \frac{n-1}{2}\right)^n.$$

Orthogonal polynomials connected with the weight $e^{-x^2/2}$ are denoted by $He_n(x)$. The relation between the polynomials $He_n(x)$ and $H_n(x)$ is

$$He_n(x) = 2^{-n/2} H_n\left(\frac{x}{\sqrt{2}}\right).$$

Sometimes, the function $e^{-x/2} H_n(x)$ is called the *Hermite function*. This function satisfies the following orthogonality relations:

$$\int_{-\infty}^{\infty} f_k(t) f_n(t)\, dt = \begin{cases} 0, & k \neq n, \\ \sqrt{\pi}\, n!\, 2^n, & k = n, \end{cases}$$

$$\int_{-\infty}^{\infty} f_k(t) f_n(t)\, dt = \begin{cases} 0, & k \neq n \pm 1, \\ \sqrt{\pi}\, n!\, 2^{n-1}, & k = n - 1. \end{cases}$$

4.9.8. Chebyshev polynomials

Chebyshev polynomials of the first and second kind are defined, correspondingly, as

$$T_n(x) = \cos(n \arccos x), \quad U_n(x) = \sin(n \arccos x), \quad (|x| \le 1),$$

or by using generating functions,

$$\frac{1 - tx}{1 - 2tx + t^2} = \sum_{n=0}^{\infty} T_n(x)\, t^n \quad (|t| < 1),$$

$$\frac{1}{1 - 2tx + t^2} = \sum_{n=0}^{\infty} U_n(x)\, t^n \quad (|t| < 1),$$

or else as solutions of the differential equation

$$(1 - x^2)y'' - xy' + n^2 y = 0,$$

where

$$y = C_1 T_n(x) + C_2 \sqrt{1 - x^2}\, U_{n-1}(x),$$

C_1, C_2 are arbitrary constants.

Chebyshev polynomials can be defined in a purely algebraic way,

$$T_n(x) = \frac{1}{2}\left(x + \sqrt{x^2 - 1}\right)^n + \frac{1}{2}\left(x - \sqrt{x^2 - 1}\right)^n.$$

The orthogonality relations for Chebyshev polynomials are:

$$\int_{-1}^{1} \frac{T_n(x) T_m(x)}{\sqrt{1 - x^2}}\, dx = \begin{cases} \frac{\pi}{2}\,\delta_{mn} & (m^2 + n^2 \ne 0), \\ \pi & (m = n = 0), \end{cases}$$

$$\int_{-1}^{1} \sqrt{1 - x^2}\, U_n(x) U_m(x)\, dx = \frac{\pi}{2}\,\delta_{mn}.$$

The power series expansions are

$$T_n(x) = \frac{n}{2} \sum_{k=0}^{\left[\frac{n}{2}\right]} \frac{(-1)^k}{n-k} \binom{n-k}{k} (2x)^{n-2k} \quad (n = 1, 2, 3, \ldots),$$

$$U_n(x) = \sum_{k=0}^{\left[\frac{n}{2}\right]} (-1)^k \binom{n-k}{k} (2x)^{n-2k} \quad (n = 0, 1, 2, \ldots).$$

These series can also be written in the following form:

$$T_n(x) = x^n - \binom{n}{2} x^{n-2}(1-x^2) + \binom{n}{4} x^{n-4}(1-x^2)^2 - \ldots,$$

$$U_n(x) = \binom{n+1}{1} x^n - \binom{n+1}{3} x^{n-2}(1-x^2) + \binom{n+1}{5} x^{n-4}(1-x^2)^2 - \ldots .$$

The power series expansion for the Chebyshev polynomials of even order satisfies

$$T_0(x) + T_2 x + T_4(x) + \ldots + T_n(x) = \frac{1}{2} + \frac{1}{2} U_n(x) \qquad (n = 0, 2, \ldots),$$

$$U_0(x) + U_2(x) + U_4(x) + \ldots + U_n(x) = \frac{1 - T_{n+2}(x)}{2(1-x^2)} \qquad (n = 0, 2, 4, \ldots),$$

and, for polynomials of odd order,

$$T_1(x) + T_3(x) + \ldots + T_n(x) = \frac{1}{2} U_n(x) \qquad (n = 1, 3, 5, \ldots),$$

$$U_1(x) + U_3(x) + \ldots + U_n(x) = \frac{x - T_{n+2}(x)}{2(1-x^2)} \qquad (n = 1, 3, 5, \ldots).$$

The first several Chebyshev polynomials are:

$$T_0(x) = 1, \quad U_0(x) = x; \quad T_1(x) = x, \quad U_1(x) = 2x; \quad T_2(x) = 2x^2 - 1, \quad U_2(x) = 4x^2 - 1;$$

$$T_3(x) = 4x^3 - 3x, \quad U_3(x) = 8x^3 - 4x; \quad T_4(x) = 8x^4 + 8x^2 + 1, \quad U_4(x) = 16x^4 - 12x^2 + 1.$$

The Chebyshev polynomials are even or odd functions depending on the order n,

$$T_n(-x) = (-1)^n T_n(x), \qquad U_n(-x) = (-1)^n U_n(x) .$$

The Chebyshev polynomials $T_n(x)$ and $U_n(x)$ have exactly n zeros,

$$T_n(\alpha_k) = 0, \quad \alpha_k = \cos \frac{2k-1}{2n} \pi, \qquad k = 1, 2, 3, \ldots, n,$$

$$U_n(\alpha_k) = 0, \quad \alpha_k = \cos \frac{k}{n+1} \pi, \qquad k = 1, 2, 3, \ldots, n,$$

see Figures 4.61, 4.62.

The Rodrigues formulas have the form:

$$T_n(x) = \frac{(-2)^n \, n!}{(2n)!} \sqrt{1-x^2} \, \frac{d^n}{dx^n} (1-x^2)^{n-1/2},$$

$$U_n(x) = \frac{(-1)^n \, (n+1)}{(2n+1)!! \, \sqrt{1-x^2}} \, \frac{d^n}{dx^n} (1-x^2)^{n+1/2}.$$

The relationship with the Gauss hypergeometric function is given by

$$T_n(x) = F\left(-n, n; \frac{1}{2}; \frac{1-x}{2}\right),$$

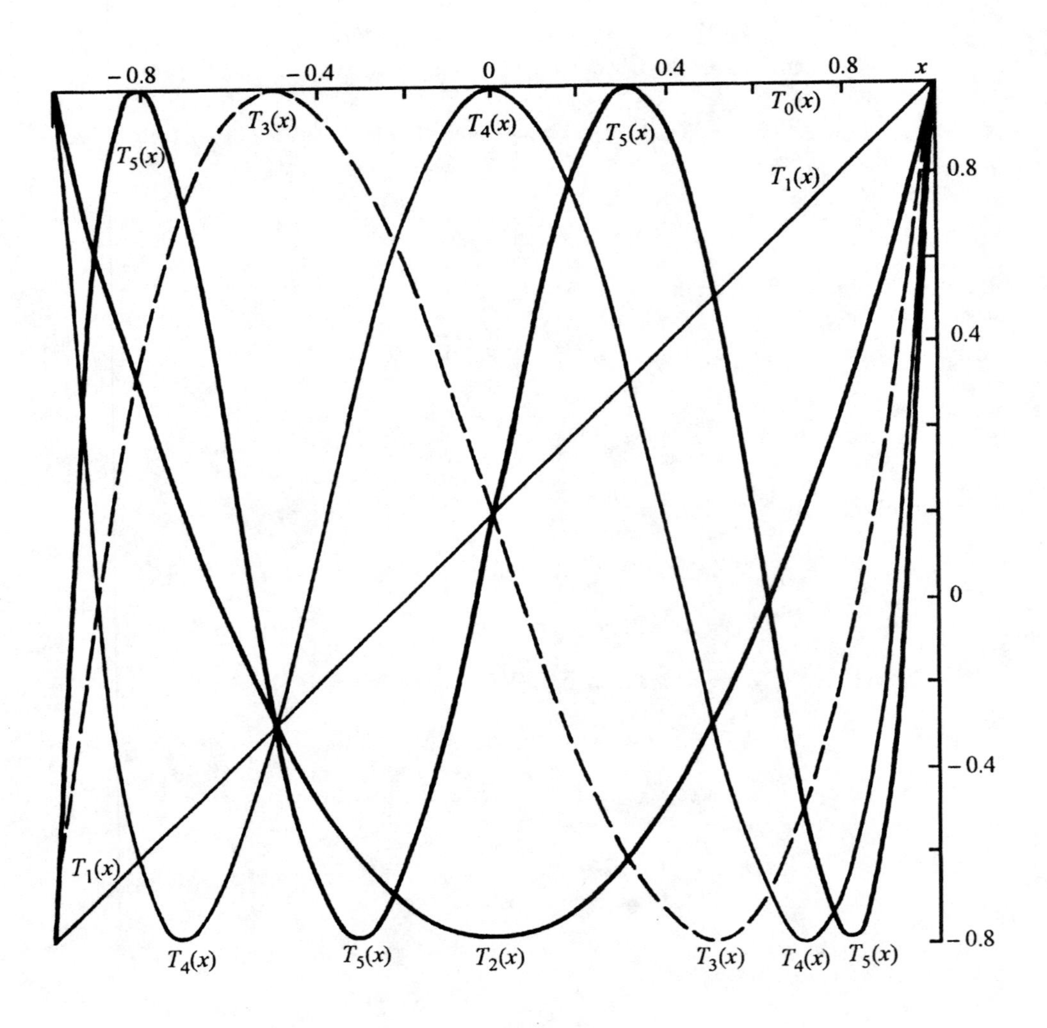

Figure 4.61. The Chebyshev polynomials of the first kind, $T_n(x)$, $n = 0,1,2,3,4,5$.

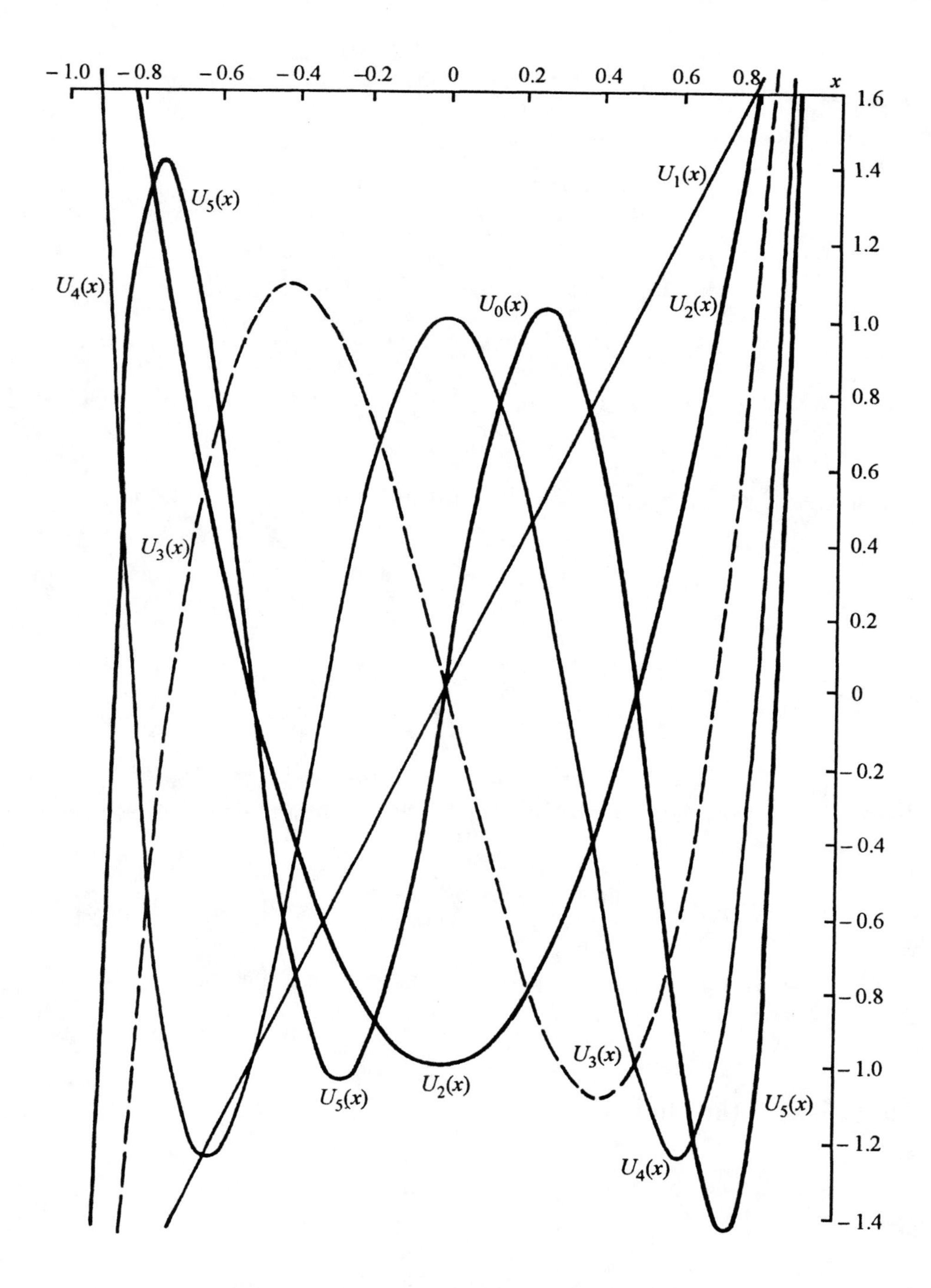

Figure 4.62. The Chebyshev polynomials of the second kind, $U_n(x)$, $n = 0,1,2,3,4,5$.

$$U_n(x) = (n+1)\,F\left(-n, n+1; \frac{3}{2}; \frac{1-x}{2}\right).$$

The polynomials satisfy the recurrence formulas,

$$T_{n+1}(x) = 2xT_n(x) - T_{n-1}(x),$$

$$U_{n+1}(x) = 2xU_n(x) - U_{n-1}(x),$$

$$2T_k(x)\,T_n(x) = T_{n+k}(x) - T_{n-k}(x),\quad n \ge k,$$

$$2(x^2-1)U_{k-1}(x)\,U_{n-1}(x) = T_{n+k}(x) - T_{n-k}(x),\quad n \ge k,$$

$$2T_k(x)\,U_{n-1}(x) = U_{n+k-1}(x) + U_{n-k-1}(x),\ n > k,$$

$$2T_n(x)\,U_{k-1}(x) = U_{n+k-1}(x) - U_{n-k-1}(x),\ n > k.$$

The relationship between the Chebyshev and Legendre polynomials is given by:

$$\left(n+\frac{1}{2}\right)\sqrt{1+x}\int_{-1}^{x}\frac{P_n(t)\,dx}{\sqrt{x-t}} = T_n(x) + T_{n+1}(x),$$

$$\left(n+\frac{1}{2}\right)\sqrt{1-x}\int_{x}^{1}\frac{P_n(t)\,dt}{\sqrt{t-x}} = T_n(x) - T_{n+1}(x).$$

Note. Apart from classical polynomials, there also exist other classes of special orthogonal polynomials, e.g. Heine polynomials, Akhiezer polynomials, orthogonal polynomials of discrete variable, e.g. Kravchuk polynomials,

$$K_n(x) = \frac{(-1)^n x!\,(N-x)!}{n!p^x q^{N-x}}\left[\frac{p^x q^{N-x+n}}{(x-n)!\,(N-x)!}\right] \quad (n = 0, 1, ..., N),$$

where $p > 0, q > 0, p + q = 1, N$ is a positive integer. There are also Charlier polynomials, Meixner polynomials, and others.

4.10. Legendre functions

Legendre functions or *spherical functions*, $P_\nu^\mu(z)$, $Q_\nu^\mu(z)$, are solutions of the differential equation

$$(1-z^2)\frac{d^2u}{dz^2} - 2z\frac{du}{dz} + \left[\nu(\nu+1) - \frac{\mu^2}{1-z^2}\right]u = 0.$$

Here ν is the power, μ — the degree, the points $x = \pm 1, \infty$ are singular (branching) points. The numbers μ, ν are called *indices*. The values of these constants can be arbitrary, real or complex.

The functions P_ν^μ and Q_ν^μ are called *Legendre functions of the first and second kind*, correspondingly. They are single-valued and regular in the complex plane cut along the real axis from 1 to ∞ such that

$$|\arg(z \pm 1)| < \pi, \quad |\arg z| < \pi.$$

The Legendre functions are connected to the Gauss hypergeometric function by the formulas

$$P_\nu^\mu(z) = \frac{2^\mu (z^2 - 1)^{-\mu/2}}{\Gamma(1 - \mu)} F\left(1 - \mu + \nu, -\mu - \nu; 1 - \mu; \frac{1 - z}{2}\right) \quad (|1 - z| < 2),$$

$$Q_\nu^\mu(z) = \frac{e^{i\mu\pi} 2^{-\nu - 1} \sqrt{\pi}}{\Gamma(\nu + \frac{3}{2})} \Gamma(\nu + \mu + 1) z^{-1 - \nu + \mu} (z^2 - 1)^{-\mu/2} \times$$

$$\times F\left(\frac{1 + \nu - \mu}{2}, 1 + \frac{\nu - \mu}{2}; \nu + \frac{3}{2}; \frac{1}{z^2}\right) \quad (|z| > 1).$$

The main relations for the Legendre functions are:

$$P_\nu^\mu(z) = P_{-\nu - 1}^\mu(z),$$

$$e^{i\mu\pi} \Gamma(\nu + \mu + 1) Q_\nu^{-\mu}(z) = e^{-i\mu\pi} \Gamma(\nu - \mu + 1) Q_\nu^\mu(z),$$

$$P_\nu^m(z) = \frac{\Gamma(\nu + m + 1)}{\Gamma(\nu - m + 1)} P_\nu^{-m}(z) \quad (m = 1, 2, 3, \ldots),$$

$$P_\nu^\mu(-z) = e^{\pm \nu\pi i} P_\nu^\mu(z) - \frac{2}{\pi} e^{-i\mu\pi} \sin[\pi(\nu + \mu)] Q_\nu^\mu(z),$$

$$Q_\nu^\mu(-z) = e^{\pm i\nu\pi} Q_\nu^\mu(z).$$

In the last two formulas, the plus or minus sign is taken depending on whether $\mathrm{Im} z \geq 0$ or not.

The *Whipple formula* is the following:

$$Q_\nu^\mu(z) = e^{i\mu\pi} \sqrt{\frac{\pi}{2}} \Gamma(\nu + \mu + 1) (z^2 - 1)^{-1/4} P_{-\mu - 1/2}^{-\nu - 1/2}\left(\frac{z}{\sqrt{z^2 - 1}}\right) \quad (\mathrm{Re}\, z > 0)$$

or

$$P_\nu^\mu(z) = \frac{i e^{i\nu\pi} \left(\frac{\pi}{2}\right)^{\frac{-1}{2}}}{\Gamma(-\nu - \mu)} (z^2 - 1)^{-1/4} Q_{-\mu - 1/2}^{-\nu - 1/2}\left(\frac{z}{\sqrt{z^2 - 1}}\right) \quad (\mathrm{Re}\, z > 0).$$

In practice, the Legendre functions are used for $z = x$, where $-1 < x < 1$. For an even integer value of the index, P_ν takes the same value at both endpoints of the interval $[-1; 1]$. In this case, it is sufficient to make the cut along the real axis from -1 to $-\infty$.

If μ is a positive integer, $\mu = m$, $m = 1,2,\ldots$, then

$$\Gamma(\nu - m + 1)\, m!\, P_\nu^m(z) = 2^{-m}\,\Gamma(\nu + m + 1)\,(z^2 - 1)^{m/2} \times$$
$$\times F\left(1 + m + \nu, m - \nu; 1 + m; \frac{1 - z}{2}\right),$$

$$\Gamma(\nu - m + 1)\, m!\, P_\nu^m(x) = (-2)^{-m}\,\Gamma(\nu + m + 1)\,(1 - x^2)^{m/2} \times$$
$$\times F\left(1 + m + \nu, m - \nu; 1 + m; \frac{1 - x}{2}\right),$$

$$P_\nu(0) = \frac{(2n - 1)!!}{(-2)^n\, n!} \quad (\nu = 2, 4, 6, \ldots, 2n, \ldots, \nu = -3, -5, -7, \ldots, -1 - 2n)$$

$$P_\nu(0) = \frac{\Gamma\left(\dfrac{1 + \nu}{2}\right)}{\sqrt{\pi}\,\Gamma\left(1 + \dfrac{\nu}{2}\right)} \cos\frac{\nu\pi}{2}; \quad -1 < \nu < 0, \; P_\nu(1) = 1,$$

$$Q_\nu(0) = \frac{-(-1)^n U(4n - 1)!!!!}{(4n + 1)!!!!}, \quad \nu = 2n + \frac{1}{2}, \; Q_\nu(1) = \infty.$$

Integral representations for Legendre functions are the following:

$$P_\nu^\mu(z) = \frac{2^{-\nu}(z^2 - 1)^{-\mu/2}}{\Gamma(-\mu - \nu)\,\Gamma(\nu + 1)} \int_0^\infty (z + \cosh t)^{-\nu - 1} \sinh^{2\nu + 1} t\, dt$$

$$(\mathrm{Re}\,(-\mu) > \mathrm{Re}\,\nu > -1),$$

$$P_\nu^\mu(\cosh\alpha) = \sqrt{\frac{2}{\pi}}\, \frac{\sinh^\mu\alpha}{\Gamma(\frac{1}{2} - \mu)} \int_0^\alpha \frac{\cosh(\nu + \frac{1}{2})t}{(\cosh\alpha - \cosh t)^{\mu + 1/2}}\, dt,$$

$$Q_\nu^\mu(\cosh\alpha) = \sqrt{\frac{\pi}{2}}\; e^{\mu\pi i}\, \frac{\sinh^\mu\alpha}{\Gamma(\frac{1}{2} - \mu)} \int_\alpha^\infty e^{-(\nu + 1/2)t} (\cosh t - \cosh\alpha)^{-\mu - 1/2}\, dt$$

$$(\alpha > 0, \mathrm{Re}\,(\nu + \mu + 1) > 0, \; \mathrm{Re}\,\mu < 1/2).$$

Zeros of the functions are determined from

$$N(P_\nu) = \begin{cases} n, & n - 1 < \nu \le n, \\ 0, & -1 \le \nu \le 0, \\ n, & -n - 1 \le \nu < -n \quad (n = 1, 2, \ldots), \end{cases}$$

and all these zeros belong to the interval $-1 < x < 1$. For the Legendre functions of the second kind, we have

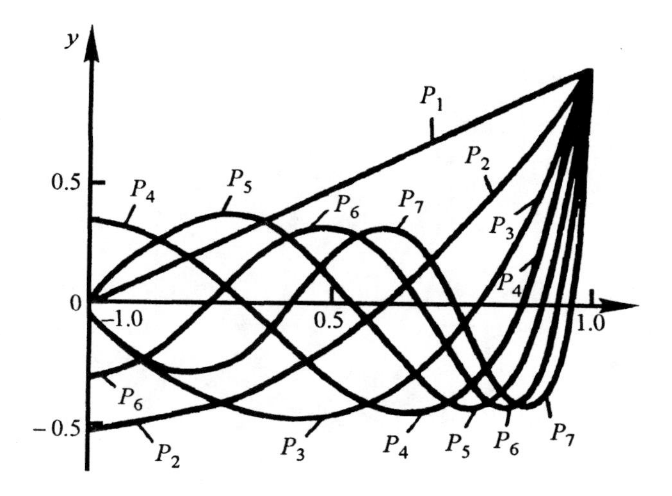

Figure 4.63. The Legendre functions of the first kind, $P_n(x)$, $n = 1,2,3,4,5,6,7$.

$$N(Q_\nu) = \begin{cases} n, & n - \frac{3}{2} < \nu \leq n - \frac{1}{2}, \\ 0, & -1 < \nu \leq -\frac{1}{2}, \\ n+1, & -\frac{1}{2} - n < \nu < -n, \\ n, & -n - 1 < \nu \leq -\frac{1}{2} - n \ (n = 1, 2, \ldots). \end{cases}$$

see Figures 4.63 – 4.67.

Recurrence relations satisfied by Legendre functions are:

$$P_\nu^{\mu+1}(z) = (z^2 - 1)^{-1/2}\left[(\nu - \mu)\, z\, P_\nu^\mu(z) - (\nu + \mu)\, P_{\nu-1}^\mu(z)\right],$$

$$(z^2 - 1)\frac{dP_\nu^\mu(z)}{dz} = (\nu + \mu)(\nu - \mu + 1)\left(z^2 - 1\right)^{\frac{1}{2}} P_\nu^{\mu-1}(z) - \mu z\, P_\nu^\mu(z),$$

$$(\nu - \mu + 1)\, P_{\nu+1}^\mu(z) = (2z + 1)\, z\, P_\nu^\mu(z) - (\nu + \mu)\, P_{\nu-1}^\mu(z),$$

$$(z^2 - 1)\frac{dP_\nu^\mu(z)}{dz} = \nu z\, P_\nu^\mu(z) - (\nu + \mu)\, P_{\nu-1}^\mu(z),$$

$$P_0(z) = 1, \quad P_0(x) = 1,$$

$$Q_0(z) = \frac{1}{2}\ln\frac{z+1}{z-1}, \quad Q_0(x) = \frac{1}{2}\ln\frac{1+x}{1-x} = x\,F\left(\frac{1}{2}, 1; \frac{3}{2}; x^2\right);$$

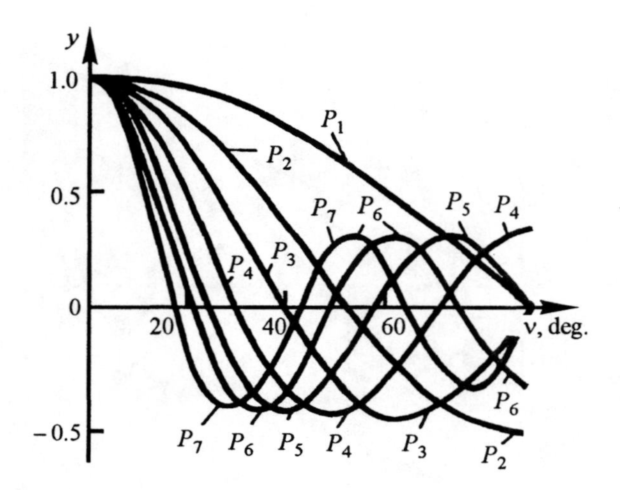

Figure 4.64. The Legendre functions of the first kind, $P_n(\cos\theta)$, $n = 1,2,3,4,5,6,7$.

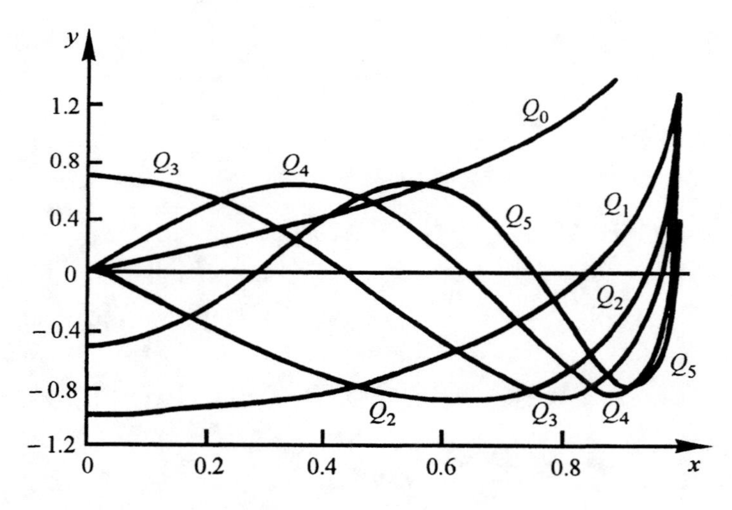

Figure 4.65. The Legendre functions of the first kind, $Q_n(x)$, $n = 0,1,2,3,4,5$.

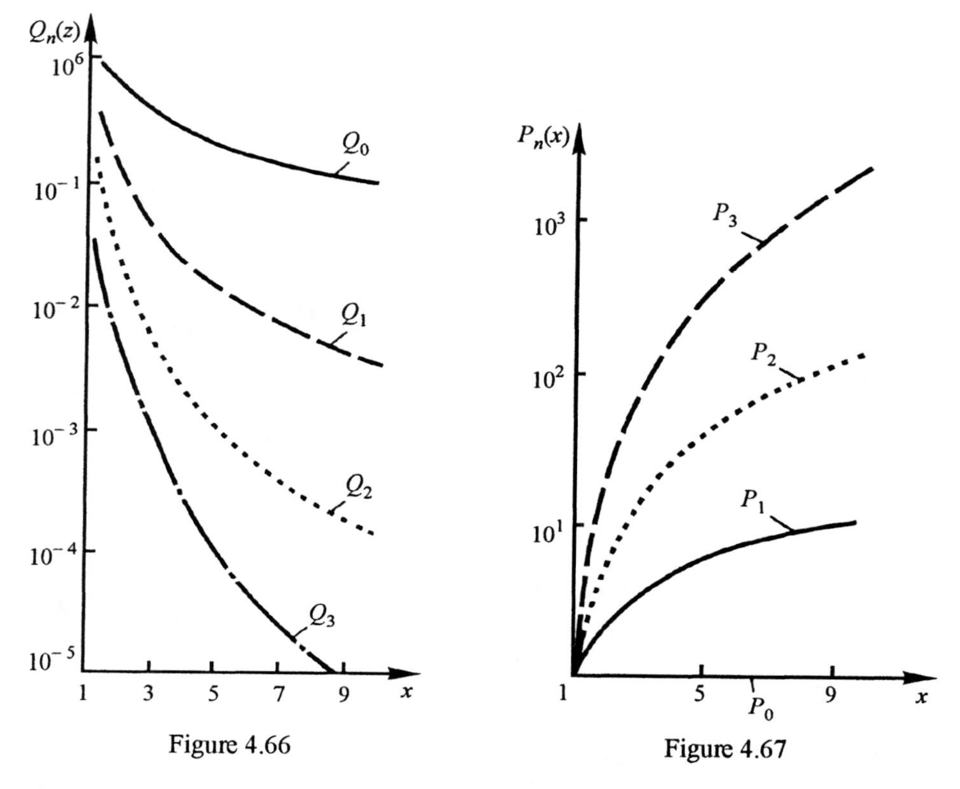

Figure 4.66

Figure 4.67

Figure 4.66. The functions $P_n(x)$, $n = 0,1,2,3$, $x \geq 1$.

Figure 4.67. The functions $Q_n(x)$, $n = 0,1,2,3$, $x > 1$.

$$P_1(z) = z;\ P_1(x) = x = \cos\theta,\quad Q_1(z) = \frac{z}{2}\ln\frac{z+1}{z-1} - 1,\quad Q_1(x) = \frac{x}{2}\ln\frac{1+x}{1-x} - 1.$$

The functions of the form $P^{\mu}_{\nu-1/2}(\cosh\eta)$, $Q^{\mu}_{\nu-1/2}(\cosh\eta)$ are called *torus functions,* and the functions $P^{\mu}_{-1/2+i\lambda}(\cos\theta)$, $Q^{\mu}_{-1/2+i\lambda}(\cos\theta)$, – *cone functions.*

4.11. Elliptic integrals and elliptic functions

Elliptic integrals are integrals of the form $\int R(x,\sqrt{P(x)}\,)dx$, where R is a rational function and $P(x)$ is a polynomial of degree three or four. Elliptic integrals can be reduced to elementary functions and standard elliptic integrals.

Elliptic integrals of the first, second, and third kind are defined by the corresponding formulas:

$$F(\varphi, k) = \int_0^{\varphi} \frac{dt}{\sqrt{1-k^2\sin^2 t}} = \int_0^{\sin\varphi} \frac{dx}{\sqrt{(1-x^2)(1-k^2x^2)}},$$

$$E(\varphi, k) = \int_0^{\varphi} \sqrt{1-k^2\sin^2 t}\; dt = \int_0^{\sin\varphi} \frac{\sqrt{1-k^2x^2}}{\sqrt{1-x^2}}\, dx,$$

$$\Pi(\varphi, n, k) = \int_0^{\varphi} \frac{dt}{(1-n\sin^2 t)\sqrt{1-k^2\sin^2 t}} = \int_0^{\sin\varphi} \frac{dx}{(1+nx^2)\sqrt{(1-x^2)(1-k^2x^2)}}.$$

Here k ($k^2 < 1$) is the modulus of the integral, n is the parameter or characteristic of the integral of the third kind, $k^1 = \sqrt{1-k^2}$ is the additional modulus, φ is the amplitude, $\Delta(\varphi) = (1 - \sin\varphi)^{1/2}$ is the delta-amplitude, m ($m = k^2$) is a parameter and m_1 ($m_1 = k^2$) is an additional parameter, $m + m_1 = 1$.

The middle parts in the above formulas are *incomplete elliptic integrals* in the normal trigonometric form; the right-hand sides are *incomplete elliptic integrals* in the normal Legendre form. The incomplete elliptic integrals are even functions of k, and odd if regarded as functions of φ.

The differences

$$F(\varphi, k) - \frac{2\varphi}{\pi}\boldsymbol{K}(k), \quad \boldsymbol{E}(\varphi, k) - \frac{2\varphi}{p}\boldsymbol{E}(k),$$

where $\boldsymbol{K}(k)$, $\boldsymbol{E}(k)$ are complete elliptic integrals, are periodic functions with period π. Note that

$$F(\varphi, 0) = E(\varphi, 0) = \varphi,$$

$$F(\varphi, 1) = \ln\tan\left(\frac{\pi}{4} + \frac{\varphi}{2}\right), \quad |\varphi| \le \frac{\pi}{2},$$

$$E(\varphi, 1) = \sin\varphi, \quad |\varphi| \le \frac{\pi}{2},$$

$$F(\varphi, \sqrt{2}) = \frac{1}{4}B\left(\frac{1}{2}, \frac{1}{4}; \sin^2 2\varphi\right) \quad \left(0 \le \varphi \le \frac{\pi}{4}\right),$$

$$F(\varphi, \sqrt{2}) = \frac{1}{4}B\left(\frac{1}{2}, \frac{3}{4}; \sin^2 2\varphi\right) \quad \left(0 \le \varphi \le \frac{\pi}{4}\right),$$

$$\Pi(\varphi, 0, k) = F(\varphi, k),$$

$$\Pi(\varphi, -1, k) = F(\varphi, k) - \frac{1}{\sqrt{1-k^2}}E(\varphi, k) + \frac{1}{\sqrt{1-k^2}}\tan\varphi\sqrt{1-k^2\sin^2\varphi},$$

$$\Pi(\varphi, n, 0) = \begin{cases} \dfrac{\arctan(\sqrt{1+n}\;\tan\varphi)}{\sqrt{1+n}}, & k > -1, \\ \tan\varphi, & k = -1, \\ \dfrac{\arctan(\sqrt{-n-1}\;\tan\varphi)}{\sqrt{-n-1}}, & k < -1, \end{cases}$$

$$\mathrm{P}\left(\frac{\pi}{2}, n, k\right) = \Pi\,(n, k).$$

The elliptic integral of the first kind is finite for all z and has singularities at the points $z = \alpha_i$ that are roots of the polynomial $\Phi(x)$. The elliptic integral of the second kind is analytic in the complex plane except for the points α_i and a finite number of poles. The elliptic integral of the third kind has a logarithmic singularity.

If the upper limit of integration is $\varphi = \pi/2$ or $\sin\varphi = 1$, then we get *complete elliptic integrals*:

$$K(k) = F\left(\frac{\pi}{2}, k\right) = K'(k') = \int_0^{\frac{\pi}{2}} \left(1 - k^2\sin^2 t\right)^{-1/2} dt = \int_0^1 \frac{dt}{\sqrt{(1-t^2)(1-k^2t^2)}} = \int_0^\infty \frac{dt}{\sqrt{(1+t^2)(1+k'^2t^2}},$$

$$E(k) = E\left(\frac{\pi}{2}, k\right) = E'(k') = \int_0^{\frac{\pi}{2}} (1 - k^2\sin^2 t)^{1/2}\, dt = \int_0^1 \frac{(1-k^2t^2)^{1/2}}{(1-t^2)^{1/2}}\, dt = \int_0^\infty \frac{\sqrt{1+k^2t^2}}{(1+t^2)^3}\, dt,$$

$$K'(k) = F\left(\frac{\pi}{2}, k'\right) K(k'),\quad E'(k) = E\left(\frac{\pi}{2}, k'\right) - E(k').$$

Figures 4.68 – 4.71 show the graphs of incomplete elliptic integrals; the graphs of complete elliptic integrals are shown in Figures 4.72 – 4.74.

The main formulas for elliptic integrals are the following:

$$K(k) = \frac{\pi}{2} F\left(\frac{1}{2}, \frac{1}{2}; 1; k^2\right) = \frac{\pi}{2k} F\left(\frac{1}{2}, \frac{1}{2}; 1; -\frac{k^2}{k'^2}\right),$$

$$E(k) = \frac{\pi}{2} F\left(-\frac{1}{2}, \frac{1}{2}; 1; k^2\right) = \frac{\pi\, k'^2}{2} = F\left(\frac{1}{2}, \frac{3}{2}; 1; k^2\right),$$

$$K\left(\frac{2\sqrt{k}}{1+k}\right) = (1+k)\,K(k),\quad k \neq 1,$$

$$E\left(\frac{2\sqrt{k}}{1+k}\right) = \frac{2E(k) - k'^2\,K(k)}{1+k}\quad (k'^2 = 1 - k^2),$$

$$K(k) = \frac{\pi}{2} \sum_{i=0}^{\infty} \left[\frac{(2i-1)!!}{(2i)!!} k^i\right]^2,$$

$$E(k) = -\frac{\pi}{2} \sum_{i=0}^{\infty} \frac{1}{2i-1} \left[\frac{(2i-1)!!}{(2i)!!} k^i\right]^2.$$

Some values of the elliptic integrals are:

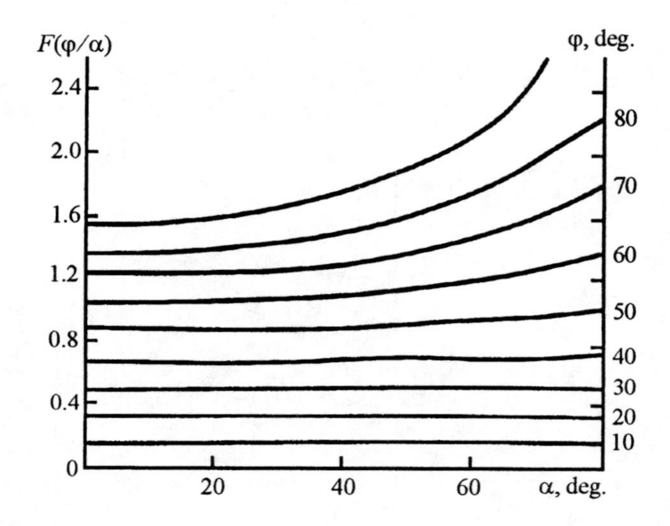

Figure 4.68. The incomplete elliptic integral of the first kind, $F(\varphi \setminus \alpha)$, α is a constant.

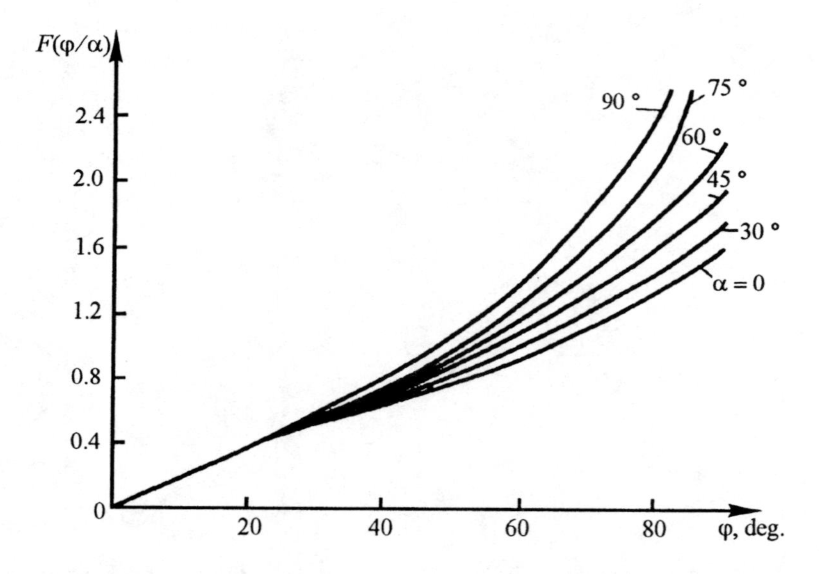

Figure 4.69. The incomplete integral of the first kind, $F(\varphi \setminus \alpha)$, α is a constant.

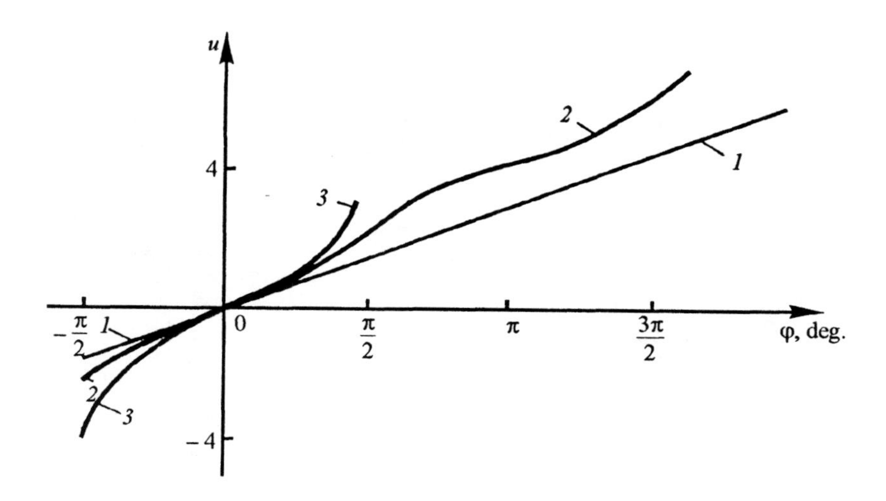

Figure 4.70. The incomplete integral of the first kind, $F(\varphi,k)$: *1.* $k^2 = 0$, *2.* $k^2 = .75$, *3.* $k^2 = 1$.

$$K(0) = \frac{\pi}{2},\ E(0) = \frac{\pi}{2},\ K(1) = \infty,\ E(1) = 1,$$

$$\frac{1}{2}\int_0^1 K\left(\sqrt{1-t^2}\right) dt = \int_0^1 E\left(\sqrt{1-t^2}\right) dt = \frac{\pi^2}{8},$$

$$\frac{d^{1/2}}{(dk^2)^{1/2}} E(k) = \frac{\sqrt{\pi}\,(1-k^2)}{2k},$$

$$\frac{d^{-1/2}}{(dk^2)^{-1/2}} K(k) = \sqrt{\pi}\ \arcsin k\,.$$

Elliptic functions. *Elliptic functions* are functions inverse to elliptic integrals. An elliptic function is a double periodic meromorphic function of complex variable. All its periods can be represented by $2m\omega_1 + 2n\omega_2$, m,n are integers, with $2\omega_2$, $2\omega_2$ principle periods. The ratio $\tau = \omega_1/\omega_2$ of the principle periods is a complex number, $\mathrm{Im}\tau > 0$. If the complex plane is covered with a lattice of period parallelograms, then a doubly periodic function takes the same value at corresponding points of the period parallelograms.

Jacobian amplitude. The function is defined by

$$\mathrm{am}(k; x) = \arcsin\{\mathrm{sn}\,(k; x)\} = \arcsin\left(\frac{\sqrt{1-\mathrm{dn}^2\,(k; x)}}{k}\right) = \varphi.$$

Other main elliptic functions are

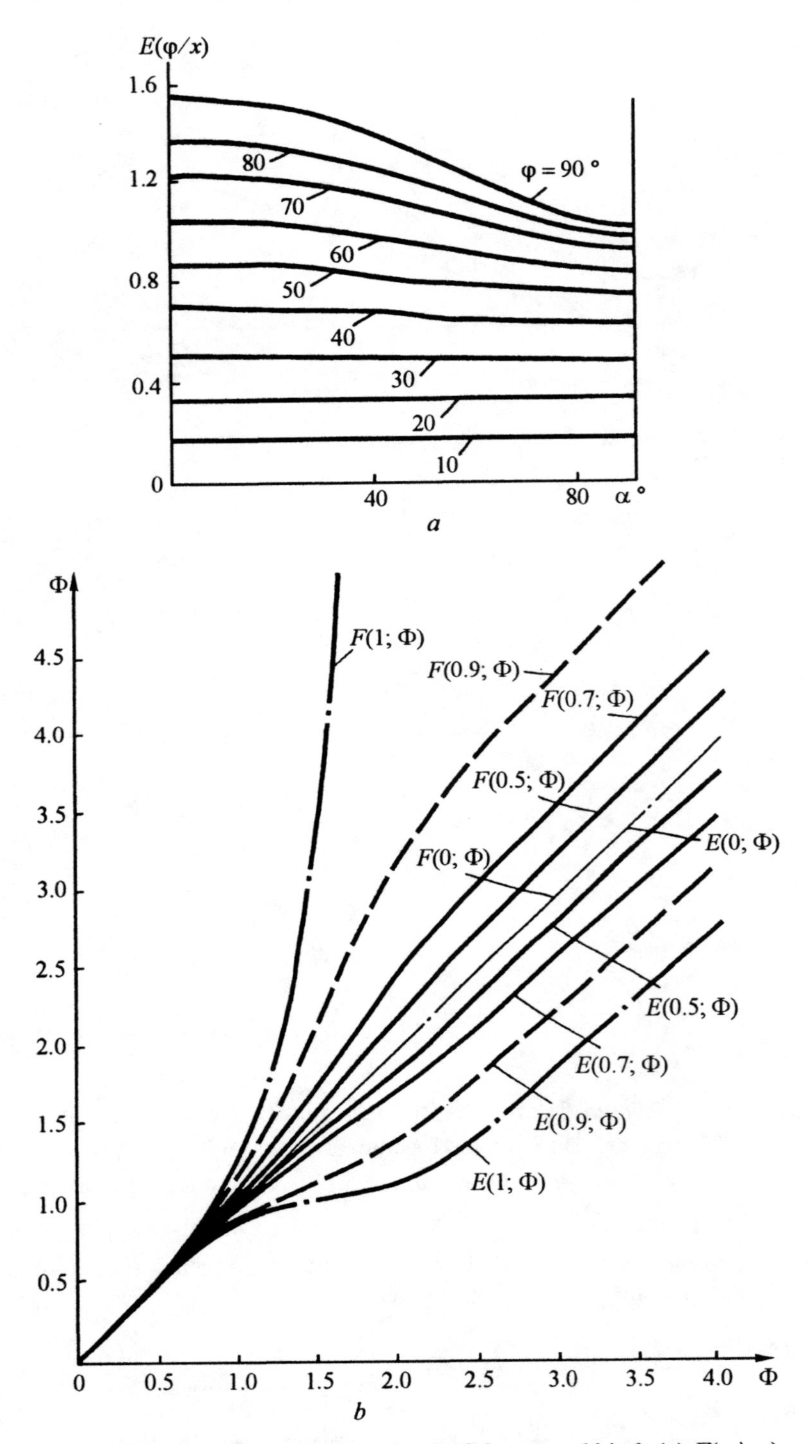

Figure 4.71. The incomplete elliptic integral of the second kind: (*a*) $F(\varphi \setminus \alpha)$, φ is a constant; (*b*) $F(k,\Phi)$, $E(k,\Phi)$.

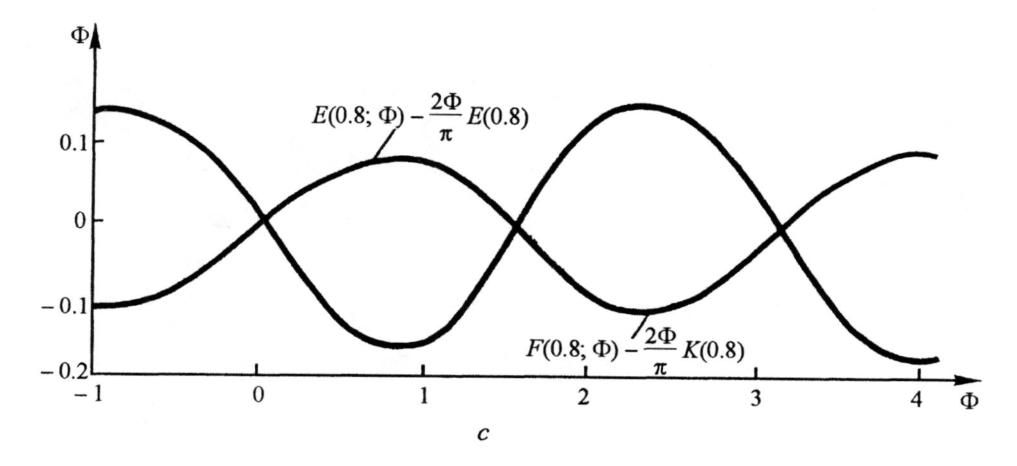

Figure 4.71. The incomplete elliptic integral of the second kind: (*c*) combinations of incomplete and complete integrals.

$$\text{sn}x = \,= \text{sn}\,(k;\,x) = \sin\,(\text{am}\,(k;\,x)),$$

$$\text{cn}x = \,= \text{cn}\,(k;\,x) = \cos\,(\text{am}\,(k;\,x)),$$

$$\text{dn}x = \,= \text{dn}\,(k;\,x) = \sqrt{1 - k^2 \sin^2\,(\text{am}\,x)}\,.$$

There are also nine functions that are used:

$$\text{ns}\,x = \frac{1}{\text{sn}\,x},\quad \text{nc}\,x = \frac{1}{\text{cn}\,x},\quad \text{nd}\,x = \frac{1}{\text{dn}\,x},$$

$$\text{cs}\,x = \frac{\text{cn}\,x}{\text{sn}\,x},\quad \text{sc}\,x = \frac{\text{sn}\,x}{\text{cn}\,x},\quad \text{sd}\,x = \frac{\text{sn}\,x}{\text{dn}\,x},$$

$$\text{ds}\,x = \frac{\text{dn}\,x}{\text{sn}\,x},\quad \text{dc}\,x = \frac{\text{dn}\,x}{\text{cn}\,x},\quad \text{cd}\,x = \frac{\text{cn}\,x}{\text{dn}\,x}.$$

These notations were suggested by Glaisher. At $x = 0$, the functions take the values: $\text{sn}\,0 = 0$, $\text{cn}\,0 = 1$, $\text{dn}\,0 = 1$; see Figures 4.75, 4.76.

All elliptic Jacobi functions can be reduced to trigonometric functions or the identity, if $k = 0$. In the same way, all elliptic Jacobi functions can be reduced to hyperbolic functions or the identity, if $k = 1$. For example,

$$\text{dc}\,(0;\,x) = \frac{1}{\text{cd}\,(0;\,x)} = \frac{1}{\cos x} = \sec x,$$

$$\text{ns}\,(1;\,x) = \frac{\text{nd}\,(1;\,x)}{\text{sd}\,(1;\,x)} = \frac{\cosh x}{\sinh x} = \coth x.$$

The following approximation formulas hold:

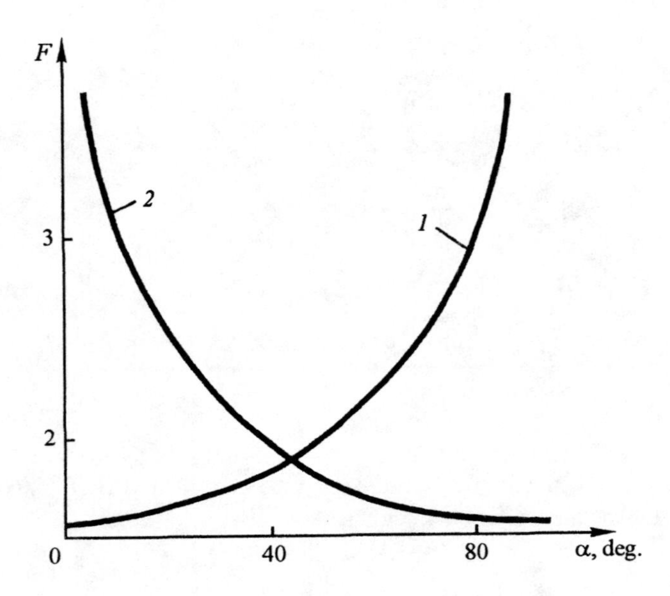

Figure 4.72. The complete elliptic integral of the first kind: *1*. $K = F(\varphi \setminus \alpha)$, *2*. $K' = F'(90° \setminus 90° - \alpha)$.

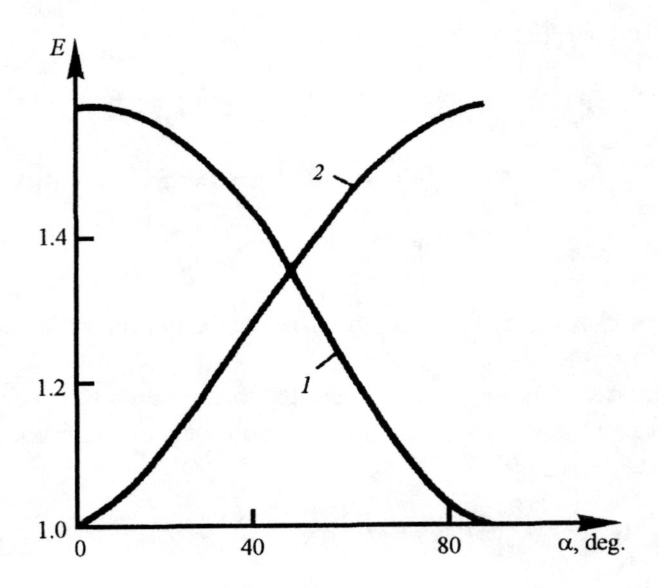

Figure 4.73. The complete elliptic integral of the second kind: *1*. $E = E(\varphi \setminus \alpha)$, *2*. $E' = E(90° \setminus 90° - \alpha)$.

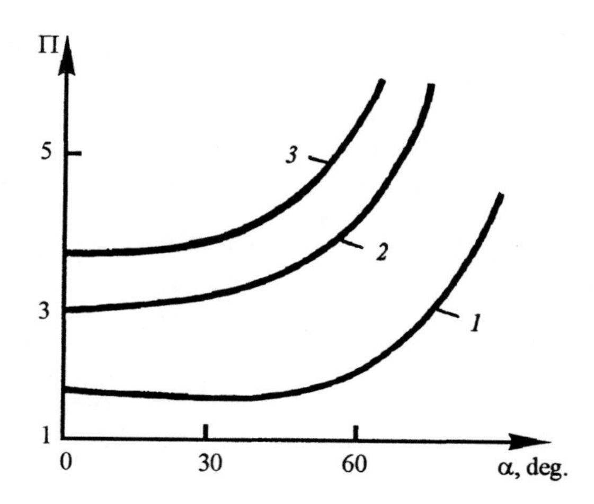

Figure 4.74. The complete elliptic integral of the second kind: *1*. $\Pi = (n, 90° \setminus \alpha)$, $n = 0$, *2*. $n = .7$, 3. $n = .8$.

$$\mathrm{cn}\,(k; x) \approx 1 - \frac{x^2}{2}, \quad \mathrm{ds}\,(k; x) \approx \mathrm{cosec}\, x, \quad \mathrm{nd}\,(k; x) \approx \cosh x.$$

We also have the following formulas:

$$\mathrm{sn}\,(-x) = -\,\mathrm{sn}\,(x), \quad \mathrm{cn}\,(-x) = \mathrm{cn}\,(x), \quad \mathrm{dn}\,(-x) = \mathrm{dn}\, x, \quad \mathrm{sn}^2 x + \mathrm{cn}^2 x = 1,$$

$$\mathrm{dn}^2 x = 1 - k^2 \mathrm{sn}^2 x = k'^2 + k^2 \mathrm{cn}^2 x = \mathrm{cn}^2 x + k'^2 \mathrm{sn}^2 x,$$

$$\mathrm{sn}^2 x = \frac{1 - \mathrm{cn}\, 2x}{1 + \mathrm{dn}\, 2x}, \quad \mathrm{cn}^2 x = \frac{\mathrm{cn}\, 2x + \mathrm{dn}\, 2x}{1 + \mathrm{dn}\, 2x},$$

$$\mathrm{dn}^2 x = \frac{\mathrm{dn}\, 2x + k^2\, \mathrm{cn}\, 2x + k'^2}{1 + \mathrm{dn}\, 2x}.$$

The Jacobi zeta function is $Z\,(\varphi \setminus \alpha) = E(\varphi \setminus \alpha) - E(\alpha) F(\alpha \setminus \alpha)/\boldsymbol{K}(\alpha)$ (Figure 4.77).

The *Haiman lambda function* has the form

$$\Lambda_0(\varphi \setminus \alpha) = \frac{F(\varphi \setminus 90° - \alpha)}{\boldsymbol{K}'(\alpha)} + \frac{2}{\pi} \boldsymbol{K}(\alpha)\, Z(\varphi \setminus 90° - \alpha),$$

$$\Lambda_0(\varphi \setminus \alpha) = \frac{2}{\pi} \left\{ \boldsymbol{K}(\alpha)\, F(\varphi \setminus 90° - \alpha) - [\boldsymbol{K}(\alpha) - E(\alpha)]\, \boldsymbol{F}(\varphi \setminus 90° - \alpha) \right\},$$

see Figure 4.78.

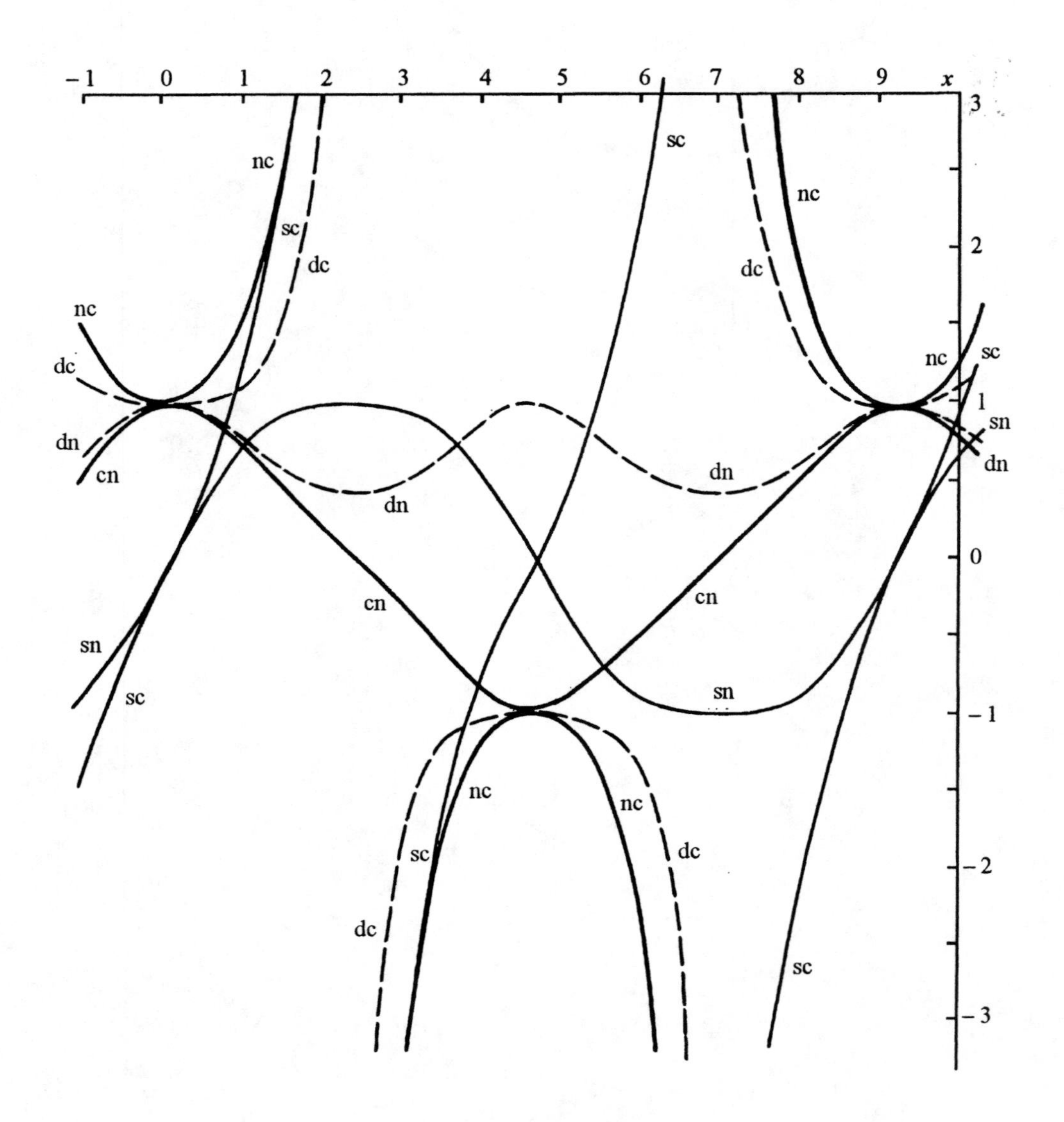

Figure 4.75. The elliptic functions, $k = .9$.

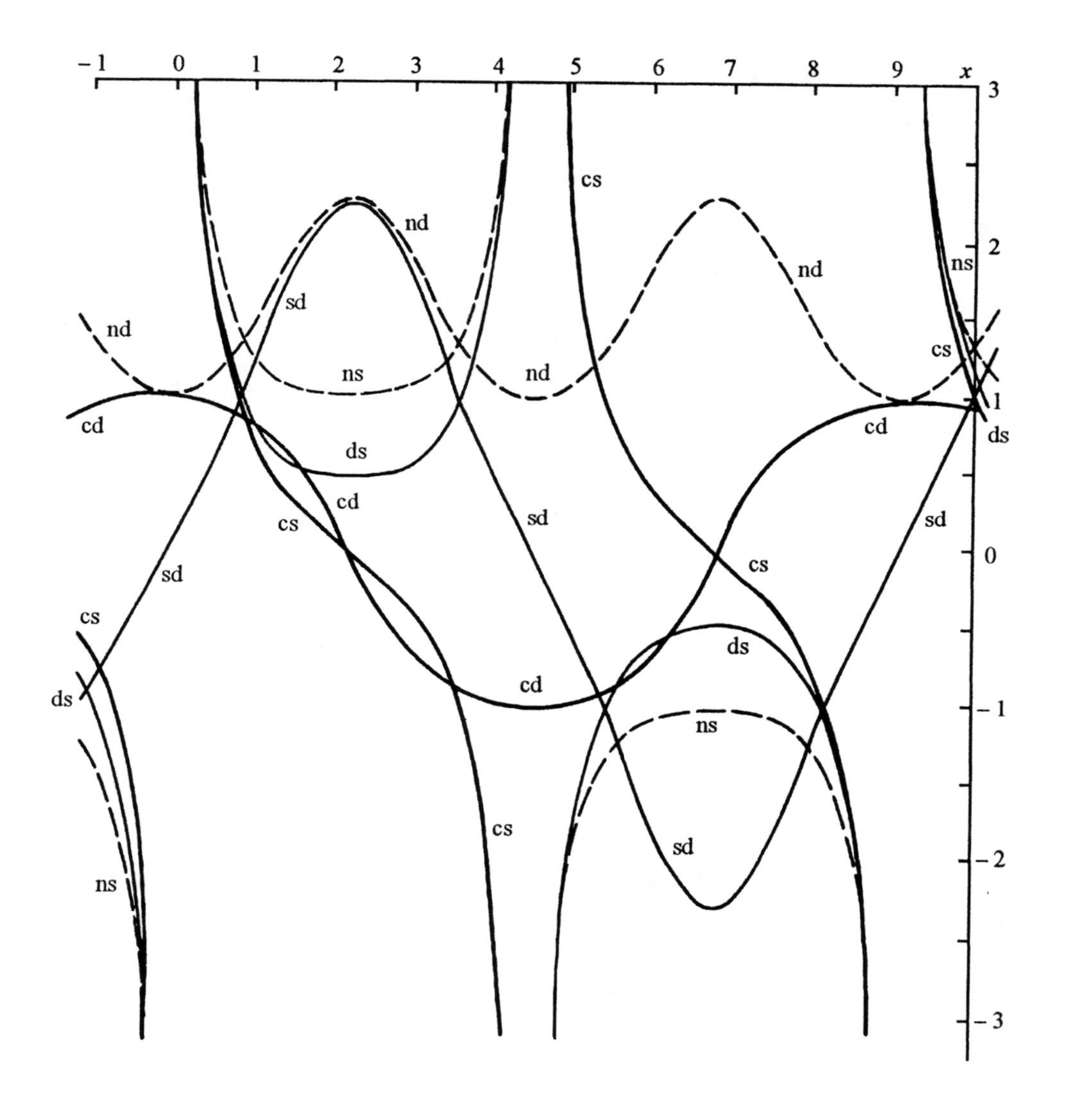

Figure 4.76. The elliptic functions, $k = .9$.

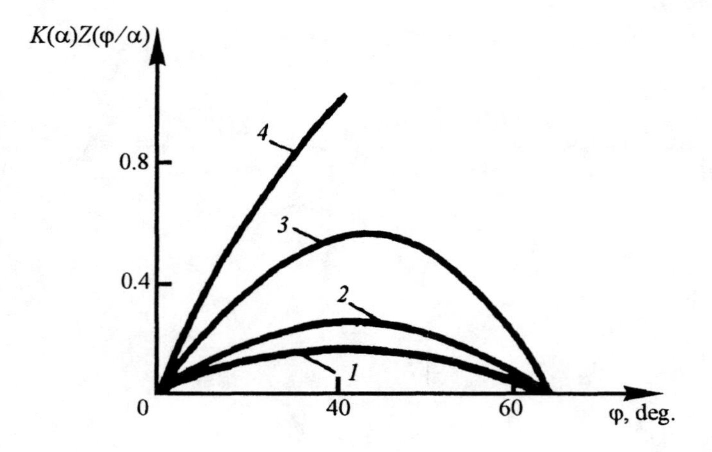

Figure 4.77. The Jacobi zeta function $K(\alpha)Z(\varphi \setminus \alpha)$: *1.* $\alpha = 30°$, *2.* $\alpha = 45°$, *3.* $\alpha = 60°$, *4.* $\alpha = 75°$.

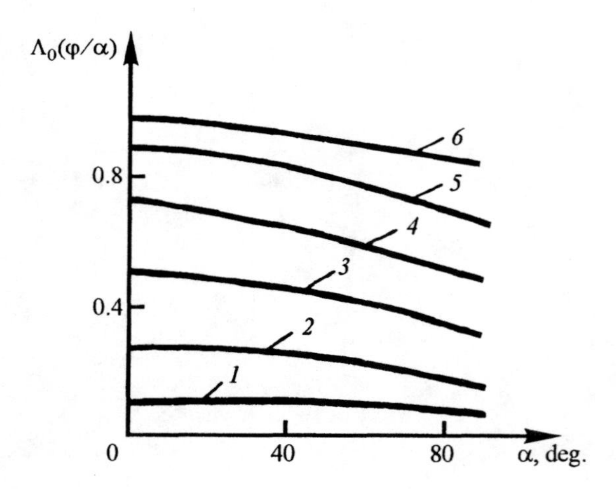

Figure 4.78. The Haiman lambda function $\Lambda_0(\varphi \setminus \alpha)$: *1.* $\varphi = 5°$, *2.* $\varphi = 15°$, *3.* $\varphi = 30°$, *4.* $\varphi = 45°$, *5.* $\varphi = 60°$, *6.* $\varphi = 75°$.

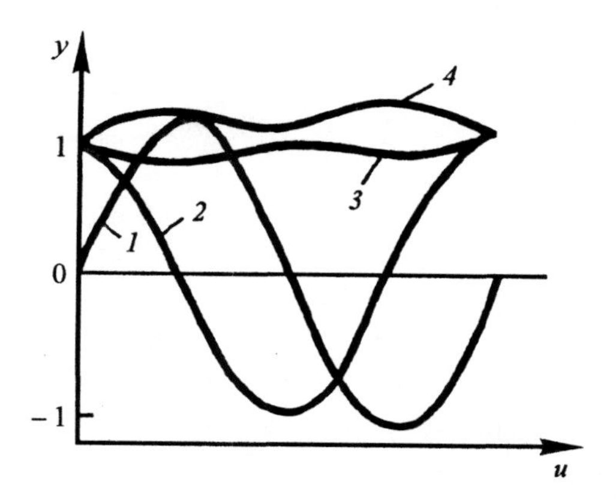

Figure 4.79. Nevill's theta functions: *1.* $\Theta_s(u)$, *2.* $\Theta_c(u)$, *3.* $\Theta_d(u)$, *4.* $\Theta_n(u)$.

Theta functions are given by

$$\Theta_1(z, q) = \Theta_1(z) = 2q^{1/4} \sum_{n=0}^{\infty} (-1)^n q^{n(n+1)} \sin (2n+1)z,$$

where $q = e^{-\pi k'/k}$ is the Jacobi parameter, $|q| < 1$,

$$\Theta_2(z, q) = \Theta_2(z) = 2q^{1/4} \sum_{n=0}^{\infty} q^{n(n+1)} \cos (2n+1)z,$$

$$\Theta_3(z, q) = \Theta_3(z) = 1 + 2 \sum_{n=0}^{\infty} q^{n^2} \cos 2nz,$$

$$\Theta_4(z, q) = \Theta_4(z) = 1 + 2 \sum_{n=1}^{\infty} (-1)^n q^{n^2} \cos 2nz.$$

Nevill's notation of the theta function is

$$\Theta_c(u) = \Theta_2(v)/\Theta_2(0); \quad \Theta_s(u) = \Theta_1(v)/\Theta_1(0) \quad (v = \pi u/2K),$$

$$\Theta_d(u) = \Theta_3(v)/\Theta_3(0); \quad \Theta_n(u) = \Theta_4(v)/\Theta_4(0),$$

see Figure 4.79.

4.12. Mathieu functions

Mathieu functions or *elliptic cylinder functions* are defined as solutions of the differential equation

$$\frac{d^2u}{dz^2} + (a - 2q\cos 2z)\, u = 0.$$

In the narrow sense, the (periodic) Mathieu functions $\varphi_m(z, q)$ are only those which are periodic solutions of this equation with period 2π. It should be noted that not all for pairs of a and q, the Mathieu equation has periodic solutions of period 2π. However, for any real value of the parameter q there is an infinite sequence of "eigen values" of the parameter a such that there exists a solution, for $q \neq 0$, and is determined up to a constant. The functions $\varphi_m(z, q)$ are entire functions of z and take real values for real $z = x$.

The Mathieu functions are split into four classes. The functions in these classes are denoted by $\mathrm{ce}_{2n}(z, q)$, $\mathrm{se}_{2n+1}(z, q)$, $\mathrm{ce}_{2n+1}(z, q)$, $\mathrm{se}_{2n+2}(z, q)$, $n = 0,1,2,\ldots$. Considered as functions of z with the parameter q being a constant, these functions are denoted by $\mathrm{ce}_{2n}(z)$, $\mathrm{se}_{2n+1}(z)$, $\mathrm{ce}_{2n+1}(z)$, $\mathrm{se}_{2n+2}(z)$.

The modified Mathieu equation is

$$\frac{d^2u}{dz^2} - (a - 2q\cosh 2z)u = 0,$$

the algebraic form of the Mathieu equation is given by

$$(1 - t^2)\frac{d^2u}{dz^2} - t\frac{du}{dt} + (a + 2q - 4qt^2)\, u = 0 \quad (\cos z = t).$$

Main relations are:

$$\mathrm{ce}_m(z) = \mathrm{ce}_m(-z),\ \mathrm{se}_m(z) = -\,\mathrm{se}_m(-z),\ \mathrm{ce}_m(k\pi + z) = \mathrm{ce}_m(k\pi - z),\ \mathrm{se}_m(k\pi + z) = -\,\mathrm{se}_m(k\pi - z)$$

$$(k \text{ is an integer}),\ \mathrm{ce}_m(\pi/2 + z) = (-1)^m\mathrm{ce}_m(\pi/2 - z),\ \mathrm{se}_m(\pi/2 + z) = (-1)^{m+1}\mathrm{se}_m(\pi/2 - z),$$

$$\varphi_m(z + \pi) = (-1)^m\varphi_m(z).$$

Symmetry of the Mathieu functions is shown in Figure 4.80.

The Mathieu functions are normalized such that

$$\frac{2}{\pi}\int_0^{\frac{\pi}{2}} \varphi_m^2(x)\, dx = \begin{cases} 1, & m = 0, \\ \frac{1}{2}, & m = 1, 2, \ldots, \end{cases}$$

if the expressions

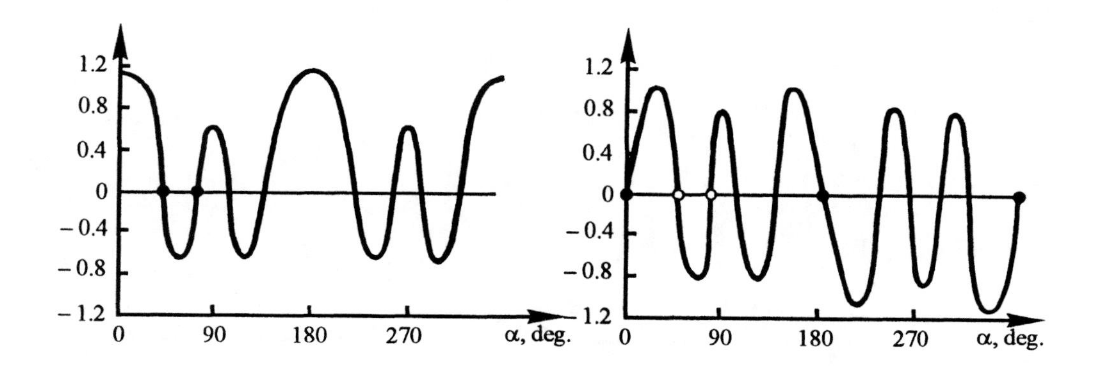

Figure 4.80. The Mathieu functions ce_1, se_5.

$$(-1)^n \mathrm{ce}_{2n}(z),\quad (-1)^n\, \mathrm{se}_{2n+1}(z),\quad (-1)^n \frac{\mathrm{ce}_{2n+1}(z)}{\cos z},\quad \frac{(-1)^{n-1}\mathrm{se}_{2n}(z)}{\cos z}$$

are positive at $z = \pi/2$. For $q \to 0$, we have:

$$\mathrm{ce}_0(z, q) \to 1,\quad \mathrm{ce}_m(z, q) \to \cos mz,\quad \mathrm{se}_m(z, q) \to \sin mz \quad (m = 1, 2, \ldots).$$

Zeros of the Mathieu functions are determined from the expressions

$$\mathrm{ce}_0(x) > 0,\quad \mathrm{se}_{2n}\, k\pi/2 = \mathrm{se}_{2n+1} k\pi = \mathrm{ce}_{2n+1}(k + 1/2)\pi = 0,$$

where k is an integer.

Each function $\mathrm{ce}_{2n}(x)$, $\mathrm{se}_{2n+1}(x)$, $\mathrm{ce}_{2n+1}(x)$, $\mathrm{se}_{2n+2}(x)$ has n real zeros that lie strictly between $x = 0$ and $x = \pi/2$. As q increases, these zeros approach $\pi/2$ (Figures 4.81 – 4.85).

The Fourier expansions of these functions have the form:

$$\mathrm{ce}_{2n}(x) = \sum_{i=0}^{\infty} A_{2n,\, 2i} \cos 2ix,\quad \mathrm{ce}_{2n+1}(x) = \sum_{i=0}^{\infty} A_{2n+1,\, 2i+1} \cos (2i + 1)\, x,$$

$$\mathrm{se}_{2n}(x) = \sum_{i=1}^{\infty} B_{2n,\, 2i} \sin 2ix,\quad \mathrm{se}_{2n+1}(x) = \sum_{i=0}^{\infty} B_{2n+1,\, 2i+1} \sin (2i + 1)\, x,$$

where the coefficients A and B satisfy the formulas: for $\mathrm{ce}_{2n}(x)$,

$$-aA_{2n,0} + 2qA_{2n,\, 2} = 0,$$

$$4qA_{2n,0} + (1 - a)A_{2n,\, 2} + 2qA_{2n,\, 4} = 0,$$

$$2qA_{2n,\, 2i-2} + (i^2 - a)A_{2n,\, 2i} + 2qA_{2n,\, 2i+2} = 0 \quad (i > 1),$$

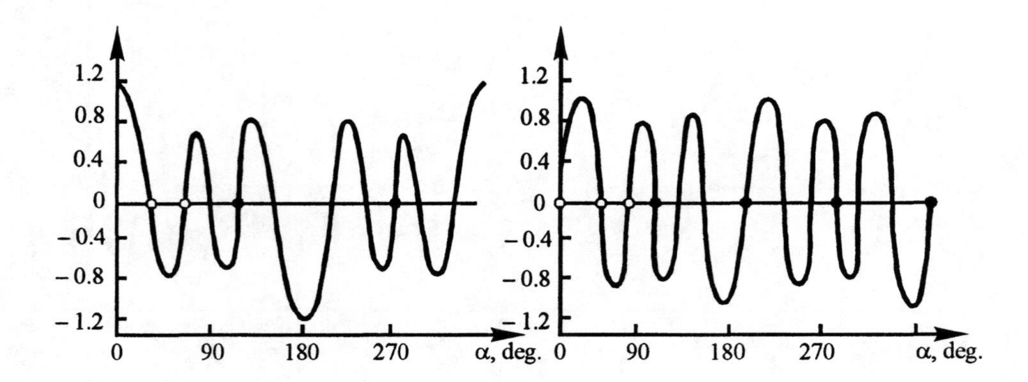

Figure 4.81. The Mathieu functions se_6.

for $se_{2n}(x)$,

$$(1-a)B_{2n,\,2}+2qB_{2n,\,4}=0,$$

$$2qB_{2n,\,2i-2}+(i^2-a)B_{2n,\,2i}+2qB_{2n,\,2i+2}=0 \quad (i>1),$$

for ce_{2n+1},

$$2qA_{2n+1,\,1}+\left(\frac{1}{4}-a\right)A_{2n+1,\,1}+2qA_{2n+1,\,3}=0,$$

$$2qA_{2n+1,\,2i-1}+\left[\left(i+\frac{1}{2}\right)^2-a\right]A_{2n+1,\,2i+1}+2qA_{2n+1,\,2i+3}=0 \quad (i>0),$$

for $se_{2n+1}(x)$,

$$-2qB_{2n+1,\,1}+\left(\frac{1}{4}-a\right)B_{2n+1,\,1}+2qB_{2n+1,\,3}=0,$$

$$2qB_{2n+1,\,2i-1}+\left[\left(i+\frac{1}{2}\right)^2-a\right]B_{2n+1,\,2i+1}+2qB_{2n+1,\,2i+3}=0 \quad (i>0).$$

Note that the normalization conditions imply that

$$2A_0^2+A_2^2+A_4^2+\ldots=B_2^2+B_4^2+\ldots=A_1^2+A_3^2+\ldots=B_1^2+B_3^2+\ldots=1$$

for any q, any index of the function, and any first index. For $ce_0(x)$, the first sum in the formula equals 2.

Solutions of the modified Mathieu equations are sometimes called *radial Mathieu functions*. Graphs of radial Mathieu functions are shown in Figures 4.86 – 4.88.

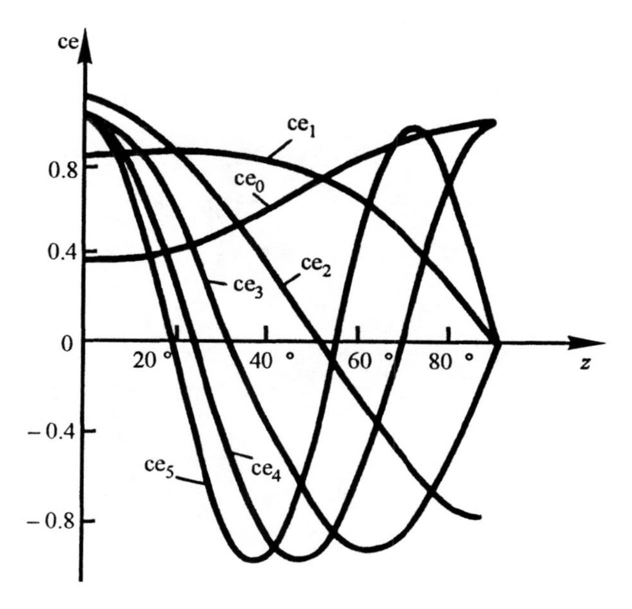

Figure 4.82. Even periodic Mathieu functions (of orders 0 — 5, $q = 1$).

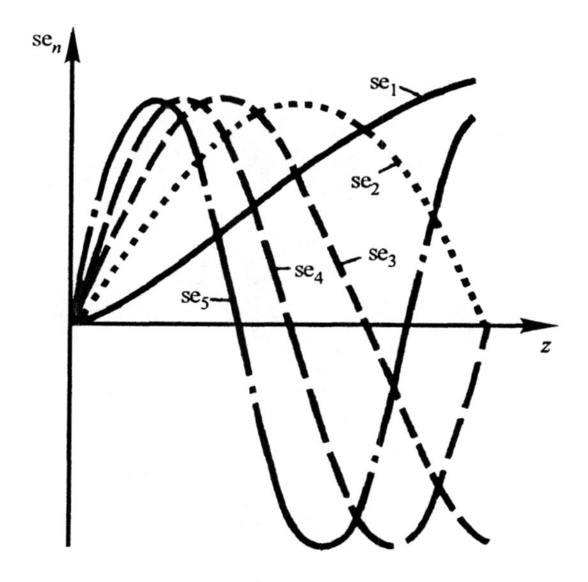

Figure 4.83. Odd periodic Mathieu functions (of orders 1 — 5, $q = 1$).

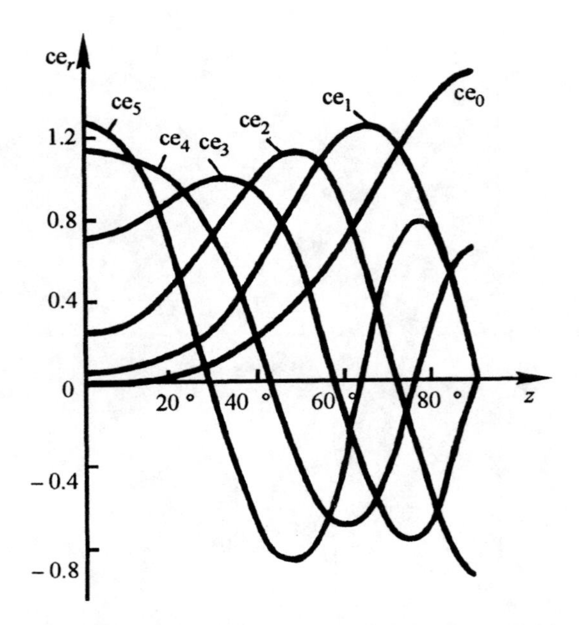

Figure 4.84. Even periodic Mathieu functions (of orders 0 — 5, $q = 10$).

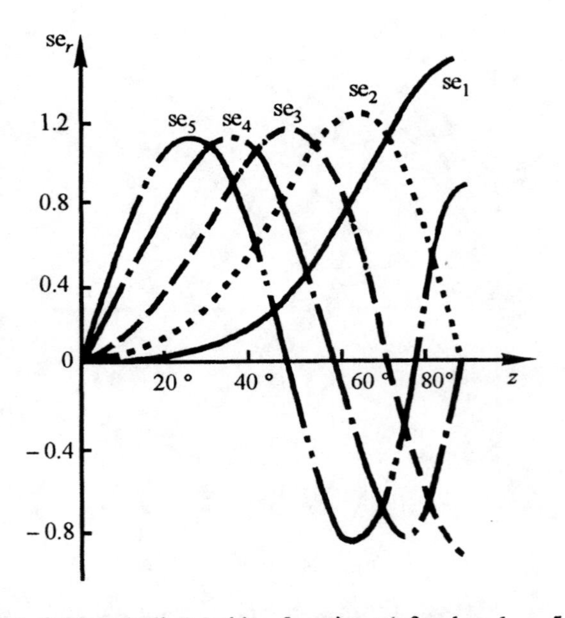

Figure 4.85. Odd periodic Mathieu functions (of orders 1 — 5, $q = 10$).

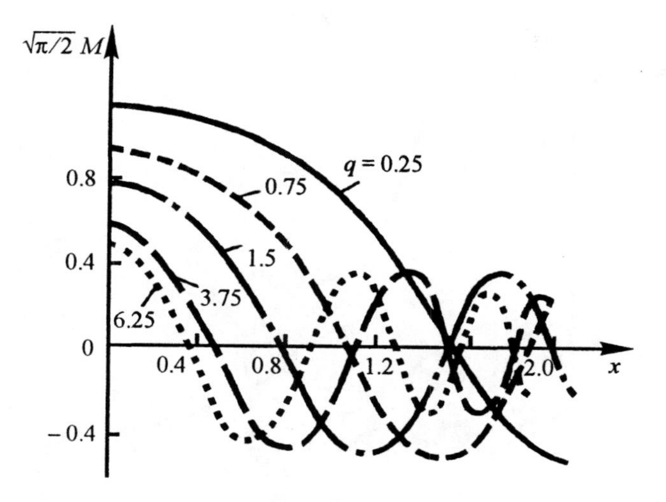

Figure 4.86. Radial Mathieu function of the first kind.

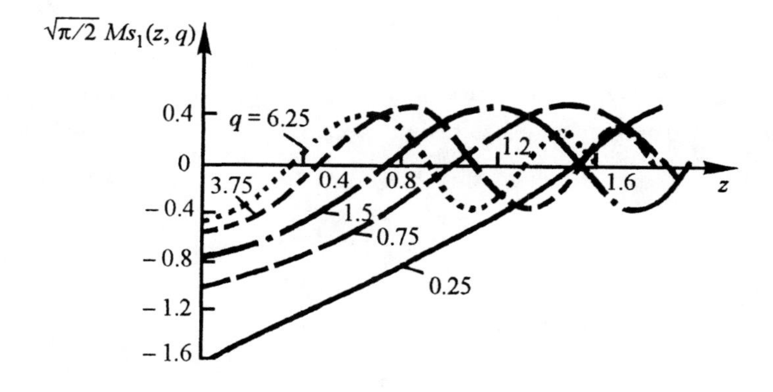

Figure 4.87. The radial Mathieu function of the second kind.

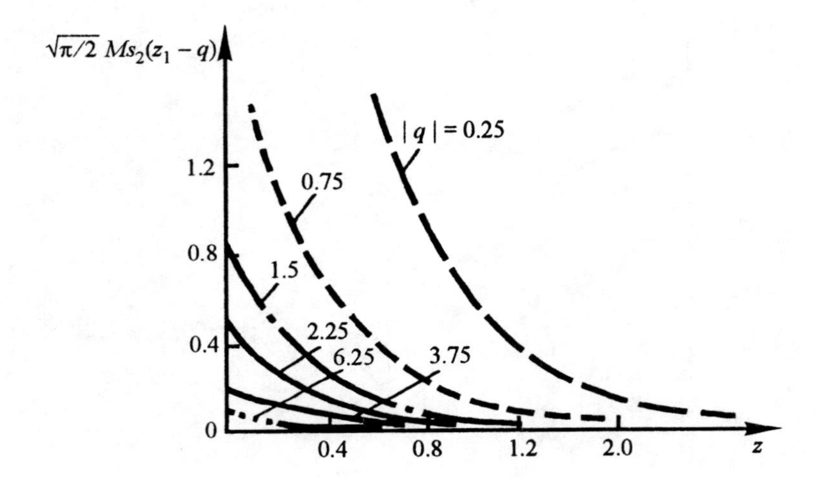

Figure 4.88. Radial Mathieu function of the third kind.

4.13. Zeta numbers and related functions

The numbers $\zeta(n)$, $\lambda(n)$, $\eta(n)$, $\beta(n)$ are called zeta, lambda, eta, and beta numbers. Here $n = 1,2,3....$ The corresponding functions, with n not necessarily being integer and positive, are defined by the integrals

$$\zeta(\nu) = \frac{1}{\Gamma(\nu)} \int_0^\infty \frac{t^{\nu-1} dt}{e^t - 1}, \quad \nu > 1,$$

$$\lambda(\nu) = \frac{1}{2\Gamma(\nu)} \int_0^\infty t^{\nu-1} \operatorname{csch} t \, dt, \quad \nu > 1,$$

$$\eta(\nu) = \frac{1}{\Gamma(\nu)} \int_0^\infty \frac{t^{\nu-1} dt}{e^t + 1}, \quad \nu > 0,$$

$$\beta(\nu) = \frac{1}{2\Gamma(\nu)} \int_0^\infty t^{\nu-1} \operatorname{sech} t \, dt, \quad \nu > 0.$$

The functions $\zeta(\nu)$, $\lambda(\nu)$ are unbounded only at $\nu = 1$ the functions $\eta(\nu)$ and $\beta(\nu)$ are always bounded. As ν increases, all these functions approach 1. The functions $\zeta(\nu)$, $\lambda(\nu)$, $\eta(\nu)$ equal zero at negative even integers, $\beta(\nu) = 0$ at $\nu = -1, -3, -5,...$ (Figure 4.89).

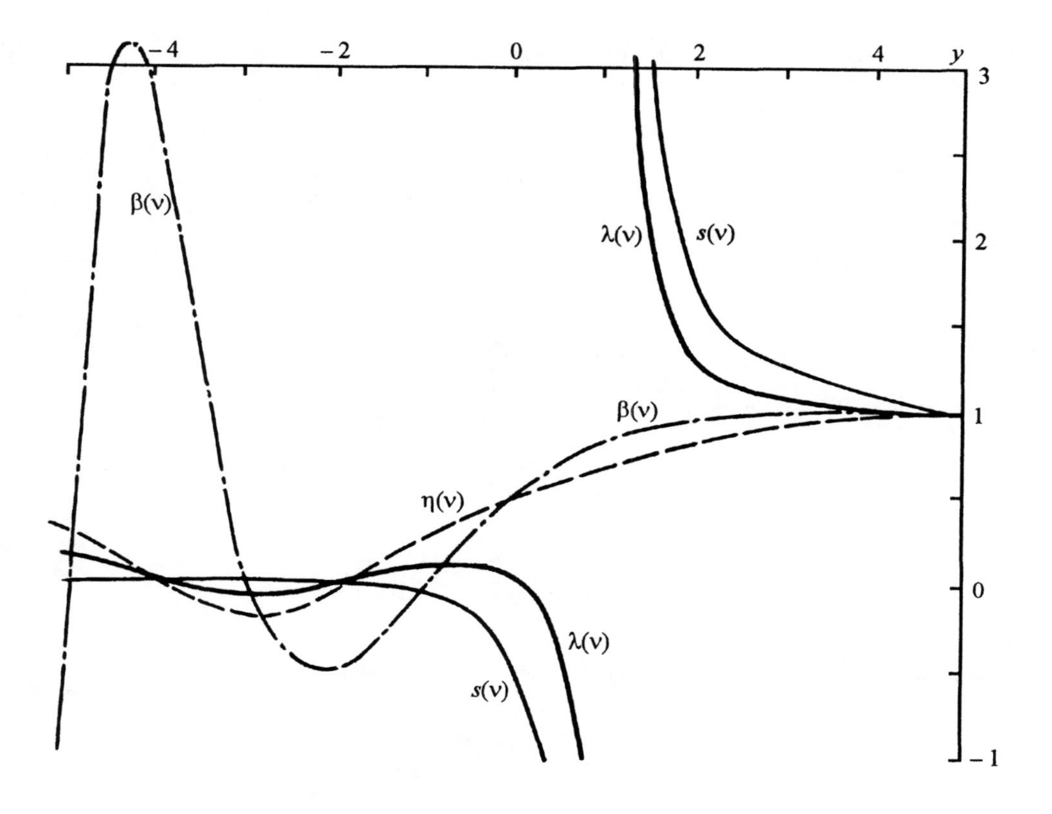

Figure 4.89. The functions $\zeta(\nu)$, $\lambda(\nu)$, $\beta(\nu)$.

The main relations are:

$$\frac{\zeta(\nu)}{2^\nu} = \frac{\lambda(\nu)}{2^\nu - 1} = \frac{\eta(\nu)}{2^\nu - 2},$$

$$\zeta(\nu) + \eta(\nu) = 2\lambda(\nu),$$

$$\zeta(1 - \nu) = \frac{2\Gamma(\nu)\,\zeta(\nu)}{(2\pi)^\nu} \cos\frac{\nu\pi}{2},$$

$$\beta(1 - \nu) = \left(\frac{2}{\pi}\right)^\nu \Gamma(\nu)\,\beta(\nu) \sin\frac{\nu\pi}{2},$$

$$\zeta(\nu) = 1 + \frac{1}{2^\nu} + \frac{1}{3^\nu} + \ldots = \sum_{k=1}^{\infty} k^{-\nu}, \quad \nu > 1,$$

$$\lambda(\nu) = 1 + \frac{1}{3^\nu} + \frac{1}{5^\nu} + \ldots = \sum_{k=1}^{\infty} (2k - 1)^{-\nu}, \quad \nu > 1,$$

$$\eta(\nu) = 1 - \frac{1}{2^\nu} + \frac{1}{3^\nu} - \ldots = \sum_{k=1}^{\infty} (-1)^{k+1} k^{-\nu}, \quad \nu > 0,$$

$$\beta(\nu) = 1 - \frac{1}{3^\nu} + \frac{1}{5^\nu} - \ldots = \sum_{k=1}^{\infty} (-1)^k (2k - 1)^{-\nu}.$$

Zeta and beta functions are related to the Bernoulli and Euler numbers via the formulas:

$$\zeta(-n) = \frac{-B_{n+1}}{n+1} \quad (n = 1, 2, 3, \ldots),$$

$$\beta(-n) = -\frac{E_n}{2} \quad (n = 0, 1, 2, \ldots),$$

$$\zeta(n) = \frac{(2\pi)^n\,|B_n|}{2n!} \quad (n = 2, 4, 6, \ldots),$$

$$\beta(n) = \left(\frac{\pi}{2}\right)^n \frac{|E_{n-1}|}{2(n-1)!} \quad (n = 1, 3, 5, \ldots).$$

The functions $\zeta(\nu)$, $\lambda(\nu)$, $\eta(\nu)$, $\beta(\nu)$ are particular cases of the Hurwitz function $\zeta(\nu; \mu)$,

$$\zeta(\nu) = \zeta(\nu; 1),$$

$$\lambda(\nu) = 2^{-\nu}\zeta\left(\nu; \frac{1}{2}\right),$$

$$\eta(\nu) = 2^{1-\nu}\,\zeta\left(\nu; \frac{1}{2}\right) - \zeta(\nu; 1) = \eta(\nu; 1),$$

$$\beta(\nu) = 4^{-\nu}\left[\zeta\left(\nu; \frac{1}{4}\right) - \zeta\left(\nu; \frac{3}{4}\right)\right] = 2^{-\nu}\,\eta\left(\nu; \frac{1}{2}\right).$$

4.14. Hurwitz function

The *Hurwitz function* can be defined in terms of the definite integral

$$\zeta(\nu; u) = \frac{1}{\Gamma(\nu)}\int_0^\infty \frac{t^{\nu-1}e^{-ut}}{1-e^{-t}}\,dt = \frac{1}{\Gamma(\nu)}\int_1^\infty \frac{\ln^{\nu-1} t}{t^{u}\,(t-1)}\,dt$$

or the Hermite integral

$$\zeta(\nu; u) = \frac{u^{-\nu}}{2} + \frac{u^{1-\nu}}{\nu-1} + 2\int_0^\infty \frac{\sin\left(\nu \arctan\frac{t}{u}\right)dt}{(u^2+t^2)^{\nu/2}\,(e^{2\pi t}-1)} \quad (u > 0),$$

see Figure 4.90.

The main relations for the Hurwitz function are:

$$\zeta(0; u) = \frac{1}{2} - u;\quad \zeta(-1; u) = \frac{-1+6u-6u^2}{12},$$

$$\zeta(-2; u) = \frac{-u+3u^2-2u^3}{6},\quad \lim_{\nu\to 1}\frac{\zeta(\nu; u)}{\Gamma(1-\nu)} = -1,$$

$$\lim_{\nu\to 1}\left[\zeta(\nu; u) - \frac{1}{\nu-1}\right] = -\psi(u),\quad \zeta(\nu; u+1) = \zeta(\nu; u) - \frac{1}{u^{\nu}} \quad (u \ge 0),$$

$$\frac{1}{u-1} = \sum_{n=2}^{\infty} \zeta(n; \nu) \quad (u > 1),\quad \frac{1}{u} = \sum_{n=2}^{\infty} (-1)^n\,\zeta(n; u) \quad (u > 0),$$

$$\psi(u) - \ln(u-1) = \sum_{n=2}^{\infty} \frac{\zeta(n; u)}{n} \quad (u > 1),$$

$$\ln u - \psi(u) = \sum_{n=2}^{\infty} (-1)^n\,\frac{\zeta(n; u)}{n} \quad (u > 0),$$

$$\zeta(\nu; u) = \sum_{k=0}^{\infty} (k+u)^{-\nu} \quad (\nu > 1),$$

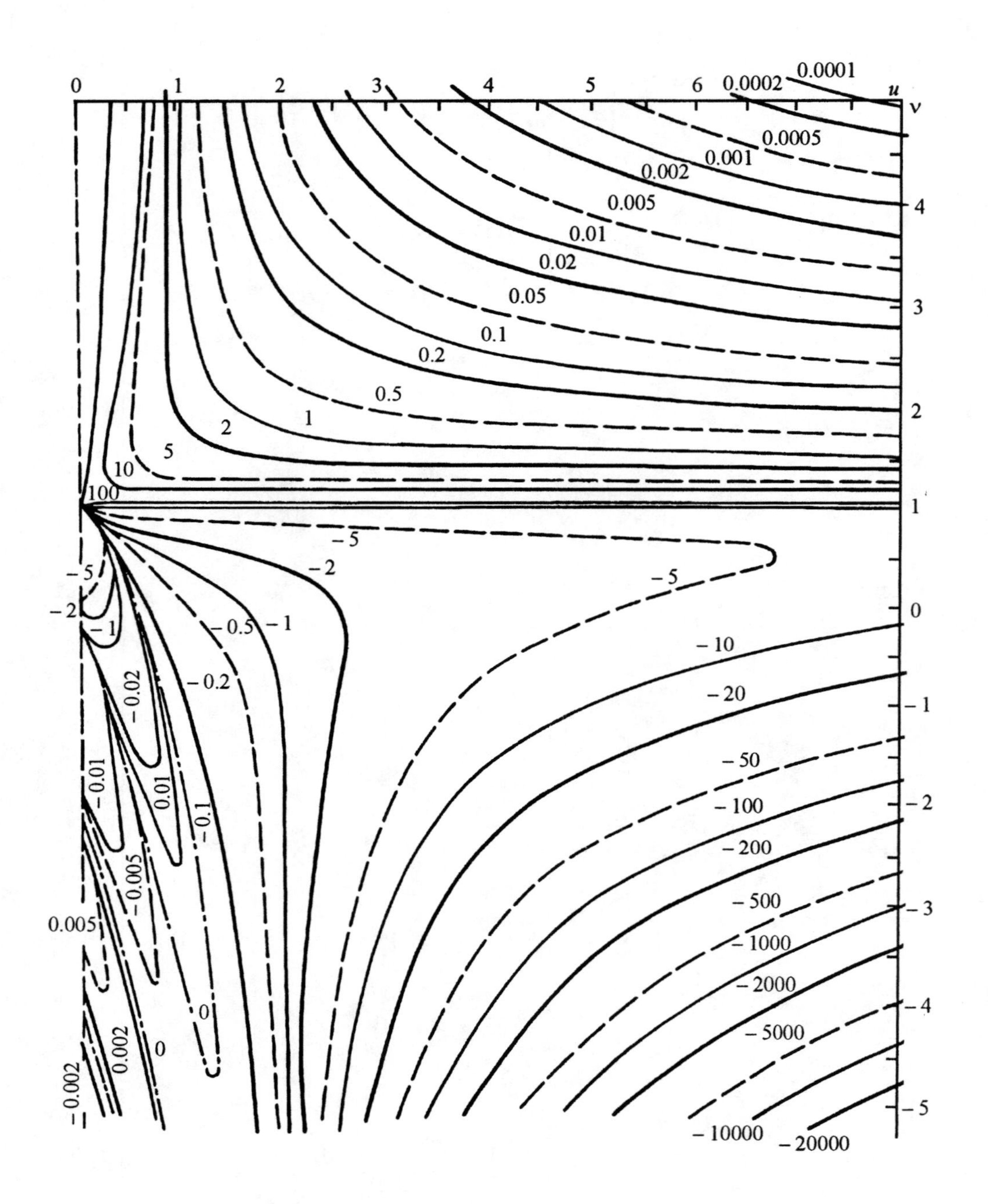

Figure 4.90. The Hurwitz functions $\zeta(\nu,u)$.

$$\zeta(\nu; u) = 2\,\Gamma(1-\nu)\sum_{k=1}^{\infty}\frac{\sin\left(2k\pi u + \dfrac{\pi\nu}{2}\right)}{(2k\pi)^{1-\nu}} \quad (\nu < 0,\ 0 \le u \le 1),$$

$$\zeta(\nu, 0) = \begin{cases}\infty, & \nu > 0,\\ \zeta(\nu), & \nu < 0,\end{cases}$$

$$\zeta(\nu; 1) = \zeta(\nu), \quad \zeta(\nu; 2) = \zeta(\nu) = 1.$$

An approximation is given by

$$\zeta(\nu; u) \sim \frac{1}{2u^{\nu}}\left(\frac{2u}{\nu-1} + 1 + \frac{1}{6u} + \ldots\right).$$

The Hurwitz function is a particular case of the Lurch function,

$$\zeta(\nu; u) = \Phi(1; \nu; u),$$

where

$$\Phi(x; \nu; u) = \frac{1}{\Gamma(\nu)}\int_0^{\infty}\frac{t^{\nu-1}e^{-ut}}{1-xe^{-t}}\,dt = \sum_{k=0}^{\infty}\frac{x^k}{(k+u)^{\nu}}.$$

Consider interesting cases of the Lurch function:

$$\Phi(x; 0; u) = \frac{1}{1-x},$$

$$\Phi(0; \nu; u) = \frac{1}{u^{\nu}}, \quad \Phi(x; -1; u) = \frac{x}{(1-x)^2} + \frac{u}{1-x},$$

$$\Phi(x; n; u) = \frac{1}{u^n}\sum_{k=0}^{\infty}\frac{(u)_k^n}{(1+u)_k^n}\,x^k.$$

$$\Phi(x; 1; 1) = -\frac{\ln(1-x)}{x}, \quad \Phi(-1; \nu; u) = \eta(\nu; u).$$

The relationship between the eta function $\eta(\nu; \mu)$ and the eta function $\eta(\nu)$ is the same as between the Hurwitz function and the zeta function.

The function $\eta(\nu; \mu)$ is defined by the series

$$\eta(\nu; u) = \sum_{k=0}^{\infty}\frac{(-1)^k}{(k+u)^{\nu}} \quad (\nu > 0,\ u \ge 0).$$

The main relations the function satisfies are:

$$\eta(\nu; u+1) = \frac{1}{u^{\nu}} - \eta(\nu; u),$$

$$\eta(\nu;\, 2u) = 2^{-\nu}\left[\zeta(\nu,\, u) - \zeta(\nu;\, \frac{1}{2} + u)\right],$$

$$\zeta(\nu,\, 0) = \begin{cases} -\eta(\nu), & \nu < 0, \\ \infty, & \nu > 0, \end{cases}$$

$$\eta(\nu;1) = \eta(\nu), \quad \eta(\nu;2) = 1 - \eta(\nu).$$

4.15. Heaviside function. Dirac's delta function

The functions $u(x-a)$ and $u(x)$ are known as *Heaviside functions*, and are defined by

$$u(x-a) = \begin{cases} 0, & x < a, \\ \frac{1}{2}, & x = a, \\ 1, & x > a. \end{cases}$$

For $f(x)u(x-a)$, we have that

$$u(x-a)\, f(x) = \begin{cases} 0, & x < a, \\ \frac{f(x)}{2}, & x = a, \\ f(x), & x > a, \end{cases}$$

see Figures 4.91, 4.92.

The earlier considered functions $\operatorname{sign} x$ and $|x|$ are connected to the function $u(x)$ by

$$\operatorname{sign} x = 2u(x) - 1, \quad |x| = x \operatorname{sign} x = 2xu(x) - x.$$

The function $\operatorname{sign} x$ can also be defined as the integral

$$\operatorname{sign} x = \frac{2}{\pi}\int_0^\infty \frac{\sin xt}{t}\, dt.$$

The main relations are:

$$u(-x) = 1 - u(x), \quad u(x-a)\, f(x) + u(a-x)\, f(x) = f(x),$$

$$u(x-a)\, f(x) = u(x-a)\sum_{k=0}^{\infty} a_k x^k = \sum_{k=0}^{\infty} a_k u(x-a)(x-a)x^k,$$

$$\frac{d}{dx}\, u(x-a) = \delta(x-a),$$

$$\frac{d}{dx}\,[u(x-a)\, f(x)] = \begin{cases} 0, & x < a, \\ \frac{df(x)}{dx}, & x > a, \end{cases}$$

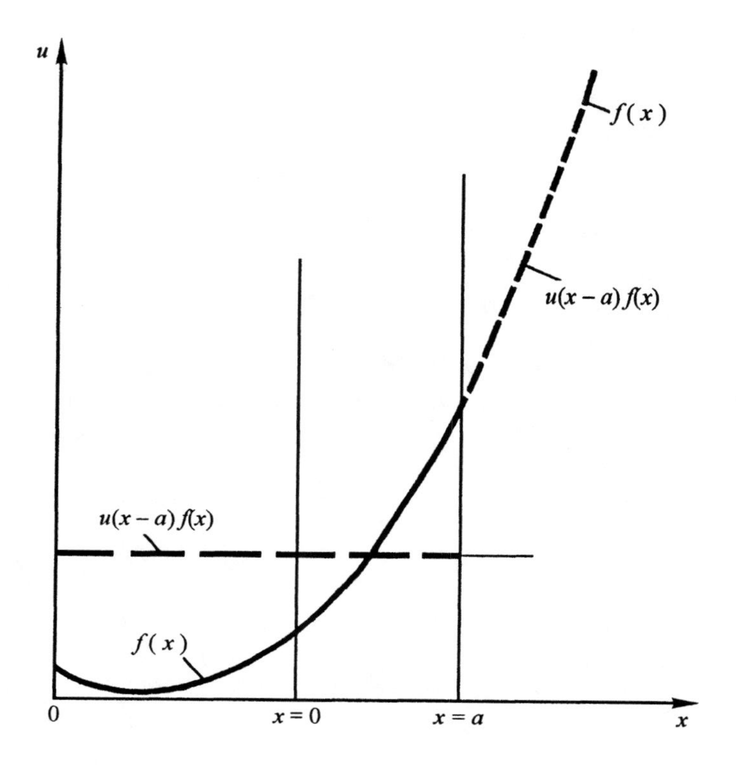

Figure 4.91. The function $u(x-a)f(x)$.

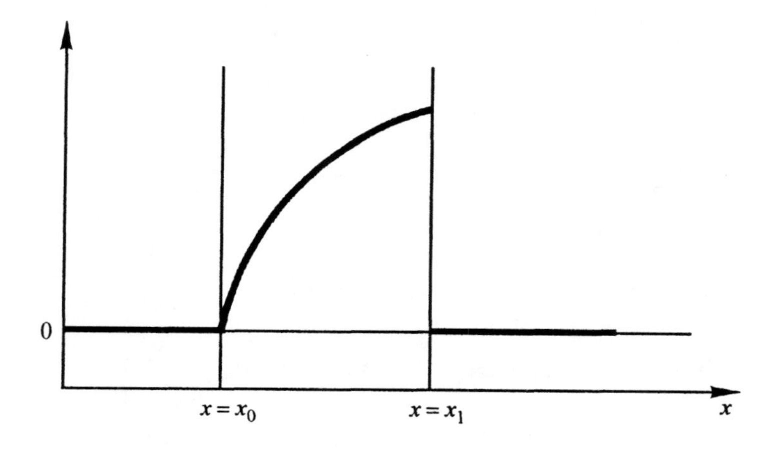

Figure 4.92. The function $[u(x-x_0) - u(x-x_1)]\,f(x)$, $x_0 < x_1$.

$$\frac{d^{\nu}u(x-a)}{\left[d(x-x_0)\right]^{\nu}}=\frac{[(x-a)]^{-\nu}}{\Gamma(1-\nu)} \quad (x_0<a<x),$$

$$\frac{d^{\nu}[u(x-a)f(x)]}{\left[d(x-x_0)\right]^{\nu}}=u(x-a)\frac{d^{\nu}f(x)}{[d(x-a)]^{\nu}} \quad (x_0<a<x),$$

$$u(x-a)\approx\frac{1}{2}[1+\tanh(bx-ba)],$$

where b is a sufficiently large positive number.

The *Dirac's delta function* $\delta(x-a)$ can be defined in many ways. For example,

$$\delta(x-a)=\lim_{b\to\infty}\sqrt{b/\pi}\,e^{-b(x-a)^2} \text{ or } \delta(x-a)=\lim_{c\to 0}\frac{1}{2c}\operatorname{sech}^2\left(\frac{x-a}{c}\right)$$

(Figure 4.93), or as $\delta(x-a)=\lim_{h\to 0}p\left(\frac{1}{h};h;x-a\right)$, where p is a pulse function (Figure 4.94) of the form

$$p(c;h;x-a)=\begin{cases}0, & x<a-\frac{h}{2},\\ c, & a-\frac{h}{2}<x<a+\frac{h}{2},\\ 0, & x>a+\frac{h}{2},\end{cases}$$

where $c=1/h$.

Still another way to define the delta function is by $\delta(x-a)=\frac{du(x-a)}{dx}$ or in terms of the integral

$$\delta(x-a)=\int_{-\infty}^{\infty}\cos 2\pi(x-a)t\,dt.$$

The main relations are:

$$\delta(a-x)=\delta(x-a),\quad \delta(\nu x)=\frac{\delta(x)}{\nu}\quad(\nu>0),$$

$$\delta(x^2-a^2)=\frac{1}{2a}[\delta(x-a)+\delta(x+a)]\quad(a>0),\quad \delta(x-a)=\begin{cases}0, & x\neq a,\\ \infty, & x=0.\end{cases}$$

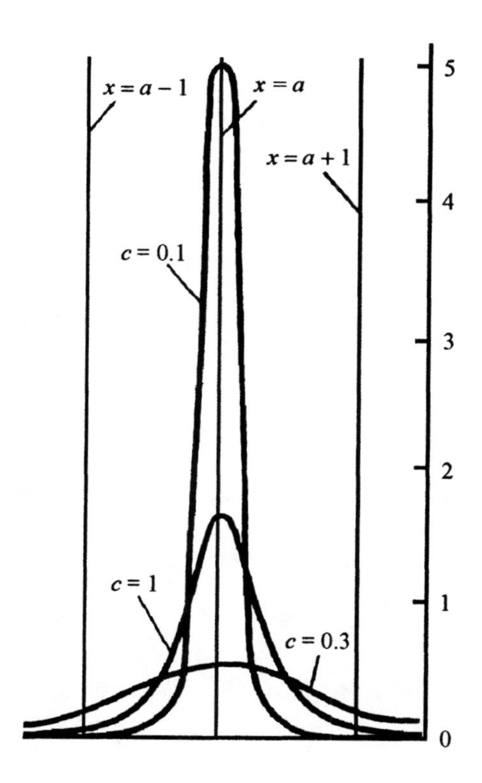

Figure 4.93. δ $(x - a)$.

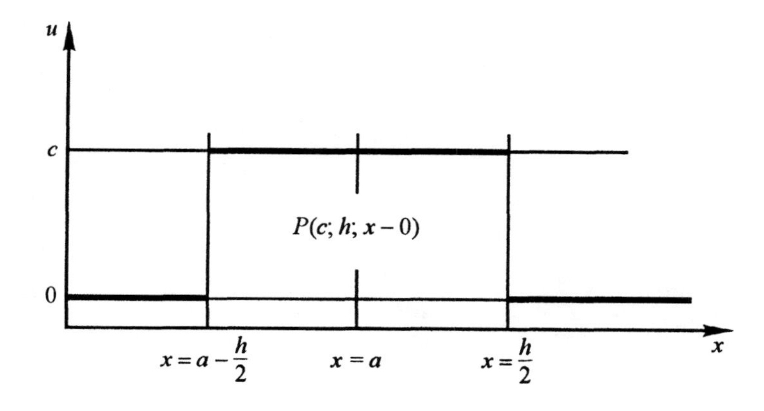

Figure 4.94. The pulse function p $(c; h; x - a)$.

Bibliography

1. Bateman, H. and Erdélyi, A. *Higher Transcendental Functions*, Vols. 1, 2, 3. McGraw-Hill, New York (1953, 1955).

2. Virchenko, N. A., Lyashko, I. I., and Shevtsov, K. I. *Graphs of Functions*, Naukova Dumka, Kiev (1981), 320 pp. (Russian).

3. Grimalyuk, V. F. (Editor). *Graphs of Curves in Polar Coordinates Handbook*, Kiev Polytechnical Institute, Kiev (1983, 1984), 36 pp. (Russian).

4. Smogorzhevskivi, A. S. and Stolova, E. S. *Theory of Curves of Degree Three Handbook*, Fizmatgiz, Moscow (1961), 263 pp. (Russian).

5. Abramowitz, M. and Stegun, I. A. (Editors). *Handbook of Mathematical Functions with Formulas, Graphs and Mathematical Tables*, National Bureau of Standards. Applied Mathematical Series, 55 (1964).

6. Shunda, N. M. *Functions and Their Graphs*, Radyans'ka Shkola, Kiev (1983), 190 pp. (Russian).

7. Janke, E. Emde, F. and Lösch. *Tafeln Höherer Funktionen*, B.G.Teubner Verlagsgesellschaft, Stuttgart (1960).

8. Spanier, J. and Oldham, K. B. *An Atlas of Functions*, Hemisphere Publishing Corporation, New York (1987), 700 pp.

Index

S

T

U

V

W